MATERIALS AND PROCESSES IN MANUFACTURING

NINTH EDITION

E. Paul DeGarmo
deceased

J T. Black
Auburn University-Emeritus

Ronald A. Kohser
University of Missouri–Rolla

with computer interactive exercises by
Barney E. Klamecki
University of Minnesota

WILEY

ACQUISITIONS EDITOR   Joseph Hayton
MARKETING MANAGER   Katherine Hepburn
SENIOR PRODUCTION EDITOR   Petrina Kulek-Ford
SENIOR DESIGNER   Dawn L. Stanley
ILLUSTRATION EDITOR   Sandra Rigby
Cover photo: Courtesy of ESAB Cutting Systems

This book was set in 10/12 Times Ten by Preparé and printed and bound by *Hamilton Printing*. The cover was printed by *Phoenix Color Corporation*.

This book is printed on acid free paper. ⊗

Library of Congress Cataloging-in-Publication Data
Materials and processes in manufacturing / E. Paul Degarmo ... [et al.].-- 9th ed.
    p. cm.
  ISBN 0-471-03306-5
   1. Manufacturing processes. 2. Materials. I. DeGarmo, E. Paul (Ernest Paul)

TS183 .M382 2002
670.42'7--dc21   2002034342

Printed in the United States of America

10 9 8 7 6 5 4 3 2 1

MATERIALS AND PROCESSES IN MANUFACTURING

CONTENTS

In the world of manufacturing, significant changes and trends are having a profound impact on our everyday lives. Whether we like it or not, we all live in a technological society, a world of manufactured goods. Every day we come in contact with hundreds of manufactured items, from the kitchen faucets, telephones, cars and planes, TVs, VCRs, furniture, clothing, and so on. These goods are manufactured in factories designed to use manufacturing technologies. What are the trends in the manufacturing world, and how do they impact manufacturing processes?

■ TRENDS IN MANUFACTURING

First, manufacturing has become a global activity with U.S. companies sending work to other countries (China, Taiwan, Mexico) to take advantage of low-cost labor, while many foreign companies are building plants in the United States, to be nearer their marketplace. These automobile manufacturers and their suppliers use just about every process described in this book and some that we do not describe, often because they are closely held secrets.

Second, many manufacturing companies are redesigning their manufacturing systems and becoming lean (or Just-in-Time) producers. Almost every plant that you can visit these days is doing something to make itself leaner. Many of them have adopted some version of the Toyota Production System and redesigned old factories to be flexible while producing superior quality, low-unit-cost goods with the shortest possible lead times. More important, when these manufacturing factories are built, they are designed with the internal customer (the workforce) in mind, so things like ergonomics and safety are key design requirements. So while this book is all about materials and processes for the product, the design of the factory cannot be ignored when it comes to making the external customer happy with the product and the internal customer satisfied with the employer.

Third, the number and variety of products and the materials from which they are made continue to proliferate, and production quantities have become smaller. Existing processes must be modified to be more flexible, and new processes must be developed (so this book gets bigger with every edition!).

Fourth, consumers want better quality and reliability, so the methods, processes, and people responsible for the quality must be continually improved. The trend toward zero defects and continuous improvement requires continuous change in the system.

Finally, the *time-to-market* for new products is continuing to decrease. Many companies are developing concurrent engineering efforts to bring product design and manufacturing closer to the customer. There are two key aspects here. First, products are designed to be easier to manufacture and assemble (called *design for manufacture/ assembly*). Second, the manufacturing system is designed to be flexible, so the company can be competitive in the global marketplace.

Basically, manufacturing is a *value-adding* activity, where the conversion of materials into products adds value to the original material. Thus, the objective of a company engaged in manufacturing is to add value and do so in the most efficient manner, using the least amount of time, material, money, space, and labor. To minimize waste and maximize efficiency, the processes and operations need to be properly selected and arranged to permit smooth and controlled flow of material through the factory and provide for product variety. Meeting these goals requires a well-designed and efficient manufacturing system.

■ PURPOSE OF THE BOOK

The purpose of this book is to give design and manufacturing engineers basic information on materials and processing. The materials section focuses on properties and behavior. Thus, aspects of smelting and refining (or other material production processes) are presented only as they affect manufacturing and manufactured products. In terms of the processes used to manufacture items (converting materials into products), this text seeks to provide a descriptive introduction to a wide variety of options, emphasizing how each process works and its relative advantages and limitations. Our goal is to present this material in a way that can be understood by individuals seeing it for the very first time. This is not a graduate text where the objective is to thoroughly understand and optimize manufacturing processes. Mathematical models and analytical equations are used only when they enhance the basic understanding of the material.

The book also serves to introduce the *language of manufacturing*. Just as there is a big difference between a gun hand and a hand gun, there is a big difference between an engine lathe and a lathe engine. Everyday English words (words like *climb*, *bloom*, *allowance*, *chuck*, *coin*, *head*, and *ironing*) have entirely different meanings on the factory floor, a place where misunderstandings can be very costly. Pity the engineer who has to go on the plant floor not knowing an engine lathe from a milling machine or what a press brake can do. This engineer quickly loses all credibility with the people who make the products (and pay the engineers' salaries). However, the modern manufacturing engineer must be able to deal with real work problem-solving techniques like Taguchi methods and six sigma and developing manufacturing cells to make product families. This requires redesign of all the elements of the manufacturing systems—the machine tools, the workholding devices, the material handling equipment and the retraining of the people who work in the system.

■ THE NINTH EDITION

Materials and Processes in Manufacturing is written to provide a solid introduction to the fundamentals of manufacturing. The book begins with a survey of engineering materials, the "stuff" that manufacturing begins with, and seeks to provide the basic information that can be used to match the properties of a material to the service requirements of a component. A variety of engineering materials are presented, along with their properties and means of modifying them. The materials section can be used in curricula that lack preparatory courses in metallurgy, materials science, or strength of materials, or where the student has not yet been exposed to those topics. In addition, various chapters in this section can be used as supplements to a basic materials course, providing additional information on topics such as heat treatment, plastics, composites, and material selection. Following the materials section, measurement, nondestructive testing, and quality control are introduced with a manufacturing perspective. Then chapters on casting, forming, powder metallurgy, material removal, and joining are all developed as families of manufacturing processes. Each section begins with a presentation of the fundamentals on which those processes are based. This is followed by a discussion of the various process alternatives, which can be selected to operate individually or be combined into an integrated system.

Although considerable effort has been made to include many of the new developments in both materials and processes, the major emphasis is on the fundamentals of production processes, for these provide an enduring basis for understanding both existing phenomena and those that are to come. Another objective of the book is to introduce the broad spectrum of manufacturing processes to individuals who will be involved in the design and manufacture of finished products. The text material is presented in a descriptive fashion, explaining the fundamental workings of the process and its capabilities, typical applications, advantages, and limitations. Furthermore, the interactions of the material and the process are emphasized throughout the text, because if one does not understand material behavior, it is difficult to understand process behavior.

Approximately 20% of the text has been revised for the Ninth Edition. We have added much new material on processes, including:

- Rapid prototyping   - Electronics   - Metalcutting dynamics

While the heart of the book is still about casting, forming, machining, and joining, we have added information on many more materials, explaining how processes are sequenced together to make goods.

In the last part of the book, we have expanded our coverage of lean manufacturing and manufacturing systems design, because the best person to design or redesign the factory is the person who best understands how things are made (process planning) and what goes into the making in terms of the cost and quality issues. In this edition, we have included more details on manufacturing cell design, with some examples of lean manufacturing cells found only in the best lean suppliers plants.

There is also in-depth material on quality control and process capability. Engineers who get involved in six sigma projects can find out what sigma really measures. There is also introductory material on surface integrity, since so many processes produce the finished surface and residual stresses in the components. The case studies have been designed to make students aware of the great importance of properly coordinating design, material selection, and manufacturing to produce a satisfactory and reliable product.

The text is intended for use by engineering (mechanical, manufacturing, and industrial) and engineering technology students, in both two- and four-year undergraduate degree programs. In addition, the book is also used by engineers and technologists in other disciplines concerned with design and manufacturing (such as aerospace and electronics). Factory personnel will find this book to be a valuable reference that concisely presents the various production alternatives and the advantages and limitations of each. Additional or more in-depth information on specific materials or processes can be found in the various references listed at the end of the text.

■ HISTORY OF THE TEXT

The first edition of *Materials and Processes in Manufacturing* was written by E. Paul DeGarmo, now deceased. It was first published in 1957 by MacMillan and soon became the emulated standard for introductory texts in manufacturing. Second, third, and fourth editions followed in 1962, 1969, and 1974. In 1977, Ronald A. Kohser wrote a letter to E. Paul DeGarmo highlighting a number of problems he found in the materials chapters. DeGarmo then invited Kohser, a professor at the University of Missouri-Rolla, to rewrite the materials section. Thus Kohser became involved with the fifth edition, published in 1979.

After publication of the fifth edition, E. Paul DeGarmo expressed a desire to complete his retirement from the University of California at Berkeley by also retiring from textbook writing. It became apparent that a person was needed who could provide expertise in areas that would complement those of Ron Kohser, and who also had the ability to present technical material in a way that first-time students could understand. After a nationwide search and screening, an offer was extended to JT. Black, who was then a professor at The Ohio State University. For the sixth edition, seventh edition, and eighth edition (published in 1984, 1988, and 1997, respectively, by Prentice Hall and 1999 by John Wiley & Sons), Ron Kohser and JT. Black have shared the responsibility for the text. The chapters on engineering materials, casting, forming, powder metallurgy, joining, and nondestructive testing have been written or revised by Ron Kohser. JT. Black has assumed the responsibility for the introduction and chapters on material removal, metrology, quality control, manufacturing systems design, and lean manufacturing.

The eighth edition was originally slated for publication in 1995 but the acquisition of the Macmillan Publishing Company by Prentice Hall, the untimely death of our Macmillan editor, and a heart attack experienced by Ron Kohser all contributed to schedule delays. In 1999, Prentice-Hall was acquired by Pearson and was forced to divest itself of certain titles in areas where they held a monopoly. DeGarmo's *Materials and Processes in Manufacturing* was acquired by John Wiley. Therefore, the eighth edition had two different publishers.

■ ACKNOWLEDGMENTS

For this ninth edition, we will be without our mentor E. Paul DeGarmo for the first time. We are forever indebted to Paul for selecting us to carry on the tradition of his book. Now we have taken on some talented assistants. Dr. Brian K. Paul, of Oregon State University, rewrote the chapter on nontraditional manufacturing processes and wrote the first drafts of the chapters on rapid prototyping and electronics. Dr. David Cochran, of MIT, helped write the three chapters on manufacturing system design and lean manufacturing. Dr. Barney Klamecki has helped with the end parts of various chapters; he wrote the solutions manual; and he created new interactive problems, case studies, and lab exercises accessible to our users via the website: *www.wiley.com/college/degarmo*. The authors also wish to acknowledge the multitude of assistance, information, and illustrations that have been provided by a variety of industries, professional organizations, and trade associations. The text has become known for the large number of clear and helpful photos and illustrations that have been graciously provided by a variety of sources. In some cases, equipment is photographed or depicted without safety guards, so as to show important details, and personnel are not wearing certain items of safety apparel that would be worn during normal operation.

Over the many editions, there have been hundreds of reviewers, faculty, and students who have made suggestions and corrections to the text. We continue to be grateful for the time and interest that they have put into this book.

As always, our wives have played a major role in preparing the manuscript. Carol Black and Barb Kohser have endured being "textbook widows" during the time when the last four editions were written. Not only did they provide loving support, but Carol and Barb also provided hours of expert proofreading, typing, and editing as the manuscript was prepared.

The authors would also like to acknowledge the contribution of Dr. Elliot Stern for the dynamics of machining section in chapter 21.

Finally special thanks to our acquisitions editor, Joseph P. Hayton, for putting up with two procrastinating professors, who tried both his patience and his abilities as he coordinated all the various activities required to produce this text as scheduled.

CHAPTER 1

INTRODUCTION TO MATERIALS AND PROCESSES IN MANUFACTURING

■ 1.1 MATERIALS, MANUFACTURING, AND THE STANDARD OF LIVING

Manufacturing is critical to a country's economic welfare and standard of living because the standard of living in any society is determined, primarily, by the *goods* and *services* that are available to its people. Manufacturing companies contribute about 20% of the GNP, employ about 18% of the workforce and account for 40% of the exports of the United States. In most cases, materials are utilized in the form of manufactured goods. Manufacturing comprises the organized activity that converts raw materials, energy, and purchased items into saleable goods. The manufactured goods are typically divided into two classes: producer goods and consumer goods. *Producer goods* are those goods manufactured for other companies to use to manufacture either producer or consumer goods. *Consumer goods* are those purchased directly by the consumer or the general public. For example, someone has to build the machine tool (a lathe) that produces the large rolls that are sold to the rolling mill to be used to roll the sheets of steel that are then formed and become the body panels of your car. Similarly, many service industries depend heavily on the use of manufactured products, just as the agricultural industry is heavily dependent on the use of large farming machines for efficient production.

Converting materials from one form to another adds value to them. The more efficiently materials can be produced and converted into the desired products that function with the prescribed quality, the greater will be the companies' productivity and the better will be the standard of living of the employees.

The history of man has been linked to his ability to work with materials, beginning with the Stone Age and ranging through the eras of copper and bronze, the Iron Age, and recently the age of steel. While ferrous materials still dominate the manufacturing world, we are entering the age of tailor-made plastics, composite materials, and exotic alloys.

A good example of this progression is shown in Figure 1-1. The early machine tools used carbon tool steels as cutting tool materials. These were replaced by steel alloys called high-speed steels because they permitted the cutting speed to double, reducing the cutting time. Over the years, many new materials have been developed to permit faster machining. In general, as materials become more complex and sophisticated, usually having greater strength (and sometimes lighter weight), they become more difficult to manufacture with existing methods, so new process technologies are being invented all the time.

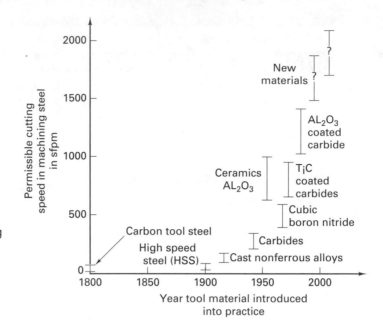

FIGURE 1-1 New cutting tool materials are continuously being introduced, resulting in faster cutting speeds (in surface feet per minute) in machine tools.

Although materials are no longer used only in their natural state, there is obviously an absolute limit to the amounts of many materials available here on earth. Therefore, as the variety of man-made materials continues to increase, resources must be used efficiently and recycled whenever possible. Of course, recycling only postpones the exhaustion date.

Like materials, processes have also proliferated greatly in the last 30 years, with new processes being developed to handle the new materials more efficiently and with less waste. Advances in manufacturing technology often account for improvements in productivity. Even when the technology is proprietary, the competition often gains access to it, usually quite quickly.

Starting with the product design, materials, men, and equipment are interactive factors in manufacturing that must be combined properly (integrated) to achieve low cost, superior quality, and on-time delivery. Typically, as shown in Figure 1-2, 40% of the selling price of a product is *manufacturing cost*. Since the selling price is determined by the customer, maintaining the profit often depends on reducing manufacturing cost. The internal customers who really make the product, called direct labor, are usually the targets of automation, but typically they account for only about 10% of the manufacturing cost even though they are the main element in increasing productivity. In Chapter 42, a manufacturing strategy is presented that attacks the materials cost, indirect costs, and general administration costs, in addition to labor costs. The material costs include the cost of storing and handling the materials within the plant. The strategy is called *lean production*.

FIGURE 1-2 Manufacturing cost is the largest cost in the selling price. The largest manufacturing cost is for materials, not direct labor.

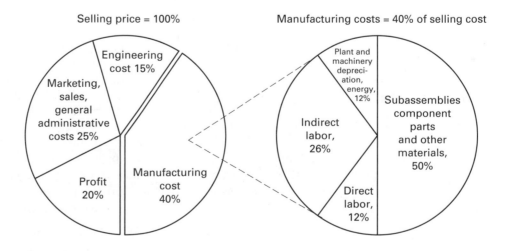

Referring again to Figure 1-2 of the total expenses (selling price less profit), about 68% of dollars are spent on people, the breakdown being about 15% for engineers; 25% for marketing, sales, general management people; 5% for direct labor, and 10% for indirect labor (55/80 = 68.75%). The average labor cost in manufacturing in the United States was around $13 per hour for the hourly workers in 1990. Reductions in direct labor will have only marginal effects on the people costs. The optimal combination of factors for producing a small quantity of a given product may be very inefficient for a larger quantity of the same product. Consequently, a systems approach, taking all the factors into account, must be used. *This requires a sound and broad understanding of materials, processes, and equipment on the part of the decision makers accompanied by an understanding of the manufacturing systems.* Materials and processes in manufacturing systems are what this book is all about.

■ 1.2 MANUFACTURING AND PRODUCTION SYSTEMS

Manufacturing is the economic term for making goods and services available to satisfy human wants. Manufacturing implies creating value by applying useful mental or physical labor.

The *manufacturing processes* are collected together to form a *manufacturing system* (MS). The manufacturing system is a complex arrangement of physical elements characterized by measurable parameters (Figure 1-3). The manufacturing system takes inputs and produces products for the external customer. The production system, or the enterprise, includes the manufacturing system as shown in Figure 1-4 and services it. In this book, a *production system* will refer to the total company and will include within it the *manufacturing system*. The production system includes the manufacturing system plus all the other functional areas of the plant for information, design, analysis, and control. These subsystems are connected by various means to each other to produce either goods or services or both. *Goods* refer to material things. *Services* are nonmaterial things that we buy to satisfy our wants, needs, or desires. *Service production systems* (SPSs) include transportation, banking, finance, savings and loan, insurance, utilities, health care, education, communication, entertainment, sporting events, and so forth. They are useful labors that do not directly produce a product. The manufacturing department has the responsibility of designing processes (sequences of operations and processes) and systems to create (make or manufacture) the product as designed. The system must be flexible to meet customer demand (volumes and mixes of products) as well as changes in product design.

As shown in Table 1-1, production terms have a definite rank of importance somewhat like rank in the army. Confusing *system* with *section* is similar to mistaking a colonel for a corporal. In either case, knowledge of rank is necessary. The terms tend to overlap because of the inconsistencies of popular usage.

An obvious problem exists here in the terminology of manufacturing and production. The same term can refer to different things. For example, *drill* can refer to the machine tool that does these kinds of operations; the operation itself, which can be done on many different kinds of machines; or the cutting tool, which exists in many different forms. It is therefore important to use modifiers whenever possible: "Use the *radial* drill

FIGURE 1-3 The manufacturing system converts inputs to outputs using processes to add value to the goods for the external customer.

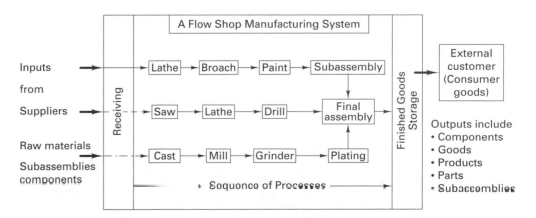

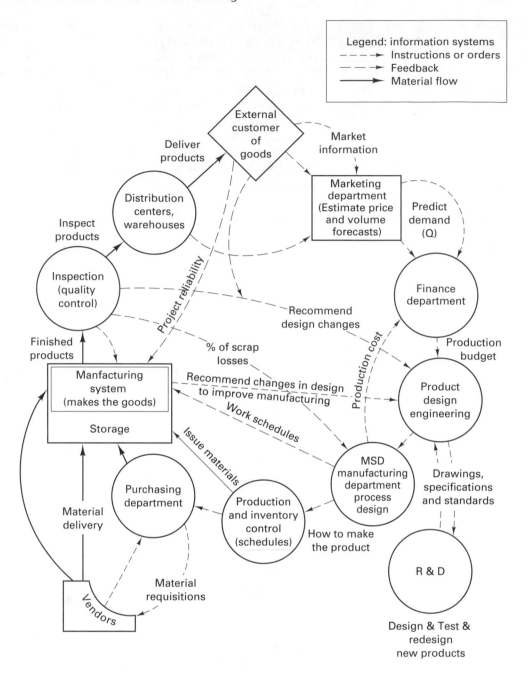

FIGURE 1-4   The functions and systems of the production system, which includes (and services) the manufacturing system. The functional departments are connected by formal and informal information systems all designed to service the manufacturing system that produces the goods.

press to drill a hole with a 1-in.-diameter spade drill." The emphasis of this book will be directed toward the understanding of the processes, machines, and tools required for manufacturing and how they interact with the materials being processed. In the last section of the book, an introduction to systems aspects is presented.

PRODUCTION SYSTEM—THE ENTERPRISE

The highest ranking term in the hierarchy is *production system*. A production system includes people, money, equipment, materials and supplies, markets, management, and the manufacturing system. In fact, all aspects of commerce (manufacturing, sales, advertising, profit, and distribution) are involved. Table 1-2 provides a partial list of production systems.

Much of the information given *for manufacturing production systems* (MPSs) is relevant to the *service production system* (SPS). Many MPSs require an SPS for proper product sales. This is particularly true in industries such as the food (restaurant) industry in which customer service is as important as quality and on-time delivery. Table 1-3 provides a short list of service industries.

TABLE 1-1.   Production Terms for Manufacturing Production Systems

Term	Meaning	Examples
Production system The enterprise	All aspects of workers, machines, and information, considered collectively, needed to manufacture parts or products; integration of all units of the system is critical (see Chapter 42).	Company that makes engines, assembly plant, glassmaking factory, foundry; sometimes called the enterprise or the business.
Manufacturing system (sequence of operations, collection of processes)	The collection of manufacturing processes and operations resulting in specific end products; an arrangement or layout of many processes, materials-handling equipment, and operators.	Rolling steel plates, manufacturing of automobiles, series of connected operations or processes, a job shop, a flow shop, a continuous process.
Machine or machine tool or manufacturing process	A specific piece of equipment designed to accomplish specific processes, often called a *machine tool*; machine tools linked together to make a manufacturing system.	Spot welding, milling machine, lathe, drill press, forge, drop hammer, die caster, punch press, grinder, etc.
Job (sometimes called a station; a collection of tasks)	A collection of operations done on machines or a collection of tasks performed by one worker at one location on the assembly line.	Operation of machines, inspection, final assembly; e.g., forklift driver has the job of moving materials.
Operation (sometimes called a process)	A specific action or treatment, often done on a machine, the collection of which makes up the job of a worker.	Drill, ream, bend, solder, turn, face, mill extrude, inspect, load.
Tools or tooling	Refers to the implements used to hold, cut, shape, or deform the work materials; called *cutting tools* if referring to machining; can refer to *jigs* and *fixtures* in workholding and *punches* and *dies* in metal forming.	Grinding wheel, drill bit, end milling cutter, die, mold, clamp, three-jaw chuck, fixture.

TABLE 1-2.   Partial List of Production Systems for Producer and Consumer Goods

Aerospace and airplanes	Foods (canned, dairy, meats, etc.)
Appliances	Footwear
Automotive (cars, trucks, vans, wagons, etc.)	Furniture
Beverages	Glass
Building supplies (hardware)	Hospital supplier
Cement and asphalt	Leather and fur goods
Ceramics	Machines
Chemicals and allied industries	Marine Engineering
Clothing (garments)	Metals (steel, aluminum, etc.)
Construction	Natural resources (oil, coal, forest, pulp and paper)
Construction materials (brick, block, panels)	Publishing and printing (books, records, newspapers)
Drugs, soaps, cosmetics	Restaurants
Electrical and microelectronics	Retail (food, department store, etc.)
Energy (power, gas, electric)	Ship building
Engineering	Textiles
Equipment and machinery (agricultural, construction and electrical products, electronics, household products, industrial machine tools, office equipment, computers, power generators)	Tire and rubber
	Tobacco
	Transportation vehicles (railroad, airline, truck, bus)
	Vehicles (bikes, cycles, ATVs, snowmobiles)

TABLE 1-3.   Types of Service Industries

Advertising and marketing
Communication (telephone, computer networks)
Education
Entertainment (radio, TV, movies, plays)
Equipment and furniture rental
Financial (banks, investment companies, loan companies)
Health care
Insurance
Transportation and car rental
Travel (hotel, motel, cruise lines)

MANUFACTURING SYSTEMS

A collection of operations and processes used to obtain a desired product(s) or component(s) is called a *manufacturing system*. The manufacturing system is therefore the design or *arrangement of the manufacturing processes*. Control of a system applies to overall control of the whole, not merely of the individual processes or equipment. The entire manufacturing system must be controlled in order to schedule and control production, inventory levels, product quality, output rates, and so forth.

ROLES OF ENGINEERS IN MANUFACTURING

Many engineers have as their function the designing of products. The products are brought into reality through the processing or fabrication of materials. In this capacity designers are a key factor in the material selection and manufacturing procedure. A *design engineer*, better than any other person, should know what the design is to accomplish, what assumptions can be made about service loads and requirements, what service environment the product must withstand, and what appearance the final product is to have. To meet these requirements, the material(s) to be used must be selected and specified. In most cases, to utilize the material and to enable the product to have the desired form, the designer knows that certain *manufacturing processes* will have to be employed. In many instances, the selection of a specific material may dictate what processing must be used. On the other hand, when certain processes must be used, the design may have to be modified in order for the process to be utilized effectively and economically. Certain dimensional sizes can dictate the processing, and some processes require certain sizes for the parts going into them. In converting the design into reality, many decisions must be made. In most instances, they can be made most effectively at the design stage. It is thus apparent that design engineers are a vital factor in the manufacturing process, and it is indeed a blessing to the company if they can *design for manufacturing*, that is, design the product so that it can be manufactured and/or *assembled* economically (i.e., low unit cost). Design for manufacturing uses the knowledge of manufacturing processes, and so the design and manufacturing engineers should work together to integrate design and manufacturing activities.

Manufacturing engineers select and coordinate specific processes and equipment to be used, or supervise and manage their use. Some design special tooling is used so that standard machines can be utilized in producing specific products. These engineers must have a broad knowledge of manufacturing processes and material behavior so that desired operations can be done effectively and efficiently without overloading or damaging machines and without adversely affecting the materials being processed. Although it is not obvious, the most hostile environment the material may ever encounter in its lifetime is the processing environment.

Industrial or *manufacturing engineers* are responsible for manufacturing systems design (or layout) of factories. They must take into account the interrelationships of the design and the properties of the materials that the machines are going to process, and the interreaction of the materials and processes. The choice of machines and equipment used in manufacturing and their arrangement in the factory are also design tasks.

Materials engineers devote their major efforts to developing new and better materials. They, too, must be concerned with how these materials can be processed and with the effects that the processing will have on the properties of the materials. Although their roles may be quite different, it is apparent that a large proportion of engineers must concern themselves with the interrelationships of materials and manufacturing processes.

As an example of the close interrelationship of design, materials selection, and the selection and use of manufacturing processes, consider the common desk stapler. Suppose that this item is sold at the retail store for $20.00. The wholesale outlet sold the stapler for $16.00 and the manufacturer probably received about $10.00 for it. Staplers typically consist of 10 to 12 parts and some rivets and pins. Thus the manufacturer had to produce and assemble the 10 parts for about $1.00 per part. Only by giving a great deal of attention to design, selection of materials, selection of processes, selection of equipment used for manufacturing (tooling), and utilization of personnel could such a result be achieved.

The stapler is a relatively simple product, yet the problems involved in its manufacture are typical of those that manufacturing industries must deal with. The elements of design, materials, and processes are all closely related, each having its effect on the others. For example, suppose the component that holds the staples is to be a machined part rather than a formed part, then entirely different processes and materials would need to be specified. Or, if a part is to be changed from metal to plastic, then a whole new set of fundamentally different materials and processes would need to come into play. Such changes would also have a significant impact on cost.

CHANGING WORLD COMPETITION

In recent years, major changes in the world of goods manufacturing have taken place. Three of these are:

1. Worldwide competition of global products
2. High-tech manufacturing or advanced technology
3. New manufacturing systems structure, strategies, and management

Worldwide (global) competition is a fact of manufacturing life and it will get stronger in the future. The goods you buy today may have been made anywhere in the world.

The second aspect, advanced manufacturing technology, usually refers to new machine tools or processes with computer-aided manufacturing. Producing machine tools is a small industry (7–8 billion tools per year) with enormous leverage. Improved processes lead to better components and more durable goods. However, the new technology is often purchased from companies that have developed the technology, so this approach is important but may not provide a unique competitive advantage if your competitors can also buy the technology, provided that they have the capital. Some companies develop their own unique process technology and try to keep it proprietary as long as they can. A good example of unique process technology was the numerical control machine tool, shown in Figure 1-5 and discussed in Chapter 32. Nowadays, computer-controlled machine tools are common to the factory floor.

The third change and perhaps the real key to success in manufacturing is to build a manufacturing system that can deliver on time to the customer, super-quality goods at

FIGURE 1-5   The same part can be made by numerical control (NC) or manual machining. The increased cost of NC can be offset by the decreased manufacturing time and improved quality.

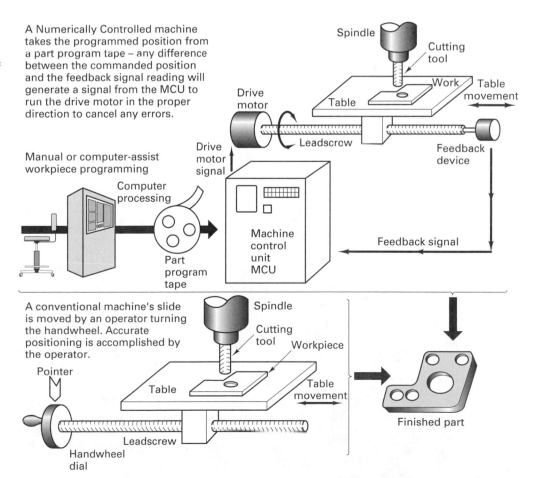

A Numerically Controlled machine takes the programmed position from a part program tape – any difference between the commanded position and the feedback signal reading will generate a signal from the MCU to run the drive motor in the proper direction to cancel any errors.

Manual or computer-assist workpiece programming

Computer processing

Part program tape

Drive motor signal

Machine control unit MCU

Feedback signal

Spindle

Cutting tool

Work

Table movement

Drive motor

Table

Leadscrew

Feedback device

A conventional machine's slide is moved by an operator turning the handwheel. Accurate positioning is accomplished by the operator.

Pointer

Spindle

Cutting tool

Workpiece

Table

Table movement

Handwheel dial

Leadscrew

Finished part

the lowest possible cost in a flexible way. This change reflects an effort to improve markedly the methodology by which goods are produced rather than simply upgrading the manufacturing-process technology.

Manufacturing system design is discussed extensively in the last section of the book and we recommend that students examine this material closely after they have gained a working knowledge of materials and processes. The next section provides a brief introduction to manufacturing system designs.

MANUFACTURING SYSTEM DESIGNS

Five manufacturing system designs can be identified: the job shop, the flow shop, the linked-cell shop, the project shop, and the continuous process. The latter system primarily deals with liquids and/or gases (such as an oil refinery) rather than solids or discrete parts.

The most common of these layouts is the *job shop*, characterized by large varieties of components, general-purpose machines, and a functional layout (Figure 1-6).

FIGURE 1-6   This rack bar machining area is functionally designed like a job shop, with lathes and broaches lined up in departments. This area is about 2 acres and the work-in-process is estimated at about 97,000 parts. The area was manned by 20 workers.

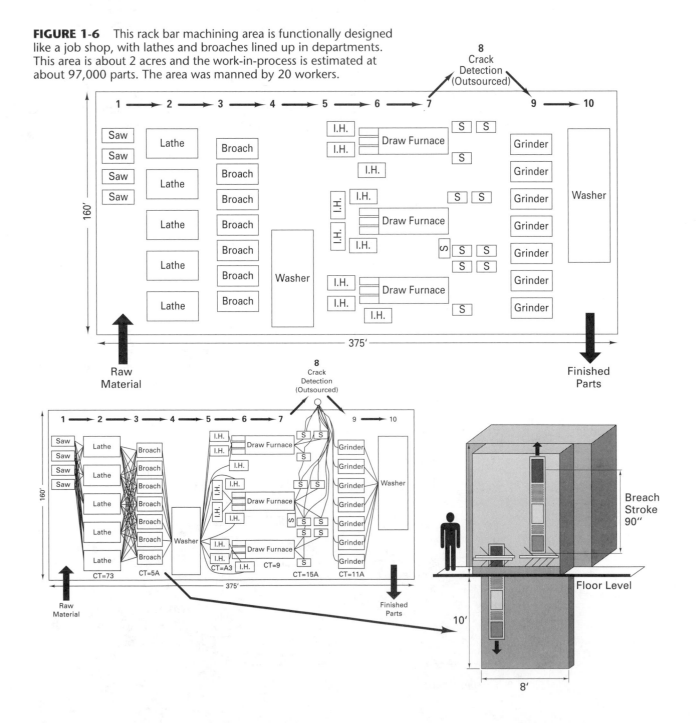

This means that machines are collected by function (all lathes together, all broaches together, all milling machines together) and the parts are routed around the shop in small lots to the various machines. The inset shows the multiple paths through the shop and a detail on one of the seven broaching machine tools. The material is moved from machine to machine in carts or containers and is called the *lot*.

Flow shops are characterized by larger lots, special-purpose machines and equipment, less variety, and more mechanization. Flow shop layouts are typically either continuous or interrupted and can be for manufacturing or assembly, as shown in Figure 1-7. If *continuous*, a production line is built that basically runs one large-volume complex item in great quantity and nothing else. The common light bulb is made this way. A transfer line producing an engine block is another typical example. If *interrupted*, the line manufactures large lots but is periodically "changed over" to run a similar but different component.

The *linked-cell* manufacturing system is composed of manufacturing cells (Figure 1-8) connected together (linked) using a unique form of inventory and information control called *kanban*. (Chapters 41 and 43.)

The *project shop* is characterized by the immobility of the item being manufactured. In the construction industry, bridges and roads are good examples. In the manufacture of goods, large airplanes, ships, large machine tools and locomotives are manufactured in project shops. It is necessary that the workers, machines, and materials come to the site. The number of end items is not very large, and therefore the lot sizes of the components going into the end item are not large. Thus the job shop usually supplies parts and subassemblies to the project shop in small lots.

Continuous processes are used to manufacture liquids, oils, gases, and powders. These manufacturing systems are usually large plants producing goods for other producers or mass-producing canned or bottled goods for consumers. The manufacturing engineer in these factories is often a chemical engineer.

FIGURE 1-7   The moving assembly line for cars is an example of the flow shop.

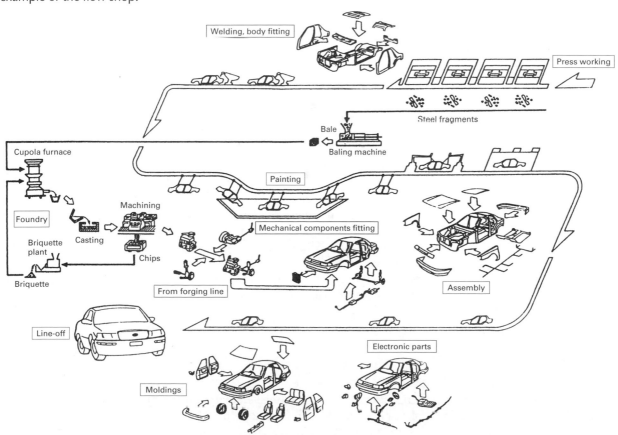

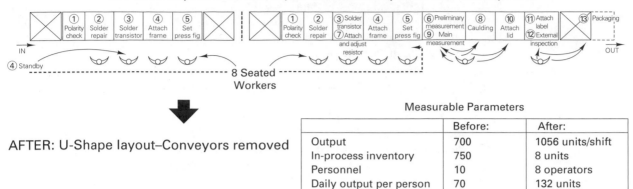

BEFORE: Layout with conveyor–Subassembly with two conveyors

AFTER: U-Shape layout–Conveyors removed

Measurable Parameters		
	Before:	After:
Output	700	1056 units/shift
In-process inventory	750	8 units
Personnel	10	8 operators
Daily output per person	70	132 units
Cycle time	0.60 minute	0.43 minute

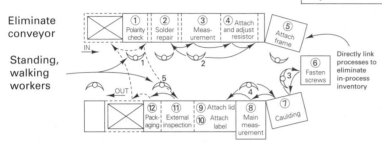

FIGURE 1-8 The traditional subassembly lines can be redesigned into U-shaped cells as part of the conversion of mass production to lean production.

Naturally, there are many hybrid forms of these manufacturing systems, but the job shop is the most common system. Because of its design, the job shop has been shown to be the least cost-efficient of all the systems. Component parts in a typical job shop spend only 5% of their time in machines and the rest of the time waiting or being moved from one functional area to the next. Once the part is on the machine, it is actually being processed (i.e., having value added to it by the changing of its shape) only about 30 to 40% of the time. The rest of the time parts are being loaded, unloaded, inspected, and so on. The advent of *numerical control* machines increased the percentage of time that the machine is making chips because tool movements are programmed and the machines can automatically change tools or load or unload parts.

However, there are a number of trends that are forcing manufacturing management to consider means by which the job shop system itself can be redesigned to improve its overall efficiency. These trends have forced manufacturing companies to convert their batch-oriented job shops into linked-cell manufacturing systems, with the manufacturing and subassembly cells structured around specific products. Another way to identify families of products with a similar set of manufacturing processes is called group technology.

Group technology (GT) can be used to restructure the factory floor. GT is a concept whereby similar parts are grouped together into part families. Parts of similar size and shape can often be processed through a *similar set of processes*. A part family based on manufacturing would have the same set or sequences of manufacturing processes. The set of processes is called a cell. Thus, with GT, job shops can be restructured into cells, each cell specializing in a particular family of parts. The parts are handled less, machine setup time is shorter, in-process inventory is lower, and the time needed for parts to get through the manufacturing system (called the throughput time) is greatly reduced.

■ 1.3 LANGUAGE OF MANUFACTURING OR PROCESS TECHNOLOGY

MANUFACTURING PROCESSES

A *manufacturing process* converts unfinished materials to finished products, often using a machine tool. For example, injection molding, die casting, progressive stamping, milling, arc welding, painting, assembling, testing, pasteurizing, homogenizing, and annealing are commonly called *processes* or *manufacturing processes*. The term *process* often implies a sequence of steps, processes, or operations for production of goods and services, as shown in Figure 1-9, which shows the processes to manufacture an Olympic type medal.

A *machine tool* is an assembly of related mechanisms on a frame or bed that together produce a desired result. Generally, motors, controls, and auxiliary devices are included. Cutting tools and workholding devices are considered separately.

A machine tool may do a single process (cutoff saw) or multiple processes, or it may manufacture an entire component. Machine sizes vary from a tabletop drill press to a 1000-ton forging press.

JOB AND STATION

In the classical system, a *job* is the total of the work or duties a worker performs. A station is the work area of a production line worker. (The multifunctional worker concept has changed these traditional definitions.)

A job is a group of related operations and tasks performed at one station or series of stations in cells. For example, the job at a final assembly station may consist of four tasks:

1. Attach carburetor.
2. Connect gas line.
3. Connect vacuum line.
4. Connect accelerator rod.

The job of a turret lathe (a semiautomatic machine) operator may include the following operations and tasks: load, start, index and stop, unload, inspect. The operator's job may also include setting up the machine (i.e., getting ready for manufacturing). Other machine operations include drilling, reaming, facing, turning, chamfering, and knurling. The operator can run more than one machine or service at more than one station.

The terms *job* and *station* have been carried over to unmanned machines. A job is a group of related operations generally performed at one station, and a *station* is a position or location in a machine (or process) where specific operations are performed. A simple machine may have only one station. Complex machines can be composed of many stations. The job at a station often includes many simultaneous operations, such as "drill all face holes" by multiple spindle drills.

OPERATION

An *operation* is a distinct action performed to produce a desired result or effect. Typical machine operations are loading and unloading. Operations can be divided into suboperational elements. For example, loading is made up of picking up part, placing part in jig, closing jig. However, suboperational elements will not be discussed here.

Operations categorized by function are:

1. *Materials handling and transporting*: change in position of the product
2. *Processing*: change in volume and quality, including assembly and disassembly; can include packaging
3. *Packaging*: special processing; may be temporary or permanent for shipping
4. *Inspecting and testing*: comparison to the standard or check process behavior
5. *Storing*: time lapses without further operations

These basic operations may occur more than once in some processes, or they may sometimes be omitted. *Remember, it is the manufacturing processes that change the value and quality of the materials.* Defective processes produce poor quality or scrap. Other operations may be necessary but do not, in general, add value, whereas operations performed by machines that do material processing usually do add value.

TREATMENTS

Treatments operate continuously on the workpiece. They usually alter or modify the product-in-process without tool contact. Heat treating, curing, galvanizing, plating, finishing, (chemical) cleaning, and painting are examples of treatments. Treatments usually do add value to the part.

These processes are difficult to include in cells because they often have long cycle times, are hazardous to the workers' health, or are unpleasant to be around because of high heat or chemicals. They are often done in large tanks or furnaces or rooms. The cycle time for these processes may *dictate* the cycle times for the entire system. These operations also tend to be material specific. Many manufactured products are given

How an olympic medal is made using the CAD/CAM process

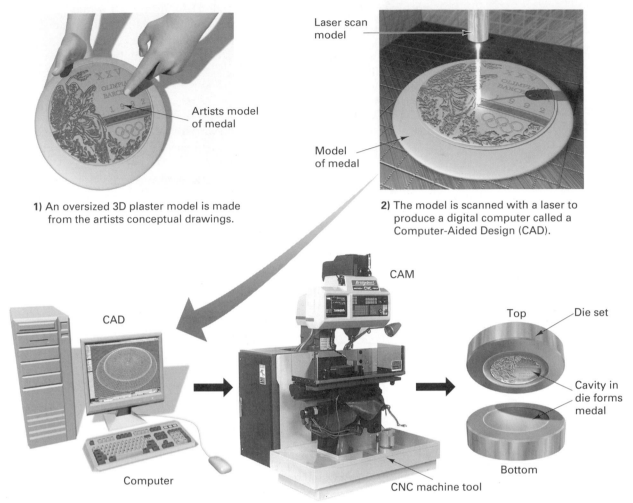

Artists model of medal

Laser scan model

Model of medal

1) An oversized 3D plaster model is made from the artists conceptual drawings.

2) The model is scanned with a laser to produce a digital computer called a Computer-Aided Design (CAD).

CAD

CAM

Top

Die set

Cavity in die forms medal

Bottom

Computer

CNC machine tool

3) The computer has software to produce a program to drive numerical control machine to cut a die set.

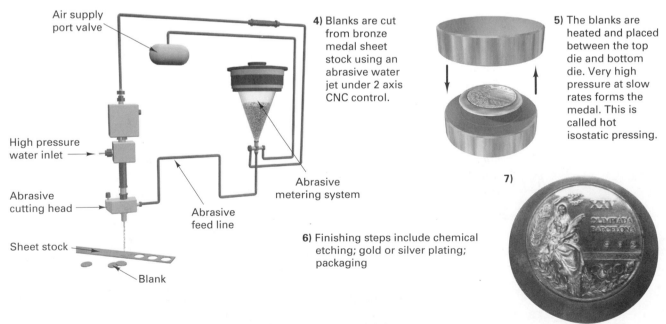

Air supply port valve

High pressure water inlet

Abrasive cutting head

Sheet stock

Blank

Abrasive feed line

Abrasive metering system

4) Blanks are cut from bronze medal sheet stock using an abrasive water jet under 2 axis CNC control.

5) The blanks are heated and placed between the top die and bottom die. Very high pressure at slow rates forms the medal. This is called hot isostatic pressing.

7)

6) Finishing steps include chemical etching; gold or silver plating; packaging

FIGURE 1-9 The manufacturing process for making Olympic medals has many steps, beginning with design and includes die making.

decorative and protective surface treatments that control the finished appearance. A customer may not buy a new vehicle because it has a visible defect in the chrome bumper, although this defect will not alter the operation of the car.

TOOLS, TOOLING, AND WORKHOLDERS

The lowest mechanism in the production term rank is the *tool*. Tools are used to hold, cut, shape, or form the unfinished product. Common hand tools include the saw, hammer, screwdriver, chisel, punch, sandpaper, drill, clamp, file, torch, and grindstone.

Basically, machines are mechanized versions of such hand tools. Most tools are for cutting (drill bits, reamers, single-point turning tools, milling cutters, saw blades, broaches, and grinding wheels). Noncutting tools for forming include extrusion dies, punches, and molds.

Tools also include workholders, jigs, and fixtures. These tools and cutting tools are generally referred to as the *tooling*, which is usually considered separate from machine tools. Cutting tools wear and fail and must be periodically replaced before parts are ruined. The workholding devices must be able to locate and secure the workpieces during processing in a repeatable, mistake-proof way.

TOOLING FOR MEASUREMENT AND INSPECTION

Measuring tools and instruments are also important for manufacturing. Common examples of measuring tools are rulers, calipers, micrometers, and gages. Precision devices that use laser optics or vision systems coupled with sophisticated electronics are becoming commonplace. Vision systems and coordinate measuring machines are becoming critical elements for achieving superior quality.

INTEGRATING INSPECTION INTO THE PROCESS

The integration of the *inspection* process into the manufacturing process or the manufacturing system is a critical step toward building products of superior quality. An example will help. Compare an electric typewriter with a computer that does word processing. The electric typewriter is flexible. It types whatever words are wanted in whatever order. It can type in Pica, Elite, or Orator but the font (disk or ball that has the appropriate type size on it) has to be changed according to the size and face of type wanted. The computer can do all of this but can also, through its software, do italics, darken the words, vary the spacing to justify the right margin, plus many other functions. It checks immediately for incorrect spelling and other defects like repeated words. The software system provides a signal to the hardware to flash the word so that the operator will know something is wrong and can make an immediate correction. If the system were designed to prevent the typist from typing repeated words, then this would be a *pokayoke*, a defect prevention. Defect prevention is better than immediate defect detection and correction. Ultimately, the system should be able to forecast the probability of a defect, correcting the problem at the source. This means that the typist would have to be removed from the process loop, perhaps by having the system type out what it is told (convert oral to written directly). Pokayoke and source inspection are keys to designing manufacturing systems that produce superior-quality products at low cost.

PRODUCTS AND FABRICATIONS

In manufacturing, material things (goods) are made to satisfy human wants. *Products* result from manufacture. Manufacture also includes conversion processes such as refining, smelting, and mining.

Products can be manufactured by fabricating or by processing. *Fabricating* is the manufacture of a product from pieces such as parts, components, or assemblies. Individual products or parts can also be fabricated. Separable discrete items such as tires, nails, spoons, screws, refrigerators, or hinges are fabricated.

Processing is also used to refer to the manufacture of a product by continuous means, or by a continuous series of operations, for a specific purpose. Continuous items such as steel strip, beverages, breakfast foods, tubing, chemicals, and petroleum are "processed." Many processed products are marketed as discrete items, such as bottles of beer, bolts of cloth, spools of wire, and sacks of flour.

Separable discrete products, both piece parts and assemblies, are fabricated in a *plant*, *factory*, or *mill*, for instance, a textile or rolling mill. Products that *flow* (liquids, gases, grains, or powders) are processed in a *plant* or *refinery*. The *continuous-process industries* such as petroleum and chemical plants are sometimes called processing industries or flow industries.

To a lesser extent, the terms *fabricating industries* and *manufacturing industries* are used when referring to fabricators or manufacturers of large products composed of many parts, such as a car, a plane, or a tractor. Manufacturing often includes continuous-process treatments such as electroplating, heating, demagnetizing, and extrusion forming.

Construction or building is making goods by means other than manufacturing or processing in factories. Construction is a form of project manufacturing of useful goods like houses, highways, and buildings. The public may not consider construction as manufacturing because the work is not usually done in a plant or factory, but it can be. There is a company in Delaware that can build a custom house of any design in their factory, truck it to the building site, and assemble it on a foundation in two or three weeks.

Agriculture, *fisheries*, and *commercial fishing* produce real goods from useful labor. Lumbering is similar to both agriculture and mining in some respects, and mining should be considered processing. Processes that convert the raw materials from agriculture, fishing, lumbering, and mining into other usable and consumable products are also forms of manufacturing.

WORKPIECE AND ITS CONFIGURATION

In the manufacturing of goods, the primary objective is to produce a component having a desired geometry, size, and finish. Every component has a shape that is bounded by various types of surfaces of certain sizes that are spaced and arranged relative to each other. Consequently, a component is manufactured by producing the surfaces that bound the shape. Surfaces may be:

1. Plane or flat
2. Cylindrical (external or internal)
3. Conical (external or internal)
4. Irregular (curved or warped)

Figure 1-10 illustrates how a shape can be analyzed and broken up into these basic bounding surfaces. Parts are manufactured by using a set or sequence of processes that will either (1) remove portions of a rough block of material (bar stock, casting, forging) so as to produce and leave the desired bounding surface, or (2) cause material to form into a stable configuration that has the required bounding surfaces (casting, forging). Consequently, in designing an object, the designer specifies the shape, size, and arrangement of the bounding surface. The part design must be analyzed to determine what materials will provide the desired properties, including mating to other components, and what processes can best be employed to obtain the end product at the most reasonable cost. This is often the job of the manufacturing engineer.

BASIC MANUFACTURING PROCESSES

It is the manufacturing processes that create or add value to a product. The manufacturing processes can be classified as:

- Casting, foundry, or molding processes
- Forming or metalworking processes
- Machining (material removal) processes
- Joining and assembly
- Surface treatments (finishing)
- Rapid prototyping
- Heat treating
- Other

These classifications are not mutually exclusive. For example, some finishing processes involve a small amount of metal removal or metal forming. A laser can be used either for joining or for metal removal or heat treating. Occasionally, we have a process such as shearing, which is really metal cutting but is viewed as a (sheet) metal-forming process. Assembly may involve processes other than joining. The categories of process types are far from perfect.

Casting and *molding* processes are widely used to produce parts that often require other follow-on processes, such as machining. Casting uses molten metal to fill a cavity. The metal retains the desired shape of the mold cavity after solidification. An important advantage of casting and molding is that, in a single step, materials can be converted from a crude form into a desired shape. In most cases, a secondary advantage is that excess or scrap material can easily be recycled. Figure 1-11 illustrates schematically some

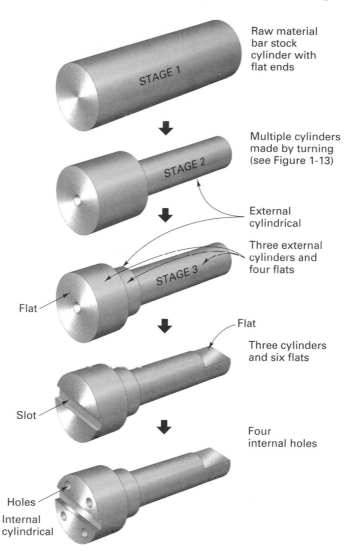

Raw material bar stock cylinder with flat ends

STAGE 1

Multiple cylinders made by turning (see Figure 1-13)

STAGE 2

External cylindrical

Three external cylinders and four flats

STAGE 3

Flat

Flat

Three cylinders and six flats

Slot

Four internal holes

Holes

Internal cylindrical

FIGURE 1-10 Pinion shaft, composed of many geometric surfaces.

of the basic steps in the lost-foam casting process, one of many processes used in the foundry industry.

Casting processes are commonly classified into two types: permanent mold (a mold can be used repeatedly) or nonpermanent mold (a new mold must be prepared for each casting made). Molding processes for plastics and composites are included in the chapters on forming processes.

Forming and *shearing* operations typically utilize material (metal or plastics) that has been previously cast or molded. In many cases the materials pass through a series of forming or shearing operations, so the form of the material for a specific operation may be the result of all the prior operations. The basic purpose of forming and shearing is to modify the shape and size and/or physical properties of the material.

Metalforming and *shearing operations* are done both "hot" and "cold," a reference to the temperature of the material at the time it is being processed with respect to the temperature at which this material can recrystallize (i.e., grow new grain structure). Figure 1-12 shows the process by which the fender of a car is made using a series of metal forming processes.

Machining or *metal removal processes* refer to the removal of certain selected areas from a part in order to obtain a desired shape or finish. Chips are formed by interaction of a cutting tool with the material being machined. Figure 1-13 shows a chip being formed by a single-point cutting tool in a machine tool called a lathe. The manufacturing engineer may be called upon to specify the cutting parameters such as cutting speed, feed, or depth of cut (DOC). The engineer may also have to select the cutting tools for the job.

Cutting tools used to perform the basic turning on the lathe are shown in Figure 1-13. The cutting tools are mounted in machine tools, which provide the required

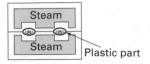

To make the foam parts, metal molds are used. Beads of polystyrene are heated and expanded in the mold to get parts.

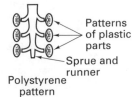

A pattern containing a sprue, runners, risers, and parts is made from single or multiple pieces of foamed polystyrene plastic.

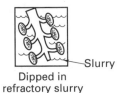

The polystyrene pattern is dipped in a ceramic slurry, which wets the surface and forms a coating about 0.005 inch thick.

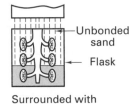

The coated pattern is placed in a flask and surrounded with loose, unbonded sand.

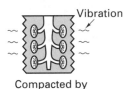

The flask is vibrated so that the loose sand is compacted around the pattern.

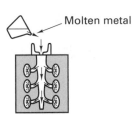

During the pouring of molten metal, the hot metal vaporizes the pattern and fills the resulting cavity.

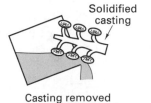

The solidified casting is removed from flask and the loose sand reclaimed.

FIGURE 1-11   Schematic of the lost-foam casting process.

movements of the tool with respect to the work (or vice versa) to accomplish the process desired. In recent years many new machining processes have been developed.

The seven basic machining processes are *shaping, drilling, turning, milling, sawing, broaching,* and *abrasive machining.* With the exception of shaping, each of the basic processes has a chapter dedicated to it. See Figure 1-14 for a schematic of the seven basic processes. Historically, eight basic types of machine tools have been developed to

Metal Forming Process for Automobile Fender

Sheet metal bending / forming

Single draw punch and die

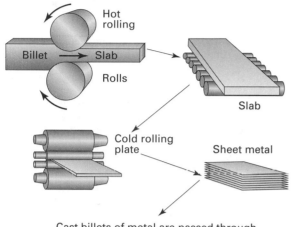

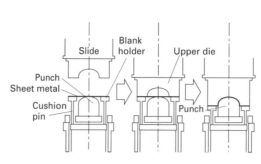

Cast billets of metal are passed through successive rollers to produce sheets of steel rolled stock.

Using sets of dies in stands of presses, the flat sheet metal is "formed" into a fender.

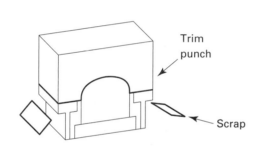

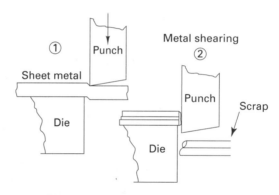

The fender is cut out of the sheet metal in the last stage using shearing processes.

Sheet metal shearing processes are like scissors cutting paper.

The sheet metal parts are welded into the body of the car.

FIGURE 1-12   The forming processes used to make a fender for a car.

accomplish the basic processes. These are shapers (and planers), drill presses, lathes, boring machines, milling machines, saws, broaches, and grinders. Most of these machine tools are capable of performing more than one of the basic machining processes. This obvious advantage has led to the development of *machining centers* specifically designed to combine many of the basic processes, plus other related processes, to be done on a single machine tool with a single workpiece setup (see Chapter 32).

Included with the machining processes are processes wherein metal is removed by chemical, electrical, electrochemical, or thermal sources. Generally speaking, these nontraditional processes have evolved to fill a specific need when conventional processes were too expensive or too slow when machining very hard materials. One of the first uses of a laser was to machine holes in ultra-high-strength metals. Lasers are being used today to drill tiny holes in turbine blades for jet engines.

In recent years a new family of processes has emerged called *rapid prototyping*. These processes produce first, or prototype, components directly from the software and specialized machines used in the computer-aided design packages. The prototypes can be field tested and modifications to the design quickly implemented. For the most part, these machines produce nonmetalic components. In contrast, the machining processes

The Machining Process
(turning on a lathe)

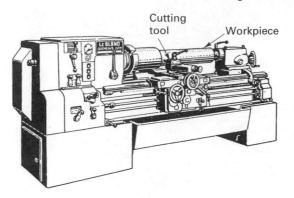

The workpiece is mounted in a machine tool (lathe) with a cutting tool.

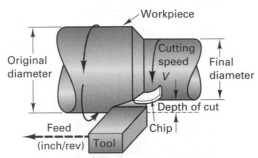

The workpiece is rotated while the tool is fed at some feed rate (inches per revolution). The desired cutting speed V determines the rpm of the workpiece.

The cutting tool interacts with the workpiece to form a chip by a shearing process. The tool shown here is an indexable carbide insert tool with a chip-breaking groove.

FIGURE 1-13 Single-point metalcutting process (turning) produces a chip while creating a new surface on the workpiece.

are recognized as having great *value-adding capability*, that is, the ability to produce components with great precision and accuracy.

Perhaps the largest collection of processes, in terms of both diversity and quantity, are the *joining processes*, which include the following:

1. Mechanical fastening
2. Soldering and brazing
3. Welding
4. Press, shrink, or snap fittings
5. Adhesive bonding
6. Assembly processes

These processes are often found in the assembly area of the plant.

Finishing processes are yet another class of processes typically employed for cleaning, removing burrs left by machining, or providing protective and/or decorative surfaces on workpieces. Surface treatments include chemical and mechanical cleaning, deburring, painting, plating, buffing, galvanizing, and anodizing.

Heat treatment is the heating and cooling of a metal for the specific purpose of altering its metallurgical and mechanical properties. Because changing and controlling these properties is so important in the processing and performance of metals, heat treatment is a very important manufacturing process. Each type of metal reacts differently to heat treatment. Consequently, a designer should know not only how a selected metal can be altered by heat treatment but, equally important, *how a selected metal will react, favorably or unfavorably, to any heating or cooling that may be incidental to the manufacturing processes.*

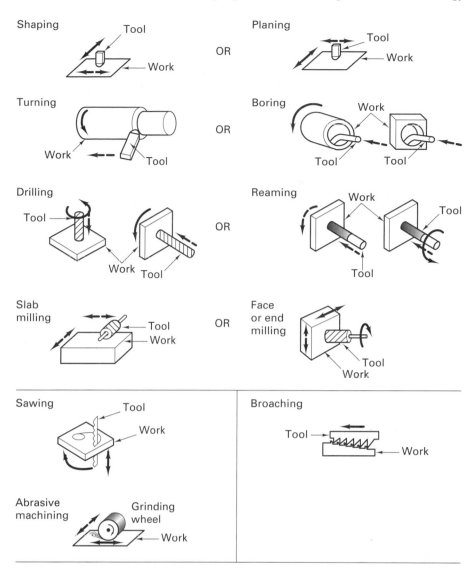

FIGURE 1-14 Schematics of the seven basic machining processes.

OTHER MANUFACTURING OPERATIONS

In addition to the processes already described, there are some other fundamental manufacturing operations that we must consider. *Inspection* determines whether the desired objectives stated by the designer in the specifications have been achieved. This activity provides feedback to design and manufacturing with regard to the process behavior. Essential to this inspection function are measurement activities.

In *testing*, a product is tried by actual function or operation or by subjection to external effects. Although a test is a form of inspection, it is often not viewed that way. In manufacturing, parts and materials are inspected for conformance to the dimensional and physical specifications, while testing may simulate the environmental or usage demands to be made on a product after it is placed in service. Complex processes may require many tests and inspections. Testing includes life-cycle tests, destructive tests, nondestructive testing to check for processing defects, wind-tunnel tests, road tests, and overload tests.

Transportation of goods in the factory is often referred to as *material handling* or *conveyance* of the goods and refers to the transporting of unfinished goods, work-in-process in the plant, and supplies to and from, between, and during manufacturing operations. Loading, positioning, and unloading are also material-handling operations. Transportation, by truck or train, is material handling between factories. Proper manufacturing system design and mechanization can reduce material handling in countless ways.

Automatic material handling is a critical part of continuous automatic manufacturing. The word *automation* is derived from automatic material handling. Material handling, a fundamental operation done by people and by conveyors and loaders, often includes positioning the workpiece within the machine by indexing, shuttle bars, slides, and clamps. In recent years, wire-guided automated guided vehicles (AGVs) and automatic storage and retrieval systems (AS/RSs) have been developed in an attempt to replace forklift trucks on the factory floor. Another form of material handling, the mechanized removal of waste (chips, trimming, and cutoffs), can be more difficult than handling the product. Chip removal must be done before a tangle of scrap chips damages tooling or creates defective workpieces.

Most texts on manufacturing processes do not mention *packaging*, yet the packaging is often the first thing the customer sees. Also, packaging often maintains the product's quality between completion and use. (Packaging is also used in electronics manufacturing to refer to placing microelectronic chips in containers for mounting on circuit boards.) Packaging can also prepare the product for delivery to the user. It varies from filling ampules with antibiotics to steel-strapping aluminum ingots into palletized loads. A product may require several packaging operations. For example, Hershey Kisses are (1) individually wrapped in foil, (2) placed in bags, (3) put into boxes, and (4) placed in shipping cartons.

Weighing, filling, sealing, and labeling are packaging operations that are highly automated in many industries. When possible, the cartons or wrapping is formed from material on rolls in the packaging machine. Packaging is a specialty combining elements of product design (styling), material handling, and quality control. Some packages cost more than their contents, for example, cosmetics and razor blades.

During *storage*, nothing happens intentionally to the product or part except the passage of time. Part or product deterioration on the shelf is called *shelf life*, meaning that items can rust, age, rot, spoil, embrittle, corrode, creep, and otherwise change in state or structure, while supposedly nothing is happening to them. Storage is detrimental, wasting the company's time and money. The best strategy is to keep the product moving with as little storage as possible. Storage during processing must be *eliminated*, not automated or computerized. Companies should avoid investing heavily in large automated systems that do not alter the bottom line. Have the outputs improved with respect to the inputs, or has storage simply increased the costs (indirectly) without improving either the quality or the throughput time?

By not storing a product, the company avoids having to (1) remember where the product is stored, (2) retrieve it, (3) worry about its deteriorating, or (4) pay storage costs. Storage is the biggest waste of all and should be eliminated at every opportunity.

UNDERSTAND YOUR PROCESS TECHNOLOGY

Understanding the process technology of the company is very important for everyone in the company. Manufacturing technology affects the design of the product and the manufacturing system, the way in which the manufacturing system can be controlled, the types of people employed, and the materials that can be processed. Table 1-4 outlines the factors that characterize a process technology. Take a process you are familiar with and think about these factors. One valid criticism of American companies is that their managers seem to have an aversion to understanding their companies' manufacturing technologies. Failure to understand the company business (i.e., its fundamental process technology) can lead to the failure of the company.

The way to overcome technological aversion is to run the process and study the technology. Only someone who has run a drill press can understand the sensitive relationship between feed rate and drill torque and thrust. All processes have these "know-how" features. Those who run the processes must be part of the decision making for the factory. The CEO who takes a vacation working on the plant floor and learning the processes will be well on the way to being the head of a successful company.

■ 1.4   PRODUCT LIFE CYCLE AND LIFE-CYCLE COST

Manufacturing systems are dynamic and change with time. There is a general, traditional relationship between a *product's life cycle* and the kind of manufacturing system used to make the product. Figure 1-15 simplifies the life cycle into these steps:

1. *Startup*. New product or new company, low volume, small company.

TABLE 1-4. Characterizing a Process Technology

Mechanics (statics and dynamics of the process)
  How does the process work?
  What are the process mechanics (statics, dynamics, friction)?
  What physically happens, and what makes it happen? (Understand the physics.)
Economics or costs
  What are the tooling costs, the engineering costs?
  Which costs are short term, which long term?
  What are the setup costs?
Time spans
  How long does it take to set up the process initially?
  What is the throughput time?
  How can these times be shortened?
  How long does it take to run a part once it is set up (cycle time)?
  What process parameters affect the cycle time?
Constraints
  What are the process limits?
  What cannot be done?
  What constrains this process (sizes, speeds, forces, volumes, power, cost)?
  What is very hard to do within an acceptable time/cost frame?
Uncertainties and process reliability
  What can go wrong?
  How can this machine fail?
  What do people worry about with this process?
  Is this a reliable, stable process?
Skills
  What operator skills are critical?
  What is not done automatically?
  How long does it take to learn to do this process?
Flexibility
  Can this process adapt easily for new parts of a new design or material?
  How does the process react to changes in part design and demand?
  What changes are easy to do?
Process capability
  What are the accuracy and precision of the process?
  What tolerances does the process meet? (What is the process capability?)
  How repeatable are those tolerances?

2. *Rapid growth.* Products become standardized and volume increases rapidly. Company's ability to meet demand stresses its capacity.

3. *Maturation.* Standard designs emerge. Process development is very important.

4. *Commodity.* Long-life, standard-of-the-industry type of product

  or

Decline. Product is slowly replaced by improved products.

    The maturation of a product in the marketplace generally leads to fewer competitors, with competition based more on price and on-time delivery than on unique product features. As the competitive focus shifts during the different stages of the product life cycle, the requirements placed on manufacturing—cost, quality, flexibility, and delivery dependability—also change. The stage of the product life cycle affects the product design stability, the length of the product development cycle, the frequency of engineering change orders, and the commonality of components, all of which have implications for manufacturing process technology.

    During the design phase of the product much of the cost of manufacturing and assembly is determined. Assembly of the product is inherently integrative as it focuses on pairs and groups of parts.

    It is crucial to achieve this integration during the design phase because about 70 percent of the life-cycle cost of a product is determined when it is designed. Design choices determine materials, fabrication methods, assembly methods, and, to a lesser degree, material-handling options, inspection techniques, and other aspects of the production system. Manufacturing engineers and internal customers can influence only a small part of the overall cost if they are presented with a finished design that does not reflect their concerns. Therefore all aspects of production should be included if product designs are to result in real function integration.

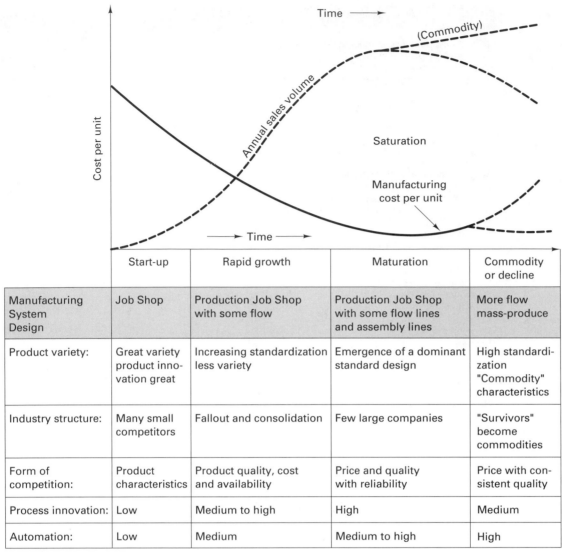

Manufacturing System Design	Job Shop	Production Job Shop with some flow	Production Job Shop with some flow lines and assembly lines	More flow mass-produce
	Start-up	Rapid growth	Maturation	Commodity or decline
Product variety:	Great variety product inno-vation great	Increasing standardization less variety	Emergence of a dominant standard design	High standardi-zation "Commodity" characteristics
Industry structure:	Many small competitors	Fallout and consolidation	Few large companies	"Survivors" become commodities
Form of competition:	Product characteristics	Product quality, cost and availability	Price and quality with reliability	Price with con-sistent quality
Process innovation:	Low	Medium to high	High	Medium
Automation:	Low	Medium	Medium to high	High

FIGURE 1-15   Product life-cycle costs change with manufacturing system designs.

Life-cycle costs include the costs of all the materials, manufacture, use, repair, and disposal of a product. Early design decisions determine about 60% of the cost, and all activities up to the start of full-scale development determine about 75%. Later deci-sions can make only minor changes to the ultimate total unless the design of the man-ufacturing system is changed.

In short, the product life-cycle concept provides a framework for thinking about the product's evolution through time and the kind of market segments that are likely to develop at various times. Life-cycle cost analysis shows that the design of the manufac-turing system determines the cost per unit, which generally decreases over time with process improvements and increased volumes. The linked-cell manufacturing system design discussed in Chapters 42 and 43 enables companies to decrease cost per unit sig-nificantly while maintaining flexibility and making smooth transitions from low-volume to high-volume manufacturing.

Low-cost manufacturing does not just happen. There is a close interdependent rela-tionship between the design of a product, the selection of materials, the selection of process-es and equipment, the design of the processes, and tooling selection and design. Each of these steps must be carefully considered, planned, and coordinated before manufacturing starts. This lead time, particularly for complicated products, may take months, or even years, and the expenditure of large amounts of money may be involved. Typically, the lead time for a completely new model of an automobile or a modern aircraft may be two to five years.

Figure 1-16 shows some of the steps involved in getting one product from the original idea stage to manufacturing. The steps are closely related to each other. For example, the

The Design / Manufacturing Process

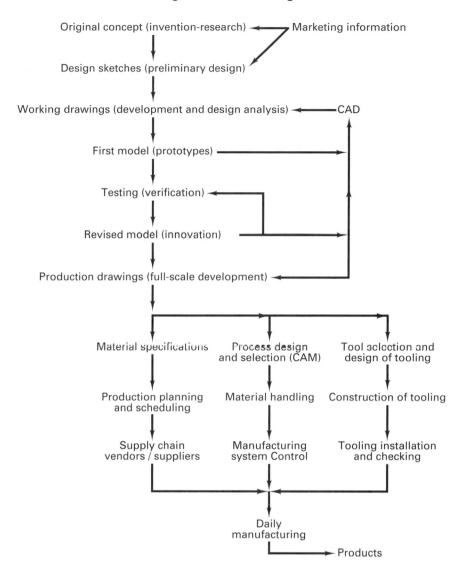

FIGURE 1-16 Traditional steps required to convert an idea into a finished product. Steps may differ depending on the complexity of the product.

design of the tooling is dependent on the design of the parts to be produced. It is often possible to simplify the tooling if certain changes are made in the design of the parts or the design of the manufacturing systems. Similarly, the material selection will affect the design of the tooling or the processes selected. Can the design be altered so that it can be produced with tooling already on hand and thus avoid the purchase of new equipment? Close coordination of all the various phases of design and manufacture is essential if economy is to result.

With the advent of computers and computer-controlled machines, the integration of the design function and the manufacturing function through the computer is a reality. This is usually called CAD/CAM (computed-aided design/computer-aided manufacturing). The key is a common database from which detailed drawings can be made for the designer and the manufacturer and from which programs can be generated to make all the tooling. In addition, extensive computer-aided testing and inspection (CATI) of the manufactured parts is taking place. There is no doubt that this trend will continue at ever-accelerating rates as computers become cheaper and smarter but at this time, the computers necessary to accomplish complete computer integrated manufacturing (CIM) are expensive and the software very complex. Implementing CIM requires a lot of manpower as well.

■ 1.5 MANUFACTURING SYSTEM DESIGN

When designing a manufacturing system, two customers must be taken into consideration: the external customer who buys the product and the internal customer who makes the product. The external customer is likely to be global and demand greater variety

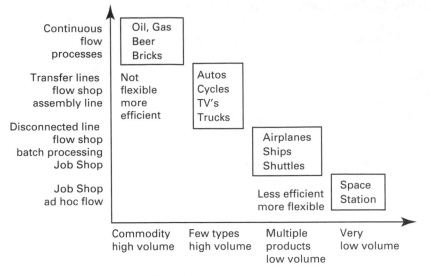

The figure to the left suggests in a general way the relationship between manufacturing systems and production volumes. The upper left represents systems with low flexibility but high efficiency compared to the lower right where volumes are low and so is efficiency. Where one is in this matrix is determined by many forces not all of which are controllable. The job of manufacturing and industrial engineers is to design and implement a system which can achieve low unit cost, superior quality with on-time delivery in a flexible way.

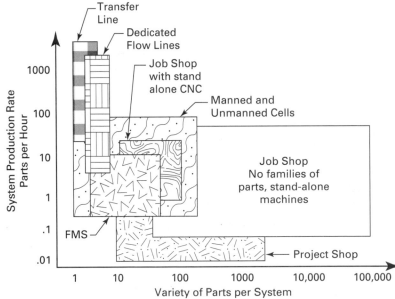

This part variety-production rate matrix shows examples of particular manufacturing systems. This matrix was developed by Black based on real factory data. Notice there is a large amount of overlap in the middle of the matrix, so the manufacturing engineer has many choices regarding which method or system to use to make his goods. This book will show the connection between the process and the manufacturing system used to produce the products, turning raw materials into finished goods.

FIGURE 1-17 Different manufacturing systems produce goods at different production rates.

with superior quality and reliability. The internal customer is often empowered to make critical decisions about how to make the products. The Toyota Motor manufacturing company is making vehicles in 25 countries. Their truck plant in Indiana has the capacity to make 150,000 vehicles per year (creating 2300 new jobs), using the Toyota production system. An appreciation of the complexity of the manufacturing system design problem is shown in Figure 1-17, where the choices in system design are reflected against volumes and variety. Clearly, there are many choices regarding which method (or system) to use to make the goods. A manufacturer never really knows how large or diverse a market will be. If a diverse and specialized market emerges, a company with a focused flow-line system may be too inflexible to meet the varying demand. If a large but homogenous market develops, a manufacturer with a flexible system may find production costs too high and the flexibility unexploitable.

NEW MANUFACTURING SYSTEMS

Manufacturing systems and production systems are discussed in Part 8. The manufacturing process technology described herein is available worldwide. Many countries have about the same level of process development when it comes to manufacturing technology. Much of the technology existing in the world today was developed in the United States. Japan and more recently Taiwan, Korea, and China are making great inroads into American markets, particularly in the automotive and electronics industries. What many people have failed to recognize is that many Japanese companies have developed and promoted a totally different kind of manufacturing production system. In future years, this new system, based on linked cells, will take its place with the Taylor system of scientific management and the Ford system for mass production (discussed in Chapters 41 through 43). The original working model for this new system is the Toyota Motor Company. The system is known worldwide as the Toyota production system or just-in-time (JIT) manufacturing.

Many American companies have successfully adopted some version of the Toyota system. The experience of dozens of these companies is amalgamated into 10 steps, which, if followed, can make any company a factory with a future. The key steps to developing the new manufacturing system are presented in Chapter 43.

For the lean production to work, 100% good units flow rhythmically to subsequent processes without interruption. In order to accomplish this, an integrated quality control (IQC) program has to be developed. The responsibility for quality has been given to manufacturing. All the employees are inspectors and are empowered to make it right the first time. There is a companywide attitude toward constant quality improvement. Make quality easy to see, stop the line when something goes wrong, and inspect things 100% if necessary to prevent defects from occurring. The results of this system are astonishing, in terms of quality, low cost, and on-time delivery of goods to the customer.

The Japanese became world-class competitors by developing superior design and process technology. Now, Japan is concentrating on new product innovation with the emphasis on high-value-added products. They are spending as much as 14% of sales dollars for research and development (R&D). The typical American company spends less than 5% for research and development. Furthermore, many Japanese firms and firms in other countries as well are now forming joint R&D ventures with U.S. companies, so the workings of the new manufacturing system is diffusing around the world.

The most important factor in economical, and successful, manufacturing is the manner in which the resources—labor, materials, and capital—are organized and managed so as to provide effective coordination, responsibility, and control. Part of the success of lean production can be attributed to a different management approach. This approach is characterized by a holistic attitude toward people and includes:

1. Working in teams or groups, improving communications
2. Consensus decision making by management
3. Decision making at the lowest possible level with strong leadership at the top
4. Mutual trust, integrity, and loyalty between workers and management
5. Innovative methods of pay for all employees, including incentive pay in the form of bonuses
6. Stable (even lifetime) employment
7. Large pool of part-time temporary workers (contract workers)

There are many companies in the United States that employ some or all of these elements and, obviously, there are many different ways a company can be organized and managed.

The real secret of successful manufacturing lies in designing a simplified system that everyone understands in terms of how it works and how it is controlled, with the decision making placed at the correct level. The engineers also must possess a broad fundamental knowledge of design, metallurgy, processing, economics, accounting, and human relations. In the manufacturing game, low-cost mass production is the result of teamwork within an integrated manufacturing production system. This is the key to producing superior quality at less cost with on-time delivery.

■ KEY WORDS

assembling	feedback	inspection	machine tool	molding	project shop
casting	filing jig	job	machining	numerical control	shearing
consumer goods	flow shop	job shop	manufacturing cost	operation	station
continuous process	forming	joining	manufacturing engineer	producer goods	tooling
design engineer	group technology	lean production	manufacturing process	product life cycle	tools
fabricating	heat treatment	linked-cell shop	manufacturing system	production system	treatments

■ REVIEW QUESTIONS

1. What role does manufacturing play relative to the standard of living of a country?
2. Are not all goods really consumer goods, depending on how you define the customer? Discuss.
3. Give examples of a job shop, flow shop, and project shop.
4. How does a system differ from a process? From a machine tool? From a job? From an operation?
5. Is a cutting tool the same thing as a machine tool?
6. What are the major classifications of basic manufacturing processes?
7. Why would it be advantageous if casting could be used to produce a complex-shaped part to be made from a hard-to-machine metal?
8. In the lost-foam casting process, what happens to the foam?
9. In making a gold medal, what do we mean by "relief image" cut into the die?
10. How is a railroad station like a job station?
11. Since no work is being done on a part when it is in storage, it does not cost you anything. True or false? Explain!
12. What forming processes are used to make a paper clip?
13. Which manufacturing system best describes your college or university? Analogize the elements in this service system to those in the manufacturing system.
14. It is acknowledged that chip-type machining is basically an inefficient process. Yet, it is probably used more than any other to produce desired shapes. Why?
15. In Figure 1-1, what mathematical equation would you use to model the data?
16. In a modern safety razor with two or three blades that sells for $1.00, what do you think the cost of the blades might be?
17. List three purposes of packaging operations.
18. Assembly is defined as "the putting together of all the different parts to make a complete machine." Think of (and describe) an assembly process. Is making a club sandwich an assembly process? What about building a house?
19. What are the physical elements in a manufacturing system?
20. In the production system, who usually figures out how to make the product?
21. In Figure 1-6, which of the physical elements are missing from this schematic? Compare to Figure 1-8.
22. Characterize the process of squeezing toothpaste from a tube (extrusion of toothpaste) using Table 1-4 as a guideline. See Chapters 17 and 18 for assistance.
23. What are the major process steps in the assembly of an automobile?
24. What difficulties might result if the step "production planning and scheduling" were omitted from the procedure shown in Figure 1-16, assuming a job-shop MS?
25. A company is considering making automobile bumpers from aluminum instead of from steel. List some of the factors it would have to consider in arriving at its decision. Many companies are critically examining the relationship of product design to manufacturing and assembly. Why do they call this concurrent engineering?
26. It has been said that low-cost products are more likely to be more carefully designed than high-priced items. Do you think this is true? Why or why not?
27. In a typical metal-cutting job using a lathe, which operations add no value to the part? Propietary processes are closely held or guarded company secrets. The chemical makeup of a lubricant for an extrusion process is a good example. Give another example of a proprietary process.
28. If the rolls for the cold-rolling mill that produce the sheet metal used in your car cost $300,000 (see Figure 1-12), how is it that your car can still cost less than $10,000?
29. Make a list of service production systems, giving an example of each and then explaining the fundamental difference between a SPS and an MPS. Table 1-3 will help you get started.
30. In the process of buying a calf, raising it to a cow, and disassembling it into "cuts" of meat for sale, where is "the value added"?
31. What kind of process is hot ISO static pressing?
32. In view of Figure 1-2, who really determines the selling price per unit?
33. What costs make up manufacturing cost (sometimes called factory cost)?
34. Look at Figures 1-10 and 1-14 and see if you can determine the machining process used to make the pinion.
35. What are major phases of a product life cycle?
36. What phases are included in life-cycle costing not shown in Figure 1-15?
37. How many different manufacturing systems might be used if you had a component with an annual projected sales of 16,000 parts per year with 10 to 12 different models (varieties)?

■ PROBLEMS

1. The Toyota Truck plant in Indiana produces 150,000 trucks per year. The plant runs one 8-hour shift, 300 days per year, and makes 500 trucks per day. About 1300 people work on the final assembly line. Each car has about 20 labor hours per car in it.
   a. Assuming the truck sells for $16,000 and workers earn $30 per hour in wages and benefits, what percentage of the cost of the truck is in direct labor?
   b. What is the production rate of the final assembly line?

   Solution: (a) $16,000 retail price
   $16,000 × 40% = $8,400 manufacturing cost
   20 hours × $30 per hour = $600 direct labor
   600/8400 ≅ 5% of manufacturing cost is direct labor

   Solution: (b) 500 trucks per 480 minute day or about 1 truck/minute or 60 trucks per hour
   Actually, Toyota uses 55 seconds per truck as the cycle time or takt time.

2. Suppose you wanted to redesign a stapler to have fewer components. (You should be able to find a stapler at a local discount store.) How much did it cost? How many parts does it have? Make up a "new parts" list and indicate which parts would have to be redesigned and which parts would be eliminated. Estimate the manufacturing cost of the stapler assuming that manufacturing costs are 40% of the selling price. What are the disadvantages of your new stapler design versus the old stapler?

PROPERTIES OF MATERIALS

■ 2.1 INTRODUCTION

When selecting a material for an engineering application, a primary concern is to assure that its properties will be adequate for the anticipated operating conditions. The various requirements of each part or component must first be estimated or determined. These may include mechanical characteristics (strength, rigidity, resistance to fracture, or the ability to withstand vibrations or impacts) and physical characteristics (weight, electrical properties, or appearance), as well as features relating to the service environment (ability to operate under extremes of temperature or resist corrosion). The selection of an appropriate engineering material is then based on a comparison of the established design requirements and the tabulated or recorded results that describe how common materials responded to various standardized tests. Test data are usually readily available, but it is important that they be used properly. It is important to consider which of the evaluated properties are significant, how the test values were determined, and what restrictions or limitations should be placed on their use. Only by being familiar with the various test procedures, their capabilities, and their limitations can one determine if the data are applicable to any particular problem.

METALLIC AND NONMETALLIC MATERIALS

Perhaps the most common classifying distinction among engineering materials is whether the material is metallic or nonmetallic. The common *metallic* materials include iron, copper, aluminum, magnesium, nickel, titanium, lead, tin, and zinc, as well as the alloys of these metals, such as steel, brass, and bronze. They possess the metallic properties of luster, high thermal conductivity, and high electrical conductivity; they are relatively ductile; and some have good magnetic properties. Some common *nonmetals* are wood, brick, concrete, glass, rubber, and plastics. Their properties vary widely, but they generally tend to be less ductile, weaker, and less dense than the metals and have poor electrical and thermal conductivities.

Although metals have traditionally been the more important of the two groups, the nonmetallic group has made great strides and new nonmetallic materials are continuously being developed. Advanced ceramics, composite materials, and engineered plastics are receiving considerable attention. In many cases, metals and nonmetals are viewed as competing materials, with selection being based on how well each is capable of providing the required properties. Where both perform adequately, total cost often becomes the deciding factor, where total cost includes both the cost of the material and the cost of fabricating the desired component.

PHYSICAL AND MECHANICAL PROPERTIES

A common means of distinguishing one material from another is through their *physical properties*. These include characteristics such as density (weight); melting point; optical properties (transparency, opaqueness, or color); the thermal properties of specific heat, coefficient of thermal expansion, and thermal conductivity; electrical conductivity; and magnetic properties. In some cases, physical properties are of prime importance when selecting a material, and several will be discussed in more detail near the end of this chapter.

More often, however, material selection is dominated by the properties that describe how a material responds to applied loads (or forces). These *mechanical properties* are usually determined by subjecting prepared specimens to standard laboratory tests. When using the results, however, it is important to remember that they apply only to the specific test conditions that were employed. Since the actual service conditions of engineered products rarely duplicate the conditions of laboratory testing, considerable caution should be exercised.

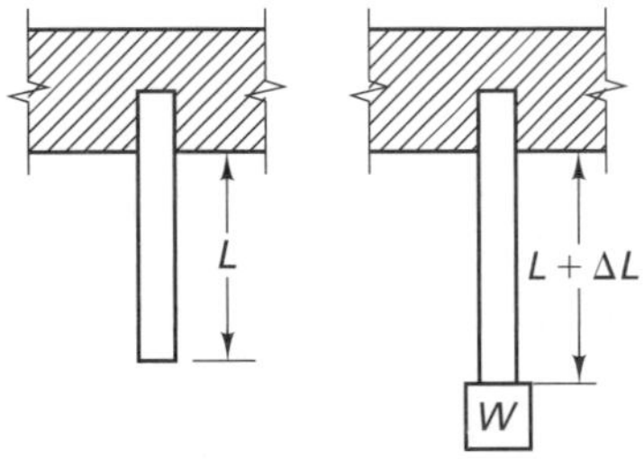

STRESS AND STRAIN

When a force or load is applied to a component, the material is deformed or distorted (*strained*), and internal reactive forces (*stresses*) are transmitted through the material. For example, if a weight, W, is suspended from a bar of uniform cross section, as in Figure 2-1, the bar will elongate by an amount ΔL. For a given weight, the magnitude of the *elongation*, ΔL, depends on the original length of the bar. The amount of elongation for each unit length, expressed as $e = \Delta L/L$, is called the *unit strain*. Although the ratio is that of a length to another length and is therefore dimensionless, it is usually expressed in terms of millimeters per meter, inches per inch, or simply as a percentage.

Application of the force also produces reactive stresses, which serve to transmit the load through the bar and on to its supports. *Stress* is defined as the force or load being transmitted divided by the cross-sectional area transmitting the load. Thus, in Figure 2-1, the stress is $S = W/A$, where A is the cross-sectional area of the supporting bar. Stress is normally expressed in megapascals (in SI units) or pounds per square inch (in the English system.)

In Figure 2-1, the weight tends to stretch or lengthen the bar, so the strain is known as a *tensile strain* and the stress as a *tensile stress*. Other types of loadings produce other types of stresses and strains (Figure 2-2). Compressive forces tend to shorten the material and produce *compressive stresses and strains*. *Shear stresses and strains* result when two forces acting on a body are offset with respect to one another.

FIGURE 2-1   Tension loading and resultant elongation.

FIGURE 2-2   Examples of tension, compression, and shear loading, and their response.

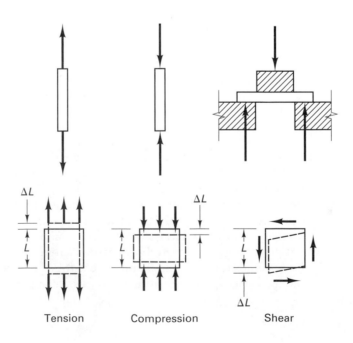

Tension   Compression   Shear

■ 2.2 STATIC PROPERTIES

When the forces that are applied to a material are constant, or nearly so, they are said to be *static*. Since static loadings are observed in many engineering applications, it is important to characterize the behavior of materials under such conditions. Consequently, a number of standardized tests have been developed to determine the *static properties* of engineering materials. The results of these tests can then be used in the selection of materials, provided that the service conditions are sufficiently similar to those of testing. Even when the service conditions differ, the results of these tests can often be used to qualitatively rate and compare the various materials.

TENSILE TEST

The most common of the static tests is the *uniaxial tensile test*. A standard specimen is loaded in tension in a testing machine like the one shown in Figure 2-3. The standard specimens ensure meaningful and reproducible results, and are designed to produce uniform uniaxial tension in the central portion of the specimen while ensuring reduced stresses in the sections that are gripped. Figure 2-4 shows two of the most common specimen designs.

Strength Properties.   A load, W, is applied and measured by the testing machine, while the elongation (ΔL) or stretch over a specified length (*gage length*) is simultaneously monitored. A plot of the coordinated load–elongation data produces a curve similar to that of Figure 2-5. Since the loads will differ for different-size specimens and the elongations will vary with different gage lengths, it is important to remove these geometric or size effects if we are to produce data that are characteristic of a given material (and not a particular specimen). If the load is divided by the *original* cross-sectional area and the elongation is divided by the *original* gage length, the size effects are eliminated and the resulting plot becomes known as an *engineering stress-engineering strain curve* (Figure 2-5). This is simply a load–elongation curve with the scales of both axes modified to remove the effects of specimen size.

   In Figure 2-5 it can be noted that the initial response is linear. Up to a certain stress, the strain and stress are directly proportional to one another. The stress at which

FIGURE 2-3   (a) Hydraulic universal (tension and compression) testing machine; (b) schematic of the load frame showing how motion of the darkened yoke can produce tension or compression with respect to the stationary (white) crosspiece. *(Courtesy of Satec Systems, Inc., Grove City, Pa.)*

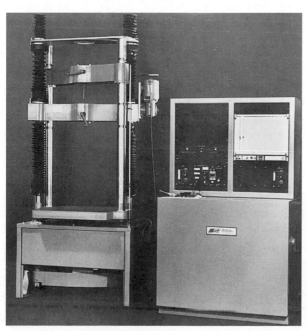

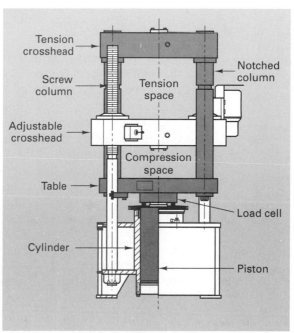

(a)   (b)

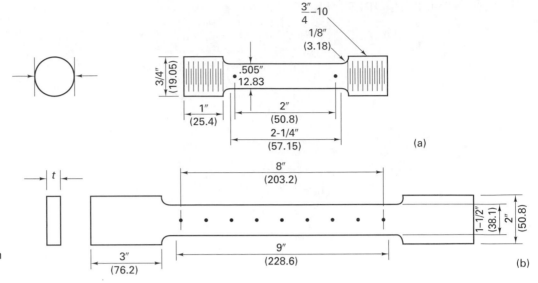

FIGURE 2-4 Two common types of standard tensile test specimens: (a) round; (b) flat. Dimensions are in inches, with millimeters in parentheses.

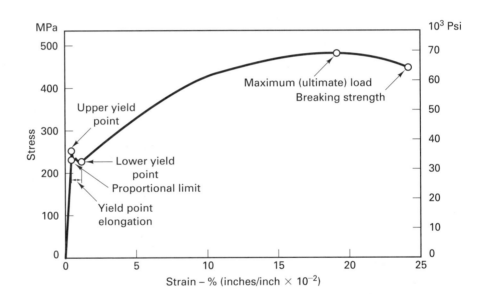

FIGURE 2-5 Engineering stress-strain diagram for a low-carbon steel.

this proportionality ceases to exist is known as the *proportional limit*. Below this value, the material obeys *Hooke's law*, which states that the strain is directly proportional to the stress. The proportionality constant, or ratio of stress to strain in this region, is known as *Young's modulus* or the *modulus of elasticity*. This is an inherent property of a given material and is of considerable engineering importance. As a measure of stiffness, it indicates the ability of a material to resist deflection or stretching when loaded and is commonly designated by the symbol E.

Up to a certain stress, if the load is removed, the specimen will return to its original length. The response is elastic, and the uppermost stress for which this behavior is observed is known as the *elastic limit*. For most materials the elastic limit and proportional limit are almost identical, with the elastic limit being slightly higher. Neither quantity should be assigned great engineering significance, however, because the determined values are often dependent on the sensitivity of the test equipment.

The amount of energy that a material can absorb while in the elastic range is called the *resilience*. The area under the load–elongation curve up to the elastic limit is the product of a force and a distance, and is equivalent to the energy absorbed by the specimen. This energy is potential energy and is released whenever the member is unloaded. If we divide the load by the original area to produce engineering stress, and the elongation by the gage length to produce engineering strain, the area beneath the elastic region of an engineering stress-strain curve corresponds to an energy per unit volume, and is known as the *modulus of resilience*.

Elongation beyond the elastic limit becomes unrecoverable and is known as *plastic deformation*. When the load is removed, the specimen retains a permanent change in shape. For most components, the onset of plastic flow represents failure, since the part dimensions will now be outside of allowable tolerances. In manufacturing, however, plastic deformation is often used to produce a desired shape, and the applied stresses must be sufficient to induce the required plastic flow. Thus, permanent deformation may be either desirable or undesirable, and it is important to determine the conditions where elastic behavior transitions into plastic flow.

When the elastic limit is exceeded, increases in strain no longer require proportionate increases in stress. For some materials, a stress value may be reached where additional strain occurs without any stress increase. This point is known as the *yield point*, or *yield-point stress*. For low-carbon steels (Figure 2-5) two distinct points are significant. The highest stress preceding extensive strain is known as the *upper yield point*, and the lower, relatively constant, "run-out" value is known as the *lower yield point*. The lower value is the one that usually appears in tabulated data.

Most materials, however, do not have a well-defined yield point, but exhibit a stress-strain curve of the form shown in Figure 2-6. For these materials, the elastic-to-plastic transition is not distinct, and must be defined through use of the *offset yield strength*, the value of the stress that will produce a given, but tolerable, amount of permanent strain. For most components, the amount of offset strain is set at 0.2%, but values of 0.1% or even 0.02% may be specified when small amounts of plastic deformation could lead to component failure. The offset yield strength is determined by drawing a line parallel to the elastic line, but displaced by the offset strain, and reporting the stress where this line intersects the actual stress-strain curve (Figure 2-6). The intersection value is reproducible and is independent of equipment sensitivity. However, it is meaningless unless it is reported in conjunction with the amount of offset strain used in its determination. If the applied stresses are then kept below the 0.2% offset yield strength of the material, the user can be guaranteed that any resulting plastic deformation will be less than 0.2% of the original dimension.

As plastic deformation is continued, the curve indicates that the material is acquiring an increased ability to bear load. If load-bearing ability is equal to material strength times cross-sectional area, and the area is decreasing as we stretch the specimen, the material must be getting stronger. (The mechanism for this phenomenon will be discussed in Chapter 3, and we will learn that the strength of a metal continues to increase with increased deformation.) During the plastic deformation portion of a tensile test, the weakest location of the specimen undergoes deformation and becomes stronger. Since it is no longer the weakest location, another location then deforms. As a consequence of this constant redistribution of deformation, the test specimen maintains its original cylindrical or rectangular geometry. As plastic deformation progresses, however, the additional increments of strength decrease in magnitude, and a point is reached when the decrease in area cancels or dominates the increase in strength. When this occurs, the load-bearing ability peaks, and the force required to continue straining the specimen begins to decrease (as shown in the curve of Figure 2-5). The stress at which the load-bearing ability peaks is known as the *ultimate strength, tensile strength*, or *ultimate tensile strength* of the material. The weakest location in the test specimen at that time continues to be the weakest location by virtue of the decrease in area, and further deformation becomes localized. This localized reduction in cross-sectional area is known as *necking*, and is shown in Figure 2-7.

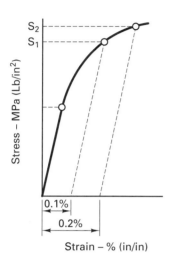

FIGURE 2-6 Stress-strain diagram for a material not having a well-defined yield point, showing the offset method for determining yield strength. S_1 is the 0.1% offset yield strength; S_2 is the 0.2% offset yield strength.

FIGURE 2-7 Standard 0.505-in.-diameter tensile specimen showing a necked region developed prior to failure.

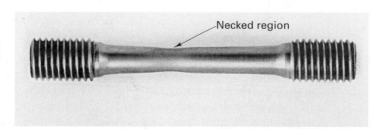

Necked region

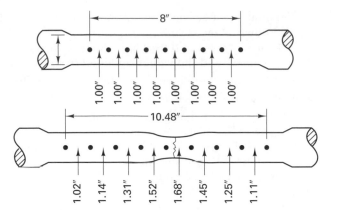

FIGURE 2-8   Final elongation in various segments of a tensile test specimen: (a) original geometry; (b) shape after fracture.

If the straining is continued, the tensile specimen will ultimately fracture. The stress at which fracture occurs is known as the *breaking strength* or *fracture strength*. For relatively ductile materials, the breaking strength is less than the ultimate tensile strength, and necking precedes fracture. For a brittle material, fracture usually terminates the stress–strain curve before necking, and possibly before the onset of plastic flow.

Ductility and Brittleness.   When evaluating the suitability of a material for certain manufacturing processes, the amount of plasticity that precedes fracture, or the *ductility*, can often be a significant property. For metal deformation processes, the greater the ductility, the more a material can be deformed without failure.

One of the simplest ways to evaluate ductility is to determine the *percent elongation* of a tensile test specimen at the time of fracture. As shown in Figure 2-8, however, ductile materials do not elongate uniformly when loaded beyond necking. If the percent change in the entire 8-in. gage length of Figure 2-8 is computed, the elongation is 31%. However, if only the center 2-in. segment is considered, the elongation of that portion is 60%. A valid comparison of material behavior, therefore, requires similar specimens with the same standard gage length.

In many cases, material "failure" is defined as the onset of localized deformation or necking. Consider a sheet of metal being formed into an automobile body panel. The operation must be performed in such a way as to maintain uniform sheet thickness if we are to assure uniform strength and corrosion resistance in the final panel. For this application, a more meaningful measure of material ductility would be the *uniform elongation* or *percent elongation prior to the onset of necking*. This value can be determined by constructing a line parallel to the elastic portion of the diagram, passing through the point of highest force or stress. The intercept where the line crosses the strain axis denotes the available uniform elongation. Since the additional deformation that occurs after necking is not considered, uniform elongation is always less than the total elongation at fracture (the generally reported elongation value).

Yet another measure of ductility is the *percent reduction in area* that occurs in the necked region of the specimen. This can be computed as

$$\text{R.A.} = \frac{A_o - A_f}{A_o} \times 100\%$$

where A_o is the original cross-sectional area and A_f is the smallest area in the necked region. Percent reduction in area, therefore, can range from 0% (brittle) to 100% (extremely plastic).

When materials fail with little or no ductility, they are said to be brittle. Brittleness, therefore, is simply the lack of significant ductility and should not be confused with a lack of strength.

Toughness.   *Toughness*, or *modulus of toughness*, is defined as the work per unit volume required to fracture a material. The tensile test can provide a measure of this property, since it corresponds to the total area under the stress–strain curve from initiation to fracture. Caution should be exercised when using toughness data, however, because the values can vary markedly with different conditions of testing. As will be seen later, variations in temperature or the rate of load application can significantly alter the stress-strain curve, and hence the toughness. Toughness is commonly associated with impact or shock loadings, and the values obtained from dynamic impact tests often fail to correlate with those obtained from the relatively static tensile test.

True Stress–True Strain Curves. The stress-strain curve in Figure 2-5 is a plot of *engineering stress*, S, versus *engineering strain*, e, where S is computed as the applied load (W) divided by the original cross-sectional area (A_o) and e is the elongation, ΔL, divided by the original gage length, L_o. As discussed previously, and illustrated in Figures 2-7 and 2-8, the cross section of the test specimen changes as the test proceeds, first uniformly and then in a nonuniform manner after necking begins. The actual stress within the specimen should be based on the instantaneous cross-sectional area, not the original, and will therefore be greater than the engineering stress plotted in Figure 2-5. *True stress*, σ, can be computed by taking simultaneous readings of the load and the minimum specimen diameter. The actual area (A) can then be computed, and true stress can be determined as

$$\sigma = W/A$$

The determination of *true strain* is somewhat more complex. In place of the change in length divided by the original length that was used to compute engineering strain, true strain is defined as the summation of the incremental strains that occur throughout the test. Thus, for a specimen that has been stretched from length L_o to length L the *true*, *natural*, or *logarithmic strain*, ε, would be:

$$\varepsilon = \int_{L_o}^{L} \frac{d\ell}{\ell} = \ln \frac{L}{L_o} = 2 \ln \frac{D_o}{D}$$

The last equality makes use of the relationship for cylindrical specimens that

$$\frac{L}{L_o} = \frac{A_o}{A} = \frac{D_o^2}{D^2}$$

and applies only up to the onset of necking.

Figure 2-9 shows the type of curve that results when the data from a uniaxial tensile test are transferred to the form of true stress versus true strain. Since the true stress is a measure of the material strength at any point during the test, it will continue to rise even after necking. Data beyond the onset of necking should be used with extreme caution, however, because the geometry of the neck transforms the stress state from uniaxial tension (stretching in one direction with compensating contractions in the other two) to triaxial tension, in which the material is stretched or restrained in all three directions. Because of the triaxial tension, voids or cracks (Figure 2-10) tend to form in the necked region and serve as a precursor to final fracture. Measurements of the external diameter no longer reflect the true load-bearing area, and the data are further distorted.

FIGURE 2-9 True stress–true strain curve for an engineering metal.

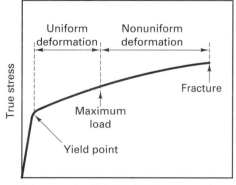

FIGURE 2-10 Section of a tensile test specimen stopped just prior to failure, showing a crack already started in the necked region. *(Courtesy of E. R. Parker.)*

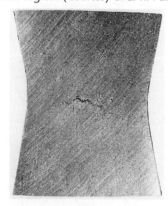

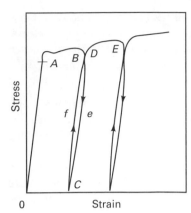

FIGURE 2-11 Stress-strain diagram obtained by unloading and reloading a specimen.

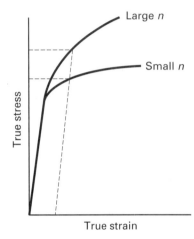

FIGURE 2-12 True stress–true strain curves for metals with large and small strain hardening. Metals with larger *n* values experience larger amounts of strengthening for a given strain.

Strain Hardening and the Strain-Hardening Exponent.   Figure 2-11 is a true stress–true strain diagram, which has been modified to show how a ductile metal (such as steel) will behave when subjected to slow loading and unloading. Loading and unloading within the elastic region will result in simply cycling up and down the linear portion of the curve between points *O* and *A*. However, if the initial loading is carried through point *B* (in the plastic region), unloading will follow the path *BeC*, which is approximately parallel to the line *OA*, and the specimen will exhibit a permanent elongation of the amount *OC*. Upon reloading from point *C*, elastic behavior is again observed as the stress follows the line *CfD*, a slightly different path from that of unloading. Point *D* is now the yield point or yield stress for the material in its partially deformed state. A comparison of points *A* and *D* reveals that plastic deformation has made the material stronger. If the test were again interrupted at point *E*, we would find a new, even higher-yield stress. Thus, within the region of plastic deformation, each of the points along the true stress–true strain curve represents the yield stress for the corresponding value of strain.

When metals are plastically deformed, they become harder and stronger, a phenomenon known as *strain hardening*. If a stress is capable of producing plastic deformation, an even greater stress will be required to continue the flow. In Chapter 3 we will discuss the atomic-scale features that are responsible for this phenomenon.

Various materials strain harden at different rates, that is, for a given amount of deformation different materials will exhibit different increases in strength. One method of describing this behavior is to mathematically fit the plastic region of the true stress–true strain curve to the equation

$$\sigma = K\varepsilon^n$$

and determine the best-fit value of *n*, the *strain-hardening exponent*.[1] As shown in Figure 2-12, a material with a high value of *n* will have a significant increase in material strength with a small amount of deformation. A material with a small *n* value will show little change in strength with plastic deformation.

Damping Capacity.   In Figure 2-11 the unloading and reloading of the specimen follow slightly different paths. The area between the two curves is proportional to the amount of energy that is converted from mechanical form to heat, and is therefore absorbed by the material. When this area is large, the material is said to exhibit good damping capacity and is able to absorb mechanical vibrations or damp them out quickly. This is an important property in applications such as crankshafts and machinery bases. Gray cast iron is used in many applications because of its high damping capacity. Materials with low damping capacity, such as brass and steel, readily transmit sound and vibrations.

COMPRESSION TESTS

When a material is subjected to compressive loadings, the relationships between stress and strain are similar to those for a tension test. Up to a certain value of stress, the material behaves elastically. Beyond this value, plastic flow occurs. In general, however, a compression test is more difficult to conduct than a standard tensile test. Test specimens must have larger cross-sectional areas to resist bending or buckling. As deformation proceeds, the material strengthens by strain hardening and the cross section of the specimen increases, combining to produce a substantial increase in required load. Friction between the testing machine surfaces and the ends of the test specimen will alter the results if not properly considered. The type of service for which the material is intended, however, should be the primary factor in determining whether the testing should be performed in tension or compression.

HARDNESS TESTING

Hardness is a very important but hard-to-define property of engineering materials. A number of tests have been developed using various phenomena. The most common of

---

[1]Taking the logarithm of both sides of the equation yields $\log \sigma = \log K + n \log \varepsilon$. This is the same form as the equation $y = mx + b$, the equation for a straight line with slope *m* and intercept *b*. Therefore, if the true stress–true strain data were plotted on a log-log scale with stress on the *y*-axis and strain on the *x*-axis, the slope of the data in the plastic region would be *n*.

the hardness tests are based on resistance to permanent deformation (indentation) under static or dynamic loading. Other tests evaluate resistance to scratching, energy absorption under impact loading, wear resistance, or resistance to cutting or drilling. Since these phenomena are not the same, the results of the various tests often do not correlate with one another. Caution should be exercised to assure that the selected test clearly evaluates the phenomena of interest.

Brinell Hardness Test.   The *Brinell hardness test* was one of the earliest accepted methods of measuring hardness. A tungsten carbide or hardened steel ball 10 mm. in diameter is pressed into the flat surface of a material by a standard load of 500, 1500, or 3000 kg, and the load is maintained for 10 to 15 seconds to permit the full amount of plastic deformation to occur. The load and ball are then removed, and the diameter of the resulting spherical indentation (usually in the range of 2 to 5 mm) is measured using a special grid or traveling microscope. The *Brinell hardness number* (BHN) is equal to the load divided by the surface area of the spherical indentation when the units are expressed as kilograms per square millimeter.

In actual practice, however, the Brinell hardness number is determined from tables that correlate the Brinell number with the diameter of the indentation produced under the various loads. Figure 2-13 shows a typical Brinell tester, along with a schematic of the testing procedure. Portable testers are available for use on pieces that are too large to be brought to a benchtop machine.

The Brinell test measures hardness over a relatively large area and is somewhat indifferent to small-scale variations in structure. It is relatively simple and easy to conduct, and is used extensively on irons and steels. On the negative side, however, the Brinell test has the following limitations:

1. It cannot be used on very hard or very soft materials.
2. The results may not be valid for thin specimens. It is best if the thickness of material is at least 10 times the depth of the indentation. Some standards specify the minimum hardnesses for which the tests on thin specimens will be considered valid.
3. The test is not valid for case-hardened surfaces.
4. The test must be conducted far enough from the edge of the material so that no edge bulging occurs.

FIGURE 2-13   (a) Brinell hardness tester; (b) Brinell test sequence showing loading and measurement of the indentation under magnification with a scale calibrated in millimeters. *[(a) Courtesy of Wilson Instruments Division, Instron Corp.]*

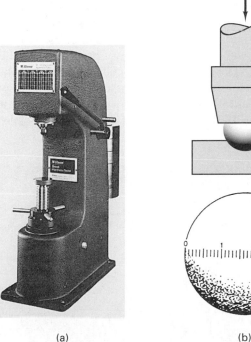

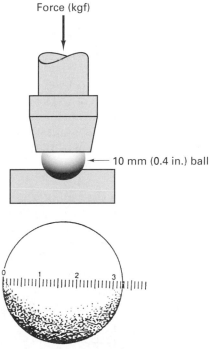

Force (kgf)

10 mm (0.4 in.) ball

(a)   (b)

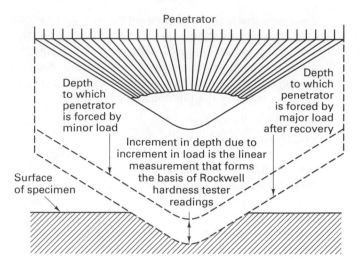

Depth to which penetrator is forced by minor load

Depth to which penetrator is forced by major load after recovery

Increment in depth due to increment in load is the linear measurement that forms the basis of Rockwell hardness tester readings

Surface of specimen

FIGURE 2-14 Operating principle of the Rockwell hardness tester. *(Courtesy of Wilson Instruments Division, Instron Corp.)*

FIGURE 2-15 Rockwell hardness tester with digital readout. *(Courtesy of MTI Corporation, Aurora, IL.)*

5. The substantial indentation may be objectionable on finished parts.

6. The edge or rim of the indentation may not be clearly defined or may be difficult to see.

The Rockwell Test.   The widely used *Rockwell hardness test* is similar to the Brinell test, with the hardness value again being determined through an indentation produced under a static load. Figure 2-14 shows the exact nature of the Rockwell test. A small indenter, either a small-diameter steel ball or a diamond-tipped 120° cone called a *brale*, is first seated firmly against the material by the application of a small but specified "minor" load. This causes a very slight elastic penetration into the surface. The indicator on the screen of the tester, like the one shown in Figure 2-15, is then set to zero, and a "major" load is applied to the indenter to produce a deeper penetration (i.e., plastic deformation). When the indicating pointer has come to rest, the major load is removed. With the minor load still applied, the tester now indicates the Rockwell hardness number on either a dial gauge or digital display. This number is really an indication of the *depth* of plastic or permanent penetration that was produced by the major load, with each unit representing a penetration depth of 2 μm.

Different combinations of major loads and indenters are available and are used for materials with various levels of strength. Table 2-1 provides a partial listing of the Rockwell scales and typical materials for which they are used. Because of the different scales, a Rockwell hardness number must be accompanied by a letter indicating the particular combination of load and indenter used in its determination. The notation $R_C 60$ (or Rockwell C 60) indicates that a brale indenter was used in combination with a major load of 150 kg, and a reading of 60 was obtained. The B and C scales are used more extensively than the others, B being common for copper and aluminum and C for steels.

Rockwell tests should not be conducted on thin materials (typically less than 1.5 mm or $\frac{1}{16}$-in.), on rough surfaces, or on materials that are not homogeneous, such as

TABLE 2-1.	Some Common Rockwell Hardness Tests		
Scale Symbol	Penetrator	Load (kg)	Typical Materials
A	Brale	60	Cemented carbides, thin steel, shallow case-hardened steel
B	$\frac{1}{16}$-in. ball	100	Copper alloys, soft steels, aluminum alloys, malleable iron
C	Brale	150	Steel, hard cast irons, titanium, deep case-hardened steel
D	Brale	100	Thin steel, medium case-hardened steel
E	$\frac{1}{8}$-in. ball	100	Cast iron, aluminum, magnesium
F	$\frac{1}{16}$-in. ball	60	Annealed coppers, thin soft sheet metals
G	$\frac{1}{16}$-in. ball	150	Hard copper alloys, malleable irons
H	$\frac{1}{8}$-in. ball	60	Aluminum, zinc, lead

gray cast iron. Because of the small size of the indentation, variations in roughness, composition, or structure can greatly influence the results. For thin materials, or where a very shallow indentation is desired (as in the evaluation of surface-hardening treatments such as nitriding or carburizing), the *Rockwell superficial hardness test* is preferred. Operating on the same Rockwell principle, this test employs smaller major and minor loads and uses a more sensitive depth-measuring device.

In comparison with the Brinell test, the Rockwell test offers the attractive advantage of direct readings in a single step. Because it requires little (if any) surface preparation and can be conducted quite rapidly, it is often used to determine if an incoming product meets specification, assure that a heat treatment was properly done, or simply monitor the quality of products undergoing mass production. It has the additional advantage of producing a small indentation that can be easily concealed on the finished product or easily removed in a later operation.

Vickers Hardness Test. The *Vickers hardness test* is also similar to the Brinell test, but uses a square-based 136° diamond pyramid as the indenter. Like the Brinell value, the Vickers hardness number is also defined as load divided by the surface area of the indentation expressed in units of kilograms per square millimeter. The advantages of the Vickers approach include the increased accuracy in determining the diagonal of a square impression as opposed to the diameter of a circle and the assurance that even light loads will produce some plastic deformation. The use of diamond as the indenter material enables the test to evaluate any material and effectively places the hardness of all materials on a single scale.

Like the other indentation (or penetration) methods, the Vickers test has a number of attractive features: (1) it is simple to conduct, (2) little time is involved, (3) little surface preparation is required, (4) the test can be done on location, (5) it is relatively inexpensive, and (6) it provides results that can be used to evaluate material strength or assess product quality.

Microhardness Tests. Various microhardness tests have been developed for applications where it is necessary to determine the hardness of a very precise area of material, or where the material or modified surface layer is exceptionally thin. Special machines, such as the one shown in Figure 2-16, have been constructed for this purpose. The location for the test is selected under high magnification. A small diamond penetrator is then loaded with a predetermined load ranging from 25 to 3600g. In the *Knoop test*, an elongated diamond-shaped indenter (long diagonal seven times the short) is used and the length of the indentation is measured with the aid of a microscope. Figure 2-17 compares the indenters for the Vickers and Knoop tests, and shows a series of Knoop indentations progressing (left-to-right) across a surface-hardened steel specimen (hardened surface to unhardened core). The hardness value, known as the *Knoop hardness number*,

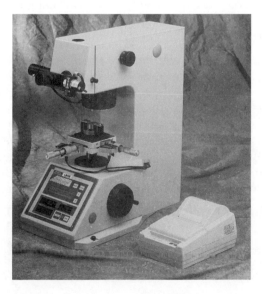

FIGURE 2-16 Microhardness tester. *(Courtesy of LECO Corporation.)*

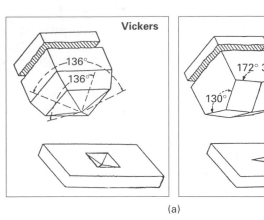 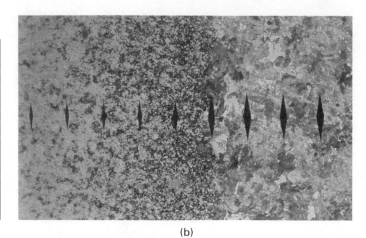

(a)   (b)

FIGURE 2-17   (a) Comparison of the diamond-tipped indenters used in the Vickers and Knoop hardness tests. (b) Series of Knoop hardness indentations progressing left-to-right across a surface hardened steel specimen (hardened surface to unhardened core). *(Courtesy Buehler Ltd., Lake Bluff, IL.)*

FIGURE 2-18   Durometer hardness tester. *(Courtesy of New Age Industries.)*

is again obtained by dividing the load in kilograms by the projected area of the indentation, expressed in square millimeters. A light-load Vickers test can also be used to determine microhardness.

A more appropriate term for microhardness tests, however, would be *microindentation* hardness testing, since it is the size of the indentation that is extremely small, not the measured value of hardness.

Other Hardness Determinations.   When testing soft, elastic materials, such as rubbers and nonrigid plastics, a *durometer* can be used. This instrument (Figure 2-18) measures the resistance of a material to elastic penetration by a spring-loaded conical steel indenter. No permanent deformation occurs. A similar test is used to evaluate the strength of molding sands used in the foundry industry.

In the *scleroscope* test, hardness is measured by the rebound of a small diamond-tipped "hammer" that is dropped from a fixed height onto the surface of the material to be tested. This test evaluates the resilience of a material, and the surface on which the test is conducted must have a fairly high polish to yield good results. Because the test is based on resilience, scleroscope hardness numbers should only be used to compare similar materials. A comparison between steel and rubber, for example, would not be valid.

Another definition of hardness is the ability of a material to resist being scratched. A crude but useful test that employs this principle is the *file test*, where one determines if a material can be cut by a simple metalworking file. The test can be either a pass–fail test using a single file, or a semiquantitative evaluation using a series of files that have been pretreated to various levels of known hardness.

Relationships among the Various Hardness Tests.   Since the various hardness tests often evaluate different material phenomena, there are no simple relationships between the different types of hardness numbers. Approximate relationships have been developed, however, by testing the same material on a variety of devices. Table 2-2 presents a correlation of hardness values for plain carbon and low-alloy steels. It may be noted that for Rockwell C numbers above 20, the Brinell values are approximately 10 times the Rockwell number. Also, for Brinell values below 320, the Vickers and Brinell values agree quite closely. Since the relationships among the various tests will differ with material, mechanical processing, and heat treatment, tables such as Table 2-2 should be used with caution.

Relationship of Hardness to Tensile Strength.   Table 2-2 and Figure 2-19 show a correlation between tensile strength and hardness for steel. For plain carbon and low-alloy steels, the tensile strength (in pounds per square inch) can be estimated by multiplying the Brinell hardness number by 500. Thus an inexpensive and quick hardness test can be used to provide a close approximation of the tensile strength of the steel.

TABLE 2-2. Hardness Conversion Table for Steels

Brinell Number	Vickers Number	Rockwell Number		Scleroscope Number	Tensile Strength	
		C	B		ksi	MPa
	940	68		97	368	2537
757[a]	860	66		92	352	2427
722[a]	800	64		88	337	2324
686[a]	745	62		84	324	2234
660[a]	700	60		81	311	2144
615[a]	655	58		78	298	2055
559[a]	595	55		73	276	1903
500	545	52		69	256	1765
475	510	50		67	247	1703
452	485	48		65	238	1641
431	459	46		62	212	1462
410	435	44		58	204	1407
390	412	42		56	196	1351
370	392	40		53	189	1303
350	370	38	110	51	176	1213
341	350	36	109	48	165	1138
321	327	34	108	45	155	1069
302	305	32	107	43	146	1007
285	287	30	105	40	138	951
277	279	28	104	39	134	924
262	263	26	103	37	128	883
248	248	24	102	36	122	841
228	240	20	98	34	116	800
210	222	17	96	32	107	738
202	213	14	94	30	99	683
192	202	12	92	29	95	655
183	192	9	90	28	91	627
174	182	7	88	26	87	600
166	175	4	86	25	83	572
159	167	2	84	24	80	552
153	162		82	23	76	524
148	156		80	22	74	510
140	148		78	22	71	490
135	142		76	21	68	469
131	137		74	20	66	455
126	132		72	20	64	441
121	121		70		62	427
112	114		66		58	

[a] Tungsten carbide ball; others standard ball.

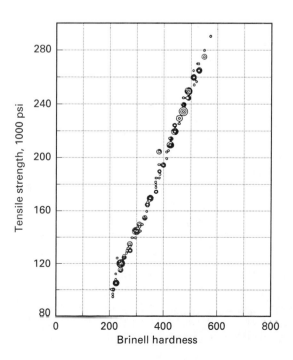

FIGURE 2-19 Relationship of hardness and tensile strength for a group of standard alloy steels. *(Courtesy of ASM International, Materials Park, Ohio.)*

For other materials, however, the relationship will be different and may exhibit too much variation to be dependable. For example, the multiplying factor for duraluminum is about 600, while for soft brass it is around 800.

■ 2.3 DYNAMIC PROPERTIES

In many engineering applications, materials are subjected to various types of dynamic loading. These may include (1) sudden impacts or loads that vary rapidly in magnitude, (2) repeated cycles of loading and unloading, or (3) frequent changes in the mode of loading, such as from tension to compression. For these operating conditions, we must be concerned with additional mechanical properties that characterize material performance under dynamic loading.

Most dynamic tests subject standard specimens to a well-controlled set of test conditions. However, the conditions of actual application rarely duplicate the conditions of a standardized test. Identical tests on different materials can indeed provide a comparison of material behavior, but for a specific set of conditions. The assumption that similar results are expected for similar conditions may not always be true. Therefore, since dynamic conditions can vary greatly, the quantitative results of standardized tests should be used with extreme caution, and one should always be aware of the limitations of each dynamic test.

IMPACT TEST

Several tests have been developed to evaluate the fracture resistance of a material when it is subjected to a rapidly applied dynamic load, or impact. Of those tests that have become common, two basic types have emerged: (1) bending impacts, which includes the standard Charpy and Izod tests, and (2) tension impacts.

The bending impact tests utilize specimens that are supported as beams. In the *Charpy test*, shown schematically in Figure 2-20, the standard specimen contains a V, keyhole, or U-shaped notch. The test specimen is supported on the ends, and an impact is applied to the center, behind the notch, to complete a three-point bending. The *Izod test* specimen, while somewhat similar in size and appearance, is supported as a cantilever beam and is impacted on the end (Figure 2-21). Impact testers, like the one shown in Figure 2-22, supply a predetermined impact energy in the form of a swinging pendulum. After breaking or deforming the specimen, the pendulum continues its upward swing with an energy equal to its original minus that absorbed by the impacted specimen. The loss of energy is measured by the angle that the pendulum attains during its upward swing.

The test specimens for bending impacts must be prepared with geometric precision to ensure consistent and reproducible results. Notch profile is extremely critical, for the test measures the energy required to both initiate and propagate a fracture. The effect of notch profile is shown dramatically in Figure 2-23. Here two specimens have been made from the same piece of steel with the same reduced cross-sectional area. The one with the keyhole notch fractures and absorbs only 43 ft-lb of energy, whereas the other specimen resists fracture and absorbs 65 ft-lb during the impact.

FIGURE 2-20   (a) Standard Charpy impact specimens and mode of loading. Illustrated are keyhole and U notches; dimensions are in millimeters with inches in parentheses. (b) Standard V-notch specimen showing the three-point bending type of impact.

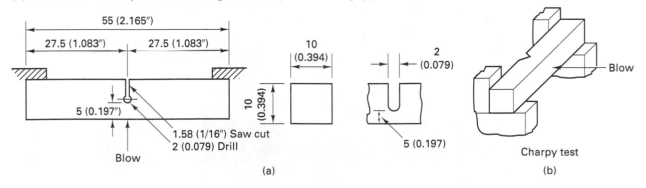

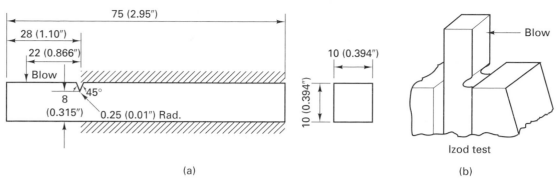

FIGURE 2-21   (a) Izod impact specimen; (b) Cantilever mode of loading in the Izod test.

Caution should also be placed on the use of impact data for design purposes. The test results apply only to standard specimens containing a standard notch. Moreover, the tests evaluate material behavior under very specific conditions. Changes in the form of the notch, minor variations in the overall specimen geometry, or faster or slower rates of loading (speed of the pendulum) can all produce significant changes in the results. Under conditions of rapid loading, wide specimens, and sharp notches, many ductile materials lose their energy-absorbing capability and fail in a brittle manner.

The results of standard impact tests, however, can be quite valuable in assessing a material's sensitivity to notches and the multiaxial stresses that exist around a notch. In addition, testing can be performed at a variety of temperatures. As will be seen in a later section of this chapter, the evaluation of how fracture resistance changes with temperature can be a valuable input when selecting engineering materials for low-temperature service.

The *tensile impact test*, illustrated schematically in Figure 2-24, eliminates the use of a notched specimen, and thereby avoids many of the objections inherent in the Charpy and Izod tests. Specimens are subjected to uniaxial impact loadings applied through drop weights, modified pendulums, or variable-speed flywheels.

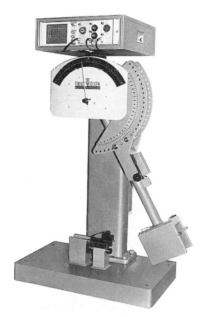

FIGURE 2-22   Impact testing machine. *(Courtesy of Tinius Olsen Testing Machine Co., Inc.)*

FATIGUE AND THE ENDURANCE LIMIT

Materials can also fail by fracture if they are subjected to repeated applications of stress, even though the peak stresses have magnitudes less than the ultimate tensile strength and usually less than the yield strength. This phenomenon, known as *fatigue*, can result from either the cyclic repetition of a particular loading cycle or entirely random variations in stress, and has been associated with almost 90% of all metallic fractures.

A periodic, sinusoidal mode of loading is often the simplest for experimental reproduction and subsequent analysis. By restricting the conditions to equal-magnitude tension–compression reversals, the test is further simplified, and data can be generated by rotating shafts under cantilever loading. If the material being evaluated in Figure 2-25 were subjected to a standard tensile test, it would require a stress in excess of 480 MPa (70,000 psi) to induce failure. Under cyclic loading with a peak stress of only 380 MPa (55,000 psi), the

FIGURE 2-23   Notched and unnotched impact specimens before and after testing. Both specimens had the same cross-sectional area.

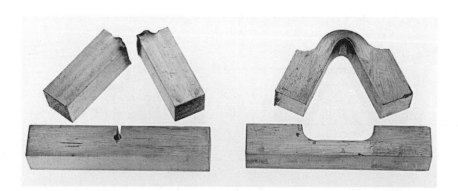

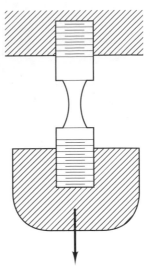

FIGURE 2-24 Tensile impact test schematic.

specimen would fail after about 100,000 cycles. If the peak stress were reduced to 350 MPa (51,000 psi), the fatigue lifetime would be extended to approximately 1,000,000 cycles. If the applied stress were reduced to a value below 340 MPa (49,000 psi), this material would not fail by fatigue, regardless of the number of stress application cycles.

Curves such as that in Figure 2-25 are known as *stress versus number of cycles*, or *S–N curves*, and summarize the results of multiple fatigue tests. Any point on the curve is the *fatigue strength*, the maximum stress that can be sustained for a specified number of loading cycles. The value of the stress below which the material will not fail regardless of the number of load cycles is known as the *endurance limit* or *endurance strength*, and may be an important criterion in many designs.

A different number of loading cycles is required to determine the endurance limit for different materials. For steels, 10 million cycles are usually sufficient. Several of the nonferrous metals, however, require 500 million cycles, and aluminum alloys often require such a great number that no endurance limit is apparent under typical test and usage conditions.

The fatigue resistance of an actual product is sensitive to a number of additional factors. One of the most important of these is the presence of stress raisers, such as sharp corners, small surface cracks, machining marks, or surface gouges. Data for the *S–N* curves are obtained from polished-surface, "flaw-free" specimens, and the reported lifetime is the cumulative number of cycles required to initiate a fatigue crack and propagate it to failure. If a part already contains a surface crack or flaw, the number of cycles required for crack initiation can be reduced significantly. In addition, the stress concentrator magnifies the stress experienced at the tip of the crack, accelerating the rate of subsequent crack growth. Great care should be taken to eliminate stress raisers and surface flaws on parts that will be subjected to cyclic loadings. Proper design and good manufacturing practices are often more important than material selection and heat treatment.

Operating temperature can also affect the fatigue performance of a material. Figure 2-26 shows *S–N* curves for Inconel 625 (a Ni–Cr–Fe alloy) determined over a range of temperatures. As temperature is increased, the fatigue strength drops significantly. Since most test data are generated at room temperature, caution should be exercised when the product application involves elevated service temperatures.

Fatigue lifetime can also be affected by changes in the environment. When metals are subjected to corrosion during the cyclic loadings, the condition is known as corrosion fatigue, and both specimen lifetime and the endurance limit can be significantly reduced. Moreover, the nature of the environmental attack need not be severe. Tests conducted in air have been shown to have shorter lifetimes than those run in a vacuum, and further lifetime reductions have been observed with increasing levels of humidity. The test results are also dependent on the frequency of the loading cycles. For slower frequencies, the environment has a longer time to act between loadings; at high frequencies,

FIGURE 2-25 Typical *S–N* or endurance-limit curve for steel. Specific numbers will vary with the type of steel and treatment.

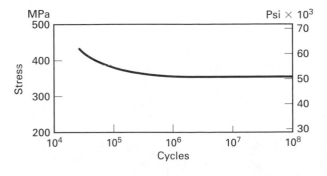

FIGURE 2-26 Fatigue strength of Inconel alloy 625 at various temperatures. *(Courtesy of Huntington Alloy Products Division, The International Nickel Company, Inc.)*

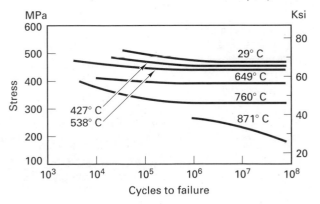

TABLE 2-3.	Ratio of Endurance Limit to Tensile Strength for Various Materials
Material	**Ratio**
Aluminum	0.38
Beryllium copper (heat-treated)	0.29
Copper, hard	0.33
Magnesium	0.38
Steel	
AISI 1035	0.46
Screw stock	0.44
AISI 4140 normalized	0.54
Wrought iron	0.63

the environmental effects may be somewhat masked. Direct application of test data to actual products, therefore, requires considerable caution.

Residual stresses can also alter fatigue behavior. If the specimen surface is in a state of compression, such as that produced from shot peening, carburizing, or burnishing, it is more difficult to initiate a fatigue crack, and so lifetime is extended. Conversely, processes that produce residual surface tension, such as welding or machining (Chapter 40), can significantly reduce the fatigue lifetime of a product.

If the magnitude of the load varies during service, the fatigue response can be quite complex. Consider the wing of a commercial airplane. The many low-stress cycles that occur as the wing vibrates during flight may be less damaging to the wing-fuselage joint than a few high-stress loadings, which might occur when the plane contacts the ground during landing. From a different perspective, however, the heavy loads may be sufficient to stretch and blunt any growing fatigue crack such that many additional small-load cycles will be required to "reinitiate" it. Evaluating how materials respond to complex patterns of loading is an area of great importance to design engineers.

Since reliable fatigue data may take a considerable time to generate, we may prefer to estimate fatigue behavior from properties that can be determined more quickly. Table 2-3 shows the approximate ratio of the endurance limit to the ultimate tensile strength for several engineering metals. For many steels the endurance limit can be approximated by 0.5 times the ultimate tensile strength as determined by a standard tensile test. For the nonferrous metals, however, the ratio is significantly lower.

FATIGUE FAILURES

Components that fail as a result of repeated or cyclic loads are commonly called *fatigue failures*. These fractures form a major part of a larger group known as progressive fractures. Consider the fracture surface shown in Figure 2-27. The two arrows identify the points of fracture initiation, which often correspond to discontinuities in the form of surface cracks, sharp corners, machining marks, or even "metallurgical notches," such as an abrupt change in metal structure. With each repeated application of load, the stress at the tip of the crack exceeds the strength of the material, and the crack grows a very small amount. Crack growth continues with each successive application of load until the remaining cross section is no longer sufficient to withstand the peak stresses. Sudden overload fracture then occurs through the remainder of the material. The overall fracture surface tends to exhibit two distinct regions: a smooth, relatively flat region where the crack was propagating by cyclic fatigue, and a coarse, ragged region, corresponding to the ductile overload tearing.

The smooth areas of the fracture often contain a series of parallel ridges radiating outward from the origin of the crack. These ridges may not be visible under normal examination, however. They may be extremely fine; they may have been obliterated by a rubbing action during the compressive stage of the repeated loading; or they may be very few in number if the failure occurred after only a few cycles of loading ("low-cycle fatigue"). Electron microscopy may be required to reveal the ridges, or *fatigue striations*, that are characteristic of fatigue failure. Figure 2-28 shows an example of these markings at high magnification.

FIGURE 2-27 Progressive fracture of an axle within a ball-bearing ring, starting at two points (arrows).

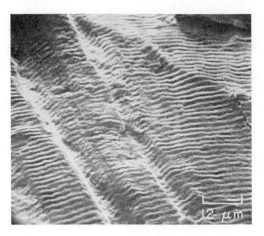

FIGURE 2-28 Fatigue fracture of AISI type 304 stainless steel viewed in a scanning electron microscope at 810×. Well-defined striations are visible. *(From "Interpretation of SEM fractographs," Metals Handbook, Vol. 9, 8th ed., ASM International, Materials Park, Ohio, 1970, p. 70.)*

For some fatigue failures, the overload area may exhibit a crystalline appearance, and the failure is sometimes attributed to the metal having "crystallized." As noted in Chapter 3, engineering metals are almost always crystalline materials. The final overload fracture simply propagated along the grain boundaries and revealed the already-existing crystalline nature of the material. The conclusion that the material crystallized is totally erroneous, and the term is a definite misnomer. This is just a form of fatigue failure.

Another common error is to classify all progressive-type failures as fatigue failures. Other progressive failure mechanisms, such as creep failure and stress–corrosion cracking, will also produce the characteristic two-region fracture. In addition, the same mechanism can produce fractures with different appearances depending on the magnitude of the load, type of loading (torsion, bending, or tension), temperature, and operating environment. Correct interpretation of a metal failure generally requires far more information than that acquired by a visual examination of the fracture surface.

A final misconception regarding fatigue failures is to assume that the failure is time dependent. The failure of materials under repeated loads below their static strengths is primarily a function of the magnitude and number of loading cycles. If the frequency of loading is increased, the time to failure should decrease proportionately. If the time does not change, the failure is dominated by one or more environmental factors, and fatigue is a secondary component.

■ 2.4 TEMPERATURE EFFECTS

The test data used in design and engineering decisions should always be obtained under conditions that simulate those of actual service. A number of engineered structures, such as aircraft, space vehicles, gas turbines, and nuclear power plants, are required to operate under temperatures as low as $-130°C$ ($-200°F$) or as high as $1250°C$ ($2300°F$). Consequently, the designer must consider the short- and long-range effects of temperature on the mechanical and physical properties of the material being considered for such applications. From a manufacturing viewpoint, the effects of temperature are equally important. Numerous manufacturing processes involve heat, and the processing may alter the material properties in both favorable and unfavorable ways. A material can often be processed successfully, or economically, only because heating or cooling can be used to change its properties.

Elevated temperatures can be quite useful in modifying the strength and ductility of a material. Figure 2-29 summarizes the changes in key properties for a medium-carbon steel. Similar effects are presented for magnesium in Figure 2-30. As noted, an increase in temperature will typically induce a decrease in strength and hardness and an increase in elongation. For manufacturing operations such as metal forming, heating to elevated temperature may be extremely attractive because the material is now both weaker and more ductile.

Figure 2-31 shows the combined effects of temperature and strain rate (speed of testing) on the ultimate tensile strength of a medium-carbon steel. These data show that the *rate of deformation* can also have a strong influence on mechanical properties. Room-temperature standard-rate tensile test data will be of little use if the application involves

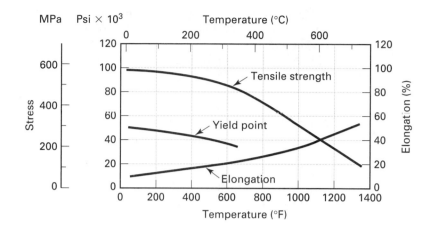

FIGURE 2-29   Some effects of temperature on the tensile properties of a medium-carbon steel.

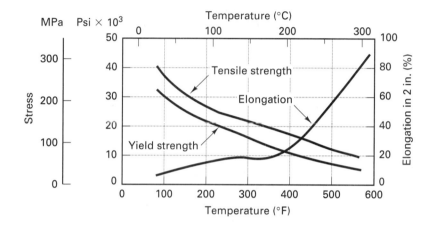

FIGURE 2-30   Effects of temperature on the tensile properties of magnesium.

a material being hot-rolled at speeds of 1300 m/min (5000 ft/min). The effects of strain rate on yield-strength follow the same trends as tensile strength.

The effect of temperature on impact properties became the subject of intense study in the 1940s when the increased use of welded-steel construction led to catastrophic failures of ships and other structures while operating in cold environments. Welding produces a monolithic (single-piece) structure, and hence a situation where cracks can propagate through a joint and continue on to other sections of the structure! Figure 2-32 shows the effect of decreasing temperature on the impact properties

FIGURE 2-31   Effects of temperature and strain rate on the tensile strength of copper. *(From A. Nadai and M. J. Manjoine, Journal of Applied Mechanics, Vol. 8, 1941, p. A82, courtesy of ASME.)*

FIGURE 2-32   Effect of temperature on the impact properties of two low-carbon steels.

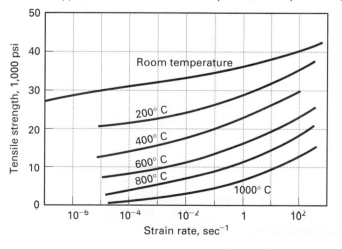

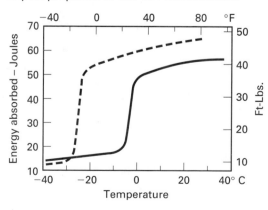

of two low-carbon steels. Although similar in form, the two curves are significantly different. The steel indicated by the solid line becomes brittle at temperatures below −4°C (25°F), while the other steel retains good fracture resistance down to −26°C (−15°F). The temperature at which the response goes from high energy absorption to low energy absorption is known as the *ductile-to-brittle transition temperature*. All steels tend to exhibit this transition when temperature is decreased. However, the temperature at which it occurs varies with carbon content and alloy. Special caution should be taken, therefore, when selecting steels for low-temperature applications.

CREEP

Long-term exposure to elevated temperatures can also lead to failure by a phenomenon known as *creep*. If a tensile-type specimen is subjected to a constant load at elevated temperature, it will elongate continuously until rupture occurs, even though the applied stress is below the yield strength of the material at the temperature of testing. While the rate of elongation is often quite small, creep can be an important consideration when designing equipment such as steam or gas turbines, power plant boilers, and other devices that operate under loads or pressures and high temperatures for long periods of time.

If a test specimen is subjected to conditions of fixed load and fixed elevated temperature, an elongation-versus-time plot can be generated, similar to the one shown in Figure 2-33. The curve contains three distinct stages: a short-lived initial stage, a rather long second stage where the elongation rate is somewhat linear, and a short-lived third stage leading to fracture. Two significant pieces of engineering data are obtained from this curve: the rate of elongation in the second stage, or *creep rate*, and the total elapsed *time to rupture*. These results are unique to the material being tested and the specific conditions of the test. Tests conducted at higher temperatures or with higher applied loads would exhibit higher creep rates and shorter rupture times.

When creep behavior is a concern, multiple tests are conducted over a range of temperatures and stresses, and the rupture time data are collected into a single *stress–rupture diagram*, like the one shown in Figure 2-34. This simple engineering tool provides an overall picture of material performance at elevated temperature. In a similar manner, creep rate data can also be plotted to show the effects of temperature and stress. Figure 2-35 presents a creep-rate diagram for a high-temperature nickel-base alloy.

FIGURE 2-33   Creep curve for a single specimen at a fixed temperature, showing the three stages of creep and reported creep rate. Note the nonzero strain at time zero due to the initial application of the load.

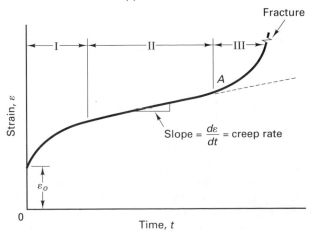

FIGURE 2-34   Stress–rupture diagram of a solution-annealed Incoloy alloy 800 (Fe–Ni–Cr alloy). *(Courtesy of Huntington Alloy Products Division, The International Nickel Company, Inc.)*

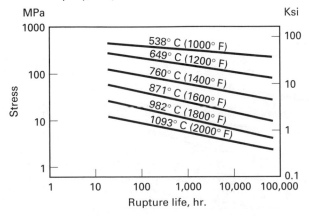

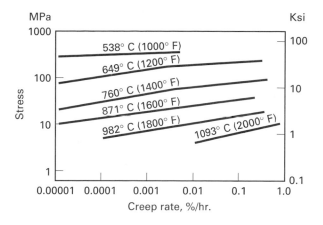

FIGURE 2-35   Creep-rate properties of a solution-annealed Incoloy alloy 800. *(Courtesy of Huntington Alloy Products Division, The International Nickel Company, Inc.)*

■ 2.5 MACHINABILITY, FORMABILITY AND WELDABILITY

While it is common to assume that the various "-ability" terms also refer to specific material properties, they actually refer to the way a material responds to specific processing techniques. As a result, they can be quite nebulous. *Machinability*, for example, depends not only on the material being machined but also on the specific machining process and the aspects of that process that are of greatest interest. Machinability ratings are generally based on relative tool life data. In certain applications, however, we may be more interested in how easy or fast a metal can be cut, and less interested in the tool life or the resulting surface finish. For other applications, surface finish or the formation of fine chips may be the most desirable feature. As a result, the term *machinability* may mean different things to different people, and it frequently involves multiple properties of a material acting in unison.

In a similar manner, *malleability*, *workability*, and *formability* all refer to a material's suitability for plastic deformation processing. Since a material often behaves differently at different temperatures, a material with good "hot formability" may have poor deformation characteristics at room temperature. Furthermore, materials that flow nicely at low deformation speeds may behave in a brittle manner when loaded at rapid rates. Formability, therefore, needs to be evaluated for a specific combination of material, process, and process conditions. The results cannot be extrapolated or transferred to other processes or process conditions. Likewise, the *weldability* of a material may also depend on the specific welding or joining process being considered.

■ 2.6 FRACTURE TOUGHNESS AND THE FRACTURE MECHANICS APPROACH

A discussion of the mechanical properties of materials would not be complete without mention of the many tests and design concepts based on the fracture mechanics approach. Instead of treating test specimens as flaw-free materials, fracture mechanics begins with the premise that *all materials contain flaws or defects of some given size*. These may be *material defects*, such as pores, cracks, or inclusions; *manufacturing defects*, in the form of machining marks, arc strikes, or contact damage to external surfaces; or *design defects*, such as abrupt section changes, excessively small fillet radii, and holes. When the specimen is subjected to loads, the applied stresses are amplified or intensified in the vicinity of these defects, leading to accelerated failure or failure under unexpected conditions.

Fracture mechanics seeks to identify the conditions under which a defect will grow or propagate to failure and, if possible, the rate of crack or defect growth. The methods concentrate on three principal quantities: (1) the size of the largest or most critical flaw, usually denoted as a; (2) the applied stress, denoted by σ; and (3) the fracture toughness, a quantity that describes the resistance of a material to fracture or crack growth, which is usually denoted by K with subscripts to signify the conditions of testing. Equations have been developed that relate these three quantities (at the onset of crack growth or propagation) for various specimen geometries, flaw locations, and flaw orientations. If nondestructive testing or quality control checks (such as those described in Chapter 11) have been applied, the size of the largest flaw that could go undetected is often known. By mathematically placing this worst possible flaw in the worst possible location and orientation, and coupling this with the largest applied stress for that location, a designer can determine the value of fracture toughness necessary to prevent that flaw from propagating during service. In a reverse manner, if the material and stress conditions were defined, the size of the maximum permissible flaw could be computed. Inspection conditions could then be selected to assure that flaws greater than this magnitude are cause for product rejection. Similarly, if a component is found to have a significant flaw and the material is known, the maximum operating stress can be determined that will assure no further growth of that flaw.

According to the philosophy of fracture mechanics, each of the flaws or defects in a material can be either *dormant* or *dynamic*. Dormant defects are those whose size remains unchanged through the lifetime of the part, and are indeed permissible. A major goal of fracture mechanics is to define the distinction between dormant and dynamic for the specific conditions of material, part geometry, and applied loading. Alternative efforts to prevent material fracture generally involve overdesign, excessive inspection, or the use of premium-quality materials.

It has already been noted that fatigue loadings account for as much as 90% of all dynamic failures. The standard method of fatigue testing applies cyclic loads to polished, flaw-free specimens, and the reported lifetime consists of both crack initiation and crack propagation. In contrast, fracture mechanics focuses on the growth of an already existing flaw. Figure 2-36 shows the *crack growth rate* (change in size per loading cycle denoted

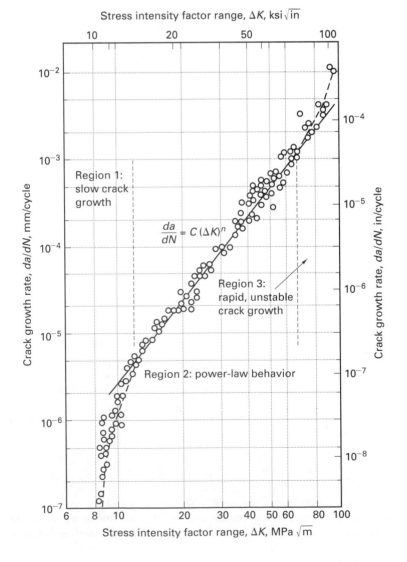

FIGURE 2-36   Plot of the fatigue crack growth rate for a typical steel using the fracture mechanics approach. Similar shape curves are obtained for most engineering metals. *(Courtesy of ASM International.)*

as *da/dN*) plotted as a function of the fracture mechanics parameter, ΔK. The fracture mechanics results provide a far more realistic guarantee of minimum service life.

Fracture mechanics is a truly integrated blend of design (applied stresses), inspection (flaw-size determination), and materials (fracture toughness). The approach has proven valuable in many areas where fractures could be catastrophic and has shown great refinement and increased acceptance in recent years.

■ 2.7 PHYSICAL PROPERTIES

For certain applications, the *physical properties* of an engineering material may be even more important than the mechanical. These include the thermal, electrical, magnetic, and optical characteristics.

We have previously discussed several ways in which the mechanical properties of materials respond to variations in temperature. In addition to these, there are some truly *thermal properties* that should be considered. The *heat capacity* or *specific heat* of a material is the amount of energy that must be added to or removed from a given mass of material to produce a 1° change in temperature. This is important in processes such as casting, where heat must be extracted rapidly to promote solidification, or heat treatment, where large quantities of material are heated and cooled. *Thermal conductivity* measures the rate at which heat can be transported through a material. While this may be tabulated separately in reference texts, it is helpful to remember that for metals, thermal conductivity is directly proportional to electrical conductivity. Metals such as copper, gold, and aluminum that possess good electrical conductivity are also good transporters of thermal energy. *Thermal expansion* is the final property of significance. Most materials expand upon heating and contract upon cooling, but the amount of expansion or contraction will vary with the material. For components that are machined at room temperature but put in service at elevated temperatures, or castings that solidify at elevated temperatures and then cool, the manufactured dimensions must be adjusted to compensate for the subsequent changes.

Electrical conductivity and *resistivity* may also be important design considerations. These properties will vary not only with the material, but also with the way the material has been processed and the temperature.

From the standpoint of *magnetic response*, materials are often classified as diamagnetic, paramagnetic, ferromagnetic, antiferromagnetic, and ferrimagnetic. These terms refer to the way in which the material responds to an applied magnetic field. Material properties, such as saturation strength, remanence, and magnetic hardness or softness, describe the strength, duration, and nature of this response.

Still other physical properties that may assume importance include *weight* or *density*, *melting* and *boiling points*, and the various *optical properties*, such as the ability to transmit, absorb, or reflect light or other electromagnetic radiation.

■ KEY WORDS

brale	electrical conductivity	impact test	Rockwell hardness test	time to rupture
breaking strength	electrical resistivity	Knoop hardness	*S–N* curve	(rupture time)
Brinell hardness number	elongation	machinability	scleroscope	toughness
Brinell hardness test	endurance limit	mechanical properties	specific heat	transition temperature
brittleness	engineering strain	metal	static properties	true strain
Charpy test	engineering stress	microhardness	strain	true stress
creep	fatigue	modulus of elasticity	strain hardening	ultimate tensile
creep rate	fatigue strength	necking	strain-hardening	strength
dormant flaw	fatigue striations	nonmetal	coefficient	uniaxial tensile test
ductile-to-brittle	formability	offset yield strength	stress	Vickers hardness test
transition temperature	fracture toughness	percent reduction in area	stress–rupture	weldability
ductility	gage length	physical properties	diagram	yield point
durometer	hardness	plastic deformation	tensile test	Young's modulus
dynamic flaw	heat capacity	proportional limit	thermal conductivity	
elastic limit	Izod test	resilience	thermal expansion	

■ REVIEW QUESTIONS

1. What are some properties commonly associated with metallic materials?
2. What are some of the more common nonmetallic engineering materials?
3. What are some of the important physical properties of materials?
4. Why should caution be exercised when applying the results from any of the standard mechanical property tests?
5. What are the standard units used to report stress and strain in the English system? In the metric or SI system?
6. Why might Young's modulus or stiffness be an important material property?
7. What are some of the tensile test properties that are used to describe or define the elastic-to-plastic transition in a material?
8. What are two tensile test properties that can be used to describe the ductility of a material?
9. Why might the uniform elongation or percent elongation prior to necking be a more meaningful measure of useful ductility?
10. Is a brittle material a weak material?
11. What is the toughness of a material?
12. What is the difference between true stress and engineering stress? True strain and engineering strain?
13. What is strain hardening or work hardening? How might this phenomenon be measured or reported? How might it be used in manufacturing?
14. What are some of the different material characteristics or responses that have been associated with the term *hardness*?
15. Describe how one would conduct a Brinell hardness test.
16. How should one report the results of a Rockwell hardness test?
17. When might a microhardness test be preferred over the standard Brinell, Rockwell, or Vickers tests?
18. Describe the reason why the various types of hardness tests often fail to agree with one another.
19. What is the relationship between penetration hardness and the ultimate tensile strength for steel?
20. Why is compression testing considered to be more difficult than tensile testing?
21. Describe several types of dynamic loading.

22. What are the two most common types of bending impact tests? How are the specimens supported and loaded during impact?
23. Why should extreme caution be used when applying impact test data?
24. What is the relationship between the fatigue strength of a metal and its yield strength as determined in a standard tensile test?
25. Fatigue strength and endurance limit are two terms that are derived from *S–N* diagrams. Define these terms and describe how they are determined from a diagram.
26. What are some of the additional factors that can alter the fatigue lifetime or fatigue behavior of a material?
27. How might the endurance limit of a steel be estimated, thereby avoiding the long time required for extensive fatigue testing?
28. What material, design, or manufacturing features may contribute to the initiation of a fatigue crack?
29. What are fatigue striations and why do they form?
30. Why is it important for a designer or engineer to know a material's properties at all possible temperatures of operation?
31. Why should one use caution when using a steel at low (below zero Fahrenheit) temperatures?
32. What are some ways to evaluate the long-term effect of elevated temperature on an engineering material?
33. How is a stress–rupture diagram developed?
34. Why are terms such as machinability, formability, and weldability considered to be poorly defined and therefore quite nebulous?
35. What is the basic premise of the fracture mechanics approach to testing and design?
36. What three principal quantities does fracture mechanics attempt to relate?
37. What are the units of fracture toughness in the English and SI systems?
38. What are the three most common thermal properties of a material, and what do they measure?
39. Describe an engineering application where the density of the selected material would be an important material consideration.

■ PROBLEMS

1. Select a product or component for which physical properties are more important than mechanical properties.
   a. Describe the product or component and its function.
   b. What are the most important properties or characteristics?
   c. What are the secondary properties or characteristics that would also be desirable?
2. Repeat Problem 1 for a product or component whose dominant required properties are of a static mechanical nature.
3. Repeat Problem 1 for a product or component whose dominant requirements are dynamic mechanical properties.
4. One of the important considerations when selecting a material for an application is to determine the highest and lowest operating temperature along with the companion properties that must be present at each extreme. The ductile-to-brittle transition temperature, discussed in Section 2-4, has been an important factor in a number of failures. An article that

summarized the features of 56 catastrophic brittle fractures that made headline news between 1888 and 1956 noted that low temperatures were present in nearly every case. The water temperature at the time of the sinking of the Titanic was −2°C, above the freezing point for salt water, but below the transition point for the steel used in construction of the hull of the ship.
   a. Which of the common engineering materials exhibits a ductile-to-brittle transition?
   b. For plain carbon and low-alloy steels, what is a typical value (or range of values) for the transition temperature?
   c. What type of material would you recommend for construction of a small vessel to transport liquid nitrogen within a building or laboratory?
   d. Figure 2P-1 summarizes the results of impact testing performed on hull plate from the RMS Titanic and similar material produced for modern steel-hulled ships. Why should

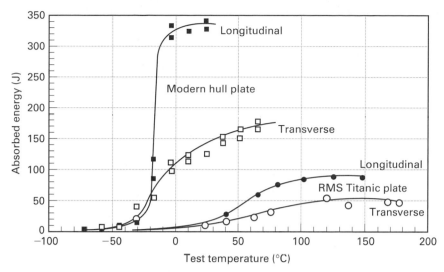

FIGURE 2P-1   Longitudinal and transverse notch toughness data: steel from the RMS Titanic versus modern steel plate. *(Courtesy I&SM, Sept. 1999, p. 33, Iron and Steel Society, Warrendale, PA.)*

there be a difference between specimens cut longitudinally (along the rolling direction) and transversely (across the rolling direction)? What advances in steel making have led to the significant improvement in low-temperature impact properties?

5. Because of the amount of handling that occurs during material production, within warehouses, and during manufacturing operations, along with the handling of loading, unloading, and shipping, material mix-ups and mixed materials are not an uncommon occurrence. Mixed materials also occur when industrial scrap is collected, or when discarded products are used as new raw materials through recycling.

   Assume that you have equipment to perform each of the tests described in this chapter (as well as access to the full spectrum of household and department store items and a small machine shop). For each of the following material combinations, determine a procedure that could permit separation of mixed materials. Use standard data-source references to help you identify distinguishable properties.

   a. Hot-rolled bars of AISI 1020 and 1040 steel.
   b. Stainless steel sheets of Type 430 ferritic stainless and Type 316 austenitic stainless.
   c. 6061-T6 aluminum and AZ91 magnesium that have become mixed in a batch of machine shop scrap.
   d. Transparent bottles of polyethylene and polypropylene (both thermoplastic polymers) that have been collected for recycling.

www.wiley.com/college/degarmo

*C*hapter 2   CASE STUDY

Overhead Conveyor for Meat Processing

A new meat-processing plant has opened and utilizes an overhead conveyor to move sides of beef (weighing approximately 300 pounds each) through the operation. The sides are suspended on hooks, which are spaced along the conveyor chain. Within the first several weeks of operation, several of the hooks broke, resulting in a 300-pound section of beef dropping from overhead. Most of the fractures occurred shortly after a particular side of beef exited the freezer unit and begun its journey through the plant. The freezer temperature is approximately 0°F,

and the hooks have been made from a low-carbon, plain-carbon steel.

a. What do you suspect is the cause of these failures?
b. What recommendations would you make regarding the failing hooks? If replacement is recommended, what material would you suggest as a better alternative?
c. Do metallurgical features play a role? Consider grain size, microstructure (i.e., heat treatment), and the type, orientation, and distribution of possible impurities.

CHAPTER 3

NATURE OF METALS AND ALLOYS

■ 3.1 STRUCTURE–PROPERTY RELATIONSHIPS

As discussed in Chapter 2, the success of many engineering activities depends on the selection of engineering materials whose properties match the specific requirements of the application. Primitive cultures are often limited to the naturally occurring materials in their environment. If the match is not a good one, compromises are required. As civilization develops, the range of engineering materials expands. Materials can be processed and their properties altered and possibly enhanced. The alloying or heat treatment of metals and the firing of ceramics are examples of techniques that can substantially alter the properties of a material. Fewer compromises are required and enhanced design possibilities emerge. Products become more sophisticated. While the early successes in altering materials were largely the result of trial and error, we now recognize that the properties of a material are a direct result of its *structure*. If we want to change the *properties*, we will most likely have to effect changes in the material structure.

Since all materials are composed of the same basic components—particles that include *protons*, *neutrons*, and *electrons*—it is amazing that so many different materials exist with such widely varying properties. This variation can be explained by the many possible combinations of these units in a macroscopic assembly. The subatomic particles combine in different arrangements to form the various elemental *atoms*, each having a nucleus of protons and neutrons surrounded by the proper number of electrons to maintain charge neutrality. The specific arrangement of the electrons surrounding the nucleus affects the electrical, magnetic, thermal, and optical properties as well as the way the atoms bond to one another. Atomic bonding then produces a higher level of structure, which may be in the form of a *molecule*, *crystal*, or *amorphous aggregate*. This structure, along with the imperfections that may be present, has a profound effect on the mechanical properties. The size, shape, and arrangement of multiple crystals, or the mixture of two or more different structures within a material, produces yet another level of structure, known as *microstructure*, because it is observed on a microscopic scale. Variations in microstructure further affect the material properties.

As a result of the ability to control structures through processing, and the ability to develop new structures through techniques such as composite materials, engineers now have at their disposal a wide variety of materials with an almost unlimited range of properties. The properties of these materials depend on all levels of their structure, from subatomic to macroscopic (Figure 3-1). This chapter will attempt to develop an understanding of the basic structure of engineering materials and how changes in that structure affect final properties.

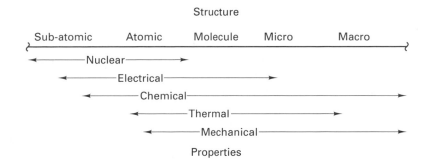

FIGURE 3-1 General relationship of the structural level to various engineering properties.

■ 3.2 THE STRUCTURE OF ATOMS

Experiments have revealed that atoms consist of a relatively dense nucleus composed of positively charged protons and neutral particles of nearly identical mass, known as neutrons. Surrounding the nucleus are the negatively charged electrons, which appear in numbers equal to the protons, so as to maintain a neutral charge balance. Distinct groupings of these basic particles produce the known elements, ranging from the relatively simple hydrogen atom to the unstable transuranium atoms over 250 times as heavy. Except for density and specific heat, however, the weight of atoms has very little influence on their engineering properties.

The light electrons that surround the nucleus play a far more significant role in determining material properties. These electrons are arranged in a characteristic structure consisting of shells and subshells, each of which can contain only a limited number of electrons. The first shell, nearest the nucleus, can contain only two. The second shell can contain eight, and the third, 32. Each shell and subshell is most stable when it is completely filled. For atoms containing electrons in the third shell and beyond, however, relative stability is achieved with eight electrons in the outermost layer or subshell.

If an atom has slightly less than the number of outer-layer electrons required for stability, it will readily accept an electron from another source. It will then have one electron more than the number of protons and becomes a negatively charged atom, or *negative ion*. Depending on the number of additional electrons, ions can have negative charges of 1, 2, 3, or more. Conversely, if an atom has a slight excess of electrons beyond the number required for stability (such as sodium, with one electron in the third shell), it will readily give up the excess electron and become a *positive ion*. The remaining electrons become more strongly bound, so the further removal of electrons becomes progressively more difficult.

The number of electrons surrounding the nucleus of a neutral atom is called the *atomic number*. More important, however, are those electrons in the outermost shell or subshell, the *valence electrons*. These are influential in determining chemical properties, electrical conductivity, some mechanical properties, the nature of interatomic bonding, atom size, and optical characteristics. Elements with similar electron configurations in their outer shells tend to have similar properties.

■ 3.3 ATOMIC BONDING

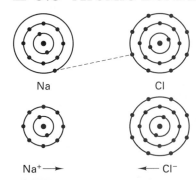

FIGURE 3-2 Ionization of sodium and chlorine, producing stable outer shells by electron transfer.

Atoms are rarely found as free and independent units, but are usually linked or bonded to other atoms in some manner as a result of interatomic forces. The electron structure of the atoms plays the dominant role in determining the nature of the bond.

Three types of *primary bonds* are generally recognized, the simplest of which is the *ionic bond*. If more than one type of atom is present, the outermost electrons can break free from atoms with excesses in their valence shell, transforming them into positive ions. The electrons are then transferred to atoms with deficiencies in their outer shell, converting them into negative ions. The positive and negative ions have an electrostatic attraction for each other, resulting in a strong bonding force. Figure 3-2 presents a crude schematic of the ionic bonding process for sodium and chlorine. The atoms, however, do not usually unite in simple pairs. All positively charged atoms attract all negatively charged atoms. Therefore, each sodium ion will attempt to surround itself with negative chlorine ions, and each chlorine ion will attempt to surround itself with positive sodium ions. Since the attraction is equal in all

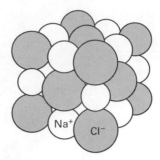

FIGURE 3-3 Three-dimensional structure of the sodium chloride crystal. Note how the various ions are surrounded by ions of the opposite charge.

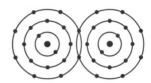

FIGURE 3-4 Formation of a chlorine molecule by a covalent bond.

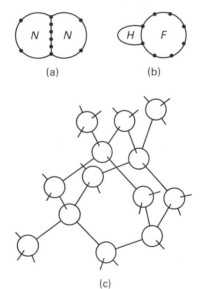

FIGURE 3-5 Examples of covalent bonding in (a) N_2, (b) HF, and (c) diamond.

directions, the result will be a three-dimensional structure, like the one shown in Figure 3-3. Since charge neutrality must be maintained within the structure, equal numbers of positive and negative charges must be present in each neighborhood. General characteristics of materials joined by ionic bonds include moderate to high strength, high hardness, brittleness, high melting point, and low electrical conductivity (electrons are captive to atoms, so charge transport requires atom—or ion—movement).

A second type of primary bond is the *covalent* type. Here the atoms in the assembly find it impossible to produce completed shells by electron transfer but achieve the same goal through electron sharing. Adjacent atoms share outer-shell electrons so that each achieves a stable electron configuration. The shared (negatively charged) electrons locate between the positive nuclei, forming a positive-negative-positive bonding link. Figure 3-4 illustrates this type of bond for a pair of chlorine atoms, each of which contains seven electrons in the valence shell. The result is a stable two-atom molecule. Stable molecules can also form from the sharing of more than one electron from each atom, as in the case of nitrogen (Figure 3-5a). The atoms in the assembly need not be identical (as in HF, Figure 3-5b), the sharing does not have to be equal, and a single atom can share electrons with more than one other atom. For atoms such as carbon and silicon, with four electrons in the valence shell, one atom may share its valence electrons with each of four neighboring atoms. The resulting structure is a three-dimensional network of bonded atoms, like the one shown in Figure 3-5c. Like the ionic bond, the covalent bond tends to produce materials with high strength and high melting point. Since atom movement within the three-dimensional structure (plastic deformation) requires the breaking of discrete bonds, covalent materials are characteristically brittle. Electrical conductivity depends on bond strength, ranging from conductive tin (weak covalent bonding), through semiconductive silicon and germanium, to insulating diamond (carbon). Ionic or covalent bonds are found in ceramic (refractories or abrasives) and polymeric materials.

A third type of primary bond is possible when a complete outer shell cannot be formed by either electron transfer or electron sharing. This bond is known as the *metallic bond* (Figure 3-6). If each of the atoms in an aggregate contains only a few valence electrons (one, two, or three), these electrons can easily be removed to produce "stable" ions. The positive ions (nucleus and inner, nonvalence electrons) then arrange in a periodic array, and are surrounded by wandering, universally shared, valence electrons, sometimes referred to as an electron cloud or electron gas. These highly mobile, free electrons account for both the high electrical and thermal conductivity values as well as the nontransparent (opaque) characteristic observed in metals (the free electrons can absorb the discrete energies of light radiation). They also provide the "cement" required for the positive-negative-positive attractions that result in bonding. Bond strength, and therefore material strength, varies over a wide range. More significant, however, is the observation that the positive ions can move within the structure without the breaking of discrete bonds. Materials bonded by metallic bonds can be deformed by atom-movement mechanisms and produce a deformed material that is every bit as strong as the original. This phenomenon is the basis of metal plasticity, ductility, and many of the shaping processes used in the fabrication of metal products.

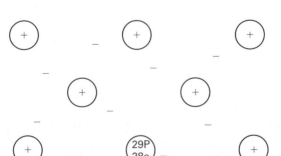

FIGURE 3-6 Schematic of the metallic bond showing the positive ions and associated electron cloud for copper. Each positive-charged ion contains a nucleus with 29 protons and stable electron shells containing 28 electrons.

■ 3.4 SECONDARY BONDS

Weak or secondary bonds, known as *van der Waals forces*, can form between molecules that possess a nonsymmetrical distribution of charge. Some molecules, such as hydrogen fluoride and water, can be viewed as electric dipoles. Certain portions of the molecule tend to be more positive or negative than others (an effect referred to as polarization). The negative part of one molecule tends to attract the positive part of another, forming a weak bond. Van der Waals forces contribute to the mechanical properties of a number of molecular polymers, such as polyethylene and polyvinyl chloride (PVC).

■ 3.5 INTERATOMIC DISTANCES AND SIZE OF ATOMS

Since the space occupied by the electron shells of an atom is far greater than the size of the actual electrons, much of the "volume" of an atom is unoccupied space. Atoms are not solid bodies, and the concept of atom size becomes somewhat nebulous. The various bonding forces tend to pull atoms together, but repelling forces exist between adjacent positive nuclei. Some equilibrium separation distance exists where the forces of attraction and repulsion are equal.

    If atoms in a solid are modeled as rigid, touching spheres, the equilibrium separation distance between neighboring atoms can be taken as the sum of the atomic radii, and atoms can be assigned a distinct size. The atomic radius is not a constant, however. Added thermal energy causes atoms to vibrate about their equilibrium positions. Since the repulsive force as atoms approach is much greater than the attractive force when they separate, the vibrations or oscillations are not symmetrical. The average location of the atoms is now at a larger-than-normal separation distance, and the result is a macroscopically observed thermal expansion of the material. Removing electrons from the outer shell will decrease the atomic radius, and adding electrons will increase it. Consequently, a negative ion is larger than the base atom, and a positive ion is smaller. Atomic radius also changes with the number of adjacent or nearest-neighbor atoms in a crystalline arrangement. With more neighbors, there is less attraction to any single neighbor atom, and the interatomic distance is increased. For example, iron has an *atomic radius* of 1.241 Å in a crystal structure where each atom has eight neighbors (body-centered cubic) and a radius of 1.269 Å in a structure with 12 adjacent neighbors (face-centered cubic).

■ 3.6 ATOM ARRANGEMENTS IN MATERIALS

As atoms bond together to form aggregates, we find that the particular arrangement of the atoms has a significant effect on the material properties. Depending on the manner of atomic grouping, materials are classified as having *molecular structures*, *crystal structures*, or *amorphous structures*.

    Molecular structures have a distinct number of atoms that are held together by primary bonds. There is only a weak attraction, however, between a given molecule and other similar groupings. Typical examples of molecules include O_2, H_2O, and C_2H_4 (ethylene). Each molecule is free to act more or less independently, so these materials exhibit relatively low melting and boiling points. Molecular materials tend to be weak, since the molecules can move easily with respect to one another. Upon changes of state from solid to liquid or liquid to gas, the molecules remain as distinct entities.

    Solid metals and most minerals have a crystalline structure. Here the atoms are arranged in a regular geometric array known as a *lattice*. Lattices are describable through a unit building block that is essentially duplicated throughout space in a repetitive manner. These blocks are known as *unit cells*. Crystalline structures will be discussed more fully in Section 3.7.

    In an amorphous structure, such as glass, the atoms have a certain degree of local order (arrangement with respect to neighbors), but when viewed as an aggregate, lack the periodically ordered arrangement of atoms that is characteristic of a crystalline solid.

■ 3.7 CRYSTAL STRUCTURES OF METALS

From a manufacturing viewpoint, metals are an extremely important class of materials. They are frequently the materials being processed, and often form both the tool and the machinery performing the processing. More than 50 of the known chemical elements are classified as metals, and about 40 have commercial importance. They are characterized by the metallic bond and possess the distinguishing characteristics of strength, good electrical and thermal conductivity, luster, the ability to be plastically deformed to a fair degree without fracturing, and a relatively high specific gravity (density) compared to nonmetals. The fact that some metals possess properties different from the general pattern simply expands their engineering utility.

When metals solidify, the atoms assume a crystalline structure; that is, they arrange themselves in a geometric lattice. Many metals exist in only one lattice form. Some, however, can exist in the solid state in two or more lattice forms, with the particular form depending on the conditions of temperature and pressure. Such metals are said to be *allotropic* or *polymorphic*, and the change from one lattice form to another is called an *allotropic transformation*. The most notable example of such a metal is iron, where the allotropic change makes it possible for heat-treating procedures that yield a wide range of final properties. It is largely because of its allotropy that iron has become the basis of our most important alloys.

There are 14 basic types of crystal structures or lattices. Fortunately, however, nearly all of the commercially important metals solidify into one of three types of lattice: body-centered cubic, face-centered cubic, or hexagonal close-packed. Table 3-1 lists the room temperature structure for a number of common metals. Figure 3-7 compares the structures to one another, as well as to the easily visualized, but rarely observed, simple cubic structure.

To begin our study of crystals, consider the *simple cubic* structure illustrated in Figure 3-7a. This structure can be constructed by placing single atoms on all corners of a cube and then linking identical cube units together. Assuming that the atoms are rigid spheres with atomic radii touching one another, computation reveals that only 52% of available space is occupied. Each atom is in direct contact with only six neighbors (plus and minus directions along the x, y, and z axes). Both of these observations are unfavorable to the metallic bond, where atoms desire the greatest number of nearest neighbors and high-efficiency packing.

The largest region of unoccupied space is in the volumetric center of the cube, where a sphere of 0.732 times the atom diameter could be inserted.[1] If the cube is expanded

FIGURE 3-7 Comparison of crystal structures: simple cubic, body-centered cubic, face-centered cubic, and hexagonal close-packed.

	Lattice structure	Unit cell schematic	Ping-pong ball model	Number of nearest neighbors	Packing efficiency	Typical metals
a	Simple cubic	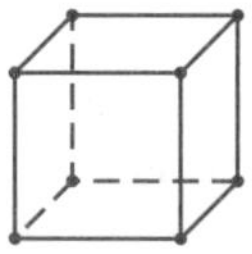		6	52%	None
b	Body-centered cubic	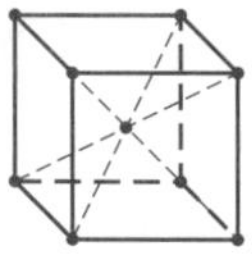		8	68%	Fe, Cr, Mn, Cb, W, Ta, Ti, V, Na, K
c	Face-centered cubic	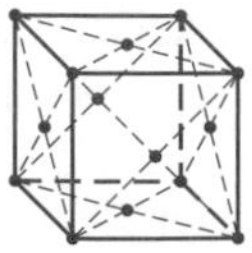		12	74%	Fe, Al, Cu, Ni, Ca, Au, Ag, Pb, Pt
d	Hexagonal close-packed	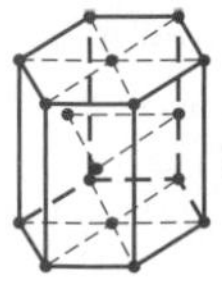		12	74%	Be, Cd, Mg, Zn, Zr

[1]The diagonal of a cube is equal to $\sqrt{3}$ times the length of the cube edge, and the cube edge is here equal to two atomic radii or one atomic diameter. Thus the diagonal is equal to 1.732 times the atom diameter and is made up of an atomic radius, open space, and another atomic radius. Since two radii equal one diameter, the open space must be equal in size to 0.732 times the atomic diameter.

TABLE 3-1.	Types of Lattices of Common Metals at Room Temperature
Metal	Lattice Type
Aluminum	Face-centered cubic
Copper	Face-centered cubic
Gold	Face-centered cubic
Iron	Body-centered cubic
Lead	Face-centered cubic
Magnesium	Hexagonal
Silver	Face-centered cubic
Tin	Body-centered tetragonal
Titanium	Hexagonal

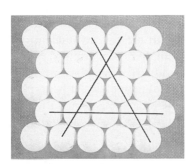

FIGURE 3-8 Close-packed atomic plane showing three directions of closest packing.

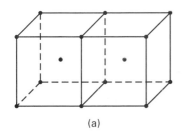

(a)

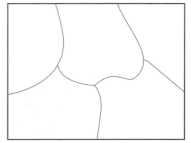

(b)

FIGURE 3-9 Growth of crystals to produce an extended lattice: (a) line schematic; (b) Ping-Pong ball model.

somewhat to allow for the insertion of an entire atom, the *body-centered-cubic (BCC)* structure results (Figure 3-7b). Each atom now has eight nearest neighbors, and 68% of the space is occupied. This structure is more favorable to metals and is observed in room-temperature iron, chromium, manganese, and the other metals listed in Figure 3-7b. Compared to materials with other structures, body-centered-cubic metals offer high engineering strength.

If Ping-Pong balls, used to simulate atoms, were placed in a box and agitated until a stable arrangement were produced, we would find that structure to consist of layered *close-packed planes*, where each layer or plane has the configuration shown in Figure 3-8. Two different structures can result, depending on the sequence in which the various layers are stacked. Both are identical in the number of nearest neighbors (12—six within the plane and three from each of the layers above and below) and the efficiency of occupying space (74%).

One of these sequences produces a structure that can also be viewed as an expanded cube with an atom inserted in the center of each of the six cube faces. This is the *face-centered-cubic (FCC)* structure shown in Figure 3-7c. It is the preferred structure for many of the engineering metals and tends to provide exceptionally high ductility (the ability to be plastically deformed without fracture).

The other sequence of stacking close-packed planes results in a structure known as *hexagonal close-packed (HCP)*, and here the individual close-packed planes can be clearly identified (Figure 3-7d). Metals having this structure tend to have poor ductility, fail in a brittle manner, and often require special processing procedures.

■ 3.8 DEVELOPMENT OF A GRAIN STRUCTURE

When a metal solidifies, a small particle of solid forms from the liquid with a lattice structure characteristic of the given material. This particle then acts like a seed or nucleus and grows as other atoms in the vicinity attach themselves. The basic crystalline unit is repeated throughout space, as illustrated in the body-centered-cubic examples of Figure 3-9.

In actual solidification, many nuclei form independently throughout the liquid and have random orientations with respect to one another. Each then grows until it begins to interfere with its neighbors, as illustrated in two dimensions in Figure 3-10. Since

FIGURE 3-10 Schematic representation of the growth of crystals to produce a polycrystalline material.

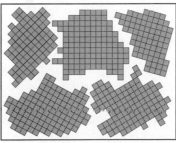
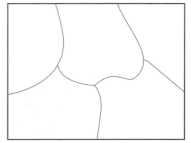

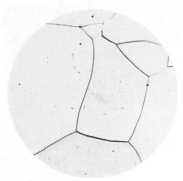

FIGURE 3-11
Photomicrograph of alpha ferrite (essentially pure iron) showing grain boundaries; 1000×. *(Courtesy of USX Corporation.)*

adjacent lattice structures have different alignments or orientations, growth cannot produce a single continuous structure. The small, continuous volumes of solid are known as crystals or *grains*, and the surfaces that divide them (i.e., the surfaces of crystalline discontinuity) are known as *grain boundaries*. The process by which a grain structure is produced upon solidification is one of *nucleation and growth*.

Grains are the smallest of the structural units in a metal that are observable with ordinary light microscopy. If a piece of metal is polished to mirror finish with a series of abrasives and then exposed to an attacking chemical for a short time (etched), the grain structure is revealed. The atoms in the grain boundaries are more loosely bonded and tend to react with the chemical more readily than those that are part of the grain interior. When viewed under reflected light, the attacked boundaries scatter light and appear dark compared to the relatively unaffected (still flat) grains (Figure 3-11). Occasionally, the individual grains are large enough to be seen by the unaided eye, as with some galvanized steels, but usually magnification is required.

The number and size of the grains in a metal vary with the rate of nucleation and the rate of growth. The greater the nucleation rate, the smaller the resulting grains. Conversely, the greater the rate of growth, the larger the grain. Because the resulting *grain structure* will influence certain mechanical and physical properties, it is an important property for an engineer to control and specify. One means of specification is through the *ASTM* (American Society for Testing and Materials) *grain size number*, defined as

$$N = 2^{n-1}$$

where N is the number of grains per square inch visible in a prepared specimen at 100X and n is the ASTM grain-size number. Low ASTM numbers mean a few massive grains, while high numbers refer to materials with many small grains.

■ 3.9 ELASTIC DEFORMATION

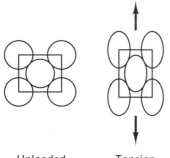

Unloaded   Tension

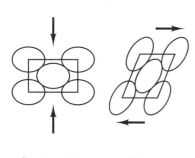

Compression   Shear

FIGURE 3-12   Distortion of a crystal lattice in response to elastic loadings.

The mechanical properties of materials are highly dependent upon the crystal structure of those materials. An understanding of mechanical behavior, therefore, begins with an understanding of the way crystals react when subjected to mechanical loads. Most studies begin with carefully prepared single crystals. Through them, we learn that the mechanical behavior is dependent on (1) the type of lattice, (2) the interatomic forces (i.e., bond strength), (3) the spacing between adjacent planes of atoms, and (4) the density of the atoms on the various planes.

If the applied loads are relatively low, the crystals respond by simply stretching or compressing the distance between atoms (Figure 3-12). The basic lattice unit does not change, and all of the atoms remain in their original positions relative to one another. The applied load serves only to disrupt the force balance of the atomic bonds, and the atoms assume new equilibrium positions with the applied load as an additional component of force. If the load is removed, the atoms return to their original positions and the crystal resumes its original size and shape. The response is *elastic* in nature, and the amount of stretch or compression is directly proportional to the applied load or stress.

Elongation or compression in the direction of loading results in an opposite change in dimensions at right angles to that direction. The ratio of lateral contraction to axial tensile strain is known as *Poisson's ratio*. This value is always less than 0.5 and is usually about 0.3.

■ 3.10 PLASTIC DEFORMATION

As the magnitude of applied load becomes greater, distortion continues to increase, and a point is reached where the atoms either (1) break bonds to produce a fracture, or (2) slide over one another in a way that would reduce the load. For metallic materials, the second phenomenon generally requires lower loads and occurs preferentially. The atomic planes shear over one another to produce a net displacement or permanent shift of atom positions, known as *plastic deformation*. Conceptually, this is similar to the distortion of a deck of playing cards

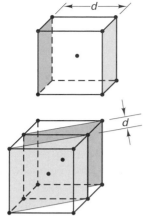

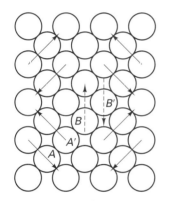

FIGURE 3-13   Schematic diagram showing crystalline planes with different atomic densities and interplanar spacings.

FIGURE 3-14   Planar schematic representing the greater deformation resistance of planes of lower atomic density and closer interplanar spacing.

when one card slides over another. The actual mechanism, however, is really a progressive one rather than all of the atoms in a plane shifting simultaneously. More significantly, this permanent change in shape occurs without a concurrent deterioration in properties.

Recalling that a crystal structure is a regular and periodic arrangement of atoms in space, it becomes possible to link the atoms into flat planes in an almost infinite number of ways. Planes having different orientations with respect to the surfaces of the unit cell will have different atomic densities and different spacing between adjacent, parallel planes (Figure 3-13). Given the choice of all possibilities, plastic deformation tends to occur along planes having the highest atomic density and greatest separation. The rationale for this can be seen in the simplified two-dimensional array of Figure 3-14. Planes A and A′ have higher density and greater separation than planes B and B′. In visualizing relative motion, the atoms of B and B′ would interfere significantly with one another, whereas planes A and A′ do not experience this difficulty.

Although Figure 3-14 represents the planes of sliding as lines, crystal structures are actually three-dimensional. Within the preferred planes are also preferred directions. If sliding occurs in a direction that corresponds to one of the close-packed directions (shown as dark lines in Figure 3-8), atoms can simply follow one another rather than each having to negotiate its own path. Plastic deformation, therefore, tends to occur by the preferential sliding of maximum-density planes (close-packed planes if present) in directions of closest packing. The specific combination of plane and direction is called a *slip system*, and the resulting shear deformation is known as *slip*.

The ease of deforming a given metal depends on the ease of shearing one atomic plane over an adjacent one and how favorably the plane is oriented with respect to the load. For example, consider the deck of playing cards. The deck does not "deform" when laid flat on the table and pressed from the top, or when stacked on edge and pressed uniformly. The cards will slide over one another only if the deck is skewed with respect to the applied load so as to induce a shear stress along the plane of sliding.

With this understanding, consider the deformation properties of the three most common crystal structures.

1. *Body-centered cubic.*   In the BCC structure, there are no close-packed planes. Slip occurs on the most favorable alternatives, which are those planes with the greatest interplanar spacing (six of which are illustrated in Figure 3-15). Within these planes, slip occurs along the directions of closest packing, which are the cube diagonals. If each specific combination of plane and direction is considered as a separate slip system, we find that the BCC materials contain 48 attractive ways to slip (plastically deform). The probability that one or more of these systems will be oriented in a favorable manner is great, but the force required to produce deformation is extremely large since there are no close-packed planes. Materials with this structure generally possess high strength with moderate ductility. (Refer to the typical BCC metals in Figure 3-7.)

2. *Face-centered cubic.*   In the FCC structure, each unit cell contains four close-packed planes, as illustrated in Figure 3-15. Each of those planes contains three close-packed directions, or face diagonals, giving 12 possible means of slip. Again, the probability that one or more of these will be favorably oriented is great, and this time, the force required to induce slip is quite low. Metals with the FCC structure are relatively weak and possess excellent ductility, as can be confirmed by a check of those listed in Figure 3-7.

FIGURE 3-15   Slip planes of the various lattice types.

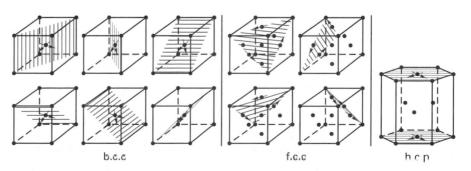

b.c.c   f.c.c   h c p

3. *Hexagonal close-packed.* The hexagonal lattice also contains close-packed planes, but only one such plane exists within the lattice. Although this plane contains three close-packed directions and the force required to produce slip is again rather low, the probability of favorable orientation to the applied load is small. As a result, metals with the HCP structure tend to have low ductility and are often classified as brittle.

■ 3.11 DISLOCATION THEORY OF SLIPPAGE

A theoretical calculation of the strength of metals based on the sliding of entire atomic planes over one another predicts yield strengths on the order of 3 million pounds per square inch or 20,000 MPa. The observed strengths of engineering metals are typically 100 to 150 times less than this value, but small laboratory crystals have been shown to exhibit the full theoretical strength.

An explanation is provided by the fact that plastic deformation does not occur by all of the atoms in one plane slipping simultaneously over all the atoms of an adjacent plane. Instead, deformation is the result of the progressive slippage of a localized disruption known as a *dislocation*. Consider a simple analogy. A carpet has been rolled onto a floor, and someone wants to move it a short distance in a given direction. One approach would be to pull on one end and try to "shear the carpet across the floor," simultaneously overcoming the frictional resistance over the entire area of contact. This would require a large force acting over a small distance. An alternative approach would be to form a wrinkle in one end of the carpet and walk it across the floor to produce a net shift in the whole carpet—a low-force-over-large-distance approach to the same task. In the region of the wrinkle, there is an excess of carpet with respect to the floor beneath it, and the motion of this excess is relatively easy.

Electron microscopes have revealed that metal crystals do not have all of their atoms in perfect arrangement, but contain a variety of localized imperfections. Two such imperfections are the *edge dislocation* and *screw dislocation* (Figure 3-16). Edge dislocations are the edges of extra half-planes of atoms. Screw dislocations correspond to partial tearing of the crystal plane. In each case the dislocation is a disruption to the regular, periodic arrangement of atoms and can be moved about with a rather low applied force. It is the motion of these atomic-scale dislocations under applied load that is responsible for the observed macroscopic plastic deformation.

All engineering metals contain dislocations, usually in abundant quantities. The ease of deformation, therefore, depends on the ease of inducing dislocation movement. Barriers to dislocation motion tend to increase the overall strength of a metal. These barriers take the form of other crystal imperfections and may be of the point type (missing atoms or *vacancies*, extra atoms or *interstitials*, or *substituting atoms* of a different variety, as may occur in an alloy), line type (another dislocation), or surface type (*crystal grain boundary* or *free surface*).

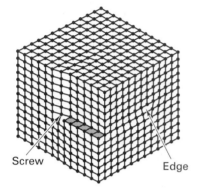

Screw   Edge

FIGURE 3-16 Schematic representation of screw and edge dislocations.

■ 3.12 STRAIN HARDENING OR WORK HARDENING

As noted in our discussion of the tensile test in Chapter 2, most metals become stronger when plastically deformed, a phenomenon known as *strain hardening* or *work hardening*. Understanding of this phenomenon can now come from our knowledge of dislocations and a further extension of the carpet analogy. Suppose that this time our goal is to move the carpet diagonally. The best way would be to move a wrinkle in one direction and then a second one 90° to the first. But suppose that both wrinkles were started simultaneously. We would find that wrinkle 1 would impede the motion of wrinkle 2, and vice versa. In essence, the device that makes deformation easy can also serve to impede the motion of other, similar devices.

In metals, plastic deformation is the result of dislocation movement. As dislocations move, they are more likely to encounter and interact with other dislocations or crystalline defects, thereby producing resistance to further motion. Moreover, mechanisms exist that markedly increase the number of dislocations in a metal during deformation, thereby enhancing the probability of interaction.

The effects of strain hardening become attractive when one considers that the mechanical working of metal is frequently performed to produce a more useful shape. Since strength can be increased substantially during deformation, a strain-hardened (deformed), inexpensive metal can often be substituted for an undeformed (such as machined to shape), costly one.

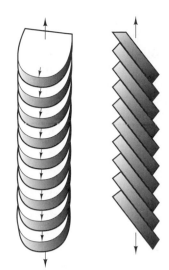

FIGURE 3-17 Schematic representation of slip and rotation resulting from deformation.

Experimental evidence has confirmed the dislocation and slippage theory of deformation. A transmission electron microscope can reveal images of the individual dislocations in a thin metal section, and studies confirm both the increase in number and interactions during deformation. Macroscopic observations also lend support. When a load is applied to a single metal crystal, deformation begins on the slip system that is most favorably oriented. The net result is often an observable slip and rotation, like that of a skewed deck of cards (Figure 3-17). Dislocation motion becomes more difficult as strain hardening produces increased resistance and rotation makes the orientation less favorable. Further deformation may then occur on alternative systems that now offer less resistance, a phenomenon known as *cross slip*.

■ 3.13 PLASTIC DEFORMATION IN POLYCRYSTALLINE METALS

Commercial metals are not single crystals, but usually take the form of polycrystalline aggregates. Within each crystal, deformation proceeds in the manner previously described. Since the various grains have different orientations, an applied load will produce different deformations within each of the crystals. This can be seen in Figure 3-18, where a metal has been polished and then deformed. The relief of the polished surface reveals the different slip planes for each of the grains.

One should note that the slip lines do not cross from one grain to another. The grain boundaries act as barriers to the dislocation motion. Therefore, metals with a finer grain structure—more grains per unit area—tend to exhibit greater strength and hardness, coupled with increased impact resistance. This near-universal improvement in properties is an attractive motivation for controlling grain size during processing.

■ 3.14 GRAIN SHAPE AND ANISOTROPIC PROPERTIES

When a metal is deformed, the grains tend to elongate in the direction of metal flow (Figure 3-19). Accompanying the nonsymmetric structure is nonsymmetry or directional variation in properties. Mechanical properties (such as strength and ductility), as well as physical properties (such as electrical and magnetic characteristics), may all exhibit directional differences. Properties that vary with direction are said to be *anisotropic*. Properties that are uniform in all directions are *isotropic*.

The directional variation of properties can be beneficial, and therefore assumes importance to both the part designer and the part manufacturer. By controlling the metal flow in processes such as forging, enhanced strength or fracture resistance can be imparted to certain locations. Caution should be exercised, however, since an improvement in one direction is generally accompanied by a decline in another. Moreover, directional variation in properties may impose serious difficulties in further processing operations, as with the further forming of rolled metal sheets.

FIGURE 3-18 Slip lines in a polycrystalline material. *(From Richard Hertzberg, Deformation and Fracture Mechanics of Engineering Materials; courtesy of John Wiley & Sons, Inc.)*

FIGURE 3-19 Deformed grains in cold-worked 1008 steel after 50% reduction by rolling; 1000×. *(From Metals Handbook, 8th ed. ASM International, Materials Park, Ohio, 1972.)*

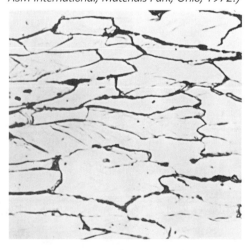

■ 3.15 FRACTURE OF METALS

If too much plastic deformation is attempted, the metal may respond by fracture. When plastic deformation precedes the break, the fracture is known as a *ductile fracture*. Fractures can also occur before the onset of plastic deformation. These sudden, catastrophic failures are known as *brittle fractures*, and are more common in metals having the BCC or HCP crystal structures. Whether the fracture is ductile or brittle, however, often depends on the specific conditions of material, temperature, state of stress, and rate of loading.

■ 3.16 COLD WORKING, RECRYSTALLIZATION, AND HOT WORKING

During plastic deformation, a portion of the deformation energy is stored within the material in the form of additional dislocations and increased grain boundary surface area.[2] If a deformed polycrystalline metal is subsequently heated to a high enough temperature, the material will seek to lower its energy. New crystals nucleate and grow to consume and replace the original structure (Figure 3-20). This process of reducing the internal energy through the formation of new crystals is known as *recrystallization*. The temperature at which recrystallization takes place is different for each metal and also varies with the amount of prior deformation. The greater the amount of prior deformation, the more stored energy, and the lower the recrystallization temperature. However, there is a lower limit below which recrystallization will not take place in a reasonable amount of time. Table 3-2 gives the lowest practical recrystallization temperatures for several materials. This temperature can often be estimated by taking 0.4 times the melting point of the metal when the melting point is expressed as an absolute temperature (Kelvin or Rankin). This is also the temperature at which atomic diffusion (atom movement within the solid) becomes significant, indicating that diffusion is an important mechanism in recrystallization.

When metals are plastically deformed at temperatures below their recrystallization temperature, the process is called *cold working*. The metal strain hardens, and the resultant structure consists of distorted grains. If deformation is continued, the metal may fracture. It is a common practice, therefore, to recrystallize the material after a certain amount of cold work. Through this *recrystallization anneal*, ductility is restored, and the material is capable of further deformation without the danger of fracture.

If the temperature of deformation is sufficiently above the recrystallization temperature, the deformation process becomes *hot working*. Deformation and recrystallization take place simultaneously, and extremely large deformations are now possible. Since a recrystallized grain structure is constantly forming, the final product will not exhibit strain hardening.

Since fine grain size is a means of imparting improved properties, recrystallization becomes an attractive property-control process. A coarse grain structure can be

TABLE 3-2.	Lowest Recrystallization Temperature of Common Metals
Metal	Temperature [°F(°C)]
Aluminum	300 (150)
Copper	390 (200)
Gold	390 (200)
Iron	840 (450)
Lead	Below room temperature
Magnesium	300 (150)
Nickel	1100 (590)
Silver	390 (200)
Tin	Below room temperature
Zinc	Room temperature

---

[2] A sphere has the least amount of surface area of any shape to contain a given volume of material. When the shape becomes altered from that of a sphere, the surface area must increase. Consider a round balloon filled with air. If the balloon is stretched or flattened into another shape, the rubber balloon is stretched further. When the applied load is removed, the balloon snaps back to its original shape, the one involving the least surface energy. Metals behave in an analogous manner. During deformation, the distortion of the crystals increases the energy of the material. Given the opportunity, the material will try to lower its energy by returning to spherical grains.

  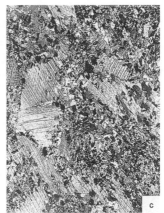 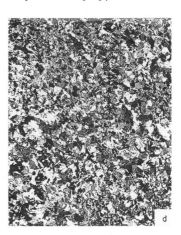

FIGURE 3-20 Recrystallization of 70–30 brass: (a) cold-worked 33%; (b) heated at 580°C (1075°F) for 3 seconds, (c) 4 seconds, and (d) 8 seconds; 45×. *(Courtesy of J.E. Burke, General Electric Company.)*

converted to a fine grain structure through recrystallization. The material must first be plastically deformed to store sufficient energy to provide the driving force. Control of the recrystallization process then establishes the final grain size in the material.

■ 3.17 GRAIN GROWTH

The recrystallization process tends to produce uniform grains of comparatively small size. If the metal is held at or above the recrystallization temperature for any appreciable time, however, the grains in the new structure will continue to increase in size. In effect, some of the grains will become larger at the expense of their small neighbors as the material seeks to further decrease the amount of grain boundary surface area. Since engineering properties tend to diminish with increased grain size, control of recrystallization is of prime importance. The material should be held at elevated temperature long enough to complete the recrystallization process. The temperature should then be decreased to prevent the further changes that would accompany *grain growth*.

■ 3.18 ALLOYS AND ALLOY TYPES

Our discussion thus far has been directed toward the nature and behavior of pure metals. For most manufacturing applications, however, metals are not used in their pure form. Instead, engineering metals tend to be *alloys*, materials composed of two or more different elements, and they tend to exhibit their own characteristic properties.

There are three ways in which a metal might respond to the addition of another element. The first, and probably the simplest, response occurs when the *two materials are insoluble in one another in the solid state*. In this case the base metal and the alloying addition each maintain their individual identities, structures, and properties. The alloy in effect becomes a composite structure, consisting of two types of building blocks in an intimate mechanical mixture.

The second possibility occurs when the *two elements exhibit some degree of solubility in the solid state*. The two materials form a *solid solution*, where the alloy element dissolves in the base metal. The solutions can be: (1) *substitutional* or (2) *interstitial*. In the substitutional solution, some atoms of the alloy element occupy lattice sites normally filled by atoms of the base metal. The replacement is totally random in nature, with the alloy atoms being distributed throughout the base lattice. In the interstitial solution, the alloy element atoms squeeze into the open spaces between the atoms in the base metal lattice.

A third possibility exists where the *elements combine to form intermetallic compounds*. In this case, the atoms of the alloying element interact with the atoms of the base metal in definite proportions and in definite geometric relationships. The bonding is primarily of the nonmetallic variety (i.e., ionic or covalent), and the lattice structures are often quite complex. Because of the type of bonding, intermetallic compounds tend to be hard, but brittle, high-strength materials.

Even though alloys are composed of more than one type of atom, their structure is still one of crystalline lattices and grains. Their behavior in response to applied loadings is similar to that of pure metals, with some features reflecting the increased level of structural complexity. Dislocation movement can be further impeded by the presence of unlike atoms. If neighboring grains have different chemistries and/or structures, they may respond differently to the same type and magnitude of load.

■ 3.19 ATOMIC STRUCTURE AND ELECTRICAL PROPERTIES

In addition to the mechanical properties, the structure of a material also influences its electrical behavior. *Electrical conductivity* refers to the net movement of charge through the material. In metals, the charge carriers are the valence electrons. The more perfect the atomic arrangement, the higher the electrical conductivity. Conversely, the greater the number of lattice imperfections or irregularities, the higher the resistance to electrical conduction.

The electrical resistance of a metal, therefore, depends largely on two factors: (1) lattice imperfections and (2) temperature. Vacant atomic sites, interstitial atoms, substitutional atoms, dislocations, and grain boundaries all act as disruptions to the regularity of a crystalline lattice. Thermal energy causes the atoms to vibrate about their equilibrium position. These vibrations cause the atoms to be out of position, which further impedes electron travel. For a metal, electrical conductivity will decrease with an increase in temperature. As the temperature is decreased, the electrical conductivity becomes primarily a function of the crystalline imperfections. The best metallic conductors, therefore, are pure metals with large grain size, at low temperature.

The electrical conductivity of a metal is due to the movement of the free electrons in the metallic bond. For covalently bonded materials, however, bonds must be broken to provide the electrons for charge transport. Therefore, the electrical properties of these materials is a function of bond strength. Diamond, for instance, is a strong insulator. Silicon and germanium have weaker bonds that are more easily broken by thermal energy. These metals are known as *intrinsic semiconductors*, since moderate amounts of thermal energy enable the materials to conduct small amounts of electricity. Continuing down Group IV of the periodic table of elements, we find that tin has such weak bonding that a high number of bonds are broken at room temperature, and the electrical behavior resembles that of a metal.

The electrical conductivity of intrinsic semiconductors can be substantially improved by a process known as *doping*. Silicon and germanium each have four valence electrons and form four covalent bonds. If one of the bonding atoms were replaced with an atom containing five valence electrons such as phosphorus or arsenic, the four covalent bonds would form, leaving an additional valence electron that was not involved in the bonding process. This extra electron would be free to move about and provide additional conductivity. Materials doped in this manner are known as *n-type extrinsic semiconductors*.

A similar effect can be created by substituting an atom with only three valence electrons, such as aluminum. An electron will be missing from one of the bonds, creating an *electron hole*. When a voltage is applied, a nearby electron can jump into this hole, creating a hole in the location that it vacated. Movement of electron holes is equivalent to a countermovement of electrons, and thus provides additional conductivity. Materials containing three-electron dopants with three valence electrons are known as *p-type semiconductors*. The ability to control the electrical conductivity of semiconductor material is the functional basis of solid-state electronics and circuitry.

■ KEY WORDS

allotropic	covalent bond	grain boundary	isotropic	slip
alloy	cross slip	grain growth	lattice	slip system
amorphous structure	crystal structure	grain size	metallic bond	solid solution
anisotropic	dislocation	hexagonal close-packed	microstructure	strain hardening
ASTM grain-size number	doping	hot work	molecular structure	substitutional atom
atomic number	ductile fracture	intermetallic compound	nucleation and growth	unit cell
atomic radius	electrical conductivity	interstitial	Poisson's ratio	vacancy
body-centered cubic	extrinsic semiconductor	intrinsic semiconductor	polymorphic	valence electrons
brittle fracture	face-centered cubic	ion	recrystallization	van der Waals forces
close-packed planes	grain	ionic bond	simple cubic	work hardening
cold work				

■ REVIEW QUESTIONS

1. Why might an engineer be concerned with controlling or altering the structure of a material?
2. What is meant by the term *microstructure*?
3. What is an ion and what are the two varieties?
4. What properties or characteristics of a material are influenced by the valence electrons?
5. What are the three types of primary bonds, and what types of atoms do they unite?
6. What are some general characteristics of ionically bonded materials?
7. What are some general properties and characteristics of covalently bonded materials?
8. What are some unique property features of materials bonded by metallic bonds?
9. What causes the bonding forces in the van der Waals type of secondary bond?
10. Why is the atomic radius of an atom often assigned different values for different crystal structures?
11. What is the difference between a crystalline material and one with an amorphous structure?
12. What are some of the general characteristics of metallic materials?
13. What is an allotropic material?
14. Why is the simple cubic crystal structure not observed in the engineering metals?
15. What are the three most common crystal structures found in metals?
16. What is the efficiency of filling space with spheres in the simple cubic structure? Body-centered-cubic structure? Face-centered-cubic structure? Hexagonal close-packed structure?
17. Describe the difference in ductility observed between the two structures formed by stacking close-packed planes.
18. What is a grain boundary?
19. What is the most common means of quantifying the grain size of a solid metal?
20. How does a metallic crystal respond to low applied loads?
21. What is plastic deformation?
22. What is a slip system in a material? What types of planes and directions tend to be preferred?
23. Based on the ease or difficulty of deformation, what is the dominant mechanical property or characteristic for each of the three most common metal crystal structures?
24. What is a dislocation? How do dislocations determine the mechanical properties of a metal?
25. What are some of the common barriers to dislocation movement that can be used to strengthen a metal?
26. What are the three major types of point defects in crystalline materials?
27. What is the mechanism (or mechanisms) responsible for the observed strain hardening of a metal?
28. Why is a fine grain size often desired in an engineering metal?
29. What is an anisotropic property? What is a possible cause?
30. What is the difference between brittle fracture and ductile fracture?
31. How does a metal increase its internal energy during plastic deformation?
32. In what ways can recrystallization be used to enable large amounts of deformation without fear of fracture?
33. What is the major distinguishing feature between hot and cold working?
34. What types of structures can be produced when an alloy is added to a base metal?
35. As a result of the ionic or covalent bonding, what types of mechanical properties are characteristic of intermetallic compounds?
36. What structural unit is responsible for the movement of electrical charge in a metal?
37. What features in a metal structure tend to impede or reduce electrical conductivity?
38. What is the difference between an intrinsic semiconductor and an extrinsic semiconductor?

*C*hapter 3 CASE STUDY

Window Frame Materials and Design

Because of your knowledge of engineering materials, a friend has asked for your assistance in evaluating various materials that are used for household window frames. He is in the process of designing a new home and wants the windows to be energy efficient and durable, and also to require low maintenance. Various suppliers have recommended wood, aluminum, and vinyl windows, each claiming their product to be superior. For each of these materials, identify its primary assets and significant limitations with regard to window-frame construction. Assume that the material will see both interior and exterior exposure. Consider ease of fabrication, strength, thermal expansion and contraction, response to moisture and humidity, durability, rigidity, ability to be finished in a variety of colors, properties at low and high extremes of temperature, and any other factors that you deem important.

Based on your performance evaluations, combined with cost, which material would you recommend for your particular location? Might your recommendation change if the location of the home were in the dry Southwest (Arizona)? New England? Alaska? Hawaii? Could you imagine some means of combining materials to produce a window that would be superior to any of the single materials?

CHAPTER 4

EQUILIBRIUM PHASE DIAGRAMS AND THE IRON–CARBON SYSTEM

■ 4.1 INTRODUCTION

As our study of engineering materials becomes more focused on specific metals and alloys, it is increasingly important that the natural characteristics and properties of the material be known. What is the basic structure of the material? Is the material uniform throughout, or is it a mixture of two or more distinct components? If there are multiple components, how much of each is present, and what are the different chemistries? Is there a component that may impart undesired properties or characteristics? What will happen if temperature is increased or decreased, pressure is changed, or chemistry is varied? The answers to these and other important questions can be obtained through the use of *equilibrium phase diagrams*.

■ 4.2 PHASES

Before we move to a discussion of equilibrium phase diagrams, it is important that we first develop a working definition of the term *phase*. As a starting definition, a phase is simply a form of material possessing a characteristic structure and characteristic properties. Uniformity of chemistry, structure, and properties is assumed throughout a phase. More rigorously, a phase has *a definable structure, a uniform and identifiable chemistry* (also known as *composition*), and distinct *boundaries* or *interfaces* that separate it from other different phases.

A phase can be continuous (like the air in a room) or discontinuous (like grains of salt in a shaker). A phase can be solid, liquid, or gas. In addition, a phase can be a pure substance or a solution, provided that the structure and composition are uniform throughout. Alcohol and water mix in all proportions and will therefore form a single phase when combined. Oil and water tend to form isolated regions with distinct boundaries and must be regarded as two distinct phases.

■ 4.3 EQUILIBRIUM PHASE DIAGRAMS

The *equilibrium phase diagram* is a graphic mapping of the natural tendencies of a material or a material system, assuming that equilibrium has been attained for all possible conditions. There are three primary variables to be considered: *temperature*, *pressure*, and *composition*. The simplest phase diagram is a pressure–temperature (*P–T*) diagram for a fixed-composition material. Areas of the diagram are assigned to the various phases, with the boundaries indicating the equilibrium conditions of transition.

As an introduction, consider the *P–T* diagram for water presented as Figure 4-1. With composition fixed, the diagram maps the stable form of water for various conditions of temperature and pressure. If pressure is held constant and temperature is

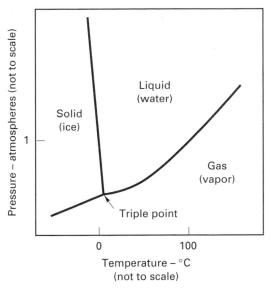

FIGURE 4-1 Pressure–temperature diagram for water.

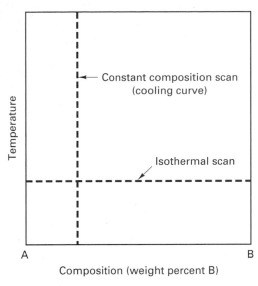

FIGURE 4-2 Mapping for a temperature–composition equilibrium phase diagram.

varied, the transition boundaries locate the melting and boiling points. For example, at 1 atmosphere pressure, water melts at 0°C and boils at 100°C. Still other uses are possible. Locate a temperature where the stable phase is liquid at atmospheric pressure. Now drop the temperature until the material goes from liquid to solid (i.e., ice). Maintain that new temperature and begin to decrease the pressure. A transition is encountered where solid goes directly to gas without melting (sublimation). The total process described above is known as *freeze drying*, and is employed in the manufacture of numerous dehydrated products. With an appropriate phase diagram, process conditions can be determined that will minimize the amount of required cooling and the pressure drop required for sublimation.

TEMPERATURE–COMPOSITION DIAGRAMS

While the *P–T* diagram for water is an excellent introduction to phase diagrams, *P–T* phase diagrams are rarely used for engineering applications. Most engineering processes are conducted at atmospheric pressure, and variations are more likely to occur in temperature and composition. The most useful mapping, therefore, would be a *temperature–composition phase diagram* at atmospheric pressure. For the remainder of the chapter, this is the form of phase diagram that will be considered.

For mapping purposes, temperature is placed on the vertical axis and composition on the horizontal. Figure 4-2 shows the axes for mapping the *A–B* system, where the left-hand vertical corresponds to pure material *A* and the percentage of *B* (usually expressed in weight percent) increases as we move toward pure *B* at the right side of the diagram. The temperature range often includes only solids and liquids, since few processes involve engineering materials in the gaseous state. Experimental investigations that will provide the details of the diagram take the form of either vertical or horizontal scans that seek to locate the transitions between phases.

COOLING CURVES

Considerable information can be obtained from vertical scans through the diagram where a fixed composition material is heated and subsequently cooled at a uniformly slow rate. Transitions in structure appear as characteristic points in the temperature-versus-time plot of the cooling history, known as a *cooling curve*.

Consider the system composed of sodium chloride (common table salt) and water. Five different cooling curves are presented in Figure 4-3. Curve (a) is for pure water being cooled from the liquid state. A decreasing-temperature line is observed for the liquid where the extraction of heat produces a concurrent drop in temperature. When the

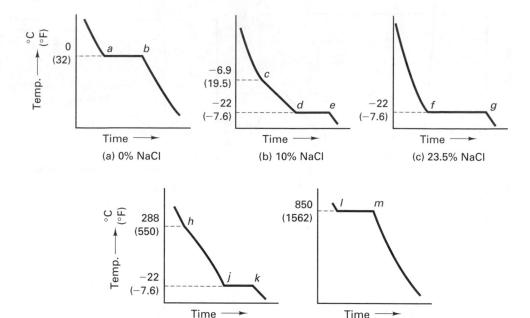

FIGURE 4-3 Cooling curves for various compositions of NaCl–H₂O solutions: (a) 0% NaCl; (b) 10% NaCl; (c) 23.5% NaCl; (d) 50% NaCl; (e) 100% NaCl.

freezing point is reached (point *a*), the material changes state and releases heat energy as part of the liquid-to-solid transition. Heat is being continuously extracted from the system, but since its source is now the change in state, there is no companion decrease in temperature. An isothermal or constant-temperature hold (*a–b*) is observed until the solidification is complete. From this point, the newly formed solid experiences a steady drop in temperature as heat extraction continues. This type of curve is characteristic of pure metals and other substances with a distinct melting point.

Curve (b) in Figure 4-3 presents the cooling curve for a solution of 10% salt in water. The liquid region undergoes continuous cooling down to point *c*, where the slope abruptly decreases. At this temperature, small particles of ice (i.e., solid) begin to form and the reduced slope is attributed to the energy released in this transition. The formation of these ice particles leaves the remaining solution richer in salt and imparts a lower freezing temperature to it. Further cooling results in the formation of additional solid, which continues to enrich the solution and further lowers the freezing point of the remaining liquid. Instead of possessing a distinct melting point or freezing point, this material is said to have a freezing range. When the temperature of point *d* is reached, the remaining liquid undergoes an abrupt reaction and solidifies into an intimate mixture of solid salt and solid water (discussed later), and an isothermal hold is observed. Further extraction of heat produces a drop in the temperature of the solidified material.

For a solution of 23.5% salt in water, a distinct freezing point is again observed, as shown in curve (c). Compositions with richer salt concentration [curve (d)] show phenomena similar to those in curve (b), but with salt being the first solid to form from the liquid. Finally, pure salt [curve (e)] exhibits behavior similar to that of pure water.

If the observed transition points are now transferred to a temperature–composition diagram, Figure 4-4 presents such a map which summarizes behavior of the system. Line *a–c–f–h–l* denotes the lowest temperature at which the material is totally liquid and is known as the *liquidus* line. Line *d–f–j* denotes a particular three-phase reaction and will be discussed later. Between the lines, two phases coexist, one being a liquid and the other a solid.

The cooling curve studies have provided some key information regarding the salt-water system, including some insight into the use of salt on highways in the winter. With the addition of salt, the freezing point of water can be lowered from 0°C (32°F) to as low as −22°C (−7.6°F). The equilibrium phase diagram can be viewed as a collective presentation of cooling curve data for an entire range of alloy compositions.

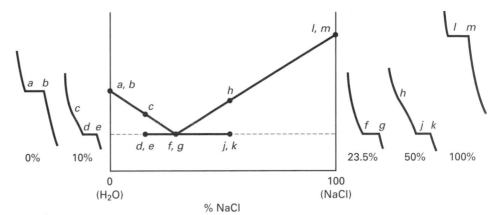

FIGURE 4-4 Partial equilibrium diagram for NaCl and H_2O derived from cooling-curve information.

SOLUBILITY STUDIES

The observant reader will note that the ends of the diagram still remain undetermined. Both pure materials have a distinct melting point, below which they appear as a pure solid. Can ice retain some salt in a single-phase solid solution? Can solid salt hold some water and remain a single phase? If so, how much, and does the amount vary with temperature? Completion of the diagram, therefore, requires several horizontal scans to determine any *solubility limits* and how they might vary with temperature.

These isothermal (constant temperature) scans usually require the preparation of specimens over a range of composition and their subsequent analysis by x-ray techniques, microscopy, or other methods to determine whether the structure and chemistry is uniform or a two-phase mixture. As we move away from the pure material, we often encounter a single-phase solid solution, in which one component is dissolved and dispersed throughout the other. If the solubility is limited, there will be a transition in the phase diagram, known as a *solvus* line, where the single-phase solid solution becomes a two-phase mixture. Figure 4-5 presents the equilibrium phase diagram for the lead–tin system, using the conventional notation in which Greek letters are used to denote the various single-phase solids. The upper portion of the diagram closely resembles the salt-water diagram, but the partial solubility of one material in the other can be observed on both ends of the solid-state region.

COMPLETE SOLUBILITY IN BOTH LIQUID AND SOLID STATES

Having developed the basic concepts of equilibrium phase diagrams, let us consider some examples, moving from the simple to the more complex. If two materials are each completely soluble in the other in both the liquid and solid states, a rather simple diagram results, like the copper–nickel diagram of Figure 4-6. The upper line is a *liquidus*

FIGURE 4-5 Lead–tin equilibrium diagram.

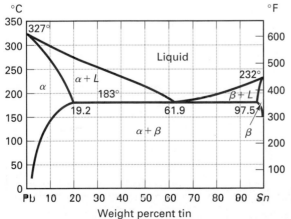

FIGURE 4-6 Copper–nickel equilibrium diagram, showing complete solubility in both liquid and solid states.

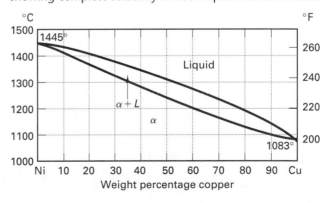

line. Above the liquidus, the two materials form a uniform-chemistry liquid solution. The lower line denotes the highest temperature at which the material is completely solid and is known as a *solidus* line. Below the solidus, the materials form a solid-state solution in which the two types of atoms are uniformly distributed throughout a single crystalline lattice. Between the liquidus and solidus is a *freezing range*, a two-phase region where liquid and solid solutions coexist.

PARTIAL SOLID SOLUBILITY

Many materials fail to exhibit complete solubility in the solid state. Each is often soluble in the other up to a certain limit or saturation point, which varies with temperature. Such a diagram has already been observed in the lead–tin system of Figure 4-5.

At the point of maximum solubility, 183°C, lead can hold up to 19.2 wt% tin in a single-phase solution and tin can hold up to 2.5% lead within its structure and still be single phase. If the temperature is decreased, however, the amount of solute that can be held in solution decreases in a continuous manner. Therefore if a saturated solution of tin in lead is cooled from 183°C, the material goes from a single-phase solution to a two-phase mixture as a tin-rich second phase precipitates from solution. This change in structure can be used to alter and control the properties in a number of engineering alloys.

INSOLUBILITY

If one or both of the components is totally insoluble in the other, the diagrams also reflect this phenomenon. Figure 4-7 illustrates the case where component *A* is completely insoluble in component *B* in both the liquid and solid states.

UTILIZATION OF DIAGRAMS

Before moving to more complex diagrams, let us return to a simple phase diagram, such as the one in Figure 4-8, and develop several useful tools. For each temperature and composition point, we can obtain three pieces of information:

1. *The phases present.* The stable phases can be determined by simply locating the point of consideration on the temperature–composition mapping and identifying the region of the diagram in which the point appears.

2. *The composition of each phase.* If the point lies in a single-phase region, the composition (or chemistry) of the phase is simply the composition of the alloy being considered. If the point lies in a two-phase region, a *tie-line* is constructed. A tie-line is an isothermal (constant-temperature) line drawn through the point of consideration, terminating at the boundaries of the single-phase regions on either side. The compositions where the tie-line intersects the neighboring single-phase regions are the compositions of those respective phases in the two-phase mixture. For example, consider point *a* in Figure 4-8. The tie-line for this point runs from S_2 to L_2. The point

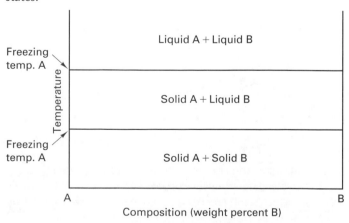

FIGURE 4-7 Equilibrium diagram of two materials, that are completely insoluble in each other in both the liquid and solid states.

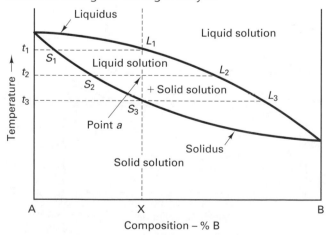

FIGURE 4-8 Equilibrium diagram showing the changes that occur during the cooling of alloy X.

S_2 is the intersection of the tie-line with the solid phase, and thus the solid in the two-phase mixture at point a has the composition of point S_2. Similarly, the liquid phase will have the composition of point L_2.

3. *The amount of each phase present.* If the point lies in a single-phase region, all of the material, or 100%, must be of that phase. If the point lies in a two-phase region, the relative amounts of the two components can be determined by a *lever-law* calculation using the previously drawn tie-line. Consider the cooling of alloy X in Figure 4-8 in a manner sufficiently slow so as to preserve equilibrium. At temperatures above t_1, the material is in a single-phase liquid state. Temperature t_1, therefore, is the lowest temperature at which the alloy is 100% liquid. If we draw a tie-line at this temperature, it runs from S_1 to L_1, and lies entirely to the left of composition X. At temperature t_3, the alloy is completely solid, and the tie-line lies completely to the right of composition X. Extrapolating these observations to the intermediate temperatures, such as temperature t_2, we predict that the fraction of the tie-line that lies to the left of point a corresponds to the fraction of the material that is liquid. This fraction can be computed as:

$$\frac{a - S_2}{L_2 - S_2} \times 100\%$$

In a similar manner, the fraction of solid corresponds to the fraction of the tie-line that lies to the right of point a. (*Note*: The mathematical relations could be rigorously derived from the conservation of either A or B atoms, as the material divides into two different compositions.) Since the calculations consider the tie-line as a lever with the phases at each end and the fulcrum at the composition line, they are called lever-law calculations.

Equilibrium phase diagrams can also be used to provide an overall picture of an alloy system, or to identify the transition points for various changes in phase. For example, the temperature required to redissolve a second phase or melt an alloy can be easily determined. The various changes that occur during the slow heating or slow cooling of a material can be predicted. In fact, most of the questions posed at the beginning of this chapter can now be answered.

SOLIDIFICATION OF ALLOY X

Let us now apply the tools we have just developed and follow the solidification of alloy X in Figure 4-8. At temperature t_1, the first minute amount of solid forms with the chemistry of point S_1. As the temperature drops, more solid forms, but the chemistries of both the solid and liquid phases shift to follow the tie-line endpoints. The chemistry of the liquid follows the liquidus line and the chemistry of the solid follows the solidus. Finally, at t_3, solidification is complete, and the composition of the single-phase solid is that of alloy X, as required.

The composition of the first solid to form is different from that of the final solid. If the cooling is sufficiently slow, such that equilibrium is maintained or approximated, the composition of the entire mass of solid shifts during cooling and follows the endpoint of the tie-line. The chemistry changes are made possible by diffusion, the process in which atoms migrate through the crystal lattice given sufficient time at elevated temperature. If the cooling rate is too rapid, however, the temperature may drop before sufficient diffusion occurs. The resultant material will have a nonuniform chemistry. The initial solid that formed will retain a chemistry that is different from the solid regions that formed later. When these variations occur on a microscopic level, the resultant structure is referred to as being *cored*. Variation on a larger scale is called *macrosegregation*.

THREE-PHASE REACTIONS

Several of the phase diagrams that were presented earlier contain a feature in which phase regions are separated by a horizontal line. These lines are further characterized by either a V intersecting from above or an inverted-V intersecting from below. The intersection of the V and the line denotes the location of a *three-phase equilibrium reaction*.

One common type of three-phase reaction, known as a *eutectic*, has already been observed in Figures 4-4, 4-5, and 4-7. It is possible to understand these reactions through use of the tie-line and lever-law concepts that have been developed. Refer to the lead–tin diagram of Figure 4-5 and consider any alloy containing between 19.2 and 97.5 wt% tin at a temperature just above the 183°C horizontal line. Tie-line and lever-law computations reveal that the material contains either a lead-rich or tin-rich solid and remaining liquid. At this temperature, any liquid that is present will have a composition of 61.9 wt% tin, regardless of the overall composition of the alloy. If we now focus on the liquid and allow it to cool to just below 183°C, a transition occurs in which liquid of composition 61.9% tin transforms to a mixture of lead-rich solid with 19.2% tin and tin-rich solid containing 97.5% tin. The relative amounts of the two components are such as to maintain the overall chemistry of 61.9% tin. The form of this eutectic transition is similar to a chemical reaction where

$$\text{liquid} \longrightarrow \text{solid}_1 + \text{solid}_2$$

Since the two solids have chemistries on either side of the intermediate liquid, a separation must have occurred within the system. Such a separation will occur in solidifying melts where the two materials are soluble in the liquid state but only partially soluble in the solid state. Separation requires atom movement, but the distances involved cannot be great. Therefore, the resulting eutectic structure is an intimate mixture of two single-phase solids. For a given reaction, the eutectic structure always forms from the same chemistry at the same temperature, and therefore has its own characteristic set of physical and mechanical properties. Alloys with the eutectic composition have the lowest melting point of all alloys within the system and are often used as casting alloys or filler material in soldering or brazing operations.

Figure 4-9 summarizes the variety of three-phase reactions that may occur in engineering systems. These include the *peritectic*, *monotectic*, and *syntectic* reactions, where the suffix *-ic* denotes that at least one of the three phases in the reaction is a liquid. If the same prefix appears with an *-oid* suffix, the reaction form is the same, but all phases involved are solids. Two such reactions are possible, the *eutectoid* and the *peritectoid*. The solid-state reactions tend to be a bit more sluggish, since all changes must occur within (usually crystalline) solids.

INTERMETALLIC COMPOUNDS

A final phase diagram feature occurs in alloy systems where the bonding attraction of the component materials is strong enough to form compounds. These compounds are single-phase solids and tend to break the diagram into recognizable subareas. If

FIGURE 4-9 Schematic summary of three-phase reactions and intermetallic compounds.

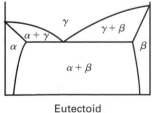

Eutectic
$(L \rightarrow S_1 + S_2)$

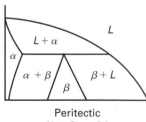

Peritectic
$(L + S_1 \rightarrow S_2)$

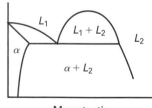

Monotectic
$(L_1 \rightarrow S_1 + L_2)$

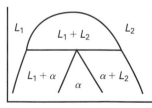

Syntectic
$(L_1 + L_2 \rightarrow S_1)$

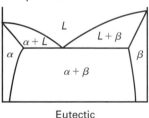

Eutectoid
$(S_1 \rightarrow S_2 + S_3)$

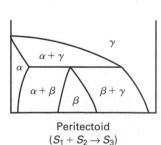

Peritectoid
$(S_1 + S_2 \rightarrow S_3)$

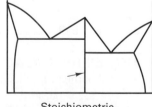

Stoichiometric
intermetallic compound

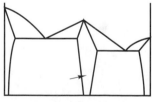

Non-stoichiometric
intermetallic compound

components *A* and *B* form a compound A_xB_y and the compound cannot tolerate any deviation from that fixed atomic ratio, the product is known as a *stoichiometric intermetallic compound* and it appears as a single vertical line in the diagram. If some degree of chemical deviation is tolerable, the vertical line expands into a single-phase region, and the compound is known as a *nonstoichiometric intermetallic compound*. Figure 4-9 shows schematic representations of both stoichiometric and nonstoichiometric compounds.

In general, intermetallic compounds tend to be hard, brittle materials, since these properties are a consequence of their ionic or covalent bonding. If they are present in large quantities or lie along grain boundaries in the form of a continuous film, the overall alloy can be extremely brittle. If the same compound is dispersed throughout the alloy in the form of small discrete particles, the result can be a considerable strengthening of the base metal.

COMPLEX DIAGRAMS

The equilibrium diagrams for actual alloy systems may be one of the basic types just discussed or some combination of them. In some cases the diagrams may appear to be quite complex and formidable. However, by focusing on a particular composition and analyzing specific points using the tie-line and lever-law concepts, even the most complex diagram can be interpreted. If the properties of the various components are known, it is then possible to predict the behavior of the resultant structures.

■ 4.4 IRON–CARBON EQUILIBRIUM DIAGRAM

Steel, composed primarily of iron and carbon, is clearly the most important of the engineering metals. For this reason, the iron–carbon equilibrium diagram assumes special importance. The diagram most frequently encountered, however, is not the full iron–carbon diagram but the iron–iron carbide diagram shown in Figure 4-10. Here, a stoichiometric intermetallic compound, Fe_3C, is used to terminate the carbon range at 6.67 wt% carbon. The names of key phases and structures, and the specific notations used on the diagram, have evolved historically and will be used in their generally accepted form.

There are four single phases within the diagram. Three of these occur in pure iron, and the fourth is the carbide intermetallic at 6.67% carbon. Upon solidification, pure iron forms a body-centered-cubic solid that is stable down to 1394°C (2541°F). Known as *delta-ferrite*, this phase is present only at extreme elevated temperatures and has little engineering importance. From 1394 to 912°C (2541 to 1674°F) pure iron assumes a

FIGURE 4-10 The iron–carbon equilibrium diagram: α, ferrite; γ, austenite; δ, δ–ferrite; Fe_3C, cementite.

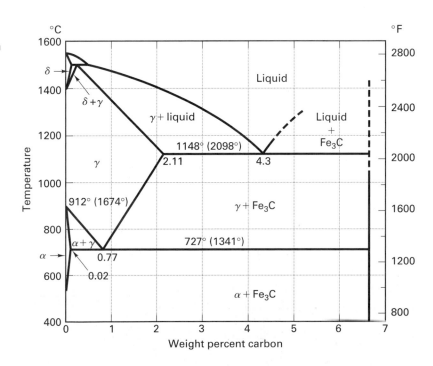

face-centered-cubic structure known as *austenite* (γ) in honor of the famed metallurgist Sir Robert Austen of England. Key features of austenite are the high formability that is characteristic of the face-centered-cubic structure and the high solubility of carbon (over 2% carbon can be dissolved in austenite). Hot forming of steel takes advantage of the high ductility and chemical uniformity of austenite. Most of the heat treatments of steel begin with the single-phase austenite structure. Alpha ferrite, or more commonly just *ferrite*, is the stable form of iron at temperatures below 912°C (1674°C). This body-centered-cubic structure can hold only 0.02 wt% carbon in solid solution and forces the creation of a two-phase mixture in most steels. The only other change to occur upon further cooling is the nonmagnetic-to-magnetic transition at the Curie point of 770°C (1418°F). Because this transition is not associated with any change in phase (but is an atomic-level transition), it does not appear on the equilibrium phase diagram.

The fourth single phase is the stoichiometric intermetallic compound, Fe_3C, which goes by the name *cementite*, or iron–carbide. Like most intermetallics, it is quite hard and brittle, and care should be exercised in controlling the structures in which it occurs. Alloys with excessive amounts of cementite, or cementite in undesirable form, tend to have brittle characteristics. Because cementite dissociates prior to melting, its exact melting point is unknown, and the liquidus line remains undetermined in the high-carbon region of the diagram.

Three distinct three-phase reactions can also be identified in the diagram. At 1495°C (2723°F), a *peritectic* occurs for alloys with a low weight percentage of carbon. Because of its high temperature and the extensive single-phase austenite region immediately below it, the peritectic reaction rarely assumes any engineering significance. A *eutectic* is observed at 1148°C (2098°F), with the eutectic composition of 4.3% carbon. All alloys containing more than 2.11% carbon will experience the eutectic reaction and are classified by the general term *cast irons*. The final three-phase reaction is a *eutectoid* at 727°C (1341°F) with the eutectoid composition of 0.77 wt% carbon. Alloys with less than 2.11% carbon miss the eutectic reaction and form a two-phase mixture when they cool through the eutectoid. These alloys are known as *steels*. Thus the point of maximum carbon solubility in iron, 2.11 wt%, forms an arbitrary separation between steels and cast irons.

■ 4.5 STEELS AND THE SIMPLIFIED IRON–CARBON DIAGRAM

If we focus on the materials normally known as steel, the diagram can be simplified considerably. Those portions near the delta phase (or peritectic) region and those with greater than 2% carbon are of little significance and can be deleted. The resulting diagram, such as the one presented as Figure 4-11, focuses on the eutectoid reaction and is quite useful in understanding the properties and processing of steel.

FIGURE 4-11 Simplified iron–carbon phase diagram.

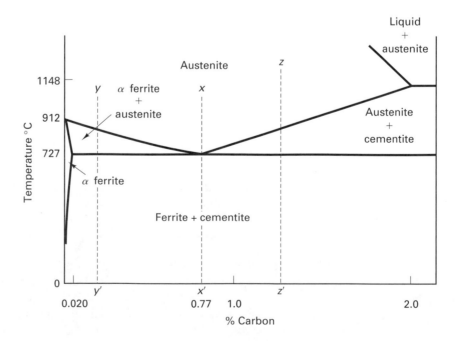

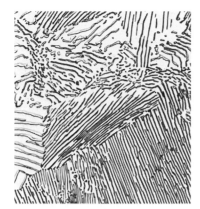

FIGURE 4-12 Pearlite; 1000×. *(Courtesy of USX Corporation.)*

FIGURE 4-13
Photomicrograph of a hypoeutectoid steel showing regions of ferrite (white) and pearlite; 500×. *(Courtesy of USX Corporation.)*

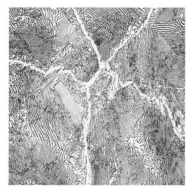

FIGURE 4-14
Photomicrograph of a hypereutectoid steel showing primary cementite along grain boundaries; 500×. *(Courtesy of USX Corporation.)*

The key transition in this diagram is the conversion of single-phase austenite (γ) to the two-phase ferrite plus carbide mixture as the temperature drops. Control of this reaction, which arises as a result of the drastically different carbon solubilities of the face-centered and body-centered structures, enables a wide range of properties to be achieved through heat treatment.

To begin to understand these processes, consider a steel of the eutectoid composition, 0.77% carbon, being slow cooled along line x x' in Figure 4-11. At the upper temperatures, only austenite is present, the 0.77% carbon being dissolved in solid solution within the face-centered structure. When the steel cools through 727°C (1341°F), several changes occur simultaneously. The iron wants to change crystal structure from the face-centered-cubic austenite to the body-centered-cubic ferrite, but the ferrite can only contain 0.02% carbon in solid solution. The surplus carbon is rejected, and forms the carbon-rich intermetallic (Fe_3C) known as cementite. The net reaction at the eutectoid, therefore, is:

$$\text{Austenite}_{0.77\%C;\ FCC} \longrightarrow \text{Ferrite}_{0.02\%C;\ BCC} + \text{Cementite}_{6.67\%C}$$

Since the chemical separation occurs entirely within crystalline solids, the resultant structure is a fine mixture of ferrite and cementite. Specimens prepared by polishing and etching in a weak solution of nitric acid and alcohol reveal a lamellar structure composed of alternating layers or plates, as shown in Figure 4-12. Since it forms from a fixed composition at a fixed temperature, this structure has its own set of characteristic properties (even though it is composed of two distinct phases) and goes by the name *pearlite* because of its resemblance to mother-of-pearl when viewed at low magnification.

Steels having less than the eutectoid amount of carbon (less than 0.77%) are called *hypoeutectoid steels* (since *hypo-* means less than). Consider the cooling of a typical hypoeutectoid alloy along line $y - y'$ in Figure 4-11. At high temperatures the material is entirely austenite. Upon cooling, however, it enters a region where the stable phases are ferrite and austenite. Tie-line and lever-law calculations show that the low-carbon ferrite nucleates and grows, leaving the remaining austenite richer in carbon. At 727°C (1341°F), the remaining austenite is of the eutectoid composition (0.77% carbon), and further cooling transforms it to pearlite. The resulting structure is a mixture of *primary* or *proeutectoid ferrite* (ferrite that forms before the eutectoid reaction) and regions of pearlite as shown in Figure 4-13.

Hypereutectoid steels (*hyper-* means greater than) are those that contain more than the eutectoid amount of carbon. When such a steel cools, as along line $z - z'$ in Figure 4-11, the process is similar to the hypoeutectoid case, except that the primary or proeutectoid phase is now cementite instead of ferrite. As the carbon-rich phase forms, the remaining austenite decreases in carbon content, again reaching the eutectoid composition at 727°C (1341°F). As before, this austenite transforms to pearlite upon slow cooling through the eutectoid temperature. Figure 4-14 is a photomicrograph of the resulting structure, which consists of primary cementite and pearlite. In this case the continuous network of primary cementite will cause the material to be extremely brittle.

It should be noted that the transitions just described are for equilibrium conditions, which can be approximated by slow cooling. Upon slow heating, the transitions will occur in the reverse manner. When the alloys are cooled rapidly, however, entirely different results may be obtained, since sufficient time may not be provided for the normal phase reactions to occur. In these cases, the equilibrium phase diagram is no longer a valid tool for engineering analysis. Since the rapid-cool processes are important in the heat treatment of steels and other metals, their characteristics will be discussed in Chapter 5, and new tools will be introduced to aid our understanding.

■ 4.6 CAST IRONS

Iron–carbon alloys with more than 2.11% carbon experience the eutectic reaction during cooling and are known as *cast irons*. The term *cast iron* applies to an entire family of metals with a wide variety of properties. Being relatively inexpensive, with good fluidity and rather low liquidus temperatures, they are readily cast and occupy an important place in engineering applications.

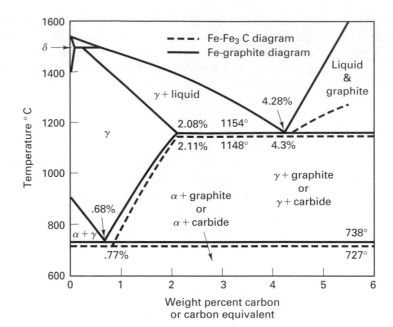

FIGURE 4-15 Iron–carbon diagram showing two possible high-carbon phases. Solid lines, iron–graphite system; dashed lines, iron–cementite (or iron–carbide).

Most commercial cast irons also contain a significant amount of silicon. A typical cast iron contains 2.0 to 4.0% carbon, 0.5 to 3.0% silicon, less than 1.0% manganese, and less than 0.2% sulfur. Silicon produces two major effects. First, it partially substitutes for carbon, so that use of the phase diagram requires replacing the weight percent carbon scale with a *carbon equivalent*. Several formulations exist to compute carbon equivalent, with the simplest being the weight percent carbon plus one-third the weight percent silicon:

$$\text{carbon equivalent} = (\text{wt\% carbon C}) + 1/3\,(\text{wt\% silicon})$$

Silicon also tends to promote the formation of graphite as the carbon-rich single phase instead of the Fe_3C intermetallic. Therefore, the eutectic reaction now has two distinct possibilities, as indicated in the modified phase diagram of Figure 4-15:

$$\text{Liquid} \longrightarrow \text{austenite} + Fe_3C$$

$$\text{Liquid} \longrightarrow \text{austenite} + \text{graphite}$$

The final microstructure of cast iron has two possible extremes: (1) all of the carbon-rich phase being Fe_3C, and (2) all of the carbon-rich phase being *graphite*. In practice, both of these extremes can be approached by controlling the chemistry and other process variables. Graphite formation is promoted by slow cooling, high carbon and silicon contents, heavy section sizes, *inoculation* practices, and the presence of sulfur, phosphorus, aluminum, magnesium, antimony, tin, copper, nickel, and cobalt. Cementite (Fe_3C) is favored by fast cooling, low carbon and silicon levels, thin sections, and alloy additions of titanium, vanadium, zirconium, chromium, manganese, and molybdenum.

Various types of cast iron are produced, depending on the chemical composition, cooling rate, and the type and amount of inoculants that are used. *Gray cast iron*, the least expensive and most common variety, is characterized by those features that promote the formation of graphite. Typical compositions range from 2.5 to 4.0% carbon, 1.0 to 3.0% silicon, and 0.4 to 1.0% manganese. The microstructure consists of three-dimensional graphite flakes (that form during the eutectic reaction) dispersed in a matrix of ferrite, pearlite, or other iron-based structure (which forms from austenite during the eutectoid reaction). The various possibilities for the matrix structure will be discussed further in Chapter 5. Figure 4-16 presents a typical section through gray cast iron, showing the graphite flakes dispersed throughout the metal matrix. Because the graphite flakes have no appreciable strength, they act essentially as voids in the structure. The pointed edges of the flakes act as preexisting notches or crack initiation sites, giving the material a characteristic brittle nature.

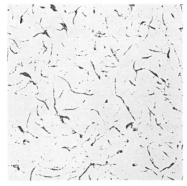

FIGURE 4-16
Photomicrograph of typical gray cast iron; 1000×. *(Courtesy of Bethlehem Steel Corporation.)*

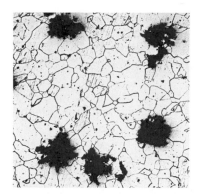

FIGURE 4-17 Photomicrograph of malleable iron showing the irregular graphite spheroids. *(Courtesy of Iron Castings Society, Rocky River, Ohio.)*

Since a large portion of any fracture follows the graphite flakes, the freshly exposed fracture surfaces have a characteristic gray appearance (Figure 4-17), and a graphite smudge can usually be obtained if one rubs a finger across the fracture. On a more positive note, the formation of the lower density graphite reduces the amount of shrinkage that occurs when the liquid goes to solid, making possible the production of more complex castings.

The size, shape, and distribution of the graphite flakes have a considerable effect on the overall properties of gray cast iron. When maximum strength is desired, small, uniformly distributed flakes are preferred with a minimum amount of intersection. A more effective means of controlling strength, however, is through control of the metal matrix structure, which is in turn controlled by the carbon and silicon contents and the cooling rate of the casting. Gray cast iron is normally sold by *class*, with the class number corresponding to the minimum tensile strength in thousands of pounds per square inch. Class 20 iron (minimum tensile strength of 20,000 psi) consists of high carbon-equivalent metal with a ferrite matrix. Higher strengths, up to class 40, are obtainable with lower carbon equivalents and a pearlite matrix. To go above class 40, alloying is required to provide solid solution strengthening, and heat-treatment practices are performed to modify the matrix. Gray cast irons can be obtained up through class 80, but in all cases the presence of the graphite flakes results in extremely low ductility.

Gray cast irons offer excellent compressive strength (compressive forces do not promote crack propagation), excellent machinability (graphite acts to break up the chips and lubricate contact surfaces), good resistance to adhesive wear and galling (graphite flakes self-lubricate), and outstanding sound and vibration damping characteristics (graphite flakes absorb transmitted energy). Table 4-1 compares the relative damping capacities of various engineering metals, and clearly shows the unique characteristic of the high-carbon-equivalent gray cast irons. High silicon contents promote good corrosion resistance and the enhanced fluidity desired for casting operations. For these reasons, coupled with low cost, gray cast iron is specified for a number of applications, including large equipment parts that are subjected to compressive loads and vibrations.

White cast iron has essentially all of its carbon in the form of iron carbide (Fe_3C) and receives its name from the white surface that appears when the material is fractured. Features promoting its formation are those that favor cementite over graphite: a low carbon equivalent (1.8 to 3.6% carbon, 0.5 to 1.9% silicon, and 0.25 to 0.8% manganese) and rapid cooling.

Because the large amount of iron carbide dominates the microstructure, white cast iron is very hard and brittle, and finds applications where high abrasion resistance is the dominant requirement. For these uses it is also common to pursue the hard, wear-resistant *martensite* structure as the metal matrix. (*Note:* This structure will be described in Chapter 5.) In this way, both the metal matrix and the high-carbon second phase contribute to the wear-resistant characteristics of the material.

White cast iron surfaces can also be formed over a base of another material. For example, mill rolls that require extreme wear resistance may have a white cast iron

TABLE 4-1.	Relative Damping Capacity of Various Metals
Material	Damping Capacity*
Gray iron (high Carbon Equivalent)	100–500
Gray iron (low Carbon Equivalent)	20–100
Ductile iron	5–20
Malleable iron	8–15
White iron	2–4
Steel	4
Aluminum	0.4

*Natural log of the ratio of successive amplitudes

surface over a steel interior. Accelerated cooling rates produced by tapered sections or metal chill bars placed in the molding sand can be used to produce white iron surfaces at selected locations of a gray iron casting. Where regions of white and gray cast iron occur in the same component, there is generally a transition region comprised of both white and gray irons, known as the *mottled zone*.

When white cast iron is exposed to an extended heat treatment at temperatures in the range of 900°C (1650°F), the cementite will dissociate into its component elements, and some or all of the carbon will be converted into irregular-shaped nodules of graphite (also referred to as clump or popcorn graphite). The product, known as *malleable cast iron*, has significantly greater ductility than that of gray cast iron because the more favorable graphite shape removes the internal notches. The rapid cooling required to produce the starting white iron structure restricts the size and thickness of malleable iron products such that most weigh less than 5 kilograms (10 lb).

Various types of malleable iron can be produced, depending on the nature of the thermal cycle. If the white iron is heated and held for a prolonged time just below the melting point, the carbon in the cementite converts to graphite (first-stage graphitization). Subsequent slow cooling through the eutectoid reaction causes the carbon-containing austenite to transform to ferrite and more graphite (second-stage graphitization). The resulting product, known as *ferritic malleable cast iron*, has a structure of irregular graphite spheroids in a ferrite matrix (Figure 4-17) and properties, such as: 10% elongation, 35 ksi (240 MPa) yield strength, 50 ksi (345 MPa) tensile strength, and excellent impact strength, corrosion resistance, and machinability. The heat-treatment times, however, are quite lengthy, often involving over 100 hours at elevated temperature.

If the material is cooled more rapidly through the eutectoid transformation, the carbon in the austenite does not form additional graphite but is retained in a pearlite or martensite matrix. The resulting *pearlitic malleable cast iron* is characterized by higher strength and lower ductility than its ferritic counterpart. Typical properties range from 1 to 4% elongation, 45 to 85 ksi (310 to 590 MPa) yield strength, and 65 to 105 ksi (450 to 725 MPa) tensile strength, with reduced machinability compared to the ferritic material.

The modified graphite structure of malleable iron provided quite an improvement in properties, but it would be even more attractive if it could be obtained directly upon solidification rather than through a prolonged heat treatment at high elevated temperature. If a high-carbon-equivalent cast iron is sufficiently low in sulfur (either by original chemistry or by desulfurization), the addition of certain materials can promote graphite formation and change the morphology (shape) of the graphite product. Inoculation with ferrosilicon will promote the formation of graphite. If magnesium (in the form of MgFeSi or MgNi alloy) is also added just prior to solidification, the graphite will form as smooth-surface spheres. The latter addition is known as a *nodulizer*, and the product is *ductile* or *nodular cast iron*. Subsequent control of cooling can produce a variety of matrix structures, with ferrite or pearlite being the most common (Figure 4-18). By controlling the matrix structure, properties can be produced that span a wide range from 2 to 18% elongation, 40 to 90 ksi (275 to 620 MPa) yield strength, and 60 to 120 ksi (415 to 825 MPa) tensile strength. The combination of good ductility, high strength, toughness, wear resistance, machinability, and low-melting-point castability makes ductile iron an attractive engineering material. Unfortunately, the costs of a nodulizer, higher-grade melting stock, better furnaces, and the improved process control required for its manufacture combine to place it among the most expensive of the cast irons.

Over the past several decades, *austempered ductile iron* has emerged as a significant engineering material. It combines the ability to cast intricate shapes with strength and wear-resistance properties that are similar to those of steel, with 8 to 10% lower density. To achieve these results, alloyed ductile iron is subjected to the austempering

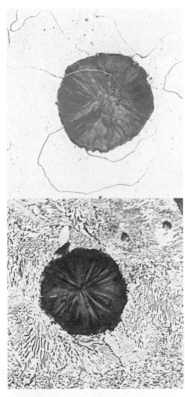

FIGURE 4-18 Ductile iron with (a) ferrite matrix and (b) pearlite matrix; 500×. Note spheroidal shape of the graphite nodule in each photo.

TABLE 4-2. Typical Mechanical Properties of Malleable, Ductile, and Austempered Ductile Cast Irons

Class or Grade	Minimum Yield Strength		Minimum Tensile Strength		Minimum Percentage Elongation	Brinell Hardness Number
	ksi	MPa	ksi	MPa		
Malleable Iron[1]						
M3210	32	224	50	345	10	156 max
M4504	45	310	65	448	4	163–217
M5003	50	345	75	517	3	187–241
M5503	55	379	75	517	3	187–241
M7002	70	483	90	621	2	229–269
M8501	85	586	105	724	1	269–302
Ductile Iron[2]						
60–40–18	40	276	60	414	18	149–187
65–45–12	45	310	65	448	12	170–207
80–55–06	55	379	80	552	6	187–248
100–70–03	70	483	100	689	3	217–269
120–90–02	90	621	120	827	2	240–300
Austempered Ductile Iron[3]						
1	80	550	125	860	10	269–321
2	100	690	165	1035	7	302–363
3	120	827	175	1205	4	363–444
4	140	965	200	1380	2	388 477
5	185		230		–	444–555

[1] ASTM Specification A–602 (Also SAE J 158)
[2] ASTM Specification A–536
[3] Ductile Iron Society

treatment that will be described in Chapter 5. Table 4-2 summarizes the typical mechanical properties of malleable, ductile and austempered ductile cast irons.

Compacted graphite cast iron (also known as vermicular graphite iron, seminodular iron, or quasi-flake iron) is also attracting considerable attention. Produced by a method similar to that used to make ductile iron (a Mg–Ce–Ti addition is made), compacted graphite iron is characterized by a graphite structure that is intermediate to the flake graphite of gray iron and the nodular graphite of ductile iron, and it tends to possess the desirable properties and characteristics of each. Table 4-3 shows how the properties of compacted graphite iron bridge the gap between gray and ductile. Strength and stiffness are greater than those of gray iron, while castability, machinability, thermal conductivity and damping capacity all exceed those of ductile. Impact and fatigue properties are good. Areas of application tend to be those where the mechanical properties of gray iron are insufficient and those of ductile iron are overkill. More specific, compacted graphite iron is attractive when the desired properties include high strength, castability, machinability, thermal conductivity, and thermal shock resistance.

TABLE 4-3. Typical Properties of Pearlitic Gray, Compacted Graphite, and Ductile Cast Irons

Property	Gray	CGI	Ductile
Tensile strength (MPa)	250	450	750
Elastic modulus (GPa)	105	145	160
Elongation (%)	0	1.5	5
Thermal conductivity (w/mk)	48	37	28
Relative damping capacity (Gray = 1)	1	0.35	0.22

■ KEY WORDS

austenite	cooling curve	graphite	liquidus	pearlite	solvus
carbon equivalent	cored structure	gray cast iron	malleable cast iron	peritectic	steel
cast iron	ductile cast iron	hypereutectoid	martensite	peritectoid	stoichiometric
cementite	equilibrium	hypoeutectoid	monotectic	phase	syntectic
compacted graphite	eutectic	inoculation	mottled zone	phase diagram	three-phase reaction
cast iron	eutectoid	intermetallic	nodular cast iron	primary phase	tie-line
complete solubility	ferrite	compound	nodulizer	solidus	white cast iron
composition	freezing range	lever law	nonstoichiometric	solubility limit	

■ REVIEW QUESTIONS

1. What are some features that are useful in defining a phase?
2. Supplement the examples provided in the text with another example of a single phase that is each of the following: continuous, discontinuous, gaseous, and a liquid solution.
3. What is an equilibrium phase diagram?
4. What three primary variables are generally considered in equilibrium phase diagrams?
5. Why is a pressure–temperature phase diagram not that useful for most engineering applications?
6. What is a cooling curve?
7. What features in a cooling curve indicate some form of change in a material's structure?
8. What is a solubility limit, and how might it be determined?
9. In general, how does the solubility of one material in another change as temperature is increased?
10. Describe the conditions of complete solubility, partial solubility, and insolubility.
11. What types of changes occur upon crossing a liquidus line? A solidus line? A solvus line?
12. What three pieces of information can be obtained for each point in an equilibrium phase diagram?
13. What is a tie-line? For what types of phase diagram regions would it be useful?
14. What points on a tie-line are used to determine the chemistry (or composition) of the component phases?
15. What tool can be used to compute the relative amounts of the component phases in a two-phase mixture? How does this tool work?
16. What is a cored structure? Under what conditions is it produced?
17. What features in a phase diagram can be used to identify three-phase reactions?
18. What is the general form of a eutectic reaction?
19. Why are alloys of eutectic composition attractive for casting and as filler metals in soldering and brazing?
20. What is a stoichiometric intermetallic compound, and how would it appear in a temperature–composition phase diagram? How would a nonstoichiometric intermetallic compound appear?
21. What type of mechanical properties would be expected for intermetallic compounds?
22. In what form(s) might intermetallic compounds be undesirable in an engineering material? In what form(s) might they be attractive?
23. What are the four single phases in the iron–iron carbide diagram? Answer with both phase diagram notation and assigned name.
24. What feature in the iron–carbon diagram is used to distinguish between cast irons and steels?
25. What features of austenite make it attractive for forming operations? What features make it attractive as a starting structure for many heat treatments?
26. Which of the three-phase reactions in the iron–carbon diagram is most important in understanding the behavior of steels? Write this reaction in terms of the interacting phases and their composition.
27. Describe the relative ability of iron to dissolve carbon in solution when in the form of austenite (the elevated temperature phase) and when in the form of ferrite at room temperature.
28. What is pearlite? Describe its structure.
29. What is a hypoeutectoid steel, and what structure will it assume upon slow cooling? What is a hypereutectoid steel and how will its structure differ from that of a hypoeutectoid?
30. In addition to iron and carbon, what other element is present in rather large amounts?
31. What are some characteristics associated with silicon in cast iron?
32. What are the two possible high-carbon phases in cast irons? What features tend to favor the formation of each?
33. Describe the microstructure of gray cast iron.
34. Which of the structural units is generally altered to increase the strength of a gray cast iron?
35. What are some of the attractive engineering properties of gray cast iron?
36. What are some of the key limitations to the engineering use of gray cast iron?
37. What is the dominant mechanical property of white cast iron?
38. What structural feature is responsible for the increased ductility and fracture resistance of malleable cast iron? How is malleable cast iron produced?
39. What is unique about the graphite that forms in ductile cast iron?
40. What requirements of ductile iron manufacture are responsible for its increased cost over materials such as gray cast iron?
41. Compacted graphite iron has a structure and properties intermediate to what two other types of cast irons?

■ PROBLEMS

1. Select a binary (two-component) phase diagram for a system not discussed in this chapter. Identify each:
   a. Single phase
   b. Three phase reaction
   c. Intermetallic compound
2. Identify at least one easily identified product or component that is currently being produced from each of the following types of cast irons:

   a. Gray cast iron
   b. White cast iron
   c. Malleable cast iron
   d. Ductile cast iron
   e. Compacted graphite cast iron

 www.wiley.com/college/degarmo

*C*hapter 4   CASE STUDY

The Blacksmith Anvils

You are an officer in the Western-America Blacksmith Association, and you have determined that a number of your members would like to have a modern equivalent of an 1870-vintage blacksmith anvil. Your objective is to replicate the design, but utilize the advantageous features of today's engineering materials. You hope ultimately to identify a producer who will make a limited number of these items for sale and distribution through your monthly magazine. The proposed design is a large forging anvil that has a total length of 20 inches. The top surfaces must be resistant to wear, deformation, and chipping.

Estimated mechanical properties call for a yield strength in excess of 70 ksi, an elongation greater than 2%, and a Brinell hardness of 200 or more on the top surface. You feel confident that you will be able to secure a minimum of 500 orders.

1. Discuss the various properties that this part must possess to adequately perform its intended task.
2. Discuss the various concerns that would influence the proposed method of fabricating the anvils.
3. Assuming that the anvils will be made from some form of ferrous metal, consider the properties of the various types of cast irons and steels with regard to this application. Which material would you recommend? Why?
4. How would you propose that a production run of 500 replica anvils be produced?

CHAPTER 5

HEAT TREATMENT

■ 5.1 INTRODUCTION

The material presented in Chapter 4 has already shown that variations in the heating or cooling of a metal can be used to produce different structures with different properties. Many engineering materials can be characterized not just by a single set of properties but by an entire spectrum that can be selected and varied at will. *Heat treatment is the controlled heating and cooling of metals for the purpose of altering their properties* and can be performed without a concurrent change in product shape. Because both physical and mechanical properties can be altered by heat treatment, it is one of the most important and widely used manufacturing processes.

While the term *heat treatment* applies only to processes where the heating and cooling are performed for the specific purpose of altering properties, heating and cooling often occur as incidental phases of other manufacturing processes, such as hot forming or welding. Material properties will be altered just as though an intentional heat treatment had been performed, and the results can be either beneficial or harmful. As a result, the designer who selects material and the engineer who specifies its processing must be fully aware of the possible changes in properties that can occur during heating and cooling. Heat treatment must be correlated with the other manufacturing processes if effective results are to be obtained. Both the theory of heat treatment and a survey of the various processes will be presented in this chapter.

■ 5.2 PROCESSING HEAT TREATMENTS

The term *heat treatment* is often associated with those thermal processes that increase the strength of a material, but the broader definition permits inclusion of another set of processes that we will call *processing heat treatments*. These are performed as a means of preparing the material for fabrication. Specific objectives may be the improvement of machining characteristics, the reduction of forming forces, or the restoration of ductility to enable further processing. Through use of the processing heat treatments, a metal can be softened for ease of fabrication and then subjected to a different heat treatment to produce a totally different set of properties for actual service.

EQUILIBRIUM DIAGRAMS AS AIDS
Most of the processing heat treatments involve rather slow cooling or extended times at elevated temperatures. These conditions tend to approximate equilibrium, and the resulting structures, therefore, can be reasonably predicted by the use of an *equilibrium phase diagram*. The diagrams can be used to indicate both the temperatures that must

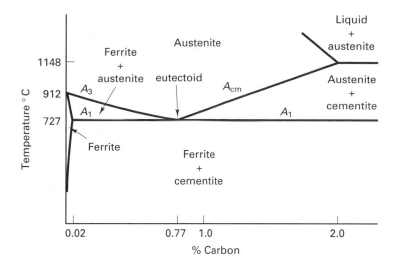

FIGURE 5-1 Simplified iron–carbon phase diagram for steels with transition lines labeled in standard notation.

be attained to produce a desired starting structure and the changes that will occur upon subsequent cooling. It should be noted, however, that the diagrams are for true equilibrium conditions, and departures from equilibrium may lead to substantially different results.

PROCESSING HEAT TREATMENTS FOR STEEL

Because many of the processing heat treatments are applied to plain-carbon and low-alloy steels, they are presented here with the simplified iron–carbon equilibrium diagram of Figure 4-11 serving as a reference guide. Figure 5-1 shows this diagram with the key transition lines labeled in standard notation. The eutectoid line is designated by the symbol A_1, and A_3 designates the boundary between austenite and ferrite + austenite.[1] The transition from austenite to austenite + cementite is designated as the A_{cm} line.

A number of process heat-treating operations are classified under the general term of *annealing*. These may be employed to reduce strength or hardness, remove residual stresses, improve toughness, restore ductility, refine grain size, reduce segregation, or alter the electrical or magnetic properties of the material. By producing a certain desired structure, characteristics can be imparted that are favorable to subsequent operations or applications. The temperature, cooling rate, and specific details of the process are determined by the material being treated and the objectives of the treatment.

In the process of *full annealing*, hypoeutectoid steels (less than 0.77% carbon) are heated to 30 to 60°C (50 to 100°F) above the A_3 temperature, held for sufficient time to convert the structure to homogeneous single-phase austenite of uniform composition and temperature, and then slowly cooled at a controlled rate to below the A_1 temperature. Cooling is usually done in the furnace by decreasing the temperature by 10 to 30°C (20 to 50°F) per hour to at least 30°C (50°F) below the A_1. At this point the metal can be removed from the furnace and air cooled to room temperature. The resulting structure is one of coarse pearlite (widely spaced lamellae) with excess ferrite in amounts predicted by the phase diagram. In this condition, the steel is quite soft and ductile.

The procedure to full-anneal a hypereutectoid alloy (greater than 0.77% carbon) is basically the same, except that the original heating is only into the austenite plus cementite region (30 to 60°C above the A_1). If the material were to be slow cooled from the all-austenite region, a continuous network of cementite may form on the grain boundaries and make the material brittle. When properly annealed, a hypereutectoid steel will have a structure of coarse pearlite with excess cementite in dispersed spheroidal form.

---

[1] Historically, an A_2 line once appeared between the A_1 and A_3. This line designated the magnetic property change known as the Curie point. This transition was later shown to be an atomic change, not a change in phase, and the line was deleted from the diagram without relabeling.

Full anneals are time consuming, and considerable energy is needed to maintain the elevated temperatures required during soaking and furnace cooling. When the maximum softness is not required and cost savings are desirable, *normalizing* may be specified. In this process, the steel is heated to 60°C (100°F) above the A_3 (hypoeutectoid) or A_{cm} (hypereutectoid) temperature, held at this temperature to produce uniform austenite, and then removed from the furnace and allowed to cool in still air. The resultant structures and properties depend on the subsequent cooling rate. Although wide variations are possible depending on the size and geometry of the metal, fine pearlite with excess ferrite or cementite is generally produced.

One should note a key difference between full annealing and normalizing. In the full anneal, the furnace imposes identical cooling conditions at all locations within the metal, which produces identical structures and properties. With normalizing, the cooling will be different at different locations. Properties will vary between surface and interior, and different thickness regions will also have different properties. When subsequent processing involves a substantial amount of machining that may be automated, the added cost of a full anneal may be justified, since it produces a product with uniform machinability at all locations.

When cold working has severely strain hardened a metal, it is often desirable to restore the ductility, either for service or to permit further processing without danger of fracture. This is often achieved through the *recrystallization* process described in Chapter 3. When the material is a low-carbon steel (<0.25% carbon), however, the specific procedure is known as a *process anneal*. The steel is heated to a temperature slightly below the A_1, held long enough to induce recrystallization of the dominant ferrite phase, and then cooled at a desired rate (usually in still air). Since the entire process is performed at temperatures within the same phase region, the process simply induces a change in phase morphology (size, shape, and distribution). The material is not heated to as high a temperature as in the full anneal or normalizing process, so a process anneal is somewhat cheaper and tends to produce less scaling.

A *stress-relief anneal* may be employed to reduce the residual stresses in large steel castings, welded assemblies, and cold-formed products. Parts are heated to temperatures below the A_1 (between 550 and 650°C or 1000 and 1200°F), held for a period of time, and then slow cooled. Times and temperatures vary with the condition of the component.

When high-carbon steels (>0.60% carbon) are to undergo extensive machining or cold-forming, a process known as *spheroidization* is often employed. Here the objective is to produce a structure in which all of the cementite is in the form of small spheroids or globules dispersed throughout a ferrite matrix. This can be accomplished by a variety of techniques, including (1) prolonged heating at a temperature just below the A_1 followed by relatively slow cooling, (2) prolonged cycling between temperatures slightly above and slightly below the A_1, or (3) in the case of tool or high-alloy steels, heating to 750 to 800°C (1400 to 1500°F) or higher and holding at this temperature for several hours, followed by slow cooling.

Although the selection of a processing heat treatment often depends on the desired objectives, steel composition strongly influences the choice. Process anneals are restricted to low-carbon steels, and spheroidization is a treatment for high-carbon material. Normalizing and full annealing can be applied to all carbon contents, but even here, preferences are noted. Since different cooling rates do not produce a wide variation of properties in low-carbon steels, the air cool of a normalizing treatment often produces acceptable uniformity. For higher carbon contents, such as the 0.4 to 0.6% range, different cooling rates can produce wide property variations, and the uniform furnace cooling of a full anneal is often preferred. Figure 5-2 provides a graphical summary of the process heat treatments.

HEAT TREATMENTS FOR NONFERROUS METALS

Most of the nonferrous metals do not have the significant phase transitions observed in the iron–carbon system, and for them, the process heat treatments do not play such a significant role. Aside from the strengthening treatment of precipitation hardening, which is discussed later, the nonferrous metals are usually heat-treated for three purposes: (1) to produce a uniform, homogeneous structure, (2) to provide stress relief, or (3) to bring about recrystallization. Castings that have been cooled too rapidly can possess a segregated

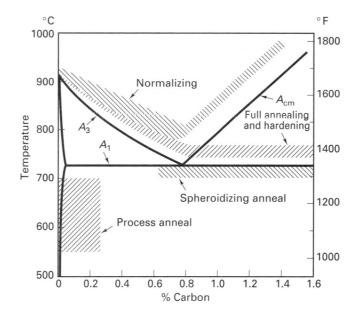

FIGURE 5-2 Graphical summary of the process heat treatments for steels on an equilibrium diagram.

solidification structure known as coring (discussed more fully in Chapter 4). *Homogenization* can be achieved by heating to moderate temperatures and then holding for a sufficient time to allow thorough diffusion to take place. Similarly, the internal stresses that are produced by forming, welding, or brazing can be reduced by heating for several hours at relatively low temperatures. *Recrystallization* (discussed in Chapter 3) is a function of the particular metal, the amount of prior straining, and the desired recrystallization time. In general, the more a metal has been strained, the lower the recrystallization temperature or the shorter the time. Without prior straining, however, recrystallization will not occur and heating will only produce undesirable grain growth.

■ 5.3 HEAT TREATMENTS USED TO INCREASE STRENGTH

Six major mechanisms are available to increase the strength of metals:

1. Solid-solution strengthening   **4.** Precipitation hardening

2. Strain hardening   **5.** Dispersion hardening

3. Grain-size refinement   **6.** Phase transformations

All of these can be induced or altered by heat treatment, but not all are applicable to any given metal.

In *solid-solution strengthening*, a base metal dissolves other atoms in solid solution, either as *substitutional solutions*, where the new atoms occupy sites in the host crystal lattice, or as *interstitial solutions*, where the new atoms squeeze into "holes" between the atoms of the base lattice. The amount of strengthening depends on the amount of dissolved solute and the size difference of the atoms involved. Since distortion of the host structure makes dislocation movement more difficult, the greater the size difference, the more effective the addition.

Strain hardening (discussed in Chapter 3) produces an increase in strength by means of plastic deformation under cold-working conditions.

Because grain boundaries act as barriers to dislocation motion, a metal with small grains tends to be stronger than the same metal with larger grains. Thus *grain-size refinement* can be used to increase strength, except at elevated temperatures, where failure is by a grain-boundary diffusion-controlled creep mechanism. It is important to note that grain-size refinement is one of the few processes that can simultaneously improve both strength and ductility.

Precipitation hardening, or *age hardening*, is a method whereby strength is obtained from a nonequilibrium structure that is produced by a three-step heat treatment. Details of this method will be provided in Section 5.4.

Strength obtained from distinct second-phase particles dispersed throughout a base material is known as *dispersion hardening*. To be effective, the dispersed particles should be stronger than the matrix, adding strength through both their reinforcing action and the additional interfacial surfaces that present barriers to dislocation movement.

Phase transformation strengthening involves those alloys that can be heated to form a single phase at elevated temperature and subsequently transform to one or more low-temperature phases upon cooling. When this feature is used to increase strength, the cooling is usually rapid and the phases that are produced are usually of a nonequilibrium nature.

■ 5.4 STRENGTHENING HEAT TREATMENTS FOR NONFERROUS METALS

All six of the mechanisms just described can be used to increase the strength of nonferrous metals. Solid-solution strengthening can impart strength to single-phase materials. Strain hardening is useful if sufficient ductility is present. Eutectic-forming alloys exhibit considerable dispersion hardening. Among all of the possibilities, however, the most effective strengthening mechanism for the nonferrous metals is precipitation hardening.

PRECIPITATION OR AGE HARDENING

In alloy systems with limited solubility that decreases with decreasing temperature, certain alloys can be heated to form an elevated temperature, single-phase solid solution, which then reverts back to two distinct phases upon cooling. If the heated single phase is cooled rapidly (quenched), however, it may be possible to produce a supersaturated single phase where the component that normally forms the second phase remains trapped within the parent lattice. Upon subsequent heating within the two-phase region, an aging process can occur in which the excess solute atoms precipitate out of the supersaturated matrix to form a controllable nonequilibrium structure.

As an example, consider the aluminum-rich portion of the aluminum–copper phase diagram presented in Figures 5-3 and 5-4. Follow the slow cooling of an alloy composed of 96% aluminum and 4% copper from a temperature of 1000°F. Upon crossing the solvus line

FIGURE 5-3 High-aluminum section of the aluminum–copper equilibrium phase diagram.

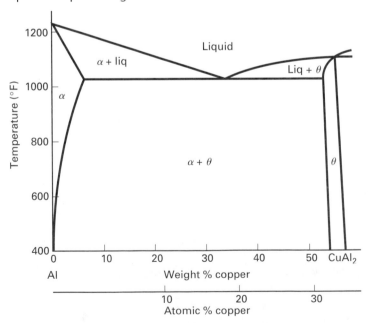

FIGURE 5-4 Enlargement of the solvus-line region of the aluminum–copper equilibrium diagram of Figure 5-3.

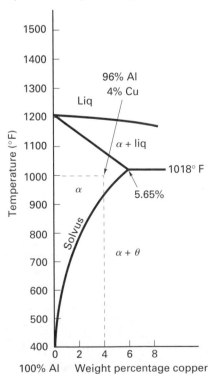

at 930° F, theta phase begins to precipitate out of the alpha-phase solid solution because the solubility of copper in aluminum decreases from 5.65% at 1018°F to less than 0.2% at room temperature. If this same alloy were rapidly cooled from 1000°F, there would not be sufficient time for the normal two-phase structure to form and the alpha phase would be retained in a highly supersaturated state. Because this supersaturated condition is not normal, the excess copper would like to precipitate out of solution and coalesce into theta-phase particles, where theta phase consists of an intermetallic compound with the formula $CuAl_2$. Atom movement or diffusion is required, however, and time at elevated temperature will have to be provided. If the temperature is now raised to around 350°F, diffusion rates increase and copper-rich clusters will begin to form. With proper control, this process can significantly increase the yield strength of the alloy.

Precipitation hardening is a three-step process. The first step, known as *solution treatment*, involves heating to a temperature above the solvus, thereby replacing the starting room-temperature structure with an elevated temperature single-phase solid solution. The solution treatment temperature should not exceed the eutectic temperature, however, or there might be melting if the original material had a cored structure. After holding for sufficient time to assure a uniform chemistry single phase, the alloy is *quenched* (rapidly cooled), usually in water, to suppress diffusion and produce a room-temperature supersaturated solid solution. In this state, the material is often soft and can be straightened, formed, or machined.

For the third step, precipitation-hardening materials can be divided into two types: (1) *naturally aging* materials, where room temperature is sufficient to move the unstable supersaturated solution toward the stable two-phase structure, and (2) *artificially aging* materials, which require elevated temperatures to provide the necessary diffusion. With natural aging materials, such as aluminum alloy rivets, some form of refrigeration may be required to retain the after-quench condition of softness. Upon removal from refrigeration, these rivets are easily headed and then progress to full strength after several days at room temperature. The properties of the artificial aging materials can be controlled by adjusting the time and temperature of the elevated temperature aging. At any time, a drop in temperature will halt diffusion and "lock in" the current structure and properties.

Aging is a continuous process that begins with the clustering of solute atoms on distinct planes within the parent lattice. Various transitions may then occur, leading ultimately to the formation of a distinct second phase with its own characteristic crystal structure.

A key concept in this sequence is that of *coherency*. If the clustered solute atoms continue to occupy lattice sites of the parent structure, the crystal planes remain continuous in all directions, and the clusters of solute atoms (which are of different size and possibly different valence from the host material) tend to distort or strain the adjacent lattice for a sizable distance in all directions. For this reason, a small cluster appears to be much larger with respect to its ability to impede dislocation motion (i.e., impart strength). When the clusters reach a certain size, however, the associated strain becomes so great that the clusters break free from the parent structure, and form distinct second-phase particles with well-defined interphase boundaries. Coherency is lost and the mechanism of strengthening reverts to *dispersion hardening*, where the particles present only their physical dimensions as effective dislocation blocks. Strength and hardness begin to decrease, and the material is said to be *overaged*. Figure 5-5 presents a family of aging curves for the 4% copper–96% aluminum alloy. For higher aging temperatures, the peak properties are achieved in a shorter time, but the peak hardness (or strength) is not as great as that for lower aging temperatures. Selection of the aging conditions (temperature and time) is a decision that is made on the basis of desired strength, available equipment, and production constraints.

The artificial aging process can be stopped at any stage by simple quenching. The structure and properties of that stage are then retained, provided that the material is not subsequently exposed to elevated temperatures that would reactivate diffusion. For example, if the 4% copper alloy of Figure 5-5 were aged for one day at 375°F and then quenched, the metal would retain a hardness of 94 Vickers (and the associated strength) throughout its useful lifetime. If higher strength were desired, a lower temperature and

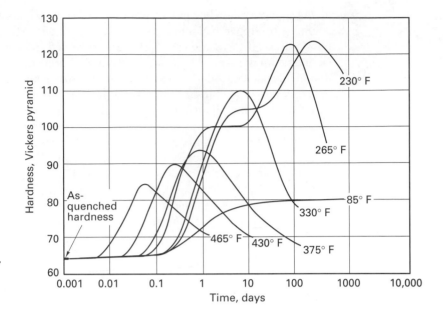

FIGURE 5-5 Aging curves for the Al–4%Cu alloy at various temperatures. (Adapted from *Journal of the Institute for Metals*, Vol. 79, p. 321, 1951.)

longer time could be selected. Artificially aging alloys are quite popular because of the ability to lock in the peak properties that occur prior to overaging.

Precipitation hardening is an extremely effective strengthening mechanism and is responsible for the attractive engineering properties of many aluminum, copper, and magnesium alloys. In many cases, the strength can be more than double that observed upon conventional cooling. Through special alloying, some age-hardenable ferrous alloys have also been produced.

■ 5.5 STRENGTHENING HEAT TREATMENTS FOR STEEL

Iron-based metals have been heat treated for centuries, and today over 90% of all heat-treatment operations are performed on steel. The striking changes that resulted from plunging red-hot steel into cold water or some other quenching medium were awe-inspiring to the ancients. Those who performed such heat treatment in the making of swords and armor were looked upon as possessing unusual powers, and much superstition arose regarding the process. Because quality was directly related to the act of quenching, great importance was placed on the quenching medium that was used. For example, urine was thought to be a superior quenching medium, and that from a red-haired boy was deemed particularly effective, as was that from a 3-year-old goat fed only ferns.

ISOTHERMAL TRANSFORMATION DIAGRAM

It has only been within the last 100 years that the art of heat treating has begun to turn into a science. One of the major barriers to understanding was the fact that the strengthening treatments were nonequilibrium in nature. Minor variations in cooling often produced major variations in structure and properties.

A useful aid to understanding the nonequilibrium processes is the *isothermal transformation* (I-T) or *time-temperature-transformation* (T-T-T) *diagram*. The information in this diagram is obtained by heating thin specimens of a particular steel to produce uniform-chemistry austenite, "instantaneously" quenching to a temperature where austenite is not the stable phase, holding for variable periods of time, and observing the resultant products via photomicrographs.

For simplicity, consider a carbon steel of eutectoid composition (0.77% carbon) and the resulting T-T-T diagram of Figure 5-6. Above 1341°F (727°C), austenite is the stable phase. Below this temperature, the face-centered austenite would like to transform to body-centered ferrite and carbon-rich cementite. Two factors control the rate of transition: (1) the motivation or driving force for the change, and (2) the ability to form the desired products (i.e., the ability to rearrange the atoms through diffusion). The region below 1341°F in Figure 5-6 can be interpreted as follows. Zero time corresponds to a sample quenched

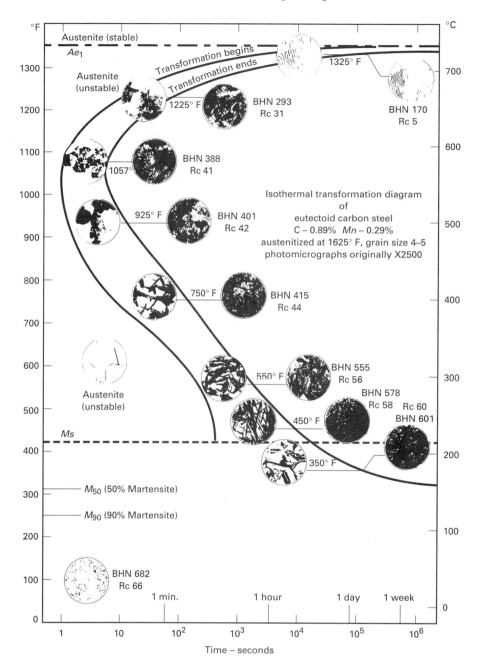

Isothermal transformation diagram
of
eutectoid carbon steel
C – 0.89% Mn – 0.29%
austenitized at 1625° F, grain size 4–5
photomicrographs originally X2500

FIGURE 5-6 Isothermal transformation diagram (T-T-T diagram) for eutectoid composition steel. Structures resulting from transformation at various temperatures are shown as insets. *(Courtesy of USX Corporation.)*

"instantaneously" to its new, lower temperature. The structure is usually unstable austenite. As time passes (moving horizontally across the diagram), a line is encountered representing the start of transformation and a second line indicating completion of the phase change. At elevated temperatures (just below 1341°F), diffusion is rapid, but the rather sluggish driving force dominates the kinetics. At a low temperature, the driving force is high but diffusion is quite limited. The phase transformation kinetics are most rapid at a compromise intermediate temperature, resulting in the characteristic C-curve shape. The portion of the C that extends farthest to the left is known as the *nose* of the T-T-T diagram.

If the transformation occurs between the A_1 temperature and the nose of the curve, the departure from equilibrium is not very great. The austenite transforms into ferrite and cementite in the layered structure known as *pearlite* (as discussed previously in the equilibrium phase diagram description of Chapter 4). Because the diffusion rates are greater at higher temperatures, the lamellar spacing (separation distance between similar layers) will be larger for pearlite produced at higher temperatures. The pearlite formed near the A_1 temperature is known as *coarse pearlite* and the structure formed near the nose is called *fine pearlite*.

If the austenite is quenched to a temperature between the nose and the temperature designated as M_s, a different structure is produced. The transformation conditions are a significant departure from equilibrium, and the amount of diffusion required to form lamellar pearlite is no longer available. The metal still has the goal of changing crystal structure from face-centered austenite to body-centered ferrite, with the excess carbon being accommodated in the form of cementite. The resulting structure, however, is not one of alternating plates but rather a dispersion of discrete cementite particles in a lathlike or needlelike matrix of ferrite. Electron microscopy may be required to resolve the carbides in this structure, which is known as *bainite*. Because of the fine dispersion of carbide, its strength exceeds that of fine pearlite, and its ductility is retained because the soft ferrite is the continuous matrix.

If austenite is quenched to below the M_s temperature, a different type of transformation occurs. The steel still wants to change from the face-centered-cubic structure to body-centered cubic, but it can no longer expel the required amount of carbon to form ferrite. Responding to the severe nonequilibrium conditions, it simply undergoes an abrupt change in crystal structure with no diffusion. The excess carbon becomes trapped, distorting the structure into a body-centered tetragonal lattice (distorted body-centered cubic), with the degree of distortion being proportional to the amount of excess carbon. The new structure (Figure 5-7) is known as *martensite*, and, with sufficient carbon, it is exceptionally strong, hard, and brittle. The highly distorted lattice effectively blocks the dislocation motion necessary for metal deformation.

As shown in Figure 5-8, the hardness and strength of steel in the martensitic condition are strong functions of the carbon content. Below 0.10% carbon, martensite is not very strong. Since no diffusion occurs during the transformation, higher-carbon steels form higher-carbon martensite, with an increase in strength and hardness and a concurrent decrease in toughness and ductility. From 0.3 to 0.7% carbon, strength and hardness increase rapidly. Above 0.7% carbon, however, the rise is far less dramatic, a feature related to the presence of retained austenite.

The amount of martensite that forms is a function of the lowest temperature that is encountered, not the time at that temperature. This feature is shown in Figure 5-9. If we return to the C curve of Figure 5-6, there is a temperature designated as M_{50}, where the structure is 50% martensite and 50% untransformed austenite. At the lower M_{90} temperature, the structure is now 90% martensite. If no further cooling occurs, the untransformed austenite can remain within the structure. This *retained austenite* can cause loss of strength or hardness, dimensional instability, and cracking or brittleness. Since most quenches are to room temperature, retained austenite becomes a significant problem when the martensite finish, or 100% martensite, temperature lies below room temperature. Higher carbon contents and

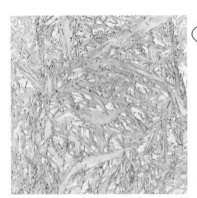

FIGURE 5-7 Photomicrograph of martensite; 1000×. *(Courtesy of USX Corporation.)*

FIGURE 5-8 Effect of carbon on the hardness of martensite.

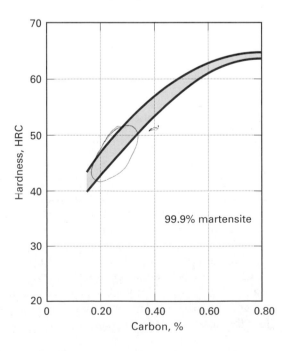

FIGURE 5-9 Schematic representation depicting the amount of martensite formed upon quenching to various temperatures from M_s through M_f.

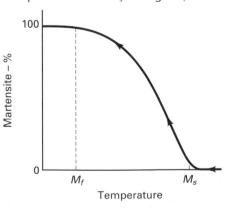

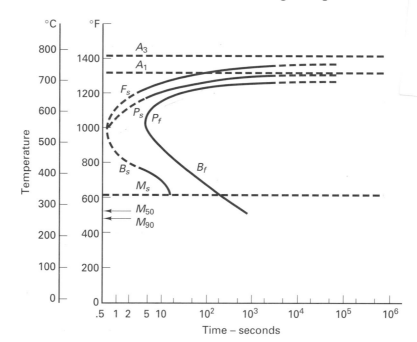

FIGURE 5-10 Isothermal transformation diagram for a hypoeutectoid steel (1050) showing additional region for primary ferrite.

alloy additions both decrease all martensite-related temperatures, and materials with these chemistries may require refrigeration or a quench in liquid nitrogen to produce full hardness.

It is important to note that all of the transformations that occur below the A_1 temperature are one-way transitions (austenite to something). The steel is simply seeking to change its crystal structure, and the various products are the result of this change. It is impossible, therefore, to convert one product to another without first reheating to above the A_1 temperature to again form the face-centered-cubic austenite.

T-T-T diagrams can be quite useful in determining the kinetics of transformation and the nature of the products. The left-hand curve shows the elapsed time at constant temperature before the transformation begins, and the right-hand curve shows the time required to complete the transformation at that temperature. If hypo- or hypereutectoid steels were considered, additional regions would have to be added to the diagram to incorporate the primary equilibrium phases that form below the A_3 or A_{cm} temperatures. These regions would not extend below the nose, however, since the nonequilibrium bainite and martensite structures can exist with variable amounts of carbon, unlike the near-equilibrium pearlite. Figure 5-10 presents the T-T-T curve for a 0.5%-carbon hypoeutectoid steel showing the additional region for the primary ferrite.

TEMPERING OF MARTENSITE

Despite its great strength, medium- or high-carbon martensite in its as-quenched form lacks sufficient toughness and ductility to be a useful engineering structure. A subsequent heating, known as *tempering*, is usually required to impart the necessary ductility and fracture resistance. As with most property-changing processes, there is a concurrent drop in other features, most notably strength and hardness.

Martensite is a supersaturated solid solution of carbon in alpha ferrite and, therefore, is a metastable structure. When heated into the range of 100 to 700°C (200 to 1300°F), the excess carbon atoms are rejected from solution, and the structure moves toward a mixture of the stable phases of ferrite and cementite. This decomposition of martensite into ferrite and cementite is a time- and temperature-dependent, diffusion-controlled phenomenon with a continuous spectrum of intermediate and transitory conditions.

Table 5-1 shows a chart-type comparison of precipitation hardening and the quench-and-temper process. Both are nonequilibrium heat treatments that involve three distinct stages. In both, the first step is an elevated temperature soaking designed to erase the prior structure and produce a uniform-chemistry, single-phase starting condition. Both treatments follow this soak with a rapid-cool quench. In precipitation hardening, the purpose of the quench is to prevent nucleation of the second phase, thereby

TABLE 5-1.   Comparison of Age Hardening with the Quench-and-Temper Process

Heat Treatment	Step 1	Step 2	Step 3
Age hardening	*Solution treatment.* Heat into the stable single-phase region (above the solvus) and hold to form a uniform-chemistry single-phase vsolid solution.	*Quench.* Rapid cool to form a nonequilibrium supersaturated single-phase solid solution (crystal structure remains unchanged, material is soft and ductile).	*Age.* A controlled reheat in the stable two-phase region (below the solvus). The material moves toward the formation of the stable two-phase structure, becoming stronger and harder. The properties can be "frozen in" by dropping the temperature to stop further diffusion.
Quench and temper for steel	*Austenize.* Heat into the stable single-phase region (above the A_3 or A_{cm}) and hold to form a uniform-chemistry single-phase solid solution (austenite).	*Quench.* Rapid cool to form a nonequilibrium supersaturated single-phase solid solution (crystal structure changes to body-centered martensite, which is hard but brittle).	*Temper.* A controlled reheat in the stable two-phase region (below the A_1). The material moves toward the formation of the stable two-phase structure, becoming weaker but tougher. The properties can be "frozen in" by dropping the temperature to stop further diffusion.

producing a supersaturated solid solution. This material is usually soft, weak, and ductile, with good toughness. Subsequent aging allows the material to move toward the formation of the stable two-phase structure, and sacrifices toughness and ductility for an increase in strength. When the proper balance is achieved, the temperature is dropped, diffusion ceases, and the current structure and properties are preserved, provided that the material is never subsequently exposed to elevated temperatures that would reactivate diffusion and permit the structure to move further toward equilibrium.

For steels, the quench induces a phase transformation as the material changes from the face-centered-cubic austenite to the distorted body-centered structure known as martensite. The product is again a supersaturated, single-phase solid solution, but the associated properties are the reverse of precipitation hardening. Martensite is strong and hard but relatively brittle. When the material is tempered, strength and hardness are sacrificed for an increase in ductility and toughness. Figure 5-11 shows the final properties

FIGURE 5-11   Properties of an AISI 4140 steel that has been austenitized, oil-quenched, and tempered at various temperatures. (Adapted from *Engineering Properties of Steel,* ASM International, Materials Park, Ohio, 1982.)

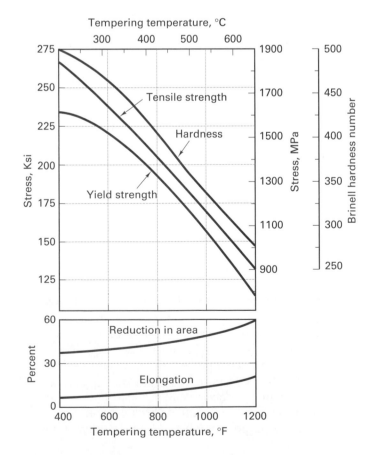

of a steel that has been tempered at a variety of temperatures. During tempering, diffusion again permits movement *toward* the stable two-phase structure, and a drop in temperature again halts diffusion and locks in properties. Therefore, by quenching steel to form 100% martensite and then tempering it at various temperatures, an infinite range of structures and corresponding properties can be produced. This is known as the *quench-and-temper process* and the product is called *tempered martensite*.

CONTINUOUS COOLING TRANSFORMATIONS

Although the T-T-T diagrams can provide useful information about the structures obtained through nonequilibrium thermal processing, they are not rigorously applicable to engineering applications because the assumptions of instantaneous cooling followed by constant temperature transformation rarely match reality. Continuous cooling from elevated temperature is far more realistic, and a diagram showing the results of continuous cooling at various rates would be far more useful. What would be the result if the temperature were to be decreased at a rate of 500°F per second, 50°F per second, or 5°F per second?

A *continuous-cooling-transformation* (C-C-T) *diagram*, like the one shown in Figure 5-12, can provide answers to these questions and numerous others. The critical cooling rates required to produce various structures can easily be determined. If the cooling is sufficiently fast, the structure will be martensite. A slow cool may produce coarse pearlite and some primary phase. Intermediate rates usually produce mixed structures, since the time at any one temperature is usually insufficient to complete the transformation. If each structure is regarded as providing a companion set of properties, the wide range of possibilities obtainable through the controlled heating and cooling of steel becomes even more evident.

JOMINY TEST FOR HARDENABILITY

The *Jominy end-quench hardenability test* and associated diagrams provide another useful tool to aid our understanding of nonequilibrium heat treatment. In this test, depicted schematically in Figure 5-13, an entire spectrum of cooling rates are produced on a single specimen by quenching a heated (i.e., austenitized) bar on one end. The quench is standardized by specifying the quench medium (water at 75°F), internal nozzle diameter ($\frac{1}{2}$ in.), water pressure (that producing a $2\frac{1}{2}$ in. vertical fountain), and the gap between the nozzle and the specimen ($\frac{1}{2}$ in.). Since the thermal conductivity of steel is essentially constant over the normal ranges of carbon and alloy additions, a characteristic cooling rate can be assigned to each position within the specimen.

FIGURE 5-12 Schematic C-C-T diagram for a eutectoid composition steel, showing several superimposed cooling curves and the resultant structures. *(Courtesy of USX Corporation.)*

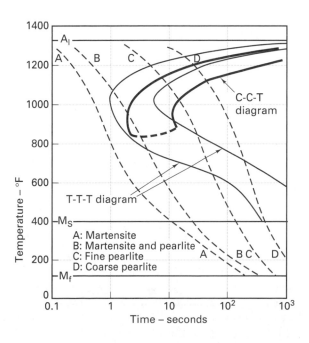

FIGURE 5-13 Schematic diagram of the Jominy hardenability test.

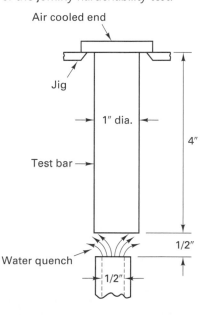

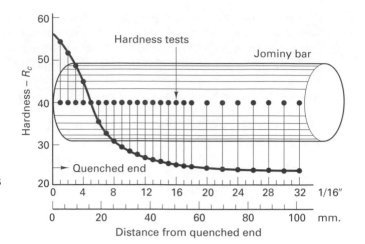

FIGURE 5-14 Typical hardness distribution along a Jominy test specimen.

After the specimen has been cooled by the quench, a flat region is ground along one side and R_C hardness readings are taken every $\frac{1}{16}$ in. along the bar. The resulting data are then plotted as shown in Figure 5-14. The resulting hardness values are correlated with position, and since the cooling rate is known for each location within the bar, the hardnesses are indirectly correlated with the cooling rate that produced them.

Application of the test assumes that for a given material, equivalent cooling conditions will produce equivalent results. If the cooling rate is known for a specific location within a part (from experimentation or theory), the properties at that location can be predicted to be those at Jominy test bar location with the equivalent cooling rate. Conversely, if specific properties are required, the required cooling rate can be easily determined. If the cooling rates are restricted by either geometry or processing limitations, various materials can be compared and a satisfactory alloy selected. Figure 5-15 shows the Jominy curves for several engineering steels.

HARDENABILITY CONSIDERATIONS

Several key effects must be considered if we are to understand the heat treatment of steel: the effect of carbon content, the effect of alloy additions, and the effect of various quenching conditions. The first two relate to the material and the third to the heat-treatment process.

Hardness is a mechanical property related to strength and is a strong function of the carbon content of a steel. *Hardenability*, however, is a measure of the depth to which full hardness can be obtained under a normal hardening cycle and is related primarily to the amounts and types of alloying elements. In Figure 5-15, all of the steels have the same carbon content, but they differ in the type and amounts of alloy elements. The maximum hardness is the same in all cases, but the depth of hardening varies considerably. Figure 5-16 shows the Jominy test results for steels containing the same alloying elements but variable amounts of carbon. Note the change in peak hardness.

The results of a heat-treatment operation depend on both the hardenability of the metal and the rate of heat extraction. The primary reason for adding alloy elements to commercial steels is to increase their hardenability, not to improve their strength. Steels with greater hardenability need not be cooled as rapidly to achieve a desired level of strength or hardness, and they can be completely hardened in thicker sections.

An accurate determination of need is required if steels are to be selected for specific applications. Strength tends to be associated with carbon content, and a general rule is to select the lowest possible level that will meet the specifications. Because heat can be extracted only from the surface of a metal, the size of the piece and the depth of required hardening set the conditions for hardenability and quench. For a given quench condition, different alloys produce different results. Because alloy additions increase the cost of a material, a general rule is to select only what is required to ensure compliance with specifications. Money is often wasted by specifying an alloy steel for an application where a plain carbon steel, or a steel with lower alloy content (less costly), would be satisfactory. Another alternative when greater depth of hardness is required

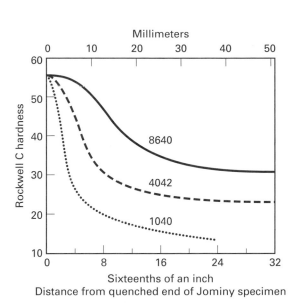

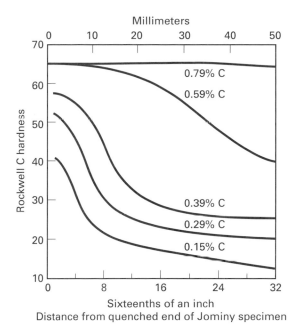

FIGURE 5-15   Jominy hardness curves for engineering steels with the same carbon content and varying types and amounts of alloy elements.

FIGURE 5-16   Jominy hardness curves for engineering steels with identical alloy conditions but variable carbon content.

is to modify the quench conditions so that a faster cooling rate is achieved. Quench changes may be limited, however, by cracking or warping problems, and other considerations relating to the size, shape, complexity, and desired precision of the part being treated.

QUENCH MEDIA

Quench media vary in their effectiveness, and one can best understand the variation by considering the three stages of quenching. When a piece of hot metal is inserted into a tank of liquid quenchant under conditions where the temperature of the metal is above the boiling point of the quenchant, the liquid adjacent to the metal will vaporize and form a gaseous layer between the metal and the liquid. Cooling is slow through this *vapor jacket* (first stage) since the gas has an insulating effect and heat transfer is largely through radiation. Bubbles soon nucleate, however, and break the jacket. New liquid contacts the metal, vaporizes (removing its heat of vaporization from the metal), and forms another bubble. As the bubbles are removed, the process continues. Because of the large quantities of heat required to vaporize a liquid, this *second stage of quenching* (or nucleate boiling phase) produces rapid rates of cooling. When the metal has cooled to the boiling point of the quenchant, vaporization can no longer occur. Heat transfer must now take place by conduction across the solid–liquid interface, aided by convection or stirring within the liquid. This conduction and convection is the *third stage of quenching*.

    Water is a fairly effective quenching medium because of its high heat of vaporization and the fact that the second stage of quenching extends down to 100°C (212°F), usually well into the temperatures for martensite formation or even below. Water is also cheap, readily available, easily stored, nontoxic, nonflammable, smokeless, and easy to filter and pump. However, with a water quench, the clinging tendency of the bubbles may cause soft spots on the metal. Agitation is recommended when using a water quench. Still other problems associated with a water quench include its oxidizing nature (i.e., its corrosiveness) and the tendency to produce excessive distortion and possible cracking.

    Brine (salt water) produces more rapid cooling than water because the salt nucleates bubbles, forcing a quick transition through the vapor jacket stage. Unfortunately, the salt in a brine quench also tends to accelerate corrosion problems unless all residue is completely removed by a subsequent rinse. Different types of salts can be used, including sodium or potassium hydroxide, and various degrees of agitation or spraying can be used to adjust the effectiveness of the quench.

When a slower cooling rate is desired, *oil* quenches are quite popular. Various oils are available that have high flash points and different degrees of quenching effectiveness. Since the boiling points can be quite high, the transition to third-stage cooling usually precedes the martensite start temperature. The slower cooling through the M_s-to-M_f martensite transformation leads to a milder temperature gradient within the piece and reduced likelihood of cracking. Problems associated with oil quenchants include water contamination, smoke, fumes, spill and disposal problems, and fire hazard. In addition, quench oils tend to be somewhat expensive.

Quite often, there is a need for a quenchant that will cool more rapidly than the oils, but slower than water or brine. To fill this gap, a number of *polymer quench* solutions (also called *synthetic quenchants*) have been developed. Tailored quenchants can be produced by varying the concentrations of the components (such as liquid organic polymers, corrosion inhibitors, and water), operating temperature, and amount of agitation. The polymer quenchants provide extremely uniform and reproducible results, are less corrosive than water and brine, and are less of a fire hazard than oils (no fires, fumes, smoke, or need for air-pollution-control apparatus). Distortion and cracking are less of a problem since the boiling point can be adjusted to be above the martensite start temperature. In addition, the polymer-rich film that forms initially on the hot metal part serves to modify the cooling rate.

When slow cooling is desired, molten salt baths can be employed to provide a medium where the quench goes directly to the third stage of cooling. Still slower cooling can be obtained by cooling in still air, burying the material in sand, or a variety of other methods.

High-pressure gas quenching uses a stream of flowing gas to extract heat, and the cooling rates can be adjusted by controlling the gas velocity and pressure. This process has already been shown to be comparable to oil quenching and has far fewer environmental and safety concerns. *Vegetable oils* may emerge as another significant industrial quenchant. They are environmentally friendly (biodegradeable), offer low toxicity, and are a renewable resource.

DESIGN CONCERNS IN THE HEAT TREATMENT OF STEEL

Product design and material selection play important roles in the satisfactory and economical heat treatment of parts. Proper consideration of these factors usually leads to simpler, more economical, and more reliable products. Failure to relate design and materials to heat-treatment procedures usually produces disappointing or variable results and may lead to a variety of service failures.

From the viewpoint of heat treatment, undesirable design features include (1) nonuniform sections or thicknesses, (2) sharp interior corners, and (3) sharp exterior corners. Since these features often find their way into the design of parts, the designer should be aware of their effect on heat treatment. Undesirable results may include nonuniform structure and properties, undesirable residual stresses, cracking, warping, and dimensional changes.

Heat can only be extracted from a piece through its exposed surfaces. Therefore, if the piece to be hardened has a nonuniform cross section, any thin region will cool rapidly and may fully harden, while thick regions may harden only on the surface, if at all. The shape that might be closest to ideal from the viewpoint of quenching would be a doughnut. The uniform cross section with high exposed surface area and absence of sharp corners is quite attractive. Since most shapes are designed to perform a function, however, compromises are usually necessary.

Residual stresses are the often-complex stresses that are present within a body, independent of any applied load. They can be induced in a number of ways, but the complex dimensional changes that can occur during heat treatment are a primary cause. Thermal expansion during heating, and contraction during cooling, is a well-understood phenomenon, but nonuniform heating or cooling can produce extremely complex results. In addition, the various phases and structures that can form within a material are usually characterized by different densities. Volume expansions or contractions accompany any phase transformation. For example, when austenite transforms to martensite, there is a volume expansion of up to 4%. Transformation to ferrite, pearlite, or other room-temperature structures also involves a volume expansion but of a smaller magnitude.

If all of the temperature changes occurred uniformly throughout a part, all of the associated dimensional changes would occur simultaneously and the resultant product would be free of residual stresses. However, most of the parts being heat-treated experience nonuniform temperatures during the cooling or quenching operation. Consider a block of hot aluminum being cooled by water sprays from top and bottom. For simplicity, let us model the block as a three-layer sandwich (see Figure 5-17). At the start of the quench, all layers are uniformly hot. However, as the water spray begins, the surface layers cool and contract. The center layer, however, does not experience the quench and remains hot. Since the part is actually one piece, the various layers must accommodate each other. The contracting surface layers exert compressive forces on the hot, weak interior, causing it to also contract, but by plastic deformation. As time passes, the interior now cools and wants to contract, but finds itself sandwiched between cold, strong surface layers. It pulls on the surface layers, placing them in compression, and the surface layers hold it back, creating tension in the interior. While the net force is zero (since there is no applied load), counterbalancing tension and compression stresses exist within the product.

Now repeat the sequence, but change the material to steel. When heated, all three layers are hot, face-centered-cubic austenite. Upon rapid cooling, the surface layers transform to martensite (the structure changes to body-centered tetragonal) and *expand*! The expanding surfaces deform the soft, weak, (and hot) austenite center, which then cools and wants to expand as it undergoes the crystal structure change to a body-centered, room-temperature product. The hard, strong surface layers hold it back, placing the center in compression, and the expanding center tries to stretch the surface layers, producing surface tension. If the tension at the surface becomes great enough, cracking can result, a phenomenon known as *quench cracking*. For the rapid quench conditions just described, aluminum will not quench crack, because the surface is in compression, but the steel might, since the residual stresses are reversed. If the cooling were not symmetrical, there might be more contraction or expansion on one side and the block might warp. With more complex shapes and nonuniform cooling, the residual stresses induced by heat treatment can be extremely complex.

The problems associated with residual stresses can be minimized if the cross sections are sufficiently uniform that the temperature differences are minimized and are not concentrated at any specific location. If this is not possible, slower cooling may be required, coupled with a material that will provide the desired properties with the slower oil or air quench. Because materials with greater hardenability are invariably more expensive, the design alternative clearly has advantages.

FIGURE 5-17   Three-layer model of a plate undergoing cooling: (a) material such as aluminum that contracts upon cooling; (b) situation for steel, which expands during the phase transformation.

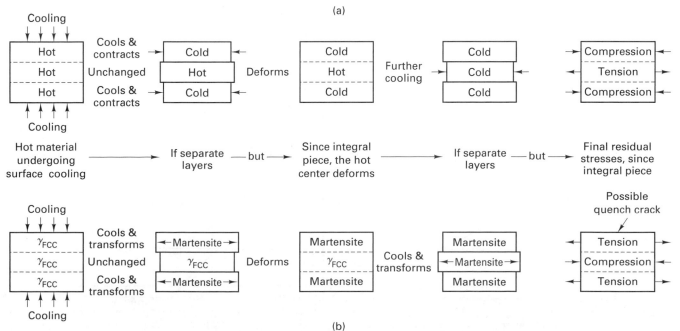

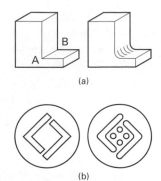

FIGURE 5-18 (a) Shape containing nonuniform sections joined by a sharp interior corner that may crack during quenching. This is improved by using a large radius to join the sections. (b) Original design containing holes, which can be modified to produce more uniform sections.

When temperature differences and the resultant residual stresses become severe or localized, cracking or distortion problems can be expected. Figure 5-18(a) shows an example where a sharp interior corner has been placed at a change in cross section. Upon quenching, stresses will concentrate along line *A–B*, and a crack is almost certain to result. If changes in cross section are required, they should be gradual, as in the redesigned version of Figure 5-18(a). Generous fillets at interior corners, radiused exterior corners, and smooth transitions all reduce problems. A material with greater hardenability and a less severe quench will also help.

Figure 5-18(b) shows the cross section of a blanking die that consistently cracked during hardening. Rounding the sharp corners and adding additional holes to provide a more uniform cross section during quenching eliminated the problem.

One of the ominous features of poor product and process design is the fact that the residual stresses may not produce immediate failure but may contribute to failure at a later time. Applied stresses add to the residual stresses already present within the part. Therefore, it is possible for applied stresses that are well within the "safe" designed limit to couple with residual stresses and produce a value sufficient to induce failure. Corrosion reactions can be accelerated in the presence of residual stresses. Dimensional changes or warping can occur when subsequent machining or grinding operations upset the equilibrium balance of the residual stresses. After the removal of some material, the remaining piece adjusts its shape to produce a new balance. Considering all the possible difficulties, it is apparent that considerable time and money can be saved if good design, material selection, and heat-treatment practices are employed. If properly performed, heat treatment can give better results with less costly materials.

TECHNIQUES TO REDUCE CRACKING

Figure 5-19(a) illustrates the cause of quench cracking using a T-T-T diagram (a misuse of the diagram, but helpful for visualization). The surface and center of a piece have different cooling rates. As a result, when the surface is transforming to martensite and expanding, the center is still hot, and therefore soft, untransformed austenite. At a later time, the center cools to the martensite transformation, and then it expands. The result is significant tension in the cold, hard surface and possible cracking.

Two variations of rapid quenching have been developed to produce strong structures with a reduced likelihood of cracking. A rapid cool is still required to prevent transformation to the softer, weaker, pearlitic structure, but instead of quenching through the martensite transformation, the component is rapidly quenched into a liquid medium that is now

FIGURE 5-19 (a) Schematic representation of the cooling paths of surface and center during a direct quench; (b) modified cooling paths experienced during the austempering and martempering processes.

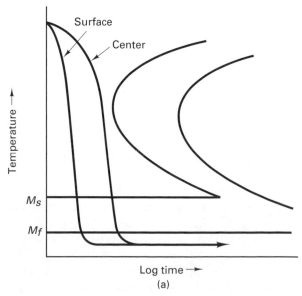

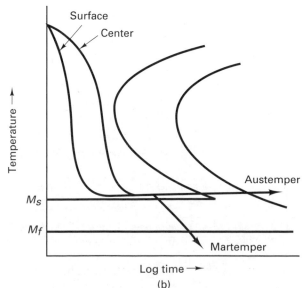

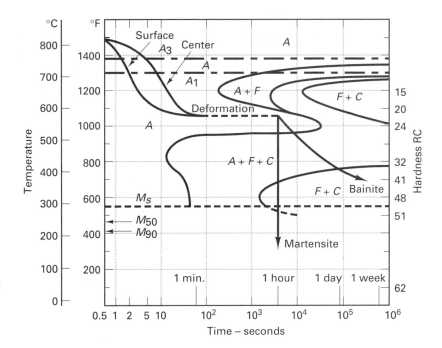

FIGURE 5-20   T-T-T diagram for 4340 steel, showing the bay and a schematic of the ausforming process. *(Courtesy of USX Corporation.)*

several degrees above the martensite start (M_s) temperature. Holding for a period of time in this bath allows the piece to return to a nearly uniform temperature. If the material is simply held at this temperature for sufficient time, the austenite will transform to bainite, and the process is known as *austempering*. If the material is brought to a uniform temperature and then slow cooled through the martensite transformation, the process is known as *martempering*, or *marquenching*. The resulting structure is martensite, which must be tempered the same as the martensite that forms directly upon quenching. Bainite usually has sufficient toughness that a temper is not required. Figure 5-19(b) illustrates the modified processes and shows that all transformations (and related volume expansions) occur at the same time, thereby eliminating the residual stresses and tendency to crack.

AUSFORMING

A process that is often confused with austempering is *ausforming*. Certain alloys tend to retard the pearlite transformation far more than the bainite reaction and produce a T-T-T curve of the shape shown in Figure 5-20. If this material is heated to form austenite and then quenched to the temperature of the "bay" between the pearlite and bainite reactions, it can retain its austenite structure for a useful period of time. Deformation can be performed on an austenite structure at a temperature where it technically should not exist. Benefits include the increased ductility of the face-centered-cubic crystal structure, the finer grain size that forms upon recrystallization at the lower temperature, and the possibility of some degree of strain hardening. Following the deformation, the metal can be slowly cooled to produce bainite or rapidly quenched to martensite, which must then be tempered. The resulting product has exceptional strength and ductility, coupled with good toughness, creep resistance, and fatigue life—properties that are far superior to those produced if the deformation and transformation processes were conducted in their normal sequence. Ausforming is an example of a growing class of *thermomechanical processes* in which deformation and heat treatment are intimately combined.

■ 5.6 SURFACE HARDENING OF STEEL

Many products require different properties at different locations. Quite frequently, this variation takes the form of a hard, wear-resistant surface coupled with a tough, fracture-resistant core. The methods developed to produce these properties can be classified into three basic groups: selective heating of the surface, altered surface chemistry, and deposition of an additional surface layer. The first two approaches will be discussed in the next sections, and platings and coatings will be described in Chapter 31.

SELECTIVE HEATING TECHNIQUES

If a steel has sufficient carbon to attain the desired surface hardness, generally greater than 0.3%, the different properties can often be obtained simply by varying the thermal histories of the various regions. Maximum hardness depends on the carbon content of the material, while the depth of that hardness depends on the depth of heating and the material's hardenability.

Flame hardening uses an oxyacetylene flame to raise the surface temperature high enough to reform austenite. The surface is then water quenched[2] and tempered to the desired level of toughness. Heat input is quite rapid and is concentrated on the surface. Slow heat transfer and short heating times leave the interior at low temperature and therefore free from any significant change.

Considerable flexibility is provided since the rate and depth of heating can be easily varied. Depth of hardening can range from thin skins to over 6 mm. ($\frac{1}{4}$ in.). Flame hardening is often used on large objects, since alternative methods tend to be limited by both size and shape. Equipment varies from crude handheld torches to fully automated and computerized units.

When using *induction hardening*, the steel part is placed inside a conductor coil, which is then energized with alternating current. The changing magnetic field induces surface currents in the steel, which heat by electrical resistance. The heating rates can be extremely rapid, and efficiency is high. Induction heating is particularly well suited to surface hardening since the rate and depth of heating can be controlled directly through the amperage and frequency of the generator.

Induction hardening is ideal for round bars and cylindrical parts but can also be adapted to more complex geometries. The process offers high quality, good reproducibility, and the possibility of automation. Figure 5-21 shows a cross section of an induction-hardened gear, where hardening has been applied to those areas expected to see high wear. Distortion during hardening is negligible since the dark areas remain cool and rigid throughout the entire process.

Laser-beam hardening has been used to produce hardened surfaces on a wide variety of geometries. An absorptive coating such as zinc or manganese phosphate is often applied to the steel to improve the efficiency of converting light energy into heat. The surface is then scanned with the laser, where beam size, beam intensity, and scanning speed have been selected to obtain the desired amount of heat input and depth of heating. It is possible for the heat to be removed simply by conductive transfer into the cool, underlying metal (autoquenching), but a water or oil quench can also be used.

Through laser-beam hardening, 0.4% carbon steel can attain surface hardnesses as high as Rockwell C 65. The process operates at high speeds, produces little distortion, induces residual compressive stresses on the surface, and can be used to harden selected surface areas while leaving the remaining surfaces unaffected. Computer software and automation can be used to control the process parameters, and conventional mirrors and optics can be used to shape and manipulate the beam.

Electron-beam hardening is similar to laser-beam hardening. Here the heat source is a beam of high-energy electrons rather than a beam of light, and the charged particles can be focused and directed by electromagnetic controls. Like laser-beam treating, the process can be readily automated, and production equipment can perform a variety of operations with efficiencies often greater than 90%. Electrons cannot travel in air, however, so the entire operation must be performed in a hard vacuum, which is the major limitation of this process. More information on laser- and electron-beam techniques, as well as other means of heating material, is provided in Chapters 36 through 39. Still other surface-heating techniques employ immersion in a pool of molten lead or molten salt (*lead pot* or *salt bath* heating).

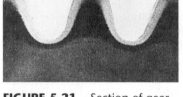

FIGURE 5-21   Section of gear teeth showing induction-hardened surfaces. *(Courtesy of TOCCO Division, Park-Ohio Industries, Inc.)*

---

[2]There is no real danger of surface cracking during the water quench. When the surface is reaustenitized, the soft austenite adjusts to the colder, stronger, underlying material. Upon quenching, the surface austenite expands during transformation. The interior is still cold and restrains the expansion, producing a surface in compression and no tendency toward cracking. This is just the opposite of the conditions that occur during the through-hardening of a furnace-soaked workpiece!

TECHNIQUES INVOLVING ALTERED SURFACE CHEMISTRY

When the steels contain insufficient carbon to achieve the desired surface properties through selective heating, an alternative approach is to alter the surface chemistry. *Carburizing*, the most common technique within this category, involves the diffusion of carbon into the elevated temperature, face-centered-cubic, austenite structure. When sufficient carbon has diffused to the desired depth, the parts are then thermally processed. Direct quenching from the carburization treatment is the simplest alternative and can often produce the desired variation in properties as a result of the different carbon contents and cooling rates. Alternative processes include a slow cool from the carburizing treatment, followed by reaustenitizing and quenching or a duplex process involving a separate surface heat treatment. These processes are more involved and more costly, but produce improved product properties. The carbon content of the surface usually varies from 0.7 to 1.2% depending on the details of the process. Case depth may range from a few thousandths of an inch to over $\frac{3}{8}$ in.

In the *pack-carburizing* process, the steel components are surrounded by a high-carbon solid material (such as carbon powder or cast iron turnings) and heated in a furnace. The hot carburizing compound produces CO gas, which reacts with the metal, releasing carbon, which is readily absorbed by the hot austenite. *Gas carburizing* replaces the solid carburizing compound with a carbon-containing gas. Compared to pack carburizing, the operation is faster and more easily controlled. Accuracy and uniformity are increased, and continuous operation is now possible.

In *liquid carburizing* the steel parts are immersed in a molten bath that supplies the carbon. At one time, liquid-carburizing baths contained cyanide, which supplied both carbon and nitrogen to the surface. Safety and environmental concerns now dictate the use of noncyanide liquid compounds. Most applications of liquid carburizing involve the production of thin cases on small parts.

Nitriding hardens the surface by producing alloy nitrides in special steels that contain nitride-forming elements like aluminum, chromium, molybdenum, or vanadium. The parts are first heat-treated and tempered at 525 to 675°C (1000 to 1250°F) prior to nitriding. After cleaning and removal of any decarburized surface material, they are heated in an atmosphere containing dissociated ammonia (nitrogen and hydrogen) for 10 to 40 hours at 500 to 625°C (950 to 1150°F). Nitrogen diffusing into the steel then forms alloy nitrides, hardening the metal to a depth of about 0.65 mm (0.025 in.). Very hard cases are formed and distortion is low. No subsequent thermal processing is required. In fact, subsequent heating should be avoided because the thermal expansions and contractions will crack the hard nitrided case. Finish grinding should also be avoided because the nitrided layer is exceptionally thin.

Ionitriding is a plasma process that has emerged as an attractive alternative to the conventional methods. Parts to be treated are placed in an evacuated "furnace" and a direct-current potential of 500 to 1000 V is applied between the parts and the furnace walls. Low-pressure nitrogen gas is introduced into the chamber and becomes ionized. The ions are accelerated toward the product surface, where they impact and generate sufficient heat to promote inward diffusion. This is the only heat associated with the process; the "furnace" acts only as a vacuum container and electrode. Advantages of the process include shorter cycle times, reduced consumption of gases, significantly reduced energy costs, reduced space requirements, and the possibility of total automation. Product quality is improved over that of conventional nitriding, and the process is applicable to a wider range of materials. *Ion carburizing* is a parallel process in which low-pressure methane is substituted for the low-pressure nitrogen, and carbon diffuses into the surface.

Ion plating and *ion implantation* are other technologies that permit modification of the surface chemistry. In Chapter 31 we expand on the surface treatment of materials and compare the processes discussed in the preceding sections to various platings, coatings, and other techniques.

■ 5.7 FURNACES

FURNACE TYPES

To facilitate production heat treatment, many styles of furnaces have been developed in a wide range of sizes. These furnaces are generally classified as batch or continuous type. *Batch furnaces* are those in which the workpiece remains stationary

FIGURE 5-22   Box electric heat-treating furnace. *(Courtesy of Lindberg, A Unit of General Signal.)*

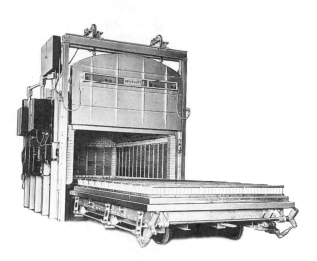

FIGURE 5-23   Car-bottom box furnace. *(Courtesy of Hevi Duty Electric Company.)*

throughout its treatment. *Continuous furnaces* move the components through the heat-treatment operation at rates selected to be compatible with the other manufacturing operations.

Horizontal batch furnaces are often called *box furnaces* because of their overall shape, and generally use gas or electricity as their source of heat. As shown in Figure 5-22, a door is provided on one end to allow the work to be inserted and removed. When large or very long workpieces are to be heated, a *car-bottom box furnace* may be employed, like the one shown in Figure 5-23. Here the work is loaded onto a refractory-topped flatcar, which can be rolled into and out of the furnace on railway rails.

In a *bell furnace*, the heating elements are contained within a bottomless "bell" that is lowered over the work. An airtight inner shell is often placed over the workpieces to contain a protective atmosphere during the heating and cooling operations. After the work is heated, the furnace unit can be lifted off and transferred to another batch, while the inner shell maintains the protective atmosphere during cooling. If extremely slow cooling is desired, an insulated cover can be placed over the heated shell.

An *elevator furnace* is an interesting modification of the bell design where the bell remains stationary and the workpieces are raised into it on a movable platform that then forms the bottom of the furnace. By placing a quench tank below the furnace, it is now possible to load the platform, raise it into the furnace and then lower it into the quench tank. This design is particularly attractive for applications where the work must be quenched as soon as possible after being removed from the furnace.

When long, slender parts are positioned horizontally, there is little resistance to sagging or warping. For these types of workpieces, a *vertical pit furnace* is prcfcrrcd. These furnaces are usually cylindrical chambers sunk into the floor with a door on top that can be swung aside to allow suspended workpieces to be lowered into the furnace. They can also be used to heat large quantities of small parts by loading them into wire-mesh baskets and stacked the baskets within the column.

Continuous furnaces are used for large production runs where the same or similar parts undergo the same thermal processing. The workpieces are moved through the furnace by some type of transfer mechanism (conveyor belt, walking beam, pusher, or monorail), and often fall into a quench tank to complete the treatment. Complex cycles of heating, holding, and quenching or cooling can be conducted in an exact and repeatable manner with low labor cost.

While all of the furnaces described in the preceding paragraphs can heat in air, most commercial furnaces can also employ *artificial gas atmospheres*. These are selected to prevent scaling or tarnishing, to prevent decarburization, or even to provide carbon or nitrogen for surface modification. Many of the artificial atmospheres are generated from natural gas, but nitrogen-based atmospheres offer reduced cost, energy savings, increased safety, and environmental attractiveness.

When a liquid heating medium is preferred, *salt bath furnaces* are a popular choice. Electrically conductive salt can be heated by passing a current between two electrodes suspended in the bath. The electrical currents also cause the bath to circulate and thereby maintain uniform temperature. Nonconductive salt can be heated by some form of immersion heater, or the containment vessels can be externally fired. The molten salt not only serves as a uniform source of heat but can also be selected to prevent scaling or decarburization. A *lead pot* is a similar furnace, where molten lead replaces salt as the heat transfer medium.

The heating rates of gas atmosphere furnaces can be significantly increased by incorporating the *fluidized-bed* concept. These furnaces consist of a bed of dry, inert particles, such as aluminum oxide, which are heated and fluidized (suspended) in a stream of flowing gas. Products introduced into the bed become engulfed in the particles, which then radiate uniform heat. Temperature and atmosphere can be altered quickly, and high heat- transfer rates, high thermal efficiency, and low fuel consumption have been observed. Since atmosphere changes can be performed in minutes, a single furnace can be used for nitriding, stress relieving, carburizing, carbonitriding, annealing, and hardening.

Electrical induction heating is another popular means of heating conductive materials, such as metal. Small parts can be through-heated and hardened as in the other methods. Long products can be heated and quenched in a continuous manner by passing them through a stationary heating coil or by having a moving coil traverse a stationary part. Localized or selective heating can also be performed at rapid production rates. Flexibility is another attractive feature, since a standard induction unit can be adapted to a wide variety of products simply by changing the induction coil and adjusting the equipment settings.

FURNACE CONTROLS

All thermal-processing operations should be conducted with rigid control if the desired results are to be obtained in a consistent fashion. Most furnaces are equipped with one or more temperature sensors. These can be coupled to a controller or computer to regulate the temperature and rate of heating or cooling. It should be remembered, however, that it is the temperature of the workpiece, not the temperature of the furnace, that controls the result, and it is this temperature that should be monitored if at all possible.

■ 5.8 HEAT TREATMENT AND ENERGY

Because of the elevated temperatures and time at those temperatures, heat treatments can consume considerable amounts of energy. However, if one considers the broader picture, heat treatment may actually prove to be an energy conservation measure. Through its use, it is possible to manufacture higher-quality, more durable products, which eliminates the need for frequent replacements. Higher strengths may also permit the use of less material in the manufacture of a product, thereby saving additional energy.

Further savings can often be obtained by integrating the manufacturing operations. For example, a direct quench and temper from hot forging may be used to replace the conventional air cool, reheat, quench, and temper sequence. One should note, however, that the integrated procedure would have greater variability in temperature, uniformity of that temperature, and austenite grain size going into the quench. If this variation is too great, additional energy may be required to conduct the reheat and soak.

■ KEY WORDS

A_1	C-C-T diagram	homogenization	pearlite	solid-solution
A_3	carburizing	induction hardening	phase transformation	strengthening
A_{cm}	coherency	interstitial solution	strengthening	solution treatment
age hardening	continuous furnaces	ionitriding	polymer quench	spheroidization
aging	dispersion hardening	isothermal transformation	precipitation	strain hardening
annealing	equilibrium phase diagram	diagram	hardening	stress-relief anneal
artificial aging	flame hardening	Jominy test	process anneal	substitutional solution
artificial gas	fluidized-bed furnaces	martempering	quench and temper	surface hardening
atmospheres	full anneal	martensite	quench cracking	T-T-T diagram
ausforming	grain-size refinement	natural aging	quenching	tempered martensite
austempering	hardenability	nitriding	recrystallization	tempering
bainite	hardness	normalize	residual stresses	thermomechanical
batch furnaces	heat treatment	overaged	retained austenite	processing

■ REVIEW QUESTIONS

1. What is heat treatment?

2. What types of properties can be altered through heat treatment?

3. Why should people performing hot forming or welding be aware of the effects of heat treatment?

4. What is the major goal of the processing heat treatments? Cite some of the specific objectives that may be sought.

5. Why might equilibrium phase diagrams be useful aids in designing and understanding the processing heat treatments?

6. What are the A_1, A_3, and A_{cm} lines?

7. What are some of the possible objectives of annealing operations?

8. While full anneals often produce the softest and most ductile structures, what may be some of the objections or undesirable features of these treatments?

9. Why are the hypereutectoid steels not furnace-cooled from the all-austenite region?

10. What is the major process difference between full annealing and normalizing?

11. What are some of the process heat treatments that can be performed without reaustenitizing the material (heating above the A_1 temperature)?

12. What types of steel would be candidates for a process anneal? Spheroidization?

13. Why is the recrystallization temperature of a metal not a well-defined temperature?

14. What are the six major mechanisms that can be used to increase the strength of a metal?

15. What is the most effective strengthening mechanism for the nonferrous metals?

16. What are the three steps in an age-hardening treatment?

17. What is the difference between natural and artificial aging? Which offers more flexibility? Over which does the engineer have more control?

18. What is the difference between a coherent precipitate and a distinct second-phase particle?

19. What is overaging?

20. What types of heating and cooling conditions are imposed in an I-T or T-T-T diagram? Are they realistic for the processing of commercial items?

21. For steels below the A_1 temperature, what are the stable equilibrium phases as predicted by the equilibrium phase diagram?

22. What are some nonequilibrium structures that appear in the T-T-T diagram for a eutectoid composition steel?

23. Which of the possible steel heat-treatment structures is the result of a diffusionless phase change?

24. What is the major factor that influences the strength and hardness of martensite?

25. Why is retained austenite an undesirable structure in heat-treated steels?

26. Why are martensitic structures usually tempered before being put into use? What properties increase during tempering? Which ones decrease?

27. In what ways is the quench-and-temper heat treatment similar to age hardening? How are the property changes different in the two processes?

28. What is a C-C-T diagram? Why is it more useful than a T-T-T diagram?

29. How do the various locations of a Jominy test specimen correlate with cooling rate?

30. What conditions are used to standardize the quench in the Jominy test?

31. What is the assumption that allows the data from a Jominy test to be used to predict the properties of various locations on a manufactured product?

32. What is hardenability?

33. What are some of the ways in which a steel product might be hardened to a greater depth?

34. What are the three stages of liquid quenching?

35. What are some of the major advantages and disadvantages of a water quench?

36. Why is an oil quench less likely to produce quench cracks than water or brine?

37. What are some of the attractive qualities of a polymer of synthetic quench?

38. What are some undesirable design features in parts that are to be heat treated?

39. Why would the residual stresses in steel be different from the residual stresses in an identically processed aluminum part?

40. What are some of the potentially undesirable effects of residual stresses?

41. What causes quench cracking to occur when steel is rapidly cooled?

42. Describe several techniques that reduce residual stresses by enabling the volumetric changes to occur simultaneously throughout the part.
43. What is thermomechanical processing?
44. What are some of the methods that can be used to selectively heat treat the surface of metal parts?
45. What are some of the attractive features of surface hardening with a laser beam?
46. What is carburizing? Why does a carburized part have to be further heat-treated after the carbon is diffused into the surface?
47. In what ways might ionitriding be more attractive than conventional nitriding or carburizing?
48. For what type of products or product mixes might a batch furnace be preferred to a continuous furnace?
49. What are some of the possible functions of artificial atmospheres used during heat treating?
50. How are parts heated in a fluidized-bed furnace? What are some of the attractive features?
51. In what ways might the heat treatment of metals actually be an energy conservation measure?

■ PROBLEMS

1. A number of heat treatments have been devised to harden the surfaces of steel and other engineering metals. Consider the processes of
   - a. Flame hardening
   - b. Induction hardening
   - c. Carburizing
   - d. Nitriding

   For each of these processes, provide information relating to
   1. A basic description of how the process works
   2. Typical materials on which the process is performed
   3. Type of equipment required
   4. Typical times, temperatures, and atmospheres required
   5. Typical depth of hardening and reasonable limits
   6. Hardness achievable
   7. Subsequent treatments or processes that might be required
   8. Information relating to distortion and/or stresses
   9. Ability to use the process to harden selective areas

2. Investigate the nine areas in Problem 1 for one of the lesser-known surface-modification treatments, such as boriding, chromizing, or similar.

3. This chapter presented four processing-type heat treatments whose primary objective is to soften, weaken, enhance ductility, or promote machinability. Consider each of the following processes as they are applied to steels:
   - a. Full annealing
   - b. Normalizing
   - c. Process annealing
   - d. Spheriodizing

   Provide information relating to
   1. A basic description of how the process works and what its primary objectives are
   2. Typical materials on which the process is performed
   3. Type of equipment used
   4. Typical times, temperatures, and atmospheres required
   5. Recommended rates of heating and cooling
   6. Typical properties achieved

4. It has been noted that *hot oil* is a more effective quench than *cold oil*. Can you explain this apparent contradiction?

www.wiley.com/college/degarmo

*C*hapter 5   CASE STUDY

A Flying Chip from a Sledgehammer

Industrial sledgehammers are used throughout JCL Industries, most having a 15-pound head made of AISI 1060 steel. To reduce tool replacement costs, the company machine shop periodically gathers hammers with heavily deformed (mushroomed) heads and grinds off the deformed segment.

A reground hammer was returned to use. Upon striking a metal plate, a chip flew from a corner of the hammer head and lodged in the eye of a nearby worker. A lawsuit resulted. Subsequent investigation revealed that the heads of new hammers had a bulk hardness between R_C44 and 55. The chip, however, had a hardness of approximately R_C65

on the surface where it fractured from the hammer. Inspection of the other hammers in the shop revealed numerous chipped regions. All of the chips, however, were on redressed hammers.

1. What do you suspect might be the problem?
2. How might you prove your suspicions?
3. How would you alter the procedures or policies of JCL Industries to eliminate a possible recurrence, yet minimize expense?

CHAPTER 6

FERROUS METALS AND ALLOYS

■ 6.1 INTRODUCTION TO HISTORY-DEPENDENT MATERIALS

Engineering materials are available with a wide range of useful properties and characteristics. Some of these are inherent to the particular material, but many others can be varied by controlling the manner of production and the details of processing. Metals are classic examples of such history-dependent materials; their final properties are clearly affected by their past processing history. The particular details of the smelting and refining process control the resulting purity and the type and nature of any influential contaminants. The solidification process imparts structural features that may be transmitted to the final product. Preliminary operations such as the rolling of sheet or plate often impart directional variations to properties, and their impact should be considered during subsequent processing and use. Thus, while it is easy to take the attitude that "metals come from warehouses," it is important to recognize that aspects of prior processing can significantly influence further operations as well as the final properties of the product. The breadth of this book does not permit full coverage of the processes and methods involved in the production of engineering metals, but certain aspects will be presented because of their role in affecting subsequent performance.

In this chapter we will introduce the major *ferrous* (iron-based) *metals* and *alloys*, summarized schematically in Figure 6-1. These materials have been the backbone of civilization and numerous varieties have been developed over the years to meet the

FIGURE 6-1 Schematic diagram depicting common ferrous metal alloys.

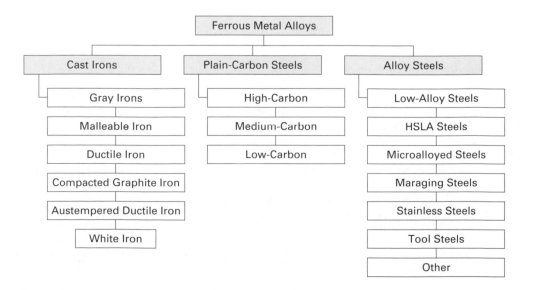

specific needs of various industries. In addition, the past two decades have seen the introduction of a number of new varieties and even classes of ferrous metals.

The availability of so many alternatives, both old and new, has often led to poor materials selection. Money can be wasted in the unnecessary specification of an expensive alloy or one that is difficult to fabricate. At other times, these materials may be absolutely necessary, and selection of a cheaper alloy would mean certain failure. Thus, it is the responsibility of the design and manufacturing engineer to be knowledgeable in the area of engineering materials and be able to make the best selection from among the available alternatives.

6.2 IRON

Iron is the fourth most plentiful element in the earth's crust and for centuries has been the most important of the basic engineering metals. The variety of metals and alloys derived from iron has played a central role in the development of civilization and will probably continue in this role in the foreseeable future. Even now, new advances in the technology of iron and iron alloys continue to expand their utility in a myriad of engineering applications.

Iron is rarely found in the metallic state but occurs in a variety of mineral compounds, known as ores. The most attractive of these ores are iron oxides coupled with companion impurities. These are first processed in a manner that breaks the iron–oxygen bonds to produce metallic iron (chemical reducing reactions). Ore, limestone, coke (carbon), and air are continuously introduced into specifically designed furnaces and molten metal is periodically withdrawn.

Within the furnace, other oxides (that were impurities in the original ore) will also be reduced. All of the phosphorus and most of the manganese will enter the molten iron. Oxides of silicon and sulfur are partially reduced, and these elements also become part of the resulting metal. Other contaminant elements, such as calcium, magnesium, and aluminum, are collected in the limestone-based slag and are removed from the system. The resulting *pig iron* tends to have roughly the following composition:

Carbon	3.0–4.5%
Manganese	0.15–2.5%
Phosphorus	0.1–2.0%
Silicon	1.0–3.0%
Sulfur	0.05–0.1%

A small portion of this iron is cast directly into final shape and is classified as cast iron. Most commercial cast iron, however, is produced by reprocessing scrap iron and steel, with the possible addition of some newly produced pig iron. The metallurgical properties of cast iron have been presented in Chapter 4, and melting and utilization in the casting process will be developed in Chapters 13 through 15. Most pig iron, however, is further processed into steel.

6.3 STEEL

The manufacture of *steel* is essentially an oxidation process that decreases the amount of carbon, silicon, manganese, phosphorus and sulfur in a mixture of molten pig iron and steel scrap. In 1856, the Kelly–Bessemer process opened up the industry by enabling the manufacture of commercial quantities of steel. The open-hearth process surpassed the Bessemer process in tonnage produced in 1908 and was producing over 90% of all steel in 1960. Currently, oxygen furnaces of a variety of types and electric arc furnaces produce most of our commercial steels.

In many of these processes, air or oxygen passes over or through the molten metal to drive a variety of exothermic refining reactions. Carbon oxidizes to form gaseous CO or CO_2, which then leaves the melt. Other elements, such as silicon and phosphorus, are similarly oxidized and rise to be collected in a removable slag. At the same time, oxygen and other elements from the reaction gases dissolve in the molten metal and may later become a cause for concern.

SOLIDIFICATION CONCERNS

Regardless of the method by which the steel is made, it must undergo a change from liquid to solid before it can become a usable product. The liquid can be converted directly into finish-shape steel castings or solidified into a form suitable for further processing. In most cases, continuous casting or the forming of discrete ingots produces the feedstock material for subsequent forging or rolling operations.

Prior to solidification, we want to remove as much contamination as possible. The molten metal is poured from the steelmaking furnaces into containment vessels, known as *ladles*. Historically, the ladles served simply as transfer and pouring containers, but they have recently emerged as the site for additional processing. *Ladle metallurgy* refers to a variety of processes designed to provide final purification and to fine-tune both the chemistry and temperature of the melt. Alloy additions can be made, carbon can be further reduced, dissolved gases can be reduced or removed, and steps can be taken to control subsequent grain size, limit inclusion content, reduce sulfur, and control the shape of any included sulfides. Stirring, degassing, reheating, and the injection of powdered alloys or cored wire can all be performed to increase the cleanliness of the steel and provide for tighter control of the chemistry and properties.

The processed liquid is then poured from these ladles into ingot molds or continuous casters, usually through some form of bottom-pouring process such as the one shown schematically in Figure 6-2. By extracting the metal from the bottom of the ladle, slag and floating matter are not transferred, and a cleaner product results.

Extending the observation that most nonmetallic contaminants will float on the molten metal, we note that a further refinement can be achieved by filling the ingot mold from the bottom. As illustrated in Figure 6-3, the bottom of several ingot molds can be connected to a central pouring ingot by ceramic tunnels. Hot metal is poured into the center ingot and is conveyed through the tunnels to fill the outer molds. Contaminants in the pour stream will tend to rise to the top of the central pouring ingot. In addition, the outer molds fill smoothly, avoiding turbulence or splashing and the additional contamination that might result as the hot material reacts with air.

Figure 6-4 schematically shows the dimensional changes of a metal undergoing cooling and solidification. The large discontinuity at the melting point is known as *solidification shrinkage* and reflects the difference in density of the liquid and solid. As a result, when metals solidify, a shrinkage void may form in the regions that held the last remaining liquid. In ingots, solidification proceeds inward from the mold walls and upward from the bottom. Shrinkage often assumes the form of a funnel-shaped pipe

FIGURE 6-2 Schematic diagram of a bottom-pouring ladle.

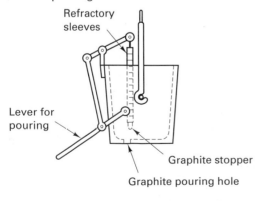

FIGURE 6-3 Pouring of ingots by the bottom-pouring process. The bottom of the center mold is connected to the bottom of the remaining molds in the cluster by ceramic channels.

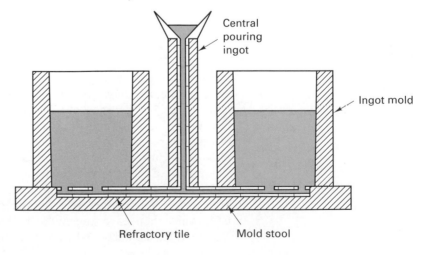

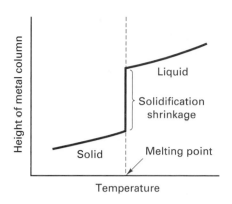

FIGURE 6-4   Height of a column of metal as a function of temperature, showing the significant shrinkage that occurs upon solidification.

coming down from the top. Oxides and contaminants form on the hot exposed surfaces, and prevent the metal from welding back together during subsequent processing.

While the amount of shrinkage is set by the change in density, the shape and location of that shrinkage can be controlled. Placing a hot block of ceramic or exothermic material on the top of the ingot, for example, can reduce the depth of pipe penetration. With additional heat at the top, the liquid present near the end of solidification is more of a uniform layer. Variation in the size and shape of the mold can further control both the geometry of solidification and the final structure of the ingot. Tapered ingots with the big end up generally produce the best quality product.

Continuous casting can be used to overcome a number of ingot-related difficulties, such as piping, entrapped slag, and structure variation along the length of the product. Figure 6-5 illustrates a typical continuous caster, in which molten metal flows from a ladle, through a tundish, into a bottomless, water-cooled mold, usually made of copper. Cooling is controlled so that the outside has solidified before the metal exits the mold. The emerging metal is then cooled by direct water sprays to produce complete solidification. The cast solid is still hot and is either bent and fed horizontally through a short reheat furnace and

FIGURE 6-5   Schematic representation of the continuous casting process for producing billets, slabs, and bars. *(Courtesy of Materials Engineering.)*

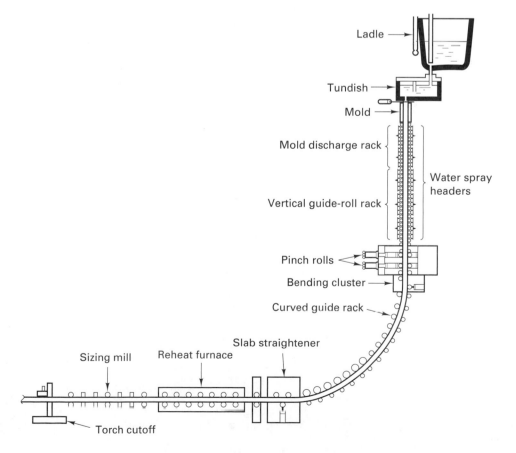

on to rolling, or is simply cut to desired lengths. By varying the shape of the mold, products can be cast with cross sections that are closer to the desired final shape.

In terms of product quality, continuous casting virtually eliminates the problems of piping and mold spatter. From a production viewpoint, it eliminates pouring into molds, stripping the molds from the solidified metal, and handling and reheating the ingots prior to rolling. Cost, energy, and scrap are all reduced significantly. In addition, the products have improved surfaces, more uniform chemical composition, and fewer oxide inclusions.

DEOXIDATION AND DEGASSIFICATION

As a result of the steelmaking process, large amounts of oxygen can be dissolved in the molten metal. During subsequent cooling and solidification, the oxygen and other gases are rejected as the solubility levels decrease significantly (Figure 6-6). The rejected oxygen frequently links with carbon to produce carbon monoxide gas, which may escape through the remaining liquid. Quite frequently, however, the bubbles become trapped, and a porous solid is produced. The bubble-induced porosity may take forms ranging from small, dispersed voids or large blowholes. The pores can often be welded shut during subsequent hot forming, if sufficient deformation is performed. Some of the voids, however, may not be fully closed, and others may not weld upon closure. Cracks and internal voids can persist into a finished product.

In many cases, potential porosity problems can be avoided by either removing the oxygen prior to solidification or making sure it does not reemerge as a gas. Aluminum, ferromanganese, or ferrosilicon can be added to molten steel to provide a material whose affinity for oxygen is higher than that of carbon. The rejected oxygen then reacts with the deoxidizer additions to produce solid metal oxides that are dispersed throughout the structure.

While deoxidizer additions can effectively tie up the dissolved oxygen, small amounts of other dissolved gases, such as hydrogen and nitrogen, can also have deleterious effects on the performance of steels. This is particularly important for alloy steels because alloy additions such as vanadium, niobium, and chromium tend to increase the solubility of these gases. Alternative degassing processes have been devised that reduce the amounts of all dissolved gases. Figure 6-7 illustrates the process of *vacuum degassing*, in which an ingot mold is placed in an evacuated chamber, and the metal stream passes through a vacuum during pouring. By creating a large amount of exposed surface during the pouring operation, the vacuum is able to extract most of the dissolved gas.

FIGURE 6-6 Solubility of gas in a metal as a function of temperature.

FIGURE 6-7 Method of vacuum degassing steel while pouring ingots.

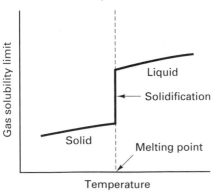

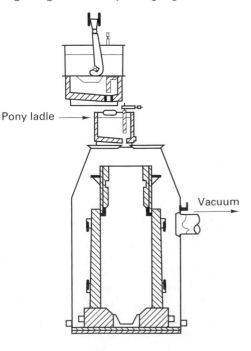

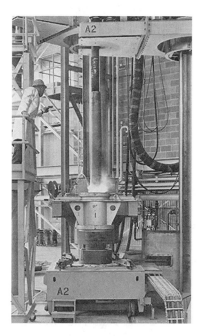

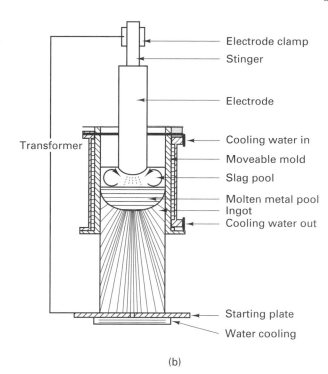

(a)

(b)

FIGURE 6-8 (a) Production of an ingot by the electroslag remelting process; (b) schematic representation of this process. *(Courtesy of Carpenter Technology Corporation.)*

An alternative to vacuum degassing is the *consumable-electrode remelting* process where an already solidified metal electrode replaces the ladle of molten metal. When the electrode is progressively remelted, the molten droplets pass through the vacuum and the extremely high surface area again provides an effective means of gas removal. If the melting is done by electric arc, the process is known as *vacuum arc remelting* (VAR). If induction heating is used, the process is known as *vacuum induction melting* (VIM). Both processes are highly effective in removing dissolved gases but are relatively ineffective in removing any nonmetallic impurities that may be present in the metal.

The *electroslag remelting* process (ESR), shown in Figure 6-8, can be used to produce extremely clean, gas-free metal. The solid electrode is again melted and recast using an electric current, but the surface of the recast liquid is now covered with a blanket of molten flux. Nonmetallic impurities float and are collected in the flux, leaving a newly solidified metal structure with much-improved quality. No vacuum is required since all of the molten material exists below the shroud of molten flux and the progressive freezing permits easy escape for the rejected gas. This process is simply a large-scale version of the electroslag welding process that will be discussed in Chapter 38.

PLAIN-CARBON STEEL

Commercial steel, while theoretically an alloy of only iron and carbon, actually contains manganese, phosphorus, sulfur, and silicon in significant and detectable quantities. When these four additional elements are present in their normal percentages, the product is referred to as *plain-carbon steel*. Its strength is primarily a function of its carbon content, increasing with increasing carbon, as can be seen in Table 6-1. Unfortunately, the ductility

TABLE 6-1.	Effect of Carbon on the Strength of Annealed Plain-Carbon Steels*		
Type of Steel	Carbon Content	Minimum MPa	Tensile Strength ksi
1020	0.20%	414	60
1030	0.30%	448	65
1040	0.40%	517	75
1050	0.50%	621	90

*Data are from ASTM Specification A732.

FIGURE 6-9 A comparison of low-carbon, medium-carbon, and high-carbon steels in terms of their balance of properties.

of plain-carbon steels decreases as the carbon content is increased, and its hardenability is quite low. In addition, the properties of ordinary carbon steels are impaired by both high and low temperatures (loss of strength and embrittlement, respectively), and they are subject to corrosion in most environments.

Plain-carbon steels are generally classed into three subgroups based on their carbon content. *Low-carbon steels* have less than 0.20% carbon and possess good formability (can be strengthened by cold work) and weldability. Their structures are usually ferrite and pearlite, and the material is generally used as it comes from the hot-forming or cold-forming processes, or in the as-welded condition. *Medium-carbon steels* have between 0.20 and 0.50% carbon, and they can be quenched to form martensite or bainite if the section size is small and a severe water or brine quench is used. The best balance of properties is obtained at these carbon levels where the high toughness and ductility of the low-carbon material is in good compromise with the strength and hardness that come with higher carbon contents. These steels are extremely popular and find numerous mechanical applications. *High-carbon steels* have more than 0.50% carbon. Toughness and formability are quite low, but hardness and wear resistance are high. Severe quenches can form martensite, but hardenability is still poor. Quench cracking is often a problem when the material is pushed to its limit. Figure 6-9 depicts the characteristic properties of low-, medium- and high-carbon steels using a balance of properties.

Compared to other engineering materials, the carbon steels offer high strength and high stiffness, coupled with reasonable toughness. They can be magnetically separated from mixed materials, and are easily recycled. Unfortunately, they also rust easily and generally require some form of surface protection, such as paint, galvanizing, or other coating. The plain-carbon steels are generally the lowest-cost steel material and should be given first consideration for many applications. Their limitations, however, may become restrictive. When improved performance is required, these steels can often be upgraded by the addition of one or more alloying elements.

ALLOY STEELS

The differentiation between plain-carbon and alloy steel is often somewhat arbitrary. Both contain carbon, manganese, and usually silicon. Copper and boron are possible additions to both classes. Steels containing more than 1.65% manganese, 0.60% silicon, or 0.60% copper are usually designated as *alloy steels*. Also, a steel is considered to be an alloy steel if a definite or minimum amount of other alloying element is specified. The most common alloy elements are chromium, nickel, molybdenum, vanadium, tungsten, cobalt, boron, and copper, as well as manganese, silicon, phosphorus, and sulfur in amounts greater than are normally present. If the steel contains less than 8% of total alloy addition, it is considered to be a *low-alloy steel*. Steels with more than 8% alloying elements are *high-alloy steels*.

EFFECTS OF THE VARIOUS ALLOYING ELEMENTS

In general, alloying elements are added to steels in small percentages (usually less than 5%) to improve strength or hardenability, or in much larger amounts (often up to 20%) to produce special properties such as corrosion resistance or stability at high or low temperatures. Additions of manganese, silicon, or aluminum may be made during

TABLE 6-2.	Effect of Carbon on the Strength of Quenched-and-Tempered Alloy Steels*		
Type of Steel	Carbon Content	Minimum MPa	Tensile Strength ksi
4130	0.30%	1030	150
4330	0.30%	1030	150
8630	0.30%	1030	150
4140	0.40%	1241	180
4340	0.40%	1241	180

*Data from ASTM Specification A732.

the steelmaking process to remove dissolved oxygen from the melt. Manganese, silicon, nickel, and copper add strength by forming solid solutions in ferrite. Chromium, vanadium, molybdenum, tungsten, and other elements increase strength by forming dispersed second-phase carbides. Nickel and copper can be added in small amounts to improve corrosion resistance. Nickel has been shown to impart increased toughness and impact resistance, and molybdenum helps resist embrittlement. Zirconium, cerium, and calcium can also promote increased toughness by controlling the shape of inclusions. Machinability can be enhanced through the formation of manganese sulfides, or by additions of lead, bismuth, selenium, or tellurium. Still other additions can be used to provide ferrite or austenite grain-size control.

Selection of an alloy steel still begins with identifying the proper carbon content. Table 6-2 shows the effect of carbon on the strength of quenched-and-tempered alloy steels. The strength values are significantly higher than those of Table 6-1, reflecting the difference between the annealed and quenched-and-tempered microstructures. The 4130 steel has about 1.2% total alloying elements, 4330 has 3.0%, and 8630 has about 1.3%, yet all have the same minimum tensile strength. These data provide confirmation that for most alloy steels the primary reason for the alloying addition is to increase *hardenability*. The most common elements for this purpose (in order of decreasing effectiveness) are manganese, molybdenum, chromium, silicon, and nickel. Boron is an extremely powerful hardenability agent. Only a few thousandths of a percent are sufficient to produce a significant effect in low-carbon steels, but the results diminish rapidly with increasing carbon content. Since no carbide formation or ferrite strengthening accompanies the addition, improved machinability and cold-forming characteristics may favor the use of boron in place of other hardenability additions. Small quantities of vanadium can also be quite effective, but the response drops off as the quantity is increased.

Table 6-3 summarizes the primary effects of the common alloying elements in steel. A working knowledge of this information may be useful in selecting an alloy steel to meet a given set of requirements. Alloying elements are often used in combination, however, resulting in the immense variety of alloy steels that are commercially available. To provide some degree of simplification, a classification system has been developed and has achieved general acceptance in a variety of industries.

AISI–SAE CLASSIFICATION SYSTEM

The most common classification scheme for alloy steels is the *AISI–SAE identification system*. This system, which classifies alloys by chemistry, was started by the Society of Automotive Engineers (SAE) to provide some standardization for the steels used in the automotive industry. It was later adopted and expanded by the American Iron and Steel Institute (AISI) and has been incorporated into the Universal Numbering System that was developed to include all engineering metals. Both plain-carbon and low-alloy steels are identified by a four-digit number, with the first number indicating the major alloying elements and the second number designating a subgrouping within the major alloy system. These first two digits can be interpreted by looking them up on a list, such as the one presented in Table 6-4. The last two digits of the number indicate the approximate amount of carbon, expressed as "points," where one point is equal to 0.01%. Thus, a

TABLE 6-3.   Principal Effects of Major Alloying Elements in Steel

Element	Percentage	Primary Function
Aluminum	0.95–1.30	Alloying element in nitriding steels
Bismuth	—	Improves machinability
Boron	0.001–0.003	Powerful hardenability agent
Chromium	0.5–2	Increase of hardenability
	4–18	Corrosion resistance
Copper	0.1–0.4	Corrosion resistance
Lead	—	Improved machinability
Manganese	0.25–0.40	Combines with sulfur to prevent brittleness
	>1	Increases hardenability by lowering transformation points and causing transformations to be sluggish
Molybdenum	0.2–5	Stable carbides; inhibits grain growth
Nickel	2–5	Toughener
	12–20	Corrosion resistance
Silicon	0.2–0.7	Increases strength
	2	Spring steels
	Higher percentages	Improves magnetic properties
Sulfur	0.08–0.15	Free-machining properties
Titanium	—	Fixes carbon in inert particles
		Reduces martensitic hardness in chromium steels
Tungsten		Hardness at high temperatures
Vanadium	0.15	Stable carbides; increases strength while retaining ductility; promotes fine grain structure

1080 steel would be a plain carbon steel with 0.80% carbon. Similarly, a 4340 steel would be a Mo–Cr–Ni alloy with 0.40% carbon. Because of the double-digit groupings, these steels are identified as a "ten eighty" and a "forty-three forty."

Letters may also be incorporated into the designation. The letter *B* between the second and third digits indicates that the base metal has been supplemented by the addition of boron. Similarly, an *L* in this position indicates a lead addition for enhanced

TABLE 6-4.   Some AISI–SAE Standard Steel Designations

AISI Number	Type	Mn	Ni	Cr	Mo	V	Other
1xxx	Carbon steels						
10xx	Plain carbon						
11xx	Free cutting (S)						
12xx	Free cutting (S) and (P)						
15xx	High manganese						
13xx	High manganese	1.60–1.90					
2xxx	Nickel steels		3.5–5.0				
3xxx	Nickel–chromium		1.0–3.5	0.5–1.75			
4xxx	Molybdenum						
40xx	Mo				0.15–0.30		
41xx	Mo, Cr			0.40–1.10	0.08–0.35		
43xx	Mo, Cr, Ni		1.65–2.00	0.40–0.90	0.20–0.30		
44xx	Mo				0.35–0.60		
46xx	Mo, Ni (low)		0.70–2.00		0.15–0.30		
47xx	Mo, Cr, Ni		0.90–1.20	0.35–0.55	0.15–0.40		
48xx	Mo, Ni (high)		3.25–3.75		0.20–0.30		
5xxx	Chromium						
50xx				0.20–0.60			
51xx				0.70–1.15			
6xxx	Chromium–vanadium						
61xx				0.50–1.10		0.10–0.15	
8xxx	Ni, Cr, Mo						
81xx			0.20–0.40	0.30–0.55	0.08–0.15		
86xx			0.40–0.70	0.40–0.60	0.15–0.25		
87xx			0.40–0.70	0.40–0.60	0.20–0.30		
88xx			0.40–0.70	0.40–0.60	0.30–0.40		
9xxx	Other						
92xx	High silicon						1.20–2.20 Si
93xx	Ni, Cr, Mo		3.00–3.50	1.00–1.40	0.08–0.15		
94xx	Ni, Cr, Mo		0.30–0.60	0.30–0.50	0.08–0.15		

machinability. A letter prefix may be used to designate the process used to produce the steel, such as *E* for electric furnace.

When hardenability is a major requirement, one might consider the H grade of AISI steels, designated by an H suffix attached to the standard designation. The chemistry specifications are somewhat less stringent, but the steel must also comply with a hardenability standard. The hardness values at specific distances from the quenched end of a Jominy specimen (see Chapter 5) must all lie within a predetermined band.

Other designation organizations, such as the American Society for Testing and Materials (ASTM) and the U.S. government (MIL and federal), have specification systems based more on specific applications. Acceptance into a given classification is generally determined by physical or mechanical properties rather than the chemistry of the metal. ASTM designations are frequently used when specifying low-carbon and structural steels.

SELECTING ALLOY STEELS

From the previous discussion it is apparent that two or more alloying elements can often produce similar effects. Thus steels with substantially different chemical compositions can possess almost identical mechanical properties. Figure 6-10 clearly demonstrates that, when properly heat-treated, steels of quite different composition (chemistry) can have almost identical property ratios. This fact is particularly important when one realizes that some alloying elements can be very costly, and others may be in short supply due to emergencies or political constraints. Overspecification has often been employed to guarantee success despite sloppy manufacturing and heat-treatment practice. The correct steel, however, is usually the least expensive one that can be consistently processed to achieve the desired properties. This usually involves taking advantage of the effects provided by all of the alloy elements.

When selecting alloy steels, it is also important to consider both use and fabrication. For one product, it might be permissible to increase the carbon content to obtain greater strength. For another application, such as one involving assembly by welding, it might be best to keep the carbon content low and use a balanced amount of alloy elements to obtain the desired strength without risking the possibility of cracking due to high-carbon

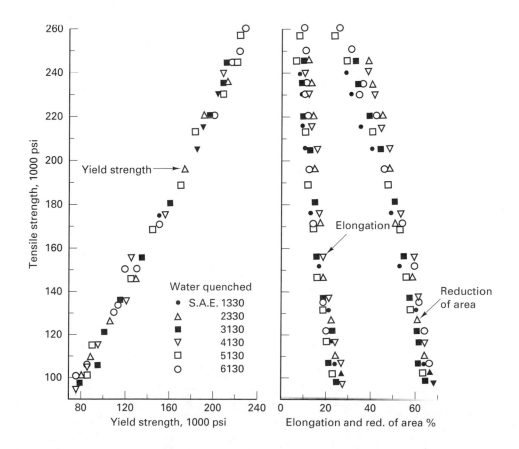

FIGURE 6-10   Relationships between the mechanical properties of a variety of properly heat-treated AISI–SAE alloy steels. *(Courtesy of ASM International, Materials Park, Ohio.)*

martensite. Steel selection, therefore, often involves defining the required properties, determining the best microstructure to provide those properties (strength can be achieved through alloying, cold work, and heat treatment, as well as combinations thereof), and selecting the steel with the best carbon content and hardenability characteristics to achieve that goal. One approach may be to purchase material on the basis of properties rather than exact chemical composition. The supplier or producer is then free to provide any material that will possess the desired characteristics, and substantial cost savings may result. To assure success, however, it is important that all of the necessary properties be specified.

HIGH-STRENGTH LOW-ALLOY STRUCTURAL STEELS

There are two general categories of alloy steels: (1) the *constructional alloys*, where the desired properties are typically developed by a separate thermal treatment and the specific alloy elements tend to be selected for their effect on hardenability, and (2) the *high-strength low-alloy* (HSLA) or microalloyed types, which rely largely on chemical composition to develop the desired properties in the as-rolled or normalized condition.

The HSLA materials often find application in large, welded structures, where the size of the product tends to preclude subsequent heat treatment. The dominant property requirements are high yield strength, good weldability, and acceptable corrosion resistance. Ductility and hardenability are somewhat limited, however.

The low-alloy structural steels often exhibit yield strengths that are nearly twice that of the plain-carbon variety. The increase in strength, and the resistance to martensite formation in a weld zone, is obtained by controlling the amounts of carbon, manganese, and silicon, and the addition of small amounts of niobium, vanadium, or other alloys. About 0.2% copper can be added to improve corrosion resistance.

Because of their higher yield strength, weight savings of 20 to 30% can often be achieved with no sacrifice to strength or safety. Rolled and welded HSLA steels are being used in automobiles, trains, bridges, and buildings. Because of their low alloy content and high-volume application, their cost is often little more than that of the ordinary plain-carbon steels. Table 6-5 presents the chemistries and properties of several of the more common types.

MICROALLOYED STEELS IN MANUFACTURED PRODUCTS

In terms of both cost and performance, *microalloyed steels* occupy a position between carbon steels and the alloy grades. These steels contain small amounts (0.05 to 0.15%) of alloying elements, such as niobium, vanadium, titanium, molybdenum, zirconium, boron, rare earth elements, or combinations thereof, and are also being used increasingly as substitutes for heat-treated steels in the manufacture of small- to medium-sized discrete parts. The primary effect of the alloy addition is to provide grain refinement and/or precipitation strengthening. Yield strengths between 500 and 750 MPa (70 and 110 ksi) can be obtained without heat treatment. Weldability can be retained or even improved if the carbon content is

TABLE 6-5.   Typical Compositions and Strength Properties of Several Groups of High-Strength Low-Alloy Structural Steels

| | | | | | | Strength Properties | | | | |
| | Chemical Compositions[a] (%) | | | | | Yield | | Tensile | | Elongation in 2 in. |
Group	C	Mn	Si	Cb	V	ksi	MPa	ksi	MPa	(%)
Columbium or vanadium	0.20	1.25	0.30	0.01	0.01	55	379	70	483	20
Low manganese–vanadium	0.10	0.50	0.10		0.02	40	276	60	414	35
Manganese–copper	0.25	1.20	0.30			50	345	75	517	20
Manganese–vanadium–copper	0.22	1.25	0.30		0.02	50	345	70	483	22

[a] All have 0.04% P, 0.05% S, and 0.20% Cu.

simultaneously decreased. In essence, these steels offer maximum strength with minimum carbon, while simultaneously preserving weldability, machinability, and formability. Compared to a quenched-and-tempered alternative, ductility and toughness are generally somewhat inferior.

Cold-formed microalloyed steels require less cold work to achieve a desired level of strength, so they tend to have greater residual ductility. Hot-formed products, such as forgings, can often be used in the air-cooled condition. By means of accurate temperature control and controlled-rate cooling, mechanical properties can be produced that approximate those of quenched-and-tempered material. Machinability can be enhanced because of the more uniform hardness and the fact that the ferrite–pearlite structure of the microalloyed steel is often more machinable than the ferrite–carbide structure of the quenched-and-tempered variety. Fatigue life and wear resistance can also be superior to those of the heat-treated counterparts.

In applications where the properties are adequate, microalloyed steels can often provide attractive cost savings. Energy savings can be substantial, straightening or stress relieving after heat treatment is no longer necessary, and quench cracking is not a problem. Due to the increase in material strength, the size and weight of finished products can often be reduced. As a result, the cost of a finished forging could be reduced by 5 to 25%.

Certain precautions must be observed, however, if these materials are to attain their optimum properties. During the elevated-temperature segments of processing, the material must be heated high enough to place all of the alloys into solution. After forming, the products should be rapidly air cooled to 540 to 600°C (1000 to 1100°F) before dropping into collector boxes. In addition, microalloyed steels tend to through-harden upon air cooling, so products fail to exhibit the lower-strength, higher-toughness interiors that are typical of the quenched-and-tempered materials.

DUAL-PHASE STEELS

Dual-phase steels use a quench with a temperature above A_1 but below A_3 to produce a structure of ferrite and high-carbon martensite in an otherwise low- or medium-carbon steel. While strengths are comparable to HSLA materials, dual-phase steels offer improved forming characteristics with no loss in weldability. A high strain-rate sensitivity means that the faster the steel is crushed, the more energy it absorbs. This feature is quite attractive for automotive structural and body applications.

FREE-MACHINING STEELS

The increased use of high-speed machining, particularly with automated machine tools, has spurred the use and development of several varieties of *free-machining steels*. These steels machine readily and form small chips when cut. The smaller chips reduce the length of contact between the chip and cutting tool, thereby reducing the associated friction and heat, and therefore, power and tool wear. The formation of small chips also reduces the likelihood of chip entanglement in the machine and makes chip removal much easier. While free-machining steels may carry a cost premium of 15 to 20% over conventional alloys, this increase may be easily recovered through higher machining speeds, larger depths of cut, and extended tool wear.

Free-machining steels are basically carbon steels that have been modified by an alloy addition of sulfur, lead, bismuth, selenium, tellurium, or phosphorus to enhance machinability. Sulfur combines with manganese to form soft manganese sulfide inclusions. These, in turn, serve as chip-breaking discontinuities within the structure. The inclusions also provide a built-in lubricant that prevents formation of a built-up edge on the cutting tool and imparts an altered cutting geometry (see Chapter 21). Insoluble lead particles work in much the same way.

The bismuth free-machining steels are an attractive alternative to the previous varieties. Bismuth is more environmentally acceptable (compared to lead), has a reduced tendency to form stringers, and can be more uniformly dispersed since its density is a better match to that of iron. Machinability is improved because the heat generated by cutting is sufficient to form a thin film of liquid bismuth that lasts for only fractions of a microsecond. Tool life is noticeably extended and the machined product is still weldable.

Use of free-machining steels is not without compromise, however. Ductility and impact properties are somewhat reduced compared to the unmodified steels. Copper-based braze joints tend to embrittle when used to join bismuth free-machining steels, and the machining additions reduce the strength of shrink-fit assemblies. If these compromises are objectionable, other methods may be used to enhance machinability. For example, the machinability of steels can be improved by cold working the metal. As the strength and hardness of the metal increase, the metal loses ductility, and subsequent machining produces chips that tear away more readily and fracture into smaller segments.

BAKE-HARDENABLE STEEL SHEET

Bake-hardenable steel sheet has assumed a significant role in automotive sheet applications. These low-carbon steels are processed in such a way that they are resistant to aging during normal storage but begin to age during sheet-metal forming. The subsequent exposure to heat during the paint-baking operation completes the aging process and adds an additional 35 to 70 MPa (5 to 10 ksi), raising the final yield strength to approximately 275 MPa (40 ksi). Since the increase in strength occurs after the forming operation, the material offers a combination of good formability and improved dent resistance in the final product. In addition, it allows weight savings to be achieved without compromising the attractive features of steel sheet, which include spot weldability, good crash energy absorption, low cost, and easy recyclability through magnetic separation.

PRECOATED STEEL SHEET

Traditional sheet metal fabrication involves the fabrication of components from bare steel, followed by the finishing of these products on a piece-by-piece basis. In this system it is not uncommon for the finishing processes to be the most expensive and time-consuming stages of manufacture.

An alternative to this procedure is to purchase mill-coated steel sheet, where the steel supplier applies the coating when the material is still in the form of a long, continuous strip. Numerous coatings can be specified, including the entire spectrum of dipped and plated metals (including aluminum, zinc, and chromium), vinyls, paints, and others. Extra caution must be exercised during fabrication to prevent damage to the coating, but the additional effort and expense are often less than the cost of finishing individual pieces.

STEELS FOR ELECTRICAL AND MAGNETIC APPLICATIONS

Soft magnetic materials can be magnetized by relatively low-strength magnetic fields, but lose almost all of their magnetism when the applied field is removed. They are widely used in products such as transformers, motors, and generators. The most common soft magnetic materials are high-purity iron, low-carbon steels, iron-silicon electrical steels, amorphous ferromagnetic alloys, iron-nickel alloys, and soft ferrites (ceramic material).

Silicon steels, with 0.5 to 5.0% silicon, have noticeably increased electrical resistivity and magnetic permeability. Increased resistivity decreases eddy-current losses, while increased permeability decreases hysteresis losses. Since silicon also embrittles steel, the amount of silicon should be kept as low as possible, especially for applications where the component will be subjected to dynamic forces.

More recently, soft magnetic *amorphous metals* have shown attractive electrical and magnetic properties. Since the material has no grains or grain boundaries: (1) the magnetic domains can move freely in response to magnetic fields, (2) the properties are the same in all directions, and (3) corrosion resistance is improved. The high magnetic strength and low hysteresis losses offer the possibility of smaller, lighter-weight magnets. When used to replace silicon steel in power transformer cores, this material has the potential of reducing core losses by as much as 50%.

Materials that exhibit permanent magnetism remain magnetized when removed from the applied field. While most permanent magnets are ceramic materials or complex metal alloys, *cobalt alloy steels* (containing up to 36% cobalt) may be specified for electrical equipment where high magnetic densities are required.

MARAGING STEELS

When super-high strength is required from a steel, the *maraging* grades become a very attractive option. These alloys contain between 15 and 25% nickel, plus significant amounts of cobalt, molybdenum, and titanium, all added to a very-low-carbon steel. A typical composition is

0.03% C	0.10% Al	4.8% Mo	0.10% Mn maximum
8.5% Ni	0.003% B	0.40% Ti	0.01% S maximum
7.5% Co	0.10% Si maximum	0.01% Zr	0.01% P maximum

This steel can be hot worked at temperatures between 750 and 1250°C (1400 and 2300°F). When air cooled from about 800°C (1500°F), it has a hardness of about $30R_C$ and a structure of soft, tough, low-carbon martensite. It is easily machined and, because of its low work-hardening rate, can be cold worked to a high degree. Aging at 500°C (900°F) for 3 to 6 hours, followed by air cooling, raises the hardness to about $52R_C$, with a yield strength in excess of 1725 MPa (250 ksi) and an elongation in excess of 11%.

Maraging alloys are very useful in applications where ultrahigh strength and good toughness are important. They can be welded, if welding is followed by the full solution and aging treatment. As might be expected from the large amount of alloy additions (over 30%) and multistep thermal processing, maraging steels are quite expensive and should be specified only when their outstanding properties are absolutely required.

STEELS FOR HIGH-TEMPERATURE SERVICE

As a general rule of thumb, plain-carbon steels should not be used at temperatures in excess of about 250°C (500°F). Conventional alloy steels extend this upper limit to around 350°C (650°F). Continued developments in areas such as missiles and jet aircraft, however, have increased the demand for metals that offer good strength characteristics, corrosion resistance, and creep resistance at operating temperatures in excess of 550°C (1000°F).

The high-temperature ferrous alloys tend to be low-carbon materials with less than 0.1% carbon. At their peak operating temperatures, 1000-hour rupture stresses tend to be quite low, often in the neighborhood of 50 MPa (7 ksi). While iron is also a major component of other high-temperature alloys, when the amounts fall below 50%, the metal is not generally classified as a ferrous material. High strength at high temperature usually requires the more expensive nonferrous materials that will be discussed in Chapter 7.

USE AND RECYCLING OF STEEL

Steel is an extremely useful engineering material. It offers strength, rigidity, and durability. From a manufacturing perspective, its formability, joinability, and paintability, as well as repairability, are all attractive. As a result, steel accounted for half of the material used in a typical 2000 model Japanese passenger car (on a per weight basis), and will likely continue at this level.

In terms of tonnage, steel is the most recycled material in commerce, nearly twice as much as paper, and far exceeding aluminum, glass, and plastics. Its magnetic properties facilitate easy recovery and separation from other materials. As a result, about two thirds of the steel production in the United States comes from the recycling of steel scrap.

■ 6.4 STAINLESS STEELS

Low-carbon steel with the addition of 4 to 6% chromium acquires good resistance to many of the corrosive media encountered in the chemical industry. This behavior is attributed to the formation of a strongly adherent iron chromium oxide. If more improved corrosion resistance and outstanding appearance are required, materials should be specified that use a superior iron chromium oxide that forms when the amount of chromium in solution (excluding chromium carbides and other forms where the chromium is no longer available to react with oxygen) exceeds 12%. When damaged, this tough, adherent, corrosion-resistant oxide heals itself, provided oxygen is present, even in very small amounts. Materials that form this oxide are referred to as the *true stainless steels*.

TABLE 6-6.	AISI Designation Scheme for Stainless Steels	
Series	Alloys	Structure
200	Chromium, nickel, manganese, or nitrogen	Austenitic
300	Chromium and nickel	Austenitic
400	Chromium and possibly carbon	Ferritic or martensitic
500	Low chromium ($<12\%$) and possibly carbon	Martensitic

Several classification schemes have been devised to categorize these alloys. The American Iron and Steel Institute (AISI) groups the metals by chemistry and assigns a three-digit number that identifies the basic family and the particular alloy within that family. This book, however, will group these alloys into microstructural families, since it is the basic structure that controls the engineering properties of the metal. Table 6-6 presents the AISI designation scheme for stainless steels and correlates it with the microstructural families.

Chromium is a ferrite stabilizer; that is, the addition of chromium tends to increase the temperature range over which ferrite is the stable structure. With sufficient chromium and a low level of carbon, a corrosion-resistant iron alloy can be produced that is ferrite at all temperatures below solidification. These alloys are known as *ferritic stainless steels*. They possess rather limited ductility or formability, but are readily weldable. No martensite can form in the welds because there is no possibility of forming the FCC austenite structure that can then transform during cooling. Since the ferritic alloys are the cheapest type of stainless steel, they should be given first consideration when a stainless alloy is required.

If increased strength is needed, the *martensitic stainless steels* should be considered. For these alloys, the chromium content is such that the material can be austenite (FCC) at high temperature and ferrite (BCC) at low. As with the standard steels, carbon can be dissolved in the face-centered-cubic austenite, which can then be quenched to trap the carbon in a martensitic structure. Variable carbon contents are available, providing a range of possible strengths and hardnesses. Caution should be taken, however, to assure more than 12% chromium in solution. Slow cools may allow the carbon and chromium to react and form chromium carbides. When this occurs, the chromium is not available to react with oxygen and form the protective oxide. As a result, the martensitic stainless steels may only be "stainless" when in the martensitic condition (when the chromium is trapped in atomic solution), and may be susceptible to red rust when annealed or normalized for machining or fabrication. The martensitic stainless steels cost about 1-1/2 times as much as the ferritic alloys, with part of the increase being due to the additional heat treatment, which generally consists of an austenitization, quench, stress relief, and temper.

Nickel is an austenite stabilizer, and with sufficient amounts of both chromium and nickel, it is possible to produce a stainless steel in which austenite is the stable structure at room temperature. Known as *austenitic stainless steels*, these alloys may cost two to three times as much as the ferritic variety, but here the added expense is attributed to the cost of the nickel and chromium alloys. Manganese and nitrogen are also austenite stabilizers and may be substituted for some of the nickel to produce a lower-cost, somewhat lower-quality austenitic stainless steel.

Austenitic stainless steels are nonmagnetic and are highly resistant to corrosion in almost all media except hydrochloric acid and other halide acids and salts. In addition, they may be polished to a mirror finish and thereby combine attractive appearance and corrosion resistance. Formability is outstanding (characteristic of the FCC crystal structure), and these steels strengthen significantly when cold worked. The response of the popular 304 alloy (also known as 18-8 because of the composition of 18% chromium and 8% nickel) to a small amount of cold work is as follows:

	Water Quench	Cold Rolled 15%
Yield strength [MPa (ksi)]	260 (38)	805 (117)
Tensile strength [MPA (ksi)]	620 (90)	965 (140)
Elongation in 2 in. (%)	68	11

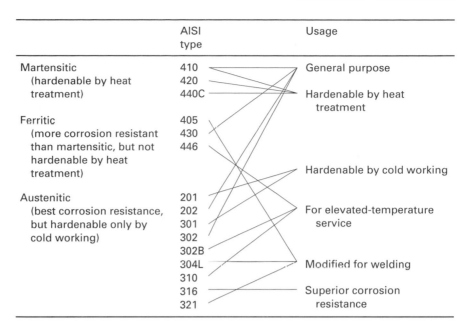

	AISI type	Usage
Martensitic (hardenable by heat treatment)	410 420 440C	General purpose / Hardenable by heat treatment
Ferritic (more corrosion resistant than martensitic, but not hardenable by heat treatment)	405 430 446	Hardenable by cold working
Austenitic (best corrosion resistance, but hardenable only by cold working)	201 202 301 302 302B 304L 310 316 321	For elevated-temperature service / Modified for welding / Superior corrosion resistance

FIGURE 6-11 Popular alloys and key properties for different types of stainless steels.

The austenitic stainless steels are often used in the water-quenched condition, where the water quench serves to retain the alloys in solid solution. No phase transformations occur during the quench since austenite is the stable phase for all of the temperatures involved.

Austenitic stainless steels are costly materials and should not be specified where the less expensive ferritic or martensitic alloys would be adequate or where a true stainless steel is not required. Figure 6-11 lists some of the popular alloys from each of the three major classifications and schematically denotes some of their key properties.

A fourth and special class of stainless steels is the *precipitation-hardening* variety. These alloys are basically martensitic or austenitic types, modified by the addition of alloying elements such as aluminum that permit the precipitation of hard intermetallic compounds at the low temperatures used to temper the martensite. With the addition of age hardening, these materials are capable of attaining properties such as a 1790-MPa (260-ksi) yield strength and 1825-MPa (265-ksi) tensile strength with a 2% elongation. However, the additional alloys and extra processing make the precipitation-hardening alloys some of the most expensive stainless steels, and they should be used only when absolutely required.

Duplex stainless steels contain between 18 and 25% chromium, 4 to 7% nickel, and up to 4% molybdenum, and are water quenched from a hot-working temperature that is between 1000 and 1050°C (1830 and 1920°F) to produce a microstructure that is approximately half ferrite and half austenite. This structure offers a higher yield strength and greater resistance to stress corrosion cracking than either the austenitic or ferritic grades.

Still other stainless alloys have been developed to meet special needs. Ordinary stainless steels are difficult to machine because of their work-hardening properties and their tendency to seize during cutting. Special *free-machining alloys* have been produced within each family, with the addition of sulfur or selenium raising the machinability to approximately that of a medium-carbon steel.

Stainless steels can lose their corrosion resistance when the amount of chromium in solution drops below 12%. Given that chromium depletion is usually caused by the formation of chromium carbides along grain boundaries (*sensitization*), and these carbides form at elevated temperatures, various means have been developed to prevent their formation. One approach is to keep the carbon content of stainless steels as low as possible, usually below 0.10%. Another method is to tie up the carbon with small amounts of *stabilizing* elements, such as titanium or niobium, that have a stronger affinity for carbon than does chromium. Rapidly cooling these metals through the carbide-forming range of 480 to 820°C (900 to 1500°F) also works to prevent carbide formation.

Another problem with high-chromium stainless steels is an embrittlement that can occur after long times at elevated temperatures. This is attributed to the formation of *sigma phase*, a brittle compound that forms at elevated temperature and coats grain boundaries, thereby producing a brittle crack path through the metal. Stainless steels used in high-temperature service should be checked periodically to detect and monitor sigma-phase formation.

■ 6.5 TOOL STEELS

Tool steels are high-carbon, high-strength, ferrous alloys that have been modified by alloy additions to provide a desired balance of strength, toughness, and wear resistance. Several classification systems have been developed, some using chemistry as a basis, while others employ hardening method or major mechanical property. The AISI–SAE system uses a letter designation to identify basic features such as quenching method, primary application, special alloy or characteristic, or specific industry involved. Table 6-7 lists the seven basic families of tool steels, the corresponding AISI–SAE letter grades, and the associated feature or characteristic. Individual alloys within the letter grades are then listed numerically to produce a letter–number identification system.

Water-hardening tool steels (W-grade) are essentially high-carbon plain-carbon steels, and account for a large percentage of all of the tool steels used. They are the least expensive variety and are used for a wide range of parts that are usually quite small and not subject to severe usage or elevated temperature. Because strength and hardness are functions of the carbon content, a wide range of properties can be achieved through composition variation. Hardenability is low, so these steels must be quenched in water to attain high hardness. They can be used only for relatively thin sections if the full depth of hardness is desired. They are also rather brittle, particularly at higher hardness.

Typical uses of the various plain-carbon steels are as follows:

0.60–0.75% carbon: machine parts, chisels, setscrews, and similar products where medium hardness is required coupled with good toughness and shock resistance

0.75–0.90% carbon: forging dies, hammers, and sledges

0.90–1.10% carbon: general-purpose tooling applications that require good balance of wear resistance and toughness, such as drills, cutters, shear blades, and other heavy-duty cutting edges

1.10–1.30% carbon: small drills, lathe tools, razor blades, and other light-duty applications in which extreme hardness is required without great toughness

In applications where improved toughness is required, small amounts of manganese, silicon, and molybdenum are often added. Vanadium additions of about 0.20% are used to

TABLE 6-7.	Basic Types of Tool Steel and Corresponding AISI–SAE Grades	
Type	AISI–SAE Grade	Significant Characteristic
1. Water-hardening	W	
2. Cold-work	O	Oil-hardening
	A	Air-hardening medium alloy
	D	High-carbon–high-chromium
3. Shock-resisting	S	
4. High-speed	T	Tungsten base
	M	Molybdenum base
5. Hot-work	H	H1–H19: chromium base
		H20–H39: tungsten base
		H40–H59: molybdenum base
6. Plastic-mold	P	
7. Special-purpose	L	Low alloy
	F	Carbon–tungsten

form strong, stable carbides that retain fine grain size during heat treating. One of the main weaknesses of the plain-carbon tool steels is their loss of hardness at elevated temperature. Prolonged exposure to temperatures over 150°C (300°F) usually results in undesired softening.

When larger parts must be hardened or distortion must be minimized, *oil-* or *air-hardening grades* (O and A designations, respectively) are often preferred. These metals contain alloy additions, and their higher hardenability permits hardening by less severe quenches. Tighter dimensional tolerances can be maintained during heat treatment, and the cracking tendency is reduced.

High-chromium tool steels are designated by the letter D and contain between 10 and 18% chromium. When properly heat-treated, these steels can retain full hardness at temperatures up to 425°C (800°F). Forging dies, die-casting die blocks, and drawing dies are often made from high-chromium tool steels.

Shock-resisting tool steels (S designation) have been developed for both hot and cold impact applications. Low carbon content (approximately 0.5% carbon) is usually specified to assure the necessary toughness, with carbide-forming alloys providing the necessary abrasion resistance, hardenability, and hot-work characteristics.

High-speed tool steels are used for cutting tools and other applications where strength and hardness must be retained at temperatures up to or exceeding red-heat (about 760°C or 1400°F). One popular member of the tungsten high-speed tool steels (T designation) is the T1 alloy, also known as 18–4–1 because the 0.7% carbon steel is alloyed with 18% tungsten, 4% chromium, and 1% vanadium. It offers a balanced combination of shock resistance and abrasion resistance, and is used for a wide variety of cutting applications. The molybdenum high-speed steels (M designation) were developed to reduce the amount of tungsten and chromium required to produce the high-speed properties. The M2 variety is quite popular since its higher carbon content and balanced analysis produce properties that are applicable to a number of high-speed applications.

Hot-work tool steels (H designation) were developed to provide strength and hardness during prolonged exposure to elevated temperature. All employ substantial additions of carbide-forming alloys. H1 to H19 are chromium-based alloys with about 5.0% chromium; H20 to H39 are tungsten-based types with 9 to 18% tungsten coupled with 3 to 4% chromium; and H20 to H59 are molybdenum-based.

Other types of tool steels include (1) the *plastic mold steels* (P designation), designed to meet the requirements of zinc die casting and plastic injection molding dies; (2) the *low-alloy special-purpose tool steels* (L designation), such as the L6 extreme toughness variety; and (3) the *carbon-tungsten type* of special-purpose tool steels (F designation), which are water hardening but substantially more wear-resistant than the plain-carbon tool steels.

Most tool steels are wrought materials, but some are designed specifically for fabrication by casting. Powder metallurgy processing has also been used to produce special compositions that are difficult or impossible to produce by wrought or cast methods.

■ 6.6 ALLOY CAST STEELS AND IRONS

The effects of alloying elements are the same regardless of the process used to produce the final shape. When the desired shape is to be made by casting, some alloys can be used to enhance process-specific features, such as fluidity and as-solidified properties. If a ferrous casting alloy contains less than about 2.0% carbon, it is considered to be a *cast steel*. Alloys with more than 2% carbon are *cast irons*.

Cast steels are usually heat-treated to produce a final quenched-and-tempered structure, and the alloy additions are selected to provide the desired hardenability and balance of properties. Similarly, if a cast iron is to be quenched, chromium, molybdenum, and nickel are often added to improve hardenability.

Most cast irons, however, are used in the as-cast condition, with the only heat treatment being a stress relief or annealing. For these applications, the alloy elements are selected for their ability to alter properties by (1) affecting the formation of graphite or cementite, (2) modifying the morphology of the carbon-rich phase, (3) strengthening the matrix material, or (4) enhancing wear resistance through the formation of alloy carbides. Nickel, for example, promotes graphite formation and tends to promote finer graphite structures.

Chromium, however, retards graphite formation and stabilizes cementite. These alloys are frequently used together in a ratio of two or three parts of nickel to one part of chromium. Between 0.5 and 1.0% molybdenum is often added to gray cast iron to impart additional strength, form alloy carbides, and help to control the size of the graphite flakes.

High-alloy cast irons have been designed to provide enhanced corrosion resistance under the elevated temperature conditions frequently encountered in the chemical industry. Within this family, the austenitic gray cast irons have become quite popular. These contain about 14% nickel, 5% copper, and 2.5% chromium. They offer good corrosion resistance to many acids and alkalis at temperatures up to about 800°C (1500°F). Alloy cast irons and cast steels are usually specified by their ASTM designation numbers.

■ KEY WORDS

AISI–SAE designation	dual-phase steels	maraging steel	sigma phase
alloy steel	duplex stainless steel	martensitic stainless steel	silicon steels
amorphous metals	electroslag remelting	microalloyed steel	solidification shrinkage
austenitic stainless steel	ferritic stainless steel	pig iron	stainless steel
bake-hardenable steel	free-machining steel	plain-carbon steel	steel
cast steel	high-strength low-alloy steel	precipitation-hardenable	tool steel
continuous casting	iron	stainless steel	vacuum arc remelting
degassification	ladle metallurgy	precoated steel	vacuum degassing
deoxidation	ladles	sensitization	vacuum induction melting

■ REVIEW QUESTIONS

1. Why might it be important to know the prior processing history of an engineering material?
2. What is a ferrous material?
3. When iron ore is reduced to metallic iron, what other elements are generally present in the metal?
4. How does steel differ from pig iron?
5. What are some of the changes that can be made to a steel by ladle metallurgy operations?
6. What is the advantage of pouring molten metal from the bottom of a ladle?
7. What is solidification shrinkage?
8. What are some of the attractive economic and processing advantages of continuous casting?
9. What are some of the techniques used to reduce the amount of dissolved oxygen in molten steel?
10. How might other gases, such as nitrogen and hydrogen, be reduced?
11. What are some of the attractive features of electroslag remelting?
12. What is plain-carbon steel?
13. What is considered a low-carbon steel? Medium-carbon? High-carbon?
14. What properties account for the high-volume use of medium-carbon steels?
15. Why should plain-carbon steels be given first consideration for applications requiring steel?
16. What are some of the common alloy elements added to steel?
17. What are some of the different reasons that alloying elements might be added to steel?
18. What alloys are particularly effective in increasing the hardenability of steel?
19. What are some of the alloy elements that tend to form stable carbides within a steel?

20. What is the significance of the last two digits in a typical four-digit AISI–SAE steel designation?
21. How are letters incorporated into the AISI–SAE designation system for steel, and what do some of the more common ones mean?
22. Why should the proposed fabrication processes enter into the considerations when selecting a steel?
23. How are the final properties usually obtained in the constructional alloy steels? In the HSLA steels?
24. What are microalloyed steels?
25. What are some of the potential benefits that may be obtained through the use of microalloyed steels?
26. What are the two phases that are present in dual-phase steels?
27. What features make dual-phase steels potentially more attractive than HSLA?
28. What are some of the various alloy additions that have been used to improve the machinability of steels?
29. What are some of the compromises associated with the use of free-machining steels?
30. What are the attractive characteristics of the bake-hardenable steels?
31. What economic factors might justify the use of precoated steel sheet?
32. Why have the amorphous metals attracted attention as potential materials for magnetic applications?
33. Under what conditions might maraging steels be required?
34. What are the typical elevated temperature limits of plain-carbon and alloy steels?
35. What features make steel an attractive material for recycling?
36. What structural feature is responsible for the observed corrosion resistance of stainless steels?
37. Why should ferritic stainless steels be given first consideration when selecting a stainless steel?

38. Which of the major types of stainless steel is likely to contain significant amounts of carbon? Why?
39. Under what conditions might a martensitic stainless steel "rust" when exposed to a hostile environment?
40. What are some of the unique properties of austenitic stainless steels?
41. What is a duplex stainless steel?
42. What is sensitization of a stainless steel, and how can it be prevented?

43. What is a tool steel?
44. How does the AISI–SAE designation system for tool steels differ from that for plain-carbon and alloy steels?
45. For what types of applications might an air-hardenable tool steel be attractive?
46. What alloying elements are used to produce the hot-worked tool steels?
47. What are some of the reasons that alloy additions are made to cast irons that will be used in their as-cast condition?

*C*hapter 6   CASE STUDY

Interior Tub of a Top-Loading Washing Machine

The interior tub of a washing machine is the container that holds the clothes during the washing and rinsing cycles, but also contains the perforations that permit removal of the water by draining and spinning. The component will see mechanical loadings from the weight of the clothes and water, and also the dynamic action of spinning water-laden fabrics. There will be exposure to a wide range of water quality, as well as the full spectrum of soaps, detergents, bleaches, and other laundry additives. The surfaces should also be resistant to the impact and abrasion of buttons, zippers, and snaps.

This part has traditionally been manufactured by the deep drawing, perforating and trimming of metal sheet, followed by some form of surface-coating treatment. For a long time, the standard material was "enameling iron"—a steel sheet with less than 0.03% carbon—which was then coated with a fired porcelain enamel. Due to the difficulties of producing ultra-low-carbon material in today's steelmaking operations, enameling iron became increasingly scarce, and manufacturers were forced to substitute the lowest-carbon, most readily available material, namely 1008 steel. This substitution further required modification of the enameling process to prevent CO and CO_2 blistering.

Your employer is presently manufacturing these tubs from 1008 steel sheet with a subsequent coating of fired porcelain enamel. Your marketing staff, however, reports that consumers tend to view a stainless steel tub to be of higher quality. As a result, your supervisor has asked you to evaluate the merits of converting to this material. You must first familiarize yourself with the current product (the base material, the forming process, and the porcelain enameling), and then determine what might be involved in converting it to stainless steel. Consider the following specific questions:

1. What are the obvious pros and cons of the present product and process? Where would you expect most problems to occur in the manufacturing process? Which aspects of fabrication are likely to be the most costly?
2. What would be the pros and cons of converting to stainless steel? In what ways would the product be superior? Are there any assets or liabilities associated with product fabrication?
3. Which stainless steel would you recommend? Begin by considering the basic types (ferritic, austenitic, and martensitic) and then refine your selection to a specific alloy if possible. Discuss the rationale for your selection.
4. Since deep drawing is a metal-deformation process, we could use cold working (strain hardening) as a strengthening mechanism. Would you find this to be attractive, or would you prefer to use a recrystallization anneal after drawing and prior to use? Why? If you elect to use cold work, might you want to at least perform a stress-relief heat treatment prior to use? Could this be done and still preserve the deformation strengthening? In deep drawing, the deformation is not uniform (increasing as we move up the sidewalls of the container), and the bottom of the tub will simply retain the properties of the starting sheet. In order to assure a minimum amount of strength at all locations, it may be desirable to begin the drawing with a partially cold-rolled sheet. Do you find this suggestion to be desirable? Why or why not?
5. After drawing and perforating, the residual drawing lubricant is removed from the part. Would any additional surface treatment be required? What would be your recommendation?

CHAPTER 7

NONFERROUS METALS AND ALLOYS

■ 7.1 INTRODUCTION

Nonferrous metals and alloys have assumed increasingly important roles in modern technology. Because of their number and the fact that their properties vary widely, they provide an almost limitless range of properties for the design engineer. While they tend to be more costly than iron or steel, these metals often possess certain properties or combinations of properties that are not available in ferrous metals, such as

1. Resistance to corrosion
2. Ease of fabrication
3. High electrical and thermal conductivity
4. Light weight
5. Strength at elevated temperatures
6. Color

Nearly all the nonferrous alloys possess at least two of the qualities just listed, and some possess nearly all. For many applications, specific combinations of these properties are highly desirable, and the availability of materials that provide them directly is a strong motivation for the use of the nonferrous alloys. Figure 7-1 presents some of the more common nonferrous metals grouped by advantageous engineering properties.

As a whole, however, the strength of the nonferrous alloys is generally inferior to that of steel. Also, the modulus of elasticity is usually lower, a fact that places them at a distinct disadvantage when stiffness is a required characteristic. Ease of fabrication is

FIGURE 7-1 Some common nonferrous metals and alloys, classified by attractive engineering property.

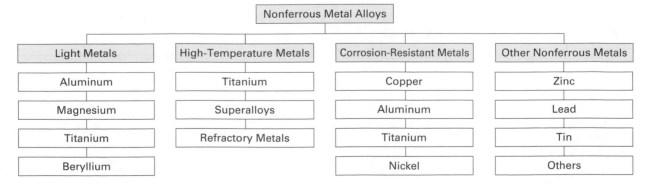

generally attractive. Those alloys with low melting points are often easy to cast in sand molds, permanent molds, or dies. Many alloys have high ductility coupled with low yield points, the ideal combination for cold working. Good machinability is also characteristic of many nonferrous alloys. The savings obtained through ease of fabrication can often overcome the higher cost of the nonferrous material and favor its use in place of steel. The one fabrication area in which the nonferrous alloys are somewhat inferior to steel is weldability. Because of a number of developments, however, it is generally possible to produce satisfactory weldments in all the nonferrous metals.

■ 7.2 COPPER AND COPPER ALLOYS

GENERAL PROPERTIES AND CHARACTERISTICS

Copper is an important engineering metal that has been in use for over 6000 years. As a pure metal, copper has been the backbone of the electrical industry. It is also the base metal of a number of highly important engineering alloys, known generically as brasses and bronzes.

The wide use of copper is based, primarily, on three important properties: its high electrical and thermal *conductivity*, its *useful strength with high ductility*, and its *corrosion resistance*. The excellent conductivity accounts for its importance to the electrical industry, and about one third of all copper produced is used in some form of electrical application, such as the commutators shown in Figure 7-2.

Pure copper in its annealed state has a tensile strength of only about 200 MPa (30 ksi), with an elongation of nearly 60%. By cold working, the tensile strength can be raised to over 450 MPa (65 ksi), with a decrease in elongation to about 5%. Its relatively low strength and high ductility make copper a very desirable metal for applications where extensive forming is required. Furthermore, the hardening effects of cold working can be easily removed, since the recrystallization temperature for copper is less than 260°C (500°F). The copper alloys also lend themselves nicely to the whole spectrum of fabrication processes, including casting, machining, joining, and surface finishing by either plating or polishing.

Unfortunately, copper is *heavier than iron*. Although the strengths can be quite high, the strength/weight ratio for the copper alloys is usually less than that for the weaker aluminum and magnesium alloys. In addition, several problems occur when copper is used at elevated temperature. If copper is stressed for a long period of time at high temperature, it is subject to intercrystalline failure at about half of its normal room-temperature strength. Material containing more than 0.3% oxygen is also subject to hydrogen embrittlement when it is exposed to reducing gases above 400°C (750°F).

COMMERCIALLY PURE COPPER

Refined copper containing between 0.02 and 0.05% oxygen (the principal impurity in copper) is called *electrolytic tough-pitch* (ETP) copper. It is often used as a base for

FIGURE 7-2   Copper and copper alloys are used for a variety of electrical applications, such as these electrical commutators. *(Courtesy of The Electric Materials Company.)*

copper alloys and may be used for electrical applications such as wire and cable, when the highest conductivity is not required. For superior conductivity, additional refining is required to reduce the oxygen content and produce *oxygen-free high-conductivity* (OFHC) copper. The better grades of conductor copper now have a conductivity rating of about 102% IACS, reflecting metallurgical improvements made since 1913, when the International Annealed Copper Standard (IACS) was established and the conductivity of pure copper was set at 100% IACS.

COPPER-BASED ALLOYS

As a pure metal, copper is not used extensively in manufactured products, except in electrical applications, and even here alloy additions such as silver, arsenic, cadmium, and zirconium are used to enhance various properties without significantly impairing conductivity. More often, copper is the base metal for some alloy, to which it imparts its good ductility, corrosion resistance, and electrical and thermal conductivity. Mechanical properties range the full spectrum from pure copper, which is soft and ductile, through manganese bronze, whose properties rival those of quenched and tempered steel.

Copper-based alloys are commonly identified through a system of numbers standardized by the Copper Development Association (CDA). Table 7-1 presents a breakdown of this system, which has been adopted by the American Society for Testing and Materials (ASTM), the Society of Automotive Engineers (SAE), and the U.S. government. Alloys numbered from 100 to 199 are mostly copper with less than 2% alloy addition. Numbers 200 to 799 are *wrought*[1] copper alloys, and the 800 and 900 series are *casting alloys.* When converted to the Unified Numbering System for metals and alloys, the three-digit numbers are converted to five digits by placing two zeros at the end, and the letter *C* is used as a prefix to denote the copper base.

COPPER–ZINC ALLOYS

Zinc is by far the most popular alloying addition to copper, with the resulting alloys being known as *brass*. If the copper content is not over 36%, the brass is a single-phase solid solution. Since this structure is identified as the alpha phase, these alloys are often called *alpha brasses*. They are quite ductile and formable, these characteristics increasing with the zinc content up to about 36%. Because of the good ductility and FCC crystal structure, the alpha brasses can be strengthened by cold working and are commercially available in various degrees of hardness. Cartridge brass, the 70% copper–30% zinc alloy, offers the best combination of strength and ductility, and has become quite popular for sheet-forming operations like deep drawing.

Above 36% zinc, the copper–zinc alloys enter a two-phase region (involving the brittle, zinc-rich, beta phase), and ductility drops markedly. Cold-working properties are rather poor for these high-zinc brasses, but deformation is rather easy when performed hot.

TABLE 7-1. Standard Designations for Copper and Copper Alloys (Copper Development Association System)

	Wrought Alloys		Cast Alloys
100–155	Commercial coppers	833–838	Red brasses and leaded red brasses
162–199	High-copper alloys	842–848	Semired brasses and leaded semired brasses
200–299	Copper–zinc alloys (brasses)	852–858	Yellow brasses and leaded yellow brasses
300–399	Copper–zinc–lead alloys (leaded brasses)	861–868	Manganese and leaded manganese bronzes
400–499	Copper–zinc–tin alloys (tin brasses)	872–879	Silicon bronzes and silicon brasses
500–529	Copper–tin alloys (phosphor bronzes)	902–917	Tin bronzes
532–548	Copper–tin–lead alloys (leaded phosphor bronzes)	922–929	Leaded tin bronzes
600–642	Copper–aluminum alloys (aluminum bronzes)	932–945	High-leaded tin bronzes
647–661	Copper–silicon alloys (silicon bronzes)	947–949	Nickel–tin bronzes
667–699	Miscellaneous copper–zinc alloys	952–958	Aluminum bronzes
700–725	Copper–nickel alloys	962–966	Copper nickels
732–799	Copper–nickel–zinc alloys (nickel silvers)	973–978	Leaded nickel bronzes

---

[1]The term *wrought* means "shaped or fabricated in the solid state" as opposed to *cast*, which is shaped as a liquid.

Many uses of brass result from the high electrical and thermal conductivity coupled with useful engineering strength. Plating characteristics are excellent and make the material an excellent base for decorative chrome or similar coatings. A unique property of alpha brass is its ability to have rubber vulcanized to it without any special treatment except thorough cleaning. As a result, brass is widely used in mechanical rubber goods.

Most brasses have good corrosion resistance. In the range of 0 to 40% zinc, the addition of a small amount of tin imparts improved resistance to seawater corrosion. Cartridge brass with tin becomes admiralty brass; Muntz metal with a tin addition is called naval brass. Brasses with 20 to 36% zinc, however, are subject to a selective corrosion, known as *dezincification*, when exposed to acidic or salt solutions. Brasses with more than 15% zinc often experience *season cracking* or *stress-corrosion cracking*. Both stress and exposure to corrosive media are required for this failure to occur (but residual stresses and atmospheric moisture may be sufficient!). As a result, cold-worked brass is usually stress relieved (to remove the residual stresses) before being placed in service.

When high machinability is required, as with automatic screw machine stock, 2 to 3% lead can be added to the brass to ensure the formation of free-breaking chips. Brass casting alloys are quite popular for use in plumbing fixtures and fittings, low-pressure valves, and a variety of decorative hardware. An alloy containing between 50 and 55% copper and the remainder zinc is often used as a filler metal in brazing. It is an effective material for joining steel, cast iron, brasses, and copper, producing joints that are nearly as strong as those obtained by welding. Decorative applications often use the range of colors (copper to white), along with the variations that can be produced through the addition of a third alloy element.

Table 7-2 lists some of the more common copper–zinc alloys and their composition, properties, and typical uses.

COPPER–TIN ALLOYS

Alloys of copper and tin, commonly called *tin bronzes*, are usually specified when they offer some form of special property. The term *bronze* is often confusing, however, since it can be used to designate any copper alloy where the major alloy addition is not zinc

TABLE 7-2. Composition, Properties, and Uses of Some Common Copper–Zinc Alloys

CDA Number	Common Name	Composition (%)					Condition	Tensile Strength		Elongation in 2 in. (%)	Typical Uses
		Cu	Zn	Sn	Pb	Mn		ksi	MPa		
200	Commercial indent bronze	90	10				Soft sheet	38	262	45	Screen wire, hardware, screws, jewelry
							Hard sheet	64	441	4	
240	Low brass	80	20				Spring	73	503	3	Drawing, architectural work, ornamental
							Annealed sheet	47	324	47	
							Hard	75	517	7	
260	Cartridge brass	70	30				Spring	91	627	3	Munitions, hardware, musical instruments, tubing
							Annealed sheet	53	365	54	
							Hard	76	524	7	
270	Yellow brass	65	35				Spring	92	634	3	Cold forming, radiator cores, springs, screws
							Annealed sheet	46	317	64	
							Hard	76	524	7	
280	Muntz metal	60	40				Hot-rolled	54	372	45	Architectural work; condenser tube
							Cold-rolled	80	551	5	
443–445	Admiralty metal	71	28	1			Soft	45	310	60	Condenser tube (salt water), heat exchangers
							Hard	95	655	5	
360	Free-cutting brass	61.5	35.3		3		Soft	47	324	60	Screw-machine parts
							Hard	62	427	20	
675	Manganese bronze	58.5	39	1		0.1	Soft	65	448	33	Clutch disks, pump rods, valve stems, high-strength propellers
							Bars, half hard	84	579	19	

or nickel. To provide clarification, the major alloy addition is usually included in the designation name.

The tin bronzes usually contain less than 12% tin. (Strength continues to increase as tin is added up to about 20%, but the high-tin alloys tend to be brittle.) Tin bronzes offer good strength, toughness, wear resistance, and corrosion resistance. They are often used for bearings, gears, and fittings that are subjected to heavy compressive loads. When the copper–tin alloys are used for bearing applications, up to 10% lead is frequently added.

The most popular wrought alloy is phosphor bronze, which usually contains from 1 to 11% tin. Alloy 521 (CDA) is typical of this class and contains 92% copper, 8% tin, and 0.15% phosphorus. Hard sheet of this material has a tensile strength of 760 MPa (110 ksi) and an elongation of 3%. Soft sheet has a tensile strength of 380 MPa (55 ksi) and 65% elongation. The material is often specified for pump parts, gears, springs, and bearings.

Alloy 905 is a bronze casting alloy containing 88% copper, 10% tin, and 2% zinc. In the as-cast condition, the tensile strength is about 310 MPa (45 ksi), with an elongation of 45%. It has very good resistance to seawater corrosion and is used on ships for pipe fittings, gears, pump parts, bushings, and bearings.

Bronzes can also be made by mixing powders of copper and tin, followed by powder metallurgy processing (described in Chapter 16). The resulting porous product can be used as filters for high-temperature or corrosive media, or can be infiltrated with oil to produce self-lubricating bearings.

COPPER–NICKEL ALLOYS

Copper and nickel exhibit complete solubility, as shown previously in Figure 4-6, and a wide range of useful alloys has been developed. Key features include high thermal conductivity, high-temperature strength, and corrosion resistance to a range of materials, including seawater. These properties, coupled with a high resistance to stress-corrosion cracking, make the copper–nickel alloys a good choice for heat exchangers, cookware, desalination apparatus, and a wide variety of coinage. *Cupronickels* contain 2 to 30% nickel. *Nickel silvers* contain no silver, but 10 to 30% nickel and at least 5% zinc. The bright silvery luster makes them attractive for ornamental applications, and they are also used for musical instruments. An alloy with 45% nickel is known as *constantan*, and the 67% nickel alloy is called *Monel*.

OTHER COPPER-BASED ALLOYS

The copper alloys discussed previously acquire their strength primarily through solid-solution strengthening and cold work. Within the copper-alloy family, alloys containing aluminum, silicon, or beryllium can strengthened by precipitation hardening.

Aluminum bronze alloys are best known for their combination of high strength and excellent corrosion resistance, and are often considered as cost-effective alternatives to stainless steel and nickel-based alloys. The wrought alloys can be strengthened through solid-solution strengthening, cold work, and the precipitation of iron- or nickel-rich phases. With less than 8% aluminum, the alloys are very ductile. When aluminum exceeds 9%, however, the ductility drops and the hardness approaches that of steel. Still higher aluminum contents result in brittle but wear-resistant materials. By varying the aluminum content and heat treatment, the tensile strength can range from about 415 to 1000 MPa (60 to 145 ksi). Typical applications include marine hardware, power shafts, sleeve bearings, and pump and valve components for handling seawater, sour mine water, and various industrial fluids. Cast alloys are available for applications where casting is the preferred means of manufacture. Since aluminum bronze exhibits large amounts of solidification shrinkage, castings made of this material should be designed with this in mind.

Silicon bronzes contain up to 4% silicon and 1.5% zinc (higher zinc contents may be used when the material is to be cast). Strength, formability, machinability, and corrosion resistance are all quite good. Tensile strengths range from a soft condition of about 380 MPa (55 ksi) through a maximum that approaches 900 MPa (130 ksi). Uses include boiler, tank, and stove applications, which require a combination of weldability, high strength, and corrosion resistance.

Copper–beryllium alloys, which ordinarily contain less than 2% beryllium, can be age hardened to produce the highest strengths of the copper-based metals, but are quite expensive to use. When annealed the material has a yield strength of 170 MPa (25 ksi), tensile strength of 480 MPa (70 ksi), and an elongation of 50%. After heat treatment, these properties can rise to 1100 MPa (160 ksi), 1250 MPa (180 ksi), and 5%, respectively. Cold work coupled with age hardening can produce even stronger material. The modulus of elasticity is about 125,000 MPa (8×10^6 psi) and the endurance limit is around 275 MPa (40 ksi). These properties make the material an excellent choice for electrical contact springs, but cost limits application to small components requiring long life and high reliability. Other applications, such as spark-resistant safety tools and spot welding electrodes, utilize the unique combination of properties: the material has the strength of steel and is nonsparking, nonmagnetic, and electrically and thermally conductive. Concerns over the toxicity of beryllium have created a demand for substitute alloys with similar properties, but no clear alternative has emerged.

■ 7.3 ALUMINUM AND ALUMINUM ALLOYS

GENERAL PROPERTIES AND CHARACTERISTICS

Although *aluminum* has only been a commercial metal for a little over 110 years, it now ranks second to steel in both worldwide quantity and expenditure and is clearly the most important of the nonferrous metals. It has achieved importance in virtually all segments of the world economy, with principal uses in transportation, construction, electrical applications, containers and packaging, consumer durables, and mechanical equipment.

A number of unique and attractive properties account for the engineering significance of aluminum. These include its workability, light weight, corrosion resistance, good electrical and thermal conductivity, optical reflectivity, and ease of recycling. Aluminum has a specific gravity of 2.7 compared to 7.85 for steel, making aluminum about one third the weight of steel for an equivalent volume. Cost comparisons are often made on the basis of cost per pound, where aluminum is at a distinct disadvantage. There are a number of applications, however, where a more appropriate comparison would be based on cost per unit volume. A pound of aluminum produces three times as many same-size parts as a pound of steel, so the cost difference becomes markedly less.

A serious weakness of aluminum from an engineering viewpoint is its relatively low modulus of elasticity, which is also about one third that of steel. Under identical loadings, an aluminum component will deflect three times as much as a steel component of the same design. Since the modulus of elasticity cannot be significantly altered by alloying or heat treatment, it is usually necessary to provide stiffness through design features such as ribs or corrugations. These can be incorporated with relative ease, however, because aluminum adapts easily to the full spectrum of fabrication processes.

COMMERCIALLY PURE ALUMINUM

In its pure state, aluminum is soft, ductile, and not very strong. In the annealed condition, pure aluminum has only about one fifth the strength of hot-rolled structural steel. Thus commercially pure aluminum is used primarily for its physical rather than its mechanical properties.

Electrical-conductor-grade aluminum is used in large quantities and has replaced copper in many applications, such as electrical transmission lines. Commonly designated by the letters EC, this grade contains a minimum of 99.45% aluminum and has an electrical conductivity that is 62% that of copper for the same size wire and 200% that of copper on an equal-weight basis.

ALUMINUMS FOR MECHANICAL APPLICATIONS

For nonelectrical applications, most aluminum is used in the form of alloys. These have much greater strength than pure aluminum, yet retain the advantages of light weight, good conductivity, and corrosion resistance. While usually weaker than steel, some alloys are now available that have tensile properties (except for ductility) that are comparable to those of the HSLA structural grades. Since alloys can be as much as 30 times stronger than pure aluminum, designers can frequently optimize their design and then tailor the material to their specific requirements. Some alloys are specifically designed for casting, while others are intended for the manufacture of wrought products.

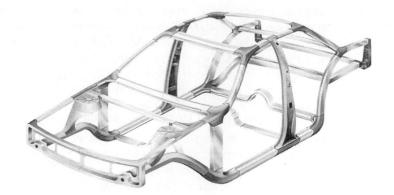

FIGURE 7-3   All-aluminum automotive space frame. *(Courtesy of Aluminum Company of America.)*

On a strength-to-weight basis, most of the aluminum alloys are superior to steel and other structural metals, but wear, creep, and fatigue properties are generally rather poor. Aluminum alloys have a finite fatigue life at all reasonable values of applied stress. In addition, aluminum alloys rapidly lose their strength as temperature is increased and should not be considered for applications involving service temperatures much above 150°C (300°F). Both the adhesive and the abrasive varieties of wear can be extremely damaging to aluminum alloys.

The selection of steel or aluminum for any given component is often a matter of cost, although considerations of light weight, corrosion resistance, low maintenance expense, or high thermal or electrical conductivity may be used to justify the added cost of aluminum. In the decade from 1991 to 2001, the use of aluminum doubled in cars and tripled in sport utility vehicles (SUVs) and light trucks. Aluminum is seeing increased use in body panels, engine blocks, manifolds, transmission housings, and wheels, where the reduced weight serves to increase fuel economy. The all-aluminum space frame shown in Figure 7-3 can reduce the weight of the structure, enhance recyclability, and reduce the number of parts going into the primary body structure. In 2001, aluminum passed plastic to become the third most-used material in automotive applications (behind steel and iron). The average North American automobile now contains over 115 kg (250 pounds) of aluminum.

CORROSION RESISTANCE OF ALUMINUM AND ITS ALLOYS

Pure aluminum is very reactive and forms a tight, adherent oxide coating on the surface as soon as it is exposed to air. This oxide is resistant to many corrosive media and serves as a corrosion-resistant barrier to protect the underlying metal. Thus, like stainless steels, the corrosion resistance of the metal is actually a property of the oxide and not the metal itself. As alloys are added to the aluminum, the oxide formation is somewhat retarded. Aluminum alloys, in general, do not have quite the superior corrosion resistance of pure aluminum.

The oxide coating on aluminum alloys also causes some difficulty relating to weldability. To produce consistent quality resistance welds, it is usually necessary to remove the oxide immediately before welding. For fusion welding, special fluxes or protective inert-gas atmospheres must be used to prevent material oxidation. Nevertheless, while welding aluminum may be more difficult than steel, suitable techniques have been developed so that high-quality, cost-effective welds can be produced by most of the welding processes.

CLASSIFICATION SYSTEM

The aluminum alloys can be divided into two major groups, wrought alloys and casting alloys, based on the method of fabrication. *Wrought alloys* are those that are shaped as solids and are therefore designed to have attractive forming characteristics, such as low yield strength, high ductility, good fracture resistance, and good strain hardening. Attractive features for the *casting alloys* include low melting point, high fluidity, and attractive as-solidified structures and properties. Clearly, these properties are distinctly different, and the alloys that have been designed to meet them are also different. As a result, separate classification systems exist for the wrought and cast aluminum alloys.

WROUGHT ALUMINUM ALLOYS

The wrought aluminum alloys can be further divided into two basic types: those that achieve strength by solid-solution strengthening and cold working and those that can be strengthened by heat treatment (age hardening). Table 7-3 lists some of the common wrought aluminum alloys in each family, using the standard four-digit designation system for aluminums. The first digit indicates the major alloy element or elements:

Major Alloying Element	
Aluminum, 99.00% and greater	1xxx
Copper	2xxx
Manganese	3xxx
Silicon	4xxx
Magnesium	5xxx
Magnesium and silicon	6xxx
Zinc	7xxx
Other element	8xxx

The second digit is usually zero. Nonzero numbers are used to indicate some form of modification to the original alloy. The last two digits simply indicate the particular alloy within the family. For example, 2024 simply means alloy number 24 within the 2xxx, or aluminum–copper, system. For the 1xxx series, the last three digits are used to denote the purity of the aluminum.

The first four digits of a wrought aluminum designation identify the chemistry of the alloy. Additional information about the alloy condition is then provided through a *temper designation*, in the form of a letter–number suffix using the following system:

-F: as fabricated
-H: strain-hardened
    -H1:   strain-hardened by working to desired dimensions; a second digit, 1 through 9, indicates the degree of hardening, 8 being commercially full-hard and 9 extrahard
    -H2:   strain-hardened by cold working, followed by partial annealing
    -H3:   strain-hardened and stabilized
-O: annealed
-T: thermally treated (heat treated)
    -T1:   cooled from hot working and naturally aged
    -T2:   cooled from hot working, cold-worked, and naturally aged
    -T3:   solution-heat-treated, cold-worked, and naturally aged
    -T4:   solution-heat-treated and naturally aged
    -T5:   cooled from hot working and artificially aged
    -T6:   solution-heat-treated and artificially aged
    -T7:   solution-heat-treated and stabilized
    -T8:   solution-heat-treated, cold-worked, and artificially aged
    -T9:   solution-heat-treated, artificially aged, and cold-worked
    -T10:   cooled from hot working, cold-worked, and artificially aged
-W: solution-heat-treated only

It can be noted from Table 7-3 that the work-hardenable alloys (those that cannot be age-hardened) are primarily those in the 1xxx (pure aluminum), 3xxx (aluminum–manganese), and 5xxx (aluminum–magnesium) series. Within these series the 1100, 3003, and 5052 alloys are quite popular.

The precipitation-hardenable alloys offer considerably higher strength. These are found primarily in the 2xxx, 6xxx, and 7xxx series. Alloy 2017, the original *duralumin*, is probably the oldest age-hardenable aluminum alloy. The 2024 alloy is stronger than 2017 and has seen considerable use in aircraft applications. An attractive feature of the 2xxx series is the fact that ductility does not significantly decrease during the strengthening heat

TABLE 7-3. Composition, Typical Properties, and Designations of Some Wrought Aluminum Alloys

Designation[a]	Composition (%) Aluminum = Balance					Form Tested	Tensile Strength		Yield Strength[b]		Elongation in 2 in. (%)	Brinell Hardness	Uses and Characteristics
	Cu	Si	Mn	Mg	Others		ksi	MPa	ksi	MPa			
Work-Hardening Alloys—Not Heat-Treatable													
1100-0	0.12				99 Al	$\frac{1}{16}$-in. sheet	13	90	5	34	35	23	Commercial Al: good forming properties
1100-H14						$\frac{1}{16}$-in. sheet	16	110	14	97	9	32	Good corrosion resistance, low yield strength
110-H18						$\frac{1}{16}$-in. sheet	24	165	21	145	5	44	Cooking utensils; sheet and tubing
3003-0	0.12		1.2			$\frac{1}{16}$-in. sheet	16	110	6	41	30	28	Similar to 1100
3003-H14						$\frac{1}{16}$-in. sheet	22	152	21	145	8	40	Slightly stronger and less ductile
3003-H18						$\frac{1}{16}$-in. sheet	29	200	27	186	4	55	Cooking utensils; sheetmetal work
5052-0				2.5		$\frac{1}{16}$-in. sheet	28	193	13	90	25	45	Strongest work-hardening alloy
5052-H32						$\frac{1}{16}$-in. sheet	33	228	28	193	12	60	High yield strength and fatigue limit
5052-H36					0.25 Cr	$\frac{1}{16}$-in. sheet	40	276	35	241	8	73	Highly stressed sheetmetal products
Precipitation-Hardening Alloys—Heat-Treatable													
2017-0	4.0	0.5	0.7	0.6		$\frac{1}{16}$-in. sheet	26	179	10	69	20	45	Duralumin, original strong alloy
2017-T4						$\frac{1}{16}$-in. sheet	62	428	40	276	20	105	Hardened by quenching and aging
2024-0	4.4	0.6		1.5		$\frac{1}{16}$-in. sheet	27	186	11	76	20	42	Stronger than 2017
2024-T4						$\frac{1}{16}$-in. sheet	64	441	45	290	19	120	Used widely in aircraft construction
2014-0	4.4	0.8	0.8	0.5		$\frac{1}{2}$-in. extruded shapes	27	186	14	97	12	45	Strong alloy for extruded shapes
2014-T6						Forgings	65	448	55	379	10	125	Strong forging alloy
2014-T6						$\frac{1}{16}$-in. sheet	70	483	60	413	8		Higher yield strength than Alclad 2024
Alclad 2014-T6	4.5	1.0	0.8	0.4	0.3 Cr	$\frac{1}{16}$-in. sheet	63	434	56	386	7		Clad with heat-treatable alloy[c]
7075-T6	1.6		0.2	2.5	5.6 Zn	$\frac{1}{16}$-in. sheet	33	228	15	103	17	60	Alloy of highest strength
7075-T6						$\frac{1}{16}$-in. sheet	76	524	67	462	11	150	Lower ductility than 2024
Alclad 7075-T6						$\frac{1}{16}$-in. sheet	76	524	67	462	11		Strongest Alclad product
Alclad 7075-T6						$\frac{1}{2}$-in. extruded shapes	80	552	70	483	6		Strongest alloy for extrusions
6061-T6	0.28			1.0	0.20 Cr	$\frac{1}{2}$-in. extruded shapes	42	290	40	276	12	95	Strong, corrosion resistant
6063-T6		0.4		0.7		$\frac{1}{2}$-in. rod extruded	35	241	31	214	12	80	Good forming properties and corrosion resistance
6151-T6		0.9		0.6	0.25 Cr	Forgings	48	331	43	297	17	90	For intricate forgings
2025-T6	4.5	0.8	0.8			Forgings	55	379	30	207	18	100	Good forgeability, lower cost
2018-T6	4			0.7	2 Ni	Forgings	55	379	40	276	10	100	Strong at elevated temperatures; forged pistons
4032-T6	0.9	12.2		1.1	0.9 Ni	Forgings	55	379	46	317	9	115	Forged aircraft pistons
2011-T3	5.5			(0.5 Bi)	0.5 Pb	$\frac{1}{2}$-in. rod	55	379	43	297	15	95	Free cutting, screw-machine products

[a] O, annealed; T, quenched and aged; H, cold-rolled to hard temper.

[b] Yield strength taken at 0.2% permanent set.

[c] Cladding alloy: 1.0 Mg, 0.7 Si, 0.5 Mn.

treatment. Within the 7xxx series are some newer alloys with strengths that approach or exceed those of the high-strength structural steels. Ductility, however, is less than that of steel, and fabrication is more difficult than for the 2024-type alloys. Nevertheless, the 7xxx series alloys have also found wide use in aircraft applications. To maintain properties, the age-hardened alloys should not be used at temperatures over 175°C (350°F). (*Note:* This temperature is comparable to the upper limit for engineering polymers.)

Because of their two-phase structure, the heat-treatable alloys tend to have poorer corrosion resistance than either pure aluminum or single-phase work-hardenable alloys. When both high strength and superior corrosion resistance are desired, wrought aluminum is often produced as *Alclad* material. A thin layer of corrosion-resistant aluminum is bonded to one or both surfaces of a high-strength alloy during rolling, and the material is further processed as a composite.

Because only moderate temperatures are required to lower the strength of aluminum alloys, extrusions and forgings are relatively easy to produce and are manufactured in large quantities. Deep drawing and other sheet-metal-forming operations can be carried out quite easily. In general, the high ductility and low yield strength of the aluminum alloys make them appropriate for almost all forming operations. Good dimensional tolerances and fairly intricate shapes can be produced with relative ease.

The machinability of aluminum-based alloys, however, can vary greatly, and special tools and techniques may be desirable if large amounts of machining are required. Free-machining alloys, such as 2011, have been developed for screw-machine work. These special alloys can be machined at very high speeds and have replaced brass screw-machine stock in many applications.

ALUMINUM CASTING ALLOYS

Although its low melting temperature tends to make it suitable for casting, pure aluminum is seldom cast. Its high shrinkage and susceptibility to hot cracking cause considerable difficulty, and scrap is high. By adding small amounts of alloying elements, however, very suitable casting characteristics can be obtained and strength can be increased. Aluminum alloys are cast in considerable quantity by a variety of processes. Many of the most popular alloys contain enough silicon to produce the eutectic reaction, which is characterized by a low melting point and high as-cast strength. Silicon also improves the fluidity of the metal, making it easier to produce complex shapes or thin sections, but high silicon also produces an abrasive, difficult-to-cut material. Copper, zinc, and magnesium are other popular alloy additions that permit the formation of age-hardening precipitates.

Table 7-4 lists some of the commercial aluminum casting alloys and uses the three-digit designation system of the Aluminum Association. The first digit indicates the alloy group as follows:

Major Alloying Element	
Aluminum, 99.00% and greater	1xx.x
Copper	2xx.x
Silicon with Cu and/or Mg	3xx.x
Silicon	4xx.x
Magnesium	5xx.x
Zinc	7xx.x
Tin	8xx.x
Other elements	9xx.x

The second and third digits identify the particular alloy or aluminum purity, and the last digit, separated by a decimal point, indicates the product form (e.g., casting or ingot). A letter before the numerical designation indicates a modification of the original alloy.

Aluminum casting alloys have been designed for both properties and process. When the strength requirements are low, as-cast properties are usually adequate. High-strength castings usually require the use of alloys that can subsequently be heat-treated. Sand casting has the fewest process restrictions. The aluminum alloys used for permanent mold casting are designed to have lower coefficients of thermal expansion (or contraction) because the molds offer restraint to the dimensional changes that occur upon cooling. Die-casting alloys require high degrees of fluidity because they are often cast in thin sections. Most of the die-casting alloys are also designed to produce high "as-cast"

TABLE 7-4. Composition, Properties, and Uses of Some Aluminum Casting Alloys

Alloy Designation[a]	Process[b]	Cu	Si	Mg	Zn	Fe	Other	Temper	Tensile Strength ksi[c]	Tensile Strength MPa	Elongation in 2 in. (%)	Uses and Characteristics
208	S	4.0	3.0		1.0	1.2		F	19	131	1.5	General-purpose sand castings, can be heat treated
242	S, P	4.0		1.6		1.0	2.0 Ni	T61	40	276	—	Withstands elevated temperatures
295	S	4.5	1.0			1.0		T6	32	221	3.0	Structural castings, heat-treatable
296	P	4.5	2.5			1.2		T6	35	241	2.0	Permanent-mold version of 295
308	P	4.5	5.5		1.0	1.0		F	24	166	—	General-purpose permanent mold
319	S, P	3.5	6.0		1.0	1.0		T6	31	214	1.5	Superior casting characteristics
354	P	1.8	9.0					—	—	—	—	High-strength, aircraft
355	S, P	1.3	5.0					T6	32	221	2.0	High strength and pressure tightness
C355	S, P	1.3	5.0					T61	40	276	3.0	Stronger and more ductile than 355
356	S, P		7.0					T6	30	207	3.0	Excellent castability and impact strength
A356	S, P		7.0					T61	37	255	5.0	Stronger and more ductile than 356
357	S, P		7.0					T6	45	310	3.0	High-strength-to-weight castings
359	S, P		9.0					—	—	—	—	High-strength aircraft usage
360	D		9.5			2.0		F	44[d]	303	2.5[d]	Good corrosion resistance and strength
A360	D		9.5			2.0		F	46[d]	317	3.5[d]	Similar to 360
380	D	3.5	8.5		3.0	2.0		F	46[d]	317	2.5[d]	High strength and hardness
A380	D	3.5	8.5		3.0	1.3		F	47[d]	324	3.5[d]	Similar to 380
383	D	1.5	10.5		3.0	1.3		F	45[d]	310	3.5[d]	High strength and hardness
384	D	3.75	11.3		1.0	1.3		F	48[d]	331	2.5	High strength and hardness
413	D	1.0	12.0			2.0		F	43[d]	297	2.5[d]	General purpose, good castability
A413	D	1.0	12.0			1.3		F	42[d]	290	3.5[d]	Similar to 413
443	D		5.25			2.0		F	33[d]	228	9.0[d]	General purpose, good castability
B443	S, P		5.25			2.0		F	17	117	3.0	General-purpose casting alloy
514	S			4.0				F	22	152	6.0	High corrosion resistance
518	D			8.0		1.8		F	45[d]	310	5.0[d]	Good corrosion resistance, strength, and toughness
520	S			10.0				T4	42	290	12.0	High strength with good ductility
535	S			6.9				F	35	241	9.0	Good corrosion resistance and machinability
712	S				5.8			F	34	234	4.0	Good properties without heat treatment
713	S, P				7.5	1.1		F	32	221	3.0	Similar to 712
771	S				7.0			T6	42	290	5.0	Aircraft and computer components
850	S, P	1.0					6.3 Sn, 1.0 Ni	T5	16	110	5.0	Bearing alloy

[a] Aluminum Association.
[b] S, sand-cast; P, permanent-mold-cast; D, die cast.
[c] Minimum figures unless noted.
[d] Typical values.

strength without heat treatment, using the rapid cooling conditions of the die-casting process to promote a fine grain size and fine eutectic structure. Tensile strengths of the aluminum permanent-mold and die-casting alloys can be in excess of 275 MPa (40 ksi).

ALUMINUM–LITHIUM ALLOYS

In the search for aluminum alloys with higher strength, greater stiffness, and lighter weight, aluminum–lithium alloys have emerged as an attractive aerospace material. Each percent of lithium (up to 4%) reduces the overall weight by 3% and increases stiffness by 6%. Alloys have already been developed that have 8 to 10% lower density, 15 to 20% greater stiffness, strengths comparable to those of existing alloys, and good resistance to fatigue crack propagation. Fracture toughness, ductility, and stress-corrosion resistance, may be poorer than for conventional alloys.

Aluminum–lithium alloys are available in both wrought and cast forms and can be fabricated just like other aluminum alloys. They are highly machinable, can be welded, and are readily adaptable to forming by forging or extrusion. Some sheet material can even be fabricated by superplastic forming. Thus, significant weight savings can be achieved without a major overhaul of manufacturing equipment that might be required with a switch to advanced composites.

In recent years, aluminum alloys comprise about 80% of the weight of commercial aircraft. Through the substitution of aluminum–lithium alloys, redesign of components to take advantage of the greater stiffness, and reduction in the size of related parts (since the material is stronger and the entire structure will be lighter), a 10 to 15% weight savings would appear to be possible in a commercial aircraft. On a more graphic basis, the weight of a full-size commercial plane could be reduced by as much as 6000 kilograms (14,000 lb). Fuel savings over the life of the airplane would more than compensate for any additional manufacturing expense. The weight of the external liquid-hydrogen tank on the U.S. Space Shuttle booster rocket was reduced by approximately 3400 kg (7500 lb) by conversion to an aluminum–lithium material.

■ 7.4 Magnesium and Magnesium Alloys

GENERAL PROPERTIES AND CHARACTERISTICS

Magnesium is the lightest of the commercially important metals, having a specific gravity of about 1.74 (30% lighter than aluminum alloys and 75% lighter than steel). Like aluminum, magnesium is relatively weak in the pure state and for engineering purposes is almost always used as an alloy. Even in alloy form, however, the metal is characterized by poor wear, creep, and fatigue properties. Strength drops rapidly when the temperature exceeds 100°C (200°F), so magnesium should not be considered for elevated-temperature service. Its modulus of elasticity is even less than that of aluminum, being between one fourth and one fifth that of steel. Thick sections are required to provide adequate stiffness, but the alloy is so light that it is often possible to use thicker sections for the required rigidity and still have a lighter structure than can be obtained with any other metal. Cost per unit volume is low, so the use of thick sections is generally not prohibitive. Moreover, since a large portion of magnesium components are cast, the thicker sections actually become a desirable feature. Ductility is frequently low, a characteristic of the HCP crystal structure, but some alloys have values exceeding 10%.

On the more positive side, magnesium alloys have a relatively high strength-to-weight ratio with some commercial alloys attaining strengths as high as 380 MPa (55 ksi). High energy absorption means good damping of noise and vibration. While many magnesium alloys require enamel or lacquer finishes to impart adequate corrosion resistance, this property has been improved markedly with the development of higher-purity alloys. In the absence of unfavorable galvanic couples, these materials resist corrosion better than steel and aluminum and have paved the way for applications in many areas, including the automotive market. The numerous limitations, however, generally restrict the use of magnesium to applications where light weight is a dominant concern. While aluminum alloys are often used for the load-bearing members of mechanical structures, magnesium alloys are best suited for those applications where lightness is the primary consideration and strength is a secondary requirement.

MAGNESIUM ALLOYS AND THEIR FABRICATION

The designation system for magnesium alloys is not as well standardized as in the case of steels or aluminums, but most producers follow a system using one or two prefix letters, two or three numerals, and a possible suffix letter. The prefix letters designate the two largest alloying metals according to the following format presented in ASTM specification B93:

A	aluminum	F	iron	M	manganese	R	chromium
B	bismuth	H	thorium	N	nickel	S	silicon
C	copper	K	zirconium	P	lead	T	tin
D	cadmium	L	beryllium	Q	silver	Z	zinc
E	rare earth						

Aluminum, zinc, zirconium, and thorium promote precipitation hardening; manganese improves corrosion resistance; and tin improves castability. Aluminum is the most common alloying element. The numerals in the designation correspond to the rounded-off whole-number percentages of the two main alloy elements and are arranged in the same order as the letters. Thus the AM60 alloy would contain approximately 6% aluminum and less than 0.5% manganese. A suffix letter is used to denote variations of the same base alloy. A temper designation suffix that is quite similar to that used with the aluminum alloys can also be used. Table 7-5 lists some of the more common magnesium alloys together with their properties and uses.

Sand, permanent mold, die, and investment casting are all well developed for magnesium alloys and take advantage of the low melting points and high fluidity. Die casting is clearly the most popular manufacturing process for magnesium, accounting for 70% of all castings. Although the magnesium alloys typically cost about twice as much as aluminum, the hot-chamber die-casting process used with magnesium is easier, more economical, and 40 to

TABLE 7-5. Composition, Properties, and Uses of Common Magnesium Alloys

Alloy	Temper	Composition (%) Al	Rare Earths	Mn	Th	Zn	Zr	Tensile Strength[a] ksi	MPa	Yield Strength[a] ksi	MPa	Elongation in 2 in. (%)	Uses and Characteristics
AM60A	F	6.0		0.13				30	207	17	117	6	Die castings
AM100A	T4	10.0		0.1				34	234	10	69	6	Sand and permanent-mold castings
AZ31B	F	3.0				1.0		32	221	15	103	6	Sheet, plate, extrusions, forgings
AZ61A	F	6.5				1.0		36	248	16	110	7	Sheet, plate, extrusions, forgings
AZ63A	T5	6.0				3.0		34	234	11	76	7	Sand and permanent-mold castings
AZ80A	T5	8.5				0.5		34	234	22	152	2	High-strength forgings, extrusions
AZ81A	T4	7.6				0.7		34	234	11	76	7	Sand and permanent-mold castings
AZ91A	F	9.0				0.7		34	234	23	159	3	Die castings
AZ92A	T4	9.0				2.0		34	234	11	76	6	High-strength sand and permanent-mold castings
EZ33A	T5		3.2			2.6	0.7	20	138	14	97	2	Sand and permanent-mold castings
HK31A	H24				3.2		0.7	33	228	24	166	4	Sheet and plates; castings in T6 temper
HM21A	T5			0.8	2.0			33	228	25	172	3	High-temperature (800°F) sheets, plates, forgings
HZ32A	T5				3.2	2.1		27	186	13	90	4	Sand and permanent-mold castings
ZH62A	T5				1.8	5.7	0.7	35	241	22	152	5	Sand and permanent-mold castings
ZK51A	T5					4.6	0.7	34	234	20	138	5	Sand and permanent-mold castings
ZK60A	T5					5.5	0.45	38	262	20	138	7	Extrusions, forgings

[a]Properties are minimums for the designated temper.

50% faster than the cold-chamber process generally required for aluminum. Wall thickness, draft angle, and dimensional tolerances are all lower than for aluminum die castings, and the shorter cycle times result in improved die life. Magnesium die castings compete well with aluminum and often replace plastic injection-molded components when improved stiffness or dimensional stability, or the benefits of electrical or thermal conductivity, are required.

Forming behavior is poor at room temperature, but most conventional processes can be performed when the material is heated to temperatures between 250 and 500°C (480 and 775°F). Since these temperatures are easily attained and generally do not require a protective atmosphere, many formed and drawn magnesium products are manufactured.

The machinability of magnesium alloys is the best of any commercial metal and, in many applications, the savings in machining costs, achieved through high cutting speeds and long tool life, more than compensate for the increased cost of the material. It is necessary, however, to keep the tools sharp and provide adequate cooling for the chips.

Magnesium alloys can be spot welded almost as easily as aluminum, but scratch brushing or chemical cleaning is necessary before forming the weld. Fusion welding is best performed with processes using an inert shielding atmosphere of argon or helium gas.

Considerable misinformation exists regarding the fire hazards when processing or using magnesium alloys. It is true that magnesium alloys are highly combustible when in a finely divided form, such as powder or fine chips, and this hazard should never be ignored. When the metal is heated above 700°C (950°F), a noncombustible, oxygen free atmosphere is required to suppress burning. Casting operations often require additional precautions due to the reactivity of magnesium with sand and water. In the form of sheet, bar, extruded product, or finished castings, however, magnesium alloys present no real fire hazard.

◾ 7.5 ZINC-BASED ALLOYS

Over 50% of all metallic *zinc* is used in the *galvanizing* of iron and steel. In this process the iron-based material is coated with a layer of zinc by one of a variety of processes that include direct immersion in a bath of molten metal (hot dipping) and electrolytic plating. The resultant coating provides excellent corrosion resistance, even when the surface is badly scratched or marred. Moreover, the corrosion resistance will persist until all of the sacrificial zinc has been depleted.

Zinc is also used as the base metal for a variety of die-casting alloys. For this purpose, zinc offers low cost, a low melting point (only 380°C or 715°F), and the attractive property of not adversely affecting steel dies when in molten metal contact. Unfortunately, pure zinc is almost as heavy as steel and is also rather weak and brittle. Therefore, when alloys are designed for die casting, the alloy elements are usually selected for their ability to increase strength and toughness in the as-cast condition while retaining the low melting point. High fluidity enables the casting of very thin sections. Good dimensional stability and the ability to be surface finished by a variety of means are additional advantages.

Two of the most popular zinc die-casting alloys are presented in Table 7-6. Alloy AG40A (also known as alloy 903 or Zamak 3) is widely used because of its excellent dimensional stability, and alloy AC41A offers higher strength and better corrosion resistance. As a whole, the zinc die-casting alloys offer a reasonably high strength and impact resistance, along with the ability to be cast to close dimensional limits with extremely thin sections. Moreover, these alloys are then machinable at a minimum of cost. Resistance to surface corrosion is adequate for a number of applications, and the material can be readily polished, plated, painted, chromated, or anodized for various applications. Energy costs are low (low melting temperature), tool life is excellent, and the zinc alloys can be efficiently recycled. While the rigidity is low compared to other metals, it is far superior to engineering plastics, and zinc die castings often compete with plastic injection moldings.

The attractiveness of zinc die casting has been further enhanced by the development of the high-aluminum zinc–aluminum casting alloys (ZA8, ZA12, and ZA27, with 8, 12, and 27% aluminum, respectively). Initially developed for sand, permanent mold, and graphite mold casting, these alloys can also be die cast to achieve higher strength, hardness, and wear resistance than are achieveable with any of the conventional alloys. As a result of their lower melting and casting costs, these materials are becoming attractive alternatives to the conventional aluminum, brass, and bronze casting alloys, as well as cast iron.

TABLE 7-6.   Composition and Properties of Some Zinc Die-Casting Alloys

Alloy	#3 SAE 903 ASTM AG40A	#5 SAE 925 ASTM AC41A	#7 ASTM AG408	ZA-8 S[1]	ZA-8 P	ZA-8 D	ZA-12 S	ZA-12 P	ZA-12 D	ZA-27 S	ZA-27 P	ZA-27 D
Composition[2]												
Aluminum	3.5–4.3	3.5–4.3	3.5–4.3	8.0–8.8			10.5–11.5			25.0–28.0		
Copper	0.25 max	0.75–1.25	0.25 max	0.8–1.3			0.5–1.2			2.0–2.5		
Zinc	balance	balance	balance	balance			balance			balance		
Properties												
Density (g/cc)	6.6	6.6	6.6		6.3			6.0			5.0	
Yield Strength (MPa)	221	228	221	200	206	290	214	269	317	372		379
(ksi)	32	33	32	29	30	42	31	39	46	54		55
Tensile Strength (MPa)	283	328	283	263	255	374	317	345	400	441		421
(ksi)	41	48	41	38	37	54	46	50	58	64		61
Elongation (% in 2 in.)	10	7	13	2	2	10	3	3	7	6		3
Impact Strength (J)	58	65	58	20		42	25		29	47		5
Modulus of Elasticity (GPa)	85.5	85.5	85.5		85.5			82.7			77.9	
Machinability[3]	E	E	E		E			VG			G	

[1] Also contains small amounts of Fe, Pb, Cd, Sn and Ni
[2] S = Sand cast   P = Permanent mold cast   D = Die cast
[3] E = Excellent   VG = Very Good   G = Good

■ 7.6 TITANIUM AND TITANIUM ALLOYS

Titanium is a strong, lightweight, corrosion-resistant metal that has been of commercial importance since about 1950. Because its properties are between those of steel and aluminum, its importance has been increasing rapidly. The yield strength of commercially pure titanium is about 415 MPa (60 ksi), but this can be raised to 1300 MPa (190 ksi) through alloying and heat treatment, a strength comparable to that of many alloy steels. Density, on the other hand, is only 56% that of steel, and the modulus of elasticity ratio is also about one half. Good mechanical properties are retained up to temperatures of 535°C (1000°F), so the metal is often considered as a high-temperature engineering material. On the negative side, titanium and its alloys suffer from high cost, fabrication difficulties, a high energy content (they require about 10 times as much energy to produce as steel), and a high reactivity at elevated temperatures (above 535°C).

Titanium alloys are generally grouped into three classes based on their microstructural features. These classes are known as alpha-, beta-, and alpha-beta-titanium alloys, the terms denoting the stable phase or phases at room temperature. Fabrication can be by casting, forging, rolling, extrusion, or welding, provided that special process modifications and controls are implemented. Many of the alloys can be heat treated to improve the overall balance of engineering properties. Advanced processing methods include powder metallurgy, mechanical alloying, rapid-solidification processing (RSP), superplastic forming, diffusion bonding, and hot-isostatic pressing (HIP).

Since titanium is significantly more expensive than either steel or aluminum, its uses relate primarily to its high strength-to-weight ratio, its good stiffness, its corrosion resistance (the result of a thin, tenacious oxide coating), and its retention of mechanical properties at elevated temperatures. Aerospace applications tend to dominate, but titanium and titanium alloys are also used in such diverse areas as chemical- and electrochemical-processing equipment, food-processing equipment, marine implements, medical implants, and sporting goods. They are often used in place of steel where weight savings are desired and to replace aluminums where high-temperature performance is necessary. Some bonding applications utilize the unique property that titanium wets glass and some ceramics. The titanium–6% aluminum–4% vanadium alloy is the most popular titanium alloy, accounting for nearly 50% of all titanium usage worldwide.

■ 7.7 NICKEL-BASED ALLOYS

Nickel-based alloys are most noted for their outstanding strength and corrosion resistance, particularly at high temperatures. *Monel* metal, containing about 67% nickel and 30% copper, has been used for years in the chemical and food-processing industries because of its outstanding corrosion characteristics. In fact, Monel probably has better corrosion resistance to more media than does any other commercial alloy. It is particularly resistant to saltwater, sulfuric acid, and even high-velocity, high-temperature steam. For the latter reason, Monel has been used for steam turbine blades. It can be polished to have an excellent appearance, similar to that of stainless steel, and is often used in ornamental trim and household ware. In its most common form, Monel has a tensile strength ranging from 500 to 1200 MPa (70 to 170 ksi), with a companion elongation ranging between 2 and 50%.

There are three special grades of Monel that contain small amounts of added alloying elements. K-Monel contains about 3% aluminum and can be precipitation hardened to a tensile strength of 1100 to 1250 MPa (160 to 180 ksi). H-Monel has 3% silicon added, and S-Monel has 4% silicon. These varieties are used for casting applications and can also be precipitation hardened. To improve the machining characteristics, a special free-machining variety, known as R-Monel, has been produced with about 0.35% sulfur.

Nickel-based alloys have also been used for electrical resistors and heating elements. These materials are primarily nickel–chromium alloys and are known by the trade name *Nichrome*. One popular alloy contains 80% nickel and 20% chromium. Another has 60% nickel, 16% chromium, and 24% iron. They have excellent resistance to oxidation while retaining useful strength at red heats.

When the nickel-based alloys are designed to provide good mechanical properties at extremely high temperatures, the materials are classified as superalloys. These alloys will be discussed along with other, similar materials in the following section.

In general, the nickel-based materials are difficult to cast, but they can be forged and hot worked. Welding operations can be performed with little difficulty.

◼ 7.8 Superalloys and Other Metals Designed for High-Temperature Service

Titanium and titanium alloys have already been cited as being useful in providing strength at elevated temperatures, but the maximum temperature for these materials is approximately 535°C (1000°F). Jet engine, gas turbine, rocket, and nuclear applications often require materials that possess high strength, creep resistance, oxidation and corrosion resistance, and fatigue resistance at temperatures up to and in excess of 1100°C (2000°F). One class of materials offering these properties is the *superalloys*, first developed extensively in the 1940s for use in turbojet aircraft. These alloys tend to be based on *nickel, iron and nickel*, or *cobalt*. Most are precipitation hardenable, and yield strengths above 700 MPa (100 ksi) are readily attained. The nickel-based alloys tend to have higher strengths at room temperature, with yield strengths up to 1200 MPa (175 ksi) and ultimate tensile strengths as high as 1450 MPa (210 ksi). The 1000-hour rupture strengths of the nickel-based alloys at 815°C (1500°F) are also higher than those of the cobalt-based material. Unfortunately, the density of the superalloy metals is significantly greater than iron, so their use is often at the expense of additional weight.

Most of the superalloys are very difficult to form or machine, so methods such as electrodischarge, electrochemical, or ultrasonic machining are often used, or the products are made to final shape as investment castings. Powder metallurgy techniques are also used extensively in the manufacture of superalloy components. Because of their ingredients, all of the alloys are quite expensive, and this limits their use to small or critical parts where the cost is not the determining factor.

A number of engineering applications have already exceeded the temperature limits of the superalloys, and still others await the development of materials that will make them feasible. One source, for example, estimates that the exhaust temperatures of future jet engines will be in excess of 1425°C (2600°F). Materials such as TD-nickel (a nickel alloy containing 2% dispersed thorium oxide) can operate at service temperatures somewhat above 1100°C (2000°F). The *refractory metals* can be used to temperatures as high as 1650°C (3000°F), provided that protective coatings are used to isolate them from gases in their operating environment. The refractory metals include *niobium, molybdenum, tantalum, rhenium*, and *tungsten*. Table 7-7 presents key properties for several of these materials. Unfortunately, all are heavier than steel, and several are significantly heavier. Figure 7-4 illustrates one high-temperature application. Figure 7-5 compares the upper limit for useful mechanical properties for a variety of engineering metals.

TABLE 7-7.   Properties of Some Refractory Metals

Metal	Melting Temperature [°F(°C)]	Density (g/cm³)	Room Temperature			Elevated Temperature [1832°F (1000°C)]	
			Yield Strength (ksi)	Tensile Strength (ksi)	Elongation (%)	Yield Strength (ksi)	Tensile Strength (ksi)
Molybdenum	4730 (2610)	10.22	80	120	10	30	50
Niobium	4480 (2470)	8.57	20	45	25	8	17
Tantalum	5430 (3000)	16.6	35	50	35	24	27
Tungsten	6170 (3410)	19.25	220	300	3	15	66

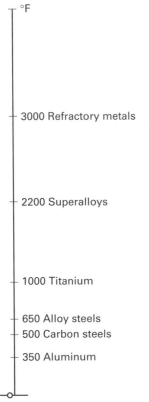

°F

3000 Refractory metals

2200 Superalloys

1000 Titanium

650 Alloy steels
500 Carbon steels

350 Aluminum

FIGURE 7-5   Temperature scale indicating the upper limit to useful mechanical properties for various engineering metals.

FIGURE 7-4   Superalloys and refractory metals are needed to withstand the high temperatures of jet engine exhaust. *(Courtesy of Northrop Grumman Corporation.)*

Other materials and technologies that offer promise for high-temperature service include directionally solidified eutectics, single crystals, intermetallic compounds, engineered ceramics, and advanced coating systems. The *intermetallic compounds* offer the potential for superior performance at very high temperature. They are hard, stiff, creep resistant, and oxidation resistant, and they have a yield strength that actually increases with temperature. In addition, the titanium and nickel aluminides are significantly lighter than the conventional superalloys. Unfortunately, the intermetallics are limited by poor ductility, fracture toughness, and fatigue resistance. Research and development efforts have overcome some of these limitations, and the intermetallics have begun to appear in commercial products.

◼ 7.9 LEAD, TIN, AND THEIR ALLOYS

The dominant properties of *lead* and lead alloys are high density coupled with strength and stiffness values among the lowest of the engineering metals. The principal uses of lead as a pure metal include storage batteries, cable cladding, and radiation-absorbing and sound-dampening shields. Lead-acid batteries are clearly the dominant product, and over 60% of the U.S. lead consumption is generated from battery recycling. Other applications utilize the properties of good corrosion resistance or low melting point. *Tin* is used primarily as a corrosion resistant coating on steel.

When in the form alloys, lead and tin are almost always used together. Bearing material and solder are the two most important uses. One of the oldest and best bearing materials is an alloy of 84% tin, 8% copper, and 8% antimony, known as genuine or tin *babbitt*. Because of the high cost of tin, however, lead babbitt, composed of 85% lead, 5% tin, 10% antimony, and 0.5% copper, is a more widely used bearing material. The tin and antimony combine to form hard particles within the softer lead matrix. The shaft rides on the harder particles with low friction, while the softer matrix acts as a cushion that can distort sufficiently to compensate for misalignment and assure a proper fit between the two surfaces. For slow speeds and moderate loads, the lead-based babbitts have proven to be quite adequate.

Soft solders are basically lead–tin alloys with a chemistry near the eutectic composition of 61.9% tin (see Figure 4-5). While the eutectic alloy has the lowest melting temperature, the high cost of tin has forced many users to specify solders with a lower-than-optimum tin content. A variety of compositions are available, each with its own characteristic melting range. Additional information on solders and soldering is provided in Chapter 39.

■ 7.10 SOME LESSER-KNOWN METALS AND ALLOYS

Several of the lesser-known metals and have achieved importance in modern technology as a result of their somewhat unique physical and mechanical properties. *Beryllium* combines a density less than aluminum with a stiffness greater than steel. *Hafnium, thorium,* and *beryllium* are used in nuclear reactors because of their low neutron-absorption characteristics. Depleted *uranium*, because of its very high density (19.1 g/cm^3), is useful in special applications where maximum weight must be put into a limited space, as in counterweights and flywheels. *Cobalt*, in addition to its use as a base metal for superalloys, is used as a binder in various powder-based components and sintered carbides, where it provides good high-temperature strength.

Zirconium is used for its outstanding corrosion resistance to most acids, chlorides, and organic acids. It offers high strength, good weldability, and fatigue resistance, and attractive neutron absorption characteristics.

Rare-earth metals have been incorporated into magnets that offer increased strength compared to the standard ferrite variety. Neodymium–iron–boron and samarium–cobalt are two common varieties.

Current efforts in aerospace, nuclear, high-temperature, and electronic applications will undoubtedly bring about still greater interest in these and a number of other uncommon metals.

■ 7.11 GRAPHITE

While technically not a metal, *graphite* is an engineering material that is showing considerable potential. It offers properties of both a metal and nonmetal, including thermal and electrical conductivity, inertness, ability to withstand high temperature, and lubricity. In addition, it possesses the unique property of increasing in strength as the temperature is elevated. Polycrystralline graphites can have mechanical strengths up to 70 MPa (10 ksi) at room temperature, which double when the temperature reaches 2500°C (4500°F).

Large quantities of graphite are used as electrodes in arc furnaces, but other uses are developing rapidly. The addition of small amounts of borides, carbides, nitrides, and silicides greatly lowers the oxidation rate at elevated temperatures and improves the mechanical strength. This makes the material highly suitable for use as rocket-nozzle inserts and permanent molds for casting various metals. It can be machined quite readily to excellent surface finishes. Graphite fibers have also found extensive use in composite materials. This application will be discussed in Chapter 8.

■ KEY WORDS

Alclad	brass	copper	intermetallic	Monel	solder	tin
aluminum	bronze	dezincification	compound	nickel	stress-corrosion cracking	titanium
babbitt	cast	galvanizing	lead	nonferrous	superalloys	wrought
beryllium	cobalt	graphite	magnesium	refractory metals	temper designation	zinc

■ REVIEW QUESTIONS

1. What types of properties do nonferrous metals possess that are not available in the ferrous metals?
2. In what respects are the nonferrous metals generally inferior to steel?
3. For what type of fabrication processes might the low-melting-point alloys be attractive?
4. What are the three properties of copper and copper alloys that account for many of their uses and applications?

5. What properties make copper attractive for cold-working processes?
6. What are some of the limiting properties of copper that might restrict its area of application?
7. What are the two primary types of commercially pure copper and how do they differ?
8. What are some of the attractive engineering properties that account for the wide use of the copper–zinc alpha brasses?
9. Why might cold-worked brass require a stress relief prior to being placed in service?
10. Why might the term *bronze* be potentially confusing when used in reference to a copper-based alloy?
11. What are some attractive engineering properties of copper–nickel alloys?
12. Describe the somewhat unique property combination that exists in heat-treated copper–beryllium alloys. What has limited its use in recent years?
13. What are some of the attractive engineering properties of aluminum and aluminum alloys?
14. How does aluminum compare to steel in terms of weight? Discuss the merits of comparing cost per unit weight versus cost per unit volume.
15. How does aluminum compare to copper in terms of electrical conductivity?
16. What features might limit the mechanical uses and applications of aluminum and aluminum alloys?
17. How is the corrosion-resistance mechanism observed in aluminum and aluminum alloys similar to that observed in stainless steels?
18. How are the wrought alloys distinguished from the cast alloys in the aluminum designation system? Why would these two groups of metals have distinctly different properties?
19. What feature in the wrought aluminum designation scheme is used to denote the condition or structure of a given alloy?
20. What is the primary strengthening mechanism in the high-strength "aircraft-quality" aluminum alloys?

21. What unique combination of properties is offered by the composite Alclad materials?
22. What specific material properties might make an aluminum casting alloy attractive for permanent mold casting? For die casting?
23. What features of aluminum–lithium alloys make them particularly attractive for aerospace applications?
24. What are some limiting properties of magnesium and magnesium alloys?
25. What particular design feature or requirement is generally necessary to justify the use of magnesium alloys?
26. Describe the designation system applied to magnesium alloys.
27. In what way can ductility be imparted to magnesium alloys so that they can be formed by conventional processes?
28. Under what conditions should magnesium be considered to be a flammable or explosive material?
29. What is the primary application of pure zinc? Of the zinc-based engineering alloys?
30. What are some of the attractive features of the zinc–aluminum casting alloys?
31. What are some of the attractive engineering properties of titanium and titanium alloys?
32. What temperature is generally considered to be the upper limit for which titanium alloys retain their useful engineering properties?
33. What property of Monel alloys dominates most of their applications?
34. What metals or combinations of metals form the bases of the superalloys?
35. What class of metals or alloys must be used when the operating temperatures exceed the limits of the superalloys?
36. Which metals are classified as refractory metals?
37. When selecting a lead–tin alloy for a solder application, why might the lead content be higher than that associated with lowest-melting-point alloy in the system?
38. What features makes beryllium a unique, lightweight metal?
39. What unique property of graphite makes it attractive for elevated-temperature applications?

*C*hapter 7   CASE STUDY

Nonsparking Wrench

A major tool-manufacturing company is considering an expansion of its line of conventional hand tools to include safety tools, capable of being used in areas such as gas leaks where the potential of explosion or fire exists. Conventional irons and steels are pyrophoric (i.e., small slivers or fragments can burn in air, forming sparks if dropped or impacted on a hard surface).

You are asked to evaluate potential materials and processes that might be used to manufacture a nonsparking pipe wrench. This product is to be produced in the same shape and range of sizes as conventional pipe wrenches and possess all of the same characteristic properties (strength in the handle, hardness in the teeth, fracture resistance, corrosion resistance, etc.). In addition, the new safety wrench must be nonsparking (or nonpyrophoric).

Your initial review of the nonferrous metals reveals that aluminum is nonpyrophoric, but lacks the strength and wear resistance needed in the teeth and jaw region of the wrench.

Copper is also nonpyrophoric, but is heavier than steel, and this may be unattractive for the larfer wrenches. Copper–2% beryllium can be age hardened to provide the strength and hardness properties equivalent to the steel that is currently being used for the jaws of the wrench, but the cost of this material is also quite high. Titanium is difficult to fabricate and may not possess the needed hardness and wear resistance. Mixed materials may create an unattractive galvanic corrosion cell. Both forging and casting appear to be viable means of forming the desired shape.

You want to produce a quality product but also wish to make the wrench in the most economical manner possible so that the new line of safety tools is attractive to potential customers.

Suggest some alternative manufacturing systems (materials coupled with companion methods of fabrication) that could be used to produce the desired wrench. What might be the advantages and disadvantages of each? Which of your alternatives would you recommend to your supervisor?

NONMETALLIC MATERIALS: PLASTICS, ELASTOMERS, CERAMICS, AND COMPOSITES

■ 8.1 INTRODUCTION

Because of their wide range of attractive properties, the nonmetallic materials have always played a significant role in manufacturing. Wood has been a key engineering material down through the centuries, and artisans have learned to select and use the various types and grades to manufacture a broad spectrum of quality products. Stone and rock continue to be key construction materials, and clay products can be traced to antiquity. Even leather has been a construction material and was used for fenders in early automobiles.

More recently, however, the *nonmetallic materials* family has expanded from the natural materials just described and now includes an extensive listing of plastics (polymers), elastomers, ceramics, and composites. Most of these are manufactured materials, so a wide variety of properties and characteristics can be obtained. New variations are being created on a continuous basis, and their uses and applications are expanding rapidly. Many observers now refer to a materials revolution as these new materials compete with steel and the other more traditional engineering metals. New products have emerged, utilizing the new properties, and existing products are continually being reevaluated for the possibility of material substitution. As the design requirements of engineering products continue to push the limits of the traditional materials, the role of manufactured nonmetallic materials will continue to expand.

Because of the breadth and number of nonmetallic materials, we will not attempt to provide information about all of them. Instead, the emphasis will be on the basic nature and properties of the various families so that the reader will be able to determine if they may be reasonable candidates for specific products and applications. For detailed information about specific materials within these families, more extensive and dedicated texts, handbooks, and compilations should be consulted.

■ 8.2 PLASTICS

It is difficult to provide a precise definition of the term *plastics*. From a technical viewpoint, the term is applied to a group of engineered materials characterized by large molecules that are built up by the joining of smaller molecules. On a more practical level, these materials are natural or synthetic resins, or their compounds, that can be molded, extruded, cast, or used as thin films or coatings. They offer low density, low tooling costs, good corrosion resistance, cost reduction, and design versatility. From a chemical viewpoint, most are organic substances containing hydrogen, oxygen, carbon, and nitrogen.

In less than a century, we have gone from a world without plastic to a world where its use and applications are limitless. The United States currently produces more plastic than steel, aluminum, and copper combined. Plastics are used to save lives in applications such as artificial organs, shatterproof glass, and bulletproof vests. They reduce the weight of cars, provide thermal insulation to our homes, and encapsulate our medicines. They form the base material in shower curtains, contact lenses, and most of our clothing and compose some of the primary components in televisions, computers, cell phones, and furniture. Even the Statue of Liberty has a plastic coating to protect it from corrosion.

MOLECULAR STRUCTURE OF PLASTICS

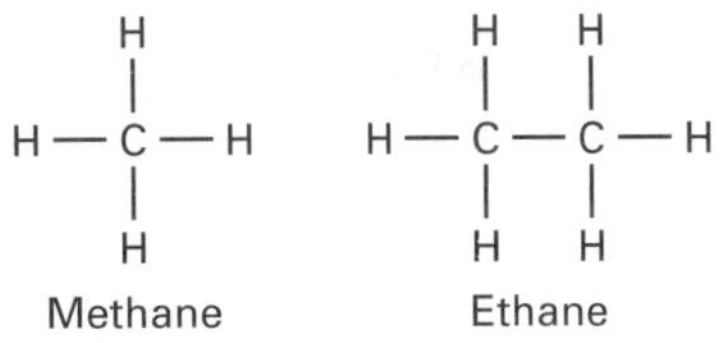

To understand the properties of plastics, it is important to first understand their molecular structure. For simplicity, let's begin with the paraffin-type hydrocarbons, in which carbon and hydrogen combine in the relationship C_nH_{2n+2}. Theoretically, the atoms can link together indefinitely to form very large molecules, extending the series depicted in Figure 8-1. The bonds between the various atoms are all single pairs of shared electrons (covalent bonds). Bonding within the molecule, therefore, is quite strong, but the attractive forces between adjacent molecules are much weaker. Because there is no provision for additional atoms to be added to the chain, these molecules are said to be *saturated*.

Carbon and hydrogen can also form molecules where the carbon atoms are held together by double or triple covalent bonds. Ethylene and acetylene are common examples (Figure 8-2). Because these molecules do not have the maximum possible number of hydrogen atoms, they are said to be *unsaturated* and are important in the polymerization process, where small molecules link to form large ones with the same constituent atoms.

In each of the previously described molecules, four electron pairs surround each carbon atom and one electron pair is shared with each hydrogen atom. Other atoms or structures can be substituted for carbon and hydrogen, however. Chlorine, fluorine, or even a benzene ring can take the place of hydrogen. Oxygen, silicon, sulfur, or nitrogen can take the place of carbon. Because of these substitutions, a wide range of organic compounds can be created.

FIGURE 8-1　Linking of hydrogen and carbon in methane and ethane molecules. Each dash represents an electron pair or covalent bond.

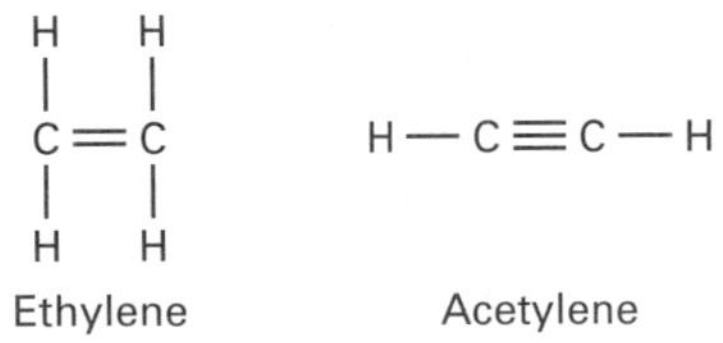

FIGURE 8-2　Double and triple covalent bonds between the carbon atoms in unsaturated ethylene and acetylene molecules.

ISOMERS

The same kind and number of atoms can also unite in different structural arrangements, known as *isomers*, and these behave as different compounds with different engineering properties. Figure 8-3 shows an example of this feature, involving propyl and isopropyl alcohol. Isomers can be considered analogous to allotropism or polymorphism in crystalline materials, where the same material possesses different properties because of the different crystal structures.

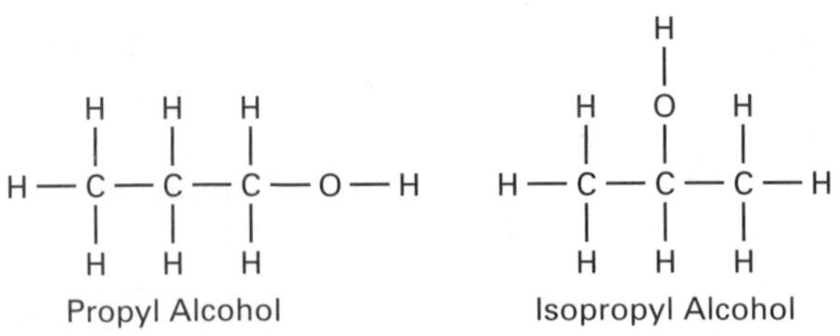

Propyl Alcohol　　Isopropyl Alcohol

FIGURE 8-3　Linking of eight hydrogen, one oxygen, and three carbon atoms to form two isomers: propyl alcohol and isopropyl alcohol.

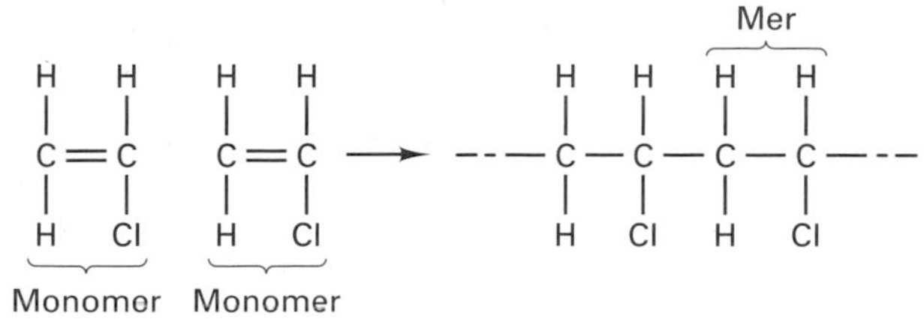

FIGURE 8-4   Polymerization by addition—the linking of identical monomers.

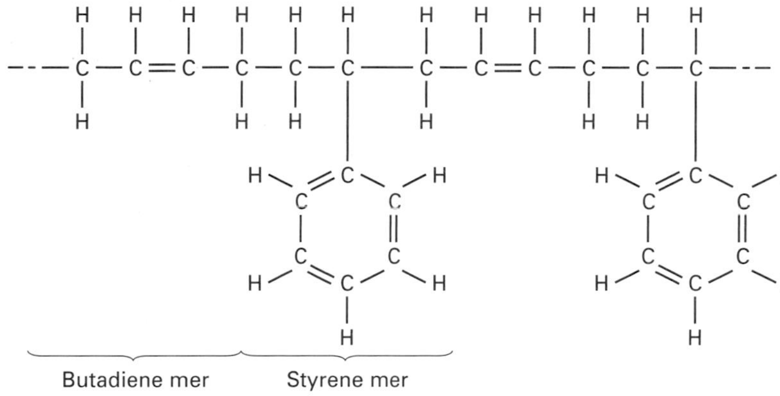

FIGURE 8-5   Polymerization by addition with two kinds of mers—the copolymerization of butadiene and styrene.

FORMING MOLECULES BY POLYMERIZATION

The polymerization process, or linking of molecules, occurs by either an *addition* or *condensation* mechanism. Figure 8-4 illustrates polymerization by addition, where a number of basic units (*monomers*) link together to form a large molecule (*polymer*) in which there is a repeated unit (*mer*). Activators or catalysts, such as benzoyl peroxide, initiate and terminate the chain. Thus, the amount of activator relative to the amount of monomer determines the average molecular weight (or average length) of the polymer chain. The average number of mers in the polymer, known as the *degree of polymerization*, ranges from 75 to 750 for most commercial plastics. *Copolymers* are a special category of polymer where two different types of mers are combined into the same addition chain. The formation of copolymers (Figure 8-5) greatly expands the possibilities of creating new types of plastics with improved physical and mechanical properties. *Terpolymers* further extend the possibilities by combining three different monomers.

In contrast to polymerization by addition, where all of the original atoms appear in the product molecule, *condensation polymerization* occurs when reactive molecules combine with one another to produce a polymer plus small, by-product molecules, such as water. Heat, pressure, and catalysts are often required to drive the reaction. Figure 8-6 illustrates the reaction between phenol and formaldehyde to form Bakelite, first performed in 1910. The structure of condensation polymers can be either linear chains or a three-dimensional framework in which all atoms are linked by strong, primary bonds.

FIGURE 8-6   The formation of phenol–formaldehyde (Bakelite) by condensation polymerization.

Formaldehyde   Phenol   Phenol   Phenol Formaldehyde   Water

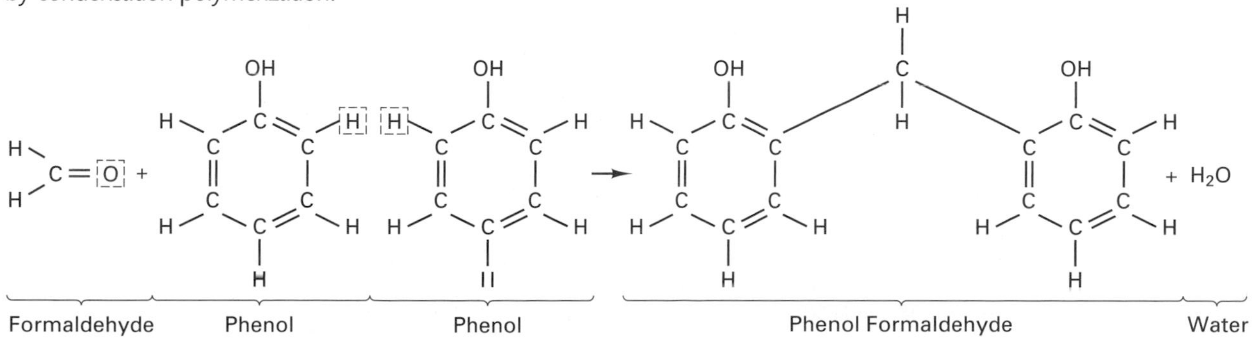

THERMOSETTING AND THERMOPLASTIC MATERIALS

The terms *thermosetting* and *thermoplastic* refer to the material's response to elevated temperature. Addition polymers (or linear condensation polymers) can be viewed as long chains of tightly bonded carbon atoms with strongly attached pendants of hydrogen, fluorine, chlorine, or benzene rings. All of the bonds within the molecules are strong covalent bonds. The attraction between neighboring molecules, however, is only by the much weaker van der Waals forces. For these materials, the intermolecular forces strongly influence the mechanical and physical properties. In general, the linear polymers tend to be flexible and tough. Because the secondary bonds are weakened by elevated temperature, plastics of this type soften with increasing temperature and become harder and stronger when cooled. The softening and hardening of these thermoplastic materials can be repeated as often as desired, and no chemical change is involved.

Because they contain molecules of different lengths, thermoplastic materials do not have a definite melting temperature but, instead, soften over a range of temperatures. Above the melting temperature, the material can be poured and cast, or formed by injection molding. Below the melting temperature, the material can retain its amorphous structure, but with companion properties that are somewhat rubbery. The application of a force produces both elastic and plastic deformation. Large amounts of permanent deformation are available and make this range attractive for molding and extrusion. At still lower temperatures, the bonds become stronger and the polymer is stiffer and somewhat leathery. Many commercial polymers (e.g., polyethylene) have useful strength in this condition. When further cooled below the glass transition temperature, the linear polymer retains its amorphous structure but becomes hard, brittle, and glasslike.

Many thermoplastics can partially *crystallize*[1] when cooled below the melting temperature. During this process the chains closely align over appreciable distances, with a companion increase in density. In addition, the polymer becomes stiffer, harder, less ductile, and more resistant to solvents and heat. The ability of a polymer to crystallize depends on the complexity of its molecules, the degree of polymerization (length of the chains), the cooling rate, and the amount of deformation during cooling.

The mechanical behavior of an amorphous (noncrystallized) thermoplastic polymer can be modeled by a common cotton ball. The individual molecules are bonded within by strong covalent bonds and are analogous to the individual fibers of cotton. The bonding forces between molecules are much weaker and are similar to the friction forces between the strands of cotton in the assembly. When pulled or stretched, plastic deformation occurs by slippage between adjacent fibers or molecular chains. Methods to increase the strength of thermoplastics, therefore, focus on restricting intermolecular slippage. Longer chains have less freedom of movement and are therefore stronger. Connecting adjacent chains to one another with primary bond cross-links, as with the sulfur links when vulcanizing rubber, can also impede deformation. Since the strength of the secondary bonds is inversely related to the separation distance of the molecules, processes such as deformation or crystallization can be used to produce a tight parallel alignment of adjacent molecules and a concurrent increase in strength, stiffness, and density. Polymers with larger side structures, such as chlorine atoms or benzene rings, may be stronger or weaker than those with just hydrogen, depending on whether the dominant effect is the impediment to slippage or the increased separation distance. Branched polymers, where the chains divide in a "Y" with primary bonds linking all segments of the chain, are often weaker since branching reduces the density and close packing of the chains.

In contrast to the thermoplastic polymers, *thermosetting plastics* usually have a highly cross-linked or three-dimensional framework structure in which all atoms are connected by strong, covalent bonds. These materials are generally produced by condensation polymerization where elevated temperature promotes an irreversible reaction, hence the term *thermosetting*. Once set, subsequent heating will not produce the softening

---

[1] It should be noted that the term *crystallize*, when applied to polymers, has a different meaning than when applied to metals and ceramics. Metals and ceramics are crystalline materials, meaning that the atoms occupy sites in a regular, periodic array, known as a lattice. In polymers, it is not the atoms that become aligned, but the molecules.

observed with the thermoplastics. Instead, thermosetting materials maintain their mechanical properties up to the temperature at which they char or burn. Since deformation requires the breaking of primary bonds, the thermosetting polymers are significantly stronger and more rigid than the thermoplastics, but they have lower ductility and poorer impact properties.

Although classification of a polymer as thermosetting or thermoplastic provides insight as to properties and performance, it also has a strong effect on fabrication. For example, thermoplastics can be easily molded. After the hot, soft material has been formed to the desired shape, however, the mold must be cooled so that the plastic will harden and be able to retain its shape upon removal. The repetitive heating and cooling cycles affect mold life, and the time required for the thermal cycles influences productivity. When a part is produced from thermosetting materials, the mold can remain at a constant temperature throughout the entire process, but the time in the mold is now determined by the setting or curing of the resins. Since the material hardens as a result of the reaction and has strength and rigidity even when hot, product removal can be performed without cooling the mold.

PROPERTIES AND APPLICATIONS

Because there are so many varieties of plastics, and new ones are being developed almost continuously, it is helpful to have a knowledge of both the general properties of plastics as well as the unique or specific properties of the various families. General properties of plastics include:

1. *Light weight.* Most plastics have specific gravities between 1.1 and 1.6, compared with about 1.75 for magnesium (the lightest engineering metal).

2. *Corrosion resistance.* Many plastics perform well in hostile, corrosive environments. Some are notably resistant to acid corrosion.

3. *Electrical resistance.* Plastics are widely used as insulating materials.

4. *Low thermal conductivity.* Plastics are relatively good thermal insulators.

5. *Variety of optical properties.* Many plastics have an almost unlimited color range, and the color goes throughout, not just on the surface. Both transparent and opaque materials exist.

6. *Formability.* Objects can frequently be produced from plastics in a single operation. Raw material can be converted to final shape through such processes as casting, extrusion, and molding. Relatively low temperatures are required for the forming of plastics.

7. *Surface finish.* Excellent surface finishes can be obtained by the same processes that produce the shape. Additional surface finishing may not be required.

8. *Comparatively low cost.* The low cost of plastics generally applies to both the material itself and the manufacturing process. Plastics frequently offer reduced tool costs and high rates of production.

9. *Low energy content.*

While the attractive features of plastics tend to be in the area of physical properties, the inferior features generally relate to mechanical strength. None of the plastics possess strength properties that approach those of the engineering metals, but their low density allows them to compete effectively on a strength-to-weight (or specific strength) basis. Many have low impact strength, although several (such as ABS, high-density polyethylene, and polycarbonate) are exceptions to this rule. The dimensional stability of plastics tends to be greatly inferior to that of metals, and the coefficient of thermal expansion is rather high. Thermoplastics are quite sensitive to heat, and their strength often drops rapidly as temperatures increase above normal environmental conditions. Thermosetting materials offer good strength retention at elevated temperature but have an upper limit of about 250°C (500°F). While the corrosion resistance of plastics is generally good, they often absorb moisture, and this, in turn, decreases strength.

Some thermoplastics can exhibit a 50% drop in tensile strength as the humidity increases from 0 to 100%. Finally, radiation, both ultraviolet and particulate, can markedly alter the properties. Many plastics used in an outdoor environment have ultimately failed due to the cumulative effect of ultraviolet radiation.

Table 8-1 summarizes the properties of a number of common plastics. By considering the information in this table along with the preceding discussion of general properties, it becomes apparent that plastics are best used in applications that require materials with low to moderate strength, light weight, low electrical and/or thermal conductivity, a wide range of available colors, and ease of fabrication into finished products. No other family of materials can offer this combination of properties. Because of their light weight, attractive appearance, and ease of fabrication, plastics have been selected for many packaging and container applications. This classification includes such items as household appliance housings, clock cases, and exteriors of electronic products, where the primary role is to contain the interior mechanisms. Applications such as insulation on electrical wires and handles for hot articles capitalize on the low electrical and thermal conductivities. Soft, pliable, foamed plastics are used extensively as cushioning material. Rigid foams are used inside sheet metal structures to provide compressive strength. Some plastics are used as adhesive or bonding agents in the assembly of products, an application discussed in Chapter 40. Others can provide inexpensive tooling for applications where pressures, temperatures, and wear requirements are not extreme.

There are many applications where only one or two of the properties of plastics are sufficient to justify their use. When special characteristics are desired that are not normally found in the commercial plastics, composite materials can often be designed that use a polymeric matrix. For example, high directional strength may be achieved by incorporating a fabric or fiber reinforcement within a plastic resin. These materials will be discussed in some detail later in the chapter.

COMMON TYPES OR FAMILIES OF PLASTICS
The following is a brief descriptive summary of the types of plastics listed in Table 8-1.

THERMOPLASTICS

ABS: contains acrylonitrile, butadiene, and styrene; low weight, good strength, and very tough; resists heat, weather, and chemicals quite well; dimensionally stable but flammable

Acrylics: highest optical clarity, transmitting over 90% of light; common trade names include Lucite and Plexiglas; high impact, flexural, tensile, and dielectric strengths; available in a wide range of colors; resist weathering; stretch rather easily

Cellulose acetate: wide range of colors; good insulating qualities; easily molded; high moisture absorption in most grades

Cellulose acetate butyrate: higher impact strength and moisture resistance than cellulose acetate; will withstand rougher usage

Ethyl cellulose: high electrical resistance and impact strength; retains toughness at low temperatures

Fluorocarbons: inert to most chemicals; high temperature resistance; very low coefficients of friction (Teflon), used for nonlubricated bearings and nonstick coatings for cooking utensils and electrical irons

Nylon (polyamides): low coefficient of friction; good strength, abrasion resistance, and toughness; excellent dimensional stability; good heat resistance; used for small gears and bearings, zip fasteners, and as monofilaments for textiles, fishing line, and ropes

Polycarbonates: high strength and outstanding toughness; good dimensional stability; transparent or easily colored

Polyethylenes: the most common polymer; inexpensive, tough, good chemical resistance and high electrical resistance; low strength; easy to shape and join; subject to weathering via ultraviolet light; flammable; used for grocery bags, milk jugs,

TABLE 8-1. Properties and Major Characteristics of Common Types of Plastics

Material	Specific Gravity	Tensile Strength (1000 lb/in²)	Impact Strength Izod (ft-lb/in. of Notch)	Top Working Temperature [°F(°C)]	Dielectric Strength[b] (V/mil)	24-Hour Water Absorption (%)	Weatherability	Colorability	Optical Clarity	Chemical Resistance	Injection Molding	Extrusions	Formable Sheet	Film	Fiber	Compression or Transfer Moldings	Castings	Reinforced Plastics	Moldings	Industrial Thermosetting Laminates	Foam
Thermoplastics																					
ABS material	1.02–1.06	4–8	1.3–10.0		300–400	0.2–0.3	0	X		0	•	•	•								
Acetal	1.4	10	1.5	250(121)	1200	0.22		X		0	•	•						•			
Acrylics	1.12–1.19	5.5–10	0.2–2.3	200(93)	400–530	0.2–0.4	X	X	X	0	•	•	•		•						
Cellulose acetate	1.25–1.50	3–8	0.75–4.0	260(127)	300–600	2.0–6.0	X	X	X		•	•	•	•	•						
Cellulose acetate butyrate	1.18–1.24	2–6	0.6–3.2	130(54)	250–350	1.8–2.1						•	•								
Cellulose propionate	1.19–1.24	1–5	0.8–9	140(60)	300	1.8–2.1		X	X		•	•	•								
Chlorinated polyether	1.4	6	3.3	300(149)	400	0.01						•		•							
Ethyl cellulose	1.16	3–6	1.8–4.0	150(66)	350	1.6–2.2		X			•	•	•								
TFE–fluorocarbon	2.1–2.3	1.5–3	2.5–4.0	500(260)	450	0	X			X		•				•					
CFE–fluorocarbon	2.1–2.15	4.5–6	3.5–3.6	390(199)	550	0	X			X		•									
Nylon	1.1–1.2	8–10	2	250(121)	385–470	0.4–5.5	X	0	0	0	•	•	•	•	•						
Polycarbonate	1.2	9.5	14	250(121)	400	0.15	X	0			•	•		•							
Polyethylene	0.96	4	10	200(93)	440	0.003		0			•	•		•							•
Polypropylene	0.9–1.27	3.4–5.3	1.02	230(110)	520–800	0.03		0	X	0	•	•		•	•						
Polystyrene	1.05–1.15	5–9	0.3–0.6	190(8)	400–600	<0.2		X	X	0	•	•	•	•							
Modified polystyrene	1.0–1.1	2.5–6	0.25–11.0	212(100)	300–600	0.03–02		X	X	X	•	•	•								
Vinyl	1.16–1.55	1–5.9	0.25–2.0	220(104)	25–500	0.2–1		X	X	X	•	•	•	•	•						•
Thermosetting plastics																					
Epoxy	1.1–1.7	4–13	0.4–1.5	325(163)	500	0.1–0.5	X	X		X						•	•	•		•	•
Melamine	1.76–1.98	5–8	0.25–5	350(177)	460	0.1		X		0								•		•	
Phenolic	1.2–1.45	5–9		300(149)	100–500	0.2–0.6				0	•					•	•	•		•	•
Polyester (other than molding compounds)	1.06–1.46	4–10	0.18–0.4	300(149)	340–570	0.5		X		0				•			•	•			
Polyester (alkyd, DAP)	1.6–1.75	3.2–8	3.6–8			0.16–0.67										•			•		
Silicone	2.0	3–5	0.2–3.0	550(288)	250–350	0.4–0.5			X					•	•	•	•	•		•	•
Urea	1.41–1.80	4–8.5	0.2–0.5	185(85)	300–600	1–3		X	0							•					•

[a] X denotes a principal reason for its use; 0 indicates a secondary reason.

[b] Short-time ASTM test.

tubes, pipes, sheeting, and electrical wire insulation; low-density polyethylene (LDPE) floats in water

PMMA (polymethyl methacrylate): hard, brittle (at room temperature), transparent or easily colored; used for items like tool handles and the interior windows of airplanes

Polypropylene: inexpensive; stronger and stiffer than polyethylene; transparent; reasonable toughness; used for beverage containers, pipes, and ropes

Polystyrenes: high dimensional stability and low water absorption; best all-around dielectric; clear, hard, and brittle at room temperature; high stiffness for a polymer; often used for rigid packaging; can be foamed to produce expanded polystyrene (trade name of Styrofoam); burns readily; softens at about 95°C

Polyvinylchloride (PVC): general-purpose thermoplastic; good resistance to ultraviolet light (good for outside applications); easily molded or extruded; uses include gas and water pipes, and window frames

Vinyls: wide range of types, from thin, rubbery films to rigid forms; tear resistant; good aging properties; good dimensional stability and water resistance in rigid forms; used for floor and wall covering, upholstery fabrics, and lightweight water hose; common trade names include Saran and Tygon

THERMOSETS

Epoxies: good strength, toughness, elasticity, chemical resistance, moisture resistance, and dimensional stability; easily compounded to cure at room temperature; used as adhesives, bonding agents, coatings, and in fiber laminates

Melamines: excellent resistance to heat, water, and many chemicals; full range of translucent and opaque colors; excellent electric-arc resistance; tableware (but stained by coffee); used extensively in treating paper and cloth to impart water-repellent properties

Phenolics: oldest of the plastics but still widely used; hard, strong, low cost, and easily molded, but rather brittle; resistant to heat and moisture; dimensionally stable; opaque, but with a wide color range; wide variety of forms: sheet, rod, tube, and laminate

Polyesters: (can be thermoplastic or thermoset) strong and resist environmental influences well; uses include boat and car bodies, pipes, vents and ducts, textiles, adhesives, coatings, and laminates

Silicones: heat and weather resistant; low moisture absorption; chemically inert; high dielectric properties; excellent sealants

Urea–formaldehyde: properties similar to those of phenolics but available in lighter colors; useful in containers and housings, but not outdoors; used in lighting fixtures because of translucence in thin sections; as a foam, may be used as household insulation

ADDITIVE AGENTS IN PLASTICS

For most uses, additional materials are incorporated into plastics to (1) improve their properties, (2) reduce their cost, (3) improve their moldability, and/or (4) impart color. These *additive constituents* are usually classified as *fillers*, *plasticizers*, *lubricants*, *coloring agents*, *stabilizers*, *antioxidants*, and *flame retardants*.

Ordinarily, *fillers* comprise a large percentage of the total volume of a molded plastic product. Their primary purpose is to improve strength, stiffness, or toughness; reduce shrinkage; reduce weight; or simply serve as an extender, providing cost-saving bulk (often at the expense of reduced moldability). To a large degree, they determine the general properties of a molded plastic. Selection tends to favor materials that are much less expensive than the plastic resin, but the various fillers can impart different properties and characteristics. Some of the most common fillers and their properties are:

1. *Wood flour* (fine sawdust): a general-purpose filler; low cost with fair strength; good moldability

2. *Cloth fibers*: improved impact strength; fair moldability

3. *Macerated cloth*: high impact strength; limited moldability

4. *Glass fibers*: high strength; dimensional stability; translucence

5. *Mica*: excellent electrical properties and low moisture absorption

6. *Calcium carbonate, silica, talc, and clay*: serve primarily as extenders

When fillers are used with a plastic resin, the resin acts as a binder, surrounding the filler material and holding the mass together. The surface of a molded part, therefore, will be almost pure resin with no exposed filler.

Coloring agents may be either dyes, which are soluble in the resins, or insoluble pigments, which impart color simply by their presence. In general, dyes are used for transparent plastics and pigments for the opaque ones. Since most fillers do not produce attractive colors, coloring agents are usually needed.

Plasticizers can be added in small amounts to improve the flow of the plastic during molding or to increase the flexibility of the thermoplastic products by reducing the intermolecular contact and strength of the secondary bonds between the polymer chains. When used for molding purposes, the amount used is governed by the intricacy of the mold. In general, it should be kept to a minimum because it is likely to affect the stability of the finished product through a gradual aging loss. When used for flexibility, plasticizers should be selected with minimum volatility, so as to impart the desired property for as long as possible.

Lubricants such as waxes, stearates, and soaps can be added to improve the moldability of plastics and to facilitate removal of parts from the mold. They are also used to keep thin polymer sheets from sticking to each other when stacked or rolled. Only a minimum amount should be used, however, because the lubricants adversely affect most engineering properties.

Heat, light (especially ultraviolet), and oxidation tend to degrade polymers. Stabilizers and antioxidants can be added to retard these effects. Flame retardants can be added when nonflammability is important. Antistatic agents may be incorporated into plastics used for applications such as electronics packaging. Antimicrobial additives can provide long-term protection from both fungus (such as mildew) and bacteria. Table 8-2 summarizes the purposes of the various additives.

ORIENTED PLASTICS

Because the intermolecular bond strength increases with reduced separation distance, processing that aligns the molecules parallel to the applied load can be used to give the long-chain thermoplastics high strength in a given direction. This orientation process can be accomplished by a forming process, such as stretching, rolling, or extrusion. The material is usually heated prior to the orienting process to aid in overcoming the intermolecular forces and is cooled immediately afterward to "freeze" the molecules in the desired orientation.

Orienting may increase the tensile strength by more than 50%, but a 25% increase is more typical. In addition, the elongation may be increased by several hundred percent.

TABLE 8-2. Additive Agents in Plastics	
Type	Purpose
Fillers	Enhance mechanical properties, reduce shrinkage, reduce weight, or provide bulk
Plasticizer	Increase flexibility, improve flow during molding, reduce elastic modulus
Lubricant	Improve moldability and extraction from molds
Coloring agents (dyes and pigments)	Impart color
Stabilizers	Retard degradation due to heat or light
Antioxidants	Retard degradation due to oxidation
Flame retardants	Reduce flammability

If the oriented plastics are reheated, they tend to deform back toward their original shape, a phenomenon known as *viscoelastic memory*. The various shrink-wrap materials are examples of this effect.

ENGINEERING PLASTICS

The standard polymers tend to be lightweight, corrosion-resistant materials with low strength and low stiffness. They are relatively inexpensive and are readily formed into a wide range of useful shapes, but are not suitable for use at elevated temperatures.

In contrast, a group of plastics has been developed with improved thermal properties (up to 350°C, or 650°F), first-rate impact and stress resistance, high rigidity, superior electrical characteristics, excellent processing properties, and little dimensional change with varying temperature and humidity. These true engineering plastics include the polyamides, polyacetals, polyarylates, polycarbonates, modified polyphenylene oxides, polybutylene terepthalates, polyketones, polysulfones, polyetherimides, and liquid crystal polymers. While the conventional plastics can be upgraded by stabilizers, fibrous reinforcements, and particulate fillers, there is usually an accompanying reduction in other properties. The engineering plastics offer a more balanced set of properties. They are usually produced in small quantities, however, and are often quite expensive.

Materials producers have also developed electroconductive polymers with tailored electrical and electronic properties and high-crystalline polymers with properties comparable to metals.

PLASTICS AS ADHESIVES

Polymeric adhesives capable of withstanding high stresses are used in many industrial applications. They are quite attractive for the bonding of dissimilar materials, such as metals to nonmetals, and have even been used to replace welding or riveting. A wide range of mechanical properties is available through variations in composition and additives, and a variety of curing mechanisms can be used. The various features of adhesive bonding are discussed in greater detail in Chapter 40.

PLASTICS FOR TOOLING

Because of their wide range of properties, their ease of conversion into desired shapes, and their excellent properties when loaded in compression, plastics have been widely used in tooling applications, such as jigs, fixtures, and a wide variety of forming-die components. Both thermoplastic and thermoset polymers (particularly the cold-setting types) have been used. By using plastics in these applications, costs can be reduced and smaller quantities of products can be economically justified. In addition, the tooling can often be produced in a much shorter time, enabling quicker production.

PLASTICS VERSUS OTHER MATERIALS

Polymeric materials have successfully competed with traditional materials in several areas. Plastics have replaced glass in a significant number of containers and flat transparent products. PVC pipe and fittings have replaced copper and brass in many plumbing applications. Plastics have even replaced ceramics in areas as diverse as sewer pipe and lavatory facilities.

While plastics and metals are often viewed as competing materials, their engineering properties are really quite different. Many of the attractive features of plastics have already been discussed. In addition to these, one may wish to add (1) the ability to be fabricated with lower tooling costs; (2) the ability to be molded at the same rate as product assembly, thereby reducing inventory; (3) a possible reduction in assembly operations and easier assembly through snap fits, friction welds, or the use of self-tapping fasteners; (4) the ability to reuse manufacturing scrap; and (5) reduced finishing costs.

Metals, on the other hand, are often cheaper and offer faster fabrication speeds and greater impact resistance. They are considerably stronger and more rigid and can withstand traditional paint cure temperatures. In addition, resistance to flames, acids, and various solvents is significantly better. Table 8-3 compares the cost per pound and elastic modulus of five engineering plastics with values for wood, steel, and aluminum. When

TABLE 8-3.	Comparison of Materials (Modulus and Cost)[a]		
Material	Modulus ($\times 10^6$ psi)	\$/pound	\$/in^3
Aluminum	10.0	0.62	0.063
Steel	30.0	0.16	0.046
Wood (oak)	1.8	0.36	0.0078
Glass/epoxy	3.0	0.92	0.066
Nylon	0.1	1.32	0.054
Polycarbonate	0.35	1.65	0.071
Polypropylene	0.2	0.32	0.010
Polystyrene	0.3	0.47	0.014

[a]Cost figures are 2001–2002 values and are clearly subject to change.

the size of the part is fixed, cost per cubic inch becomes a more valid comparison, and here the plastics become more attractive because of their low density.

The automotive industry is a good indication of the expanding use of plastics. Polymeric materials now account for about 250 pounds of a typical vehicle, compared to only 25 pounds in 1960, 105 in 1970, and 195 in 1980. In addition to the traditional application areas of dashboards, interiors, body panels, and trim, plastics are now being used for bumpers, intake manifolds, valve covers, fuel tanks, and fuel lines and fittings. If we include clips and fasteners, there are over 1000 plastic parts in a typical automobile.

RECYCLING OF PLASTICS

Because of the wide variety of types and compositions, all with similar physical properties, the recycling of mixed plastics is far more difficult than the recycling of mixed metals. These materials must be sorted not only on the basis of resin type, but also by type of filler and color.

If the various types of resins can be identified and kept separate, many of the thermoplastic materials can be readily recycled into useful products. Packaging is the largest single market for plastics, and there is currently a well-established network to collect and recycle PET (the polyester used in soft-drink bottles) and high-density polyethylene (the plastic used in milk, juice, and water jugs). The properties generally deteriorate with recycling, however, so applications must often be downgraded with reuse. PET is being recycled into new bottles, fiber-fill insulation, and carpeting. Recycled polyethylene is used for new containers, plastic bags, and recycling bins. Polystyrene has been recycled into cafeteria trays and videocassette cases. Plastic "lumber" offers weather and insect resistance and a reduction in required maintenance, but at higher cost than traditional wood.

When thermoplastics and thermosets are mixed in varying amounts, the material is often regarded more as an alternative fuel (competing with coal and oil) than as a resource for recycling into quality products. On an equivalent weight basis, polystyrene and polyethylene have heat contents greater than fuel oil and far in excess of paper and wood. As a recycling alternative, decomposition processes can be used to break polymers down into useful building blocks. Hydrolysis (exposure to high-pressure steam) and pyrolysis (heating in the absence of oxygen) methods can be used to convert plastics into simple petrochemical materials, but even these processes require some control of the input material. As a result, only about one-third of all plastic now finds a second life.

■ 8.3 ELASTOMERS

Elastomers are a special class of linear polymers that display an exceptionally large amount of elastic deformation when a force is applied. Many can be stretched to several times their original length. Upon release of the force, the deformation can be completely recovered, as the material quickly returns to its original shape. In addition, the cycle can be repeated numerous times with identical results, as with the stretching of a rubber band.

The elastic properties of most engineering materials are the result of a change in the distance between adjacent atoms (i.e., bond length) when loads are applied. Hooke's law is commonly obeyed, wherein twice the force will produce twice the stretch. When the applied load is removed, the interatomic forces return all of the atoms to their original position and the elastic deformation is recovered completely.

In the elastomeric polymers, the linear chain-type molecules are twisted or curled, much like a coil spring. When a force is applied, the polymer stretches by uncoiling. When the load is removed, the molecules recoil as the bond angles return to their original, unloaded, values, and the material returns to its original size and shape. The relationship between force and stretch does not, however, follow Hooke's Law.

In reality, the behavior of elastomers is a bit more complex. While the chains indeed uncoil when placed under load, they also tend to slide with respect to one another to produce a small degree of viscous deformation. When the load is removed, the molecules recoil, but the viscous deformation is not recovered and the elastomer retains some permanent change in shape.

By linking the coiled molecules to one another by strong covalent bonds, a process known as *cross-linking*, it is possible to restrict the viscous deformation while retaining the large elastic response. The elasticity or rigidity of the product can be determined by controlling the number of cross-links within the material. Small amounts of cross-linking leave the elastomer soft and flexible, as in a rubber band. Additional cross-linking further restricts the uncoiling, and the material becomes harder, stiffer, and more brittle, like the rubber used in bowling balls. Since the cross-linked bonds can only be destroyed by extremely high temperatures, the engineering elastomers can be tailored to possess a wide range of stable properties and stress-strain characteristics.

If placed under constant strain, however, even highly cross-linked material will exhibit some viscous flow over time. Consider a rubber band stretched between two nails. While the dimensions remain fixed, the force or stress being applied to the nails will continually decrease. This phenomenon is known as *stress relaxation*. The rate of this relaxation depends on the material, the force, and the temperature.

RUBBER

Natural rubber, the oldest commercial elastomer, is made from the processed sap of a tropical tree. In its crude form it is an excellent adhesive, and many cements can be made by dissolving it in suitable solvents. Its use as an engineering material dates from 1839, when Charles Goodyear discovered that it could be vulcanized (cross-linked) by the addition of about 30% sulfur followed by heating to a suitable temperature. The cross-linking restricts the movement of the molecular chains and imparts strength. Subsequent research found that the properties could be further improved by various additives (such as carbon black), which act as stiffeners, tougheners, and antioxidants. Accelerators have been found that speed up the vulcanization process. These have enabled a reduction in the amount of sulfur, such that most rubber compounds now contain less than 3% sulfur. Softeners can be added to facilitate processing, and fillers can be used to add bulk.

Rubber can now be compounded to provide a wide range of characteristics, ranging from soft and gummy to extremely hard. When additional strength is required, textile cords or fabrics can be coated with rubber. The fibers carry the load and the rubber serves as a matrix to join the cords while isolating them from one another to prevent chafing. For severe service, steel wires can be used as the load-bearing medium. Vehicle tires and heavy-duty conveyor belts are examples of this technology.

Natural rubber compounds are outstanding for their flexibility, good electrical insulation, low internal friction, and resistance to most inorganic acids, salts, and alkalies. However, they have poor resistance to petroleum products, such as oil, gasoline, and naphtha. In addition, they lose their strength at elevated temperatures, so it is advisable that they not be used at temperatures above 80°C (175°F). Unless they are specially compounded, they also deteriorate fairly rapidly in direct sunlight.

ARTIFICIAL ELASTOMERS

Seeking to overcome some of these limitations, as well as the uncertainty in the supply and price of natural rubber, a number of synthetic or artificial elastomers have been developed and have come to assume great commercial importance. While some are a bit inferior to natural rubber, others offer distinctly different and, frequently, superior properties. Polyisoprene offers the same molecular structure as that of natural rubber, with equal or superior properties. Silicone rubbers are based on a linear chain of silicon and oxygen atoms and can operate in service environments as hot as 230°C (450°F). Various mixes and blends offer retention of physical properties at elevated temperatures; flexibility at low temperatures; resistance to acids, bases, and other aqueous and organic fluids; resistance to flex fatigue; ability to absorb energy and provide damping; good weatherability; ozone resistance; and availability in a variety of different hardnesses. Thus, elastomeric materials can now be selected and used for a wide range of engineering applications.

Table 8-4 lists some of the more common artificial elastomers, along with natural rubber for comparison, and gives their properties and some typical uses. It should be remembered, however, that the properties of these materials can vary widely, depending on the details of their compounding and processing.

ELASTOMERS FOR TOOLING APPLICATIONS

When an elastomer is confined, it acts like a fluid, transmitting force uniformly in all directions. For this reason, elastomers can be substituted for one half of a die set in sheet-metal-forming operations. Elastomers are also used to perform bulging and form reentrant sections that would be impossible to form with rigid dies except through the use of costly multipiece tooling. Because they can be compounded to range from very soft to very hard; hold up well under compressive loading; are impervious to oils, solvents, and other similar fluids; and can be made into a desired shape quickly and economically, the engineering elastomers have become increasingly popular as tool materials. In addition, the elastomeric tooling will not mark or damage highly polished or prepainted surfaces. The urethanes are currently the most popular elastomer for tooling applications.

■ 8.4 CERAMICS

The first materials used by humans were natural materials such as wood and stone. The discovery that certain clays could be mixed, shaped, and hardened by firing led to what was probably the first human-made material—a ceramic. Traditional ceramic products, such as bricks and pottery, have continued to be part of our lives. More recently, *ceramic materials* have played an important role in the electrical industry because of their high electrical resistivity, and they have come to be associated with a wide variety of other engineering applications. Most of these utilize their outstanding physical properties, including their ability to withstand high temperatures (refractories and refractory coatings), provide a variety of electrical and magnetic properties (solid-state electronics), and resist wear (coated cutting tools). In general, ceramics are hard, brittle, high-melting-point materials with low electrical and thermal conductivity, low thermal expansion, good chemical and thermal stability, good creep resistance, high elastic modulus, and high compressive strengths. More recently, a family of structural ceramics has emerged, and these materials now provide enhanced mechanical properties that make them attractive for many load-bearing applications.

Glass and glass products account for about half of the ceramic materials market. Advanced ceramic materials (including the structural ceramics, electrical and magnetic ceramics, and fiber-optic material) compose another 20%. Whiteware and porcelain enameled products (such as household appliances) account for about 10% each, while refractories and structural clay products make up most of the difference.

NATURE AND STRUCTURE OF CERAMICS

Ceramic materials are compounds of metallic and nonmetallic elements (often in the form of oxides, carbides, and nitrides) and exist in a wide variety of compositions and forms. Most have crystalline structures, but unlike metals, the bonding electrons are generally captive in strong ionic or covalent bonds. The absence of free electrons makes the ceramic materials poor electrical conductors and results in many being transparent in thin sections. Because of the strength of the primary bonds, most ceramics have high melting temperatures, high rigidity, and high compressive strength.

The crystal structures of ceramic materials can be quite different from those observed in metals. In many ceramics, atoms of significantly different size must be accommodated within the same structure (as with the sodium chloride crystal of Figure 3-3), and the interstitial sites, therefore, become extremely important. Charge neutrality must be maintained

TABLE 8-4. Properties and Uses of Common Elastomers

Elastomer	Specific Gravity	Durometer Hardness	Tensile Strength (psi) Pure Gum	Black	Elongation (%) Pure Gum	Black	Service Temperature [°F(°C)] Min.	Max.	Resistance to:[a] Oil	Water Swell	Tear	Typical Application
Natural rubber	0.93	20–100	2500	4000	750	650	−65 (−54)	180 (82)	P	G	G	Tires, gaskets, hose
Polyacrylate	1.10	40–100	350	2500	600	400	0 (−18)	300 (149)	G	P	F	Oil hose, O-rings
EDPM (ethylene propylene)	0.85	30–100	1	3		500	−40 (−40)	300 (149)	P	G	G	Electric insulation, footwear, hose, belts
Chlorosulfonated polyethylene	1.10	50–90	4	2		400	−65 (−54)	250 (121)	G	E	G	Tank lining, chemical hose, shoes, soles and heels
Polychloroprene (neoprene)	1.23	20–90	3500	4000	800	550	−50 (−46)	225 (107)	G	G	G	Wire insulation, belts, hose, gaskets, seals, linings
Polybutadiene	1.93	30–100	1000	3000	800	550	−80 (−62)	212 (100)	P	P	G	Tires, soles and heels, gaskets, seals
Polyisoprene	0.94	20–100	3000	4000		600	−65 (−54)	180 (82)	P	G	G	Same as natural rubber
Polysulfide	1.34	20–80	350	1000	600	400	−65 (−54)	180 (82)	E	G	G	Seals, gaskets, diaphragms, valve disks
SBR (styrene-butadiene)	0.94	40–100	2			1200	−65 (−54)	225 (107)	P	G	G	Molded mechanical goods, disposable pharmaceutical items
Silicone	1.1	25–90		1200		450	−120 (−84)	450 (232)	F	E	P	Electric insulation, seals, gaskets, O-rings
Epichlorohydrin	1.27	40–90		2		325	−50 (−46)	250 (121)	G	G	G	Diaphragms, seals, molded goods, low-temperature parts
Urethane	0.85	62–95	5000		700		−54 (−65)	212 (100)	E	F	E	Caster wheels, heels, foam padding
Fluoroelastomers	1.65	60–90	1	3		400	−40 (−40)	450 (232)	E	E	F	O-rings, seals, gaskets, roll coverings

[a] P, poor; F, fair; G, good; E, excellent.

throughout the ionic structures. Covalent materials must have structures with a limited number of nearest neighbors, set by the number of shared-electron bonds. These features often dictate a less efficient packing, and hence lower densities, than those observed for metallic materials. As with metals, the same chemistry material can often exist in more than one structural arrangement (polymorphism). Silica (SiO_2), for example, can exist in three forms—quartz, tridymite, and crystobalite—depending on the conditions of temperature and pressure.

Ceramic materials can also exist in the form of chains, similar to the linear molecules in plastics. As with the polymeric materials having this structure, the bonds between the chains are not as strong as those within the chains. Consequently, when forces are applied, cleavage or shear can occur between the chains. In other ceramics, the atoms bond in the form of sheets to produce layered structures. Relatively weak bonds exist between the sheets, and these surfaces become the preferred sites for fracture. Mica is a good example of such a material.

A noncrystalline structure is also possible in solid ceramics. This *amorphous* condition is referred to as the *glassy state*, and the materials are known as *glasses*.

CLAY AND WHITEWARE PRODUCTS

Many ceramic products are still based on *clay*, to which various amounts of quartz and feldspar have been added. Selected proportions are mixed with water, shaped, dried, and fired to produce the structural clay products of brick, roof and structural tiles, drainage pipe, and sewer pipe, as well as the *whiteware* products of sanitary ware (toilets, sinks, and bathtubs), dinnerware, china, decorative floor and wall tile, pottery, and other artware.

REFRACTORY MATERIALS

Refractory materials are ceramics that have been designed to provide acceptable mechanical or chemical properties at high operating temperatures. They may take the form of bricks and shaped products, bulk materials (often used as coatings), and insulating ceramic fibers. Most are based on stable oxide compounds, where the coarse oxide particles are bonded by finer refractory material. Various carbides, nitrides, and borides can also be used in refractory applications.

Three distinct classes of refractories can be identified: *acidic*, *basic,* and *neutral*. Common acidic refractories are based on silica (SiO_2) and alumina (Al_2O_3) and can be compounded to provide high-temperature resistance along with high hardness and good mechanical properties. (*Note:* Machinable silica ceramics have been used as insulating tiles on the U.S. space shuttle.) Magnesium oxide (MgO) is the core material for most basic refractories. These are generally more expensive than the acidic materials but are often required in metal-processing applications to provide compatibility with the metal. Neutral refractories, containing chromite (Cr_2O_3), are often used to separate the acidic and basic materials since they tend to attack one another. The combination is often attractive when a basic refractory is necessary on the surface for chemical reasons, and the cheaper, acidic material is used beneath to provide strength and insulation. Figure 8-7 shows a variety of high-strength alumina components.

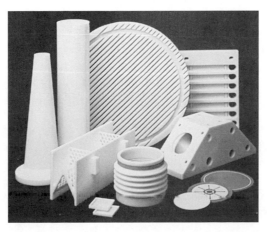

FIGURE 8-7   A variety of high-strength alumina (acid refractory) components, including a filter for molten metal. *(Courtesy of Wesgo Division, GTE.)*

ABRASIVES

Because of their high hardness, ceramic materials, such as silicon carbide and aluminum oxide (alumina), are often used for abrasive applications, such as grinding. Materials such as manufactured diamond and cubic boron nitride have such phenomenal properties that they are often termed *superabrasives*. Materials used for abrasive applications are discussed in greater detail in Chapter 27.

CERAMICS FOR ELECTRICAL AND MAGNETIC APPLICATIONS

Ceramic materials also have a variety of useful electrical and magnetic properties. Some ceramics, such as silicon carbide, are used as resistors and heating elements for electric furnaces. Others have semiconducting properties and are used for thermistors and rectifiers. Dielectric, piezoelectric, and ferroelectric behavior can also be utilized in many applications. Barium titanate, for example, is used in capacitors and transducers. High-density clay-based ceramics and aluminum oxide make excellent high-voltage insulators. The magnetic ferrites have been used in a number of magnetic applications. Considerable attention has also been directed toward the ceramic superconductors.

GLASSES

When some molten ceramics are cooled at a rate that exceeds a critical value, the material solidifies into a rigid, noncrystalline (i.e., amorphous) solid, known as a *glass*. Most commercial glasses are based on silica (SiO_2) with additives to alter the structure or reduce the melting point. Various chemistries can be used to optimize optical properties, thermal stability, and resistance to thermal shock.

Glass is soft and moldable when hot, making shaping rather straightforward. When cool and solid, glass is strong in compression, but brittle and weak in tension. Traditional applications include automotive and window glass, bottles and other containers, light bulbs, and fiberglass insulation. There is also a wide variety of specialty applications, including glass fiber for fiber-optic communications, glass fiber to reinforce composites, cookware, TV tubes and monitors, and a variety of medical and biological products.

GLASS CERAMICS

These materials are first shaped as a glass and then heat treated to promote devitrification or recrystallization of the material, resulting in a structure that contains large amounts of crystalline material within an amorphous base. The crystalline phase helps to retard creep at high temperatures. Strength is greater than with the traditional glasses, and the thermal expansion coefficient is near zero, thereby providing good resistance to thermal shock. The white Pyroceram (trade name) material commonly found in Corningware is a common example of a glass ceramic.

CERMETS

Cermets are combinations of metals and ceramics (usually oxides, carbides, nitrides, or carbonitrides), united into a single product by the procedures of powder metallurgy. This usually involves pressing mixed powders at pressures ranging from 70 to 280 MPa (10 to 40 ksi) followed by sintering in a controlled-atmosphere furnace at about 1650°C (3000°F). Cermets combine the high hardness and refractory characteristics of ceramics with the toughness and thermal shock resistance of metals. They are used as crucibles, jet engine nozzles, and aircraft brakes, and in other applications requiring hardness, strength, and toughness at elevated temperature. One example of the latter is the cutting tool, where cermets enable higher cutting speeds than those achievable with high-speed tool steel, tungsten carbide, or the coated carbides. See Chapter 22.

CERAMICS FOR MECHANICAL APPLICATIONS: THE STRUCTURAL AND ADVANCED CERAMICS

Because of the strong ionic or covalent bonding and high shear resistance, ceramic materials tend to have low ductility and high compressive strength. Theoretically, ceramics could also have high tensile strengths. However, because of their high melting points and lack of ductility, most ceramics are processed in the solid state, and products are made from powdered material. After high-pressure compaction, voids remain between the powder particles, and a portion of these persist through sintering. Contamination can also occur on particle surfaces and become part of the internal structure of the product. As a result, small cracks, pores, and impurities tend to be an integral part of most ceramics and act as stress concentrators. When loads are applied, the effect of these flaws cannot be reduced through plastic flow, and the result is a brittle fracture. Applying the principles of fracture mechanics, we find that ceramics are sensitive to very small flaws. Failures typically occur at tensile stress values between 20 and 210 MPa (3 and 30 ksi), more than an order of magnitude less than the corresponding compressive strength.

Since the size, number, shape, and location of the flaws are likely to differ from part to part, ceramic parts produced from identical material by identical methods often fail at very different applied loads. As a result, the mechanical properties of ceramic products tend to follow a statistical spread that is much less predictable than for metals. This feature tends to limit the use of ceramics in critical high-strength applications.

If the various flaws and defects could be eliminated or reduced to very small size, however, high and consistent tensile strengths could be obtained. Hardness, wear resistance, and strength at elevated temperatures would be attractive properties, along with light weight (specific gravities of 2.3 to 3.85), high stiffness, dimensional stability, low thermal conductivity, corrosion resistance, and chemical inertness. Reliability might still be low, however, and failure would still occur by brittle fracture with little, if any, prior warning. Because of the poor thermal conductivity, thermal shock may be a problem. The cost of these "flaw-free" or "restricted flaw" materials would be rather high. Joining to other engineering materials and machining would be extremely difficult, so products would have to be fabricated through the use of net-shape processing.

Advanced or structural ceramics is an emerging technology with a broad base of current and potential applications and products that are characterized by high strength, high fracture toughness, fine grain size and little or no porosity. The base materials currently include silicon nitride, silicon carbide, partially stabilized zirconia, transformation-toughened zirconia, alumina, sialons, boron carbide, boron nitride, titanium diboride, and ceramic composites (such as ceramic fibers in a glass, glass–ceramic, or ceramic matrix). Applications include a wide variety of wear-resistant parts (including bearings, seals, valves, and dies), cutting tools, punches, dies, and engine components, as well as use in heat exchangers and furnaces. Porous products have been used as substrate material for catalytic converters and as filters for streams of molten metal. Biocompatible ceramics have been used as substitutes for joints and bones, and as dental implants.

Alumina (or aluminum oxide) ceramics are the most common for industrial applications. They are relatively inexpensive and offer high hardness and abrasion resistance, low density, and high electrical resistivity. Alumina is strong in compression and retains useful properties at high temperatures, but it is limited by low toughness and low tensile strength. Due to its high melting point, it is generally processed in a powder form.

Silicon carbide and silicon nitride offer excellent strength and wear resistance with moderate toughness. They work well in high-stress, high-temperature applications, such as turbine blades, and may well replace nickel- or cobalt-based superalloys. Figure 8-8 shows gas-turbine

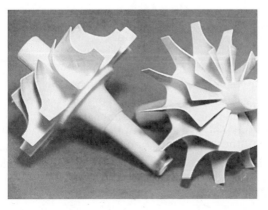

FIGURE 8-8 Gas-turbine rotors made of silicon nitride. The lightweight material (one-half the weight of stainless steel) offers strength at elevated temperature as well as excellent resistance to corrosion and thermal shock. *(Courtesy of Wesgo Division, GTE.)*

TABLE 8-5.   Properties of Some Structural Ceramics

Material	Density (g/cm³)	Tensile Strength (ksi)	Compressive Strength (ksi)	Modulus of Elasticity (10^6 psi)	Fracture Toughness (ksi $\sqrt{\text{in.}}$)
Al_2O_3	3.98	30	400	56	5
Sialon	3.25	60	500	45	9
SiC	3.1	25	560	60	4
ZrO_2 Partially stabilized	5.8	65	270	30	10
ZrO_2 Transformation toughened	5.8	50	250	29	11
Si_3N_4 (hot pressed)	3.2	80	500	45	5

FIGURE 8-9   A variety of components manufactured from silicon nitride, including an exhaust valve and turbine blade. *(Courtesy of Wesgo Division, GTE.)*

rotors made from injection-molded silicon nitride. They are designed to operate at 1250°C (2300°F), where the material retains over half of its room-temperature strength and does not require external cooling. Figure 8-9 shows some additional silicon nitride products. Table 8-5 compares the mechanical properties of some of the structural ceramics.

Sialon (a *si*licon–*al*uminum–*o*xygen–*n*itrogen structural ceramic) is really a solid solution of alumina and silicon nitride, and it bridges the gap between them. More aluminum oxide enhances hardness, while more silicon nitride improves toughness. The resulting material is stronger than steel, extremely hard; and as light as aluminum. It has good resistance to corrosion, wear, and thermal shock; is an electrical insulator; and retains good tensile and compressive strength up to 1400°C (2550°F). It has excellent dimensional stability, with a coefficient of thermal expansion that is only one-third that of steel and one-tenth that of plastic. When overloaded, however, it exhibits the ceramic property of failure by brittle fracture.

Partially stabilized zirconia combines the resistance to thermal shock, wear, and corrosion; the low thermal conductivity; and the low friction coefficient of zirconia with the enhanced strength and toughness brought about by doping the material with oxides of calcium, yttrium, or magnesium. Transformation-toughened zirconia has even greater toughness as a result of dispersed second phases throughout the ceramic matrix. When a crack approaches the metastable phase, it transforms to a more stable structure, increasing in volume to compress and stop the crack.

The high cost of the structural ceramics continues to be a barrier to their widespread acceptance. High-grade ceramics are currently several times more expensive than their metal counterparts. Even factoring in enhanced lifetime and improved performance, there is still a need to reduce cost. Work continues, however, toward the development of a low-cost, high-strength, high-toughness ceramic with a useful temperature range. Parallel efforts are under way to assure flaw detection in the range of 10 to 50 mm. If these efforts are successful, ceramics could compete where tool steels, powdered metals, coated materials, and tungsten carbide are now being used. Potential applications include engines, turbochargers, gas turbines, bearings, pump and valve seals, and other products that operate under high-temperature, high-stress environments.

ADVANCED CERAMICS AS CUTTING TOOLS

The high hardness, retention of hardness at elevated temperature, and low reactivity with metals makes ceramic materials attractive for cutting applications, and cutting tools have improved significantly through advances in ceramic technology. Silicon carbide is a common abrasive in many grinding wheels. Cobalt-bonded tungsten carbide has been a popular alternative to high-speed tool steels for many tool and die applications. Many carbide tools are now enhanced by a variety of vapor-deposited ceramic coatings. Thin layers of titanium carbide, titanium nitride, and aluminum oxide can stop the interaction between the metal being cut and the binder phase of the carbide. This results in a significant reduction in friction and wear and enables faster rates of cutting. Silicon nitride, boron carbide, cubic boron nitride, and polycrystalline diamond cutting tools now offer even greater tool life, higher cutting speeds, and reduced machine downtime. With advanced tool materials, cutting speeds

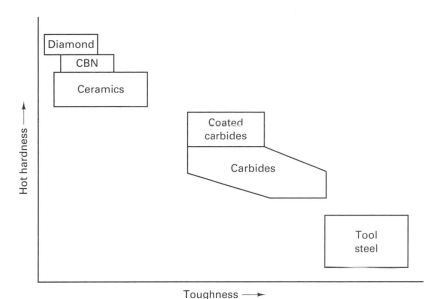

FIGURE 8-10 Graphical mapping of the combined toughness and hardness for a variety of cutting-tool materials.

can be increased from 60 to 1500 m/min (200 to 5000 ft/min). Use of these ultrahigh-speed materials, however, will require companion developments in the area of machine tools. High-speed spindles must be perfectly balanced, and workholding devices must withstand the high centrifugal forces. Chip-removal methods must be developed that can remove the chips as fast as they are formed.

With the enactment of more stringent environmental legislation, dry machining may be pursued to reduce or eliminate coolant and lubricant-disposal problems. Ceramic materials are currently the best materials for dry operation. Yet another application of ceramic tools is direct machining of materials that once required grinding, a process sometimes called *hard machining*. Figure 8-10 shows the combination of toughness and hardness for a variety of cutting-tool materials.

■ 8.5 COMPOSITE MATERIALS

A *composite material* is a nonuniform solid consisting of two or more different materials that are mechanically or metallurgically bonded together. Each of the various components retains its identity in the composite and maintains its characteristic structure and properties. There are recognizable interfaces between the materials. The composite material, however, generally possesses characteristic properties (or combinations of properties), such as stiffness, strength, weight, high-temperature performance, corrosion resistance, hardness, and conductivity, that are not possible with the individual components by themselves. Analysis of these properties shows that they depend on (1) the properties of the individual components; (2) the relative amounts of the components; (3) the size, shape, and distribution of the discontinuous components; (4) the orientation of the various components; and (5) the degree of bonding between the components. The materials involved can be organics, metals, or ceramics. Hence a wide range of freedom exists, and composite materials can often be designed to meet a desired set of engineering properties and characteristics.

There are many types of composite materials and several methods of classifying them. One method is based on geometry and consists of three distinct families: laminar or layered composites, particulate composites, and fiber-reinforced composites.

LAMINAR OR LAYERED COMPOSITES

Laminar composites are those having distinct layers of material bonded together in some manner and include thin coatings, thicker protective surfaces, claddings, bimetallics, laminates, sandwiches, and others. Plywood is probably the most common engineering material in this category and is an example of a laminate material. Layers of wood veneer are adhesively bonded with their grain orientations at various angles to one another.

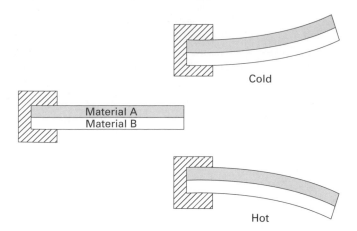

FIGURE 8-11 Schematic of a bimetallic strip where material A has the greater coefficient of thermal expansion.

Strength and fracture resistance are improved, properties are somewhat uniform within the plane of the sheet, swelling and shrinkage tendencies are minimized, and large pieces are available at reasonable cost. Safety glass is another laminate in which a layer of polymeric adhesive is placed between two pieces of glass and serves to retain the fragments when the glass is broken. *Ar*amid–*Al*uminum *L*aminates (Arall) consist of thin sheets of aluminum bonded with woven adhesive-impregnated aramid fibers. The combination offers light weight coupled with high fracture, impact, and fatigue resistance. Other laminates have become popular in decorative items, such as Formica countertops, imitation hardwood flooring, and furniture.

Bimetallic strip is a laminate of two metals with significantly different coefficients of thermal expansion. Changes in temperature produce flexing or curvature in the product, which may be employed in thermostat and other heat-sensing applications. Figure 8-11 illustrates this effect. Still other laminar composites are designed to provide enhanced surface characteristics while retaining a low-cost, high-strength, or lightweight core. Many clad materials fit this description. Alclad metal, for example, consists of high-strength, heat-treatable aluminum with an exterior cladding of one of the more corrosion-resistant, non-heat-treatable aluminum alloys. U.S. coinage is another laminate, designed to conserve the more costly, high-nickel-content material while providing a lustrous, corrosion-resistant surface. Other laminates have surface layers that have been selected primarily for enhanced wear resistance or improved appearance.

Sandwich material is a laminar structure composed of a thick, low-density core placed between thin, high-density surfaces. Corrugated cardboard is an example of a sandwich structure. Other engineering sandwiches incorporate cores of a polymer foam or honeycomb structure, to produce a lightweight, high-strength, high-rigidity composite.

PARTICULATE COMPOSITES

Particulate composites consist of discrete particles of one material surrounded by a matrix of another material. Concrete is a classic example, consisting of sand and gravel particles surrounded by cement. Asphalt consists of similar aggregate in a matrix of bitumin, a thermoplastic polymer. In both of these examples, the particles are rather coarse. Other particulate composites involve extremely fine particles and include many of the multi-component powder metallurgy products, specifically those where the dispersed particles do not diffuse into the matrix material.

Dispersion-strengthened materials are particulate composites where a small amount of hard, brittle, small-size particles (typically, oxides or carbides) are dispersed throughout a softer, more ductile metal matrix. Pronounced strengthening can be induced, which decreases gradually as temperature is increased. Creep resistance, therefore, is improved significantly. Examples of dispersion-strengthened materials include sintered aluminum powder (SAP)[2], which consists of an aluminum matrix strengthened by up to 14% aluminum oxide, and TD-nickel[3], a nickel alloy containing 1 to 2 wt% thoria (ThO_2).

---

[2]Sintered aluminum powder
[3]Thoria-dispersed nickel

Other types of particulate composites, known as *true particulate composites*, contain large amounts of coarse particles and are usually designed to produce some desired combination of properties rather than increased strength. Cemented carbides, for example, consist of hard ceramic particles, such as tungsten carbide, tantalum carbide, or titanium carbide, embedded in a metal matrix, which is usually cobalt. Although the hard, stiff carbide could withstand the high temperatures and pressure of cutting, it is extremely brittle. Toughness is imparted by combining the carbide particles with cobalt powder, pressing the material into the desired shape, heating to melt the cobalt, and then resolidifying the compacted material. Varying levels of toughness can be imparted by varying the amount of cobalt in the composite.

Grinding and cutting wheels are often formed by bonding abrasives, such as alumina (Al_2O_3), silicon carbide (SiC), cubic boron nitride (CBN), or diamond, in a matrix of glass or polymeric material. As the hard particles wear, they fracture or pull out of the matrix, exposing new cutting edges. By combining tungsten powder and powdered silver or copper, electrical contacts can be produced that offer both high conductivity and resistance to wear and arc erosion. Foundry molds and cores are often made from sand (particles) and an organic or inorganic binder (matrix). Particulate-type metal-matrix composites are made by introducing a variety of ceramic or glass particles into aluminum or magnesium matrices. Particulate-toughened ceramics using zirconia and alumina matrices are being used as bearings, bushings, valve seats, die inserts, and cutting-tool inserts. In addition, many plastics might be considered to be particulate composites because the additive fillers and extenders can be viewed as dispersed particles. Designation as a composite, however, is usually reserved for polymers where the particles are added for the primary purpose of property modification. One such example is the combination of granite particles in an epoxy matrix currently being used in some machine tool bases. This unique material offers high strength and a vibration damping capacity that exceeds that of gray cast iron.

Because of their unique geometry, the properties of particulate composites are usually *isotropic*, that is, uniform in all directions. This may be particularly important in engineering applications.

FIBER-REINFORCED COMPOSITES

The most popular type of composite material is the *fiber-reinforced composite* geometry, where continuous or discontinuous thin fibers of one material are embedded in a matrix of another. The matrix supports and transmits forces to the fibers, protects them from environments and handling, and provides ductility and toughness, while the fibers carry most of the load and impart enhanced stiffness. Wood and bamboo are two naturally occurring fiber composites, consisting of cellulose fibers in a lignin matrix. Bricks of straw and mud may well have been the first human-made material of this variety, dating back to near 800 B.C. Automobile tires now use fibers of nylon, rayon, aramid (Kevlar), or steel in various numbers and orientations to reinforce the rubber and provide added strength and durability. Steel-reinforced concrete is actually a double composite, consisting of a particulate matrix reinforced with steel fibers.

Glass-fiber reinforced resins, the first of the modern fibrous composites, were developed shortly after World War II in an attempt to produce lightweight materials with high strength and high stiffness. Glass fibers about 10μm in diameter are bonded in a variety of polymers, generally epoxy or polyester resins. Between 30 and 60% by volume is made up of fibers of either E-type borosilicate glass (tensile strength of 500 ksi and elastic modulus of 10.5×10^6 psi) or the stronger, stiffer, high-performance S-type magnesia–alumina–silicate glass (with tensile strength of 670 ksi and elastic modulus of 12.4×10^6 psi).

It is important to note that a fiber of material tends to be stronger than the same material in bulk form because the size of any flaw is limited to the diameter of the fiber. Moreover, the complete failure of a given fiber does not propagate through the assembly, as would a flaw in an identical bulk material.

Glass fibers are still the most widely used reinforcement, primarily because of their lower cost and adequate properties for many applications. Current uses of glass-fiber-reinforced plastics include sporting goods, boat hulls and bathtubs. Limitations of

TABLE 8-6. Properties and Characteristics of Some Common Reinforcing Fibers

Fiber Material	Specific Strength[a] (10^6 in.)	Specific Stiffness[b] (10^6 in.)	Density ($lb/in.^3$)	Melting Temperature[c] (°F)
Al_2O_3 whiskers	21.0	434	0.142	3600
Boron	4.7	647	0.085	3690
Ceramic fiber (Mullite)	1.1	200	0.110	5430
E-type glass	5.6	114	0.092	<3140
High-strength graphite	7.4	742	0.054	6690
High-modulus graphite	5.0	1430	0.054	6690
Kevlar	10.1	347	0.052	—
SiC whiskers	26.2	608	0.114	4890

[a]Strength divided by density.   [b]Elastic modulus divided by density.   [c]Or maximum temperature of use.

the glass-fiber material are generally related to strength and stiffness. Alternative fibers have been developed for applications requiring enhanced properties. Boron–tungsten fibers (boron deposited on a tungsten core) offer an elastic modulus of 380,000 MPa (55×10^6 psi) with tensile strengths in excess of 2750 MPa (400 ksi). Silicon carbide filaments (SiC on tungsten) have an even higher modulus of elasticity.

Graphite and aramid (Kevlar) are other popular reinforcing fibers. Graphite fibers can be either the PAN type, produced by the thermal pyrolysis of synthetic organic fibers, primarily polyacrylonitrile, or pitch type, made from petroleum pitch. They have low density and a range of high tensile strengths (4150 to 5175 MPa or 600 to 750 ksi) and high elastic moduli (40,000 to 65,000 MPa or 40 to 65×10^6 psi). Graphite's negative thermal-expansion coefficient can also be used to offset the positive values of most matrix materials. Kevlar is an organic aramid fiber with a tensile strength up to 4500 MPa (650 ksi), 185,000 MPa (27×10^6 psi) elastic modulus, a density approximately one half that of aluminum, and good toughness. In addition, it is flame retardant and transparent to radio signals, making it attractive for a number of military and aerospace applications where the service temperature is not excessive. Table 8-6 lists some of the key engineering properties for several of the common reinforcing fibers.

Within the composite, the reinforcing fibers can be arranged in a variety of orientations. Fiberglass, for example, contains short, randomly oriented fibers. Unidirectional fibers can be used to produce highly directional properties, with the fiber directions being tailored to the direction of loading. Woven fabrics or tapes can be produced and then layered in various orientations to produce a plywood-like product. The layered materials can be stitched together to add a third dimension to the weave, and complex three-dimensional shapes can be woven from fibers and later injected with a matrix material.

The properties of fiber-reinforced composites depend strongly on several characteristics: (1) the properties of the fiber material; (2) the volume fraction of fibers; (3) the aspect ratio of the fibers, that is, the length-to-diameter ratio; (4) the orientation of the fibers; (5) the degree of bonding between the fiber and the matrix; and (6) the properties of the matrix. The matrix materials should be strong, tough, and ductile so that they can transmit the loads to the fibers and prevent cracks from propagating through the composite. Both thermosetting and thermoplastic polymeric resins have been used. Popular thermosets include epoxies, polyesters, bismaleimides, and polyimides. From a manufacturing viewpoint, it is often easier and faster to heat and cool a thermoplastic than to cure a thermoset, and the thermoplastics are tougher and more tolerant to damage. Attractive thermoplastic matrix materials are usually those with good high-temperature and chemical-resistant properties, which includes the thermoplastic polyimides, polyphenylene sulfide (PPS), polyether ether ketone (PEEK), and the liquid-crystal polymers. The matrix material should be selected to match the temperature of operation. Polymeric materials can only be used for temperatures below 315°C (600°F). Above this temperature, metal or ceramic matrices should be considered.

While the attractive properties of the fiber-reinforced composites are usually in the area of performance, the limitations or weaknesses often relate to fabrication. The processing of epoxy resins and fibers requires special precautions against toxic fumes, fiber fragments and fire. The materials are extremely difficult to shape and join and nearly impossible to recycle.

ADVANCED FIBER-REINFORCED COMPOSITES

Advanced composites are materials that have been developed for applications requiring exceptional combinations of strength, stiffness, and light weight. Fiber content generally exceeds 50% (by weight), and the modulus of elasticity is typically greater than 110 GPa (16×10^6 psi). Superior creep and fatigue resistance, low thermal expansion, low friction and wear, vibration-damping characteristics, and environmental stability are other properties that may also be required in these materials.

There are four basic types of advanced composites where the matrix material is matched to the fiber and application:

1. The advanced *organic or resin-matrix composites* frequently use high-strength, high-modulus fibers of graphite, aramid (Kevlar), or boron. Properties can be put in desired locations or orientations at about one-half the weight of aluminum (or one-sixth that of steel). Thermal expansion can be designed to be low or even negative. Unfortunately, these materials have a maximum service temperature of about 315°C (600°F) because the polymeric matrix loses strength when heated. Table 8-7 compares the properties of some of the common resin-matrix composites with those of several of the lightweight or low-thermal-expansion metals. Typical applications include sporting equipment (tennis rackets, skis, golf clubs, and fishing poles), lightweight armor plate, and a myriad of low-temperature aerospace components.

2. *Metal-matrix composites* (MMCs) can be used for operating temperatures up to 1250°C (2300°F), where the conditions require high strength and/or high stiffness, good thermal conductivity, exceptional wear resistance, and good ductility and toughness. The ductile matrix material can be aluminum, copper, magnesium, titanium, nickel, superalloy, or even intermetallic compound, and the reinforcing fibers may be graphite, boron carbide, alumina, or silicon carbide. Fine whiskers (tiny needlelike single crystals of 1 to 10 μm in diameter) of sapphire, silicon carbide, and silicon nitride have also been used as the reinforcement, as well as wires of titanium, tungsten, molybdenum, beryllium, and stainless steel. The reinforcing fibers may be either continuous or discontinuous, and typically comprise between 10 and 60% of the composite by volume.

   Compared to the engineering metals, these composites offer higher stiffness and strength (especially at elevated temperatures), a lower coefficient of thermal expansion, and enhanced resistance to fatigue, abrasion, and wear. Compared to the organic matrix materials, they offer higher heat resistance, as well as improved electrical and thermal conductivity. They are nonflammable, do not absorb water or gases, and are corrosion resistant to fuels and solvents. Unfortunately, these materials are quite expensive, the

---

TABLE 8-7.   Properties of Several Fiber-Reinforced Composites (in the Fiber Direction) Compared to Those of Lightweight or Low-Thermal-Expansion Metals

Material	Specific Strength[a] (10^6 in.)	Specific Stiffness[b] (10^6 in.)	Density (lb/in.^{3})	Thermal Expansion Coefficient [in./(in.-°F)]	Thermal Conductivity [Btu/(hr-ft-°F)]
Boron–epoxy	3.3	457	0.07	2.2	1.1
Glass–epoxy (woven cloth)	0.7	45	0.065	6	0.1
Graphite–epoxy: high modulus (unidirectional)	2.1	700	0.063	−0.5	75
Graphite–epoxy: high strength (unidirectional)	5.4	400	0.056	−0.3	3
Kevlar–epoxy (woven cloth)	1	80	0.5	1	0.5
Aluminum	0.7	100	0.10	13	100
Beryllium	1.1	700	0.07	7.5	120
Invar[c]	0.2	70	0.29	1	6
Titanium	0.8	100	0.16	5	4

[a] Strength divided by density.   [b] Elastic modulus divided by density.   [c] A low-expansion metal containing 36% Ni and 64% Fe.

vastly different thermal expansions of the components may lead to debonding, and the assemblies may be prone to degradation through interdiffusion or galvanic corrosion.

Graphite-reinforced aluminum can be designed to have near-zero thermal expansion in the fiber direction. Aluminum oxide-reinforced aluminum has been used in automotive connecting rods to provide stiffness and fatigue resistance with lighter weight. Aluminum reinforced with silicon carbide has been fabricated into automotive drive shafts, cylinder liners, and brake drums, as well as aircraft wing panels, all offering significant weight savings. Fiber-reinforced superalloys may well become a preferred material for applications such as turbine blades.

3. *Carbon–carbon composites* (graphite fibers in a graphite or carbon matrix) offer the possibility of a heat-resistant material that could operate at temperatures above 2000°C (3600°F), with a strength that is 20 times that of conventional graphite, a density that is 30% lighter (1.38 g/cm^3), and a low coefficient of thermal expansion. Not only does this material withstand high temperatures, it actually gets stronger when heated. Companion properties include good toughness, good thermal and electrical conductivity, and resistance to corrosion and abrasion. For temperatures over 540°C (1000°F), however, the composite requires some form of coating to protect it from oxidizing. Various coatings can be used for different temperature ranges. Current applications include the nose cone and leading edge of the space shuttle, aircraft and racing car disc brakes, automotive clutches, aerospace turbines and jet engine components, rocket nozzles, and surgical implants.

4. *Ceramic-matrix composites* (CMCs) offer light weight, high-temperature strength and stiffness, and good dimensional and environmental stability. The matrix provides high-temperature resistance. Glass matrices can operate at temperatures as high as 1500°C (2700°F). The crystalline ceramics, usually based on alumina, silicon carbide, silicon nitride, boron nitride, titanium diboride, or zirconia, can be used for even hotter conditions. The fibers add directional strength, increase fracture toughness, improve thermal shock resistance, and can be incorporated in unwoven, woven, knitted, and braided form. Typical reinforcements include carbon fiber, glass fiber, fibers of the various matrix materials, and ceramic whiskers. Composites with discontinuous fibers tend to be used primarily for wear applications, such as cutting tools, forming dies, and automotive parts such as valve guides. Other applications include lightweight armor plate and radomes. Continuous-fiber ceramic composites are used more for applications involving the combination of high temperatures and high stresses and have been shown to fail in a noncatastrophic manner. Application examples include gas-turbine components, high-pressure heat exchangers, and high-temperature filters. Unfortunately, the cost of ceramic–ceramic composites ranges from high to extremely high, so applications will be restricted to those where the benefits are quite attractive.

HYBRID COMPOSITES

Hybrid composites involve two or more different types of fibers in a common matrix. The particular combination of fibers is usually selected to balance strength and stiffness, provide dimensional stability, reduce cost, reduce weight, or improve fatigue and fracture resistance. Types of hybrid composites include (1) interply (alternating layers of fibers), (2) intraply (mixed strands in the same layer), (3) interply–intraply, (4) selected placement (where the more costly material is used only where needed), and (5) interply knitting (where plies of one fiber are stitched together with fibers of another type).

DESIGN AND FABRICATION

The design of composite materials involves the selection of the component materials; the determination of the relative amounts of each component; the determination of size, shape, distribution, and orientation of the components; and the selection of an appropriate fabrication method. Many of the possible fabrication methods have been specifically developed for use with composite materials. For example, fibrous composites can be manufactured into useful shapes through compression molding, filament winding, pultrusion (where bundles of coated fibers are drawn through a heated die), cloth lamination, and autoclave curing (where pressure and elevated temperature are applied simultaneously). Fiber-reinforced plastics can be injection-molded to produce products that compete with zinc die castings. Sheets of glass-fiber-reinforced plastic (sheet-molding compounds) can be press formed to provide lightweight, corrosion-resistant products that are similar to those made from sheet metal. One current application of this technology is the cargo beds for pickup trucks, where the composite product offers reduced weight coupled with resistance to dents, scratches, and corrosion. With a variety of materials, geometries, and processes, it is now possible to tailor a composite material product for a specific application.

A significant portion of Chapter 20 is devoted to a more complete description of the fabrication methods that have been developed for composite materials.

ASSETS AND LIMITATIONS

Figure 8-12 graphically presents the strength-to-weight ratios of various aerospace materials as a function of temperature. The superiority of the various advanced composites over the conventional aerospace metals is clearly evident. The weight of a graphite–epoxy composite I-beam is less than one-fifth that of steel, one-third that of titanium, and one-half that of aluminum. Its ultimate tensile strength equals or exceeds that of the other three materials, and it possesses an almost infinite fatigue life. The greatest limitations of this and other composites are their relative brittleness and the high cost of both materials and fabrication.

While there has been considerable advancement in the field, manufacturing with composites can still be quite labor intensive, and there is a persistent lack of trained designers, established design guidelines and data, information about fabrication costs, and reliable methods of quality control and inspection. It is often difficult to predict the interfacial bond strength, the strength of the composite and its response to impacts, and the probable modes of failure. Defects can involve delaminations, voids, missing layers, contamination, fiber breakage, and (hard-to-detect) improperly cured resin. There is often concern about heat resistance. Many composites with polymeric matrices are sensitive to moisture, acids, chlorides, organic solvents, oils, and ultraviolet radiation, and tend to cure forever, causing continually changing properties. In addition, most composites have limited ability to be repaired if damaged, preventive maintenance procedures are not well established, and recycling is often extremely difficult. Assembly operations with composites generally require the use of industrial adhesives.

FIGURE 8-12   Graphical depiction of the strength/weight ratio of various aerospace materials as a function of temperature. Note the role of the various advanced fiber-reinforced composites. *(Adapted with permission of DuPont Company.)*

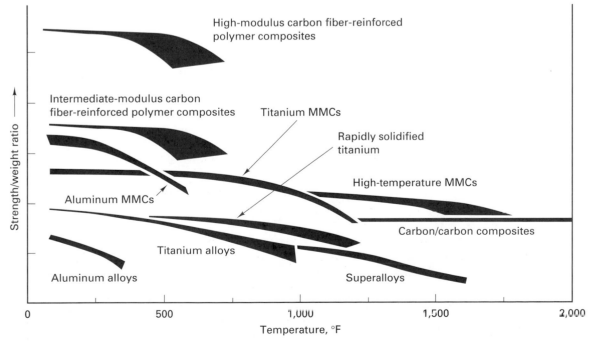

On the positive side, the availability of a corrosion-resistant material with strength and stiffness greater than those of steel at only one-fifth the weight may be sufficient to justify some engineering compromises. In addition, products can often be designed to significantly reduce the number of parts, number of fasteners, assembly time, and cost.

AREAS OF APPLICATION

Many composite materials are stronger than steel, lighter than aluminum, and stiffer than titanium. They can also possess low thermal conductivity, good heat resistance, good fatigue life, low corrosion rates, and adequate wear resistance. For these reasons they have become well established in several areas.

Aerospace applications frequently require light weight, stiffness, and fatigue resistance, and composites may well account for a considerable fraction of the weight of a current airplane design. Figure 8-13 shows a schematic of the F-22 Raptor, proposed to replace the F-15 as the U.S. Air Force fighter in 2004. Traditional materials, such as aluminum and steel, make up about 20% of the F-22 structure by weight. However, its higher speed, longer range, greater agility, and reduced detectability are made possible through the use of 42% titanium and 24% composite material.

Sports are highly competitive, and victories are often decided by fractions of a second or tenths of a millimeter. As a result, both professionals and amateurs are willing to invest in equipment that will improve performance. The materials of choice have evolved from naturally occurring wood, twine, gut, and rubber to a wide variety of high-technology metals, polymers, ceramics, and composites. Golf club shafts, baseball bats, fishing rods, tennis rackets, bicycle frames, and skis are now available in a wide variety of fibrous composites. Figure 8-14 shows several of these applications.

Potential automotive uses, in addition to body panels, include drive shafts, springs, and bumpers. Weight savings compared to existing parts is generally 20 to 25%. Truck manufacturers now use fiber-reinforced composites for cab shells and bodies, oil pans, fan shrouds, instrument panels, and engine covers.

FIGURE 8-13   Schematic diagram showing the materials used in the various sections of the F-22 Raptor. Traditional materials, such as aluminum and steel, comprise only 20% by weight. Titanium accounts for 42%, and 24% is composite material. The plane is capable of flying at Mach 2. *NOTE:* RTM is resin-transfer molding. *(Reprinted with permission of ASM International.)*

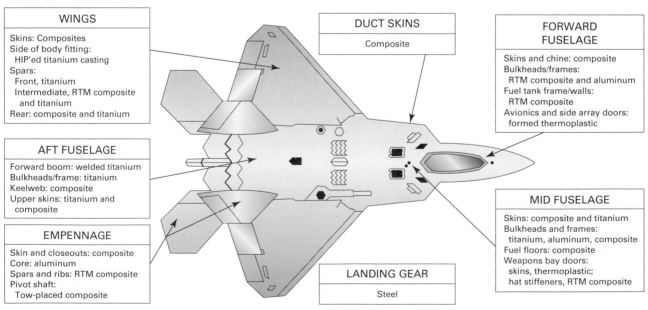

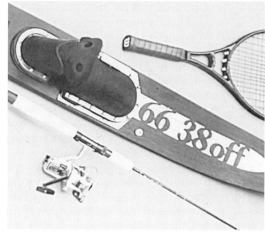

FIGURE 8-14 Composite materials are often used in sporting goods to improve performance through light weight, high stiffness, and high strength, and also to provide attractive styling. *(Courtesy of Fiberite Corporation.)*

■ KEY WORDS

abrasive	clay	elastomer	metal-matrix composite	terpolymer
addition polymerization	composite	fiber-reinforced	oriented plastics	thermoplastic
additive agents	condensation polymerization	composite	particulate composite	thermosetting
advanced ceramic	copolymer	fillers	plastic	unsaturated
amorphous	cross-linked	glass	plasticizer	monomer
carbon–carbon composite	crystallize	hybrid composite	polymer	whiteware
ceramic	degree of polymerization	isomer	refractory material	
ceramic-matrix composite	dispersion-strengthened	laminar composite	saturated monomer	
cermet	material	mer	superabrasive	

■ REVIEW QUESTIONS

1. What are some naturally occurring nonmetallic materials that have been used for engineering applications?
2. What are some material families that would be classified under the general term *nonmetallic engineering materials*?
3. How might plastics be defined from the viewpoints of chemistry, structure, fabrication, and processing?
4. What is the primary type of atomic bonding within polymers?
5. What is the difference between a saturated and an unsaturated molecule?
6. What is an isomer?
7. Describe and differentiate the two means of forming polymers: addition and condensation.
8. What is degree of polymerization?
9. Describe and differentiate thermoplastic and thermosetting plastics.
10. What does it mean when a polymer "crystallizes"?
11. What are some of the ways that a thermoplastic polymer can be made stronger?
12. Why are thermosetting polymers characteristically brittle?
13. How do thermosetting polymers respond to subsequent heating?
14. Describe how thermoplastic or thermosetting characteristics affect productivity during the fabrication of a molded part.
15. What are some attractive engineering properties of polymeric materials?
16. What are some limiting properties of plastics, and in what general area do they fall?

17. What are some environmental conditions that might adversely affect the engineering properties plastics?
18. What are some reasons that additive agents are incorporated into plastics?
19. What are some functions of a filler material?
20. What are some of the more common filler materials used in plastics?
21. What is the role of a stabilizer or antioxidant?
22. What is the primary engineering benefit of an oriented plastic?
23. What are some properties and characteristics of the "engineering plastics"?
24. What are some products for which plastics have captured segments of markets traditionally held by other materials?
25. Which type of plastic is most easily recycled?
26. Why is the recycling of mixed plastics more difficult than the recycling of mixed metals?
27. What is the unique mechanical property of elastomeric materials, and what structural feature is responsible for it?
28. How can cross-linking be used to control the engineering properties of elastomers?
29. What are some of the limitations of natural rubber?
30. What are some outstanding physical properties of ceramic materials?
31. Why are the crystal structures of ceramics frequently more complex than those observed for metals?
32. What is the common name given for ceramic material in the noncrystalline, or amorphous, state?

33. What is the dominant property of refractory ceramics?
34. What is the dominant property of ceramic abrasives?
35. How are glass products formed or shaped?
36. What are cermets, and what properties or combination of properties do they offer?
37. Why do most ceramic materials fail to possess their theoretically high tensile strength?
38. Why do the mechanical properties of ceramics generally show a wider statistical spread than the same properties of metals?
39. If all significant flaws or defects could be eliminated from the structural ceramics, what features might still limit their possible applications?
40. What are some specific materials that are classified as structural ceramics?
41. What are some attractive and limiting properties of sialon (one of the structural ceramics)?
42. What are some ceramic materials that are currently being used for cutting-tool applications, and what features or properties make them attractive?
43. What is a composite material?
44. What are the basic structural features of a composite material that influence and determine its properties?
45. What are the three primary geometries of composite materials?

46. What feature in a bimetallic strip makes its shape sensitive to temperature?
47. What is the attractive aspect to the strength that is induced by the particles in a dispersion-strengthened particulate composite material?
48. Which of the three primary composite geometries is most likely to possess isotropic properties?
49. What is the primary role of the matrix in a fiber-reinforced composite? Of the fibers?
50. What are some of the more popular fiber materials used in fiber-reinforced composite materials?
51. What are some possible fiber orientations or arrangements in a fiber-reinforced composite material?
52. What are some features that influence the properties of fiber-reinforced composites?
53. In what ways are metal-matrix composites superior to straight engineering metals? To organic-matrix composites?
54. What features might be imparted by the fibers in a ceramic-matrix composite?
55. What are some major limitations to the extensive use of composite materials in engineering applications?
56. What are some properties of composite materials that make them attractive for aerospace applications?

■ PROBLEMS

1. a. One of Leonardo da Vinci's sketchbooks contains a crude sketch of an underwater boat (or submarine). Leonardo did not attempt to develop or refine this sketch further, possibly because he recognized that the engineering materials of his day (wood, stone, and leather) were inadequate for the task. What properties would be required for the body of a submersible vehicle? What materials might you consider?
   b. Another of Leonardo's sketches bears a crude resemblance to a helicopter—a flying machine. What properties would be desirable in a material that would be used for this type of application?
   c. Try to identify a possible engineering product that would require a material with properties that do not exist among today's engineering materials. For your application, what are the demanding features or requirements? If a material were to be developed for this application, from what family or group do you think it would emerge? Why?
2. Select a product (or component of a product) that can reasonably be made from materials from two or more of the basic materials families (metals, polymers, ceramics, and composites).
   a. Describe briefly the function of the product or component.
   b. What properties would be required for this product or component to perform its function?
   c. What two materials groups might provide reasonable candidates for your product?
   d. Select a candidate material from the first of your two families and describe its characteristics. In what ways does it meet your requirements? How might it fall short of the needs?

   e. Repeat part (d) for a candidate material from the second material family.
   f. Compare the two materials to one another. Which of the two would you prefer? Why?
3. Coatings have been applied to cutting tool materials since the early 1970s. Desirable properties of these coatings include high-temperature stability, chemical stability, low coefficient of friction, high hardness for edge retention, and good resistance to abrasive wear. Consider the ceramic material coatings of titanium carbide (TiC), titanium nitride (TiN), and aluminum oxide (Al_2O_3) and compare them with respect to the conditions required for deposition and the performance of the resulting coatings.
4. Ceramic engines continue to constitute an area of considerable interest and are frequently discussed in the popular literature. If perfected, they would allow higher operating temperatures with a companion increase in engine efficiency. In addition, they would lower sliding friction and permit the elimination of radiators, fan belts, cooling system pumps, coolant lines, and coolant. The net result would be reduced weight and a more compact design. Estimated fuel savings could amount to 30% or more.
   a. What are the primary limitations to the successful manufacture of such a product?
   b. What types of ceramic materials would you consider to be appropriate?
   c. What methods of fabrication could produce a product of the required size and shape?
   d. What types of special material properties or special processing might be required?

www.wiley.com/college/degarmo

*C*hapter 8 CASE STUDY

Two-Wheel Dolly Handles

The items illustrated in Figure CS-8 are the handle grips for an industrial-quality, pneumatic-tire, two-wheel dolly. They are designed to be bolted onto box-channel tubular sections using four bolt-holes, which are sized to accommodate $\frac{3}{8}$-inch diameter bolts. The major service requirements are strength, durability, fracture resistance, reasonable appearance, and possibly light weight.

Your employer is currently marketing such a dolly using handles that are made as permanent-mold aluminum castings. The firm is in the process of updating its line and is reevaluating the design and manufacture of each of its products. The dolly is part of your assignment, and you have been asked specifically to determine whether the handles should be replaced by an alternative material, such as a polymer or low-cost composite.

Investigate the properties and cost* of alternative materials, including means of fabricating the desired shape, and make your recommendation. Since the existing design was for cast metal, you might want to make minor modifications. Make sure the alternative materials possess adequate properties in the bolt-hole region, and if not, recommend some form of reinforcement.

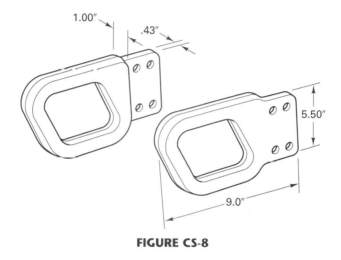

FIGURE CS-8

---

* Since the size of the part will remain relatively unchanged, material costs should be compared on the basis of \$/in³ or \$/cm³ and not \$/lb or \$/kg.

CHAPTER 9

MATERIAL SELECTION

■ 9.1 INTRODUCTION

Manufacturing operations are targeted toward the production of products or components that will adequately perform their intended task. Meeting this objective implies the manufacture of components from selected engineering materials, with the required geometrical shape and precision, and with companion material structures that are optimized for the service environment that the component must withstand. The ideal product is one that will just meet all requirements. Anything better will usually incur added cost through higher-grade materials, enhanced processing, or improved properties that may not be necessary. Anything worse will likely cause product failure, dissatisfied customers, and possible unemployment.

It was not all that long ago that each of the materials groups had their own well-defined uses and markets. Metals were specified when strength, toughness, and durability were the primary requirements. Ceramics were generally limited to low-value applications where heat or chemical resistance was required and any loadings were compressive. Glass was used for its optical transparency, and plastics were relegated to low-value applications where low cost and light weight were attractive features, and performance properties were secondary.

Such clear delineations no longer exist. Many of the metal alloys in use today did not exist as little as 30 years ago, and the common alloys that have been in use for a century or more have been much improved due to advances in metallurgy and production processes. New on the scene are amorphous metals, oxide dispersion-strengthened alloys produced by powder metallurgy, mechanical alloyed product, and directionally solidified materials. Ceramics, polymers, and composites are now available with specific properties that often transcend the traditional limits and boundaries. Advanced structural materials offer higher strength and stiffness, strength at elevated temperature, light weight, and resistance to corrosion, creep, and fatigue. Other materials have enhanced thermal, electrical, optical, magnetic, and chemical properties.

As a result of these dynamic changes, the selection of engineering materials has become extremely important and the process now requires constant reevaluation. New materials are continually being developed, others may no longer be available, and prices are always subject to change. Concerns regarding environmental pollution, recycling, and worker health and safety continue to impose new constraints. Desires for weight reduction, energy savings, or improved corrosion resistance may well motivate a change in engineering material. Pressures from domestic and foreign competition, increased demand for quality and serviceability, and negative customer feedback can all prompt a reevaluation. Finally, the proliferation of product liability actions, many of which are the result of improper material use, has further emphasized the need for constant reevaluation of the engineering materials in a product.

The automotive industry alone consumes approximately 60 million metric tons of engineering materials worldwide every year—primarily steel, cast iron, aluminum, copper, glass, lead, polymers, rubber, and zinc. In recent years, the drive toward lighter weight has led to an increase in the use of the lighter-weight metals and high-strength steels, as well as plastics and composites.

FIGURE 9-1 The Raytheon Premier six-passenger business jet has a fuselage constructed from high strength composite material—a carbon fiber/epoxy honeycomb. *(Courtesy of Raytheon Aircraft Company, Wichita, KS.)*

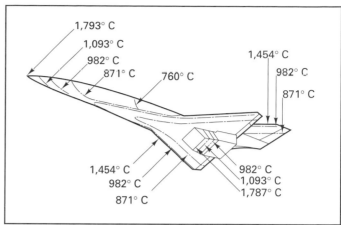

FIGURE 9-2 Predicted maximum temperatures of a transatmospheric hypersonic aircraft. The extreme temperatures coupled with the desire for light weight may well require radically new engineering materials.

A million metric tons of engineering materials go into aerospace applications every year. The principal materials tend to be aluminum, magnesium, titanium, superalloys, polymers, rubber, steel, metal-matrix composites, and polymer-matrix composites. Competition is intense, and materials substitutions are frequent. Aluminum-lithium alloys have replaced conventional aluminums. The use of advanced composite materials in aircraft construction has risen from less than 2% in 1970 to the point where they will now account for one-quarter of the weight of the U.S. Air Force's new F-22 Advanced Tactical Fighter. Titanium is used extensively for applications that include the exterior skins surrounding the engines, as well as the engine frames. The Raytheon Premier six-passenger business jet, shown in Figure 9-1, has a fuselage that is constructed from carbon fiber/epoxy honeycomb composite. Radically new materials will be needed to endure the temperatures that will be encountered by hypersonic aircraft. Figure 9-2 shows some of these predicted temperatures.

At one time, bicycle frames were constructed almost exclusively from welded steel tubing. Now, companies offer frames in a wide range of engineering materials, including aluminum alloys, titanium, and various fiber-reinforced composites. One top-of-the-line carbon fiber frame now weighs only 2.5 lb!

Window frames were once made almost exclusively from wood. While wood remains a competitive material, a trip to any building supply will reveal a selection that includes anodized aluminum in a range of colors, as well as frames made from colored vinyl and other polymers. Each has its companion advantages and limitations. Auto bodies were fabricated from steel sheet, and assembled by welding. Designers now select from steel, aluminum, and polymeric sheet molding compounds, and may use adhesive bonding to produce the joints.

The vacuum cleaner assembly shown in Figure 9-3 is typical of many engineering products, where a variety of materials are used for the various components. Table 9-1 lists the material changes that were recommended in just one revision of the appliance. The materials for 12 components were changed completely, and that for a thirteenth was modified. Eleven different reasons were given for the changes. An increased emphasis on lighter weight has brought about even further changes.

The list of engineering materials now includes metals and alloys, ceramics, glasses, plastics, elastomers, concrete, composite materials, and others. It is not surprising, therefore, that a single person might have difficulty making the necessary decisions concerning the materials in even a simple manufactured product. More frequently, the design engineer or design team will work in conjunction with various materials specialists to select the materials that will be needed to convert today's designs into tomorrow's reality.

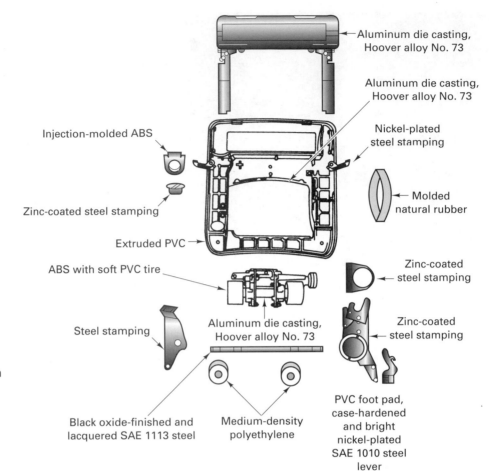

FIGURE 9-3 Materials used in various parts of a vacuum cleaner assembly. *(Courtesy of Metals Progress.)*

TABLE 9-1. Examples of Material Selection and Substitution in the Redesign of a Vacuum Cleaner

Part	Former Material	New Material	Benefits
Bottom plate	Assembly of steel stampings	One-piece aluminum die casting	More convenient servicing
Wheels (carrier and caster)	Molded phenolic	Molded medium-density polyethylene	Reduced noise
Wheel mounting	Screw-machine parts	Preassembled with a cold-headed steel shaft	Simplified replacement, more economical
Agitator brush	Horsehair bristles in a die-cast zinc or aluminum brush back	Nylon bristles stapled to a polyethylene brush back	Nylon bristles last seven times longer and are now cheaper than horsehair
Switch toggle	Bakelite molding	Molded ABS	Breakage eliminated
Handle tube	AISI 1010 lock-seam tubing	Electric seam-welded tubing	Less expensive, better dimensional control
Handle bail	Steel stamping	Die-cast aluminum	Better appearance, allowed lower profile for cleaning under furniture
Motor hood	Molded cellulose acetate (replaced Bakelite)	Molded ABS	Reasonable cost, equal impact strength, much improved heat and moisture resistance: eliminated warpage problems
Extension-tube spring latch	Nickel-plated spring steel, extruded PVC cover	Molded acetal resin	More economical
Crevice tool	Wrapped fiber paper	Molded polyethylene	More flexibility
Rug nozzle	Molded ABS	High-impact styrene	Reduced costs
Hose	PVC-coated wire with a single-ply PVC extruded covering	PVC-coated wire with a two-ply PVC extruded covering separated by a nylon reinforcement	More durability, lower cost
Bellows, cleaning-tool nozzles, cord insulation, bumper strips	Rubber	PVC	More economical, better aging and color, less marking

Source: Metal Progress, by permission.

■ 9.2 MATERIAL SELECTION AND MANUFACTURING

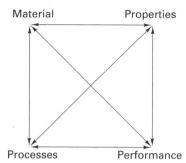

FIGURE 9-4 Schematic showing the interrelation between material, properties, processing, and performance.

The interdependence between materials and their processing must also be recognized. New processes are frequently associated with new materials, and their implementation can often cut production costs and improve product quality. Therefore, a change in material may well require a change in the manufacturing process, and, conversely, improvements in processes will often lead to a reevaluation of the materials being processed. Improper processing of a well-chosen material may well result in a defective product. If satisfactory results are to be achieved, considerable care must be exercised in selecting both the engineering materials and the manufacturing processes used to produce the product.

Many textbooks on materials and manufacturing processes spend considerable time discussing the interrelationships between the structure and properties of engineering materials, the processing used to produce a product, and its subsequent performance. As Figure 9-4 attempts to depict, each of these aspects is related to all of the others. An engineering material may possess different properties depending upon its structure. Processing of that material will alter the structure, which in turn will alter the properties. Altered properties will certainly alter performance. The objective of manufacturing, therefore, is to devise an optimized system to produce the desired product.

■ 9.3 THE DESIGN PROCESS

The first step in the manufacturing process is *design*—the determining in rather precise detail what to produce, what properties it must possess, what to make it out of, how to make it, how many to make, and how it will ultimately be used.

Design usually takes place in several distinct stages: (1) conceptual, (2) functional, and (3) production. During the *conceptual-design* stage, the designer is concerned primarily with the functions that the product is to fulfill. Several concepts are often considered, and a determination is made that the concept is either not practical, or is sound and should be developed further. Here the only concern about materials is that materials exist that can provide the desired properties. If such materials are not available, consideration is given to whether there is a reasonable prospect that new ones could be developed within the limitations of cost and time.

At the *functional- or engineering-design* stage, a workable design is developed, including a detailed plan for manufacturing. Geometric features are determined and dimensions are specified, along with allowable tolerances. Specific materials are selected for each component. Consideration is given to appearance, cost, reliability, producibility, and serviceability, in addition to the various functional factors. Often, a *prototype* or working model is constructed to permit a full evaluation of the product. It is possible that the prototype evaluation will show that some changes have to be made in either the design or material before the product can be advanced to production. This should not be taken, however, as an excuse for not doing a thorough job of material selection. There is much merit to the practice that all prototypes be built with the same materials that will be used in production and, where possible, with the same manufacturing techniques. It is of little value to have a perfectly functioning prototype that cannot be manufactured economically in the desired volume, or one that is substantially different from what the production units will be like.

It is important to have a complete understanding of the functions and performance requirements of each component and to follow this with a thorough materials analysis, selection, and specification. If these decisions are deferred, they may end up being made by individuals who are less knowledgeable about all of the functional aspects of the product.

In the *production-design* stage, we consider whether the specified materials are compatible with the manufacturing processes and equipment. Can they be processed economically, and are they available in the necessary quantities and quality?

As actual manufacturing begins, changes in both the materials and processes may be suggested. In most cases, however, changes made after the tooling and machinery have been placed in production tend to be quite costly. Good up-front material selection and thorough product evaluation can do much to eliminate the need for change.

The subsequent availability of new materials and new processes, however, may well present possibilities for cost reduction or improved performance. Before adopting new materials, however, they should be evaluated very carefully to ensure that all of their characteristics related to both processing and performance are well established. Remember that it is indeed rare that as much is known about the properties and reliability of a new material as an established one. Numerous product failures and product liability cases have resulted from new materials being substituted before their long-term properties were fully known.

■ 9.4 PROCEDURE FOR MATERIAL SELECTION

The selection of an appropriate material and its subsequent conversion into a useful product with desired shape and properties can be a rather complex process. Nearly every engineered item goes through the following sequence of activities: design → material selection → process selection → production → evaluation → and possible redesign or modification. Numerous engineering decisions must be made along the way.

Several methods have been developed for approaching a design and selection problem. The *case-history method* is one of the simplest. Evaluating what has been done in the past, or what is currently being done by a competitor, can yield important information. Use it as a starting base, and either duplicate or modify the details of that solution, that is, engineering material and method of manufacture. The basic assumption is that similar requirements can be met with similar solutions.

This approach is quite useful, and many manufacturers continually examine and evaluate their competitors' products for just this purpose. The real issue here, however, is "how similar is similar." A minor variation in service requirement, such as operating temperature or a corrosive environment, may be sufficient to justify a totally different material and manufacturing method. Moreover, this approach tends to preclude the use of new materials, new technology, and manufacturing advances that may have occurred since the formulation of the original solution. It is equally unwise, however, to totally ignore the benefits of past experience.

Other design and selection activities involve the *modification of an existing product*, generally in an effort to reduce cost, improve quality, or overcome a problem or defect that has been encountered. A customer may have requested a product like the current one, but capable of operating at higher temperatures, or in an acidic environment, or at higher pressure. Efforts here generally begin with an evaluation of the current product and its present method of manufacture. The most frequent pitfall is to overlook one of the original design requirements and recommend a change that in some way compromises the total performance of the product. Examples of such oversights are provided in Section 9.8.

The safest and most comprehensive approach to part manufacture is to approach the task as though it were the *development of an entirely new product*. Here the full sequence of design, material selection, and process selection will be followed prior to production.

The first step in any material selection problem is to define the needs of the product—describe the target that should ultimately be hit. Without prior biases about material or method of fabrication, a clear picture of all the characteristics necessary for this part to adequately perform its intended function can be developed. These requirements will fall into three major areas: (1) shape or geometry considerations, (2) property requirements, and (3) manufacturing concerns. By formulating these requirements, we will then be in a better position to evaluate candidate materials and companion methods of fabrication.

A dimensioned sketch can begin to answer many of the questions about the size, shape, and complexity of the part, and these *shape considerations* have a strong influence on decisions relating to the proposed method or methods of fabrication. While many features of part geometry are somewhat obvious, geometric considerations are often more complex than first imagined. Typical questions might include:

1. What is the relative size of the component?
2. How complex is its shape? Are there any axes or planes of symmetry? Are there any uniform cross sections? Could the component be divided into several simpler shapes that might be easier to manufacture?

3. How many dimensions must be specified?

4. How precise must these dimensions be? Are all precise? How many are restrictive, and which ones?

5. How does this component interact geometrically with other components? Are there any restrictions imposed by this interaction?

6. What are the surface finish requirements? Must all surfaces be finished? Which ones do not?

7. How much can a dimension change by wear or corrosion and the part still function adequately?

8. Could a minor change in part geometry increase the ease of manufacture or improve the performance (fracture resistance, fatigue resistance, etc.) of the part?

Producing the right shape is only part of the desired objective. If the part is to perform adequately, it must also possess the necessary mechanical and physical properties, as well as the ability to endure anticipated environments for a specified period of time. These requirements will certainly influence the selection of both the material and the method of fabrication and their consideration and specification are an important part of component design.

MECHANICAL PROPERTIES

1. How much static strength is required?

2. If the part is accidentally overloaded, is it permissible to have a sudden brittle fracture, or is plastic deformation and distortion a desirable precursor to failure?

3. How much can the material bend, stretch, or compress under load and still function properly?

4. Are any impact loadings anticipated? If so, of what type, magnitude, and velocity?

5. Can you envision vibrations or cyclic loadings? If so, of what type, magnitude, and frequency?

6. Is wear resistance desired? Where? How much? How deep?

7. Will all of the above requirements be needed over the entire range of operating temperature? If not, which properties are needed at the lowest extreme? At the highest?

PHYSICAL PROPERTIES (ELECTRICAL, MAGNETIC, THERMAL, AND OPTICAL)

1. Are there any electrical requirements? Conductivity? Resistivity?

2. Are any magnetic properties desired?

3. Are thermal properties significant? Thermal conductivity? Changes in dimension with change in temperature?

4. Are there any optical requirements?

5. Is weight a significant factor?

6. How important is appearance? Is there a preferred color, texture, or feel?

SERVICE ENVIRONMENT

Another important area to consider is the interaction of the product with its *service environment* throughout the entire product lifetime. Consideration should be given to shipping, storage, and use!

1. What are the lowest, highest, and normal temperatures the product will see? Will temperature changes be cyclic? How fast will temperature changes occur?

2. What is the most severe environment that is anticipated as far as corrosion or deterioration of material properties is concerned?

3. What is the desired service lifetime for the product?

4. What is the anticipated level of inspection and maintenance?

5. What is the potential liability if the product should fail?

6. Should the product be manufactured with disassembly or *recyclability* in mind? Are there any disposal concerns?

MANUFACTURING CONCERNS

A final area of concern is a variety of factors that will directly influence the method of manufacture. Some of these *manufacturing concerns* are:

1. How many of the components are to be produced? At what rate? (*Note:* One-of-a-kind parts are rarely made by processes that require dedicated patterns, molds or dies, since the expense of the tooling is hard to justify.)

2. What is the desired level of quality compared to similar products on the market?

3. What are the quality control and inspection requirements?

4. Are there any assembly (or disassembly) concerns? Any key relationships or restrictions with respect to mating parts?

5. What are the largest and smallest section thicknesses?

6. Have standard sizes and shapes been specified wherever possible (both as finished shapes and as starting raw material)?

7. Has the design addressed the requirements that will facilitate ease of manufacture (machinability, castability, formability, weldability, hardenability)?

The considerations just mentioned are only a sample of the many questions that must be addressed when precisely defining what it is that we want to produce. While there is a natural tendency to want to jump to an answer, that is, a material and method of manufacture, time spent determining the various requirements will be well rewarded. Collectively, the requirements will direct and restrict material and process selections. It is possible that several families of materials, and numerous members within those families, all appear to be adequate, and selection then becomes a matter of preference. It is also possible that one or more of the requirements emerges as a dominant restrictor (such as the need for ultrahigh strength, superior wear resistance, the ability to function at high operating temperature, or the ability to withstand extremely corrosive environments), and selection then becomes focused on those materials offering that specific characteristic.

It is important that *all* factors are listed and *all* service conditions and uses be considered. Many failures and product liability claims have resulted from simple engineering oversights or failure to consider the entire spectrum of conditions that a product might experience in its lifetime. Consider the failure of several large electric power transformers where fatigue cracks formed around several horizontal cooling fins that had been welded to the exterior of the casing. The loss of cooling oil through the cracks led to overheating and failure of the transformer coils. Since transformers operate under static conditions, fatigue had not been considered in the original design and material selection. When the horizontal fins were left unsupported during shipping, however, the resulting vibrations were sufficient to induce the fatal cracks. It is also not uncommon for the most severe corrosion environment to be experienced during shipping or storage as opposed to normal operation. Products also encounter unexpected service conditions. Numerous parts failed on earthmoving equipment that was used to construct the Alaskan pipeline. When this equipment had been designed, extreme subzero temperatures had not been considered as part of the operating conditions.

Assume that a complete and thorough evaluation has been made of required properties; it may be helpful to assign a relative importance to the various needs. Some requirements may be *absolutes*, while others may be *relative*. Absolute requirements are those where there can be no compromise. The consequence of not meeting them is product failure. As a result, materials that fall short of absolute requirements can be automatically eliminated. For example, if a component must possess good electrical conductivity, plastics and ceramics would not be appropriate. Relative or compromisable properties are those that frequently differentiate good, better, and best, where all would be deemed to be adequate.

■ 9.5 ADDITIONAL FACTORS TO CONSIDER

When evaluating candidate materials, an individual is often directed to handbook-type data that has been obtained through standardized materials characterization tests. It is important to note the conditions of these tests in comparison with those of the proposed application. Significant variations in factors such as temperature, rates of loading, or surface finish can lead to major changes in a material's behavior. In addition, one should keep in mind that the handbook values often represent an average or mean and that actual material properties may vary to either side of that value. Where vital information is missing or the data may not be applicable to the proposed use, one is advised to consult with the various materials producers or qualified materials engineers.

At this point it is probably appropriate to introduce *cost* as an additional selection factor. Because of competition and marketing pressures, economic considerations are often as important as technological considerations. However, we have adopted the philosophy that cost should not be considered until a material has been shown to meet the necessary property requirements. If acceptable candidates have been identified, cost becomes an important part of the selection process, and both material cost and the cost of fabrication should be considered. Often, the final decision involves a compromise between cost, ease of fabrication, and performance or quality. Numerous questions might be asked, such as:

1. Is the material too expensive to meet the marketing objectives?

2. Is a more expensive material justifiable if it offers improved performance?

3. How much additional expense might be justified to gain ease of fabrication?

In addition, it is important that the appropriate cost figures are considered. Material costs are most often reported in the form of dollars per pound, or some other form of cost per unit weight. If the product has a fixed size, however, material comparisons should probably be based on cost per unit volume. For example, aluminum has a density about one-third that of steel. For products where the size is fixed, 1 lb of aluminum can be used to produce three times as many parts as 1 lb of steel. As long as the "per pound" cost of aluminum is less than three times that of steel, aluminum will be the cheaper material. When the density of materials is quite different, as with magnesium and stainless steel, the relative rankings based on cost per pound and cost per cubic inch can be radically different.

Material availability is another important consideration. The material selected may not be available in the size, quantity, or shape desired, or may not be available in any form at all. The reliability of supply may be another consideration, especially when the material or a major component of it is imported from a limited number of sources, and political events may drastically alter that availability. In these cases one should be prepared to recommend alternative materials, provided that they, too, are feasible candidates for the specific use.

Still other factors to be considered when making material selections include:

1. Are there possible misuses of the product that should be considered? If the product is to be used by the general public, one should definitely anticipate the worst. Screwdrivers are routinely used as chisels and pry bars (different forms of loading than that of the torsional twist intended). Scissors may be used as wire cutters. Other products are similarly misused.

2. Have there been any failures of this or similar products? If so, what were the identified causes? Failure analysis results should definitely be made available to the designers, who can directly benefit from them.

3. Has the material (or class of materials) being considered established a favorable or unfavorable performance record? Under what conditions was unfavorable performance noted?

4. Has an attempt been made to benefit from material standardization, whereby multiple components are manufactured from the same material or by the same manufacturing process? Although function, reliability, and appearance should not be sacrificed, one should not overlook the potential for savings and simplification that standardization has to offer.

■ 9.6 CONSIDERATION OF THE MANUFACTURING PROCESS

The overall value of an engineering material depends not only on its physical and mechanical properties but also on our ability to shape it into useful objects in an economical and timely manner. Without the necessary shape, parts cannot perform, and without economical production, the material will be limited to a few high-value applications. For this reason, our material selection should be further refined by considering the possible fabrication processes and the suitability of each "prescreened" material to each process. Familiarity with the various manufacturing alternatives is a necessity, together with a knowledge of the associated limitations, economics, product quality, surface finish, precision, and so on. All processes are not compatible with all materials. Steel, for example, cannot be fabricated by die casting. Titanium can be forged successfully by isothermal techniques but generally not by conventional drop hammers. Wrought alloys cannot be cast, and casting alloys are not attractive for forming.

Certain fabrication processes have distinct ranges of product size, shape, and thickness, and these should be compared with the requirements of the product. Each process has its characteristic precision and surface finish. Since secondary operations, such as machining, grinding, and polishing, all require the handling, positioning, and processing of individual parts, they can add significantly to manufacturing cost. Usually it is best to hit the target with as few operations as possible. Some processes require prior heating or subsequent heat treatment. Still other considerations include production rate, production volume, desired level of automation, and the amount of labor required, especially if it is skilled labor. All of these concerns will be reflected in the cost of fabrication. There may also be additional constraints, such as the need to design a product so that it can be produced with existing equipment or facilities.

It is not uncommon for a certain process to be implied by the geometric details of a component design, such as the presence of cored holes, the magnitude of draft allowances, or the recommended surface finish. The designer often specifies these features prior to consultation with manufacturing experts. It is best, therefore, to consider all possible methods of manufacture and, where appropriate, recommend a change in design to accommodate a more attractive means of production.

■ 9.7 ULTIMATE OBJECTIVE

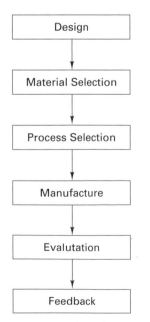

FIGURE 9-5 Sequential flow chart showing activities leading to the production of a part or product.

The real objective of this activity is to arrive at a combination of material and process (or sequence of processes) that is the best solution for a given product. Figure 9-5 depicts a series of activities that move from a well-defined set of needs and objectives through the manufacture and evaluation of a product. Numerous decisions will be made along the way and most are judgmental in nature and highly interrelated. An example of such decision making is the purchase of a new toaster at a discount department store. Will the two-slice model be adequate, or do you see a need the four-slice model? Is the wide-slice feature (to accommodate waffles and bagels) desirable or necessary? How important is appearance? What level of quality is preferred? If your specific needs can be met by models that are "good," "better," and "best," is the added expense of "better" or "best" a worthwhile investment? Various heating-element designs may be used, ranging from wide-spaced wires or strips wrapped around an insulator to closer-spaced wrappings or even fine coils. Do you feel that this feature reflects quality or has an effect on performance? If so, do you have a preference? Ultimately, how much are you willing to spend, and which one do you want to buy? The many decisions related to the purchase of a new car can be used to provide a similar analogy.

While Figure 9-5 depicts the various considerations in developing an optimized manufacturing system as having a definite, sequential manner, one should be aware that they are often rearranged and are almost always interrelated. Figure 9-6 shows a modified form, where material selection and process selection have been moved to be parallel instead of sequential. It is not uncommon for one of the two selections to be dominant, and the other to become dependent or secondary. For example, the production of a large quantity of small, intricate parts with thin walls, precise dimensions, and

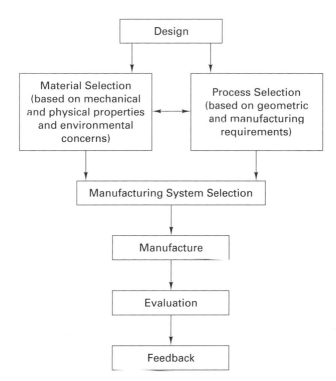

FIGURE 9-6   Alternative flow chart showing parallel selection of material and process.

smooth surfaces is an ideal candidate for diecasting. Material selection, therefore, may be limited to die castable materials—assuming feasible alternatives are available. In a converse example, highly restrictive material properties, such as the ability to endure extreme elevated temperatures or severe corrosive environments, may significantly limit the material options. Fabrication will tend to be limited to those processes that are compatible with the candidate materials.

In both models, decisions in one area generally impose restrictions or limitations in another. As shown in Figure 9-7, selection of a material may limit processes and selection

FIGURE 9-7   Compatibility chart of materials and processes. Selection of a material may restrict possible processes. Selection of a process may restrict possible materials.

Process \ Material	Irons	Steel	Aluminum	Copper	Magnesium	Nickel	Refractory Metals	Titanium	Zinc
Sand Casting	X	X	X	X	X	X			0
Permanent Mold Cast	X	0	X	0	X	0			0
Die Casting			X	0	X				X
Investment Casting		X	X	X	0	0			
Closed-die Forging		X	0	0	0	0	0	0	
Extrusion		0	X	X	X	0	0	0	
Cold Heading		X	X	X		0			
Stamping, Deep Draw		X	X	X	0	X		0	0
Screw Machine	0	X	X	X	0	X	0	0	0
Powder Metallurgy	X	X	0	X		0	X	0	

Key: X = Routinely performed
      0 = Performed with difficulty, caution, or some sacrifice (such as die life)
  Blank = Not recommended

of a process may limit material. Each material has its own set of performance characteristics. The various fabrication methods impart characteristic properties to the material, and all of these may not be beneficial. Processes designed to modify certain properties (such as heat treatment) may adversely affect others. Economics, environment, energy, efficiency, recycling, inspection, and serviceability all tend to influence decisions.

On rare occasions, a single solution will emerge as the obvious choice. More likely, several combinations of materials and processes will all meet the specific requirements, each with its own strengths and limitations. Compromise, opinion, and judgment all enter into the final decision making, and our desire is to achieve the best solution while not overlooking a major requirement. Listing and ranking the required properties will help ensure that all of the necessary factors were considered and weighed in making the ultimate decision. If no material-process combination meets the requirements or the compromises appear to be too severe, it may be necessary to redesign the product, compromise the requirements, or develop new materials or processes.

The individuals making materials and manufacturing decisions must understand the product, the materials, the manufacturing processes, and all of the various interrelations. This often requires multiple perspectives and diverse expertise, and it is not uncommon to find the involvement of an entire team of individuals. Design engineers ensure that each of the requirements is met, and that any compromise or adjustment in those requirements is acceptable. Materials specialists bring expertise in candidate materials and the effects of various processing. Manufacturing personnel know the capabilities of various processes, the equipment available, and the cost of associated tooling. Quality and environmental specialists add their perspective and expertise. Failure analysis personnel can provide valuable experience gained from past unsuccessful efforts. Customer representatives or marketing specialists may also be consulted for their opinions. Clear and open communication is vital to the making of sound decisions and compromises.

The design and manufacture of a successful product is an iterative and evolving process. The development of new materials, technological advances in processing methods, increased restrictions in environment and energy, and the demand for enhanced performance of existing products provide a continuing challenge. A change in material may well require a change in manufacturing process, and improvements in processing may warrant a reevaluation of material. The next topic considers such a change.

■ 9.8 MATERIALS SUBSTITUTION

As new technology is developed or market pressures arise, it is not uncommon for new materials to be substituted into an existing design or manufacturing system. Quite often, the substitution brings about improved quality, reduced cost, ease of manufacturing, simplified assembly, or enhanced performance. When making a *material substitution*, however, it is also possible to overlook certain requirements and cause more harm than good.

Consider the efforts related to the production of lighter-weight, more-fuel-efficient automobiles. The development of high-strength low-alloy steel sheets (HSLA) provided the opportunity to match the strength of the traditional body panels with thinner-gage material. Having overcome some of the early fabrication problems, the substitution appeared to be a natural one. However, it is important to step back, consider the total picture, and become aware of any possible compromises. While strength was indeed increased, corrosion resistance and elastic stiffness (rigidity) remained essentially unaltered. Thinner sheets would corrode in a shorter time and previously unnoticed vibrations could become a significant problem. Measures to retard corrosion and design modifications to reduce vibration would probably be necessary before the new material could be effectively substituted.

Aluminum castings might be considered as an alternative to cast iron for transmission housings. Corrosion resistance would be enhanced and weight savings would be substantial. However, the mechanical properties must be assured to be adequate, and consideration would also have to be given to the area of noise and vibration. Gray cast iron has excellent damping characteristics, and effectively eliminates these undesirable features. Aluminum, however, transmits noise and vibration, and its use would probably require the addition of some form of sound isolation material. Likewise, cast iron engine blocks do an excellent job of damping vibrations. When aluminum was first used for this application, vibrations became an immediate problem, and a companion redesign of the support system was required.

Polymeric materials have been used successfully for body panels, bumpers, fuel tanks, pumps, and housings. Composite-material drive shafts have been used in place of metal. Cast crankshafts have replaced forgings. Cast metal, powder metallurgy products, and composite materials have all been used for connecting rods. Ceramic and reinforced plastic components have been used on engines. Numerous other examples could be cited. When making materials substitution in a successful product, it is important to first consider all of the design requirements. Approaching a design or material modification as thoroughly as one approaches a new problem may well avoid costly errors.

■ 9.9 EFFECT OF PRODUCT LIABILITY ON MATERIALS SELECTION

Product liability actions, court awards, and rising insurance costs have made it imperative that designers and manufacturers employ the very best procedures in selecting and processing materials. Although many individuals feel that the situation has grown to absurd proportions, there have also been many instances where sound procedures were not used in selecting materials and methods of manufacture. In today's business and legal climate, such negligence cannot be tolerated.

An examination of recent product liability claims has revealed that the five most common causes have been:

1. Failure to know and use the latest and best information about the materials being specified.

2. Failure to foresee, and account for, all reasonable uses of the product. The designer must also consider the entire spectrum of possible product misuse, since a number of product liability cases have been lodged where the claimant was inappropriately using a product.

3. Use of materials for which there were insufficient or uncertain data, particularly with regard to long-term properties.

4. Inadequate and unverified quality control procedures.

5. Material selection made by people who were completely unqualified.

An examination of the faults just listed reveals that there is no good reason for them to exist. Consideration of each of them, however, is good practice when seeking to assure the production of a good-quality product, and can greatly reduce the number and magnitude of product liability claims.

■ 9.10 AIDS TO MATERIAL SELECTION

From the discussion in this chapter, it is apparent that those who select materials should have a broad, basic understanding of the nature and properties of materials and their processing characteristics. Providing this background is a primary purpose of this book. However, the number of engineering materials is so great, and the mass of information that is both available and useful is so large, that a single book of this type and size cannot be expected to furnish all that is required. Anyone who does much work in material selection needs to have ready access to many sources of data.

It is almost imperative that one has access to the information contained in the various volumes of *Metals Handbook*, published by ASM International. This multivolume series contains a wealth of information about both engineering metals and associated manufacturing processes. The one-volume *Metals Handbook Desk Edition* provides the highlights of this information in a less voluminous, more concise format. A parallel *ASM Engineered Materials Handbook* series provides information for the nonmetallic engineering materials. Various volumes address composites, plastics, adhesives, and ceramics.

ASM also offers a one-volume *ASM Metals Reference Book* that provides extensive data about metals and metalworking in tabular or graphic form. The *ASM Engineered*

Materials Reference Book provides similar information for ceramics, plastics, composites, and electronic materials. Additional handbooks are available for specific materials, such as titanium alloys, stainless steels, and tool steels. Various technical magazines often provide annual issues that serve as information databooks. Some of these include *Modern Plastics*, *Industrial Ceramics*, and ASM's *Advanced Materials and Processes*.

Persons selecting materials and processes should have available several of the handbooks published by various materials organizations, technical societies, and trade associations. These may be material related (such as the Aluminum Association's *Aluminum Standards and Data* and the Copper Development Association's *Standards Handbook: Copper, Brass, and Bronze*), process related (such as the *Steel Castings Handbook* by the Steel Founder's Society of America and the *Heat Treater's Guide* by ASM International), or profession related (such as the *SAE Handbook* by the Society of Automotive Engineers, *ASME Handbook* by the American Society for Mechanical Engineers, and the *Tool and Manufacturing Engineers Handbook* by the Society for Manufacturing Engineers). These may be supplemented further by a variety of supplier-provided information. While the latter is excellent and readily available, the user should recognize that supplier information might not provide a truly objective viewpoint.

It is also important to have accurate information on the cost of various materials. Since these tend to fluctuate, it may be necessary to consult a daily or weekly publication such as the *American Metal Market* newspaper or on-line service. Costs associated with various processing operations are more difficult to obtain and can vary greatly from one company to another. These costs may be available from within the firm or may have to be estimated from outside sources. A variety of texts and software packages are available.

With the evolution of high-speed computers with large volumes of searchable memory, computerized materials selection is now a reality. Most of the textbook and handbook references are now available on CDs, and all of the information in an entire handbook series can be accessed almost instantaneously. Programs have been written to utilize information databases and actually perform materials selection. The various property requirements can be specified and the entire spectrum of engineering materials can be searched to identify possible candidates. Search parameters can then be tightened or relaxed so as to produce a desired number of candidate materials. In a short period of time, a wide range of materials can be considered, far greater than could be considered in a manual selection. Process simulation packages can then be used to verify the likelihood of producing a successful product.

While the capabilities of computers and computer software are indeed phenomenal, the knowledge and experience of trained individuals should not be overlooked. Experienced personnel should reevaluate the final materials and manufacturing sequence to assure full compliance with the needs of the product.

■ KEY WORDS

absolute requirement	design	material availability	product liability	recyclability
case history	functional design	material selection	production design	relative requirements
conceptual design	manufacturing concerns	material substitution	prototype	service environment
cost				

■ REVIEW QUESTIONS

1. What are the possible undesirable features of significantly exceeding the requirements of a product?
2. In a manufacturing environment, why should the selection and use of engineering materials be a matter of constant reevaluation?
3. What are some of the recent shifts in the materials used in a family automobile? How do the materials in use today compare to those used in cars 50 or 75 years ago?
4. What are some of the new materials that have found use in the aerospace industry?

5. Discuss the interrelation between engineering material and the fabrication processes used to produce the desired shape and properties.

6. What is design?

7. What are the three primary phases of product design, and how does the consideration of materials differ in each?

8. What is the benefit of requiring prototype products to be manufactured from the same materials that will be used in production and by the same manufacturing techniques?

9. What precautions should be exercised when considering a new material or process for the manufacture of a product?

10. What are some of the possible pitfalls in the case-history approach to materials selection?

11. What is the most frequent pitfall when seeking to improve an existing product?

12. What should be the first step in any materials selection problem?

13. In what ways does the concept of shape considerations go beyond a dimensioned sketch?

14. How might temperature enter into the specification of mechanical properties?

15. What are some of the important aspects of the service environment to be considered when selecting an engineering material?

16. What are some of the possible manufacturing concerns that should be specified?

17. Why is it important to resist jumping to the answer and first perform a thorough evaluation of product needs and requirements?

18. Why is it important for all factors and all service conditions to be considered?

19. What is the difference between an absolute and relative requirement?

20. What are some possible pitfalls when using handbook data to assist in materials selection?

21. Why might it be appropriate to defer cost considerations until after evaluating the performance capabilities of various engineering materials?

22. Give an example of a product or component where material cost should be compared on a cost per pound basis. Give a contrasting example where cost per unit volume would be more appropriate.

23. In what way might failure analysis data be useful in a material selection decision?

24. Why should consideration of the various fabrication process possibilities be included in material selection?

25. Why might material selection and process selection be better performed in a parallel fashion as opposed to sequential?

26. Why is it likely that multiple individuals will be involved in the material and process selection activity?

27. Give an example of a materials substitution that has led to a subsequent problem.

28. What are some of the most common causes of product liability losses?

■ PROBLEMS

1. One simple tool that has been developed to assist in materials selection is a comparison chart, such as the one shown in Figure 9-A. The various desired properties are weighted as to their significance, and candidate materials are evaluated on a scale such as 1 to 5 or 1 to 10 with regard to their ability to provide that property. A rating number is then computed by multiplying the property rating by its weighted significance and summing the results. Potential materials can then be compared in a uniform, unbiased manner, and the best candidates can often be identified. In addition, by placing all the requirements on a single sheet of paper, the designer is less likely to overlook a major requirement. Finalist materials should then be reevaluated to assure that no key requirement has been overlooked or excessively compromised.

   Three materials, X, Y, and Z, are available for a certain use. Any material selected must have good weldability. Tensile strength, stiffness, stability, and fatigue strength have also been identified as key requirements. Fatigue strength is considered the most important of these requirements, and stiffness is least important. The three materials can be rated as follows:

	X	Y	Z
Weldability	Excellent	Poor	Good
Tensile strength	Good	Excellent	Fair
Stiffness	Good	Good	Good
Stability	Good	Excellent	Good
Fatigue strength	Fair	Good	Excellent

   Use a rating chart such as that in Figure 9-A to determine which material you would recommend.

2. The chalk tray on a classroom chalk board has very few performance requirements. As a result, it can be made from a wide spectrum of materials. Wood, aluminum, and even plastic have been used in this application. Discuss the performance and durability requirements and the pros and cons of the three listed materials. Chalk trays have a continuous cross section, but the processes used to produce such a configuration may vary with material. Discuss how a chalk tray might be mass-produced from each of the three materials classifications.

3. Examine the properties of wood, aluminum, and extruded vinyl as they relate to household window frames. Discuss the pros and cons of each, considering both ease of manufacture and performance. Which would be your preference for your particular location? Might your preference change if you were located in Alaska or Hawaii?

4. The individual turbine blades used in the exhaust region of jet engines must withstand high temperatures, high stresses, and highly corrosive operating conditions. These demanding conditions severely limit the material possibilities, and most jet engine turbine blades have been manufactured from one of the high-temperature superalloys. The fabrication processes are limited to those that are compatible with both the material and the desired geometry. Through the 1960s and early 1970s the standard method of production was investment casting and the resultant product was a polycrystalline solid with thousands of polyhedral crystals. In the 1970s production shifted to unidirectional solidification, where elongated crystals ran the entire length of the blade. More recently, production now offers single crystal products. Investigate this product to determine how the various material and processing conditions produce products with differing performance characteristics.

Rating chart for selecting materials

Material	Go-No-Go** screening			Relative rating number (†rating number x *weighting factor)								Material rating number
	Corrosion	Weldability	Brazability	Strength (5)*	Toughness (5)	Stiffness (5)	Stability (5)	Fatigue (4)	As-welded strength (4)	Thermal stresses (3)	Cost (1)	Σ rel rating no. / Σ rating factors

*Weighting factor = 1 lowest to 5 most important

† Range = 1 poorest to 5 best

**Code = S = satisfactory

U = unsatisfactory

FIGURE 9-A　Rating chart for comparing materials for a specific application.

www.wiley.com/college/degarmo

Chapter 9　CASE STUDY

Material Selection

This study is designed to get you to question why parts are made from a particular material and how they could be fabricated to their final shape. For one or more of the products listed below, write a brief evaluation that addresses the following questions.

QUESTIONS:

1. What are the normal use or uses of this product or component? What are its normal operating conditions in terms of temperatures, loadings, impacts, corrosive media, etc.?
2. What are the major properties or characteristics that the material must possess in order for the product to function?
3. What material (or materials) would you suggest and why?
4. How might you propose to fabricate this product?
5. Would the product require heat treatment For what purpose? What kind of treatment?
6. Would this product require any surface treatment or coating? For what purpose? What would you recommend?
7. Would there be any concerns relating to environment? Recycling? Product liability?

PRODUCTS:

A. The head of a carpenter's claw hammer
B. The lid for a top-loading washing machine
C. Residential interior door knob
D. A paper clip
E. A thumb tack
F. A pair of scissors
G. A moderate-to-high-quality household cookpot
H. Case for a jeweler-quality wrist watch
I. Jet engine turbine blade to operate in the exhaust region of the engine
J. Standard open-end wrench
K. A socket-wrench socket to install and remove spark plugs
L. The frame of a 10-speed bicycle
M. Interior panels of a microwave oven
N. Handle segments of a retractable blade utility knife with internal storage for additional blades
O. The outer skin of an automobile muffler
P. The interior crank handle for an automobile window
Q. The basket section of a grocery-store shopping cart
R. The body of a child's toy wagon
S. Decorative handle for a kitchen cabinet
T. Automobile engine block
U. The motor housing for a chain saw
V. Nails for the installation of aluminum siding on homes
W. Household dinnerware (knife, fork, and spoon)
X. The blades on a high-quality cutlery set
Y. A shut-off valve for a $\frac{1}{2}$-in. household water line
Z. The base plate (with heating element) for an electric steam iron

MEASUREMENT AND INSPECTION

■ 10.1 INTRODUCTION

Measurement, the act of measuring or being measured, is the fundamental activity of inspection. The intent of *inspection* is to ensure that what is being manufactured will conform to the specifications of the product. Most products are manufactured to standard sizes and shapes. For example, the base of a 60-W light bulb has been standardized so that when one bulb burns out, the next will also fit the socket in the lamp. The socket in the lamp has also been designed and made to accept the standard bulb size. Christmas tree light bulbs are made to a different standard size. Standardization is a necessity for interchangeable parts and is also important for economic reasons. A 69-W light bulb cannot be purchased because that is not a standard wattage. Light bulbs are manufactured only in standard wattages so that they can be mass produced in large volumes by high-speed automated equipment. This results in a low unit cost.

Large-scale manufacturing based on the principles of standardization of sizes and interchangeable parts became common practice early in the twentieth century. Size control must be built into machine tools and workholding devices through the precision manufacture of these machines and their tooling. The output of the machines must then be checked carefully (1) to determine the capability of specific machines and (2) for the control and maintenance of the quality of the product. A designer who specifies the dimensions and tolerances of a part often does so to enhance the function of the product, but the designer is also determining the machines and processes needed to make the part. Frequently, the design engineer has to alter the design or the specifications to make the product easier or less costly to manufacture, assemble, or inspect (or all of these). Designers should always be prepared to do this provided that they are not sacrificing functionality, product reliability, or performance.

ATTRIBUTES VERSUS VARIABLES

The examination of the product during or after manufacture, manually or automatically, falls in the province of *inspection*. Basically, inspection of items or products can be done in two ways:

1. By attributes, using gages to determine if the product is good or bad, resulting in a yes or no, go or not-go decision.
2. By variables, using calibrated instruments to determine the actual dimensions of the product for comparison with the size desired.

In an automobile, a speedometer and oil pressure gage are variable types of measuring instruments, and an oil pressure light is an attributes type of gage. As is typical of an attributes gage, the driver does not know *what* the pressure actually is if the light goes on, only that it is not good.

On the factory floor, *measurement* is the generally accepted industrial term for inspection by variables. *Gaging* (or *gauging*) is the term for determining whether the dimension or characteristic is larger or smaller than the established standard or is within some range of acceptability. Variable types of inspection generally take more time and are more expensive than attribute inspection, but yield more information because the magnitude of the characteristic is known in some standard unit of measurement.

■ 10.2 STANDARDS OF MEASUREMENT

The four fundamental measures on which all others depend are *length*, *time*, *mass*, and *temperature*. Three of these basic measures are defined in terms of material constants, as shown in Table 10-1, along with the original definitions. These four measures, along with the *ampere* and the *candela*, provide the basis for all other units of measurement, as shown in Figure 10-1, along with the original definitions. Most mechanical measurements involve combinations of units of mass, length, and time. Thus, the newton, a unit of force, is derived from Newton's second law of motion ($f = ma$) and is defined as the force that gives an acceleration of 1 m/sec/sec to a mass of 1 kilogram. Figure 10-2 and Table 10-2 provide basic metric-English conversions.

LINEAR STANDARDS

When people first sought a unit of length, they adopted parts of the human body, mainly the hands, arms, or feet. Such tools were not very satisfactory because they were not universally standard in size. Satisfactory measurement and gaging must be based on a reliable, preferably universal, standard or standards. These have not always existed. For example, although the musket parts made in Eli Whitney's shop were interchangeable, they were not interchangeable with parts made by another contemporary gun maker *from the same drawings*, because the two gunsmiths *had different foot rulers*. Today, the entire industrialized world has adopted the *international meter* as the standard of linear measurement. The inch, used by both the United States and Great Britain, has been defined officially as 2.54 centimeters. The U.S. standard inch is 41,929.399 wavelengths of the orange-red light from krypton-86.

Although *officially* the United States is committed to conversion to the metric (SI) system of measurement, which uses millimeters for virtually all linear measurements in manufacturing, the English system of feet and inches is still being used by many manufacturing plants, and its use will probably continue for some time.

TABLE 10-1. International System of Units, Founded on Seven Base Quantities on Which All Others Depend

Quantity	Name of Base	Symbol	Definition or Comment
Length	Meter (or metre)	m	Original: 1/10,000,000 of quadrant of earth's meridian passing through Barcelona and Dunkirk. Present: 1,650,763.73 wavelengths in vacuum of transition between energy levels $2p_{10}$ and $5d_5$ of krypton-86 atoms, excited at triple point of nitrogen ($-210°C$).
Mass	Kilogram	kg	Original: Mass of 1 cubic decimeter (1000 cubic centimeters) of water at its maximum density (4°C). Present: Mass of Prototype Kilogram No. 1 kept at International Bureau of Weights and Measures at Sèvres, France.
Time	Second	s	Original: 1/86,400t of mean solar day. Present: 9,192,631,770 cycles of frequency associated with transition between two hyperfine levels of isotope cesium-133.
Electric current	Ampere		Present: The rate of motion of charge in a circuit is called the *current*. The unit of current is the *ampere*. One ampere exists when the charge flows at a rate of 1 coulomb per second.
Thermodynamic	Degree Celsius	°C (K)	Present: 1/273.16 of the thermodynamic temperature of the triple point of temperature (Kelvin) water (0.01°C).
Amount of substance	Mole	mol	Present: A mole is an artificially chosen number ($N_0 = 6.02 \times 10^{23}$) that measures the number of molecules.
Luminous intensity	Candle		Present: One lumen per square foot is a footcandle.

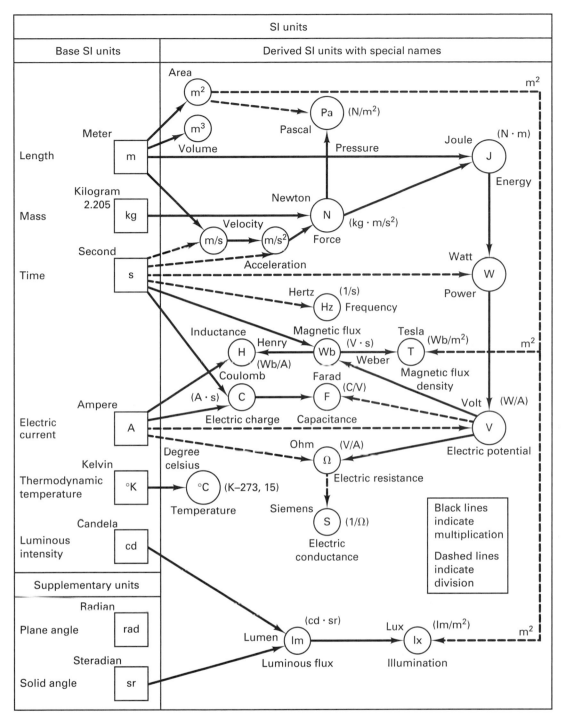

FIGURE 10-1 Relationship of secondary physical quantities to basic SI units. Solid lines, multiplication; dashed lines, division.

LENGTH STANDARDS IN INDUSTRY

Gage blocks provide industry with linear standards of high accuracy that are necessary for everyday use in manufacturing plants. The blocks are small, rectangular, square, or round in cross section and are made from steel or carbide with two very flat and parallel surfaces that are certain specified distances apart (see Figure 10-3). These gage blocks were first conceived by Carl E. Johansson in Sweden just before 1900. By 1911 he was able to produce sets of such blocks on a very limited scale, and they came into limited but significant use during World War I. Shortly after the war, Henry Ford recognized the importance of having such gage blocks generally available. He arranged for Johansson to come to the United States, and through facilities provided by the Ford Motor

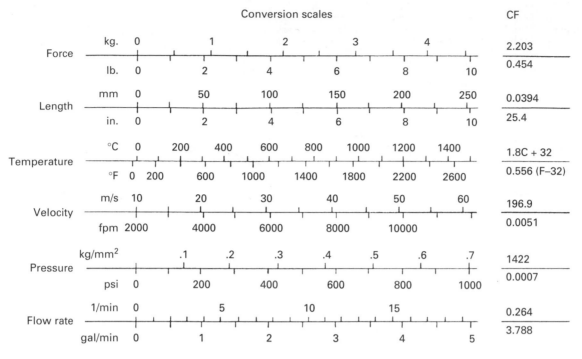

FIGURE 10-2   Metric to English conversion of some widely used measures in engineering.

Company, methods were devised for the large-scale production of gage block sets. Today, gage block sets of excellent quality are produced by a number of companies in this country and abroad.

Steel gage blocks are made of alloy steel, hardened to R_c65 and carefully heat-treated (*seasoned*) to relieve internal stresses and to minimize subsequent dimensional change. Carbide gage blocks provide extra wear resistance. The measuring surfaces of each block are surface-ground to approximately the required dimension and are then lapped and mirror polished to bring the block to the final dimension and to produce a very flat and smooth surface. The surface finish is 0.4 millionths of an inch (0.01 μm). (Surface finish is discussed in Section 10.10.)

Gage blocks are commonly made to conform to National Bureau of Standards No. 731/222 131, having Grades of 0.5, 1, 2, and 3 with tolerances as follows (ANSI/ASME B89.1OM-1984):

Grade		Inches	Millimeters	Recalibration Period
0.5	(laboratory)	±0.000,001	±0.000 03	Annually
1	(laboratory)	±0.000,002	±0.000 05	Annually
2	(precision)	+0.000,004	+0.000 10	Monthly to semiannually
		−0.000,002	−0.000 05	
3	(working)	+0.000,008	+0.000 20	Monthly to quarterly
		−0.000,004	−0.000 10	

Blocks up to 1 in. in length have absolute accuracies as stated, whereas the tolerances are per inch of length for blocks larger than 1 in. Some companies supply blocks in AA quality (which corresponds to Grade 1) and in A+ quality (which corresponds to Grade 2).

Grade 0.5 (grand-master) blocks are used as a basic reference standard in calibration laboratories. Grade 1 (laboratory-grade) blocks are used for checking and calibrating other grades of gage blocks. Grade 2 (precision-grade) blocks are used for

TABLE 10-2. Metric-English Conversions

Measurement	Metric Symbol	Metric Unit	English Conversion
Linear dimensions	m	metre	1 in. = 0.0254 m[a]
			1 ft = 0.348 m[a]
	cm	centimetre	1 in. = 2.54 cm[a]
	mm	millimetre	1 in. = 25.4 mm[a]
	μm	micrometre	1 μin. = 0.0254 μm[a]
Area	m²	square metre	1 ft² = 0.093 m²[b]
	cm²	square centimetre	1 in.² = 6.45 cm²[b]
	mm²	square millimetre	1 in.² = 645.16 mm²[a]
Volume and capacity	m³	cubic metre	1 ft³ = 0.028 m³[c]
	cm³	cubic centimetre	1 in³ = 16.39 cm³
	mm³	cubic millimetre	1 in³ = 16.387.06 mm³[b]
	L	litre	1 ft³ = 28.32 L[c]
			1 U.S. gal = 3.79 L[c]
Velocity, acceleration and flow	m/s	metres per second	1 ft/s = 0.3048 m/s[a]
	m/min	metres per minute	1 ft/min = 0.3048 m/mm[a]
	m/s²	metres per second squared	1 in/s² = 0.0254 m/s²[c]
	L/mm	litres per minute	1 ft³/min = 28.3 l/min[c] and
			1 gallon (U.S. liquid)/min = 3.785 l/min[c]
Mass	g	gram	1 oz = 28.36g[c]
	kg	kilogram	1 lb = 0.45 kg[c]
	t	metric ton	1 short ton (2000 lb) = 0.7072t[d]
Force	N	newton	1 lb-force = 4.448 N[c]
	kN	kilonewton	1 short ton-force (2000 lb) = 8.896 kN[c]
Bending moment or torque	N·m	newton-metre	1 oz-force/in. = 0.007 N·m[c]
			1 lb-force/in. = 0.113 N·m[c]
			1 lb-force/ft = 1.3558 N·m[c]
Pressure	Pa	pascal	1 lb/ft² = 47.88 Pa[c]
	kPa	kilopascal	1 lb/in² = 6.895 kPa[c]
Energy, work, or quantity of heat	J	joule	1 Btu[d] = 1055.056[c]
	kJ	kilojoule	1 Btu[d] = 1.055 kJ[c]
Power	W	watt 1	1 hp (550 ft-lb/s) = 745.7 W[c]
			1 hp (electric) = 746 W[a]
	kW	kilowatt	1 hp (550 ft-lb/s) = 0.7457 kW[b]
Temperature	C	Celsius	degrees C = $\dfrac{\text{degrees F} - 32}{1.8}$
Frequency	Hz	hertz	1 cycle per second = 1 Hz
	kHz	kilohertz	1000 cycles per second = 1 kHz
	MHz	megahertz	1,000,000 cycles per second = 1 MHz

[a] Comments on the metric system:

   The *re* spelling of *metre* conforms with the such standards as ANZI Z210.1, ISO Standard 1000, and recommendations of the Society of Manufacturing Engineers Metric Advisory Committee. The use of metre also avoids confusion with meter and micrometer measuring instruments.

   Metric symbols are usually presented the same way in singular and in plural (1 mm, 100 mm), and periods are not used after symbols, except at the end of sentences. Degree (instead of radian) continues to be used for plane angles, but angles are expressed with decimal subdivisions rather than minutes and seconds. Surface finishes are specified in micrometres.

   Accuracy of conversion. Multiplying an English measurement by an exact metric conversion factor often provides an accuracy not intended by the original value. In general, use one less significant digit to the right of the decimal point than was given in the original value; 0.032 in. (0.81) in. (0.008 mm), etc. Fractions of an inch are converted to the nearest tenth of a millimetre: $\frac{1}{8}$ in. (3.2 mm).

   Weight, mass, and force. Confusion exists in the use of the term *weight* as a quantity to mean either force or mass. In nontechnical circles, the term *weight* nearly always means mass, and this use will probably persist. Weight is a force generated by mass under the influence of gravity. Mass is expressed in gram (g), kilogram (kg), or metric ton (t) units, and force in newton (N) or kilonewton (kN) units.

   Pressure. Kilopascal (kPa) is the recommended unit for fluid pressure. Absolute pressure is specified either by using the identifying phrase "absolute pressure" or by adding the word *absolute* after the unit symbol, separating the two by a comma or a space.

[b] *Exact.*   [c] *Approximate.*   [d] *International Table.*

checking Grade 3 blocks and master gages. Grade 3 (B or working-grade) blocks are used to calibrate or check routine measuring devices such as micrometers, or in actual gaging operations.

The dimensions of individual blocks are established by light-beam interferometry, with which it is possible to calibrate these blocks routinely with an uncertainty as low as one part per million.

Gage blocks usually come in sets containing various numbers of blocks of various sizes, such as those shown in Figure 10-3. By "wringing the blocks together" in various combinations, as shown in Figure 10-4, any desired dimension can be obtained. For

FIGURE 10-3   Standard set of rectangular gage blocks with ±0.000050-in. accuracy, and three individual blocks shown.

Lapped and mirror polished to a very low micro-surface finish.

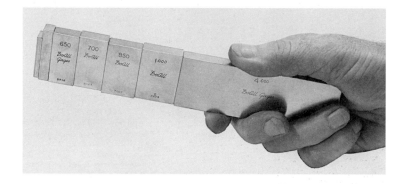

FIGURE 10-4   Seven gage blocks wrung together to build up a desired dimension. *(Courtesy of DoALL Company.)*

example, suppose the last two blocks on the stack are 0.100 and 0.05 in., what is the total length of the wrung together stack of gage blocks? Here is the breakdown of blocks that one would have in an 81 block set of English #2 gage blocks.

9 Blocks	0.1001	through 0.1009 in.	in steps of 0.0001 in.
49 Blocks	0.101	through 0.149 in.	in steps of 0.001 in.
19 Blocks	0.050	through 0.950 in.	in steps of 0.050 in.
4 Blocks	1.000	through 4.000 in.	in steps of 1.000 in.

Gage blocks are wrung together by sliding one past another using hand pressure. They will adhere to one another with considerable force and must not be left in contact for extended periods of time. Gage blocks are available in different shapes (squares, angles, rounds, and pins), so standards of high accuracy can be obtained to fill almost any need. In addition, various auxiliary clamping, scribing, and base block attachments are available that make it possible to form very accurate gaging devices, such as the setup shown in Figure 10-5.

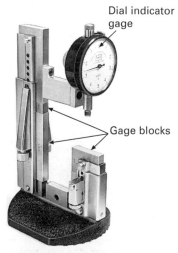

Dial indicator gage

Gage blocks

FIGURE 10-5   Wrung-together gage blocks in a special holder, used with a dial gage to form an accurate comparator. *(Courtesy of DoALL Company.)*

STANDARD MEASURING TEMPERATURE

Because all the commonly used metals are affected dimensionally by temperature, a standard measuring temperature of 68°F (20°C) has been adopted for precision-measuring work. All gage blocks, gages, and other precision-measuring instruments are calibrated at this temperature. Consequently, when measurements are to be made to accuracies greater than 0.0001 in. (0.0025 mm), the work should be done in a room in

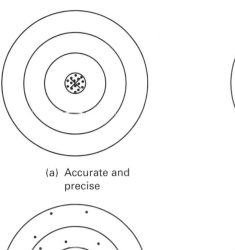
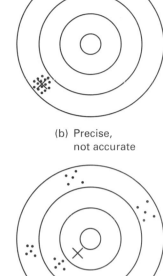
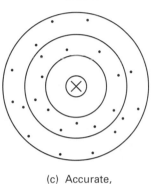

(a) Accurate and precise

(b) Precise, not accurate

(c) Accurate, not precise

(d) Precise within sample
Not precise between samples
Not accurate over all or within sample

FIGURE 10-6 *Accuracy versus precision. Dots in targets represent location of shots. Cross (×) represents the location of the average position of all shots. (a) Accurate and precise; (b) precise, not accurate; (c) accurate, not precise; (d) precise within sample, not precise between samples, not accurate over all or within sample.*

which the temperature is controlled at standard. Although it is true that to some extent both the workpiece and the measuring or gaging device *may* be affected to about the same extent by temperature variations, one should not rely on this. Measurements to even 0.0001 in. (0.0025 mm) should not be relied on if the temperature is very far from 68°F (20°C).

ACCURACY VERSUS PRECISION IN PROCESSES

It is vitally important that the difference between accuracy and precision be understood. *Accuracy* refers to the ability to hit what is aimed at (the bull's-eye of the target). *Precision* refers to the repeatability of the process. Suppose that five sets of five shots are fired at a target from the same gun. Figure 10-6 shows some of the possible outcomes. In Figure 10-6a, inspection of the target shows that this is a good process—accurate and precise. Figure 10-6b shows precision (repeatability) but poor accuracy. The agreement with a standard is not good. In Figure 10-6c the process is on the average quite accurate, as the X is right in the middle of the bull's-eye, but the process has too much scatter or variability; it does not repeat. Finally, in Figure 10-6d, a failure to repeat accuracy between samples with respect to time is observed; the process is not stable. These four outcomes are typical but not all-inclusive of what may be observed.

In measuring instruments used in the factory, precision or repeatability is critical because the processes must be repeatable as well as accurate. These are the two main characteristics of what is called *process capability*, which is discussed in Chapter 12.

■ 10.3 ALLOWANCE AND TOLERANCE

Two factors, allowance and tolerance, must be specified if one is to obtain the desired fit between mating parts. *Allowance* is the intentional, desired difference between the dimensions of two mating parts. It is the difference between the dimension of the largest interior-fitting part (shaft) and that of the smallest exterior-fitting part (hole). Figure 10-7 shows shaft A designed to fit into the hole in block B. This difference (0.0535 − 0.5025) thus determines the condition of *tightest* fit between mating parts. Allowance may be specified so that either *clearance* or *interference* exists between the mating parts. In the

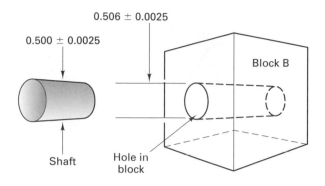

0.500 ± 0.0025

0.506 ± 0.0025

Block B

Shaft

Hole in block

FIGURE 10-7 When mating parts are designed, each shaft must be smaller than each hole for a clearance fit.

case of a shaft and mating hole, it is the difference in diameters of the largest shaft and the smallest hole. With clearance fits, the largest shaft is smaller than the smallest hole, whereas with interference fits, the hole is smaller than the shaft.

Tolerance is an undesirable but permissible deviation from a desired dimension. There is variation in all process, and no part can be made *exactly* to a specified dimension, except by chance. Furthermore, such exactness is neither necessary nor economical. Consequently, it is necessary to permit the actual dimension to deviate from the desired theoretical dimension (called the *nominal*) and to control the degree of deviation so that satisfactory functioning of the mating parts will still be ensured.

Now we can see that the objective of *inspection*, by means of measurement techniques, is to provide feedback information on the actual size of the parts with reference to the size specified by the designer on the part drawing.

The manufacturing processes that make the shaft are different from those that make the hole, but both the hole and the shaft are subject to deviations in size because of variability in the processes and the materials. Thus, while the designer wishes ideally that all the shafts would be exactly 0.500 (Figure 10-8a) and all the holes 0.506, the reality of processing is that there will be deviations in size around these nominal or ideal sizes.

Most manufacturing processes result in products whose measurements of the geometrical features and sizes are distributed normally (Figure 10-8b). That is, most of the measurements are clustered around the average dimension, $\overline{X}$. $\overline{X}$ will be equal to the nominal dimension only if the process is 100% accurate, that is, perfectly centered. More likely, parts will be distributed on either side of the average, and the process might be described (modeled) with a normal distribution. In normal distributions, as shown in Figure 10-8c, 99.73% of the measurements (X_i) will fall within plus or minus 3 standard deviations ($\pm 3\sigma$) of the mean, 95.46% will be within $\pm 2\sigma$, and 68.26% within $\pm 1\sigma$, where

$$\sigma = \sqrt{\frac{\sum\limits_{i=1}^{n} \left(X_i - \overline{X}\right)^2}{n}} \qquad (10\text{-}1)$$

and

$$\overline{X} = \frac{\sum\limits_{i=1}^{n} X_i}{n} \qquad \text{for } n \text{ items} \qquad (10\text{-}2)$$

In general, the designer applies nominal values to the mating parts according to the desired fit between the parts. Tolerances are added to those nominal values in recognition of the fact that all processes have some natural amount of variability.

Assume that the data for both the hole and the shaft are normally distributed. The $\pm 3\sigma$ added to the mean (μ) gives the upper and lower natural tolerance limits

$$\mu + 3\sigma = \text{UNTL}$$

$$\mu - 3\sigma = \text{LNTL}$$

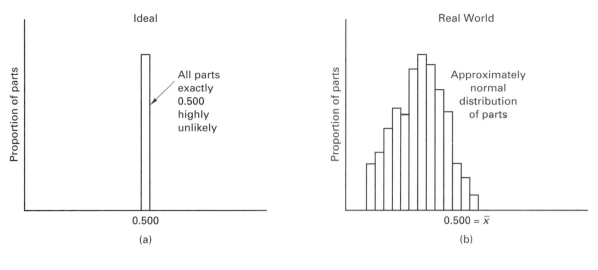

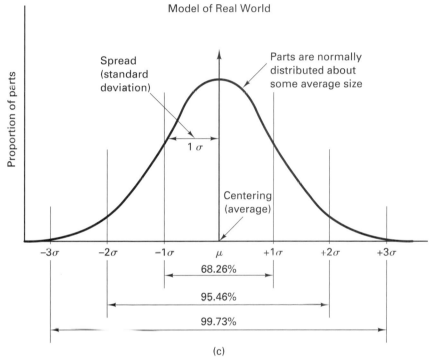

FIGURE 10-8   (a) In the ideal situation, the process would make all parts exactly the same size. (b) In the real world of manufacturing, parts have variability in size. (c) The distribution of sizes can often be modified with a normal distribution.

As shown in Figure 10-9a, the average fit of two mating parts is equal to the difference between the mean of the shaft distribution and the mean of the hole distribution. The *range of fit* would be the difference between the minimum diameter shaft and the maximum diameter hole. The minimum *clearance* would be the difference between the smallest hole and the largest shaft. The tighter the distributions (i.e., the more precise the process), the better the fit between the parts.

If tool wear is considered, the diameter of the shaft will tend to get larger as the tool wears. However, the diameter of the hole in block B of Figure 10-7 decreases in size with tool wear. If no corrective action is taken, the means of the distributions will shift toward each other, and the fit will become increasingly tight (the clearance will be decreased). If the hole and the shaft distributions overlap by 2 standard deviations, as shown in Figure 10-9b, 6 parts out of every 10,000 could not be assembled. As the shaft and the hole distributions move closer together, more interference between parts will occur, and the fit will become tighter, eventually becoming an interference fit.

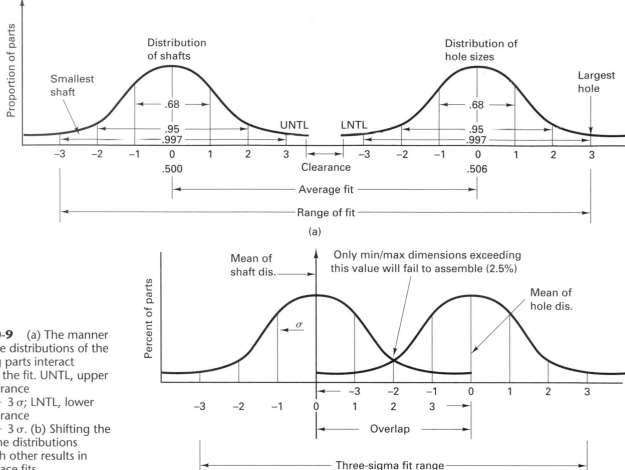

FIGURE 10-9 (a) The manner in which the distributions of the two mating parts interact determines the fit. UNTL, upper natural tolerance limit = $\mu + 3\sigma$; LNTL, lower natural tolerance limit = $\mu - 3\sigma$. (b) Shifting the means of the distributions toward each other results in some interface fits.

The designer must specify the tolerances according to the function of the mating parts. Suppose that the mating parts are the cap and the body of an ink pen. The cap must fit snugly but must be easily removed by hand. A snug fit would be too tight for a dead bolt in a door lock. A snug fit is also too tight for a high-speed bearing for rotational parts but is not tight enough for permanently mounting a wheel on an axle. In the next section we introduce the manner in which tolerances and allowances are specified, but design engineers are expected to have a deeper understanding of this topic.

SPECIFYING TOLERANCE AND ALLOWANCES

Tolerance can be specified in four ways: bilateral, unilateral, limits, and geometrically. *Bilateral* tolerance is specified as a plus or minus deviation from the nominal size, such as 2.000 ± 0.002 in. More modern practice uses the *unilateral* system, where the deviation is in one direction from the basic size, such as

$$
50.8 \text{ mm} \quad \begin{array}{c} +0.1 \text{ mm} \\ \\ -0.0 \text{ mm} \end{array} \quad \text{or} \quad 2.000 \text{ in.} \quad \begin{array}{c} +0.004 \text{ in.} \\ \\ -0.000 \text{ in.} \end{array} \quad \text{in metric}
$$

In the first case, that of bilateral tolerance, the dimension of the part could vary between 1.998 and 2.002 in., a total tolerance of 0.004 in. For the example of unilateral tolerance, the dimension could vary between 2.000 and 2.004 in., again a tolerance of 0.004 in. Obviously, to obtain the same maximum and minimum dimensions with the two systems, different basic sizes must be used. The maximum and minimum dimensions that result from application of the designated tolerance are called *limit dimensions*, or *limits*. (*Geometric* tolerances are discussed in the next section.)

There can be no rigid rules for the amount of clearance that should be provided between mating parts; the decision must be made by the designer, who considers how the parts are to function. The American National Standards Institute, Inc. (ANSI) has established eight classes of fits that serve as a useful guide in specifying the allowance and tolerance for typical applications and that permit the amount of allowance and tolerance to be determined merely by specifying a particular class of fit. These classes are as follows:

CLASS 1: *Loose fit:* large allowance. Accuracy is not essential.

CLASS 2: *Free fit:* liberal allowance. For running fits where speeds are above 600 rpm and pressures are 4.1 MPa (600 psi) or above.

CLASS 3: *Medium fit:* medium allowance. For running fits under 600 RPM and pressures less than 4.1 MPa (600 psi) and for sliding fits.

CLASS 4: *Snug fit:* zero allowance. No movement under load is intended, and no shaking is wanted. This is the tightest fit that can be assembled by hand.

CLASS 5: *Wringing fit:* zero to negative allowance. Assemblies are selective and not interchangeable.

CLASS 6: *Tight fit:* slight negative allowance. An interference fit for parts that must not come apart in service and are not to be disassembled or are to be disassembled only seldom. Light pressure is required for assembly. Not to be used to withstand other than very light loads.

CLASS 7: *Medium force fit:* An interference fit requiring considerable pressure to assemble; ordinarily assembled by heating the external member or cooling the internal member to provide expansion or shrinkage. Used for fastening wheels, crank disks, and the like, to shafting. The tightest fit that should be used on cast iron external members.

CLASS 8: *Heavy force and shrink fits:* Considerable negative allowance. Used for permanent shrink fits on steel members.

The allowances and tolerances that are associated with the ANSI classes of fits are determined according to the theoretical relationship shown in Table 10-3. The actual resulting dimensional values for a wide range of basic sizes can be found in tabulations in drafting and machine design books.

In the ANSI system, the hole size is always considered basic, because the majority of holes are produced through the use of standard-size drills and reamers. The internal member, the shaft, can be made to any one dimension as readily as to another. The allowance and tolerances are applied to the basic hole size to determine the limit dimensions of the mating parts. For example, for a basic hole size of 2 in. and a Class 3 fit, the dimensions would be:

Allowance		0.0014 in.
Tolerance		0.0010 in.
Hole		
	Maximum	2.0010 in.
	Minimum	2.0000 in.

TABLE 10-3.   ANSI Recommended Allowances and Tolerances

Class of Fit	Allowance	Average Interference	Hole Tolerance	Shaft Tolerance
1	$0.0025\sqrt[3]{d^2}$	—	$+0.0025\sqrt[3]{d}$	$-0.0025\sqrt[3]{d}$
2	$0.0014\sqrt[3]{d^2}$	—	$+0.0013\sqrt[3]{d}$	$-0.0013\sqrt[3]{d}$
3	$0.0009\sqrt[3]{d^2}$	—	$+0.0008\sqrt[3]{d}$	$-0.0008\sqrt[3]{d}$
4	0	—	$+0.0006\sqrt[3]{d}$	$-0.0004\sqrt[3]{d}$
5	—	0	$+0.0006\sqrt[3]{d}$	$+0.0004\sqrt[3]{d}$
6	—	$0.00025d$	$+0.0006\sqrt[3]{d}$	$+0.0006\sqrt[3]{d}$
7	—	$0.0005d$	$+0.0006\sqrt[3]{d}$	$+0.0006\sqrt[3]{d}$
8	—	$0.001d$	$+0.0006\sqrt[3]{d}$	$+0.0006\sqrt[3]{d}$

Shaft

Maximum	1.9986 in.
Minimum	1.9976 in.

It should be noted that for both clearance and interference fits, the permissible tolerances tend to result in a looser fit.

The *ISO System of Limits and Fits* (Figure 10-10a) is widely used in a number of leading metric countries. This system is considerably more complex than the ANSI system just discussed. In this system each part has a *basic size*. Each limit of size of a part, high and low, is defined by its *deviation* from the basic size, the magnitude and sign being obtained by subtracting the basic size from the limit in question. The difference between the two limits of size of a part is called *tolerance*, an absolute amount without sign.

There are three classes of fits: (1) *clearance fits*; (2) *transition fits* (the assembly may have either clearance or interference); and (3) *interference fits*. Either a *shaft-* or *hole-basis system* may be used (Figure 10-10b). For any given basic size, a range of tolerances and deviations may be specified with respect to the line of zero deviation, called the *zero line*. The tolerance is a function of the basic size and is designated by a number symbol, called the *grade* (e.g., the *tolerance grade*). The *position* of the tolerance with respect to the zero line, also a function of the basic size, is indicated by a letter symbol (or two letters), a capital letter for holes, and a lowercase letter for shafts, as illustrated in (Figure 10-10c).[1] Thus the specification for a hole and a shaft having a basic size of 45 mm might be 45 H8/g7.

Eighteen standard grades of tolerances are provided, called IT 01, IT 0, and IT 1 through IT 16, providing numerical values for each nominal diameter, in arbitrary steps up to 500 mm (i.e., 0–3, 3–6, 6–10, . . . , 400–500 mm). The valve of the tolerance unit, i, for grades 5–16 would be

$$i = 0.45\sqrt{D} + 0.001\,D$$

where i is in micrometers and D in millimeters.

Standard shaft and hole deviations are provided by similar sets of formulas. However, for practical application, both tolerances and deviations are provided in three sets of rather complex tables. Additional tables give the values for basic sizes above 500 mm, and for "commonly used shafts and holes" in two categories: "general purpose" and "fine mechanisms and horology" (horology is the art of making timepieces).

GEOMETRIC TOLERANCES

Geometric tolerances state the maximum allowable deviation of a form or a position from the perfect geometry implied by a drawing. These tolerances specify the diameter or the width of a tolerance zone necessary for a part to meet its required accuracy. Figure 10-11 shows the various symbols used to specify the required geometric characteristics of dimensioned drawings. A modifier is used to specify the limits of size of a part when applying geometric tolerances. The *maximum material condition* (MMC) indicates that a part is made with the largest amount of material allowable (i.e., a hole at its smallest permitted diameter or a shaft at its largest permitted diameter). The least material condition (LMC) is the converse of the maximum material condition. *Regardless of feature size* (RFS) indicates that tolerances apply to a geometric feature for any size it may be. Many geometric tolerances or *feature control symbols* are stated with respect to a particular datum or reference surface. Up to three datum surfaces can be given to specify a tolerance. Datum surfaces are generally designated by a letter symbol.

Figure 10-11b gives examples of the symbols used for datum planes and feature control symbols.

There are four tolerances that specify the permitted variability of forms: *flatness*, *straightness*, *roundness*, and *cylindricity*. Form tolerances describe how an actual feature may vary from a geometrically ideal feature.

---

[1] It will be recognized that the "position" in the ISO system essentially provides the "allowance" of the ANSI system.

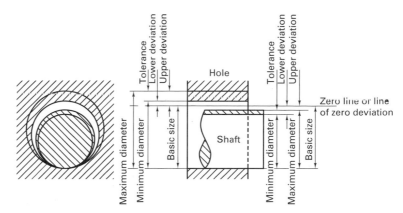

(a) Basic size, deviation and tolerance in the ISO system.

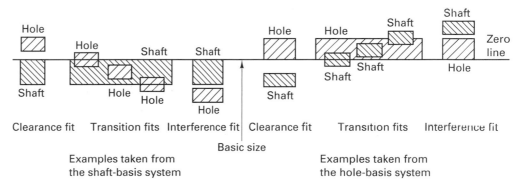

(b) Shaft-basis and hole-basis system for specifying fits in the ISO system.

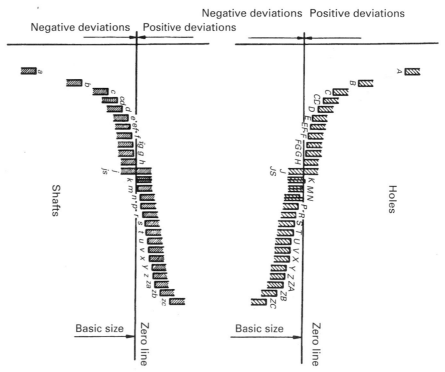

(c) Position of the various tolerance zones for a given diameter in the ISO system.

FIGURE 10-10 The ISO System of Limits and Fits (By permission from Recommendations R286-1962, *System of Limits and Fits,* copyright 1962, American Standards Institute, New Jersey).

	Tolerance	Characteristic	Symbol
Individual features	Form	Straightness	—
		Flatness	▱
		Circularity	○
		Cylindricity	⌀̸
Individual or related features	Profile	Line	⌒
		Surface	⌓
Related features	Orientation	Angularity	∠
		Perpendicularity	⊥
		Parallelism	//
	Location	Position	⊕
		Concentricity	◎
	Runout	Circular runout	↗
		Total runout	↗↗

Notes	⌀ DIA	Ⓜ MMC	Ⓛ LMC	Ⓢ RFS

(a)

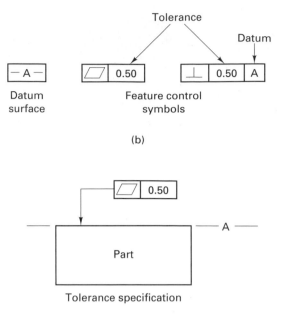

(b)

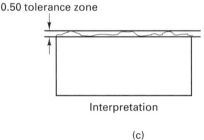

Tolerance specification

0.50 tolerance zone

Interpretation

(c)

FIGURE 10-11 (a) Geometric tolerancing symbols; (b) feature control symbols for part drawings; (c) example of use of geometric tolerancing—form tolerancing for flatness.

The surface of a part is ideally flat if all its elements are coplanar. The *flatness* specification describes the tolerance zone formed by two parallel planes that bound all the elements on a surface. A 0.5-mm tolerance zone is described by the feature control symbol in Figure 10-11c. The distance between the highest point on the surface to the lowest point on the surface may not be greater than 0.5 mm.

In Section 10.7, on coordinate measuring machines, additional examples of geometric tolerances can be found.

■ 10.4 INSPECTION METHODS FOR MEASUREMENT

The field of *metrology*, even limited to geometrical or dimensional measurements, is far too large to cover here. The remainder of this chapter concentrates on basic linear measurements and the attributes and variables devices most commonly found in a company's metrology or quality control facility. At a minimum, such labs would typically contain optical flats: one or two granite measuring tables; an assortment of indicators, calipers, micrometers, and height gages; an optical comparator; a set or two of Grade 1 gage blocks; a coordinate measuring machine; a laser scanning device; a laser interferometer; a toolmaker's microscope; and pieces of equipment specially designed to accommodate the company's products.

Table 10-4 provides a summary of inspection methods, listing five basic kinds of devices: air, light optical and electron optical, electronic, and mechanical. The variety seems to be endless, but digital electronic readouts connected to any of the measuring devices are becoming the preferred method.

The discrete digital readout on a clear liquid-crystal display (LCD) eliminates reading interpretations associated with analog scales and can be entered directly into dedicated microprocessors or computers for permanent recording and analysis. The added

TABLE 10-4.	Five Basic Kinds of Inspection Method		
Method	Typical Accuracy	Major Applications	Comments
Air	0.5–10 μ.in. or 2 to 3% of scale range	Gaging holes and shafts using a calibrated difference in air pressure or airflow, with magnifications of 20,000–40,000 to 1; also used for machine control, sorting, and classifying.	High precision and flexibility; can measure out-of-round, taper, concentricity, camber, squareness, parallelism, and clearance between mating parts; noncontact principle good for delicate parts.
Optical light energy	0.2–2 μ.in. or better with laser interferometry 0.5–1 second of arc in autocollimation optical comparators	Interferometry; checking flatness and size of gage blocks; finding surface flaws; measuring spherical shapes, flatness of surface plates, accuracy of rotary index tables; includes all light microscopes and devices common on plant floor	Largest variety of measuring equipment; autocollimators are used for making precision angular measurements; lasers are used to make precision in-process measurements; laser scanning.
Optical electron energy	100 Å	Precision measurement in scanning electron microscopes of microelectronic circuits and other small precision parts.	Part size restricted by vacuum chamber size; electron beam can be used for processing and part testing of electronic circuits.
Electronics	0.5–10 μ.in.	Widely used for machine control, on-line inspection, sorting, and classification; ODs, IDs, height, surface, and geometrical relationships, profile tracing for roundness, surface roughness, contours, etc.; most devices are comparators with movement of stylus or spindle producing an electronic signal that is amplified electronically; commonly connected to microprocessors and minicomputers for process adjustment.	Electronic gages come in many forms but usually have a sensory head or detector combined with an amplifier; capable of high magnification with resolution limited by size or geometry for sensory head; readouts commonly have multiple magnification steps; solid-state digital electronics make these devices small, portable, stable, and extremely flexible, with extremely fast response time.
Mechanical	1–10 μ.in.	Large variety of external and internal measurements using dial indicators, micrometers, calipers, and the like; commonly used for bench comparators for gage calibration work.	Moderate cost and ease of use make many of these devices and workhorses of the shop floor; highly dependent on workers' skills and often subject to problems of linkages.

speed and ease of use for this type of equipment have allowed it to be routinely used on the plant floor instead of in the metrology lab. In summary, the trend toward tighter tolerances (greater precision) and accuracy associated with the need for superior quality and reliability has greatly enhanced the need for improved measurement methods.

FACTORS IN SELECTING INSPECTION EQUIPMENT

Many inspection devices use electronic output to communicate directly to microprocessors. Inspection devices are being built into the processes themselves and are often computer-aided. In-process inspection generates feedback sensory data from the process or its output to the computer control of the machine, which is the first step in making the processes responsive to changes (Adaptive Control). In addition to in-process inspection, many other quality checks and measurements of parts and assemblies are needed. In general, six factors should be considered when selecting equipment for an inspection job by measurement techniques.

1. *Gage Capability*. The measurement device (or working gage) should be 10 times more precise than the tolerance to be measured. This is known on the factory floor as "the rule of 10." The rule actually applies to all stages in the inspection sequence, as shown in Figure 10-12. The master gage should be 10 times more precise than that of the inspection device. The reference standard used to check the master gage should be 10 times more precise than the master gage. The application of the rule greatly reduces the probability of rejecting good parts or accepting bad components and performing additional work on them. Additional discussion of gage capability is found in Chapter 12.

2. *Linearity*. This factor refers to the calibration accuracy of the device over its full working range. Is it linear? What is its degree of nonlinearity? Where does it become nonlinear, and what, therefore, is its real linear working region?

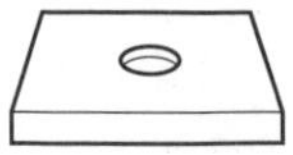 Tolerance needed on part ± 0.001 on hole diameter	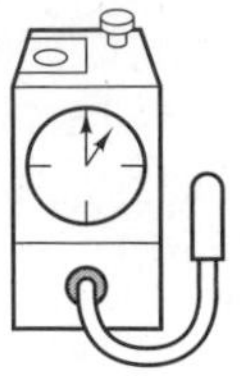 Precision needed on gage ± 0.0001 in.	 To check and set the air gage, needs to be ± 0.00001 in.	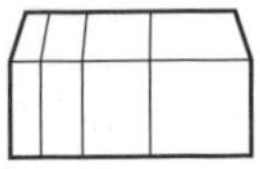 In the manufacture of the master gage, a standard of precision of at least ± 0.000001 in. is needed
Workpiece	Air gage or working gage	Master gage	Reference end standard

FIGURE 10-12 The rule of 10 states that for reliable measurements each successive step in the inspection sequence should have 10 times the *precision* of the preceding step.

3. *Repeat accuracy.* How repeatable is the device in taking the same reading over and over on a given standard?

4. *Stability.* How well does this device retain its calibration over a period of time? Stability is also called *drift.* As devices become more accurate, they often lose stability and become more sensitive to small changes in temperature and humidity.

5. *Magnification.* The amplification of the output portion of the device over the actual input dimension. The more accurate the device, the greater must be its magnification factor, so that the required measurement can be read out (or observed) and compared with the desired standard. Magnification is often confused with resolution, but they are not the same thing.

6. *Resolution.* This is sometimes called *sensitivity* and refers to the smallest unit of scale or dimensional input that the device can detect or distinguish. The greater the resolution of the device, the smaller will be the things it can resolve and the greater will be the magnification required to expand these measurements up to the point where they can be observed by the naked eye.

Some other factors of importance in selecting inspection devices include the type of measurement information desired, the range or the span of sizes the device can handle versus the size and geometry of the workpieces, the environment, the cost of the device, and the cost of installing, training, and using the device. The last factor depends on the speed of measurement, the degree to which the system can be automated, and the functional life of the device in service.

■ 10.5 MEASURING INSTRUMENTS

Because of the great importance of measuring in manufacturing, a great variety of instruments are available that permit measurements to be made routinely, ranging in accuracy from $\frac{1}{64}$ to 0.00001 in. and from 0.5 to 0.0003 mm. Machine-mounted measuring devices (probes and lasers) for automatically inspecting the workpiece during manufacturing are beginning to compete with postprocess gaging and inspection, in which the part is inspected, automatically or manually, after it has come off the machine. In-process inspection for automatic size control has been used for some years in grinding to compensate for the relatively rapid wear of the grinding wheel. Today, touch trigger probes, with built-in automatic measuring systems, are being used on machine tools to determine cutting tool offsets and tool wear. These systems are discussed in Chapter 32.

For manually operated analog instruments, the ease of use, precision, and accuracy of measurements can be affected by (1) the least count of the subdivisions on the instrument, (2) line matching, and (3) the parallax in reading the instrument. Elastic deformation of the instrument and workpiece and temperature effects must be considered. Some instruments are more subject to these factors than others. In addition, the skill of the person making the measurements is very important. Digital readout devices in measuring instruments lessen or eliminate the effect of most of these factors, simplify many measuring problems, and lessen the chance of making a math error.

LINEAR MEASURING INSTRUMENTS

Linear measuring instruments are of two types: direct reading and indirect reading. *Direct-reading instruments* contain a line-graduated scale so that the size of the object being measured can be read directly on this scale. *Indirect-reading instruments* do not contain line graduations and are used to transfer the size of the dimension being measured to a direct-reading scale, thus obtaining the desired size information indirectly.

The simplest and most common direct-reading linear measuring instrument is the *machinist's rule*, shown in Figure 10-13. Metric rules usually have two sets of line graduations on each side, with divisions of $\frac{1}{2}$ and 1 mm; English rules have four sets, with divisions of $\frac{1}{16}, \frac{1}{32}, \frac{1}{64}$, and $\frac{1}{100}$ in. Other combinations can be obtained in each type.

The machinist's rule is an end- or line-matching device. For the desired reading to be obtained, an end and a line, or two lines, must be aligned with the extremities of the object or the distance being measured. Thus, the accuracy of the resulting reading is a function of the alignment and the magnitude of the smallest scale division. Such scales are not ordinarily used for accuracies greater than $\frac{1}{64}$ in. (0.01 in.) or about $\frac{1}{2}$ mm.

Several attachments can be added to a machinist's rule to extend its usefulness. The *square head* (Figure 10-14) can be used as a miter or trisquare or to hold the rule in an upright position on a flat surface for making height measurements. It also contains a small bubble-type level so that it can be used by itself as a level. The *bevel protractor* permits the measurement or layout of angles. The *center head* permits the center of cylindrical work to be determined.

The *vernier caliper* (Figure 10-15) is an end-measuring instrument, available in various sizes, that can be used to make both outside and inside measurements to theoretical accuracies of 0.01 mm or 0.001 in. End-measuring instruments are more

FIGURE 10-13 Machinist's rules: (a) metric; (b) inch graduations; 10ths and 100ths on one side, 32nds and 64ths on the opposite side. *(Courtesy of L.S. Starrett Company.)*

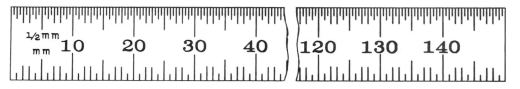

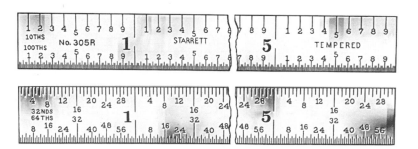

FIGURE 10-14 Combination set. *(Courtesy of MTI Corporation.)*

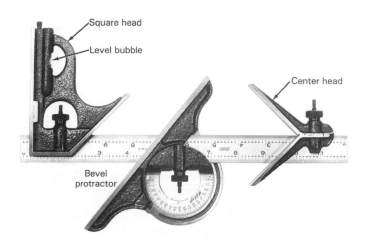

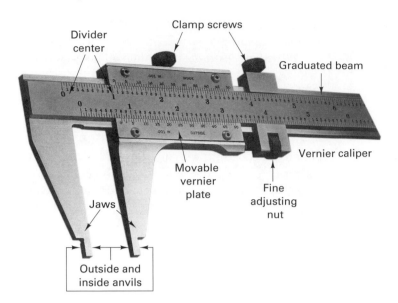

FIGURE 10-15   This vernier caliper can make measurements using both inside (for holes) and outside (shafts) anvils.

accurate and somewhat easier to use than line-matching types because their jaws are placed against either end of the object being measured, so any difficulty in aligning edges or lines is avoided. However, the difficulty remains in obtaining uniform contact pressure, or "feel," between the legs of the instrument and the object being measured.

A major feature of the vernier caliper is the auxiliary scale (Figure 10-16). The caliper shown has a graduated beam with a metric scale on the top, a metric vernier plate, and an English scale on the bottom with an English vernier. The manner in which readings are made is explained in the figure.

FIGURE 10-16   Vernier caliper graduated for English and metric (direct) reading. The metric reading is 27 + 0.42 = 27.42 mm.

Refer to the upper bar graduations and metric vernier plate. Each bar graduation is 1.00 mm. Every tenth graduation is numbered in sequence–10 mm, 20 mm, 30 mm, 40 mm, etc. over the full range of the bar. This provides for direct reading in millimeters.

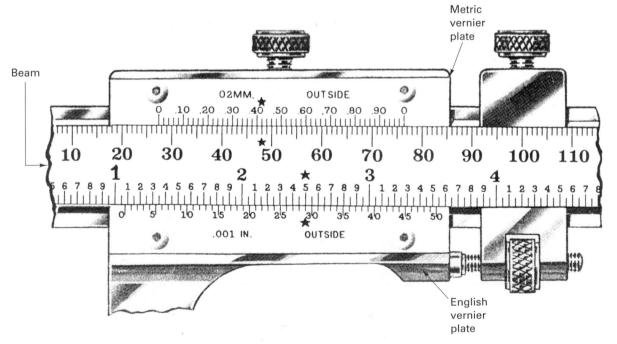

In the picture, the vernier plate zero line is one inch (1.000") plus one twentieth (0.50") beyond the zero line on the bar, or 1.050". The 29th graduation on the vernier plate coincides with a line on the bar (as indicated by stars). 29 × 0.001 (.029") is therefore added to the 1.050" bar reading, and the total is 1.079".

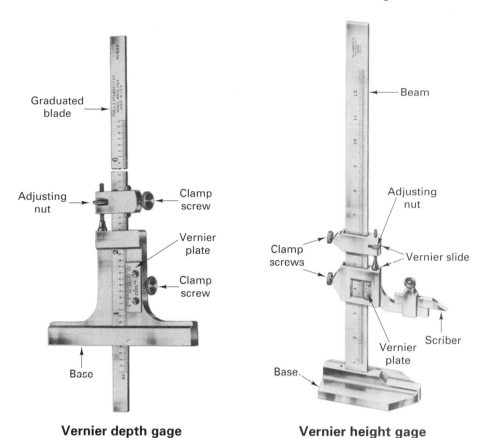

FIGURE 10-17 Variations in the vernier caliper design result in other basic gages.

Vernier depth gage   **Vernier height gage**

Figure 10-17 also shows a vernier depth gage for measuring the depth of holes or the length of shoulders on parts and a vernier height gage for making height measurements. Figure 10-18 shows calipers that have a dial indicator or a digital readout that replace the vernier. The latter two calipers are capable of making inside and outside measurements as well as depth measurements, as shown in Figure 10-19, which shows typical applications for this end-measuring device. The digital device can be zeroed at any position, as shown in the left-hand sketch of applications A through D, which greatly speeds the inspection processes and improves reading accuracy (eliminates math errors).

The *micrometer caliper*, more commonly called a *micrometer*, is one of the most widely used measuring devices. Until recently, the type shown in Figure 10-20 was virtually standard. It consists of a fixed anvil and a movable spindle. When the thimble is rotated on the end of the caliper, the spindle is moved away from the anvil by means of an accurate screw thread. On English types, this thread has a lead of 0.025 in., and one revolution of the thimble moves the spindle this distance. The barrel, or sleeve, is calibrated in 0.025-in. divisions, with each $\frac{1}{10}$ of an inch being numbered. The circumference at the edge of the thimble is graduated into 25 divisions, each representing 0.001 in.

A major difficulty with this type of micrometer is making the reading of the dimension shown on the instrument. To read the instrument, the division on the thimble that coincides with the longitudinal line on the barrel is added to the largest reading exposed on the barrel.

Micrometers graduated in ten-thousandths of an inch are the same as those graduated in thousandths, except that an additional vernier scale is placed on the sleeve so that a reading of ten-thousandths is obtained and added to the thousandths reading. The vernier consists of 10 divisions on the sleeve, shown in B, which occupy the same space as 9 divisions on the thimble. Therefore, the difference between the width of one of the 10 spaces on the vernier and one of the 9 spaces on the thimble is one-tenth of a division on the thimble, or one-tenth of one-thousandth, which is one ten-thousandth. To read a ten-thousandths micrometer, first obtain the thousandths reading, then see which of the lines on the vernier coincides with a line on the thimble. If it is the line marked "1," add one ten-thousandths; if it is the line marked "2," add two ten-thousandths; and so on.

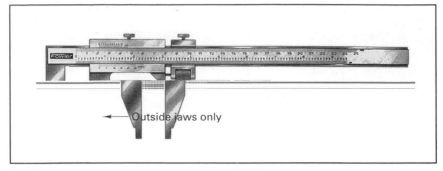

Vernier caliper with inch or metric scales and 0.001 in. accuracy

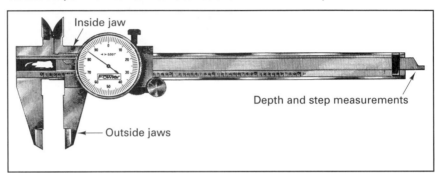

Dial caliper with 0.001 in. accuracy

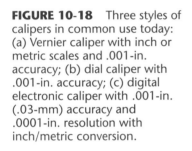

FIGURE 10-18   Three styles of calipers in common use today: (a) Vernier caliper with inch or metric scales and .001-in. accuracy; (b) dial caliper with .001-in. accuracy; (c) digital electronic caliper with .001-in. (.03-mm) accuracy and .0001-in. resolution with inch/metric conversion.

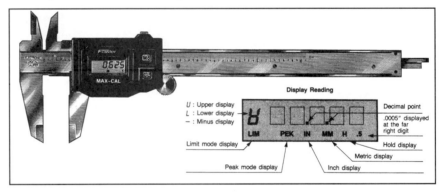

Digital electronic caliper with 0.001 in. (0.03 mm) accuracy and 0.0001 in. resolution with inch/metric conversion.

EXAMPLE: REFER TO INSETS A AND B IN FIGURE 10-20.

The "2" line on sleeve is visible, representing	0.200 in.
Two additional lines, each representing .025″	2 × .025″ = 0.050 in.
Line "O" on the thimble coincides with the reading line on the sleeve, representing	0.000 in.
The "O" lines on the vernier coincide with lines on the thimble, representing	0.000 in.
The micrometer reading is	0.2500 in.

Now you try to read inset C.

The "2" line on sleeve is visible, representing	0.200 in.
Two additional lines, each representing .025″	2 × .025″ = 0.050 in.
The reading line on the sleeve lies between the "O" and "1" on the thimble, so ten-thousandths of an inch is to be added as read from the vernier.	
The "7" line on the vernier coincides with a line on the thimble, representing	7 × .0001″ = .0007 in.
The micrometer reading is	0.2507 in.

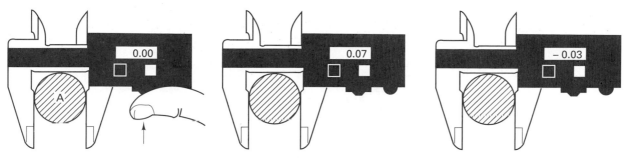

(a) Deviation from reference size (A = reference size). Use outside jaws.

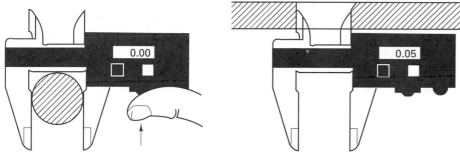

(b) Comparison, e.g., between plug and hole. Use outside + inside jaws.

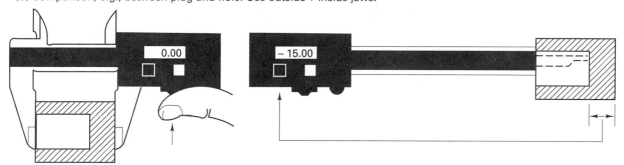

(c) Measurement of wall thickness. Use outside jaws + depth rod.

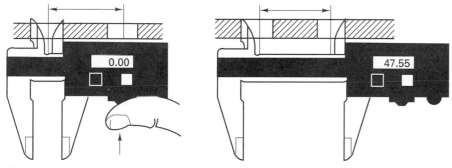

(d) Measurement of center of distance between two identical holes. Use inside jaws.

FIGURE 10-19 Four typical applications of a digital caliper. (a) Deviation from reference size (A, reference size). Use outside jaws. (b) Comparisons, e.g., between plug and hole. Use outside + inside jaws. (c) Measurement of wall thickness. Use outside jaws + depth rod. (d) Measurement of center of distance between two identical holes. Use inside jaws. Reading reset to zero (0.000) in each left-hand figure.

However, owing to the lack of pressure control, micrometers can seldom be relied on for accuracy beyond 0.0005 in., and such vernier scales are not used extensively. On metric micrometers the graduations on the sleeve and thimble are usually 0.5 mm and 0.01 mm, respectively (see the problems at the end of chapter).

Many errors have resulted from the ordinary micrometer being misread, the error being ±0.025 in. or ±0.5 mm. Consequently, direct-reading micrometers have been developed. Figure 10-21 shows a digital outside micrometer that reads to 0.001 in. on the

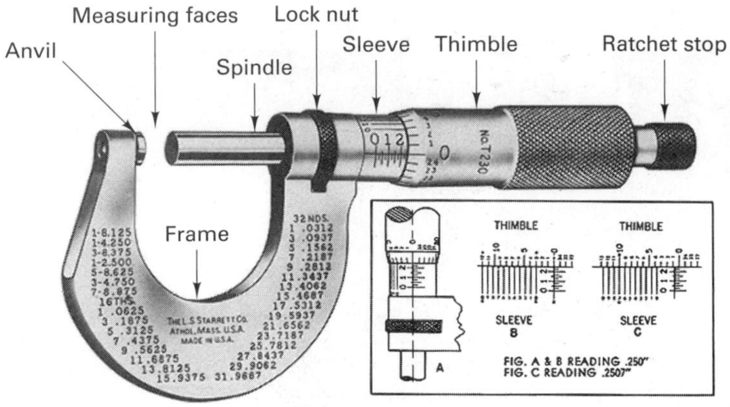

FIGURE 10-20 Micrometer caliper graduated in ten-thousandths of an inch with insets A, B, and C showing two example readings *(Courtesy Starrett Bulletin No. 1203.)*

FIGURE 10-21 Digital micrometer for measurements from 0 to 1 in., in .0001-in. graduations.

digit counter and 0.0001 in. on the vernier on the sleeve. The range of a micrometer is limited to 1 in. Thus a number of micrometers of various sizes are required to cover a wide range of dimensions. To control the pressure between the anvil, the spindle, and the piece being measured, most micrometers are equipped with a ratchet or a friction device, as shown in Figures 10-20 and 10-21. Calipers that do not have this device may be overtightened and sprung by several thousandths by applying excess torque to the thimble. Micrometer calipers should not usually be relied on for measurements of greater accuracy than 0.01 mm or 0.001 in., unless they are of the new digital design.

Micrometer calipers are available with a variety of specially shaped anvils and/or spindles, such as point, balls, and disks, for measuring special shapes, including screw threads. Micrometers are also available for inside measurements, and the micrometer principle is also incorporated into a *micrometer depth gage*.

Bench micrometers with direct readout to data processors are becoming standard inspection devices on the plant floor (see Figure 10-22). The data processor provides a record of measurement as well as control charts and histograms. Direct-gaging height gages, calipers, indicators, and micrometers are also available with statistical analysis capability. See Chapter 12 for more discussion on control charts and process capability.

Larger versions of micrometers, called *supermicrometers*, are capable of measuring 0.0001 in. when equipped with an indicator that shows that a selected pressure between the anvils has been obtained. The addition of a digital readout permits the device to measure to ±0.00005 in. (0.001 mm) directly when it is used in a controlled-temperature environment.

The toolmaker's microscope, shown in Figure 10-23, is a versatile instrument that measures by optical means; no pressure is involved. Thus it is very useful for making accurate measurements on small or delicate parts. The base, on which the special microscope is mounted, has a table that can be moved in two mutually perpendicular, horizontal directions (X and Y) by means of accurate micrometer screws that can be read to 0.0001 in., or, if so equipped, by means of the digital readout. Parts to be measured are mounted on the table, the microscope is focused, and one end of the desired part feature is aligned with the cross-line in the microscope. The reading is then noted, and the table is moved until the other extremity of the part coincides with the cross-line. From the final reading, the desired measurement can be determined. In addition to a wide

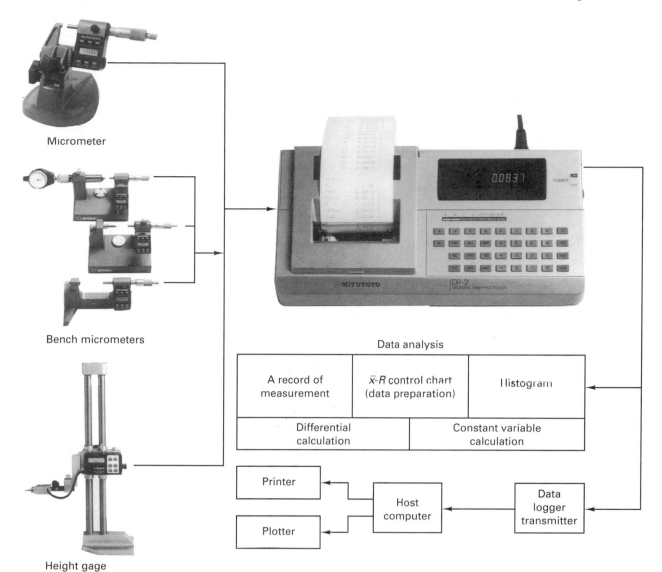

Micrometer

Bench micrometers

Height gage

FIGURE 10-22   Direct gaging system for process control and statistical analysis of inspection data. *(Courtesy of MITUTOYO.)*

variety of linear measurements, accurate angular measurements can also be made by means of a special protractor eyepiece. These microscopes are available with digital readouts.

The *optical projector* or *comparator* (Figure 10-24) is a large optical device on which both linear and angular measurements can be made. As with the toolmaker's microscope, the part to be measured is mounted on a table that can be moved in X and Y directions by accurate micrometer screws. The optical system projects the image of the part on a screen, magnifying it from 5 to more than 100 times. Measurements can be made directly by means of the micrometer dials, the digital readouts, or the dial indicators, or on the magnified image on the screen by means of an accurate rule. A very common use for this type of instrument is the checking of parts, such as dies and screws. A template is drawn to an enlarged scale and is placed on the screen. The projected contour of the part is compared to the desired contour on the screen. Some projectors also function as low-power microscopes by providing surface illumination.

MEASURING WITH LASERS

One of the earliest and most common metrological uses of low-power lasers has been in interferometry. The interferometer uses light interference bands to determine distance and thickness of objects (Figure 10-25). First, a beam splitter divides a beam of light into a measurement beam and a reference beam. The measurement beam travels to a reflector (optical glass plate A) resting on the part whose distance is to be measured, while the reference beam is directed at fixed reflector B. Both beams are reflected back through the beam splitter, where they are recombined into a single beam before traveling

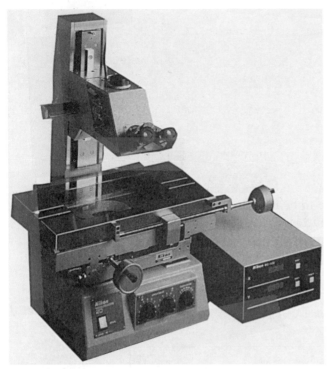

FIGURE 10-23 Toolmaker's microscope with digital readouts for *X* and *Y* table movements. *(Courtesy of Nikon.)*

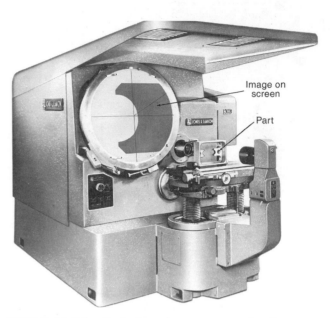

FIGURE 10-24 Optical comparator, measuring the contour on a workpiece. Digital indicators with in./mm conversions add to the utility of optical comparators.

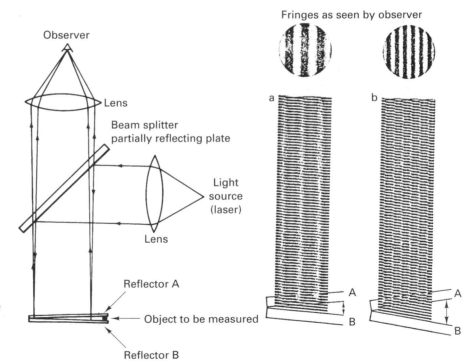

FIGURE 10-25 Interference bands can be used to measure the size of objects to great accuracy. *(Based on the Michelson interferometer invented in 1882.)*

to the observer. This recombined beam produces interference fringes, depending on whether the waves of the two returning beams are in phase (called *constructive interference*) or out-of-phase (termed *destructive interference*). In-phase waves produce a series of bright bands, and out-of-phase waves produce dark bands. The number of fringes can be related to the size of the object, measured in terms of light waves of a given frequency. The following example will explain the basics of the method.

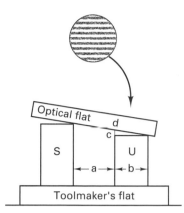

FIGURE 10-26 Method of calibrating gage block by light-wave interference.

To determine the size of object U in Figure 10-26, a calibrated reference standard S, plus an optical flat and a toolmaker's flat, are needed along with a monochromatic light source. *Optical flats* are quartz or special glass disks, from 2 to 10 in. (50 to 250 mm) in diameter and about $\frac{1}{2}$ to 1 in. (12 to 25 mm) thick, whose surfaces are very nearly true planes and nearly parallel. Flats can be obtained with the surfaces within 0.000001 in. (0.00003 mm) of true flatness. It is not essential that both surfaces be accurate or that they be exactly parallel, but one must be certain that only the accurate surface is used in making measurements. A *toolmaker's flat* is similar to an optical flat but is made of steel and usually has only one surface that is accurate. A *monochromatic light source*, light of a single wavelength, must be used. Selenium, helium, or cadmium light sources are commonly used along with helium-neon lasers.

The block to be measured is U and the calibrated block is S. Distances a and b must be known but do not have to be measured with great accuracy. By counting the number of *interference bands* shown on the surface of block U, the distance c–d can be determined. Because the difference in the distances between the optical flat and the surface of U is one-half wavelength, each dark band indicates a change of one-half wavelength in the elevation. If a monochromatic light source having a wavelength of 23.2 μin. (0.589 μin.) is used, each interference band represents 11.6 μin. (0.295 μm). Then, by simple geometry, the difference in the heights of the two blocks can be computed. The same method is applicable for making precise measurements of other objects by comparing them with a known gage block.

Accurate measurement of distances greater than a few inches was very difficult until the development of laser interferometry, which permits accuracies of ±0.5 part per million over a distance of 6.1 m with 0.01 μm resolution. Such equipment is particularly useful in checking the movement of machine tool tables, aligning and checking large assembly jigs, and making measurements of intricate machined parts such as tire-tread molds.

The Hewlett-Packard laser interferometer (Figure 10-27) uses a helium-neon laser beam split into two beams, each of different frequency and polarized. When the beams are recombined, any relative motion between the optics creates a Doppler shift in the frequency. This shift is then converted into a distance measurement. The laser light has less tendency to diverge (spread out) and is also monochromatic (of the same wavelength). A process that has been largely confined to the optical industry and the metrology lab is now suitable for the factory, where its extremely precise distance-measuring capabilities have been applied to the alignment and calibration of machine tools.

The company's first two-frequency interferometer calibration system was introduced in 1970 to overcome workplace contamination by thermal gradients, air turbulence, oil mist, and so on, which affect the intensity of light. Doppler laser interferometers are relatively insensitive to such problems. The system can be used to measure linear distances, velocities, angles, flatness, straightness, squareness, and parallelism in machine tools.

Lasers provide for accurate machine tool alignment. Large, modern machine tools can move out of alignment in a matter of months, causing production problems often attributed to the cutting tools, the work holders, the machining conditions, or the numerical control part program.

Light interference also makes it possible to determine easily whether a surface is exactly flat. The achievement of interference fringes is largely dependent on the coherence of the light used. The availability of highly coherent *laser* light (in-phase light of a single frequency) has made interferometry practical in far less restrictive environments than in the past. The sometimes arduous task of extracting usable data from a close-packed series of interference fringes has been taken over by microprocessors.

The most widely used laser technique for inspection and in-process gaging is known as *laser scanning*. At its most basic level, the process consists of placing an object between the source of the laser beam and a receiver containing a photodiode. A microprocessor then computes the object's dimensions based on the shadow that the object casts (Figure 10-28).

The noncontact nature of laser scanning makes it well suited to in-process measurement, including such difficult tasks as the inspection of hot-rolled, or extruded material, and its comparative simplicity has led to the development of highly portable systems. The bench gage versions can measure to resolutions of 0.0001 mm.

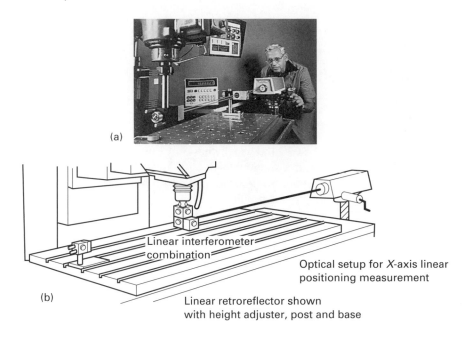

(a)

(b)

Linear interferometer combination

Optical setup for *X*-axis linear positioning measurement

Linear retroreflector shown with height adjuster, post and base

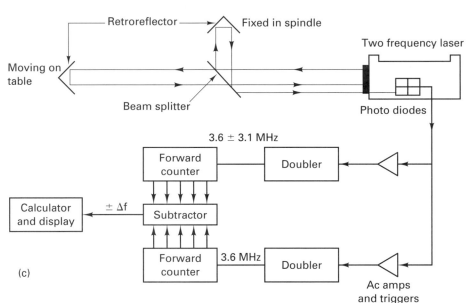

(c)

FIGURE 10-27
(Top) Calibrating the *X*-axis linear table displacement of a vertical spindle milling machine; (middle) schematic of optical setup; (bottom) schematic of components of a two frequency laser interferometer. *(Courtesy of Hewlett-Packard.)*

■ 10.6 VISION SYSTEMS FOR MEASUREMENT

If a picture is worth a thousand words, then vision systems are the tome of inspection methods (see Figure 10-29). Machine vision is used for visual inspection, for guidance and control, or for both. Normal TV image formation on photosensitive surfaces or arrays is used, and the video signals are analyzed to obtain information about the object. Each picture frame represents the object at some brief interval of time. Each frame must be dissected into picture elements (called *pixels*). Each pixel is *digitized* (has binary numbers assigned to it) by fixing the brightness or gray level of each pixel to produce a *bit-map* of the object (Figure 10-30). That is, each pixel is assigned a numerical value based on its shade of gray. Image preprocessing improves the quality of the image data by removing unwanted detail. The bit map is stored in a buffer memory. By analyzing and processing the digitized and stored bit map, the patterns are extracted, edges located, and dimensions determined.

Sophisticated computer algorithms using artificial intelligence have greatly reduced the computer operations needed to achieve a result, but even the most powerful video-based systems currently require one to two seconds to achieve a measurement. This

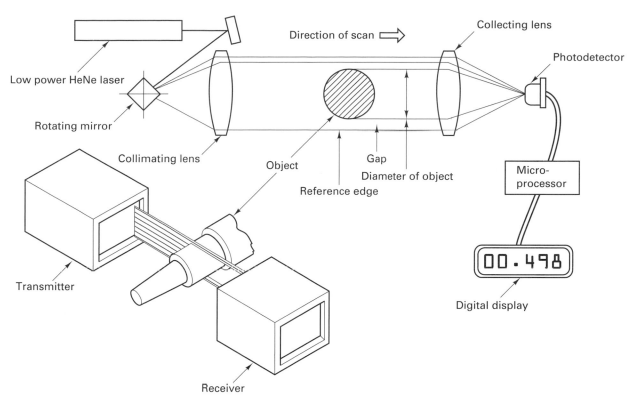

FIGURE 10-28 Scanning laser measuring system. *(Courtesy of ZYGO Corporation.)*

FIGURE 10-29 Schematic of element of a machine vision system.

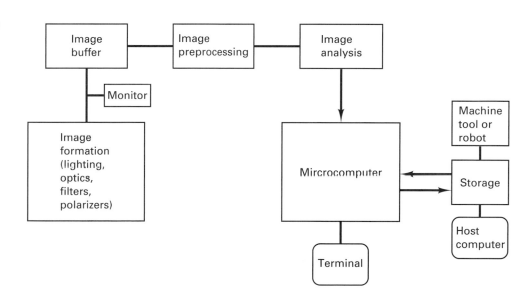

FIGURE 10-30 Vision systems use a gray scale to identify objects. (a) Object with three different gray values. (b) One frame of object (pixels). (c) Each pixel assigned a gray scale number.

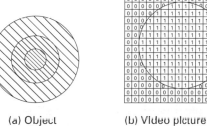

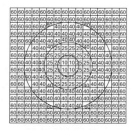

(a) Object (b) Video picture frame (c) Digitized frame

TABLE 10-5.   Laser Scanning versus Vision Systems

Variable	Laser-Scanning Systems	Video-Based Systems
Ambient lighting	Independent	Dependent
Object motion	Object usually stationary	Multiple cameras or strobe lighting may be required
Adaptability to robot systems	Readily adapted; some limitations on robot motion speed or overall system operation	Readily adapted; image-processing delays may delay system operation
Signal processing	Simple; computers often not required	Requires relatively powerful computers with sophisticated software
Cycle time	Very fast	Seconds of computer time may be needed
Applicability to simple tasks	Readily handled; edges and features produce sharp transitions in signal	Requires extensive use of sophisticated software algorithms to identify edges
Sizing capability	Can size an object in a single scan per axis	Can size on horizontal axis in one scan; other dimensions require full-frame processing
Three-dimensional capability	Limited three dimensionality; needs ranging capability	Uses two views of two cameras with sophisticated software or structured light
Accuracy and precision	Submicrometer 0.001 to 0.0001 in. or better accuracy; highly repeatable	Depends on resolution of cameras and distance between camera and object; systems with 0.004-in. precision and 0.006-in. accuracy are typical

may be too long a time for many on-line production applications. Table 10.5 provides a comparison of vision systems to laser scanning.

With the recent emphasis on quality and 100% inspection, applications for inspection by machine vision have increased markedly. Vision systems can check hundreds of parts per hour for multiple dimensions. Resolutions of ±0.01 in. have been demonstrated but 0.02 in. is more typical for part location. Machine vision is useful for robot guidance in material handling, welding, and assembly but nonrobotic inspection and part location applications are still more typical. The use of vision systems in inspection, quality control, sorting, and machining tool monitoring will continue to expand. Systems can cost $100,000 or more to install and must be justified on the basis of improved quality rather than labor replacement.

■ 10.7 COORDINATE MEASURING MACHINES

Precision measurements in three-dimensional Cartesian coordinate space can be made with *coordinate measuring machines* (CMM) of the design shown in Figure 10-31. The parts

FIGURE 10-31 Coordinate measuring machine with inset showing probe and a part being measured.

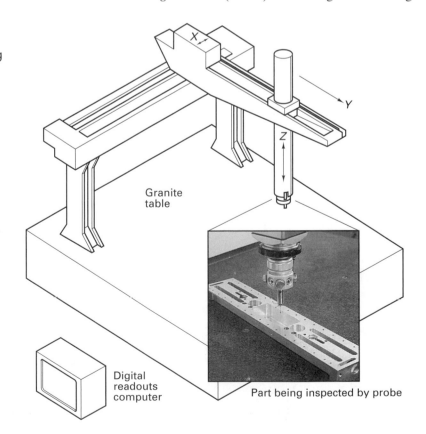

Granite table

Digital readouts computer

Part being inspected by probe

are placed on a large granite flat or the table. The vertical arm carries a probe which can be precisely moved in *x–y–z* directions to produce 3D measurements. In this design, the vertical column rides on a bridge beam and carries a touch-trigger probe. Such machines use digital readouts, air bearings, computer controls, and granite tables to achieve accuracies of the order of 0.0002 to 0.0004 in. over spans of 10 to 30 in. or more. These systems may have computer routines that give the best fit to feature measurements and that provide the means of establishing geometric tolerances discussed earlier in this chapter. Figure 10-32 gives a partial listing of the results one can achieve with these machines.

FIGURE 10-32 Examples of geometric form tolerances developed by probing surface with a CMM.

Straightness

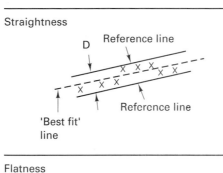

Straightness
Measured or previously calculated points may be used to determine a 'best fit' line. The form routine establishes two reference lines which are parallel to the 'best fit' line, and which just contain all of the measured or calculated points.
Straightness is defined as the distance D between these two reference lines.

Flatness

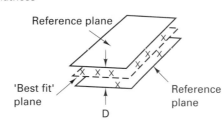

Flatness
Measured or previously calculated points may be used to determine the 'best fit' plane. The form routine establishes two reference planes which are parallel to the 'best fit' plane, and which just contain all of the measured or calculated points.
Flatness is defined as the distance D between these two reference planes.

Roundness

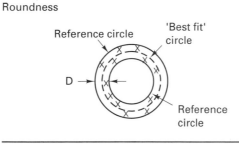

Roundness
Measured or previously calculated points may be used to determine the 'best fit' circle. The form routine establishes two reference circles which are concentric with the 'best fit' circle, and which just contain all of the measured or calculated points.
Roundness is defined as the difference D in radius of these two reference circles.

Cylindricity

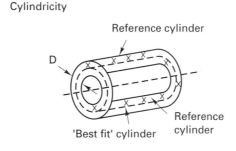

Cylindricity
Measured or previously calculated points may be used to determine the 'best fit' cylinder. The form routine establishes two reference cylinders which are co-axial to the 'best fit' cylinder, and which just contain all of the measured or calculated points.
Cylindricity is the difference D in radius of these two reference cylinders. Also applicable to stepped cylinders.

Conicity

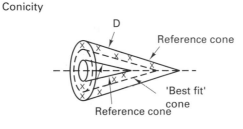

Conicity
Measured or previously calculated points may be used to determine the 'best fit' cone. The form routine establishes two reference cones which are co-axial with and similar to the 'best fit' cone, and which just contain all of the measured or calculated points.
Conicity is defined as the distance D between the side of these two reference cones.

(continues on next page)

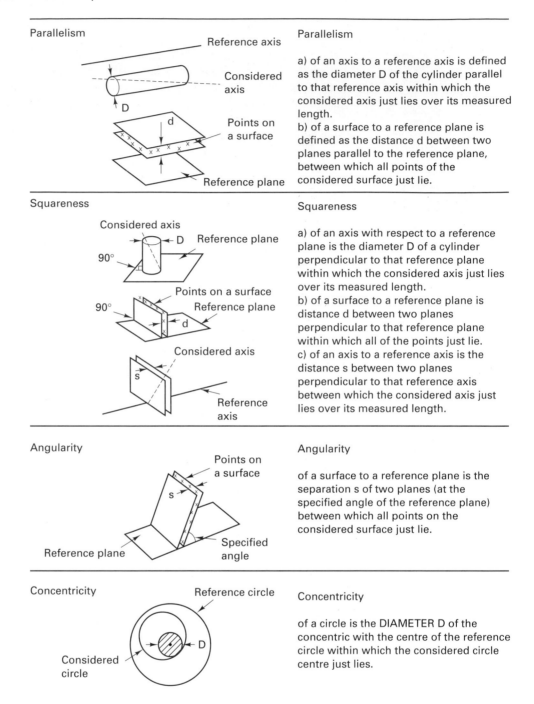

Parallelism

a) of an axis to a reference axis is defined as the diameter D of the cylinder parallel to that reference axis within which the considered axis just lies over its measured length.
b) of a surface to a reference plane is defined as the distance d between two planes parallel to the reference plane, between which all points of the considered surface just lie.

Squareness

a) of an axis with respect to a reference plane is the diameter D of a cylinder perpendicular to that reference plane within which the considered axis just lies over its measured length.
b) of a surface to a reference plane is distance d between two planes perpendicular to that reference plane within which all of the points just lie.
c) of an axis to a reference axis is the distance s between two planes perpendicular to that reference axis between which the considered axis just lies over its measured length.

Angularity

of a surface to a reference plane is the separation s of two planes (at the specified angle of the reference plane) between which all points on the considered surface just lie.

Concentricity

of a circle is the DIAMETER D of the concentric with the centre of the reference circle within which the considered circle centre just lies.

FIGURE 10-32 *(continued)*

■ 10.8 ANGLE-MEASURING INSTRUMENTS

Accurate angle measurements are usually more difficult to make than linear measurements. Angles are measured in degrees (a degree is 1/360 part of a circle) and decimal subdivisions of a degree (or in minutes and seconds of arc). The SI system calls for measurements of plane angles in radians, but degrees are permissible. The use of degrees will continue in manufacturing, but with minutes and seconds of arc possibly being replaced by decimal portions of a degree.

The bevel protractor (Figure 10-33) is the most general angle-measuring instrument. The two movable blades are brought into contact with the sides of the angular part, and the angle can be read on the vernier scale to 5 minutes of arc. A clamping device is provided to lock the blades in any desired position so that the instrument can be used for both direct measurement and layout work. As indicated previously, an angle attachment on the combination set can also be used to measure angles, similar to the way a bevel protractor is used but usually with somewhat less accuracy.

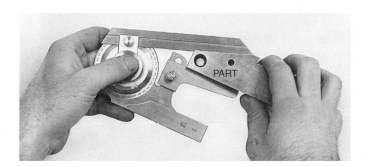

FIGURE 10-33 Measuring an angle on a part with a bevel protractor. *(Courtesy of Brown & Sharpe Mfg. Co.)*

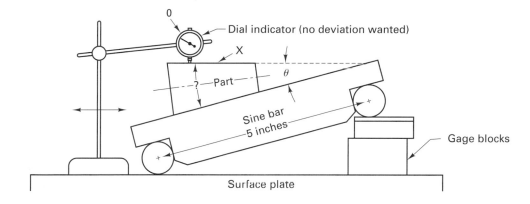

FIGURE 10-34 Setup to measure an angle on a part using a sine bar. The dial indicator is used to determine when the part surface *X* is parallel to the surface plate.

The toolmaker's microscope is very satisfactory for making angle measurements, but its use is restricted to small parts. The accuracy obtainable is 5 minutes of arc. Similarly, angles can be measured on the optical contour projector. Angular measurements can also be made by means of an angular interferometer with the laser system.

A *sine bar* may be used to obtain accurate angle measurements if the physical conditions will permit. This device (Figure 10-34) consists of an accurately ground bar on which two accurately ground pins of the same diameter are mounted an exact distance apart. The distances used are usually either 5 or 10 in., and the resulting instrument is called a 5- or 10-in. sine bar. Sine bars also are available with millimeter dimensions. Measurements are made by using the principle that the sine of a given angle is the ratio of the opposite side to the hypotenuse of the right triangle.

The object being measured is attached to the sine bar, and the inclination of the assembly is raised until the top surface is exactly parallel with the surface plate. A stack of gage blocks is used to elevate one end of the sine bar, as shown in Figure 10-34. The height of the stack directly determines the difference in height of the two pins. The difference in height of the pins can also be determined by a dial indicator gage or any other type of gage. The difference in elevation is then equal to either 5 or 10 times the sine of the angle being measured, depending on whether a 5- or 10-in. bar is being used. Tabulated values of the angles corresponding to any measured elevation difference for 5- or 10-in. sine bars are available in various handbooks. Several types of sine bars are available to suit various requirements.

Accurate measurements of angles to 1 second of arc can be made by means of *angle gage blocks*. These come in sets of 16 blocks that can be assembled in desired combinations. Angle measurements can also be made to ±0.001° on rotary indexing tables having suitable numerical control.

■ 10.9 GAGES FOR ATTRIBUTES MEASURING

In manufacturing, particularly in mass production, it may not be necessary to know the exact dimensions of a part, only that it is within previously established limits. Limits can often be determined more easily than specific dimensions by the use of attribute-type instruments called *gages*. They may be of either fixed type or deviation type, may be used for both linear and angular dimensions, and may be used manually or mechanically (automatically).

FIXED-TYPE GAGES

Fixed-type gages are designed to gage only one dimension and to indicate whether it is larger or smaller than the previously established standard.

They do not determine how much larger or smaller the measured dimension is than the standard. Because such gages fulfill a simple and limited function, they are relatively inexpensive and are easy to use.

Gages of this type are ordinarily made of hardened steel of proper composition and are heat treated to produce dimensional stability. Hardness is essential to minimize wear and maintain accuracy. Because steels of high hardness tend to be dimensionally unstable, some fixed gages are made of softer steel with a hard chrome plating on the surface to provide surface hardness. Chrome plating can also be used for reclaiming some worn gages. Where gages are to be subjected to extensive use, they may be made of tungsten carbide at the wear points.

The *plug gage* is one of the most common types of fixed gages. As shown in Figure 10-35, plug gages are accurately ground cylinders used to gage internal dimensions, such as holes. The gaging element of a *plain plug gage* has a single diameter. To control the minimum and maximum limits of a given hole, two plug gages are required. The smaller, or *go gage*, controls the minimum because it must go (slide) into any hole that is larger than the required minimum. The larger, or *no-go gage*, controls the maximum dimension because it will not go into any hole unless that hole is over the maximum permissible size. The go and no-go plugs are often designed with two gages on a single handle for convenience in use. The no-go plug is usually much shorter than the go plug; it is subjected to little wear because it seldom slides into any holes. Figure 10-36 shows a *step-type go/no-go gage* that has the go and not-go diameters on the same end of a single plug, the go portion being the outer end. The user knows that the part is good if the *go* gage goes into the hole but the *no-go* gage does not go. Such gages require careful use and should never be forced into (or onto) the part. Obviously these plug gages were specially designated and made for checking a specific hole on a part.

In designing plug and snap ring gages, the key principle is: *It is better to reject a good part than declare a bad part to be within specifications*. All gage design decisions are made with this principle in mind. Gages must have tolerances like any manufactured components. All gages are made with gage and wear tolerances. Gage tolerance allows for the permissible variation in the manufacture of the gage. It is typically 5 to 20% (depending on the industry) of the tolerance on the dimension being gaged. Wear tolerances compensate for the wear of the gage surface as a result of repeated use. Wear tolerance is applied only to the go side of the gage because the no-go side should seldom see contact with a part surface. It is typically 5 to 20% of the dimensional tolerance.

Plug-type gages are also made for gaging shapes other than cylindrical holes. Three common types are *taper plug gages, thread plug gages, and spline gages*. Taper plug gages gage both the angle of the taper and its size. Any deviation from the correct angle is indicated by looseness between the plug and the tapered hole. The size is indicated by the depth to which the plug fits into the hole, the correct depth being denoted by a mark on the plug. Thread plug gages come in go and no-go types. The go gage must screw into the threaded holes, and the no-go gage must not enter.

FIGURE 10-35   Plain plug gage having the go member on the left end (1.1250-in. diameter) and not-go member on the right end. *(Courtesy of Sheffield.)*

FIGURE 10-36   Step type plug gage with go and no-go elements on the same end. *(Courtesy of Sheffield.)*

FIGURE 10-37   Go and no-go (on right) ring gages for checking a shaft. *(Courtesy of Automation and Measurement Division, Bendix Corporation.)*

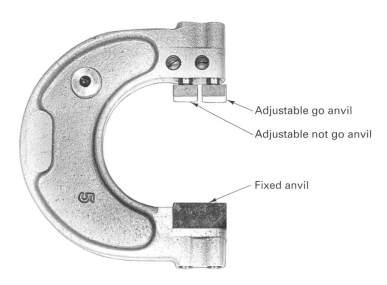

Adjustable go anvil

Adjustable not go anvil

Fixed anvil

FIGURE 10-38   Adjustable go and not-go snap gage. *(Courtesy of Bendix Corporation, Automation and Measurement Division.)*

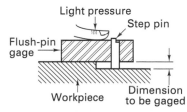

Light pressure

Step pin

Flush-pin gage

Workpiece

Dimension to be gaged

FIGURE 10-39   Flush-pin gage being used to check height of step.

Ring gages are used to check shafts or other external round members. These are also made in go and no-go types as shown in Figure 10-37. Go ring gages have plain knurled exteriors, whereas no-go ring gages have a circumferential groove in the knurling, so that they can easily be distinguished. *Ring thread gages* are made to be slightly adjustable because it is almost impossible to make them exactly to the desired size. Thus they are adjusted to exact, final size after the final grinding and polishing have been completed.

Snap gages are the most common type of fixed gage for measuring external dimensions. As shown in Figure 10-38, they have a rigid, U-shaped frame on which are two or three gaging surfaces, usually made of hardened steel or tungsten carbide. In the adjustable type shown, one gaging surface is fixed, and the other(s) may be adjusted over a small range and locked at the desired position(s). Because in most cases one wishes to control both the maximum and the minimum dimensions, the *progressive* or *step-type snap gage* (Figure 10-39) is used most frequently. These gages have one fixed anvil and two adjustable surfaces to form the outer go and the inner no-go openings, thus eliminating the use of separate go and no-go gages.

Snap gages are available in several types and a wide range of sizes. The gaging surfaces may be round or rectangular. They are set to the desired dimensions with the aid of gage blocks.

Many types of special gages are available or can be constructed for special applications. The *flush-pin gage* (Figure 10-39) is an example for gaging the depth of a shoulder. The main section is placed on the higher of the two surfaces, with the movable step pin resting on the lower surface. If the depth between the two surfaces is sufficient but not too great, the top of the pin, but not the lower step, will be slightly above the top surface of the gage body. If the depth is too great, the top of the pin will be below the surface. Similarly, if the depth is not great enough, the lower step on the top of the pin will

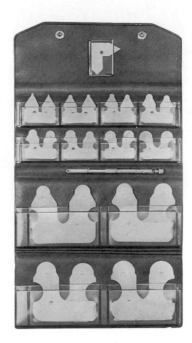

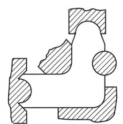

FIGURE 10-40 Set of radius gages, showing how they are used. *(Courtesy of MTI Corporation.)*

FIGURE 10-41 Thread pitch gages. *(Courtesy of L.S. Starrett Company.)*

be above the surface of the gage body. When a finger, or fingernail, is run across the top of the pin, the pin's position with respect to the surface of the gage body can readily be determined.

Several types of *form gages* are available for use in checking the *profile* of various objects. Two of the most common types are *radius gages* (Figure 10-40) and *screw-thread pitch gages*, shown in Figure 10-41.

DEVIATION-TYPE GAGES

A large amount of gaging, and some measurement, is done through the use of *deviation-type gages*, which determine the amount by which a measured part deviates, plus or minus, from a standard dimension to which the instrument has been set. In most cases, the deviation is indicated directly in units of measurement, but in some cases, the gage shows only whether the deviation is within a permissible range. A good example of a deviation-type gage is a flashlight battery checker, which shows if the battery is good (green), bad (red), or borderline (yellow), but not how much voltage or current is generated. Such gages use mechanical, electrical, or fluidic amplification techniques so that very small linear deviations can be detected. Most are quite rugged, and they are available in a variety of designs, amplifications, and sizes.

Dial indicators, as shown in Figure 10-5, are a widely used form of deviation-type gage. Movement of the gaging spindle is amplified mechanically through a rack and pinion and a gear train and is indicated by a pointer on a graduated dial. Most dial indicators have a spindle travel equal to about $2\frac{1}{2}$ revolutions of the indicating pointer and are read in either 0.001 or 0.0001 in. (or 0.02 or 0.002 mm).

The dial can be rotated by means of the knurled bezel ring to align the zero point with any position of the pointer. The indicator is often mounted on an adjustable arm to permit its being brought into proper relationship with the work. It is important that the axis of the spindle be aligned exactly with the dimension being gaged if accuracy is to be achieved. Digital dial indicators are also readily available (Figure 10-42).

Dial indicators should be checked occasionally to determine if their gage capability has been lost through wear in the gear train. Also, it should be remembered that the pressure of the spindle on the work varies because of spring pressure as the spindle moves into the gage. This spring pressure normally causes no difficulty unless the spindles are used on soft or flexible parts.

Linear variable-differential transformers (LVDT) are used as sensory elements in many electronic gages, usually with a solid-state diode display or in automatic inspec-

FIGURE 10-42 Digital dial indicator with 1 in. range and .0001-in. accuracy. *(Courtesy of CDI.)*

tion setups. These devices can frequently be combined into multiple units for the simultaneous gaging of several dimensions. Ranges and resolutions down to 0.0005 and 0.00001 in. (0.013 and 0.00025 mm, respectively) are available.

Air gages have special characteristics that make them especially suitable for gaging holes or the internal dimensions of various shapes. A typical gage of this type, shown earlier in Figure 10-12, indicates the clearance between the gaging head and the hole by measuring either the volume of air that escapes or the pressure drop resulting from the airflow. The gage is calibrated directly in 0.0001-in. or 0.02-mm divisions. Air gages have an advantage over mechanical or electronic gages for this purpose in that they detect not only linear size deviations but also out-of-round conditions. Also, they are subject to very little wear because the gaging member is always slightly smaller than the hole and the airflow minimizes rubbing. Special types of air gages can be used for external gaging.

■ 10.10 SURFACE ROUGHNESS MEASUREMENT

The material removal processes discussed in Part 5 generate a wide variety of surfaces textures, generally referred to as *surface finish*. The cutting processes leave a wide variety of surface patterns on the materials. *Lay* is the term used to designate the direction of the predominate surface pattern produced by the machining process. In addition, certain other terms and symbols have been developed and standardized for specifying the surface quality. The most important terms are *surface roughness*, *waviness*, and *lay* (Figure 10-43). *Roughness* refers to the finely spaced surface irregularities. It results from machining operations in the case of machined surfaces. *Waviness* is surface irregularity of greater spacing than in roughness. It may be the result of warping, vibration, or the work being deflected during machining.

A variety of instruments are available for measuring surface roughness and surface profiles. The majority of these devices use a diamond stylus that is moved at a constant rate across the surface, perpendicular to the lay pattern. The rise and fall of the stylus is detected electronically (often by an LVDT device), is amplified and recorded on a strip-chart, or is processed electronically to produce average or root-mean-square readings for a meter (Figure 10-44). The unit containing the stylus and the driving motor may be handheld or supported by skids that ride of the workpiece or some other supporting surface.

Roughness is measured by the height of the irregularities with respect to an average line. These measurements are usually expressed in micrometers or microinches. In most cases, the arithmetical average (AA) is used. In terms of the measurements indicated in Figure 10-44, the AA would be as follows:

$$AA = \frac{\sum_{i=1}^{n} y_i}{n}$$

Cutoff refers to the sampling length used for the calculation of the roughness height. When it is not specified, a value of 0.030 inch (0.8 mm) is assumed. In the previous equation, y_i is a vertical distance from the centerline and n is the total number of vertical measurements taken within a specified cutoff distance. This average roughness value is also called R_a. Occasionally used is the *root-mean-square* (rms) valve, which is defined as

$$rms = \sqrt{\frac{\sum_{i=1}^{n} y_i^2}{n}}$$

The instrument shown in Figure 10-45 is capable of making a series of parallel offset traces on the surface, providing the two-dimensional profile maps shown in the bottom of the figure. Notice how the different machining processes create different roughness profiles. Areas of from 0.005 × 0.005 in. (0.13 × 0.13 mm) up to 2 × 2 in. (50.8 × 50.8 mm), depending on the magnification selected, can be profiled.

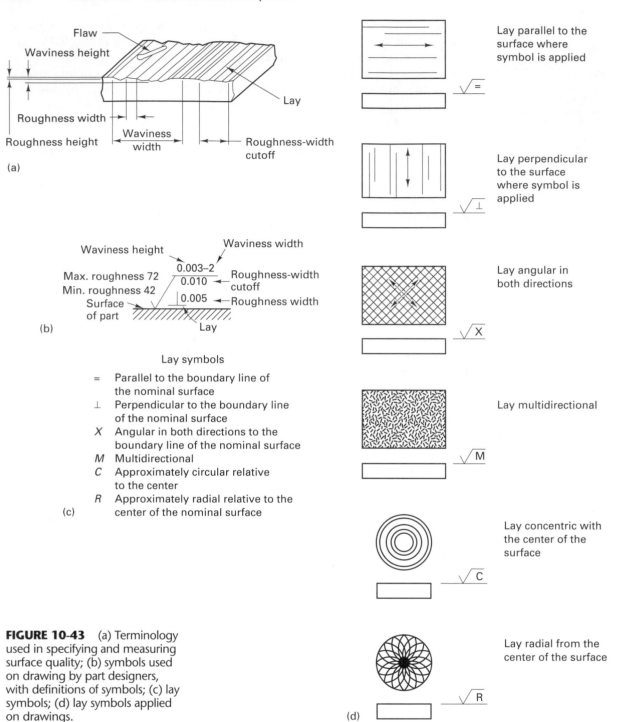

(a)

(b)

Lay symbols

= Parallel to the boundary line of the nominal surface
⊥ Perpendicular to the boundary line of the nominal surface
X Angular in both directions to the boundary line of the nominal surface
M Multidirectional
C Approximately circular relative to the center
R Approximately radial relative to the center of the nominal surface

(c)

Lay parallel to the surface where symbol is applied

Lay perpendicular to the surface where symbol is applied

Lay angular in both directions

Lay multidirectional

Lay concentric with the center of the surface

Lay radial from the center of the surface

(d)

FIGURE 10-43 (a) Terminology used in specifying and measuring surface quality; (b) symbols used on drawing by part designers, with definitions of symbols; (c) lay symbols; (d) lay symbols applied on drawings.

The resolution of stylus profile devices is determined by the radius or the diameter of the tip of the stylus. When the magnitude of the geometric features begins to approach the magnitude of the tip of the stylus, great caution should be used in interpreting the output from these devices. As a case in point, Figure 10-46 shows a scanning electron micrograph of a face-milled surface on which has been superimposed (photographically) a scanning electron micrograph of the tip of a diamond stylus (tip radius of 0.0005 in.). Both micrographs have the same final magnification. Surface flaws of the same general size as the roughness created by the machining process are difficult to resolve with the stylus-type device, where both these features are about the same size as the stylus tip.

This example points out the difference between resolution and detection. Stylus tracing devices often can detect the presence of a surface crack, step, or ridge on the

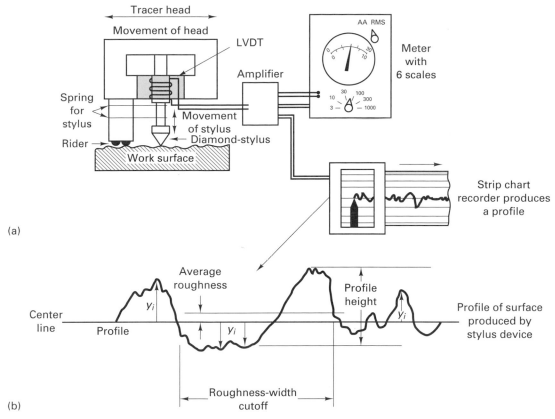

(a)

(b)

FIGURE 10-44   (a) Schematic of stylus profile device for measuring surface roughness and surface profile with two readout devices shown: a meter for AA or rms values and a strip chart recorder for surface profile. (b) Profile enlarged.

FIGURE 10-45

(Top) Microtopographer, a stylus profile device used to measure and depict surface roughness and character (surface profile); (bottom) some typical surface-roughness profiles. *(Courtesy of Measurement Systems Division Gould, Inc.)*

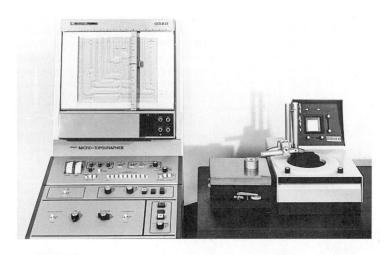

| Blanchard ground X & Y 200X Z 50X | Milled X & Y 50X Z 200 X | Ground (rust spots) X & Y 50X Z 200X | Bead blasted X & Y 50X Z 200X | EDM machined X & Y 200X Z 200X |

FIGURE 10-46 Typical machined steel surface as created by face milling and examined in the SEM. A micrograph (same magnification) of a 0.00005-in. stylus tip has been superimposed at the top.

part but cannot resolve the geometry of the defect when the defect is of the same order of magnitude as the stylus tip or smaller.

Another problem with these devices is that they produce a reading (a line on the chart) where the stylus tip is not touching the surface, as is demonstrated in Figure 10-47(a), which shows the *s* from the word *trust* on a U.S. dime. The SEM micrograph was made after the topographical map of Figure 10-47(b) had been made. Both figures are at about the same magnification. The tracks produced by the stylus tip are easily seen in the micrograph. Notice the difference between the features shown in the micrograph and the trace, indicating that the stylus tip was not in contact with the surface many times during its passage over the surface (left no track in the surface), yet the trace itself is continuous.

Figure 10-48 shows a schematic for a new laser-based instrument capable of measuring surface roughness. This instrument has the advantage of not needing to contact the surface to obtain a reading. The parameter s_n is a mean value from the scattered light distribution and has an entirely different meaning from R_a and is not (yet) an established standard. The noncontact feature, coupled with the measurement speed (20 readings per second) and the ability to read small parts easily are strong advantages. Clearly, such systems will be incorporated into machines in the future for in-process control of surface roughness.

The range of surface roughnesses that are typically produced by various manufacturing processes is indicated in Figure 10-49. This is a very general picture of typical ranges

FIGURE 10-47 (a) SEM micrograph of a U.S. dime, showing the *S* in the word *TRUST* after the region has been traced by a stylus-type machine. (b) Topographical map of the *S* region of the word *TRUST* from a U.S. dime. (Compare to part (a).)

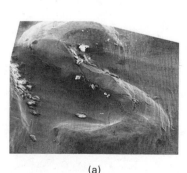

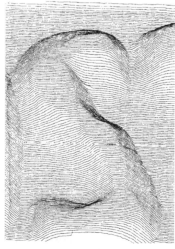

(a)   (b)

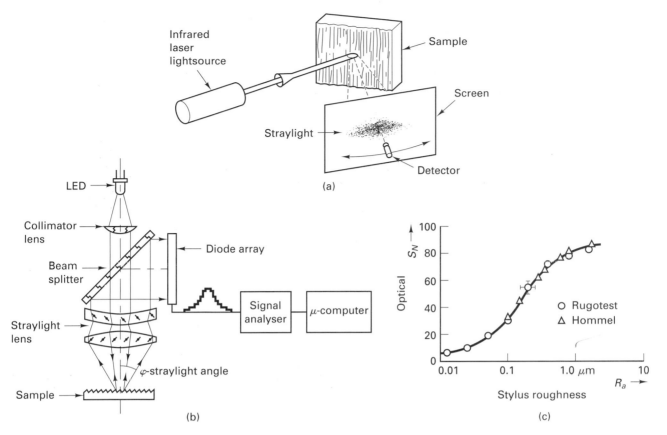

FIGURE 10-48   (a) Laser light scattering on rough surface; (b) schematic of roughness measuring system; (c) good correlation between optical and stylus roughness in the low RMS ranges. *(Courtesy of Rodenstock.)*

associated with these processes. However, one can usually count on it being more expensive to generate a fine finish (low roughness). To aid designers, samples of metals with various levels of surface roughness have been produced. See Figure 10-50 for an example.

It is difficult to visualize a surface having a given roughness, inasmuch as the same value of roughness may reflect different surface characteristics when produced by different processes. Clearly, such systems will be incorporated into machines in the future for in-process control of surface roughness.

■ KEY WORDS

accuracy	geometric tolerances	magnification	ring gage	time
allowance	interference bands	mass	rule of 10	tolerance
ampere	interference fit	metrology	sine bar	toolmaker's flat
attributes	laser interferometer	micrometer caliper	snap gage	toolmaker's microscope
candela	lay	optical comparator	stability	variables
clearance fit	length	plug gage	super micrometer	vernier caliper
coordinate measuring machine	linearity	precision	surface roughness	vision system
drift	machinist's rule	resolution	temperature	waviness
gage blocks				

■ REVIEW QUESTIONS

1. What are some of the advantages to the consumer of standardization and of interchangeable parts?
2. Why is it important to interface the manufacturing engineering requirements with the design phase as early as possible?
3. Explain the difference between attributes and variables inspection.

4. Why have so many variable-type devices in autos been replaced with attribute-type devices?
5. What are the four basic measures upon which all others depend?
6. What is a Pascal, and how is it made up of the basic measures?
7. What are the different grades of gage blocks, and why do they come in sets?

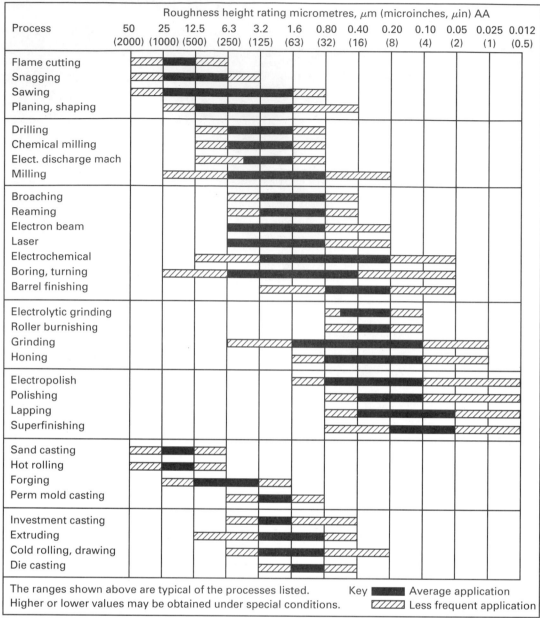

Process	Roughness height rating micrometres, μm (microinches, μin) AA												
	50 (2000)	25 (1000)	12.5 (500)	6.3 (250)	3.2 (125)	1.6 (63)	0.80 (32)	0.40 (16)	0.20 (8)	0.10 (4)	0.05 (2)	0.025 (1)	0.012 (0.5)
Flame cutting													
Snagging													
Sawing													
Planing, shaping													
Drilling													
Chemical milling													
Elect. discharge mach													
Milling													
Broaching													
Reaming													
Electron beam													
Laser													
Electrochemical													
Boring, turning													
Barrel finishing													
Electrolytic grinding													
Roller burnishing													
Grinding													
Honing													
Electropolish													
Polishing													
Lapping													
Superfinishing													
Sand casting													
Hot rolling													
Forging													
Perm mold casting													
Investment casting													
Extruding													
Cold rolling, drawing													
Die casting													

The ranges shown above are typical of the processes listed.
Higher or lower values may be obtained under special conditions.

Key ▬▬ Average application
▨▨ Less frequent application

Extracted from General Motors Drafting Standards, June 1973 revision

FIGURE 10-49 Comparison of surface roughness produced by common production processes. *(Courtesy of American Machinist.)*

FIGURE 10-50 Set of surface-roughness standards being used in a drafting room. *(Courtesy of Surface Checking Gage Co.)*

8. What keeps gage blocks together when they are "wrung together"?
9. What is the difference between tolerance and allowance?
10. What type of fit would describe the following situations?
    a. The cap of a ball-point pen
    b. The lead in a mechanical lead pencil, at the tip
    c. The bullet in a barrel of a gun
11. Here is a table that provides a description of fits from clearance to interference. Give an example of each of these fits.

	ISO Symbol			Example
	Hole Basis	Shaft Basis		
Clearance Fits	H11/c11	C11/h11	*Loose-running fit:* for wide commercial tolerances or allowances on external members	
	H9/d9	D9/h9	*Free-running fit:* not for use where accuracy is essential, but good for large temperature variations	
	H8/f7	F8/h7	*Close-running fit:* for running on accurate machines and for accurate location at moderate speeds and journal pressures	
	H7/g6	G7/h6	*Sliding fit:* not intended to run freely, but to move and turn freely and locate accurately	
	H7/h6	H7/h6	*Locational-clearance fit:* provides snug fit for locating stationary parts, but can be freely assembled and disassembled	
Transition Fits	H7/k6	K7/h6	*Locational-transition fit:* for accurate location; a compromise between clearance and interference	
	H7/n6	N7/h6	*Locational-transition fit:* for more accurate location where greater interference is permissible	
Interference Fits	H7/p6	P7/h6	*Locational-interference fit:* for parts requiring rigidity and alignment with prime accuracy of location, but without special bore pressure requirements	
	H7/s6	S7/h6	*Medium-drive fit:* for ordinary steel parts or shrink fits on light sections; the tightest fit usable with cast iron	
	H7/u6	U7/h6	*Force fit:* for highly stressed parts or for shrink fits where the heave pressing forces required are impractical	

12. Why might you use a shrink fit rather than welding to join two steel parts? What does the word *shrink* imply?
13. Explain the difference between accuracy and precision.
14. Into which of the five basic kinds of inspection does interferometry fall?
15. What factors should be considered in selecting measurement equipment?
16. Explain how you could determine if your ordinary bathroom scale is linear and has good repeat accuracy, assuming the scale is analog.
17. Design and describe a simple experiment that demonstrates the difference between magnification and resolution.
18. Explain what is meant by the statement that usable magnification is limited by the resolution of the device.
19. What is parallax? (Why do linesmen in tennis sit looking down the line?)
20. What is the rule of 10?
21. What is the principle of vernier calipers?
22. What are the two most likely sources of error in using micrometer calipers?
23. What is the major disadvantage of a micrometer caliper as compared with a vernier caliper? The advantages?
24. What would be the major difficulty in obtaining an accurate measurement with a micrometer depth gage if it were not equipped with a ratchet or friction device for turning the thimble?
25. Suppose you had a 2-ft steel bar in your supermicrometer. Could you detect a length change if the temperature of the bar changed 20°F?
26. Why is the toolmaker's microscope particularly useful for making measurements on delicate parts?
27. In what two ways can linear measurements be made using an optical projector?
28. What type of instrument would you select for checking the accuracy of the linear movement of a machine tool table through a distance of 50 inches?
29. What are the chief disadvantages of using a vision system for measurement compared to laser scanning?
30. What is a CMM (coordinate measuring machine)?
31. What is the principle of a sine?
32. How can the not-go member of a plug gage be easily distinguished from the go member?
33. What is the primary precaution that should be observed in using a dial gage?
34. What tolerances are added to gages when they are being designed?
35. Explain how a go/not-go ring gage works for check a shaft.
36. Why are air gages particularly well suited for gaging the diameter of a hole?
37. Explain the principle of measurement by light-wave interference.
38. How does a toolmaker's flat differ from an optical flat?
39. Two surfaces can have the same microinch roughness but be different in appearance. Explain!
40. Why are surface-finish blocks often used for specifying surface finish rather than microinch values?
41. What limits the resolution of a stylus-type surface-measuring device in finding profiles?
42. What is the general relationship between surface roughness and tolerance? Between tolerance and cost to produce the surface and/or tolerance?
43. What is the main disadvantage of the laser-based instrument for surface measurement?

■ PROBLEMS

1. Read the 25-division vernier graduated in English (Figure 10-A).
2. Read the 25-division vernier graduated in metric (direct reading) (Figure 10-B).
3. Convert the larger of the two readings to units of the smaller and subtract.
4. Suppose in Figure 10-34, the height of the gage blocks are 3.2500 in. What is the angle θ assuming that the dial indicator is reading 0.0 ± 0.001 in.?
5. What is the estimated error in this measurement, given that Grade 3 working blocks are being used?
6. In Figure 10-C, the sleeve-thimble region of three micrometers graduated in thousandths of an inch are shown. What are the readings for these three micrometers? (*Hint:* Think of the various units as if you were making change from a ten dollar bill. Count the figures on the sleeve as dollars, the vertical lines on the sleeve as quarters, and the divisions on the thimble as cents. Add up your change, and put a decimal point instead of a dollar sign in front of the figures.)

7. Figure 10-D shows the sleeve-thimble region of two micrometers graduate in thousandths of an inch with a vernier for an additional ten-thousandths. What are the readings?
8. In Figure 10-E, two examples of a metric vernier micrometer are shown. The micrometer is graduated in hundredths of a millimeter (0.01 mm), and an additional reading in two-thousandths of a millimeter (0.002 mm) is obtained from vernier on the sleeve. What are the readings?
9. In checking a 1-in.^{2} gage block by means of a helium light source, five dark bands were observed. There was a 2-in. distance between the front edges of the two blocks. What was the difference in height between the two blocks?
10. In Figure 10-F, the angle θ on the part needs to be inspected. The setup used is shown in Figure 10-G. (No sine plate was available.) Determine the angle θ from the part drawing and the valve of X for the stack of gage blocks.
11. Figure 10-H shows a section of a vernier caliper. What is the reading for the outside caliper?

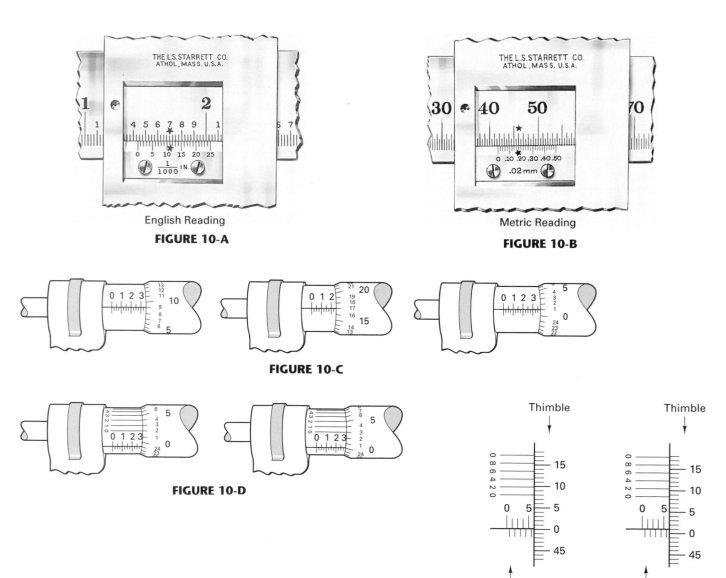

English Reading

FIGURE 10-A

Metric Reading

FIGURE 10-B

FIGURE 10-C

FIGURE 10-D

FIGURE 10-E

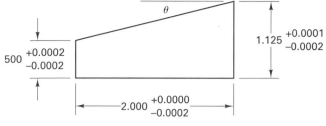

FIGURE 10-F

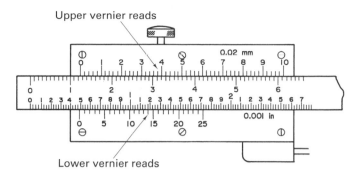

FIGURE 10-H

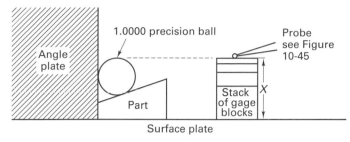

FIGURE 10-G

www.wiley.com/college/degarmo

Chapter 10 CASE STUDY

Machining Accuracy Over the Last Century

Progress in machine tool technology over the last 100 years has led to the continual redefinition of precision as shown in Figure CS-10, developed by Taniguchi. The trend here is very clear—that precision in machining continues to improve over time and approaches some limit. Discuss this figure, addressing such issues as:

1. What is the limit in machining precision?

2. What is nanoprocessing? Give some examples.
3. What is the current industrial level of precision?
4. What is the correct title for the vertical axis on this figure?
5. What processes might be grouped into the nanotechnology field? For example, what level of precision is needed in a CD player or an artificial joint?

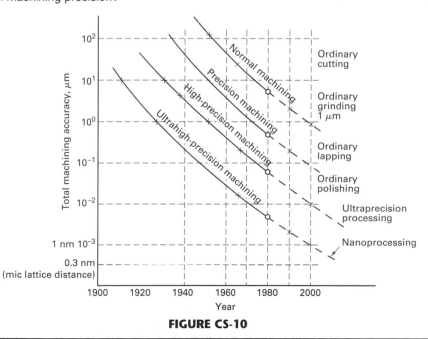

FIGURE CS-10

CHAPTER 11

NONDESTRUCTIVE INSPECTION AND TESTING

■ 11.1 DESTRUCTIVE VERSUS NONDESTRUCTIVE TESTING

The objective of most manufacturing operations is the manufacture of high-quality products, and this objective implies the absence of any defects that might cause poor performance or product failure. Care in product design, material selection, fabrication of the desired shape, heat treatment, and surface treatment, as well as consideration of all possible service conditions, can do much to assure the manufacture of quality products. However, it is also important that we confirm that our efforts have been successful—that the product is indeed free from any harmful flaws or defects.

A variety of tests have been developed to evaluate product quality and ensure the absence of any performance-impairing flaws. *Destructive testing* provides one such means of product assessment. Components or assemblies are selected and then subjected to conditions that induce failure. Determining the specific conditions where failure occurs can provide insight into the performance characteristics and quality of the remaining products. Statistical methods are used to determine the probability that the remaining products would exhibit similar behavior. For example, assume that 100 parts are produced and a randomly selected one is tested to failure with satisfactory results. Is it safe to assume that the remaining 99 would also be acceptable? A satisfactory test of another randomly selected part (or better yet, the first and last of the 100 parts) would further increase our confidence in the remaining 98. Additional tests would enhance this confidence, but the cost of destroying each of the tested (i.e., destroyed) products must be borne by the remaining quantity. Regardless of the amount of testing, there will still be some degree of uncertainty since none of the remaining products has actually been subjected to any form of property assessment.

Proof testing is another means of assuring product quality. Here a product is subjected to a load or pressure of some determined magnitude (generally equal to or greater than the designed capacity or the condition expected during operation). If the part remains intact, there is reason to believe that it will subsequently perform in an adequate fashion, provided it is not subjected to abuse or service conditions that exceed its rated level. Proof tests can be conducted under laboratory conditions or at the site of installation or assembly, as with large manufactured assemblies such as pressure vessels.

In some situations, *hardness tests* can be used to provide insight into the quality of a product. With the correct material and proper heat treatment, the resulting hardness values should fall within a well-defined range of values. Abnormal results usually indicate some form of manufacturing error, such as improper material, missed operations, or poorly controlled processes. Hardness tests can be performed quickly, and the surface indentations are often small enough that they can be concealed or easily removed from a product. The results, however, relate only to the surface strength of the product and bear no correlation to defects such as cracks or voids.

Table 11-1 provides a summary of the advantages and limitations of destructive testing and compares that approach with *nondestructive testing*. In nondestructive testing,

TABLE 11-1.   Advantages and Limitations of Destructive and Nondestructive Testing

Destructive Testing

ADVANTAGES

1. Provides a direct and reliable measurement of how a material or component will respond to service conditions.
2. Provides quantitative results, useful for design.
3. Does not require interpretation of results by skilled operators.
4. Usually find agreement as to meaning and significance of test results.

DISADVANTAGES

1. Applied only to a sample; must show that the sample is representative of the group.
2. Tested parts are destroyed during testing.
3. Usually cannot repeat a test on the same item or use the same specimen for multiple tests.
4. May be restricted for costly or few-in-number parts.
5. Hard to predict cumulative effect of service usage.
6. Difficult to apply to parts in use; if done, testing terminates their useful life.
7. Extensive machining or preparation of test specimens is often required.
8. Capital equipment and labor costs are often high.

Nondestructive Testing

ADVANTAGES

1. Can be performed directly on production items without regard to cost or quantity available.
2. Can be performed on 100% of production lot (when high variability is observed) or a representative sample (if sufficient similarity is noted).
3. Different tests can be applied to the same item, and a test can be repeated on the same specimen.
4. Can be performed on parts that are in service; the cumulative effects of service life can be monitored on a single part.
5. Little or no specimen preparation is required.
6. The test equipment is often portable.
7. Labor costs are usually low.

DISADVANTAGES

1. Results often require interpretation by skilled operators.
2. Different observers may interpret the test results differently.
3. Properties are measured indirectly and results are often qualitative or comparative.
4. Some test equipment requires a large capital investment.

the product is examined in a manner that retains its usefulness for future service. Tests can be performed on parts during or after manufacture, or even on parts that are already in service. An entire production lot can be inspected, or representative samples can be taken. Different tests can be applied to the same item, either simultaneously or sequentially, and the same test can be repeated on the same specimen for additional verification. Little or no specimen preparation is required, and the equipment is often portable, permitting on-site testing in most locations.

Nondestructive tests can have a variety of objectives, including the detection of internal or surface flaws, the measurement of a product's dimensions, the determination of a material's structure or chemistry, or the evaluation of a material's physical or mechanical properties. In general, nondestructive tests incorporate the following aspects: (1) some means of probing a material or product; (2) a means by which a flaw, defect, material property, or specimen feature interacts with or modifies whatever is probing; (3) a sensor to detect the response; (4) a device to indicate or record the response; and (5) a way to interpret and evaluate quality.

How you choose to look at a material or product generally depends on what you are looking at, what you wish to see, and in how much detail you wish to examine it. Each of the various inspection processes has characteristic advantages and limitations. Some can be performed only on certain types of materials (such as electrical conductors or ferromagnetic materials). Many are limited in the type, size, and orientation of flaws that they can detect. There may be geometric restrictions relating to part size, part complexity, or the accessibility of critical surfaces or locations. The availability of required equipment, the cost of the operation, the need for a skilled operator or technician, and the possibility of producing a permanent test record are additional considerations when selecting a test procedure.

Regardless of the specific method, nondestructive testing can be a vital element in good manufacturing practice. Its potential value has been widely recognized as productivity and production rates increase, consumers demand higher-quality products, and product liability continues to be a concern. Rather than being an added manufacturing cost, nondestructive testing can actually expand profit by ensuring product reliability and customer satisfaction. In addition to its role in quality control, nondestructive testing can also be used as an assessment aid in product design. Periodic testing can provide a means of controlling a manufacturing process, reducing overall manufacturing costs by preventing the continued manufacture of out-of-specification, defective, or poor-quality parts.

The remainder of this chapter consists of an overview of the various nondestructive test methods. Each is presented along with a discussion of its underlying principle, associated advantages and limitations, compatible materials, and typical applications.

■ 11.2 VISUAL INSPECTION

Probably the simplest and most widely used nondestructive testing method is *visual inspection*, summarized in Table 11-2. The human eye is a very discerning instrument and, with training, the brain can readily interpret the signals. Optical aids such as mirrors, magnifying glasses, and microscopes can expand the capabilities of this system. Video cameras and computer systems, such as digital image analyzers, can be used to automate the inspection and perform quantitative geometrical evaluations. Bore scopes and similar tools can provide accessibility to otherwise inaccessible locations. Only the surfaces of a product can be examined, but that is often sufficient to reveal corrosion, contamination, surface finish flaws, and a wide variety of surface discontinuities.

■ 11.3 LIQUID PENETRANT INSPECTION

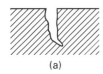

(a)

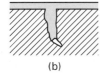

(b)

(c)

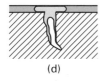

(d)

FIGURE 11-1 Liquid-penetrant testing: (a) initial surface with open crack; (b) penetrant is applied and is pulled into the crack by capillary action; (c) excess penetrant is removed; (d) developer is applied, some penetrant is extracted, and the product inspected.

Liquid penetrant testing, also called dye penetrant inspection, is an effective method of detecting surface defects in metals and other nonporous materials and is illustrated schematically in Figure 11-1. The piece to be tested is first subjected to a thorough cleaning and is dried prior to the test. Then a *penetrant*, a liquid material capable of wetting the entire surface and being drawn into fine openings, is applied to the surface of the workpiece by dipping, spraying, or brushing. Sufficient time is given for capillary action to draw the penetrant into any surface discontinuities, and the excess penetrant liquid is then removed by wiping, water wash, or solvent. The surface is then coated with a thin film of *developer*, an absorbent material capable of drawing traces of penetrant from the defects back onto the surface. Brightly colored dyes or fluorescent materials that glow under ultraviolet light are generally added to the penetrant to make these traces more visible, and the developer is often selected to provide a contrasting background. Radioactive tracers can also be added and used in conjunction with photographic paper to produce a permanent image of the defects. Cracks, laps, seams, lack of bonding, pinholes, gouges, and tool marks can all be detected. After inspection, the developer and residual penetrant are removed by a second cleaning operation.

To be successful, the inspection for surface defects must be correlated with the manufacturing operations. If previous processing involved techniques that might have induced the flow of surface material, such as shot peening, honing, burnishing, machining, or various forms of cold working, a chemical etching may be required to remove material that might be covering critical flaws. An alternative procedure is to penetrant test

---

TABLE 11-2. Visual Inspection

Principle: Illuminate the test specimen and observe the surface. Can reveal a wide spectrum of surface flaws and geometric discontinuities. Use of optical aids or assists (such as magnifying glass, microscopes, illuminators, and mirrors) is permitted. While most inspection is by human eye, video cameras and computer-vision systems can be employed.

Advantages: Simple, easy to use, relatively inexpensive.

Limitations: Depend on skill and knowledge of inspector. Limited to detection of surface flaws.

Material limitations: None.

Geometrical limitations: Any size or shape providing viewing accessibility of surfaces to be inspected.

Permanent record: Photographs or videotapes are possible. Inspectors' reports also provide valuable records.

Remarks: Should always be the initial and primary means of inspection and is the responsibility of everyone associated with parts manufacture.

Principle: A liquid penetrant containing fluorescent material or dye is drawn into surface flaws by capillary action and subsequently revealed by developer material in conjunction with visual inspection.
Advantages: Simple, inexpensive, versatile, portable, easily interpreted, and applicable to complex shapes.
Limitations: Can only detect flaws that are open to the surface; surfaces must be cleaned before and after inspection; deformed surfaces and surface coatings may prevent detection; and the penetrant may be wiped or washed out of large defects. Cannot be used on hot products.
Material limitations: Applicable to all materials with a nonporous surface.
Geometrical limitations: Any size or shape permitting accessibility of surfaces to be inspected.
Permanent record: Photographs, videotapes, and inspectors' reports provide the most common records.

before any surface-finishing operations, when significant defects will still be open and available for detection. Penetrant inspection systems can range from aerosol spray cans of cleaner, penetrant, and developer (for portable applications), to automated, mass-production equipment using sophisticated computer vision systems. Table 11-3 shows a summary of the process and its advantages and limitations.

■ 11.4 MAGNETIC PARTICLE INSPECTION

Magnetic particle inspection, summarized in Table 11-4, is based on the principle that ferromagnetic materials (such as the alloys of iron, nickel, and cobalt), when magnetized, will have distorted magnetic fields in the vicinity of material defects. As shown in Figure 11-2, surface and subsurface flaws, such as cracks and inclusions, will produce magnetic anomalies that can be mapped with the aid of magnetic particles on the specimen surface.

As with the previous method, the specimen must be cleaned prior to inspection. A suitable magnetic field is then established in the part. As shown in Figure 11-3, orientation can be quite important. For a flaw to be detected, it must produce a significant disturbance of the magnetic field at or near the surface. If a bar of steel is placed within an energized coil, a magnetic field will be produced whose lines of flux travel along the axis of the bar. Any defect perpendicular to this axis will significantly alter the field. If the perturbation is sufficiently large and close enough to the surface, the flaw can be detected.

Principle: When magnetized, ferromagnetic materials will have a distorted magnetic field in the vicinity of flaws and defects. Magnetic particles will be strongly attracted to regions where the magnetic flux breaks the surface.
Advantages: Relatively simple, fast, easy-to-interpret; portable units exist; can reveal both surface and subsurface flaws and inclusions (as much as 1/4-in or 6-mm deep) and small, tight cracks.
Limitations: Parts must be relatively clean; alignment of the flaw and the field affects the sensitivity so that multiple inspections with different magnetizations may be required; can only detect defects at or near surfaces; must demagnetize part after test; high current source is required; some surface processes can mask defects; postcleaning may be required.
Material limitations: Must be ferromagnetic; nonferrous metals such as aluminum, magnesium, copper, lead, tin, and titanium and the ferrous (but not ferromagnetic) austenitic stainless steels cannot be inspected.
Geometrical limitations: Size and shape are almost unlimited, most restrictions relate to the ability to induce uniform magnetic fields within the piece. Hard to use on rough surfaces.
Permanent record: Photographs, videotapes, and inspectors' reports are most common. In addition, the defect pattern can be preserved on the specimen by an application of transparent lacquer, or transferred to a piece of transparent tape that has been applied to the specimen and peeled off.

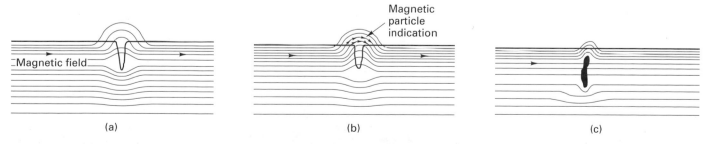

(a)   (b)   (c)

FIGURE 11-2   (a) Magnetic field showing disruption by a surface crack; (b) magnetic particles are applied and are preferentially attracted to field leakage; (c) subsurface defects can also produce surface-detectable disruptions if they are sufficiently close to the surface.

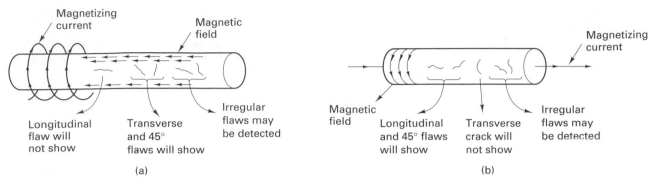

FIGURE 11-3 (a) A bar placed within a magnetizing coil will have an axial magnetic field. Defects parallel to this field may go unnoticed while those that disrupt the field and are sufficiently close to a surface are likely to be detected. (b) When magnetized by a current passing through it, the bar has a circumferential magnetic field and the geometries of detectable flaws are reversed.

However, if the flaw is in the form of a crack aligned with the specimen axis, there will be little perturbation of the lines of flux and the flaw is likely to go undetected.

If the cylindrical specimen is then magnetized by passing a current through it, a circumferential magnetic field will be produced. Any axial defect now becomes a significant perturbation, and a defect perpendicular to the axis will likely go unnoticed. To fully inspect a product, therefore, a series of inspections may be required using various forms of magnetization. Passing a current between various points of contact is a popular means of inducing the desired fields. Electromagnetic coils of various shapes and sizes are also used. Alternating-current methods are most sensitive to surface flaws, while direct-current inspections are better for detecting subsurface defects, such as nonmetallic inclusions.

Once the specimen has been subjected to a magnetic field, magnetic particles are applied to the surface in the form of either a dry powder or a suspension in a liquid carrier. These particles are attracted to places where the lines of magnetic flux break the surface, revealing anomalies that can then be interpreted. To better reveal the orientation of the lines of flux, the particles are often made in an elongated form. They can also be treated with a fluorescent material to enhance observation under ultraviolet light or coated with a lubricant to prevent oxidation and enhance their mobility. Figure 11-4 shows a component of a truck front-axle assembly: as manufactured, under straight magnetic particle inspection, and under ultraviolet light with fluorescent particles.

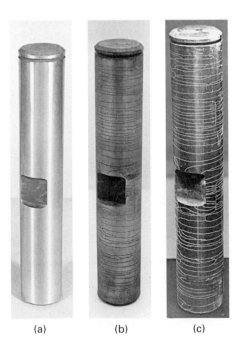

FIGURE 11-4 Front-axle king pin for a truck. (a) As manufactured and apparently sound; (b) inspected under conventional magnetic particle inspection to reveal numerous grinding-induced cracks; (c) fluorescent particles and ultraviolet light make the cracks even more visible. *(Courtesy of Magnaflux Corporation.)*

After magnetic particle inspection, it is not uncommon for some residual magnetization to be retained by the parts. It is usually necessary to demagnetize the inspected parts before further processing or before placing them in use. One common means of demagnetization is to place the parts inside a coil powered by alternating current and then gradually reduce the current to zero. A final cleaning operation generally completes the process.

In addition to in-process and final inspection of parts during manufacture, magnetic particle inspection is used extensively during the maintenance and overhaul of equipment and machinery. The testing equipment ranges from small, portable units to complex automated systems.

■ 11.5   ULTRASONIC INSPECTION

Sound has long been used to provide an indication of product quality. A cracked bell will not ring true, but a fine crystal goblet will have a clear ring when lightly tapped. Striking an object and listening to the characteristic ring is an ancient art but is limited to the detection of large defects because the wavelength of audible sound is rather large compared to the size of most defects. By reducing the wavelength of the signal to the ultrasonic range, typically between 100,000 and 25 million hertz, ultrasonic inspection can be used to detect rather small defects and flaws.

As shown in Table 11-5, *ultrasonic inspection* involves sending high-frequency waves through a material and observing the response. Within the specimen, sound waves can be affected by voids, impurities, changes in density, delaminations, interfaces with materials having a different speed of sound, and other imperfections. At any interface, part of the ultrasonic wave will be reflected and part will be transmitted. If the incident beam is at an angle to an interface where materials change, the transmitted portion of the beam will be bent to a new angle by the phenomenon of refraction. By receiving and interpreting either transmitted or reflected signals, ultrasonic inspection can be used to detect flaws within the material, measure thickness from only one side, or characterize metallurgical structure.

An ultrasonic inspection system begins with a pulsed oscillator and *transducer*, a device that transforms electrical energy into mechanical vibrations. The pulsed oscillator generates a burst of alternating voltage, with a characteristic principal frequency, duration, profile, and repetition rate. This burst is then applied to a sending transducer, which uses a piezoelectric crystal to convert the electrical oscillations into mechanical vibrations. Because air is a poor transmitter of ultrasonic waves, an acoustic *coupling medium*—generally a liquid such as oil or water—is required to link the transducer to the piece to be inspected and transmit the vibrations into the part. The pulsed vibrations then propagate through the part with a velocity that depends on the density and elasticity of the test material. A receiving transducer is then used to convert the transmitted or reflected vibrations back into electrical signals. The receiving transducer is often identical to the sending unit, and the same transducer can actually perform both functions. A receiving unit then amplifies, filters, and processes the signal for display, possible recording, and final interpretation. An electronic clock is generally integrated into the system to time the responses and provide reference signals for comparison purposes.

Depending on the test objectives and part geometry, several different inspection methods can be employed:

TABLE 11-5.   Ultrasonic Inspection

Principle: High-frequency sound waves are propagated through a test specimen and the transmitted or reflected signal is monitored and interpreted.
Advantage: Can reveal internal defects; high sensitivity to most cracks and flaws; high-speed test with immediate results; can be automated and recorded; portable; high penetration in most important materials (up to 60 ft in steel); indicates flaw size and location; access to only one side is required; can also be used to measure thickness, Poisson's ratio, or elastic modulus; presents no radiation or safety hazard.
Limitations: Difficult to use with complex shapes; external surfaces and defect orientation can affect the test (may need dual transducer or multiple inspections); a couplant is required; the area of coverage is small (inspection of large areas requires scanning); trained, experienced, and motivated technicians may be required.
Material limitations: Few—can be used on metals, plastics, ceramics, glass, rubber, graphite, and concrete, as well as joints and interfaces between materials.
Geometric limitations: Small, thin, or complex-shaped parts or parts with rough surfaces and nonhomogeneous structure pose the greatest difficulty.
Permanent record: Ultrasonic signals can be recorded for subsequent playback and analysis. Strip charts can also be used.

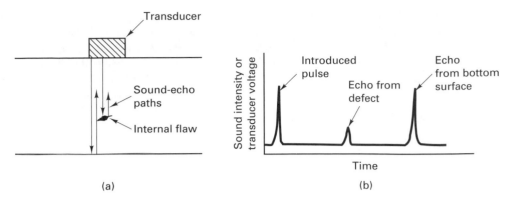

FIGURE 11-5 (a) Ultrasonic inspection of a flat plate with a single transducer; (b) plot of sound intensity or transducer voltage versus time showing the initial pulse and echoes from the bottom surface and intervening defect.

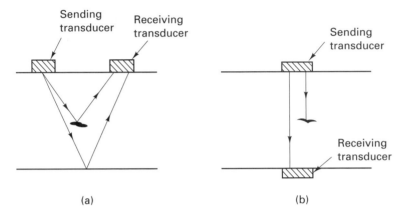

FIGURE 11-6 (a) Dual transducer ultrasonic inspection in the pulse-echo mode; (b) dual transducers in through-transmission configuration.

1. In the *pulse-echo technique*, an ultrasonic pulse is introduced into the piece to be inspected, and the echoes from opposing surfaces and any intervening flaws are detected by the receiver. The time interval between the initial emitted pulse and the various echoes can be displayed on the horizontal axis of a display screen. Defects are identified by the position and amplitude of the various echoes. Figure 11-5 shows a schematic of a single-transducer pulse-echo inspection and the companion signal as it would appear on a display. Figure 11-6a depicts a dual-transducer pulse-echo examination. Both cases require access to only one side of the specimen.

2. The *through-transmission technique* requires separate sending and receiving transducers. As shown in Figure 11-6b, a pulse is emitted by the sending transducer and detected by a receiver on the opposite surface. Flaws in the material decrease the amplitude of the transmitted signal because of back-reflection and scattering.

3. *Resonance testing* can be used to determine the thickness of a plate or sheet from one side of the material. Input pulses of varying frequency are fed into the material. When resonance is detected by an increase in energy at the transducer, the thickness can be calculated from the speed of sound in the material and the time of traverse. Ultrasonic thickness gages can be calibrated to provide direct digital readout of the thickness of a material.

Reference standards—specimens of known thickness or containing various types and sizes of machined "flaws"—are often used to ensure consistent results and aid in interpreting any indications of internal discontinuities.

■ 11.6 RADIOGRAPHY

Radiographic inspection, summarized in Table 11-6, employs the same principles and techniques as those of medical X-rays. A shadow pattern is created when certain types of radiation (X-rays, gamma rays, or neutron beams) penetrate an object and are differentially absorbed due to variations in thickness, density, or chemistry, or the presence of defects in the specimen. The transmitted radiation is registered on a photographic film that provides a permanent record and a means of analyzing the component. Fluorescent screens can provide direct conversion of radiation into visible light and enable

TABLE 11-6.   Radiography

Principle: Some form of radiation (X-ray, gamma ray, or neutron beam) is passed through the sample and is differentially absorbed depending on the thickness, type of material, and the presence of internal flaws or defects.

Advantages: Probes the internal regions of a material; provides a permanent record of the inspection; can be used to determine the thickness of a material; very sensitive to density changes.

Limitations: Most costly of the NDT methods (involves expensive equipment); radiation precautions are necessary (potentially dangerous to human health); the defect must be at least 2% of the total section thickness to be detected (thin cracks can be missed if oriented perpendicular to the beam); film processing requires time, facilities, and care; the image is a two-dimensional projection of a three-dimensional object, so the location of an internal defect requires a second inspection at a different angle; complex shapes can present problems; a high degree of operator training is required.

Material limitations: Applicable to most engineering materials.

Geometric limitations: Complex shapes can present problems in setting exposure conditions and obtaining proper orientation of source, specimen, and film. Two-side accessibility is required.

Permanent record: A photographic image is part of the standard test procedure.

---

fast and inexpensive viewing without the need for film processing. The fluorescent image, however, usually does not offer the sensitivity of the photographic methods.

Various types of radiation can be used for inspection. X-rays are an extremely short wavelength form of electromagnetic radiation that are capable of penetrating many materials that reflect or absorb visible light. They are generated by high-voltage electrical apparatus—the higher the voltage, the shorter the X-ray wavelength and the greater the energy and penetrating power of the beam. Gamma rays, another useful form of electromagnetic radiation, are emitted during the disintegration of radioactive nuclei. Various radioactive isotopes can be selected as the radiation source. Neutron beams for radiography can be obtained from nuclear reactors, nuclear accelerators, or radioisotopes. For most applications it is necessary to moderate the energy and collimate the beam before use.

The absorption of X-rays and gamma rays depends on the thickness, density, and atomic structure of the material being inspected. The higher the atomic number, the greater the attenuation of the beam. Figure 11-7 shows a radiograph of the historic Liberty Bell. The famous crack is clearly visible, along with the internal spider (installed to support the clapper in 1915) and the steel beam and bolts installed in the wooden

FIGURE 11-7   Radiograph of the Liberty Bell. The photo reveals the famous crack, as well as the iron spider installed in 1915 to support the clapper and the steel beam and supports, which were set into the yoke in 1929. *(Courtesy of Eastman Kodak Company.)*

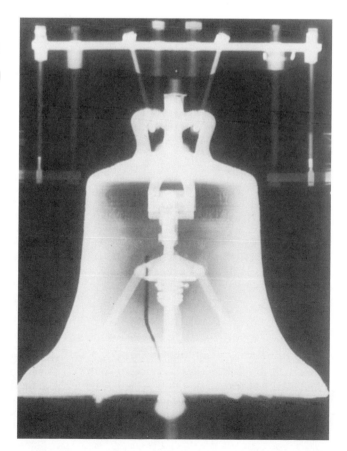

yoke in 1929. Other radiographs disclosed previously unknown shrinkage separations and additional cracks in the bell, as well as a crack in the bell's clapper.

In contrast to X-ray absorption, neutron absorption varies widely from atom to atom, with no pattern in terms of atomic number. Unusual contrasts can be obtained that would be impossible with other inspection methods. For example, hydrogen has a high neutron absorption. The presence of water in a product can be easily detected by neutron radiography. X-rays, on the other hand, are readily transmitted through water, and its presence could be missed.

When a radiation beam is passed through an object, part of the radiation is scattered in all directions. This scatter produces an overall "fogging" of the radiograph, reducing the contrast and sharpness of the image. The thicker the material, the more troublesome the scattered radiation becomes. Photographic considerations relating to the exposure time and development also affect the quality of the radiographic image. Image enhancing computer software can help reveal the subtle but important variations in photographic density.

A standard test piece, or *penetrameter*, is often included in a radiographic exposure. Penetrameters are made of the same or similar material as the specimen and contain features with known dimensions. The image of the penetrameter is compared to the image of the product being inspected. Regions of similar intensity are considered to be of similar thickness.

Radiography is not inexpensive, however. Many users, therefore, recommend extensive use only during the development of a new product or process, followed by spot-checks and statistical methods during subsequent production.

■ 11.7 EDDY-CURRENT TESTING

When an electrically conductive material is exposed to an alternating magnetic field such as that generated by a coil of wire carrying an alternating current, small electric currents are induced on or near the surface of the material (Figure 11-8). These induced *eddy currents*, in turn, generate their own opposing magnetic field, which then reduces the strength of the field from the coil. This change in magnetic field causes a change in the *impedance* of the coil, which in turn changes the magnitude of the current flowing through it. By monitoring the impedance of the exciting coil, or a separate indicating coil, eddy-current testing can be used to detect any condition that would affect the current-carrying ability (or conductivity) of the test specimen. Figure 11-9 shows

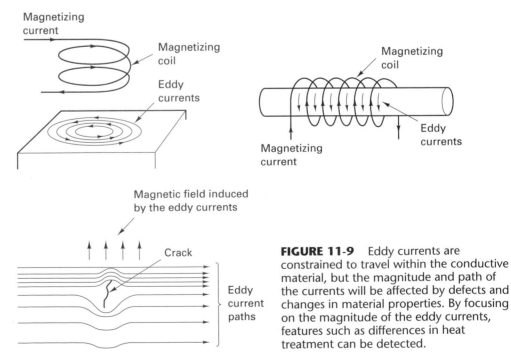

FIGURE 11-8 Relation of the magnetizing coil, magnetizing current, and induced eddy currents. The magnetizing current is actually an alternating current, producing a magnetic field that forms, collapses, and reforms in the opposite direction. This dynamic magnetic field induces the eddy currents, and the changes in the eddy currents produce a secondary magnetic field that interacts with the sensor coil or probe.

FIGURE 11-9 Eddy currents are constrained to travel within the conductive material, but the magnitude and path of the currents will be affected by defects and changes in material properties. By focusing on the magnitude of the eddy currents, features such as differences in heat treatment can be detected.

TABLE 11-7.   Eddy-Current Testing

Principle: When an electrically conductive material is brought near an alternating-current coil that produces an alternating magnetic field, surface currents (eddy currents) are generated in the material. These surface currents generate their own magnetic field, which interacts with the original, modifying the impedance of the originating coil. Various material properties and/or defects can affect the magnitude and direction of the induced eddy currents and can be detected by the electronics.

Advantages: Can detect both surface and near-surface irregularities; applicable to both ferrous and nonferrous metals; versatile—can detect flaws, variations in alloy or heat treatment, variations in plating or coating thickness, wall thickness, and crack depth; intimate contact with the specimen is not required; can be automated; electrical circuitry can be adjusted to select sensitivity and function; pass/fail inspection is easily conducted; high speed; low cost; no final cleanup is required.

Limitations: Response is sensitive to a number of variables, so interpretation may be difficult; sensitivity varies with depth, and depth of inspection depends on the test frequency; reference standards are needed for comparison; trained operators are generally required.

Material limitations: Only applicable to conductive materials, such as metals; some difficulties may be encountered with ferromagnetic materials.

Geometric limitations: Depth of penetration is limited; must have accessibility of coil or probe; constant separation distance between coils and specimen is required for good results.

Permanent record: Electronic signals can be recorded using devices such as strip-chart recorders.

how the eddy-current paths would be forced to alter around a crack, thereby changing the characteristics of the induced magnetic field in that vicinity.

Eddy-current testing, summarized in Table 11-7, can be used to detect surface and near-surface flaws, such as cracks, voids, inclusions, and seams. Stress concentrations, differences in metal chemistry, or variations in heat treatment (i.e., microstructure and hardness) will all affect the magnetic permeability and conductivity of a metal and therefore alter the eddy-current characteristics. Material mix-ups and processing errors can therefore be detected. Specimens can be sorted by hardness, case depth, residual stresses, or any other structure-related property. Thickness (or variation in thickness) of platings, coatings, or even corrosion can be detected and measured.

Eddy-current test equipment can range from simple, portable units with handheld probes to fully automated systems with computer control and analysis. Each system, however, includes:

1. A source of changing magnetic field capable of inducing eddy currents in the part being tested. This source generally takes the form of a coil (or coil-containing probe) carrying alternating current of a specific frequency and amplitude. (*Note:* High frequency gives shallow penetration and a higher eddy current density on the surface. Low frequency gives deeper penetration.) Various coil geometries are used for different-shaped specimens.

2. A means of sensing the field changes caused by the interaction of the eddy currents with the original magnetic field. Either the exciting coil itself or a secondary sensing coil can be used to detect the impedance changes. Differential testing can be performed using two oppositely wound coils wired in series. In this method, only differences in the signals between the two coils are detected as one or both coils are scanned over the specimen.

3. A means of measuring and interpreting the resulting impedance changes. The simplest method is to measure the induced voltage of the sensing coil, a reading that evaluates the cumulative effect of all variables affecting the eddy-current field. Phase analysis can be used to determine the magnitude and direction of the induced eddy-current field. Familiarity with characteristic impedance responses or comparison with signals from other parts or other locations can be used to identify or interpret the desired features in the specimen.

When comparing alternative techniques, eddy current is usually not as sensitive as penetrant testing in detecting small, open flaws, but it requires none of the cleanup operations and is noticeably faster. In a similar manner, it is not as sensitive as magnetic particle inspection to small subsurface flaws but it can be applied to all metals (ferromagnetic and nonferromagnetic alike). In addition, eddy-current testing offers capabilities that cannot be duplicated by the other methods, such as the ability to differentiate between various chemistries and heat treatments.

TABLE 11-8. Acoustic Emission Monitoring

Principle: Almost all materials will emit high-frequency sound (acoustic emissions) when stressed, deformed, or undergoing structural changes, such as the formation or growth of a crack or defect. These emissions can now be detected and provide an indication of dynamic change within the material.

Advantages: The entire structure can be monitored with near-instantaneous detection and response; continuous surveillance is possible; defects inaccessible to other methods can be detected; inspection can be in harsh environments; and the location of the emission source can be determined.

Limitations: Only growing or "active" flaws can be detected (the mere presence of defects is not detectable); background signals may cause difficulty; there is no indication of the size or shape of the flaw; expensive equipment is required, and experience is required to interpret the signals.

Material limitations: Virtually unlimited, provided that they are capable of transmitting sound.

Geometric limitations: Requires continuous sound-transmitting path between the source and the detector. Size and shape of the component affect the strength of the emission signals that reach the detector.

■ 11.8 ACOUSTIC EMISSION MONITORING

Materials experiencing the dynamic events of deformation or fracture emit stress waves in frequencies as high as 1 MHz. While these sounds are inaudible to the human ear, they are detectable through the use of sophisticated electronics. Transducers, amplifiers, filters, counters, and computers can be used to isolate and analyze the sonic emissions of a cracking or deforming material. Much like the warning sound of ice cracking underneath boots or skates, the acoustic emissions of materials can be used to provide a warning of impending danger. They can detect deformations as small as 10^{-12} in./in. (that occur in short intervals of time), initiation or propagation of cracks (including stress-corrosion cracking), delamination of layered materials, and fiber failure in composites. By using multiple sensors, it is possible to accurately pinpoint the source of these sounds by a triangulation method similar to that used to locate seismic sources (earthquakes) in the earth.

Acoustic emission monitoring, summarized in Table 11-8, involves listening for indications of failure. Temporary monitoring can be used to detect the formation of cracks in materials during production operations, such as welding and subsequent cooling of the weld region. Monitoring can also be employed to assure the absence of plastic deformation during preservice proof testing. Continuous surveillance may be used when the product or component is particularly critical, as with bridges and nuclear reactor pressure vessels. The sensing electronics can be coupled to an alarm and safety system to protect and maintain the integrity of the structure.

In contrast to the previous inspection methods, acoustic emission cannot detect an existing defect in a static product. Instead, it is a monitoring technique designed to detect a dynamic change in the material, such as the formation or growth of a crack or defect, or the onset of plastic deformation.

■ 11.9 OTHER METHODS OF NONDESTRUCTIVE TESTING AND INSPECTION

LEAK TESTING

Leak testing is a form of nondestructive testing designed to determine the existence or absence of leak sites and the rate of material loss through the leaks. Various testing methods have been developed, ranging from the rather crude bubble-emission test (pressurize, immerse, and look for bubbles), through simple pressure drop tests with either air or liquid as the pressurized media, to advanced techniques involving tracers, detectors, and sophisticated apparatus. Each has its characteristic advantages, limitations, and sensitivity. Selection should be on the basis of cost, sensitivity, reliability, and compatibility with the specific product to be tested.

THERMAL METHODS

Temperature-sensing devices (including thermometers, thermocouples, pyrometers, temperature-sensitive paints and coatings, liquid crystals, infrared scanners, infrared film, and others) can also be used to evaluate the soundness of engineering materials and components. Parts can be heated and then inspected during cool-down to reveal abnormal temperature distributions that are the result of faults or flaws. The identification of "hot spots" on an operating component is often an indication of a flaw or defect, and may provide advanced warning of impending failure. For example, faulty electrical components tend to be hotter than defect-free devices. Composite materials (difficult to inspect by many standard techniques) can be subjected to brief pulses of intense heat and then inspected to reveal the temperature pattern produced by the subsequent thermal conductivity. Thermal anomalies tend to appear in areas where the bonding between the components is poor or incomplete. In another technique, ultrasonic waves are used to produce heat at internal defects, which are then detected by infrared examination.

STRAIN SENSING

Although used primarily during product development, strain-sensing techniques can also be used to provide valuable insight into the stresses and stress distribution within a part. Brittle coatings, photoelastic coatings, or electrical resistance strain gages can be applied to the external surfaces of a part, which are then subjected to an applied stress. The extent and nature of cracking, the photoelastic pattern produced, or the electrical resistance changes then provide insight into the strain at various locations. X-ray diffraction methods and extensometers have also been used.

ADVANCED OPTICAL METHODS

Although visual inspection is often the simplest and least expensive of the nondestructive inspection methods, there are also several advanced optical methods. Monochromatic laser light can be used to detect differences in the backscattered pattern from a part and a master. The presence or absence of geometrical features such as holes or gear teeth is readily detected. Holograms can provide three-dimensional images of an object, and holographic interferometry can detect minute changes in the shape of an object under stress.

RESISTIVITY METHODS

The *electrical resistivity* of a conductive material is a function of its chemistry, processing history, and structural soundness. Measurement of resistivity can therefore be used for alloy identification, flaw detection, or the assurance of proper processing. Tests can be developed to evaluate the effects of heat treatment, the amount of cold work, the integrity of welds, or the depth of case hardening. The development of sensitive micro-ohm-meters has greatly expanded the possibilities in this area.

COMPUTED TOMOGRAPHY

While X-ray radiography provides a single image of the X-ray intensity being transmitted through an object, X-ray *computed tomography* (CT) is an inspection technique that provides a cross-sectional view of the interior of an object along a plane parallel to the X-ray beam. This is the same technology that has revolutionized medical diagnostic imaging (CAT scans), with the process parameters (such as the energy of the X-ray source) being adapted to permit the nondestructive probing of industrial products. Basic systems include an X-ray source, an array of detectors, a mechanical system to move and rotate the test object, and a dedicated computer system. The intensity of the received signal is recorded at each of the numerous detectors with the part in a variety of orientations. Complex numerical algorithms are then used to construct an image of the interior of the component. Internal boundaries and surfaces can be determined clearly, enabling inspection and dimensional analysis of a product's interior. The presence of cracks, voids, or inclusions can also be detected and their precise location can be determined.

CT inspections are slow and costly, so they are currently used only when the component is critical and the more standard inspection methods prove to be inadequate due to features such as shape complexity, thick walls, or poor resolution of detail. The video images of the CT technique also permit easy visualization and interpretation.

Acoustic holography is another computer reconstruction technique, this time based on ultrasound reflections from within the part.

CHEMICAL ANALYSIS AND SURFACE TOPOGRAPHY

While nondestructive inspection is usually associated with the detection of flaws and defects, various nondestructive techniques can also be employed to determine the chemical and elemental analysis of surface and near-surface material. These techniques include Auger electron spectroscopy (AES), energy-dispersive X-ray analysis (EDX), electron spectroscopy for chemical analysis (ESCA), and various forms of secondary-ion mass spectroscopy (SIMS). Because of its large depth of focus, the scanning electron microscope has become an extremely useful tool for observing the surfaces of materials. More recently, the atomic-force microscope and scanning tunneling microscope have extended this capability and can now provide information about surface topography with resolution to the atomic scale.

■ 11.10 DORMANT VERSUS CRITICAL FLAWS

There was a time when the detection of a flaw was considered to be sufficient cause for rejecting a material or component, and material specifications often contained the term *flaw-free*. Such a criterion, however, is no longer practical, because the sensitivity of detection methods has increased dramatically. If materials were rejected upon detection of a flaw, we would find ourselves rejecting nearly all commercial engineering materials.

If a defect is sufficiently small, it is possible for it to remain dormant throughout the useful lifetime of a product, never changing in size or shape. Such a defect is clearly allowable. Larger defects, or defects of a more undesirable geometry, may grow or propagate under the same conditions of loading, often causing sudden or catastrophic failure. These flaws would be clearly unacceptable. The objective (or challenge), therefore, is to identify the conditions below which a flaw remains *dormant* and above which it becomes *critical* and a cause for rejection. This issue is addressed in the section on "Fracture Toughness and the Fracture Mechanics Approach" in Chapter 2.

■ KEY WORDS

acoustic emission	eddy-current testing	magnetic particle	proof test	tomography
acoustic holography	electrical resistivity	inspection	pulse-echo method	transducer
computed tomography	flaw-free	nondestructive testing	radiographic	ultrasonic inspection
coupling medium	hardness testing	(also nondestructive	inspection	visual inspection
critical flaw	impedance	inspection)	resonance testing	
destructive testing	leak testing	penetrameter	through transmission	
dormant flaw	liquid penetrant testing	penetrant	technique	

■ REVIEW QUESTIONS

1. Why must destructive testing be performed on a statistical basis?
2. What is a proof test, and what assurance does it provide?
3. What quality-related features can a hardness test reasonably ensure?
4. What exactly is nondestructive testing, and what are some attractive features of the approach?
5. What are some possible objectives of nondestructive testing?
6. What are some factors that should be considered when selecting a nondestructive testing method?
7. How might the costs of nondestructive testing actually be considered as an asset rather than a liability?
8. Why should visual inspection be considered as the initial and primary means of inspection?
9. What is the primary limitation of a visual inspection?
10. What types of defects can be detected in a liquid penetrant test?
11. What is the primary materials-related limitation of magnetic particle inspection?
12. Describe how the orientation of a flaw with respect to a magnetic field can affect its detectability during magnetic particle inspection.
13. What is the major limitation of sonic testing, where one listens to the characteristic ring of a product in an attempt to detect defects?
14. What is the role of a coupling medium in ultrasonic inspection?
15. What are three types of ultrasonic inspection methods?
16. What types of radiation can be used in radiographic inspection of manufactured products?
17. What are penetrameters, and how are they used in radiographic inspection?
18. While radiographs offer a graphic image that looks like the part being examined, the technique has some significant limitations. What are some of these limitations?
19. Why would we not expect eddy-current examination to be useful with ceramics or polymeric materials?
20. What types of detection capabilities are offered by eddy-current inspection that cannot be duplicated by the other methods?
21. Why can't acoustic emission methods be used to detect the presence of an existing but static defect?
22. How can acoustic emission be used to determine the location of a flaw or defect?
23. How can temperature be used to reveal defects?
24. What kinds of product features can be evaluated by electrical resistivity methods?
25. What type of information can be obtained through computed tomography?
26. What are some of the techniques that can be used to determine the chemical composition of surface and near-surface material?
27. Why is it important to determine the distinction between allowable and critical flaws, as opposed to rejecting all materials that contain detectable flaws?

■ PROBLEMS

1. A manufacturing company routinely specifies X-ray radiography to assure the absence of cracks in its cast metal products. The primary reason for selecting radiography is the availability of a hard-copy record of each inspection for use in any possible liability litigation. Discuss the pros and cons of their selection. What other processes might you want to consider? If some form of permanent record is desirable, discuss how these might be obtained for the various alternative processes.

2. For each of the inspection methods listed below, cite one major limitation to its use.
   a. Visual inspection
   b. Liquid penetrant inspection
   c. Magnetic particle inspection
   d. Ultrasonic inspection
   e. Radiography
   f. Eddy-current testing
   g. Acoustic emission monitoring

3. Which of the major nondestructive inspection methods might you want to consider if you want to detect (1) surface flaws and (2) internal flaws in products made from each of the following materials?
   a. Ceramics
   b. Polymers
   c. Fiber-reinforced composites with (i) polymer matrix and (ii) metal matrix (Consider various fiber materials.)

4. Discuss the application of nondestructive inspection methods to powder metallurgy (metallic) products with low, average, and high density.

5. It has been said, "Total Quality Control is superior to 100% inspection because inspection, like any other process, has its own defect rate. Some substandard product, therefore, is always likely to slip through." Do you agree or disagree? Discuss.

6. The pulse-echo ultrasonic technique can be used to determine the thickness of a part or structure. By accurately measuring the time it takes for a short ultrasonic pulse to travel through the thickness of a material, reflect from the back or inside surface, and return to the transducer, the distance can be calculated by:

$$d = Vt/2$$

Where:
   d is the thickness of the test piece.
   V is the speed of sound in the material being tested.
   t is the measured round-trip time.

   What are some alternative means of measuring thickness? Briefly discuss their relative pros and cons. Consider such features as geometric and material constraints, and the ability to measure from one side only.

7. If V for a particular metal is 5000 m/sec and a part made of that material is 3 mm thick, what is the transit time for the pulse to cross the material and reflect back to the source/receptor? (*Note:* The transit time is a consideration in evaluating equipment capabilities and may well influence cost!)

8. With nondestructive inspection methods using wave phenomena, the detection limit or resolution is dependent upon the wavelength being used. Compare the wavelengths of:
   a. Ultrasonic waves of frequency 500,000 Hz
   b. X-rays
   c. Acoustic emission stress waves of frequency 1 MHz

   Note: Velocity = frequency × wavelength

www.wiley.com/college/degarmo

*C*hapter 11   CASE STUDY

Portable Failure Analysis Kit

You are a member of a corporate failure analysis group, working out of the home office of a large petroleum corporation with production sites throughout North America. These sites generally contain a variety of equipment, including pumps, valves, pipelines, and storage tanks, and these occasionally experience failure under both normal and abnormal production conditions. Much of this equipment is large and cumbersome. Therefore, when a failure occurs, you must frequently travel to the site of the failure to conduct your investigation. Your primary objectives during these visits are to collect information, inspect the site, and acquire specimens and samples that can be taken back to your base laboratory for further investigation and study.

   Your present assignment is to design and equip a portable failure analysis kit for on-site investigation. This kit will be in a suitcase that can be stored in your laboratory. Upon notification of a problem, you can simply pick up the kit and be on your way to that location.

1. Your overall container is restricted to the size of a typical suitcase and must be easily carried by a single individual (suggested container weight of less than 50 pounds). Moreover, you are being asked to purchase and equip the case for less than $2000.00. (It is hoped that your case can be duplicated and distributed to other individuals within the corporation.) With these constraints, develop a list of the items that you would include in your case. For each item, discuss briefly: a description of the item, its intended use or uses, and justification of the cost (if over $200.00), the weight (if over five pounds), and the size (if it occupies over 20% of the suitcase).

2. If additional funds could be provided to upgrade your suitcase (up to $5000), but the size and weight restrictions were maintained, what additional or upgraded items would you want to include? Discuss each in the manner used in part 1.

CHAPTER 12

PROCESS CAPABILITY AND QUALITY CONTROL

■ 12.1 INTRODUCTION

All manufacturing processes display some level of variability, referred to as inherent capability or some inherent uniformity. For example, suppose that we view "shooting at a metal target" as a "process" for putting holes in a piece of metal. I hand you the gun and tell you to take nine shots at the bull's-eye. Figure 12-1 shows some possible results. You are the operator of the process. To measure the *process capability* (PC), that is, your ability to consistently hit the bull's-eye you are aiming at, the target is inspected after you have finished shooting. So, the capability of manufacturing processes is determined by measuring the output of the process. In *quality control* (QC), the product is examined to determine whether or not the processing accomplished was what was specified by the designer in the design, usually the nominal size and the tolerance.

FIGURE 12-1 The concepts of accuracy (aim) and precision (repeatability) are shown in the four target outcomes.

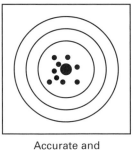

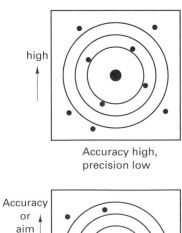

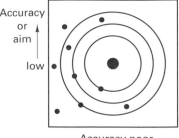

high

Accuracy high, precision low

Accurate and precise

Accuracy or aim

low

Accuracy poor, precision poor

Accuracy poor, precision good

PC and QC studies both use statistical and analytical tools, but PC studies are directed at the machines used in the processing rather than the output or products from the processes. Going back to our example, a PC study would quantify the inherent accuracy and precision in the shooting process. Accuracy is reflected in your aim (the average of all your shots), whereas precision reflects the repeatability of the process. The QC program is designed to root out problems that can cause defective products during production. Traditionally, the objective of QC studies has been to find defects in the process. The more progressive point of view is to inspect the process to prevent the problems that can cause defective products from occurring during production.

12.2 DETERMINING PROCESS CAPABILITY

The *nature of the process* refers to both the *variability* (or inherent uniformity) and the aim of the process. Thus in the target-shooting example, a perfect process would be capable of placing nine shots right in the middle of the bull's-eye, one right on top of the other. The process would display no variability with perfect *accuracy*. Such performance would be very unusual in a real industrial process. The variability may have assignable causes and may be correctable if the cause can be found and eliminated. That variability to which no cause can be assigned and which cannot be eliminated is said to be inherent in the process and is therefore its nature.

Some examples of assignable causes of variation in processes include multiple machines for the same components, operator blunders, defective materials, or progressive wear in the tools during machining. Sources of inherent variability in the process include variation in material properties, operator variability, vibrations and chatter, and the wear of the sliding components in the machine, perhaps resulting in poorer operation of the machine. These kinds of variations, which occur naturally in processes, usually display a random nature and often cannot be eliminated. In QC terms, these are referred to as *chance causes*. Sometimes, the causes of assignable variation cannot be eliminated because of cost. Almost every process has multiple causes of variability occurring simultaneously, so it is extremely difficult to separate the effects of the different sources of variability during the analysis.

MAKING PC STUDIES BY THE TRADITIONAL METHOD

The object of the PC study is to determine the inherent nature of the process as compared to the desired specifications. The output of the process must be examined under normal conditions, or what is typically called *hands-off conditions*. The inputs (e.g., materials, setups, cycle times, temperature, pressure, and operator) are fixed or standardized. The process is allowed to run without tinkering or adjusting, while the output (i.e., the product or units or components) is documented with respect to (1) time, (2) source, and (3) order of production. A sufficient number of data have to be taken to ensure confidence in the statistical analysis of the data. The capability of the gage (its precision) used to measure the products must exceed the expected tolerance on the part by one order of magnitude. (See the discussion of the rule of 10 in Chapter 10.)

Prior to any data collection, these steps must be taken:

1. Design the PC experiment (standard method):

    Use normal or hands-off process conditions; specify machine settings for speed, feed, volume, pressure, material, temperature, operator, and so on.

2. Define the inspection method and the inspection means (the procedure and the instrumentation). In selecting the gage, consider these aspects:

    a. Features that the gage will be checking

    b. Speed or rate of operation

    c. Level of accuracy and precision

    d. Skill of the operator

    e. Portability of gages or part, or both

    f. Environment (clean and stable, cutting fluids)

g. Workpiece (clean, lubricants present)

h. Cost (initial, maintenance, daily)

3. Decide how many items (measurements) will be needed to perform the statistical analysis.

4. For a standard PC study, use homogeneous input material, and try to contrast it with normal (more variable) input material.

5. Data sheets must be designed to record date, time, source, order of production, and all the process parameters being used (or measured) while the data are being gathered.

6. Assuming that the standard PC study approach is being used, the process is run, and the parts are made and measured.

Let us assume that the designer specified the part to be 1.000 ± 0.005 in. After some units have been manufactured according to the process plan without any adjustment of the process, each unit is measured, and the data are recorded on the data sheet. A frequency distribution, in the form of a *histogram*, or a *run chart* (Figure 12-2) is developed. This histogram shows the raw data and the desired value, along with the upper and lower *specification limits*, where LSL represents lower specification limit and USL the upper specification limit. The statistical data are used to estimate the mean and the standard deviation of this distribution. The run chart shows the same data, but here the data are plotted against time.

The mechanics of this statistical analysis are shown in Figure 12-3. The true mean of the distribution, designated $\overline{X}'$ ("X bar prime") is to be compared with the nominal value. The estimate of the true *standard deviation*, designated σ' ("sigma prime"), is used to determine how the process compares with the desired tolerance. *The purpose of the analysis is to obtain estimates of $\overline{X}'$ and σ' values, the true process parameters, because*

FIGURE 12-2 The process capability study compares the part as made by the process to the specifications called out by the designer. Histogram and run charts are commonly used.

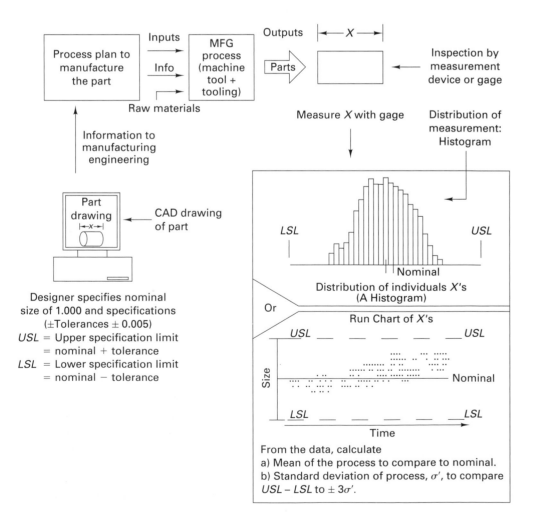

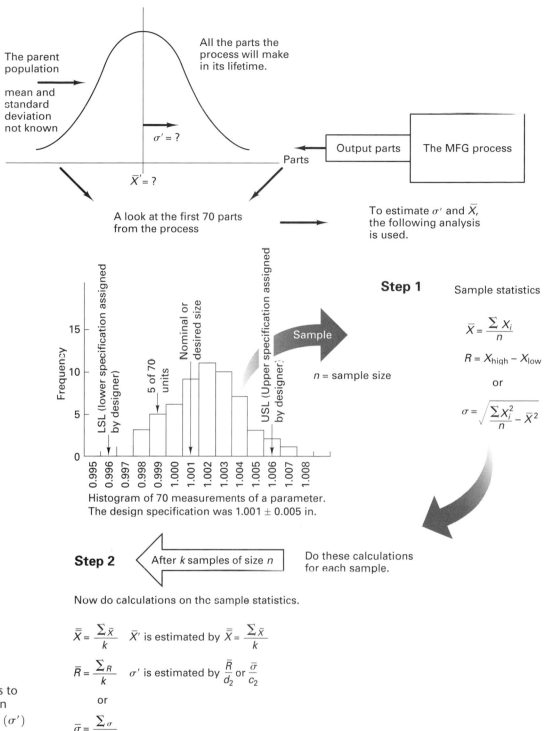

FIGURE 12-3   Calculations to obtain estimates of the mean $\left(\bar{X}'\right)$ and standard deviation (σ') of a process.

they are not known. A sample size of five was used in this example, so $n = 5$. Fourteen groups of samples were drawn from the process, so $k = 14$. For each sample, the *sample mean* $\bar{X}$ and the *sample range R* are computed. For large samples $(n > 12)$, the standard deviation of each sample should be computed rather than the range. Next, the average of the sample averages, $\bar{\bar{X}}$, is computed. This is sometimes called the *grand average.* This is used to estimate the mean of the process, $\bar{X}'$. The standard deviation of the process, which is a measure of the spread or variability of the process, is estimated from either the average of the sample ranges, $\bar{R}$, or the average of the sample standard deviations, $\bar{\sigma}$, using either $\bar{R}/d_2$ or $\bar{\sigma}/c_2$. The factors d_2 and c_2 depend on the sample size *n* and are given in the table in Figure 12-13. The process capability is defined by $\pm 3\sigma'$ or $6\sigma'$. Thus, $\overline{X}' \pm 3\sigma'$ defines the natural capability limits of the process, assuming the process is approximately normally distributed. Note that a distinction is made between a sample and a population. A sample is of a specified, limited size and is drawn from the population. The population is the large source of items, which can include all the items the process will ever produce under the specified conditions. Our calculations assume that this distribution was normal or bell-shaped. Figure 10-8 shows a typical normal curve and the areas under the curve as defined by the standard deviation. Other distributions are possible, but the histogram in Figure 12-3 clearly suggested that this process can best be described by a normal probability distribution. Now it remains for the process engineer and the operator to combine their knowledge of the process with the results from the analysis in order to draw conclusions about the ability of this process to meet specifications.

HISTOGRAMS

A histogram is a representation of a frequency distribution that uses rectangles whose widths represent class intervals and whose heights are proportional to the corresponding frequencies. The frequency histogram is a type of diagram in which data are grouped into cells (or intervals), and the frequency of observations falling into each interval can be noted. All the observations within a cell are considered to have the same value, which is the midpoint of the cell. So, a histogram is a picture that describes the variation in a process. It is good to have this visual impression of the distribution of values, along with the mean and standard deviation. Histograms are used in many ways in QC, for example,

- To determine the process capability (central tendency and dispersion)
- To compare the process with the specifications
- To suggest the shape of the population (e.g., normality)
- To indicate discrepancies in data, such as gaps

There are several types of histograms. A histogram shows either absolute frequency (actual occurrence) or relative frequency (percentage). Cumulative histograms show cumulative frequency and relative cumulative frequency. Each type has its own advantages and is used in different situations. Figure 12-4 shows frequency versus location for 45 measurements. The aim (accuracy) of the process is low, but all the data are within

FIGURE 12-4 Histogram shows the output mean from the process versus nominal and the tolerance specified by the designer versus the spread as measured by the standard deviation, σ.

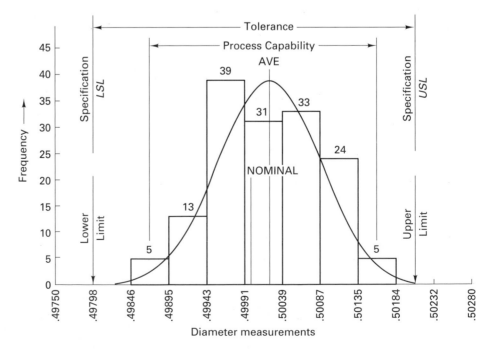

the tolerances. The disadvantage of the histogram is that it does not show trends and does not take time into account. We can take the data from the histogram and spread them out over time to create a run chart or run diagram.

RUN CHART OR DIAGRAM

A run diagram is a plot of a quality characteristic as a function of time. It provides some idea of general trends and degree of variability. Run charts reveal information that histograms cannot, such as certain trends over time or at certain times of day. Individual measurements (not samples) are taken at regular time intervals, and the points are plotted on a connected line graph as a function of time. The graph can be used to find obvious trends in the data. Figure 12-5 shows a run diagram with measurements made every hour over four shifts.

Run diagrams are very important at startup to identify the basic nature of a process. Without this information, one may use an inappropriate tool in analyzing the data. For example, a control chart or histogram might hide tool wear if frequent tool changes and adjustments are made between groups of observations. As a result, run diagrams (with 100% inspection where feasible) should always precede the use of control charts for averages and ranges.

WHAT PC STUDIES TELL ABOUT THE PROCESS

PC studies provide the answers to these questions:

1. Does the process have the ability to meet specifications?

To answer this question, a process capability index, C_p, is often computed, where

$$C_p = \frac{\text{Tolerance spread}}{6\sigma'} = \frac{USL - LSL}{6\sigma'} \tag{12-1}$$

A value of $C_p \geq 1.33$ is considered good.

FIGURE 12-5   An example of a run chart or graph.

The lengths of manufactured steering rack bars are measured. A run diagram is constructed to determine how the process is behaving. During 32 hours, measurements are made every hour (60 minutes) in the order that rack bars are produced. (Length is in inches.)

First shift	35	40	27	30	30	34	26	31
Second shift	24	23	28	15	23	17	16	21
Third shift	15	13	28	8	20	9	5	11
First shift	16	5	9	13	16	10	9	10

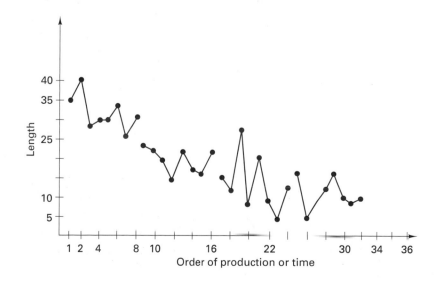

2. Is the process well centered with respect to the desired nominal specification? To answer this question, the amount of mismatch, D, is computed, where

$$D = \frac{\text{Estimated process mean} - \text{Nominal}}{1/2(\text{Tolerance spread})} = \frac{\overline{X}' - \text{Nominal}}{1/2(USL - LSL)} \quad (12\text{-}2)$$

The condition for not producing rejects is

$$\overline{X}' - \text{Nominal} + 3\sigma' < 1/2(USL - LSL) \quad (12\text{-}3)$$

The capability index combines both factors and is defined as

$$C_{pk} = \frac{\text{MIN}\left\{\left|USL - \overline{X}'\right|, \left|LSL - \overline{X}'\right|\right\}}{3\sigma'} \quad (12\text{-}4)$$

In response to the first question, the width of the histogram is compared with the specifications (Figure 12-6). The natural spread of the process, $6\sigma'$, is computed and is then compared with the upper and lower tolerance limits. Three situations exist:

FIGURE 12-6 Histograms comparing measurements from a process to designer specifications.

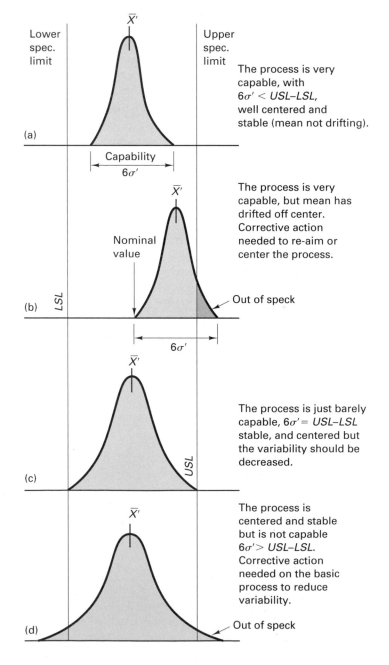

1. $6\sigma' < USL - LSL$ or $C_p > 1$, or process variability less than tolerance spread, see Figure 12-6a.

2. $6\sigma' = USL - LSL$ or $C_p = 1$, or process variability equal to tolerance spread, see Figure 12-6c.

3. $6\sigma' > USL - LSL$ or $C_p < 1$, or process variability greater than tolerance spread, see Figure 12-6d.

In Situation 1, the machine is capable of meeting the tolerances applied by the designer. Generally speaking, if process capability is on the order of two-thirds to three-fourths of the design tolerance, there is a high probability that the process will produce all good parts over a long period of time. If the PC is on the order of one-half or less of the design tolerance, it may be that the selected process is too good; that is, the company may be producing ball bearings when what is called for is marbles. In this case, it may be possible to trade off some precision in this process for looser specifications elsewhere, resulting in an overall economic gain. Quality in well-behaved processes can be maintained by checking the first, middle and last part of a lot or production run. If these parts are good, then the lot is certain to be good. This is called $N = 3$. Naturally, if the lot size is 3 or less, this is 100% inspection. Sampling and control charts are also used under these conditions to maintain the process aim and variability.

When the process is not capable of meeting the design specifications, there are a variety of alternatives, including:

a. Shifting this job to another machine with greater process capability.

b. Getting a review of the specifications to see if they may be relaxed.

c. Sorting the product, to separate the good from the bad. This entails 100% inspection of the product, which may not be a feasible economic alternative unless it can be done automatically. Automatic sorting of the product on a 100% basis can ensure near-perfect quality of all the accepted parts. The automated station shown in Figure 12-7 checks parts for the proper diameter with the aid of a linear variable differential transformer (LVDT). As a part approaches the inspection station on a motor-driven conveyor system, a computer-based controller activates a clamping device. Embedded in the clamp is an LVDT position sensor with which the control computer can measure the diameter of the part. Once the measurement has been made, the computer releases the clamp, allowing the part to be carried away. If the diameter of the part is within a given tolerance, a solenoid-actuated gate operated by the computer lets the part pass. Otherwise, the part is ejected into a bin. With the fast-responding LVDT, 100% of manufactured parts can be automatically sorted quickly and economically.

Automated sorting does not determine what caused the defects, so this example is an "automated defect finder." How would one change this inspection system to make it "inspect to prevent" the defect from occurring?

FIGURE 12-7 A linear variable differential transformer (LVDT) is a key element in an inspection station checking part diameters. Momentarily clamped into the sensor fixture, a part pushed the LVDT armature into the device winding. The LVDT output is proportional to the displacement of the armature. The transformer makes highly accurate measurements over a small displacement range.

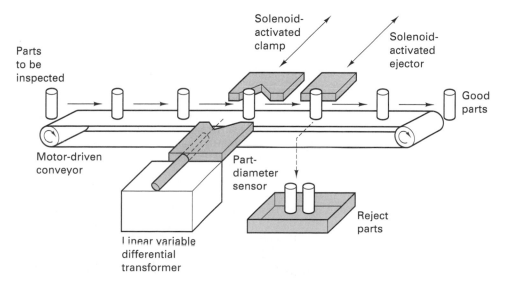

d. Determining whether the *precision* of the process can be improved by:
1. Switching cutting tools, workholding devices or materials.
2. Overhauling the existing process and/or developing a preventive maintenance program.
3. Finding and eliminating the causes of variability, using cause-and-effect diagrams.
4. Combinations of (1), (2), and (3).
5. Using a factorial (Taguchi) procedure, that is, designing experiments to reduce the variability of the process.

In Figure 12-6c, the process capability is almost exactly equal to the assigned tolerance spread, so if the process is not perfectly centered, defective pcoducts will always result. Thus, situation 2 should be treated like situation 3 unless the process can be perfectly centered and maintained. Tool wear, which causes the distribution to shift, must be negligible. Then the situation can be treated as that in 1, particularly if a small percentage of parts just outside tolerance is acceptable.

The second question deals with the ability of the process to maintain centering so that the average of the distribution comes as close as possible to the desired nominal value, see Figure 12-6b. This process needs to be recentered so that the mean of the process distribution is at or near the nominal value. Most processes can be reaimed. Poor accuracy is often due to assignable causes, which can be eliminated.

In addition to direct information about the accuracy and the precision of the process, PC studies can also tell the manufacturing engineer how pilot processes compare with production processes, and vice versa. If the source and time of the manufacture of each product are carefully recorded, information about the instantaneous reproducibility can be found and compared with the repeatability of the process with respect to time (time-to-time variability). More important, since almost all processes are duplicated, PC studies generate information about machine-to-machine variability. Going back to our target-shooting example, suppose that nine different guns were used, all of the same make and type. The results would have been different, just as having nine marksmen use the same gun would have resulted in yet another outcome. Thus, PC studies generate information about the homogeneity and the differences in multiple machines and operators.

It is quite often the case in such studies that one variable dominates the process. Target shooting viewed as a process is "operator"-dominated in that the outcome is highly dependent on the skill (the capability) of the "worker." Processes that are not well engineered nor highly automated, or in which the worker is viewed as "highly skilled," are usually operator-dominated. Processes that change or shift uniformly with time but that have good repeatability in the short run are often machine dominated. For example, the mean of a process $\left(\overline{X}'\right)$ will usually shift after a tool change, but the variability may decrease or remain unchanged. Machines tend to become more precise (to have less variability within a sample) after they have been broken in (i.e., the rough contact surfaces have smoothed out because of wear) but will later become less precise (will have less repeatability) due to poor fits between moving elements (called *backlash*) of the machine under varying loads. Other variables that can dominate processes are setup, input components, and even information.

In many machining processes in use today, the task of tool setting has been replaced by an automatic tool positioning capability (see Chapter 32), which means that one source of variability in the process has been eliminated and the process becomes more repeatable. In the same light, it will be very important in the future for manufacturing engineers to know the process capability of robots they want to use in the workplace.

The discussion to this point has assumed that the *parent population* is normally distributed, that is, has the classic bell-shaped distribution in which the percentages shown in Figure 10-8 are dictated by the number of standard deviations from the central value or mean. The shape of the histogram may reveal the nature of the process to be skewed to the left or the right (unsymmetrical), often indicating some natural limit in the process. Drilled holes exhibit such a trend as the drill tends to make the hole oversize. Another possibility is a bimodal distribution (two distinct peaks), often caused by

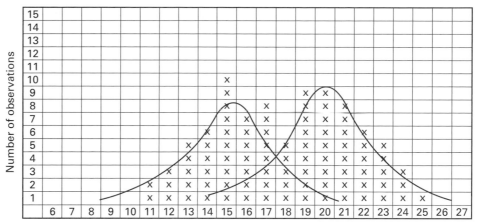

FIGURE 12-8 Example of a check sheet for gathering data on a process.

two processes being mixed together. Suppose you had a manufacturing cell and you were using a check sheet to gather data on one of the operations. See Figure 12-8. The time to perform an assembly task is recorded. Why do you think the data are bimodal? In this cell design, the operation was shared by two operators. The data would suggest that they do not do the assembly process the same way since there appears to be a 5- or 6-second difference in their average time to complete this task. Clearly further study is needed to determine what is going on here. For the next check sheet, use a different symbol for each worker.

The check sheet is an excellent way to view data while they are being collected. It can be constructed using predetermined parameters based on experience with the cell or system. The appropriate interval is checked as the data are being collected. This often allows the central tendency and the spread of the data to be seen. The check sheet can provide basically the same information as the histogram, but it is easier to build (once the check sheet is formatted). The possibilities are endless and require a careful recording of all the sources of the data to track down the factors that result in loss of precision and accuracy in the process. Rapid feedback on quality is perhaps the most important factor, so these data-gathering tasks are done right on the factory floor.

CAUSE-AND-EFFECT DIAGRAMS

One of the most effective methods for improving quality is the cause-and-effect diagram, also known as a fishbone diagram because of its structure. Initially developed by Kaorw Ishikowa in 1943, this diagram organizes theories about the probable cause of a problem. On the main line is a quality characteristic that is to be improved or the quality problem being investigated. Fishbone lines are drawn from the main line. These lines organize the main factors that could have caused the problem (see Figure 12-9). Branching from each of these factors are even more detailed factors. Everyone taking part in making a diagram gains new knowledge of the process. When a diagram serves as a focus for the discussion, everyone knows the topic, and the conversation does not stray. The diagram is often structured around four branches; the machine tools (or processes), the operators (workers), the method, and the material being processed. Another version of the diagram is called the CEDAC, the cause-and-effect diagram with the addition of cards. The effect is often tracked with a control chart (see Figure 12-16). The possible causes of the defect or problem are written on 3 × 5-inch cards and inserted in slots in the charts.

The three main applications of C&E diagrams:

I. *Cause enumeration:* listing of every possible cause and subcause.

   a. *Visual presentations:* one of the most widely used graphical techniques for QC.

   b. Better understanding of the relationships within the process yielding a better understanding of the process as a whole.

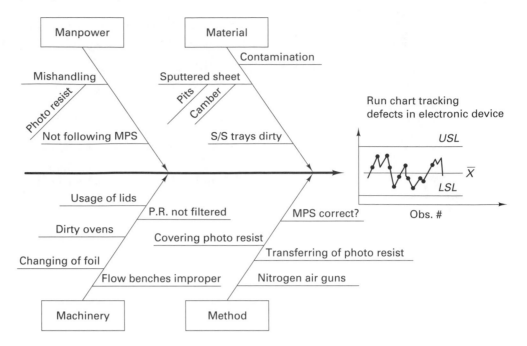

FIGURE 12-9 Example of a fishbone diagram using a run chart to show effects.

II. *Dispersion analysis:* causes grouped under similar headings; the 4 Ms are men, machines, materials, and methods.

   a. Each *major* cause is thoroughly analyzed.

   b. Possibility of not identifying root cause (may not fall into main categories).

III. *Process analysis:* similar to creating a flow diagram.

   a. Each part of the process listed in the sequence in which operations are performed.

INTRODUCTION TO TAGUCHI METHODS

Foreign competition has forced American manufacturers to take a second look at quality, as evidenced by the major emphasis (reemphasis) on statistical process control (SPC) in American industry. This drive toward superior quality has led to the introduction of Taguchi methods for improvement in products, product design, and processes. Basically, SPC looks at processes and control, the latter loosely implying "improvement." Taguchi methods, however, span a much wider scope of functions and include the design aspects of products and processes, areas that were seldom if every formally treated from the quality standpoint. Another threshold has been reached in quality control, witnessed by an expanding role of quality in the production of goods and services. The consumer is the central focus of attention on quality, and the methods of quality design and controls have been incorporated into all phases of production.

The Taguchi methods incorporate the following general features:

1. Quality is defined in relation to the total loss to the consumer (or society) from less-than-perfect quality of the product. The methods include placing a monetary value on quality loss. Anything less than perfect is waste.

2. In a competitive society, continuous quality improvement and cost reduction are necessary for staying in business.

3. Continuous quality improvement requires continuous reduction in the variability of product performance characteristics with respect to their target values.

4. The quality and cost of a manufactured product are determined by the engineering designs of the product and its manufacturing system.

5. The variability in product and process performance characteristics can be reduced by exploiting the nonlinear (interactive) effects of the process or product parameters on the performance characteristics.

6. Statistically planned (Taguchi) experiments can be used to determine settings for processes and parameters that reduce the performance variation.

7. Design and improvement of products and processes can make them *robust*, or insensitive to uncontrollable or difficult-to-control variations, called *noise* by Taguchi.

In the Taguchi approach, specified combinations of all of the input parameters, at various levels, that are believed to influence the quality characteristics being measured are used. These combinations should be run with the objective of selecting the best level for individual factors. For example, speed levels may be high, normal, and low; and operators may be fast or slow. For a Taguchi approach, material is often an input variable specified at different levels: normal, homogeneous, and highly variable. If a material is not controllable, it is considered a noise factor.

Taguchi methods can be used as an alternative approach to making a PC study. The Taguchi approach uses a truncated experiment design (called an *orthogonal array*) to determine which process inputs have the greatest effect on process variability (i.e., precision) and which have the least. Those inputs that have the greatest influence are set at levels that minimize their effect on process variability. As shown in Figure 12-10 factors A, B, C, and D all have an effect on process variability V. By selecting a high level of A and low levels of B, C, and D, the inherent variability of the process can be reduced. Those factors that have little effect on the process variable V are used to adjust or recenter the process aim. In other words, Taguchi methods seek to minimize or dampen the effect of the causes of variability and thus to reduce the total process variability. This is the goal of the six sigma method, which is discussed next.

The methods are, however, more than just mechanical procedures. They infuse an overriding new philosophy into manufacturing management that basically makes quality the primary issue in manufacturing. The manufacturing world is rapidly becoming aware that the consumer is the ultimate judge of quality. Continuous quality improvement toward perfect quality is the ultimate goal. Finally, it is recognized that the ultimate

FIGURE 12-10 The use of Taguchi methods can reduce the inherent process variability as shown in the upper figure. Factors A, B, C, and D versus process variable V shown in lower figure.

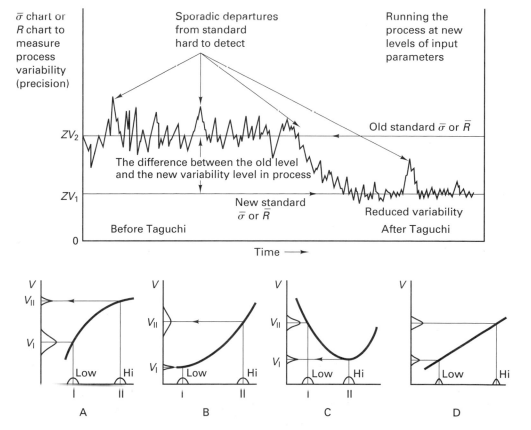

quality and lowest cost of a manufactured product are determined to a large extent by the engineered designs of

1. The product
2. The manufacturing process technology and the sequences of processes
3. The manufacturing system (integration of the product and the process)

So, a new understanding of quality has emerged. Process variability is not fixed. It can be improved! The noise level of a process can be reduced by exploring the nonlinear effects of the products (or process) parameters on the performance characteristics.

In Chapter 32 process capability is addressed again in the context of machining centers, programmable (NC) machines. In machine tools, accuracy and precision in processing are affected by machine alignment, the setup of the workholder, the design and rigidity (accuracy) of the workholder, the accuracy of the cutting tools, the design of the product, the temperature, and the operating parameters. The Taguchi methods provide a means of determining which of the input parameters are most influential in product quality.

MOTOROLA'S SIX SIGMA

In order to meet the quality challenge of the Japanese an American company, Motorola, developed the six sigma concept. The concept is shown in Figure 12-11 in terms of four sigma and six sigma capability. Most people do not know how sigma prime (σ') is

FIGURE 12-11   To move to six sigma capability from four sigma capability requires that the process capability (variability) be greatly improved (σ reduced). The curves in these figures represent histograms or curves fitted to histograms.

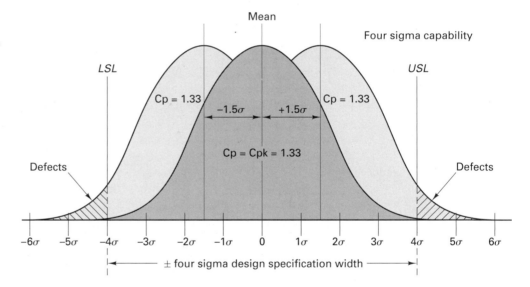

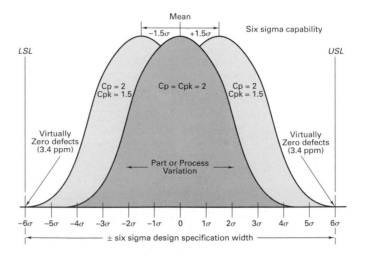

determined, (σ' is estimated from $\bar{R}/d_2$ or σ/C_2 sample data), that a sigma is a standard deviation, or what sigma measures (sigma measures the repeatability or variability or lack of precision in a process.)

In essence, the six sigma concept calls for the process to be improved to the place where there are 12 standard deviations between *USL* and *LSL*. As the variability of a process changes, so does sigma. A reduction in sigma (a reduction in spread) reflects an improvement in process or an improvement in precision (better repeatability). As the process is improved, sigma decreases. So the question is, "How do I improve the process?" This is the essence of process capability studies, reviewed in the first part of this chapter.

Here is an example. A foundry was having a problem with cores breaking in the molds during pouring. A PC study determined that the core strength was widely variable and it was the low strength cores that broke when the molten metal hit them during filling. Increasing the resin content in the cores and changing the gating in the mold did not eliminate the problem, so a Taguchi study was run that revealed that core strength was highly dependent on the grain size of the sand, which was also highly variable. The sand-preparation process was revised to yield more uniform sized grains, which, when used in the core-packing process, reduced the variability in the strength characteristic and eliminated the core breakage problem.

■ 12.3 Inspection and Quality Control

In virtually all manufacturing, it is extremely important that the dimensions and quality of individual parts be known and maintained. This is of particular importance where large quantities of parts, often made in widely separated plants, must be capable of interchangeable assembly. Otherwise, difficulty may be experienced in subsequent assembly or in service, and costly delays and failures may result. In recent years, defective products resulting in death or injury to the user have resulted in expensive litigation and damage awards against manufacturers. Inspection is the function that controls the quality (e.g., the dimensions, the performance, and the color) manually, by using operators or inspectors, or automatically, with machines, as discussed previously.

The economics-based question, "How much should be inspected?" has three possible answers.

1. *Inspect every item being made.* 100% Inspect every item being made; 100% checking with prompt execution of feedback and immediate corrective action can ensure perfect quality.
2. *Sample.* Inspect some of the product by sampling and make decisions about the quality of the process based on the sample.
3. *None.* Assume that everything made is acceptable or that the product is inspected by the consumer, who will exchange it if it is defective. (This is not a recommended procedure).

The reasons for not inspecting all of the product (i.e., for sampling) include:

1. Everything has not yet been manufactured—the process is continuing to make the item—so we have to look at some before we are done with all.
2. The test is destructive.
3. There is too much product for all of it to be inspected.
4. The testing takes too much time or is too complex or too expensive.
5. It is not economically feasible to inspect everything even though the test is simple, cheap, and quick.

Some characteristics are nondissectible; that is, they cannot be measured during the manufacturing process because they do not exist until after a whole series of operations has taken place. The final edge geometry of a razor blade is a good example, as is the yield strength of a rolled bar of steel.

Sampling (looking at some percentage of the whole) requires the use of statistical techniques that permit decisions about the acceptability of the whole based on the quality found in the sample. This is known as *statistical process control* (SPC).

STATISTICAL PROCESS CONTROL (SPC)

Looking at some (sampling) and deciding about the behavior of the whole (the parent population) is common in industrial inspection operations. The basic SPC techniques are the histogram and control charts. The histogram, shown earlier in this chapter, explains how the natural tolerance limits of the distribution can be compared with the engineering specifications to determine if the process is centered at the *nominal* value and to estimate the spread compared to tolerance values (or product specifications).

As an example, examine the data in Figure 12-12. These are measurements of the gaps in 125 retainers. They are supposed to have a nominal size of 0.70 mm. The population is assumed to be normal and only chance variations are occurring. What does the

FIGURE 12-12   Example of $\bar{X}$ and R charts and the data set of 25 samples ($k = 12$) of size 5($n = 5$) *(Courtesy: Ford)*

VARIABLES CONTROL CHART X&R
Averages & Ranges

Part/Asm. Name *Retainer*	Operation *Bend Clip*	Specification .50 – .90 mm	Nominal Size .70 mm
Part No. 1234567	Department 105	Gage *Depth Gage Micrometer*	
Parameter *Gap. Dim. "A"*	Machine 030	Sample Size/Frequency 5/2 *Hours*	

Date		6/8			6/9				6/10				6/11				6/12				6/15				6/16
Time of day		8	10	12	2	8	10	12	2	8	10	12	2	8	10	12	2	8	10	12	2	·8	10	12	2	8
Operator																										
Sample Measurements / Value of X	1	.65	.75	.75	.60	.70	.60	.75	.60	.65	.60	.80	.85	.70	.65	.90	.75	.75	.75	.65	.60	.50	.60	.80	.65	.65
	2	.70	.85	.80	.70	.75	.75	.80	.70	.80	.70	.75	.75	.70	.70	.80	.80	.70	.70	.65	.60	.55	.80	.65	.60	.70
	3	.65	.75	.80	.70	.65	.75	.65	.80	.85	.60	.90	.85	.75	.85	.80	.75	.75	.60	.85	.65	.65	.65	.75	.65	.70
	4	.65	.85	.70	.75	.85	.85	.75	.75	.85	.80	.60	.65	.75	.75	.75	.80	.70	.70	.65	.60	.80	.65	.65	.60	.60
	5	.85	.65	.75	.65	.80	.70	.70	.75	.75	.65	.80	.70	.70	.60	.85	.65	.80	.60	.70	.65	.80	.75	.65	.70	.65
Sum		3.50	3.85	3.80	3.40	3.75	3.65	3.65	3.60	3.90	3.35	3.75	3.80	3.60	3.55	4.10	3.75	3.80	3.35	3.50	3.10	3.30	3.45	3.50	3.20	3.30
Average $\bar{X}$			.77	.76	.68	.75	.73	.73	.72	.78	.67	.75	.76	.72	.71	.82	.75	.76	.67	.70	.62	.66	.69	.70	.64	.66
Range R		.20	.20	.10	.15	.20	.25	.15	.20	.20	.20	.40	.20	.05	.25	.15	.15	.15	.15	.20	.05	.30	.20	.15	.10	.10

$\bar{\bar{X}} = (.70 + .77 + \ldots + .64 + .66)/25 = 17.90/25 = .716$ $UCL_{\bar{X}} = .716 + (.58 \times .178) = .819$

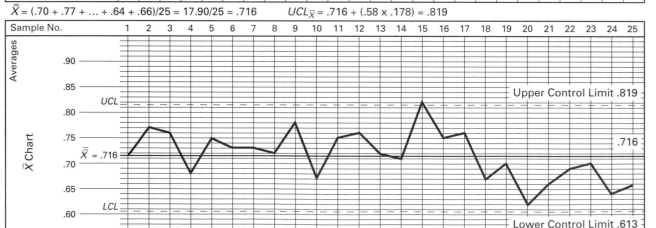

$\bar{R} = 4.45/25 = .176$ $UCL_R = 2.11 \times .178 = .376$ $LCL_R = 0$

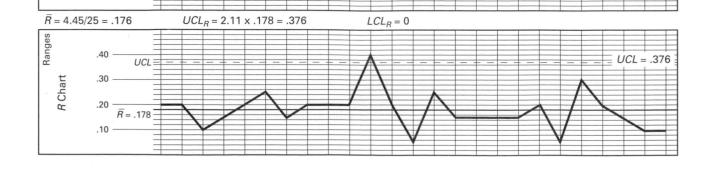

histogram of these 125 measurements look like? The mean for the entire population, μ, is obtained by

$$\mu = \frac{\sum\limits_{i=1}^{n} X_i}{n} = \frac{\sum\limits_{i=1}^{125} X_i}{125} \tag{12-5}$$

where X_i is an individual measurement and n is the number of items or measurements. The mean is a measure of the central tendency around which the individual measurements tend to group. The variability of the individual measurements about the average may be indicated by the standard deviation, σ, where

$$\sigma = \sqrt{\frac{\sum\limits_{i=1}^{n} (X_i - \mu)^2}{n}} \tag{12-6}$$

The standard deviation is of particular interest. For normal distributions, 68.26% of all measurement values will lie within the $\pm 1\sigma$ range from the average, 95.46% within the $\pm 2\sigma$ range, and 99.73% within the $\pm 3\sigma$ range; refer to Chapter 10. If the standard deviation for a population of items is known, then not over 0.07% of the items will fall outside the $\mu \pm 3\sigma$ *as long as only chance variations occur.* A process would be selected that produced a distribution of parts whose $6\sigma'$ value is less than the total part tolerance.

Note that these values may not be the same values as obtained earlier in the control chart study. Here we are finding the mean and standard deviation of 125 items, the entire population. In the QC study, the true mean and standard deviation of the population was being *estimated* from sample data, which is more typical of the situation one has to deal with.

QUALITY CONTROL CHARTS

Control charts for variables are used to monitor the output of a process by sampling (looking at some), by measuring selected quality characteristics, by plotting the sample data on the chart and then by making decisions about the performance of the process.

Figure 12-13 shows the basic structure of two charts commonly used for variables types of measurements. The $\overline{X}$ chart tracks the aim (accuracy) of the process. The R chart (or σ chart) tracks the precision or variability of the process. Usually, only the $\overline{X}$ chart and the R chart are used unless the sample size is large, and then σ charts are used in place of R charts. Because some sample statistics tend to be normally distributed about their own mean, $\overline{X}$ values are normally distributed about $\overline{\overline{X}}$, R values are normally distributed about $\overline{R}$, and σ values are normally distributed about $\bar{\sigma}$.

Quality control charts are widely used as aids in maintaining quality and in achieving the objective of detecting trends in quality variation before defective parts are actually produced. These charts are based on the previously discussed concept that if only chance causes of variation are present, the deviation from the specified dimension or attribute will fall within predetermined limits.

When sampling inspection is used, the typical sample sizes are from 3 to about 12 units. The $\overline{X}$ chart tracks the sample averages ($\overline{X}$ values). The R chart plots the range values (R values). The σ chart plots the standard deviation values and is used alternately with the R chart, usually when the sample size is large. Let us assume that the data shown in Figure 12-12 now represent the results of 125 samples of size 5 taken from a large number of retainers but not the entire parent population. The sample data will be used to prepare the control charts shown in Figures 12-12 and 12-14.

The center line of the $\overline{X}$ chart was computed prior to actual usage of the charts in control work,

$$\overline{\overline{X}} = \sum_{i=1}^{k} \overline{X}_i/k \tag{12-7}$$

where $\overline{X}_i$ was a sample average and k was the number of sample averages. The horizontal axis for the charts is time, thus indicating *when* the sample was taken. $\overline{\overline{X}}$ serves as an estimate for X', the true center of the process distribution and the center line of the $\overline{X}$ chart. The upper and lower control limits are commonly based on three standard error units, $3\sigma_{\overline{X}}$ (i.e., standard deviations for $\overline{X}$).

Area under a normal curve

- 99.7%
- 95.5%
- 68%

$$-3^{\sigma} \quad -2^{\sigma} \quad -1^{\sigma} \qquad +1^{\sigma} \quad +2^{\sigma} \quad +3^{\sigma}$$

Calculate for each sample

(Mean) $\bar{X} = \dfrac{\sum X}{n}$

n = Sample size

(Range) R = Hi-Low

or

(Sigma) $\sigma = \sqrt{\dfrac{\sum(X_i - \bar{X})^2}{n-1}}$

k = number of samples
and
n = sample size

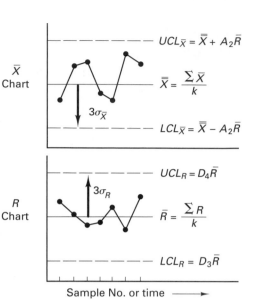

$\bar{X}$ Chart

$UCL_{\bar{X}} = \bar{\bar{X}} + A_2\bar{R}$

$\bar{\bar{X}} = \dfrac{\sum \bar{X}}{k}$

$3\sigma_{\bar{X}}$

$LCL_{\bar{X}} = \bar{\bar{X}} - A_2\bar{R}$

R Chart

$UCL_R = D_4\bar{R}$

$3\sigma_R$

$\bar{R} = \dfrac{\sum R}{k}$

$LCL_R = D_3\bar{R}$

Sample No. or time ⟶

Out of Control

- RUN — 7 pts. on one side of the central line
- TREND — 6 successive increasing or decreasing points
- POINT outside control limits

Calculate the average range $\bar{R}$ and the process average $\bar{\bar{X}}$ after k samples

$$\bar{R} = \dfrac{R_1 + R_2 + \dots R_k}{k} \qquad \bar{\bar{X}} = \dfrac{\bar{X}_1 + \bar{X}_2 + \dots \bar{X}_k}{k}$$

$UCL_{\bar{X}} = \bar{\bar{X}} + A_2\bar{R} = \bar{\bar{X}} + 3\sigma_{\bar{X}}$

$LCL_{\bar{X}} = \bar{\bar{X}} - A_2\bar{R} = \bar{\bar{X}} - 3\sigma_{\bar{X}}$

$UCL_R = D_4\bar{R} \qquad LCL_R = D_3\bar{R}$

$3\sigma_{\bar{X}} = 3\dfrac{\sigma'}{\sqrt{n}} = 3\dfrac{\bar{R}}{d_2\sqrt{n}} = A_2\bar{R}$

Sample Size	$\bar{X}$	R Chart		Est.	σ'
(n)	A_2	D_3	D_4	d_2	c_2
2	1.88	.00	3.27	1.13	.56
3	1.02	.00	2.57	1.69	.72
4	0.73	.00	2.28	2.06	.80
5	0.58	.00	2.11	2.33	.84
6	0.48	.00	2.00	2.53	.87
7	0.42	.08	1.92	2.70	.89
8	0.37	.14	1.86	2.85	.90
9	0.34	.18	1.82	2.97	.91
10	0.31	.22	1.78	3.08	.92

FIGURE 12-13   Quality control chart calculations.

Thus,

$$UCL_{\bar{X}} = \text{upper control limit on } \bar{X} \text{ chart}$$

$$= \bar{\bar{X}} + 3\sigma_{\bar{X}}$$

$$= \bar{X}' + A_2\bar{R} \tag{12-8}$$

$$LCL_{\bar{X}} = \text{lower control limit, } \bar{X} \text{ chart} = \bar{\bar{X}} - 3\sigma_{\bar{X}}$$

$$= \bar{X}' - A_2\bar{R}$$

where $\sigma_{\bar{X}} = \dfrac{\sigma'}{\sqrt{n}}$ (see Figure 12-13 for A_2 values). The standard deviation for the distribution of sample averages, $\sigma_{\bar{X}}$, is determined directly from the estimate of the standard deviation of the parent population, σ'. The $\bar{X}$ chart is used to track the central tendency (aim) of the process. In this example, assume that samples are being taken hourly and that the average of *each sample* (not individual values) is plotted.

The R chart is used to track the variability or dispersion of the process. A σ chart could also be used. R is computed for each sample ($X_{\text{HIGH}} - X_{\text{LOW}}$). The value of $\bar{R}$ was determined previously in PC study from

$$\bar{R} = \dfrac{\sum\limits_{i=1}^{k} R}{k} \tag{12-9}$$

where $\bar{R}$ represents the average range of k range values. The range values may also be normally distributed about $\bar{R}$, with standard deviation σ_R. To determine the upper and lower control limits for the charts, the following relationships are used.

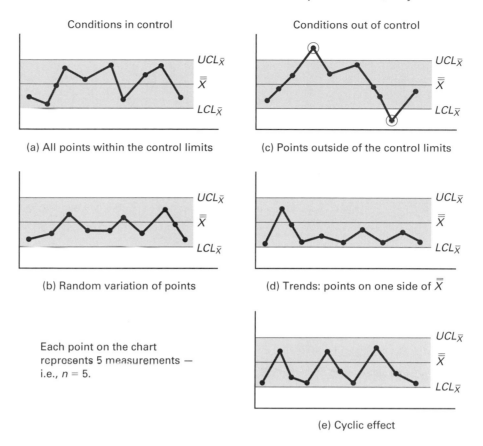

FIGURE 12-14 Examples of X bar control charts showing control versus out-of-control conditions.

$$UCL_R = \text{upper control limit } R \text{ chart,} = \bar{R} + 3\sigma_R = D_2\bar{R} = 2.11\bar{R} \text{ for } n = 5$$
$$UCL_R = \text{lower control limit} = R - 3\sigma_R = D_3\bar{R} = 0 \tag{12-10}$$

where D_4 and D_3 are constants and are given in Figure 12-13. For small values of n, the distance between center line R and LCL_R is more than $3\sigma_R$, but LCL_R cannot be negative, as negative range values are not allowed, by definition. Hence, $D_3 = 0$ for values of n up to 6.

After control charts have been established, and the average and range values have been plotted for each sample group, the chart acts as a control indicator for the process. If the process is operating under chance cause conditions, the data will appear random (will have no trends or pattern), see Figure 12-15. If $\bar{X}$, R or σ values fall outside the control limits or if nonrandom trends (run or cyclic effects) appear, an assignable cause or change may have occurred, and some action should be taken to correct the problem.

Trends in the control charts often indicate the existence of an assignable cause factor before the process actually produces a point outside the control limit. In grinding operations, wheel wear (wheel undersize) results in the parts becoming oversized and corrective action should be taken. (Redress and reset wheel or replace with new wheel.) Note that defective parts can be produced even if the points on the charts are in control. That is, it is possible for something to change in the process, causing defective parts to be made and the sample point still to be within the control limits. Since no corrective action was suggested by the charts, a type II error was made. Subsequent operations will then involve performing additional work on products already defective. Thus the effectiveness of the SPC approach in improving quality is often deterred by the lag in time between the discovery of an abnormality and the corrective action.

With regard to control charts in general, it should be kept in mind that the charts are only capable of indicating that something has happened, not what happened, and that a certain amount of detective work will be necessary to find out what has occurred to cause a break from the random, normal pattern of sample points on the charts. Keeping careful track of when and where the sample was taken will be very helpful in such investigations, but the best procedure is to have the operator do the inspection and run the chart. In this way, quality feedback is very rapid and the causes of defects readily found.

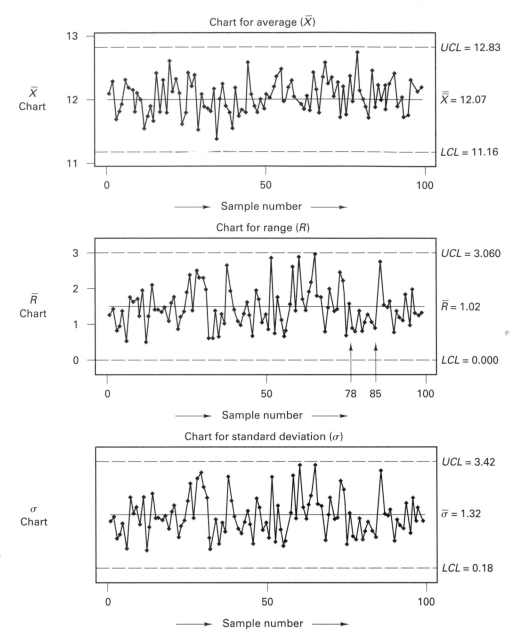

FIGURE 12-15 Examples of $\bar{X}$, R and σ for 100 samples showing random nature.

■ 12.4 DETERMINING CAUSES FOR PROBLEMS IN QUALITY

The best way to quickly isolate quality problems is to make everyone an inspector. This means every worker, foreman, supervisor, engineer, manager, and so forth, is responsible for making it right the first time and every time. One very helpful tool in this effort is the fishbone diagram. As shown in Figure 12-16, the fishbone diagram can be used in conjunction with the control chart to root out the causes of problems. Also, the problem (bad welds) can have multiple causes, but in general, the cause will lie in the process, operators, materials, or method (i.e., the four main branches on the chart). Every time a quality problem is caused by one of these events, it is noted by the observer, and corrective action is taken. As before, experimental design procedures can help identify causes that affect performance.

In summary, the data gathered to develop the PC study can be used to prepare the initial control charts from the process after the removal of all assignable causes for variability and proper setting of the process average. After the charts have been in place for some time and a large amount of data have been obtained from the process output,

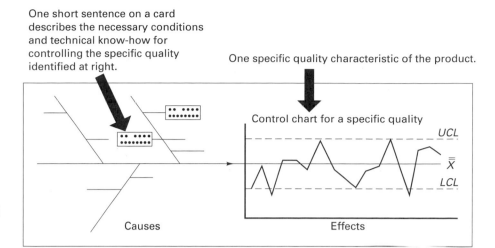

FIGURE 12-16 A cause and effect diagram with cards is used with a control chart to analyze the process.

the PC study can be redone to obtain better estimates of the natural spread of the process during actual production. On-line or in-process methods such as SPC and off-line (Taguchi) methods are key elements in a total quality control program.

SAMPLING ERRORS

It is important to understand that in sampling, two kinds of decision errors are always possible; see Figure 12-17. Suppose that the process is running perfectly, but the sample data indicate that something is wrong. You decide to stop the process to make adjustments. This is a type I error. Suppose that the process was not running perfectly and was making defective products. However, the sample data did not indicate that anything was wrong. You did not stop the process and set it right. This is a type II error. Both types of errors are possible in sampling. For a given sample size, reducing the chance of one type of error will increase the chance of the other. Increasing the sample size or the frequency of sampling reduces the probability of errors but increases the cost of the inspection. Many companies determine the size of the errors they are willing to accept according to the overall cost of making the errors plus the cost of inspection. If, for example, a type II error is very expensive in terms of product recalls or legal suits, the company may be willing to make more type I errors, to sample more, or even to go to 100% inspection on very critical items to ensure that the company is not accepting defective materials as good and passing them on to the customer. The inspection should take place immediately after the processing.

As mentioned earlier, in any continuing manufacturing process, variations from established standards are of two types: (1) *assignable cause variations*, such as those due to malfunctioning equipment or personnel, or to defective material, or to a worn or broken

FIGURE 12-17 When you look at some of the output from a process (sample) and decide about the quality of the process, you can make two kinds of errors.

		The process is running It really has	
		Changed (making defects)	Not changed
The sample data from the process is	Has changed	No error	Type II β error
Plotted on a control chart. You decide that the process	Has not changed	Type I α error	No error

tool; and (2) *normal chance variations*, resulting from the inherent nonuniformities that exist in materials and in machine motions and operations. Deviations due to assignable causes may vary greatly. Their magnitude and occurrence are unpredictable, and thus should be prevented. However, if the assignable causes of variation are removed from a given operation, the magnitude and frequency of the chance variations can be predicted with great accuracy. Thus if one can be assured that only chance variations will occur, the quality of the product will be better known, and manufacturing can proceed with assurance about the results. By using statistical process control procedures, one may detect the presence of an assignable-cause variation and remove the cause, often before it causes quality to become unacceptable. Also, the astute application of statistical experimental design methods (Taguchi experiments) can help to identify some assignable causes.

To sum up, use PC analysis and process improvements to get the process into the condition described in Figure 12-11. Use control charts to keep the process centered ($\overline{X}$ chart) with no increases in variability (R chart or σ chart).

GAGE CAPABILITY

The instrument (gage) used to measure the process will also have some inherent precision and accuracy, often referred to as gage capability. In other words, the observed variation in the component part being measured is really composed of the actual process variation plus the variation in the measured system; see Figure 12-18. The measuring system will display:

Bias: poor accuracy or aim
Linearity: accuracy changes over the span of measurements
Stability: accuracy changes over time
Repeatability: loss of precision in the gage (variability)
Reproducibility: variation due to different operators

Different operators have different means and different variation about the mean when performing a measurement. Determining the capability of the gage is called an *R and R* study. Space does not permit a full discussion here of *R and R* studies, but detailed descriptions are found in the references cited at the end of this chapter. In particular, students involved in ISO9000 studies should examine the *Measurement System Analysis Reference Manual*, published by the Automotive Industry Action Group.

TOTAL QUALITY CONTROL (TQC)

The phrase *total quality control* (TQC) was first used by A. V. Feigenbaum in *Industrial Quality Control* in May 1957. TQC means that all departments of a company must participate in quality control (Table 12-1). Quality control is the responsibility of workers at every level in every department, all of whom have studied quality control. It begins at the product design stages and carries through the manufacturing system, where the emphasis is on making it right the first time. It is surprising how few companies have embarked on implementing Taguchi methods in order to reduce the inherent variability in their processes. This is probably because many manufacturing (mechanical) engineers have not had any coursework in this area.

Also, a change in company culture is often needed to give the responsibility for quality to the worker, along with the authority to stop the process when something goes wrong. An attitude of defect prevention and a habit of constant improvement of quality are fundamental to lean production (see Chapter 42). Companies such as Toyota have accomplished TQC (they call it company-wide quality control) by extensive education of the

FIGURE 12-18   Gage capability (variation) contributes to the total observed variation in a measurement of a part.

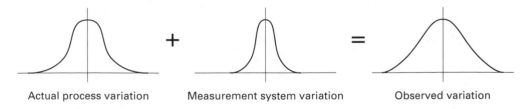

Actual process variation   Measurement system variation   Observed variation

TABLE 12-1. Quality Control: Concepts and Categories

TQC Category	TQC Concept
1. Organization	• Manufacturing engineering has responsibility for quality—quality circles
2. Goals	• Habit of improvement for everyone everywhere in the manufacturing system. • Perfection—zero defects—not a program, a goal
3. Basic principles	• Process control—defect prevention, not detection • Easy-to-see quality—quality on display so customers can see and inspect processes—easy to understand quality • Insist on compliance with maintenance • Line stop when something goes wrong • Correct your own errors • 100% check; MO-CO-MOO • Project-by-project improvements

4. Facilitating concepts	• QC department as facilitator
	   Audit suppliers
	   Help in quality improvement projects
	   Training workers, supervisors, suppliers
	• Small lot sizes through rapid changeover
	• Housekeeping
	• Less-than-full-capacity scheduling
	• Check machines daily, use checklists like airplane pilot
	• Total preventive maintenance (TPM)
	• 8-4-8-4 two-shift scheduling

5. Techniques and aids	• Expose problems, solve problems
	• Defect prevention, pokayokes for checking 100% of parts
	• $N = 2$, for checking first and last item in a lot ($N = 3$)
	• Analysis tools
	   Cause-and-effect diagrams
	   Histograms and run charts, check sheets
	   Control charts ($\overline{X}, R, \sigma$)
	   Scatter diagrams
	   Pareto charts
	   Process flowcharts
	   Taguchi and DOE methods

TABLE 12-2. Typical Problems Tackled by Quality or Production Teams

Product quality
Paperwork
Hardware
Communications
Service
Processes and methods
Scrap reduction
Productivity
Cost reductions
Production delays

workers, giving them the analysis tools they need (control charts with cause-and-effect diagrams) to find and expose the problems. Workers are encouraged to correct their own errors, and 100% inspection (often done automatically) is the rule. Passing defective products on to the next process is not allowed. The goal is perfection. *Quality circles*, now popular in the United States, are just one of the methods used by Japanese industries to achieve perfection. See Table 12-2 for typical methods tackled by quality teams.

LINE STOP

A pair of yellow and red lights hanging above the workers on the assembly line can be used to alert everyone in the area to the status of the processes. Many companies use Andon boards, which hang above the aisles. The number on the board reflects stations on the line. A worker can turn on a yellow light when assistance is needed, and nearby workers will move to assist the worker having a problem. The line keeps moving, however, until the product reaches the end of the station. Only then is a red light turned on if the problem cannot be solved quickly and the line needs to be stopped. When the problem is solved and everyone is ready to go again, the red light goes off and everyone starts back to work, all in synch.

Every worker should be given the authority to stop the production line to correct quality problems. In systems using pokayoke or autonomation, devices may stop the line automatically. The assembly line or manufacturing cell should be stopped immediately and started again only when the necessary corrections have been made. Although stopping the line takes time and money, it is advantageous in the long run. Problems can be found immediately, and the workers have more incentive to be attentive because they do not want to be responsible for stopping the line.

IMPLEMENTING QUALITY COMPANY-WIDE

The basic idea of integration is to shift functions that were formerly done in the staff organization (called the production system) into the manufacturing system. What happens to the quality control department? The department serves as the facilitator and therefore acts to promote quality concepts throughout the plant. In addition, its staff educates and trains the workers in statistical and process control techniques and provides engineering assistance on visual and automatic inspection installations. Its most important functions will be training the entire company in quality control.

Another important function of the QC department will be to work with and audit the vendors. The vendor's quality must be raised to the level at which the buyer does not need to inspect incoming material, parts, or subassemblies. The vendor simply becomes an extension of the buyer's plant. Ultimately, each vendor will deliver to the plant perfect materials that need no incoming inspection. Note that this means the acceptable quality level (AQL) of incoming material is 0%. Perfection is the goal. For many years this country has lived with the unwritten rule that 2 or 3% defective was about as good as you could get: Better quality just costs too much. For the mass production systems, this was true. To achieve the kinds of quality that Toyota, Honda, Sanyo, and many others have demonstrated, a company has to eliminate the job shop (a functional manufacturing system) and restructure the production system, integrating the quality function directly into the linked-cell manufacturing system; see Chapter 43 for details.

The quality control department also performs complex or technical inspections, total performance checks (often called end item inspection), chemical analysis, X-ray analysis, destructive tests, or tests of long duration.

MAKING QUALITY VISIBLE

Visual display on quality should be placed throughout manufacturing facilities to make quality evident. These displays tell workers, managers, customers, and outside visitors what quality factors are being measured, what the current quality improvement projects are, and who has won awards for quality. Examples of visible quality are signs showing quality improvements, framed quality awards presented to or by the company, and displays of high-precision measuring equipment.

These displays have several benefits. When customers visit your plant to inspect your processes, they want to see measurable standards of quality. Highly visible indicators of quality such as control charts and displays should be posted in every department. Everyone is informed on current quality goals and the progress being made. Displays and quality awards are also an effective way to show the workforce that the company is serious about quality.

POKAYOKES; SOURCE, SELF, AND SUCCESSIVE CHECKS

Many companies have developed an extensive QC program based on having many inspections. However, inspections can only find defects, not prevent them. Adding more inspectors and inspections merely uncovers more defects, but does little to prevent them. Clearly, the least costly system is one that produces no defects. But is this possible? Yes, it can be accomplished through two methods—pokayoke and source inspection.

Many people do not believe that the goal of zero defects is possible to reach, but many companies have achieved this goal or have reduced their defect level to virtually zero using pokayokes and source inspection.

Pokayoke is a Japanese word for defect prevention. Pokayoke devices and procedures are often devised mainly for preserving the safety of operations. The idea is to develop a method, mechanism, or device that will prevent the defect from occurring rather than to find the defect after it has occurred. Pokayokes can be attached to machines to automatically check the products or parts in a process. Pokayokes differ from source inspections in that they are usually attributes inspections. The production of a bad part is prevented by the device. Some devices may automatically shut down a machine if a defect is produced, preventing the production of an additional defective part. The pokayoke system uses 100% inspection to guard against unavoidable human error. Figure 12-19 illustrates a pokayoke device. This device ensures that the workers remember to apply labels to the products, thus preventing defective products.

Such devices work very well when physical detection is needed, but many items can be checked only by sensory detection methods, such as the surface finish on a bearing

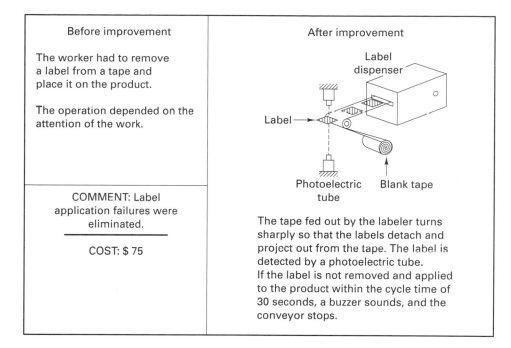

Before improvement	After improvement
The worker had to remove a label from a tape and place it on the product. The operation depended on the attention of the work.	*(diagram: Label dispenser, Label, Photoelectric tube, Blank tape)*
COMMENT: Label application failures were eliminated. ——— COST: $ 75	The tape fed out by the labeler turns sharply so that the labels detach and project out from the tape. The label is detected by a photoelectric tube. If the label is not removed and applied to the product within the cycle time of 30 seconds, a buzzer sounds, and the conveyor stops.

FIGURE 12-19 Example of pokayoke device for preventing defects. *(Shingo, 1986).*

race or the flatness of a glass plate. For such problems a system of self-checks and successive checks can be used.

Source inspection looks for errors before they become defects, and either stops the system or makes corrections or automatically compensates or corrects for the error condition to prevent a defective item from being made. The common term for source inspection in manufacturing processes is adaptive control (A/C).

There are two ways to look at source inspections: vertically and horizontally. Vertical source inspections try to control upstream processes that can be the source or the cause of defects downstream. It is always necessary to examine source processes because they may have a much greater impact on quality than the processes being examined. Finding the source of a problem requires asking why at every opportunity. Here is an example.

Some steel bars were being cylindrically ground. After grinding, about 10% of the bars warped (bent longitudinally) and were rejected. The grinding process was studied extensively and no sure solution was found. Looking upstream, a problem with the heat-treating process that preceded cylindrical grinding was detected. About 10% of the bars were not getting a complete, uniform heating prior to quench. Asking why, it was found that these bars were always lying close to the door of the oven, and it was found that the door was not properly sealed, which resulted in a temperature gradient inside. Quenching of the bars induced a residual stress that was released by the grinding and caused the warping. Asking why five times uncovered the source of the problem.

Horizontal source inspections detect defect sources within the processes and then introduce corrections to keep from turning errors into defects. This is commonly called adaptive control for preventing defects or in-process quality control.

TEAMS (AKA QUALITY CIRCLES)

Several popular programs are built upon the concept of participative management, such as quality circles, improvement teams, and task groups. These programs have been very successful in many companies, but have failed miserably in others. The difference is often due to the way management implemented the program. Programs must be integrated and managed within the context of a lean manufacturing system design strategy. For example, asking an employee for a suggestion that management does not use (or cannot explain why it does not use) defeats a suggestion system. Management must learn to trust the employees' ideas and decisions and move the decision making to the factory floor.

TABLE 12-3.   Quality Circle Objectives
1. Develop workers' skill and knowledge. 2. Introduce a team effort among workers, supervisors, and managers. 3. Unlock the creativity inherent in workers. 4. Improve quality consciousness. 5. Create a more harmonious workforce, leading to higher morale. 6. Encourage commitment and contribution to corporate goals of better quality and higher productivity. 7. Encourage leadership qualities in circle leaders. 8. Improve communications and extend recognition.

A quality circle is usually a group of employees within the same department or factory floor area. Meetings are held to work on problems. An organization structure is usually composed of members, a team leader, a facilitator, a manufacturing engineer, and a steering committee.

Quality circles usually have the following main objectives: provide all workers with a chance to demonstrate their ideas; raise employee morale; and encourage and develop workers' knowledge, quality control techniques, and problem-solving methods. They also unify company-wide QC activities, clarify managerial policies, and develop leadership and supervisory capabilities (see Table 12-3).

Quality circles have been implemented in U.S. companies with limited success when they are not part of a lean manufacturing strategy. It is possible for quality circles to work in the United States, but they must be encouraged and supported by management. Everyone must be taught the importance and benefits of integrated quality control.

SUPERIOR QUALITY IN MANUFACTURING/ASSEMBLY CELLS

The cells are designed for a "make one–check one–move one on" (MO-CO-MOO) strategy. The part receives successive checks after each processing or assembly step. For *successive checks* to be successful, several rules should be followed. All the possible variables and attributes should not be measured, because this would eventually lead to errors and confusion in the inspection process. The part should be analyzed so that only one or two points are inspected after each step in the process. This is the heart of MO-CO-MOO. Only the most important elements just produced are inspected.

Another important rule is that the immediate feedback of a defect leads to immediate action. Since the parts are produced in an integrated manufacturing system, this will be very effective in preventing the production of more defective parts. Suppose the cell has only one or two workers and they are not in a position to directly check each other's work after each step. Here is where the decouplers can play a role by proving automatic successive checking of the parts' critical features before proceeding to the next step. Only perfect parts are pulled from one process to the next through the decoupler. See Chapter 42 for additional discussions.

In assembly lines, a worker may inspect each part immediately after producing it. This is called *self-checking*. There is an immediate feedback to the worker on quality. However, it would be difficult for many workers not to allow a certain degree with on bias to creep into their inspection, whether they were aware of it or not, since they are inspecting their own work. Within cells operated by multiple workers, the operator of the downstream station or process can inspect the parts produced by the upstream operator. If there is a problem with the parts, the defective item is immediately passed back to the worker at the previous station. There the defect is verified and the problem corrected. Action is immediately taken to prevent any more defective parts. While this is going on, the line is shut down.

■ 12.5 SUMMARY

The designer of the product must have quality in mind during the quality design phase, seeking the least costly means to assure the quality of the desired functional characteristics. Major factors that can be handled during the early stages of the product design cycle include temperature, humidity, power variations, and deterioration of materials and tools. Compensation for these factors is difficult or even impossible to implement after the product is in production. The distinction between superior- and poor-quality products can be seen in their variability in the face of internal and external causes. This is where Taguchi parameter design methods can be important.

The secret to successful process control is putting the control of quality in the hands of the workers. Many companies in this country are currently engaged in SQC (statistical quality control) but they are still inspecting to *find* defects. The number of defects will not be reduced merely by making the inspection stage better or faster or automated. You are simply more efficient at discovering defects. The trick is to *inspect* to *prevent* defects. How can this be done? Here are the basic ideas: Use source inspection techniques that control quality at the stage where defects originate. Use 100% inspection with immediate feedback rather than sampling. Make every worker an inspector. Minimize the time it takes to carry out corrective action. Remember that people are human and not infallible. Devise methods and devices to prevent them from making errors. Can you think of such a device? Does your car have a procedure that prevents you from locking the ignition keys inside the car? That is a pokayoke.

Concentrate on making processes efficient, not simply on making the operators and operations more proficient. Continuous improvement requires that you redesign the manufacturing system continuously, reducing the time required for products to move through the system (i.e., the throughput time). This approach seems to be the American stumbling block. Industrial engineers can do operations improvement work like buying a better machine or improving the ergonomics of a task. However, they need to do more systems improvement work. Too often fancy, complex computerized solutions are devised to solve complex manufacturing process problems. Why not simplify the manufacturing system so that the need for complex solutions disappears? For more information on manufacturing system design, see Chapters 41, 42, and 43.

■ KEY WORDS

accuracy	fishbone	nominal	process capability (PC)	sample mean	statistical process control (SPC)
control	diagram	parent population	quality control (QC)	specification limit	Taguchi methods
chart	histogram	precision	range (R)	standard deviation (σ)	variability

■ REVIEW QUESTIONS

1. Define a process capability study in terms of accuracy or precision.

2. What does the *nature of the process* refer to?

3. Suppose you have a "pistol-shooting" process that is accurate and precise. What might the target look like if, occasionally while shooting, a sharp gust of wind blew left to right?

4. Review the steps required to making a PC study of a process.

5. Why don't standard tables exist detailing the natural variability of a given process, like rolling, extruding, or turning?

6. What are Taguchi or factorial experiments, and how might they be used to do a process capability study?

7. How does the Taguchi approach differ from the experimental method outlined in this chapter?

8. Why are Taguchi experiments so important compared to classical experiments?

9. Here are some common, everyday processes with which you are familiar. What variable or aspect to the process might dominate the process, in terms of quality, not output?

   a. Baking a cake (from scratch; from a cake mix).

   b. Mowing the lawn.

   c. Washing dishes in a dishwasher.

10. Explain why the diameter measurements for holes produced by the process of drilling could have a skewed rather than a normal distribution.

11. What are some common manufactured items that may

    a. Receive 100% inspection.

    b. Receive no final inspection.

    c. Receive some final inspection, that is, sampling?

12. What are common reasons for sampling inspection rather than 100% inspection?

13. Fill in this table with one of the following statements: no error—the process is good, no error—the process is bad, type I or alpha error, type II or beta error.

		In reality, if we looked at everything the process made, we would know that it had:	
		Changed	Not changed
The sample suggested that the process had:	Changed		
	Not changed		

14. Now explain why when we sample, we cannot avoid making type I and type II errors?

15. Which error can lead to legal action from the consumer for a defective product that caused bodily injury?

16. Define and explain the difference between
    a. σ' and $\sigma_{\bar{X}}$
    b. σ' and $\bar{\sigma}$
    c. $\sigma_{\bar{X}}, \sigma_R, \sigma_\sigma$

17. What is C_p, and why is a value of 0.80 not good? How about a value of 1.00? 1.3?

18. Explain what is measured by the mismatch factor, D.

19. What are some of the alternatives available to you when you have the situation where $6\sigma' > USL - LSL$?

20. C_{pk} is also a process capability index. How does it differ from C_p?

21. In a sigma chart, are σ values for the samples normally distributed about $\bar{\sigma}$? Why or why not?

22. What is an assignable cause, and how is it different from a chance cause?

23. Why is the range used to measure variability when the standard deviation is really a better statistic?

24. How is the standard deviation of a distribution of sample means related to the distribution from which the samples were drawn?

25. In the last two decades, the quality in automobiles has significantly improved. What do *you* think is the main cause for this marked quality improvement?

■ PROBLEMS

1. For the items listed in the following chart, obtain a quantity of 48. Measure the indicated characteristics and determine the process mean and standard deviation. Use a sample size of 4, so that 12 samples are produced.

Item	Characteristic(s) You Can Measure
Flat washer	Weight, width, diameter of hole, outside diameter
Paper clip	Length, diameter of wire
Coin (penny, dime)	Diameter, thickness at point, weight
Your choice	Your choice

2. Perform a process capability study to determine the PC of the process that makes M&M candy. You will need to decide what characteristics you want to measure, (weight, diameter, thickness, etc.), how you will measure it (use rule of 10), and what kind of M&Ms you want to inspect (how many bags of M&Ms you wish to sample). Take sample size of $4(n = 4)$. Make a histogram of the individual data and estimate $\bar{X}'$ and σ as outlined in the chapter. If you decide to measure the weight characteristics, you can check your estimate of $\bar{X}'$ by weighing all the M&Ms together and dividing by the total number of M&Ms.

3. For the data given in Figure 12-3, compute C_p, D, and C_{pk}, making any assumptions needed to perform the calculations.

4. For the data given in Figure 12-4, compute C_p, D, and C_{pk}.

5. Calculate $\bar{\bar{X}}$ and $\bar{R}$ and the control limits for the $\bar{X}$ and R control charts shown in Figure 12-A. The sample mean, $\bar{X}$ and range for the first nine subgroups and the data for each sample are given in the bottom of the figure. There are 25 samples of size 4. Therefore $K = 25, n = 4$. Complete the bottom part

of the table and then compute the control limits for both charts. Construct the charts plotting $\bar{\bar{X}}$ and $\bar{R}$ bar as solid lines and control limits as dashed lines as shown in Figure 12-12. The first four data points have been plotted and the points connected, but are they all correctly plotted? Replot any points that are incorrectly plotted. Plot the rest of the data on the charts and comment on your findings.

6. For the data given in Figure 12-A, estimate the mean and standard deviation for the process from which these samples were drawn (i.e., the parent population) and discuss the process capability in terms of C_p, C_{pk} and D. The USL and LSL for this dimension are 0.9 and 0.5, respectively, and the nominal is 0.7.

7. Compare the results from your analysis of Figure 12-A with the findings of problem 4.

8. Figure 12-B contains data from a process that produces holes (drilling) with limits of 6.00 to 6.70 mm. The control charts for $\bar{X}$ and R using $n = 5$ and $k = 25$ are shown in the figure, also. (*Note*: The numbers in the body of the table are 6.47, 6.19, 6.19, 6.29, etc.)
    a. Using the data for $n = 5, k = 25$, develop the process capability indexes C_p and C_{pk}, and discuss the capability of this process.
    b. Using the data $n = 5, k = 25$, develop the σ control chart and use $\bar{\sigma}$ to estimate σ' for the process capability indexes C_p and C_{pk}.
    c. Develop $\bar{X}, R$, and σ charts for samples sizes of 4 or 3 by ignoring X_5 or X_3 and X_5 (or any combination of individual values). Use the charts to perform a process capability study.

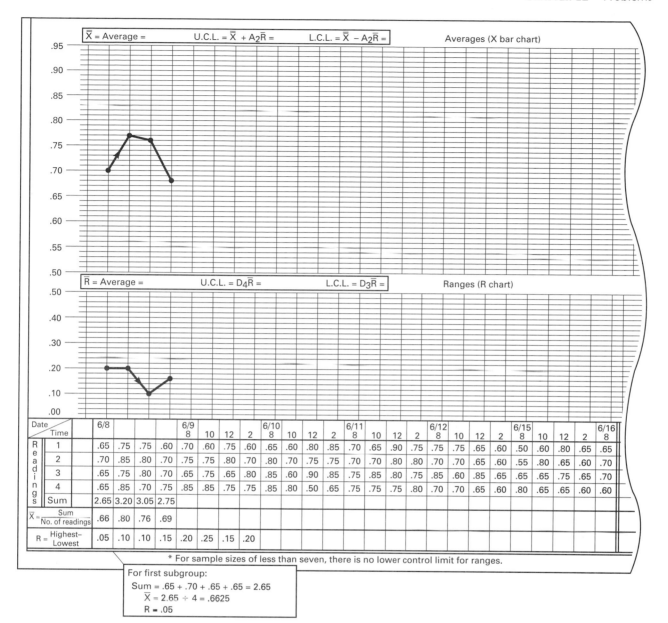

$\overline{X}$ = Average =   U.C.L. = $\overline{\overline{X}}$ + A$_2$$\overline{R}$ =   L.C.L. = $\overline{\overline{X}}$ − A$_2$$\overline{R}$ =   Averages (X bar chart)

$\overline{R}$ = Average =   U.C.L. = D$_4$$\overline{R}$ =   L.C.L. = D$_3$$\overline{R}$ =   Ranges (R chart)

Date / Time		6/8				6/9 8	10	12	2	6/10 8	10	12	2	6/11 8	10	12	2	6/12 8	10	12	2	6/15 8	10	12	2	6/16 8
Readings	1	.65	.75	.75	.60	.70	.60	.75	.60	.65	.60	.80	.85	.70	.65	.90	.75	.75	.75	.65	.60	.50	.60	.80	.65	.65
	2	.70	.85	.80	.70	.75	.75	.80	.70	.80	.70	.75	.75	.70	.70	.80	.80	.70	.70	.65	.60	.55	.80	.65	.60	.70
	3	.65	.75	.80	.70	.65	.75	.65	.80	.85	.60	.90	.85	.75	.85	.80	.75	.85	.60	.85	.65	.65	.65	.75	.65	.70
	4	.65	.85	.70	.75	.85	.85	.75	.75	.85	.80	.50	.65	.75	.75	.75	.80	.70	.70	.65	.60	.80	.65	.65	.60	.60
	Sum	2.65	3.20	3.05	2.75																					
$\overline{X}$ = Sum / No. of readings		.66	.80	.76	.69																					
R = Highest− Lowest		.05	.10	.10	.15	.20	.25	.15	.20																	

* For sample sizes of less than seven, there is no lower control limit for ranges.

For first subgroup:
Sum = .65 + .70 + .65 + .65 = 2.65
$\overline{X}$ = 2.65 ÷ 4 = .6625
R = .05

FIGURE 12-A

Product name	Cylinder		Sample size	3, 4 or 5	Period		1996, 10.15	$\overline{X}$ Control chart		R Chart
Quality characteristic	Hole diameter	Samples	Timing of taking samples	Daily			1975, 10.30	Center line $\overline{\overline{X}}$ = 6.299		Center line $\overline{R}$ = 0.274
Limits of allowable range	Max. 6.70 mm	Section		00 00	Person in charge		C. Black	$UCL\overline{x}$ + 0.58 $\overline{R}$ = 6.458		UCL 2.11 × $\overline{R}$ = 0.578
	Min. 6.00 mm	Measuring instrument serial number		103037	Person in charge of inspection		Pogi Bear	$LCL\overline{x}$ − 0.58 $\overline{R}$ = 6.140		LCL = 0

	Lot No.	1	2	3	4	5	6	7	8	9	10	11	12	13	14	15	16	17	18	19	20	21	22	23	24	25	Total	Mean
Measured values	X_1	47	19	19	29	28	40	15	35	27	23	28	31	22	37	25	7	38	35	31	12	52	20	29	28	42		
	X_2	32	37	37	29	12	35	30	44	37	45	44	25	37	32	40	31	0	12	20	27	42	31	47	27	34		
	X_3	44	31	31	42	45	11	12	32	26	26	40	24	19	12	24	23	41	29	35	38	52	15	41	22	15		
	X_4	35	25	25	59	36	38	33	11	20	37	31	32	47	38	50	18	40	48	24	40	24	3	32	32	29		
	X_5	20	34	34	38	25	33	26	38	35	32	18	22	14	30	19	32	37	20	47	31	25	28	22	54	21		
	Total	178	146	101	197	146	157	116	160	145	163	161	134	139	149	158	111	156	144	157	148	195	97	171	163	141		
	Mean $\overline{x}$	35.6	29.2	20.2	39.4	29.2	31.4	23.2	32.0	29.0	32.6	32.2	26.8	27.8	29.8	31.6	22.2	31.2	28.8	31.4	29.6	39.0	19.4	34.2	32.6	28.2	746.6	29.86
	Range R	27	18	33	30	33	29	21	33	17	22	26	10	33	26	31	25	41	36	27	28	28	28	25	32	27	686	27.44

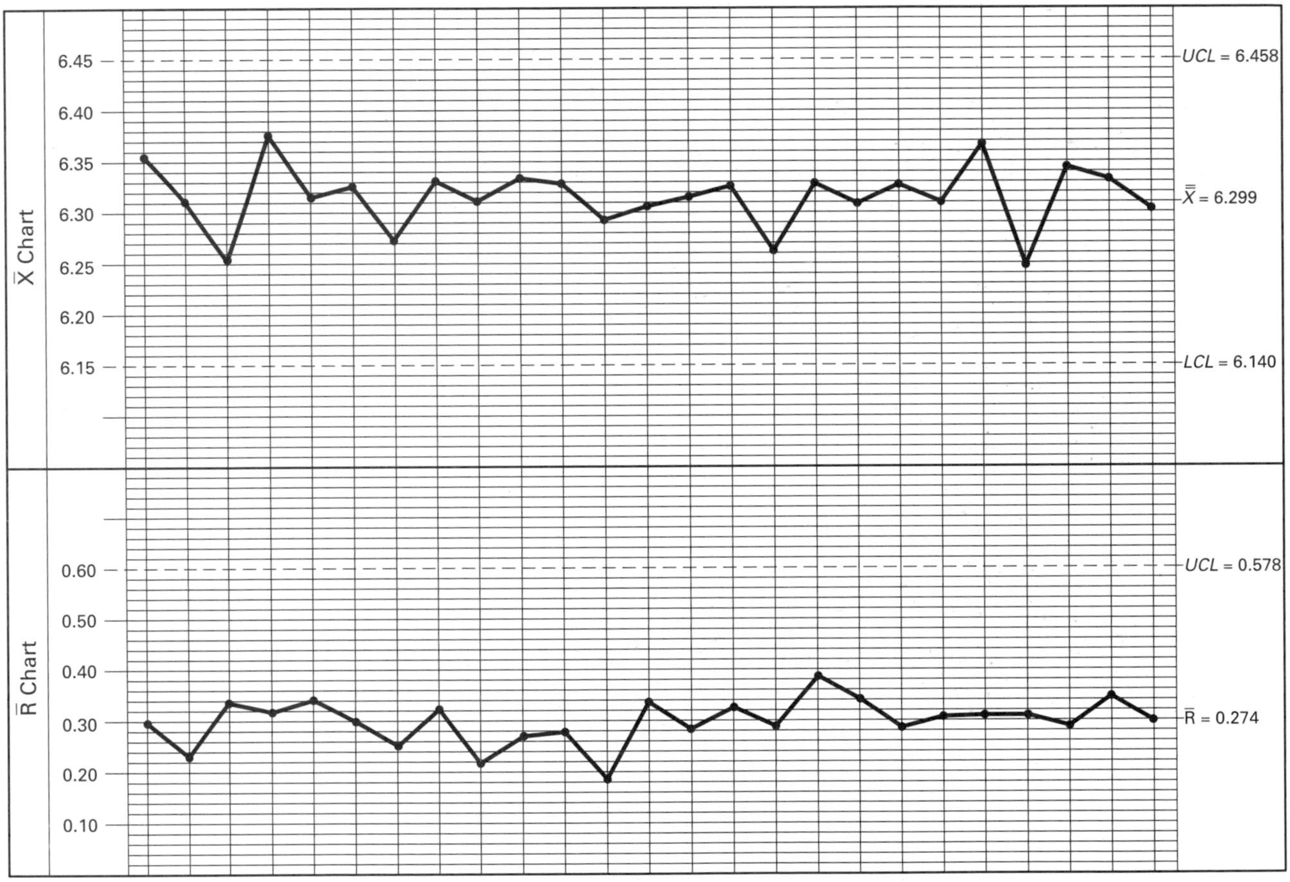

FIGURE 12-B

Chapter 12 CASE STUDY

Boring QC Chart Blunders

You have recently been hired by the BRC Company as the quality engineer. Your new boss, Abigail Crawford, brought in one of the leading textbooks in manufacturing engineering (NOT DeGarmo) to show you this problem. After reading the discussion and studying the figure, Abigail asked you the following questions.

1. What kind of control chart is this?

2. What do you estimate the standard deviation (σ') of the parent population to be, assuming that the control limits on this chart are based on three sigma.

3. The dashed line in the figure labeled "mean" is not the mean or the average—the mean is $\overline{\overline{X}}$ and is equal to 0.00017. What is the line really called?

4. From the information given in the discussion and in the figure, determine the values of C_p and C_{pk}.

5. What is the most glaring error in the figure (*Hint:* something one never does with control charts) and why is it so wrong?

Example: Maintaining accuracy in boring using control charts.

The workpiece shown in Figure CS-12 is made of gray cast iron and is bored to the tolerances indicated (5.5125/5.5115 in.). These parts were bored on a chucking machine. Each of the 19 points plotted on the vertical axis of the control chart represents the average of bore diameter measurements made on four parts (sample size). The horizontal broken lines at +0.0005 in. and −0.0005 in. represent upper and lower specified limits, respectively. The solid line $\overline{\overline{X}}$ = 0.00017 in. is the estimate of the process capability based on a study of several samples bored on the machine. The upper and lower control limits are then calculated from $\overline{\overline{X}}$. We note that samples 4–9 show a definite trend toward undersized bored holes. If the operation had been continued without any changes, the successive bored holes very likely would have been out of tolerance. To avoid this situation (out of control), the boring tools were reset toward the upper control limit before parts in sample 10 and the rest were bored. *(Source: ASM International.)*

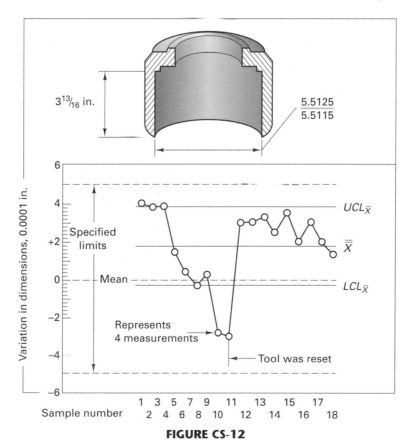

FIGURE CS-12

CHAPTER 13

FUNDAMENTALS OF CASTING

■ 13.1 INTRODUCTION TO MATERIALS PROCESSING

Materials processing is the science and technology by which a material is converted into a useful shape with a structure and properties that are optimized for the proposed service environment. A less technical definition of materials processing might be "all that is done to convert stuff into things."

Numerous processes and operations can be involved in the manufacture of products and components, and many of them are associated with the production of a desired shape. These shape-producing processes are often grouped into four basic "families," as indicated in Figure 13-1. *Casting processes* exploit the fluidity of a liquid as it flows, assumes the shape of a prepared container, and solidifies upon cooling. The *material removal processes* produce a desired shape by removing segments from an initially oversized piece. While these processes have often been referred to as *machining*, that term is generally used to describe the mechanical cutting of materials. The more general term *material removal* is used to incorporate material removal by all means, including chemical, thermal, and physical processes. *Deformation processes* exploit the ductility of certain materials, most notably metals, and produce the desired shape by mechanically moving or rearranging the solid through a phenomenon known as plasticity. *Consolidation processes* put pieces together, and include welding, brazing, soldering, adhesive bonding, and mechanical fasteners. *Powder metallurgy* is the manufacture of a desired shape from particulate material and involves aspects of casting, forming, and consolidation.

Each of the various families has distinct advantages and limitations, and the various processes within the families have their own unique characteristics. For example, cast products can have extremely complex shapes, but also possess structures that are produced by solidification and are therefore subject to such defects as shrinkage and porosity. Material removal processes are capable of outstanding dimensional precision, but produce scrap when material is cut away to produce the desired shape. Deformation processes can have high rates of production, but generally require powerful equipment and dedicated tools or dies. Complex products can often be assembled from simple shapes, but the joint areas are often affected by the joining process and may possess characteristics different from the original base material.

In addition to product design and material selection, the manufacture of any product or component also involves selection of the process or processes to be used in obtaining the desired shape and achieving the desired properties. Decisions should be made with the knowledge of all available alternatives and their associated assets and limitations. A large portion of this book is dedicated to presenting the various processes that can be applied to engineering materials. They are grouped according to the four basic categories, with powder metallurgy being included at the end of this section. The emphasis is on process fundamentals, descriptions of the various alternatives, and an assessment of associated assets and limitations. We will begin with a survey of the casting processes.

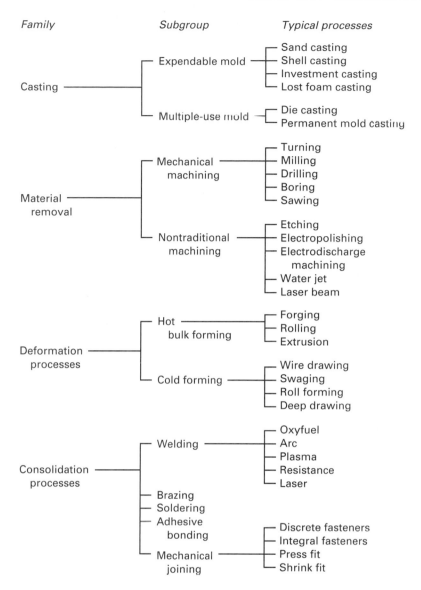

Family	Subgroup	Typical processes

FIGURE 13-1 The four materials processing families, with subgroups and typical processes.

■ 13.2 INTRODUCTION TO CASTING

In the *casting* processes, a solid material is first melted, heated to proper temperature, and sometimes treated to modify its chemical composition. The molten material, generally metal, is then poured into a cavity or mold that contains it in the desired shape during subsequent cool-down and solidification. Thus, in a single step, simple or complex shapes can be made from any material that can be melted. The product can have virtually any configuration the designer desires. In addition, the resistance to working stresses can be optimized, directional properties can be controlled, and a pleasing appearance can be produced.

Cast parts range in size from a fraction of a centimeter and a fraction of a gram (such as the individual teeth on a zipper), to over 10 meters and many tons (such as the huge propellers and stern frames of ocean liners). Moreover the casting processes have distinct advantages when the production involves complex shapes, parts having hollow sections or internal cavities, parts that contain irregular curved surfaces (except those made from thin sheet metal), very large parts, or parts made from metals that are difficult to machine.

It is almost impossible to design a part that cannot be cast by one or more of the commercial casting processes. However, as with all manufacturing techniques, the best results and lowest cost are only achieved if the designer understands the various options and tailors the design to use the most appropriate process in the most efficient manner. The various casting processes are distinguished primarily by the mold material (whether sand, metal, or other material) and pouring method (gravity, vacuum, low pressure, or high pressure). All share the requirement that the material should solidify in a manner that will maximize the properties and avoid the formation of defects, such as shrinkage voids, gas porosity, and trapped inclusions.

BASIC REQUIREMENTS OF CASTING PROCESSES

Six basic requirements are associated with most casting processes:

1. A *mold cavity*, having the desired shape and size, must be produced with due allowance for shrinkage of the solidifying material. Any geometrical feature desired in the finished casting must exist in the cavity. Consequently, the mold material must be able to reproduce the desired detail and also have a refractory character so that it will not contaminate the molten material that it will contain. Either a new mold must be prepared for each casting (single-use molds) or the mold must be made from a material that can withstand repeated use. The *multiple-use molds* are generally made of metal or graphite. Since they tend to be quite costly, their use is generally restricted to products where large quantities are desired. The more economical *single-use molds* are usually preferred for the production of smaller quantities.

2. A *melting process* must be capable of providing molten material not only at the proper temperature, but also in the desired quantity, with acceptable quality, and at a reasonable cost.

3. A *pouring technique* must be devised to introduce the molten metal into the mold. Provision should be made for the escape of all air or gases present in the cavity prior to pouring, as well as those generated by the introduction of the hot metal. The molten material is then free to fill the cavity, producing a high-quality casting that is fully dense and free of defects.

4. The *solidification process* should be properly designed and controlled. Castings should be designed so that solidification and solidification shrinkage can occur without producing internal porosity or voids. In addition, the molds should not provide excessive restraint to the shrinkage that accompanies cooling. If they do, the casting may crack when it is still hot and its strength is low.

5. It must be possible to remove the casting from the mold (i.e., *mold removal*). With single-use molds that are broken apart and destroyed after each casting, mold removal presents no serious difficulty. With multiple-use molds, however, the removal of a complex-shaped casting may be a major design problem.

6. After the casting is removed from the mold, various *cleaning, finishing, and inspection* operations may be required. Extraneous material is usually attached where the metal entered the cavity, excess material may be present along mold parting lines, and mold material often adheres to the casting surface. All of these must be removed from the finished casting.

Each of these requirements will be considered in more detail as we move through the chapter. The fundamentals of solidification, pattern design, gating, and risering will be developed. Various defects will also be considered, together with their causes and cures.

■ 13.3 CASTING TERMINOLOGY

Before proceeding with the process fundamentals, it is helpful to first become familiar with a variety of casting terms. Figure 13-2 shows the cross section of a two-part sand mold and incorporates many features of a typical casting process. The process starts with the construction of a *pattern*, an approximate duplicate of the final casting. *Molding material* is then packed around the pattern and the pattern is removed. The *flask* is the rigid metal or wood frame that holds the molding aggregate. In a horizontally-parted two-part mold, the *cope* is the name given to the top half of the pattern, flask, mold, or core. The *drag* refers to the bottom half of any of these features. A *core* is a sand (or metal) shape that is inserted into a mold to produce the internal features of a casting, such as holes

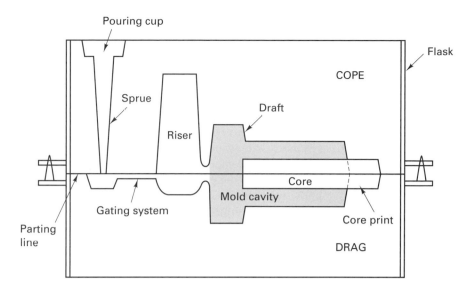

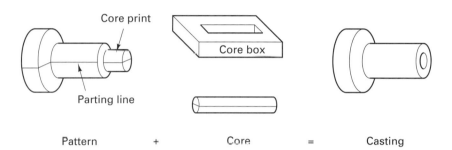

FIGURE 13-2 Cross section of a typical two-part sand mold, indicating various mold components and terminology.

or passages for water cooling. A *core print* is a region that is added to the pattern, core, or mold and is used to locate and support the core within the mold. The mold material and the core then combine to produce the *mold cavity*, the shaped hole into which the molten metal is poured and solidified to produce the desired casting. A *riser* is an extra void created in the mold that will also fill with molten metal. It provides a reservoir of liquid that can flow into the mold cavity to compensate for any shrinkage that occurs during solidification. If the riser contains the last material to solidify, shrinkage voids should be located in the riser and not the final casting.

The *gating system* is the network of connected channels used to deliver the molten metal to the mold cavity. The *pouring cup* (or pouring basin) is the portion of the gating system that initially receives the molten metal from the pouring vessel and controls its delivery to the rest of the mold. From the pouring cup, the metal travels down a *sprue* (the vertical portion of the gating system), then along horizontal channels, called *runners*, and finally through controlled entrances, or *gates*, into the mold cavity. Additional channels, known as *vents*, may be included in a mold or core to provide an escape for the gases that are originally present in the mold or are generated during the pour. (These and additional features will be discussed later in the chapter and are illustrated in Figure 13-9.)

The *parting line* or *parting surface* is the interface that separates the cope and drag halves of a mold, flask, or pattern and also the halves of a core in some core-making processes. *Draft* is the taper on a pattern or casting that permits it to be withdrawn from the mold. The mold or die used to produce casting cores is known as a *core box*. Finally, the term *casting* is used to describe both the process and the product when molten metal is poured and solidified in a mold.

■ 13.4 THE SOLIDIFICATION PROCESS

Casting is a *solidification process* where the molten material is poured into a mold and then allowed to freeze into the desired final shape. Many of the structural features that ultimately control product properties are set during solidification. Furthermore, many casting defects, such as *gas porosity* and *solidification shrinkage*, are solidification phenomena, and they can be reduced or eliminated by controlling the solidification process.

Solidification occurs in two stages, nucleation and growth, and it is important to control both of these processes. *Nucleation* occurs when a stable particle of solid forms from within the molten liquid. As the material changes state, its internal energy is reduced since at lower temperatures the solid phase is more stable than the liquid. At the same time, however, interface surfaces are created between the new solid and the parent liquid. Formation of these surfaces requires a positive contribution of energy. As a result, nucleation generally occurs at a temperature somewhat below the equilibrium melting point (the temperature where the internal energies of the liquid and solid are equal). The difference between the melting point and the actual temperature of nucleation is known as the amount of *undercooling*.

In most practical situations, the nucleation process occurs on some form of existing surface since solidification no longer requires the creation of a full, surrounding interface. These surfaces are usually present in the form of mold or container walls, or solid impurity particles contained within the molten liquid.

Each nucleation event will then produce a crystal or grain in the final casting. Since fine-grained materials (many small grains) possess enhanced mechanical properties, efforts to promote nucleation tend to be beneficial to the final product. It is not uncommon, therefore, to intentionally introduce impurities into the liquid before pouring into the mold. These small particles of solid provide numerous sites for nucleation and promote formation of a uniform, fine-grained product. The practice of intentionally introducing impurities is known as *inoculation* or *grain refinement*.

The second step in the solidification process is *growth*, which occurs as the heat of fusion is extracted from the liquid material. The direction, rate, and type of growth can be controlled by the way in which the heat is removed. *Directional solidification*, in which the solidification interface sweeps continuously through the material, can be used to assure the production of a sound casting. The molten material on the liquid side of the interface can flow into the mold to continuously compensate for the shrinkage that occurs as the material changes from liquid to solid. Faster rates of cooling generally produce products with finer grain size and superior mechanical properties.

COOLING CURVES

Cooling curves, such as those introduced in Chapter 4, can provide one of the most useful tools for studying the solidification process. By inserting thermocouples into a casting and monitoring the temperature versus time, one can obtain valuable insight into what is happening in the various regions.

Figure 13-3 shows a typical cooling curve for a pure or eutectic-composition material (one with a distinct melting point) and is useful for depicting many of the principal features and terms. The *pouring temperature* is the temperature of the liquid metal

FIGURE 13-3 Cooling curve for a pure metal or eutectic-composition alloy (metals with a distinct freezing point), indicating major features related to solidification.

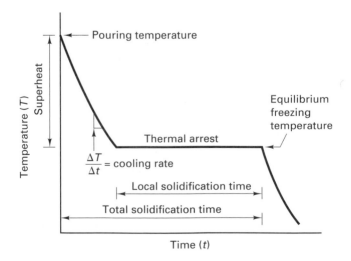

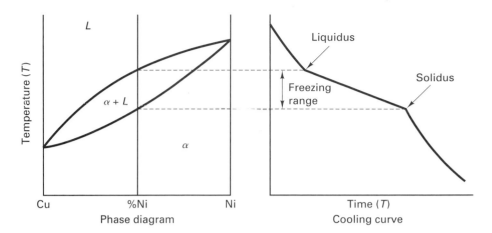

FIGURE 13-4 Phase diagram and companion cooling curve for an alloy with a freezing range. The slope changes indicate the onset and termination of solidification.

when it first enters the mold cavity. *Superheat* is the difference between the pouring temperature and the freezing temperature of the material. The higher the superheat, the more time is given for the material to flow into the intricate details of the mold cavity before it begins to freeze. The *cooling rate* is the rate at which the liquid or solid is cooling and can be viewed as the slope of the cooling curve at any given point. A *thermal arrest* is the plateau in the cooling curve that occurs during the solidification of a material with fixed melting point. At this temperature, the heat being removed from the mold comes from the latent heat of fusion that is released during the solidification process. The time from the start of pouring to the end of solidification is known as the *total solidification time*. The time from the start of solidification to the end of solidification is called the *local solidification time*.

If an alloy is used that does not have a distinct melting point, such as the one shown in Figure 13-4, the difference between the *liquidus* and *solidus* temperatures is known as the *freezing range*. The onset and termination of solidification now appear as slope changes in a cooling curve. If undercooling was required to induce the initial nucleation, the subsequent solidification may release enough heat to cause an increase in temperature back to the melting point. This increase in temperature, known as *recalescence*, is shown in Figure 13-5.

The specific form of a cooling curve depends on the type of material being poured, the nature of the nucleation process, and the rate and means of heat removal from the mold. Analysis of experimental cooling curves can provide valuable insight into both the process and the product. Fast cooling rates and short solidification times lead to finer structures and improved mechanical properties.

FIGURE 13-5 Cooling curve depicting undercooling and subsequent recalescence.

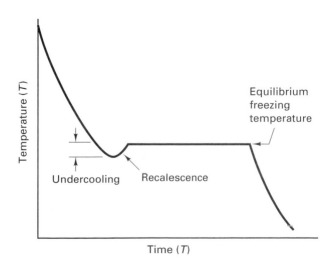

PREDICTION OF SOLIDIFICATION TIME: CHVORINOV'S RULE

The amount of heat that must be removed from a casting to cause it to solidify is dependent upon both the amount of superheating and the volume of metal in the casting. Conversely, the ability to remove heat from a casting is directly related to the amount of exposed surface area through which the heat can be extracted and the environment surrounding the molten material (i.e., the mold and mold surroundings). These observations are reflected in *Chvorinov's rule*[1], which states that t_s, the total solidification time, can be computed by

$$t_s = B(V/A)^n \qquad \text{where } n = 1.5 \text{ to } 2.0$$

The total solidification time, t_s, is the time from pouring to the completion of solidification; V is the volume of the casting; A is the surface area; and B is the *mold constant*, which incorporates the characteristics of the metal being cast (its density, heat capacity, and heat of fusion), the mold material (its density, thermal conductivity, and heat capacity), the mold thickness, and the amount of superheat.

Test specimens can be cast to determine B for a given mold material, casting material, and condition of casting. This value can then be used to compute the solidification times for other castings made under the same conditions. Since a riser and casting both lie within the same mold and fill with the same metal under the same conditions, Chvorinov's rule can be used to compare the solidification times of each, and thereby assure that the casting will solidify before the riser. This condition is absolutely essential if the liquid within the riser is to effectively feed the casting and compensate for solidification shrinkage. Aspects of riser design, including the use of Chvorinov's rule, will be developed later in this chapter.

Different cooling rates and solidification times can produce substantial variation in the structure and properties of the resulting casting. Die casting, for example, uses metal molds, and the faster cooling produces higher-strength castings than sand casting, which uses a more insulating mold material. Even the various types and condition of sand can produce different cooling rates. Sands with high moisture contents extract heat faster than ones with low moisture.

THE CAST STRUCTURE

The structures that result when molten metal is poured into a mold and permitted to solidify may have as many as three distinct regions or zones. The *chill zone* is a narrow band of randomly oriented crystals that forms on the surface of a casting. Rapid nucleation occurs here due to the presence of the mold walls and the relatively rapid surface cooling. As additional heat is removed from the surfaces, the grains of the chill zone begin to grow inward, and the rate of heat extraction and solidification decreases. Since most crystals have directions of rapid growth, a selection process then begins. Those crystals whose rapid-growth direction is perpendicular to the casting surface grow fast and shut off adjacent grains whose rapid-growth direction is at some intersecting angle. The favorably oriented crystals continue to grow, producing the long, thin columnar grains of a *columnar zone*. The companion properties are highly directional, since the selection process has converted the purely random structure of the surface into one of aligned parallel crystals. Figure 13-6 shows a cast structure containing both chill and columnar zones.

In many materials, new crystals can nucleate in the interior of the casting and then grow to produce another region of spherical, randomly oriented crystals, known as the *equiaxed zone*. Low pouring temperatures, alloy additions, or the addition of inoculants can be used to promote the formation of this region, which is far more desirable than columnar grains. Isotropic properties (uniform in all directions) are characteristic of this region of the casting.

MOLTEN METAL PROBLEMS

Castings begin with molten metal and the many reactions that occur between molten metal and its surroundings can lead to defects in the final casting. Oxygen and

FIGURE 13-6 Internal structure of a cast metal bar showing the chill zone at the periphery, columnar grains growing toward the center, and a central shrinkage cavity.

---

[1]N. Chvorinov, "Theory of Casting Solidification," *Giesserei*, Vol. 27, 1940, pp. 177–180, 201–208, 222–225.

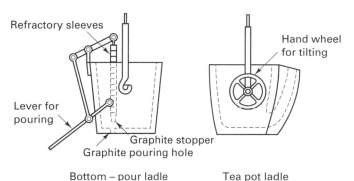

FIGURE 13-7 Two types
of ladles used to pour castings.
Note how each avoids pouring
the impure material from the top
of the molten pool.

Refractory sleeves

Hand wheel
for tilting

Lever for
pouring

Graphite stopper
Graphite pouring hole

Bottom – pour ladle   Tea pot ladle

molten metal often react to produce metal oxides (i.e., a nonmetallic or ceramic material), which can then be carried with the molten metal during pouring and filling of the mold. Known as *dross* or *slag*, this material can become trapped in the casting and impair surface finish, machinability, and mechanical properties. Material eroded from the linings of furnaces and pouring ladles and loose sand particles that flow freely in the mold contribute to the dross or slag.

Dross and slag can be controlled by using special precautions during melting and pouring and by good design practice. Fluxes can be used to cover and protect molten metal during melting, or melting and pouring can be performed under a vacuum or protective atmosphere. Measures can be taken to agglomerate the dross and cause it to float to the surface of the metal, where it can be skimmed off prior to pouring. Special ladles can be used which pour from beneath the surface, such as those depicted in Figure 13-7. Gating systems can often be designed to trap any dross that might enter the mold and keep it from flowing into the mold cavity. In addition, ceramic filters can be inserted into the feeder channels of the mold.

Liquid metals can also contain significant amounts of dissolved gas. When these materials solidify, the solid structure cannot accommodate the gas, and the rejected atoms tend to form bubbles or *gas porosity* within the casting. Figure 13-8 shows the maximum solubility of hydrogen in aluminum as a function of temperature. Note the large decrease that occurs as the material goes from liquid to solid.

Several techniques can be used to prevent the formation of gas porosity. One approach is to prevent the gas from initially dissolving in the molten metal. Melting can be performed under vacuum, in an environment of low-solubility gases, or under a protective flux that excludes contact with the air. Superheat temperatures can be kept low to

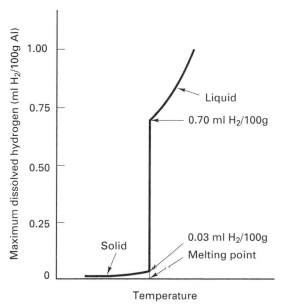

FIGURE 13-8 Maximum solubility of
hydrogen in aluminum as a function
of temperature.

Maximum dissolved hydrogen (ml H_2/100g Al)

1.00

0.75

0.50

0.25

0

Liquid

0.70 ml H_2/100g

0.03 ml H_2/100g

Melting point

Solid

Temperature

minimize gas solubility. In addition, careful handling and pouring can do much to streamline the flow of molten metal and minimize the turbulence that brings air and molten metal into contact.

Other methods attempt to remove the gas from the molten metal before it is poured into castings. *Vacuum degassing* sprays the molten metal through a low-pressure environment. Under these conditions, the amount of dissolved gas is reduced as the material seeks to establish an equilibrium with its new surroundings. (See a discussion of Sievert's law in any basic chemistry text.) Passing small bubbles of inert or reactive gas through the melt, known as *gas flushing*, can also be effective. In seeking equilibrium, the dissolved gas enters the flushing gas and is carried away. Bubbles of nitrogen or chlorine, for example, are particularly effective in removing hydrogen from molten aluminum.

Still another approach is to react the dissolved gas with something to produce a low-density compound. These compounds then float to the surface and can be removed with the dross or slag. Oxygen can be removed from copper by the addition of phosphorus. Steels can be deoxidized with additions of aluminum or silicon. The resulting phosphorus, aluminum, or silicon oxides are then removed by skimming, or left on the top of the container as the remaining high-quality metal is poured from beneath the surface.

FLUIDITY

When molten metal is poured to produce a casting, it should first *flow* into all regions of the mold cavity and then *freeze* into this new shape. It is vitally important that these two functions occur in the proper sequence. If the metal begins to freeze before it has completely filled the mold, defects known as *misruns* and *cold shuts* are produced.

The ability of a metal to flow and fill a mold is known as *fluidity*. Fluidity affects the minimum section thickness that can be cast, the maximum length of a thin section, the fineness of detail, and the accuracy of filling mold extremities. Although no single method has been accepted as a means of measuring fluidity, various "standard molds" have been developed where the results are sensitive to metal flow. One popular approach produces castings in the form of a long, thin spiral that progresses outward from a central sprue. The length of the final casting provides a good measure of fluidity.

POURING TEMPERATURE

Fluidity is dependent on the composition, freezing temperature, and freezing range of the metal or alloy, as well as the surface tension of oxide films. The most important controlling factor, however, is usually the *pouring temperature* or the amount of *superheat*. The higher the pouring temperature, the higher the fluidity. Excessive temperatures should be avoided, however. At high pouring temperatures, metal–mold reactions are accelerated. In addition, the fluidity may be so great as to permit *penetration*, a defect where the metal not only fills the mold cavity but also fills the small voids between the particles of a sand mold. The surface of the resultant casting would then contain small particles of embedded sand.

GATING SYSTEM

As molten metal is poured into a mold, the gating system conveys the material and delivers it to all sections of the mold cavity. The speed or rate of metal movement is important as well as the degree of cooling that occurs while it is flowing. Slow filling and high loss of heat can result in misruns and cold shuts. Rapid rates of filling, on the other hand, can produce erosion of the gating system and mold cavity and might result in the entrapment of mold material in the final casting. The cross-sectional areas of the various channels can be selected to regulate flow. In addition, the shape and length of the channels are influential in controlling temperature loss. When heat loss is to be minimized, short channels with round or square cross sections are the most desirable. The gates are usually attached to the thickest or heaviest sections of a casting to control shrinkage and to the bottom of the casting to minimize turbulence and splashing. For large castings, multiple gates and runners may be used to introduce metal to more than one point of the mold cavity.

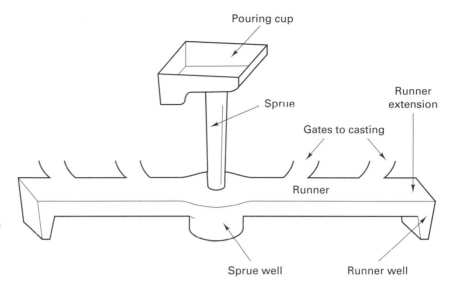

FIGURE 13-9   Typical gating system for a horizontal parting plane mold, showing key components involved in controlling the flow of metal into the mold cavity.

Gating systems should be designed to minimize *turbulent flow*, which tends to promote absorption of gases, oxidation of the metal, and erosion of the mold. Figure 13-9 shows a typical gating system for a mold with a horizontal parting line and can be used to identify some of the key components that can be optimized to promote the smooth flow of molten metal. Short sprues are desirable, since they minimize the distance that the metal must fall when entering the mold and the kinetic energy that the metal acquires during that fall. Rectangular pouring cups prevent the formation of a vortex or spiraling funnel, which tends to suck gas and oxides into the sprue. Tapered sprues also prevent vortex formation. A large *sprue well* can be used to dissipate the kinetic energy of the falling stream. Finally, the *choke*, or smallest cross-sectional area, serves to control the rate of metal flow. If the choke is located near the base of the sprue, flow through the runners and gates is slowed and flow is rather smooth. If the choke is moved to the gates, the metal might enter the mold cavity with a fountain effect, an extremely turbulent mode of flow, but the small connecting area would make separation of the casting and gating system easier.

Gating systems can also be designed to trap dross and sand particles and keep them from entering the mold cavity. Given sufficient time, the lower-density particles will rise to the top of the molten metal. Long, flat runners can be beneficial (but these promote cooling of the metal), as well as gates that exit from the lower portion of the runners. In addition, the first metal to enter the mold is most likely to contain the foreign matter (dross from the top of the pouring ladle and loose particles washed from the walls of the gating system). *Runner extensions and wells* (see Figure 13-9) can be used to catch and trap this first metal and keep it from entering the mold cavity. These are particularly effective with aluminum castings since aluminum oxide has approximately the same density as molten aluminum. Screens or ceramic filters can also be inserted into the gating system to trap the foreign material.

The specific details of a gating system often depend on the metal being cast. Turbulent-sensitive metals (such as aluminum and magnesium) and alloys with low melting points generally employ gating systems that concentrate on eliminating turbulence and trapping dross. Turbulent-insensitive alloys (such as steel, cast iron, and most copper alloys), and alloys with a high melting point, generally use short, open gating systems that provide for quick filling of the mold cavity.

SOLIDIFICATION SHRINKAGE

Once they are in the mold cavity and begin to cool, most metals and alloys undergo a noticeable volumetric contraction. There are three principal stages of *shrinkage*: (1) *shrinkage of the liquid*, (2) *solidification shrinkage* as the liquid turns into solid, and (3) *solid metal contraction* as the solidified material cools to room temperature. The amount of liquid metal contraction depends on the coefficient of thermal contraction

TABLE 13-1.	Solidification Shrinkage of Some Common Engineering Metals (Expressed in Percent)
Aluminum	6.6
Copper	4.9
Magnesium	4.0
Zinc	3.7
Low-carbon steel	2.5–3.0
High-carbon steel	4.0
White cast iron	4.0–5.5
Gray cast iron	−1.9

(a property of the metal being cast) and the amount of superheat. Liquid contraction, however, is rarely a problem in casting production because the metal in the gating system continues to flow into the mold cavity as the metal in the cavity cools and contracts.

As the metal cools between the liquidus and solidus temperatures and changes state from liquid to solid, significant amounts of shrinkage can occur, the amount of which varies from alloy to alloy, as indicated by the data in Table 13-1. Not all metals contract upon solidification, however. Some actually expand, such as gray cast iron, where low-density graphite flakes form as part of the solid structure.

When solidification shrinkage does occur, it is important to control the form of the resulting void. Metals and alloys with short freezing ranges, such as pure metals and eutectic alloys, tend to form large cavities or pipes. These can be avoided by designing the casting to have directional solidification. Freezing begins farthest away from the feed gate or riser and moves progressively toward it. As the metal solidifies and shrinks, the shrinkage void is continually filled with additional liquid metal. The final shrinkage void forms when the liquid metal is exhausted and is located external to the desired casting in either the riser or the gating system.

Alloys with large freezing ranges have a period of time when the material is in a slushy (liquid plus solid) condition. As the material cools between the liquidus and solidus, the relative amount of solid increases and tends to trap small isolated pockets of liquid. It is almost impossible for additional liquid to feed into the shrinkage areas, and the resultant structure will contain small but numerous shrinkage pores dispersed throughout. This type of shrinkage is far more difficult to prevent by means of gating and risering, and a porous product may be inevitable. If a gas- or liquid-tight product is desired, the castings may need to be impregnated (the pores filled with a resinous material or lower-melting-temperature metal) in a subsequent operation. Castings with dispersed porosity tend to have poor ductility, toughness, and fatigue life.

After solidification is complete, the casting will contract further as it cools to room temperature. This solid metal contraction is often called patternmaker's contraction, since compensation for these dimensional changes should be made when the mold cavity or pattern is designed. Additional concern arises, however, if the casting is to be produced in a rigid mold, such as the metal molds used in die casting. If the mold provides constraint during the time of contraction, tensile forces can be generated within the casting and cracking can occur. It may be desirable, therefore, to eject the hot castings as soon as solidification is complete.

RISERS AND RISER DESIGN

Risers are added reservoirs designed to feed liquid metal to the solidifying casting as a means of compensating for solidification shrinkage. To effectively perform this function, the risers must solidify after the casting. If the reverse were true, liquid metal would flow from the casting toward the solidifying riser and the casting shrinkage would be even greater. Hence, castings should be designed to produce directional solidification that sweeps from the extremities of the mold cavity toward the riser. In this way, the riser can continuously feed molten metal, and will compensate for the solidification shrinkage of the entire mold cavity. If this type of solidification is not possible, multiple risers may be necessary with various sections of the casting each solidifying toward their respective risers.

The risers should also be designed to conserve metal. If we define the *yield* of a casting as the casting weight divided by the total weight of metal poured (sprue, gates, risers, and casting), it is clear that there is a motivation to make the risers as small as possible, yet still able to perform their task. This is usually done through proper consideration of riser size, shape, and location, as well as the type of connection between the riser and casting.

According to Chvorinov's rule, a good shape for a riser would be one that has a long freezing time (i.e., a small surface area per unit volume). While a sphere would make the most efficient riser, this shape presents considerable difficulty to both the patternmaker and the moldmaker, who must remove the pattern from the mold. As a result, the most popular shape for a riser is a cylinder, where the height-to-diameter ratio is varied depending upon the nature of the alloy, location of the riser, the size of the flask, and other variables.

Risers should be located so that directional solidification occurs from the extremities of the mold cavity back toward the riser. Since the thickest regions of a casting will be the last to freeze, the risers should feed directly into these locations. Various types of risers are possible. A *top riser* is one that sits on top of a casting. Because of their location, top risers have shorter feeding distances and occupy less space within the flask, thereby providing more freedom for the layout of the pattern and gating system. *Side risers* are located adjacent to the mold cavity, displaced horizontally along the parting line. Figure 13-10 depicts both a top and a side riser. If the riser is contained entirely within the mold, it is known as a *blind riser*; if it is open to the atmosphere, it is called an *open riser*. Blind risers are usually larger than open risers because of the additional heat loss that occurs where the top of the riser is in contact with mold material.

Live risers (also known as hot risers) receive the last hot metal that enters the mold and generally do so at a time when the metal in the mold cavity has already begun to cool and solidify. Thus they can be smaller than *dead (or cold) risers*, which fill with metal that has already flowed through the mold cavity. As shown in Figure 13-10, top risers are almost always dead risers. Risers that are part of the gating system are generally live risers.

The minimum size of a riser can be calculated from Chvorinov's rule by setting the total solidification time for the riser to be greater than the total solidification time for the casting. Since both cavities receive the same metal and are in the same mold, the mold constant, B, will be the same for both regions. Assuming that $n = 2$, and a safe difference in solidification time is 25% (the riser takes 25% longer to solidify than the casting), we can write this condition as

$$t_{\text{riser}} = 1.25 t_{\text{casting}} \qquad (13\text{-}1)$$

or

$$(V/A)^2_{\text{riser}} = 1.25(V/A)^2_{\text{casting}} \qquad (13\text{-}2)$$

Calculation of the riser size then requires selection of a riser geometry, which is generally cylindrical. For a cylinder of diameter D and height H, the volume and surface area can be written as

$$V = \pi D^2 H/4$$
$$A = \pi D H + 2(\pi D^2/4)$$

FIGURE 13-10   Schematic of a sand casting mold, showing an open-type top riser (left) and a blind-type side riser (right). The side riser is a live riser, receiving the last hot metal to enter the mold. The top riser is a dead riser, receiving metal that has flowed through the mold cavity.

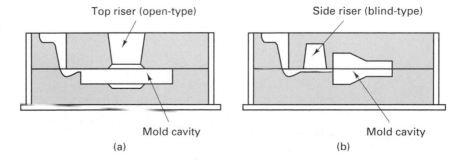

Top riser (open-type)   Side riser (blind-type)

Mold cavity   Mold cavity

(a)   (b)

Selecting a specific height-to-diameter ratio then enables the V/A ratio for the riser to be written as a simple expression with one unknown, D. The V/A ratio for the casting is calculated for its particular geometry. Substitution into equation (13-2) produces a relation with one unknown, that is, the size of the required riser. One should note that if the riser and casting share a surface, as with a blind top riser, the area of the common surface should be subtracted from both components since it will not be a surface of heat loss to either. There are actually a number of methods to calculate riser size, but the Chvorinov's rule method will be the only one presented here.

A final aspect of riser design is the connection between the riser and the casting. Since the riser must ultimately be separated from the casting, it is desirable that the connection area be as small as possible. On the other hand, the connection area must be large enough that the link does not freeze before solidification of the casting is complete. Risers should be placed close to the casting with relatively short connections. Short connections are most desirable. The mold material surrounding the link will then receive heat from both the casting and the riser. It will heat rapidly and remain hot throughout the cast, thereby preventing solidification of the metal in the channel.

RISERING AIDS

Various methods have been developed to assist the risers in performing their job. Some are intended to promote directional solidification, while others seek to reduce the number and size of the risers, and thereby increase the yield of a casting. These techniques generally work by either speeding the solidification of the casting (chills) or retarding the solidification of the riser (sleeves or toppings).

External chills are masses of high-heat-capacity, high-thermal-conductivity material that are placed in the mold, adjacent to the casting, to accelerate the cooling of various regions. Chills can effectively promote directional solidification or increase the effective feeding distance of a riser. They can often be used to reduce the number of risers required for a casting.

Internal chills are pieces of metal that are placed within the mold cavity to absorb heat and promote more rapid solidification. Since some of this metal will melt during the operation, it will absorb not only the heat-capacity energy needed to raise its temperature, but also some heat of fusion. Internal chills ultimately become part of the final casting, so they must be made from the same alloy as that being cast.

Ways to slow the cooling of risers include (1) switching from a blind riser to an open riser, (2) placing *insulating sleeves and toppings* around the riser, and (3) surrounding the sides or top of the riser with *exothermic material* that supplies added heat to just the riser segment of the mold. These techniques generally seek to reduce the riser size rather than promote directional solidification.

Finally, it is important to note that risers are not always necessary. For alloys with large freezing ranges, the risers would not be particularly effective, and one generally accepts the fine, dispersed porosity. For processes such as die casting, low-pressure permanent molding, and centrifugal casting, the positive pressures associated with the process provide the feeding action that is required to compensate for solidification shrinkage.

■ 13.5 PATTERNS

Casting processes can be divided into two basic categories: those for which a new mold must be created for each casting (the *expendable-mold processes*) and those that use a permanent, *reusable mold*. Most of the expendable mold processes begin with some form of reusable *pattern*—a duplicate of the part to be cast, modified dimensionally to reflect both the casting process and the material being cast.

The modifications that are incorporated into a pattern are called *allowances*, and the most important of these is the *shrinkage allowance*. Following solidification, a casting continues to contract as it cools to room temperature, the amount of this contraction being as much as 2% or $\frac{1}{4}$ in./ft. To produce the desired final dimensions, the pattern must be slightly larger than the room temperature casting. The exact amount of this compensation depends on the metal that is being cast. Typical allowances for some common engineering metals are:

Cast iron	0.8–1.0%	$\left(\dfrac{1}{10}-\dfrac{1}{8}\ \text{in./ft}\right)$
Steel	1.5–2.0%	$\left(\dfrac{3}{16}-\dfrac{1}{4}\ \text{in./ft}\right)$
Aluminum	1.0–1.3%	$\left(\dfrac{1}{8}-\dfrac{5}{32}\ \text{in./ft}\right)$
Magnesium	1.0–1.3%	$\left(\dfrac{1}{8}-\dfrac{5}{32}\ \text{in./ft}\right)$
Brass	1.5%	$\left(\dfrac{3}{16}\ \text{in./ft}\right)$

Shrinkage allowances are often incorporated into a pattern through use of special *shrink rules*—measuring devices that are larger than a standard rule by the appropriate shrink allowance. For example, a shrink rule for brass would designate 1 ft at a length that is actually 1 ft $\frac{3}{16}$ in. A pattern made to shrink rule dimensions would then produce a proper-size casting after cooling.

Some caution should be exercised when using shrink rules, however, for thermal contraction may not be the only factor affecting the final dimensions. The various phase transformations discussed in Chapter 4 are often accompanied by significant expansions or contractions. Examples include eutectoid reactions, martensitic reactions, and graphitization.

In many casting processes, mold material is formed around the pattern and the pattern is then removed to create the mold cavity. To facilitate pattern removal, molds are often made in two or more sections. Consideration must then be given to the location of the *parting line*, the surface where one section of the mold mates with the other section or sections. A flat parting line is usually preferred, but the casting design or molding practice may dictate the use of irregular parting surfaces.

If the pattern contains surfaces that are perpendicular to the parting line (parallel to the direction of pattern withdrawal), the friction between the pattern and the mold, or any horizontal movement of the pattern during extraction, would tend to damage the mold. This damage would be particularly severe at the corners where the mold cavity intersects the parting surface. This difficulty can be minimized by incorporating a slight taper, or *draft*, on all surfaces parallel to the direction of withdrawal. As soon as the pattern is withdrawn a slight amount, it is free from the mold material on all surfaces, and it can be withdrawn further without damaging the mold. Figure 13-11 illustrates the use of parting lines and draft.

The required amount of draft is determined by the size and shape of the pattern, the depth of the mold cavity, the method used to withdraw the pattern, the pattern material, the mold material, and the molding procedure. Draft is seldom less than 1° or $\frac{1}{8}$ in./ft, with

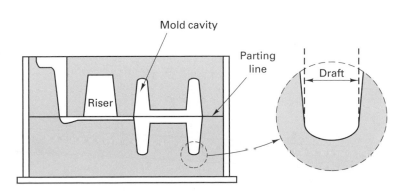

FIGURE 13-11   Two-part mold showing the parting line and the incorporation of a draft allowance on vertical surfaces.

a minimum taper of about $\frac{1}{16}$ in. over the length of any surface. For interior surfaces where the opening is small, such as a hole in the center of a hub, the draft should be increased to about $\frac{1}{2}$ in./ft. Since draft allowances tend to increase the size of a pattern and thus the size and weight of a casting, it is generally desirable to keep them to the minimum that will permit satisfactory pattern removal. Modern molding procedures, which provide higher strength to the molding material before the pattern is withdrawn, and the use of molding machines that incorporate mechanical pattern withdrawal, have permitted substantial reductions in draft allowances. These improvements have enabled the production of lighter castings with thinner sections, thereby saving both weight and the amount of subsequent machining.

When machined surfaces must be provided on castings, it is often necessary to add a *machining allowance*, or *finish allowance*, to the pattern. This allowance depends to a great extent on the casting process and the mold material. Ordinary sand castings have rougher surfaces than those of shell-mold castings. Die castings are sufficiently smooth so that very little or no metal has to be removed, and investment castings frequently require no additional machining. Consequently, the designer should relate the finishing allowance to the casting process and also remember that draft may provide part or all of the extra metal needed for machining.

If a core is to be used to form a hole or interior cavity, it too must be made oversized to compensate for shrinkage (all of the metal surrounding the hole will contract, making the hole smaller). In addition, core prints should be added to the pattern to produce the added segments of the mold cavity. However, if a machining allowance is to be incorporated in a core, its value should be subtracted from the core dimensions (rather than added) because machining will increase the size of the hole.

Some casting shapes require yet an additional allowance for *distortion*. For example, the arms of a U-shaped section may be restrained by the mold, while the base of the U is free to shrink. This restraint will result in a final casting with outwardly sloping arms. However, by designing the arms to originally slope inward, they will distort to a straight shape upon cooling. Long, horizontal sections tend to sag in the center unless some form of ribbing provides adequate support. Distortion depends greatly on the particular configuration of the casting, and the designer must use experience and judgment to provide the appropriate distortion allowance. Fillets on inside corners and radiused edges on outside corners will help to eliminate sites of localized slow or rapid cooling.

Figure 13-12 illustrates the manner in which the various allowances are incorporated into a casting pattern. Since allowances increase both the weight of a casting and the amount of metal that has to be removed by machining, efforts are generally made to reduce them to the lowest value possible.

If the casting is to be made directly in a reuseable metal mold, all of the "pattern allowances" should be incorporated directly into the mold cavity. In addition to the allowances discussed above, the change in mold dimensions caused by the heating of the mold from room temperature to its elevated operating temperature should be included as an additional correction.

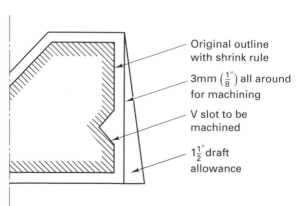

FIGURE 13-12   Various allowances incorporated into a casting pattern.

Original outline with shrink rule

3mm $\left(\frac{1''}{8}\right)$ all around for machining

V slot to be machined

$1\frac{1}{2}°$ draft allowance

■ 13.6 Design Considerations in Castings

To produce the best-quality product at the lowest possible cost, it is important that the designers of castings give careful attention to several process requirements and, if possible, work closely with the producing foundry. It is not uncommon for minor and readily permissible changes in design to greatly facilitate and simplify the casting of a component and also reduce the number and severity of defects.

One of the first features that must be considered by a designer is the *location of the parting plane*, an important part of all processes that use segmented or separable molds. The location of the parting plane can affect each of the following: (1) the number of cores, (2) the method of supporting the cores, (3) the use of effective and economical gating, (4) the weight of the final casting, (5) the final dimensional accuracy, and (6) the ease of molding.

In general, it is desirable to minimize the use of cores. A change in the location or orientation of the parting plane can often assist in this objective. Note that the change illustrated in Figure 13-13 not only eliminates the need for a core, but also reduces the weight of the casting by eliminating the need for draft. Figure 13-14 shows another example of how a core can be eliminated by a simple design change.

Certain design features can also dictate the location of the parting line. Figure 13-15 shows how the specification of round edges can restrict the location of the parting plane. The specification of draft can also fix the parting plane, as indicated in Figure 13-16. This figure also shows that considerable freedom can be provided simply by noting the need to provide for draft or by letting it be an option of the foundry. Since mold closure may not always be consistent, consideration should also be given to the fact that dimensions across the parting plane are subject to more variation than those that lie entirely within a given segment of the mold.

Controlling the solidification process is of prime importance in obtaining quality castings, and this control is also related to design. Those portions of a casting that have a high

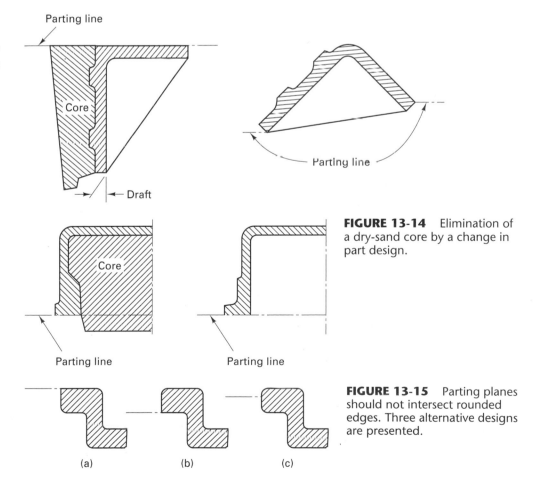

FIGURE 13-13 Elimination of a core by changing the location or orientation of the parting plane.

FIGURE 13-14 Elimination of a dry-sand core by a change in part design.

FIGURE 13-15 Parting planes should not intersect rounded edges. Three alternative designs are presented.

(a)   (b)   (c)

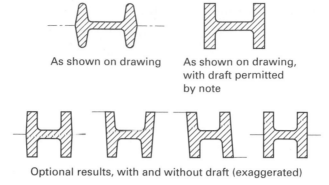

As shown on drawing

As shown on drawing, with draft permitted by note

Optional results, with and without draft (exaggerated)

FIGURE 13-16 (*Top left*) Design where the location of the parting plane is specified by the draft. (*Top right*) Part with draft unspecified. (*Bottom*) Various options to produce the top right part, including a no-draft design.

FIGURE 13-17 Typical guidelines for section change transitions in castings.

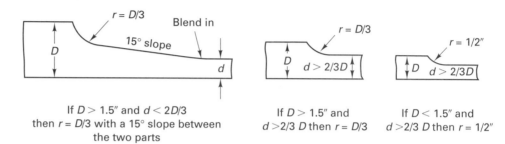

If $D > 1.5''$ and $d < 2D/3$ then $r = D/3$ with a 15° slope between the two parts

If $D > 1.5''$ and $d > 2/3\ D$ then $r = D/3$

If $D < 1.5''$ and $d > 2/3\ D$ then $r = 1/2''$

ratio of surface area to volume will experience more rapid cooling and will be stronger and harder than the other regions. Heavier sections will cool more slowly and, unless special precautions are observed, may contain shrinkage cavities and porosity or have weaker, large-grain-size structures. Ideally, a casting should have uniform thickness in all directions. In most cases, however, this is not possible. When the section thickness must change, it is best if these changes are gradual, as indicated in the recommendations of Figure 13-17.

When sections of castings intersect, two problems can arise. The first of these is *stress concentration*. Generous fillets (inside radii) at all interior corners can help to minimize potential problems. Excessive fillets, however, can augment the second problem, known as *hot spots*. Figure 13-18 shows that localized thick sections tend to exist where sections of castings intersect. These thick sections cool more slowly than the other locations and tend to be sites of localized, abnormal shrinkage. When the differences in sections are large, like those illustrated in Figure 13-19, the hot-spot areas are likely to contain objectionable defects, such as porosity or shrinkage cavities.

Defects, such as voids, porosity, and cracks, can be sites of subsequent failure and should be prevented if at all possible. Where heavy sections must exist, an adjacent riser is often used to feed the section during solidification and shrinkage. If the riser is designed properly, the shrinkage cavity will lie totally within the riser, as illustrated in Figure 13-20, and can be removed when the riser is cut off.

When sections intersect to form continuous ribs, contraction occurs in opposite directions as the various ribs contract. As a consequence, cracking frequently occurs at the intersections. By staggering the ribs, as shown in Figure 13-21, there is opportunity

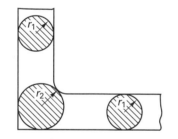

FIGURE 13-18 The "hot spot" at section r_2 is caused by intersecting sections.

FIGURE 13-19 Hot spots often result from intersecting sections of various thickness.

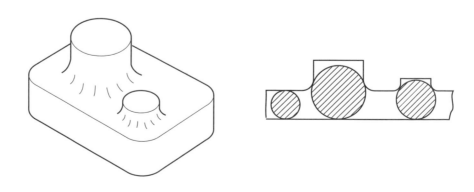

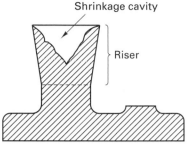

FIGURE 13-20 Use of a riser to keep the shrinkage cavity external to the casting.

TABLE 13-2.	Recommended Minimum Section Thicknesses for Various Engineering Metals and Casting Processes				
Material	Casting Process	Minimum		Desirable	
		mm	in.	mm	in.
Steel	Sand	4.76	$\frac{3}{16}$	6.35	$\frac{1}{4}$
Gray iron	Sand	3.18	$\frac{1}{8}$	4.76	$\frac{3}{16}$
Malleable iron	Sand	3.18	$\frac{1}{8}$	4.76	$\frac{3}{16}$
Aluminum	Sand	3.18	$\frac{1}{8}$	4.76	$\frac{3}{16}$
Magnesium	Sand	4.76	$\frac{3}{16}$	6.35	$\frac{1}{4}$
Zinc alloys	Die	0.51	0.020	0.76	0.030
Aluminum alloys	Die	1.27	0.050	1.52	0.060
Magnesium alloys	Die	1.27	0.050	1.52	0.060

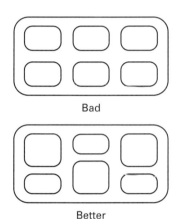

FIGURE 13-21 Method of using staggered ribs to prevent cracking during cooling.

for distortion to occur, thereby providing relaxation to the high residual stresses that would otherwise induce cracking.

Large unsupported areas should be avoided in all types of casting, since these regions tend to warp during cooling. While the warpage may not be detrimental to performance, it generally disrupts the smooth surface appearance that is often desired in a casting.

The location of the parting line may be another appearance consideration. A small amount of fin, or flash, is often present at this location. When the flash is removed (or left in place if it is small enough), a region of surface imperfection is created. If this location is in the middle of a flat surface, it will be clearly visible. However, if the parting line is moved to coincide with a corner, the parting line "defect" will go largely unnoticed.

Minimum section thickness should also be considered when designing castings. Specific values are rarely given because they tend to vary with the shape and size of the casting, the type of metal, the method of casting, and the practice of the individual foundry. Table 13-2 presents average values of both the minimum and desirable section thicknesses for some combinations of common foundry materials and casting processes.

■ KEY WORDS

allowance	directional solidification	growth	parting line (parting surface)	solidification shrinkage
blind riser	draft	hot spot	pattern	solidus
casting	drag	inoculation	penetration	sprue
chill	dross	internal chill	pouring cup	sprue well
chill zone	equiaxed zone	liquidus	pouring temperature	stress concentrators
choke	expendable-mold process	live riser	recalescence	superheat
Chvorinov's rule	external chill	local solidification time	riser	thermal arrest
cold shut	flask	machining allowance	runner	top riser
columnar zone	fluidity	misruns	runner extension	toppings
cooling curve	freezing range	mold cavity	shrink rule	total solidification time
cooling rate	gas flushing	mold constant	shrinkage	turbulent flow
cope	gas porosity	mold material	side riser	undercooling
core	gate	multiple-use mold	single-use mold	vacuum degassing
core box	gating system	nucleation	slag	vent
core print	grain refinement	open riser	sleeves	yield
dead riser				

■ REVIEW QUESTIONS

1. What is materials processing?
2. What are the four basic families of shape-production processes? Cite one advantage and one limitation of each family.
3. Describe the capabilities of the casting process in terms of size and shape of the product.
4. How might the desired production quantity influence the selection of a single- or multiple-use molding process?
5. Why is it important to provide a means of venting gases from the mold cavity?

6. What types of problems can arise if the mold material provides too much restraint to the solidifying and cooling metal?
7. What is a casting pattern? Flask? Core? Mold cavity? Riser?
8. What are some of the components that combine to make up the gating system of a mold?
9. What is a parting line or parting surface?
10. What is draft and why is it used?
11. What are the two stages of solidification, and what occurs during each?
12. Why is it that most solidification does not begin until the temperature falls somewhat below the equilibrium melting temperature (i.e., undercooling is required)?
13. Why might it be desirable to promote nucleation in a casting through inoculation or grain refinement processes?
14. Heterogeneous nucleation begins at preferred sites within a mold. What are some probable sites for heterogeneous nucleation?
15. Why might directional solidification be desirable in the production of a cast product?
16. Describe some of the key features observed in the cooling curve of a pure metal.
17. What is superheat?
18. What is the freezing range for a metal or alloy?
19. Discuss the roles of casting volume and surface area as they relate to the total solidification time and Chvorinov's rule.
20. What characteristics of a specific casting process are incorporated into the mold constant, B, of Chvorinov's rule?
21. What is the correlation between cooling rate and final properties of a casting?
22. What is the chill zone of a casting, and why does it form?
23. Which of the three regions of a cast structure is least desirable? Why are its properties highly directional?
24. How can we promote the formation of an equiaxed zone and minimize the size of the columnar zone?
25. What is dross or slag, and how can it be prevented from becoming part of a finished casting?
26. What are some of the possible approaches that can be taken to prevent the formation of gas porosity in a metal casting?
27. What is fluidity, and how can it be measured?
28. What is a misrun and what causes them to form?
29. What defect can form in sand castings if the pouring temperature is too high and fluidity is too great?

30. Why is it important to design the geometry of the gating system to control the rate of metal flow as it travels from the pouring cup into the mold cavity?
31. What are some of the undesirable consequences that could result from turbulence of the metal in the gating system and mold cavity?
32. What is a choke and how does its placement affect metal flow?
33. What features can be incorporated into the gating system to aid in trapping dross and loose mold material that is flowing with the molten metal?
34. What features of the metal being cast tend to influence whether the gating system is designed to minimize turbulence and reduce dross, or promote rapid filling to minimize temperature loss?
35. What are the three stages of contraction or shrinkage as a liquid is converted into a finished casting?
36. Why is it more difficult to prevent shrinkage voids from forming in metals or alloys with large freezing ranges?
37. What type of flaws or defects form during the cooling of an already-solidified casting?
38. Why is it desirable to design a casting to have directional solidification sweeping from the extremities of the mold to the riser?
39. Based on Chvorinov's rule, what would be an ideal shape for a casting riser? A desirable shape from a practical perspective?
40. Define the following riser-related terms: top riser, side riser, open riser, blind riser, live riser, and dead riser.
41. What assumptions were made when using Chvorinov's rule to calculate the size of a riser in the manner presented in the text?
42. What is the purpose of a chill? Of an insulating sleeve? Of exothermic material?
43. What types of modifications or allowances are generally incorporated into a casting pattern?
44. What is a shrink rule, and how does it work?
45. What is the purpose of a draft or taper on pattern surfaces?
46. Why is it desirable to make the pattern allowances as small as possible?
47. What are some of the features of the casting process that are directly related to the location of the parting plane?
48. What types of problems can occur when sections of a casting intersect?

■ PROBLEMS

1. Using Chvorinov's rule as presented in the text with $n = 2$, calculate the dimensions of an effective riser for a casting that is a 2 in. by 4 in. by 6 in. rectangular plate. Assume that the casting and riser are not connected, except through a gate and runner, and that the riser is a cylinder of height/diameter ratio $H/D = 1.5$. The finished casting is what fraction of the combined weight of the riser and casting?
2. Reposition the riser in Problem 1 so that it sits directly on top of the flat rectangle, with its bottom circular surface being part of the surface of the casting, and recompute the size and yield fraction. Which approach is more efficient?
3. A rectangular casting having the dimensions 3 in. by 5 in. by 10 in. solidifies completely in 11.5 minutes. Using $n = 2$ in Chvorinov's rule, calculate the mold constant B. Then compute the solidification time of a 0.5 in. by 8 in. by 8 in. casting poured under the same conditions.

www.wiley.com/college/degarmo

Chapter 13 CASE STUDY

The Cast Oil-Field Fitting

A cast iron, T-type fitting is being produced for the oil drilling industry, using an air-set or no-bake sand for both the mold and the core. A silica sand has been used in combination with a catalyzed alkyd-oil/urethane binder. Figure CS-13 shows a cross section of the mold with the core in place (part a), and a cross section of the finished casting (part b). The final casting contains several significant defects. Gas bubbles are observed in the bottom section of the horizontal tee. A penetration defect is observed near the bottom of the inside diameter, and there is an enlargement of the casting at location *C*.

1. What is the most likely source of the gas bubbles? Why are they present only at the location noted? What might you recommend as a solution?

2. What factors may have caused the penetration defect? Why is the defect present on the inside of the casting, but not on the outside? Why is the defect near the bottom of the casting, but not near the top?

3. What factors led to the enlargement of the casting at point *C*? What would you recommend to correct this problem?

4. Another producer has noted penetration defects on all surfaces of his castings, both interior and exterior. What would be some possible causes? What could you recommend as possible cures?

5. Could these molds and cores be reclaimed (i.e., recycled) after breakout? Discuss.

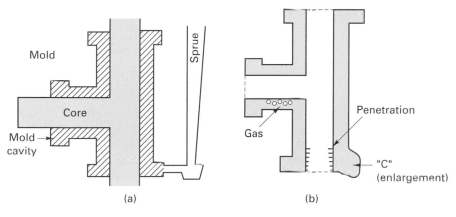

FIGURE CS-13

CHAPTER 14

EXPENDABLE-MOLD CASTING PROCESSES

■ 14.1 INTRODUCTION

The versatility of metal casting is made possible by a number of distinctly different processes, each with its own set of characteristic advantages and benefits. Selection of the best process requires a familiarization with the various options and capabilities, and an understanding of the needs of the specific product. Some influential factors include the quality of the cast surface, the desired dimensional precision, the number of castings desired, the type of pattern and corebox that are needed, the cost of making the required mold or die, and restrictions imposed by the selected material.

As we begin to survey the various casting processes used to produce manufactured items, it is helpful to have some form of process classification. One approach focuses on the molds and patterns and utilizes the following three categories:

1. Single-use molds with multiple-use patterns
2. Single-use molds with single-use patterns
3. Multiple-use molds

Categories 1 and 2 are often combined under the more general heading "expendable-mold casting processes," and these will be presented in this chapter. Sand, plaster, ceramics, or other refractory materials are combined with binders to form the mold material. Those processes where a mold can be used multiple times will be presented in Chapter 15. The multiple-use molds are usually made from metal.

Since the casting processes are associated primarily with the production of metal products, the emphasis of the casting chapters will be on metal casting. The metals most frequently cast are iron, steel, aluminum, brass, bronze, magnesium, certain zinc alloys, and nickel-based superalloys. Of these, cast iron and aluminum are dominant, primarily because of their low cost, good fluidity, adaptability to a variety of processes, and the wide range of possible properties. The processes used to fabricate products from polymers, ceramics (including glass), and composites will be discussed in Chapter 20.

■ 14.2 SAND CASTING

Sand casting is by far the most common and possibly the most versatile of the casting processes. Granular refractory material (such as silica, zircon, olivine, or chromite sand) is mixed with small amounts of other materials, such as clay and water, and is then packed around a pattern that has the shape of the desired casting. Because the grains can pack into thin sections and can be economically used in large quantities, products spanning a wide range of sizes and detail can be made by this method. If the pattern must be removed before pouring, the mold is usually made in two or more pieces. An opening called a *sprue hole* is cut from the top of the mold through the sand and connected to a system of channels called *runners*. The molten metal is poured down the sprue hole, flows through the runners, and enters the mold cavity through one or more openings, called *gates*. Gravity flow is the most common means of introducing the metal into the mold. After solidification, the mold is broken and the finished casting is removed. Because the mold is destroyed, a new mold must be made for each casting. Figure 14-1 shows the essential steps and basic components of a sand casting process.

FIGURE 14-1 Sequential steps in making a sand casting. (a) A pattern board is placed between the bottom (drag) and top (cope) halves of a flask, with the bottom side up. (b) Sand is then packed into the drag half of the mold. (c) A bottom board is positioned on top of the packed sand, and the mold is turned over, showing the top (cope) half of pattern with sprue and riser pins in place. (d) The cope half of the mold is then packed with sand. (e) The mold is opened, the pattern board is drawn (removed), and the runner and gate are cut into the parting surface of the sand. (e') The parting surface of the cope half of the mold is shown with the pattern and pins removed. (f) The mold is reassembled with the pattern board removed, and molten metal is poured through the sprue. (g) The contents are shaken from the flask and the metal segment is separated from the sand, ready for further processing.

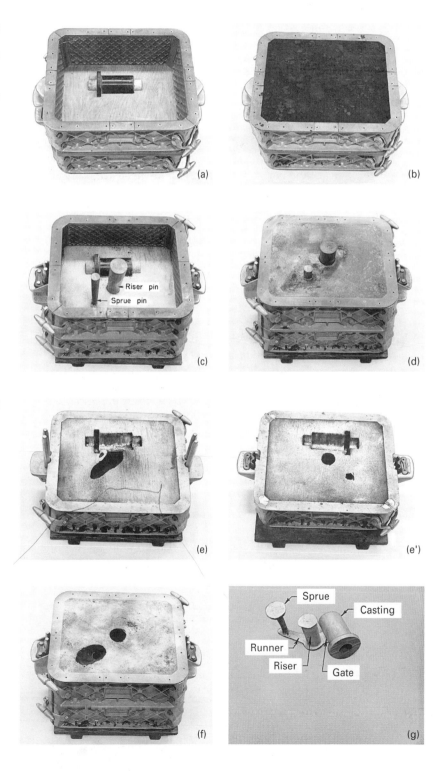

PATTERNS AND PATTERN MATERIALS

The first step in making a sand casting is the design and construction of a *pattern*. This is a duplicate of the part to be cast, modified in accordance with the requirements of the casting process, the metal being cast, and the particular molding technique that is being used. The pattern material is determined primarily by the number of castings to be made, but is also influenced by the size and shape of the casting, the desired dimensional precision, and the molding process. Wood patterns are relatively easy to make and are frequently used when small quantities of castings are required. Wood, however, is not very dimensionally stable. It may warp or swell with changes in humidity, and it tends to wear with repeated use. Metal patterns are more expensive but are more stable and durable. Hard plastics, such as urethanes, offer another alternative, and are often preferred with processes that use strong, organically-bonded sands that tend to stick to other pattern materials. In the full-mold and lost-foam processes, expanded polystyrene (EPS) is used, and investment casting uses patterns made from wax. In the latter processes, both the pattern and the mold are single-use, that is, destroyed each time a casting is produced.

TYPES OF PATTERNS

Many types of patterns are used in the foundry industry. Selection is usually based on the number of duplicate castings required and the complexity of the part.

One-piece or *solid patterns*, such as the one shown in Figure 14-2, are the simplest and often the least expensive type to make. This type of pattern is essentially a duplicate of the part to be cast, modified only by the various allowances discussed in Chapter 13 and by the possible addition of core prints. They are relatively cheap to construct, but the subsequent molding process is usually slow. One-piece patterns, therefore, are generally used when the shape is relatively simple and the number of duplicate castings is rather small.

If a one-piece pattern is simple in shape and contains a flat surface, it can be placed directly on a *follow board*. The entire mold cavity will be in one segment of the mold, and the follow board will create the parting surface. If the shape is more complex or the parting plane is to be more centrally located, it may be necessary for the molder to hand-cut an irregular parting surface. This is a time-consuming process and requires a skilled worker. More often, special follow boards are used with inset cavities designed so that the one-piece pattern is positioned at the correct depth for the parting line. The follow board again determines the parting surface, as illustrated in Figure 14-3.

Split patterns are used when moderate quantities of a casting are desired. The pattern is divided into two segments along a single parting plane, which corresponds to the parting plane of the mold. The bottom segment of the pattern is positioned in a flask and the lower (*drag*) portion of the mold is produced. The flask is then inverted, and the upper segment of the pattern is attached. Tapered pins in the cope half of the pattern align with holes in the drag segment to assure proper positioning. With the full pattern now in place, mold material is then packed into the upper (*cope*) flask. The two segments of the flask are separated and the pattern is removed to produce the mold cavity. Sprues and runners are cut and the mold is then ready to be poured. Figure 14-4 shows a split pattern that also contains several core prints (lighter color).

Match-plate patterns, such as the one shown in Figure 14-5, further simplify the process, and can be coupled with modern molding machines to produce large quantities of duplicate castings. The cope and drag segments of a split pattern are permanently

FIGURE 14-2 Single-piece pattern for a pinion gear.

FIGURE 14-3 Method of using a follow board to position a single-piece pattern and locate a parting surface. The final figure shows the flask of the previous operation (the drag segment) inverted for construction of the upper portion of the mold (cope segment).

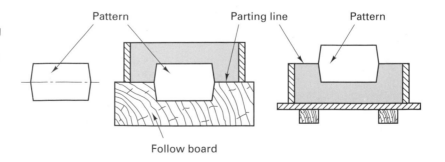

Pattern   Parting line   Pattern

Follow board

FIGURE 14-4   Split pattern, showing the two sections together and separated. The light-colored portions are core prints.

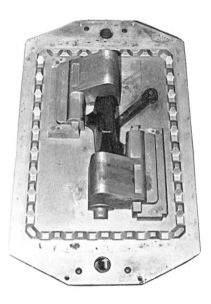 

FIGURE 14-5   Match-plate pattern used to produce two identical parts in a single flask. (*Left*) Cope side; (*right*) drag side. (*Note:* The views are opposite sides of a single pattern board.)

fastened to opposite sides of a wood or metal *match plate*. Using holes that align with pins on the flask, the match plate is positioned between the upper and lower segments of the flask. Mold material is then packed on both sides of the match plate to complete both the cope and drag segments of a mold. The mold sections are then separated and the match-plate pattern is removed. When the mold is reassembled using the pins and guide holes, the cavities in the cope and drag join in proper alignment.

In most cases, the necessary gates and runners are also incorporated on the match plate. This eliminates the need for hand cutting and guarantees that these features will be uniform and of the proper size in each mold, thereby reducing the possibility of defects. The gate and runner system can be seen on the cope side of the match-plate pattern in Figure 14-5 (the dark center section), and also on the cope section of the pattern in Figure 14-6. These patterns also include core prints and risers, and further illustrate the common practice of including more than one pattern on a match plate.

When large quantities are to be produced, or when the casting is large, it may be desirable to have the cope and drag halves of split patterns attached to separate match plates to produce mating *cope-and-drag patterns*. This enables independent molding of the cope and drag segments of a mold. Large molds can be handled more easily in separate segments, and small molds can be made at a faster rate. Figure 14-6 shows the mating pieces of a cope-and-drag pattern.

FIGURE 14-6 Cope-and-drag pattern for molding two heavy parts. (*Left*) Cope section; (*right*) drag section. (*Note:* These are two separate pattern boards.)

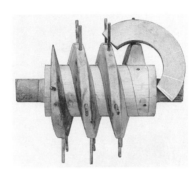

FIGURE 14-7 Loose piece pattern for molding a large worm gear. After sufficient sand has been packed around the pattern to hold the pieces in position, the wooden pins are withdrawn. The mold is completed and the pieces of the pattern are removed in a designated sequence.

When the product has a geometry such that a one-piece or split pattern could not be removed from the molding sand, a *loose-piece pattern* can sometimes be developed. Separate pieces are held to a primary pattern segment by beveled grooves or pins (Figure 14-7). After molding, the primary segment of the pattern is withdrawn. The hole that is created then permits the remaining segments to be moved in the directions necessary for their extraction. In some loose-piece patterns, a single sliding pin is often used to hold all the segments in place. After the sand is compacted around the pattern, the pin is removed to permit the individual segments to be withdrawn. Loose-piece patterns are expensive. They require careful maintenance, slow the molding process, and increase molding costs. They do, however, enable the sand casting of complex shapes that would otherwise require the full-mold, lost-foam, or investment processes. Whenever possible, however, design changes should be considered that would eliminate the need for split-piece patterns.

In all foundry patterns it is important that internal corners contain a small radius, called a *fillet*, rather than meet in a distinct line. Fillets prevent shrinkage cracks from forming at the intersections and reduce the stress concentrations in the finished product. Designers should provide generous fillets whenever possible. Radii of 6 and 3 mm ($\frac{1}{4}$ and $\frac{1}{8}$ in.) are the most common. Fillets are often added to patterns through the use of pre-radiused wax, leather, or plastic strips that are glued to the pattern or pressed into place with a heated fillet tool.

SANDS AND SAND CONDITIONING

The sand used to make molds must be carefully prepared if it is to provide satisfactory and uniform results. Ordinary silica (SiO_2), zircon, olivine, or chromite sands are compounded with additives to meet four requirements:

1. *Refractoriness*: the ability to withstand high temperatures
2. *Cohesiveness* (also referred to as *bond*): the ability to retain a given shape when packed into a mold
3. *Permeability*: the ability to permit gases to escape through it
4. *Collapsibility*: the ability to permit the metal to shrink after it solidifies and ultimately to free the casting by disintegration of the surrounding mold

Refractoriness is provided by the basic nature of the sand. Cohesiveness, bond, or strength is obtained by coating the sand grains with clays, such as bentonite, kaolinite, or illite that become cohesive when moistened. Collapsibility is sometimes enhanced by adding cereals or other organic materials, such as cellulose, that burn out when they come in contact with the hot metal. The combustion reduces both the volume and strength of the restraining sand. Permeability is a function of the size of the sand particles, the amount and type of clay or bonding agent, the moisture content, and the compacting pressure.

Good molding sand always represents a compromise between conflicting factors. The size of the sand particles, the amount of bonding agent (such as clay), the moisture content, and the organic additives are all selected to obtain an acceptable compromise of the four requirements. The composition must be carefully controlled to assure

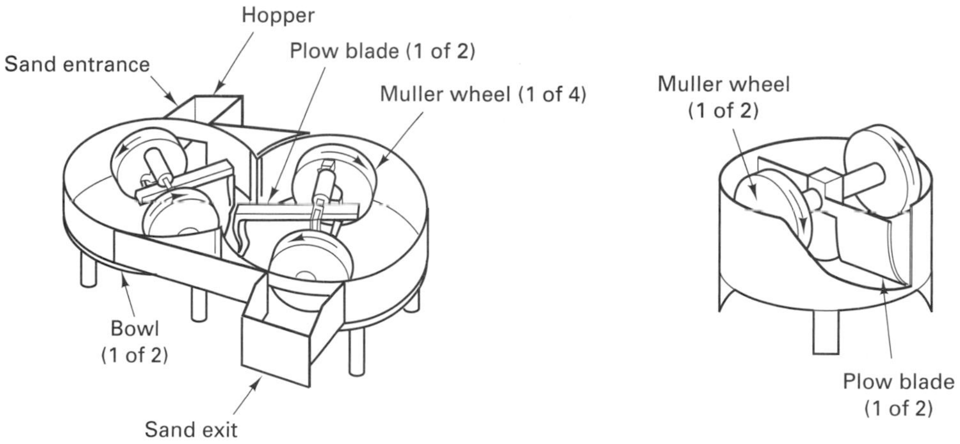

FIGURE 14-8   Schematic diagram of a continuous (*left*) and batch-type (*right*) sand muller. Plow blades move and loosen the sand and the muller wheels mix the components. *(Courtesy of ASM International.)*

satisfactory and consistent results. A typical green-sand mixture contains about 88% silica sand, 9% clay, and 3% water. Since molding material is often reclaimed and recycled, the temperature of the mold during pouring and solidification is also important. If organic materials have been incorporated into the mix to provide collapsibility, a portion will burn during the pour. Adjustments will be necessary, and ultimately some or all of the mold material may have to be discarded and replaced with new.

To achieve good molding, it is important for each grain of sand to be coated uniformly with the additive agents. This is achieved by putting the ingredients through a *muller*, a device that kneads, rolls, and stirs the sand. Figure 14-8 shows both a continuous and batch-type muller, that utilize blades and wheels to produce the mixing. After mulling, the sand is often discharged through an aerator, which fluffs it so that it does not pack too hard during handling.

SAND TESTING

Maintaining consistent sand quality may be of little concern to the casting designer, but it is a significant matter to the foundry that is expected to produce consistent, high-quality products. Sands can be characterized by grain shape, surface smoothness, density, and contaminants. In addition, standard tests and procedures have been developed to evaluate *grain size, moisture content, clay content,* and *compactability,* as well as *mold hardness, permeability,* and *strength.*

Grain size is determined by shaking a known amount of clean, dry sand downward through a set of 11 standard sieves of decreasing mesh size. After shaking for 15 minutes, the amount remaining on each sieve is weighed, and the weights are converted into an AFS (American Foundrymen's Society) grain fineness number.

Moisture content is usually determined by a special device that measures the electrical conductivity of a small sample of sand that is compressed between two prongs. Another method is to measure the weight lost from a 50-g sample after it has been subjected to a temperature of about 110°C (230°F) for sufficient time to drive off all the water.

Clay content can be determined by washing the clay from a 50-g sample of molding sand, using water that contains sufficient sodium hydroxide to make it alkaline. Several cycles of agitation and washing may be required to fully remove the clay. The remaining sand is then dried and weighed to determine the amount of clay removed from the original sample.

Permeability and strength tests are conducted on a *standard rammed specimen.* A sufficient amount of sand is placed into a 2-in.-diameter steel tube so that after a 14-lb weight is dropped three times from a height of 2 in., the final height of the specimen is within $\frac{1}{32}$ in. of two inches.

Permeability is a measure of how easily gases can pass through the narrow voids between the sand grains. Air in the mold before pouring, plus the steam that is produced when the hot metal contacts the moisture in the sand, must be allowed to escape, rather than prevent mold filling or be trapped in the casting as porosity or blow holes. During the permeability test, shown schematically in Figure 14-9, a sample tube containing the

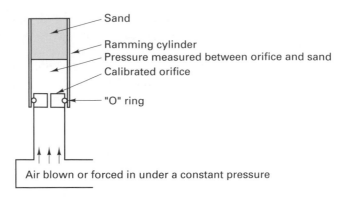

Sand

Ramming cylinder

Pressure measured between orifice and sand

Calibrated orifice

"O" ring

Air blown or forced in under a constant pressure

FIGURE 14-9 Schematic of a permeability tester in operation. The standard sample in a metal sleeve is sealed by an O-ring onto the top of the unit. *(Courtesy of Harry W. Dietert Company.)*

rammed specimen is subjected to an air pressure of 10 g/cm². By means of either a flow rate determination or measurement of the steady-state pressure between the orifice and the sand, an *AFS permeability number*[1] is determined. Most test devices are calibrated to provide a direct readout of the permeability number.

The molding material must have sufficient strength to maintain the integrity of the mold cavity as the mold is handled between molding and pouring. In addition, the material must withstand the erosion of the liquid metal as it flows into the mold, and the pressures induced by a column of molten metal. The *compressive strength* of the sand (also referred to as green compressive strength) is a measure of these properties. It is determined by removing the rammed specimen from the compacting tube and placing it in a mechanical testing device. A compressive load is then applied until the specimen breaks, which usually occurs in the range of 10 to 30 psi (0.07 to 0.2 MPa). When there is too little moisture in the sand, the grains are poorly bonded and strength is poor. If there is excess moisture, the extra water acts as a lubricant and strength again is poor. In between, there is a maximum strength and an optimum water content that will vary with the content of other materials in the mix. A similar optimum also applies to permeability, since unwetted clay blocks vent passages, as does excess water. Sand coated with a uniform thin film of moist clay provides the best molding properties. A ratio of 1 part water to 3 parts clay (by weight) is often a good starting point.

The *hardness* of the compacted sand can provide a quick indication of mold strength and give additional insight into the strength and permeability characteristics. Hardness can be determined by the resistance of the sand to penetration by a 0.2-in. (5.08-mm)-diameter spring-loaded steel ball. A typical test instrument is shown in Figure 14-10.

Compactibility is determined by sifting loose sand into a steel cylinder, leveling off the column, striking it three times with a standard weight (as in making a standard rammed specimen), and then measuring the final height. The *percent compactibility* is the change in height divided by the original height, times 100%. This value can often be correlated with the moisture content of the sand, where a compactibility of around 45% indicates a proper level of moisture. A low compactibility is usually associated with too little moisture.

FIGURE 14-10 Mold hardness tester. *(Courtesy of Harry W. Dietert Company.)*

SAND PROPERTIES AND SAND-RELATED DEFECTS

The characteristics of the sand granules can be very influential in determining the properties of the molding material. Round grains give good permeability and minimize the amount of clay required because of their low surface area. Angular sands give better green strength because of the mechanical interlocking of the grains. Large grains provide good permeability and better resistance to high temperature melting and expansion,

---

[1] The AFS permeability number is defined as follows:

$$\text{AFS Number} = (V \times H)/(P \times A \times T)$$

Where V is the volume of air (2000 cm³), H is the height of the specimen (5.08 cm), P is the pressure (10 g/cm²), A is the cross section area of the specimen (20.268 cm²), and T is the time in seconds to pass a flow of 2000 cm³ of air through the specimen. Substituting the constants, the permeability number is equal to $3000.2/T$.

while fine-grained sands produce a better surface finish on the final casting. Uniform-size sands give good permeability, while a distribution of sizes enhances surface finish.

Silica sand is cheap and light weight, but when hot metal is poured into a silica sand mold, the sand becomes hot, undergoes one or more phase transformations, and has a substantial expansion in volume. Because sand is a poor thermal conductor, only the sand that is adjacent to the mold cavity becomes hot and expands. The remaining material stays fairly cool, does not expand, and provides a high degree of mechanical restraint. Because of this uneven heating, the sand at the surface of the mold cavity may buckle or fold. Castings having large, flat surfaces are more prone to *sand expansion defects* since a considerable amount of expansion must occur in a single, fixed direction.

Sand expansion defects can be minimized in a number of ways. Certain particle geometries permit the sand grains to slide over one another, thereby relieving the expansion stresses. Excess clay can be added to absorb the sand expansion, or volatile additives, such as cellulose, can be added to the mix. When the sand becomes hot, the cellulose burns, creating voids that then accommodate the sand expansion. Another alternative is the use of olivine or zircon sand in place of silica. Since these sands do not undergo phase transformations upon heating, their expansion is only about half that of silica sand. Unfortunately, both types are much more expensive and notably heavier in weight.

Voids can form in castings when the molten metal remains in contact with trapped or evolved gas. The most common cause is low sand permeability and/or large amounts of gas evolution caused by high moisture or excessive amounts of volatiles. If adjustments to the mold composition are not sufficient to eliminate the voids, vent passages may have to be cut, a procedure that adds significantly to the mold-making cost.

Molten metal can also penetrate between the sand grains, causing the mold material to become embedded in the surface of the casting. *Penetration* can be the result of high pouring temperatures (excess fluidity), high metal pressure (possibly due to excessive cope height or pouring from too high an elevation above the mold), or the use of high-permeability sands with coarse, uniform particles. Fine-grained materials, such as silica flour, can be used to fill the voids, but this reduces permeability and increases the likelihood of gas and expansion defects.

Hot tears or *cracks* can form in castings made from metals or alloys with large amounts of solidification shrinkage. During solidification, as the metal tries to contract, it may find itself restrained by a strong mold or core. Tensile stresses can develop while the metal is still partially liquid, or fully solidified, but still hot and weak. If the stresses become great enough, the casting will crack. Hot tears are often attributed to a lack of collapsibility, the ability of the sand to break down and crumble after the casting has been poured and solidified. Sand additives, such as cellulose, can be particularly helpful in improving collapsibility. Table 14-1 summarizes the desirable properties of a sand-based molding material.

THE MAKING OF SAND MOLDS

Hand ramming is often the preferred method of mold making when only a few castings are to be made from any given design. In most cases, however, the sand molds are made

TABLE 14-1.   Desirable Properties of a Sand-Based Molding Material

1. Is inexpensive in bulk quantities
2. Retains properties through transportation and storage
3. Uniformly fills a flask or container
4. Can be compacted or set by simple methods
5. Has sufficient elasticity to remain undamaged during pattern withdrawal
6. Can withstand high temperatures and maintains its dimensions until the metal has solidified
7. Is sufficiently permeable to allow the escape of gases
8. Is sufficiently dense to prevent metal penetration
9. Is sufficiently cohesive to prevent wash-out of mold material into the pour steam
10. Is chemically inert to the metal being cast
11. Can yield to solidification and thermal shrinkage, thereby preventing hot tears and cracks
12. Has good collapsibility to permit easy removal and separation of the casting
13. Can be recycled

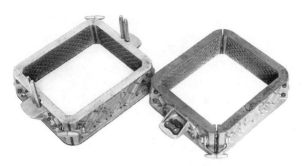

FIGURE 14-11 Bottom (*left*) and top halves of a tapered cam-latch snap flask, drag flask closed and cope (*right*) open for removal.

using specially designed molding machines. The various methods differ in the type of flask required, the way sand is packed within the flask, whether mechanical assistance is provided to turn or handle the mold, and whether a flask is even required. In all cases, however, the molding machines greatly reduce the labor and skill required, and lead to castings with better dimensional accuracy and consistency.

Molding usually begins with a pattern, such as the match-plate pattern discussed earlier, and a *flask*. The flasks may be either straight-walled containers with guide pins or removable jackets, and they are generally constructed of lightweight aluminum or magnesium. Figure 14-11 shows a snap flask, designed to open slightly to permit removal after the mold material has been packed in place.

The sand can be packed in the flask by one or more basic techniques. A *sand slinger* uses the centrifugal force of a rotating impeller to throw sand against the pattern. The slinger is manipulated over the pattern and flask to build a mold by depositing layers of flung sand. Sand slinging is used primarily with large molds and large castings.

In a method known as *jolting*, a flask is placed over the pattern, filled with sand, and the pattern, flask, and sand are then lifted and dropped several times, as shown in Figure 14-12. The weight and kinetic energy of the sand produces optimum packing around the pattern. Jolting machines can be used on the first half of a match-plate pattern or on both halves of a cope-and-drag operation.

Squeezing machines use an air-operated squeeze head, a flexible diaphragm, or small individually activated squeeze heads to compact the sand. Squeezing provides firm packing near the squeeze head, but the density diminishes as you move farther into the mold. High-pressure machines with a flexible diaphragm, commonly called Taccone machines, can produce a more uniform density around all sections of an irregular pattern. Figure 14-13 illustrates the squeezing process, and Figure 14-14 compares squeezing with a flat plate and squeezing with a flexible diaphragm.

A combination of jolting and squeezing is often used to produce a more uniform density throughout the mold. Here a match-plate pattern is positioned between the cope and drag sections of a flask, and the assembly is placed upside down on the molding machine. A parting compound is sprinkled on the pattern, and the drag section of the flask is filled with sand. The entire assembly is then jolted a specified number of times to pack the sand around the drag side of the pattern. A squeeze head is then swung into place, and pressure is applied to complete the drag portion of the mold. The flask is then

FIGURE 14-12 Jolting a mold section.

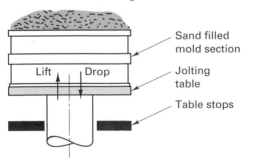

Sand filled mold section

Jolting table

Table stops

Lift Drop

FIGURE 14-13 Squeezing a sand-filled mold section.

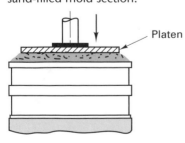

Platen

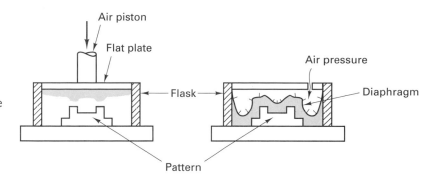

FIGURE 14-14 Schematic diagram showing relative sand densities obtained by flat-plate squeezing where all areas get vertically compressed by the same amount of movement (*left*) and by flexible-diaphragm squeezing where all areas flow to the same resisting pressure (*right*).

inverted and a squeezing operation is performed on the cope half. (*Note:* Jolting here might cause the already-compacted sand to break free of the inverted drag section of the pattern!) If the cope and drag segments are made on separate machines using separate cope-and-drag patterns, both jolting and squeezing can be performed on each part of the mold.

While some patterns include the sprue hole, it is most often cut by hand. (Hand cutting should be performed before removal of the pattern, so loose sand does not fall into the mold cavity.) The pouring basin may also be hand cut, or it may be formed by a protruding shape on the squeeze board. The gates and runners are usually included on the pattern. Because hand labor is costly and time consuming, most patterns are designed to minimize the amount of hand working.

After the mold is complete, the tapered molding flask may be removed to prevent possible damage to it during the pour. An inexpensive metal band called a *slip jacket* may be positioned around the mold to hold the sand in place. Heavy metal weights are often placed on top of the molds to prevent the sections from separating when the hydrostatic pressure of the molten metal presses upward on the cope. The slip jackets and added weights are needed only during pouring and for a few minutes afterward, while solidification occurs. They are then removed and placed on other molds, thereby reducing the amount of equipment needed in the operation.

For mass-production molding, a number of automatic mold-making devices have been developed. These include automatic match-plate machines, automatic cope-and-drag machines, and machines that produce some form of stacked segments. Figure 14-15 depicts a *vertically parted flaskless molding machine*, where the cope-and-drag patterns are incorporated into opposing sides of a vertical mold. Sand is deposited between the patterns and squeezed with a horizontal motion. The patterns are withdrawn, cores are set, and the mold block is joined to those that were previously molded. Since each block contains the right-hand cavity of one mold and the left-hand cavity of another, an entire mold is made with each cycle of the machine. (Previous techniques required two separate molding operations to produce the individual cope and drag segments of a mold.) A vertical gating system is usually included on the pattern, and the vertically parted molds are usually poured individually. If a common runner is used to connect multiple mold segments, the

FIGURE 14-15 Vertically parted flaskless molding with inset cores. Note how one mold block contains both the cope and drag impressions. (*Courtesy of Belens Corporation.*)

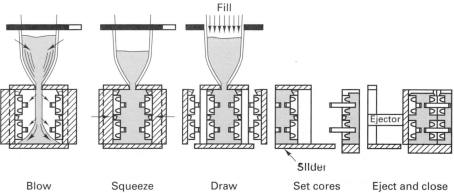

Blow   Squeeze   Draw   Set cores   Eject and close

method is known as the *H-process*. Since metal cools as it travels through long runners, the individual cavities of the H-process often fill with different temperature metal. To assure product uniformity, most producers reject the H-process and prefer to pour their vertically-parted molds on an individual basis.

In *stack molding*, sections containing a cope impression on the bottom and a drag impression on the top are piled vertically on top of one another. Metal is poured down a common sprue, which is connected to a horizontal gating system at each of the parting planes. For molds that are too large to be made by either hand ramming or by one of the many types of molding machines, large flasks can be placed on the foundry floor. Various types of mechanical aids can then be used to add and pack the sand. For example, a *sand slinger* can be used to produce large molds with uniform compaction throughout. Additional tamping can be done with a pneumatic rammer.

Extremely large molds can be constructed in sunken pits. Because of the size, the complexity, and the need for strength, pit molds are frequently assembled using smaller sections of baked or dried sand. Added binders may be required to provide the necessary strength.

GREEN-SAND, DRY-SAND, AND SKIN-DRIED MOLDS

Green-sand casting is the most widely used process for casting ferrous and nonferrous metals, and accounts for nearly 90% of all castings made in the United States. The mold material is composed of sand with a binder of clay, water, and additives. Tooling costs are low, and the entire process is one of the least expensive casting processes. Almost any metal can be cast, and there are few limits on the size, shape, weight, and complexity of the products. The process has evolved from a manually intensive operation, to a mechanized and automated system capable of producing over 300 molds per hour. As a result, it can be economically applied to both small and large production runs.

Design limitations are usually related to the rough surface finish, poor dimensional accuracy, and the resulting need for subsequent machining. Still other problems can be attributed to the low strength of the mold material and the moisture that is present in the binder. Table 14-2 provides a process summary for green-sand molding, and Figure 14-16 shows a variety of parts that can be produced by conventional sand casting.

Some of the problems associated with the green-sand process can be reduced by heating the mold to a temperature of 300°F or higher, and baking until most of the moisture is driven off. This strengthens the mold and reduces the amount of gases generated when the hot metal enters the cavity. These *dry-sand molds* are very durable and may be stored for a relatively long period of time. They are not very popular, however, because of the long times required for drying, the added cost of that operation, and the availability of practical alternatives. An attractive compromise is to produce a *skin-dried mold*, drying only the sand that is adjacent to the mold cavity. Torches are often used to perform the drying, and the water is usually removed to a depth of about 13 mm $\left(\frac{1}{2}\text{ in.}\right)$.

Molds used for the casting of large, heavy parts from steel are almost always skin-dried, because the pouring temperatures are significantly higher than those for cast iron. These molds may also be given a high-silica wash prior to drying to increase the refractoriness of the surface, or the more thermally stable zircon sand can be used as a facing. Additional binders, such as molasses, linseed oil, or corn flour, may be added to the facing sand to enhance the strength of the skin-dried segment.

TABLE 14-2. Green-Sand Casting

Process: Sand, bonded with clay and water, is packed around a wood or metal pattern. The pattern is removed and molten metal is poured into the cavity. When the metal has solidified, the mold is broken and the casting is removed.
Advantages: Almost no limit on size, shape, weight or complexity; low cost; almost any metal can be cast.
Limitations: Tolerances and surface finish are poorer than in other casting processes; some machining is often required; relatively slow production rate.
Common metals: Cast iron, steel, stainless steel, and casting alloys of aluminum, copper, magnesium, and nickel.
Size limits: 30 g to 3000 kg (1 oz to 6000 lb)
Thickness limits: As thin as 0.25 cm (3/32 in.), with no maximum
Typical tolerances: 0.8 mm for first 15 cm ($\frac{1}{32}$ in. for first 6 in.), 0.003 cm for each additional cm; additional increment for dimensions across the parting line
Draft allowance: 1–3°
Surface finish: 2.5–25 microns (100–1000 μin.) rms

FIGURE 14-16 A variety of sand cast aluminum parts. *(Courtesy of Bodine Aluminum Inc.)*

SODIUM SILICATE–CO₂ MOLDING

Molds (and cores) can also be made from sand that receives its strength from the addition of 3 to 4% *sodium silicate*, a liquid inorganic binder that is also known as *water glass*. The sand can be mixed with the liquid sodium silicate in a standard muller and can be packed into flasks by any of the methods discussed previously. It remains soft and moldable until it is exposed to a flow of CO_2 gas. Then, it hardens in a matter of seconds by the reaction:

$$Na_2SiO_3 + CO_2 \longrightarrow Na_2CO_3 + SiO_2 \text{ (colloidal)}$$

The CO_2 gas is nontoxic and odorless, and no heating is required to initiate or drive the reaction. The hardened sands, however, have poor collapsibility, making shakeout and core removal difficult. Unlike most other sands, the heating that occurs as a result of the pour makes the mold even stronger (a phenomenon similar to the firing of a ceramic material). Additives that will burn out during the pour are often used to enhance the collapsibility of sodium silicate molds. In addition, care must be taken to prevent the carbon dioxide in the air from hardening the sand before the mold-making process is complete.

A modification of the CO_2 process can be used when certain portions of a mold require higher strength, better accuracy, thinner sections, or deeper draws than can be achieved with ordinary molding sand. Sand mixed with sodium silicate is packed around a metal pattern to a thickness of about 1 in., followed by regular molding sand as a backing material. After the mold is fully rammed, CO_2 is introduced through vents in the metal pattern. This hardens the adjacent sand, and the pattern can now be withdrawn with less possibility of damaging the mold.

NO-BAKE, AIR-SET, OR CHEMICALLY BONDED SANDS

One alternative to the sodium silicate process involves the use of organic resin binders that cure by room temperature chemical reactions that occur with liquid catalysts. Two or more binder components are mixed with the sand just prior to the molding operation, and the curing reactions begin immediately. Because the mix is workable for only a short period of time, the molds (or cores) must be made in a reasonably rapid fashion. After a few minutes to a few hours (depending on the specific binder and curing agent), the sands harden enough to be removed from the pattern without distortion. After time for further curing, the molds are ready to pour.

Various *no-bake sand systems* are available, with selection being based on the metal being poured, the cure time desired, the complexity and thickness of the casting, and the desire for sand reclamation. Each system is based on organic resin binders, curing agents or catalysts, and various additives and modifiers. Like the molds produced by the sodium silicate process, no-bake offers high dimensional accuracy, good hot strength, and high resistance to mold-related casting defects. Patterns can incorporate thinner sections and deeper draws. In contrast to the sodium silicate material, however, the no-bake molds decompose readily after the metal has been poured, providing excellent shakeout characteristics.

SHELL MOLDING

Many molds are now being made by the *shell-molding process*, which offers better surface finish than can be obtained with ordinary sand molding, better dimensional accuracy, and a higher production rate with reduced labor requirements. In many cases, the process can be completely mechanized and adapted for mass production.

Figure 14-17 illustrates the six basic steps of the shell process:

1. Fine silica sand, in which the individual grains have been precoated with a thin layer of thermosetting phenolic resin and liquid catalyst, is dumped, blown, or shot onto a metal pattern (usually some form of cast iron) that has been heated to 230 to 315°C (450 to 600°F). It is allowed to stand for a few minutes while the heat from the pattern partially cures (polymerizes and crosslinks) a layer of material, forming a strong, solid-bonded region adjacent to the pattern. The actual thickness depends on the pattern temperature and the time of contact, but typically ranges between 10 and 20 mm (0.4 to 0.8 in.).

FIGURE 14-17 Schematic of the shell-molding process. (a) A heated pattern is placed over a dump box containing a sand and resin mixture. (b) The box is inverted and a shell partially cures around the pattern. (c) The box is righted, the top is removed, and placed in an oven to further cure the shell. (d) The shell is stripped from the pattern. (e) Matched shells are then joined and supported in a flask ready for pouring.

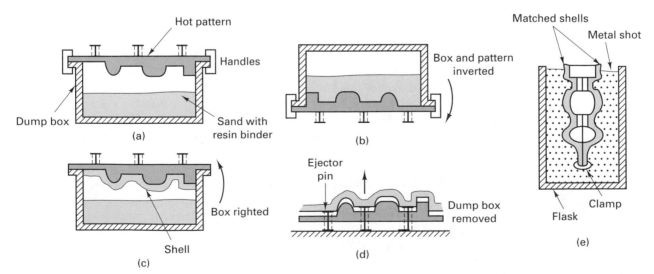

2. The pattern and sand mixture are then inverted. All of the excess sand drops free, leaving only the layer of partially cured material that adhered to the pattern.

3. The box is righted and the pattern/shell is removed, and placed in an oven where the partially cured "shell," which is heated for a few minutes to complete the curing process.

4. The hardened shell (with tensile strength between 350 and 450 psi) is then stripped from the pattern.

5. Two or more shells are then clamped or glued together with a thermoset adhesive to produce a mold, which may be poured immediately or stored almost indefinitely.

6. Shell molds are often placed in a pouring jacket and surrounded with metal shot or sand to provide extra support during the pour.

Because the sand is compounded for almost no shrinkage and a metal pattern is used, the shell has excellent dimensional accuracy. Tolerances of 0.08 to 0.13 mm (0.003 to 0.005 in.) are quite common. Shell-mold sand is finer than ordinary foundry sand and, in combination with the plastic resin, produces a very smooth casting surface. Cleaning, machining, and other finishing costs can be reduced significantly. In addition, the shell-mold process offers an excellent level of product consistency.

Figure 14-18 shows a set of metal patterns, the two shells before clamping, and the resulting shell-mold casting. Machines for making shell molds vary from simple ones for small operations, to large, completely automated devices for mass production. The cost of a metal pattern is often rather high and its design must include the gate and runner system since these cannot be cut after molding. Large amounts of expensive binder are required, but the amount of material actually used to form a thin shell is not that great. High productivity, low labor costs, smooth surfaces, and a level of precision that reduces that amount of required machining all combine to make the process economical for even moderate quantities. The thin shell provides for the easy escape of the gases that evolve

FIGURE 14-18   (*Top*) Two halves of a shell-mold pattern. (*Bottom*) The two shells before clamping, and the final shell-mold casting with attached pouring basin, runner and riser. (*Courtesy of Shalco Systems, an Acme-Cleveland Company.*)

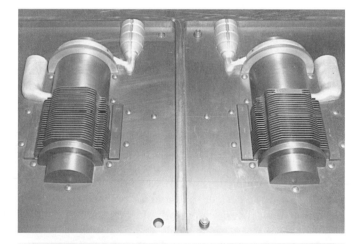

TABLE 14-3.   Shell-Mold Casting

Process: Sand coated with a thermosetting plastic resin is dropped onto a heated metal pattern, which cures the resin. The shell segments are stripped from the pattern and assembled. When the poured metal solidifies, the shell is broken away from the finished casting.
Advantages: Faster production rate than sand molding; high dimensional accuracy with smooth surfaces.
Limitations: Requires expensive metal patterns. Plastic resin adds to cost; part size is limited.
Common metals: Cast irons and casting alloys of aluminum and copper
Size limits: 30 g (1 oz) minimum; usually less than 10 kg (25 lb); mold area usually less than 0.3 m^2 (500 in^2)
Thickness limits: Minimums range from 0.15 to 0.6 cm ($\frac{1}{16}$ to $\frac{1}{4}$ in.), depending on material
Typical tolerances: Approximately 0.005 cm/cm or in/in.
Draft allowance: $\frac{1}{4}$ to $\frac{1}{2}$ degree
Surface finish: $\frac{1}{3}$–4.0 microns (50–150 μin.) rms

during the pour, and the volume of evolved gas is rather low because of the absence of moisture in the mold material. When the shell becomes hot, some of the resin binder burns out, providing excellent collapsibility and shakeout characteristics. Table 14-3 summarizes the features of shell molding.

OTHER SAND-BASED MOLDING METHODS

Over the years, a variety of processes have been proposed to overcome some of the limitations of the more traditional methods. While few have become commercially significant, several are included here to illustrate the nature of these efforts.

In one method, known as the *V-process* or *vacuum molding*, a vacuum performs the role of the sand binder. Figure 14-19 depicts the production sequence, which begins by draping a thin sheet of heat-softened plastic over a special vented pattern (usually made from another plastic). The sheet is then drawn tightly to the pattern surface by a vacuum within the pattern. A vacuum flask is then placed over the pattern, the flask is filled with sand, a sprue and pouring cup are formed, and a second sheet of plastic is placed over the mold. A vacuum is then drawn on the flask itself, compacting the sand and providing the necessary strength and hardness. The pattern vacuum is released, and the pattern is withdrawn. The other segment of the mold is made in a similar fashion, and the two mold halves are assembled. The mold is then poured while a vacuum is maintained in both the cope and drag segments of the flask. While the thin layer of plastic film vaporizes almost instantaneously, the vacuum is still sufficient to hold the sand in shape until the metal has cooled and solidified. When the vacuum is released, the sand reverts to a loose, unbonded state, and falls away from the casting.

Advantages of the vacuum process include the total absence of moisture-related defects. Since no binder is used, binder cost is eliminated and the loose, dry sand is completely and directly reusable. Since there is no clay, water, or other binder to impair permeability, finer sands can be used, resulting in better surface finish in the resulting castings. No fumes (binders burning up) are generated during the pouring operation. Shakeout characteristics are exceptional, since the mold virtually collapses when the vacuum is released. Unfortunately, the process is relatively slow because of the additional steps and the time required

FIGURE 14-19   Schematic of the V-process or vacuum molding. (a) A vacuum is pulled on a pattern, drawing a shrink-wrap plastic sheet tightly against it. (b) A vacuum flask is placed over the pattern, filled with dry unbonded sand, a pouring basin and sprue are formed, the remaining sand is leveled, a second sheet is placed on top, and a mold vacuum is drawn. (c) The pattern vacuum is then broken and the pattern is withdrawn. With the mold vacuum being maintained, the cope and drag segments are assembled, and the molten metal is poured.

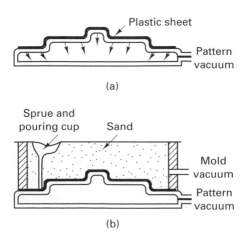

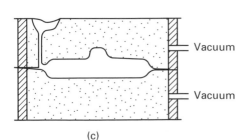

to pull a sufficient vac̲ ̲ ̲ ̲ ̲ ̲ It i̲ ̲ ̲ used primarily for the production of prototype, frequently modified, or l̲ ̲ ̲ ̲ ̲ ̲ ̲ olume parts (more than 10 but less than 15,000).

In the *Eff-set process*, sand with a small amount of clay and quite a bit of water is first packed around a pattern. The pattern is removed and liquid nitrogen is sprayed onto the mold surface. The ice that forms becomes the binder, and molten metal is poured into the mold while the surface is in its frozen condition. As with the V-process, binder cost is low and shakeout is excellent.

■ 14.3 CORES AND CORE MAKING

Casting processes are unique in their ability to incorporate internal cavities or reentrant sections with relative ease. To produce these features, however, it is often necessary to use *cores* as part of the mold. Figure 14-20 shows an example of a product that could not be made by any process other than casting with cores. While these cores constitute an added cost, they do much to expand the capabilities of the process, and good design practice can often facilitate and simplify their use.

Consider the simple belt pulley shown schematically in Figure 14-21. Various methods of fabrication are suggested in the four sketches, beginning with the casting of a solid form and the subsequent machining of the through-hole for the drive shaft. A large volume of metal would have to be removed through a substantial amount of costly machining. A more economical approach would be to make the pulley with a cast-in hole of the approximate final size. Figure 14-21b depicts an approach where each half of the

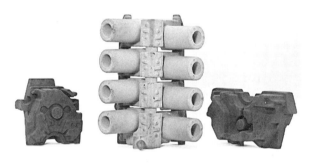
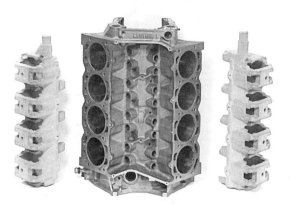

FIGURE 14-20   V-8 engine block (*bottom center*) and the five dry-sand cores that are used in the construction of its mold. *(Courtesy of General Motors Corporation.)*

FIGURE 14-21   Four methods of making a hole in a cast pulley. Three show the use of a core.

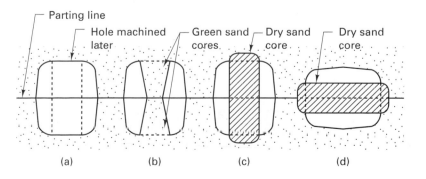

pattern includes a tapered hole, which receives the same green sand that is used for the remainder of mold. While these protruding sections are an integral part of the mold, they are also known as *green-sand cores*. Unfortunately, green-sand cores have a relatively low level of strength. If the protrusions are narrow or long, it might be difficult to withdraw the pattern without breaking them, or they may not have enough strength to even support their own weight. For long cores, a considerable amount of machining may still be required to remove the draft that must be provided on the pattern. For more complex shapes, it is often impossible to withdraw the pattern, and green-sand cores would no longer be an option.

Dry-sand cores can be used to overcome some of the cited difficulties. These cores are made independent from the remainder of the mold and are then inserted into core prints that hold them in position. The sketches in Figure 14-21c and Figure 14-21d show dry-sand cores in the vertical and horizontal positions. Dry-sand cores can be made in several ways. In each, the sand, mixed with some form of binder, is packed into a wood or metal core box that contains a cavity of the desired shape. A *dump core box* such as the one shown in Figure 14-22 offers a very simple approach. Sand is packed into the box and scraped level with the top surface (which acts like the parting line in a traditional mold). A wood or metal plate is then placed over the top of the box, and the box is turned over and lifted upward, leaving the molded sand resting on the plate. After baking or hardening, the core segments are joined with hot-melt glue or some other bonding agent. Rough spots along the parting line are removed with files or sanding belts, and the final product may be given a thin coating to provide a smoother surface or greater resistance to heat. Graphite, silica, or mica can be sprayed or brushed onto the surface.

Single-piece cores are often made in a *split core box*. The two halves of the core box are clamped together, and an opening is provided in one or both ends through which sand is introduced and rammed. After the sand is compacted, the halves of the box are separated to permit removal of the core. A core-extruding machine, similar to the familiar meat grinder, can be used to produce cores that have a uniform cross section. The individual cores are simply cut to the desired length as the product emerges from the machine, and they are then placed in core supports for hardening. More complex cores can be made in core-blowing machines that use separating dies and receive the sand in a manner similar to injection molding or diecasting.

Cores are often the most fragile part of a mold assembly. To provide the necessary strength, a variety of processes have been developed, many using some form of special binder. The oldest of these processes, the *core-oil process*, uses a vegetable or synthetic oil binder, and water with cereal or clay to develop green strength. The wet sand is blown or rammed into a relatively simple corebox at room temperature. The uncured cores are then transferred to flat plates or special supports and placed in convection

FIGURE 14-22 (*Upper right*) A dump-type core box; (*bottom*) Two core halves ready for baking; and (*upper left*) a completed core made by gluing two opposing halves together.

ovens at 200 to 260°C (400 to 500°F) for curing. The heat causes the binder to cross-link or polymerize, producing a strong organic bond between the grains of sand. The process is simple and the materials are inexpensive, but the dimensional accuracy of the resultant cores is often difficult to maintain.

In the *hot-box method*, sand containing a liquid thermosetting binder and catalyst is packed into a core box that has been heated to around 230°C (450°F). When the sand contacts the hot surface, the catalyst is heated, and the initial stages of curing occur within 10 to 30 seconds. The core can then be removed from the pattern and will hold its shape during further handling. For some materials, the cure completes through an exothermic curing reaction. Others require further baking to complete the process.

Room temperature curing is the dominant characteristic of the *cold-box process*. Binder-coated sand is first blown into the core box, which is then sealed and a gas or vaporized catalyst is passed through the permeable core to polymerize the resin. Hollow cores can be produced by introducing small amounts of the curing gas through holes in the core-box pattern. The uncured sand in the center can be dumped free and reused. It is not uncommon for the required gases to be either toxic (an amine gas) or odorous (SO_2), making special handling of both incoming and exhaust gas a process requirement. Because of the low temperatures, core-box tooling can be made from wood, metal, or even plastic.

Room-temperature cores can also be made with the *air-set* or *no-bake sands*. These systems eliminate the gassing operation of the cold-box process through the simultaneous use of an organic resin binder and a curing catalyst. As discussed previously, there is only a short time period to form the core after the components are mixed. *Shell molding* is another core-making alternative, producing hollow cores with excellent permeability.

Selecting the actual method of core production is usually based on a number of considerations, including production quantity, production rate, required precision, required surface finish, and the metal being poured. Certain metals may be sensitive to gases that are emitted from the cores when they come into contact with the hot metal. Other materials may have low pouring temperatures that are inadequate to break down the binder and permit collapsibility and removal from the final casting.

To function properly, cores must have the following characteristics:

1. Sufficient strength before hardening to permit handling in the "green" condition.
2. Sufficient hardness and strength after baking or hardening to withstand handling and the forces of the molten metal. Compressive strength should be between 100 and 500 psi (0.7 to 3.5 MPa).
3. Adequate permeability to permit the escape of gases. Since cores are largely surrounded by molten metal, they should possess exceptionally good permeability.
4. Collapsibility. After pouring, the cores must be weak enough to permit shrinkage of the solidified casting as it cools, thereby preventing cracking. In addition, they must be easily removable from the interior of the finished product via shakeout.
5. Adequate refractoriness. Since the cores are largely surrounded by hot metal, they can become quite a bit hotter than the adjacent mold material.
6. A smooth surface.
7. Minimum generation of gases when heated by the pour.

Various techniques have been developed to enhance the natural properties of cores and core materials. Internal wires or rods can be used to impart additional strength. Collapsibility can be enhanced by making the cores hollow or by placing a material such as straw in the center. Enhanced collapsibility is particularly important in steel castings, where a large amount of shrinkage is observed. All but the smallest of cores must be vented to permit the escape of trapped and evolved gases. Vent holes can be produced by pushing small wires into the core. Coke or cinders are sometimes placed in the center of large cores to improve permeability.

Since the core material must be removed from the finished casting, the cores must be connected to the outer surfaces of the mold cavity. Recesses at these connection

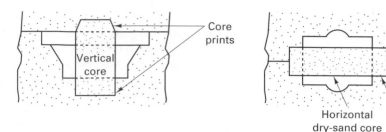

FIGURE 14-23 Molds cavities containing core prints to hold and position a vertical core (*left*) and horizontal core (*right*).

points, known as *core prints*, are used to support the cores and hold them in proper position during mold filling. Figure 14-23 shows the cross section of several molds where core prints are used to hold vertical and horizontal cores.

If the cores do not pass completely through the casting, or additional measures are necessary to support the weight of the core or keep it from being moved or floated by the molten metal, small metal supports, called *chaplets*, can be positioned between the core and the surfaces of the mold cavity, as shown in Figure 14-24. The use of chaplets should be minimized, however, since they become an integral part of the finished casting and may cause defects or be a location of weakness. Chaplets should be of the same, or at least comparable, composition as the casting material. They should be large enough that they do not completely melt and permit the core to move, but small enough that their surface melts and fuses with the metal being cast.

Core sections can also be used to facilitate the molding of shapes that contain reentrant angles or sections. Consider a round pulley with a recessed groove around its perimeter. Figure 14-25 produces the part by using a third segment of flask, called a *cheek*. By adding a second parting plane, the entire mold can be made by green-sand molding around withdrawable patterns. Additional molding operations are required, but this may be an attractive approach when only a few products are to be made, since it eliminates the need for a special core box.

If we want to produce a substantial number of identical pulleys, a simple green-sand mold can be used in conjunction with a ring-shaped core, as shown in Figure 14-26. Rapid machine molding can be used, with the green sand pattern providing a seat for the separately manufactured core. Molding time is reduced at the expense of a core box and a separate core-making operation.

FIGURE 14-24 (*Left*) Typical chaplets. (*Right*) Method of supporting a core by use of chaplets (relative size of the chaplets is exaggerated).

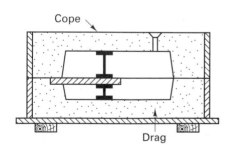

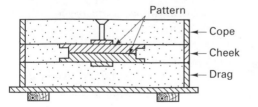

FIGURE 14-25 Method of making a reentrant angle or inset section by using a three-piece flask.

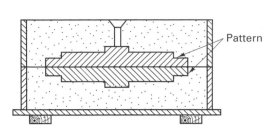
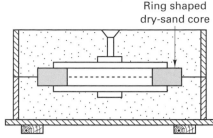

Ring shaped
dry-sand core

Pattern

FIGURE 14-26   Molding an inset section using a dry-sand core.

■ 14.4 OTHER EXPENDABLE-MOLD PROCESSES WITH MULTIPLE-USE PATTERNS

PLASTER MOLD CASTING

In *plaster molding* the mold material is plaster of paris (also known as calcium sulfate or gypsum), with various additions to improve green strength, dry strength, permeability, and castability. Talc or magnesium oxide helps to prevent cracking and reduces the setting time, lime or cement helps to control expansion during baking, glass fibers can be added to improve strength, and sand is often used as a filler.

The mold material is first mixed with water, and the creamy slurry is poured over a pattern and allowed to set. Hydration of the plaster produces a hard solid that can then be stripped from the pattern, which is usually made from metal. (*Note:* When complex angular surfaces or reentrant angles are required, flexible rubber patterns can be used. Because of the strength of the plaster, the rubber patterns can generally be removed without damage to the mold cavity.) After the mold is removed from the pattern, it is baked to remove the excess water, assembled, and poured.

The metal patterns, coupled with the plaster mold material, provide an excellent surface finish and good dimensional accuracy. Because of the low heat capacity and low thermal conductivity of the mold, cooling is slow, so the poured metal can flow into and replicate thin sections and fine detail. Unfortunately, only the lower-melting-temperature nonferrous alloys (such as aluminum, copper, magnesium, and zinc) can be cast in plaster molds. At the high temperatures of ferrous casting, the plaster would first undergo a phase transformation and then melt. Table 14-4 summarizes the features of plaster casting.

CERAMIC MOLD CASTING

Ceramic mold casting (summarized in Table 14-5) is similar to plaster mold casting, except that the mold can now withstand the higher-melting-point metals. Cope-and-drag molds are formed around withdrawable patterns (which can be metal or wood), using a ceramic slurry as the mold material. As with the plaster process, ceramic molding can produce thin sections, fine detail, and smooth surfaces, thereby eliminating a considerable amount of finish machining. These advantages, however, must be weighed against

TABLE 14-4.   Plaster Casting

Process: A slurry of plaster, water, and various additives is poured over a pattern and allowed to set. The pattern is removed and the mold is baked to remove excess water. After pouring and solidification, the mold is broken and the casting is removed.

Advantages: High dimensional accuracy and smooth surface finish; can reproduce thin sections and intricate detail to make net- or near-net-shaped parts.

Limitations: Lower-temperature nonferrous metals only; long molding time restricts production volume or requires multiple patterns; mold material is not reusable; maximum size is limited.

Common metals: Primarily aluminum and copper

Size limits: As small as 30 g (1 oz) but usually less than 7 kg (15 lb)

Thickness limits: Section thickness as small as 0.06 cm (0.025 in.)

Typical tolerances: 0.01 cm on first 5 cm (0.005 in. on first 2 in.), 0.002 cm per additional cm (0.002 in. per additional in.)

Draft allowance: $\frac{1}{2}$–1 degree

Surface finish: 1.3–4 microns (50–125 μin.) rms

TABLE 14-5.	Ceramic Mold Casting

Process: Stable ceramic powders are combined with binders and gelling agents to produce the mold material.
Advantages: Intricate detail, close tolerances, and smooth finish.
Limitations: Mold material is costly and not reusable.
Common metals: Ferrous and high-temperature nonferrous metals are most common; can also be used with alloys of aluminum, copper, magnesium, titanium, and zinc.
Size limits: 100 grams to several thousand kilograms (several ounces to several tons)
Thickness limits: As thin as 0.13 cm (0.050 in.); no maximum
Typical tolerances: 0.01 cm on the first 2.5 cm (0.005 in. on the first in.), 0.003 cm per each additional cm (0.003 in. per each additional in.)
Draft allowance: 1° preferred
Surface finish: 2–4 microns (75–150 μin.) rms

the greater cost of the mold material. For large molds, the ceramic material can be used to produce a facing around the pattern, which is then backed up by a material such as fireclay. This back up material is not only less expensive, but it can also be reused.

One of the most popular of the ceramic molding techniques is the *Shaw process.* A reusable pattern is placed inside a slightly tapered flask, and a slurry like mixture of refractory aggregate, hydrolyzed ethyl silicate, alcohol, and a gelling agent is poured on top. This mixture sets to a rubbery state that permits removal of the pattern and the flask, and the mold surface is then ignited with a torch. During "burn-off," most of the volatiles are consumed, and a three-dimensional network of microscopic cracks (microcrazing) forms in the ceramic. The gaps are small enough to prevent metal penetration but large enough to provide venting of air and gas (permeability) and to accommodate both the thermal expansion of the ceramic particles during the pour and the subsequent shrinkage of the solidified metal (i.e., provide collapsibility). A subsequent baking operation removes all of the remaining volatiles, making the mold hard and rigid. Before pouring, the ceramic molds are often preheated to ensure proper filling and to control the solidification characteristics of the metal. Figure 14-27 shows a set of intricate cutters that were produced by this process. Figure 14-28 shows how the Shaw process can be combined with the lost-wax process (discussed in Section 14.5) to produce complex-geometry products. Cores can also be used in this technique in much the same way as with sand casting.

EXPENDABLE GRAPHITE MOLDS

For metals such as titanium, which tend to react with many of the more common mold materials, powdered *graphite* can be combined with cement, starch, and water and compacted around a pattern. The pattern is then removed and the mold is fired at 1000°C (1800°F) to consolidate the graphite. After pouring, the mold is broken to remove the metal casting.

FIGURE 14-27 Group of cutters produced by ceramic mold casting. *(Courtesy of Avnet Shaw Division of Avnet, Inc.)*

FIGURE 14-28 Method of combining ceramic mold casting and wax pattern casting to produce the complex vanes of an impeller. *(Left)* A wax pattern is added to a metal pattern. After the metal pattern is withdrawn and the wax pattern is melted out *(center)*, the complex casting can be poured *(right)*. *(Courtesy of Avnet Shaw Division of Avnet, Inc.)*

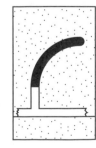
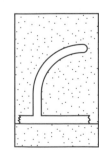

RUBBER-MOLD CASTING

Several types of artificial elastomers are available that can be compounded in liquid form and then poured over a pattern to form a semirigid mold. The molds are sufficiently flexible to permit stripping from an intricate pattern or patterns with reentrant surfaces. Unfortunately, rubber molds are suitable only for small castings and low-melting-point materials. Wax patterns for investment casting can be made in this manner, as well as finished castings of plastics and metals which can be poured at temperatures below 250°C (500°F).

■ 14.5 EXPENDABLE-MOLD PROCESSES USING SINGLE-USE PATTERNS

INVESTMENT CASTING

While *investment casting* is actually a very old process and has been performed by dentists and jewelers for a number of years, it was not until the end of World War II that it attained any degree of industrial importance. Developments and demands in the aerospace industry, such as rocket components and jet engine turbine blades, required high-precision complex shapes from high-melting-point metals that are not readily machinable. Investment casting offers almost unlimited freedom in both the complexity of shapes and types of materials that can be cast. As experience and capabilities developed, so did the market, and at present, millions of investment castings are produced each year.

Investment casting uses the same type of binder and molding aggregate as the ceramic molding process, and typically involves the following steps:

1. *Produce a master pattern.* The pattern is a modified replica of the desired product made from metal, wood, plastic, or some other easily worked material.

2. *From the master pattern, produce a master die.* This can be made from low-melting-point metal, steel, or possibly even wood. If low-melting-point metal is used, the die can often be cast directly from the master pattern. Steel dies may be machined directly, eliminating the need for step 1. Rubber molds can also be made from the master pattern.

3. *Produce wax patterns.* Patterns are made by pouring molten wax into the master die, or injecting it under pressure, and allowing it to harden. Plastic and frozen mercury have also been used as pattern material. When cores are required, they can be made from soluble wax or ceramic. The soluble wax cores are dissolved from the patterns prior to investment coating, while the ceramic cores remain as part of the wax pattern and are ultimately removed from the cast part during cleaning.

4. *Assemble the wax patterns onto a common wax sprue.* The individual wax patterns are attached to a central sprue and runner system by means of heated tools and melted wax. In some cases, several pattern pieces may first be united to form a complex, single pattern that if made in one piece, could not be withdrawn from a master die. The result of this assembly is a pattern cluster, or *tree*.

5. *Coat the cluster with a thin layer of investment material.* This step is usually accomplished by dipping the cluster into a watery slurry of finely ground refractory material. A thin but very smooth layer of investment material is deposited onto the wax pattern, ensuring a smooth surface and good detail in the final product.

6. *Produce the final investment around the coated cluster.* After the initial layer is formed, the cluster can be redipped, but this time the wet ceramic is coated with a layer of sand and allowed to dry. This process can be repeated until the investment coating is the desired thickness (typically 5 to 15 mm or $\frac{3}{16}$ to $\frac{5}{8}$ in.). As an alternative, the single-coated cluster can be placed upside down in a flask and liquid investment material poured around it. The flask is then vibrated to permit the escape of entrapped air and to settle the investment material around the cluster.

7. *Allow the investment to fully harden.*

8. *Melt or dissolve the wax pattern to remove it from the mold.* This is generally accomplished by placing the molds upside down in an oven, where the wax melts and runs out, and any residue subsequently vaporizes. This step is the most distinctive feature of the process, because it enables a complex pattern to be removed from a single-piece mold. Extremely complex shapes with reentrant sections can be readily cast.

   (*Note:* In the early years of the process, only small parts were cast. When the molds were placed in the oven, the molten wax was absorbed into the porous investment. Because the wax "disappeared," the process was called the *lost-wax process*, and the name is still used occasionally.)

9. *Preheat the mold in preparation for pouring.* Heating to 550 to 1100°C (1000 to 2000°F) ensures complete removal of the mold wax, cures the mold to give added strength, and allows the molten metal to retain its heat and flow more readily into all of the thin sections. It also gives better dimensional control because the mold and the metal will shrink together during cooling.

10. *Pour the molten metal.* Various methods, beyond simple pouring, can be used to ensure complete filling of the mold, especially when complex, thin sections are involved. Among these methods are the use of positive air pressure, evacuation of the air from the mold, and some form of centrifugal pouring process.

11. *Remove the casting from the mold.* This is accomplished by breaking the mold away from the casting. Techniques include mechanical vibration and high-pressure water.

Figure 14-29 depicts the investment procedure where the investment material fills the entire flask. Figure 14-30 shows the shell-investment method. Table 14-6 summarizes the features of investment casting.

Investment casting is a complex process and tends to be rather expensive. Its unique advantages, however, often justify its use, and many of the steps can be automated. Extremely complex shapes can be cast as a single piece. Thin sections, down to 0.40 mm (0.015 in.), can be produced. Excellent dimensional tolerances can be obtained in combination with very smooth surfaces. Machining can often be completely eliminated or greatly reduced. Where machining is required, allowances of as little as 0.4 to 1 mm (0.015 to 0.040 in.)

FIGURE 14-29 Investment casting steps for the flask-cast method. (*Courtesy of Investment Casting Institute, Dallas, Texas.*)

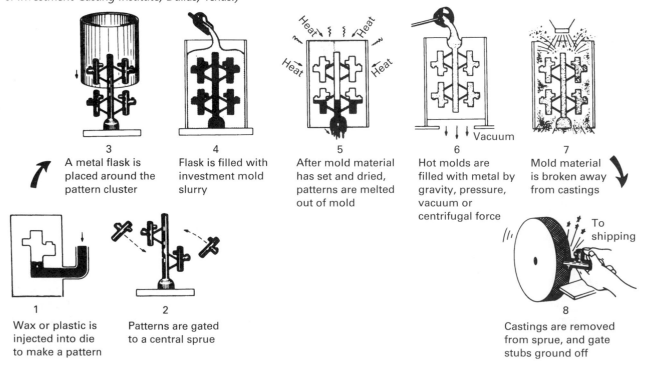

3
A metal flask is placed around the pattern cluster

4
Flask is filled with investment mold slurry

5
After mold material has set and dried, patterns are melted out of mold

6
Hot molds are filled with metal by gravity, pressure, vacuum or centrifugal force

7
Mold material is broken away from castings

1
Wax or plastic is injected into die to make a pattern

2
Patterns are gated to a central sprue

8
Castings are removed from sprue, and gate stubs ground off

To shipping

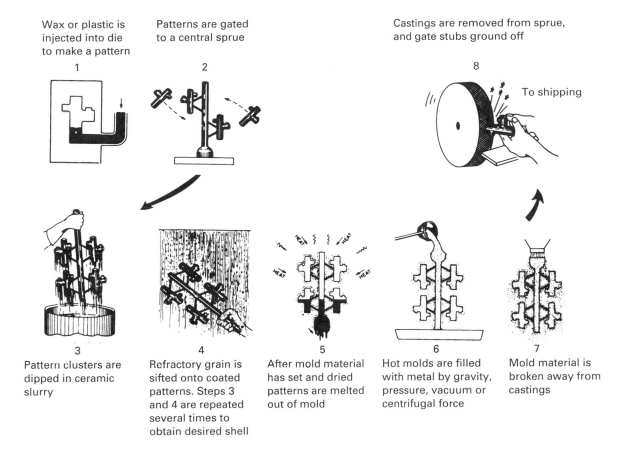

Wax or plastic is injected into die to make a pattern

Patterns are gated to a central sprue

Castings are removed from sprue, and gate stubs ground off

To shipping

3 Pattern clusters are dipped in ceramic slurry

4 Refractory grain is sifted onto coated patterns. Steps 3 and 4 are repeated several times to obtain desired shell

5 After mold material has set and dried patterns are melted out of mold

6 Hot molds are filled with metal by gravity, pressure, vacuum or centrifugal force

7 Mold material is broken away from castings

FIGURE 14-30 Investment casting steps for the shell-casting procedure. *(Courtesy of Investment Casting Institute, Dallas, Texas.)*

TABLE 14-6.	Investment Casting

Process: A refractory slurry is formed around a wax or plastic pattern and allowed to harden. The pattern is then melted out and the mold is baked. Molten metal is poured into the mold and solidifies. The mold is then broken away from the casting.

Advantages: Excellent surface finish; high dimensional accuracy; almost unlimited intricacy; almost any metal can be cast; no flash or parting line concerns.

Limitations: Costly patterns and molds; labor costs can be high; limited size.

Common metals: Just about any castable metal. Aluminum, copper, and steel dominate; also performed with stainless steel, nickel, magnesium, and the precious metals.

Size limits: As small as 3 g $\left(\frac{1}{10} \text{ oz}\right)$ but usually less than 5 kg (10 lb)

Thickness limits: As thin as 0.06 cm (0.025 in.), but less than 7.5 cm (3.0 in.)

Typical tolerances: 0.01 cm for the first 2.5 cm (0.005 in. for the first inch) and 0.002 cm for each additional cm (0.002 in. for each additional in.)

Draft allowance: None required

Surface finish: 1.3–4 microns (50 to 125 μin.) rms

---

are usually ample. As-cast surfaces are generally very smooth. These advantages are especially attractive when making products from the difficult-to-machine metals.

While most investment castings are less than 10 cm (4 in.) in size and weigh less than $\frac{1}{2}$ kg (1 lb), castings up to 1 m (36 in.) and 35 kg (80 lb) have been produced. Products ranging from stainless steel and titanium golf club heads to superalloy turbine blades have become quite routine. Some typical investment castings are shown in Figure 14-31. From the figure, one will note that a high degree of shape complexity is a common characteristic.

COUNTER-GRAVITY INVESTMENT CASTING

Counter-gravity investment casting turns the pouring process upside down. In one variation of the process, a ceramic shell mold is placed in a mold chamber with the sprue end down. The open end of the chamber is set against a seal, and the open end of the sprue is lowered into the molten pool. A vacuum is then induced within the chamber. As the

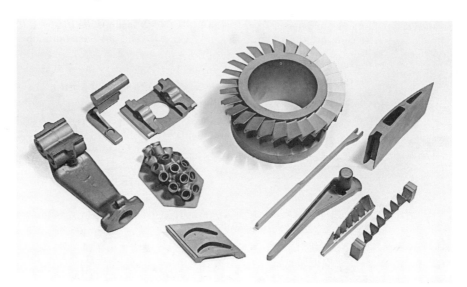

FIGURE 14-31 Typical parts produced by investment casting. *(Courtesy of Haynes Stellite Company.)*

air is withdrawn from both the chamber and the shell mold, the vacuum draws metal through the central sprue and into the mold. After the castings have solidified, the vacuum is released, and the excess molten metal flows back into the furnace. This technique is often called the *Hitchiner* process since it was developed by the Hitchiner Manufacturing Company in Milford, New Hampshire.

Similar results can be achieved by using a low-pressure inert gas to drive the molten metal upward into the mold. For a more complete picture of this approach, see the discussion on low-pressure permanent mold casting in Chapter 15.

The counter-gravity processes have a number of distinct advantages. Because the molten metal flows back from the sprue into the melt, there is a more efficient use of molten metal. The metal that flows into the mold is withdrawn from below the surface and flows with little turbulence. Since the gating system does not need to control turbulence, simpler gating systems can be used, further reducing the amount of metal that does not become product. In the counter-gravity process, between 60 and 95% of the withdrawn metal becomes cast product, compared to a 15 to 50% level for gravity poured castings. This metal entering the mold is free of slag and dross, and has a very low level of inclusions. The pressure differential allows the metal to flow into thinner sections. In addition, lower melting temperatures can be used, resulting in improved grain structure and companion properties.

FULL-MOLD AND LOST-FOAM CASTING

Several limitations are common to most of the casting processes that have been presented. Some form of pattern is usually required, and this pattern may be costly to design and fabricate. The pattern cost may be hard to justify when the number of identical castings is rather small. In addition, the pattern is usually withdrawn from the mold, and this withdrawal often requires some form of design modification or compromise, division into multiple pieces, or special molding procedures. Investment casting overcomes the withdrawal limitations through the use of patterns that can be removed by melting and vaporization. Unfortunately, this process also has its set of limitations, including a large number of individual operations and the need to remove the investment (i.e., mold) material from the finished castings.

In the *full-mold* and *lost-foam* processes, the pattern is made of *expanded polystyrene*, and remains in the mold during the pouring of the metal. When the molten metal is poured, the polystyrene melts and burns, and the metal fills the space that was occupied by the pattern.

When small quantities are required, the pattern can be hand cut or machined from pieces of foamed polystyrene (a material similar to that used in Styrofoam drinking cups). This material is extremely light in weight (19 g/l or 1.2 lb/ft^3) and can be cut by a number of methods, including ones as simple as an electrically heated wire. Foamed material in the form of a pouring basin, sprue, runner segments, and risers can be attached with hot melt glue to form a complete gating and pattern assembly. Small products can be assembled into clusters or trees, similar to investment casting.

When larger quantities are desired, a metal mold or die is used to mass-produce the single-use patterns. Hard beads of polystyrene are first preexpanded and stabilized. The preexpanded beads are then injected into a preheated metal die or mold (usually made from aluminum) at low pressure, where a steam cycle causes them to further expand, fill the die, and fuse. After cooling within the mold, the result is a pattern that is about 2.5% polymer and 97.5% air. Pattern dies can be very complex and large quantities of patterns can be produced accurately and rapidly. If size or complexity is great, the pattern can be divided into multiple segments, which are then assembled by gluing.

Once the polystyrene pattern has been produced and a polystyrene gating system has been attached, there are various options for the completion of the mold. In the *full-mold process*, shown schematically in Figure 14-32, green sand or some type of chemically bonded sand is compacted around the pattern and gating system, taking care not to crush or distort it. In the *lost-foam* technique, shown in Figure 14-33, the polystyrene assembly is first dipped into a water-based ceramic that wets the surface and forms a thin refractory coating (rigid enough to prevent mold collapse during pouring, but thin and sufficiently permeable to permit the escape of the molten and gaseous pattern material). After the coating dries, the pattern is then suspended in a flask and surrounded by fine unbonded sand that is vibrated into place. During the pour, the molten metal vaporizes and replaces the expanded polystyrene pattern, while the coating contains the metal, isolating it from the loose, unbonded sand. After cooling, the sand is then dumped from the flask, leaving the casing and attached gating system. Figure 14-34 shows the series of operations used in producing a rather complex lost-foam casting.

The full-mold and lost-foam processes (known collectively as *evaporative-pattern casting*) can be used for castings of any size, and can employ both ferrous and nonferrous metals (although nonferrous is most common). Because of the reduced pattern cost, small quantities can often be produced in an economical manner. Since the pattern need not be withdrawn, no draft is required in the design, and the processes are attractive for complex shapes that would ordinarily require cores, loose-piece patterns, or

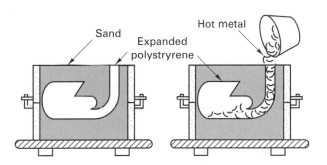

FIGURE 14-32 Schematic of the full mold process. (*Left*) An uncoated expanded polystyrene pattern is surrounded by bonded sand to produce a mold. (*Right*) Hot metal progressively vaporizes the expanded polystyrene pattern and fills the resulting cavity.

FIGURE 14-33 Schematic of the lost-foam casting process. In this process, the polystyrene pattern is dipped in a ceramic slurry, and the coated pattern is then surrounded with loose, unbonded sand.

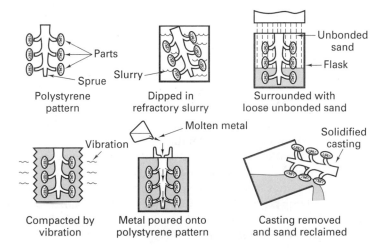

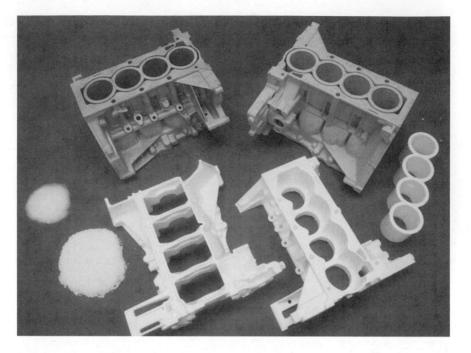

FIGURE 14-34 The stages of a lost-foam casting, proceeding counterclockwise from the lower left: polystyrene beads → expanded polystyrene pellets → three foam pattern segments → an assembled and dipped polystyrene pattern → a finished metal casting which is a metal duplicate of the polystyrene pattern. *(Courtesy of Saturn Corporation, Spring Hill, TN.)*

TABLE 14-7.   Lost-Foam Casting

Process: A pattern containing a sprue, runners, and risers is made from single or multiple pieces of foamed plastic, such as polystyrene. It is dipped in a ceramic material, dried, and positioned in a flask, where it is surrounded by loose sand. Molten metal is poured directly onto the pattern, which vaporizes and is vented through the sand.
Advantages: Almost no limits on shape and size; most metals can be cast; no draft is required and no flash is present (no parting lines).
Limitations: Pattern cost can be high for small quantities; patterns are easily damaged or distorted because of their low strength.
Common metals: Aluminum, iron, steel, and nickel alloys; also performed with copper and stainless steel.
Size limits: 0.5 kg to several thousand kg (1 lb to several tons)
Thickness limits: As small as 2.5 mm (0.1 in.) with no upper limit
Typical tolerances: 0.003 cm/cm (0.003 in./in.) or less
Draft allowance: None required
Surface finish: 2.5–25 microns (100–1000 μin.) rms

extensive finish machining. Because of the high precision and smooth surface finish, machining and finishing operations can be reduced or totally eliminated. Cores and parting lines are not required, eliminating the companion surface lines or fins. Metal yield (product weight versus the weight of poured metal) is high, and the back-up sand is often directly reusable. For many castings, risers are not required. As the metal moves into the pattern, it loses heat due to the melting and volatilizing of the foam. Thus the material farthest from the gate is the coolest, and solidification proceeds in a directional manner back to the gate (eliminating the need for a riser). The process can produce parts with high internal complexity, and is an attractive means of replacing multicomponent assemblies with a single casting. For these and other reasons, evaporative-pattern casting is growing rapidly in popularity and use. Table 14-7 summarizes the process and its capabilities. See also Figure 1-11.

■ 14.6 SHAKEOUT, CLEANING, AND FINISHING

In each of the casting processes presented in this chapter, the final step involves separating the castings from the molds and mold material. *Shakeout* operations are designed to separate the molds and sand from the flasks (i.e., containers), separate the castings from the molding sand, and separate or remove the cores from the castings. Punchout machines have been designed to force the entire contents of a flask (molding sand and casting) from the container. Vibratory machines can operate on entire flasks, or the extracted contents, and are available in a range of styles, sizes, and vibratory frequencies. Rotary separators can be used to separate the sand from castings, by placing the mold contents inside a slow-turning, large-diameter, rotating drum. The tumbling action breaks the gates and runners from the castings, crushes lumps of sand, extracts the cores, and empties internal pockets of sand. Because of possible damage to lightweight or thin-sectioned castings, rotary tumbling is usually restricted to cast iron, steel, and brass castings of reasonable thickness.

Processes, such as blast cleaning, can be used to remove adhering sand, oxide scale, and parting line burrs. Compressed air or centrifugal force is used to propel abrasive particles against the surfaces of the casting. The propelled abrasive media can be metal shot (usually iron or steel), fine aluminum oxide, glass beads, or naturally occurring quartz or silica. The blasting action is often combined with some form of tumbling or robotic manipulation. Additional finishing operations include grinding, trimming, and various forms of machining.

■ 14.7 SUMMARY

A number of processes have been developed to utilize the fluidity of a liquid and subsequent solidification as a means of producing a desired shape. Each has its own characteristic set of capabilities, advantages, and limitations, and the selection of the best method for a given application requires an understanding of all possible alternatives. Chapter 14 presented those processes that utilize a single-use (expendable) mold. Chapter 15 will supplement this knowledge with a survey of multiple-use mold processes.

■ KEY WORDS

ceramic mold	dry-sand mold	hot-box method	permeability	squeezing
chaplets	Eff-set process	hot tears	plaster mold	split pattern
cheek	evaporative pattern casting	investment casting	refractoriness	sprue
cohesiveness	expanded polystyrene	jolting	runner	stack molding
cold-box process	expendable mold	loose-piece pattern	sand expansion defects	standard rammed
collapsibility	flask	lost-foam casting	sand slinger	specimen
compactibility	follow board	lost-wax process	shakeout	V-process
compressive strength	full-mold casting	match-plate	Shaw process	vacuum molding
cope and drag pattern	gates	muller	shell molding	vertically parted
core	graphite mold	no-bake sand	skin-dried mold	flaskless molding
core-oil process	green-sand molding	one-piece pattern	slip jacket	water glass
core prints	hand ramming	pattern	sodium silicate–CO_2	
counter-gravity investment casting	hardness	penetration	molding	

■ REVIEW QUESTIONS

1. What are some of the factors that influence the selection of a specific casting process as a means of making a product?
2. What are the three basic categories of casting processes when classified by molds and patterns?
3. What metals are frequently cast into products?
4. Which type of casting is the most common and most versatile?
5. What is a casting pattern?
6. What are some of the materials used in making casting patterns? What features should be considered when selecting a pattern material?
7. What is the simplest and least expensive type of casting pattern?
8. What is a match plate and how does it aid molding?
9. How is a cope-and-drag pattern different from a matchplate pattern? When might this be attractive?
10. For what types of products might a loose-piece pattern be required?
11. What are the four primary requirements of a molding sand?
12. In what ways might a molding sand be a compromise material?

13. What is a muller, and what function does it perform?
14. What are some of the properties or characteristics of foundry sands that are evaluated by standard tests?
15. What is a standard rammed specimen for evaluating foundry sands, and how is it produced?
16. What is permeability, and why is it important in molding sands?
17. How does the ratio of water to clay affect the compressive strength of green sand?
18. How does the size and shape of the sand grains relate to molding sand properties?
19. What is a sand expansion defect, and what is its cause?
20. How can sand expansion defects be minimized?
21. What features can cause the penetration of molten metal between the grains of the molding sand?
22. What are hot tears and what can cause them to form?
23. Describe the distribution of sand density after compaction by jolting, squeezing, and a jolt–squeeze combination.
24. What is a slip jacket and how is it used?

25. How can the use of vertically parted flaskless molding reduce the number mold sections required to produce a series of castings?
26. How might extremely large molds be made?
27. What are some of the limitations or problems associated with green sand as a mold material?
28. What restricts the use of dry sand molding?
29. What are some of the advantages and limitations of the sodium silicate–CO_2 process?
30. What is the primary feature of no-bake sands?
31. What material serves as the binder in the shell-molding process, and how is it cured?
32. Why do shell molds have excellent permeability and collapsibility?
33. What is the sand binder in the V-process? The Eff-set process?
34. What types of geometric features might require the use of cores?
35. What is the primary limitation of green-sand cores?
36. What is the sand binder in the core-oil process, and how is it cured?
37. What is the binder in the hot-box core-making process?
38. What is the primary attraction of the cold-box coremaking process?
39. What is an attractive feature of shell-molded cores?
40. Why is it common for greater permeability, collapsibility, and refractoriness to be required of cores than for the base molding sand?

41. Why is it important that chaplets not completely melt during the pouring and solidification of a casting?
42. Why are plaster molds only suitable for the lower-melting-temperature nonferrous metals and alloys?
43. What is the primary performance difference between plaster and ceramic molds?
44. For what materials might a graphite mold be required?
45. What materials are used to produce the patterns for investment casting?
46. Why are investment casting molds generally preheated prior to pouring?
47. Why are investment castings sometimes called "lost-wax" castings?
48. What are some of the advantages of counter-gravity investment casting over the conventional gravity pour approach?
49. What are some of the benefits of not having to remove the pattern from the mold (as in investment casting, full-mold casting, and lost-foam casting)?
50. Since both use expanded polystyrene as a pattern, what is the primary difference between full-mold and lost-foam casting?
51. What are some of the attractive features of the evaporative pattern processes?
52. What are some of the objectives of a shakeout operation?
53. How might castings be cleaned after shakeout?

Chapter 14 CASE STUDY

Moveable and Fixed Jaw Pieces for a Heavy-duty Bench Vise

Figure CS-14 presents a cut-away sketch of the moveable and fixed jaw pieces of a heavy-duty vise that might see use in vocational schools, factories and machine shops. The vise is intended to have a rated maximum clamping force of 15 tons. The slide of the moving jaw has been designed to be a 2-in. box channel. The jaw width is 5 in., the maximum jaw opening is 6 in., and the depth of the throat is 4 in. The designer has elected to use replaceable, serrated jaws, and suggests that the material used for the receiving jaw pieces have a yield strength in excess of 35 ksi, with at least 15% elongation in a uniaxial tensile test (to assure that an overload or hammer impact would not produce brittle fracture).

1. Determine some possible combinations of material and process that could fabricate the desired shapes with the required properties. Of the alternatives presented, which would you prefer and why?

2. Would the components require some form of subsequent heat treatment? Consider the possibilities of stress relief, homogenization, or the establishment of desired final properties. What would you recommend?

3. One of your colleagues has suggested that the slides be finished with a coat of paint. Do you think a surface treatment is necessary or desirable for your selected material and process? If so, what would you recommend? If not, is a different surface treatment preferred, and what might that be?

Note: shaded surfaces have been produced by cross-sectional cuts

FIGURE CS-14

MULTIPLE-USE-MOLD CASTING PROCESSES

■ 15.1 INTRODUCTION

In each of the expendable-mold casting processes discussed in Chapter 14, a separate mold must be created for each pour. Variations in mold consistency, mold strength, moisture content, pattern removal, and other factors contribute to dimensional and property variation from casting to casting. In addition, the need to create and then destroy a separate mold for each pour results in rather low rates of production.

The multiple-use-mold casting processes overcome many of these limitations, but they, in turn, have their own assets and liabilities. Since the molds are generally made from metal, many of the processes are restricted to the casting of the lower-melting-point nonferrous metals and alloys. Part size is often limited, and the dies or molds can be rather costly.

■ 15.2 PERMANENT MOLD CASTING

In the *permanent mold casting* process, a reusable mold is machined from gray cast iron, steel, bronze, graphite, or other material. The molds are usually made in segments, which are often hinged to permit rapid and accurate opening and closing. After a preheating, a refractory or mold coating is applied to the preheated mold, sand or metal cores are positioned, the mold is clamped shut, and molten metal is poured in and flows through the feeding system into the mold cavity by simple gravity flow. After solidification, the mold is opened and the product is removed. Since the heat from the previous cast is usually sufficient to maintain mold temperature, the process can be immediately repeated, with a single refractory coating serving for several pouring cycles. Aluminum-, magnesium-, zinc-, lead-, and copper-based alloys are the metals most frequently cast. If graphite is used as the mold material, iron and steel castings can also be made by the permanent mold process.

Numerous advantages can be cited for the permanent mold process. The mold is reusable, and a good surface finish is obtained if the mold is in good condition. Dimensional accuracy can often be held to within 0.25 mm (0.010 in.). Directional solidification can be promoted by selectively heating or chilling various portions of the mold or by varying the thickness of the mold wall. The result is a sound, relatively defect-free casting with good mechanical properties. The faster cooling rates of the metal mold produce stronger products than would result from a sand casting process. Expendable sand cores or *retractable metal cores* can be used to increase the complexity of the casting. Multiple cavities can often be included in a single mold.

On the negative side, the process is generally limited to the lower-melting-point alloys, and the high tooling (mold) costs make low production runs prohibitively expensive. When applied to steels or cast irons, mold life tends to be extremely short. For the low-temperature metals, the mold life is limited because of erosion by the molten metal and thermal fatigue, but typically ranges from 10,000 to 120,000 cycles. The actual mold life varies with:

1. *Alloy being cast.* The higher the melting point, the shorter the mold life.

2. *Mold material.* Gray cast iron has about the best resistance to thermal fatigue and machines easily. Thus it is used most frequently for permanent molds.

3. *Pouring temperature.* Higher pouring temperatures reduce mold life, increase shrinkage problems, and induce longer cycle times.

4. *Mold temperature.* If the temperature is too low, misruns are produced and high temperature differences form in the mold. If the temperature is too high, excessive cycle times result and mold erosion is aggravated.

5. *Mold configuration.* Differences in section sizes of either the mold or the casting can produce temperature differences within the mold and reduce its life.

Mold complexity is often restricted because the rigid cavity offers no collapsibility to compensate for shrinkage of the casting. As a best alternative, it is common practice to open the mold and remove the casting immediately after solidification. This prevents the formation of hot tears that may form if the product is restrained during the shrinkage that accompanies cooldown.

Permanent molds are usually heated at the beginning of a run and continuous operation then maintains the mold at a fairly uniform elevated temperature. This minimizes the degree of thermal fatigue, facilitates metal flow, and controls the cooling rate of the metal being cast. Since the mold temperature rises when a casting is produced, it may be necessary to provide a mold-cooling delay before the cycle is repeated. Refractory washes or graphite coatings can be applied to the mold walls to prevent the casting from sticking and prolong the mold life. When pouring cast iron, an acetylene torch is often used to apply a coating of carbon black to the mold.

Since the molds are not permeable, special provision must be made for *venting*. This is usually accomplished through the slight cracks between mold halves or by very small vent holes that permit the escape of trapped air but not the passage of molten metal. Since gravity is the only means of inducing metal flow, risers must still be employed to compensate for shrinkage, and yields are generally less than 60%. Both sand and retractable metal cores can be used to increase part complexity.

High-volume production is usually required to justify the high cost of the metal molds. Automated machines are frequently used to coat the mold, pour the metal, and remove the casting. Figure 15-1 shows a variety of automobile and truck pistons that were manufactured by the permanent mold process, which is summarized in Table 15-1.

FIGURE 15-1 Truck and car pistons, mass-produced by the millions using permanent mold casting. *(Courtesy of General Motors Corporation.)*

TABLE 15-1. Permanent Mold Casting

Process: Mold cavities are machined into mating metal die blocks, which are then preheated and clamped together. Molten metal is then poured into the mold and enters the cavity by gravity flow. After solidification, the mold is opened and the casting is removed.

Advantages: Good surface finish and dimensional accuracy; metal mold gives rapid cooling and fine-grain structure; multiple-use molds (up to 120,000 uses); metal cores or collapsible sand cores can be used.

Limitations: High initial mold cost; shape, size, and complexity are limited; yield rate rarely exceeds 60%, but runners and risers can be directly recycled; mold life is very limited with high-melting-point metals such as steel.

Common metals: Alloys of aluminum, magnesium, and copper are most frequently cast; irons and steels can be cast into graphite molds; alloys of lead, tin, and zinc are also cast.

Size limits: 100 grams to 75 kilograms (several ounces to 150 pounds).

Thickness limits: Minimum depends on material but generally greater than 3 mm $\left(\frac{1}{8}\text{ in.}\right)$; maximum thickness about 50 mm (2.0 in.).

Geometric limits: The need to extract the part from a rigid mold may limit certain geometric features. Uniform section thickness is desirable.

Typical tolerances: 0.4 mm for the first 2.5 cm (0.015 in. for the first inch) and 0.02 mm for each additional centimeter (0.002 in. for each additional inch); 0.25 mm (0.01 in.) added if the dimension crosses a parting line

Draft allowance: 2°–3°

Surface finish: 2.5 to 7.5 μm (100–250 μin.) rms

SLUSH CASTING

Hollow castings can be produced by a variant of permanent mold casting known as *slush casting.* Hot metal is poured into the metal mold and is allowed to cool until a shell of the desired thickness has formed. The mold is then inverted and the remaining liquid is poured out. The resulting casting is a hollow shape with good surface detail but variable wall thickness. Common applications include the casting of ornamental objects such as candlesticks, lamp bases, and statuary from the low-melting-temperature metals.

LOW-PRESSURE PERMANENT MOLD CASTING

In *low-pressure permanent mold* (LPPM) casting, illustrated in Figure 15-2, a low-pressure gas (3 to 15 psi) acts on the surface of a molten metal bath. In response, molten metal rises up a refractory tube and enters the metal mold or gating system from the bottom. The metal is exceptionally clean, since it flows from the center of the melt and is fed directly into the mold (a distance of about 10 cm or 3 to 4 in.), never passing through the atmosphere. (*Note:* This is particularly attractive for maintaining the cleanliness of rapidly oxidizing metals such as aluminum.) By controlling the pressure, the mold can be filled in a controlled, nonturbulent manner, which further minimizes gas porosity and dross formation.

FIGURE 15-2 Schematic of the low-pressure permanent mold process. *(Courtesy of Amsted Industries.)*

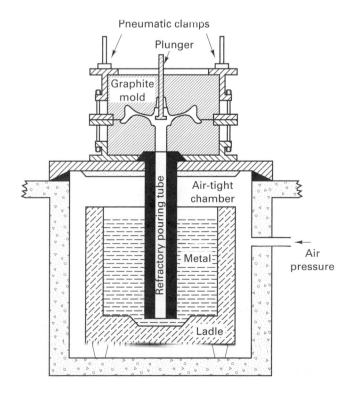

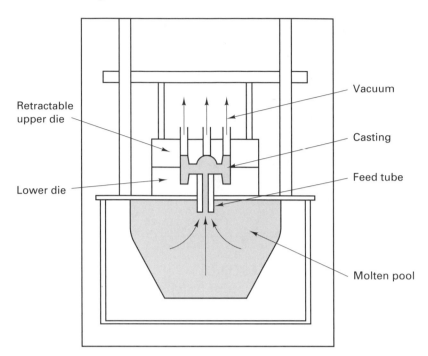

FIGURE 15-3 Schematic illustration of vacuum permanent mold casting.

Mold cooling is designed to promote directional solidification from the top down, and the applied pressure continually feeds molten metal to compensate for shrinkage. When solidification is complete and the pressure is released, the unused metal in the feed tube simply drops back into the crucible. Since the pressurized metal in the feed tube acts as a riser, no additional risers are required. This, coupled with the immediate reuse of the molten metal remaining in the feed tube, leads to yields that are generally greater than 85%. Nearly all low-pressure permanent mold castings are made from aluminum or magnesium, but some copper-base alloys are also used. Mechanical properties are typically 5% better than those of conventional permanent mold castings. Compared to conventional permanent molding, cycle times are somewhat longer, however.

VACUUM PERMANENT MOLD CASTING

Figure 15-3 depicts yet another variation of permanent mold casting, where a vacuum is used to draw material into the mold cavity. All of the benefits and features of the low-pressure process are retained, including the subsurface extraction of molten metal from the melt, the bottom feed to the mold, the minimal metal disturbance during pouring, the self-risering action, and the downward directional solidification. Thin-walled castings can be produced with high metal yield and excellent surface quality. Because of the vacuum, the cleanliness of the metal and the dissolved gas content are both superior to that of the low-pressure process. Final castings typically range from 0.2 to 5 kg (0.4 to 10 lb), and have mechanical properties that are 10 to 15% better than those of conventional permanent mold products.

■ 15.3 DIE CASTING

In the *die-casting* process, molten metal is forced into metal molds under pressures of several thousand pounds per square inch (tens of MPa) and held under this pressure during solidification. Because of the combination of metal molds or dies and high pressure, fine sections and excellent detail can be achieved, together with long mold life. Most die castings are made from nonferrous metals and alloys, and special zinc-, copper-, and aluminum-based alloys have been designed to have excellent properties when die cast. Ferrous-metal die castings are also possible, but are still considered somewhat uncommon. Production rates are high, the products exhibit good strength, shapes can be quite intricate, and dimensional precision and surface qualities are excellent. There is almost a complete elimination of subsequent machining. While targeted to the production of small to medium sized parts, the size and weight of die castings are increasing constantly. Parts weighing up to 10 kg (20 lb) and measuring up to 600 mm (24 in.) can now be made routinely.

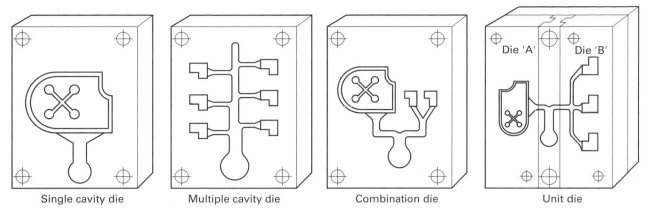

Single cavity die   Multiple cavity die   Combination die   Unit die

FIGURE 15-4 Some common types of die-casting dies. *(Courtesy of American Die Casting Institute, Inc., Des Plaines, Illinois.)*

Since the dies used in die casting are usually made from hardened hot-work tool steels (because cast iron cannot withstand the high pressures), they tend to be rather expensive. As shown in Figure 15-4, the dies may be relatively simple, containing only one or two mold cavities, or they may be complex, containing multiple cavities of the same or different products. To permit the removal of the casting, the rigid dies must separate into at least two pieces. It is not uncommon, however, for die-casting dies to be far more complicated, containing multiple sections that open and close in several different directions. In addition, the various die sections often contain water-cooling passages, retractable cores, and moving pins to knock out or eject the finished casting.

Die life is usually limited by wear (or erosion), which is strongly dependent on the temperature of the molten metal. Surface cracking can also occur in response to the large number of heating and cooling cycles applied to the die surfaces. If the rate of temperature change is the dominant feature, the problem is called *heat checking*. If the number of cycles is the primary cause, the problem is called *thermal fatigue*.

In the die-casting process, the water-cooled dies are usually lubricated and then clamped tightly together. The molten metal is then injected under pressure. Since high injection pressure tends to cause turbulence and air entrapment, the specified values of pressure and time of application vary considerably. There has been a trend toward the use of larger gates and lower injection pressure, followed by higher pressure after the mold has been filled completely and the metal has started to solidify. This cycle tends to reduce both the porosity and inclusion content of the finished casting. When solidification is complete, the dies separate, and ejector pins extract the finished casting along with its attached runners and sprues.

There are two basic types of die-casting machines. Figure 15-5 schematically illustrates the *hot-chamber*, or *gooseneck*, design. A gooseneck is partially submerged in a reservoir of molten metal. To begin the cycle, molten metal flows through an open port

FIGURE 15-5 Principal components of a hot-chamber die-casting machine. *(Courtesy of Noranda Sales Corp. Ltd.)*

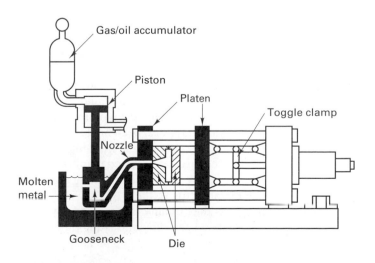

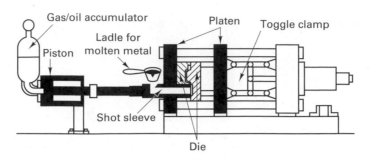

FIGURE 15-6 Principal components of a cold-chamber die-casting machine. *(Courtesy of Noranda Sales Corp. Ltd.)*

to fill the gooseneck. A mechanical plunger or air-injection system then forces the metal out of the gooseneck and into the die, where it rapidly solidifies.

Hot-chamber die-casting machines offer fast cycling times (up to about 15 cycles per minute) and the added advantage that the molten metal is injected from the same chamber in which it is melted (i.e., there is no handling or transfer of molten metal). Unfortunately, the hot chamber design cannot be used for the higher-melting-point metals, and it is unattractive for aluminum since there is a tendency for molten aluminum to pick up some iron as a result of the extended time in contact with the casting equipment. Hot-chamber machines, therefore, see primary use with zinc-, tin-, and lead-based alloys.

Cold-chamber machines are usually employed for the die casting of materials that are not suitable for the hot-chamber design. These include alloys of aluminum, magnesium, copper, and high-aluminum zinc. As illustrated in Figure 15-6, the metal is melted in a separate furnace, then transported to the die-casting machine, where a measured quantity is fed into an unheated shot chamber (or injection cylinder) and subsequently driven into the die by a hydraulic or mechanical plunger. The pressure is then maintained or increased during solidification. Since molten metal must be transferred to the chamber for each cycle, the cold-chamber process has a longer operating cycle than that of the hot-chamber machines. Nevertheless, productivity is still high.

Die-casting dies fill with metal so fast that there is little time for air in the mold cavity to escape, and the metal molds offer no permeability. Trapped air causes problems in the form of blow holes, porosity, or misruns. To minimize these defects, it is crucial that the dies be properly vented, usually by wide, thin (0.13 mm or 0.005 in.) vents along the parting line. The air must escape through the vents before they fill with molten metal, because the thin geometry results in rapid freezing and a plugging of the hole. The metal that solidifies in the vents must be trimmed off after the casting has been ejected. This can be done with special trimming dies that also serve to remove the sprues and runners.

No risers are used in the die-casting process since the high pressures ensure continuous feed of molten metal from the gating system into the casting. However, die castings frequently contain porosity, which was induced due to entrapped air or the turbulent mode of die filling. This porosity tends to be confined to the center of the casting, and smooth metal flow, good venting, and proper application of pressure can do much to minimize its formation. The rapidly solidified surface is usually harder and stronger than the slower-cooled interior, and is usually sound and suitable for plating or decorative applications.

In the *pore-free casting* process, air problems have been reduced by introducing oxygen into the mold before each die-casting shot. When the molten metal is injected, it reacts with the oxygen to form fine, dispersed oxide particles, virtually eliminating gas porosity. The products can be welded and heat-treated, and they possess greater strength than conventional die castings. Pore-free casting has been performed on aluminum, zinc, and lead alloys.

Zinc die castings have also been made by a process known as *heated-manifold direct-injection die casting* (also known as direct-injection die casting or runnerless die casting). Molten zinc is forced through a heated manifold and then through heated mininozzles directly into the die cavity, thereby eliminating the need for sprues, gates, and runners. Less cooling of the metal occurs, so the product surface is improved. Energy is conserved, scrap is reduced, and product quality is increased. Existing die-casting machines can be used with the addition of a heated manifold and modification of the various dies.

Sand cores cannot be used in die casting because the high pressures and flow rates cause the cores to either disintegrate or have excessive metal penetration. As a result, metal cores are required. Provisions must be made for their retraction, usually before the die is opened for removal of the casting. A close fit must be maintained between the halves of the die and also between the moving cores and die sections if metal is to be prevented from flowing into the gap. Since the cores are rigid, retraction motions must be either in a straight line or a circular arc. Loose core pieces (still metal) can be inserted into the die at the beginning of each cycle and then removed from the casting after it has been ejected from the die. This procedure permits more complex shapes to be cast, such as holes with internal threads, but production rate is slowed and costs increase.

The die-casting process also permits the incorporation of cast-in *inserts*. These include prethreaded bosses, electrical heating elements, threaded studs, and high-strength bearing surfaces that are positioned in the die before the lower-melting-temperature metal is injected. Suitable recesses must be provided in the die for positioning and support, and the casting cycle tends to be slowed by the additional operations.

Smooth surfaces and excellent dimensional accuracy are attractive features of die casting. For aluminum-, magnesium-, zinc-, and copper-based alloys, linear tolerances of 3 mm/m (0.003 in./in.) are not uncommon. Thinner sections can be cast than with either sand or permanent mold casting. The actual value of the minimum section thickness and draft depend on the type of metal, as follows:

Metal	Minimum Section	Minimum Draft
Aluminum alloys	0.89 mm (0.035 in.)	1 : 100 (0.010 in./in.)
Brass and bronze	1.27 mm (0.050 in.)	1 : 80 (0.015 in./in.)
Magnesium alloys	1.27 mm (0.050 in.)	1 : 100 (0.010 in./in.)
Zinc alloys	0.63 mm (0.025 in.)	1 : 200 (0.005 in./in.)

Because of the features in the preceding table, most die castings require no finish machining except for the removal of the small amount of excess metal fin, or flash, around the parting line and the possible drilling or tapping of holes. Production rates are high, and a set of dies can produce many thousands of castings without significant change in dimensions. While die casting is most economical for large production volumes, quantities as low as 2000 can be justified if extensive secondary machining or surface finishing can be eliminated.

Thin-wall zinc die casting is now considered to be a significant competitor to plastic injection molding. The die castings are stronger, stiffer, more dimensionally stable, and more heat resistant. In addition, the metal parts are more resistant to ultraviolet radiation, weathering, and stress caracking when exposed to various reagents.

Table 15-2 summarizes the features of the die-casting process. Figure 15-7 presents a variety of aluminum and zinc die castings. Table 15-3 presents a comparison of the properties of die cast alloys with the properties of other engineering materials.

TABLE 15-2.   Die Casting

Process: Molten metal is injected into closed metal dies under pressures ranging from 10 to 175 MPa (1500–25,000 psi). Pressure is maintained during solidification, after which the dies separate and the casting is ejected along with its attached sprues and runners. Cores must be simple and retractable and take the form of moving metal segments.

Advantages: Extremely smooth surfaces and excellent dimensional accuracy; rapid production rate; product tensile strengths as high as 415 MPa (60 ksi).

Limitations: High initial die cost; limited to high-fluidity nonferrous metals; part size is limited; porosity may be a problem; some scrap in sprues, runners, and flash, but this can be directly recycled.

Common metals: Alloys of aluminum, zinc, magnesium, and lead; also possible with alloys of copper and tin.

Size limits: Less than 30 grams (1 oz) up through about 7 kg (15 lb) most common.

Thickness limits: As thin as 0.75 mm (0.03 in.), but generally less than 13 mm $\left(\frac{1}{2}\text{ in.}\right)$.

Typical tolerances: Varies with metal being cast; typically 0.1 mm for the first 2.5 cm (0.005 in. for the first inch) and 0.02 mm for each additional centimeter (0.002 in. for each additional inch).

Draft allowances: 2°

Surface finish: 1–2.5 μm (40–100 μ.in.) rms.

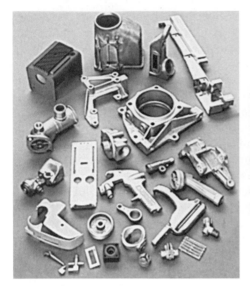

FIGURE 15-7   Variety of aluminum (*left*) and zinc (*right*) die castings. *(Courtesy of Yoder Die Casting Corporation.)*

TABLE 15-3.   Comparison of Properties (Die Cast Metals vs. Other Engineering Materials)

Material	Yield Strength		Tensile Strength		Elastic Modulus	
	MPa	ksi	MPa	ksi	GPa	10^6 psi
Die Cast Alloys						
360 Aluminum	170	25	300	44	71	10.3
380 Aluminum	160	23	320	46	71	10.3
AZ91D Magnesium	160	23	230	34	45	6.5
Zamak 3 zinc (AG4OA)	221	32	283	41	—	—
Zamak 5 zinc (AC41A)	269	39	328	48	—	—
ZA-8 (zinc–aluminum)	283–296	41–43	365–386	53–56	85	12.4
ZA-27 (zinc–aluminum)	359–379	52–55	407–441	59–64	78	11.3
Other Metals						
Steel sheet	172–241	25–35	276	40	203	29.5
HSLA steel sheet	414	60	414	60	203	29.5
Powdered iron	483	70		—	120–134	17.5–19.5
Plastics						
ABS	—	—	55	8	7	1.0
Polycarbonate	—	—	62	9	7	1.0
Nylon 6[1]	—	—	152	22	10	1.5
PET[1]	—	—	145	21	14	2.0

[1]30% glass reinforced

■ 15.4   SQUEEZE CASTING AND SEMISOLID CASTING

In the *squeeze-casting* process, molten metal is introduced into a metal mold die cavity through large gate areas and at slower metal velocities to avoid turbulence. When the cavity has filled, high pressure (20 to 175 MPa, or 3,000 to 25,000 psi) is then applied and maintained during solidification. Intricate shapes can be produced at lower pressures than would normally be required for hot or cold forging. Both retractable and disposable cores can be used to create holes and internal passages. Gas and shrinkage porosity are substantially reduced, and mechanical properties are enhanced. The process is most commonly applied to aluminum and magnesium castings. A further adaptation of the process can be used to produce metal–matrix composites by forcing the pressurized liquid (usually aluminum or magnesium) around or through foamed or fiber reinforcements that have been positioned in the mold.

For most alloy compositions, there is a freezing range where liquid and solid coexist. Another die-casting modification involves the injection of this *semisolid* material into reuseable metal dies under high injection pressures. A special class of semisolids, *thixotropic material*, can be handled mechanically, like a solid, but flows like a liquid when agitated or squeezed. Bars of stirred, cast metal are cut to prescribed lengths, reheated to the semisolid state (where the material is about 40% liquid and 60% solid), transferred to the shot chamber of a cold-chamber die casting machine, and injected under pressure. An alternative is to begin with liquid metal and cool it to the semisolid state while stirring or shearing the material. The shear forces break up the dendrites, which results in a slurry of rounded particles of solid in a liquid melt that can be directly injected into a casting die. Because the slurry contains no superheat and is already partially solidified, it freezes quickly. In addition, the solid component increases the viscosity and makes it more difficult to flow material into the die. Successful castings are possible with about 30% solid content in the slurry.

The absence of turbulent flow during the casting operation minimizes gas pickup and entrapment. Since the material is already partially solid and completes its solidification and subsequent cooling while under pressure, solidification shrinkage and related porosity are reduced. The result is a high-quality intricate part with good finish and precision and reduced warpage or distortion. Because of the absence of entrapped porosity, high-temperature heat treatments can be used to enhance strength, such as the T6 age-hardening treatment in aluminums.

■ 15.5 CENTRIFUGAL CASTING

In *centrifugal casting* (a category that includes true centrifugal casting, semicentrifugal casting, and centrifuging), the inertial forces of rotation or spinning are used to distribute the molten metal into the mold cavity or cavities. In *true centrifugal casting*, a dry-sand, graphite, or metal mold rotates about either a horizontal or vertical axis at speeds of 300 to 3000 rpm. As the molten metal is introduced, it is flung to the surface of the mold, where it solidifies into some form of hollow product. The exterior profile is usually round (as with pipes and gun barrels), but hexagons and other symmetrical shapes are also possible.

No core or mold is needed to shape the interior, which will always have a round profile because the molten metal is uniformly distributed by the centrifugal forces. When rotation is about the horizontal axis, as illustrated in Figure 15-8, the inner surface is always cylindrical. If the mold is oriented vertically, gravitational forces cause the inner surface to become a section of a parabola. As shown in Figure 15-9, the exact shape is a function of the speed of rotation. Wall thickness can be controlled by varying the amount of metal that is introduced into the mold.

During the rotation, the metal is forced against the outer walls of the mold with considerable force, and solidification begins at the outer surface. Centrifugal force continues to feed molten metal as solidification progresses inward. Since the process compensates for shrinkage, no risers are required. The final product has a strong, dense exterior with all the lighter impurities (including dross and pieces of the refractory mold coating) collecting on the inner surface of the casting. For some applications, this surface may be removed by a light boring operation.

FIGURE 15-8 Schematic representation of a horizontal centrifugal casting machine. *(Courtesy of American Cast Iron Pipe Company.)*

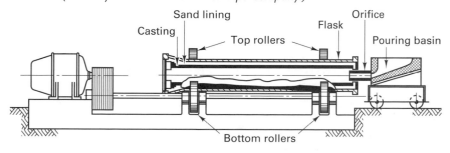

FIGURE 15-9 Vertical centrifugal casting, showing the effect of rotational speed on the shape of the inner surface. Paraboloid A results from fast spinning, whereas slow spinning will produce paraboloid B.

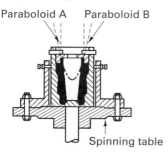

FIGURE 15-10 Electrical products (collector rings, slip rings, and rotor end rings) that have been centrifugally cast from aluminum and copper. *(Courtesy of The Electric Materials Company.)*

TABLE 15-4. Centrifugal Casting

Process: Molten metal is introduced into a rotating sand, metal, or graphite mold and held against the mold wall by centrifugal force until it is solidified.
Advantages: Can produce a wide range of cylindrical parts, including ones of large size; good dimensional accuracy, soundness, and cleanliness.
Limitations: Shape is limited; spinning equipment can be expensive.
Common metals: Iron, steel, stainless steel, and alloys of aluminum, copper, and nickel.
Size limits: Up to 3 m (10 ft) in diameter and 15 m (50 ft) in length.
Thickness limits: Wall thickness 2.5 to 125 mm (0.1–5 in.).
Typical tolerances: O.D. to within 2.5 mm (0.1 in.); I.D. to about 4 mm (0.15 in.).
Draft allowance: 10 mm/m $\left(\frac{1}{8}\,\text{in./ft}\right)$
Surface finish: 2.5–12.5 μm (100–500 μin.) rms.

Pipe (up to 12 m, or 40 ft, in length), pressure vessels, cylinder liners, brake drums, and the starting material for bearing rings and the parts illustrated in Figure 15-10 can all be manufactured by centrifugal casting. The equipment is rather specialized and can be quite expensive for large castings. The permanent molds can also be expensive, but they offer a long service life, especially when coated with some form of refractory dust or wash. Since no sprues, gates, or risers are required, yields can be greater than 90%. Composite products can be made by the centrifugal casting of a second material on the inside surface of an already cast alloy. Table 15-4 summarizes the features of the centrifugal casting process.

■ 15.6 SEMICENTRIFUGAL CASTING

In *semicentrifugal casting* (Figure 15-11) the centrifugal force assists the flow of metal from a central reservoir to the extremities of a rotating symmetrical mold, which may be either expendable or multiple-use. The rotational speeds are usually lower than for

FIGURE 15-11 Schematic of a semicentrifugal casting process.

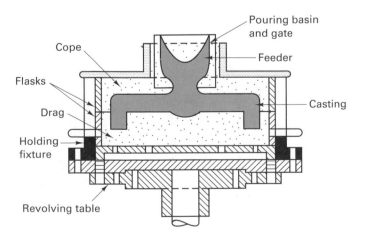

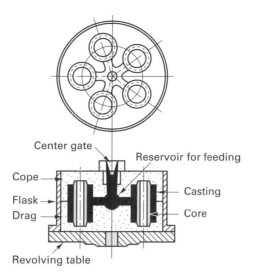

FIGURE 15-12 Schematic of a centrifuging process. Metal is poured into the central pouring sprue and spun into the various mold cavities. *(Courtesy of American Cast Iron Pipe Company.)*

true centrifugal casting, and it is not uncommon for several molds to be stacked on top of one another, being fed by a common pouring basin and sprue.

The central reservoir acts as a riser and must be large enough to ensure that it will be the last material to freeze. Since the lighter impurities concentrate in the center, the process is best used for castings where the central region will ultimately be hollow. Cores can be used to increase the complexity of the product.

■ 15.7 CENTRIFUGING

Centrifuging (Figure 15-12) uses centrifugal action to force the metal from a central pouring reservoir, through runners, into separate mold cavities that are offset from the axis of rotation. Relatively low rotational speeds are required to produce sound castings with thin walls and intricate shapes. Centrifuging is often used to assist in the pouring of investment casting trees.

In an adaptation of the centrifuging process, pewter, zinc, or wax can be cast into spinning rubber molds to produce products with close tolerances, smooth surfaces, and excellent detail. These can be finished products or the low-melting-point patterns for subsequent assembly onto investment casting trees.

■ 15.8 CONTINUOUS CASTING

As discussed in Chapter 6 and depicted in Figure 6-5, *continuous casting* is usually employed in the solidification of basic shapes that become the feedstock for deformation processes such as rolling and forging. By producing a special mold, continuous casting can also be used to produce long lengths of complex cross-section product, such as the one depicted in Figure 15-13. Since each product is simply a cutoff section of the continuous

FIGURE 15-13 Gear produced by continuous casting. (*Left*) As-cast material; (*right*) after machining. *(Courtesy of American Smelting and Refining Company.)*

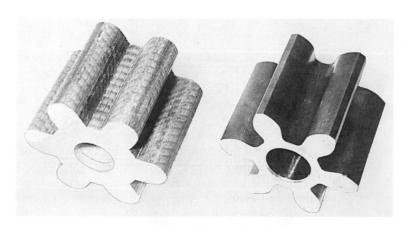

strand, a single mold is all that is required to produce a large number of pieces. Quality is high as well, since the metal can be protected from contamination during melting and pouring, and only a minimum of handling is required.

■ 15.9 ELECTROMAGNETIC (OR LEVITATION) CASTING

In *electromagnetic casting* the metal is contained and solidified in an electromagnetic field designed to counteract the gravitational forces and hydrostatic pressure of the metal column. Since there is no container, the surfaces of the molten metal are completely exposed and open to the direct impingement of water jets or spray. The intense stirring induced by the electromagnetic field helps to create a final structure that is homogeneous, equiaxed, and fine grained. The exterior surface is quite smooth, and the process can be automated and adapted for continuous casting.

■ 15.10 MELTING AND POURING

All casting processes begin with molten metal. Ideally, the molten metal should be available in an adequate amount, at the desired temperature, with the desired chemistry and minimum contamination. The melting furnace should be capable of holding material for an extended period of time without deterioration of quality, be economical to operate, and be capable of being operated without contributing to the pollution of the environment. Except for experimental or very small operations, virtually all foundries use cupolas, air furnaces, (also known as direct fuel-fired furnaces), electric-arc furnaces, electric resistance furnaces, or electric-induction furnaces. In locations such as fully integrated steel mills, molten metal may be taken directly from a steelmaking furnace and poured into casting molds. This practice is usually reserved for exceptionally large castings. For small operations, gas-fired crucible furnaces are common, but these have rather limited capacities.

Selection of the most appropriate melting procedure depends on such factors as (1) the temperature needed to melt and superheat the metal, (2) the alloy being melted and the form of available charge material, (3) the desired melting rate or quantity of metal, (4) the desired quality of the metal, (5) the availability and cost of various fuels, (6) the variety of metals or alloys to be melted, (7) whether melting is to be batch or continuous, (8) the required level of emission control, and (9) the various capital and operating costs.

The feedstock entering the melting furnace may take several forms. While prealloyed ingot may be purchased for remelt, it is not uncommon for the starting material to be a mix of commercially pure primary metal and commercial scrap, which also includes recycled gates, runners, sprues, and risers, as well as defective castings. Alloy elements can be added as either pure materials or master alloys that are high in a particular element but are designed to have both a lower melting point than the pure material and a density that allows for good mixing. Preheating the metal being charged is another common practice, and it can increase the melting rate by as much as 30%.

CUPOLAS

A significant amount of gray, nodular, and white cast iron is still melted in *cupolas*, although many foundries have converted to electric induction furnaces. A cupola is a refractory-lined, vertical steel shell into which alternating layers of coke, iron (pig iron and/or scrap), limestone or other flux, and possible alloy additions are charged and melted under forced-air draft. The operation is similar to that of a blast furnace, with the molten metal collecting at the bottom of the cupola to be tapped off either continuously or at periodic intervals.

Cupolas are simple and economical, can be obtained in a wide range of capacities, and can produce cast iron of excellent quality if the proper raw materials are used and good control is practiced. They are used exclusively for the melting of cast iron and can be operated either continuously or in a batch-type mode. Control of temperature and chemistry is somewhat difficult, however. The nature of the charged materials and the reactions that occur within the cupola can all affect the product chemistry. Moreover, by the time the final chemistry is determined through analysis of the tapped product, a substantial charge of material is already working its way through the furnace. Final chemistry adjustments, therefore, are often performed in the ladle, using the various techniques of ladle metallurgy discussed in Chapter 6.

Various methods can be used to increase the melting rate and improve the economy of operation. In a hot-blast cupola, the stack gases are put through a heat exchanger, enabling the incoming air to be preheated. Oxygen-enriched blasts can be used to further increase the temperature and accelerate the rate of melting. Plasma torches can be used to melt the iron scrap. The resulting melting rates can be quite high, such that it is not uncommon for a continuously operating cupola to produce as much as 120 tons of hot metal per hour.

INDIRECT FUEL-FIRED FURNACES (OR CRUCIBLE FURNACES)

Although limited in size and melting rate, *indirect fuel-fired furnaces* are often used to melt small batches of nonferrous metal. They usually take the form of crucibles or holding pots whose outer surface is heated by an external flame. Stirring action, temperature control, and chemistry control are poor, but these furnaces offer low capital cost. The containment crucibles are generally made from clay and graphite, silicon carbide, cast iron, or steel.

Crucible furnaces heated by electrical resistance heating can often provide better control of temperature and chemistry.

DIRECT FUEL-FIRED FURNACES OR REVERBERATORY FURNACES

Direct fuel-fired furnaces, also known as *reverberatory furnaces*, are similar to small open-hearth furnaces but are less sophisticated. As illustrated in Figure 15-14, a fuel-fired flame is passed directly over the pool of molten metal. Heat is transferred through radiant heating from the refractory roof and walls and convective heating from the hot gases. Capacity is much larger than the crucible furnace, but the operation is still limited to the batch melting of nonferrous metals and the holding of cast iron that has been previously melted in a cupola. The rate of heating and melting and the temperature and composition of the molten metal are all easily controlled.

ARC FURNACES

Arc furnaces are the preferred method of melting in many foundries because of (1) their rapid melting rates, (2) their ability to hold the molten metal for any desired period of time, and (3) the greater ease of incorporating pollution control equipment.

The basic features and operating cycle of a *direct-arc furnace* can be described with the aid of Figure 15-15. The top of the wide, shallow unit is first lifted or swung aside to permit the introduction of charge material. The top is then replaced, and the electrodes

FIGURE 15-14 Cross section of an air furnace. Hot combustion gases pass across the surface of a molten metal pool.

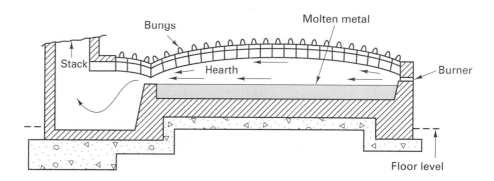

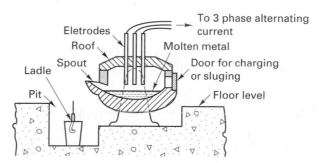

FIGURE 15-15 Schematic diagram of a three-phase electric arc furnace.

FIGURE 15-16 Electric arc furnace, tilted for pouring. *(Courtesy of Pittsburgh Lectromelt Furnace Corporation.)*

are lowered to create an arc between the electrodes and the metal charge. The path of the heating current is usually through one electrode, across an arc, through the metal charge, and back through another arc to another electrode.

Fluxing materials are usually added to provide a protective cover to the pool of molten metal. Reactions between the slag and the metal are efficient because of the large interface area and the fact that the slag is as hot as the metal. Because the metal is covered and can be maintained at a given temperature for long periods of time, arc furnaces can be used to produce high-quality metal of almost any desired composition. They are available in sizes up to about 200 tons (but capacities of 25 tons or less are most common), and up to 50 tons per hour can be melted conveniently in batch operations. Arc furnaces are generally used with ferrous alloys, especially steel, and provide good mixing and homogeneity to the molten bath. Unfortunately, the noise and level of particle emissions can be rather high, and the consumption of electrodes, refractories, and power results in high operating costs. Figure 15-16 shows the pouring of an electric arc furnace. Note the still-glowing electrodes at the top of the furnace.

INDUCTION FURNACES

Because of their very rapid melting rates and the relative ease of controlling pollution, electric *induction furnaces* have become another popular means of melting metal. There are two basic types of induction furnaces. The *high-frequency*, or *coreless* units, shown schematically in Figure 15-17, consist of a crucible surrounded by a water-cooled coil of copper tubing. A high-frequency electrical current passes through the coil, creating an alternating magnetic field. The varying magnetic field induces secondary currents in the metal being melted, which bring about a rapid rate of heating.

Coreless induction furnaces are used for virtually all common alloys, the maximum temperature being limited only by the refractory and the ability to insulate against heat loss. They provide good control of temperature and composition and are available in a range of capacities up to about 65 tons. Because there is no contamination from the heat source, they produce very pure metal. Operation is generally on a batch basis.

Low-frequency or *channel-type* induction furnaces are also seeing increased use. As shown in Figure 15-18, only a small channel is surrounded by the primary (current-carrying) coil. A secondary coil is formed by a loop, or channel, of molten metal, and all the metal is free to circulate through the loop and gain heat. Enough molten metal must be placed into the furnace to fill the secondary coil, with the remainder of the charge taking a variety of forms. The heating rate is very high and the temperature can be accurately controlled. As a result, channel-type furnaces are often preferred as holding furnaces, where the molten metal is maintained at a constant temperature for an extended period of time. Capacities can be quite large, up to about 250 tons.

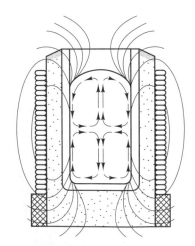

FIGURE 15-17 Schematic showing the basic principle of a coreless induction furnace.

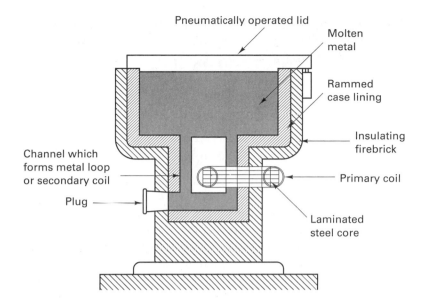

FIGURE 15-18 Cross section showing the principle of the low-frequency or channel-type induction furnace.

Labels in figure:
- Pneumatically operated lid
- Molten metal
- Rammed case lining
- Insulating firebrick
- Primary coil
- Laminated steel core
- Channel which forms metal loop or secondary coil
- Plug

■ 15.11 POURING PRACTICE

To transfer the metal from the melting furnace to the molds, some type of pouring device, or ladle, is usually required. The primary considerations for this operation are (1) to maintain the metal at the proper temperature for pouring, and (2) to assure that only high-quality metal is introduced into the molds. The specific type of *pouring ladle* is determined largely by the size and number of castings to be poured. In small foundries, a handheld, shank-type ladle is used for manual pouring. In larger foundries, either bottom-pour or teapot-type ladles are used, like the ones illustrated in Figure 13-7. These are often used in conjunction with a conveyor line that moves the molds past the pouring station. By extracting metal from beneath the surface, slag and other impurities that float on top of the melt are not permitted to enter the mold.

High-volume, mass-production foundries often use automatic pouring systems, such as the one shown in Figure 15-19. Molten metal is transferred from the main melting furnace to a holding furnace by overhead crane. A programmed amount of molten metal is further transferred into individual pouring ladles and is then poured automatically into the corresponding molds. Laser-based control units position the pouring ladle over the sprue and control the flow rate into the sprue cup.

FIGURE 15-19 Machine for automatic pouring of molds on a conveyor line. *(Courtesy of Roberts Corporation.)*

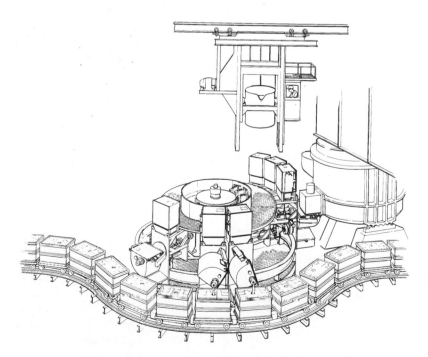

■ 15.12 CLEANING, FINISHING, AND HEAT TREATING OF CASTINGS

CLEANING AND FINISHING

After solidification and removal from molds, most castings require some additional cleaning and finishing. Specific operations may include all or several of the following:

1. Removing cores
2. Removing gates and risers
3. Removing fins, flash, and rough spots from the surface
4. Cleaning the surface
5. Repairing any defects

Cleaning and finishing operations can be quite expensive, so consideration should be given to their minimization when designing the product and selecting the specific method of casting. In addition, consideration should also be directed toward the possibility of automating the cleaning and finishing.

Sand cores can usually be removed by mechanical shaking. At times, however, they must be removed by chemically dissolving the core binder. On small castings, gates and risers can sometimes be knocked off. For larger castings, a cutting operation is usually required. Most nonferrous metals and cast irons can be cut with an abrasive cutoff wheel, power hacksaw, or band saw. Steel castings frequently require an oxyacetylene torch. Plasma arc cutting can also be used to remove sprues, gates, and risers.

The specific method of cleaning often depends on the size and complexity of the casting. After the gates and risers have been removed, small castings are often processed through tumbling barrels to remove fins, flash, and sand that may have adhered to the surface. Tumbling may also be done to remove cores and, in some cases, gates and risers. Metal shot or abrasive material is often added to the barrel to aid in the cleaning. Conveyors can be used to pass larger castings through special cleaning chambers were they are subjected to blasts of abrasive or cleaning material. Extremely large castings usually require manual finishing, using pneumatic chisels, portable grinders, and manually directed blast hoses.

While defect-free castings are always desired, flaws such as cracks, voids, and laps are not uncommon. In some cases, especially when the part is large and the production quantity is small, it may be more attractive to repair the part rather than change the pattern, die, or process. If the material is weldable, repairs are often made by removing the defective region (usually by chipping or grinding) and filling the created void with deposited weld metal. Porosity that is at or connected to free surfaces can be filled with resinous material, such as polyester, by a process known as *impregnation*. (See Chapter 16 for a further discussion of this process.)

HEAT TREATMENT AND INSPECTION OF CASTINGS

Heat treatment is an attractive means of altering properties while retaining the shape of the product. Steel castings are frequently given a full anneal to reduce the hardness (and brittleness) of rapidly cooled, thin sections and to reduce the internal stresses that result from uneven cooling. Nonferrous castings are often heat-treated to provide stress relief and prepare them for subsequent machining. For final properties, virtually all of the treatments discussed in Chapter 5 can be applied. Ferrous-metal castings often undergo a quench-and-temper treatment, and many nonferrous castings are age hardened to impart additional strength. The variety of heat treatments is largely responsible for the wide range of properties and characteristics available in cast metal products.

Virtually all of the nondestructive *inspection techniques* that were presented in Chapter 11 can be applied to cast metal products. X-ray radiography, liquid penetrant inspection, and magnetic particle inspection are extremely common.

■ 15.13 ROBOTS IN FOUNDRY OPERATIONS

Many of the operations that are performed in a foundry are ideally suited for robotic automation since they tend to be dirty, dangerous, and dull. Robots can dry molds, coat cores, vent molds, and clean or lubricate dies. They can tend stationary, cyclic equipment, such as die-casting machines, and if the machines are properly grouped, one robot can often service two or three machines. In the finishing room, robots can be equipped with plasma cutters or torches to remove sprues, gates, and runners. They can perform grinding and blasting operations, as well as various functions involved in the heat treatment of castings. In the investment casting process, robots can be used to dip the wax patterns into the refractory slurry and produce the desired molds. In a similar manner, they have been used to dip full-mold Styrofoam patterns in their refractory coating and hang them on conveyors to dry. In a fully automized operation, robots could be used to position the pattern, fill the flask with sand, pour the metal, and use a torch to remove the sprue.

■ 15.14 PROCESS SELECTION

As shown in the individual process summaries that have been included throughout Chapters 14 and 15, each of the casting processes has its own characteristic set of capabilities, assets, and limitations. The requirements of the particular product (such as size, complexity, required dimensional precision, desired surface finish, total quantity to be made, and desired rate of production) often limit the number of processes that should be considered as production candidates. Further selection is often based on cost.

Some aspects of product cost, such as the cost of the material and the energy required to melt it, are somewhat independent of the specific process. The cost of other features, such as patterns, molds, dies, melting and pouring equipment, scrap material, cleaning, inspection, and all related labor, can vary markedly and be quite dependent on the process. For example, pattern and mold costs for sand casting are quite a bit less than the cost of die-casting dies. Die casting, on the other hand, offers high production rates and a high degree of automation. When a small quantity of parts is desired, the cost of the die or tooling must be distributed over the total number of parts, and unit cost (or cost per casting) is high. When the total quantity is large, the tooling cost is distributed over many parts, and the cost per piece decreases.

Figure 15-20 shows the relationship between unit cost and production quantity for both sand and die casting. Sand casting is an expendable mold process. Since an individual mold is made for each pour, increasing quantity does not lead to a significant drop in unit cost. Die casting involves a multiple-use mold, and the cost of the die can be distributed over the total number of parts. As shown, sand casting is often less expensive for small production runs, and processes such as die casting are preferred for large quantities. (*Note:* While the die-casting curve in Figure 15-20 is a smooth line, it is not uncommon for an actual curve to contain abrupt discontinuities. If the lifetime of a set of tooling is 50,000 casts, the cost per part for 45,000 pieces, using one set of tooling, would actually be less than 60,000, since the latter would require a second set of dies.)

In most cases, more than one or two processes are possible candidates for production, and curves for all of the options should be included. The final selection is then based on a combination of economic and technical considerations.

Table 15-5 presents a comparison of casting processes, including green-sand casting, chemically bonded sand molds (shell, sodium silicate, and air-set), ceramic mold and investment casting, permanent mold casting, and die casting. The processes are compared on the basis of cost for both small and large quantities, thinnest section, dimensional precision, surface finish, ease of casting a complex shape, ease of changing the design while in production, and range of castable materials.

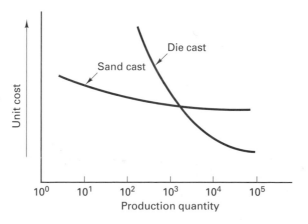

FIGURE 15-20 Typical unit cost of castings for both sand casting and die casting. Note how the large cost of a die casting die diminishes as it is spread over a larger quantity of parts.

TABLE 15-5.   Comparison of Casting Processes

Property or Characteristic	Green-Sand Casting	Chemically Bonded Sand (Shell, Sodium Silicate, Air-Set)	Ceramic Mold and Investment Casting	Permanent Mold Casting	Die Casting
Relative cost for small quantity	Lowest	Medium high	Medium	High	Highest
Relative cost for large quantity	Low	Medium high	Highest	Low	Lowest
Thinnest section (inches)	$\frac{1}{10}$	$\frac{1}{10}$	$\frac{1}{16}$	$\frac{1}{8}$	$\frac{1}{32}$
Dimensional precision ($+/-$ in inches)	0.01–0.03	0.005–0.015	0.01–0.02	0.01–0.05	0.001–0.015
Relative surface finish	Fair to good	Good	Very good	Good	Best
Ease of casting complex shape	Fair to good	Good	Best	Fair	Good
Ease of changing design while in production	Best	Fair	Fair	Poor	Poorest
Castable metals	Unlimited	Unlimited	Unlimited	Low-melting-point metals	Low-melting-point metals

■ KEY WORDS

arc furnace	electromagnetic casting	indirect fuel-fired furnace	pore-free casting	slush casting
centrifugal casting	heat checking	induction furnace	pouring ladle	squeeze casting
centrifuging	heat treatment	inserts	retractable metal cores	thermal fatigue
cold-chamber machines	heated-manifold direct-injection die casting	inspection	reverberatory furnaces	thixotropic material
continuous casting		low-pressure permanent mold casting	semicentrifugal casting	true centrifugal casting
cupola	hot-chamber machines	permanent mold casting	semisolid casting	vacuum permanent mold casting
die casting				venting
direct fuel-fired furnace	impregnation			

■ REVIEW QUESTIONS

1. What are some of the major disadvantages of the expendable-mold casting processes?
2. What are some possible limitations of multiple-use molds?
3. What are some common mold materials for permanent mold casting? What are some of the metals more commonly cast?
4. Describe some of the process advantages of permanent mold casting.
5. Why might low production runs be unattractive for permanent mold casting?
6. What features affect the life of a permanent mold?
7. Why are permanent mold castings generally removed from the mold immediately after solidification has been completed?
8. How is venting provided in the permanent mold process?
9. What types of products would be possible candidates for manufacture by slush casting?
10. How does low-pressure permanent mold casting differ from the traditional gravity-pour process?
11. What are some of the attractive features of the low-pressure permanent mold process?
12. What are some additional advantages of vacuum permanent mold casting over the low-pressure process?
13. Contrast the feeding pressures on the molten metal in low-pressure permanent molding and die casting.
14. Contrast the materials used to make dies for gravity permanent mold casting and die casting. Why is there a notable difference?
15. Why might the pressure on the molten metal vary during the die casting cycle?
16. For what types of materials would a hot-chamber die-casting machine be appropriate?
17. What metals are routinely cast with cold-chamber die-casting machines?
18. How does the air in the mold cavity escape in the die-casting process?
19. Are risers employed in die casting? Can sand cores be used?
20. What are some of the advantages of the heated-manifold direct-injection die casting of zinc?
21. What are some of the attractive features of die casting compared to alternative casting methods?
22. When might low quantities be justified for the die-casting process?
23. Describe the squeeze-casting process.
24. What is a thixotropic material? How does it provide an attractive alternative to squeeze casting?
25. What are some of the attractive features of semisolid casting?
26. Contrast the structure and properties of the outer and inner surfaces of a centrifugal casting.
27. What is the difference between semicentrifugal casting and centrifuging?

28. What are some of the attractive features of electromagnetic casting?
29. What are some of the factors that influence the selection of a furnace type or melt procedure in a casting operation?
30. What types of metals are commonly melted in cupolas?
31. In cupola-melting operations, why are many of the chemistry control operations performed in the ladle, via ladle metallurgy?
32. What are some of the ways that the melting rate of a cupola can be increased?
33. What are some of the pros and cons of indirect fuel-fired furnaces?
34. What are some of the attractive features of arc furnaces in foundry applications?
35. Why are channel induction furnaces attractive for metal-holding applications where molten metal must be held at a specified temperature for long periods of time?

36. What are the primary functions of a pouring operation?
37. What are some of the typical cleaning and finishing operations that are performed on castings?
38. What are some common ways to remove cores from castings? To remove runners, gates, and risers?
39. How might defective castings be repaired to permit successful use in their intended applications?
40. What are some of the ways that industrial robots can be employed in metal-casting operations?
41. Describe some of the features that affect the cost of a cast product. Why might the cost vary significantly with the quantity to be produced?

Chapter 15   CASE STUDY

Baseplate for a Household Steam Iron

The item depicted in Figure CS–15 is the baseplate of a high-quality household steam iron. It is rated for operation at up to 1200 watts and is designed to provide both steady steam and burst of steam features. Incorporated into the design is an integral electrical resistance heating "horseshoe" that must be thermally coupled to the baseplate but remain electrically insulated. (This component often takes the form of a resistance heating wire, surrounded by ceramic insulation, all encased in a metal tube.) There is a complex series of channels designed to receive the steam and distribute it evenly through a number of small vent holes in the base, each about $\frac{1}{16}$ in. in diameter. There are about a dozen larger threaded recesses, about $\frac{1}{8}$ in. in diameter, that are used in assembling the various components.

1. Discuss the various features that this component must possess in order to function in an adequate fashion. Consider strength, impact resistance, thermal conductivity, corrosion resistance, weight, and other factors.

2. What material or materials would appear to be strong candidates?

3. What are some possible means of producing the desired shape? Which would you prefer? Could the heating element assembly be incorporated during manufacture, or does it have to be added as a secondary operation? What are the major advantages of the method you propose?

4. Could all of the design features (holes, webs, and recesses) be incorporated in the initial manufacturing operation, or would secondary processing be required? If secondary processing is required, for what features, and how would you recommend that they be produced?

5. Some commercial irons have baseplates for which the bottom surfaces have been finished by a simple buff and polish, while others have a teflon coating or have been anodized. If your desire is to produce a high-quality product, what form of surface finishing would you recommend?

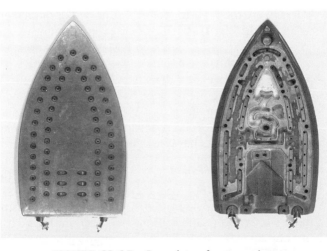

FIGURE CS-15   Baseplate of a steam iron.

CHAPTER 16

POWDER METALLURGY

■ 16.1 INTRODUCTION

Powder metallurgy is the name given to the process by which fine powdered materials are blended, pressed into a desired shape (compacted), and then heated (sintered) in a controlled atmosphere to bond the contacting surfaces of the particles and establish desired properties. The process, commonly designated as P/M, readily lends itself to the mass production of small, intricate parts of high precision, often eliminating the need for additional machining or finishing. There is little material waste; unusual materials or mixtures can be utilized; and controlled degrees of porosity or permeability can be produced. Major areas of application tend to be those for which the P/M process has strong economical advantage or where the desired properties and characteristics would be difficult to obtain by any other method. Because of its level of manufacturing maturity, powder metallurgy should actually be considered as a possible means of manufacture for any part where the geometry and production quantity are appropriate.

While a crude form of iron powder metallurgy existed in Egypt as early as 3000 B.C., and the ancient Incas made jewelry and other artifacts from precious metal powders, mass manufacturing of P/M products did not begin until the mid- or late-nineteenth century. At this time, powder metallurgy was used to produce copper coins and medallions, platinum ingots, and tungsten wires, the primary material for light bulb filaments early in the twentieth century. By the 1920s the tips of tungsten carbide cutting-tools and nonferrous bushings were being produced. Self-lubricating bearings and metallic filters were other early products. A period of rapid technological development occurred after World War II based primarily on automotive applications, and iron and steel replaced copper as the dominant P/M material. Aerospace and nuclear developments created accelerated demand for refractory and reactive materials where powder processing is quite attractive. Full-density products emerged in the 1960s, and high-performance superalloy components, such as aircraft turbine engine parts, were a highlight of the 1970s. Developments in the 1980s and 90s included the commercialization of rapidly-solidified and amorphous powders as well as P/M injection molding technology.

Recent years have been ones of rapid growth for the P/M industry. From 1960 to 1980, the consumption of iron powder increased tenfold. A similar increase occurred between 1980 and 1990, and the exponential growth continued through most of the 1990s. While most products are still under 50 mm (2 in.) in size, some have been produced with weights up to 45 kg (100 lb) with linear dimensions up to 500 mm (20 in.).

Automotive applications now account for nearly 70% of the powder metallurgy market. In 1990, the average U.S. automobile contained about 10 kg (21 lb) of P/M parts. By 1995, the amount had increased to over 13.6 kg (30 lb), and by 2000, had risen to

16.3 kg (36 lb). It is expected that by 2003, this number will reach 20 kg (nearly 45 lb). Other areas where powder metallurgy products are used extensively include household appliances, recreational equipment, hand tools, hardware items, office equipment, industrial motors, and hydraulics. Areas of rapid growth include aerospace applications, advanced composites, electronic components, magnetic materials, metalworking tools, and a variety of biomedical and dental applications. Iron and low-alloy steels now account for 85% of all P/M usage, with copper and copper-based powders comprising about 7%. Stainless steel, high-strength and high-alloy steels, and aluminum and aluminum alloys are also high-volume materials.

■ 16.2 BASIC PROCESS

The powder metallurgy process generally consists of four basic steps: (1) powder manufacture, (2) mixing or blending, (3) compacting, and (4) sintering. Optional secondary processing often follows to obtain special properties or enhanced precision. Figure 16-1 presents a simplified flowchart of the conventional die-compaction P/M process.

■ 16.3 POWDER MANUFACTURE

The properties of powder metallurgy products are highly dependent on the characteristics of the starting powders that are used. Some important properties and characteristics include *chemistry and purity*, *particle size*, *size distribution*, *particle shape*, and the *surface texture* of the particles. Several processes can be used to produce powdered material, with each imparting distinct properties and characteristics to the powder and hence to the final product.

Over 80% of all commercial powder is produced by some form of melt *atomization*, where a liquid is fragmented into molten droplets which then solidify into particles. Various forms of energy have been used to form the droplets. In the method illustrated in Figure 16-2a, molten metal emerges from an orifice where it is atomized by a stream of impinging gas or liquid. In Figure 16-2b, an electric arc impinges on a rapidly rotating electrode, all contained within a chamber purged with inert gas. Centrifugal force causes the molten droplets to fly from the surface of the electrode and freeze in flight.

Regardless of the process details, atomization is an extremely useful means of producing *prealloyed powders*. By starting with an alloyed melt or prealloyed electrode, each powder particle has the desired alloy composition. Powders of aluminum alloys, copper alloys, stainless steel, nickel-based alloys (such as Monel), titanium alloys, cobalt-based alloys, and various low-alloy steels have all been commercially produced. The size and shape of the powder particles can be varied and depend on process features such

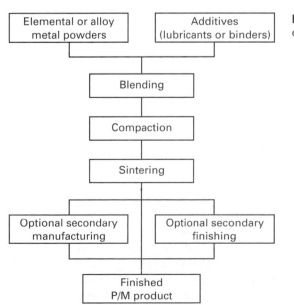

FIGURE 16-1 Simplified flowchart of the powder metallurgy process.

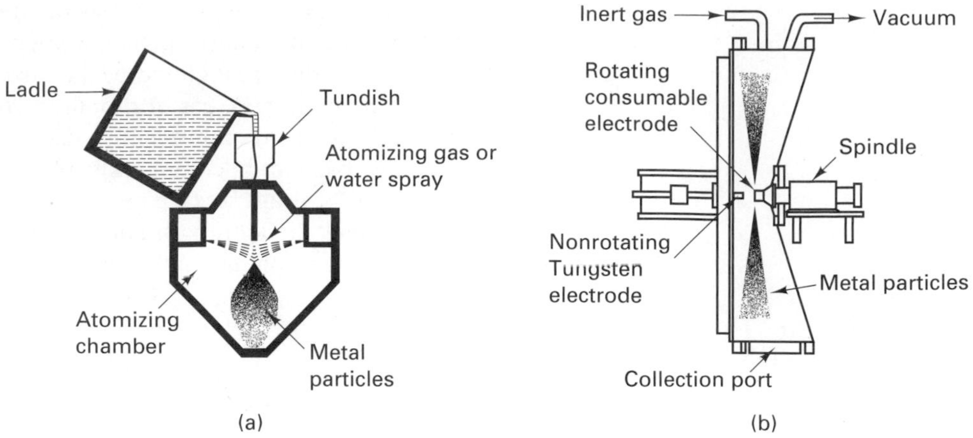

FIGURE 16-2 Two methods for producing metal powders: (a) melt atomization, (b) atomization from a rotating consumable electrode.

as the velocity and media of the atomizing jets or the speed of electrode rotation, the starting temperature of the liquid (which affects the time that surface tension can act on the individual droplets prior to solidification), and the environment provided for cooling. When cooling is slow (such as in gas atomization) and surface tension is high, spherical shapes can form before solidification. With more rapid cooling, such as with water atomization, irregular shapes tend to be produced.

Other methods of powder manufacture include *chemical reduction of particulate compounds* (generally crushed oxides or ores), *electrolytic deposition* from solutions or fused salts, *pulverization* or *grinding* of brittle materials (comminution), *thermal decomposition of hydrides or carbonyls*, *precipitation from solution*, and *condensation of metal vapors*.

Almost any metal, metal alloy, or nonmetal (ceramic, polymer, or wax or graphite lubricant) can be converted into powder form by one or more of the powder production methods. Some methods can produce only elemental powder (often of high purity), while others can produce prealloyed particles. Further operations, such as drying or heat treatment, may be performed prior to further processing.

■ 16.4 RAPIDLY SOLIDIFIED POWDER (MICROCRYSTALLINE AND AMORPHOUS)

Increasing the cooling rate of the liquid material can result in the formation of an ultrafine or microcrystalline grain size. In these materials, a large percentage of the atoms are located in grain boundary regions, giving unusual properties (such as high diffusivity), expanded alloy possibilities, and good formability. If the cooling rate approaches or exceeds $10^{6\circ}$C/sec, metals can solidify without becoming crystalline. These *amorphous* or glassy metals can also exhibit unusual or unique properties, which include high strength, improved corrosion resistance, and reduced energy to induce and reverse a magnetization. Amorphous metal transformer cores lose 60 to 70% less energy in magnetization than conventional silicon steels. As a result, it is estimated that over half of all new power distribution transformers purchased in the United States will utilize amorphous metal cores.

Production of amorphous material, however, requires immensely high cooling rates and hence ultra-small dimensions. Atomization with rapid cooling and the "splat quenching" of a metal stream onto a cool surface to produce a continuous ribbon are two prominent methods. Since much of the ribbon material is further fragmented into powder, powder metallurgy is the primary means of fabricating useful products from amorphous material.

■ 16.5 POWDER TESTING AND EVALUATION

In addition to the evaluation of bulk chemistry, surface chemistry, particle size and size distribution, particle shape, surface texture, and internal structure, metal powders must also be evaluated for their suitability for further processing. *Flow rate* is a measure of the ease by which powder can be fed and distributed into a die. Poor flow characteristics can result in nonuniform die filling and in nonuniform density and properties in a product.

Associated with the flow characteristics is the *apparent density*, a measure of a powder's ability to fill available space without the application of external pressure. A low apparent density means that there is a large fraction of unfilled space in the loose-fill powder. *Compressibility* tests evaluate the effectiveness of applied pressure in raising the density of the powder, and *green strength* is used to describe the strength of the pressed powder immediately after compacting. It is well established that higher product density correlates with superior mechanical properties, such as strength and fracture resistance. Good green strength is required to maintain smooth surfaces, sharp corners, and intricate details during ejection from the compacting die or tooling and the subsequent transfer to the sintering operation.

■ 16.6 POWDER MIXING AND BLENDING

It is rare that a single powder will possess all of the characteristics desired in a given process and product. Most likely, the starting material will be a mixture of various grades or sizes of powder, or powders of different compositions, with additions of *lubricants* or *binders*.

The final product chemistry is often obtained by combining pure metal or non-metal powders, rather than using prealloyed material. Sufficient diffusion must then occur during the sintering operation to produce a uniform chemistry and structure in the final product. Unique *composites* can also be produced, such as the distribution of an immiscible reinforcement material in a matrix, or the combination of metals and nonmetals in a single product such as a tungsten carbide–cobalt matrix cutting tool for high-temperature service.

Some powders, such as graphite, can even play a dual role, serving as a lubricant during compacting and a source of carbon as it alloys with iron during sintering to produce steel. Lubricants such as graphite or stearic acid improve the flow characteristics and compressibility at the expense of reduced green strength. Binders produce the reverse effect. Most lubricants or binders are not wanted in the final product and are removed (volatilized or burned off) in the early stages of sintering, leaving holes that are reduced in size or closed during subsequent heating.

Blending or *mixing* operations can be done either dry or wet, where water or other solvent is used to enhance particle mobility, reduce dusting, and lessen explosion hazards. Large lots of powder can be homogenized with respect to both chemistry and distribution of components, sizes, and shapes. Quantities up to 16,000 kg (35,000 lb) have been blended in single lots to ensure uniform behavior during processing and the production of a consistent product.

■ 16.7 COMPACTING

One of the most critical steps in the P/M process is *compacting*. Loose powder is compressed and densified into a shape known as a green compact, usually at room temperature. High product density and the uniformity of that density throughout the compact are generally desired characteristics. In addition, the compacts should possess sufficient green strength for in-process handling and transport to the sintering furnace.

Most compacting is done with mechanical presses and rigid tools, but hydraulic and hybrid (combinations of mechanical, hydraulic, and pneumatic) presses can also be used. Figure 16-3 shows a typical mechanical press for compacting powders and a removable set of compaction tooling. The removable die sets allow the time-consuming alignment and synchronization of tool movements to be set up while the press is producing parts with another die set. Compacting pressures generally range between 40 and 1650 MPa (3 to 120 tons/in^2) depending on material and application (see Table 16-1), with the range of 140 to 690 MPa (10 to 50 tons/in^2) being the most common. While most P/M presses have total capacities of less than 100 tons (9×10^5N), increasing numbers are being purchased with higher capacity. Because of press capacity, powder metallurgy products are often limited to cross section areas or pressing areas of less than 6500 mm^2 (10 in^2) have become more common. Some P/M presses now have capacities up to 3000 tons (27 MN) and are capable of pressing areas up to 65,000 mm^2 (100 in^2). Where larger products are desired, compaction

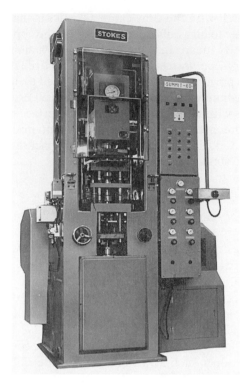
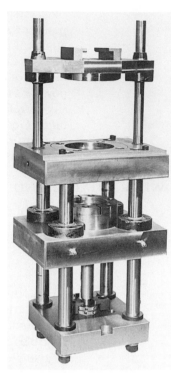

FIGURE 16-3 (*Left*) Typical press for the compacting of metal powders. A removable die set (*right*) allows the machine to be producing parts with one die set while another is being fitted to produce a second product. *(Courtesy of Sharples-Stokes Division, Pennwalt Corporation.)*

TABLE 16-1.	Typical Compacting Pressures for Various Applications	
Application	Compaction tons/in²	Pressures MPa
Porous metals and filters	3–5	40–70
Refractory metals and carbides	5–15	70–200
Porous bearings	10–25	146–350
Machine parts (medium density iron & steel)	20–50	275–690
High-density copper and aluminum parts	18–20	250–275
High-density iron and steel parts	50–120	690–1650

FIGURE 16-4 Typical compaction sequence for a single-level part, showing the functions of the feed shoe, die, core rod, and upper and lower punches. Loose powder is shaded; compacted powder is solid black.

can be performed by dynamic methods, such as use of an explosively induced shock wave to provide the compaction pressure. Metalforming processes, such as rolling, forging, extrusion, and swagging, have also been adapted to compact powders.

Figure 16-4 shows the typical compaction sequence for a mechanical press. With the bottom punch in its fully raised position, a feed shoe moves into position over the die.

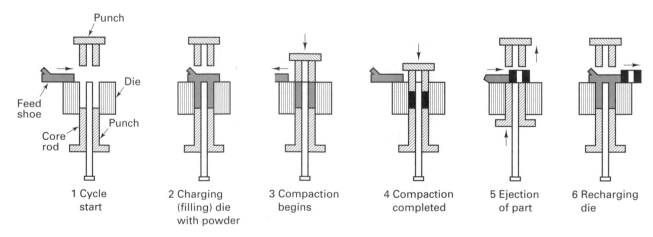

| 1 Cycle start | 2 Charging (filling) die with powder | 3 Compaction begins | 4 Compaction completed | 5 Ejection of part | 6 Recharging die |

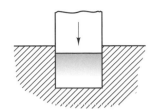

FIGURE 16-5   Compaction with a single punch, showing the resultant nonuniform density (*shaded*).

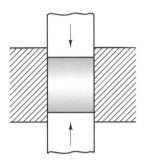

FIGURE 16-6   Density distribution obtained with a double-acting press and two moving punches. Note the increased uniformity compared to Figure 16-5.

The feed shoe is an inverted container filled with powder, connected to the powder supply by a flexible feed tube. With the feed shoe in position, the bottom punch descends to a preset fill depth, and the shoe retracts, leveling the powder. The upper punch then descends and compacts the powder as it penetrates the die. The upper punch retracts and the bottom punch rises to eject the green compact. As the die shoe advances for the next cycle, its forward edge clears the compacted product from the press, and the cycle repeats.

During compacting, the powder particles move primarily in the direction of the applied force. Since the loose-fill dimensions are two to two and a half times the pressed dimensions, the amount of particle travel in the pressing direction can be substantial. The amount of lateral flow, however, is quite limited. In fact, it is rare to find a particle in the compacted product that has moved more than three particle diameters off of its original axis of pressing. Thus, the powder does not flow like a liquid but simply compresses until an equal and opposing force is created. This opposing force is probably a combination of (1) resistance by the bottom punch, and (2) friction between the particles and the die surfaces. As illustrated in Figure 16-5, when the pressure is applied by only one punch, maximum density occurs below the punch and decreases as one moves down the column. It is very difficult to transmit uniform pressures and produce uniform density throughout a compact. By use of a double-action press, where pressing movements occur from both top and bottom (Figure 16-6), more uniform density can be obtained and thicker products can be compacted. Since sidewall friction is a key factor in compaction, the resulting density shows a strong dependence on both the thickness and width of the part being pressed. For good, uniform compaction, the ratio of thickness/width should be kept below 2.0 whenever possible. Products with ratios above 2.0 tend to exhibit considerable variation in density.

As shown in Figure 16-7, the average density of the compact depends on the amount of pressure that is applied, with the particular response being strongly dependent upon the characteristics of the powder being compressed (its size, shape, surface texture, mechanical properties, etc.). The final density may be reported as either an absolute density in units such as grams per cubic centimeter, or as a percentage of the theoretical density, where the difference between this number and 100% is the amount of void space left within the compact.

Figure 16-8 shows that a single displacement will produce different degrees of compaction in different thicknesses of powder. Therefore, it is impossible for a single punch to produce uniform density in a multithickness part. When the product requires nonuniform thickness, more complicated presses or compaction methods must be employed. Figure 16-9 illustrates two methods to compact a dual-thickness part. By providing different amounts

FIGURE 16-7   Effect of compacting pressure on green density (the density after compaction but before sintering) for several commercial powders.

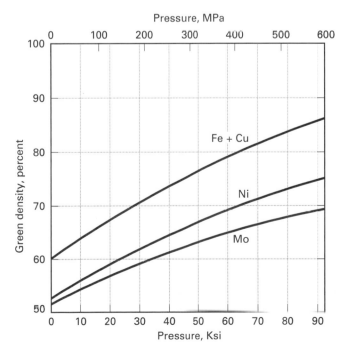

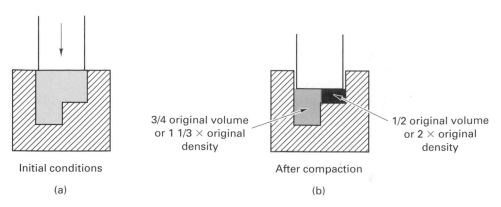

FIGURE 16-8 Two-thickness part with only one moving punch. (*Left*) Initial conditions; (*right*) after compaction by the upper punch, showing drastic difference in compacted density.

3/4 original volume or 1 1/3 × original density

1/2 original volume or 2 × original density

Initial conditions

(a)

After compaction

(b)

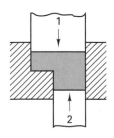

Single lower punch

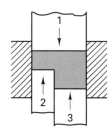

Double lower punch

FIGURE 16-9 Two methods of compacting two thickness parts to near-uniform density. Both involve the controlled movement of two or more punches.

of motion to the various punches and synchronizing these movements to provide simultaneous compaction, a uniformly compacted product can be produced.

Because the complexity of the part often dictates the complexity of compaction equipment, powder metallurgy components have been grouped into classes. Class 1 components are the simplest and easiest to produce. They are thin, single-level parts that can be pressed with a force from one direction. The thickness is generally less than 6.35 mm ($\frac{1}{4}$ in.). Class 2 parts are single-level parts of any thickness that require pressing from two directions. As discussed above, these are usually thicker parts. Class 3 parts are double-level parts and require pressing from two directions. Class 4 parts are the most complex of those produced by rigid die compaction. They are multilevel parts that require two or more pressing motions. These classes are summarized in Table 16-2 and Figure 16-10.

When extremely complex shapes are desired, the powder is generally encapsulated in a flexible mold and immersed in a pressurized gas or liquid, a process known as *isostatic* (uniform pressure) *compaction*. Production rates in this process are extremely low, but parts up to several hundred pounds can be compacted effectively.

TABLE 16-2.	Classes for Conventional Press-and-Sinter Parts	
Class	Levels	Press Actions
1	1	Single
2	1	Double
3	2	Double
4	More than 2	Double or multiple

FIGURE 16-10 Sample geometries of the four basic classes of press-and-sinter powder metallurgy parts. Note the increase in the complexity of the pressing operation as class increases.

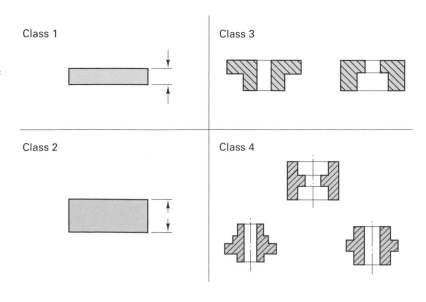

Class 1

Class 3

Class 2

Class 4

Warm compaction emerged as a common practice in the 1990s. By preheating the powder prior to pressing, the metal is softened and responds better to applied pressures. Better compaction results in improved properties, both as-compacted and after final processing.

Another means of enhancing compaction is to increase the amount of lubricant in the powder. This reduces the friction between the powder and the die wall and enhances the transmission of compaction pressure through the powder. Unfortunately, additional lubricant may reduce the green strength to the point where it is insufficient for part ejection and handling, or reduce final properties to the point where they are unacceptable.

While pressing rates vary widely, small mechanical presses can typically produce up to 100 pieces per minute. By means of mechanisms such as bulk movement of particles, deformation of individual particles, and particle fracture or fragmentation, mechanical compaction can raise the density of loose powder to about 80% of an equivalent cast or forged metal. Sufficient strength can be imparted by this compaction to hold the shape and permit a reasonable amount of handling. This strength is referred to as *green strength*. In addition, the compaction process sets both the nature and distribution of the porosity remaining in the product.

COMPACTION TOOLING (PUNCHES AND DIES)

Because powder particles tend to be somewhat abrasive and high pressures are involved during compaction, wear of the tool components is a major concern. Consequently, compaction tools are usually made of hardened tool steel. For particularly abrasive powders, or for high-volume production, cemented carbides may be employed. Die surfaces should be highly polished and the dies should be heavy enough to withstand the high pressing pressures. Lubricants are also used to reduce die wear.

METAL INJECTION MOLDING (MIM) OR POWDER INJECTION MOLDING (PIM)

For many years, injection molding has been used to produce small, complex-shaped components from plastic. A thermoplastic resin is heated to impart the necessary degree of fluidity and is then pressure injected into a die, where it cools and hardens. Die casting is a similar process for metals but is restricted to alloys with relatively low melting temperature (lead-, zinc-, aluminum-, and copper-based materials). Small, complex-shaped products of the higher-melting-point metals are generally made by more costly processes, such as investment casting, machining directly from metal stock, or conventional powder metallurgy. *Metal injection molding* is a rather recent extension of conventional powder metallurgy that offers the shape forming capability of plastics, the precision of die casting, and the materials flexibility of powder metallurgy.

Since powdered material does not flow like a fluid, complex shapes are produced by first mixing ultrafine (usually in the range of 2–20 μm) spherical-shaped metal, ceramic, or carbide powder with a low molecular weight thermoplastic or wax material in amounts up to 50% powder by volume. This mixture is frequently produced in the form of pellets, which become the feedstock for the injection process. After heating to a paste-like consistency (about 260°C or 500°F), the material is injected into a mold cavity under sufficient pressure (about 70 MPa or 10,000 psi) and temperature to assure die filling. After cooling and ejection, the binder material is removed by either solvent extraction, controlled heating to above the volatilization temperature, or a catalytic debindering process in which nitric acid is added to a flow of nitrogen carrier gas to break down the binder molecules and carry away the decomposition products. The parts then undergo conventional sintering, where they shrink 15 to 25% and the density increases from about 60% up to as much as 99% of ideal. The sintering process also sets the final properties of the material. Secondary processes generally take the form of surface cleaning or finishing, plating, machining, and heat treating. Figure 16-11 summarizes this sequence of activities.

Removing the binder is currently the most expensive and time consuming part of the process. Heating rates, temperatures, and debinding times must be carefully controlled and adjusted for part thickness. Different variations of P/M injection molding use different binders or fluidizers and different means of removal.

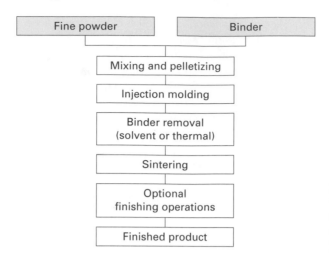

FIGURE 16-11 Schematic flowchart of the metal injection molding process (MIM) used to produce small, intricate shaped parts from metal powder.

While the size of conventional P/M products is generally limited by press capacity, the size of P/M injection moldings is more limited by economics (cost of the fine powders) and binder removal. The best candidates for P/M injection molding are complex-shaped metal parts with thicknesses of less than 6.3 mm ($\frac{1}{4}$ in.) and weights under 60 grams (2 oz.). Section thicknesses as small as 0.25 mm (0.010 in.) are possible because of the fineness of the powder. Despite the high cost of ultrafine powders and extensive processing, P/M injection molding has become an attractive method for producing parts that would either be impossible to compact in a conventional press or would require significant amounts of secondary machining.

Large production volumes (more than 2000 to 5000 identical parts) are generally required to justify the cost of die design and manufacture. The relatively high final density (95 to 99% of ideal compared to 75 to 90% for conventional P/M parts), the uniformity of that density, the close tolerances (0.3 to 0.5%), and excellent surface finish all combine to make the process attractive for many applications. Parts can be made from a wide selection of metal alloys, including steels, stainless steel, tool steel, brass, copper, titanium, tungsten, ceramics, and many specialty materials. The final properties are generally close to those of wrought or cast equivalents.

■ 16.8 SINTERING

In the *sintering* operation, the pressed-powder compacts are heated in a controlled atmosphere environment to a temperature below the melting point but high enough to permit solid-state diffusion, and held for sufficient time to permit bonding of the particles. Most metals are sintered at temperatures of 70 to 80% of their melting point, while certain refractory materials may require temperatures near 90%. When the product is composed of more than one material, the sintering temperature may even be above the melting temperature of one or more components. The lower-melting-point materials then melt and flow into the voids between the remaining particles, and the process becomes *liquid-phase sintering*.

Most sintering operations involve *three stages*, and many sintering furnaces employ three corresponding zones. The *first operation*, the *burn-off* or *purge*, is designed to combust any air, volatilize and remove lubricants or binders that would interfere with good bonding, and slowly raise the temperature of the compacts in a controlled manner. Rapid heating would produce high internal pressure from air entrapped in closed pores or volatilizing lubricants, and would result in swelling or fracture of the compacts. When the compacts contain appreciable quantities of volatile materials, their removal creates additional *porosity* and *permeability* within the pressed shape. The manufacture of products such as metal filters is designed to take advantage of this feature. When the products are load-bearing components, however, high amounts of porosity are undesirable, and the amount of volatilizing lubricant is kept to an optimized minimum. The *second, or high-temperature stage* is where the desired solid-state diffusion and bonding between the powder particles take place. Atoms move toward the points of contact between the particles; the areas of contact become larger; and the part becomes a solid mass with small pores of various sizes and shapes. The mechanical bonds of compaction become metallurgical bonds. The time in this stage must be sufficient to produce the desired density and final properties, and usually varies from 10 minutes to several hours. Finally, a *cooling period* is required to lower the temperature of the products while retaining them in a controlled atmosphere. This feature serves to prevent oxidation that would occur upon direct discharge into air as well as possible thermal shock from rapid cooling. Both batch and continuous furnaces are used for sintering.

All three stages of sintering must be conducted in the oxygen-free conditions of a *vacuum* or *protective atmosphere*. This is critical because the compacted shapes typically have 10 to 25% residual porosity, and some of the internal voids are connected to exposed surfaces. At elevated temperatures, rapid oxidation would occur and significantly impair the quality of interparticle bonding. *Reducing atmospheres*, commonly based on hydrogen, dissociated ammonia, or cracked hydrocarbons, are preferred since they can reduce any oxide already present on the particle surfaces and combust harmful gases that are liberated during the sintering. *Inert gases* cannot reduce existing oxides but will prevent the formation of any additional contaminants. *Vacuum* sintering is frequently employed with stainless steel, titanium, and the refractory metals. *Nitrogen atmospheres* are also common.

During the sintering operation, a number of changes occur in the compact. Metallurgical bonds form between the powder particles as a result of solid-state atomic diffusion, and strength, ductility, toughness, and electrical and thermal conductivities all increase. If different chemistry powders were blended, interdiffusion will promote the formation of alloys or intermetallic phases. In addition, there will be a concurrent increase in density and contraction in product dimensions. To meet final tolerances, the dimensional shrinkage will have to be compensated through the design of oversized compaction dies. During sintering, not all of the porosity is removed, however. Conventional pressed-and-sintered P/M products generally contain between 5 and 25% residual porosity.

Sinter brazing is the process in which two or more separate pieces are joined by brazing while they are also being sintered. The individual pieces are compacted separately, and are assembled with the braze metal positioned so it will flow into the joint. When the assembly is heated for sintering, the braze metal melts and flows between the joint surfaces to create the bond. As sintering continues, much of the braze metal diffuses into the surrounding metal, producing a final joint that is often stronger than the materials being joined.

■ 16.9 HOT ISOSTATIC PRESSING

In conventional press-and-sinter powder metallurgy, the pressing or compaction is usually performed at room temperature, and the sintering, at atmospheric pressure. *Hot isostatic pressing* (HIP) combines powder compaction and sintering into a single operation that involves gas-pressure squeezing at elevated temperature. While this may seem to be an improvement over the two-step approach, it should be noted that heated powders may need to be "protected" and the pressurizing media must be prevented from entering the voids between the particles. One approach to hot isostatic pressing begins by sealing the powder in a flexible, airtight, evacuated container, which is then subjected to a high-temperature, high-pressure environment. Conditions for processing irons and steels involve pressures around 70 to 100 MPa (10,000 to 15,000 psi) coupled with temperatures in the neighborhood of 1250°C (2300°F). For the nickel-based superalloys, refractory metals, and ceramic powders, the equipment must be capable of 310 MPa (45,000 psi) and 1500°C (2750°F). Multiple pieces, totaling up to several tons, can now be processed in a single cycle that typically lasts several hours.

After processing, the products emerge at full density with uniform, isotropic properties that are often superior to those of other processes. Near-net shapes are possible, thereby eliminating material waste and costly machining operations. Since the powder is totally isolated and compaction and sintering occur simultaneously, the process is attractive for reactive or brittle materials, such as beryllium, uranium, zirconium, and titanium. Since die compaction is not required, large parts are now possible, and shapes can be produced that would be impossible to eject from rigid compaction dies. Hot isostatic pressing has also been employed to densify existing parts (such as those that have been conventionally pressed and sintered), heal internal porosity in castings, and seal internal cracks in a variety of products. The elimination or reduction of defects yields startling improvements in strength, toughness, fatigue resistance, and creep life.

Several aspects of the HIP process make it expensive and unattractive for high-volume production. The first is the high cost of *canning* the powder in a flexible isolating medium that can resist the subsequent temperatures and pressures, and then later removing this material from the product (*decanning*). Sheet metal containers are most common, but glass and even ceramic molds have been used. The second problem involves the relatively long time for the HIP cycle. While process advances have reduced cycle times from 24 hours to 6 to 8 hours, production is still limited to several loads a day, and the number of parts per load is limited by the ability to produce and maintain uniform temperature throughout the pressure chamber.

The *sinter-HIP* process and *pressure-assisted sintering* are techniques that have been developed to produce full-density powder products without the expense of canning and decanning. Conventionally compacted P/M parts are placed in a pressurizable chamber and sintered (heated) under vacuum for a time that is sufficient to seal the surface and isolate all internal porosity. (*Note:* This generally requires a product density greater than 92–95% of theoretical.) While maintaining the elevated temperature, the vacuum is broken and high pressure is then applied for the remainder of the process. The sealed surface produced during the vacuum sintering acts as an isolating can during the high-pressure stage. These processes start with as-compacted powder parts, and eliminate the additional heating and cooling cycle that would be required if parts were first compacted and sintered in the conventional manner and then subjected to the HIP process for final densification.

■ 16.10 OTHER TECHNIQUES TO PRODUCE HIGH-DENSITY P/M PRODUCTS

High-density P/M parts can also be produced by using the high-temperature forming processes. Sheets of sintered powder (produced by roll compaction and sintering) can be reduced in thickness and further densified by hot rolling in the process depicted in Figure 16-12. Rods, wires, and small billets can be produced by the hot extrusion of encapsulated powder or pressed-and-sintered slugs. Forging can be applied to form complex shapes from canned powder or simple-shaped sintered preforms. By using powdered material, these processes offer the combined benefits of powder metallurgy and the respective forming process, such as the production of fabricated shapes with uniform fine grain size, and uniform chemistry or unusual alloy composition.

The *Ceracon process* is another method of raising the density of conventional pressed-and-sintered P/M products without requiring encapsulation or canning.

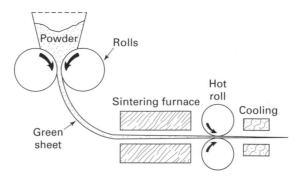

FIGURE 16-12 Method of producing continuous sheet products from powdered feedstock.

A heated preform is first surrounded by hot granular material, usually smooth-surface ceramic particles. When the assembly is then compacted in a conventional hydraulic press, the granular material transmits pressure in a somewhat uniform manner. Since no gas or liquid is involved in the pressurization, encapsulation is not required to prevent the pressurizing medium from entering the material. At the end of the cycle, the part and the pressurizing medium separate freely, and the pressure-transmitting granules are then reheated and reused.

Still another means of producing a high-density shape from fine particles is *in-situ compaction* or *spray forming* (also known as the *Osprey process*). Consider an atomizer similar to that of Figure 16-2a, in which jets of inert or harmless gas (nitrogen or carbon dioxide) propel molten droplets down into a collecting container. If the droplets solidify before impact, the container fills with loose powder. If the droplets remain liquid during their flight, the container fills with molten metal and then solidifies into a conventional casting. However, if the cooling of the droplets is controlled so that they are semisolid (and computers can provide the necessary process control), they act as "slush balls" and flatten upon impact. The remaining freezing occurs quickly, and the product is a uniform chemistry, fine grain size, high-density (in excess of 98% of theoretical) solid. Depending on the shape of the collecting container, the spray-formed product can be a finished part, a strip or plate, a deposited coating, or a preform for subsequent operations, such as forging. Both ferrous and nonferrous products can be produced with deposition rates as high as 200 kilograms (400 pounds) per minute. Simultaneous deposition of two or more materials, injecting secondary particles into the stream, and promoting in-stream reactions can all be used to produce unique composites.

■ 16.11 SECONDARY OPERATIONS

Powder metallurgy products are often ready to use when they emerge from the sintering furnace. Many P/M products, however, utilize one or more secondary operations to provide enhanced precision, improved properties, or special characteristics.

During sintering, product dimensions shrink due to densification. In addition, warping or distortion may occur during cooldown from elevated temperature. As a result, a second pressing operation, known as *repressing*, *coining*, or *sizing*, may be required to restore or improve dimensional precision. The part is placed in a die and subjected to pressures equal to or greater than the initial pressing pressure. A small amount of plastic flow takes place, resulting in high dimensional accuracy, sharp detail, and improved surface finish. The associated cold working and increase in part density may combine to increase part strength by 25 to 50%. (*Note:* Because of the shrinkage that occurs during sintering, repressing cannot be performed with the same set of tooling that was used for the original powder compaction.)

If massive metal deformation takes place in the second pressing, the operation is known as *P/M forging*. Conventional powder metallurgy is used to produce a preform, which is one forging operation removed from the finished shape. The normal forging sequence of billet or bloom production, shearing, reheating, and sequential deformation to the desired shape is replaced by the manufacture of a comparatively simple-shaped powder metallurgy preform followed by a final hot-forging operation. The forging stage produces the complex shape, adds precision, provides the benefits of metal flow, and increases the density (often up to 99% of theoretical). The increase in density is accompanied by a significant improvement in mechanical properties. While protective atmospheres or coatings are required to prevent oxidation of the powder perform during heating and hot forging, the P/M process can often provide a significant reduction in scrap or waste. (By controlling preform weight to within 0.5%, flash-free forging can often be performed.) Forged products can benefit from the improved properties of powder metallurgy, such as the absence of segregation, the uniform fine grain size, and the use of novel alloys or unique composites. The conventional powder metallurgy process can be expanded to larger sizes and increased complexity. The tolerance requirements of cams, splines, and gears can usually be

FIGURE 16-13 Comparison of conventional forging and the forging of a powder metallurgy preform to produce a gear blank (or gear). Moving left to right, the top sequence shows the sheared stock, upset section, forged blank, and exterior and interior scrap associated with conventional forging. The finished gear is generally machined from the blank with additional generation of scrap. The bottom pieces are the powder metallurgy preform and forged gear produced without scrap by P/M forging. *(Courtesy of GKN Forging Limited.)*

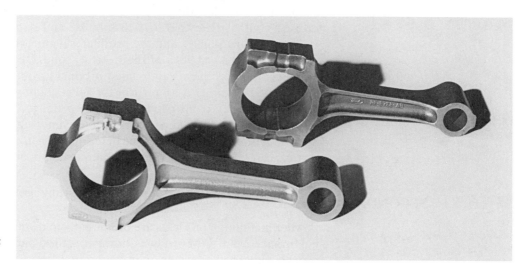

FIGURE 16-14 P/M forged connecting rods have been produced by the millions. *(Courtesy of Metal Powder Industries Federation.)*

met without subsequent machining. Figure 16-13 illustrates the reduction in scrap by comparing the same part made by conventional forging and the P/M forge approach. Figure 16-14 shows several P/M forged connecting rods. Millions have been produced and are currently being used by a number of automotive manufacturers.

Impregnation or infiltration are secondary processes that utilize the interconnected porosity or permeability of low-density P/M products. *Impregnation* refers to the forcing of oil or other liquid, such as a polymeric resin, into the porous network. This can be done by immersing the part in a bath and applying pressure or by a combination vacuum-pressure process. The most common application is that of oil-impregnated bearings. After impregnation, the bearing material contains from 10 to 40% oil by volume, which will provide lubrication over an extended lifetime of operation. In a similar manner, P/M parts can be impregnated with fluorocarbon resin (such as Teflon) to produce products offering a combination of high strength and low friction.

When the presence of pores is undesirable, P/M products may be subjected to metal *infiltration*. In this process a molten metal or alloy with a melting point lower than the P/M constituent flows into the interconnected pores of the product under pressure or by capillary action. Steel parts are often infiltrated with copper, for example. After infiltration, the engineering properties such as strength and toughness are improved to a level where they are generally comparable to those of solid metal products. Infiltration can also be used to seal pores prior to plating, improve machinability or corrosion resistance, or make the components gas- or liquidtight. Additional heating after infiltration can cause interdiffusion between the infiltrant and base metal, further enhancing mechanical properties.

Powder metallurgy products can also be subjected to the more conventional finishing operations such as *heat treatment*, *machining*, or *surface treatment*. If the part is of high density (<10% porosity) or has been metal impregnated, conventional processing is employed. Special precautions must be taken, however, when processing low-density P/M products. During heat treatment, protective atmospheres must again be used and certain liquid quenchants should be avoided. When machining, speeds and feeds must be adjusted and care taken to avoid pickup of lubricant or coolant. In general, P/M products should be machined using sharp tools, light cuts, and high feed rates. When large amounts of machining are required, special machinability-enhancing additions may be incorporated into the initial powder blend. Nearly all common methods of surface finishing can be applied, including platings and coatings, diffusion treatments, surface hardening, and steam treatment (which is used to produce a hard, corrosion-resistant oxide on ferrous parts). As with the other secondary processes, some process modifications may be required if the part has a reasonable amount of porosity or permeability. Since most parts are small and are produced in large quantity, barrel tumbling is another common means of cleaning, deburring, and surface modification.

■ 16.12 PROPERTIES OF P/M PRODUCTS

Because the properties of powder metallurgy products depend on so many variables—type and size of powder, amount and type of lubricant, pressing pressure, sintering temperature and time, finishing treatments, and so on—it is difficult to provide generalized information. Products can range all the way from low-density, highly porous parts with tensile strengths as low as 70 MPa (10 ksi) up to high-density pieces with tensile strengths of 1250 MPa (180 ksi) or more.

As shown in Figure 16-15, most mechanical properties show a strong dependence on product density, with the fracture-limited properties of toughness, ductility, and fatigue life being more sensitive than strength and hardness. The voids in the P/M part act as stress concentrators and assist in starting and propagating fractures. The yield strength of P/M products made from the weaker metals is often equivalent to the same material in wrought form. If higher-strength materials are used or the fracture-related tensile strength is specified, the properties of the P/M product tend to fall below those of wrought equivalents by varying but usually substantial amounts. Table 16-3 shows the properties of a few powder metallurgy materials compared with those of wrought material of similar composition. When larger presses or processes such as P/M forging or hot isostatic pressing are used to produce higher density, the strength of the P/M products approaches that of the wrought material. If the processing results in full density with fine grain size, P/M parts can actually have properties that exceed their wrought or cast equivalents. Since the mechanical properties of powder metallurgy

FIGURE 16-15   Mechanical properties versus as-sintered density for two iron-based powders. Properties depicted include yield strength, tensile strength, Charpy impact energy (shown in ft-lbs), and elongation in a 1-in. gage length.

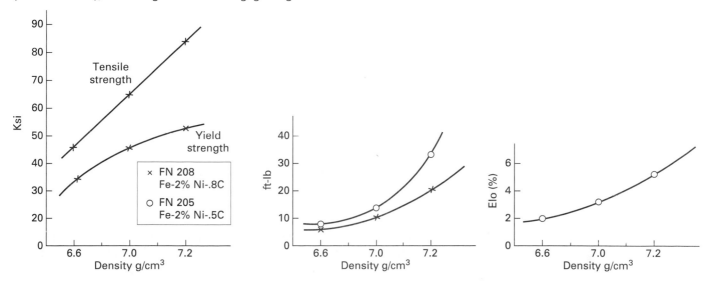

TABLE 16-3.   Comparison of Properties of Powder Metallurgy Materials and Equivalent Wrought Metals

Material[a]	Form and Composition	Condition[b]	Theoretical Density (%)	Tensile Strength		Elongation in 2 in. (%)
				10^3 psi	MPa	
Iron	Wrought	HR	—	48	331	30
	P/M—49% Fe min	As sintered	89	30	207	9
	P/M—99% Fe min	As sintered	94	40	276	15
Steel	Wrought AISI 1025	HR	—	85	586	25
	P/M—0.25%C, 99.75% Fe	As sintered	84	34	234	2
Stainless steel	Wrought Type 303	Annealed	—	90	621	50
	P/M Type 303	As sintered	82	52	358	2
Aluminum	Wrought 2014	T6	—	70	483	20
	P/M 201 AB	T6	94	48	331	2
	Wrought 6061	T6	—	45	310	15
	P/M 601 AB	T6	94	36.5	252	2
Copper	Wrought OFHC	Annealed	—	34	234	50
	P/M Copper	As sintered	89	23	159	8
		Repressed	96	35	241	18
Brass	Wrought 260	Annealed	—	44	303	65
	P/M 70% Cu-30% Zn	As sintered	89	37	255	26

[a]Equivalent wrought metal shown for comparison.   [b]HR, hot rolled; T6, age hardened.

products are so dependent upon density, *it is important that P/M products be designed and materials selected so that the final properties will be achieved with the anticipated amount of final porosity.*

Physical properties can also be affected by porosity. Corrosion resistance tends to be reduced due to the presence of entrapment pockets and fissures. Electrical, thermal, and magnetic properties all vary with density, usually decreasing in the presence of pores. The presence of porosity, however, increases the ability to damp both sound and vibration, and many P/M parts have been designed to take advantage of this feature.

■ 16.13 DESIGN OF POWDER METALLURGY PARTS

Powder metallurgy is a manufacturing system whose ultimate objective is to economically produce products for specific engineering applications. Success begins with good design and follows with good material and proper processing. In designing parts that are to be made by a powder metallurgy, it must be remembered that P/M is a special manufacturing process and provision should be made for a number of unique factors. Products that are converted from other manufacturing processes without modification in design rarely perform as well as parts designed specifically for manufacture by powder metallurgy. Some basic rules for the design of P/M parts are:

1. The shape of the part must permit ejection from the die. Sidewall surfaces should be parallel to the direction of pressing. Holes or recesses should have uniform cross section with axes and sidewalls parallel to the direction of punch travel.

2. The shape of the part should be such that powder is not required to flow into small cavities such as thin walls, narrow splines, or sharp corners.

3. The shape of the part should permit the construction of strong tooling.

4. The shape of the part should be within the thickness range for which P/M parts can be adequately compacted.

5. The part should be designed with as few changes in section thickness as possible.

6. Parts can be designed to take advantage of the fact that certain forms and properties can be produced by P/M which are impossible, impractical, or uneconomical to obtain by any other method.

7. If necessary, the design should be consistent with available equipment. Pressing areas should match press capability, and the number of thicknesses should be consistent with the number of available press actions.

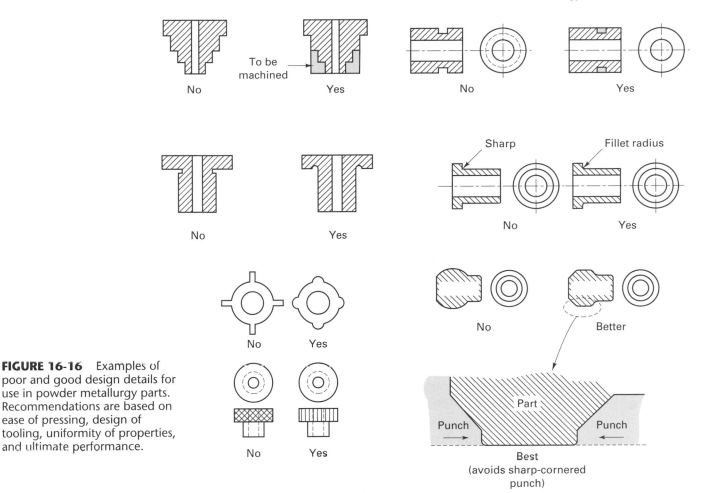

FIGURE 16-16 Examples of poor and good design details for use in powder metallurgy parts. Recommendations are based on ease of pressing, design of tooling, uniformity of properties, and ultimate performance.

8. Consideration should also be made for product tolerances. Higher precision and repeatability is observed for dimensions in the radial direction (set by the die) than for those in the axial or pressing direction (set by punch movement).

9. Finally, design should consider and compensate for the dimensional changes that will occur after pressing, such as the shrinkage that occurs during sintering.

The ideal powder metallurgy part, therefore, has a uniform cross section and a single thickness that is small compared to the cross-sectional width or diameter. More complex shapes are indeed possible, but it should be remembered that uniform strength and properties require uniform density. Holes that are parallel to the direction of pressing are easily accommodated. Holes at angles to this direction, however, must be made by secondary processing. Multiple-stepped diameters, reentrant holes, grooves, and undercuts should be eliminated whenever possible. Abrupt changes in section, narrow deep flutes, and internal angles without generous fillets should also be avoided. Straight serrations can be readily molded, but diamond knurls cannot. Punches should be designed to eliminate sharp points or thin sections that could easily wear or fracture. Figure 16-16 illustrates some of these design recommendations and restrictions.

■ 16.14 POWDER METALLURGY PRODUCTS

The products that are commonly produced by powder metallurgy can generally be classified into six groups.

1. *Porous or permeable products, such as bearings, filters, and pressure or flow regulators.* Oil-impregnated bearings, made from either iron or copper alloys, constitute a large volume of P/M products. They are widely used in home appliance and automotive applications since they require no lubrication or maintenance during their service life.

P/M filters can be made with pores of almost any size, some as small as 0.0025 mm (0.0001 in.). Unlike many alternative filters, powder metallurgy filters can withstand conditions of elevated temperature, high applied stresses, and corrosive environments.

2. *Products of complex shapes that would require considerable machining when made by other processes.* Because of the dimensional accuracy and fine surface finish that are characteristic of the P/M process, many parts require no further processing and others require only a small amount of finish machining. Tolerances can generally be held to within 0.1 mm (0.005 in.). Large numbers of small gears are currently being made by the powder metallurgy process. Other complex shapes, such as pawls, cams, and small activating levers, can be made quite economically.

3. *Products made from materials that are difficult to machine or with high melting points.* Some of the first modern uses of powder metallurgy were the production of tungsten lamp filaments and tungsten carbide cutting tools.

4. *Products where the combined properties of two or more metals (or metals and non-metals) are desired.* This unique capability of the powder metallurgy process is applied to a number of products. In the electrical industry, copper and graphite are frequently combined in applications like motor or generator brushes where copper provides the current-carrying capacity and graphite provides lubrication. Bearings have been made of graphite combined with iron or copper, or from mixtures of two metals, such as tin and copper, where the harder material provides wear resistance and the softer material deforms in a way that better distributes the load. Electrical contacts often combine copper or silver with tungsten, nickel, or molybdenum. Here, the copper or silver provides high conductivity, while the high melting temperature material provides resistance to fusion when the contacts experience arcing and subsequent closure.

5. *Products where the powder metallurgy process produces clearly superior properties.* The development of processes that produce full density has resulted in P/M products that are clearly superior to those produced by competing techniques. In areas of critical importance such as aerospace applications, the additional cost of the processing may be justified by the enhancement of properties. As another example, consider the production of P/M magnets. A magnetic field can be used to align the particles prior to sintering, resulting in a product with extremely high flux density.

6. *Products where the powder metallurgy process offers definite economic advantage.* Consideration of the process advantages described in the next section reveals features that may make powder metallurgy the most economical among two or more alternative ways to produce an equivalent part.

Figure 16-17 shows an array of typical powder metallurgy products.

FIGURE 16-17 Typical parts produced by the powder metallurgy process. *(Courtesy of PTX-Pentronix, Inc.)*

■ 16.15 ADVANTAGES AND DISADVANTAGES OF POWDER METALLURGY

Like all other manufacturing processes, powder metallurgy has distinct advantages and disadvantages that should be considered if the technique is to be employed economically and successfully. Among the important advantages are:

1. *Elimination or reduction of machining.* The *dimensional accuracy* and *surface finish* of P/M products are such that subsequent machining operations can be totally eliminated for many applications. If unusual dimensional accuracy is required, simple coining or sizing operations can often give accuracies equivalent to those of most production machining.

2. *High production rates.* All steps in the P/M process are simple and readily automated. Labor requirements are low, and product uniformity and reproducibility are among the highest in manufacturing.

3. *Complex shapes can be produced.* Subject to the limitations discussed previously, complex shapes can be produced, such as combination gears, cams, and internal keys. It is often possible to produce parts by powder metallurgy that cannot be machined or cast economically.

4. *Wide variations in compositions are possible.* Parts of very high purity can be produced. Metals and ceramics can be intimately mixed. Immiscible materials can be combined, and solubility limits can be exceeded. In most cases the chemical homogeneity of the product exceeds that of all competing techniques.

5. *Wide variations in properties are available.* Products can range from low-density parts with controlled permeability to high-density parts with properties that equal or exceed those of equivalent wrought counterparts. Damping of noise and vibration can be tailored into a P/M product. Magnetic properties, wear properties, and others can all be designed to match the needs of a specific application.

6. *Scrap is eliminated or reduced.* Powder metallurgy is the only common manufacturing process in which no material is wasted. In casting, machining, and press forming, the scrap can often exceed 50% of the starting material. This is particularly important where expensive materials are involved and may make it possible to use more costly materials without increasing the overall cost of the product. An example of such a product would be the rare-earth magnets.

The major disadvantages of the powder metallurgy process are:

1. *Inferior strength properties.* Because of the residual porosity, powder metallurgy parts generally have mechanical properties that are inferior to wrought or cast products of the same material. Their use may be limited when high stresses are involved. The required strength and fracture resistance, however, can often be obtained by using different materials or by employing alternate or secondary processing techniques that are unique to powder metallurgy.

2. *Relatively high tooling cost.* Because of the high pressures and severe abrasion involved in the process, the P/M dies must be made of expensive materials and be relatively massive. Because of the need for part-specific tooling, production quantities of less than 10,000 identical parts are normally not practical.

3. *High material cost.* On a unit weight basis, powdered metals are considerably more expensive than wrought or cast stock. However, the absence of scrap and the elimination of machining can often offset the higher cost of the starting material. In addition, powder metallurgy is usually employed for rather small parts where the material cost per part is not very great.

4. *Design limitations.* The powder metallurgy process is simply not feasible for many shapes. Parts must be able to be ejected from the die. The thickness/diameter (or thickness/width) ratio is limited. Thin vertical sections are difficult, and the overall size must be within the capacity of available presses. Few parts exceed 150 cm^2 (25 in^2) in pressing area.

5. *Density variations produce property variations.* The nonuniform product density that is frequently produced in compacting operations generally results in property variations throughout the part. For some products, these variations may be unacceptable.

6. *Health and safety hazards.* Many metals, such as aluminum, titanium, magnesium, and iron, are pyrophoric—they can ignite or explode when in particle form with large surface/volume ratios. Fine particles can also remain airborne for long times and can be inhaled by workers. To minimize the health and safety hazards, the handling of metal powders frequently requires the use of inert atmospheres, dry boxes, and hoods, as well as special cleanliness of the working environment.

■ 16.16 PROCESS SUMMARY

For many years, powder metallurgy products carried the stigma of "low strength" or "inferi-or mechanical properties." This label was largely the result of comparisons where "identical" parts were made of the same material, but by various methods of manufacture. In essence, the size, shape *and material* were all specified. In such a comparison, the product with 10 to 25% residual porosity would naturally be inferior to fully dense products made by casting, forming, or machining processes. Unfortunately, it is this type of comparison that is frequently made when one considers converting an existing design or existing part to P/M manufacture.

A far more valid comparison can be obtained by specifying size, shape, *and desired mechanical properties.* Each process can then be optimized by the selection of *both* mate-rial and process conditions. Powder metallurgy can now use unique materials, such as iron copper blends for which there are no cast or wrought equivalents. The P/M products can be designed to provide the targeted properties while containing the typical amounts of residual porosity. Since all products will then possess the targeted mechanical properties, process comparison can then be based on economic factors, such as total production cost. On this basis, powder metallurgy has emerged as a significant manufacturing process, and its products no longer carry the stigma of "inferior mechanical properties."

Table 16-4 summarizes some of the important manufacturing features of four pow-der processing methods. Note the variations in product size, production rate, produc-tion quantity, mechanical properties, and cost.

TABLE 16-4.   Comparison of Four Powder Processing Methods

Characteristic	Conventional Press and Sinter	Metal Injection Molding (MIM)	Hot Isostatic Pressing (HIP)	P/M Forging
Size of workpiece	Intermediate <5 pounds	Smallest <1/4 pound	Largest 1–1000 pounds	Intermediate <5 pounds
Shape complexity	Good	Excellent	Very good	Good
Production rate	Excellent	Good	Poor	Excellent
Production quantity	>5000	>5000	1–1000	>10,000
Dimensional precision	Excellent ±0.001 in/in	Good ±0.003 in/in	Poor ±0.020 in/in	Very good ±0.0015 in/in
Density	Fair	Very good	Excellent	Excellent
Mechanical properties	80–90% of wrought	90–95% of wrought	Greater than wrought	Equal to wrought
Cost	Low $0.50–5.00/lb	Intermediate $1.00–10.00/lb	High >$100.00/lb	Somewhat low $1.00–5.00/lb

■ KEY WORDS

amorphous	composites	liquid-phase sintering	porosity	sinter-HIP
apparent density	compressibility	lubricant	powder metallurgy	sintering
atomization	flow rate	metal injection molding	prealloyed powder	size distribution
binder	green strength	(MIM)	pressure-assisted	sizing
blending	hot isostatic pressing	mixing	sintering	spray forming
burn-off	(HIP)	P/M forging	protective atmosphere	(Osprey)
canning	impregnation	particle shape	rapidly solidified powder	surface texture
coining	infiltration	particle size	repressing	warm compaction
compacting	isostatic compaction	permeability	sinter brazing	

■ REVIEW QUESTIONS

1. What type of product would be considered to be a prospect for powder metallurgy manufacture?
2. What were some of the earliest powder metallurgy products?
3. What are some of the primary market areas for P/M products?
4. Which metal family currently dominates the powder metallurgy market?
5. What are the four basic steps that are usually involved in making products by powder metallurgy?
6. What are some of the important properties and characteristics of metal powders to be used in powder metallurgy?
7. What is the most common method of producing metal powders?
8. What are some of the other techniques that can be employed to produce particulate material?
9. Which of the powder manufacturing processes are likely to be restricted to the production of elemental (unalloyed) metal particles?
10. Why is powder metallurgy a key process in producing products from amorphous or rapidly solidified powders?
11. Why is flow rate an important powder characterization property?
12. What is apparent density, and how is it related to the final density of a P/M product?
13. What is green strength, and why is it important to the manufacture of high-quality P/M products?
14. What are some of the objectives of powder mixing or blending?
15. How does the addition of a lubricant affect compressibility? Green strength?
16. How might the use of a graphite lubricant be fundamentally different from the use of wax or stearates?
17. What types of composite materials can be produced through powder metallurgy?
18. What are some of the objectives of the compacting operation?
19. What limits the cross-sectional area of most P/M parts to several square inches or less?
20. Describe the movement of powder particles during compaction. What feature is responsible for the fact that powder does not flow and transmit pressure like a liquid?
21. For what conditions might a double-action pressing be more attractive than compaction with a single moving punch?
22. How is the final density of a P/M product typically reported?
23. Describe the four common classes of conventional powder metallurgy products.
24. What is isostatic compaction? For what product shapes might it be preferred?
25. What is the benefit of warm compaction?
26. How is the metal powder used in metal injection molding (MIM) different from the metal powder used in a conventional press-and-sinter production?
27. What are some of the ways that the binder can be removed from metal injection molded parts?
28. Why are P/M injection molding products injection molded to sizes that are considerably larger than the desired product?
29. Describe the ideal geometry for a metal injection molded product.
30. For what types of parts is P/M injection molding an attractive manufacturing process?
31. What are the three stages associated with most P/M sintering operations?
32. How is the sintering temperature usually related to the melting temperature of the material being sintered?
33. Why is it necessary to raise the temperature of P/M compacts slowly to the temperature of sintering?
34. Why is a protective atmosphere required during sintering?
35. What are some of the changes that occur to the compact during sintering?
36. What is the purpose of the sinter brazing process?
37. What are some of the attractive properties of hot isostatic pressed products?
38. What are some of the major limitations of the HIP process?
39. What is the attractive feature of the sinter-HIP and pressure-assisted sintering processes?
40. What are some of the other methods that can produce high-density P/M products?
41. Describe the spray forming process, and the unique feature that enables production of high-density, fine grain size products.
42. What is the purpose of repressing, coining, or sizing operations?
43. Why can we not use the original compaction tooling to perform repressing?
44. What is the major difference between repressing and P/M forging?
45. What is the difference between impregnation and infiltration? How are they similar?
46. What role does product density play when determining how to heat treat, machine, or surface treat a powder metallurgy product?
47. The properties of P/M products are strongly tied to density. Which properties show the strongest dependence?
48. What advice would you want to give to a person who is planning to convert the manufacture of a component from die casting to powder metallurgy?
49. What is the shape of an "ideal" powder metallurgy product?
50. What are some P/M products that have been intentionally designed to use the porosity or permeability features of the process?
51. Give an example of a product where two or more materials are mixed to produce a composite P/M product with a unique set of properties.
52. Why is finish machining such an expensive component in parts manufacture?
53. Describe some of the materials that can be made into P/M parts that could not be used for processes such as casting and forming.
54. Why is P/M not attractive for parts with low production quantities?
55. What features of the P/M process often compensate for the higher cost of the starting material?
56. How might you respond to the criticism that P/M parts have inferior strength?

■ PROBLEMS

1. When specifying the starting material for casting processes, the primary variables are chemistry and purity. Any structure of the starting material will be erased by the melting. For forming processes, the material remains in the solid state, so the principal concerns relating to the starting material are chemistry and purity, ductility, yield strength, strain-hardening characteristics, grain size, and so on. What are some of the characteristics that should be specified for the starting powder to assure the success of a powder metallurgy process? In what ways are these similar or different from those mentioned for casting and forming processes?

2. In conventional powder metallurgy manufacture, the material is compacted with applied pressure at room temperature and then sintered by elevated temperature at atmospheric pressure. With P/M hot pressing, the loose powder is subjected to pressure while it is also at elevated temperature. It would appear, therefore, that hot pressing could produce a finished part in a single operation and would be a more economical and attractive manufacturing process. What features have been overlooked in this argument that would tend to favor the press-and-sinter sequence for conventional manufacture?

www.wiley.com/college/degarmo

Chapter 16 CASE STUDY

Impeller for an Automobile Water Pump

The component pictured in Figure CS-16 is the impeller of a water pump used by a major automotive manufacturer. The outer diameter of the component is 2.75 in. and the total height of the six curved vanes is 0.75 in. (with a tolerance of 0.005 in.). The inner diameter of the center hole is 0.625 in., and the flat base is 0.187 in. thick. Vibration and balance considerations require accurate positioning and uniform thickness of the six curved vanes. A relatively smooth surface finish is desirable for good fluid flow.

The maximum operating temperature has been estimated at 300°F, and the contact fluid should be a water/antifreeze mixture with corrosion resistant additives. The designer has provided a target tensile strength of 30,000 psi, and notes that a minimum amount of fracture resistance is also desirable. Since there should be no direct metal-to-metal rubbing, enhanced wear resistance does not appear to be necessary. This is a high volume component, however, so low total cost (material plus manufacturing) would appear to be a prime objective.

Similar components have been sand cast from cast iron, with a grinding operation being required to maintain controlled height. The manufacturer is interested in improving quality and lowering cost.

1. Is a ferrous material needed to provide the desired properties, or might a nonferrous metal be acceptable?
2. What processes would you want to consider to mass produce such a shape? Are there more attractive casting processes? Is this a candidate for metal forming, and if so, which process or processes? Is powder metallurgy a possibility for this product? If so, can the desired properties be achieved with the densities that are common for a traditional press-and-sinter operation?
3. Investigate the various material-process combinations that would be candidates for production. Select and defend your "best choice."
4. Might this part be a candidate for manufacture from a nonmetal, such as molded nylon, some other polymer, or even some form of reinforced composite material? How would you suggest producing the desired shape if one or more of these materials were considered? Would you have to compromise on any of the performance requirements?

FIGURE CS-16 An automobile water pump impeller.

CHAPTER 17

FUNDAMENTALS OF METAL FORMING

■ 17.1 INTRODUCTION

The casting processes described in Chapters 13 through 15, and the powder metallurgy techniques presented in Chapter 16, have included a variety of methods for producing a desired shape from an engineering material. Each method had its characteristic set of capabilities, advantages, and limitations. Selection of the best method to make a given product, however, requires not only considerable information about the part to be made, but also a complete understanding of *all* of the available techniques for shape production and their related characteristics.

To further our knowledge of shape production alternatives, we will continue our survey by considering the family of *deformation processes*. These processes have been designed to exploit a remarkable property of some engineering materials (most notably metals) known as *plasticity*, the ability to flow as solids without deterioration of their properties. Since all processing is done in the solid state, there is no need to handle molten material or deal with the complexities of solidification. Since the material is simply moved (or rearranged) to produce the shape, as opposed to the cutting away of unwanted regions, the amount of waste is reduced substantially. Unfortunately, the forces required are often high. Machinery and tooling can be quite expensive, and large production quantities may be necessary to justify the approach.

The overall usefulness of metals is due largely to the ease by which they can be formed into useful shapes. Nearly all metal products undergo metal deformation at some stage of their manufacture. By rolling, cast ingots, strands, and slabs are reduced in size and converted into basic forms such as sheets, rods, and plates. These forms then undergo further deformation to produce wire, or the myriad of finished products formed by processes such as forging, extrusion, sheet metal forming, and others. The deformation may be *bulk flow* in three dimensions, simple *shearing*, simple or compound *bending*, or complex combinations of these. The stresses producing these deformations can be tension, compression, shear, or any of the other varieties included in Table 17-1. As shown in Table 17-2, the specific processes are numerous and varied. In addition, a wide range of speeds, temperatures, tolerances, surface finishes, and deformation amounts are possible.

TABLE 17-1. Classification of States of Stress

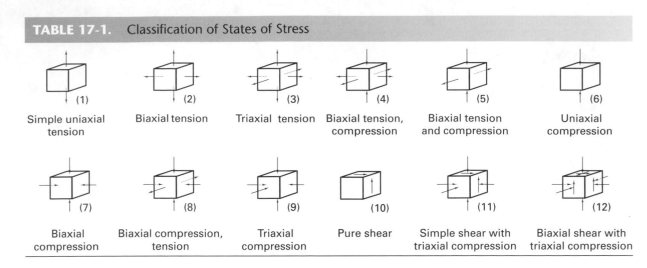

(1) Simple uniaxial tension	(2) Biaxial tension	(3) Triaxial tension
(4) Biaxial tension, compression	(5) Biaxial tension and compression	(6) Uniaxial compression
(7) Biaxial compression	(8) Biaxial compression, tension	(9) Triaxial compression
(10) Pure shear	(11) Simple shear with triaxial compression	(12) Biaxial shear with triaxial compression

■ 17.2 FORMING PROCESSES: INDEPENDENT VARIABLES

Forming processes tend to be *complex systems* consisting of independent variables, dependent variables, and independent-dependent interrelations. *Independent variables* are those aspects of the process over which the engineer has direct control, and they are generally selected or specified when setting up a process. Consider some of the independent variables in a typical forming process:

1. *Starting material.* When specifying the starting material, we may define not only the chemistry of that material, but also its condition. In so doing, we define the properties and characteristics of the material to be deformed. These may be chosen entirely for ease of fabrication, or the starting properties may be restricted by the desire to achieve the required final properties directly through the deformation process.

2. *Starting geometry of the workpiece.* The starting geometry may be dictated by previous processing, or it may be selected from a variety of available shapes. Economic considerations often influence this decision.

3. *Tool or die geometry.* This is an area of major significance and has many aspects, such as the diameter and profile of a rolling mill roll, a bend radius in a sheet-forming operation, the die angle in wire drawing or extrusion, and the cavity details when forging. Since the *tooling* will induce and control the metal flow as the material goes from starting shape to finished product, success or failure of a process often depends on tool geometry.

4. *Lubrication.* It is not uncommon for friction to account for more than 50% of the power supplied to a deformation process. In addition, lubricants can also act as coolants, thermal barriers, corrosion inhibitors, and parting compounds. Hence, their selection is an aspect of great importance to the success of an operation. Specification includes type of lubricant, amount to be applied, and the method of application.

5. *Starting temperature.* Since material properties can vary greatly with temperature, temperature selection and control is often key to the success or failure of a metal forming operation. Specification of starting temperatures may well include the temperatures of both the workpiece and the tooling.

6. *Speed of operation.* Most deformation processing equipment can be operated over a range of speeds. Since speed can directly influence the forces required for deformation (see Figure 2-31), the lubricant effectiveness, and the time available for heat transfer, its selection affects far more than the production rate.

7. *Amount of deformation.* While some processes control this variable through the design of tooling, others, such as rolling, may permit its adjustment at the discretion of the operator.

TABLE 17-2. Classification of Some Forming Operations

Process	Schematic Diagram	State of Stress in Main Part During Forming[a]
Rolling		7
Forging		9
Extrusion		9
Shear spinning		12
Tube spinning		9
Swaging or kneading		7
Deep drawing		In flange of blank, 5 In wall of cup, 1
Wire and tube drawing	(a) (b)	8
Stretching		2
Straight bending		At bend, 2 and 7
Contoured flanging	(a) Convex	At outer flange, 6 At bend, 2 and 7
	(a) Concave	At outer flange, 1 At bend, 2 and 7

[a] Numbers correspond to those in parentheses in Table 17-1

■ 17.3 DEPENDENT VARIABLES

After specification of the independent variables, the process in turn determines the nature and values for a second set of variables. Known as *dependent variables*, these, in essence, are the consequences of the independent variable selection. Examples of dependent variables include:

1. *Force or power requirements.* To convert a selected material from a starting shape to a final shape, with a specified lubricant, tooling geometry, speed, and starting temperature, will require a certain amount of force or power. A change in any of the independent variables will bring about a change in the force or power required, but the effect is indirect. We cannot directly specify the force or power; we can only specify the independent variables and then experience the consequences of that selection.

   It is important, however, that we be able to predict the forces or powers that will be required for an operation. Without a reasonable estimate of forces or power, we would be unable to specify the equipment for the process, select appropriate tool or die materials, compare various die designs or deformation methods, and ultimately optimize a process.

2. *Material properties of the product.* While we can easily specify the properties of the starting material, the combined effects of deformation and temperature experienced during the deformation process will certainly change them. While the starting properties of the material may be of interest to the manufacturer, the customer is far more concerned with our ability to produce both the desired final shape and the desired final properties. While the starting properties may be selected for ease of processing, it is important to know how they will be altered by the shape-producing process.

3. *Exit (or final) temperature.* Deformation generates heat within the material. Hot workpieces will cool when in contact with cold tooling. Lubricants can break down or decompose when overheated, or may react with the workpiece. The engineering properties of a material are altered by both the mechanical and thermal aspects of a deformation process. Therefore, if we are to control a process and produce quality products, it is important to know and control the temperature of the material throughout the deformation. (*Note:* The fact that temperature may vary from location to location within both the deforming and finished product further adds to the complexity of this variable.)

4. *Surface finish and precision.* Both *surface finish* and *dimensional precision* are characteristics of the resultant product that are dependent on the specific details of the process.

5. *Nature of the material flow.* Deformation processes generally exert forces or pressures on material surfaces and control the movement of the external surfaces through dies or tooling. The obvious objective of an operation is the production of a desired shape.

What may not be obvious, however, is the fact that the internal flow of material may be of equal importance. As will be shown later in this chapter, product properties can be significantly affected by the details of material flow, and material flow depends on all of the details of a process. Customer satisfaction requires not only the production of a desired geometric shape, but also on that shape possessing the right set of companion properties, without any surface or internal defects.

■ 17.4 INDEPENDENT–DEPENDENT RELATIONSHIPS

Figure 17-1 serves to illustrate a major problem facing metal-forming personnel. On the left side are the *independent variables, those aspects of the process for which control is direct and immediate.* On the right side are the *dependent variables, those aspects for which control is totally indirect.* Unfortunately, it is the dependent variables that we want to control, but the dependent variables are determined by the process, as complex consequences of the independent variable selection. If we want to change a dependent variable, we must determine which independent variable (or combination of independent variables) is to be changed, in what manner, and by how much. To make appropriate decisions, it is important for us to develop an understanding of the *independent variable–dependent variable interrelations.*

Independent variables	Links	Dependent variables
Starting material		Force or power requirements
Starting geometry	-Experience-	
Tool geometry		Product properties
Lubrication	-Experiment-	Exit temperature
Starting temperature		Surface finish
Speed of deformation	-Modeling-	Dimensional precision
Amount of deformation		Material flow details

FIGURE 17-1   Schematic of the metalforming system showing independent variables, dependent variables, and the various means of linking the two.

The link between independent and dependent variables is truly the most important area of knowledge for a person in metalforming. Unfortunately, such links are often difficult to obtain. Metalforming processes are complex systems composed of the material being deformed, the tooling performing the deformation, lubrication at surfaces and interfaces, and various other process parameters such as temperature and speed. The number of different forming processes (and variations thereof) is quite large. In addition, various materials often behave differently in the same process and multitudes of different lubricants exist. Some processes are sufficiently complex that they may have 15 or more interacting independent variables.

Information on the interdependencies of the independent and dependent variables can be obtained in three distinct ways:

1. *Experience.* Unfortunately, this requires long-time exposure to a process and is generally limited to the specific materials, equipment, and products encountered in the realm of past contact. Young employees are not likely to have the experience necessary to solve production problems. In addition, a single change in an area such as material, temperature, speed, or lubricant may render all past experience irrelevant.

2. *Experiment.* While possibly the least likely to be in error, direct experiment is both time consuming and costly. Size and speed of deformation are often reduced when conducting laboratory studies. Unfortunately, lubricant performance and heat transfer behave differently at different speeds and sizes, and their effects are generally altered. The most valid experiment, therefore, is one conducted under full-size and full-speed production conditions—generally too costly to consider to any great degree. While laboratory experiments can provide valuable insight, caution should be exercised when extrapolating lab-scale results to more realistic production conditions.

3. *Process modeling.* Here one approaches the problem with a high-speed computer and one or more mathematical models of the process. Numerical values are provided for the various independent variables and the models are used to compute predictions for the dependent variables. Most techniques rely on the applied theory of plasticity with various simplifying assumptions. Alternatives vary from crude, first-order approximations to sophisticated, computer-based solutions, such as the finite element method. Solutions may be algebraic relations that describe the process and reveal trends between the variables, or simply a numerical answer based on the specific input values.

■ 17.5 PROCESS MODELING

Metalforming simulations using the finite element modeling method became common in the 1980s, but generally required high-power minicomputers or engineering workstations. By the mid-1990s, the rapid increase in computing power had made it possible to model complex processes on desktop personal computers. Process simulations are now quick, inexpensive, and quite accurate. As a result, modeling is being used in all areas of manufacturing, including part design, manufacturing process design, heat treatment and surface treatment optimization, and others. Models can predict how a material will respond to a rolling process, fill a forging die, flow through an extrusion die, or solidify in a casting. Entire heat treatments can be simulated, including cooling rates in various quenchants. Models can even predict the strain distribution, residual stresses, microstructure, and final properties at all locations within a product.

Advanced simulation techniques can provide a clear and thorough understanding of a process, eliminating costly trial-and-error development cycles. Product design and manufacturing methods can be optimized for quality and reliability, while reducing production costs and minimizing lead times. When coupled with appropriate sensors, the same models can be used to determine the type of adjustments needed to provide on-line process control. In addition, process models can serve as laboratory tools to explore new ideas or new products. New employees can become familiar with what works and what doesn't in a quick and inexpensive manner.

It is important to note, however, that the accuracy of any model can be no better than that of the input variables. For example, when modeling a metal forming operation, the mechanical properties of the deforming material (i.e., yield strength, ductility, etc.) must be known for the specific conditions of temperature, strain (amount of prior deformation), and strain rate (speed of deformation) being considered. The mathematical descriptions of material behavior as a function of the process conditions are known as *constitutive relations*. The development of such relationships is not an easy task, however, because the same material may respond differently to the same conditions if its microstructure is different. A 1040 steel that has been annealed (ferrite and pearlite) will not have the same properties as a quenched and tempered (tempered martensite) steel of the same chemistry. Microstructure and its effects on properties are difficult to describe in quantitative terms that can be handled by a computer.

Another rather elusive variable is the friction between the tool and the workpiece. Studies have shown the value of friction to depend on contact pressure, contact area, surface finish, lubricant, speed, and the mechanical properties of the two contacting materials. We know that these parameters often vary from location to location and also change with time during a process, but many models tend to describe friction with a single variable of constant magnitude. Any variations with time and location are simply ignored in favor of mathematical simplicity or because of a lack of any better information.

At first glance, problems such as those just discussed appear to be a significant barrier to the use of mathematical models. It should be noted, however, that the same lack of knowledge applies to the person trying to document, characterize, and extrapolate the results of experience or experiments. Process modeling often reveals features that might otherwise go unnoticed and can be quite useful when attempting to prevent or eliminate defects, optimize performance, or extend a process into a previously unknown area.

■ 17.6 GENERAL PARAMETERS

While much metalforming knowledge is specific to a given process, there are certain features that are common to all processes, and these will be presented here.

It is extremely important to characterize the *material being deformed*. What is its strength or resistance to deformation at the relevant conditions of temperature, speed of deformation, and amount of prior straining? What are the formability limits and conditions of anticipated fracture? What is the effect of temperature or variations in temperature? To what extent does the material strain harden? What are the recrystallization kinetics? Will the material react with various environments or lubricants? These and many other questions must be answered to assess the suitability of a material to a given deformation process. Since the properties of engineering materials vary widely, the details will not be presented at this time. The reader is referred to the various chapters on engineering materials, as well as the more in-depth references cited in Chapter 9.

Another general parameter is the *speed of deformation* and the various related effects. Some rate-sensitive materials may shatter or crack if impacted, but will deform plastically when subjected to slow-speed loadings. Other materials appear to be stronger when deformed at higher speeds. For these *speed-sensitive materials*, more energy is needed to produce the same result if we wish to do it faster, and stronger tools may be required. Mechanical data obtained from slow strain rates in tensile tests may be totally useless if the deformation process operates at a significantly greater rate of deformation. Speed sensitivity is also greatest when the material is at elevated temperature, a condition that is frequently encountered in metalforming operations.

In addition to the changes in mechanical properties, faster deformation speeds tend to promote improved lubricant efficiency. Fast speeds also reduce the time for heat transfer and cooling. During hot working, workpieces stay hotter and less heat is transferred to the tools.

Other general parameters include *friction and lubrication* and *temperature*. Both of these are of sufficient importance that they will be discussed in some detail.

■ 17.7 FRICTION AND LUBRICATION UNDER METALWORKING CONDITIONS

High forces or high pressures are applied through tools to induce deformation or flow of a material. Because of the relative motion between the workpiece and the tool, an important consideration in metal deformation processes is the friction that exists at this interface. For some processes, more than 50% of the input energy is spent in overcoming friction. Changes in lubrication can alter the mode of material flow during forming, create or eliminate defects, alter the surface finish and dimensional precision of the product, and modify product properties. Production rates, tool design, tool wear, and process optimization all depend on the ability to determine and control friction between the tool and workpiece.

In most cases, we want to economically reduce the effects of friction. However, some deformation processes, such as rolling, can only operate when sufficient friction is present. Regardless of the process, friction effects are hard to measure. As previously noted, the specific friction conditions depend on a number of variables, including contact area, contact pressure, surface finish, speed, lubricant, and temperature. Because of the many variables, the effects of friction are extremely difficult to scale down for laboratory testing or extrapolate from laboratory tests to production conditions.

It should be noted that friction under metalworking conditions is significantly different from the friction encountered in most mechanical devices. The friction conditions of gears, bearings, journals, and similar components generally involve (1) two surfaces of similar material and similar strength, (2) under elastic loads such that neither body undergoes permanent change in shape, (3) with wear-in cycles that produce surface compatibility, and (4) generally low-to-moderate temperatures. Metalforming operations, on the other hand, involve a hard, nondeforming tool interacting with a soft workpiece at pressures sufficient to cause plastic flow in the weaker material. Only a single pass is involved as the tool induces deformation; the workpiece is often at elevated temperature; and the contact area is frequently changing as the workpiece deforms.

Figure 17-2 shows the relationship between frictional resistance and contact pressure. For light, elastic loads, friction is directly proportional to the applied pressure, with

FIGURE 17-2   Effect of contact pressure on the frictional resistance between two surfaces.

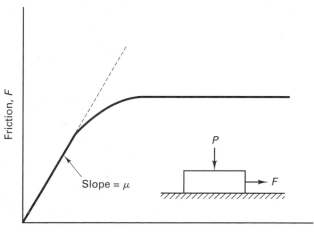

Contact pressure, P

the proportionality constant, μ, being known as the *coefficient of friction*, or more specifically, the Coulomb coefficient of friction. At high pressures, friction becomes independent of contact pressure and is more closely related to the strength of the weaker material.

An understanding of these results can be obtained from modern friction theory, whose primary premise is that "flat surfaces are not flat" but have some degree of roughness. When two irregular surfaces interact, sufficient contact is established to support the applied load. At the lightest of loads, only three points of contact may be necessary to support a plane. As the load is increased, the contacting points deform and the contact area increases, initially in a linear fashion. As load increases, more area comes into contact, and at some high value of load, there is full contact between the surfaces. Additional loads can no longer bring additional area into contact, and friction can now be described by a constant, independent of pressure.

Friction is the resistance to sliding along an interface. From a mechanistic viewpoint, this resistance can be attributed to (1) abrasion, the force necessary to plow the peaks of a harder material through a softer one, and/or (2) adhesion, the force necessary to rip apart microscopic weldments that form between the two materials. Since the weldment tears generally occur in the weaker of the two materials, it is reasonable to assume that the resistance attributed to both features is proportional to the strength of the weaker material and to the actual area of metal-to-metal contact. The curve depicted in Figure 17-2, a plot of friction versus contact pressure, can also be viewed as a plot of actual contact area at the interface versus the contact pressure. Figure 17-2 and the associated theory unfortunately applies only to unlubricated metal-to-metal contact. The addition of a lubricant, as well as any variation in its type or amount, can significantly alter the response.

Surface deterioration or *wear* is another phenomenon that is directly related to friction. Since the workpiece only interacts with the tooling during a single forming operation, any wear experienced by the workpiece is usually not objectionable. In fact, a shiny, fresh-metal surface produced by wear is often viewed as desirable. Manufacturers whose processes retain most or all of the original dull finish may be accused of selling old or substandard products. Wear on the tooling, however, is quite the reverse. Tooling is expensive and it is expected to shape many workpieces. Tooling wear generally means that the dimensions of the workpiece will change. Tolerance control is lost, and at some point, the tools will have to be replaced. Other consequences of tool wear include increased frictional resistance (increased required power and decreased process efficiency), poor surface finish on the product, and loss of production during tool changes.

Lubrication is a key to success in many metalforming operations. While lubricants are generally selected for their ability to reduce friction and suppress tool wear, secondary considerations may include the ability to act as a thermal barrier, keeping heat in the workpiece and away from the tooling; the ability to act as a coolant, removing heat from the tools; and the ability to retard corrosion if left on the formed product. Other influencing factors include ease of application and removal; lack of toxicity, odor, and flammability; reactivity or lack of reactivity with material surfaces; adaptability over a useful range of pressure, temperature, and velocity; surface wetting characteristics; cost; availability; and the ability to flow or thin and still function as a lubricant. Lubricant selection is further complicated by the fact that lubricant performance may change with any change in the interface conditions. The exact response is often dependent on such factors as the finish of both surfaces, the area of contact, the applied load, the speed, the temperature, and the amount of lubricant.

The ability to select an appropriate lubricant can be a critical factor in determining whether a process is successful or unsuccessful, efficient or inefficient. For example, if a lubricant layer can prevent mechanical contact between the tool and the workpiece (full-fluid or solid layer separation), the forces and power required may decrease by as much as 30 to 40%, and tool wear becomes almost nonexistent. Considerable effort, therefore, has been directed to the study of friction and lubrication, a subject known as *tribology*, as it applies to both general metalworking conditions and specific metalforming processes. A substantial information base has been developed that can aid in optimizing the use of lubricants in metalworking.

■ 17.8 TEMPERATURE CONCERNS

In metalworking operations, workpiece temperature can be one of the most important process variables. The role of temperature in altering the properties of a material has been discussed in Chapter 2. In general, an increase in temperature brings about a decrease in strength, an increase in ductility, and a decrease in the rate of strain hardening—all effects that would tend to promote ease of deformation.

Forming processes tend to be classified as hot working, cold working, or warm working based on both the temperature and the material being formed. In hot working, the deformation is performed under conditions of temperature and strain rate where recrystallization occurs simultaneously with the deformation. To achieve this, the temperature of deformation is usually in excess of 0.6 times the melting point of the material on an absolute temperature scale (Kelvin or Rankine). Cold working is deformation under conditions where the recovery processes are not active. Here the working temperatures are usually less than 0.3 times the workpiece melting temperature. Warm working is deformation under the conditions of transition (i.e., a working temperature between 0.3 and 0.6 times the melting point).

HOT WORKING

Hot working is defined as the plastic deformation of metals at a temperature above the recrystallization temperature. It is important to note, however, that the recrystallization temperature varies greatly with different materials. Tin is near hot-working conditions at room temperature; steels require temperatures near 2000°F; and tungsten does not enter the hot-working regime until about 4000°F. Thus the term hot working does not necessarily correlate with high or elevated temperature, although such is often the case.

As shown in Figures 2-29 and 2-30, elevated temperatures bring about a decrease in the yield strength of a metal and an increase in ductility. At the temperatures of hot working, recrystallization eliminates the effects of strain hardening, so there is no significant increase in yield strength or hardness, or corresponding decrease in ductility. The true stress–true strain curve is essentially flat once we exceed the yield point, and deformation can be used to drastically alter the shape of a metal without fear of fracture and without the requirement of excessively high forces. In addition, the elevated temperatures promote diffusion that can remove or reduce chemical inhomogeneities; pores can be welded shut or reduced in size during the deformation; and the metallurgical structure can often be altered through recrystallization to improve the final properties. An added benefit is observed for steels, where hot working involves the deformation of weak, ductile, face-centered-cubic austenite, as opposed to the stronger body-centered-cubic ferrite that is the stable structure at lower temperatures.

From a negative perspective, the high temperatures of hot working may promote undesirable reactions between the metal and its surroundings. Tolerances are poorer due to thermal contractions and possible warping or distortion that results from nonuniformity in the cooling. The metallurgical structure may also be nonuniform, since the final grain size depends on the amount of deformation, temperature at last deformation, cooling history after the deformation, and other factors, all of which may vary throughout a workpiece.

While recrystallization sets the minimum temperature for hot working, the upper boundary for hot working is usually determined by factors such as excess oxidation, grain growth, or undesirable phase transformations. To keep the forming forces as low as possible and enable hot deformation to be performed for a reasonable amount of time, the starting temperature of the workpiece is usually set at or near the highest temperature for hot working.

Structure and Property Modification by Hot Working. When metals solidify, particularly in the large sections that are typical of ingots or continuously cast slabs or strands, coarse structures tend to form with a certain amount of chemical segregation. The size of the grains is usually not uniform, and undesirable grain shapes can be quite common, such as the columnar grains that have been revealed in Figure 17-3. Small gas cavities or shrinkage porosity can also form during solidification.

FIGURE 17-3 Cross section of a 4-in. diameter cast copper bar showing the as-cast grain structure.

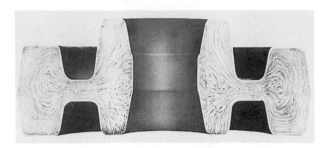

FIGURE 17-4 Flow structure of a hot-formed (forged) transmission gear blank. *(Courtesy of Bethlehem Steel Corporation.)*

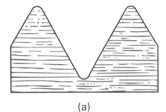

(a)

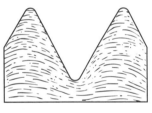

(b)

FIGURE 17-5 Schematic comparison of the grain flow in a machined thread (a) and a rolled thread (b). The rolling operation further deforms the axial structure produced by the previous wire- or rod-forming operations, while machining simply cuts through it.

If a cast metal is reheated without prior deformation, it will simply experience grain growth and the accompanying deterioration in engineering properties. However, if the metal has experienced a sufficient amount of deformation, the distorted structure will be rapidly replaced by new strain-free grains. This *recrystallization* is then followed by either (1) grain growth, (2) additional deformation and recrystallization, or (3) a drop in temperature that will terminate diffusion and "freeze in" the recrystallized structure. The structure in the final product is that formed by the last recrystallization and the thermal history that followed. By replacing the starting structure with one of fine, spherical-shaped grains, it is possible to produce an increase not only in strength but also in ductility and toughness—a somewhat universal enhancement of properties.

Engineering properties can also be improved through the reorientation of inclusions or impurity particles that are present in the metal. With normal melting and cooling, many impurities tend to locate along grain boundary interfaces. If these are unfavorably oriented or intersect surfaces, they can initiate a crack or assist its propagation through a metal. When a metal is plastically deformed, the impurities tend to flow along with the base metal, or fracture into rows of fragments (*stringers*) that are aligned in the direction of working. These nonmetallic impurities do not recrystallize with the base metal but retain their distorted shape and orientation. The product exhibits a *flow structure*, like the one shown in Figure 17-4, and properties tend to vary in different directions. Through proper design of the deformation, the impurities can often be reoriented into a "crack-arrestor" configuration where they are perpendicular to the direction of crack propagation. For example, the outer lobe of the forging in Figure 17-4 has excellent fracture resistance since all flow lines are parallel to the external surface. The impurities appear as crack initiators or crack propagators only at the top and bottom of the inner lobe, which hopefully is a low-stress or noncritical location.

Figure 17-5 schematically compares a machined thread and a rolled thread in a threaded fastener. By reorienting the axial defects in the starting wire or rod to be parallel to the thread surfaces, the rolled thread offers improved strength and fracture resistance.

Temperature Variations. The success or failure of a hot deformation process often depends on the ability to control the temperatures within the workpiece. Over 90% of the energy imparted to a deforming workpiece will be converted into heat. If the deformation process is sufficiently rapid, the temperature of the workpiece may actually increase. More common, however, is the cooling of the workpiece in its lower-temperature environment. Heat is lost through the workpiece surfaces, with the majority of the loss occurring where the workpiece is in direct contact with lower-temperature tooling. Nonuniform temperatures are produced, and flow of the hotter, weaker, interior may well result in cracking of the colder, less ductile, surfaces. Thin sections cool faster than thick sections, and this may further complicate the flow behavior.

To minimize problems, it is desirable to keep the workpiece temperatures as uniform as possible. Heated dies can reduce the rate of heat transfer, but die life tends to be compromised. For example, dies are frequently heated to 325 to 450°C (600 to 850°F) when used in the hot forming of steel. Tolerances could be improved and contact times could be increased if the tool temperatures could be raised to 550 to 650°C (1000 to 1200°F), but tool life drops so rapidly that these conditions become quite unattractive.

A final concern is the cool-down from the temperatures of hot working. Nonuniform cooling can introduce significant amounts of *residual stress* in hot-worked products. Associated with these stresses may be warping or distortion, and possible cracking.

COLD WORKING

The plastic deformation of metals below the recrystallization temperature is known as *cold working*. Here, the deformation is usually performed at room temperature, but mildly elevated temperatures may be used to provide increased ductility and reduced strength. From a manufacturing viewpoint, cold working has a number of distinct advantages, and the various cold-working processes have become quite prominent. Recent advances have expanded their capabilities, and a trend toward increased cold working appears likely to continue.

When compared to hot working, the advantages of cold working include:

1. No heating is required.
2. Better surface finish is obtained.
3. Superior dimensional control is achieved since the tooling sets dimensions at room temperature. As a result, little, if any, secondary machining is required.
4. Products possess better reproducibility and interchangeability.
5. Strength, fatigue, and wear properties are all improved through strain hardening.
6. Directional properties can be imparted.
7. Contamination problems are minimized.

Some disadvantages associated with cold-working processes include:

1. Higher forces are required to initiate and complete the deformation.
2. Heavier and more powerful equipment and stronger tooling are required.
3. Less ductility is available.
4. Metal surfaces must be clean and scale-free.
5. Intermediate anneals may be required to compensate for the loss of ductility that accompanies strain hardening.
6. The imparted directional properties may be detrimental.
7. Undesirable residual stresses may be produced.

The strength levels induced by *strain hardening* are often comparable to those produced by the strengthening heat treatments. Even when the precision and surface finish of cold working are not required, it may be cheaper to produce a product by cold working a less expensive alloy (achieving the strength by strain hardening) than by heat-treating parts that have been hot formed from an alloy that will respond to the strengthening heat treatment. In addition, better and more ductile metals and an improved understanding of plastic flow have done much to reduce the difficulties often experienced during cold forming. As an added benefit, most cold working processes eliminate or minimize the production of waste material and the need for subsequent machining—a significant feature with today's emphasis on conservation and materials recycling.

Because the cold-forming processes require powerful equipment and product-specific tools or dies, they are best suited for large-volume production of precision parts, where the quantity of products can justify the cost of the equipment and tooling. Considerable effort has been devoted to developing and improving cold-forming machinery along with methods that enable these processes to be economically attractive for modest production quantities. By grouping products made from the same starting material and using quick-change tooling, cold-forming processes can often be adapted to small-quantity or just-in-time manufacture.

Metal Properties and Cold Working.   The suitability of a metal for cold working is determined primarily by its tensile properties, and these are a direct consequence of its metallurgical structure. Cold working will alter that structure, thereby altering the tensile properties of the resulting product. It is important for both sets of properties to be considered by the designer when selecting metals that are to be processed by cold working.

Figure 17-6 presents the true stress–true strain curves for both a low- and a high-carbon steel. Focusing on the low-carbon material, we note that plastic deformation cannot occur until the strain exceeds X_1, the strain associated with the elastic limit, point *a*,

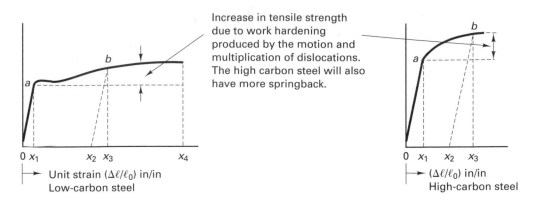

FIGURE 17-6 Use of true stress–true strain diagrams to assess the suitability of two metals for cold working.

on the stress–strain curve. Plastic deformation continues until the strain reaches the value X_4, where the metal ruptures. From the viewpoint of cold working, two features are significant: (1) the magnitude of the yield-point stress, which determines the force required to initiate permanent deformation, and (2) the extent of the strain region from X_1 to X_4, which indicates the amount of plastic deformation (or ductility) that can be achieved without fracture. If a considerable amount of deformation is desired, a material like the low-carbon steel is more desirable than the high-carbon variety. Greater ductility would be available and less force would be required to initiate and continue the deformation. The curve on the right, however, has a higher strain-hardening coefficient (see Chapter 2 for discussion). If strain hardening is being used to impart strength, this material would have a greater increase in strength for the same amount of cold work. In addition, the material on the right would be more attractive for shearing operations and may be easier to machine (see Chapter 21).

Springback is another cold-working phenomenon that can be explained with the aid of a stress-strain diagram. When a metal is deformed by the application of a load, part of the resulting deformation is elastic. For example, if a metal is stretched to point X_1 in Figure 17-6 and the load is removed, it will return to its original size and shape because all of the deformation is elastic. If, on the other hand, the metal is stretched by an amount X_3, corresponding to point b on the stress-strain curve, the total strain is made up of two parts, a portion that is elastic and another that is plastic. When the deforming load is removed, the stress relaxation will follow line bX_2, and the final strain will only be X_2. The decrease in strain, $X_3 - X_2$, is known as *elastic springback*.

In cold-working processes, springback can be extremely important. If a desired size is to be achieved, the deformation must be extended beyond that point by an amount equal to the springback. Since different materials have different elastic moduli, the amount of springback from a given load will change from one material to another. A substitution in material may well require adjustments in the forming process. Fortunately, springback is a predictable phenomenon, and most difficulties can be prevented by proper design procedures.

Preparing Metals for Cold Working. The quality of the starting material is often key to the success or failure of a cold-working operation. To obtain a good surface finish and maintain dimensional precision, the starting material must be clean and free of oxide or scale that might cause abrasion and damage to the dies or rolls. Scale can be removed by pickling, a process in which the metal is dipped in acid and then washed. In addition, sheet metal and plate is sometimes given a light cold rolling prior to the major deformation. The rolling operation not only assures uniform starting thickness but also produces a smooth starting surface.

The light cold-rolling pass can also serve to remove the *yield-point phenomenon* and the associated problems of nonuniform deformation and surface irregularities in the product. Figure 17-7 presents a blow-up of the left-hand region of Figure 2-5 or Figure 17-6, a stress-strain curve that is typical of many low-carbon steels. After loading to the upper yield point, the material exhibits a *yield-point runout* wherein the material

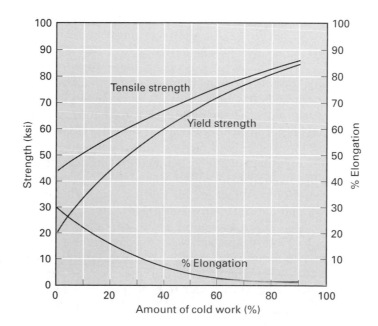

FIGURE 17-7 (*Left*) Stress-strain curve for a low-carbon steel showing the commonly observed yield-point runout; (*Right*) Luders bands or stretcher strains that form when this material is stretched to an amount less than the yield-point runout.

can strain up to several percent with no additional force being required. Consider a piece of sheet metal that is to be formed into an automotive body panel. If a segment of that panel were to receive a total stretch less than the yield-point value, a stress equal to the yield-point stress would have to be applied. Under this stress, the material is free to not deform at all, to deform the entire amount of the yield-point runout, or to select some point in between. It is not uncommon for some regions to deform the entire amount and thin correspondingly while adjacent regions resist deformation and retain the original thickness. The resulting ridges and valleys, shown in Figure 17-7, are referred to as *Luders bands* or *stretcher strains* and are very difficult to remove or conceal. By first cold rolling the material to a strain near or past the yield-point runout, subsequent forming occurs in a smooth-line region of the curve where a well-defined strain corresponds to each value of stress. When the body panel is shaped from this material, the deformation and thinning are now uniform across the piece.

Figure 17-8 shows the effect of cold working on the mechanical properties of pure copper. Individual tensile tests were conducted on specimens that had experienced progressively greater amounts of cold work. As the graph shows, yield strength and tensile strength increase with increased deformation. While not presented on the graph, hardness generally follows tensile strength. Elongation (or ductility) decreases, with the

FIGURE 17-8 Mechanical properties of pure copper as a function of the amount of cold work (expressed in percent).

amount of cold working generally being limited by the onset of fracture. Reduction in area, another measure of ductility, shows a similar decline, along with electrical conductivity and corrosion resistance.

An *annealing* heat treatment is often applied to a metal prior to cold working as a means of maximizing the amount of starting ductility. If the required amount of deformation exceeds the fracture limit, *intermediate anneals* may be performed to restore ductility (set the amount of cold work back to zero), thereby enabling further working without danger of fracture. Often the desired final properties coincide with a given amount of cold work. If the last anneal is properly positioned in the deformation cycle, the desired shape can be produced along with the mechanical properties that accompany the amount of cold work imparted since that anneal. For example, if the last anneal were moved earlier in a deformation sequence, a product that used to contain 10% cold work may now contain 40% cold work and be stronger but less ductile. In all annealing operations care should be exercised to control the grain size of the resulting material. Grain sizes that are too large or too small can both be detrimental.

Cold working, like hot working, also produces an anisotropic structure—one whose properties vary with direction. Here, the *anisotropy* is related to the distorted crystal structure, and is not simply a function of the nonmetallic inclusions. Also associated with cold working is the generation of *residual stresses*. Anisotropy and residual stresses can be beneficial, and they can also be quite harmful. In any case, they should be considered to be a consequence of cold working, and worthy of consideration as to their effect on performance.

WARM FORMING

Deformation produced at temperatures intermediate to hot and cold forming is known as *warm forming*. Compared to cold forming, warm forming offers the advantages of reduced loads on the tooling and equipment, increased material ductility, and a possible reduction in the number of anneals due to a reduction in the amount of strain hardening. The use of higher forming temperatures can often expand the range of materials and geometries that can be formed by a given process or piece of equipment. High-carbon steels can often be formed without a prior spheroidization treatment. Compared to hot forming, the lower temperatures of warm working produce less scaling and decarburization, and enable production of products with better dimensional precision and smoother surfaces. Finish machining is reduced and less material is converted into scrap. Because of the finer structures and the presence of some strain hardening, the as-formed properties may be adequate for many applications, enabling the elimination of final heat treatment operations. The warm regime generally requires less energy than hot working due to the decreased energy in heating the workpiece (lower temperature), energy saved through higher precision (less material being heated), and the possible elimination of postforming heat treatments. Tools last longer, for while they must exert 25 to 60% higher forces, there is less thermal shock and thermal fatigue.

When energy was cheap, metal forming was usually conducted in either the hot- or cold-working regimes, and warm working was largely ignored. Even today, material behavior is less well characterized for the warm working temperatures (the warm-working temperatures for steel are between 550 and 800°C or 1000 and 1500°F). Lubricants are not as fully developed for the warm working temperatures and pressures, and die design technology is not as well established. Nevertheless, the pressures of energy and material conservation, coupled with the other cited benefits, strongly favor the continued development of warm working. Cold forming is still the preferred method for fabricating small components, but warm forming is considered to be attractive for larger parts (up to about 10 lb) and steels with more than 0.35% carbon and/or high alloy content.

Hot working and warm forming are usually applied to the bulk forming processes, like forging and extrusion. For sheet material, the surface-to-volume ratio is generally large enough that the workpiece is prone to rapid changes in temperature. As the auto manufacturers seek to further increase fuel efficiency (while maintaining performance, comfort, affordability and safety), there has been an increased interest in

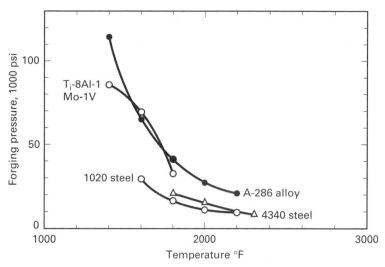

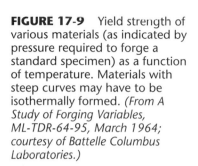

FIGURE 17-9 Yield strength of various materials (as indicated by pressure required to forge a standard specimen) as a function of temperature. Materials with steep curves may have to be isothermally formed. *(From A Study of Forging Variables, ML-TDR-64-95, March 1964; courtesy of Battelle Columbus Laboratories.)*

aluminum sheet applications. Unfortunately, the formability of high-strength aluminum sheet is much lower than similar strength low-carbon steels. If the steel is simply converted to aluminum, and the design and tooling remain unchanged, fracture usually occurs in the more heavily worked regions. If the material, die, and blank holder are all heated to 200–300°C (400–575°F), however, many aluminum alloys show a significant increase in formability, and satisfactory parts can be produced.

ISOTHERMAL FORMING

Figure 17-9 shows the relationship between yield strength (or forging pressure) and temperature for several engineering metals. The 1020 and 4340 steels show a moderate increase in strength with decreasing temperature. In contrast, the strength of the titanium alloy and the A-286 nickel-based superalloy shows a much stronger dependence on temperature. Within the realm of typical hot-working temperatures, cooling of as little as 100°C (200°F) can produce a doubling in strength. During hot forming, cooler surfaces surround a hotter interior, and the variations in strength can result in nonuniform deformation and cracking of the less ductile surface.

To successfully deform temperature-sensitive materials, deformation may have to be performed under *isothermal* (constant temperature) conditions. The dies or tooling must be heated to the same temperature as the workpiece, sacrificing die life for product quality. Deformation speeds must be slowed so that any heat generated by deformation can be removed in a manner that would maintain a uniform and constant temperature. Inert atmospheres may be required because of the long times at elevated temperature. Although such methods are indeed costly, they are often the only means of producing satisfactory products from certain materials. As a result of the unique conditions of forming, isothermally-formed components generally exhibit close tolerances, low residual stresses, and fairly uniform metal flow.

■ KEY WORDS

abrasion
adhesion
anisotropy
annealing
bending
bulk flow
coefficient
    of friction
cold working

constitutive
    relation
deformation
    processes
dependent variable
dimensional
    precision
dislocation
elastic springback

flow structure
friction
hot working
independent
    variable
intermediate
    anneal
isothermal forming
lubrication

Luders bands
oriented struc-
    ture
plasticity
recrystallization
shearing
speed sensitivity
springback
strain hardening

stretcher strains
stringers
surface finish
tooling
tribology
warm working
wear
yield-point
    runout

■ REVIEW QUESTIONS

1. What is plasticity?
2. What are some of the general assets of the metal deformation processes? Some general liabilities?
3. Why might large production quantities be necessary to justify metal deformation as a means of manufacture?
4. What is an independent variable in a metalforming process?
5. What is the significance of tool and die geometry in designing a successful metalforming process?
6. Why is lubrication often a major concern in metalforming?
7. What are some of the possible roles of a lubricant in addition to reducing friction?
8. What are some of the secondary effects that may occur when the speed of a metalforming process is varied?
9. What is a dependent variable in a metalforming process?
10. Why is it important to be able to predict the forces or powers required to perform specific forming processes?
11. Why is it important to know and control the thermal history of a metal as it undergoes deformation?
12. Why is it often difficult to determine the specific relationships between independent and dependent variables?
13. What are the three distinct ways of determining the interrelation of independent and dependent variables?
14. What features limit the value of laboratory experiments in modeling metalforming processes?
15. What features have contributed to the expanded use of process modeling?
16. What features may limit the accuracy of a mathematical model?
17. What are some of the uses or applications of process models?
18. What is a constitutive relation for an engineering material?
19. What simplifying assumptions are often made regarding friction between the tool and workpiece?
20. What type of information about the material being deformed may be particularly significant to a metalforming engineer?
21. Why is friction such an important parameter in metalworking operations?
22. What are several ways in which the friction conditions during metalworking differ from the friction conditions found in most mechanical equipment?
23. According to modern friction theory, frictional resistance can be attributed to what two physical phenomena?
24. Discuss the significance of wear in metalforming: wear on the workpiece and wear on the tooling.
25. Lubricants are often selected for properties in addition to their ability to reduce friction. What are some of these additional properties?
26. What are some of the benefits that can be obtained by fully separating a tool and workpiece by an intervening layer of lubricant?
27. If the temperature of a material is increased, what changes in properties might occur that would promote the ease of deformation?
28. Define the various regimes of cold working, warm working, and hot working in terms of the melting point of the material being formed.
29. What is an acceptable definition of hot working?
30. What are some of the attractive manufacturing and metallurgical features of hot-working processes?
31. What are some of the negative aspects of hot working?
32. Describe how hot working can be used to improve the grain structure of a metal.
33. If the deformed grains recrystallize during hot working, how can the process impart an oriented or flow structure (and directionally dependent properties) to the product?
34. Why are heated dies or tools often employed in hot-working processes?
35. What generally restricts the upper temperature to which dies or tooling is heated?
36. Compared to hot working, what are some of the advantages of cold-working processes?
37. What are some of the disadvantages of cold-forming processes?
38. How could cold working be used to reduce the cost of a moderate-to-high-strength product?
39. How can the tensile test properties of a metal be used to assess its suitability for cold forming?
40. Why is elastic springback an important consideration in cold-forming processes?
41. What are Luders bands or stretcher strains, and what causes them to form?
42. What engineering properties are likely to decline during the cold working of a metal?
43. How can the selective placement of the final intermediate anneal be used to establish desired final properties in a cold formed product?
44. What are some of the advantages of warm forming compared to cold forming? Compared to hot forming?
45. What features have contributed to the slow development of warm working?
46. What material feature is considered to be the driving force for isothermal forming?
47. Why is isothermal forming considerably more expensive than conventional hot forming?

■ PROBLEMS

1. Copper rod is being reduced from a hot-rolled 3/8-in. diameter to a final diameter of 0.100 in. by wire drawing through a series of dies. The final wire should have a yield strength in excess of 50,000 psi and an elongation greater than 10%. Use Figure 17-8 to determine a desirable amount of final cold work. Describe the drawing sequence and the placement of the last intermediate anneal, so that the final product has both the desired size and the desired properties.
2. a. List and discuss the various economic factors that should be considered when evaluating a possible switch from cold forming to warm forming.

   b. Repeat part (a) for a possible conversion from hot forming to warm forming.
3. An advertisement for automobile spark plugs cited the superiority of rolled threads over machined threads. Figure 17-5 shows such a comparison for hot forming. The spark plug threads, however, were cold rolled. Discuss the assets and liabilities of the cold rolling of threads compared to thread formation by conventional machining.

*C*hapter 17 CASE STUDY

Repairs to a Damaged Propeller

The propeller of a moderately large pleasure boat has been cast from a nickel-aluminum-bronze alloy that contains 82% Cu, 9% Al, 4% Ni, 4% Fe, and 1% Mn. It is approximately 13 inches in diameter with three 10-pitch blades, and has been designed for both fresh- and salt-water usage.

1. One of the blades has struck a rock and is badly bent. A replacement propeller is quite expensive and cannot be obtained for several weeks. An attractive alternative, therefore, may be to repair the existing piece. Would you recommend such a repair and how would you proceed? Can it simply be hammered back into shape? Would you recommend any additional processing, either before or after the repair? What is the rationale for your recommendations?

2. A second propeller, identical to the one above, has also been damaged by an impact. This time, however, the damage is in the form of a crack at the base of one of the blades, as shown in Figure CS-17. Since the crack does not penetrate into the hub, it is proposed that a repair be made using some form of welding or brazing process. Would you recommend such a repair? If so, how would you suggest the repair be made? Explain the rationale for your recommendations and outline the procedure that should be followed. Would there be any sacrifice in quality or performance with the repaired propeller?

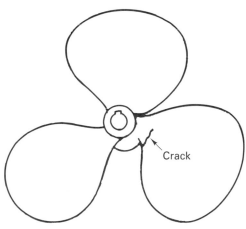

FIGURE CS-17

CHAPTER 18

HOT-WORKING PROCESSES

■ 18.1 INTRODUCTION

The shaping of metal by deformation is as old as recorded history. The Bible, in the fourth chapter of Genesis, introduces Tubal-cain and cites his ability as a worker of metal. While we do not know of his equipment, it is well established that metal forging was practiced before written records. Processes such as rolling and wire drawing were common in the Middle Ages and probably date back much further. In North America, by 1680 the Saugus Iron Works near Boston had an operating drop forge, rolling mill, and slitting mill.

Although the basic concepts of many forming processes have remained largely unchanged throughout history, the details and equipment have evolved considerably. Manual processes were converted to machine processes during the industrial revolution. The machinery then became bigger, faster, and more powerful. Waterwheel power was replaced by steam and then by electricity. More recently, computer-controlled, automated operations have become the norm.

■ 18.2 CLASSIFICATION OF DEFORMATION PROCESSES

A wide variety of processes have been developed to mechanically shape material, and a number of methods have been proposed to provide classification. One approach divides the processes into *primary* and *secondary*. Primary processes reduce a cast material into intermediate shapes, such as slabs, plates, or billets. Secondary processes further convert these shapes into finished or semifinished products. Unfortunately, some processes clearly fit both categories, depending on the particular product being made.

A more useful division focuses on the size and shape of the workpiece and how that size and shape is changed. *Bulk deformation processes* are those where the surface area of the workpiece changes significantly. Thicknesses or cross sections are reduced or shapes are changed. Since the volume of the material remains constant, other dimensions must change in proportion. Thus the enveloping surface area is altered, usually increasing as the product lengthens or the shape becomes more complex. In contrast, *sheet-forming operations* involve the deformation of a material where the thickness and surface area remain relatively constant. Even here, the division is not without confusion. Coining, for example, begins with sheet material but alters the thickness in a complex manner that is essentially bulk deformation.

In Chapters 18 and 19 we present a survey of metal deformation processes, where the division of processes is based on the workpiece temperature. Processes that are normally performed "hot" are presented in Chapter 18, and processes normally performed "cold" are deferred to Chapter 19. Even here, the division is somewhat blurred, especially in view of the increased emphasis on energy conservation, the growth of "warm working," and new advances in technology. Processes that were traditionally performed hot are now being performed cold, and cold-forming processes can often be aided by some degree of heating. Since sheet material has such a large surface-to-volume ratio, it tends to lose heat rapidly. As a result, most sheet-forming operations are performed cold.

■ 18.3 HOT-WORKING PROCESSES

The hot-working processes possess features (discussed in Chapter 17) that often make them an attractive means of producing a desired shape. At elevated temperatures, metals weaken and become more ductile. With continual recrystallization, massive deformation can take place without fear of fracture due to diminished ductility. In steels, hot forming involves the deformation of the weaker, austenite structure, which then cools and transforms to the stronger, room-temperature ferrite, or much stronger nonequilibrium structures, such as martensite.

Some of the hot-working processes that are of major importance in modern manufacturing are:

1. Rolling   3. Extrusion   5. Pipe welding
2. Forging   4. Hot drawing   6. Piercing

■ 18.4 ROLLING

As shown in Figure 18-1, *rolling* is usually the first process that is used to convert material into a finished wrought product. Thick starting stock can be rolled into blooms, billets, or slabs, or these shapes can be obtained directly from continuous casting. A *bloom* has a square or rectangular cross section, with a thickness greater than 15 cm (6 in.) and a width

FIGURE 18-1 Schematic flowchart for the production of various finished and semifinished steel shapes. Note the abundance of rolling operations. *(Courtesy of American Iron and Steel Institute, Washington, D.C.)*

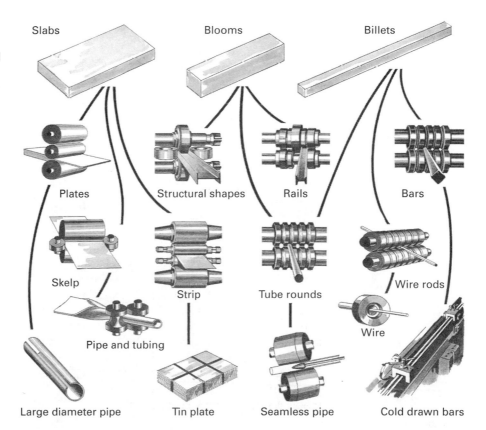

no greater than twice the thickness. A *billet* is usually smaller than a bloom and has a square or circular cross section. Billets are usually produced by some form of deformation process, such as rolling or extrusion. A *slab* is a rectangular solid where the width is greater than twice the thickness. Slabs can be further rolled to produce *plate*, *sheet*, and *strip*. These hot-worked products often form the starting material for subsequent processing using techniques such as cold forming or machining. Sheet and strip can be fabricated into products or further cold rolled into thinner, stronger material, or even into foil. Blooms and billets can be further rolled into finished products, such as *structural shapes* or railroad rail, or they can be processed into semifinished shapes, such as *bar*, *rod*, *tube*, or *pipe*.

From a tonnage viewpoint, hot rolling is clearly predominant among all manufacturing processes, and hot-rolling equipment and practices are sufficiently advanced that standardized, uniform-quality products can be produced at relatively low cost. Because shaped rolls are both massive and costly, hot-rolled products can normally be obtained only in standard shapes and sizes for which there is sufficient demand to permit economical production.

BASIC ROLLING PROCESS

As shown in Figure 18-2, heated metal is passed between two rolls that rotate in opposite directions, the gap between the rolls being somewhat less than the thickness of the entering metal. Because the rolls rotate with a surface velocity that exceeds the speed of the incoming metal, friction along the contact interface acts to propel the metal forward. The metal is squeezed and elongates to compensate for the decrease in thickness or cross-sectional area. The amount of deformation that can be achieved in a single pass between a given pair of rolls depends on the friction conditions along the interface. If too much is demanded, the rolls cannot advance the material and simply skid over its surface. If too little deformation is taken, the operation will be successful, but the additional passes required to produce a part will increase the production cost.

ROLLING TEMPERATURES

In hot rolling, as with all hot-working processes, temperature control is a requirement for success. The starting material should be heated to a uniform elevated temperature. If the temperature is not uniform, the subsequent deformation will not be uniform. Consider a piece being reheated for additional rolling. If the soaking time is insufficient, the hotter exterior will flow in preference to the cooler, stronger interior. If a part is removed from the furnace and cools prior to working, or has cooled during previous working

FIGURE 18-2 Schematic representation of the hot-rolling process, showing the deformation and recrystallization of the metal being rolled.

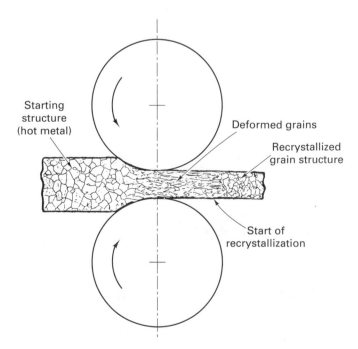

Starting structure (hot metal)

Deformed grains

Recrystallized grain structure

Start of recrystallization

operations, the cooler surfaces will tend to resist deformation. Cracking and tearing of the surface may result as the hotter, weaker interior tries to deform.

It is not uncommon for high-volume producers to begin with continuous-cast feedstock. The cooling from solidification is controlled so as to enable direct insertion into a hot-rolling operation without additional handling or reheating. For smaller operations or secondary processing, the starting material is often a room-temperature solid, such as an ingot, slab, or bloom. This material must first be brought to the desired rolling temperature, usually in gas- or oil-fired soaking pits or furnaces. For plain-carbon and low-alloy steels, the soaking temperature is usually about 1200°C (2200° F). For smaller cross sections, induction coils may be used to heat the material for rolling.

Hot rolling is usually terminated when the temperature falls to about 50 to 100°C (100 to 200°F) above the recrystallization temperature of the material. Such a *finishing temperature* ensures the production of a uniform fine grain size and prevents the possibility of unwanted strain hardening. Before additional deformation can be performed, a period of reheating is required to reestablish desirable hot-working conditions.

ROLLING MILL CONFIGURATIONS

As illustrated in Figure 18-3, rolling mill stands are available in a variety of roll configurations. Early reductions, often called primary, roughing, or breakdown passes, usually employ a two- or three-high configuration with 60 to 140-cm (24- to 55-in.) diameter rolls. The *two-high nonreversing mill* is the simplest design, but the material can only pass through the mill in one direction. The *two-high reversing mill* permits back-and-forth rolling, but the rolls must be stopped, reversed, and brought back to rolling speed between each pass. The *three-high mill* eliminates the need for roll reversal but requires some form of elevator on each side of the mill to raise or lower the material and mechanical manipulators to turn or shift the product between passes.

As shown in Figure 18-4, smaller-diameter rolls produce less length of contact for a given reduction and therefore require lower force and less energy to produce a given change in shape. The smaller cross section, however, provides reduced stiffness, and the rolls are prone to flex elastically since they are supported on the ends and pressed apart by the metal passing through the middle (a condition of three-point bending). *Four-high* and *cluster* arrangements use backup rolls to support the smaller work rolls. These configurations are used in the hot rolling of wide plate and sheets and in cold rolling, where even small deflections in the roll would result in an unacceptable variation in product thickness. Foil is almost always rolled on *cluster mills* since the small thickness requires small-diameter rolls. In a cluster mill, the roll in contact with the work can be as small

FIGURE 18-3 Various roll configurations used in rolling operations.

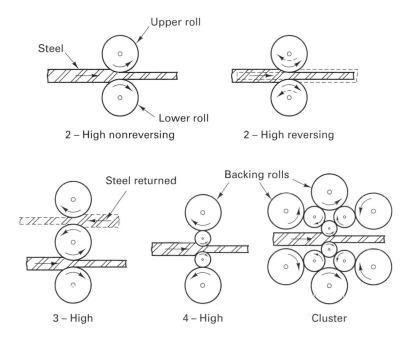

2 – High nonreversing

2 – High reversing

3 – High

4 – High

Cluster

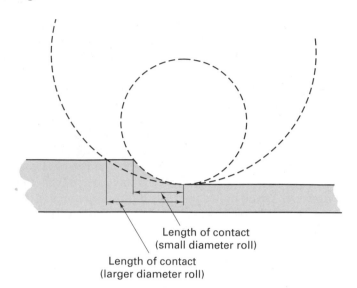

FIGURE 18-4 Schematic showing the effect of roll diameter on length of contact for a given reduction.

Length of contact (small diameter roll)

Length of contact (larger diameter roll)

as 6 mm ($\frac{1}{4}$ in.) in diameter. To counter the need for even smaller rolls, some foils are produced by *pack rolling*, a process where two or more layers of metal are rolled simultaneously as a means of providing a thicker input material. Household aluminum foil is usually rolled as a double sheet, as evidenced by the one shiny side (in contact with the roll) and one dull side (in contact with the other piece of foil).

In the rolling of nonflat or shaped products, such as structural shapes and railroad rail, the sets of rolls contain contoured grooves that sequentially form the desired shape and cross section and control the metal flow. Figure 18-5 shows some typical roll-pass sequences used in the production of structural shapes.

CONTINUOUS ROLLING MILLS

When the volume of a product justifies the investment, rolling may be performed on a continuous rolling mill. Billets, blooms, or slabs are heated and fed through an integrated series of nonreversing rolling mill stands. Continuous mills for the hot rolling of steel strip, for example, often consist of a roughing train of approximately four four-high mill stands and a finishing train of six or seven additional four-high stands. In a continuous structural mill, the rolls in each stand contain only one set of shaped grooves, in contrast to the multigrooved rolls used when the product is produced by back-and-forth passes through a single stand.

FIGURE 18-5 Typical roll-pass sequences used in producing various structural shapes.

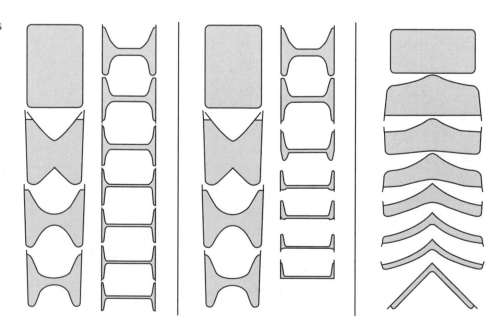

If a single piece of material is in multiple rolling stations at the same time, it is imperative that the same amount of material pass through each stand in the same amount of time. If the cross section is reduced, the speed must be increased proportionately. Therefore, as a material is reduced in size, the rolls of each successive stand must turn faster than those of the preceding one. If a subsequent stand is running too slow, material will accumulate between stands. If the demand for incoming material exceeds the output of the previous stand, the material is placed in tension, and may tear or rupture.

The synchronization of six or seven mill stands is not an easy task, especially when key variables such as temperature and lubrication may vary during a single run, and the product may be exiting the final stand at speeds in excess of 110 kilometers per hour (70 miles per hour). Computer control is basic to successful rolling, and modern mills are equipped with numerous sensors to provide the needed information. When continuous casting units feed directly into continuous rolling mills, the time lapse from final solidification to finished rolled product is often a matter of a few minutes or less.

RING ROLLING

Ring rolling is a special rolling process where one roll is placed through the hole of a thick-walled ring, and a second roll presses in from the outside (Figure 18-6). As the rolls squeeze and rotate, the wall thickness is reduced and the diameter of the ring increases. Shaped rolls can be used to produce a wide variety of cross-section profiles. The resulting seamless rings have a circumferential grain orientation, and find application in products such as rockets, turbines, airplanes, pipelines, and pressure vessels. Diameters can be as large as 8 m (25 ft) with face heights as great as 2 m (80 in.).

CHARACTERISTICS, QUALITY, AND PRECISION OF HOT-ROLLED PRODUCTS

Because hot-rolled products are formed and finished above the recrystallization temperature, they have little directionality in their properties and are relatively free of deformation-induced residual stresses. These characteristics may vary, however, depending on the thickness of the product and the presence of complex sections. Nonmetallic inclusions do not recrystallize, so they may impart some degree of directionality. In addition, substantial residual stresses can be induced during nonuniform cooling from the temperatures of hot working. Thin sheets often show some definite directional characteristics, whereas thicker plate (such as that above 20 mm or 0.8 in.) will usually have very little. Because of the high residual stresses in the rapidly cooled edges, a complex shape, such as an I- or H-beam, may warp in a noticeable fashion if a portion of one flange is cut away.

As a result of the hot deformation and the good control that is maintained during processing, hot-rolled products are normally of uniform and dependable quality, and considerable reliance can be placed on them. It is quite unusual to find any voids, seams, or laminations when these products are produced by reliable manufacturers. Of course, the surfaces of hot-rolled products are usually a bit rough and are originally covered with a tenacious high-temperature oxide, known as *mill scale*. This is usually removed by an acid pickling operation, resulting in a surprisingly smooth surface finish.

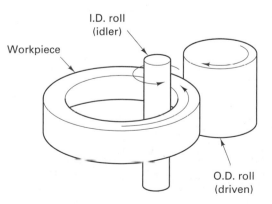

FIGURE 18-6 Schematic of a horizontal ring rolling operation. As the thickness of the ring is reduced, its diameter will increase.

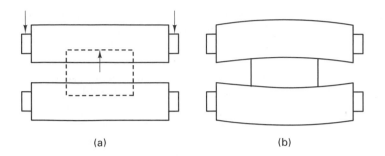

FIGURE 18-7 Loading on a rolling mill roll. The top roll is pressed upward in the center while being supported on the end. The bottom figure shows the elastic response to the three-point bending.

(a)   (b)

The dimensional tolerances of hot-rolled products vary with the kind of metal and the size of the product. For most products produced in reasonably large tonnages, the tolerances are within 2 to 5% of the specified dimension (either height or width).

FLATNESS CONTROL AND ROLLING DEFECTS

If we are rolling a flat product with uniform thickness, the gap between the rolls must be a uniform one. Attaining such an objective, however, may be difficult. Consider the upper roll in a set that is rolling sheet or plate. As shown schematically in Figure 18-7, the material presses upward in the middle of the roll, and the roll is held in place by bearings that are mounted on either end and supported in the mill frame. The roll is loaded in three-point bending and tends to flex in a manner that produces a thicker center and thinner edge. If a roll is always used to reduce the same material at the same temperature by the same amount, the forces and deflections can be predicted, and the roll can be designed to have a specified amount of crowning, or barrel-shape deviation from a cylindrical profile. When the roll is subjected to the specified load, it will deflect into flatness. This feature is illustrated in Figure 18-8.

If the applied load is not of the designed magnitude, however, the resulting profile will not be flat and defects may result. For example, if the correction is insufficient, the edges will still be thinner than the center. The thinner material will try to become longer but must remain attached to the thicker, and therefore shorter, center. The result may be either wavy edges or fractures in the center. If the correction is excessive, the center becomes thinner and longer, and the result can be a wavy center or cracking of the edges.

Since roll deflections are proportional to the forces applied to the rolls, product flatness can also be improved by measures that reduce these forces. Heating the workpiece generally makes it weaker, so increased temperature will reduce the force on the rolls. Horizontal tensions can also be applied to the piece as it is being rolled (strip tension in sheet metal rolling). Since these tensions combine with the vertical compression to deform the piece (stretching while squeezing), the roll forces and associated deflections are less. Other techniques to improve flatness include an increase in the elastic modulus of the rolls themselves through material selection, or to provide some form of backup support that resists deflection.

THERMOMECHANICAL PROCESSING AND CONTROLLED ROLLING

As with most deformation processes, rolling is generally viewed as being a means of changing the shape of a material. While heat may be used to reduce forces and promote plasticity, the thermal processes that produce or control mechanical properties (heat treatments)

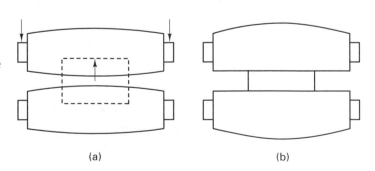

FIGURE 18-8 Use of a "crowned" roll to compensate for roll flexure. When the roll flexes in three-point bending, the crowned roll flexes into flatness.

(a)   (b)

are usually performed as subsequent operations. *Thermomechanical processing*, of which *controlled rolling* is an example, consists of integrating deformation and thermal processing into a single process that will produce not only the desired shape, but also the desired properties, such as strength and toughness. The heat for the property modification is the same heat used in the rolling operation, and subsequent heat treatment is unnecessary.

A successful thermomechanical operation begins with process design. The starting material must be specified and the composition closely maintained. Then a time–temperature–deformation system must be developed to achieve the desired objective. Possible goals include production of a uniform fine grain size; controlling the nature, size, and distribution of the various transformation products (such as ferrite, pearlite, bainite, and martensite in steels); controlling the reactions that produce solid solution strengthening or precipitation hardening; and producing a desired level of toughness. Starting structure (controlled by composition and prior thermal treatments), deformation details, temperature during the various stages of deformation, and the conditions of cooldown from the working temperature must all be specified and controlled. Moreover, the attainment of uniform properties requires uniform temperatures and deformations throughout the product. Computer-controlled facilities are an absolute necessity if thermomechanical processing is to be successfully performed.

Possible benefits of thermomechanical processing include improved product properties; substantial energy savings (by eliminating subsequent heat treatment); and the possible substitution of a cheaper, less-alloyed metal for a highly alloyed one that responds to heat treatment.

■ 18.5 FORGING

Forging is the term applied to a family of processes where the deformation is induced by localized compressive forces. The equipment can take the form of hammers, presses, or special forging machines. While the deformation can be done in the hot, cold, warm, or isothermal mode, the term *forging* usually implies hot forging done above the recrystallization temperature.

Forging is the oldest known metalworking process. From the days when prehistoric peoples discovered that they could heat sponge iron and beat it into a useful implement by hammering with a stone, forging has been an effective method of producing many useful shapes. Modern forging has developed from the ancient art practiced by the armor makers and the immortalized village blacksmith. High-powered hammers and mechanical presses have replaced the strong arm, the hammer, and the anvil, and modern metallurgical knowledge has supplemented the art and skill of the craftsman in controlling the heating and handling of the metal. Parts can range in size from ones whose largest dimension is less than 2 cm (1 in.) to others weighing more than 170 metric tons (450,000 lb).

A variety of forging processes have been developed that offer a wide range of capabilities. A single piece can be economically fashioned by some methods, while others can mass produce thousands of identical parts. The metal may be (1) *drawn out* to increase its length and decrease its cross section, (2) *upset* to decrease the length and increase the cross section, or (3) *squeezed in closed impression dies* to produce multidirectional flow. As indicated in Table 17-2, the state of stress in the work is primarily uniaxial or multiaxial compression.

Common forging processes include:

1. Open-die drop-hammer forging   **5.** Automatic hot forging

2. Impression-die drop-hammer forging   **6.** Roll forging

3. Press forging   **7.** Swaging

4. Upset forging

OPEN-DIE DROP-HAMMER FORGING

In concept, *open-die hammer forging* is the same type of forging done by the blacksmith of old, but massive mechanical equipment is now used to impart the repeated blows. The metal is first heated to the proper temperature by gas, oil, or electric furnaces or by electrical induction heating. The impact is then delivered by some type of mechanical hammer, the simplest type being a *gravity drop* or *board hammer*. Here the hammer is attached to the lower end of a hardwood board, which is raised by being gripped between two rotating, roughened rollers, which then separate to release it for free-fall. Few of these are still in use. *Steam* or *air hammers*, which use pressure to both raise and propel the hammer, are far more common. These give higher striking velocities, more control of striking force, easier automation, and the ability to shape pieces up to several tons. *Programmable, computer-controlled hammers* can provide blows of differing impact speed (energy) for each of the various stages of an operation. Their use can greatly increase the efficiency of the process and also minimize the amount of noise and vibration, which are the most common outlets for the excess energy not absorbed in the deformation of the workpiece. Figure 18-9 shows a large double-frame hammer as well as a simple schematic of its basic tooling.

Open-die forging does not fully control the flow of metal. The operator must obtain the desired shape by orienting and positioning the workpiece between blows. The hammer may contact the workpiece directly, or specially shaped tools can be inserted to assist in making simple shapes (such as round, concave, or convex surfaces), forming holes, or performing a cutoff operation. Manipulators may be used to position larger workpieces, which may weigh several tons. While some finished parts can be made by this technique, open-die forging is usually employed to preshape metal in preparation for further operations. For example, consider such massive parts as turbine rotors and generator shafts. These parts may have dimensions up to 20 m (70 ft) in length and up to 1 m (3 ft) in diameter. Open-die forging is used to induce oriented plastic flow and minimize the amount of subsequent machining. Figure 18-10 schematically depicts the unrestricted flow of material and the formation of both a multidiameter cylindrical shaft and a seamless ring by open-die forging.

FIGURE 18-9 Double-frame drop hammer and a schematic of the basic tooling. *(Courtesy of Erie Press Systems, Erie, Pa.)*

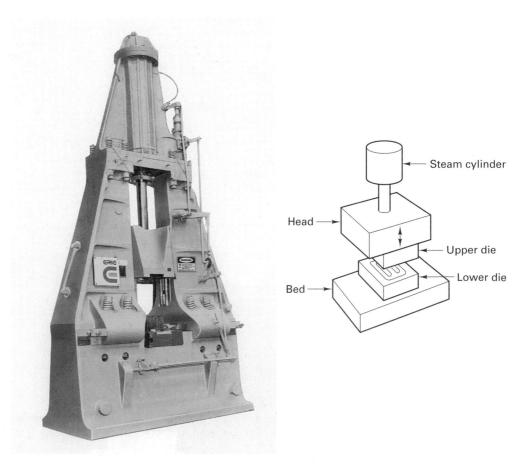

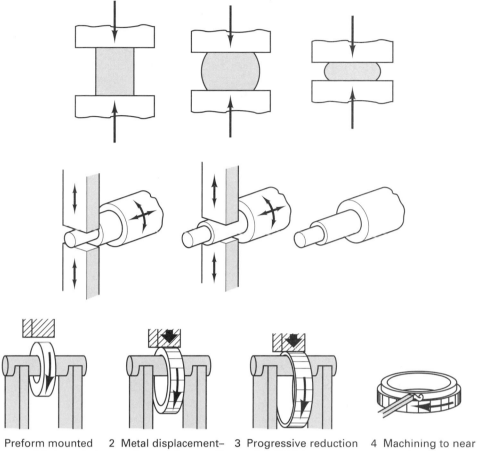

1 Preform mounted   2 Metal displacement–   3 Progressive reduction   4 Machining to near
   on saddle/mandrel.   reduce preform wall   of wall thickness to   net shape.
                            thickness to increase   produce ring
                            diameter.   dimensions.

FIGURE 18-10   (*Top*) Schematic of the unrestrained flow of material in open-die forging. Note the barrel shape that forms due to friction between the die and material. (*Middle*) Open-die forging of a multidiameter shaft. (*Bottom*) Forging of a seamless ring by the open-die method. (*Courtesy of Forging Industry Association, Cleveland, Ohio.*)

IMPRESSION-DIE DROP-HAMMER FORGING

Open-die hammer forging (or smith forging, as it has been called) is a simple and flexible process, but it is not practical for large-scale production. It is a slow operation, and the size, shape, and dimensional precision of the resulting workpiece is dependent on the skill of the operator. *Impression-die* or *closed-die forging* overcomes these difficulties by using shaped dies to control the flow of metal, as shown in Figure 18-11. Figure 18-12 shows a typical set of dies. The upper piece attaches to the hammer and the lower piece to the anvil. The heated metal is positioned in the lower cavity and struck one or more blows by the upper die. This hammering causes the metal to flow and completely fill the die cavity. Excess metal is squeezed out around the periphery of the cavity to form a *flash*. This material cools rapidly, increases in strength, and effectively blocks the formation of additional flash. By trapping

FIGURE 18-11   Schematic of the impression-die forging process showing partial die filling and the beginning of flash formation in the center sketch, and the final shape with flash in the right-hand sketch.

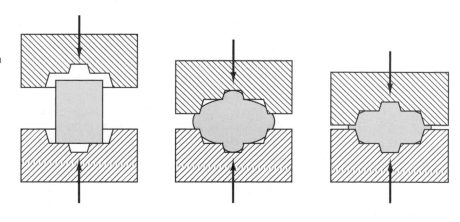

FIGURE 18-12 Impression drop-forging dies and the product resulting from each impression. The flash is trimmed from the finished connecting rod in a separate trimming die. The sectional view shows the grain flow resulting from the forging process. *(Courtesy of Forging Industry Association, Cleveland, Ohio.)*

material within the die, the flash ensures the filling of all of the cavity details. The flash is ultimately trimmed from the part as the final forging operation.

In *flashless forging*, also known as true closed-die forging, the metal is deformed in a cavity that provides total confinement. Accurate workpiece sizing is required since complete filling of the cavity must be assured with no excess material. Accurate workpiece positioning is also necessary, along with good die design and control of lubrication. The major advantage of this approach is the elimination of the scrap generated during flash formation, an amount that is typically between 20 and 45% of the starting material.

Most conventional forgings are impression-die with flash and are produced in dies with a series of cavities where one or more blows of the hammer are used for each step in the sequence. The first impression is often an *edging*, *fullering*, or *bending* impression to distribute the metal roughly in accordance with the requirements of the later cavities. Intermediate impressions are for *blocking* the metal to approximately its final shape, with generous corner and fillet radii. For small production lots, the cost of further cavities may not be justified, and the blocker-type forgings are further finished by machining. More often, the final shape and size are set by additional forging in a *final* or *finisher impression*. Figure 18-12 shows an example of these steps and the shape of the part at the conclusion of each. Since every part is shaped in the same die cavities, each mass-produced part is a close duplicate of all the others.

The shape of the various cavities controls the flow of material, and the flow, in turn, imparts the oriented structure discussed in Chapter 17. (Grain flow that follows the outline of the component is in the crack arrestor orientation, improving strength, ductility and resistance to impact and fatigue.) Through forging, we can also control the size and shape of various cross sections, so the metal can be distributed as needed to resist the applied loads. Couple these factors with a fine recrystallized grain structure (hot working) and the absence of voids (compressive forming stresses), and we see why forgings often have about 20% higher strength/weight ratios compared with cast or machined parts of the same material.

Board hammers, steam hammers, and air hammers are all used in impression die forging. An alternative to the hammer and anvil arrangement is the *counterblow machine*,

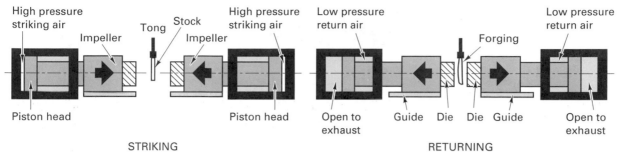

STRIKING   RETURNING

FIGURE 18-13 Schematic diagram of an impactor in the striking and returning modes. *(Courtesy of Chambersburg Engineering Company.)*

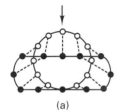

(a)

Conventional forged disc with paths of flow

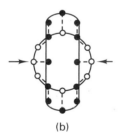

(b)

Disc formed by impacter with paths of flow

FIGURE 18-14 A comparison of the metal flow in conventional forging and impacting.

or *impactor*, illustrated in Figure 18-13. These machines have two horizontal hammers that simultaneously impact a workpiece that is positioned between them. Excess energy simply becomes recoil, in contrast to the hammer and anvil arrangement, where energy is lost to the machine foundation and a heavy machine base is required. Impactors also operate with less noise and less vibration and produce distinctly different flow patterns, as illustrated in Figure 18-14.

Conventional closed-die forging begins with a simple hot-rolled shape and utilizes reheating and working to progressively convert it into a more complex geometry whose metallurgical structure and properties are superior to cast or machined equivalents. A variety of approaches can be used, each of which can be further automated through the use of induction heating, mechanical feeding, positioning and manipulation, and the direct heat treatment of parts as they emerge from the machine.

Several alternative processes have been developed to conserve energy while simultaneously producing a product that is somewhere between a conventional forging and a conventional casting. In one approach, a forging preform is cast from liquid metal, removed from the mold while still hot, and then finish-forged in a single-cavity die. The flash is then trimmed and the part is quenched to room temperature. Forging preforms can also be produced by the spraying deposition of metal droplets into shaped collectors, as with the *Osprey process* described in Chapter 16. The preforms are removed from the mold, and the final shape and properties are imparted by a final forging. Still another approach is semisolid forging, discussed in Chapter 15.

Design of Impression-Die Forgings and Associated Tooling.   Forging dies are usually made of high-alloy or tool steel and can be expensive to design and construct. Impact resistance, wear resistance, strength at elevated temperature, and the ability to withstand cycles of rapid heating and cooling must all be outstanding. In addition, considerable care is required to produce and maintain a smooth and accurate cavity and parting plane. Better and more economical results will be obtained if the following rules are observed:

1. The dies should part along a single, flat plane if at all possible. If not, the parting plane should follow the contour of the part.

2. The parting surface should be a plane through the center of the forging and not near an upper or lower edge.

3. Adequate draft should be provided—at least 3° for aluminum and 5 to 7° for steel.

4. Generous fillets and radii should be provided.

5. Ribs should be low and wide.

6. The various sections should be balanced to avoid extreme differences in metal flow.

7. Full advantage should be taken of fiber flow lines.

8. Dimensional tolerances should not be closer than necessary.

The various design details, such as the number of intermediate steps, the shape of each, the amount of excess metal required to assure die filling, and the dimensions of the flash at each step, are often a matter of experience. Each component is a new design entity and brings its own unique challenges. Computer-aided design has made notable

TABLE 18-1.	Thickness Tolerances for Steel Drop Hammer Forgings				
Mass of Forging		Minus		Plus	
lb	kg	in.	mm	in.	mm
1	0.45	0.006	0.15	0.018	0.48
2	0.91	0.008	0.20	0.024	0.61
5	2.27	0.010	0.25	0.030	0.76
10	4.54	0.011	0.28	0.033	0.84
20	9.07	0.013	0.33	0.039	0.99
50	22.68	0.019	0.48	0.057	1.45
100	45.36	0.029	0.74	0.087	2.21

advances, however, and the development and accessibility of high-speed, immense-memory computers have enabled the accurate modeling of many complex shapes.

Good dimensional accuracy is a characteristic of impression-die forging. With reasonable care, the dimensions for steel products can be maintained within the tolerances shown in Table 18-1. It should be noted, however, that the dimensions across the parting plane are affected by closure of the dies, and are therefore dependent on die wear and the thickness of the final flash. Dimensions contained entirely within a single die segment can be maintained at a significantly greater level of accuracy. Draft angles can sometimes be reduced, occasionally approaching zero, but this is not recommended for general practice.

Selection of a lubricant is also critical to successful forging. The lubricant not only affects the friction and wear and associated metal flow, but it may also be expected to act as a thermal barrier (restricting heat flow from the workpiece to the dies) and a parting compound (preventing the part from sticking in the cavities).

PRESS FORGING

In hammer or impact forging, the metal flows to dissipate the energy imparted in the hammer–workpiece collision. Speeds are high, so the forming time is short. Contact times under load are on the order of milliseconds. There is little time for heat transfer and cooling of the workpiece, and the adiabatic heating that occurs during deformation helps to minimize chilling. It is possible, however, that all of the energy can be dissipated by deformation of just the surface of the metal (coupled with additional absorption by the anvil and foundation), and the interior of the workpiece remains essentially undeformed. For example, consider the deformation of a metal wood-splitting wedge after it has been struck repeatedly by a sledge hammer. The top is usually "mushroomed," whereas the remainder retains the original geometry and taper.

If large pieces or thick products are to be formed, *press forging* may be required. The deformation is now analyzed in terms of forces or pressures (rather than energy), and the slow squeezing action penetrates completely through the metal, producing a more uniform deformation and flow. New problems can arise, however, because of the longer time of contact between the dies and the workpiece. As the surface of the workpiece cools, it becomes stronger and less ductile and may crack if deformation is continued. Heated dies are generally used to reduce heat loss, promote surface flow, and enable the production of finer details and closer tolerances. Periodic reheating of the workpiece may also be required.

Forging presses are of two basic types, mechanical and hydraulic, and are usually quite massive. *Mechanical presses* use means such as cams, cranks, or toggles to produce a preset and reproducible stroke. Because of their mechanical drives, different forces are available at the various stroke positions. Production presses are quite fast, capable of up to 50 strokes per minute, and are available in capacities ranging from 300 to 18,000 tons (3 to 160 MN). *Hydraulic presses* move in response to fluid pressure in a piston and are generally slower, more massive, and more costly to operate. On the positive side, hydraulic presses are much more flexible and can have greater capacity. Since motion is in response to flow of pressurized drive fluids, hydraulic presses can be programmed to have different strokes for different operations and even different speeds within a stroke. Machines with capacities up to 50,000 tons (445 MN) are currently in operation in the United States.

Presses can be used to perform all types of forging, including open-die and impression-die. Impression-die press forgings usually require less draft than drop forgings and have higher dimensional accuracy. In addition, press forgings can often be completed in a single closing of the dies, and the process can be readily automated.

UPSET FORGING

Upset forging involves increasing the diameter of a material by compressing its length. In terms of the number of pieces produced, it is the most widely used of all forging processes. Parts can be upset forged both hot and cold, with the operation generally being performed on special high-speed machines where the forging motion is horizontal and the workpiece is rapidly moved from station to station. The starting stock is usually wire or rod, but some machines can forge bars up to 25 cm (10 in.) in diameter.

Upset forging generally employs split dies that contain multiple positions or cavities. A typical die set is shown in Figure 18-15. The dies separate enough for the bar to advance between them and move into position. They are then clamped together and a heading tool or ram moves longitudinally against the bar, upsetting it into the cavity. Separation of the dies then permits transfer to the next position or removal of the product. If a new piece is started with each die separation, and an operation is performed in each cavity simultaneously, a finished product can be made with each cycle of the machine. By including a shearing operation as the initial piece moves into position, the process can operate with continuous coil or long-length rod as its incoming feedstock.

Upset-forging machines are often used to form heads on bolts and other fasteners, and to shape valves, couplings, and many other small components. The following three rules, illustrated in Figure 18-16, should be followed in designing parts that are to be upset-forged:

1. The length of unsupported metal that can be gathered or upset in one blow without injurious buckling should be limited to three times the diameter of the bar.

FIGURE 18-15   Set of upset forging dies and punches. The product resulting from each of the four positions is shown along the bottom. *(Courtesy of Ajax Manufacturing Company.)*

FIGURE 18-16   Schematics illustrating the rules governing upset forging. *(Courtesy of National Machinery Company.)*

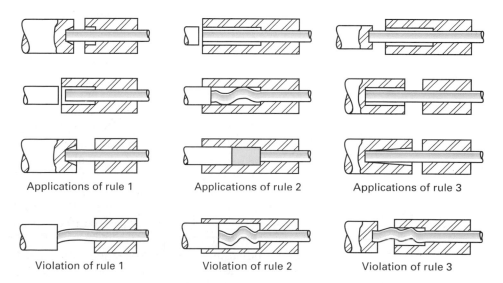

Applications of rule 1   Applications of rule 2   Applications of rule 3

Violation of rule 1   Violation of rule 2   Violation of rule 3

FIGURE 18-17   Typical parts made by upsetting and related operations. *(Courtesy of National Machinery Company.)*

2. Lengths of stock greater than three times the diameter may be upset successfully provided that the diameter of the upset is not more than $1\frac{1}{2}$ times the diameter of the bar.

3. In an upset requiring stock length greater than three times the diameter of the bar, and where the diameter of the cavity is not more than $1\frac{1}{2}$ times the diameter of the bar (the conditions of rule 2), the length of unsupported metal beyond the face of the die must not exceed the diameter of the bar.

Figure 18-17 illustrates the variety of parts that can be produced by upsetting and subsequent piercing, trimming, and machining operations.

AUTOMATIC HOT FORGING

Several equipment manufacturers now offer highly automated upset equipment in which mill-length steel bars (typically 7 m or 24 ft long) are fed into one end at room temperature and hot-forged products emerge from the other end at rates of up to 180 parts per minute (86,400 parts per 8-hour shift). These parts can be solid or hollow, round or symmetrical, up to 6 kg (12 lb) in weight, and up to 18 cm (7 in.) in diameter.

The process begins with the lowest-cost steel bar stock: hot-rolled and air-cooled carbon or alloy steel. The bar is first heated to 1200 to 1300°C (2200 to 2350°F) in under 60 seconds as it passes through high-power induction coils. It is then descaled by rolls, sheared into individual blanks, and transferred through several successive forming stages, during which it is upset, preformed, final-forged, and pierced (if necessary). Small parts can be produced at up to 180 parts per minute, with rates for larger pieces on the order of 90 parts per minute. Figure 18-18 shows a typical sequence and a variety of ferrous products.

The *automatic hot forging* process has a number of attractive features. Low-cost input material and high production speeds have already been cited. Minimum labor is required, and since no flash is produced, material savings can be as much as 20 to 30% over conventional forging. With a consistent finishing temperature near 1050°C (1900°F), an air cool can often produce a structure suitable for machining, eliminating the need for an additional anneal or normalizing treatment. Tolerances are generally ±0.3 mm (±0.012 in.), surfaces are clean, and draft angles need only be $\frac{1}{2}$ to 1° (as opposed to the conventional 3 to 5°). Tool life is nearly double that of conventional forging because the contact times are only on the order of 6/100 of a second.

Automatic hot formers can also be coupled with high-rate, cold-forming operations. Preform shapes can be hot formed at rates that approach 180 parts per minute. These products can then be cold formed to final shape on machines that operate at speeds near 90 parts per minute. The benefits of the combined operations include high-volume production at low cost, coupled with the precision, surface finish, and strain hardening that are characteristic of a cold-finished material.

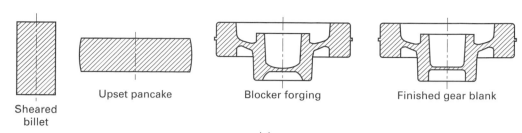

Sheared billet   Upset pancake   Blocker forging   Finished gear blank

(a)

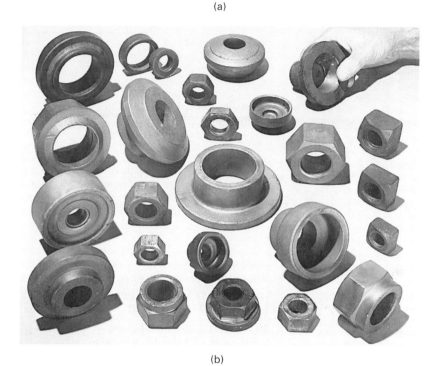

FIGURE 18-18 (*Top*) Typical four-step sequence to produce a spur-gear forging. The sheared billet is progressively shaped into an upset pancake, blocker forging, and finished gear blank. (*Bottom*) Samples of ferrous parts produced by automatic hot forging at rates between 90 and 180 parts per minute. (*Courtesy of National Machinery Company.*)

(b)

To justify an automatic hot forging operation, however, large quantities of a given product must be required. A single production line may well require an initial investment in excess of $10 million.

ROLL FORGING

In *roll forging*, round or flat bar stock is reduced in thickness and increased in length to produce such products as axles, tapered levers, and leaf springs. As illustrated in Figure 18-19, roll forging is performed on machines that have two cylindrical or semicylindrical rolls,

FIGURE 18-19 (*Left*) Roll-forging machine in operation. (*Right*) Rolls from a roll-forging machine and the various stages in roll forging a part. (*Courtesy of Ajax Manufacturing Company.*)

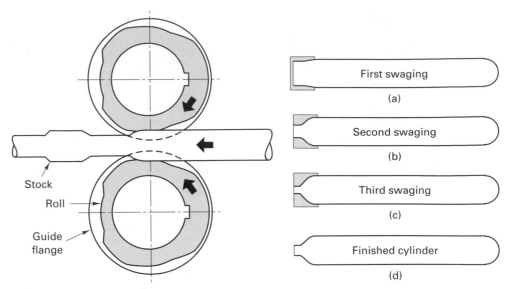

FIGURE 18-20 Schematic of the roll-forging process showing the two shaped rolls and the stock being formed. *(Courtesy of Forging Industry Association, Cleveland, Ohio.)*

FIGURE 18-21 Steps in swaging a tube to form the neck of a gas cylinder. *(Courtesy of USX Corporation.)*

each containing one or more shaped grooves. A heated bar is inserted between the rolls. When the bar encounters a stop, the rolls rotate, and the bar is progressively shaped as it is rolled out toward the operator. The piece is then transferred to the next set of grooves (or rotated and reinserted in the same groove) and the process repeats until the desired size and shape is produced. Figure 18-20 shows the cross section of one set of grooves and a piece being formed. In most cases there is no flash, and the oriented structure imparts favorable properties.

SWAGING

The *swaging* process involves the hammering of a rod or tube to reduce its diameter where the die itself acts as the hammer. Repeated blows are delivered from various angles, causing the metal to flow inward and assume the contour of the die. Since most swaging is usually performed cold, it will be discussed in more detail in Chapter 19.

The term *swaging* is also applied to a process where material is forced into a confining die to reduce its diameter, and this process is often performed hot. Figure 18-21 shows a sequence of hot swaging being used to reduce and form the end of a pressurized gas cylinder.

NET-SHAPE AND NEAR-NET-SHAPE FORGING

As much as 80% of the cost of a forged gear can be incurred during the machining operations that follow forging, and a finished aerospace wing spar may contain as little as 4% of the original billet (the remaining 96% being lost as scrap in the forging and subsequent machining operations). To minimize both the expense and waste, considerable effort has been made to develop processes that can form parts close enough to final dimensions that little or no final machining is required. These are known as *net-shape*, or *near-net-shape*, operations, and may also be referred to as *precision forging*. Cost savings often result from the reduction or elimination of secondary machining (and the associated handling, positioning, and fixturing), the companion reduction in scrap, and an overall decrease in the amount of energy required to produce the product.

Precision or near-net-shape forgings can now be produced with draft angles of less than 1° (or even zero draft). Complex shapes can be forged with such close tolerances that little or no finish machining is required. Since the design and implementation of net-shape processing can be rather expensive, application is usually reserved for parts where a significant cost reduction can be achieved.

■ 18.6 EXTRUSION

In the *extrusion* process, metal is compressed and forced to flow through a suitably shaped die to form a product with reduced but constant cross section. Although extrusion may be performed either hot or cold, hot extrusion is commonly employed for many metals to reduce the forces required, eliminate cold-working effects, and reduce directional properties. Basically, the extrusion process is like squeezing toothpaste out of a tube. In the case of metals, a common arrangement is to have a heated billet placed inside a confining chamber. A ram advances from one end, causing the billet to first upset and conform to the confining chamber. As the ram continues to advance, the pressure builds until the material flows plastically through the die *and extruds*. See Figure 18-22. The stress state within the material is one of triaxial compression.

Aluminum, magnesium, copper, lead, and alloys of these metals are commonly extruded, taking advantage of the relatively low yield strengths and low hot-working temperatures. Steels, stainless steels, nickel-based alloys, and titanium are far more difficult to extrude. Their yield strengths are high, and the metals have the tendency to weld to the walls of the die and confining chamber under the required conditions of temperature and pressure. With the development and use of phosphate-based and molten glass lubricants, however, hot extrusions can be routinely produced from these high-strength, high-temperature metals. These lubricants are able to withstand the required temperatures and adhere to the billet, flowing and thinning in a way that prevents metal-to-metal contact throughout the process.

As shown in the left-hand segment of Figure 18-23, almost any cross-sectional shape can be extruded from the nonferrous metals. Size limitations are few because presses are now available that can extrude any shape that can be enclosed within a 75-cm (30-in.) diameter circle. In the case of steels and the other high-strength metals, the shapes and

FIGURE 18-22 Direct extrusion schematic showing the various equipment components. *(Courtesy of Wean United, Inc., Hydraulic Machinery Division.)*

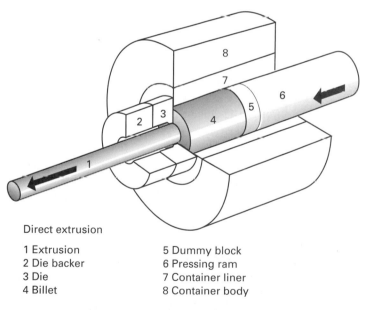

Direct extrusion

1 Extrusion
2 Die backer
3 Die
4 Billet
5 Dummy block
6 Pressing ram
7 Container liner
8 Container body

FIGURE 18-23 Typical shapes produced by extrusion. *(Left)* Aluminum products. *(Courtesy of Aluminum Company of America.)* *(Right)* Steel products. *(Courtesy of Allegheny Ludlum Steel Corporation.)*

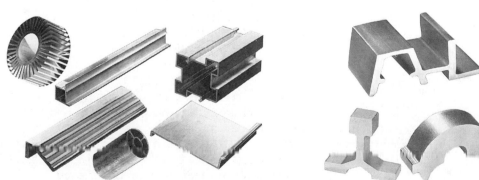

sizes are a bit more limited, but, as the right-hand segment of Figure 18-23 shows, considerable freedom still exists.

Extrusion has a number of attractive features. Many shapes can be produced as extrusions that are not possible by rolling, such as ones containing reentrant angles or longitudinal holes. No draft is required, so extrusions can offer savings in both metal and weight. Since the deformation is compressive, the amount of reduction in a single step is limited only by the capacity of the equipment. Billet-to-product cross-sectional area ratios can be in excess of 100-to-1 for the weaker metals. In addition, extrusion dies can be relatively inexpensive, and one die may be all that is required to produce a given product. Conversion from one product to another requires only a single die change, so small quantities of a desired shape can be produced economically. The major limitation of the process is the requirement that the cross section be uniform for the entire length of the product.

Extruded products have good surface finish and dimensional precision. For most shapes, tolerances of ±0.003 cm/cm or in./in., with a minimum of ±0.075 mm (±0.003 in.) are easily attainable. Grain structure is typical of other hot-worked metals, but strong directional properties (longitudinal versus transverse) are usually observed. Standard product lengths are about 6 to 7 m (20 to 24 ft), but lengths in excess of 12 m (40 ft) have been produced.

EXTRUSION METHODS

Extrusions can be produced by various techniques and equipment configurations. Hot extrusion is usually done by either the direct or indirect method, both of which are illustrated in Figure 18-24. In *direct extrusion*, a solid ram drives the entire billet to and through a stationary die and must provide additional power to overcome the frictional resistance between the surface of the moving billet and the confining chamber. In *indirect extrusion*, a hollow ram pushes the die back through a stationary, confined billet. Since there is no relative motion, friction between the billet and the chamber is eliminated. The required force is lower and longer billets can be used with no penalty in power or efficiency. Figure 18-25 shows the ram force versus ram position curves for both direct and indirect extrusion. The areas below the lines have units of Newton-meters or foot-pounds, and are therefore proportional to the work or power required to produce the part. The area between the two curves is the power required to overcome the billet–chamber friction during direct extrusion, an amount that can be saved by converting to indirect extrusion. Unfortunately, the added complexity of the indirect process (applying force through a hollow ram, extracting the product through the hollow, and removing residual billet material at the end of the stroke) serves to increase the purchase price and maintenance cost of the required equipment.

With either process, the speeds of hot extrusion are usually rather fast, so as to minimize the cooling of the billet within the chamber. Extruded products can emerge at rates up to 300 m/min (1000 ft/min). The extrusion speed may be restricted, however, by the large amounts of heat that are generated by the massive deformation and the associated rise in temperature. Sensors are often used to monitor the temperature of the emerging product and feed this information back to a control system. For materials whose properties are not sensitive to strain rate, ram speed may be maintained at the highest level that will keep the product temperature below some predetermined value.

Lubrication is another important area of concern. If the reduction ratio (cross section of billet to cross section of product) is 100, the product will be 100 times longer than the starting billet. If the product has a complex cross section, its perimeter can be significantly greater than a circle of equivalent area. Since the surface area of the product is the length times the perimeter, this value can easily be an order of magnitude

FIGURE 18-24 Direct and indirect extrusion. In direct extrusion, the ram and billet both move and friction with the chamber opposes forward motion of the billet. For indirect extrusion, the billet is stationary. There is no billet-chamber friction, since there is no relative motion.

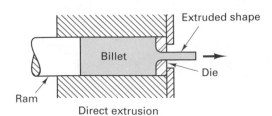

Direct extrusion

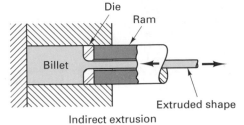

Indirect extrusion

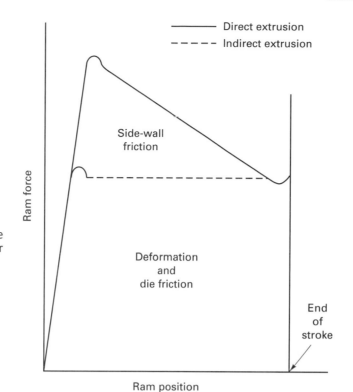

FIGURE 18-25   Diagram of the ram force versus ram position for both direct and indirect extrusion of the same product. The area under the curve corresponds to the amount of work (force × distance) performed. The difference between the two curves is attributed to billet-chamber friction.

greater than the surface area of the original billet. A lubricant that is applied to the starting piece must thin considerably as the material passes through the die and is converted to product. An acceptable lubricant is expected to reduce friction and act as a barrier to heat transfer at all stages of the process.

Impact extrusion, *hydrostatic extrusion*, and *continuous extrusion* are usually performed cold and will be discussed in Chapter 19.

EXTRUSION OF HOLLOW SHAPES

Hollow shapes, and shapes with more than one longitudinal cavity, can be extruded by several methods. For tubular products, the stationary or moving *mandrel* processes of Figure 18-26 are quite common. The die forms the outer profile, while the mandrel shapes and sizes the interior.

FIGURE 18-26   Two methods of extruding hollow shapes using internal mandrels. In the upper schematic the mandrel and ram have independent motions; in the lower sequence they move as a single unit.

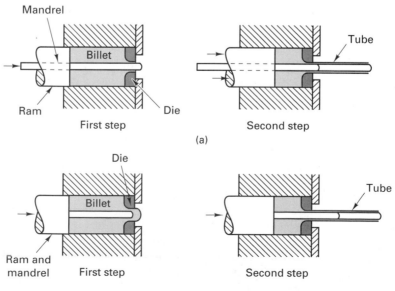

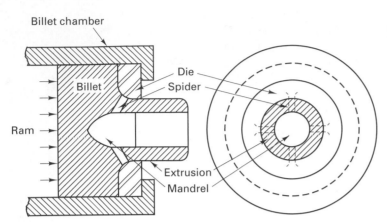

FIGURE 18-27   Hot extrusion of a hollow shape using a spider-mandrel die. Note the four arms connecting the die and the mandrel.

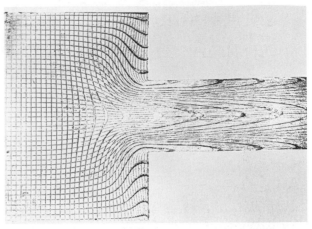

FIGURE 18-28   Grid pattern showing the metal flow in a direct extrusion. The billet was sectioned and the grid pattern was engraved prior to extrusion.

For products with multiple or more-complex cavities, a *spider-mandrel* die (also known as a porthole, bridge, or torpedo die) may be required. As illustrated in Figure 18-27, hot metal flows around the arms of a "spider," and a further reduction then forces the material back together. Since the metal is never exposed to contamination, perfect welds result. Unfortunately, lubricants cannot be used since they will contaminate the surfaces to be welded. The process is therefore limited to materials that can be extruded without lubrication and can also be easily pressure welded.

Since additional tooling is required, hollow extrusions will obviously cost more than solid ones, but a wide variety of continuous cross-section shapes can be produced that cannot be made economically by any other process.

METAL FLOW IN EXTRUSION

The flow of metal during extrusion is often complex, and some care must be exercised to prevent surface cracks, interior cracks, and other flow-related defects. Metal near the center of the chamber can often pass through the die with little distortion, while metal near the surface undergoes considerable shearing. In direct extrusion, friction between the forward-moving billet and the stationary chamber and die serves to further impede surface flow. The result is often a deformation pattern similar to the one shown in Figure 18-28. If the surface regions of the billet undergo excessive cooling, the deformation is further impeded and cracks tend to form on the product surface. If quality is to be maintained, process control must be exercised in the areas of design, lubrication, extrusion speed, and temperature.

■ 18.7 HOT DRAWING OF SHEET AND PLATE

Drawing is a plastic deformation process in which a flat sheet or plate is formed into a recessed, three-dimensional part with a depth more than several times the thickness of the metal. The metal assumes the desired configuration as a punch descends into a mating die, or the die moves upward over a punch. Hot drawing is used for forming relatively thick-walled parts of simple geometries, usually cylindrical. Because the material is hot, there is often considerable thinning as it passes through the dies. In contrast, cold drawing uses relatively thin metal, changes the thickness very little or not at all, and produces parts in a wide variety of shapes.

Hot drawing is illustrated in the upper left-hand schematic of Figure 18-29. A heated sheet or plate is positioned over a female die. A punch then descends, pushing the metal through the die, converting the circular blank to a cylindrical cup. The height of the cup walls is determined by the difference between the diameter of the original blank and the diameter of the punch. This dimension is limited by the onset of several defects. Wrinkles can appear in the cup walls as the circumference is reduced, or the punch can act as a piercing tool, tearing the blank around the punch perimeter.

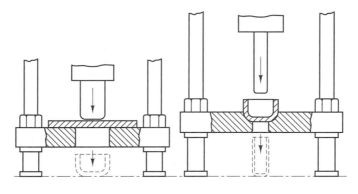

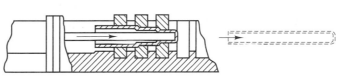

FIGURE 18-29 Methods of cup forming or hot drawing. (*Upper left*) First draw. (*Upper right*) Redraw operation. (*Lower*) Multiple-die drawing. (*Courtesy of USX Corporation.*)

There are several ways to produce cups with taller walls. If the gap between the punch and the die is less than the thickness of the incoming material, the cup wall is thinned and elongated by a process called *ironing* or *wall ironing*. If the thinning is objectionable, the initial drawing can be followed by a further reduction in diameter and concurrent increase in wall height. This *redrawing* uses a smaller punch and die, as shown in the upper right-hand segment of Figure 18-29. The lower segment of this figure illustrates still another alternative, where the cup is pushed through a series of dies with a single punch.

Some drawn products are designed to utilize part of the original disk as a flange around the top of the cup. For these products, the punch does not push the material completely through the die, but descends to a predetermined depth and then retracts. The partially drawn cup is then ejected upward, and the perimeter of the remaining flange is trimmed to the desired size and shape.

■ 18.8 PIPE WELDING

Large quantities of steel pipe are made by two processes that use hot forming of steel strip coupled with the deformation-induced welding of its free edges. Both of these processes, *butt welding* of pipe and *lap welding* of pipe, utilize steel in the form of *skelp*—long strips with specified width, thickness, and edge configuration. Because the skelp has been produced by hot rolling, and the welding process produces further compressive working and recrystallization, pipe welded by these processes tends to be very uniform in quality.

BUTT-WELDED PIPE

In the butt-welding process for making pipe, steel skelp is heated to a specified hot-working temperature by passing it through a furnace. Upon exiting the furnace, it is pulled through forming rolls that shape it into a cylinder and bring the free ends into contact. The pressure exerted between the edges of the skelp is sufficient to upset the metal and produce a welded seam. Additional sets of rollers then size and shape the pipe and it is cut to standard, preset lengths. Product diameters range from 3 mm ($\frac{1}{8}$ in.) to 75 mm (3 in.), and speeds can approach 150 m/min (500 ft/min).

LAP-WELDED PIPE

The lap-welding process for making pipe differs from the butt-welding technique in that the skelp now has beveled edges and the rolls form the weld by forcing the lapped edges down against a supported mandrel. This process is used primarily for larger sizes of pipe, with diameters from about 50 mm (2 in.) to 400 mm (14 in.). Because the product is driven over a supported mandrel, product length is limited to about 6 to 7 m (20 to 25 ft).

■ 18.9 PIERCING

Thick-walled *seamless tubing* can be made by *rotary piercing*, a process illustrated in Figure 18-30. A heated billet is fed longitudinally into the gap between two large, convex-tapered rolls. These rolls are rotated in the same direction, but the axes of the rolls are off-set from the axis of the billet by about 6°, one to the right and the other to the left. The clearance between the rolls is preset at a value less than the diameter of the incoming billet. As the billet is caught by the rolls, it is simultaneously rotated and driven forward. The reduced clearance between the rolls forces the billet to deform into a rotating ellipse. As shown in the right-hand segment of Figure 18-28, rotation of the elliptical section causes the metal to shear about the major axis. A crack tends to form down the center axis of the billet, and the cracked material is then forced over a pointed mandrel that enlarges and shapes the opening to create a seamless tube. The result is a short length of thick-walled seamless tubing, which can then be passed through sizing rolls to reduce the diameter and/or wall thickness. Seamless tubes can also be expanded in diameter by passing them over an enlarging mandrel. As the diameter and circumference increase, the walls correspondingly thin.

The *Mannesmann mills* commonly used in hot piercing can be used to produce tubing up to 300 mm (12 in.) in diameter. Larger-diameter tubes can be produced on *Stiefel mills*, which use the same principle but replace the convex rolls of the Mannesmann mill with larger-diameter conical disks.

FIGURE 18-30   (*Left*) Principle of the Mannesmann process of producing seamless tubing. (*Courtesy of American Brass Company.*) (*Right*) Mechanism of crack formation in the Mannesmann process.

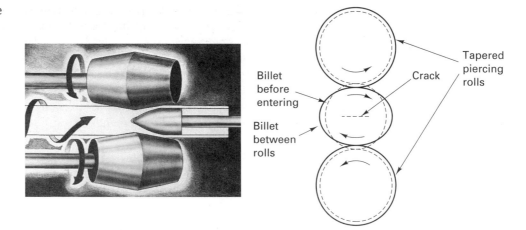

■ KEY WORDS

automatic hot forging	counterblow machine	impactor	near-net-shape	press forging	slab
bar	direct extrusion	impression-die	forging	redrawing	spider-mandrel die
billet	drawing	forging	net-shape	ring rolling	Stiefel mill
blocking	drop-hammer forging	indirect extrusion	forging	rod	strip
bloom	extrusion	ironing	open-die	roll forging	structural shape
bulk deformation	finishing temperature	lap-welding	forging	rolling	swaging
processes	flash	mandrel	pack rolling	rotary piercing	thermomechanical
butt-welding	flashless forging	Mannesmann	piercing	seamless tubing	processing
closed-die forging	forging	mill	pipe	sheet	tube
cluster mill	hammer	mechanical press	pipe welding	sheet forming	upset
controlled rolling	hydraulic press	mill scale	plate	skelp	upset forging

■ REVIEW QUESTIONS

1. Briefly describe the evolution of forming equipment from ancient to modern.
2. What are some of the possible means of classifying metal-deformation processes?
3. Why is the division of forming processes into hot working and cold working a somewhat nebulous classification?
4. What are some features of metals at elevated temperature that

make hot forming an attractive technique to alter shape?
5. What are some of the common terms applied to the various shapes of rolled products?
6. Why are hot-rolled products generally limited to standard shapes and sizes?
7. Why is it undesirable to minimize friction between the work-piece and tooling in a rolling operation?

8. Why is it important to control the finishing temperature of a hot-rolling operation?

9. Discuss the relative advantages and typical uses of two-high rolling mills with large-diameter rolls, three-high mills, and four-high mills.

10. Why is foil almost always rolled on a cluster mill?

11. Why is speed synchronization of the various rolls so vitally important in a continuous or multistand rolling mill?

12. What types of products are produced by ring rolling?

13. Explain how hot-rolled products can have directional properties and residual stresses.

14. Discuss the problems in maintaining uniform thickness in a rolled product and some of the associated defects.

15. Why is a "crowned" roll always designed for a specific operation on a specific material?

16. What is thermomechanical processing, and what are some of its possible advantages?

17. Why are steam or air hammers more attractive than board hammers for hammer forging?

18. What is the difference between open-die and impression-die forging?

19. Why is open-die forging not a practical technique for large-scale production of identical products?

20. What additional controls must be exercised to perform flashless forging satisfactorily?

21. What is a blocker impression in a forging sequence?

22. What attractive features are offered by counterblow forging equipment, or impactors?

23. Why are different tolerances usually applied to dimensions contained within a single die cavity and dimensions across the parting plane?

24. For what types of forging products or conditions might a press be preferred over a hammer?

25. Why are heated dies generally employed in hot-press forging operations?

26. Describe some of the primary differences among hammers, mechanical presses, and hydraulic presses.

27. What is upset forging?

28. What are some of the typical products produced by upset-forging operations?

29. What are some of the attractive features of automatic hot forging? What is a major limitation?

30. How does roll forging differ from a conventional rolling operation?

31. What is meant by the term *swaging*?

32. What are some possible objectives of near-net-shape forging?

33. What are some of the attractive features of the extrusion process?

34. What is the primary shape limitation of the extrusion process?

35. What is the primary attraction of indirect extrusion?

36. What property of a lubricant is critical in extrusion that might not be required for processes such as forging?

37. Why can lubricants not be used in conjunction with a spider-mandrel extrusion die?

38. How can tall, thin cups be produced by hot drawing? (*Note:* The wall height is greater than can be produced in a single drawing operation.)

39. What is meant by the term *ironing* in hot-drawing operations?

40. What two hot-forming operations can be used to produce pipe from steel strip?

41. What limits the length of seamless pipe that can be produced by hot-piercing operations?

■ PROBLEMS

1. Some snack foods, such as rectangular corn chips, are often formed by a rolling-type operation and are subject to the same types of defects common to rolled sheet and strip. Obtain a bag of such a snack and examine the chips to identify examples of rolling-related defects such as those discussed in Section 18.4 and shown in Figure 18-P1.

FIGURE 18-P1 Some typical defects that occur during rolling: wavy edges, edge cracking, and center cracking.

2. Consider the extrusion of a cylindrical billet, and compute the following.
   a. Assume the starting billet to have a length of 0.3 m and a diameter of 15 cm. This is extruded into a cylindrical product that is 3 cm in diameter and 7.5 m long (a reduction ratio of 25). Neglecting the areas on the two ends, compute the ratio between the product surface area (wraparound cylinder) and the surface area of the starting billet.
   b. How would this ratio change if the product were a square with the same cross-sectional area as that of the 3 cm diameter circle?
   c. Consider a cylinder-to-cylinder extrusion with a reduction ratio of R. Derive a general expression of the relative surface areas of product to billet as a function of R. (*Hint:* Start with a cylinder with length and diameter both equal to 1 unit. Since the final area will be $1/R$ times the original, the final length will be R units and the final diameter will be proportional to $1/\sqrt{R}$).

d. If the final product had a more complex cross-sectional shape than a cylinder. Would the final area be greater than or less than that computed in part c?

3. The force required to compress a cylindrical solid between flat parallel dies (See Figure 18-10) has been estimated (by a theory of plasticity analysis) to be

$$\text{force} = \pi R^2 \sigma_o \frac{1 + 2\,mR}{3\sqrt{3}\,T}$$

where

R = radius of the cylinder
T = thickness of the cylinder
σ_o = yield strength
m = friction factor

An engineering student is attempting to impress his date by demonstrating some of the neat aspects of metal forming. He places a shiny penny between the platens of a 60,000-lb capacity press and proceeds to apply pressure. Assume that the coin has a $\frac{3}{4}$-in. diameter and is $\frac{1}{16}$ in. thick. The yield strength is estimated as 50,000 psi, and since no lubricant is applied, friction is that of complete sticking, or $m = 1.0$.

a. Compute the force required to induce plastic deformation.

b. If this force is greater than the capacity of the press (60,000 lb), compute the pressure when the full-capacity force of 60,000 lb is applied.

c. If the press surfaces are made from thick plates of a material with a yield strength of 120,000 psi, describe the results of the demonstration.

4. Mathematical analysis of the rolling of flat strip reveals that the roll-separation force (the squeezing force required to deform the strip) is directly proportional to the term

$$1 + \frac{K_1 mL}{t_{av}}$$

where

K_1 = geometric constant
m = friction factor
L = length of contact
t_{av} = avarage thickness of the strip in the roll bite

Since L is proportional to the roll radius, R, K_1L can be re-placed by K_2R, so the force becomes proportional to the term:

$$1 + \frac{K_2 mR}{t_{av}}$$

If K_2 and m are both positive numbers, how will the roll-separation force change as the strip becomes thinner? How can this effect be minimized? Relate your observations to the types of rolling mills used for various thicknesses of product.

5. If the area under the indirect extrusion curve of Figure 18-25 is proportional to the power required to extrude a product without billet-chamber frictional resistance, how could the relative regions of the direct extrusion curve be used to determine a crude measure of the mechanical "efficiency" of direct extrusion?

www.wiley.com/college/degarmo

*C*hapter 18 CASE STUDY

Outboard Motor Brackets

The components depicted in Figure CS-18 are two styles of mounting brackets used to attach outboard motors to the stern plates of small boats. The sketches show the parts as unfinished castings or forgings. Finish machining will add a threaded hole through the bottom of the short leg to accommodate a tightening turn screw. Holes will also be drilled through the long leg to permit attachment of the motor.

Since outboard motors are frequently carried to the boat, weight is a definite concern. The bracket will also operate in a wet environment, possibly even saltwater, and will be subjected to impacts and vibration, in addition to its steady-state force. The designer has recommended a minimum yield strength of 345 MPa (50 ksi), coupled with a 5% minimum elongation in a uniaxial tensile test to assure adequate resistance to brittle fracture.

1. Briefly discuss the major service requirements for these components. Which of the various engineering materials would you recommend for this application?

2. For your recommended material, what are some of the possible means of producing the desired shapes? Which of the possible alternatives would you recommend? Why?

3. Would a heat treatment be necessary to establish the required properties? If so, what type of treatment should be used?

4. Would some form of surface treatment be desirable? What type of treatment would you recommend?

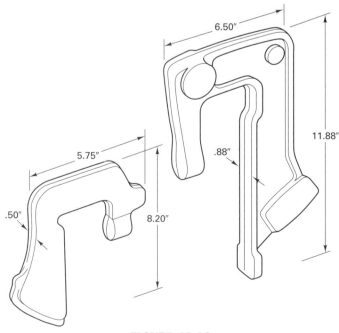

FIGURE CS-18

CHAPTER 19

COLD-WORKING PROCESSES

■ 19.1 INTRODUCTION

A number of attractive features are associated with the cold deformation of a metal. Strain hardening acts to increase the strength, so cheaper, weaker starting materials may be selected, or costly heat treatment of the finished product may be eliminated. By proper positioning of the last anneal, a final product can be produced with a specified amount of cold work and the associated properties that have been selected to meet specific needs. Surface finish and dimensional precision are quite good. In addition, substantial energy savings are possible. While greater forces are required to induce deformation, we no longer need to heat the workpiece to the forming temperature.

As a result of these features, a number of cold-working processes have been developed to perform a wide variety of deformations. These processes can be classified into four basic categories: *squeezing*, *bending*, *shearing*, and *drawing*. Table 19-1 lists some of the cold-working processes according to their primary form of deformation.

■ 19.2 SQUEEZING PROCESSES

Most of the squeezing-type cold-working processes have hot-working counterparts or are extensions of hot-working processes. The primary reason for cold deformation is to obtain better dimensional accuracy and improved surface finish. In many cases, the equipment is basically the same, except that it must be more powerful to deform the higher-strength starting material and overcome the additional resistance caused by strain hardening. If power is limited, compromise may have to be made in either the size of the workpiece or the amount of deformation.

TABLE 19-1.   Classification of the Major Cold-Working Operations

Squeezing		Bending	
1. Rolling	7. Staking	1. Angle	6. Flanging
2. Swaging	8. Coining	2. Roll	7. Straightening
3. Cold forging	9. Peening	3. Draw and	
4. Extrusion	10. Burnishing	compression	
5. Sizing	11. Hubbing	4. Roll-forming	
6. Riveting	12. Thread rolling	5. Seaming	
Shearing		Drawing	
1. Shearing; slitting	5. Shaving	1. Bar and tube	6. Sheet metal drawing
2. Blanking	6. Trimming	drawing	7. Ironing
3. Piercing; lancing;	7. Cutoff	2. Wire drawing	8. Superplastic forming
perforating	8. Dinking	3. Spinning	
4. Notching;		4. Embossing	
nibbling		5. Stretch forming	

COLD ROLLING

Cold rolling is clearly the dominant cold-working process in terms of product tonnage. Sheets, strips, bars, and rods are cold-rolled into products that have smooth surfaces and accurate dimensions. Because of the smaller size and higher strength of the material (compared to hot rolling) most cold rolling is performed on four-high or cluster-type rolling mills.

Cold-rolled *sheet* and *strip* can be obtained in various conditions, including *skin-rolled*, *quarter-hard*, *half-hard*, and *full-hard*. Skin-rolled metal is subjected to only a $\frac{1}{2}$ to 1% reduction to produce a smooth surface and uniform thickness, and to remove or reduce the yield-point phenomenon (i.e., prevent formation of Luders bands upon further forming). This material is well suited for subsequent cold-working operations where good ductility is required. Quarter-hard, half-hard, and full-hard sheet and strip experience greater amounts of cold reduction, up to 50%. Their yield points are higher, properties have become directional, and ductility has decreased. Quarter-hard steel can be bent back on itself across the grain without breaking. Half-hard and full-hard can be bent back 90° and 45°, respectively, about a radius equal to the material thickness.

COLD-ROLLED SHAPES

If a product has a uniform (or nearly uniform) cross section and relatively small transverse dimensions (less than about 5 cm or 2 in.), cold rolling of rod or bar may be an attractive alternative to extrusion or machining. Strain hardening can be employed to provide up to 20% additional strength to the material, and the process offers smooth surfaces and high dimensional precision. Like the hot forming of structural shapes discussed in Chapter 18, the cold rolling of shapes generally requires a series of shaping operations. Separate passes (and roll grooves) may be required for sizing, breakdown, roughing, semiroughing, semifinishing, and finishing. A minimum order of several tons of product may be required to justify the cost of the required tooling.

THREAD ROLLING

Thread rolling is a more specialized cold-rolling operation that is an alternative to the cutting of threads. See the discussion in Chapter 30.

SWAGING

Swaging (also known as rotary swaging or radial forging) uses external hammering to reduce the diameter, taper, or point round bars or tubes, as shown in Figure 19-1. Figure 19-2 shows the internal components of the machine, and can be used to explain the process. The dies, located in the center of the apparatus, consist of two blocks of hardened tool steel, with a central hole that has a conical input transitioning to a cylinder. An external motor drives a large, massive flywheel, which is connected to the central spindle of the machine. High-speed rotation of the central unit generates centrifugal force, which causes the matching die

FIGURE 19-1 Tube being reduced in a rotary swaging machine. *(Courtesy of Torrington Company.)*

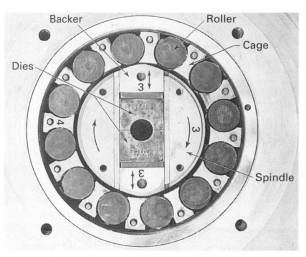

FIGURE 19-2 Basic components and motions of a rotary swaging machine. *Note:* The cover plate has been removed to reveal the interior workings. *(Courtesy of Torrington Company.)*

segments and backing blocks to separate. As the spindle rotates, the backing blocks are driven into opposing rollers that have been mounted in a massive machine housing. To pass beneath these rollers, the backer blocks must squeeze the dies tightly together. Once the assembly clears the rollers, the dies once again separate and the cycle repeats, generating as many as 2000 blows per minute.

With the machine in motion, the operator simply inserts a rod or tube between the dies and advances it during the periods of die separation. Because the dies rotate, the repeated closures hammer the workpiece from a variety of angles, reducing the diameter and increasing the length. Since the rotating spindle is usually hollow, the workpiece can be passed through the machine, or withdrawn after a preset length has been reduced.

Swaging operations can also be used to form products with internal shapes of constant cross section. A shaped mandrel is inserted into a tube (or closed-end workpiece), and the metal is collapsed around it to simultaneously shape and size both the interior and exterior of the product. If the parts are extremely long, they can be fed over a short, stationary mandrel that is positioned between the dies. Swaging over a mandrel can be used to form parts with internal gears, splines, recesses, and sockets. Figure 19-3 shows a variety of swaged products, many containing shaped holes or recesses.

FIGURE 19-3 A variety of swaged parts, some with internal details. *(Courtesy of Cincinnati Milacron, Inc.)*

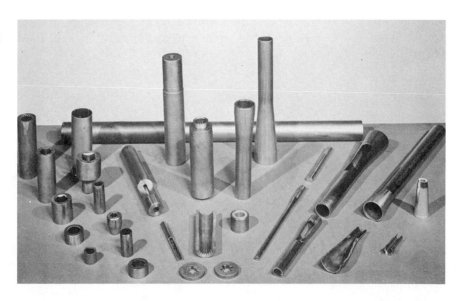

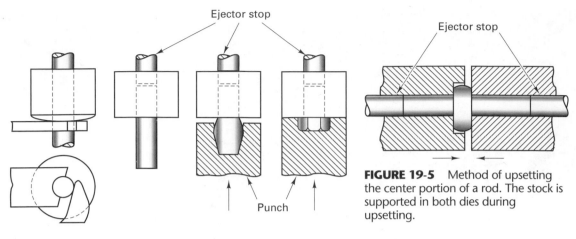

FIGURE 19-4 Typical steps in a shearing and cold-heading operation.

FIGURE 19-5 Method of upsetting the center portion of a rod. The stock is supported in both dies during upsetting.

COLD FORGING (COLD FORMING)

Large quantities of products are now being made by *cold forging* (or *cold forming*), a family of processes in which slugs of material are squeezed into shaped die cavities to produce finished parts of precise shape and size. *Cold heading*, illustrated schematically in Figure 19-4, is used for making enlarged sections on the ends of rod or wire, such as the heads of nails, bolts, rivets, or other fasteners. Two variations of the process are common. In the sequence illustrated, a piece of rod is first sheared to a preset length and then transferred to a holder-ejector assembly. Heading punches then strike one or more blows on the exposed end to perform the upsetting. If intermediate shapes are required, the piece is transferred from station to station, or the various heading punches sequentially rotate into position. When the heading is completed, the ejector stop advances and expels the product.

In the second variation, a continuous rod (or wire) is fed forward to produce a preset extension, clamped, and the head is formed. The rod is then advanced to a second preset length and sheared, and the cycle repeats. This procedure is particularly attractive for producing nails, since the point can be formed in the shearing or cutoff operation.

Enlarged sections can also be produced at locations other than the ends of a rod or wire, as illustrated in Figure 19-5. To produce the deformation and control final dimensions, ejector stops must be provided in both segments of the opposing tooling.

By using various types of dies and combining high-speed operations such as heading, upsetting, extrusion, bending, coining, thread rolling, and knurling, a wide variety of relatively complex parts can be cold formed to close tolerances. Figure 18-17 has already presented a display of typical products. The larger parts are generally hot formed and machined, while the smaller ones are cold formed. Since cold forming is a chipless manufacturing process, producing parts by deformation that would otherwise be machined from bar stock or hot forgings, the material is used more efficiently and waste is reduced.

Figure 19-6 compares the manufacture of a spark-plug body by machining from hexagonal bar stock with manufacture by cold forming. Material is saved, machining

FIGURE 19-6 Manufacture of a spark plug body: (*left*) by machining from hexagonal bar stock; (*right*) by cold forming. Note the reduction in waste. (*Courtesy of National Machinery Co.*)

Cutting (74% waste)   Cold forming (6% waste)

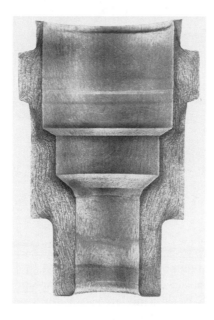

FIGURE 19-7 Section of the cold-formed spark-plug body of Figure 19-6, etched to reveal the flow lines. The cold-formed structure produces an 18% increase in strength over the machined product. *(Courtesy of National Machinery Co.)*

time and cost are reduced, and the product is stronger, due to cold work, and tougher, as illustrated by the flow lines revealed in Figure 19-7. By converting from screw machining to cold forming, a manufacturer of cruise-control housings reduced material usage by 65%, while simultaneously increasing production rate by a factor of 5.

Cold forming is generally associated with the manufacture of small parts from the weaker nonferrous metals, but the process is now used extensively on steel, and parts can be up to 45 kg (100 lb) in weight and 18 cm (7 in.) in diameter. At the small end of the scale, microformers are now cold forming extremely small electronic components with dimensional accuracies within 0.005 mm (0.0002 in.). Shapes are usually axisymmetric or those with relatively small departures from symmetry. Production rates are high, dimensional tolerances and surface finish are excellent, and the amount of machining can generally be reduced (material savings). Strain hardening can provide additional strength, and favorable grain flow can enhance toughness and fatigue life. Unfortunately, the cost of the required tooling, coupled with the high production speed, generally requires large-volume production.

IMPACT EXTRUSION (OR COLD EXTRUSION)

Chapter 18 has already presented the processes of direct and indirect extrusion, where the workpiece is usually deformed hot. Great advances have also been made in cold extrusion, a family of processes more commonly known as *impact extrusion*. Figure 19-8 illustrates the basic principles of several variations, *forward* and *backward* using both

FIGURE 19-8 Backward and forward extrusion with open and closed dies.

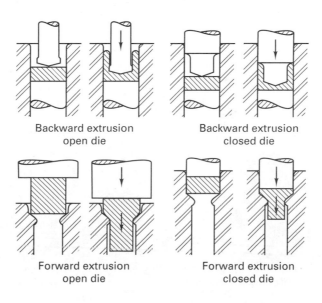

Backward extrusion
open die

Backward extrusion
closed die

Forward extrusion
open die

Forward extrusion
closed die

FIGURE 19-9 Steps in the forming of a bolt by cold extrusion, cold heading, and thread rolling. *(Courtesy of National Machinery Co.)*

open and closed dies. These processes were first used to shape low-strength metals such as lead, tin, zinc, and aluminum into products such as collapsible tubes for toothpaste, medications, and other creams; small "cans" for shielding electronic components; zinc cases for flashlight batteries; and larger cans for food and beverages.

In recent years, impact extrusion has been applied to the forming of mild steel parts. The process is often used in combination with cold heading, as shown in the example of Figure 19-9. When heading alone is used, there is a definite limit to the ratio of the head and stock diameters (as presented in Figure 18-16 and related discussion). The combination of forward extrusion and cold heading overcomes this limitation by using an intermediate starting diameter. The shank diameter is reduced by forward extrusion while upsetting is used to increase the diameter of the head. Figure 19-10 illustrates another cold-forming operation, this time involving two extrusions and a central upset. In both of these cases, considerable metal is saved compared to the machining of parts from larger-diameter stock.

FIGURE 19-10 Cold-forming sequence involving cutoff, squaring, two extrusions, an upset, and a trimming operation. Also shown are the finished part and the trimmed scrap. *(Courtesy of National Machinery Co.)*

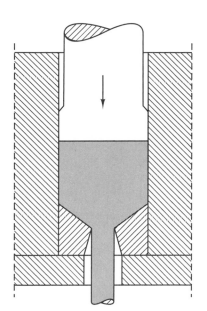
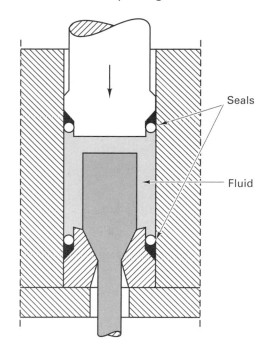

Seals

Fluid

FIGURE 19-11 Comparison of conventional (*left*) and hydrostatic (*right*) extrusion. Note the addition of the pressurizing fluid and the O-ring and miter-ring seals on both the die and ram.

HYDROSTATIC EXTRUSION

Another type of cold extrusion, known as *hydrostatic extrusion*, is illustrated schematically in Figure 19-11. Here high-pressure fluid surrounds the workpiece, and applies the force necessary to extrude it through the die. The product emerges into either atmospheric pressure or a lower-pressure fluid-filled chamber. The process resembles direct extrusion, but the fluid pressure surrounding the billet prevents any upsetting. Since the billet does not come into contact with the surrounding chamber, billet-chamber friction is eliminated, and the pressurized fluid can further act as a lubricant between the billet and the die.

While process efficiency can be significantly greater than most other extrusion processes, there are problems related to the fluid and the associated high pressures, which typically range between 900 and 1700 MPa (125 to 250 ksi). Temperatures are limited since the fluid acts as a heat sink and the pressurizing fluids (typically light hydrocarbons and oils) burn or decompose at moderately low temperatures. Seals must be designed to contain the pressurized fluid without leaking, and measures must be taken to prevent the complete ejection of the product, often referred to as *blowout*. Because of these features, hydrostatic extrusion is usually employed only where the process offers unique advantages that cannot be duplicated by the more conventional methods.

Pressure-to-pressure extrusion is one of these unique capabilities. In this variant, the product emerges from one pressurized chamber into a second high-pressure chamber. In effect, the metal deformation is performed in a highly-compressed environment. Crack formation begins with void formation, void growth and void coalescence. Voids are suppressed in a compressed environment, resulting in a phenomenon known as *pressure-induced ductility*. Relatively brittle materials such as molybdenum, beryllium, tungsten, and various intermetallic compounds can be plastically deformed without fracture, and materials with limited ductility become highly formable. Products can be made that could not be otherwise produced, and materials can be considered that would have been rejected because of their limited ductility at room temperature and atmospheric pressure.

CONTINUOUS EXTRUSION

Conventional extrusion is a discontinuous process, converting finite-length billets into finite-length products. If the pushing force could be applied to the periphery of the feedstock, rather than the back, continuous feedstock could be converted into continuous product, and the process could become one of *continuous extrusion*. The first continuous extrusion of solid metal feedstock was performed in 1970. Since then a number of techniques have been proposed with varying degrees of success. In terms of commercial application, the most significant is probably the *Conform process*, illustrated schematically

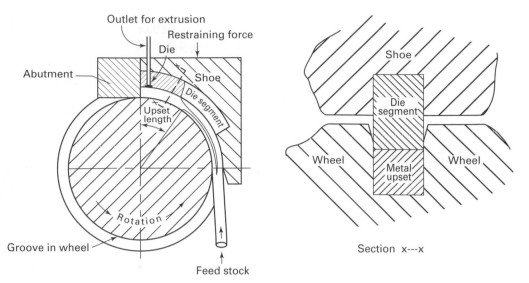

FIGURE 19-12 Cross-sectional schematic of the Conform continuous extrusion process, showing bar feed stock or particle feed. The material upsets at the abutment and extrudes. Section x---x shows the segment in the shoe.

in Figure 19-12. Continuous feedstock is inserted into a grooved wheel and is driven by surface friction into a chamber created by a mating die segment. Upon impacting a protruding abutment, the material upsets to conform to the chamber, and the increased wall contact further increases the driving friction. Upsetting continues until the pressure reaches a value sufficient to extrude the material through the die opening. At this point, the rate of material entering the machine equals the rate of product emerging, and a steady-state continuous process is established.

Since surface friction is the propulsion force, the feedstock can take a variety of forms, including solid rod, metal powder, punchouts from other forming operations, or chips from machining. Metallic and nonmetallic powders can be intimately mixed and coextruded. Rapidly solidified material can be extruded without exposure to the elevated temperatures that would harm the properties. Polymeric materials and even fiber-reinforced plastics have been successfully extruded. The most common feed, however, is coiled aluminum or copper rod.

Continuous extrusion complements and competes with wire drawing and shape rolling as a means of producing nonferrous products with small, but uniform, cross sections. It is particularly attractive for complex profiles and cross sections that contain one or more holes. Since extrusion operations can perform massive reductions through a single die, one Conform operation can produce an amount of deformation equivalent to ten conventional drawing or cold-rolling passes. In addition, sufficient heat can be generated by the deformation that the product will emerge in an annealed condition, ready for further processing without intermediate heat treatment.

ROLL EXTRUSION

Thin-walled cylinders can be produced from thicker-wall material by the *roll-extrusion* process. In the variant depicted in Figure 19-13a, internal rollers increase the internal diameter as they squeeze the rotating material against an external confining ring. The tube elongates as the wall thickness is reduced. In Figure 19-13b, the internal diameter is maintained as external rollers squeeze the material against a rotating mandrel. Although the process has been used to produce cylinders from 2 cm to 4 m (0.75 to 156 in.) in diameter, most products have diameters between 7.5 and 50 cm (3 and 20 in.).

SIZING

Sizing involves squeezing all or selected regions of forgings, ductile castings, or powder metallurgy products, to achieve a prescribed thickness or enhanced dimensional precision. By incorporating sizing, designers can make the initial tolerances of a part more liberal, enabling the use of less-costly production methods. Those dimensions that must be precise are then set by one or more sizing operations that are usually performed on simple, mechanically driven presses.

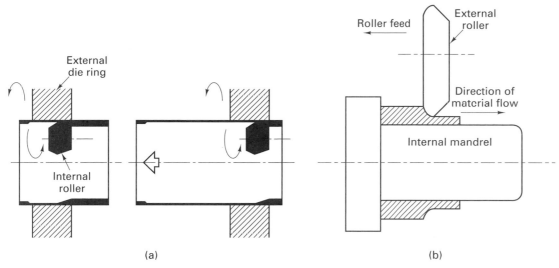

(a) (b)

FIGURE 19-13 The roll extrusion process: (a) with internal rollers expanding the inner diameter; (b) with external rollers reducing the outer diameter.

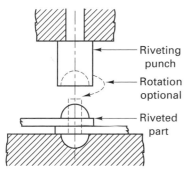

FIGURE 19-14 Joining components by riveting.

RIVETING

In *riveting*, an expanded head is formed on the shank end of a fastener to permanently join sheets or plates of material. Although riveting is usually done hot in structural applications, in manufacturing, it is almost always done cold. Where there is access to both sides of the work, the method illustrated in Figure 19-14 is commonly used. The shaped punch may be driven by a press or contained in a special, hand-held riveting hammer. When a press is used, the rivet is usually headed in a single squeezing action, although the heading punch may also rotate so as to to shape the head in a progressive manner, an approach known as *orbital forming*. Special riveting machines, like those used in aircraft assembly, can punch the hole, place the rivet in position, and perform the heading operation, all in about 1 second.

It is often desirable to use riveting in situations where there is access to only one side of the assembly. Figure 19-15 shows two types of special rivets that can be used for one-side access applications. The shank on the "blind" side of an *explosive rivet* expands to form a retaining head when a heated tool is touched against the exposed segment and detonates the charge. In the pull type, or *pop-rivet*, a pull-up pin is used to expand a tubular shank. After performing its function, the pull pin breaks or is cut off flush with the head.

STAKING

Staking is a method of permanently joining parts together when a segment of one part protrudes through a hole in the other. As shown in Figure 19-16, a shaped punch is driven into the exposed end of the protruding piece. The deformation causes radial expansion, mechanically locking the two pieces together. Because the tooling is simple

FIGURE 19-15 Rivets for use in "blind" riveting: (*left*) explosive type; (*right*) shank-type pull-up. (*Courtesy of Huck Manufacturing Company.*)

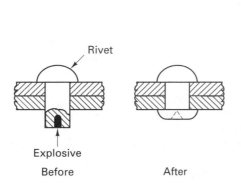

Rivet

Before After

Explosive

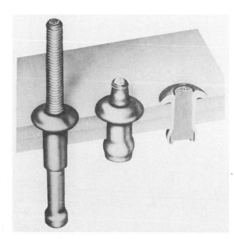

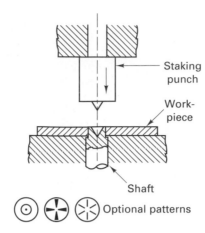

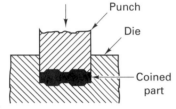

FIGURE 19-16 Permanently attaching a shaft to a plate by staking.

and the operation can be completed with a single stroke of a press, staking is a convenient and economical method of fastening when permanence is desired and the appearance of the punch mark is not objectionable. Figure 19-16 includes some of the decorative punch designs that are commonly used.

COINING

The term *coining* refers to the cold squeezing of metal while all of the surfaces are confined within a set of dies. The process, illustrated schematically in Figure 19-17, is used to produce coins, medals, and other products where exact size and fine detail are required, and thickness varies about a well-defined average. Because of the total confinement (there is no possibility for excess metal to escape from the die), the input material must be accurately sized to avoid breakage of the dies or press. Coining pressures may be as high as 1400 MPa or 200,000 psi.

HUBBING

Hubbing[1] is a cold-working process that is used to plastically form recessed cavities in a workpiece. As shown in Figure 19-18, a male hub (or master) is made with the reverse profile of the desired cavity. After hardening, the hub is pressed into an annealed block (usually by a hydraulic press) until the desired impression is produced. (Production of the cavity can often be aided by machining away some of the metal in regions where large amounts of material would be displaced.) The hub is withdrawn, and the displaced metal

FIGURE 19-17 The coining process.

FIGURE 19-18 Hubbing a die block in a hydraulic press. Inset shows close-up of the hardened hub and the impression in the die block. The die block is contained in a reinforcing ring. The upper surface of the die block is then machined flat to remove the bulged metal.

---

[1] This process should not be confused with "hobbing," a machining process used for cutting gears.

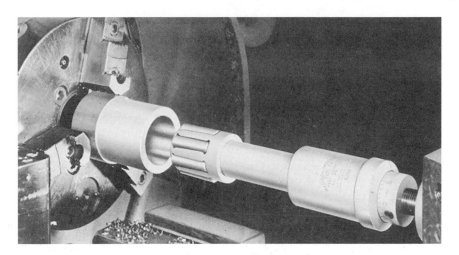

FIGURE 19-19 Tool for roller burnishing. The burnishing rollers move outward by means of a taper. *(Courtesy of Madison Industries, Inc.)*

is removed by a facing-type machining cut. The workpiece, which now contains the desired cavity, is then hardened by heat treatment.

Hubbing is often more economical than die sinking (machining the cavity), especially when multiple impressions are to be produced. One hub can be used to form a number of identical cavities, and it is generally easier to machine a male profile (with exposed surfaces) than a female cavity (where you are cutting in a hole).

SURFACE IMPROVEMENT BY COLD WORKING

Cold-working processes can also be used to improve or alter the surfaces of metal products. *Peening* is the mechanical working of surfaces by repeated blows of impelled shot or a round-nose tool. The highly localized impacts flatten and broaden the metal surface, but the underlying material restricts spread, resulting in a surface with residual compression. Since the net loading on a material surface is the applied load minus the residual compression, peening tends to enhance the fracture resistance and fatigue life of tensile-loaded components. For this reason, shot impellers are frequently used to peen shafting, crankshafts, connecting rods, gear teeth, and other cyclic-loaded components.

Manual or pneumatic hammers are frequently used to peen the surfaces of metal weldments. Solidification shrinkage and thermal contraction produce surfaces with residual tension. Peening can reduce or cancel this effect, thereby reducing associated distortion and preventing cracking.

Burnishing involves rubbing a smooth, hard object (under considerable pressure) over the minute surface irregularities that are produced during machining or shearing. The edges of sheet metal stampings can be burnished by pushing the stamped parts through a slightly tapered die having its entrance end a little larger than the workpiece and its exit slightly smaller. As the part rubs along the sides of the die, the pressure is sufficient to smooth the slightly rough edges that are characteristic of a blanking operation (see Figure 19-36).

Roller burnishing, illustrated in Figure 19-19, can be used to improve the size and finish of internal and external cylindrical and conical surfaces. The hardened rolls of a burnishing tool press against the surface and deform the protrusions to a more-nearly-flat geometry. The resulting surfaces possess improved wear and fatigue resistance, since they have been cold worked and are now in residual compression.

■ 19.3 BENDING

As we approach the subject of bending, it is important to first establish a few definitions. *Bending* is the plastic deformation of metals about a linear axis with little or no change in the surface area. Multiple bends can be made simultaneously, but to be classified as true bending, and treatable by simple bending theory, each axis must be linear and independent of the others. If multiple bends are made with a single die, the process is often called *forming*. When the axes of deformation are not linear, or are not independent, the process is known as *drawing* and/or *stretching*, not bending, and these operations will be treated later in the chapter.

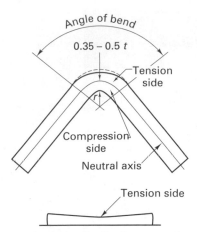

FIGURE 19-20 (*Top*) Nature of a bend in sheet metal showing tension on the outside and compression on the inside; (*bottom*) The upper portion of the bend region, viewed from the side, shows how the center portion will thin more than the edges.

As shown in Figure 19-20, simple bending causes the metal on the outside to be stretched while that on the inside is compressed. The location that is neither stretched nor compressed is known as the *neutral axis* of the bend. Since the yield strength of metals in compression is somewhat higher than the yield strength in tension, the metal on the outer side yields first, and the neutral axis is displaced from the midpoint between the two surfaces. The neutral axis is generally located between one-third and one-half of the way from the inner surface, depending on the bend radius and the material being bent. Because of this lack of symmetry and the dominance of tensile deformation, the metal is generally thinned at the bend. In a linear bend, the thinning is greatest in the center of the sheet, and less at the free edges where a pulling in of the free edge provides some compensation.

On the inner side of a bend, the compressive stresses can induce upsetting and a companion thickening of material. While this thickening somewhat offsets the thinning of the outer section, the upsetting can also produce an outward expansion of the free edges. This contraction of the tensile segment and expansion of the compression segment can produce significant distortion of edge surfaces that terminate a linear bend. This effect is particularly pronounced when bends are produced across the width of thick but narrow plates.

Still another consequence of the combined tension and compression is the elastic recovery that occurs when the bending load is removed. The stretched region contracts and the compressed region expands, resulting in a small amount of "unbending," known as *springback*. To produce a product with a specified angle, the metal must be overbent by an amount equal to the subsequent springback.

ANGLE BENDING (BAR FOLDER AND PRESS BRAKE)

Machines like the *bar folder*, shown in Figure 19-21, can be used to make angle bends up to 150° in sheet metal under 1.5 mm ($\frac{1}{16}$ in.) thick. The workpiece is inserted under the folding leaf and aligned in the proper position. Raising the handle then actuates a cam, causing the leaf to clamp the sheet. Further motion of the handle bends the metal to the desired angle. These machines are manually operated and can be used to produce linear bends up to about 3.5 m (12 ft) in length.

Bends in heavier sheet, or more complex bends in thin material, are generally made on *press brakes*, like the one shown in Figure 19-22. These are mechanical or hydraulic presses with a long, narrow bed and short, adjustable strokes. The metal is bent between interchangeable dies that are attached to both the bed and the ram. As illustrated in Figures 19-22 and 19-23, the different dies can be used to produce many types of bends. The metal can be repositioned between strokes to produce complex

FIGURE 19-21 Phantom section of a bar folder, showing position and operation of internal components. (*Courtesy of Niagara Machine and Tool Works.*)

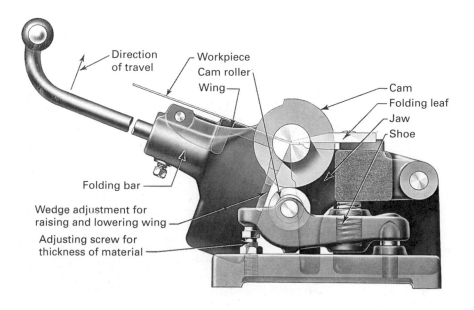

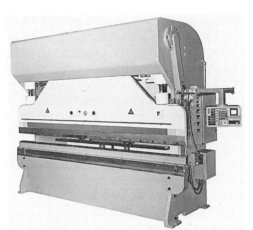

FIGURE 19-22 (*Left*) Press brake with CNC gauging system. *(Courtesy of DiAcro Division, Houdaille Industries, Inc.)* (*Right*) Close-up view of press brake dies forming corrugations. *(Courtesy of Cincinnati Incorporated.)*

FIGURE 19-23 Press brake dies can form a variety of angles and contours. *(Courtesy of Cincinnati Incorporated.)*

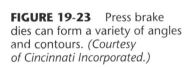

contours or repeated bends, such as corrugations. Figure 19-24 shows how a roll bead can be formed with repeated strokes, repositioning, and multiple sets of tooling. Seaming, embossing, punching, and other operations can also be performed with press brakes, but these operations can usually be done more efficiently on other types of equipment.

The upper and lower tools and support structures on a press brake are essentially beams that are loaded in the same three-point bending previously discussed for rolling mill rolls. Elastic deflections can cause a variety of bend deviations and defects, and a number of means have been developed to minimize the problems.

FIGURE 19-24 Dies and operations used in the press brake forming of a roll bead. *(Courtesy of Cincinnati Incorporated.)*

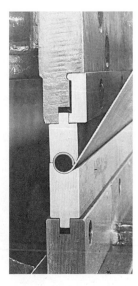

DESIGN FOR BENDING

Several factors must be considered when designing parts that are to be shaped by bending. One of the primary concerns is determining the smallest bend radius that can be formed without metal cracking (i.e., the *minimum bend radius*). This value is dependent on both the ductility of the metal (as measured by the percent reduction in area observed in a standard tensile test) and the thickness of the material being bent. Figure 19-25 shows how the ratio of the minimum bend radius R to the thickness of the material t varies with material ductility. As this plot reveals, an extremely ductile material is required if we wish to produce a bend with radius less than the thickness of the metal. If possible, bends should be designed with large bend radii. This permits easier forming and allows the designer to select from a wider variety of engineering materials.

If the punch radius is large and the bend angle is shallow, large amounts of springback are often encountered. The sharper the bend, the more likely the surfaces will be stressed beyond the yield point. Less severe bends have large amounts of elastically-stressed material, and large amounts of springback. In general, when the bend radius is greater than four times the material thickness, the tooling or process must provide springback compensation.

If the metal has experienced previous cold work or has marked directional properties, these features should be considered when designing the bending operation. Whenever possible, it is best to make the bend axis perpendicular to the direction of previous working, as shown in Figure 19-26. The explanation for this behavior has little to do with the grain structure of the metal, but is more closely related to the mechanical loading applied to the weak inclusions. Cracks can easily start along tensile-loaded inclusions and propagate to full cracking of the bend. If perpendicular bends are required, it is often best to place each at 45° to the rolling direction, rather than have one longitudinal and one transverse.

Another design concern is determining the dimensions of a flat blank that will produce a bent part of the desired precision. As discussed earlier in the chapter, metal tends to thin and lengthen when it is bent. The amount of lengthening is a function of both the stock thickness and the bend radius. Figure 19-27 illustrates one method that has been found to give satisfactory results in determining the blank length for bent products. In addition, the minimum length of any protruding leg should be at least equal to the bend radius plus $1\frac{1}{2}$ times the thickness of the metal.

FIGURE 19-25   Relationship between the minimum bend radius, R, (relative to thickness) and the ductility of the metal being bent (as measured by the reduction in area in a uniaxial tensile test).

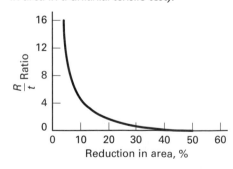

FIGURE 19-26   Bends should be made with the bend axis perpendicular to the rolling direction. When intersecting bends are made, both should be at angle to the rolling direction, as shown.

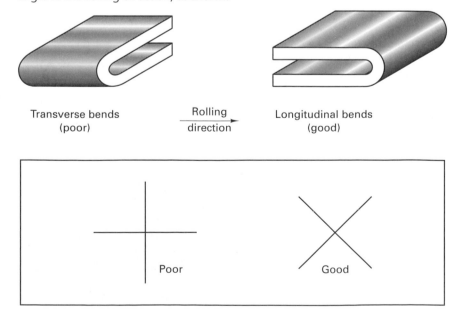

FIGURE 19-27 One method of determining the starting blank size (*L*) for several bending operations. Due to thinning, the product will lengthen during forming. ℓ_1, ℓ_2, and ℓ_3 are the desired product dimensions. See table to determine *D* based on size of radius *R* where *t* = stock thickness.

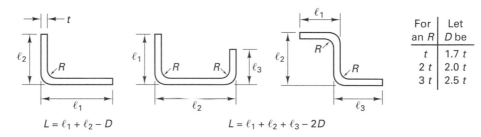

For an *R*	Let *D* be
t	1.7 *t*
2 *t*	2.0 *t*
3 *t*	2.5 *t*

$$L = \ell_1 + \ell_2 - D$$

$$L = \ell_1 + \ell_2 + \ell_3 - 2D$$

Whenever possible, the tolerance on bent parts should not be less than 0.8 mm ($\frac{1}{32}$ in.). Bends of 90° or greater should not be specified without first determining whether the material and bending method will permit them. Parts with multiple bends should be designed with most (or preferably all) of them to be of the same bend radius. This will reduce setup time and tooling costs. Consideration should also be given to providing regions for adequate clamping or support during manufacture. Bending near the edge of a material will distort the edge. If an undistorted edge is required, additional material must be included and a trimming operation performed after bending.

AIR-BEND, BOTTOMING, AND COINING DIES

Yet another design decision is the use of air-bend, bottoming, or coining dies. As shown in Figure 19-28, *bottoming dies* contact and compress the full area within the tooling. The angle of the resulting bend is set by the geometry of the tooling, and subsequent springback. If the results are outside specifications, or the material is changed and produces a different amount of springback, the geometry of the tooling will have to be modified. Once the geometry of the tool is successfully set, however, reproducibility of the bend geometry is excellent, provided there is consistency within the size and properties of the material being bent.

In contrast, *air-bend dies* produce the bend geometry by simple three-point bending. Since the resulting angle is controlled by the bottoming position of the upper die, a single set of tooling can produce a range of bend geometries from 180° through the included angle of the die, and the pressure required to form the bends is the lowest of the three options. Product reproducibility depends on the ability to control the stroke of the press. Adaptive control and on-the-fly corrections are frequently used with air-bend tooling.

If bottoming dies continue to move beyond the bottoming position, the material between the upper and lower dies is plastically deformed, and the operation extends into *coining*. Springback can be significantly reduced, and more consistent results can be achieved with materials having variation in structure and thickness. Unfortunately, the loading is greatly increased on both the press and the tools.

Mechanical presses are generally used for bottom bending and coining, while hydraulic presses are preferred for air bending.

FIGURE 19-28 Comparison of air-bend (*left*) and bottoming (*right*) press brake dies. With the air-bend die, the amount of bend is controlled by the bottoming position of the upper die.

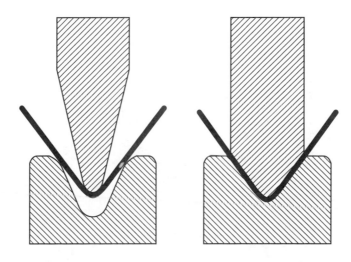

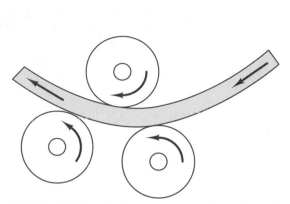

FIGURE 19-29 (*Left*) Schematic of the roll-bending process; (*right*) the roll bending of an I-beam section. Note how the material is continuously subjected to three-point bending. (*Courtesy of Buffalo Forge Company.*)

ROLL BENDING

Roll bending is a continuous form of three-point bending where plates, sheets, beams, pipe, and even rolled shapes and extrusions are bent to a desired curvature using forming rolls. As shown in Figure 19-29, roll bending machines usually have three rolls in the form of a triangle. The two lower rolls are driven and the position of the upper roll is adjustable to control the degree of curvature in the product. The rolls on the machine in Figure 19-29 are supported on only one end. When wide material is being formed, the longer long rolls usually require support on both ends. The support frame on one end can often be swung clear to permit the removal of closed circular shapes or partially-rolled pieces. Because of the variety of applications, roll bending machines are available in a wide range of sizes, some being capable of bending plate up to 25 cm (10 in.) thick.

DRAW BENDING, COMPRESSION BENDING, AND PRESS BENDING

Bending machines can also utilize a clamp and pressure tool to bend material against a form block. In *draw bending*, illustrated in Figure 19-30, the workpiece is clamped against a bending form and the entire assembly is rotated to draw the workpiece under a stationary pressure tool. In *compression bending*, also illustrated in Figure 19-30, the bending form remains stationary and the pressure tool moves along the surface of the workpiece.

Press bending, also shown in Figure 19-30, utilizes a downward descending bend die, which pushes into the center of material that is supported on either side by wing dies. As the ram descends, the wing dies pivot up, bending the material around the form on the ram. The flexibility of each of the above processes is somewhat limited because a certain length of the product must be used for clamping.

FIGURE 19-30 (a) Draw bending, in which the form block rotates; (b) compression bending, in which a moving tool compresses the workpiece against a stationary form, (c) press bending, where the press ram moves the bending form.

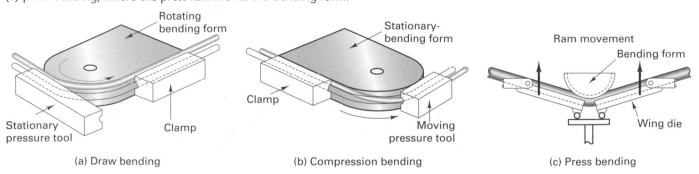

(a) Draw bending   (b) Compression bending   (c) Press bending

TUBE BENDING

Quite often the material being bent is a tube or pipe, and this geometry presents additional problems. Key parameters are the outer diameter of the tube, the wall thickness, and the radius of the bend. Small diameter, thick-wall tubes usually present little difficulty. As the diameter increases, wall thickness decreases, or the bend radius becomes smaller, the outside of the tube tends to pull to the center, flattening the tube, and the inside may wrinkle. For many years, the common method of overcoming these problems was to pack the tube with wet sand, produce the bend, and then remove the sand from the interior. Flexible mandrels have now replaced the sand, and are currently available in a wide variety of styles and sizes.

ROLL FORMING

The continuous *roll forming* of flat strip into complex sections has become a highly developed forming technique that competes directly with press brake forming, extrusion, and stamping. As shown in Figures 19-31 and 19-32, the process involves the progressive bending of metal strip as it passes through a series of forming rolls at speeds up to 80 m/min (270 ft/min). Only bending takes place, and all bends are parallel to one another. The thickness of the starting material is preserved, except for thinning at the bend radii.

FIGURE 19-31   (a) Schematic representation of the cold roll-forming process being used to convert sheet or plate into tube. (b) Some typical shapes produced by roll forming.

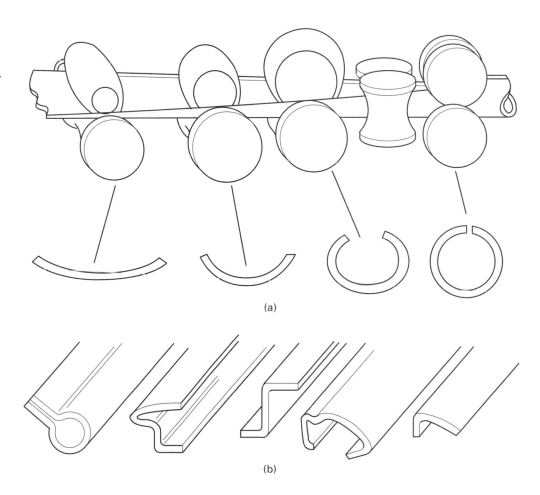

(a)

(b)

FIGURE 19-32   Eight-roll sequence for the roll forming of a box channel. *(Courtesy of the Aluminum Association, New York.)*

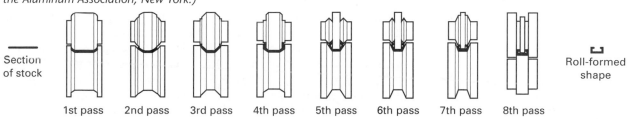

Section of stock   1st pass   2nd pass   3rd pass   4th pass   5th pass   6th pass   7th pass   8th pass   Roll-formed shape

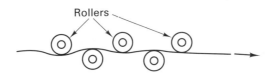

FIGURE 19-33 Various types of seams used on sheet metal.

Rollers

FIGURE 19-34 Method of straightening rod or sheet by passing it through a set of straightening rolls. For rods, another set of rolls is used to provide straightening in the transverse direction.

Any material that can be bent can be roll formed, including cold rolled, hot rolled, polished, prepainted, coated, and plated metals in thicknesses ranging from 0.1 through 20 mm (0.005 through $\frac{3}{4}$ in.). A variety of moldings, channeling, gutters and downspouts, automobile beams and bumpers, and other shapes of uniform wall thickness and uniform cross section are now being formed.

By changing the rolls, a single roll forming machine can produce a wide variety of different shapes. However, changeover, setup, and adjustment may take several hours, so a production run of at least 3000 m (10,000 ft) is usually required for any given product. To produce pipe or tubular products, a resistance welding unit or seaming operation is often integrated with the roll forming.

SEAMING AND FLANGING

Seaming is a bending operation that can be used to join the ends of sheet metal in some form of mechanical interlock. Figure 19-33 shows several of the more common seam designs that can be formed by a series of small rollers. Seaming machines range from small hand-operated types to large automatic units capable of producing hundreds of seams per minute. Common products include cans, pails, drums, and other similar containers.

Flanges can be rolled on sheet metal in essentially the same manner as seams. In many cases, however, the forming of both flanges and seams is a drawing operation, since the bending occurs along a curved axis.

STRAIGHTENING

The objective of *straightening* or *flattening* is the opposite of bending, and these operations are often performed before subsequent cold forming to assure the use of flat or straight material that is reasonably free of residual stresses. *Roll straightening* or *roller leveling*, illustrated in Figure 19-34, subjects the material to a series of reverse bends. The rod, sheet, or wire is passed through a series of rolls with progressively decreased offsets from a straight line. As the material is bent, first up and then down, the surfaces are stressed beyond their elastic limit, replacing any permanent set with a flat or straight profile. Tension applied along the length of the product can help induce the required deformation.

Sheet may also be straightened by a process called *stretcher leveling*. Here the material is gripped mechanically and stretched beyond the elastic limit to produce the desired flatness.

■ 19.4 SHEARING OPERATIONS

Shearing is the mechanical cutting of materials without the formation of chips or the use of burning or melting. It is often used to prepare material for subsequent operations, and its success helps to assure the accuracy and precision of the finished product. When the two cutting blades are straight, the process is called *shearing*. When the blades are curved, the processes have special names, such as *blanking*, *piercing*, *notching*, and *trimming*. In terms of tool design and material behavior, all are shearing-type operations.

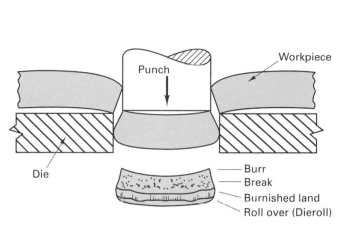

FIGURE 19-35   Simple blanking with a punch and die.

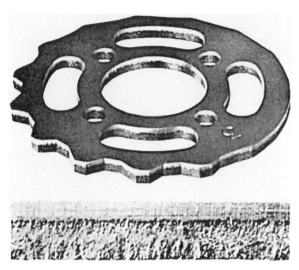

FIGURE 19-36   (*Top*) Conventionally sheared surface showing the distinct regions of deformation and fracture, and (*bottom*) magnified view of the sheared edge. (*Courtesy of American Feintool, Inc.*)

A simple type of shearing operation is illustrated in Figure 19-35. As the punch (or upper blade) pushes the workpiece, the metal responds by plastically flowing into the die (or over the lower blade). Because the clearance between the two tools is small, usually between 5 and 10% of the thickness of the metal being cut (depending on the material being cut), the deformation is in the form of highly localized shear. The punch pushes downward on the metal, the material flows into the die, and the opposite surface bulges slightly. When the penetration is between 15 and 60% of the metal thickness, the actual amount depending on the material ductility and strength, an instability arises. The applied stress exceeds the shear strength and the metal tears or ruptures through the remainder of its thickness. These two distinct stages of the shearing process, deformation and fracture, are often visible on the edges of sheared parts, as shown in Figure 19-36.

Because of the normal inhomogeneities in a metal and the possibility of nonuniform clearance between the shear blades, the final shearing does not occur in a uniform manner. Fracture and tearing begin at the weakest point and proceed progressively or intermittently to the next-weakest location. The result is usually a rough and ragged edge.

If the punch and die (or upper and lower shearing blades) have proper clearance and are maintained in good condition, sheared edges can be produced that have sufficient smoothness to permit use without further finishing. The quality of the sheared edge can often be improved by clamping the starting stock firmly against the die (from above), maintaining proper clearance and alignment between the punch and the die, and restraining the movement of the piece by a plunger or rubber die cushion that applies opposing pressure from below the workpiece. Each of these measures causes the shearing to take place more uniformly around the perimeter of the cut.

If the entire shearing operation is performed in a compressive environment, fracture is suppressed and the relative amount of smooth edge (produced by deformation) is increased. Above a certain pressure, no fracture occurs and the entire edge is smooth, deformed metal. Figure 19-37 shows one method of producing a compressive environment. In a process known as *fineblanking*, a V-shaped protrusion is incorporated into the hold-down or pressure plate at a location slightly external to the contour of the cut. As pressure is applied to the hold-down or pressure plate, the protrusion is driven into the material compressing the region to be cut. Matching upper and lower punches then grip the material and descend in unison, extracting the desired segment. The sheared edges

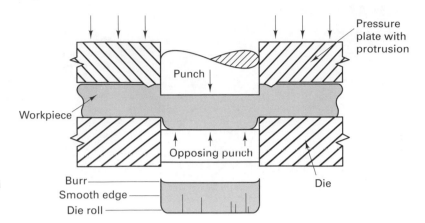

FIGURE 19-37 Method of obtaining a smooth edge in shearing by using a shaped pressure plate to put the metal into localized compression and a punch and opposing punch descending in unison.

are now smooth and square, as shown in Figure 19-38. Fineblanked parts are usually less than 6 mm ($\frac{1}{4}$ in.) in thickness and typically have complex-shaped perimeters. Dimensional accuracy is often within 0.05 mm (0.002 in.), and holes, slots, bends, and semi-pierced projections can often be incorporated as part of the fineblanking operation. Secondary edge finishing can often be eliminated, and the work hardening that occurs during the shearing process enhances wear resistance.

In fineblanking, the clearance between the punch and the die is reduced, and a triple action press is generally required. The fineblanking force is about 40% greater than conventional blanking of the same contour, and the extra material required for the impinging protrusion often forces a greater separation between nested parts.

Figure 19-39 illustrates another means of shearing under compression. Bar stock is pressed against the closed end of a feed hole, placing the stock in a state of compression. A transverse punch then shears the material into smooth-surface, burr-free slugs, ready for further processing.

SIMPLE SHEARING

When sheets of metal are to be sheared along a straight line, *squaring shears*, like the one shown in Figure 19-40, are frequently used. As the upper ram descends, a clamping bar or set of clamping fingers presses the sheet of metal against the machine table to hold it firmly in position. A moving blade then comes down across a fixed blade and shears the metal. On larger shears, the moving blade is often set at an angle or "rocks" as it descends, so the cut is made in a progressive fashion from one side of the material to the other, much like a pair of household scissors. This action significantly reduces the amount of cutting force required, although the total energy expended is still the same. Since work is equal to force times distance, a low force–long stroke operation can often be used in place of a high force and short stroke.

FIGURE 19-38 Fine-blanked surface of the same component as shown in Figure 19-36. *(Courtesy of American Feintool, Inc.)*

FIGURE 19-39 Method of smooth shearing rod by putting it into compression during shearing.

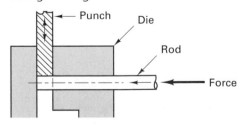

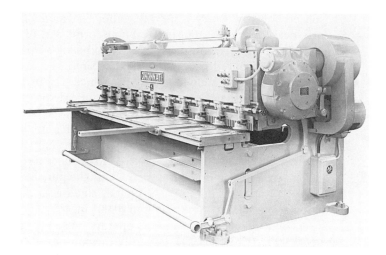

FIGURE 19-40   A 3-m (10 ft) power shear for 6.5 mm ($\frac{1}{4}$ in.) steel. *(Courtesy of Cincinnati Incorporated.)*

The direction of descent of the upper blade may also be inclined about $\frac{1}{2}$ to $2\frac{1}{2}$ degrees with respect to the lower blade and move along this line of inclination. While squareness and edge quality may be compromised, this action helps to ensure that the sheared material does not become wedged between the blades.

SLITTING

Slitting is the length-wise shearing process used to cut rolls of sheet metal into several rolls of narrower width. Here the shearing blades take the form of cylindrical rolls with circumferential mating grooves. The raised ribs of one roll match the recessed grooves on the other. The process is now continuous and can be performed rapidly and economically. Moreover, since the distance between adjacent shearing edges is fixed, the resultant strips have accurate and constant width, more consistent than that obtained from alternative cutting processes.

PIERCING AND BLANKING

Piercing and *blanking* are shearing operations where the shear blades are closed, curved lines along the edges of a punch and die. Both involve the same basic cutting action, the primary difference being one of definition. Figure 19-41 shows that in blanking, the piece being punched out becomes the workpiece and any major burrs or undesirable features should be left on the remaining strip. In piercing, the punch-out is the scrap and the remaining strip is the workpiece. Piercing and blanking are usually done on some form of mechanical press.

There are a number of variations of piercing and blanking and some have come to acquire specific names. *Lancing* is a piercing operation that forms either a line cut (slit) or an actual hole in the metal, like those shown in the left-hand portion of Figure 19-42. The purpose of lancing is to permit the adjacent metal to flow more readily in subsequent forming operations. In the case illustrated, the lancing makes it easier to form the radial grooves, which were shaped before the ashtray was blanked from the strip stock and shallow drawn to final shape. *Perforating* consists of piercing a large number of closely

FIGURE 19-41   Schematic showing the difference between piercing and blanking.

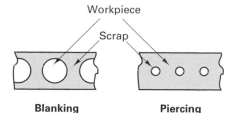

FIGURE 19-42   (*Left to right*) Piercing, lancing, and blanking precede the forming of the final ashtray. The small round holes assist positioning and alignment.

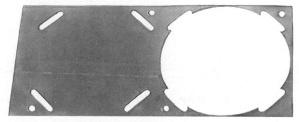

FIGURE 19-43 Shearing operation being performed on a nibbling machine. *(Courtesy of Tech-Pacific.)*

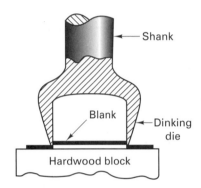

FIGURE 19-44 The dinking process.

spaced holes. *Notching* is used to remove segments from along the edge of an existing product. The edge of the strip or blank forms part of the punch-out perimeter.

In *nibbling*, a contour is cut by producing a series of overlapping slits or notches, as shown in Figure 19-43. In this manner, simple tools can be used to cut a complex shape from sheets of metal up to 6 mm ($\frac{1}{4}$ in.) thick. The process is widely used on parts where the quantities do not justify the expense of a dedicated blanking die. Edge smoothness is determined by the shape of the tooling and the degree of overlap in successive cuts.

Shaving is a finishing operation in which a small amount of metal is sheared away from the edge of an already blanked part. Its primary use is to obtain greater dimensional accuracy, but it may also be employed to produce a squared or smoother edge. Because only a small amount of metal is removed, the punches and dies must be made with very little clearance. Blanked parts, such as small gears, can be shaved to produce dimensional accuracies within 0.025 mm (0.001 in.).

A *cutoff* is a punch and die operation used to separate a stamping or other product from a strip of stock. Frequently, the contour of the cutoff completes the periphery of a workpiece. Cutoff operations are quite common in progressive die sequences, like several to be presented shortly.

Dinking is a modified shearing operation that is used to blank shapes from low-strength materials, such as rubber, fiber, or cloth. As illustrated in Figure 19-44, the shank of a die is either struck with a hammer or mallet or the entire die is driven downward by some form of mechanical press.

TOOLS AND DIES FOR PIERCING AND BLANKING

As shown in Figure 19-45, the basic components of a piercing and blanking die set are a *punch*, a *die*, and a *stripper plate*, which is attached above the die to keep the strip material from ascending with the retracting punch. The position of the stripper plate and the size of its hole should be such that it does not interfere with either the horizontal motion of the strip as it feeds into position or the vertical motion of the punch.

Theoretically, the punch should fit within the die with a uniform clearance that approaches zero. On its downward stroke, it should not enter the die but should stop just as its base aligns with the top surface of the die. In general practice, the clearance is between 5 to 7% of the stock thickness and the punch enters slightly into the die cavity.

If the face of the punch is normal to the axis of motion, the entire perimeter is cut simultaneously. By tilting the punch face on angle, a feature known as *shear* or *rake angle*, the cutting force can be reduced substantially. As shown in Figure 19-46, the periphery is now cut in a progressive fashion, similar to the action of a pair of scissors or the opening of a "pop-top" beverage can. Variation in the shear angle controls the length of cut that is made at any given time and the total stroke that is necessary to complete the operation. Adding shear to a punch reduces the force but increases the stroke. It is an attractive way to cut thicker or stronger material on an existing piece of equipment.

FIGURE 19-45 The basic components of piercing and blanking dies.

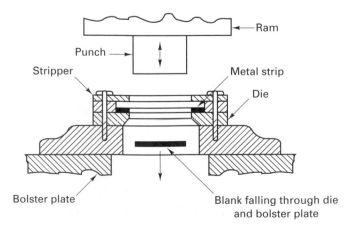

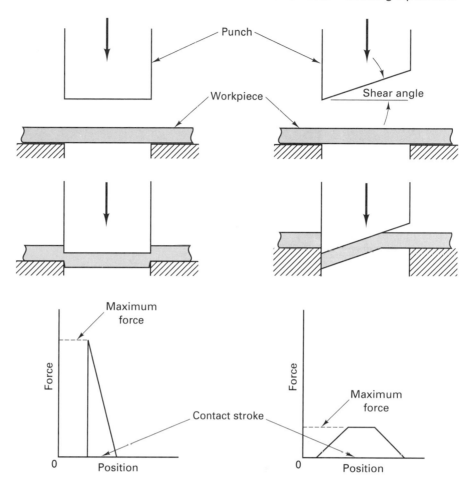

FIGURE 19-46 Blanking with a square-faced punch (*left*) and one containing angular shear (*right*). Note the difference in maximum force and contact stroke. The total work (the area under the curve) is the same for both processes.

FIGURE 19-47 Typical die set having two alignment guideposts. *(Courtesy of Danly Machine Specialties, Inc.)*

Punches and dies should also be in proper alignment so that a uniform clearance is maintained around the entire periphery. The die is usually attached to the bolster plate of the press, which, in turn, is attached to the main press frame. The punch is attached to the movable ram, enabling motion in and out of the die with each stroke of the press. Punches and dies can also be mounted on a separate *punch holder* and *die shoe*, like the one shown in Figure 19-47, to create an *independent die set*. The holder and shoe are permanently aligned and guided by two or more guide pins. By aligning a punch and die, and fastening them to the die set, the entire unit can be inserted into a press without having to set or check the tool alignment. This can significantly reduce the amount of production time lost during tool change. Moreover, when a given punch and die are no longer needed, they can be removed and new tools attached to the shoe and holder assembly.

In most cases the punch holder is attached directly to the ram of the press, and ram motion acts to both raise and lower the punch. On smaller die sets, springs can be incorporated to provide the upward motion. The press ram contacts the top of the punch holder and forces it downward. As the ram retracts, the springs cause the punch to return to its starting position. This form of construction makes the die set fully self-contained. It can be rapidly inserted and removed from a press, thereby reducing setup time.

A wide variety of standardized, self-contained die sets have been developed. Known as *subpress dies* or *modular tooling*, these can often be assembled and combined on the bed of a press to pierce or blank large parts that would otherwise require large and costly complex die sets. Figure 19-48 shows an assembly of subpress dies where a piece of sheet is inserted between the tooling and the downward motion of the press produces a variety of holes and slots, all in proper relation to one another.

Punches and dies are usually made from low-distortion or air-hardenable, tool steel so they can be hardened after machining with minimal warpage. The die profile is maintained for a depth of about 3 mm ($\frac{1}{8}$ in.) from the upper face, beyond which an angular clearance or back relief is generally provided (see Figure 19-45) to reduce friction

FIGURE 19-48 A piercing and blanking setup using self-contained subpress tool units. *(Courtesy of Strippit Division, Houdaille Industries, Inc.)*

between the part and the die and to permit the part to fall freely from the die after being sheared. The 3 mm depth provides adequate strength and sufficient metal so that the die can be resharpened by grinding a few thousandths of an inch from its face.

Dies can be made in a single piece, or they can be made in component sections that are assembled on the punch holder and die shoe. The latter procedure simplifies production and enables the replacement of single sections in the event of wear or fracture. Complex dies like the one shown in Figure 19-49 can often be assembled from the many standardized punch and die components that are available. Substantial savings can often be achieved by modifying the design of parts to enable the use of standard die components. An added advantage of this approach is that when the die set is no longer needed, the components can be removed and used to construct tooling for another product.

Another technique that can be used to cut metal and a wide variety of softer materials (such as plastics, wood, cork, felt, fabrics, and cardboard) is the "steel-rule" or "cookie-cutter" die. Here the cutting die is fashioned from hardened steel strips, known as steel rule, that are mounted on edge and held in position by some means, such as saw-cut grooves in a piece of plywood. The mating piece of tooling may be either a flat piece of hardwood or steel, a male shape that conforms to the part profile, or a set of matching grooves into

FIGURE 19-49 A progressive piercing, forming, and cutoff die set built up mostly from standard components. The part produced is shown at the bottom. *(Courtesy of Oak Manufacturing Company.)*

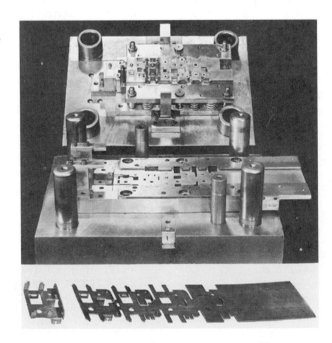

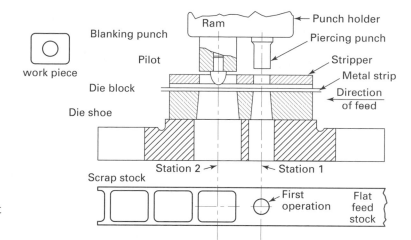

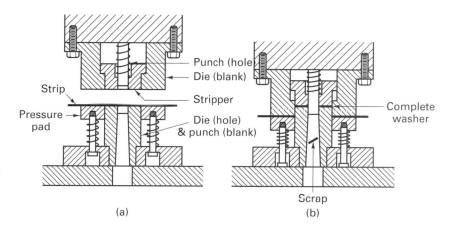

FIGURE 19-50 Progressive piercing and blanking die for making a square washer. Note that the punches are of different lenght.

FIGURE 19-51 Method for making a simple washer in a compound piercing and blanking die. Part is blanked (a) and subsequently pierced (b). The blanking punch contains the dic for piercing.

which the protruding rule can descend. Rubber pads are usually inserted between the strips to replace the stripper plate. During the compression stroke, the rubber compresses and allows the cutting action to proceed. As the ram ascends, the rubber then expands to push the blank free of the steel-rule cavity. Steel-rule dies are less expensive to construct than solid dies and are more attractive for producing small quantities of parts.

Many parts require multiple cutting-type operations, and it is often desirable to perform these with a single cycle of a press. Multiple operations can be accomplished with the types of dies shown in Figures 19-50 and 19-51. For simplicity, their operation is discussed in terms of manufacturing simple, flat washers from a continuous strip of metal.

The *progressive die* set, depicted in Figure 19-50, is the simpler of the two types. Basically, it consists of two or more sets of punches and dies mounted in tandem. The strip stock is fed into the first die, where a hole is pierced as the ram descends. When the ram raises, the stock is advanced and the pierced hole is positioned under the blanking punch. As the ram descends on the second stroke, a pilot on the bottom of the blanking punch enters the hole that was pierced on the previous stroke to ensure accurate alignment. Further descent of the punch blanks the completed washer from the strip, and at the same time, the first punch pierces the hole for the next washer. As the process continues, a finished part is completed with each stroke of the press.

Progressive dies can be used for many combinations of piercing, blanking, forming, lancing, and drawing, as shown by the examples in Figures 19-42, 19-49 and 19-52. They

FIGURE 19-52 The various stages of an 11-station progressive die. *(Courtesy of the Minster Machine Company.)*

are relatively simple to construct and are economical to maintain and repair, since a defective punch or die does not require replacement of the entire die set. However, if accurate alignment of the various operations is required, compound dies should be specified.

The material moves through a progressive die in the form of a continuous strip. As the products are shaped, they must remain attached to the strip or carrier until the final cutoff operation. While the attachment restricts some of the forming operations and prevents part reorientation between steps, it enables the quick and accurate positioning of material in each of the die segments.

If individual parts are mechanically moved from die to die within a single press, the dies are known as *transfer dies*. Part handling automation operates in harmony with the press motions to move, orient, and position the parts as they travel from operation to operation.

In *compound dies* like the one shown schematically in Figure 19-51, piercing and blanking, or other combinations of operations, occur sequentially during a single stroke of the ram. While dies of this type are more precise, they are usually more expensive to construct and also more susceptible to breakage.

When many holes of varying sizes and shapes are to be placed in sheet components, numerically controlled *turret-type punch presses* may be specified. In these machines, as many as 60 separate punches and dies are contained within a turret that can quickly be rotated to provide the specific tooling required for an operation. Between operations, the workpiece is repositioned through numerically controlled X–Y movements of the worktable.

DESIGN FOR PIERCING AND BLANKING

The construction, operation, and maintenance of piercing and blanking dies can be greatly facilitated if designers of the parts to be fabricated keep a few simple rules in mind:

1. Diameters of pierced holes should not be less than the thickness of the metal, with a minimum of 0.3 mm (0.025 in.). Smaller holes can be made, but with difficulty.

2. The minimum distance between holes, or between a hole and the edge of the stock, should be at least equal to the metal thickness.

3. The width of any projection or slot should be at least 1 times the metal thickness and never less than 2.5 mm ($\frac{3}{32}$ in.).

4. Keep tolerances as large as possible. Tolerances below about ±0.075 mm (±0.003 in.) will require shaving.

5. Arrange the pattern of parts on the strip to minimize scrap.

■ 19.5 DRAWING AND SHEET METAL FORMING

Cold drawing is a term that can refer to two somewhat different operations. If the starting stock is sheet metal, cold drawing refers to the forming of parts where plastic flow occurs over a curved axis. This is one of the most important of all cold-working operations because a wide range of shapes, from small cups to large automobile body panels, can be readily fabricated. Cold drawing is similar to hot drawing (discussed briefly in Chapter 18), but the higher deformation forces, thinner metal, limited ductility, and closer dimensional tolerances create some distinctive problems.

If the starting stock is wire, rod, or tubing, cold drawing processes reduce the cross section of the material by pulling it through a die. These processes are similar to extrusion, but the applied forces are tensile, pulling on the product rather than pushing on the workpiece.

ROD, BAR, AND TUBE DRAWING

One of the simplest cold-drawing operations is *rod* or *bar drawing*, illustrated schematically in Figure 19-53. One end of a rod is reduced or pointed, so that it can pass through a die of somewhat smaller cross section. The protruding material is then placed in grips and pulled in tension, drawing the remainder of the rod through the die. The rods reduce in section, elongate, and become stronger (strain harden). Since the product cannot be readily bent or coiled, straight-pull *draw benches* are generally employed with finite-length feedstock. Hydraulic cylinders can be used to provide the pull for short-length products, while chain drives, as depicted in Figure 19-54, can be used to draw products up to 30 m (100 ft) in length.

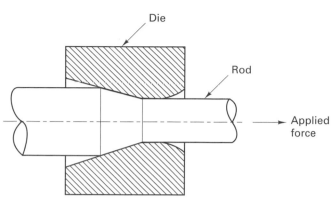

FIGURE 19-53   Schematic diagram of the rod or bar-drawing process.

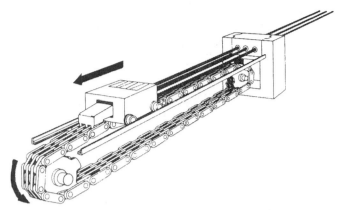

FIGURE 19-54   Schematic diagram of a chain-driven multiple-die-draw bench used to produce finite lengths of straight rod or tube. *(Courtesy of Wean United, Inc.)*

The reduction in area is usually restricted to between 20 and 50%, since higher values require higher pulling forces that may exceed the tensile strength of the reduced product. To produce a desired size or shape, multiple draws may be required through a series of progressively smaller dies. Intermediate anneals may also be required to restore ductility and enable further deformation.

Tube drawing can be used to produce high-quality tubing where the product requires the smooth surfaces, thin walls, accurate dimensions, and added strength (from the strain hardening) that are characteristic of cold forming. Internal mandrels are often used to control the inside diameter of tubes, which range from about 12 to 250 mm ($\frac{1}{2}$ to 10 in.) in diameter. As shown in Figure 19-55, these mandrels are inserted through the incoming stock and are held in place during the drawing operation.

Thick-walled tubes and those less than 12 mm ($\frac{1}{2}$ in.) in diameter are often drawn without a mandrel in a process known as *tube sinking*. Precise control of the inner diameter is sacrificed in exchange for process simplicity and the ability to draw long lengths of product. If a controlled internal diameter must be produced in a long-length product, it is possible to utilize a *floating plug*, like the one shown in Figure 19-56. This plug must be designed for the specific conditions of material, reduction, and friction. If the friction on the plug surface is too great, the flowing tube will pull it too far forward, pinching off or fracturing the tube wall. If the amount of friction is insufficient, the plug will chatter or vibrate within the tube and will not assume a stable position. If properly designed, the floating plug will assume a stable position within the die, and size the internal diameter while the external die shapes and sizes the outside of the tube.

The drawing of bar stock can also be used to make products with shaped cross sections. By using cold drawing instead of hot extrusion, the material emerges with precise

FIGURE 19-55   Cold-drawing smaller tubing from larger tubing. The die sets the outer dimension while the stationary mandrel sizes the inner diameter.

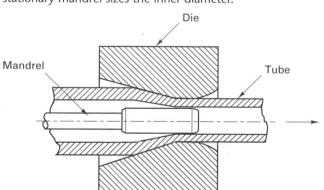

FIGURE 19-56   Tube drawing with a floating plug.

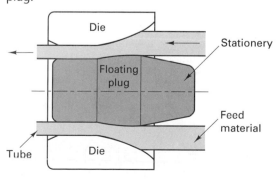

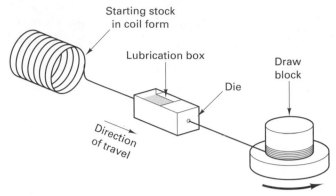

FIGURE 19-57   Schematic of wire drawing with a rotating draw block. The rotating motor on the draw block provides a continuous pull on the incoming wire.

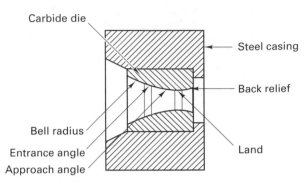

FIGURE 19-58   Cross section through a typical carbide wire drawing die showing the characteristic regions of the contour.

dimensions and excellent surface finish. Inexpensive materials strengthened by strain hardening can often replace stronger alloys or ones that would require additional heat treatment. Small parts with complex but constant cross sections can be economically made by sectioning long lengths of cold-drawn shaped bars to produce the individual products. Steels, copper alloys, and aluminum alloys have all been cold drawn into shaped bars.

WIRE DRAWING

Wire drawing is essentially the same process as bar drawing except that it involves smaller-diameter material. Because the material can be coiled, the process can now be conducted in a somewhat continuous manner on rotating draw blocks, like the one illustrated schematically in Figure 19-57. Wire drawing usually begins with large coils of hot-rolled rod stock approximately 9 mm ($\frac{3}{8}$ in.) in diameter. After descaling or other forms of surface preparation, one end of the coil is pointed, fed through a die, gripped, and the drawing process begins.

Wire dies generally have a configuration similar to the one shown in Figure 19-58. The contact regions are usually made of wear-resistant tungsten carbide or polycrystalline, manufactured diamond. Single-crystal diamonds can be used for the drawing of very fine wire, and wear-resistant and low-friction coatings can be applied to the various die material substrates. Lubrication boxes often precede the individual dies to help reduce friction drag and prevent wear of the dies.

Because the tensile load is applied to the already reduced product, the amount of reduction is severely limited. Multiple draws are usually required to affect any significant change in size. To convert hot-rolled rod stock to the fine wire that is used in household telephone lines requires passes through as many as 20 or 30 individual dies. To minimize handling and labor, these operations are usually performed on tandem machines, like the one shown schematically in Figure 19-59. Between 3 and 12 dies are mounted in a single machine, and the material moves continuously from one station to another in a synchronized manner that prevents any localized accumulation or tension that might induce fracture.

FIGURE 19-59   Schematic of a multistation synchronized wire drawing machine. To prevent accumulation or breakage, it is necessary to assure that the same volume of material passes through each station in a given time. The loops around the sheaves between the stations use wire tensions and feedback electronics to provide the necessary speed control.

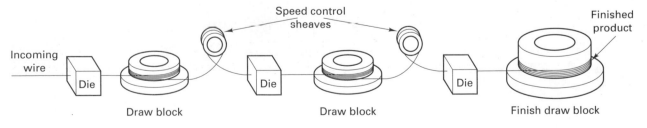

After passing through all the dies in a tandem machine, the material usually requires an intermediate anneal before it can be subjected to further deformation. By controlling the placement of the last anneal so the final product has a selected amount of cold work, wires can be made with a wide range of strengths (or tempers). When maximum ductility and conductivity is desired, the wire should be annealed in controlled-atmosphere furnaces after the final draw.

SPINNING

Spinning is a cold-forming operation where a rotating disk of sheet metal is progressively shaped over a male form, or mandrel, to produce rotationally symmetrical shapes, such as cones, hemispheres, cylinders, bells, and parabolas. Localized pressure is applied through a simple round-ended wooden or metal tool or small roller, which traverses the entire surface of the part to produce the progressive deformation depicted in Figure 19-60.

The form block possessing the shape of the desired part is attached to a rotating spindle, such as the drive section of a simple lathe. A disk of metal is centered on the small end of the form and held in place by a pressure pad attached to the tailstock of the lathe. As the disk and form rotate, the operator applies localized pressure against the metal, causing it to flow progressively against the form, as shown in the two stages of Figure 19-61. Because the final diameter of the part is less than that of the starting disk, the circumferential length decreases. This decrease must be compensated by either an increase in thickness, a radial elongation, or circumferential buckling. Control of the process is highly dependent on the skill of the operator.

During spinning, the form block sees only localized compression, and the metal does not move across it under pressure. As a result, a form block can often be made of hardwood or even plastic. Its major requirement is simply to replicate the shape with a smooth surface. Tooling cost can be extremely low, making spinning an attractive process for producing making small quantities of a single part. With automation, however, spinning can also be used to mass-produce such high-volume items as lamp reflectors, cooking utensils, bowls, and the bells of some musical instruments. When large quantities are being produced, a metal form block is generally preferred.

Spinning is usually considered for simple shapes that can be directly withdrawn from a one-piece form. More complex shapes, such as those with reentrant angles, can be spun over multipiece or offset forms. Complex form blocks can also be made from frozen water, which is melted out of the product after spinning.

SHEAR FORMING OR FLOW TURNING

Cones, hemispheres, and similar shapes are often formed by *shear forming* or *flow turning*, a modification of the spinning process in which each element of the blank maintains

FIGURE 19-60   Progressive stages in the spinning of a sheet metal product.

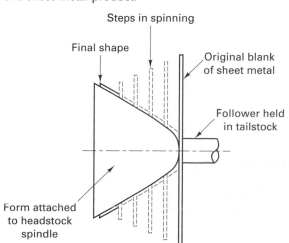

FIGURE 19-61   Two stages in the spinning of a metal reflector. *(Courtesy of Spincraft, Inc.)*

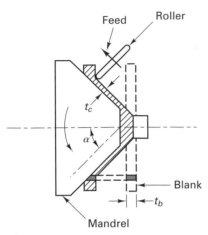

FIGURE 19-62 Schematic representation of the basic shear-forming process.

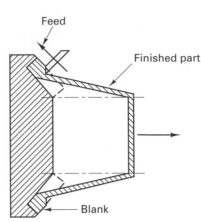

FIGURE 19-63 Forming a conical part by reverse shear forming.

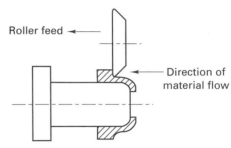

FIGURE 19-64 Shear forming a cylinder by the direct process.

its distance from the axis of rotation. Because there is no circumferential shrinkage, the metal flow is entirely by shear and no compensating stretch has to occur. As shown in Figure 19-62, the wall thickness of the product, t_c, will vary with the angle of the particular region according to the relationship: $t_c = t_b \sin \alpha$, where t_b is the thickness of the starting blank. If α is less than 30°, it may be necessary to complete the forming in two stages with an intermediate anneal in between. Reductions in wall thickness as high as 8:1 are possible, but the limit is usually set at about 5:1 or 80%.

Conical shapes are usually shear formed by the direct process depicted in Figure 19-62. The bottom of the product is held against the face of the form block or mandrel, while the material being formed moves in the same direction as the rollers. Products can also be formed by a reverse process, like that illustrated in Figure 19-63. By controlling the position and feed of the forming rollers, the reverse process can be used to shape concave, convex, or conical parts without a matching form block or mandrel.

Cylinders can be shear formed by both the direct and reverse processes. As shown in Figure 19-64, the direct process restricts the length of the product to the length of the mandrel. No schematic is provided for the reverse process, because it is essentially the same as the roll extrusion process, depicted in Figure 19-13.

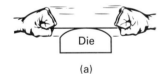

(a)

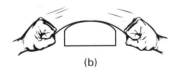

(b)

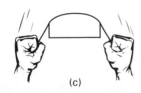

(c)

FIGURE 19-65 Schematic representation of the motions and steps involved in stretch-wrap forming.

STRETCH FORMING

Stretch forming, illustrated in principle in Figure 19-65, is an attractive means of producing large sheet metal parts in low or limited quantities. A sheet of metal is gripped by two or more sets of jaws that stretch it and wrap it around a single form block which serves as the die. Various combinations of stretching, wrapping, and motion of the block can be employed, depending on the shape of the part.

Because most of the deformation is induced by the tensile stretching, the forces on the form block are far less than those normally encountered in bending or forming. Consequently, there is very little springback, and the workpiece conforms very closely to the shape of the tool. Since stretching accompanies bending or wrapping, wrinkles are pulled out before they occur. Because the forces are so low, the form blocks can often be made of wood, low-melting-point metal, or even plastic.

Stretch forming, or *stretch-wrap forming* as it is often called, is quite popular in the aircraft industry and is frequently used to form aluminum and stainless steel into cowlings, wing tips, scoops, and other large panels. Low-carbon steel can be stretch formed to produce large panels for the automotive and truck industry.

If mating male and female dies are used to shape the metal while it is being stretched, the process is known as *stretch-draw forming*.

SHEET METAL DRAWING (SHALLOW DRAWING AND DEEP DRAWING)

The forming of closed-bottom cylindrical or rectangular containers from metal sheet is one of the most important and widely used manufacturing processes. When the depth of the product is less than its diameter (or the smallest dimension of its opening), the process is considered to be *shallow drawing*. When the depth is greater than the diameter, it is known as *deep drawing*.

Consider the simple operation of converting a circular disk of sheet metal into a flat-bottom cylindrical cup. Figure 19-66 shows the blank positioned over a die opening, and a circular punch descending to perform the operation. The material beneath the punch remains largely unaffected and becomes the bottom of the cup. The cup wall is formed by pulling the remainder of the disk inward and over the radius of the die, as shown in Figure 19-67. As the material is pulled inward, the circumference decreases. Since the volume of material must remain constant, the decrease in circumferential dimension must be compensated by an increase in another dimension, such as thickness. Since the material is thin, an alternative response is to relieve the circumferential compression by a buckling or wrinkling. One means of suppressing the wrinkles is to compress the sheet between the die and a blankholder surface during forming, hopefully forcing an increase in either radial length or thickness.

In single-action presses, where there is only one movement that is available, springs or air pressure are often used to clamp the metal between the die and pressure ring. When multiple-actions are available (two or more independent motions), as shown in Figure 19-68, the hold-down force can now be applied in a manner that is independent of the punch position. This restraining force can also be varied during the drawing operation. For this reason, multiple-action presses are usually specified for the drawing of more complex parts, while single-action presses can be used for the simpler operations.

Key variables in the deep drawing process include the blank diameter and the punch diameter, which combine to determine the *draw ratio*, the die radius, the punch radius, the clearance between the punch and the die, the thickness of the blank, lubrication, and the hold-down pressure. Once the process has been designed and the tooling manufactured, the primary variable for process adjustment is the *hold-down pressure* or *blankholder force*. If the force or pressure is too low, wrinkling may occur at the start of the stroke. If it is too high, there is too much restraint, and the descending punch will simply tear the disk or some portion of the already-formed cup wall.

When drawing a shallow cup, there is little change in circumference, and a small area is being confined by the blankholder. As a result, the tendency to wrinkle or tear is low. As cup depth increases, there is an increased tendency for forming both of the defects.

FIGURE 19-66   Schematic of the deep drawing process.

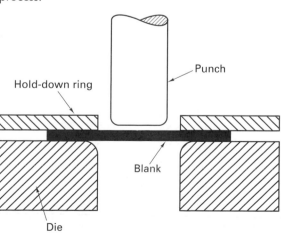

FIGURE 19-67   Flow of material during deep drawing. Note the circumferential compression as the radius is pulled inward.

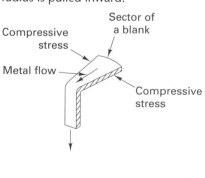

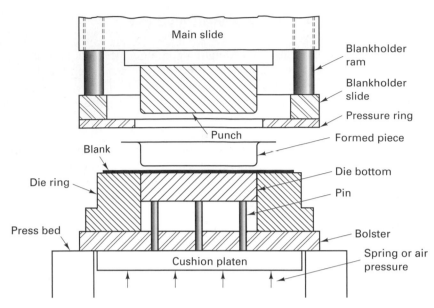

FIGURE 19-68 Drawing on a double-action press, where the blankholder uses the second press action.

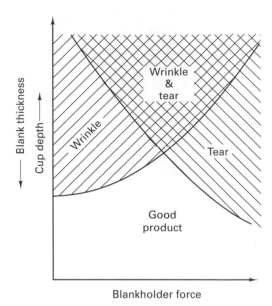

FIGURE 19-69 Defect formation in deep drawing as a function of blankholder force, blank thickness, and cup depth.

In a similar manner, thin material is more likely to wrinkle or tear than thick material. Figure 19-69 summarizes the effects of cup depth, blank thickness, and blankholder pressure.

The limitations of wrinkling and tearing may require deep drawn parts to be made in multiple operations. Figure 19-70 shows two ways in which drawn parts can be further formed into deeper cups. When part geometry becomes more complex, such as rectangular or asymmetric parts, it is best if the surface area and thickness of the material can remain relatively constant. Different regions may need to be differentially constrained. One technique that can provide variable constraint is the use of *draw beads*, rib-like projections and matching grooves in the die and blankholder. The added force of bending and unbending restricts the flow of material, and the degree of constraint can be varied by adjusting the height shape and size of the bead and bead cavity.

Because of prior rolling and other metallurgical and process variables, the flow of sheet metal is generally not uniform, even in the simplest drawing operation. Excess material may be required to assure final dimensions, and *trimming* may be required to establish both the size and uniformity of the final part. Figure 19-71 shows a shallow-drawn part before and after trimming. Trimming obviously adds to the production cost because it not only converts some of the starting material to scrap and but also adds another operation to the manufacturing process. Since the shape has already been produced, trimming requires accurate positioning or manipulation of the workpiece or the construction of a separate trimming die.

FIGURE 19-70 Cup redrawing to further reduce diameter and increase wall height. (*Left*) forward redraw, (*right*) reverse redraw.

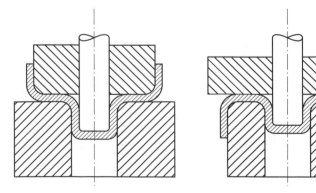

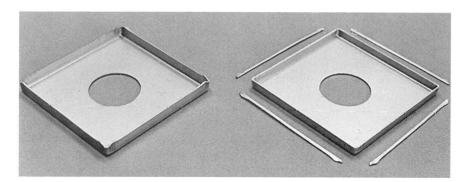

FIGURE 19-71 Pierced, blanked, and drawn part before and after trimming.

The amount of trimming scrap can often be reduced or even eliminated through the use of complex-shaped blanks and improved metallurgical and process control. Because of the complex shape of the starting blank, additional scrap is likely to be generated during the blanking operation. The choice of simple blank and blanking scrap or complex blank and reduced trimming is usually dictated by process economics and depends on such factors as the complexity of the part and the quantity to be produced.

FORMING WITH RUBBER TOOLING OR FLUID PRESSURE

Blanking and drawing operations usually require mating male and female, or upper and lower die sets, and such tooling can be quite expensive. In addition, process setup requires that the various tool components be precisely positioned and aligned. Consequently, numerous methods have been developed to reduce tooling cost and/or setup time and expense. Still other approaches seek to extend the amount of deformation that can be performed with a single set of tools, thereby eliminating the need for multiple operations and associated tooling, as well as any intermediate anneals. Although most of these methods have distinct limitations, such as complexity of shape or types of metal that can be formed, they also have definite areas of application. Some alternative processes will now be presented.

Several forming methods replace either the male or female member of the die set with rubber or fluid pressure. The *Guerin process* (also known as rubber-die forming), depicted in Figure 19-72, is based on the phenomenon that rubber of the proper consistency, *when totally confined*, acts as a fluid and transmits pressure uniformly in all directions. Blanks of sheet metal are placed on top of form blocks, which can be made of wood, Bakelite, polyurethane, epoxy, or low-melting-point metal. The upper ram contains a pad of rubber 20 to 25 cm (8 to 10 in.) thick mounted within a steel container. As the ram descends, the rubber pad becomes confined and transmits force to the metal, causing it to bend to the desired shape. Since no female die is used and inexpensive form blocks replace the male die, the total tooling cost is quite low. There are no mating tools to align, process flexibility is quite high (different shapes can even be formed at the same time), wear on the material and tooling is low, and the surface quality of the workpiece is easily maintained, a feature that makes the process attractive for forming prepainted or specially-coated sheet. When reentrant sections are produced (as in the bottom portion of Figure 19-72), it must be possible to slide the parts lengthwise from the form blocks or to disassemble a multipiece form from within the product.

FIGURE 19-72 The Guerin process for forming sheet metal products.

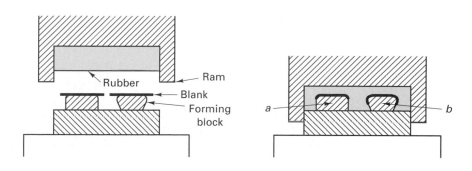

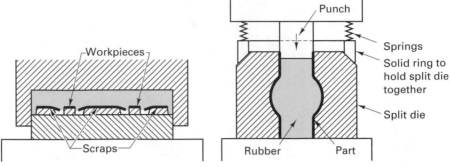

FIGURE 19-73 Method of blanking sheet metal using the Guerin process.

FIGURE 19-74 Method of bulging tubes with rubber tooling.

The Guerin process was developed by the aircraft industry, where the production of small numbers of duplicate parts clearly favors the low cost of tooling. It can be used on aluminum sheet up to 3 mm ($\frac{1}{8}$ in.) thick and on stainless steel up to 1.5 mm ($\frac{1}{16}$ in.). Magnesium sheet can also be formed if it is heated and shaped over heated form blocks.

Most of the forming done with the Guerin process is multiple-axis bending, but some shallow drawing can also be performed. The process can also be used to pierce or blank thin gages of aluminum, as illustrated in Figure 19-73. For this application the blanking blocks are shaped the same as the desired workpiece, with a face, or edge, of hardened steel. Round-edge supporting blocks are positioned a short distance from the blanking blocks to support the scrap skeleton and permit the metal to bend away from the shearing edges.

In *bulging*, fluid or rubber is used to transmit the pressure required to expand a metal blank or tube outward against a split female mold or die. For simple shapes, rubber tooling can be inserted, compressed, and then easily removed, as shown in Figure 19-74. For more complicated shapes, fluid pressure is used to form the bulge. More complex equipment is now required since pressurized seals must be formed and maintained while still enabling the easy insertion and removal of material that is required for mass production.

SHEET HYDROFORMING

Sheet hydroforming is really a family of processes in which a rubber bladder backed by oil pressure, or pockets of pressurized liquid, replace either the solid punch or female die of the traditional tool set. In a variant known as *high-pressure flexible die forming*, or *Flexforming*, the rubber pad of the Guerin process is replaced by a flexible rubber diaphragm backed by controlled hydraulic pressures at values between 140 and 200 MPa (20,000 and 30,000 psi). As illustrated in Figure 19-75, the solid punch presses the sheet against the resisting medium, whose pressure is adjusted throughout the stroke.

In a variation of this process that shares similarity to stretch forming, the periphery of the workpiece is held in place by a hold-down, while the punch is in a retracted position. The fluid is then pressurized, causing the workpiece to balloon toward the punch. Since the pressure is uniformly distributed over the workpiece, the sheet is uniformly stretched and uniformly thinned. The punch then descends, causing the pre-stretched metal to conform to its profile.

The flexible membrane can also be used to replace the hardened male punch, as shown in Figure 19-76. Here the ballooning action causes the material to conform fully to the female die, which may be made of epoxy or other low cost material. *Parallel-plate hydroforming*, or *pillow forming*, extends the process to the simultaneous production of upper and lower contours. As shown in Figure 19-77, two sheet metal blanks are laser welded around the periphery or are strongly clamped between upper and lower dies. Pressurized fluid is then injected between the sheets, simultaneously forming both upper and lower profiles. This may be a more attractive means of producing complex sheet metal containers, since the manufacturer no longer has to cope with the problems of aligning and welding two separately-formed complex-shaped pieces.

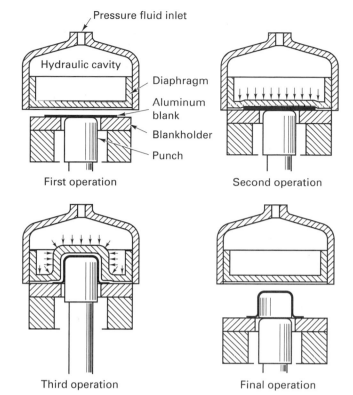

FIGURE 19-75 High-pressure flexible-die forming, showing (1) the blank in place with no pressure in the cavity; (2) press closed and cavity pressurized; (3) ram advanced with cavity maintaining fluid pressure; and (4) pressure released and ram retracted. *(Courtesy of Aluminum Association, New York.)*

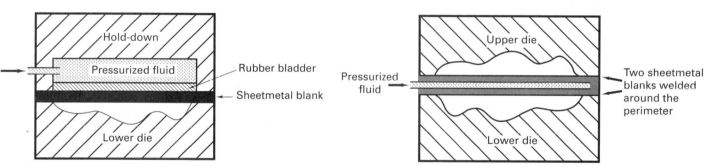

FIGURE 19-76 Schematic of one form of sheet hydroforming.

FIGURE 19-77 Two-sheet hydroforming or "pillow forming."

While the most attractive feature of sheet hydroforming is probably the reduced cost of tooling, there are other attractive features. By producing a more uniform distribution of strain, materials exhibit greater formability. Drawing limits are generally about $1\frac{1}{2}$ times those of conventional deep drawing. Deeper parts can therefore be formed without fracture. Complex shapes that required multiple operations can be formed in a single pressing, with excellent surface finish. Part dimensions are more accurate and more consistent.

Cycle times for sheet hydroforming are on the order of 1 to 3 minutes, but the reduced tool costs make the process attractive for prototype manufacturing and low-volume production (up to about 10,000 identical parts) like that encountered in the aerospace industry. More recently, sheet hydroforming has attracted attention within the automotive community because of its ability to not only produce low-volume parts in an economical manner, but also to successfully shape lower-formability materials, such as alloyed aluminum sheet and high-strength steels.

TUBE HYDROFORMING

Tube hydroforming, shown schematically in Figure 19-78, has emerged as a significant process for manufacturing strong, lightweight, one-piece automotive components, which frequently replace an assembly of welded stampings. Typical parts include engine cradles, frame rails, roof headers, radiator supports, and exhaust components, as shown in Figure 19-79.

In elementary terms, a tubular blank, either straight or preshaped, is placed in an encapsulating die, and the ends are sealed. A fluid is then introduced through one of the end plugs, achieving sufficient pressure to expand the material to the shape of the die. At the same time, the end closures may move inward to help compensate or overcome the thinning that would otherwise accompany radial expansion. In actual operation, the process may use combinations or even sequences of internal pressure and axial motion to control the flow and final thickness of the material.

In *low-pressure tube hydroforming*, the tube is first filled with fluid and then the dies are closed around the tube. The primary purpose of the fluid is to act as a liquid mandrel that prevents collapsing as the tube is bent to the contour of the die. The cross section shape can be changed, but the shapes must be simple, corner radii must be large, and there must be minimal expansion of the tube diameter. In *high-pressure tube hydroforming*, an internal pressure between 100 and 700 MPa (15,000 and 100,000 psi) is used to expand the diameter of the tube, forming tight corner radii and significantly altered cross sections. *Pressure-sequence hydroforming* begins by applying low internal pressure as the die is closing. This supports the inside wall of the tube and allows it to conform to the cavity. When the die is fully closed, high pressure is applied to complete the forming of the tube walls.

Attractive features of tube hydroforming include the ability to use lightweight, high-strength materials, the increase in strength that results from strain hardening, and the ability to utilize designs with varying thickness or varying cross section. Disadvantages include the long cycle time (low production rate) and relatively high cost of tooling and process setup.

FIGURE 19-78 Tube hydroforming.

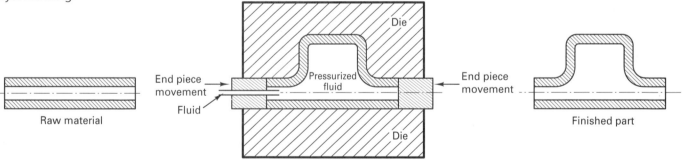

FIGURE 19-79 Use of hydroformed tubes in automotive applications. *(Courtesy of MetalForming, a publication of PMA Services, Inc., for the Precision Metalforming Association, Independence, OH.)*

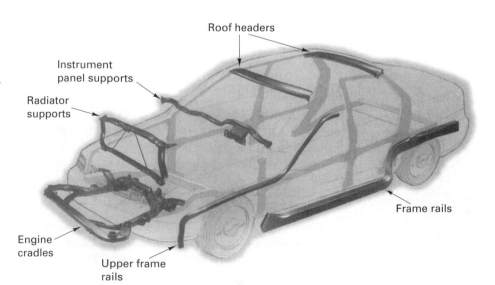

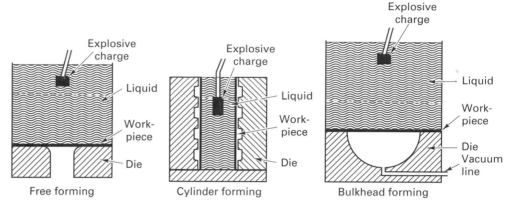

FIGURE 19-80   Three methods of high-energy-rate-forming with explosive charges. *(Courtesy of Materials Engineering.)*

HIGH-ENERGY-RATE FORMING

A number of methods have been developed to form metals through the application of large amounts of energy in a very short time interval (high strain rate). These are known as *high-energy-rate forming* processes and often go by the abbreviation *HERF*. Many metals tend to deform more readily under the ultrarapid load application rates used in these processes. As a consequence, HERF makes it possible to form large workpieces and difficult-to-form metals with less-expensive equipment and cheaper tooling than would otherwise be required.

Another advantage of HERF is that there is less difficulty related to springback. This is probably associated with two factors: (1) high compressive stresses are set up in the metal when it is forced against the die, and (2) some slight elastic deformation of the die occurs under the ultrahigh pressure. The latter results in a slight overforming of the workpiece and the appearance that no springback has occurred.

High-energy-release rates can be obtained by five distinct methods: (1) underwater explosions, (2) underwater spark discharge (electrohydraulic techniques), (3) pneumatic-mechanical means, (4) internal combustion of gaseous mixtures, and (5) the use of rapidly formed magnetic fields (electromagnetic techniques). Specific processes have been developed around each of these approaches.

Figure 19-80 illustrates three commonly used procedures involving the use of *explosive* charges: free forming, cylinder forming, and bulkhead forming. While these procedures can be used for a wide range of products, they are particularly suited for parts of thick material like the 10-ft-diameter elliptical dome shown in Figure 19-81. The only equipment that is required is a tank of water with about 2 m (6 ft) of water above the workpiece. The female die can be made of an inexpensive material such as wood, plastic, or low-melting-temperature metal.

The *spark-discharge* method uses the energy of an electrical discharge to shape the metal. Electrical energy is stored in large capacitor banks and is then released in a controlled discharge, either between two electrodes or across an exploding bridgewire. High-energy shockwaves propagate through a pressure-transmitting medium impacting and deforming the workpiece material. The spark-discharge methods are most often used for bulging operations in small parts. Compared to explosive forming, the discharge techniques are easier and safer, use smaller tanks, and do not have to be performed in remote areas.

The *pneumatic-mechanical* and *internal combustion* techniques are preferred when the HERF methods are applied to mass production of forged-type products. In a pneumatic-mechanical press a sudden application of high pressure gas propels the upper die onto the workpiece. Internal combustion presses operate on the same principle as that of an automobile engine. The explosion of a gaseous mixture propels the upper die. Die velocities can be up to 15 meters per second (50 fps) and cycle rates may be as high as 60 strokes per minute. Single or repeated blows can be used to form the part.

FIGURE 19-81   Explosively formed elliptical dome 10 ft in diameter being removed from a forming die. *(Courtesy of NASA.)*

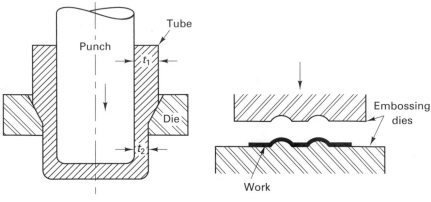

FIGURE 19-82 Schematic of the ironing process.

FIGURE 19-83 Embossing.

In *electromagnetic forming*, a conductive coil is positioned within a cylinder of conductive material, around the cylinder, or adjacent to a flat sheet. When a large capacitor bank is discharged, producing a current surge through the coiled conductor, the electromagnetic forces cause the material to be repelled from the coil. The deformation is very rapid and can be used to expand or contract tubing, or to permanently assemble component parts.

IRONING

Ironing is the name given to the process that thins the walls of a drawn cylinder by passing it between a punch and die whose separation is less than the original wall thickness. As shown in Figure 19-82, the walls thin and lengthen, while the thickness of the base remains unchanged. The most common example of an ironed product is the thin-walled beverage can.

EMBOSSING

Embossing, shown in Figure 19-83, is a pressworking process in which raised lettering or other designs are impressed in sheet material. Basically, it is a very shallow drawing operation where the depth of the draw is limited to one to three times the thickness of the metal, and the material thickness remains largely unchanged. An example of an embossed product is the patterned or textured industrial stair tread.

SUPERPLASTIC SHEET FORMING

Conventional metals and alloys typically exhibit tensile elongations in the range of 10 to 30%. By producing sheet materials with ultrafine-grain-size, and performing the deformation at low strain rates and elevated temperatures, elongation can exceed 100% and may be as high as 2000 to 3000%. This *superplastic* behavior can be used to form material into large, complex-shaped products with compound curves. Deep or complex shapes can be made as single-piece, single-operation pressings rather than multistep conventional pressings or multipiece assemblies.

At the elevated temperatures required to promote superplasticity (about 900°C for titanium, between 450 and 520°C for aluminum, or generally above half of the melting point of the material on an absolute scale), the strength of material is sufficiently low so that many of the superplastic forming techniques are adaptations of processes used to form thermoplastics (discussed in Chapter 20). The tooling doesn't have to be exceptionally strong, so form blocks can often be used in place of die sets. In thermoforming, a vacuum or pneumatic pressure causes the sheet to conform to a heated male or female die. Blow forming, vacuum forming, deep drawing, and combined superplastic forming and diffusion bonding are other possibilities. Precision is excellent and fine details or surface textures can be reproduced accurately. Springback and residual stresses are almost nonexistent, and the products have a fine, uniform grain size.

The major limitation to superplastic forming is the low forming rate that is required to maintain superplastic behavior. Cycle times may range from 2 minutes to as much as 2 hours per part, compared to the several seconds that is typical of conventional presswork. As a result, applications tend to be limited to low-volume products such as those common to the aerospace industry. By making the products larger and eliminating assembly operations, the weight of products can often be reduced, there are fewer fastener holes to initiate fatigue cracks, tooling and fabrication costs are reduced, and there is a shorter production lead time.

DESIGN AIDS FOR SHEET METAL FORMING

A majority of sheet metal failures occur due to thinning or fracture, and both are the result of excessive deformation in a given region. A quick and economical means of evaluating the severity of deformation in a formed part is to use *strain analysis* and a *forming limit diagram*. A pattern or grid, such as the one in Figure 19-84, is placed on the surface of a sheet by scribing, printing, or etching. The ability to detect point-to-point variations in strain distribution generally requires circle diameters between 2.5 and 5 mm (0.1 to 0.2 in.). The sheet is then deformed, converting the circles into ellipses, and the distorted pattern is then measured and evaluated. Regions where the enclosed area has expanded are locations of sheet thinning and possible failure. Regions where the area has contracted have undergone sheet thickening and may be sites of possible buckling or wrinkles.

Using the ellipses on the deformed grid, the major strains (strain in the direction of the largest radius or diameter) and the associated minor strains (strain 90° from the major) can be determined for a variety of locations, and the values can be plotted on a forming-limit diagram such as the one shown in Figure 19-84. If both major and minor strains are positive (right-hand side of the diagram), the deformation is known as *stretching*, and the sheet metal will definitely decrease in thickness. If the minor strain is negative, this contraction may partially or wholly compensate any positive stretching in the major direction. The combination of tension and compress is known as *drawing*, and the thickness may decrease, increase, or stay the same, depending on the relative magnitudes of the two strains. Regions where both strains are negative do not appear on the diagram, since its purpose is to reveal locations of possible fracture, and fractures only occur in a tensile environment.

FIGURE 19-84   (*Left*) Typical pattern for sheet metal deformation analysis; (*right*) forming limit diagram used to determine whether a metal can be shaped without risk of fracture. Fracture is expected when strains fall above the lines.

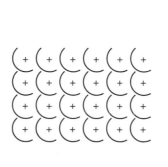

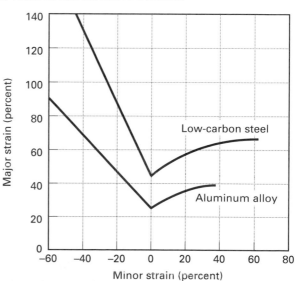

Those strains that fall above the forming limit line indicate regions of probable fracture. Possible corrective actions include modification of the lubricant, change in the die design, or variation in the clamping or hold-down pressure. Strain analysis can also be used to determine the best orientation of blanks relative to the rolling direction, compare the effectiveness of various lubricants, or assist in the design of dies for complex-shaped products.

■ 19.6 ALTERNATIVE METHODS OF PRODUCING SHEET-TYPE PRODUCTS

ELECTROFORMING

Several manufacturing processes have been developed to produce sheet-type products by directly depositing metal onto preshaped forms or mandrels. In a process known as *electroforming*, the metal is deposited by plating. Nickel, iron, copper, or silver can be deposited in thicknesses up to 16 mm ($\frac{5}{8}$ in.). When the desired thickness has been attained, plating is stopped and the product is stripped from the mandrel.

A wide variety of sizes and shapes can be made by electroforming, and the fabrication of a product requires only a single pattern or mandrel. Low production quantities can be made in an economical fashion, with the principal limitation being the need to strip the product from the mandrel. Dimensional tolerances are good, and the surface finishes of the replicated interior can be outstanding. For most applications, such as the production of multiple molds from a single master pattern, the interior surface is the critical one, and the wall thickness serves only to provide the necessary strength. Irregularities in the exterior are not critical, and various types of backup material can be employed to provide additional support. For other applications, the exterior dimensions are also important, and uniform deposition is desired.

SPRAY FORMING

Similar parts can also be formed by *spray deposition*. One approach is to inject powdered material into a plasma torch (a stream of hot ionized gas with temperatures up to 11,000°C or 20,000°F). The particles melt, and are propelled onto a shaped form or mandrel. Upon impact, the droplets flatten and undergo rapid solidification to produce a dense, fine-grained product. Multiple layers can be deposited to build up a desired size, shape, and thickness. Because of the high adhesion, the mandrel or form is usually removed by machining or chemical etching. Most applications of plasma spray forming involve the fabrication of specialized products from difficult-to-form or ultrahigh-melting-point materials.

The *Osprey process*, described in Chapter 16, can also be adapted to produce thin, spray-formed products. Here, molten metal flows through a nozzle, where it is atomized and carried by high velocity nitrogen jets. The semisolid particles ultimately impact and solidify onto a target or substrate. Tubes, plates, and simple forms can be produced from a variety of materials. Layered structures can also be produced by sequenced deposition.

■ 19.7 PRESSES

CLASSIFICATION OF PRESSES

The primary tool for performing many of the cold-working operations discussed in this chapter is some form of press, and successful forming often depends on using the right kind of press. When selecting a press for a given application, consideration should be given to the capacity required, the type of power (manual, mechanical, or hydraulic), the number of slides or drives, the type of drive, the stroke length for each drive, and the type of frame or construction. Table 19-2 lists some of the major types of presses and groups them by press drive.

Manually operated presses such as foot-operated or *kick presses* are generally used for very light work such as shearing small sheets.

TABLE 19-2.	Classification of the Various Drive Mechanisms of Commercial Presses	
Manual	Mechanical	Hydraulic
Kick presses	Crank	Single-slide
	Single	Multiple-slide
	Double	
	Eccentric	
	Cam	
	Knuckle joint	
	Toggle	
	Screw	
	Rack and pinion	

Moving toward larger equipment, we find that *mechanical drives* tend to provide fast motion and positive control of displacement. Once built, however, the flexibility of a mechanical press is limited, since the length of the stroke is set by the design of the drive. The available force usually varies with position, so mechanical presses are preferred for operations that require the maximum pressure near the bottom of the stroke, such as cutting, shallow forming, drawing (up to about 10 cm) and progressive and transfer die operations. Typical capacities range up to about 9000 metric tons.

Figure 19-85 depicts some of the basic types of mechanical press drive mechanisms. *Crank-driven* presses are the most common type because of their simplicity. They are used for most piercing and blanking operations and for simple drawing. Double-crank presses offer a means of actuating blank holders or operating multiple-action dies. *Eccentric or cam drives* are used where only a short ram stroke is required. Cam action can also provide a dwell at the bottom of the stroke and is often the preferred method of actuating the blank holder in deep-drawing processes. *Knuckle-joint drives* provide a very high mechanical advantage along with fast action. They are often preferred for coining, sizing, and Guerin forming. *Toggle mechanisms* are used principally in drawing presses to actuate the blank holder, and *screw-type drives* offer great mechanical advantage coupled with an action that resembles a drop hammer (but slower and with less impact). For this reason, screw presses have become quite popular in the forging industry.

FIGURE 19-85 Schematic representation of the various types of press drive mechanisms.

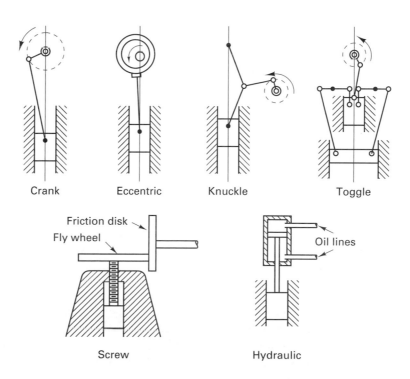

Crank   Eccentric   Knuckle   Toggle

Friction disk
Fly wheel

Oil lines

Screw   Hydraulic

In contrast to the mechanical presses, *hydraulic presses* produce motion as the result of piston movement, and longer or variable-length strokes can be programmed within the limitations of the cylinder (which may be as long as 250 cm or 100 in.). Forces and pressures are more accurately controlled and full pressure is available throughout the entire stroke. Speeds can be programmed to vary or remain constant during an operation. Since position is varied through fluid displacement, the reproducibility of position will have greater variation than a mechanical press. Hydraulic presses are available in capacities exceeding 50,000 metric tons and are preferred for operations requiring a steady pressure throughout a substantial stroke (such as deep drawing), operations requiring wide variation in stroke length, and operations requiring high or widely variable forces. In general, hydraulic presses tend to be slower than the mechanical variety, but some are available that can provide up to 600 strokes per minute in a high-speed blanking operation. By using multiple hydraulic cylinders, programmed loads can be applied to the main ram, while a separate force and timing are used on the blank holder.

TYPES OF PRESS FRAME

Another matter of importance in selecting a press is determining the type of frame. See Table 19-3. Frame design often imposes limitations on the size and type of work that can be accommodated, how that work is fed and unloaded, the overall stiffness of the machine, and the die exchange time. Table 19-3 presents a classification of basic frame types.

Presses that have their frames in the shape of an arch (arch-frame presses) are seldom used today, except with screw drives for coining operations. *Gap-frame presses*, whose frames are in the shape of the letter *C*, are among the most versatile and commonly preferred presses. They provide unobstructed access to the dies from three directions and permit large workpieces to be fed into the press. Gap-frame presses are available in a wide variety of sizes, from small bench types of about 1 metric ton up to 300 metric tons or more.

Several popular design features include open back, inclinability, adjustable bed, and sliding bolster. *Open-back presses* allow for ejection of the products or scrap through an opening in the back of the press frame. *Inclinable presses* can be tilted, so that ejection can be assisted by gravity or compressed air jets. As a result of these features, the open-back inclinable (OBI) presses are the most common form of gap-frame press. The addition of an *adjustable bed* allows the base of the machine to raise or lower to accommodate different workpieces. Finally, the OBI press shown in Figure 19-86 contains a *sliding bolster*, a feature that permits a second die to be set up on the press while another is in operation. Die changeover then requires only a few minutes to unclamp the punch segment of the active die, move the second die set into position, clamp the new set to the press ram, and resume operation.

A *horn press* is a special type of upright gap-frame press where a heavy cylindrical shaft or "horn" appears in place of the usual bed. Curved or cylindrical workpieces can be placed over the horn for such operations as seaming, punching, and riveting, as illustrated in Figure 19-87. On some presses, both a horn and a bed are provided, with provision for swinging the horn aside when not needed.

FIGURE 19-86 Inclinable gap-frame press with sliding bolster to accommodate two die sets for rapid change of tooling. *(Courtesy of Niagara Machine & Tool Works.)*

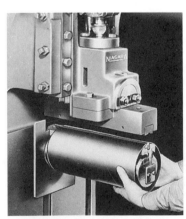

FIGURE 19-87 Making a seam on a horn press. Note the protruding "horn" that replaces the lower press bed. *(Courtesy of Niagara Machine & Tool Works.)*

TABLE 19-3.	Classification of Presses According to Type of Frame	
Arch	Gap	Straight-Sided
Crank or eccentric Percussion	Foot	Many variations, but all with straight-sided frames
	Bench	
	Vertical	
	Inclinable	
	Inclinable	
	Open back	
	Horn	
	Turret	

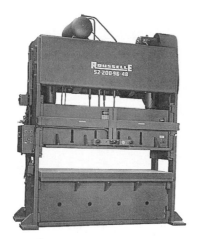

FIGURE 19-88 A 200-ton (1800-kN) straight-side press. *(Courtesy of Rousselle Corporation.)*

Turret presses are especially useful in the production of sheet metal parts with numerous holes or slots that vary in size and shape. They usually employ a modified gap-frame structure, and add upper and lower turrets that carry a number of punches and dies. The two turrets are geared together so that any desired tool set can be quickly rotated into position. Another type uses a single turret that carries a number of subpress dies.

Straight-sided presses have frames that consist of a crown, two uprights, a base or bed, and one or more moving slides. Accessibility is generally from the front and rear, but openings are often provided in the side uprights to permit feeding and unloading of workpieces. Straight-sided presses are available in a wide variety of sizes and designs and are the preferred design for most hydraulic, large-capacity, or specialized mechanical-drive presses. As an added benefit, any elastic deflections tend to be uniform across the working surface, as opposed to the angular deflections that are typical of the gap-frame design. Figure 19-88 shows a typical straight-sided press.

SPECIAL TYPES OF PRESSES

A number of presses have been designed to perform specific types of operations. *Transfer presses* have a long moving slide that enables multiple operations to be performed simultaneously in a single machine. Multiple die sets are mounted side by side along the slide. After the completion of each stroke, a continuous strip or individual workpieces are automatically and progressively advanced (transferred) to the next station by a mechanism like the one shown in Figure 19-89. Transfer presses can be used to perform blanking, piercing, forming, trimming, drawing, flanging, embossing, and coining. Figure 19-90 illustrates the production of a part that incorporates a variety of these operations.

FIGURE 19-89 Schematic showing the arrangement of dies and the transfer mechanism used in transfer presses. *(Courtesy of Verson Allsteel Press Company.)*

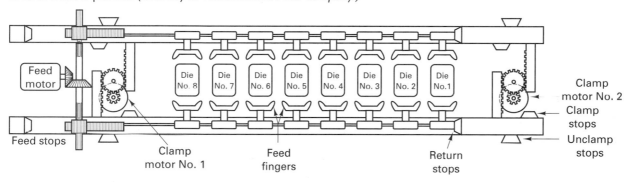

FIGURE 19-90 Various operations that can be performed during the production of stamped and drawn parts on a transfer press. *(Courtesy of U.S. Baird Corporation, Stratford, Conn.)*

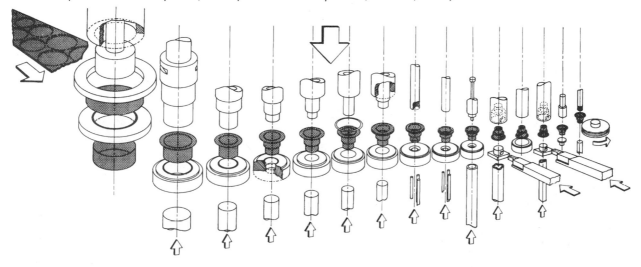

FIGURE 19-91 Multislide machine with guards and covers removed. *(Courtesy of U.S. Baird Corporation, Stratford, Conn.)*

By using a single press to perform multiple operations, transfer presses offer high production rates, high flexibility, and reduced costs (attributed to the reduced labor, floor space, energy, and maintenance). Since production is usually between 500 and 1500 parts per hour, these machines are usually restricted to operations where 4000 or more identical parts are required daily, each involving three or more separate operations. A total production run of 30,000 or more identical parts is generally desired between major changes in tooling. As a result, transfer presses are used primarily in industries such as automotive and appliances, where large numbers of identical products are being produced.

Four-slide or *multislide* machines like the one shown in Figure 19-91 are extremely versatile presses that are designed to produce small, intricately shaped parts from wire or coil feeds. The basic machine has four power-driven slides (or motions) set 90° apart. The attached tooling is controlled by cams and designed to operate in a progressive cycle. In sheet metal forming, coiled strip stock is fed into the machine, where it is straightened and progressively pierced, notched, bent, and cut off at the various slide stations. Figure 19-92 schematically presents the operating mechanism of one such machine. As the material moves

FIGURE 19-92 Schematic of the operating mechanism of a multislide machine. The material enters on the right and progresses toward the left as operations are performed. *(Courtesy of U.S. Baird Corporation, Stratford, Conn.)*

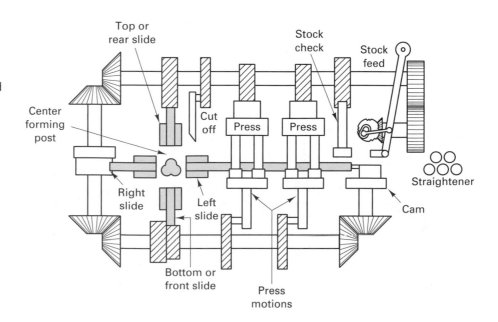

FIGURE 19-93   Example of the piercing, blanking, and forming operations performed on a multislide machine. *(Courtesy of U.S. Baird Corporation, Stratford, Conn.)*

from right to left, it undergoes a straightening, two successive pressing operations, various operations from all four directions, and a final cutoff. Figure 19-93 shows the carrier strip and the successive operations as flat strip is pierced, blanked, and formed into a folded sheet metal product. The strip stock may be up to 75 mm (3 in.) wide and 2.5 mm ($\frac{3}{32}$ in.) thick. Wires up to about 3 mm ($\frac{1}{8}$ in.) in diameter are also commonly processed. Products such as hinges, links, clips, and razor blades can be formed at very high rates on these machines. Setup times are long, so large production runs are preferred.

PRESS-FEEDING DEVICES

Although hand feeding may still be used in some press operations, operator safety and the desire to increase productivity have motivated a strong shift to feeding by some form of mechanical device. When continuous strip is used it can be fed automatically by double-roll feeds mounted on the side of the press. Discrete products can be moved and positioned in a wide variety of ways. Dial-feed mechanisms enable an operator to insert workpieces into the front holes of a dial. The dial then indexes with each stroke of the press to move the parts progressively into proper position between the punch and die. Lightweight parts can be fed by suction-cup mechanisms, vibratory-bed feeders, and similar devices. Robots are being used in increasing numbers to place parts into presses and remove them after forming. See chapter 32. Technology and equipment now exist to replace almost all manual feeding if such a transition is truly desired.

■ KEY WORDS

air-bend die	draw bending	kick press	rake angle	steel-rule die
backward extrusion	draw ratio	lancing	riveting	strain analysis
bar folder	drawing	mechanical press	roll bending	stretch forming
bending	electroforming	minimum bend radius	roll extrusion	stretcher leveling
blankholder force	electromagnetic forming	modular tooling	roll forming	stretching
blanking	embossing	multislide press	roll straightening	strip
bottoming die	explosive forming	neutral axis	roller burnishing	stripper plate
bulging	explosive rivet	nibbling	rubber tooling	subpress die
burnishing	fineblanking	notching	seaming	superplastic
coining	flanging	orbital forming	shallow drawing	forming
cold forging	floating plug	Osprey process	shaving	swaging
cold heading	flow turning	parallel-plate hydroforming	shear	thread rolling
cold rolling	forming	peening	shear forming	transfer die
cold working	forming limit diagram	perforating	shearing	transfer press
compound die	forward extrusion	piercing	sheet	trimming
compression bending	Guerin process	pillow forming	sheet hydroforming	tube bending
Conform process	high-energy-rate forming	pneumatic-mechanical	sizing	tube drawing
continuous extrusion	hold-down force	pop rivet	slitting	tube hydroforming
cutoff	hubbing	press bending	spark discharge	tube sinking
deep drawing	hydraulic press	press brake	spinning	turret-type punch
die	hydrostatic extrusion	pressure induced ductility	springback	press
dinking	impact extrusion	pressure-to-pressure extrusion	squaring shears	upsetting
draw bead	independent die set	progressive die	spray forming	wire drawing
draw bench	ironing	punch	staking	

■ REVIEW QUESTIONS

1. What are some of the attractive features of cold-working processes?
2. Why might cold-working equipment have to be more powerful than that used for hot working?
3. What benefits can be obtained by subjecting sheet metal to a skin-rolling pass?
4. Why would the cold rolling of shaped product not be an attractive means to produce a small amount of prototype product?
5. How can the swaging process impart different sizes and shapes to an interior cavity and the exterior of a product?
6. How might cold forging be used to substantially reduce material waste?
7. What are some of the attractive properties or characteristics of cold forging or cold-forged products?
8. If a product contains a large-diameter head and a small-diameter shank, how can the processes of cold extrusion and cold heading be combined to save metal?
9. What are some of the unique capabilities and special limitations of hydrostatic extrusion?
10. What is the unique capability provided by pressure-to-pressure hydrostatic extrusion?
11. How is the feedstock pushed through the die in continuous extrusion processes?
12. What type of products can be made by the roll extrusion process?
13. What types of rivets can be used when there is access to only one side of a joint?
14. Why might hubbing be an attractive way to produce a number of identical die cavities?
15. How might a peening operation increase the fracture resistance of a product?
16. What is burnishing?
17. When making bends in sheet metal, what is the distinction between bending, forming, and drawing?
18. Why does a metal usually become thinner in the region of a bend?
19. What is springback, and why is it a concern during bending?
20. What types of operations can be performed on a press brake?
21. What factors determine the minimum bend radius for a material?
22. If a right-angle bend is to be made in a cold-rolled sheet, should it be made with the bend lying along or perpendicular to the direction of previous rolling?
23. From a manufacturing viewpoint, why is it desirable for all bends in a product (or component) to have the same radius?
24. What is the difference between air-bend and bottoming dies? Which is more flexible? Which produces more reproducible bends?
25. What type of products are produced by roll bending?
26. How can we prevent flattening or wrinkling when bending a tube?
27. What type of product geometry can be produced by cold-roll forming? Is the process appropriate for making short lengths of specialized products?
28. What are some methods for straightening or flattening rod or sheet?
29. What is a definition of shearing?
30. Why are sheared or blanked edges generally not smooth?
31. What measures can be employed to improve the quality of a sheared edge?
32. Why are fineblanking presses more complex than those used in conventional blanking?
33. Why might a long shearing cut be made in a progressive fashion?
34. What is a slitting operation?
35. What are the differences between piercing and blanking?
36. Why would you make the piercing punch shorter than the blanking punch in the progressive die set in Figure 19-50?
37. What are some types of blanking or piercing operations that have come to acquire specific names?
38. What is the purpose of having a shear angle on a punch?
39. Why is it important that a blanking punch and die be in proper alignment?
40. What is the major benefit of mounting punches and dies on independent die sets?
41. What is the major benefit of assembling a complex die set from standard subpress dies?
42. What is the benefit of making dies as a multipiece assembly?
43. What is a progressive die set?
44. What is the difference between progressive dies and transfer dies?
45. How do compound dies differ from progressive dies?
46. What is the attractive feature of a turret-type punch press?
47. What two distinctly different processes are often referred to as cold drawing?
48. What is the difference between tube drawing and tube sinking?
49. For what types of products might a floating plug be employed?
50. Why are rods generally drawn on draw benches, while wire is drawn on draw block machines?
51. Why are multiple passes usually required in wire-drawing operations?
52. Why is the tooling cost for a spinning operation relatively low?
53. How is shear forming different from spinning?
54. For what types of products would stretch forming be an appropriate manufacturing technique?
55. What is the distinction between shallow drawing and deep drawing?
56. What is the function of the pressure ring or hold-down in a deep-drawing operation?
57. Explain why thin material may be difficult to draw into a defect-free cup.
58. What are draw beads and what function do they perform?
59. Why is a trimming operation often included in a deep-drawing manufacturing sequence?
60. How does the Guerin process reduce the cost of tooling?
61. How can fluid pressure or rubber tooling be used to perform bulging?
62. What is sheet hydroforming?
63. What explanation can be given for the greater formability observed during sheet hydroforming?
64. What is the purpose of the inward movement of the end plugs during tube hydroforming?
65. What are some of the basic methods that have been used to achieve the high energy-release rates needed in the HERF processes?
66. Why is springback rather minimal in high-energy-rate forming?
67. What are some well-known products that have been produced by processes including ironing? By embossing?
68. What material and process conditions are associated with superplastic forming?
69. What is the major limitation of the superplastic forming of sheet metal? What are some of the more attractive features?
70. How can strain analysis be used to determine locations of possible defects or failure in sheet metal components?
71. What is a forming limit diagram?

72. Explain the key difference between the right- and left-hand sections of a forming limit diagram. These sections correspond to stretching and drawing.
73. Describe two alternative methods of producing complex-shaped thin products without requiring sheet metal deformation techniques.

74. What are the primary assets and limitations of mechanical press drives? Of hydraulic drives?
75. What are some of the common types of press frames?
76. What is the purpose of inclining or tilting a press?
77. Describe how multiple operations are performed simultaneously in a transfer press.

■ PROBLEMS

1. Compare the forming processes of wire drawing, conventional extrusion, and continuous extrusion with respect to continuity, reduction in area possible in a single operation, possible materials, speeds, typical temperatures, and other important processing variables.
2. Consider the various means of producing tubular products, such as extrusion, seam welding, butt-welding during forming, piercing, and the various drawing operations. Describe the advantages, limitations, and typical applications of each.

3. Tube and sheet hydroforming have been undergoing rapid growth. Investigate new applications for these processes in automotive and other fields.
4. What are some of the techniques for minimizing the amount of springback in sheet forming operations?
5. Select a cold forming process and investigate the residual stresses that typically accompany or result from that process.

www.wiley.com/college/degarmo

Chapter 19 CASE STUDY

Diesel Engine Fuel Metering Lever

The component pictured in Figure CS-19 is a fuel-metering device for a large diesel engine. The long dimension of the part is approximately 6.3 cm ($2\frac{1}{2}$ in.), and the width is 2.2 cm ($\frac{7}{8}$ in.). The large hole through the center is 0.93 cm (0.368 in.) in diameter, and the two small holes are tapped to accept 0.24 cm ($\frac{3}{32}$ in.) threaded bolts. All horizontal surfaces are parallel, and a realistic production run would be for about 20,000 identical pieces.

1. Consider the geometry of the piece and identify at least three possible means of producing the required shape. For each, discuss briefly the relative pros and cons.

2. If the part were to require mechanical properties of 380 MPa (55 ksi) tensile strength, a surface hardness of R_B 55, and an elongation of at least 1%, what types of materials would seem to be appropriate for this part?
3. Discuss the compatibility of the materials selected in Part 2 and the processes identified in Part 1. Are there any significant limitations?
4. What do you feel would be the best material-and-process solution from among the identified possibilities? For this solution, in what form would you purchase the starting material? Would a heat treatment be required to achieve the necessary final properties?

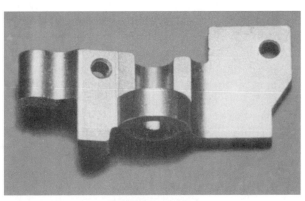

FIGURE CS-19

CHAPTER 20

FABRICATION OF PLASTICS, CERAMICS, AND COMPOSITES

■ 20.1 INTRODUCTION

In Chapters 6, 7, and 8, *plastics, ceramics*, and *composites* were shown to be substantially different from metals in both structure and properties. It is reasonable to expect, therefore, that the principles of material selection and product design, as well as the fabrication processes, will also be somewhat different. In addition, there will also be some similarities. The specific material will still be selected for its ability to provide the required properties; and the fabrication processes, for their ability to produce the desired shape in an economical and practical manner.

In terms of differences, plastics, ceramics, and composites tend to be used closer to their design limits, and many of the fabrication processes convert the raw material into a finished product in a single operation. Large, complex shapes can often be formed as a single unit, eliminating the need for multipart assembly operations. Materials in these classes can often provide integral and variable color, and the processes used to manufacture the shape can frequently produce the desired finish and precision. As a result, finishing operations are often unnecessary—an attractive feature since, for many of these materials, altering the final dimensions or surface would be both difficult and costly. The joining and fastening operations used with these materials also tend to be quite different from those used with metals.

As with metals, the properties of these materials are affected by the processes used to produce the shape. Thus fabrication of an acceptable product, therefore, involves the selection of both (1) an appropriate material and (2) a companion method of processing, such that the resulting combination can provide the desired shape, properties, precision, and finish.

■ 20.2 FABRICATION OF PLASTICS

Chapter 8 presented material about the wide variety of plastics or polymers that are currently used as engineering materials. As we move our attention to the fabrication of parts and shapes, we find that there is also a variety of processes from which to choose. Determination of the preferred method depends on the desired size, shape, and quantity, as well as whether the polymer is a *thermoplastic, thermoset*, or *elastomer*. Thermoplastic polymers can be heated to produce either a soft, formable solid or a liquid. The pre-polymerized material can then be cast, injected into a mold, or forced into or through dies to produce a desired shape. Thermosetting polymers have far fewer options, because once the polymerization has occurred, the framework structure is established, and no further deformation can occur. Thus the polymerization reaction must take place during the shape-forming operation. Elastomers are sufficiently unique that they will be treated in a separate section of this chapter.

Casting, blow molding, compression molding, transfer molding, cold molding, injection molding, reaction injection molding, extrusion, thermoforming, rotational molding, and *foam molding* are all processes that are used to shape polymers. Each has its distinct set of advantages and limitations that relate to part design, compatible materials, and production cost. To make optimum selections, we must be familiar with the shape capabilities of a process as well as how the process affects the properties of the material.

CASTING

Casting is the simplest of the shape forming processes because no fillers are used and no pressure is required. While not all plastics can be cast, there are a number of castable thermoplastics, including acrylics, nylons, urethanes, and PVC plastisols. The thermoplastic polymer is simply melted, and the liquid is poured into a container having the shape of the desired part. Several variations of the process have been developed. Small products can be cast directly into shaped molds. Plate glass can be used as a mold to cast individual pieces of thick plastic sheet. Continuous sheets and films can be produced by injecting the liquid polymer between two moving belts of highly polished stainless steel, the width and thickness being set by resilient gasket strips on either end of the gap. Thin sheets can be made by ejecting molten liquid from a gap-slot die onto a temperature-controlled chill roll. The molten plastic can also be spun against a rotating mold wall (centrifugal casting) to produce hollow or tubular shapes.

Some thermosets (such as phenolics, polyesters, epoxies, silicones, and urethanes) can also be cast, as well as any resin that will polymerize at low temperatures and atmospheric pressure. Because of the need for curing, the casting of thermoset resins usually involves additional processing, often some form of heating while in the mold. Figure 20-1 depicts a process where a steel pattern is dipped into molten lead, withdrawn, and allowed to cool. A thin lead sheath is produced when the pattern is removed, and this becomes the mold for the plastic resin. Curing occurs, either at room temperature or by heating for long times at temperatures in the range of 65 to 95°C (150 to 200°F). After curing, the product is removed, and the lead sheaths can be reused.

Since cast plastics contain no fillers, they have a distinctly lustrous appearance, and a wide range of transparent and translucent colors are available. Since the product is shaped as a liquid, fiber or particulate reinforcement can be easily incorporated. The process is relatively inexpensive because of the comparative lack of costly dies, equipment, and controls. Typical products include sheets, plates, films, rods, and tubes, as well as small objects, such as jewelry, ornamental shapes, gears, and lenses. While dimensional precision can be quite high, quality problems can occur because of inadequate mixing, air entrapment, gas evolution, and shrinkage.

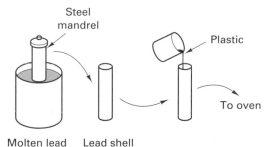

Steel mandrel

Plastic

Molten lead   Lead shell

To oven

FIGURE 20-1   Steps in the casting of plastic parts using a lead shell mold.

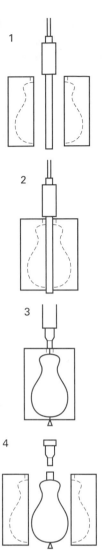

FIGURE 20-2 Steps in blow molding plastic parts: (1) a tube of heated plastic is placed in the open mold, (2) the mold closes over the tube, (3) air expands the tube against the sides of the mold, and (4) after sufficient cooling, the mold opens to release the product.

BLOW MOLDING

A variety of *blow molding* processes have been developed, the most common being used to convert thermoplastic polyethylene, polyvinyl chloride (PVC), polypropylene, and PEEK resins into bottles and other hollow-shape containers. A round, solid-bottom, hollow-tube preform, known as a *parison*, is made from heated plastic by either extrusion or injection molding. The heated preform is then positioned between the halves of a split mold, the mold closes, and the preform is expanded against the mold by air or gas pressure. The mold is then cooled, the halves separated, and the product is removed. Any flash is then trimmed for direct recycling. Figure 20-2 depicts a schematic of this process, which has currently been expanded to include the engineering thermoplastics and has been used to produce products as varied as automotive fuel tanks, seat backs, ductwork, and bumper beams.

Variations of blow molding have been designed to provide both axial and radial expansion of the plastic (for enhanced strength) as well as to produce multilayered products. In one process, a sheet of heated plastic is placed between upper and lower cavities, the lower one having the shape of the product. Both cavities are then pressurized to 2 to 4 MPa (300 to 600 psi) with a nonreactive gas such as argon. When the pressure in the lower segment is then vented, the gas in the upper segment "blows" the material into the lower die cavity.

Because the thermoplastics must be cooled before removal from the mold, the molds for blow molding must contain the desired cavity as well as a cooling system, venting system, and other design features. The mold material must provide thermal conductivity and durability while being inexpensive and compatible with the resins being processed. Beryllium copper, aluminum, tool steels, and stainless steels are all popular.

COMPRESSION MOLDING OR HOT-COMPRESSION MOLDING

In *compression molding*, illustrated schematically in Figure 20-3, solid granules or preformed tablets of *unpolymerized* plastic are introduced into an open, heated cavity. A heated plunger then descends to close the cavity and apply pressure. As the material melts and becomes fluid, it is driven into all portions of the cavity. The heat and pressure are maintained until the material has "set" (i.e., cured or polymerized). The mold is then opened and the part is removed. A wide variety of heating systems and mold materials are used, and multiple cavities can be placed within a mold to produce more than one part in a single pressing. The process is simple and used primarily with the thermosetting polymers, although recent developments permit the shaping of thermoplastics and composites. Cycle times are set by the rate of heat transfer and the reaction or curing rate of the polymer. They typically range from under 1 minute to as much as 20 minutes or more.

The tool and machinery costs for compression molding are often lower than for competing processes, and the dimensional precision and surface finish are high, thereby reducing or eliminating secondary operations. Compression molding is most economical when it is applied to small production runs of parts requiring close tolerances, high impact strength, and low mold shrinkage. It is a poor choice when the part contains thick sections (the cure times become quite long) or when large quantities are desired. Most products have relatively simple shapes because the flow of material is rather

FIGURE 20-3 Schematic representation of the hot-compression molding process: (1) solid granules or a preform pellet is placed in a heated die, (2) a heated punch descends and applies pressure, and (3) after curing (thermosets) or cooling (thermoplastics), the mold is opened and the part is removed.

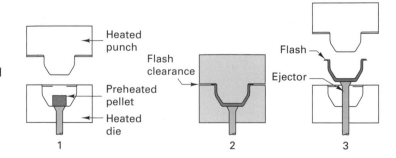

limited. Typical compression-molded parts include gaskets, seals, exterior automotive panels, aircraft fairings, and a wide variety of interior panels.

More recently, compression molding has been used to form fiber-reinforced plastics, both thermoplastics and thermosets, into parts with properties that rival the engineering metals. In the thermoset family, polyesters, epoxies, and phenolics can be used as the base of fiber-containing sheet-molding compound, bulk-molding compound, or sprayed-up reinforcement mats. These are introduced into the mold and shaped and cured in the normal manner. Cycle times range from about 1 to 5 minutes per part, and typical products include wash basins, bathtubs, equipment housings, and various electrical components.

If the starting material is a fiber-containing thermoplastic, precut blanks are first heated in an infrared oven to produce a soft, pliable material. The blanks are then transferred to the press, where they are shaped and cooled in specially designed dies. Compared to the thermosets, cycle times are reduced and the scrap is often recyclable. In addition, the products can be joined or assembled using the thermal "welding" processes applied to plastics.

Compression molding equipment is usually rather simple, typically consisting of a hydraulic or pneumatic press with parallel platens that apply the heat and pressure. Pressing areas range from 15 cm^2 (6 in^2) to as much as 2.5 m^2 (8 ft^2), and the force capacities range from 6 to 9,000 metric tons. The molds are usually made of tool steel and are polished or chrome plated to improve material flow and product quality. Mold temperatures typically run between 150 and 200°C (300 and 400°F) but can go as high as 650°C (1200°F). They are heated by a variety of means, including electric heaters, steam, oil, and gas.

TRANSFER MOLDING

Transfer molding is sometimes used to reduce the turbulence and uneven flow that can result from the high pressures of hot-compression molding. As shown in Figure 20-4, the *unpolymerized* raw material is now placed in a plunger cavity, where it is heated until molten. The plunger then descends, forcing the molten plastic through channels or runners into adjoining die cavities. Temperature and pressure are maintained until the thermosetting resin has completely cured. To shorten the cycle and extend the lifetime of the cavity, plunger, runner, and gates, the charge material may be preheated before being placed in the plunger cavity.

Because the material enters the die cavities as a liquid, there is little pressure until the cavity is completely filled. Thin sections, excellent detail, and good tolerances and finish are all characteristics of the process. In addition, inserts can be incorporated into the products of transfer molding. They are simply positioned within the cavity and maintained in place as the liquid resin is introduced around them.

As a process, transfer molding combines elements of both compression molding and injection molding (to be discussed) and enables some of the advantages of injection molding to be utilized with thermosetting polymers. The thermosetting resins can be reinforced with fillers, such as cellulose, glass, silica, alumina, or mica, to improve the mechanical or electrical properties and reduce shrinkage or warping. Common products include electrical switchgear and wiring devices, parts of household appliances that require heat resistance, structural parts that require hardness and rigidity under load, under-hood automotive parts, and parts that require good resistance to chemical attack.

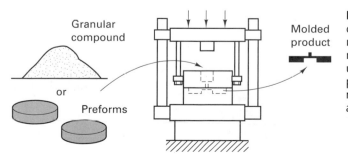

FIGURE 20-4   Schematic diagram of the transfer molding process. Molten material is formed in the upper heated cavity. A plunger then drives the molten material into an adjacent die.

COLD MOLDING

In *cold molding*, the uncured thermosetting material is pressed to shape while cold and is then removed from the mold and cured in a separate oven. While the process is faster and more economical, the resulting products generally lack good surface finish and dimensional precision.

INJECTION MOLDING

Injection molding is the most widely used process for high-volume production of thermoplastic parts. Figure 20-5 illustrates one approach to the process, where granules of raw material are fed by gravity from a hopper into a cavity that lies ahead of a moving plunger. As the plunger advances, the material is forced through a preheating chamber and on through a torpedo section, where it is melted and superheated to 200 to 300°C (400 to 600°F). The material exits the torpedo section through a nozzle that seats against a mold. Other types of injection units use screws that have both rotation and axial movements, or combinations of screws and plungers, to control the flow of material and generate the injection pressure. Alternative methods of heating the material include heated barrels and the shearing action of the screws.

Sprues and runners then channel the molten material into one or more closed-die cavities. Since the dies remain cool, the plastic solidifies almost as soon as the mold is filled. Premature solidification would cause defective parts, so the material must be rapidly forced into the mold cavities by pressures in the range 35 to 140 MPa (5 to 20 ksi), which are maintained during solidification. The mold halves must clamp tightly together during molding and then be easily separated for part ejection. Impact forces should be minimized during die closure, since they can adversely affect die life. Various types of clamping designs have been developed, including toggle, hydraulic, and hydromechanical.

Control systems coordinate all of the functions of the process, including the time required for cooling within the mold. By heating the material for the next part as the mold is separating for part ejection, a complete molding cycle can be completed in 1 to 30 seconds. The process is quite similar to the die casting of molten metal, and the result is usually a finished product needing no further work before assembly or use.

Some injection molding machines incorporate a hot runner distribution system to transfer the material from the injection nozzle to the mold cavities. If the runners are cold, the material in the runner solidifies with each cycle and needs to be ejected and reprocessed or disposed. With hot runners, the thermoplastic material is maintained in a liquid state until it reaches the gate. The material in the runners can be used in the subsequent shot, thereby reducing shot size and cycle time, since less material must be heated. Quality is improved, since all material enters the mold at the same temperature, recycled sprues and runners are not incorporated into the charge, and there is less turbulence since pressurized material is not injected into empty runners. Hot runners do add an additional degree of complexity to the design, operation, and control of the system, so the additional cost must be weighed against the benefits cited.

Injection molding can also be applied to the thermosetting materials, but the process must be modified to provide the temperature and pressure required for curing. The mold is now heated, and the time in the heated mold must be sufficient to complete

FIGURE 20-5 Schematic diagram of the injection molding process. A moving plunger advances material through a heating region (in this case, through a heated manifold and over a heated torpedo) and further through runners into an opening and closing mold where the molten thermoplastic cools and solidifies.

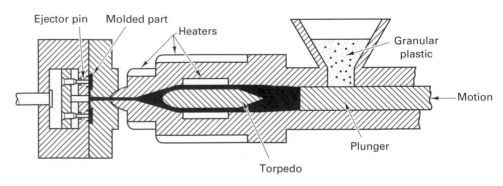

the curing process. The major deterrent to the injection molding of the thermosets is the relatively long cycle times that are required.

REACTION INJECTION MOLDING

Figure 20-6 depicts the *reaction injection molding* process, in which two or more liquid monomers are metered into a unit where they are intimately mixed by the impingement of liquid streams that have been pressurized to a value between 13 and 20 MPa (2000 and 3000 psi). The combined material flows through a pressure-reducing chamber and exits the mixhead directly into a mold. An exothermic chemical reaction takes place between the two components, resulting in thermoset polymerization. Since no heating is required, the production rates are set primarily by the curing time of the polymer, which is often less than 1 minute. Molds are made from steel, aluminum, or nickel shell, with selection being made on the basis of number of parts to be made and the desired quality. The molds are generally clamped in low-tonnage presses.

At present, the dominant materials for reaction injection molding are polyurethanes, polyamides, and composites containing short fibers or flakes. Properties can span a wide range, depending on the combination and percentage of base chemicals and the additives that are used. Different formulations can result in elastomeric or flexible, structural foam (foam core with a hard, solid outer skin), solid (no foam core), or composite products. Part size can range from $\frac{1}{2}$ to 50 kg (1 to 100 lb), shapes can be quite complex (with variable wall thickness), and surface finish is excellent. Automotive applications include steering wheels, airbag covers, instrument panels, door panels, armrests, headliners, and center consoles, as well as body panels, bumpers, and wheel covers. Rigid polyurethanes are also used in such products as household refrigerators, water skis, hot-water heaters, and picnic coolers.

From a manufacturing perspective, reaction injection molding has a number of attractive features. The low processing temperatures and low injection pressures make the process attractive for molding large parts, and the large size can often enable parts consolidation. Thermoset parts can generally be fabricated with less energy than the injection molding of thermoplastics, with similar cycle times and a similar degree of automation. The metering and mixing equipment and related controls tend to be quite sophisticated and costly, but the lower temperatures and pressures enable the use of cheaper molds, which can be quite large.

EXTRUSION

Long plastic products with uniform cross sections can be readily produced by the *extrusion* process depicted in Figure 20-7. Thermoplastic pellets or powders are fed through a hopper into the barrel chamber of a screw extruder. A rotating screw propels the material through a preheating section, where it is heated, homogenized, and compressed, and then forces it through a heated die and onto a conveyor belt. To preserve its newly imparted shape, the material is cooled and hardened by jets of air or sprays of water. It continues to cool as it passes along the belt and is then either cut into lengths

FIGURE 20-6   The reaction injection molding process. (*Left*) Measured amounts of reactants are mixed in the mixing head and injected into the split mold. (*Right*) After sufficient curing, the mold is opened and the component is ejected.

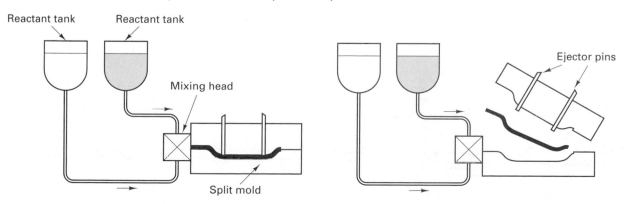

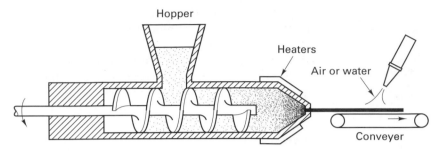

FIGURE 20-7 A screw extruder producing thermoplastic product. Some units may have a changeable die at the exit to permit production of different shaped parts.

or coiled, depending on whether the material is rigid or flexible and the desires of the customer. The process is continuous and provides a cheap and rapid method of molding. Common production shapes include a wide variety of constant cross section profiles, as well as tubes, pipes, and even coated wires and cables. Thermoplastic foam shapes can also be produced. If an emerging tube is expanded by air pressure, allowed to cool, and then rolled, the product can be a double layer of sheet or film.

While single-screw extruders are common and versatile, the twin-screw design has become quite popular, especially for the processing of the more complex plastics. Two parallel screws are located within a single barrel, and can be counterrotating or co-rotating, intermeshing or nonintermeshing. In the co-rotating, intermeshing design, the roots and flanks of one screw continuously wipe the crests of the other. The material follows a figure-eight path as it advances down the heated barrel.

THERMOFORMING

In the *thermoforming* process, thermoplastic sheet material is heated to a working temperature and then formed into a finished shape by heat, pressure, or vacuum. The starting material can be either discrete sheets or a continuous roll of material. If continuous material is used, it is usually heated by passing through an oven or other heating device. The material emerges over a male or female mold and is formed by the application of vacuum and/or pressure. Cooling occurs upon contact with the mold, and the product hardens in its new shape. After sufficient cooling, the part is removed from the mold and trimmed. The unused strip material is diverted for recycling.

Figure 20-8 shows the process using a female mold cavity and discrete sheets of material. Here the material is placed directly over the die or pattern and heated in place. Pressure and/or vacuum is then applied, causing the material to draw into the cavity. Male form blocks or mating male and female dies can also be used. An entire cycle requires only a few minutes. Typical products tend to be simple-shaped, thin-walled parts, such as panels for light fixtures or pages of Braille text for the blind.

ROTATIONAL MOLDING

Rotational molding can be used to produce hollow, seamless products of a wide variety of sizes and shapes, including storage tanks, bins and refuse containers, doll parts,

FIGURE 20-8 One method of molding thermoplastic sheets using a combination of heat and vacuum, a type of thermoforming.

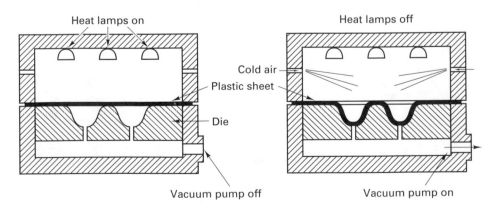

footballs, helmets, and even boat hulls. The process begins with a closed mold or cavity that has been filled with a premeasured amount of thermoplastic powder or liquid. The molds are either preheated or placed in a heated oven and are then rotated simultaneously about two perpendicular axes. Other designs rotate the mold about one axis while tilting or rocking about another. In either case, the resin melts and is distributed in the form of a uniform-thickness coating over all of the surfaces of the mold. The mold is then transferred to a cooling chamber, where the motion is continued and air or water is used to slowly drop the temperature. After the material has solidified, the mold is opened and the hollow product is removed. The lightweight rotational molds are frequently made from cast aluminum, but sheet metal is often used for larger parts, and electroformed or vaporformed nickel can also be used to reproduce fine detail.

FOAM MOLDING

Foamed plastics have become an important and widely used form of polymer product. In foam molding, a foaming agent is mixed with the plastic resin and releases gas or volatilizes when the material is heated during molding. The materials expand to 2 to 50 times their original size, resulting in products with densities ranging from 32 to 640 g/L (2 to 40 lb/ft³). *Open-cell foams* have interconnected pores that permit the permeability of gas or liquid. *Closed-cell foams* have the property of being gas- or liquid-tight.

Both rigid and flexible foams have been produced using both thermoplastic and thermosetting materials. The rigid type is useful for structural applications (including housings for computers and business machines), packaging, and shipping containers; as patterns for the full-mold and lost-foam casting processes (see Chapter 14); and for injection into the interiors of thin-skinned metal components, such as aircraft fins and stabilizers. Flexible foams are used primarily for cushioning.

Variations of the conventional molding processes can be used to produce a variety of unusual products. By introducing foaming material into the interior of a mold that has partly filled, parts can be produced with a solid outer skin and a rigid foam core. Figure 20-9 shows the cross section of a plastic gear with this type of dual structure.

OTHER PLASTIC-FORMING PROCESSES

In the *calendering* process, a mass of dough-like thermoplastic is forced between and over two or more counterrotating rolls to produce thin sheets or films of polymer, which are then cooled to induce hardening. Product thicknesses generally range between 0.3 and 1.0 mm (0.01 to $\frac{1}{16}$ in.), but can be reduced further to as low as 0.05 mm (0.002 in.) by subsequent stretching. Embossed designs can be incorporated into the rolls to produce products with textures or patterns.

Conventional *drawing* can be used to produce fibers, and *rolling* can be performed to change the shape of plastic extrusions. In addition to changing the product dimensions, these processes can also serve to induce crystallization or produce a preferred orientation to the thermoplastic polymer chains.

Filaments, fibers, and yarns can be produced by *spinning*, a modified form of extrusion. Molten thermoplastic polymer is forced through a die containing many small

FIGURE 20-9   A plastic gear having a solid outer skin and a rigid foam core. *(From American Machinist.)*

holes. Where multistrand yarns or cables are desired, the dies can rotate or spin to produce the twists and wraps.

It is not uncommon to find the various plastic-forming processes combined in either sequential or integrated forms to produce specific products. For example, the closed-bottom parison for blow molding can be formed as a separate injection molding, or as an integrated extrusion operation that is then followed by blow molding. Another classic example is the manufacture of the thin plastic bags that are dispensed from a roll in the produce sections of most grocery stores. Polymer granules flow through a hopper and enter the barrel of a screw extruder. As the screw drives the material forward, it is melted and driven through an open-ended metal die that forms a thin-walled plastic tube. Air flows through the center of the die, causing the diameter of the tube to expand substantially as the tube emerges from the die constraint. Air jets around the circumference of the expanded tube then cool the thin plastic material, after which it is passed through rolls to flatten the expanded tube. The flattened tube is then periodically seam welded and perforated, and wound on a roll.

MACHINING OF PLASTICS

Plastics can be milled, sawed, drilled, and threaded much like metals, but their properties are so variable that it is impossible to give instructions that are exactly correct for all. It is important, however, to remember some of the general characteristics of plastics that will affect their machinability. Since plastics tend to be poor thermal conductors, little of the heat that results from chip formation will be conducted into the material or be carried away in the chips. The thermoplastics tend to soften and swell, and occasionally bind or clog the cutting tool. In addition, considerable elastic flexing can occur, and this, coupled with material softening, reduces the precision of final dimensions. Because of their higher rigidity and reduced softening, the thermosetting materials are easier to machine to precise dimensions.

The high temperatures that develop at the point of cutting also cause the tools that are machining plastic to run very hot and they may fail more rapidly than when cutting metal. Carbide tools may be preferred over high-speed tool steels if the cuts are of moderate duration or if high-speed cutting is performed.

The tools that are used to machine plastics should also be kept sharp at all times. Drilling is best done by means of straight-flute drills or by "dubbing" the cutting edge of a regular twist drill to produce a zero rake angle. Such configurations are shown in Figure 20-10. Rotary files, saws, and milling cutters should be run at high speeds to improve cooling but with the feed carefully adjusted to avoid clogging the cutter.

Coolants can often be used advantageously if they do not discolor the plastic or induce gumming. Water, soluble oil and water, and weak solutions of sodium silicate have been used effectively. Filled and laminated plastics can be quite abrasive, and the fine dust that is produced during machining can also be a health hazard.

Laser machining is an attractive alternative to mechanical cutting. By vaporizing the material instead of forming chips, precise cuts can be achieved. Minute holes can be drilled, such as those in the nozzles of aerosol cans.

FIGURE 20-10 Straight-flute drill (*top*) and "dubbed" drill (*bottom*) used for drilling plastics.

OTHER FINISHING AND ASSEMBLY OPERATIONS

One of the advantages of polymeric materials is the possibility of integral color, and it is quite possible that the surface of as-formed products is adequate for final use. Some of the finishing processes that can be applied to plastics include printing, hot stamping, vacuum metallizing, electroplating, and painting. Some of these are described in detail in Chapter 31.

Thermoplastics can often be joined by heating the relevant surfaces or regions. Heat can be applied through a stream of hot gases or by a tool like a soldering iron. Joining heat can also be generated through ultrasonic vibrations. The welding techniques that are applied to plastics are presented in Chapter 38. Adhesive bonding is another popular means of joining plastic and is presented in Chapter 40. Because of the low modulus of elasticity, plastics can be easily flexed, and *snap-fits* are another popular means of assembling plastic components. Because of the softness of some polymeric materials, self-tapping screws can also be used.

DESIGNING FOR FABRICATION

The primary objective of any manufacturing activity is the production of satisfactory components or products, and this involves the selection of an appropriate material or materials. When polymers are selected as the material of construction, it is usually as a result of one or more of their somewhat unique properties, which include light weight, corrosion resistance, good thermal and electrical insulation, and the possibility of integral color. While these properties are indeed attractive, one should also be aware of the more common limitations, such as softening or burning at elevated temperatures, poor dimensional stability, or the deterioration of properties with age. The basic properties and characteristics of polymeric materials have been developed in Chapter 8.

A second area of manufacturing concern is selecting the process or processes to be used in producing the shape and establishing the desired properties. Each of the wide variety of fabrication processes has distinct advantages and limitations, and efforts should be made to utilize the unique features. In addition, the production of quality products also requires an awareness of all of the various aspects of a given process. For example, consider a molding process in which a liquid or semifluid polymer is introduced into a mold cavity and allowed to harden. The proper amount of material must be introduced and caused to flow in such a way as to completely fill the cavity. Air that originally occupied the cavity needs to be vented and removed. Shrinkage will occur during solidification and/or cooling and may not occur in a uniform manner. Heat transfer must be provided to control the cooling and/or solidification. Finally, a means must be provided for part removal or ejection from the mold. Surface finish and appearance, the resultant engineering properties, and the ultimate cost of production are all dependent on the good design and execution of the molding process.

In all molded products, it is important to provide adequate fillets between adjacent sections to ensure smooth flow of the plastic into all sections of the mold and eliminate stress concentrations at sharp interior corners. These fillets also make the mold less expensive to produce and reduce the danger of mold fracture during use. Even the exterior edges should be rounded where possible. A radius of 0.25 to 0.40 mm (0.010 to 0.015 in.) is scarcely noticeable but will do much to prevent an edge from chipping. Sharp corners should also be avoided in products that will be used for electrical applications, since they tend to increase voltage gradients, which can lead to product failure.

Wall or section thickness is also very important, since the hardening or curing time is determined by the thickest section. If possible, sections should be kept nearly uniform in thickness, since nonuniformity can lead to serious warpage and dimensional control problems. As a general rule, one should use the minimum thickness that will provide satisfactory end-use performance. The specific value will be determined primarily by the size of the part and, to some extent, the process and the type of plastic being used. Recommended minimum thicknesses for molded plastics are as follows:

Small parts	1.25 mm (0.050 in.)
Average-sized parts	2.15 mm (0.085 in.)
Large parts	3.20 mm (0.125 in.)

Thick corners should also be avoided because they can lead to gas pockets, undercuring, or cracking. When extra strength is needed in a corner, it can usually be provided by incorporating ribs into the design.

Economical production is also facilitated by appropriate dimensional tolerances. A minimum tolerance of ±0.08 mm (±0.003 in.) should be allowed in directions that are parallel to the parting line of the mold or contained within a mold segment. In directions that cross a parting surface, a minimum tolerance of ±0.25 mm (±0.010 in.) is desirable. In both cases, increasing these values by about 50% can simultaneously reduce both manufacturing difficulty as well as cost.

Since most molds are reusable, careful attention should be given to the removal of the part. Rigid, metal molds should be designed so that they can be easily opened and closed. A small amount of unidirectional taper should be provided to facilitate part withdrawal. Undercuts should be avoided whenever possible, since they will prevent part removal unless additional mold sections are used. These must move independently of the major segments of the mold, adding to the costs of mold production and maintenance and slowing the rate of production.

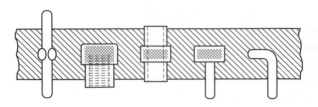

FIGURE 20-11 Typical metal inserts used to provide threaded cavities, holes, and alignment pins in plastic parts.

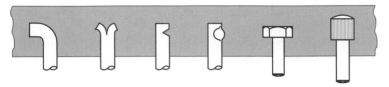

FIGURE 20-12 Various ways of anchoring metal inserts in plastic parts (*left to right*): bending, splitting, notching, swaging, noncircular head, and grooves and shoulders. Knurling is depicted in Figure 20-11.

INSERTS

Metal *inserts* are often incorporated into plastic products to provide enhanced performance or unique features. Since molded threads are difficult to produce, machined threads require additional processing, and both types tend to chip or deform, threaded inserts are frequently used when assemblies require considerable strength or when frequent disassembly and reassembly is anticipated. Figure 20-11 depicts one form of threaded insert, along with other types that serve to provide alignment or mounting pins and holes. These metal inserts are usually made from brass or steel.

The successful use of inserts requires careful attention to design since they are generally held in place by only a mechanical bond that must resist both rotation and pullout. Knurling or grooving is often required to provide suitable sites for gripping. A medium or coarse knurl is usually adequate to resist torsional loads and moderate axial forces. Circumferential grooves are excellent for axial loads but offer little resistance to torsional rotation. Axial grooves resist rotation, but do little to prevent pullout. Other means of anchoring include bending, splitting, notching, and swaging. Headed parts with noncircular heads may be used as formed. Combinations of notches, grooves, and shoulders are also common. Figure 20-12 provides a summary of the various means of insert attachment.

Consideration should also be given to minimizing the stresses that occur around a cast-in insert as a result of the differential contraction during cooling. Preheating the inserts can be beneficial; the plastic can be selected to minimize shrinking; and nonmetallic inserts can be considered.

If an insert is to act as a boss for mounting or serve as an electrical terminal, it should protrude slightly above the surface of the plastic. This permits a firm connection to be made without creating an axial load that would tend to pull the insert from its surroundings. If the insert serves to hold two mating parts together, it should be flush with the surface. In this way, the parts can be held together snugly without danger of loosening the insert. In all cases, the wall thickness of the surrounding plastic must be sufficient to support any load that may be transmitted by the insert. For small inserts, the wall thickness should be at least half the diameter of the insert. For inserts larger than 13 mm ($\frac{1}{2}$ in.) in diameter, the wall thickness should be at least 6.5 mm ($\frac{1}{4}$ in.).

DESIGN FACTORS RELATED TO FINISHING

Because plastics are frequently used where consumer acceptance is of great importance, special attention should be given to finish and appearance. In many cases, plastic parts can be designed to require very little finishing or decorative treatment. Fins and rough spots can often be removed by a barrel tumbling with suitable abrasives or polishing agents. Smoothing and polishing occur in the same operation. By etching the surfaces of a mold, decorations or letters can be produced that protrude approximately 0.01 mm (0.004 in.) above the surface of the plastic. When higher relief is required, the mold can be engraved, but this adds significantly to mold cost.

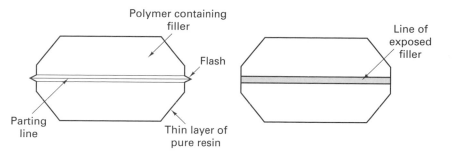

FIGURE 20-13 Trimming the flash from a plastic part ruptures the thin layer of pure resin and creates a line of exposed filler.

Whenever possible, depressed letters or designs should be avoided. These features, when transferred to the mold, become raised above the surrounding surface. Mold making then requires a considerable amount of intricate machining as the surrounding material is cut away from the design or letters. When recessed features are required, manufacturing cost can be reduced if they can be incorporated into a small area that is raised above the primary surface.

A prime objective in the design of plastic parts is often the elimination of secondary machining, especially on surfaces that would be exposed to the customer. Even when fillers are used (as they are in most plastics), the surfaces of molded parts have a thin film of pure resin. This film provides the high luster that is characteristic of polymeric products. Machining cuts through the surface, exposing the underlying filler. The result is a poor surface appearance, as well as a site for the absorption of moisture.

One location that frequently requires machining, however, is the parting line that is produced where the mold segments come together. Perfect mating is difficult to achieve, especially when the parting surfaces are not flat. As a result, a small fin, or "flash," is produced around the part perimeter, as illustrated in Figure 20-13. When the flash is trimmed off, the resulting line of exposed filler may be objectionable. By locating the parting line along a sharp corner, it is easier to maintain satisfactory mating of the mold sections, and the exposed filler that is created by flash removal will be confined to a corner, where it is less noticeable.

Because plastics have a low modulus of elasticity, large flat areas are not rigid and should be avoided whenever possible. Ribbing or doming, like that illustrated in Figure 20-14, can be used to provide the required stiffness. In addition, flat surfaces tend to reveal flow marks from the molding operation, as well as scratches that occur during service. External ribbing then serves the dual function of increasing strength and rigidity while masking the surface flaws. Dimpled or textured surfaces can also be used to provide a pleasing appearance and conceal scratches.

Holes that are formed by pins that protrude from the mold often require special consideration. In compression molding, these pins can be subjected to considerable bending during mold closure and filling. When these pins are supported only at one end, their length should not exceed twice the diameter. In processes with reduced filling pressures, the length can be as much as five times the diameter without excessive problems.

Holes that are to be threaded or used for self-tapping screws should be countersunk. This not only assists in starting the tap or screw but also reduces chipping at the outer edge of the hole. If the threaded hole is less than 6.5 mm ($\frac{1}{4}$ in.) in diameter, it is best to cut the threads after molding, using some form of thread tap. For diameters greater than 6.5 mm, the threads can be molded or an insert should be used. If the threads are molded, however, special provisions must be made to remove the part from the mold. These additional operations are generally uneconomical, since they extend the molding time and reduce productivity.

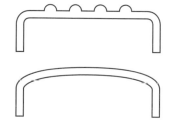

FIGURE 20-14 Stiffness can be imparted to large surfaces of plastic parts through the use of ribbing or doming.

■ 20.3 PROCESSING OF RUBBER AND ELASTOMERS

Rubber and elastomeric products can also be produced by a variety of fabrication processes. Probably the simplest of these is *dipping*, a process used to produce relatively thin parts with uniform wall thickness, such as boots, gloves, and fairings. A master form is first produced, usually from some type of metal. This form is then immersed into a

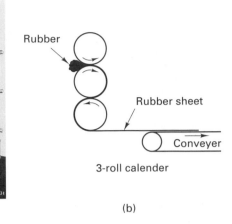

(a)

(b)

3-roll calender

FIGURE 20-15 *(Left)* Three-roll calender used for producing rubber or plastic sheet. *(Courtesy of Farrel-Birmingham Company, Inc.)* *(Right)* Schematic diagram showing the method of making sheets of rubber with a three-roll calender.

liquid preparation or compound (usually based on natural rubber, neoprene, or silicone), then removed and allowed to dry. With each dip, a certain amount of the liquid adheres to the surface, and repeated dips are used to produce a final desired thickness. After vulcanization, usually in steam, the products can be stripped from the molds.

Dipping can be accelerated by using electrostatic charges. A negative charge is introduced to the latex particles, and the form or mold is charged positively, either through an applied voltage or by a coagulant coating that releases positive ions when dipped into the solution. The attraction and neutralization of the charges causes the elastomeric particles to be deposited on the form at a faster rate and in thicker layers than the basic process. With electrostatic deposition, many products can be made in a single immersion.

When products are to be made of solid elastomers, the first step is the compounding of the elastomeric resin, vulcanizers, fillers, antioxidants, accelerators, and other pigments. This is usually done in some form of mixer, which blends the components to form a homogeneous mass. Adaptations of the processes previously discussed for plastics are frequently used to produce the desired shapes. Injection, compression, and transfer molding are used, along with special techniques for foaming. Urethanes and silicones can also be directly cast to shape.

Rubber compounds can be made into sheets using *calenders*, like that shown in Figure 20-15. The sheet coming from the calender is often rolled with a fabric liner to prevent the material from sticking. Three- or four-roll calenders can also be used to place a rubber or elastomer covering over cord or woven fabric. In the three-roll geometry, only one side of the fabric can be coated in each pass. The four-roll arrangement, shown schematically in Figure 20-16, enables both sides to be coated simultaneously.

Products such as inner tubes, garden hoses, tubing, and strip moldings can be produced by the *extrusion* process. The compounded material is forced through a die by a screw device similar to that described for plastics.

Adhesives have been developed that enable the bonding of rubber or artificial elastomers to metal, usually brass or steel. Only moderate pressures and temperatures are required to obtain excellent adhesion.

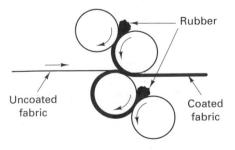

4-roll calender

FIGURE 20-16 Arrangement of the rolls, fabric, and coating material for coating both sides of a fabric in a four-roll calender.

■ 20.4 PROCESSING OF CERAMICS

The fabrication processes applied to ceramic materials generally fall into two distinct classes, based on the properties of the material. *Glasses* can be manufactured into useful articles by first heating the material to produce a molten or viscous state, shaping the material by means of *viscous flow*, and then cooling the material to produce a solid product. *Crystalline ceramics* have a characteristically brittle behavior and are normally manufactured into useful components by pressing moist aggregates or powder into a shape, followed by drying, and then bonding by one of a variety of mechanisms, which include chemical reaction, *vitrification* (cementing with a liquefied material), and *sintering* (solid-state diffusion).

FABRICATION TECHNIQUES FOR GLASSES

Glass is generally shaped at elevated temperatures where the viscosity can be controlled. A number of the processes begin with material in the liquid or molten condition. Sheet and plate glass is formed by processes such as rolling through water-cooled rolls or floating on a bath of molten tin. Glass shapes can also be produced by pouring the molten material directly into a mold. The cooling rate is then controlled, usually as slow as possible, to minimize residual stresses and the tendency for cracking. Glass fibers can be produced by forcing liquid glass through multiple openings in a metal-extrusion die.

Other glass-forming processes begin with viscous masses and use mating male and female die members to press the material into the desired shape, as illustrated in Figure 20-17.

A process similar to the *blow molding* of plastics can be used to expand cup-shaped pieces of viscous material against the outside of heated dies. This process is illustrated in Figure 20-18.

Special heat treatments can also be applied to glass material. By applying forced cooling to the exposed surfaces, a residual stress pattern of surface compression can be induced. (The surface layers cool and contract. This is followed by the cooling of the interior, which then tries to contract but is restrained by the surface, creating the surface compression.) This glass, called *tempered glass*, is stronger and more fracture resistant, since cracks tend to initiate on free surfaces. *Annealing* can be used to reduce unfavorable residual stresses that might lead to cracking.

Glass-ceramics form a unique class of materials that are part crystalline and part glass. They are fabricated into shape as a glass, and are then subjected to a special heat treatment (*devitrification*) that controls the nucleation and growth of the crystalline component. The final properties include good strength and toughness, along with low thermal expansion. Typical products include cookware (such as the white CorningWare products), ceramic stove tops, and materials used in electrical and computer components.

FABRICATION OF CRYSTALLINE CERAMICS

Crystalline ceramics are hard, brittle materials with high melting points. As a result, they cannot be formed by techniques requiring either plasticity (i.e., forming methods) or melting (i.e., casting methods). Instead, these materials are generally processed in the solid state by techniques that utilize particles or aggregates and resemble those used in powder metallurgy. Dry powders can be pressed into useful shapes either at environmental or elevated temperatures. *Dry pressing, isostatic pressing*, and *hot-isostatic pressing (HIP)* are all common techniques and exhibit features and limitations similar to those discussed in Chapter 16.

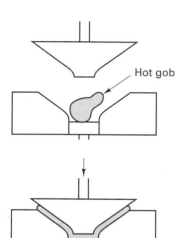

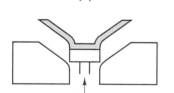

FIGURE 20-17   Viscous glass can be easily shaped by mating male and female die members.

FIGURE 20-18   Thin-walled glass shapes can be produced by a combination of pressing and blow molding.

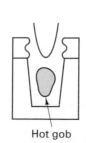
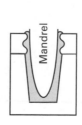
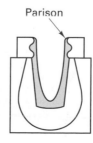
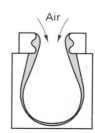

In the *slip casting* process, ceramic powder is mixed with a liquid to form a slurry, which is then cast into a mold containing very fine pores. Capillary action pulls the liquid from the slurry, allowing the ceramic particles to arrange into a "green" body with sufficient strength for subsequent handling. Hollow shapes can be produced by pouring out the remaining slurry once a desired thickness of solid has formed on the mold walls.

Clay products are based on special types of ceramics blended with water and various additives to produce a material that can be shaped by most of the traditional forming methods. *Plastic forming* can also be applied to other ceramics if the ceramic particles are combined with additives, such as plastic, that impart plasticity when subjected to pressure and heat. *Extrusion* can produce products with constant cross sections. After the shape is produced, additive materials are usually removed by dissolving in a solvent or by controlled heating or drying, and the remaining ceramic is fused together by a firing operation.

Injection molding was discussed earlier in this chapter as a means of forming plastics, and metal injection molding (MIM) was presented in Chapter 16 as a way of producing small, complex-shaped metal parts. An adaptation of injection molding can also be used to form complex, three-dimensional shapes from ceramic materials. Ceramic powder is mixed with wax or wax-polymer material at about 90°C (200°F), and the high-viscosity, coated particles are then injected into an aluminum die under low pressure, on the order of 0.7 MPa (100 psi). The mixture cools, and after about 30 seconds, it is sufficiently hard to permit ejection from the die. The parts then undergo a dewaxing, which can involve thermal, solvent, catalytic, or wicking methods, and subsequent firing to desired final density. As with metal injection molding, the die forms a part that is considerably oversized, and controlled shrinkage during firing produces the final dimensions. Most parts' major dimensions are less than 10 cm (4 in.); wall thickness, less than 6 mm ($\frac{1}{4}$ in.); and tolerances, on the order of ±1% or ±0.1 mm (0.005 in.), whichever is greater. Most parts are made from the oxide ceramics, such as alumina or zirconia, but the process has also been used with silicon carbide and silicon nitride.

Table 20-1 summarizes the primary processes used to fabricate shapes from crystalline ceramics.

PRODUCING STRENGTH IN PARTICULATE CERAMICS

The processes described previously can all be used to produce useful shapes from ceramic materials, but useful strength generally requires a subsequent heating operation, known as *firing* or *sintering*. Slurry-type materials must first be dried in a controlled manner that is designed to control dimensional changes and minimize stresses, distortion, and cracking. The material is then heated to temperatures between 0.5 and 0.8 times the absolute melting point, where diffusion processes act to fuse the particles together and impart the desired mechanical and physical properties. The temperature and time are selected to control the resulting grain size, pore size, and pore shape. In some firing operations, surface melting (*liquid-phase sintering*) or component reactions (*reaction sintering*) can produce a substantial amount of liquid material (*vitrification*). The liquid then flows to produce a glassy bond between the ceramic particles and either solidifies or crystallizes.

TABLE 20-1.	Processes Used to Form Products From Crystalline Ceramics		
Process	Starting material	Advantages	Limitations
Dry axial pressing	Dry powder	Low cost; can be automated	Limited cross sections; density gradients
Isostatic processing	Dry powder	Uniform density; variable cross sections; can be automated	Long cycle times; small number of products per cycle
Slip casting	Slurry	Large sizes; complex shapes; low tooling cost	Long cycle times; labor-intensive
Injection molding	Ceramic–plastic blend	Complex cross sections; fast; can be automated; high volume	Binder must be removed; high tool cost
Forming processes (e.g., extrusion)	Ceramic–binder blend	Low cost; variable shapes (such as long lengths)	Binder must be removed; particles oriented by flow
Clay products	Clay, water, and additives	Easily shaped by forming methods; wide range of size and shape	Requires controlled drying

Cementation is an alternative method of producing strength that does not require elevated temperature. A liquid binder material is used to coat the ceramic particles, and a subsequent chemical reaction converts the liquid to a solid, forming strong, rigid bonds.

Prototypes or small production quantities of ceramic products have also been made by the *laser sintering* of ceramic powders. Parts are made by building up successive layers of material by laser sintering (or laser melting) of thin layers of heat-fusible powder. For ceramic parts, the powder particles are actually coated with a very thin thermoplastic polymer binder. The laser then acts on the polymer coating to produce the bond. After the laser bonding, the parts then undergo conventional debinding and sintering to about 55 to 65% of theoretical density. Isostatic pressing prior to sintering can raise the final density to 90 to 99% of theoretical.

MACHINING OF CERAMICS

Most ceramic materials are brittle, and the techniques used to cut metals will generally produce uncontrolled or catastrophic cracks. In addition, ceramics are typically hard materials. Since ceramics are often used as abrasives or coatings on cutting tools, the tools needed to cut them have to be even harder.

Direct production to the desired final shape is clearly the most attractive alternative, but there are times when a material removal operation is necessary. Such machining can be performed before or after the final firing. Before firing, the material tends to be rather weak and fragile. While fracture is always a concern, a more significant consideration might be the dimensional changes that will occur upon subsequent firing. Shrinkage may be as much as 30%, so it may be difficult to achieve or maintain close final tolerances. For this reason, machining before firing, known as *green machining*, is usually rough machining designed to reduce the amount of finishing that will be required after firing.

When machining is performed after firing, the processes are generally ones we might consider to be nonconventional. Grinding, lapping, and polishing with diamond abrasives, drilling with diamond-tipped tooling, cutting with diamond saws, ultrasonic machining, laser and electron-beam machining, water-jet machining, and chemical etching have all been used. When mechanical forces are applied, material support is quite critical (since ceramic materials are almost always brittle). Because of the hardness of the ceramic, the tools must be quite rigid. Selection and use of coolants are also important issues.

Materials producers have developed "machinable" ceramics that lend themselves to precision shaping by more traditional machining operations. It should be noted, however, that these are indeed special materials and not characteristic of ceramics as a whole.

JOINING OF CERAMICS

When we consider joining operations, the unique properties of ceramics once again introduce fabrication limitations. Brittle ceramics cannot be joined by fusion welding or deformation bonding, and threaded assemblies should be avoided whenever possible. Therefore, most joining utilizes some form of adhesive bonding, brazing, diffusion bonding, or special cements. Even with these methods, the stresses that develop on the surfaces can lead to premature failure. As a result, most ceramic products are designed to be monolithic (single-piece) structures rather than multipart assemblies.

DESIGN OF CERAMIC COMPONENTS

Since ceramics are brittle materials, special care should be taken to minimize bending and tensile loading as well as design stress raisers. Sharp corners and edges should be avoided where possible. Outside corners should be chamfered to reduce the possibility of edge chips. Inside corners should have fillets of sufficient radius to minimize crack initiation. Undercuts are difficult to produce and should be avoided. Specifications should generally use the largest possible tolerances, since these can often be met with products in the as-fired condition. Extremely precise dimensions usually require hand grinding, and costs can escalate significantly. In addition, consideration should be given to surface finish requirements, since grinding, polishing, and lapping operations can increase production cost substantially.

■ 20.5 FABRICATION OF COMPOSITE MATERIALS

As shown in Chapter 8, composite materials can be designed to offer a number of attractive properties. In some market areas, such as aerospace and sporting goods, their acceptance and growth have been phenomenal. Use can only occur, however, if the material can be produced in useful shapes at an acceptable cost and rate of production. Many of the manufacturing processes designed for composites are slow, and some require extensive amounts of hand labor. There is often a high amount of variability between nominally identical products, and inspection and quality control methods are not well developed. While these limitations may be acceptable for certain applications, they provide definite barriers to the use of composites for high-volume, mass-produced items. Faster production speeds, increased use of automation, reduced variability, and integrated quality control continue to be important as we seek to expand the use of composite materials.

In Chapter 8, composite materials were classified by their basic geometry as particulate, laminar, and fiber-reinforced. Since the fabrication processes are often unique to a specific type of composite, they will also be grouped by these three basic geometries.

FABRICATION OF PARTICULATE COMPOSITES

Particulate composites usually consist of a fracture-resistant metallic or polymeric matrix and dispersed particles of a second material. Their fabrication, however, rarely requires processes unique to composite materials. Instead, the particles are simply dispersed in the matrix by introduction into a liquid melt or slurry, or by blending the various components as solids, using powder metallurgy methods. Subsequent processing generally follows the conventional methods of casting or forming, or utilizes techniques that are common to powder metallurgy. These have been presented elsewhere in the text and will not be repeated here.

Reinforcement particles have also been blended into the highly viscous slurries of rheocast material, the semisolid mixtures that are viscous when agitated but retain their shape when static. Various reinforced products have also been made by spray forming multicomponent feeds.

FABRICATION OF LAMINAR COMPOSITES

Laminar composites include coatings and protective surfaces, claddings, bimetallics, laminates, and a host of other materials. Their production generally involves processes designed to form a high-quality bond between distinct layers of different materials. When the layers are metallic, as in claddings and bimetallics, the composites can be produced by hot or cold *roll bonding*. Sheets of the various materials are passed simultaneously through the rolls of a conventional rolling mill. If the amount of deformation is great enough, surface oxides and contaminants are broken up and dispersed, metal-to-metal contact is established, and the two surfaces become joined by a solid-state bond. U.S. coinage is a common example of a roll-bonded material.

Explosive bonding is another practical means of bonding layers of metal. A sheet of explosive material progressively detonates above the layers to be joined, causing a pressure wave to sweep across the interface. A small open angle is maintained between the two surfaces. As the pressure wave propagates, any surface films are liquefied or scarfed off and are jetted out the open interface. Clean metal surfaces are then forced together at high pressures, forming a solid-state bond with a characteristically wavy configuration at the interface. Wide plates (too wide to roll bond conveniently) and dissimilar materials with large differences in mechanical properties are attractive candidates for explosive bonding.

Adhesive bonding is another attractive means of joining the various layers and can be applied to both metallic and nonmetallic materials. The lamination of polymer matrix composites (to be discussed later in this chapter when each ply is a fiber-reinforced or woven layer) often utilizes films of unpolymerized resin that are introduced between the layers. Pressing at elevated temperature then cures the resin and completes the bond. Brazing can be employed to join layers of metallic material and form composites that can withstand moderate elevated temperatures.

In *sandwich structures*, such as corrugated cardboard or the honeycomb shown in Figure 20-19, thin layers of facing material are bonded, usually by adhesive, to a

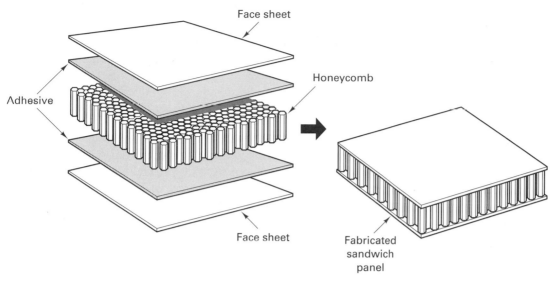

FIGURE 20-19 Fabrication of a honeycomb sandwich structure using adhesive bonding to join the facing sheets to the lightweight honeycomb filler. *(Courtesy of ASM International.)*

lightweight filler material. Special fabrication methods may be employed to produce the foam, corrugated, or honeycomb filler.

FABRICATION OF FIBER-REINFORCED COMPOSITES

A number of processes have been developed to produce and shape the fiber-reinforced composites. Variations are based primarily on the orientation of the fibers, the length of continuous filaments, and the geometry of the final product. Each seeks to embed the *fibers* in a selected *matrix* with the proper alignment and spacing necessary to produce the desired properties. Discontinuous fibers can be combined with a matrix to provide either a random or a preferred orientation. Continuous fibers are normally aligned in a unidirectional fashion in rods or tapes, woven into fabric layers, wound around a mandrel, or woven into a three-dimensional shape.

Production of Reinforcing Fibers.   A number of processes have been developed to produce the various types of reinforcement fibers used in composites. Metallic fibers, glass fibers, and many polymeric fibers (including the popular Kevlar) can be produced by variations of conventional wire drawing and extrusion. Boron, carbon, and ceramic fibers such as silicon carbide are too brittle to be produced by the deformation methods. Boron fibers are produced by chemical vapor deposition around a tungsten filament. Carbon (graphite) fibers can be made by carbonizing (decomposing) an organic material that is more easily formed to the desired shape.

The individual fine filaments are often bundled into *yarns* (twisted assemblies of filaments), *tows* (untwisted assemblies of fibers), and *rovings* (untwisted assemblies of yarns or tows). Fibers can also be chopped into short lengths, usually 12 mm ($\frac{1}{2}$ in.) or less, for incorporation into the various sheet or bulk molding compounds. In these materials the fibers usually assume a random orientation.

Processes Designed to Combine Fibers and a Matrix.   A variety of processes have been developed to combine the fiber and the matrix into a unified material suitable for further processing. If the matrix material can be liquefied and the temperature is not harmful to the fibers, casting-type processes can be an attractive means of coating the reinforcement. The pouring of concrete around steel reinforcing rod is a crude example of this method. In the case of the high-tech, fiber-reinforced plastics and metals, the liquid can be introduced between the fibers by means of *capillary action, vacuum infiltration*, or *pressure casting*. In a modification of *centrifugal casting*, the resin is introduced into the center of a rotating mold and is then uniformly forced against and into the reinforcing material. Yet another alternative is to draw the fibers through a bath of molten material and combine them into aligned bundles before the liquid solidifies.

Prepregs involve the formation of a woven fabric that has been infiltrated with a matrix material. *Mats* are sheets of nonwoven, randomly oriented fibers in a matrix. When the matrix is a polymeric material, the resin in the prepreg or mat is usually only partially cured. Later fabrication then involves the stacking of layers and the application of heat and pressure to further cure the resin and bond the layers into a continuous solid matrix. Prepreg layers can be stacked in various orientations to provide various directional properties.

Individual filaments can be coated with a matrix material by drawing through a molten bath, plasma spraying, vapor deposition, electrodeposition, or other techniques. These coated fibers can then be used, either individually or in various assemblies. The coated fibers can also be wound around a mandrel with a specified spacing, and then cut to produce *tapes* that contain continuous, unidirectionally aligned filaments. These tapes are generally one fiber diameter in thickness and can be up to 1.2 m (48 in.) wide.

When the temperatures of the molten matrix become objectionable or potentially damaging, the matrix can often be bonded to the fibers by means of either diffusion or deformation bonding (hot pressing or rolling). A common arrangement is to position aligned or woven fibers between sheets of foil material. Loosely woven fibers can also be infiltrated with a particulate matrix, which is then compacted at high pressures and sintered to form a solid mass.

Sheet-molding compounds (SMC) are composed of chopped fibers (usually glass) and partially cured resin, along with fillers, pigments, catalysts, thickeners, and other additives, in sheets approximately 2.5 mm (0.1 in.) thick. With strengths in the range of 35 to 70 MPa (5 to 10 ksi) and the ability to be press-formed in heated dies, these materials offer a feasible alternative to sheet metal in applications where light weight, corrosion resistance, and integral color are attractive features.

After initial compounding and a few days of curing, sheet-molding compounds generally take on the consistency of leather, making them easy to handle and mold. When placed in a heated mold, the viscosity is quickly reduced and the material flows easily under pressures of about 7 MPa (1000 psi). The elevated temperatures accelerate the chemical reactions and final curing can often be completed in less than 60 seconds. As an added benefit, sheet-molding compounds can be easily recycled. One possible disadvantage, however, is that polymer flow may orient the reinforcing fibers, making the final orientation nonrandom and difficult to predict and control.

Bulk-molding compounds (BMC) are fiber-reinforced, thermoset, molding materials, where the short fibers are distributed in random orientation. The starting material is usually a bulk material with the consistency of putty or modeling clay, although pellets and granules are also possible. The final shape is usually produced by compression molding in heated dies, but transfer molding and injection molding are other possibilities.

Fabrication of Final Shapes from Fiber-Reinforced Composites.

A number of processes have been developed for the production of finished products from fiber-reinforced material. Many are simply extensions or adaptations of processes that are used to shape the matrix material (usually metals or polymers). Others are unique to the family of fiber-reinforced composites. The dominant techniques will be discussed individually in the sections that follow.

PULTRUSION *Pultrusion* is a continuous process that is used to produce relatively simple shapes of uniform cross section, such as round, rectangular, tubular, plate, sheet, and structural products. As shown in Figure 20-20, bundles of continuous reinforcing fibers are drawn through a bath of thermoset polymer resin, and the impregnated material is then gathered to produce a desired cross-sectional shape. This material is then pulled

FIGURE 20-20 Schematic diagram of the pultrusion process. The heated dies cure the thermoset resin.

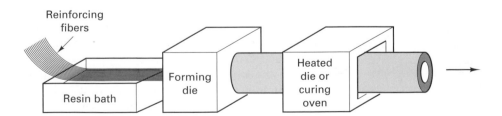
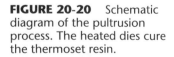

Reinforcing fibers

Resin bath

Forming die

Heated die or curing oven

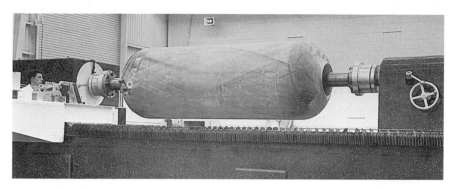

FIGURE 20-21   A large tank being made by filament winding. *(Courtesy of Rohr Corporation.)*

through one or more heated dies, which further shapes the product and cures the resin. Upon emergence from the heated dies, the product is cooled by air or water, cut to length, and further fabricated into products such as fishing poles, golf club shafts, and ski poles. Extremely high strengths are possible (since the reinforcement can be as much as 75% of the final structure), with densities about 20% that of steel or 60% that of aluminum. Cross sections can be as much as 1.5 m (60 in.) wide and 0.3 m (12 in.) thick.

FILAMENT WINDING   Resin-coated or resin-impregnated, high-strength, continuous filaments, bundles, or tapes made from fibers of glass, graphite, boron, Kevlar, or similar materials can be used to produce cylinders, spheres, cones, and other container-type shapes that have exceptional strength-to-weight ratios. The filaments are wound over a form or mandrel, using longitudinal, circumferential, or helical patterns, or a combination of these, designed to take advantage of their highly directional strength properties. By adjusting the density of the filaments in various locations and selecting the orientation of the wraps, products can be designed to have strength where needed and lighter weight in less critical regions. After winding, the part and mandrel are placed in an oven for curing, after which the product is stripped from the form. The matrix, often an epoxy-type polymer, binds the structure together and transmits the stresses to the fibers.

Figure 20-21 shows a large tank being produced by filament winding. Products such as pressure tanks and rocket motor casings can be made in virtually any size, some as large as 4.5 m (15 ft) in diameter and 20 m (65 ft) long. Moderate production quantities are feasible, and because the process can be highly mechanized, uniform quality can be maintained. A new form block is all that is required to produce a new size or design. Because the tooling is so inexpensive, the process offers tremendous potential for cost savings and flexibility. With advancements in equipment and control, parts no longer need to be axisymmetric. Filament-wound products can now be made with changing surfaces, nonsymmetric cross sections, and compound curvatures.

LAMINATION AND LAMINATION-TYPE PROCESSES   In the lamination process, prepregs, mats, or tapes are stacked to produce a desired thickness and cured under pressure and heat. The resulting products possess unusually high strength properties as a result of the integral fiber reinforcement. Because the surface is a thin layer of pure resin, laminates usually possess a smooth, attractive appearance. If the resin is transparent, the fiber material is visible and can impart a variety of decorative effects. Other decorative laminates use a separate patterned face sheet that is bonded to the laminate structure.

Laminated materials can be produced as sheets, tubes, and rods. Flat sheets can be made using the method illustrated in Figure 20-22. Prepreg sheets or reinforcement sheets saturated in resin are stacked and then compressed under pressures on the order of 7 MPa (1000 psi). Figure 20-23 depicts the technique used to produce rods or tubes. For tubing, the impregnated stock is wound around a mandrel of the desired internal diameter. Solid rods are made by using a small-diameter mandrel, which is removed prior to curing, or by wrapping the material tightly about itself. Sheet laminating can also be a continuous process. Multiple reinforcement sheets are passed through a resin bath, faced with a nonstick sheet, and passed through squeeze rolls. In all cases, the final operation is a curing, usually involving elevated temperature and possibly applied pressure.

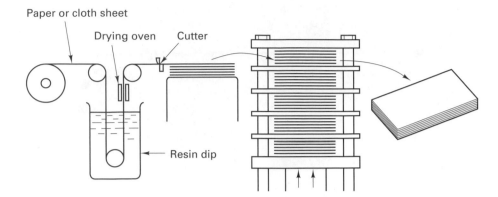

FIGURE 20-22 Method of producing multiple sheets of laminated plastic material.

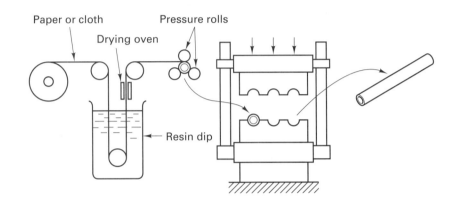

FIGURE 20-23 Method of producing laminated plastic tubing. In the final operation, the rolled tubes are cured by being held in heated tooling.

Because of their excellent strength properties, plastic laminates find a wide variety of uses. Some sheets can be easily blanked and punched. Gears machined from thick laminated sheets have unusually quiet operating characteristics when matched with metal gears.

Many laminated products are not flat but contain relatively simple curves and contours. Manufacturing processes that require zero to moderate pressures and relatively low curing temperatures can be used to produce boat bodies, automobile body panels, aerospace panels, safety helmets, and similar products. In one technique, the only tooling is a female mold or male form block that can be made from metal, hardwood, or even particleboard. The layers of prepreg or resin-dipped fabric are stacked in various orientations until the desired thickness is obtained. Care must be taken to avoid the entrapment of air bubbles and ensure that no impurities (such as oil, dirt, or other contaminants) are introduced between the layers. The entire assembly (mold and material) is then placed in a nonadhering, flexible bag, and the contained air is evacuated. In the *vacuum-bag molding* process, air pressure holds the laminate against the mold while the resin cures. Curing generally occurs at room temperature, but moderately elevated temperatures may be used, which can be obtained through a medium such as live steam. In a variation known as *pressure-bag molding*, a flexible membrane is positioned over the female mold cavity and pressurized to force the individual plies together and drive out entrapped air and excess resin. Pressures usually range from 0.2 to 0.4 MPa (30 to 50 psi) but can be as high as 2 MPa (250 psi). The pressure is coupled with room- or low-temperature curing. Pressure-bag molding has been used to produce extremely large components, such as the skins of military aircraft, large air-deflectors for tractor-trailers, and body panels for trucks.

Higher heats and pressures can be used when the part is cured in an *autoclave*. The supporting molds and vacuum-bagged layups are placed inside a heated pressure vessel where curing occurs under elevated temperatures and pressures in the range 0.4 to 0.7 MPa (50 to 100 psi). Denser, void-free moldings are produced, and the properties can be further enhanced through the use of matrix resins that require higher-temperature cures. The size of the autoclave limits the size of the product.

When production quantities are large and quality needs to be high, matched metal dies can be substituted for the mold and bag. The process then becomes a modification of *compression molding*. A sheet-molding compound, bulk-molding compound, or preformed mat is placed on the press, and heat and pressure are applied. Temperatures typically range from 110 to 160°F (225 to 325°F), and pressures range from 1 to 7 MPa (250 to 1000 psi). With heated dies, the thermoset curing occurs during the compression operation, and cycles range from 1 to 5 minutes.

Resin-transfer molding is a low-pressure process that is intermediate to the slow, labor-intensive layup processes and the faster compression molding or injection molding processes, which generally require more expensive tooling. Continuous fiber mat or woven material (usually employing glass fiber) is positioned dry in the bottom half of a matching mold, which is then closed and clamped. Because of the low pressures employed in the process, the mold tooling does not need to be steel but can be electroformed nickel shells, epoxy composite, or aluminum. In addition, low-capacity presses can be used to clamp the mold segments, and inflatable bags can be used to produce simple holes or hollows, in much the same way that cores are used in conventional casting. A low-viscosity catalyzed resin is then injected into the mold, where it displaces the air, permeates the reinforcement, and subsequently cures at low temperatures. The resulting products can have excellent surfaces on both sides, since both mold surfaces can be precoated with a pigmented gel. Large parts can often be made as a single unit with a relatively low capital investment. Cycle times range from a few minutes to a few hours, depending on the part size and the resin system being used. The aerodynamic hood and fender assembly for Ford Motor Company's AeroMax heavy-duty truck (Figure 20-24) is an example of a large resin transfer molding.

When the quality demands are not as great and the reinforcement-to-resin ratio is not exceptionally high, pressing operations can often be eliminated. The layers of pliable resin-coated cloth are simply placed in an open mold or draped over a form in a process known as *hand layup* or *open mold processing*. Squeegees or rollers are used to manually ensure good contact and remove any entrapped air, and the assembly is then allowed to cure, generally at room temperature. If prepreg layers are not used, a layer of mat, cloth, or woven roving can be put in place, and a layer of resin brushed, sprayed, or poured on. The process can then be repeated to build the desired thickness.

While the hand layup process is slow and labor intensive and has part-to-part and operator-to-operator variability, the tooling costs are sufficiently low that single items or small quantities become economically feasible. Molds or forms can be made from wood, plaster, plastics, aluminum, or steel, so design changes and the associated tool modifications are rather inexpensive, and manufacturing lead time can be quite short. In addition, large parts can be produced as single units, significantly reducing the amount of assembly, and various types of reinforcement can be incorporated into a single product, expanding design options. High-quality surfaces can be produced by applying a pigmented gel coat to the mold before the layup.

FIGURE 20-24 Aerodynamic styling and smooth surfaces characterize the hood and fender of Ford Motor Company's AeroMax truck. This panel was produced as a resin transfer molding by Rockwell International. *(Courtesy of ASM International.)*

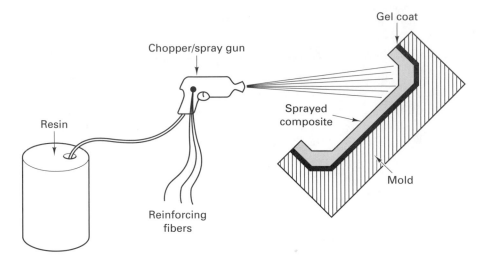

FIGURE 20-25 Schematic diagram of the spray forming of chopped-fiber-reinforced polymeric composite.

SPRAY MOLDING   When continuous or woven fibers are not required to produce the desired properties, sheet-type parts can be produced by mixing chopped fibers and catalyzed resin and spraying the combination into or onto a mold form, as shown in Figure 20-25. Rollers or squeegees can be used to remove entrapped air and work the resin into the reinforcement. Room-temperature curing is usually preferred, but elevated temperatures are sometimes used to accelerate the cure. As with the hand layup process, an initial gel coat can be used to produce a smooth, pigmented surface.

SHEET STAMPING   Thermoplastic sheets that have been reinforced with nonwoven fiber can often be heated and press-formed in a manner similar to conventional sheet metal forming. Precut blanks are heated and placed between the halves of a matched metal mold that is mounted in a vertical press. Ribs, bosses, and contours can be formed in parts with essentially uniform thickness. Cycle times range from 25 to 50 seconds for most parts.

INJECTION MOLDING   The injection molding of fiber-reinforced plastics is a process that competes with metal die castings and offers comparable properties at considerably reduced weight. In the simplest variation, chopped or continuous fibers are placed in a mold cavity that is then closed and injected with resin. An improved method utilizes chopped fibers, up to 6 mm ($\frac{1}{4}$ in.) in length, which are premixed with the heated thermoplastic (often nylon) prior to injection. Another variation uses a feedstock of discrete pellets that have been manufactured by slicing continuous-fiber pultruded rods. The benefits of adding fiber reinforcement (compared to conventional plastic molding) include increased rigidity and impact strength, reduced possibility of brittle failure during impact, better dimensional stability at elevated temperatures and in humid environments, improved abrasion resistance, and better surface finish due to the reduced dimensional contraction and absence of related sink marks. The molding process is quite rapid, and the final parts can be both precise and complex.

BRAIDING, THREE-DIMENSIONAL KNITTING, AND THREE-DIMENSIONAL WEAVING   The primary causes of failure in lamination-type composites are interlaminar cracking and delamination (layer-separation) upon impact. To overcome these problems, the high-strength reinforcing fibers can also be interwoven into three-dimensional preforms by processes that include weaving, braiding, and stitching through the thickness of stacked two-dimensional performs. Resin is then injected into the assembly and the resultant product is cured for use. Complex shapes can be produced with the fiber orientations selected for optimum properties. Computers can be used to design and control the weaving, making the process less expensive than many of the more labor-intensive techniques.

Fabrication of Fiber-Reinforced Metal–Matrix Composites.   Continuous-fiber *metal–matrix composites* can be produced by variations of filament winding, extrusion, and pultrusion. Fiber-reinforced sheets can be produced by electroplating, plasma spray deposition coating, or vapor deposition of metal onto a fabric or mesh, which are then shaped and bonded. Diffusion bonding of foil–fabric sandwiches, roll bonding, and coextrusion are other means of producing fiber-reinforced metal products. Various casting processes have been adapted to place liquid metal around the fibers by means of capillary action, gravity, pressure (die casting and squeeze casting), or vacuum (countergravity casting). Products that incorporate discontinuous fibers can also be produced by powder metallurgy or spray-forming techniques and further fabricated by hot pressing, superplastic forming, forging, or some types of casting. In general, efforts are made to reduce or eliminate the need for finish machining, which would require the use of diamond or carbide tools, or methods such as EDM.

In terms of properties, graphite-reinforced aluminum has been shown to be twice as stiff as steel and one-third to one-fourth the weight, with practically zero thermal expansion. Aluminum reinforced with silicon carbide exhibits increased strength (tension, compression, and shear at both room and elevated temperature), as well as increased hardness, fatigue strength, and elastic modulus. Thermal creep and thermal expansion are both reduced, but ductility, thermal conductivity, and electrical conductivity are also decreased. Magnesium, copper, and titanium alloys as well as the superalloys have also been used as the matrix in fiber-reinforced metal–matrix composites.

Fabrication of Fiber-Reinforced Ceramic–Matrix Composites.

Unlike polymeric– or metal–matrix composites, where failures originate in or along the reinforcement fibers, *ceramic–matrix composites* often fail due to flaws in the matrix. If the reinforcement is bonded strongly to the matrix, a matrix crack will propagate right through to the fibers. To impart toughness, therefore, it is often desirable to promote a weak bond between the fiber and matrix. Cracks are then deflected along the fiber–matrix interface rather than through the fiber.

The matrix materials and reinforcement fibers for ceramic–ceramic composites have been discussed in Chapter 8, along with some of the unique property combinations that can be achieved. Fabrication techniques for the ceramic–matrix composites are often quite different from the other composite families. One approach is to pass the fibers or mats through a slurry mixture that contains the matrix material. The impregnated material is then dried, assembled, and fired. Other techniques include the chemical vapor deposition or chemical vapor infiltration of a coated fiber base, where the coating serves to weaken an otherwise strong bond. Silicon nitride matrices can be formed by reaction bonding. The reinforcing fibers are dispersed in silicon powder, which is then reacted with nitrogen. Hot-pressing techniques can also be used with the various ceramic matrices. When the matrix is a glass, the heated material behaves much like a polymer, and the processing methods are often similar to those used for polymer–matrix composites.

Secondary Processing and Finishing of Fiber-Reinforced Composites.

The various fiber-reinforced composites can often be processed further with conventional equipment (sawed, drilled, routed, tapped, threaded, turned, milled, sanded, and sheared), but special considerations should be exercised. Cutting some materials may be like cutting multilayer cloth, and precautions should be used to prevent the formation of splinters and cracks as well as frayed or delaminated edges. Sharp tools, high speeds, and low feeds are generally required. Cutting debris should be removed quickly to prevent the cutters from becoming clogged.

In addition, many of the reinforcing fibers are extremely abrasive and quickly dull most conventional cutting tools. Diamond or polycrystalline diamond tooling may be required to achieve realistic tool life. Abrasive slurrys can be used in conjunction with rigid tooling to assure the production of smooth surfaces. Lasers and water-jets are alternative cutting tools. Lasers, however, can burn or carbonize the material or produce undesirable heat-affected zones. Water-jets can create moisture problems with some plastic resins.

When fiber-reinforced materials must be joined, the major concern is the lack of continuity of the fibers in the joint area. Thermoplastics can be softened and welded by applying pressure with heated tools, combining pressure and ultrasonic vibration, or using pressure and induction heating. Thermoset materials generally require the use of mechanical joints or adhesives, with each method having its characteristic advantages and limitations. Metal–matrix composites are often brazed.

■ KEY WORDS

adhesive bonding
annealing
autoclave
blow molding
braiding
bulk-molding
  compound
calendering
casting
cementation
ceramics
ceramic–matrix
  composites
clay products

cold molding
composites
compression molding
crystalline ceramics
devitrification
dipping
dry pressing
elastomer
explosive bonding
extrusion
fibers
filament winding
firing
foam molding

glass
glass ceramic
hand layup
hot-isostatic pressing
injection molding
inserts
isostatic pressing
lamination
laser sintering
mats
metal–matrix composites
open-mold processing
parison
plastics

prepregs
pressure-bag molding
pultrusion
reaction injection
  molding
resin-transfer
  molding
roll bonding
rotational molding
rovings
sandwich structures
sheet-molding
  compound
sintering

slip casting
spinning
spray molding
tapes
tempered glass
thermoforming
thermoplastic polymer
thermosetting polymer
tows
transfer molding
vacuum-bag molding
viscous flow
vitrification
yarns

■ REVIEW QUESTIONS

1. Why are the fabrication processes applied to plastics, ceramics, and composites often different from those applied to metals? What are some of the key differences?
2. How does the fabrication of a thermoplastic polymer differ from the processing of a thermosetting polymer?
3. What are some of the ways that plastic sheet, plate, and tubing can be cast?
4. Why do cast plastic resins typically have a lustrous appearance?
5. What types of polymers are most commonly blow molded?
6. Why do blow molding molds typically contain a cooling system?
7. For what types of parts and production volumes would compression molding be an appropriate process?
8. What are typical mold temperatures for compression molding? What is the most common mold material?
9. What are some of the attractive features of the transfer molding process?
10. What is the most widely used process for the fabrication of thermoplastic materials (in terms of number of parts produced)?
11. In what ways is injection molding of plastic similar to the die casting of metal?
12. What is the benefit of a hot runner distribution system in plastic injection molding?
13. Why is the cycle time for the injection molding of thermosetting polymers significantly longer than that for the thermoplastics?
14. How are the individual components mixed in the reaction injection molding process?
15. What are some of the attractive consequences of the low temperatures and low pressures of the reaction injection molding process?
16. What are some of the typical production shapes that are produced by the extrusion of plastics?
17. What are some attractive features of a twin-screw extruder design?
18. For what types of materials and products might thermoforming be considered to be attractive?
19. What types of products are produced by rotational molding?
20. What is the difference between open-cell and closed-cell foamed plastics?
21. What are some typical applications for rigid-type foamed plastics?
22. What type of products are produced by the spinning process?
23. What are some of the general properties of plastics that affect their machinability?
24. What property of plastics is responsible for making snap-fit assembly a popular alternative for plastic products?
25. What are some of the attractive properties of plastics that favor their selection? What are some of the common limitations?
26. Why should adequate fillets be included between adjacent sections of a mold? What is a major benefit of rounding exterior corners?
27. Why is it most desirable to have uniform wall thickness in plastic products?
28. Why are product dimensions less precise when they cross a mold parting line?
29. Why might threaded inserts be preferred over other means of producing threaded holes in a plastic component?
30. What are some of the ways in which metal inserts are held in place in a plastic part?
31. When designing a decorative surface (design or lettering) on a plastic product, why is it desirable that the details be raised on the product rather than depressed?
32. Why does locating a parting line on a sharp corner make that feature less noticeable?
33. What is the benefit of countersinking holes that are to be threaded or used for self-tapping screws?
34. What types of products can be produced from elastomeric materials using the dipping process?
35. What process or equipment is used to form rubber compounds into sheets?
36. What are the two basic classes of ceramic materials, and how does their processing differ?
37. What are some of the special heat treatment operations performed on glass products?
38. What are glass-ceramics?
39. What are some of the techniques that can be used to impart some degree of plasticity to crystalline ceramic materials?
40. Describe the differences between the injection molding of plastics and the injection molding of ceramics.
41. What is the purpose of the firing or sintering operations in the processing of crystalline ceramic products?
42. What are the benefits and limitations of machining ceramic materials before firing versus after firing?
43. Why are joining operations usually avoided when fabricating products from ceramic materials?
44. Discuss some of the design guidelines that relate to the production of parts from ceramic material.

45. Why are the processes used to fabricate particulate composites essentially the same as those used for conventional material?
46. What are some of the processes that can be used to produce a high-quality bond between the layers of a laminar composite?
47. What are some of the forms in which reinforcement fibers can be used in composite materials?
48. What is a prepreg?
49. What are sheet-molding compounds (SMC)? Bulk-molding compounds (BMC)?
50. In what way is pultrusion similar to wire drawing?
51. What are some typical products that are made by filament winding? Why might this process be attractive for the production of small quantities of large parts?

52. What are some of the various molding processes that can be used to shape products from laminated sheets of woven fibers?
53. What type of reinforcing fibers can be incorporated in the spray molding process? Injection molding?
54. What is the major benefit of three-dimensional fiber reinforcement?
55. Describe some of the ways in which a metal matrix can be introduced into a fiber-reinforced composite.
56. Why might it be desirable to weaken the bond between a reinforcing fiber and a ceramic–matrix material?
57. Discuss some of the techniques used to cut fiber-reinforced composites.
58. What is the major concern when considering the joining of fiber-reinforced composites?

■ **PROBLEMS**

1. Consider some of the more prominent sporting goods that are fabricated from composite materials, such as skis, tennis rackets, golf club shafts, and body panels for racing cars. For two products, identify appropriate composite materials and companion shape-producing fabrication methods.

www.wiley.com/college/degarmo

*C*hapter 20   CASE STUDY

Fabrication of Lavatory Wash Basins

Lavatory wash basins (bathroom sinks) have been successfully made from a variety of engineering materials, including cast iron, steel, stainless steel, ceramics, and polymers (such as melamine). Your company, Diversified Household Products, Inc., is considering a possible entrance into this market and has assigned you the tasks of (1) assessing the competition and (2) recommending the "best" approach toward producing this product.

1. For each of the materials (or families of materials), describe the material properties that are attractive for a washbasin application. What are the primary limitations or disadvantages?
2. For each of the materials (or families of materials), describe possible means of fabricating lavatory wash basins. Consider sheet metal forming, casting, molding, joining, and other types of fabrication processes. If multiple options exist, which one do you consider to be most attractive? Comment on the attractive features of

the proposed system (materials and process) as well as the relative quality and cost.
3. Wash basins generally require a surface that is nonporous and stain resistant, scratch resistant, corrosion resistant, and attractive (and possibly available in a variety of colors). One approach to providing these properties on a steel or cast iron substrate is a coating of porcelain enamel. For each of the systems discussed in Question 2, discuss the need for additional surface treatment. What type of treatment would you recommend?
4. Most sinks contain an overflow feature that diverts excess water to the drain at a location beneath the stoppered basin. Discuss how this feature can be incorporated into each of your material-process manufacturing systems.
5. If your company were to consider producing lavatory wash basins on a competitive basis, which of the alternative manufacturing systems (material and manufacturing process) would you recommend? What features make it the most attractive?

CHAPTER 21

FUNDAMENTALS OF MACHINING/ ORTHOGONAL MACHINING

■ 21.1 INTRODUCTION

Machining is the process of removing unwanted material from a workpiece in the form of chips. If the workpiece is metal, the process is often called *metal cutting* or *metal removal*. U.S. industries annually spend $60 billion to perform metal removal operations because the vast majority of manufactured products require machining at some stage in their production, ranging from relatively rough or nonprecision work, such as cleanup of castings or forgings, to high-precision work involving tolerances of 0.0001 in. or less and high-quality finishes. Thus machining undoubtedly is the most important of the basic manufacturing processes.

Beginning with the work of F. W. Taylor at Midvale steel in the 1880s, the process has been the object of considerable research and experimentation that have led to improved understanding of the nature of both the process itself and the surfaces produced by it. While this research effort led to marked improvements in machining productivity, the complexity of the process has resulted in slow progress in obtaining a complete theory of chip formation.

What makes this process so unique and difficult to analyze?

- Prior workhardening greatly affects the process.
- Different materials behave differently.
- The process is asymmetrical and unconstrained, bounded only by the cutting tool.
- The level of strain is very large.
- The strain rate is very high.
- The process is sensitive to variations in tool geometry, tool material, temperature, environment (cutting fluids), and process dynamics (chatter and vibration).

The objective of this chapter is to put all this in perspective for the practicing engineer.

■ 21.2 FUNDAMENTALS

The process of metal cutting is complex because it has such a wide variety of inputs which are listed in Figure 21-1. The variables are:

- The machine tool selected to perform the process
- The cutting tool selected (geometry and material)
- The properties and parameters of the workpiece
- The cutting parameters selected (speed, feed, depth of cut)
- The workpiece holding devices or fixtures or jigs

As we can see from Figure 21-1, the wide variety of inputs creates a host of outputs most of which are critical to satisfactory performance of the component and product.

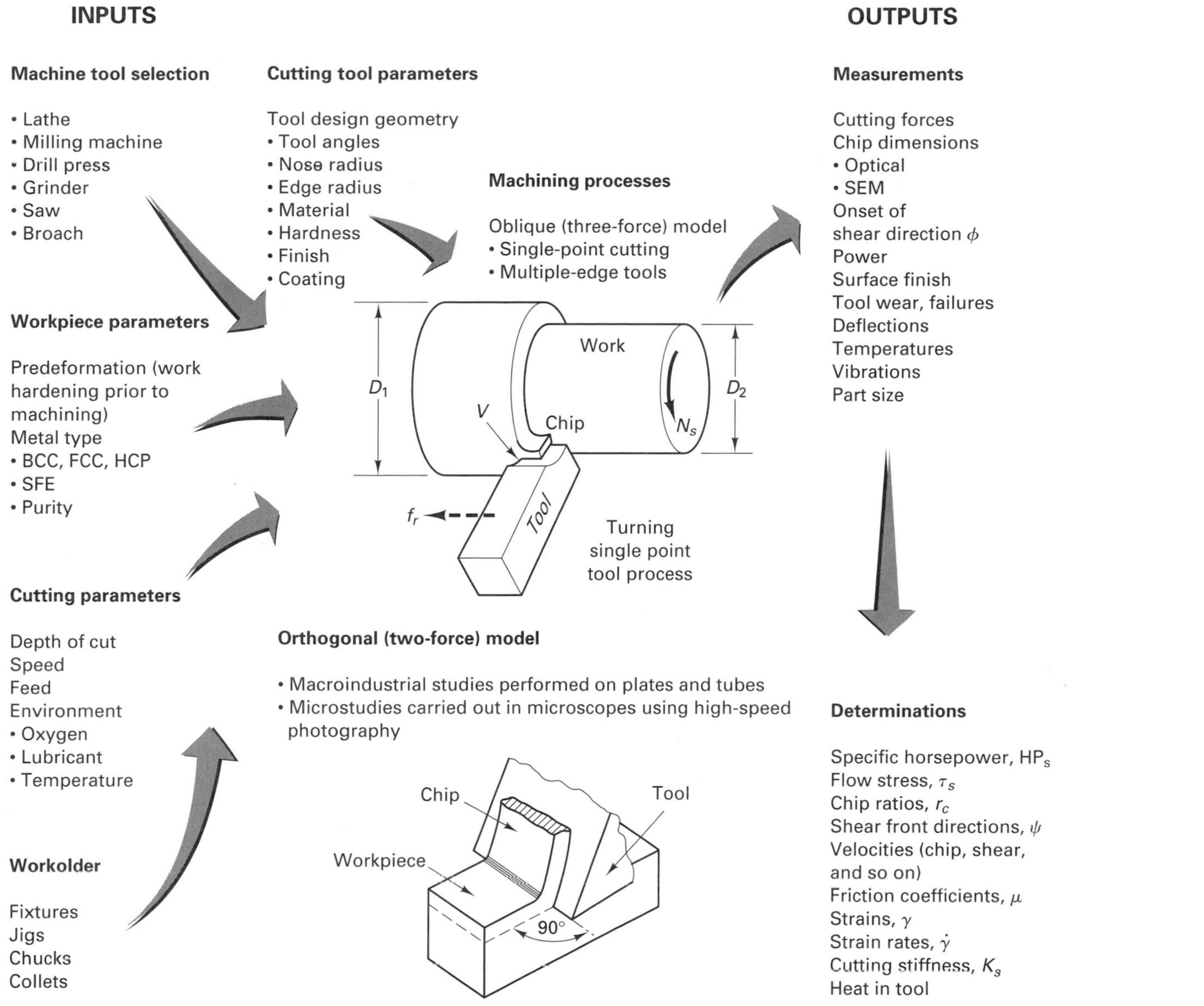

INPUTS

Machine tool selection

- Lathe
- Milling machine
- Drill press
- Grinder
- Saw
- Broach

Workpiece parameters

Predeformation (work hardening prior to machining)
Metal type
- BCC, FCC, HCP
- SFE
- Purity

Cutting parameters

Depth of cut
Speed
Feed
Environment
- Oxygen
- Lubricant
- Temperature

Workolder

Fixtures
Jigs
Chucks
Collets

Cutting tool parameters

Tool design geometry
- Tool angles
- Nose radius
- Edge radius
- Material
- Hardness
- Finish
- Coating

Machining processes

Oblique (three-force) model
- Single-point cutting
- Multiple-edge tools

Orthogonal (two-force) model

- Macroindustrial studies performed on plates and tubes
- Microstudies carried out in microscopes using high-speed photography

OUTPUTS

Measurements

Cutting forces
Chip dimensions
- Optical
- SEM
Onset of shear direction ϕ
Power
Surface finish
Tool wear, failures
Deflections
Temperatures
Vibrations
Part size

Determinations

Specific horsepower, HP_s
Flow stress, τ_s
Chip ratios, r_c
Shear front directions, ψ
Velocities (chip, shear, and so on)
Friction coefficients, μ
Strains, γ
Strain rates, $\dot{\gamma}$
Cutting stiffness, K_s
Heat in tool

FIGURE 21-1 The fundamental inputs and outputs to machining processes.

There are seven basic chip formation processes (see Figure 21-2): *turning, milling, drilling, sawing, broaching, shaping (planing), and grinding (abrasive machining)* discussed in Chapters 23–27. Part 2 of the book describes work materials and Chapter 22 will provide additional insights on cutting tool materials and tool geometry. Usually the workpiece material is determined by the design engineer to meet the functional requirements of the part in service. The manufacturing engineer will often have to select cutting tool parameters, and workholder parameters and then cutting parameters based on that work material decision. Let us begin with the assumption that the workpiece material has been selected and you have decided to use a high-speed steel cutting tool for a turning operation.

For all metal-cutting processes, it is necessary to distinguish between speed, feed, and depth of cut. The turning process will be used to introduce these terms. See Figure 21-3. In general, *speed* (V) is the primary cutting motion, which relates the velocity of the cutting tool relative to the *workpiece*. It is generally given in units of surface feet per minute (sfpm), inches per minute (in./min), or meters per minute (m/m) or meters per second (m/s). *Speed* (V) is shown with the heavy dark arrow. *Feed* (f_r) is the amount of material removed per revolution or per pass of the tool over the workpiece. In turning, feed is in inches per revolution, and the tool feeds parallel to the rotational axis of the workpiece. Depending

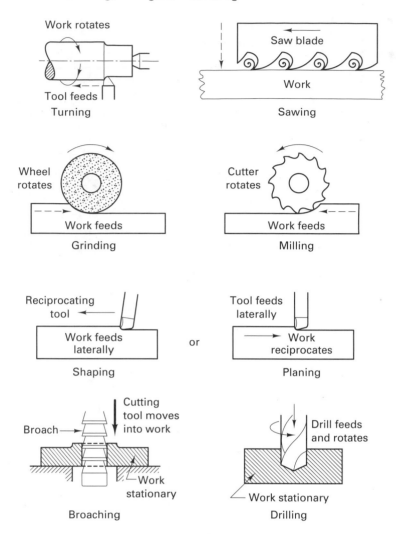

FIGURE 21-2 The seven basic machining processes used in chip formation.

on the process, feed units are inches per revolution, inches per cycle, inches per minute, or inches per tooth. Feed is shown with dashed arrows. The *depth of cut* (DOC), represents the third dimension. In turning, it is the distance the tool is plunged into the surface. It is half the difference in the diameter D_1, the initial diameter, and D_2, the final diameter:

$$\text{DOC} = \frac{D_1 - D_2}{2} = d \tag{21-1}$$

The selection of the cutting speed V determines the surface speed of the rotating part that is related to the outer diameter of the workpiece.

$$V = \frac{\pi D_1 N_s}{12} \tag{21-2}$$

where D_1 is in inches, V is speed in surface feet per minute, and N_s is the revolutions per minute (rpm) of the workpiece. The input to the lathe will be in revolutions per minute of the spindle.

Figure 21-3 shows a typical *machine tool* for the turning process, a lathe. Workpieces are held in *workholding devices*. (See Chapter 29 for details on the design of workholders.) In this example, a three-jaw chuck is used to hold the workpiece and rotate it against the tool. The chuck is attached to the spindle, which is driven through gears by the motor. The *cutting tool* is used to machine the workpiece and is the most critical component. The geometry of a single point (single cutting edge) of a typical high-speed steel tool used in turning is found in Chapter 22. Various tool geometry is usually ground onto HSS blanks, depending on what material is being machined. Cutting tool material and geometry must be selected before speed and feed can be determined, as shown in Figure 21-4. This table, taken from *Metcut's Machinability Data Handbook*, gives the MfE/IE starting values for cutting speed

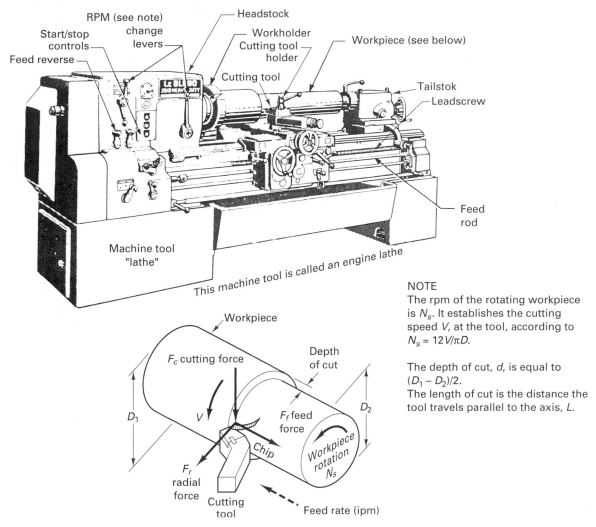

FIGURE 21-3   Turning a cylindrical workpiece on a lathe requires you select the cutting speed, feed, and depth of cut.

NOTE
The rpm of the rotating workpiece is N_s. It establishes the cutting speed V, at the tool, according to $N_s = 12V/\pi D$.

The depth of cut, d, is equal to $(D_1 - D_2)/2$.
The length of cut is the distance the tool travels parallel to the axis, L.

(sfpm or m/min) and feed (ipr or mm/r) for a given depth of cut, a given work material (hardness), and a given process (turning). Notice how speed decreases as DOC or feed increases, and cutting speeds increase with carbide and coated carbide tool materials.

To process different metals, the input parameters to the machine tools must be determined. For the lathe, the input parameters are DOC, the feed, and the rpm value of the spindle. The rpm value depends on the selection of the cutting speed V. Rewriting equation for N_s:

$$N_s = \frac{12V}{\pi D_1} \cong \frac{3.8V}{D_1} \tag{21-3}$$

Cutting speed, feed, and DOC selection depend on many factors, and a great deal of experience and experimentation are required to find the best combinations. A good place to begin is by consulting tables of recommended values as shown in Figure 21-4. Most tables are arranged according to the process being used, the material being machined, the hardness, and the cutting tool material. The table given is a sample, to be used only for solving turning problems in the book. For industrial calculations, standard references listed in the appendix or cutting tool manufacturers should be consulted.

This table is for turning processes only. The amount of metal removed per pass determines the DOC. In practice, roughing cuts are heavier than finishing cuts in terms of feed and DOC and are run at a lower surface speed. Note that this table provides recommendations of V and f_r in both English and metric units based on the DOC needed to perform the job. Table values are usually conservative and should be considered starting points for determining the operational parameters for a process.

Turning, Single Point and Box Tools

Material	Hard-ness Bhn	Condition	Depth of Cut* in / mm	HSS Speed fpm/m/min	HSS Feed ipr/mm/r	HSS Tool Material AISI/ISO	Uncoated Brazed fpm/m/min	Uncoated Indexable fpm/m/min	Uncoated Feed ipr/mm/r	Uncoated Tool Grade C/ISO	Coated Speed fpm/m/min	Coated Feed ipr/mm/r	Coated Tool Grade C/ISO
1. FREE MACHINING CARBON STEELS, WROUGHT (cont.) Medium Carbon Leaded (cont.) (materials listed on preceding page)	225 to 275	Hot Rolled, Normalized, Annealed, Cold Drawn or Quenched and Tempered	.040	160	.008	M2, M3	500	610	.007	C-7	925	.007	CC-7
			.150	125	.015	M2, M3	390	480	.020	C-6	600	.015	CC-6
			.300	100	.020	M2, M3	310	375	.030	C-6	500	.020	CC-6
			.625	80	.030	M2, M3	.240	290	.040	C-6	—	—	—
			1	**49**	**.20**	**S4, S5**	**150**	**185**	**.18**	**P10**	**280**	**.18**	**CP10**
			4	**38**	**.40**	**S4, S5**	**120**	**145**	**.50**	**P20**	**185**	**.40**	**CP20**
			8	**30**	**.50**	**S4, S5**	**95**	**115**	**.75**	**P30**	**150**	**.50**	**CP30**
			16	**24**	**.75**	**S4, S5**	**73**	**88**	**1.0**	**P40**	—		
	275 to 325	Hot Rolled, Normalized, Annealed or Quenched and Tempered	.040	135	.007	T15, M42†	460	545	.007	C-7	825	.007	CC-7
			.150	105	.015	T15, M42†	350	425	.020	C-6	525	.015	CC-6
			.300	85	.020	T15, M42†	275	380	.030	C-6	425	020	CC-6
			.625	—	—	—	—	—	—	—	—	—	—
			1	**41**	**.18**	**S9, S11†**	**140**	**165**	**.18**	**P10**	**250**	**.18**	**CP10**
			4	**32**	**.40**	**S9, S11†**	**105**	**130**	**.50**	**P20**	**160**	**.40**	**CP20**
			8	**26**	**.50**	**S9, S11†**	**84**	**100**	**.75**	**P30**	**130**	**.50**	**CP30**
			16	—	—	—	—	—	—		—		
	325 to 375	Quenched and Tempered	.040	100	.007	T15, M42†	390	480	.007	C-7	725	.007	CC-7
			.150	80	.015	T15, M42†	300	375	.020	C-6	475	.015	CC-6
			.300	65	.020	T15, M42†	230	290	.030	C-6	375	.020	CC-6
			.625	—	—	—	—	—	—	—	—	—	—
			1	**30**	**.18**	**S9, S11†**	**120**	**145**	**.18**	**P10**	**220**	**.18**	**CP10**
			4	**24**	**.40**	**S9, S11†**	**90**	**115**	**.50**	**P20**	**145**	**.40**	**CP20**
			8	**20**	**.50**	**S9, S11†**	**70**	**88**	**.75**	**P30**	**115**	**.50**	**CP30**
			16	—	—	—	—	—	—		—		
	375 to 425	Quenched and Tempered	.040	70	.007	T15, M42†	325	400	.007	C-7	600	.007	CC-7
			.150	55	.015	T15, M42†	250	310	.020	C-6	400	.015	CC-6
			.300	45	.020	T15, M42†	200	240	.030	C-6	325	.020	CC-6
			.625	—	—	—	—	—	—	—	—	—	—
			1	**21**	**.18**	**S9, S11†**	**100**	**120**	**.18**	**P10**	**185**	**.18**	**CP10**
			4	**17**	**.40**	**S9, S11†**	**76**	**95**	**.50**	**P20**	**120**	**.40**	**CP20**
			8	**14**	**.50**	**S9, S11†**	**60**	**73**	**.75**	**P30**	**100**	**.50**	**CP30**
			16	—	—	—	—	—	—		—		
2. CARBON STEELS, WROUGHT Low Carbon 1005 1010 1020 1006 1012 1023 1008 1015 1025 1009 1017	85 to 125	Hot Rolled, Normalized, Annealed or Cold Drawn	.040	185	.007	M2, M3	535	700	.007	C-7	1050	.007	CC-7
			.150	145	.015	M2, M3	435	540	.020	C-6	700	.015	CC-6
			.300	115	.020	M2, M3	340	420	.030	C-6	550	.020	CC-6
			.625	90	.030	M2, M3	265	330	.040	C-6	—	—	—
			1	**56**	**.18**	**S4, S5**	**165**	**215**	**.18**	**P10**	**320**	**.18**	**CP10**
			4	**44**	**.40**	**S4, S5**	**135**	**165**	**.50**	**P20**	**215**	**.40**	**CP20**
			8	**35**	**.50**	**S4, S5**	**105**	**130**	**.75**	**P30**	**170**	**.50**	**CP30**
			16	**27**	**.75**	**S4, S5**	**81**	**100**	**1.0**	**P40**	—	—	—
	125 to 175	Hot Rolled, Normalized, Annealed or Cold Drawn	.040	150	.007	M2, M3	485	640	.007	C-7	950	.007	CC-7
			.150	125	.015	M2, M3	410	500	.020	C-6	625	.015	CC-6
			.300	100	.020	M2, M3	320	390	.030	C-6	500	.020	CC-6
			.625	80	.030	M2, M3	245	305	.040	C-6	—	—	—
			1	**46**	**.18**	**S4, S5**	**150**	**195**	**.18**	**P10**	**290**	**.18**	**CP10**
			4	**38**	**.40**	**S4, S5**	**125**	**150**	**.50**	**P20**	**190**	**.40**	**CP20**
			8	**30**	**.50**	**S4, S5**	**100**	**120**	**.75**	**P30**	**150**	**.50**	**CP30**
			16	**24**	**.75**	**S4, S5**	**75**	**95**	**1.0**	**P40**	—	—	—
	175 to 225	Hot Rolled, Normalized, Annealed or Cold Drawn	.040	145	.007	M2, M3	460	570	.007	C-7	850	.007	CC-7
			.150	115	.015	M2, M3	385	450	.020	C-6	550	.015	CC-6
			.300	95	.020	M2, M3	300	350	.030	C-6	450	.020	CC-6
			.625	75	.030	M2, M3	235	265	.040	C-6	—	—	—
			1	**44**	**.18**	**S4, S5**	**140**	**175**	**.18**	**P10**	**260**	**.18**	**CP10**
			4	**35**	**.40**	**S4, S5**	**115**	**135**	**.50**	**P20**	**170**	**.40**	**CP20**
			8	**29**	**.50**	**S4, S5**	**90**	**105**	**.75**	**P30**	**135**	**.50**	**CP30**
			16	**23**	**.75**	**S4, S5**	**72**	**81**	**1.0**	**P40**	—	—	—
	225 to 275	Annealed or Cold Drawn	.040	125	.007	M2, M3	410	510	.007	C-7	750	.007	CC-7
			.150	95	.015	M2, M3	360	400	.020	C-6	500	.015	CC-6
			.300	75	.020	M2, M3	285	315	.030	C-6	400	.020	CC-6
			.625	60	.030	M2, M3	220	240	.040	C-6	—	—	—
			1	**38**	**.18**	**S4, S5**	**125**	**155**	**.18**	**P10**	**230**	**.18**	**CP10**
			4	**29**	**.40**	**S4, S5**	**110**	**120**	**.50**	**P20**	**150**	**.40**	**CP20**
			8	**23**	**.50**	**S4, S5**	**87**	**95**	**.75**	**P30**	**120**	**.50**	**CP30**
			16	**18**	**.75**	**S4, S5**	**67**	**73**	**1.0**	**P40**	—	—	—

See section 15.1 for Tool Geometry.
*Caution: Check Horsepower requirements on heavier depths of cut.

See section 16 for Cutting Fluid Recommendations.
†Any premium HSS (T15, M33, M41–M47) or (S9, S10, S11, S12).

FIGURE 21-4 Examples of a table for selection of speed and feed for turning. (*Source: Metcut's Machinability Data Handbook.*)

Once cutting speed V has been selected, Equation (21-3) is used to determine the spindle rpm, N_s. The speed and feed can be used with the DOC to estimate the metal removal rate for the process, or MRR. For turning, the MRR is

$$\text{MRR} \cong 12Vf_r d \qquad (21\text{-}4)$$

This is an approximate equation for MRR. For turning, MRR values can range from 0.1 to 600 in^3/min. The MRR can be used to estimate the horsepower needed to perform a cut, as will be shown later. For most processes, the MRR equation can be viewed as the volume of metal removed divided by the time needed to remove it.

$$\text{MRR} = \frac{\text{volume of cut}}{T_m}$$

where T_m is the cutting time in minutes. For turning, the cutting time depends upon the length of cut L divided by the rate of traverse of the cutting tool past the rotating workpiece $f_r N_s$ as shown in Figure 21-5. Therefore,

$$T_m = \frac{L + \text{allowance}}{f_r N_s} \qquad (21\text{-}5)$$

An allowance is usually added to the L term to allow for the tool to enter and exit the cut.

Turning is an example of a single-point tool process, as is shaping. Milling and drilling are examples of multiple-point tool processes. Figures 21-5 through 21-9 show the basic process schematically. Speed (V) is shown in these figures with a dark heavy arrow. Feed (f) is the amount of material removed per pass of the tool over the workpiece and is shown as a dashed arrow.

FIGURE 21-5 Relationship of speed, feed, and depth of cut in turning, boring, facing, and cutoff operations typically done on a lathe.

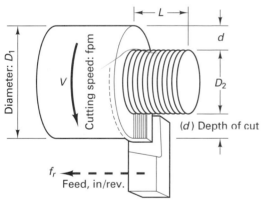

Turning

Speed, stated in surface feet per minute (sfpm), is the peripheral speed at the cutting edge. Feed per revolution in turning is a linear motion of the tool parallel to the rotating axis of the workpiece. The depth of cut reflects the third dimension.

L = length of cut

$T_m = \dfrac{L + A}{f_r N_s}$

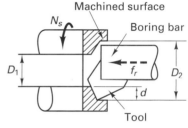

Boring

Enlarging hole of diameter D_1 to diameter D_2. Boring can be done with multiple cutting tools. Feed in inches per revolution, f_r.

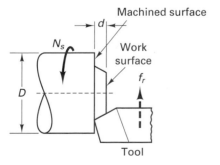

Facing

Tool feeds to center of workpiece so $L = D/2$. The cutting speed is decreasing as the tool approaches the center of the workpiece.

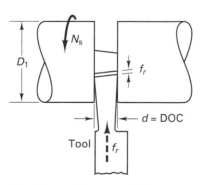

Grooving, parting or cutoff

Tool feed perpendicular to the axis of rotation. The width of the tool produces the depth of cut (DOC).

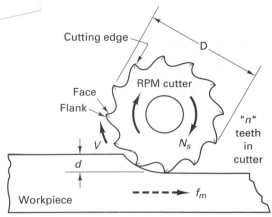

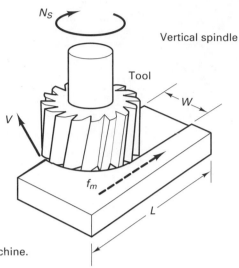

Slab milling – multiple tooth

Slab milling is usually performed on a horizontal milling machine. Equations for T_m and MRR derived in Chapter 25.

The tool rotates at rpm N_s. The workpiece translates past the cutter at feed rate f_m, the table feed. The length of cut, L, is the length of workpiece plus allowance, L_A,

$$L_A = \sqrt{\frac{D^2}{4} - \left(\frac{D}{2} - d\right)^2} = \sqrt{d(D-d)} \text{ inches}$$

$$T_m = (L + L_A)/f_m$$

The MRR = Wdf_m where W = width of the cut and d = depth of cut.

Face milling
Multiple tooth cutting

Given a selected cutting speed V and a feed per tooth f_t, the rpm of the cutter is $N_s = 12V/\pi D$ for a cutting of diameter D. The table feed rate is $f_m = f_t n N_s$ for a cutter with n teeth.
The cutting time, $T_m = (L + L_A + L_o)/f_m$
where $L_o = L_A = \sqrt{W(D - W)}$ for $W < D/2$
  or $L_o = L_A = D/2$ for $W \geqslant D/2$.
The MRR = Wdf_m where d = depth of cut.

FIGURE 21-6   Basics of milling processes (slab, face and end milling) including equations for cutting time and metal-removal rate (MRR).

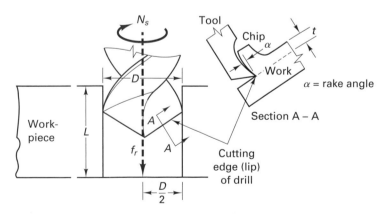

Drilling-multiple edge tool

Select cutting spaced V, fpm and feed, f_r, in./rev.
D = diameter of the drill which rotates 2 cutting edges at rpm N_s. V = velocity of outer edge of the lip of the drill.
$N_s = 12V/\pi D$. T_m = cutting time = $(L + A)/f_r N_s$ where f_r is the feed rate in in. per rev. The allowance $A = D/2$.
The MRR = $(\pi D^2/4) f_r N_s$ in.^{3} /min which is approximately $3DVf_r$.

FIGURE 21-7   Basics of the drilling (hole-making) processes, including equations for cutting time T_m and metal-removal rate (MRR).

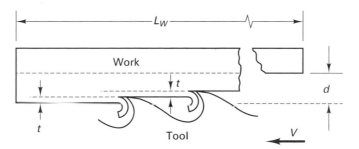

The T_m for broaching is $T_m = L/12V$. The MRR (per tooth) is $12tWV$ in³/min where $V =$ cutting velocity in fpm, W is the width of cut, t – rise per tooth.

FIGURE 21-8 Process basics of broaching. Equations for cutting time and metal-removal rate (MRR), developed in Chapter 26.

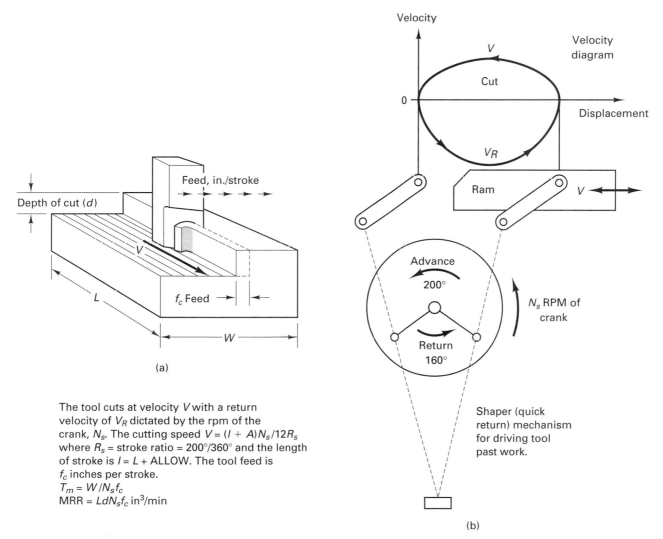

The tool cuts at velocity V with a return velocity of V_R dictated by the rpm of the crank, N_s. The cutting speed $V = (l + A)N_s/12R_s$ where R_s = stroke ratio = 200°/360° and the length of stroke is $l = L + $ ALLOW. The tool feed is f_c inches per stroke.

$T_m = W/N_s f_c$

MRR $= LdN_s f_c$ in³/min

Shaper (quick return) mechanism for driving tool past work.

FIGURE 21-9 (a) Basics of the shaping process, including equations for cutting time (T_m) and metal-removal rate (MRR). (b) The relationship of the crank rpm N_s to the cutting velocity V.

TABLE 21-1.	Shop Formulas for Turning, Milling, Drilling, and Broaching (English Units)			
Parameter	Turning	Milling	Drilling	Broaching
Cutting speed, fpm	$V = 0.262 \times D_1 \times \text{rpm}$	$V = 0.262 \times D_m \times \text{rpm}$	$V = 0.262 \times D_d \times \text{rpm}$	V
Revolutions per minute, N_s	$\text{rpm} = 3.82 \times V_c/D_t$	$\text{rpm} = 3.82 \times V_c/D_m$	$\text{rpm} = 3.82 \times V_c/D_d$	—
Feed rate, in./min	$f_m = f_r \times \text{rpm}$	$f_m = f_t \times \text{rpm}$	$f_m = f_r \times \text{rpm}$	—
Feed per rev tooth pass, in./rev	f_r	f_t	f_r	—
Cutting time, min, T_m	$T_m = L/f_m$	$T_m = L/f_m$	$T_m - L/f_m$	$T_m = L/12V$
Rate of metal removal, in³/min	$\text{MRR} = 12 \times d \times f_r \times V_c$	$\text{MRR} = w \times d \times f_m$	$\text{MRR} = \pi D^2 d/4 \times f_m$	$\text{MRR} = 12 \times w \times d \times V$
Horsepower required at spindle	$\text{hp} = \text{MRR} \times \text{HP}_s$	$\text{hp} = \text{MRR} \times \text{HP}_s$	$\text{hp} = \text{MRR} \times \text{HP}_s$	—
Horsepower required at motor	$\text{hp}_m = \text{MRR} \times \text{HP}_s/E$	$\text{hp}_m = \text{MRR} \times \text{HP}_s/E$	$\text{hp}_m = \text{MRR} \times \text{HP}_s/E$	$\text{hp}_m = \text{MRR} \times \text{HP}_s/E$
Torque at spindle	$t_1 = 63{,}030 \, \text{hp/rpm}$	$t_1 = 63{,}030 \, \text{hp/rpm}$	$t_1 = 63{,}030 \, \text{hp/rpm}$	—
Symbols	D_1 = Diameter of workpiece in turning, inches D_m = Diameter of milling cutter, inches D_d = Diameter of drill, inches d = Depth of cut, inches E = Efficiency of spindle drive f_m = Feed rate, inches per minute f_r = Feed, inches per revolution f_t = Feed, inches per tooth hp_m = Horsepower at motor		hp = horsepower at spindle L = Length of cut, inches n = Number of teeth in cutter HP_s = Unit power, horsepower per cubic inch per minute, specific horsepower MRR = Revolution per minute of work or cutter, N_s t_s = Torque at spindle, inches-pound T_m = Cutting time, minutes V = Cutting speed, feet per minute w = Width of cut, inches	

Values for specific horsepower (unit power) are given in Table 21-4

For many of the basic processes, the equations for T_m and MRR are given. These equations are commonly referred to as *shop equations* and are as fundamental as the processes themselves, so the student should be as familiar with them as with the basic processes. If one keeps track of the units and visualizes the process, the equations are, for the most part, straightforward. See Table 21-1 for summary.

In addition to turning, other operations can be performed on the lathe. For example, as shown in Figure 21-5, a flat surface on the rotating part can be produced by facing or a cutoff operation. Boring can produce an enlarged hole, and grooving puts a slot in the workpiece.

The process of milling requires two figures because it takes different forms depending upon the selection of the machine tool and the cutting tool. Milling, a multiple-tooth process, has two feeds: the amount of metal an individual tooth removes, called the feed per tooth f_t, and the rate at which the table translates past the rotating tool, called the table feed rate f_m, in inches per minute. It is calculated from

$$f_m = f_t n N_s \tag{21-6}$$

where n is the number of teeth in a cutter and N_s is the rpm value of the cutter. Just as was shown for turning, standard tables of speeds and feeds for milling provide values for the recommended cutting speeds and feeds per tooth, f_t.

Figure 21-10 and Table 21-2 provide a summary of the basic machining processes in terms of typical machine tools which can perform the process, the typical sizes (min–max), the production rates (part/hour), tolerances (precision or repeatability) and surface finish (roughness). Milling has pretty much replaced shaping and planing, although gear shaping is still a viable process. Milling combined with other rotational multiple-edge tool processes (drilling or reaming) is often performed in machining centers rather than on milling machines. The turret lathe has been replaced by CNC turning centers with multiple turrets in many factories (see Chapter 32).

Operation	Block diagram	Most commonly used machines	Machines less frequently used	Machines seldom used
Turning	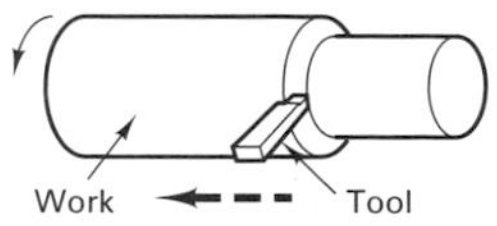	Lathe NC lathe machining center	Boring mill	Turret lathe
Grinding	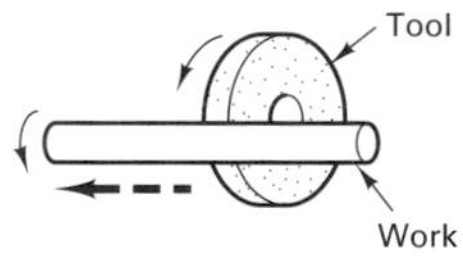	Cylindrical grinder		Lathe (with special attachment)
Sawing (of plates and sheets)	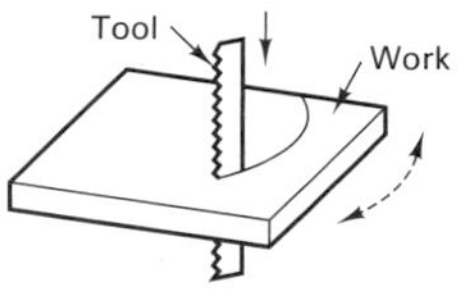	Contour or band saw	Laser Flame cutting Plasma arc	
Drilling	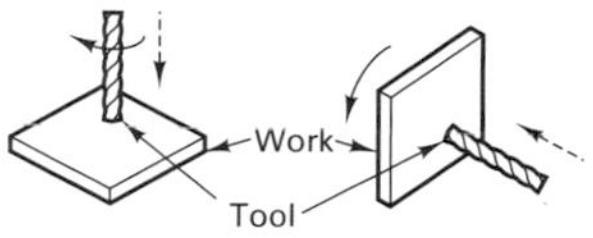	Drill press Machining center (nc) Vert. milling machine	Lathe Horizontal boring machine	Horizontal milling machine Boring mill
Boring	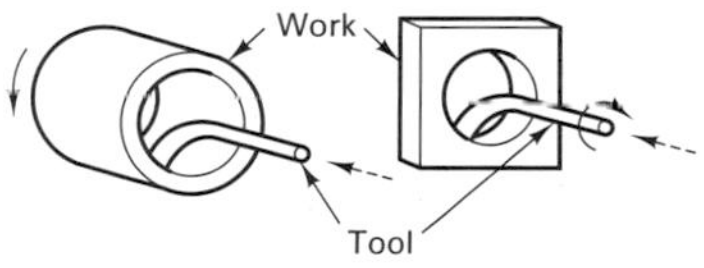	Lathe Boring mill Horizontal boring machine Machining center		Milling machine Drill press
Reaming	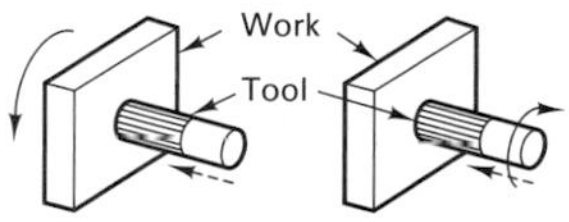	Lathe Drill press Boring mill Horizontal boring machine Machining center	Milling machine	
Grinding	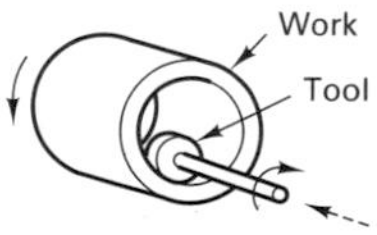	Cylindrical grinder		Lathe (with special attachment)
Sawing	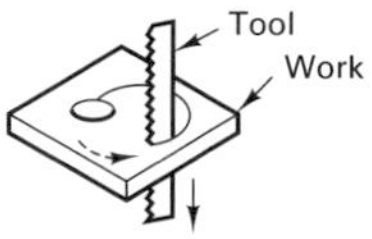	Contour or band saw		
Broaching	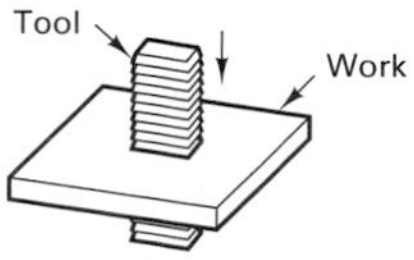	Broaching machine	Arbor press (keyway broaching)	

(continues on next page)

FIGURE 21-10 Operations and machines for machining cylindrical and/or flat surfaces.

Operation	Block diagram	Most commonly used machines	Machines less frequently used	Machines seldom used
Facing	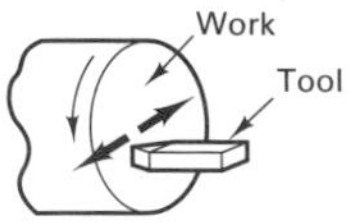	Lathe	Boring mill	
Broaching	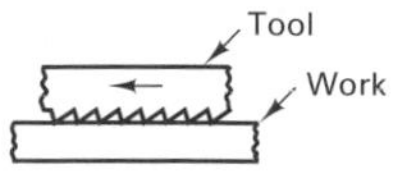	Broaching machine		Turret broach
Grinding	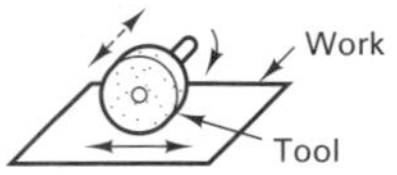	Surface grinder		Lathe (with special attachment)
Sawing	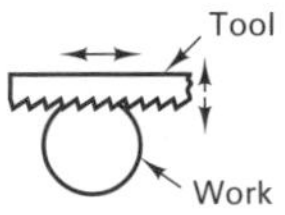	Cutoff saw	Contour saw	
Shaping	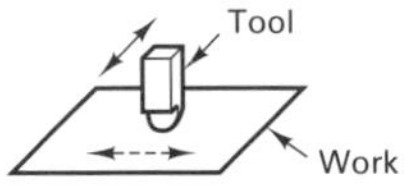	Horizontal shaper	Vertical shaper	
Planing	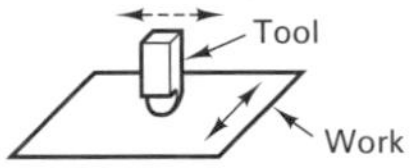	Planer		
Milling	slab milling 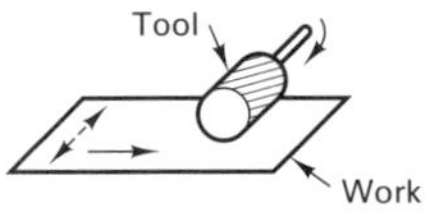	Milling machine	Lathe with special milling tools	
	face milling 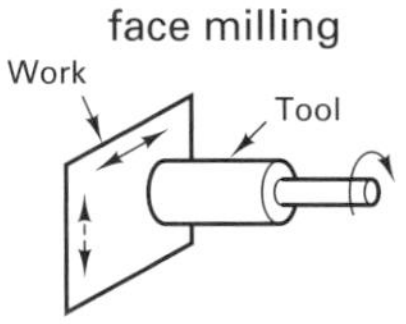	Milling machine Machining center	Lathe with special milling tools	Drill press (light cuts)

FIGURE 21-10   *(continued)*

■ 21.3 ENERGY AND POWER IN MACHINING

All of the process described to this point are examples of oblique, or three-force, cutting and were shown in Figure 21-2. The cutting force system in a conventional, oblique-chip formation process is shown schematically in Figure 21-11. Oblique cutting has three components:

1. F_c: Primary cutting force acting in the direction of the cutting velocity vector. This force is generally the largest force and accounts for 99% of the power required by the process.

2. F_f: Feed force acting in the direction of the tool feed. This force is usually about 50% of F_c but accounts for only a small percentage of the power required because feed rates are usually small compared to cutting speeds.

3. F_r: radial or thrust force acting perpendicular to the machined surface. This force is typically about 50% of F_f and contributes very little to power requirements because velocity in the radial direction is negligible.

The oblique cutting geometry shown in Figures 21-1 and 21-3 is repeated in Figure 21-11, which shows the general relationship between these forces and speed, feed, and depth of cut. Note that these figures cannot be used to determine forces for a specific process.

TABLE 21-2. Basic Machining Processes

Applicable Process	Raw Material Form	Size Maximum	Size Minimum	Typical Production Rate	Material Choice	Typical Tolerance	Typical Surface Roughness
Turning (engine lathes)	Cylinders, preforms, castings, forgings	78 in. dia. × 73 in. long	$\frac{1}{64}$ in. typical	1–10 parts/hour	All ferrous and nonferrous material considered machinable	±0.002 in. on dia. common; ±0.001 in. obtainable	125–250
Turning (CNC)	Bar, rod, tube, preforms	36 in. dia. × 93 in. long	$\frac{1}{64}$ in. dia.	1–2 parts/minute to 1–4 parts/hour	Any material with good machinability rating	±0.001 in. on dia. where needed; ±0.0005 in. possible	63 or better
Turning (automatic screw machine)	Bar, rod	Generally 2 in. dia × 6 in. long	$\frac{1}{16}$ in. dia. and less, weight less than 1 ounce	10–30 parts/minute	Any material with good machinability rating	±0.0005 in. possible; ±0.001 to ±0.003 in. common	63 average
Turning (Swiss automatic machining)	Rod	Collets adapt to $\frac{1}{2}$ in. dia.	Collets adapt to less than $\frac{1}{2}$ in.	12–30 parts/minute	Any material with good machinability rating	±0.0002 in. to ±0.001 in. common	63 and better
Boring (vertical)	Casting, preforms	98 in. × 72 in.	2 in. × 12 in.	2–20 hours/piece	All ferrous and nonferrous	±0.0005 in.	90–250
Milling	Bar, plate, rod, tube	4–6 ft long	Limited usually by ability to hold part	1–100 parts/hour	Any material with good machinability rating	±0.0005 in. possible; ±0.001 in. common	63–250
Hobbing (milling gears)	Blanks, preforms, rods	10-ft-dia. gears 14-in. face width	0.100 in. dia.	1 part/minute	Any material with good machinability rating	±0.001 in. or better	63
Drilling	Plate, bar, preforms	$3\frac{1}{2}$-in.-dia. drills (1-in.-dia. normal)	0.002-in. drill dia.	2–20 second/hole after setup	Any unhardened material; carbides needed for some case-hardened parts	±0.002–±0.010 in. common; ±0.001 in. possible	63–250
Sawing	Bar, plate, sheet	2-in. armor plate ($\frac{1}{6}$ in. is preferred)	0.010 in. thick	3–30 parts/hour	Any nonhardened material	±0.015 in. possible	250–1000
Broaching	Tube, rod, bar, plate	74 in. long	1 in.	300–400 parts/minute	Any material with good machinability rating	±0.0005–±0.001 in.	32–125
Grinding	Plate, rod, bars	36 in. wide × 7 in. dia.	0.020 in. dia.	1–1000 pieces/hour	Nearly all metallic materials plus many nonmetallic	0.0001 in. and less	16
Shaping	Bar, plate, casting	3 ft × 6 ft	Limited usually by ability to hold part	1–4 parts/hour	Low- to medium-carbon steels and nonferrous metals best; no hardened parts	±0.001–±0.002 in. (larger parts) ±0.0001–±0.0005 in. (small–medium parts)	63–250
Planing	Bar, plate, casting	42 ft wide × 18 ft high × 76 ft long	Parts too large for shaper work	1 part/hour	Low- to medium-carbon steels or nonferrous materials best	±0.001–±0.005 in.	63–125
Gear shaping	Blanks	120-in.-dia. gears 6-in. face width	1 in. dia.	1–60 parts/hour	Any material with good machinability rating	±0.001 in. or better at 200 D.P. to 0.0065 in. at 30 D.P.	63

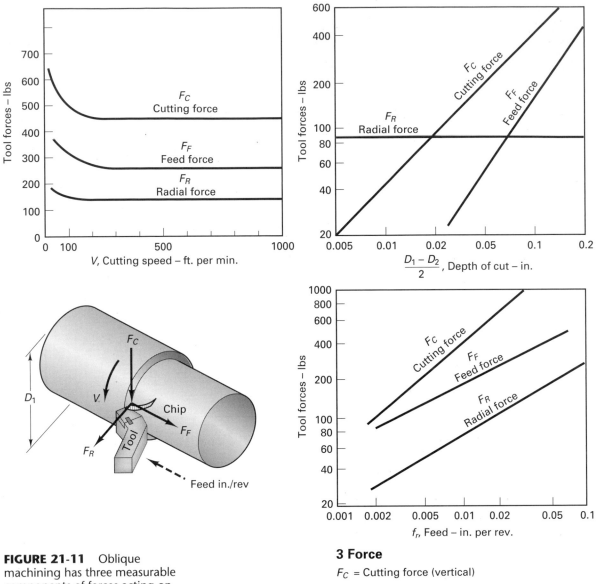

FIGURE 21-11   Oblique machining has three measurable components of forces acting on the tool. The forces vary with speed, depth of cut, and feed.

3 Force

F_C = Cutting force (vertical)

F_R = Radial force (thrust)

F_F = Feed force

The power required for cutting is

$$P = F_c V \,(\text{ft-lb/min}) \qquad (21\text{-}7)$$

The horsepower at the spindle of the machine is therefore

$$\text{hp} = \frac{F_c V}{33,000} \qquad (21\text{-}8)$$

In metal cutting a very useful parameter is called the unit, or specific, horsepower HP_s, which is defined as

$$\text{HP}_s = \frac{\text{hp}}{\text{MRR}} \,(\text{hp/in}^3/\text{min}) \qquad (21\text{-}9)$$

In turning, for example, where MRR $\cong 12 V f_r d$, then

$$\text{HP}_s = \frac{F_c}{396,000 f_r d} \qquad (21\text{-}10)$$

Thus this term represents the approximate power needed at the spindle to remove a cubic inch of metal per minute.

Values for specific horsepower HP_s, which is also called unit power, are given in Table 21-3. These values are obtained through orthogonal metal-cutting experiments described later in this chapter.

Specific horsepower is related to and correlates well with shear stress τ_s for a given metal, which will be derived later. Unit power is sensitive to material properties (e.g., hardness), rake angle, depth of cut, and feed, whereas τ_s is sensitive to material properties only.

Specific power can be used in a number of ways. First, it can be used to estimate the motor horsepower required to perform a machining operation for a given material. HP_s values from the table are multiplied by the approximate MRR for the process. The motor horsepower, HP_m, is then

$$HP_m = \frac{HP_s \times MRR \times CF}{E} \tag{21-11}$$

TABLE 21-3. Values for Unit Power and Specific Energy (cutting stiffness)

Material (Hardness)	Unit Power (hp-min./in^3) HP_s	Specific Energy (in.-lb/in^3) K_s or U
Steel (120 Bhn)	1.12	443,000
Steel (120 Bhn)	0.86	347,000
Steel (120 Bhn)	0.76	301,000
Steel (120 Bhn)	0.64	254,000
Steel (120 Bhn)	0.54	214,000
Steel (160 Bhn)	1.25	495,000
Steel (160 Bhn)	0.59	234,000
Steel (200 Bhn)	1.50	594,000
Steel (200 Bhn)	0.73	290,000
Steel (300 Bhn)	1.87	740,000
Steel (300 Bhn)	0.92	364,000
SAE-302	0.72	285,000
SAE-350	1.20	475,000
SAE-410	0.75	297,000
Gray CI (130 Bhn)	0.29–0.33	127,000
Meehanite	0.55–0.76	262,000
K-Monel	0.80	317,000
Inconel 700	1.40	554,000
High-Temperature Alloy A 286	1.20	475,000
High-Temperature Alloy S 816	1.25	495,000
Titanium A-55	0.65–0.76	281,000
Titanium C-130	0.81–0.93	345,000
Titanium (250–275 BHN)	1.8–2.0	
Aluminum 2014-T6, 2014-T4	0.24	95,100
Aluminum 6064-TO	0.34	125,000
Aluminum 3003-O	0.16	63,400
Aluminum 108 (55 BHN)	0.15	49,400
Muntz Metal	0.55	218,000
Phosphor Bronze	0.33	131,000
Cartridge Brass	0.48	190,000
Copper Alloys (10–80 R_B)	0.5–0.6	
Copper (50 R_B)	0.9–1.0	
Magnesium (40–90 BHN at 500 kg)	0.16	
Tungsten, Tantilum (210–320 BHN)	2.6–2.8	
Nickel Alloys (280–360 BHN)	1.8–2.0	
Nickel/Cobalt Alloys (200–360 BHN)	2.0–2.5	

Values assume normal feed ranges and sharp tools. Multiply values by 1.25 for a dull tool.

Calculation of Unit Power (HP_s)

$HP = F_c V / 33000$

$HP_s = HP/MRR$ Where

$MRR = 12Vtw$ for tube turning

$HP_s = F_c V / 12Vtw \times 33000 = F_c/tw \times 396000$

Calculation of specific energy (U)

$U = F_c V / Vtw = F_c/tw$ for tube turning

where E is the efficiency of the machine. The E factor accounts for the power needed to overcome friction and inertia in the machine and drive moving parts. Usually, 80% is used. Correction factors (CFs) may also be used to account for variations in cutting speed, feed, and rake angle. There is usually a tool wear correction factor of 1.25 used to account for the fact that dull tools use more power than sharp tools.

The primary cutting force F_c can be roughly estimated according to

$$F_c \cong \frac{\text{HP}_s \times \text{MRR} \times 33,000}{V} \tag{21-12}$$

This type of estimate of the major force F_c is useful in analysis of deflection and vibration problems in machining and in the proper design of workholding devices, because these devices must be able to resist movement and deflection of the part during the process.

In general, increasing the speed, the feed, or the depth of cut will increase the power requirement. Doubling the speed doubles the horsepower directly. Doubling the feed or the depth of cut doubles the cutting force F_c. In general, increasing the speed does not increase the cutting force F_c, a surprising experimental result. However, speed has a strong effect on tool life because most of the input energy is converted into heat, which raises the temperature of the chip, the work, and the tool, to the latter's detriment. Tool life (or tool death) is discussed in Chapter 22.

Equation 21-12 can be used to estimate the maximum depth of cut, d, for a process as limited by the available power.

$$d_{\text{max}} = \frac{\text{HP}_m \times E}{12\text{HP}_s V f_r (CF)} \tag{21-13}$$

Another handbook value useful in chatter or vibration calculations is cutting stiffness K_s. In this text, the term *specific energy* U will be used interchangeably with cutting stiffness K_s.

It is interesting to compute the total specific energy in the process and determine how it is distributed between the primary shear and the secondary shear that occurs at the interface between the chip and the tool. It is safe to assume that the majority of the input energy is consumed by these two regions.

Therefore,

$$U = U_s + U_f \tag{21-14}$$

where specific energy (also called cutting stiffness) is

$$U = \frac{F_c V}{V f_r d} = \frac{F_c}{f_r d} = K_s (\text{turning}). \tag{21-15}$$

The specific shear energy is

$$U_s = \frac{F_s V_s}{V f_r d} \tag{21-16}$$

where V_s is the shear velocity and F_s is the shear force.

Specific friction energy is

$$U_f = \frac{F V_c}{V f_r d} = \frac{F r_c}{f_r d} \tag{21-17}$$

where V_c is the chip velocity and r_c is the chip thickness ratio. See equation 21-18.

Usually, 30 to 40% of the total energy goes into friction and 60 to 70% into the shear process.

Typical values for U are given in Table 21-3. This is experimental data developed by the orthogonal machining experiment described in the next section.

■ 21.4 ORTHOGONAL MACHINING (TWO FORCE)

In order to understand this complex process, the tool geometry is simplified from the three-dimensional (oblique) geometry, which typifies most processes, to a two-dimensional (orthogonal) geometry. Low speed orthogonal plate machining as shown in Figure 21-12 uses a flat plate setup in a milling machine. The workpiece is moving past

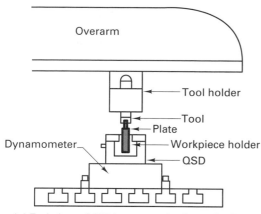

(a) End view of OPM setup on horizontal mill.
Table feed is used for cutting speed.

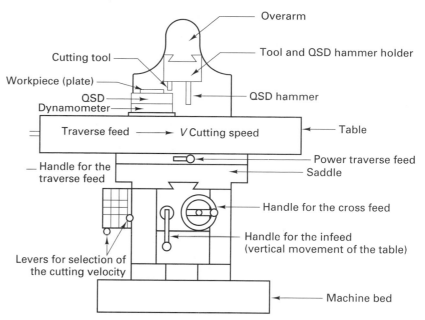

(b) Horizontal milling setup for OPM using QSD. Front view.

FIGURE 21-12   Schematics of the orthogonal plate machining setups: (a) End view of table, QSD, and plate being machined for OPM. (b) Front view of horizontal milling machine. (c) Orthogonal plate machining with fixed tool, moving plate. The feed mechanism of the mill used to produce low cutting speeds.

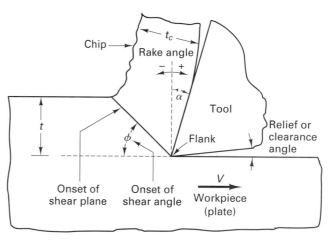

(c) Orthogonal plate machining with fixed tool, moving plate.

the tool at velocity V. The feed of the tool is now called t, the uncut chip thickness. The DOC is the width of the plate w. The cutting edge of the tool is perpendicular to the direction of motion V. The angle that the tool makes with respect to a vertical from the workpiece is called the *back rake angle* α. A positive angle is shown in the schematic. The chip is formed by *shearing*. The *onset of shear* occurs at an angle ϕ with respect to the horizontal. This model is sufficient to allow us to consider the behavior of the work material during chip formation, the influence of the most critical elements of the tool geometry (the edge radius of the cutting tool and the back rake angle α), and the interactions that occur between the tool and the freshly generated surfaces of the chip against the rake face and the new surface as rubbed by the flank of the tool.

Basically, the chip is formed by a localized shear process that takes place over a very narrow zone. This large-strain, high-strain-rate, plastic deformation evolves out of a radial compression zone that travels ahead of the tool as it passes over the workpiece. This radial compression zone has, like all plastic deformations, an elastic compression region that changes into a plastic compression region when the yield strength of the material is exceeded. The plastic compression generates dislocation tangles and networks in annealed metals. The applied stress level increases as the material approaches the tool where the material has no recourse but to shear. The onset of the shear process takes place along the lower boundary of the shear zone defined by the shear angle ϕ. The shear lamella (microscopic shear planes) lie at the angle ψ to the shear plane.

This can be seen in the videograph in Figure 21-13 and the schematic made from the videograph (see Figure 21-14). The videograph was made by videotaping the orthogonal machining of an aluminum plate at over 100× with a high-speed videotaping machine capable of 1000 frames per second. By machining at low speeds ($V = 8.125$ ipm), the behavior of the process was captured and then observed at playback at very slow frame rates. The uncut chip thickness was $t = 0.020$. The termination of the shear process as defined by ψ cannot be observed in the still videograph but can easily be seen in the videos.

FIGURE 21-13 Videograph made from the orthogonal plate machining process.

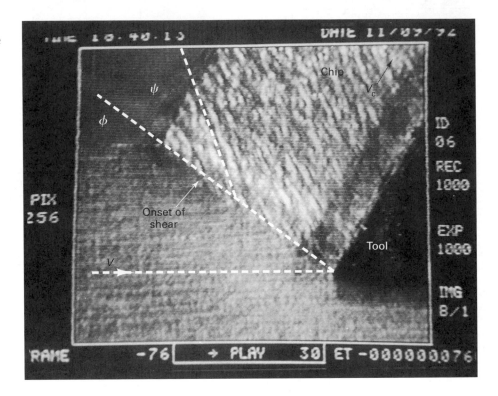

FIGURE 21-14 Schematic representation of the material flow, i.e., the chip-forming shear process.

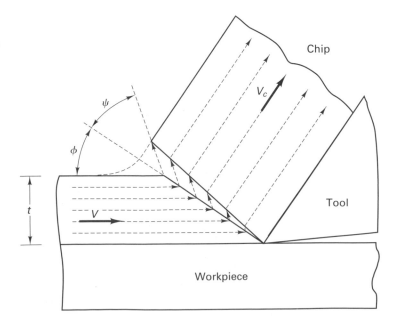

If the work material has hard second-phase particles dispersed in it, they can act as barriers to the shear front dislocations, which cannot penetrate the particle. The dislocations create voids around the particles. If there are enough particles of the right size and shape, the chip will fracture through the shear zone, forming segmented chips. *Free-machining steels*, which have small percentages of hard second-phase particles added to them, use this metallurgical phenomena to break up the chips for easier chip handling.

Orthogonal machining is done to test machining mechanics and theory. Orthogonal machining (measuring two forces) can be obtained in laboratory practice by:

1. Machining a plate as shown in Figures 21-12, 21-13, 21-14, and 21-15.
2. End-cutting a tube wall in a turning setup, Figures 21-15 and 21-16.
3. End-cutting a plate feeding in a facing direction, Figure 21-15.

In *oblique* machining, as in shaping, drilling, and single-point turning, the cutting edge and the cutting motion are not perpendicular to each other. In the orthogonal case, the cutting velocity vector and the cutting edge are perpendicular. As shown in Figure 21-16, OTT can be done on solid cylinders that have had a groove machined on the end to form a tube wall *w*, or a tubular workpiece can be used. The tubular workpieces can be mounted in a lathe and normal cutting speeds developed for the machining experiment. This setup has the

FIGURE 21-15 Three ways to perform orthogonal machining. (a) Orthogonal plate machining on a horizontal milling machine, good for low speed cutting. (b) Orthogonal tube turning, on a lathe, high speed cutting (see Figure 21-16). (c) Orthogonal disk machining on a lathe, very high speed machining with tool feeding (ipr) in the facing direction.

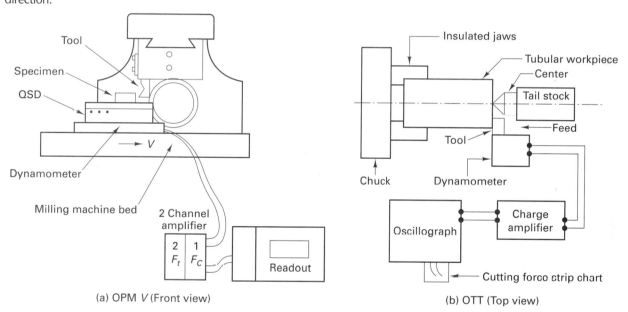

(a) OPM *V* (Front view)

(b) OTT (Top view)

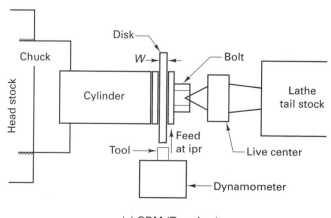

(c) ODM (Top view)

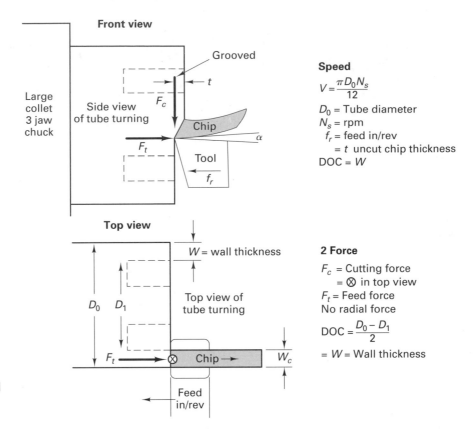

Speed

$$V = \frac{\pi D_0 N_s}{12}$$

D_0 = Tube diameter
N_s = rpm
f_r = feed in/rev
   = t uncut chip thickness
DOC = W

2 Force

F_c = Cutting force
   = $\otimes$ in top view
F_t = Feed force
No radial force

$$DOC = \frac{D_0 - D_1}{2}$$

= W = Wall thickness

FIGURE 21-16 Orthogonal tube turning (OTT) produces a two-force cutting operation at speeds equivalent to those used in oblique machining model/Merchant's.

advantage of being very easy to modify so that cutting temperature experiments can be performed, using the tool/chip thermocouple method. The orthogonal case is more easily modeled for temperature experiments. It will be used here to describe the process further.

■ 21.5 MERCHANT'S MODEL

For the purpose of modeling chip formation, assume that the shear process takes place on a single narrow plane rather than on the set of shear fronts that actually comprise a narrow shear zone. Further, assume that the tool's cutting edge is perfectly sharp and no contact is being made between the flank of the tool and the new surface. The workpiece passes the tool with velocity V, the cutting speed. The uncut chip thickness is t. Ignoring the plastic compression, chips having thickness t_c are formed by the shear process. The chip has velocity V_c. The shear process then has velocity V_s and occurs at the onset of shear angle ϕ. The tool geometry is given by the back rake angle α and the clearance angle γ. The velocity triangle for $V, V_c,$ and V_s is also shown (Figure 21-17). The chip makes contact with the rake face of the tool over length l_c. The plate thickness is w.

FIGURE 21-17 Velocity diagram associated with Merchant's orthogonal machining model.

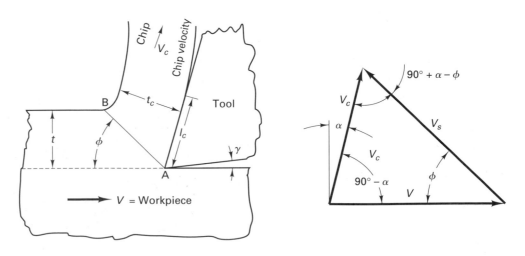

From orthogonal machining experiments, the chip thickness is measured and used to compute the shear angle from the *chip thickness ratio*, r_c, defined as t/t_c;

$$r_c = \frac{t}{t_c} = \frac{AB \sin \phi}{AB \cos(\phi - \alpha)} \tag{21-18}$$

where AB is the length of the shear plane from the tool tip to the free surface.

Equation (21-18) may be solved for the *shear angle* ϕ as a function of the measurable chip thickness ratio by expanding the cosine term and simplifying:

$$\tan \phi = \frac{r_c \cos \alpha}{1 - r_c \sin \alpha} \tag{21-19}$$

There are numerous other ways to measure chip ratios and obtain shear angles both during (dynamically) and after (statically) the cutting process. For example, the ratio of the length of the chip L_c to the length of the cut L can be used to determine r_c. Many researchers use the chip compression ratio, which is the reciprocal of r_c, as a parameter. See Problem 2 at the end of the chapter for another method. The shear angle can be measured statically by instantaneously interrupting the cut through the use of *quick-stop devices*. These devices disengage the cutting tool from the workpiece while cutting is in progress, leaving the chip attached to the workpiece. Optical and scanning electron microscopy is then used to observe the direction of shear. Figure 21-18 was made using a quick-stop device. High-speed motion pictures and high-speed video graphic systems have also been used to observe the process at frame rates as high as 30,000 frames per second. Figure 21-14 is a high-speed videograph. Machining stages have been built that allow the process to be performed inside a scanning electron microscope and recorded on videotapes for high-resolution, high-magnification examination of the deformation process. Using sophisticated electronics and slow-motion playback, this technique can be used to measure the shear velocity. The vector sum of V_s and V_c equals V.

For consistency of volume, we observe that

$$r_c - \frac{t}{t_c} = \frac{\sin \phi}{\cos(\phi - \alpha)} = \frac{V_c}{V} \tag{21-20}$$

indicating that the chip ratio (and therefore the onset of shear angle) can be determined dynamically if a reliable means to measure V_c can be found.

The ratio of V_s to V is

$$\frac{V_s}{V} = \frac{\cos \alpha}{\cos(\phi - \alpha)} \tag{21-21}$$

These velocities are important in power calculations, heat and temperature calculations, and vibration analysis associated with chatter in chip formation.

FIGURE 21-18 Three characteristic types of chips. (*Left* to *right*) discontinuous, continuous, and continuous with built-up edge. Chip samples produced by quick-stop technique. (*Courtesy of Cincinnati Milacron, Inc.*)

■ 21.6 MECHANICS OF MACHINING (STATICS)

Orthogonal machining has been defined as a two-force system. Consider Figure 21-19, which shows a free-body diagram of a chip that has been separated at a shear plane. It is assumed that the resultant force R acting on the back of the chip is equal and opposite to the resultant force R' acting on the shear plane. The resultant R is composed of the *friction force F* and the normal force N acting on the tool/chip interface contact area. The resultant force R' is composed of a *shear force F_s* and normal force F_n acting on the shear plane area A_s. Since neither of these two sets of forces can usually be measured, a third set is needed, which can be measured using a dynamometer (force transducer) mounted either in the workholder or the tool holder. Note that this set has resultant R, which is equal in magnitude to all the other resultant forces in the diagram. The resultant force R is composed of a *cutting force F_c* and a tangential (normal) force F_t. Now it is necessary to express the desired forces (F_s, F_n, F, N) in terms of the measured dynamometer components, F_c and F_t, and appropriate angles. To do this, a circular force diagram is developed in which all six forces are collected in the same force circle (Figure 21-20). The only symbol in this figure as yet undefined is β, which is the angle between the normal force N and the resultant R. It is called friction angle β and is used to describe the friction coefficient μ on the tool–chip interface area, which is defined as F/N, so that

$$\beta = \tan^{-1}\mu = \tan^{-1}\frac{F}{N} \tag{21-22}$$

FIGURE 21-19   Free-body diagram of orthogonal chip formation process, showing equilibrium condition between resultant forces R and R'.

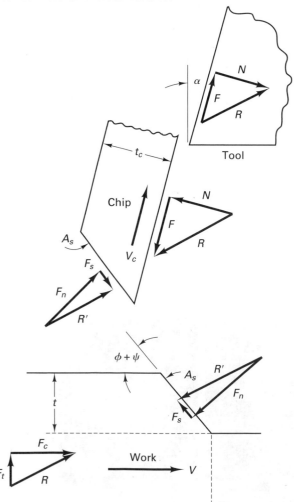

FIGURE 21-20   Merchant's circular force diagram used to derive equations for F_s, F_n, F, and N as functions of F_c, F_t, ϕ, α, and β.

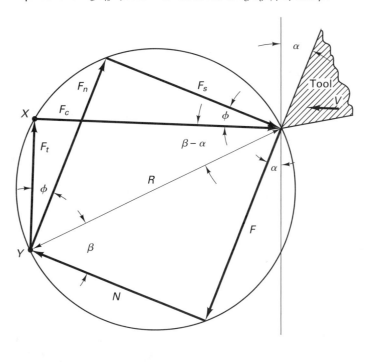

The friction force F and its normal N can be shown to be

$$F = F_c \sin \alpha + F_t \cos \alpha \tag{21-23}$$

$$N = F_c \cos \alpha - F_t \sin \alpha \tag{21-24}$$

and the resultant R is

$$R = \sqrt{F_c^2 + F_t^2} \tag{21-25}$$

Notice that in the special situation where the back rake angle is zero, $F = F_t$ and $N = F_c$, so that in this orientation, the friction force and its normal can be directly measured by the dynamometer.

The forces parallel and perpendicular to the shear plane can be shown from the force circle diagram (Figure 21-20) to be

$$F_s = F_c \cos \phi - F_t \sin \phi \tag{21-26}$$

$$F_n = F_c \sin \phi + F_t \cos \phi \tag{21-27}$$

F_s is of particular interest, because it is used to compute the shear stress on the shear plane. This shear stress is defined as

$$\tau_s = \frac{F_s}{A_s} \tag{21-28}$$

where

$$A_s = \frac{tw}{\sin \phi} \tag{21-29}$$

recalling that t was the uncut chip thickness and w was the width of the workpiece. The *shear stress (flow stress)* is, therefore,

$$\tau_s = \frac{F_c \sin \phi \cos \phi - F_t \sin^2 \phi}{tw} \text{ psi} \tag{21-30}$$

For a given polycrystalline metal, this shear stress is a material constant, not sensitive to variations in cutting parameters, tool material, or the cutting environment. Figure 21-21 gives some typical values for the flow stress for a variety of metals, plotted against hardness.

FIGURE 21-21 Shear stress τ_s variation with the Brinell hardness number for a group of steels and aerospace alloys. Data of some selected fcc metals are also included. (Adapted with permission from S. Ramalingham and K. J. Trigger, *Advances in Machine Tool Design and Research*, 1971, Pergamon Press.)

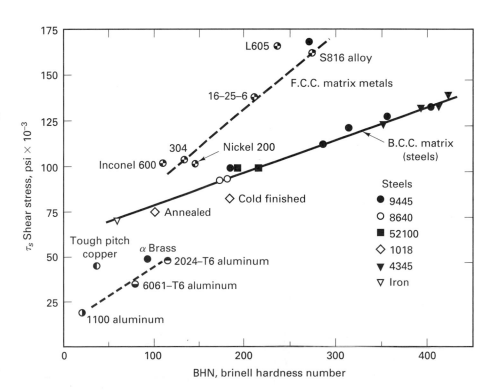

■ 21.7 SHEAR STRAIN γ AND SHEAR FRONT ANGLE ψ

Based on the experimental findings and using the Merchant chip formation bubble, a revised "stack of cards" model, as shown in Figure 21-22, can be developed. From this model, strain is expressed as

$$\gamma = \cos\alpha / [\sin(\phi + \psi)\cos(\phi + \psi - \alpha)] \tag{21-31}$$

where

$\phi =$ the angle of the onset of the shear plane

$\psi =$ the shear front angle

Using the available machining data, ψ decreases, reaches a minimum, and rises again, for all rake angles for a given metal of some hardness.

The minimum energy principle has been reported to have use in various fields such as physics, metal forming processes, and machining processes.

The specific shear energy (shear energy/volume) equals shear stress × shear strain

$$U_s = \tau \times \gamma \tag{21-32}$$

In order to find an equation for γ, the minimum energy principle is used so that ψ will take on values to reduce shear energy to a minimum.

That is

$$dU_s/d\psi = 0 \tag{21-33}$$

The shear front angle is obtained by

$$\psi = 45° - \phi + \alpha/2 \tag{21-34}$$

FIGURE 21-22   The revised "stack of cards" model for calculating shear strain in metal cutting. (Inset) Merchant's bubble model for chip formation.

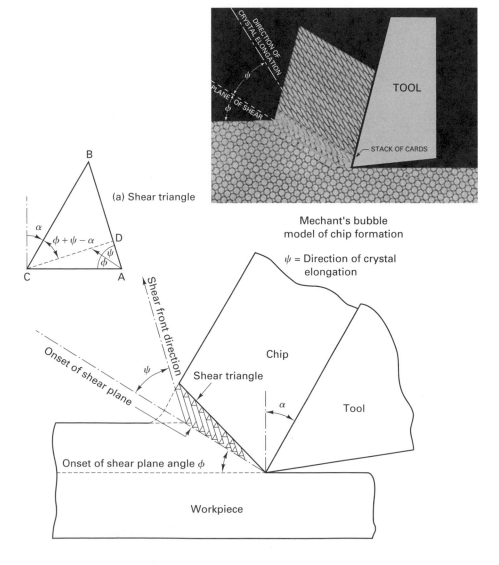

Substituting ψ in Equation (21-34) into Equation (21-31), the shear strain can be expressed as

$$\gamma = \cos\alpha/(1 + \sin\alpha) \qquad (21\text{-}35)$$

which shows that the *shear* strain is dependent only on the rake angle α. Generally speaking, metal-cutting strains are quite large compared to other plastic deformation processes, on the order of 1 to 2 in./in.

The agreement between the predicted shear strain from this model and measured shear strain obtained from metal-cutting experiments is exceptionally good.

This large strain occurs, however, over very narrow regions resulting in extremely high shear strain rates, $\dot{\varepsilon}$, typically in the range of 10^4 to 10^8 in./in. per second. It is this combination of large strains and high strain rates operating within a process constrained only by the rake face of the tool that results in great difficulties in theoretical analysis of this process.

Metal cutting experiments in copper with a hardness gradient ranging from dead soft to full hard have experimentally verified Equation (21-34) to 99% confidence.

- The material begins to shear at the lower boundary of the shear zone, defined by the angle ϕ. As the hardness of a material increases, ϕ increases, while ψ decreases, so $\psi + \phi = 45° + \alpha/2$ is maintained.

- The material in the shear zone shears at an inclination angle ψ to the plane of the onset of shear plane ϕ for aluminum and steel.

- Shear strain and shear front angle can be determined by

$$\gamma = 2\cos\alpha/(1 + \sin\alpha)$$

$$\psi + \phi = 45° + \alpha/2$$

where ϕ and ψ vary with hardness.

■ 21.8 MECHANICS OF MACHINING (DYNAMICS)

Machining is a dynamic process of large strain and high strain rates. All the process variables are dependent variables. The process is intrinsically a closed-loop interactive process as shown in Figure 21-23.

FIGURE 21-23 Machining dynamics is a closed loop interactive process which creates a force-displacement response.

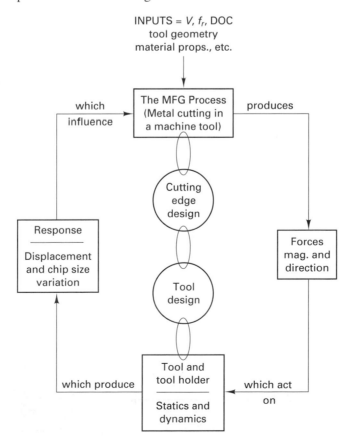

This means:

- The initial inputs determine the chip load on the tool.
- The chip load determines the forces.
- The forces produce the deflection and alter the chip load on the tool.
- The altered chip load determines the new forces.
- The force-displacement response repeats.

Remember that plastic deformation is always proceeded by elastic deformation, which behaves like a big spring. The mechanism by which a process dissipates energy is called chatter or vibration. In machining, it has long been observed in practice that rotational speed may greatly influence process stability and chatter. Experienced operators commonly listen to machining noise and interactively modify the speed when optimizing a specific application. In addition, experience demonstrates that the performance of a particular tool may vary significantly based on the machine tool employed and other characteristics such as the workpiece, fixture holder, and the like. Today more than ever, the manufacturing industry is more competitive and responsive, characterized by both high volume and small batch production seeking economies of scale. High productivity is achieved by increased machine and tooling capabilities along with the elimination of all non-value-added activities. Few companies can afford lengthy trial-and-error approaches to machining-process optimization or additional processes to treat the effect of chatter.

In metal cutting, chatter is a self-excited vibration that is caused by the closed-loop force-displacement response of the machining process. The process-induced variations in the cutting force may be caused by changes in the cutting velocity, chip cross section (area), tool/chip interface friction, built-up edge, workpiece variation, or most commonly, process modulation resulting in regeneration of vibration. When more energy is input into the dynamic machining system than can be dissipated by mechanical work, damping, and friction, equilibrium (the state of minimum potential energy) is sought by the machining system through the generation of chatter vibration.

The proper classification of the type of vibration is the first step in identifying and solving the cause of unwanted vibration (see Figure 21-24).

FIGURE 21-24   There are three types of vibration in machining.

• **Free Vibration** The response to an initial condition or sudden change. The amplitude of the vibration decreases with time and occurs at the natural frequency of the system often produced by interrupted machining. Often appears as lines or shadows following a surface discontinuity.

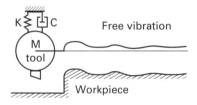

• **Forced Vibration** The response to a periodic (repeating with time) input. The response and input occur at the same frequency. The amplitude of the vibration remains constant for a set input condition and is nonlinearly related to speed. Unbalance, misalignment, tooth impacts and resonance of rotating systems are the most common examples.

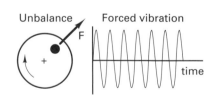

• **Self-Excited Vibration** The periodic response of the system to a constant input. The vibration may grow in amplitude (unstable) and occurs near the natural frequency of the system regardless of the input. Chatter due to the regeneration of surface waviness is the most common metal cutting example.

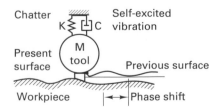

- *Free Vibration* is the response to any initial condition or sudden change. The amplitude of the vibration decreases with time and occurs at the natural frequency of the system. Interrupted machining is an example that often appears as lines or shadows following a surface discontinuity.
- *Forced Vibration* is the response to a periodic (repeating with time) input. The response and input occur at the same frequency. The amplitude of the vibration remains constant for set input conditions and is linearly related to speed. Unbalance, misalignment, tooth impacts, and resonance of rotation systems are the most common examples.
- *Self-Excited Vibration* is the periodic response of the system to a constant input. The vibration may grow in amplitude (become unstable) and occurs near the natural frequency of the system regardless of the input. Chatter due to the regeneration of waviness in the machined surface is the most common metal cutting example.

How do we know chatter exists? Listen and look! Chatter is characterized by the following:

1. The sudden onset of vibration (a screech or buzz or whine) that rapidly increases in amplitude until a maximum threshold (saturation) is reached.
2. The frequency of chatter remains very close to a natural frequency (critical frequency) of the machining system and changes little with variation of process parameters. The largest force-displacement response occurs at resonance and therefore the greatest energy dissipation.
3. Chatter often results in unacceptable surface finish exhibited by a helical or angular pattern (pearled or fish scaled) superimposed over normal feed marks.
4. Visible surface undulations are found in the feed direction and corresponding wavy or serrated chips with variable thickness.

Figure 21-25 shows some typical examples of chatter visible in the surface finish marks.

There are several important factors that influence the stability of a machining process:

- Cutting stiffness of the workpiece material (related to the machinability), K_s
- Cutting-process parameters (speed, feed, DOC, total width of chip)
- Cutter geometry (rake and clearance angles, edge prep, insert size and shape)
- Dynamic characteristics of the machining process (tooling, machine tool, fixture, and workpiece).

K_s, cutting stiffness, is closely aligned with flow stress but simpler to calculate in that ϕ is not used. Like flow stress, cutting stiffness can be viewed as a material property of the workpiece, dependent on hardness.

CHIP FORMATION AND REGENERATIVE CHATTER

In machining, the chip is formed due to the shearing of the workpiece material over the chip area (A = thickness × width = $t \times w$), which results in a cutting force F_c. The magnitude of the resulting cutting force is predominantly determined by the material cutting stiffness K_s and the chip area such that $F_c = K_s \times t \times w$. The direction of the cutting force F_c is influenced mainly by the geometries of the rake and clearance angles as well as the edge prep.

Machining operations require an overlap of cutting paths that generate the machined surface (see Figure 21-26). In single-point operations, the overlap of cutting paths does not occur until one complete revolution. In milling or drilling, overlap occurs in a fraction of a revolution depending on the number of cutting edges on the tool.

The cutting force causes a relative displacement X between the tool and workpiece, which affects the uncut chip thickness t and, in turn, the cutting force. This coupled relationship between displacement in the Y direction (modulation direction) and the resulting cutting force forms a closed-loop response system. The modulation direction is normal to the surface defining the chip thickness.

A phase shift ϵ between subsequent overlapping surfaces results in a variable chip thickness and modulation of the displacement, causing chatter vibration. The phase shift between overlapping cutting paths is responsible for producing chatter. However, there is a preferred speed that corresponds to a phase-locked condition ($\epsilon = 0$) that results in a constant chip thickness t. A constant chip thickness results in a steady cutting force and the

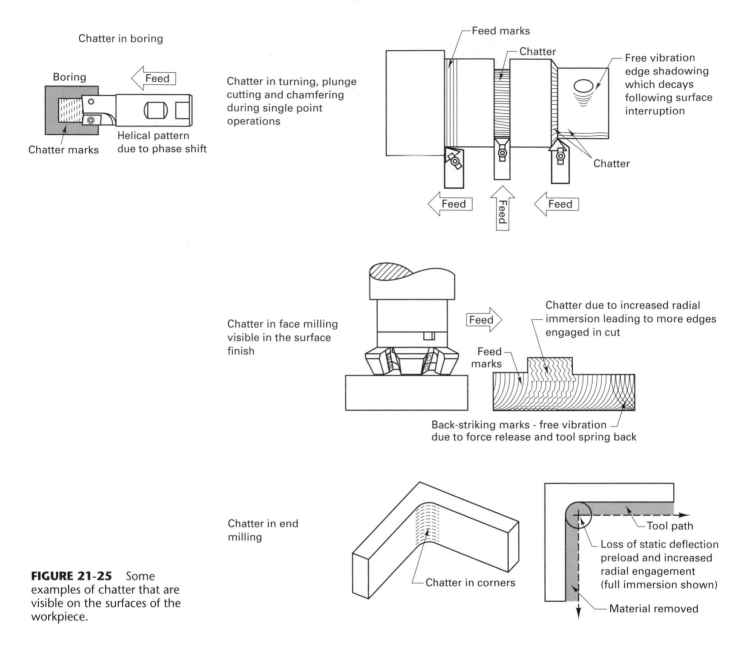

Chatter in boring

Boring

Feed

Chatter marks

Helical pattern due to phase shift

Chatter in turning, plunge cutting and chamfering during single point operations

Feed marks

Chatter

Free vibration edge shadowing which decays following surface interruption

Chatter

Feed Feed Feed

Chatter in face milling visible in the surface finish

Feed

Chatter due to increased radial immersion leading to more edges engaged in cut

Feed marks

Back-striking marks - free vibration due to force release and tool spring back

Chatter in end milling

Chatter in corners

Tool path

Loss of static deflection preload and increased radial engagement (full immersion shown)

Material removed

FIGURE 21-25 Some examples of chatter that are visible on the surfaces of the workpiece.

FIGURE 21-26 When the overlapping cuts get out of phase with each other, a variable chip thickness is produced, resulting in a change in F_c on the tool or workpiece.

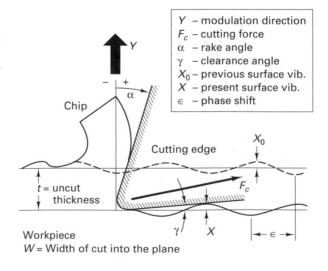

Y – modulation direction
F_c – cutting force
α – rake angle
γ – clearance angle
X_0 – previous surface vib.
X – present surface vib.
ϵ – phase shift

Y

Chip

$-$ $+$

α

Cutting edge

X_0

t = uncut thickness

F_c

γ X ϵ

Workpiece
W = Width of cut into the plane

elimination of the feedback mechanism responsible for regenerative chatter. This what the operators are trying to achieve when they vary cutting speeds (see Figure 21-27).

HOW DO THE IMPORTANT FACTORS INFLUENCE CHATTER?

- *Cutting stiffness K_s:* This is a material property related to shear flow stress, hardness and work hardening and is often described in a relative sense of the machinability of materials. Materials such as steel and titanium require much greater shear forces than aluminum or cast iron; therefore, the corresponding larger cutting forces lead to greater displacement in the Y direction and less machining stability.

- *Speed:* The process parameters are the easiest factors to change chatter and its amplitude. The rotational *speed* of the tool affects the phase shift between overlapping surfaces and the regeneration of vibration. A handheld speed analyzer[1] that produces dynamically preferred speed recommendations is commercially available. When applied to processes exhibiting a relative rotational motion between the cutting tool and workpiece, it recommends a speed to eliminate chatter.

The most successful applications are in

- Milling, boring, and turning
- Multipoint tools
- Machining aluminum and cast iron
- High speed machining
- Thin-chip, high speed die machining

FIGURE 21-27   Regenerative chatter in turning and milling produced by variable uncut chip thickness.

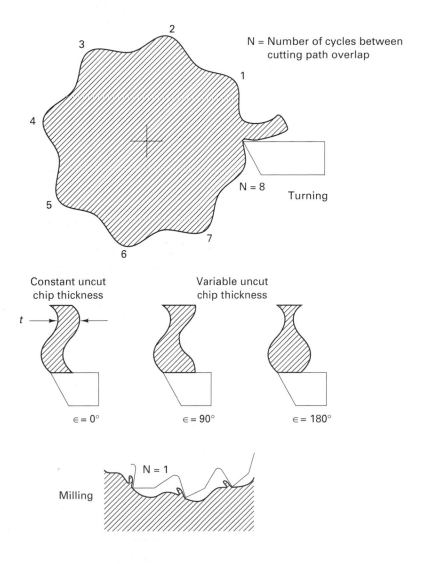

---

[1] *Best Speed* by Design Manufacturing Inc., Tampa, Florida.

At slow speeds (relative to the vibration frequency) process stability is mainly due to increased frictional losses occurring between the tool clearance angle (defined by γ) and the present surface vibration X. This interference and friction dissipates energy in the form of heat and is called *process damping*. As machining speeds are increased, the wavelength of the surface vibration also increases, which reduces the slope of the surface and eliminates process induced damping. Additionally, chatter becomes more significant as speeds increase, because existing forces approach the natural frequencies of the machining system. The analyzer measures and identifies the vibrational frequencies of the chatter noise and determines which speeds will most closely result in $\epsilon = 0$. A zero phase shift between overlapping surfaces eliminates the variation of the chip thickness and eliminates the modulation resulting in chatter.

- *Feed:* The *feed* per tooth defines the average uncut chip thickness t and influences the magnitude of the cutting force. The feed does not greatly influence the stability of the machining process (i.e., whether chatter occurs) but does control the severity of the vibration. Because no cutting force exists if the vibration in the Y direction results in the loss of contact between the tool and workpiece, the maximum amplitude of chatter vibration is limited by the feed.

- *DOC:* The *depth of cut* is the primary cause and control of chatter. The DOC defines the chip width and acts as the feedback gain in the closed-loop machining process. The stability limit (or borderline between stable machining and chatter) may be experimentally determined by incrementally increasing the DOC until the onset of chatter. It can also be analytically predicted based on a thorough understanding of the machining system dynamics and material cutting stiffness.

- *Total Width of Chip:* The *total width of chip* is equal to the DOC times the number of cutting edges engaged in the cut. The total width of cut directly influences the stability of the process. At a fixed DOC that corresponds to the stability limit, increasing the number of engaged cutting edges will result in chatter. The number of engaged teeth in the cut may be increased by adding inserts (using a fine-pitch cutter), or increasing the radial immersion of a milling cutter. Conversely, reducing the number of edges in the cut will have a stabilizing effect on the process.

  The *cutting tool geometry* influences the magnitude and direction of the cutting force, specifically the amount of the force component in the modulation direction Y. A greater projection of force in the Y direction results in increased displacement and vibration normal to the surface, leading to potential chatter.

- As the *back rake angle* α increases (becomes more positive), the length of the onset of shear plane decreases, which reduces the magnitude of the cutting force, F_c. A more positive rake also directs the cutting force to be more tangential and reduces the force component in the Y direction. In general, a more positive cutting geometry increases process stability especially at higher speeds. An insufficient feed compared to the edge radius results in less efficient machining, greater tool deflection, and poorer machining stability.

- A *reduced clearance angle* γ, which increases the frictional contact between the tool and workpiece, may produce process damping. The stabilizing effect is due to energy dissipation in the form of heat, which potentially decreases tool life and may thermally distort the workpiece or increase the heat-affected zone in the workpiece. The initial wear of a new cutting edge may have a stabilizing effect on chatter.

- The *size* (nose radius), *shape* (diamond, triangular, square, round) and *lead angle* of the insert all influence the chip area shape and the corresponding Y direction (see Figure 21-28). In milling, the feed direction is transverse to the tool axis (i.e., radial), and the DOC is defined by the axial immersion of the tool. In milling, as the lead angle of the cut increases, or the shape of the insert becomes rounded (same effect as a large nose radius compared to a small DOC), the Y orientation is directed away from the more flexible radial tool direction and toward the stiffer axial direction. The orientation of the modulation direction Y toward a dynamically more rigid direction results in decreased vibrational response and therefore greater process stability (less tendency for chatter). In boring, the feed direction is axial, and the DOC is determined

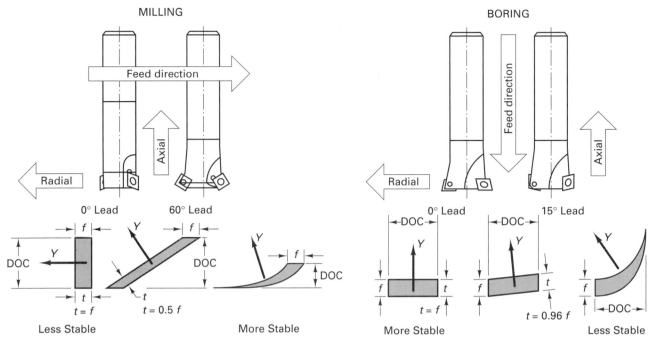

FIGURE 21-28 Milling and boring operations can be made more stable by correct selection of insert geometry.

by the radial direction. Therefore, in boring, a reduced lead angle or a less round (and smaller nose radius compared to the DOC) insert maintains a more axial (stiffer tool direction) orientation of Y, leading to greater stability.

Because the stability of the machining process is a direct result of the dynamic *force-displacement characteristics* between the tool and workpiece, all components of the machining system (tool, spindle, workpiece, fixture, machine tool) may, to varying degrees, influence chatter. Maximizing the dynamics (the product of the static stiffness and damping) of the machining system leads to increased process stability. For example, the static stiffness of an overhung (cantilever) tool with circular cross section varies nonlinearly with the diameter D and the unsupported length L. Machining stability is increased by having the tool with the largest possible diameter tool with the minimum overhang. The frequency of chatter occurs near the most flexible vibrational mode of the machining system.

STABILITY LOBE DIAGRAM

A stability lobe diagram (Figure 21-29) relates the total width of cut that can be machined to the rotational speed of the tool with a specified number of cutting edges. If the total width of cut is maintained below a minimum level (although this may be of limited

FIGURE 21-29 Dynamic analysis of the cutting process produces a stability lobe diagram which defines speeds which produce stable and unstable cutting conditions.

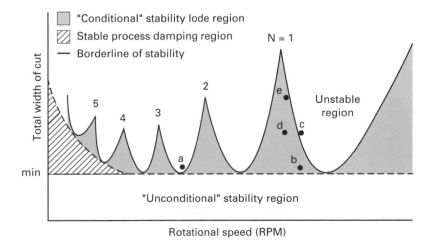

practical value for some machining systems), then the process stability exhibits speed independence or "unconditional" stability. At slow speeds, increased stability may be achieved within the process damping region. The "conditional" stability lobe regions allow increased total width of cut (DOC × number of edges engaged in the cut) at dynamically preferred speeds at which the phase shift ϵ between overlapping or consecutive cutting paths approaches zero. The stability lobe number N indicates complete cycles of vibration that exist between overlapping surfaces. As can be seen by the diagram, the higher speeds correspond to lower lobe numbers and provide the greatest potential increase in the total width of cut and material removal rate (due to greater lobe height and width). If the total width of cut exceeds the borderline of stability, even if the process is operating at a preferred speed, chatter occurs. The greater the total width of cut above the stability limit, the more unstable and violent will be the chatter vibration.

When a chatter condition occurs, such as at point a on the stability lobe diagram, the rotational speed is adjusted to the first recommended speed ($N = 1$), which results in stable machining at point b on the diagram. The DOC may be incrementally increased until chatter again occurs as the stability border is crossed at point c. Using the analyzer again while chattering under the new operating conditions will result in a modified speed recommendation corresponding to point d. If desired, the DOC may again be incrementally increased (conservative steps promote safety) to point e. In general, do not attempt to maintain the DOC (and total width of cut) right up to the borderline of stability because workpiece variation affecting K_s, speed errors, or small changes in the dynamic characteristics of the machining system may result in crossing the stability limit into severe chatter. The amplitude of chatter vibration may be more safely limited by temporary reduction of the feed per tooth until a preferred speed and stable depth of cut have been established.

HEAT AND TEMPERATURE IN METAL CUTTING

In metal cutting, the power put into the process ($F_c V$) is largely converted to heat, elevating the temperatures of the chip, the workpiece, and the tool. These three elements of the process, along with the environment (which includes the cutting fluid), act as the heat sinks. Figure 21-30 shows the distribution of the heat to these three sinks as a function of cutting speed. As speed increases, a greater percentage of the heat ends up in the chip to the point where the chips can be cherry red or even burn at high cutting speeds.

FIGURE 21-30 Distribution of heat generated in machining to the chip, tool, and workpiece. Heat going to the environment is not shown. Figure based on the work of A. O. Schmidt.

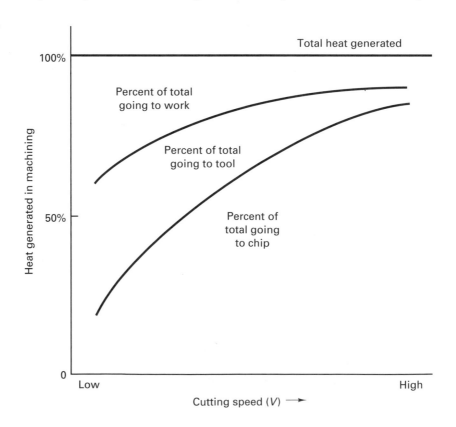

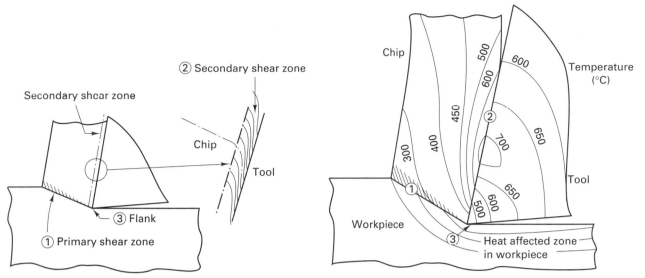

FIGURE 21-31   There are three main sources of heat in metal cutting. (1) Primary shear zone. (2) Secondary shear zone-tool/chip (T/C) interface. (3) Tool flank. The peak temperature occurs at the center of the T/C interface.

There are three main *sources* of heat. Listed in order of their heat-generating capacity, they are shown in Figure 21-31.

1. The shear front itself, where plastic deformation results in the major heat source. Most of this heat stays in the chip.
2. The tool/chip interface contact region, where additional plastic deformation takes place in the chip, and considerable heat is generated due to sliding friction.
3. The flank of the tool, where the freshly produced workpiece surface rubs the tool.

There have been numerous experimental techniques developed to measure cutting temperatures and some excellent theoretical analyses of this "moving" multiple-heat-source problem. Space does not permit us to explore this problem in depth. Figure 21-32 shows the effect of cutting speed on the tool/chip interface temperature. The rate of wear of the tool at the interface can be shown to be directly related to temperature (see Figure 21-32b).

FIGURE 21-32   Typical relationships of temperature to cutting speed and crater wear for machining steels.

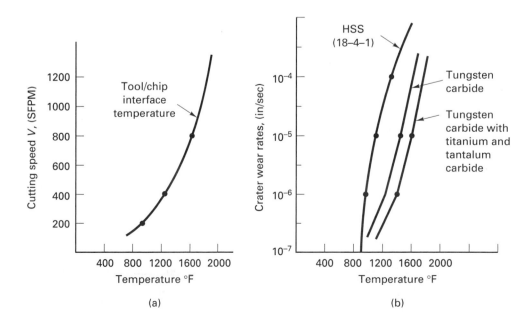

Because cutting forces are concentrated in small areas near the cutting edge, these forces produce large pressures. The tool material must be hard (to resist wear) and tough (to resist cracking and chipping). Tools used in interrupted cutting, such as milling, must be able to resist impact loading as well. Tool materials must sustain their hardness at elevated temperatures. The challenge to manufacturers of cutting tools has always been to find materials that satisfy these severe conditions. Cutting tool materials that do not lose hardness at the high temperatures associated with high speeds are said to have "hot hardness." Obtaining this property usually requires a trade-off in toughness, as hardness and toughness are generally opposing properties. In Chapter 22 cutting tool materials will be addressed in more depth.

■ 21.9 SUMMARY

In this chapter, the basics of the machining processes have been presented. Chapters 23 through 30 provide additional information on the various operations and machine tools. Modern machine tools can often perform several basic operations. As described in Chapter 32, machining centers are now widely used. These chapters on basic processes must be carefully studied. The relationship between the basic processes and the machine tools that can be used to perform these processes has been presented. In Chapter 32, an anatomy for automation will be presented, which will show that most of the machines described in this chapter will be of the A(2) or A(3) level of automation. Machining centers, which are numerical control machines, are A(4). Machining centers have automatic tool-change capability and are usually capable of milling, drilling, boring, reaming, tapping (hole threading), and other minor machining processes. For particular machines, you will need to become familiar with new terminology, but in general all machining processes will need inputs concerning rpm (given that you selected the cutting speed), feeds, and depths of cut. Note also from these tables that the same process can be performed on two or more different machine tools. There are many ways to produce flat surfaces, internal and external cylindrical surfaces, and special geometries in parts. Generally, the quantity to be made is the driving factor in the selection of processes, as we shall explain later.

This chapter also introduced the student to orthogonal machining, a laboratory machining process used to determine experimentally values for specific horsepower, cutting stiffness, and flow stress for machining.

As noted previously, the properties of the work material are important in chip formation. High-strength materials require larger forces than materials of lower strength, causing greater tool and work deflection, increased friction, heat generation, and operating temperatures and requiring greater work input. The structure and composition also influence metal cutting. Hard or abrasive constituents, such as carbides in steel, accelerate tool wear.

Work hardness prior to machining is an important factor because it controls the onset of shear. Onset is delayed by increased hardness, so ϕ increases, as does τ_s. Highly ductile materials not only permit extensive plastic deformation of the chip during cutting, which increases heat generation and temperature, but also result in longer, "continuous" chips that remain in contact longer with the tool face, thus causing more frictional heat. Chips of this type are severely deformed and have a characteristic curl. On the other hand, materials that are already heavily work hardened or brittle, such as gray cast iron, lack the ductility necessary for appreciable plastic deformation. Consequently, the compressed material ahead of the tool fails in brittle fracture, sometimes along the shear front, producing small fragments. Such chips are termed *discontinuous* or *segmented*.

A variation of the continuous chip, often encountered in machining ductile materials, is associated with a *built-up edge* (BUE) formation on the cutting tool. The local high temperature and extreme pressure in the cutting zone cause the work material to adhere or pressure-weld to the cutting edge of the tool forming the built-up edge, rather like a dead metal zone in the extrusion process. Although this material protects the cutting edge from wear, it modifies the geometry of the tool. BUEs are not stable and will break off periodically, adhering to the chip or passing under the tool and remaining on the machined surface. Built-up edge formation can be eliminated or minimized by reducing the depth of cut, altering the cutting speed, using positive rake tools, applying a coolant, or changing cutting-tool materials.

■ KEY WORDS

back rake angle	cutting stiffness	grinding (abrasive machining)	sawing	specific horsepower
boring	cutting tool	machine tool	self-excited vibration	speed
broaching	depth of cut	metal cutting	shaping	stability lobe diagram
chatter	drilling	milling	shear angles	turning
built-up edge	dynamics	oblique machining	shear velocity	vibration
chip ratio	feed	orthogonal machining	shear strain	workholding device
chip velocity	friction force	regenerative chatter	shear force (flow stress)	workpiece
cutting force				

■ REVIEW QUESTIONS

1. Why has the metal-cutting process resisted theoretical solution for so many years?
2. What variables must be considered in understanding a machining process?
3. Which of the basic chip formation processes are single point, and which are multiple point?
4. How is feed related to speed in machining operations such as turning?
5. Milling has two feeds. What are they, and which one is an input parameter to the machine tool?
6. What is the fundamental mechanism of chip formation?
7. What are the implications of Figure 21-13, given that this videograph was made at a very low cutting speed?
8. What is the difference between oblique machining and orthogonal machining?
9. Note that the units for the approximate equation for MRR for turning are not correct. When is the approximate equation not very good (yields a large error in MRR values)?
10. For orthogonal machining, the cutting edge radius is assumed to be small compared to the uncut chip thickness. Why?

11. How does the magnitude of the strain and strain rate values of metal cutting compare to tensile testing?
12. Why is titanium such a difficult metal to machine? (Note its high value of HP_s.).
13. Explain why you get segmented or discontinuous chips when you machine cast iron.
14. Why is shear stress so important?
15. Which of the three cutting forces in oblique cutting consumes most of the power?
16. How is the energy in a machining process typically consumed?
17. Where does the energy consumed ultimately go?
18. State two ways of estimating the primary force F_c.
19. How is cutting speed related to tool wear?
20. What is the relationship between hardness and temperature in metal-cutting tool materials?
21. Why doesn't the cutting force F_c increase with increased speed V?
22. Why does the cutting force increase with increased feed or DOC?

■ PROBLEMS

1. Suppose you have the following data obtained from a metal-cutting experiment (orthogonal machining). Compute the shear angle, the shear stress, the specific energy, the shear strain, and the coefficient of friction at tool/chip interface. How does your HP_s and τ_s values compare with the value found in Chapter 21?

2. Suppose you weighed a short chip fragment. You measured the length of the short chip fragment L_c. Assume that it weighs W. The density of this metal is p lb/in^3. Can you obtain the chip thickness ratio r_c?

3. For the data in Problem 1, determine the specific shear energy and the specific friction energy.

Machining data for 1020 steel, as-received, in air with a K3H carbide tool, orthogonally (tube cutting on lathe) with tube OD = 2.875. The cutting speed was 530 fpm. The tube wall thickness was 0.200 in. The back rake angle was zero for all cuts.

			Data					
Run Number	F_c	F_t	Feed ipr ×1/1000	Chip Ratio r_c	ϕ	τ_s	μ	HP_s
1	330	295	4.89	0.331				
2	308	280	4.89	0.381				
3	410	330	7.35	0.426				
4	420	340	7.35	0.426				
5	510	350	9.81	0.458				
6	540	395	9.81	0.453				

4. Derive equations for F and N using the circular force diagram. (*Hint:* Make a copy of the diagram. Extend a line from point X intersecting force F perpendicularly. Extend a line from point Y intersecting the previous line perpendicularly. Find the angle α made by these constructions.)

5. Derive equations for F_s and F_n using the circular force diagram. (*Hint:* Construct a line through X parallel to vector F_n. Extend vector F_s to intersect this line. Construct a line from X perpendicular to F_n. Construct a line through point Y perpendicular to the line through X. Construct a line through point Y perpendicular to the line through X.)

6. For the data in Problem 1, calculate the shear strain and compare it to $1/r_c$. Comment on the comparison, $1/r_c = t_c/t = \dfrac{L}{L_c}$ assuming that $W = W_c$.

7. A manufacturing engineer needs an estimate of the cutting force F_c to estimate the loss of accuracy of a machining process due to deflection. The material being machined is Inconal 600 with a BHN value of 100. The cutting speed was 250 ft/min, the feed was 0.020 in./rev, and the depth of cut was 0.250 in. The chip from the process measured 0.080 in. thick. Estimate the cutting force F_c, assuming that $F_t = \dfrac{F_c}{2}$.

8. Using Table 21-1, determine the maximum and minimum MRR values for rough machining (turning) a 1020 carbon steel with a BHN value of 200. Repeat for finish machining assuming a DOC value equal to 10% of the roughing DOC.

9. Estimate the horsepower needed to remove metal at 550 in^3/min with a feed of 0.005 in./rev at a DOC value of 0.675 in. The cutting force F_c was measured at 10,000 lb. Comment on these values.

10. For a turning process, the horsepower required was 24 hp. The metal removal rate was 550 in^3/min. Estimate the specific horsepower and compare to published values for 1020 steel at 200 BHN.

www.wiley.com/college/degarmo

CUTTING TOOLS FOR MACHINING

■ 22.1 INTRODUCTION

Success in metal cutting depends upon the selection of the proper cutting tool (material and geometry) for a given work material. A wide range of cutting tool materials is available with a variety of properties, performance capabilities, and cost. These include high carbon steels and low/medium alloy steels, high-speed steels, cast cobalt alloys, cemented carbides, cast carbides, coated carbides, coated high speed steels, ceramics, cermets, whisker-reinforced ceramics, sialons, sintered polycrystalline cubic boron nitride (CBN), sintered polycrystalline diamond, and single-crystal natural diamond. Figure 22-1 shows the common tool materials in a matrix of tool materials ranked by the cutting speeds used to machine a unit volume of steel materials, assuming equal tool lives. As the speed (feed rate and DOC) increases, so does the metal removal rate. The time required to remove a given unit volume of material therefore decreases. Notice the five-fold increase in speed that the $TiC/Al_2O_3/TiN$-coated carbide has over the WC/Co tool (250 → 1200 sfpm). Today, approximately 85% of carbide tools are coated, almost exclusively by the *chemical vapor deposition* (CVD) process. The cutting tool is the most critical part of the machining system.

The cutting tool material, cutting parameters, and tool geometry selected directly influence the productivity of the machining operation. Figure 22-2 outlines the input variables that influence the tool material selection decision. The elements which influence the decision are:

- Work material characteristics (chemical and metallurgical state, hardness)
- Part characteristics (geometry, accuracy, finish, and surface-integrity requirements)

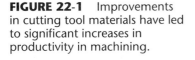

FIGURE 22-1 Improvements in cutting tool materials have led to significant increases in productivity in machining.

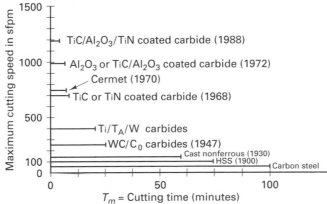

Time required to remove a unit volume of metal using various tool materials machining 1018 steel.

$TiC/Al_2O_3/TiN$ coated carbide (1988)

Al_2O_3 or TiC/Al_2O_3 coated carbide (1972)

Cermet (1970)

TiC or TiN coated carbide (1968)

$Ti/T_A/W$ carbides

WC/C_0 carbides (1947)

Cast nonferrous (1930)

HSS (1900)

Carbon steel

T_m = Cutting time (minutes)

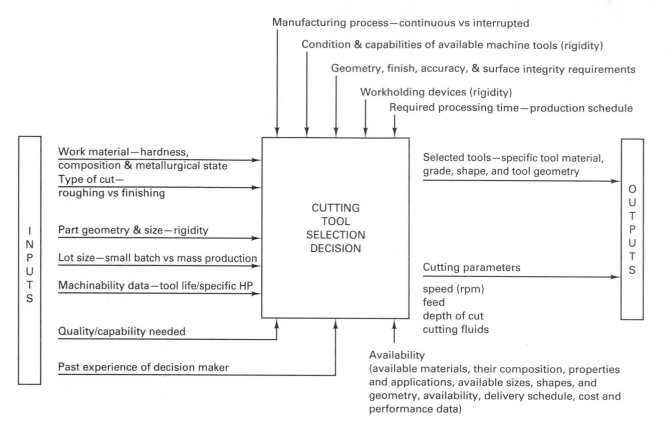

FIGURE 22-2   The selection of the cutting tool material and geometry followed by the selection of cutting conditions for a given application depends upon many variables.

- Machine tool characteristics, including the workholders (adequate rigidity with high horsepower, and wide speed and feed ranges)
- Support systems (operator's ability, sensors, controls, method of lubrication and chip removal)

Tool material technology is advancing rapidly, enabling many difficult-to-machine materials to be machined at higher removal rates and/or cutting speeds with greater performance reliability. Higher speed and/or removal rates usually improve productivity. Predictable tool performance is essential when machine tools are computer controlled and have minimal operator interaction. Long tool life is desirable when machines are placed in cellular manufacturing systems.

The cutting tool is subjected to severe conditions. Tool temperatures of 1000°C, severe friction and high local stresses require that the tool have these characteristics.

1. High hardness (Figure 22-3)
2. High hardness temperature, *hot hardness* (refer to Figure 22-3)
3. Resistance to abrasion, wear, chipping of the cutting edge
4. High toughness (impact strength) (refer to Figure 22-4)
5. Strength to resist bulk deformation
6. Good chemical stability (inertness or negligible affinity with the work material)
7. Adequate thermal properties
8. High elastic modulus (stiffness)
9. Consistent tool life
10. Correct geometry and surface finish

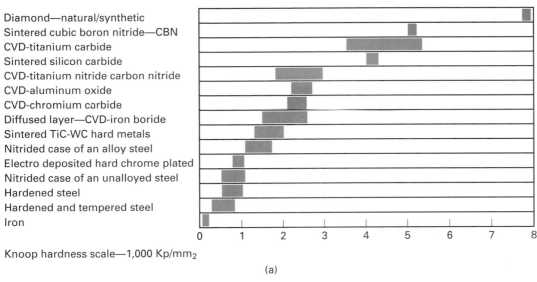

Diamond—natural/synthetic
Sintered cubic boron nitride—CBN
CVD-titanium carbide
Sintered silicon carbide
CVD-titanium nitride carbon nitride
CVD-aluminum oxide
CVD-chromium carbide
Diffused layer—CVD-iron boride
Sintered TiC-WC hard metals
Nitrided case of an alloy steel
Electro deposited hard chrome plated
Nitrided case of an unalloyed steel
Hardened steel
Hardened and tempered steel
Iron

Knoop hardness scale—1,000 Kp/mm$_2$

(a)

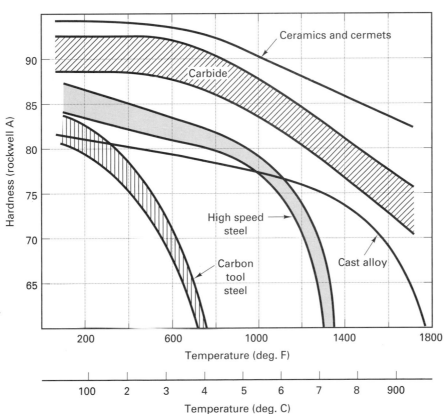

(b)

FIGURE 22-3 (a) Hardness of cutting materials and (b) decreasing hardness with increasing temperature, called hot hardness. Some materials display a more rapid drop in hardness above some temperatures. *(From Metal Cutting Principles, 2nd ed. Courtesy of Ingersoll Cutting Tool Company.)*

Table 22-1 compares these properties for various cutting tool materials. Figure 22-3 compares various tool materials on the basis of hardness, the most critical characteristic, and hot hardness (hardness decreases slowly with temperature). Figure 22-4 compares hot hardness with toughness or the ability to take impacts during interrupted cutting. Naturally, it would be most wonderful if these materials were also easy to fabricate, readily available, and inexpensive, since cutting tools are routinely replaced but this is not usually the case. Obviously, many of the requirements conflict and therefore tool selection will always require trade-offs.

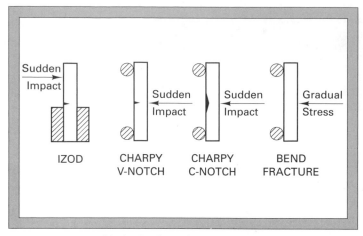

Methods of toughness testing

Toughness

Toughness (as considered for tooling materials) is the relative resistance of a material to breakage, chipping, or cracking under impact or stress. Toughness may be thought of as the opposite of brittleness. Toughness testing is not the same as standardized as hardness testing. It may be difficult to correlate the results of different test methods. Common toughness tests include Charpy impact tests and bend fracture tests.

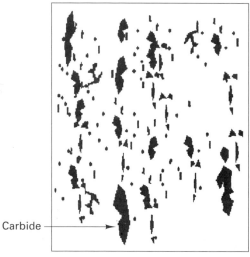

Conventional tool steel microstructure

Wear Resistance

Alloy elements (Cr, V, W, Mo) form hard carbide particles in tool steel microstructures. Amount & type present influence wear resistance.

Hardness of carbides:

• Hardened steel	• 60/65 HRC
• Chromium carbides	• 66/68 HRC
• Moly, tungsten carbides	• 72/77 HRC
• **Vanadium carbides**	• **82/84 HRC**

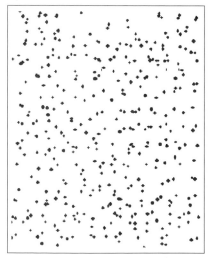

P/M tool steels microstructure

Microstructure of P/M tool steel versus conventional tool steels shows the fine carbide distribution, uniformly distributed.

FIGURE 22-4 The most important properties of tool steels are:

1. Hardness—resistance to deforming and flattening
2. Toughness—resistance to breakage and chipping
3. Wear resistance—resistance to abrasion and erosion.

TABLE 22-1. Salient Properties of Cutting Tool Materials[a]

Property	Carbon and Low/Medium-Alloy Steels	High-Speed Steels	Sintered Cemented Carbides	Coated HSS	Coated Carbides	Ceramics	Polycrystalline CBN	Diamond
Toughness	◄——————————————————— Decreasing ———————————————————►							
Hot hardness	◄——————————————————— Increasing ———————————————————►							
Impact strength	◄——————————————————— Decreasing ———————————————————►							
Wear resistance	◄——————————————————— Increasing ———————————————————►							
Chipping resistance	◄——————————————————— Decreasing ———————————————————►							
Cutting speed	◄——————————————————— Increasing ———————————————————►							
Depth of cut	Light to medium	Light to heavy	Light to heavy	Light to heavy	Light to heavy	Light to heavy	Light to heavy	Very light for single-crystal diamond
Finish obtainable	Rough	Rough	Good	Good	Good	Very good	Very good	Excellent
Method of manufacture	Wrought	Wrought cast, HIP sintering	Cold pressing and sintering, PM	PVD[b] after forming	CVD[c]	Cold pressing and sintering or HIP sintering	High-pressure–high-temperature sintering	High-pressure–high-temperature sintering
Fabrication	Machining and grinding	Machining and grinding	Grinding	Machining and grinding, coating	Grinding before coating	Grinding	Grinding and polishing	Grinding and polishing
Thermal shock resistance	◄——————————————————— Increasing ———————————————————►							
Tool material cost	◄——————————————————— Increasing ———————————————————►							

[a] Overlapping characteristics exist in many cases. Exceptions to the rule are very common. In many classes of tool materials a wide range of composition and properties is obtainable.
[b] Physical vapor deposition.
[c] Chemical vapor disposition.

■ 22.2 TOOL COATING PROCESSES

The two most effective coating processes for improving the life and performance of tools are the chemical vapor deposition (CVD) and physical vapor deposition (PVD) of titanium nitride (TiN) and titanium carbide (TiC). The CVD process, used to deposit a protective coating onto carbide inserts, has been benefitting the metal removal industry for many years and is now being applied with equal success to steel. The PVD processes have quickly become the preferred TiN coating processes for high-speed steel and carbide tipped cutting tools.

CVD

Chemical vapor deposition (CVD) is an atmosphere controlled process carried out at temperatures in the range of 950 to 1050°C (1740 to 1920°F). Figure 22-5 shows a schematic of the CVD process.

Cleaned tools ready to be coated are staged on precoated graphite work trays (shelves) and loaded onto a central gas distribution column (tree). The tree loaded with parts to be coated is placed inside the retort of the CVD reactor. The tools are heated under an inert atmosphere until the coating temperature is reached. The coating cycle is initiated by the introduction of titanium tetrachloride ($TiCl_4$), hydrogen, and methane (CH_4) into the reactor. $TiCl_4$ is a vapor and is transported into the reactor via a hydrogen carrier gas; CH_4 is introduced directly. The chemical reaction for the formation of TiC is:

$$TiCl_4 + CH_4 \longrightarrow TiC + 4HCl \qquad (22\text{-}1)$$

To form titanium nitride, a nitrogen/hydrogen gas mixture is substituted for methane. The chemical reaction for TiN is:

$$2TiCl_4 + 2H_2 + N_2 \longrightarrow 2TiN + 4HCl \qquad (22\text{-}2)$$

PVD

There are three different PVD methods currently in use: reactive sputtering, reactive ion plating, and arc evaporation. In each of the methods, the TiN coating is formed by reacting free titanium ions with nitrogen away from the surface of the tool, and relying upon a physical means to transport the coating onto the tool surface.

The PVD processes are carried out under a hard vacuum, with the workpieces heated to temperatures in the range of 200 to 485°C (400 to 900°F). Substrate heating enhances the adhesion of the coatings.

Of the three, PVD/ARC, shown in Figure 22-6, is the most recent development. The plasma sources are from several arc evaporators located on the sides and top of the vacuum chamber. Each evaporator generates plasma from multiple arc spots. In this way a highly localized electrical arc discharge causes minute evaporation of the material of the cathode and a self-sustaining arc is produced that generates a high energy and concentrated plasma.

The kinetic energy of deposition is much greater than that found in any other PVD method. During coating, this energy is of the order of 150 electron volts and more. Therefore, the plasma is highly reactive and the greater percentage of the vapor is atomic and ionized.

FIGURE 22-5 Chemical vapor deposition (CVD) is used to apply layers, (TiC, TiN, etc.) to carbide cutting tools.

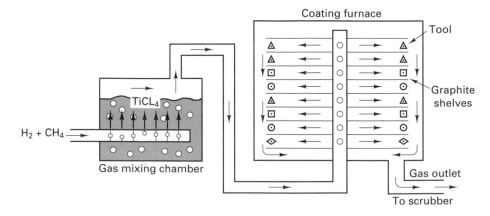

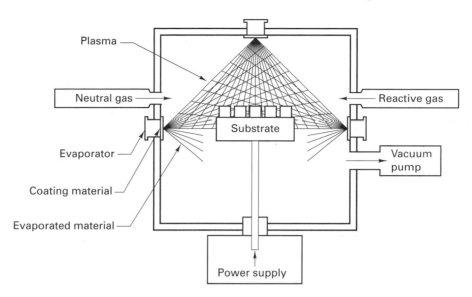

FIGURE 22-6 Schematic of PVC arc evaporation process.

Coating temperatures can be selected and controlled so that metallurgy is preserved. This enables a coating of a wide variety of sintered carbide tools, for example, brazed tools, solid carbide tools such as drills, end mills, form tools, and inserts.

The PVD arc evaporation process will preserve substrate metallurgy, surface finish, edge sharpness, geometrical straightness, and dimensions.

CVD AND PVD—COMPLEMENTARY PROCESSES

CVD and PVD are complementary coating processes. The differences between the two processes and resultant coatings dictate which coating process to use on different tools.

Since CVD is done at higher temperatures, the adhesion of these coatings tends to be superior to a PVD-CVD deposited coating. CVD coatings are normally deposited thicker than PVD coatings (6 to 9 μm for CVD, 1 to 3 μm for PVD).

With CVD multiple coatings, layers may be readily deposited but the tooling materials are restricted. CVD coated tools must be heat treated after coating. This limits the application to loosely toleranced tools. However, the CVD process, being a gaseous process, results in a tool which is coated uniformly all over; this includes blind slots and blind holes.

Since PVD is mainly a line-of-sight process, all surfaces of the part to be coated may be masked. PVD also requires fixturing of each part in order to effect the substrate bias.

APPLICATIONS

Applications for the two different process are

CVD

- loosely toleranced tooling
- piercing and blanking punches, trim dies, phillips punches, upsetting punches
- AISI A, D, H, M, and air hardening and tool steel parts
- solid carbide tooling

PVD

- all HSS, solid carbide and carbide-tipped cutting tools
- fine blanking punches, dies (.001 in. tolerance or less)
- noncomposition dependent process; virtually all tooling materials, including mold steels and bronze

■ 22.3 CUTTING TOOL MATERIALS

In nearly all machining operations, cutting speed and feed are limited by the capability of the tool material. Speeds and feeds must be kept low enough to provide for an acceptable tool life. If not, the time lost changing tools may outweigh the productivity gains from increased cutting speed.

Coated and uncoated carbides are currently the most extensively used tool materials. Coated tools cost only about 15 to 20% more than uncoated tools, so a modest improvement in performance can justify the added cost. About 15 to 20% of all tool steels are coated, mostly by the *physical vapor deposition* (PVD) processes. Diamond and CBN are used for applications in which, despite higher cost, their use is justified. Cast cobalt alloys are being phased out because of the high raw material cost and the increasing availability of alternate tool materials. New ceramic materials called *cermets* (ceramic material in a metal binder) are being introduced that will have significant impact on future manufacturing productivity.

Tool requirements for other processes that use noncontacting tools, as in electrodischarge machining (EDM) and electrochemical machining (ECM), or *no tools at all* (as in laser machining), are discussed in Chapter 28. Grinding abrasives will be discussed in Chapter 27.

TOOL STEELS

Carbon steels and low/medium alloy steels, called *tool steels,* were once the most common cutting-tool materials. Plain-carbon steels of 0.90 to 1.30% carbon when hardened and tempered have good hardness and strength and adequate toughness and can be given a keen cutting edge. However, tool steels lose hardness at temperatures above 400°F because of tempering and have largely been replaced by other materials for metal cutting.

The most important properties for tool steels are hardness, hot hardness and toughness. Low/medium alloy steels have alloying elements such as Mo and Cr, which improve hardenability, and W and Mo, which improve wear resistance. These tool materials also lose their hardness rapidly when heated to about their tempering temperature of 300 to 650°F and they have limited abrasion resistance. Consequently, low/medium alloy steels are used in relatively inexpensive cutting tools (e.g., drills, taps, dies, reamers, broaches, and chasers) for certain low-speed cutting applications when the heat generated is not high enough to reduce their hardness significantly. High-speed steels, cemented carbides, and coated tools are also used extensively to make these kinds of cutting tools. Although more expensive, they have longer tool life and improved performance. These steels greatly benefit from PM manufacturing due to uniformly distributed carbides.

HIGH-SPEED STEELS

First introduced in 1900 by F. W. Taylor and White, high-alloy steel was superior to tool steel in that it retains its cutting ability at temperatures up to 1100°F, exhibiting good "red hardness." Compared with tool steel, it can operate at about double or triple cutting speeds to about 100 sfpm with equal life, resulting in its name *high-speed steel*, often abbreviated HSS.

Today's high-speed steels contain significant amounts of W, Mo, Co, V, and Cr besides Fe and C. W, Mo, Cr, and Co in the ferrite as a solid solution provide strengthening of the matrix beyond the tempering temperature, thus increasing the hot hardness. Vanadium (V), along with W, Mo, and Cr, improves hardness (R_c 65–70) and wear resistance. Extensive solid solutioning of the matrix also ensures good hardenability of these steels.

Although many formulations are used, a typical composition is that of the 18-4-1 type (tungsten 18%, chromium 4%, vanadium 1%), called T1. Comparable performance can also be obtained by the substitution of approximately 8% molybdenum for the tungsten, referred to as a tungsten equivalent (W_{eq}). High-speed steel is still widely used for drills and many types of general-purpose milling cutters and in single-point tools used in general machining. For high-production machining it has been replaced almost completely by carbides, coated carbides, and coated HSS.

HSS main strengths are

- great toughness—superior transverse rupture strength
- easily fabricated
- best for sever applications where complex tool geometry is needed (gear cutters, taps, drills, reamers, dies)

High-speed steel tools are fabricated by three methods: cast, wrought, and sintered (using the powder metallurgy technique). Improper processing of cast and wrought products can result in carbide segregation, formation of large carbide particles and significant variation of carbide size, and nonuniform distribution of carbides in the matrix.

The material will be difficult to grind to shape and will cause wide fluctuations of properties, inconsistent tool performance, distortion and cracking.

To overcome some of these problems, a powder metallurgy technique has been developed which uses the hot isostatic pressing (HIP) process on atomized, prealloyed tool steel mixtures. Because the various constituents of the PM alloys are "locked" in place by the compacting procedure, the end product is a more homogeneous alloy, Figure 22-4. PM high-speed steel cutting tools exhibit better grindability, greater toughness, better wear resistance, and higher red (or hot) hardness, and they perform more consistently. They are about double the cost of regular HSS.

TiN-COATED HIGH-SPEED STEEL

Coated high-speed steel (HSS) does not provide as dramatic improvements in cutting speeds as do coated carbides, with increases of 10 to 20% being typical. First introduced in 1980 for gear cutters (hobs) and in 1981 for drills, TiN-coated HSS tools have demonstrated their ability to more than pay for the extra cost of the coating process.

In addition to hobs, gear-shaper cutters, and drills, HSS tooling coated by TiN now includes reamers, taps, chasers, spade-drill blades, broaches, bandsaw and circular saw blades, insert tooling, form tools, end mills, and an assortment of other milling cutters.

Physical vapor deposition (PVD) has proved to be the most viable process for coating HSS, primarily because it is a relatively low-temperature process that does not exceed the tempering point of HSS. Therefore, no subsequent heat treatment of the cutting tool is required. Films 0.0001 to 0.0002 in. in thickness adhere well and withstand minor elastic, plastic, and thermal loads. Thicker coatings tend to fracture under the typical thermomechanical stresses of machining.

There are many variations to the physical vapor deposition (PVD) process, as outlined in Table 22-2. The process usually depends on gas pressure and is performed in a vacuum chamber. PVD processes are carried out with the workpieces heated to temperatures in the 400 to 900°F. Substrate heating enhances coating adhesion and film structure.

Because surface pretreatment is critical in PVD processing, tools to be coated are subject to a vigorous cleaning process. Precleaning methods typically involve degreasing, ultrasonic cleaning, and Freon drying. Deburring, honing, and more active cleaning methods are also used.

The advantages of TiN-coated HSS tooling include reduced tool wear. Less tool wear results in less stock removal during tool regrinding, thus allowing individual tools to be reground more times. For example, a TiN hob can cut 300 gears per sharpening; the uncoated tool would cut only 75 parts per sharpening. Therefore the cost per gear is reduced from 20 cents to 2 cents. Naturally, reduced tool wear means longer tool life.

Higher hardness, with typical values for the thin coatings, "equivalent" to R_c 80–85, as compared to R_c 65–70 for hardened HSS, means reduced abrasion wear. Relative inertness (i.e., TiN does not react significantly with most workpiece materials) results in greater tool life through a reduction in adhesion. TiN coatings have a low coefficient of friction. This results in an increase in the shear angle, which reduces the cutting forces, spindle power, and heat generated by the deformation processes. PVD coatings generally fail in high-stress applications such as cold extrusion, piercing, roughing, and high speed machining.

CAST COBALT ALLOYS

Cast cobalt alloys, popularly known as *stellite tools*, are cobalt-rich, chromium-tungsten-carbon cast alloys having properties and applications in the intermediate range between high-speed steel and cemented carbides. Although comparable in room-temperature hardness to high-speed steel tools, cast cobalt alloy tools retain their hardness to a much higher temperature. Consequently, they can be used at higher cutting speeds (25% higher) than HSS tools. Cast cobalt alloys are hard as cast, and cannot be softened or heat treated.

Cast cobalt alloys contain a primary phase of Co-rich solid solution strengthened by Cr and W and dispersion hardened by complex hard, refractory carbides of W and Cr. Other elements added include V, B, Ni, and Ta. The casting provides a tough core and elongated grains normal to the surface. The structure is not, however, homogeneous.

Tools of cast cobalt alloys are generally cast to shape and finished to size by grinding. They are available only in simple shapes, such as single-point tools and saw

TABLE 22-2. Surface Treatments for Cutting Tools

Process	Method	Hardness[a] and Depth	Advantages	Limitations
Black Oxide	HSS cutting tools are oxidized in a steam atmosphere at 1000°F	No change in prior steel hardness	Prevents built-up edge formations in machining of steel.	Strictly for HSS tools
Nitriding Case Hardening	Steel surface is coated with nitride layer by use of cyanide salt at 900°F to 1600°F, or ammonia, gas, or N_2 ions.	To 72 R_c; Case depth: 0.0001 to 0.100 in.	High production rates with bulk handling. High surface hardness. Diffuses into the steel surfaces. Simulates strain hardening.	Can only be applied to steel. Process has embrittling effect because of greater hardness. Post-heat treatment needed for some alloys.
Electrolytic Electroplating	The part is the cathode in a chromic acid solution; anode is lead. Hard chrome plating is the most common process for wear resistance.	70–72 R_c; 0.0002 to 0.100 in.	Low friction coefficient, antigalling. Corrosion resistance. High hardness.	Moderate production; pieces must be fixtured. Part must be very clean. Coating does not diffuse into surface, which can affect impact properties.
Vapor Deposition Chemical Vapor Deposition (CVD)	Deposition of coating material by chemical reactions in the gaseous phase. Reactive gases replace a protective atmosphere in a vacuum chamber. At temperatures of 1800°F to 1200°F, a thin diffusion zone is created between the base metal and the coating.	To 84 R_c; 0.0002 to 0.0004 in.	Large quantities per batch. Short reaction times reduce substrate stresses. Excellent adhesion, recommended for forming tools. Multiple coatings can be applied (TiN, TiC, AL_2O_3). Line-of-sight not a problem.	High temperatures can affect substrate metallurgy, requiring post heat treatment, which can cause dimensional distortion (except when coating sintered carbides). Necessary to reduce effects of hydrogen chloride on material properties, such as impact strength. Usually not diffused. Tolerances of +.001 required for HSS tools.
Physical Vapor Desposition (PVD sputtering)	Plasma is generated in a vacuum chamber by ion bombardment to dislodge particles from a target made of the coating material. Metal is evaporated and is condensed or attracted to substrate surfaces.	To 84 R_c; To 0.0002 in. thick	A useful experimental procedure for developing wear surfaces. Can coat substrates with metals, alloys, compounds, and refactories. Applicable for all tooling.	Not a high production method. Requires care in cleaning. Usually not diffused.
PVD (Electron beam)	A plasma is generated in vacuum by evaporation from a molten pool which is heated by an electron beam gun.	To 84 R_c; To 0.0002 in. thick	Can coat reasonable quantities per batch cycle. Coating materials are metals, compounds, alloys, and refactories. Substrate metallurgy is preserved. Very good adhesion. Fine particle deposition. Applicable for all tooling.	Parts require fixturing and orientation in line-of-sight process. Ultra cleanliness required.
PVD/ARC	Titanium is evaporated in a vacuum and reacted with nitrogen gas. Resulting titanium nitride plasma is ionized and electrically attracted to the substrate surface. A high energy process with multiple plasma guns.	To 85 R_c; To 0.0002 in. thick	Process at 900°F preserves substrate metallurgy. Excellent coating adhesion. Controllable deposition of grain size and growth. Dimensions, surface finish, and sharp edges are preserved. Can coat all high speed steels without distortion.	Parts must be fixtured for line-of-sight process. Parts must be very clean. No by-products formed in reaction. Usually only minor diffusion.

[a] Rockwell hardness values above 68 are estimates.

blades, because of limitations in the casting process and expense involved in the final shaping (grinding). The high cost of fabrication is primarily due to the high hardness of the material in the as-cast condition. Materials machinable with this tool material include plaincarbon steels, alloy steels, nonferrous alloys, and cast iron.

Cast cobalt alloys are currently being phased out for cutting-tool applications because of increasing costs, shortages of strategic raw materials (Co, W, and Cr), and the development of other, superior tool materials at lower cost.

CARBIDE OR SINTERED CARBIDES
Carbide cutting tool inserts are traditionally divided into two primary groups:

1. Straight tungsten grades, which are used for machining cast irons, austenitic stainless steel, and nonferrous and nonmetallic materials.

2. Grades containing major amounts of titanium, tantalum, and/or columbium carbides, which are used for machining ferritic workpieces. There are also the titanium carbide grades, which are used for finishing and semifinishing ferrous alloys.

The classification of carbide insert grades employs a C-classification system in the United States and ISO P and M classification system in Europe and Japan. These classifications are based on application, rather than composition or properties. Each cutting tool vendor can provide proprietary grades and recommended applications.

Carbides, which are nonferrous alloys, are also called *sintered* (or cemented) carbides because they are manufactured by powder metallurgy techniques. The P/M process is outlined in Figure 22-7. See Chapter 19 for details on powder metallurgy processes. These materials became popular during World War II, as they afforded a four- or five-fold increase in cutting speeds. The early versions had tungsten carbide as the major constituent, with a cobalt binder in amounts of 3 to 13%. Most carbide tools in use today are either straight WC or multicarbides of W-Ti or W-Ti-Ta, depending upon the work material to be machined. Cobalt is the binder. These tool materials are much harder, and chemically more stable, have better hot hardness, high stiffness, and lower friction, and operate at higher cutting speeds than HSS. They are more brittle and more expensive and use strategic metals (W, Ta, Co) more extensively.

Cemented carbide tool materials based on TiC have been developed primarily for auto industry applications using predominantly Ni and Mo as a binder. These are used for higher speed (>1000 ft/min) finish machining of steels, and some malleable cast irons.

Cemented carbide tools are available in insert form in many different shapes: squares, triangles, diamonds, and rounds. They can be either brazed or mechanically clamped onto the tool shank. Mechanical clamping (Figure 22-8) is more popular because when one edge or corner becomes dull, the insert is rotated or turned over to expose a new cutting edge. Mechanical inserts can be purchased in the "as-pressed" state or the insert can be ground to closer tolerances. Naturally, precision ground inserts cost more. Any part tolerance less than ±0.003 in. normally cannot be manufactured without radial adjustment of the cutting tool, even with ground inserts. If no radial adjustment is performed, precision-ground inserts should be used only when the part tolerance is between ±0.003 and ±0.006. Pressed inserts have an application advantage as the cutting edge is unground and thus does not leave grinding marks on the part after machining. Ground inserts can break under heavy cutting loads because the grinding marks on the

FIGURE 22-7 PM process for making cemented carbide insert tools.

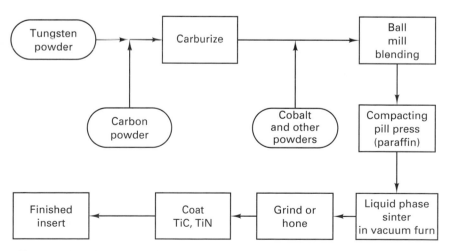

Tungsten is carburized in a high-temperature furnace, mixed with cobalt and blended in large ball mills. After ball milling, the powder is screened and dried. Paraffin is added to hold the mixture together for compacting. Carbide inserts are compacted using a pill press. The compacted powder is sintered in a high-temperature vacuum furnace. The solid cobalt dissolves some tungsten carbide, then melts and fills the space between adjacent tungsten carbide grains. As the mixture is cooled, most of the dissolved tungsten carbide precipitates onto the surface of existing grains. After cooling, inserts are finish ground and honed or used in the pressed condition.

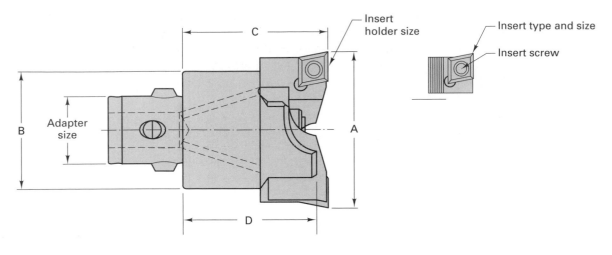

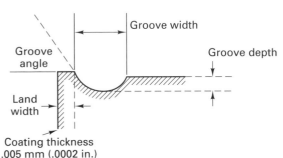

FIGURE 22-8 Boring head with carbide insert cutting tools. These inserts have a chip groove which can cause the chips to curl tightly and break into small, easily disposed lengths.

insert produce stress concentrations that result in brittle fracture. Diamond grinding is used to finish carbide tools. Abusive grinding can lead to thermal cracks and premature (early) failure of the tool. Brazed tools have the carbide insert brazed to the steel tool shank. These tools will have a more accurate geometry than the mechanical insert tools but they are more expensive. Since cemented carbide tools are relatively brittle, a 90° corner angle at the cutting edge is desired. To strengthen the edge and prevent edge chipping, it is rounded off by honing, or an appropriate chamfer or a negative land (a T-land) on the rake face is provided. The preparation of the cutting edge can affect tool life. The sharper the edge (smaller edge radius), the more likely the edge is to chip or break. Increasing the edge radius will increase the cutting forces, so a trade-off is required. Typical edge radius values are 0.001 to 0.003 in.

A *chip groove* (Figure 22-8) with a positive rake angle at the tool tip may also be used to reduce cutting forces without reducing the overall strength of the insert significantly. The groove also breaks up the chips for easier disposal by causing them to curl tightly.

For very low speed cutting operations, the chips tend to weld to the tool face and cause subsequent microchipping of the cutting edge. Cutting speeds are generally in the 150 to 600 ft/min range. Higher speeds (>1000 ft/min) are recommended for certain less difficult-to-machine materials (such as aluminum alloys), and much lower speeds (100 ft/min) for more difficult-to-machine materials (such as titanium alloys). In interrupted cutting applications, it is important to prevent edge chipping by choosing the appropriate cutter geometry and cutter position with respect to the workpiece. For interrupted cutting, finer grain size and higher cobalt content improve toughness in straight WC-Co grades.

After use, carbide inserts (called disposable or throwaway inserts) are generally recycled in order to reclaim the Ta, WC, and Co. This recycling not only conserves strategic materials, but also reduces costs. A new trend is to regrind these tools for future use where the actual size of the insert is not of critical concern.

COATED CARBIDE TOOLS

Beginning in 1969 with TiC coated WC, coated tools became the norm in the metalworking industry because coating can consistently improve tool life 200 or 300% or more. In cutting tools, material requirements at the surface of the tool need to be abrasion-resistant, hard, and chemically inert to prevent the tool and the work material from interacting chemically with each other during cutting. A thin, chemically stable, hard refractory coating of TiC, TiN, or AL_2O_3 accomplishes this objective. The bulk of the tool is a tough, shock-resistant carbide which can withstand high-temperature plastic deformation and resist breakage. The result is a composite tool as shown in Figure 22-9.

To be effective, the coatings should be hard, refractory, chemically stable, and chemically inert to shield the constituents of the tool and the workpiece from interacting chemically under cutting conditions. The coatings must be fine grained, free of binders and porosity. Naturally, the coatings must be metallurgically bonded to the substrate. Interface coatings are graded to match the properties of the coating and the substrate. The coatings must be thick enough to prolong tool life but thin enough to prevent brittleness.

Coatings should have a low coefficient of friction so the chips do not adhere to the rake face. TiC-coated tools were introduced in 1969. Coating materials now include single coatings of TiC, TiN, Al_2O_3, HfN, or HfC. Multiple coatings are used, with each layer imparting its own characteristic to the tool. The most successful combinations are: TiN/TiC/TiCN/TiN and TiN/TiC/Al_2O_3. Chemical vapor deposition (CVD) is used to obtain coated carbides. The coatings are formed by chemical reactions that take place only on or near the substrate. Like electroplating, chemical vapor deposition is a process in which the deposit is built up atom by atom. It is therefore capable of producing deposits of maximum density and of closely reproducing fine detail on the substrate surface.

Control of critical variables such as temperature, gas concentration, and flow pattern is required to assure adhesion of the coating to the substrate. The coating-to-substrate adhesion must be better for cutting tool inserts than for most other coatings applications to survive the cutting pressure and temperature conditions without flaking off. Grain size and shape are controlled by varying temperature and/or pressure.

The purpose of multiple coatings is to tailor the coating thickness for prolonged tool life. Multiple coatings allow a stronger metallurgical bond between the coating and

FIGURE 22-9   Triple-coated carbide tools provide resistance to wear and plastic deformation in machining of steel, abrasive wear in cast iron, and built-up edge formation.

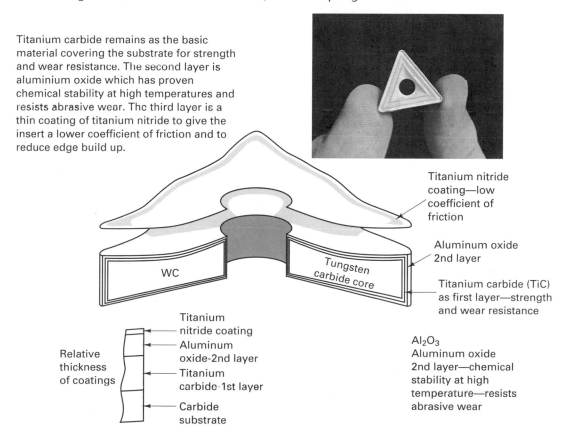

Titanium carbide remains as the basic material covering the substrate for strength and wear resistance. The second layer is aluminium oxide which has proven chemical stability at high temperatures and resists abrasive wear. The third layer is a thin coating of titanium nitride to give the insert a lower coefficient of friction and to reduce edge build up.

Titanium nitride coating—low coefficient of friction

Aluminum oxide 2nd layer

Titanium carbide (TiC) as first layer—strength and wear resistance

WC

Tungsten carbide core

Relative thickness of coatings

Titanium nitride coating
Aluminum oxide-2nd layer
Titanium carbide-1st layer
Carbide substrate

Al_2O_3 Aluminum oxide 2nd layer—chemical stability at high temperature—resists abrasive wear

the substrate and provide a variety of protection processes for machining different work materials, thus offering a more general-purpose tool material grade. A very thin final coat of TiN coating (5 μm) can effectively reduce crater formation on the tool face by one to two orders of magnitude relative to uncoated tools.

Coated inserts of carbides are finding wide acceptance in many metal-cutting applications. Coated tools have two or three times the wear resistance of the best uncoated tools with the same breakage resistance. This results in a 50 to 100% increase in speed for the same tool life. Because most coated inserts cover a broader application range, fewer grades are needed, and therefore inventory costs are lower. Aluminum oxide coatings have demonstrated excellent crater wear resistance by providing a chemical/diffusion reaction barrier at the tool/chip interface, permitting a 90% increase in cutting speeds in machining some steels.

Coated carbide tools have progressed to the place where in the United States about 80–90% of the carbide tools used in metalworking are coated.

CERAMICS

Ceramics are made of pure *aluminum oxide*, Al_2O_3, or Al_2O_3 used as a metallic binder. Using P/M, very fine particles are formed into cutting tips under a pressure of 20 to 28 tons/in^2 (267 to 386 MPa) and sintered at about 1800°F (1000°C). Unlike the case with ordinary ceramics, sintering occurs without a vitreous phase.

Ceramics usually are in the form of disposable tips. They can be operated at two to three times the cutting speeds of tungsten carbide. They almost completely resist cratering, run with no coolant, and have about the same tool life at their higher speeds as tungsten carbide does at lower speeds. As shown in Table 22-3, ceramics are usually as hard as carbides but are more brittle (lower bend strength) and therefore require more rigid toolholders and machine tools in order to take advantage of their capabilities. Their hardness and chemical inertness make ceramics a good material for high-speed finishing and/or high-removal-rate machining applications of superalloys, hard-chill cast iron, and high-strength steels. Because ceramics have poor thermal and mechanical shock resistance, interrupted cuts and interrupted application of coolants can lead to premature tool failure. Edge chipping is usually the dominant mode of tool failure. Ceramics are not suitable for aluminum, titanium, and other materials that react chemically with alumina-based ceramics. Recently, whisker-reinforced ceramic materials that have greater transverse rupture strength have been developed. The whiskers are made from silicon carbide.

CERMETS

Cermets are a new class of tool materials best suited for finishing. Cermets are ceramic TiC, nickel, cobalt, and tantalum nitrides. TiN and other carbides are used for binders. Cermets have superior wear resistance, longer tool life, and can operate at higher cutting speeds with superior wear resistance. Cermets have higher hot hardness and oxidation resistance than cemented carbides. The better finish imparted by a cermet is due to its low level of chemical reaction with iron [less cratering and *built-up-edge* (BUE)]. Compared to carbide, the cermet has less toughness, lower thermal conductivity, and greater thermal expansion, so thermal cracking can be a problem during interrupted cuts.

Cermets are usually cold-pressed and proper processing techniques are required to prevent insert cracking. New cermets are designed to resist thermal shocking during

TABLE 22-3. Properties of Cutting Tool Materials Compared for Carbides, Ceramics, HSS, and Cast Cobalt[a]

	Hardness Rockwell A or C	Transverse Rupture (bend) Strength ($\times 10^3$ psi)	Compressive Strength ($\times 10^3$ psi)	Modulus of Elasticity (e) ($\times 10^6$ psi)
Carbide C1–C4	90–95 R_A	250–320	750–860	89–93
Carbide C5–C8	91–93 R_A	100–250	710–840	66–81
High Speed Steel	86 R_A	600	600–650	30
Ceramic (oxide)	92–94 R_A	100–125	400–650	50–60
Cast cobalt	46–62 R_C	80–120	220–335	40

[a]Exact properties depend upon materials, grain size, bonder content, volume.

milling by using high nitrogen content in the titanium carbonitride phase (produces finer grain size) and adding WC and TaC to improve shock resistance. PVD coated cermets have the wear resistance of cermets and the toughness range of a coated carbide, and perform well with a coolant.

Figure 22-10 shows a comparison of speed/feed coverage of typical cermets compared to ceramics, carbides, and coated carbides. The values illustrate that cermets can clearly cover a wide range of important metalcutting applications.

FIGURE 22-10 Comparison of cermets with various cutting tool materials.

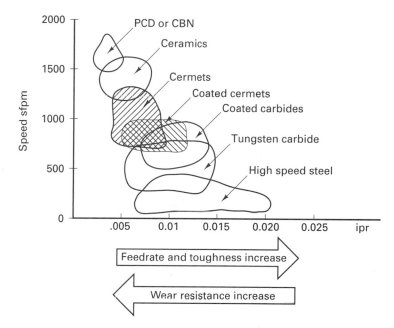

Tool Material Group	General Applications	Versus Cermet
PCD Polycrystal diamond	High-speed machining of aluminum alloys, nonferrous metals and nonmetals.	Cermets can machine same materials, but at lower speeds and significantly less cost per corner.
CBN Cubic boron nitride	Hard workpieces and high-speed machining on cast irons.	Cermets cannot machine the harder workpieces that CBN can. Cermets cannot machine cast iron at the speeds CBN can. The cost per corner of cermets is significantly less.
Ceramics (Cold press)	High-speed turning and grooving of steels and cast iron.	Cermets are more versatile and less expensive than cold press ceramics, but cannot run at the higher speeds.
Ceramics (Hot press)	Turning and grooving of hard workpieces, high-speed finish machining of steels and irons.	Cermets cannot machine the harder workpieces or run at the same speeds on steels and irons but are more versatile and less expensive.
Ceramics (Silicon nitride)	Rough and semi-rough machining of cast irons in turning and milling applications at high speeds and under unfavorable conditions.	Cermets cannot machine cast iron at the high speeds of silicon nitride ceramics, but in moderate speed applications cermets may be more cost effective.
Coated Carbide	General-purpose machining of steels, stainless steels, cast iron, etc.	Cermets can run at higher cutting speeds and provide better tool life at less cost for semi-roughing to finishing applications.
Carbides	Tough material for lower speed applications on various materials.	Cermets can run at higher speeds, provide better surface finishes and longer tool life for semi-roughing to finishing applications.

DIAMONDS

Diamond is the hardest material known. Industrial diamonds are now available in the form of polycrystalline compacts, which are finding industrial application in the machining of aluminum, bronze, and plastics, greatly reducing the cutting forces as compared to carbides. Diamond machining is done at high speeds with fine feeds for finishing and produces excellent finishes. Recently, *single-crystal* diamonds, with a cutting-edge radius of 100 Å or less, have been used for precision machining of large mirrors. However, single-crystal diamonds have been used for years to machine brass watch faces, thus eliminating polishing. They have also been used to slice biological materials into thin films for viewing in transmission electron microscopes. (This process, known as ultramicrotomy, is one of the few industrial versions of orthogonal machining in common practice.)

The salient features of diamond tools include high hardness, good thermal conductivity, the ability to form a sharp edge of cleavage (single-crystal, natural diamond), very low friction, nonadherence to most materials, the ability to maintain a sharp edge for a long period of time, especially in machining soft materials such as copper and aluminum, and good wear resistance.

To be weighed against these advantages are some shortcomings, which include a tendency to interact chemically with elements of group IVB to group VIII of the periodic table. In addition, diamond wears rapidly when machining or grinding mild steel. It wears less rapidly with high-carbon alloy steels than with low-carbon steel, and has occasionally machined grey cast iron (which has high carbon content) with long life. Diamond has a tendency to revert at high temperatures (700°C) to graphite and/or to oxidize in air. Diamond is very brittle and is difficult and costly to shape into cutting tools, the process for doing the latter being a tight held industry practice.

Limited supply, increasing demand, and high cost of natural diamonds have led to the ultrahigh-pressure (50 Kbar), high-temperature (1500°C) synthesis of diamond from graphite at the General Electric Company in the mid-1950s, and the subsequent development of *polycrystalline* sintered diamond tools in the late 1960s.

Polycrystalline diamond (PCD) tools consist of a thin layer (0.5 to 1.5 mm) of fine-grain-size diamond particles sintered together and metallurgically bonded to a cemented carbide substrate. A high-temperature/high-pressure process, using conditions close to those used for the initial synthesis of diamond, is needed. Fine diamond powder (1 to 30 μm) is first packed on a support base of cemented carbide in the press. At the appropriate sintering conditions of pressure and temperature in the diamond stable region, complete consolidation and extensive diamond-to-diamond bonding take place. Sintered diamond tools are then finished to shape, size, and accuracy by laser cutting and grinding. See Figure 22-11. The cemented carbide provides the necessary elastic support for the hard and brittle diamond layer above it. The main advantages of sintered polycrystalline tools over natural single-crystal tools are better quality, greater toughness, and improved wear resistance, resulting from the random orientation of the diamond grains and the lack of large cleavage planes.

Diamond tools offer dramatic performance improvements over carbides. Tool life is often greatly improved as is control over part size, finish, and surface integrity.

Positive rake tooling is recommended for the vast majority of diamond tooling applications. If BUE is a problem, increasing cutting speed and using more positive rake angles may eliminate it. If edge breakage and chipping are problems, one can reduce the feed rate. Coolants are not generally used in diamond machining unless, as in the machining of plastics, it is necessary to reduce airborne dust particles. Diamond tools can be reground.

There is much commercial interest in being able to coat HSS and carbides directly with diamond, but getting the diamond coating to adhere reliably has been difficult. Diamond-coated inserts would deliver roughly the same performance as PCD tooling when cutting nonferrous materials but could be given more complex geometries and chip breakers while reducing the cost per cutting edge.

POLYCRYSTALLINE CUBIC BORON NITRIDES

Polycrystalline cubic boron nitride (PCBN) is a man-made tool material widely used in the automotive industry for machining hardened steels and superalloys. It is made in a

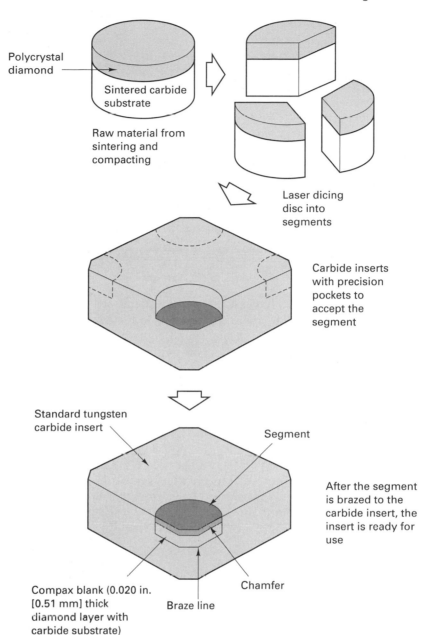

Polycrystal diamond

Sintered carbide substrate

Raw material from sintering and compacting

Laser dicing disc into segments

Carbide inserts with precision pockets to accept the segment

Standard tungsten carbide insert

Segment

After the segment is brazed to the carbide insert, the insert is ready for use

Compax blank (0.020 in. [0.51 mm] thick diamond layer with carbide substrate)

Braze line

Chamfer

FIGURE 22-11 Polycrystalline diamond tools are carbides with diamond inserts. They are restricted to simple geometries.

compact form for tools by a process quite similar to that used for sintered polycrystalline diamonds. It retains its hardness at elevated temperatures (Knoop 4700 at 20°C, 4000 at 1000°C) and has low chemical reactivity at the tool/chip interface. This material can be used to machine hard aerospace materials like Inconel 718 and René 95 as well as chilled cast iron.

Although not as hard as diamond, PCBN is less reactive with such materials as hardened steels, hard-chill cast iron, and nickel- and cobalt-based superalloys. PCBN can be used efficiently and economically to machine these difficult-to-machine materials at higher speeds (fivefold) and with a higher removal rate (fivefold) than cemented carbide, and with superior accuracy, finish, and surface integrity. PCBN tools are available in basically the same sizes and shapes as sintered diamond and are made by the same process. The cost of an insert is somewhat higher than either cemented carbide or ceramic tools, but the tool life may be five to seven times that of a ceramic tool. Therefore, to see the economy of using PCBN tools, it is necessary to consider all the factors.

TABLE 22-4. Cost Comparison for Machining Liner Bores in 1500 Engine Blocks[a]

	Ceramic TNG-433	PCBN BTNG-433
Cost per insert	$14.90	$208.00
Edges per insert	6	3
Cost per edge	$2.48	$69.33
Time per index (6 tools)	.25 hr	.25 hr
Cost per index at $45 per hour	$11.25	$11.25
Indexes per 1500 blocks	43	3
Indexing cost (Indexes × $11.25)	$483.75	$33.75
Insert cost for 6 spindles	$638.34	$1248.00
Labor and tool cost	$1122.09	$1281.00
Cost per bore	$.125	$.142
Total number of toolchanges	43	3
Downtime for 1500 blocks	×.25 hr	×.25 hr
	10.75 hr	.75 hr

[a] To see the economy of using PCBN cutting tools, it is important to consider all factors of the operation, especially downtime for tool changing.

Here is an industrial example of analysis of tooling economics, where a comparison is being made between two tool materials (insert tools). A manufacturer of diesel engines is producing an in-line six-cylinder engine block which is machined on a transfer line. Each cylinder hole must be bored to accept a sleeve liner. This operation has a depth of cut of 0.062 in. per side, for a total of 0.125 in. stock removal. The tolerance on this bore is ±0.001 in. and the spindle is operating at 2000 sfpm. Ceramic inserts are used on this operation, but with these inserts, wear was severe enough to require indexing after only 35 pieces. The ceramic insert was replaced with PCBN inserts made of a high-content BCN. Both inserts had a 0.001 to 0.002 in. radius hone for edge preparation.

Table 22-4 is a cost comparison between the ceramic and PCBN insert. The PCBN insert used in the application is a full-top PCBN insert, meaning that the entire top of the insert is a layer of PCBN material. At first glance the PCBN tool appears to be extremely expensive. The insert cost $208.00 and provides only three usable edges, whereas the ceramic insert costs $14.90 and provides six usable edges.

The ceramic tool must be indexed every 35 pieces. The PCBN tool is indexed every 500 pieces. The cost per bore, including insert cost and the cost of labor to perform indexing, comes to $0.125 per bore for the ceramic tool and $0.142 per bore for the PCBN tool. This appears to make the ceramic tool more cost-effective, but downtime for indexing has not been accounted for. The ceramic insert required 10.75 hours of downtime for indexing, whereas the PCBN tool required only 0.75 hour of downtime for indexing. Use of the PCBN cutting tool will significantly reduce the total cost per piece by eliminating 10 hours of downtime and increasing the productivity of the machine. Later in this chapter the economics of machining will be addressed again.

The two predominant wear modes of PCBN tools are notching at the *depth-of-cut-line* (DCL) and *microchipping*. In some cases, the tool will exhibit flank wear of the cutting edge. These tools have been used successfully for heavy interrupted cutting and for milling white cast iron and hardened steels using negative lands and honed cutting edges.

Since diamond and PCBN are extremely hard but brittle materials, new demands are being placed on the machine tools and on machining practice in order to take full advantage of the potential of these tool materials. These demands include:

- Use of more rigid machine tools, and machining practices involving gentle entry and exit of the cut in order to prevent microchipping
- Use of high-precision machine tools, because these tools are capable of producing high finish and accuracy
- Use of machine tools with higher power, because these tools are capable of higher metal removal rates and faster spindle speeds

Table 22-5 shows the application of cutting tool materials to workpiece materials.

TABLE 22-5.   Application of Cutting Tool Materials to Workpiece Materials

Workpiece Material	Applicable Tool Material			
	Carbide Coated Carbide	Ceramic, Cermet	Cubic Boron Nitride	Diamond Compacts
Cast irons, carbon steels	X	Uninterrupted finishing cuts X		
Alloy steels, alloy cast iron	X	X	X	
Aluminum, brass	X	X		X
High silicon aluminum	X			X
Nickel-based	X	X	X	
Titanium	X			
Plastic composites	X		X	

■ 22.4 TOOL GEOMETRY

Figure 22-12 shows the cutting tool geometry for a single point tool (HSS) used in turning. The *back rake angle* affects the ability of the tool to shear the work material and form the chip. It can be positive or negative. Positive rake angles reduce the cutting forces,

FIGURE 22-12   Standard terminology to describe the geometry of single-point tools.
a) orthogonal view of tool
b) oblique view of tool from cutting edge
c) top view
d) oblique view from shank end

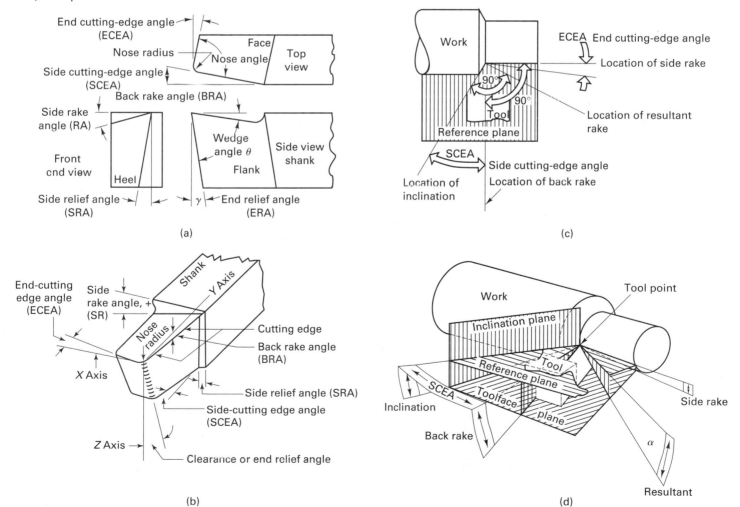

resulting in smaller deflections of the workpiece, toolholder, and machine. In machining hard work materials, the back rake angle must be small, even negative for carbide and diamond tools. Generally speaking, the higher the hardness of the workpiece the smaller the back rake angle. For high-speed steels, back rake angle is normally chosen in the positive range, depending on the type of tool (turning, planing, end milling, face milling, drilling, and so on) and the work material.

For carbide tools, inserts for different work materials and toolholders can be supplied with several standard values of back rake angle; −6 to +6°. The side rake angle and the back rake angle combine to form the effective rake angle. This is also called the true rake angle or resultant rake angle of the tool.

True rake inclination of a cutting tool has a major effect in determining the amount of chip compression and the shear angle. A small rake angle causes high compression, tool forces, and friction, resulting in a thick, highly deformed, hot chip. Increased rake angle reduces the compression, the forces, and the friction, yielding a thinner, less-deformed and cooler chip. Unfortunately, it is difficult to take much advantage of these desirable effects of larger positive rake angles, since they are offset by the reduced strength of the cutting tool, due to the reduced tool section, and by its greatly reduced capacity to conduct heat away from the cutting edge.

To provide greater strength at the cutting edge and better heat conductivity, zero or negative rake angles commonly are employed on carbide, ceramic, polydiamond, and PCBN cutting tools. These materials tend to be brittle, but their ability to hold their superior hardness at high temperatures results in their selection for high-speed and continuous machining operations. While the negative rake angle increases tool forces, it keeps the tool in compression and provides added support to the cutting edge. This is particularly important in making intermittent cuts and in absorbing the impact during the initial engagement of the tool and work.

In general, the power consumption is reduced by approximately 1% for each 1° in α. γ is the end relief angle. The wedge angle θ determines the strength of the tool and its capacity to conduct heat and depends on the values of α and γ. The relief angles mainly affect the tool life and the surface quality of the workpiece. To reduce the deflections of the tool and the workpiece and to provide good surface quality, larger relief values are required. For high-speed steel, relief angles in the range of 5 to 10° are normal, with smaller values being for the harder work materials. For carbides, the relief angles are lower to give added strength to the tool.

The side and end-cutting edge angles define the nose angle and characterize the tool design. The nose radius has a major influence on surface finish. Increasing the nose radius usually decreases tool wear and improves surface finish.

Tool nomenclature varies with different cutting tools, manufacturers, and users. Many terms are still not standard because of all this variety. The most common tool terms will be used in later chapters to describe specific cutting tools.

The introduction of coated tools has spurred the development of improved tool geometries. Specifically, *low-force groove* (LFG) geometries have been developed which reduce the total energy consumed, and break up the chips into shorter segments. These grooves effectively increase the rake angle, which increases the shear angle and lowers the cutting force and power. This means that higher cutting speeds or lower cutting temperatures (and better tool lives) are possible.

As a chip breaker, the groove deflects the chip at a sharp angle and causes it to break into short pieces that are easier to remove and are not so likely to become tangled in the machine and possibly cause injury to personnel. This is particularly important on high-speed, mass-production machines.

The shapes of cutting tools used for various operations and materials are compromises, resulting from experience and research so as to provide good overall performance. For coated tools, edge strength is an important consideration. A thin coat enables the edge to retain high strength, but a thicker coat exhibits better wear resistance. Normally, tools for turning have a thickness of 6 to 12 μm. Edge strength is higher for multilayer coated tools. The radius of the edge should be 0.0005 to 0.005 in.

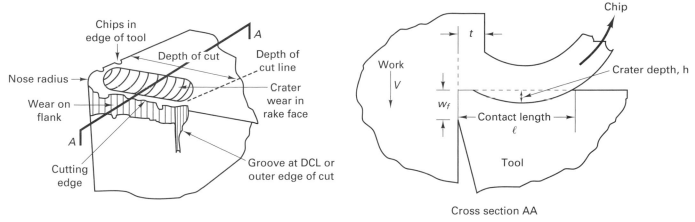

FIGURE 22-13 Sketch of a worn tool showing various wear elements resulting during oblique cutting: W_f, flank wear land length; t, uncut chip thickness.

■ 22.5 TOOL FAILURE AND TOOL LIFE

In metal cutting, the failure of the cutting tool can be classified into two broad categories, according to the failure mechanisms that caused the tool to die (or fail):

1. *Slow-death mechanisms*: gradual tool wear on the flank(s) of the tool below the cutting edge (called flank wear) or wear on the rake face of the tool (called *crater wear*) or both
2. *Sudden-death mechanisms*: rapid, usually unpredictable and often catastrophic failures resulting from abrupt, premature death of a tool

Figure 22-13 shows a sketch of a "worn" tool, showing *crater wear* and *flank wear*, along with wear of the tool nose radius and an outer diameter groove (the DCL groove). As the tool wears, its geometry changes. This geometry change will influence the cutting forces, the power being consumed, the surface finish obtained, the dimensional accuracy, and even the dynamic stability of the process. Worn tools are duller creating greater cutting forces, often resulting in chatter in processes that otherwise are usually relatively free of vibration. The actual wear mechanisms active in this high-temperature environment are abrasive, adhesion, diffusion, or chemical interactions. It appears that in metal cutting, any or all of these mechanisms may be operative at a given time in a given process.

The sudden-death mechanisms are more straightforward but less predictable. These mechanisms are categorized as plastic deformation, brittle fracture, fatigue fracture, or edge chipping. Here again it is difficult to predict which mechanism will dominate and result in a tool failure in a particular situation. What can be said is that tools, like people, die (or fail) from a great variety of causes under widely varying conditions. Therefore, tool life should be treated as a random variable, or probabilistically, and not as a deterministic quantity.

■ 22.6 FLANK WEAR

During machining, the tool is performing in a hostile environment wherein high contact stresses and high temperatures are commonplace, and therefore tool wear is always an unavoidable consequence. At lower speeds and temperatures, the tool most commonly wears on the flank. In Figure 22-14 four characteristic tool wear curves (average values) are shown for four different cutting speeds, V_1 through V_4. V_4 is the fastest cutting speed and therefore generates the fastest wear rates. Such curves often have three general regions, as shown in the figure. The central region is a steady-state region (or the region of secondary wear). This is the normal operating region for the tool. Such curves are typical for both flank wear and crater wear. When the amount of wear reaches the value W_f, the permissible tool wear on the flank, the tool is said to be "worn out." W_f is typically set at 0.025 to 0.030 in. for flank wear. For crater wear, the depth of the crater is used to determine tool failure.

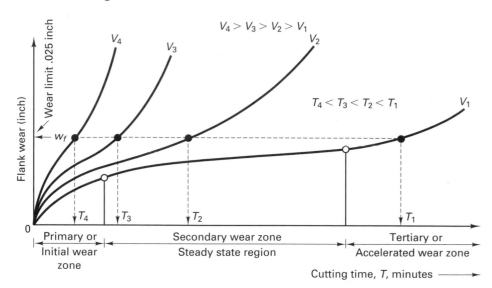

FIGURE 22-14   Typical tool wear curves for flank wear at different velocities.

Suppose that the tool wear experiment were to be repeated 15 times without changing any of the input parameters. The result would look like Figure 22-15, which depicts the variable nature of tool wear and shows why tool wear must be treated as a random variable. In Figure 22-15 the average time is denoted as μ_t and the standard deviation as σ_t, where the wear limit criterion was 0.025 in. At a given time during the test, 35 minutes, the tool displayed flank wear ranging from 0.013 to 0.021 in. with an average of $\mu_w = .00175$ in. with standard deviation σ_w. Now here is how the next figure known as a Taylor Tool Life plot is developed.

Using the empirical tool wear data shown in Figure 22-14, which used the values of T (time in minutes) associated with V (cutting speed) for a given amount of tool wear, W_f, (see the dashed-line construction) Figure 22-16 was developed. When V and T are plotted on log-log scales, a linear relationship appears described by the equation

$$VT^n = \text{constant} = K \qquad (22\text{-}3)$$

This equation is called the Taylor Tool Life equation because in 1907, F. W. Taylor published his now-famous paper "On the Art of Cutting Metals" in ASME Transactions, wherein tool

FIGURE 22-15   Tool wear on the flank displays a random nature, as does tool life.

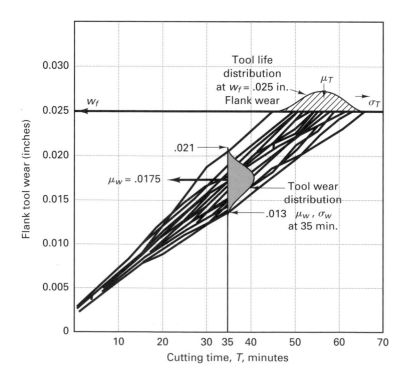

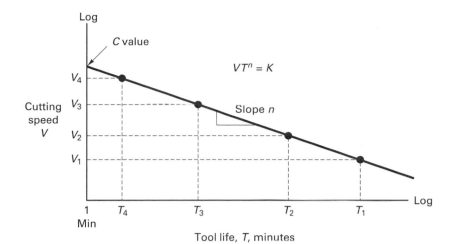

FIGURE 22-16 Construction of the Taylor Tool Life curve using data from deterministic tool wear plots like those of Figure 22-14.

life (T) was related to cutting speed (V) and feed (f). This equation had the form

$$T = \frac{\text{constant}}{f_x V_y} \tag{22-4}$$

which over the years took the more widely published form

$$VT^n = K$$

where n is an exponent which depends mostly on tool material but is affected by work material, cutting conditions, and environment and K is a constant that depends on all the input parameters, including feed. Table 22-6 provides some data on Taylor Tool Life constants.

Figure 22-17 shows typical tool life curves for one tool material and three work materials. Notice that all three plots have about the same slope, n. Typical values for n are 0.14 to 0.16 for HSS, 0.21 to 0.25 for uncoated carbides, 0.30 for TiC inserts, 0.33 for polydiamonds, 0.35 for TiN inserts and 0.40 for ceramic-coated inserts.

It takes a great deal of experimental effort to obtain the constants for the Taylor equation, as each combination of tool and work material will have different constants. Note that for a tool life of 1 minute, $K = V$, or the cutting speed that yields about 1 minute of tool life for this tool. A great deal of research has gone into developing more sophisticated versions of the Taylor equation, wherein constants for other input parameters (typically feed, depth of cut, and work material hardness) are experimentally determined: for example:

$$VT^n F^m d^p = K' \tag{22-5}$$

where n, m, and p are exponents and K' is a constant. Equations of this form are also deterministic and determined empirically.

The problem has been approached probabilistically in the following way. Since T depends upon speed, feed, materials, and so on, one writes

$$T = \frac{K^{1/n}}{V^{1/n}} = \frac{K}{V^m} \tag{22-6}$$

where K is now a random variable which represents the effects of all unmeasured factors and V^m is an input variable.

The sources of tool life variability include factors such as:

1. Variation in work material hardness (from part to part and within a part)
2. Variability in cutting-tool materials, geometry, and preparation
3. Vibrations in machine tool, including rigidity of work and tool-holding devices
4. Changing surface characteristics of workpieces

The examination of the data from a large number of tool life studies wherein a variety of steels were machined has shown that regardless of the tool material or process,

TABLE 22-6. Tool Life Information for Various Materials and Conditions

Source	Tool Material	Geometry	Workpiece Material	Size of Cut. in. Depth	Feed	Cutting Fluid	$VT^n = K$ n	K
1	High carbon steel	8.14, 6.6, 6.15, 3/64	Yellow brass (.60 Cu, 40 Zn, .85 NI, .006 Pb)	.050 .100	.0255 .0127	Dry Dry	.081 .096	242 299
1	High carbon steel	8.14, 6.6, 6.15, 3/64	Bronze (.9 Cu, .1.5n)	.050 .100	.0255 .0127	Dry Dry	.086 .111	190 232
1	HSS-18-4-1	8.14, 6.6, 6.15, 3/64	Cast Iron 160 Bhn Cast iron, nickel, 164 Bhn Cast iron, NI-Cr, 207 Bhn	.050 .050 .050	.0255 .0255 .0255	Dry Dry Dry	.101 .111 .088	172 186 102
1	HSS-18-4-1	8.14, 6.6, 6.15, 3/64	Stell, SAE B1113 C.D. Stell, SAE B1112 C.D. Stell, SAE B1120 C.D. Stell, SAE B1120 + Pb C.D. Stell, SAE B1035 C.D. Stell, SAE B1035 + Pb C.D.	.050 .050 .050 .050 .050 .050	.0127 .0127 .0127 .0127 .0127 .0127	Dry Dry Dry Dry Dry Dry	.080 .105 .100 .060 .110 .110	260 225 270 290 130 147
1	HSS-18-4-1	8.14, 6.6, 6.15, 3/64 8.14, 6.6, 6.13, 3/66 8.14, 6.6, 6.15, 3/64 8.14, 6.6, 6.15, 3/64	Stell, SAE 1045 C.D. Stell, SAE 2340 185 Bhn Stell, SAE 2345 198 Bhn Stell, SAE 3140 190 Bhn	.100 .100 .050 .100	.0127 .0125 .0255 .0125	Dry Dry Dry Dry	.110 .147 .105 .160	192 143 126 178
1	HSS-18-4-1	8.14, 6.6, 6.15, 3/64	Stell, SAE 4350 363 Bhn Stell, SAE 4350 363 Bhn Stell, SAE 4350 363 Bhn Stell, SAE 4350 363 Bhn Stell, SAE 4350 363 Bhn	.0125 .0125 .0250 .100 .100	.0127 .0255 .0255 .0127 .0255	Dry Dry Dry Dry Dry	.080 .125 .125 .110 .110	181 146 95 78 46
1	HSS-18-4-1	8.14, 6.6, 6.15, 3/64	Stell, SAE 4140 230 Bhn Stell, SAE 4140 271 Bhn Stell, SAE 6140 240 Bhn	.050 .050 .050	.0127 .0127 .0127	Dry Dry Dry	.180 .180 .150	190 159 197
1	HSS-18-4-1	8.22, 6.6, 6.15, 3/64	Monel metal 215 Bhn	.100 .150 .100 .100	.0127 .0255 .0127 .0127	Dry Dry Em SMO	.080 .074 .080 .105	170 127 185 189
1	Stellite 2400	0.0, 6.6, 6.0, 3/32	Steel, SAE 3240 annealed	.187 .125 .062 .031	.031 .031 .031 .031	Dry Dry Dry Dry	.190 .190 .190 .190	215 240 270 310
1	Stellite No. 3	0.0, 6.6, 6.0, 3/32	Cast iron 200 Bhn	.062	.031	Dry	.150	205
1	Carbide (T 64)	6.12, 5.5, 10.45	Steel, SAE 1040 annealed Steel, SAE 1060 annealed Steel, SAE 1060 annealed Steel, SAE 1060 annealed Steel, SAE 1060 annealed Steel, SAE 1060 annealed Steel, SAE 1060 annealed Steel, SAE 2340 annealed	.062 .125 .187 .250 .062 .062 .062 .062	.025 .025 .025 .025 .021 .042 .062 .025	Dry Dry Dry Dry Dry Dry Dry Dry	.156 .167 .167 .167 .167 .164 .162 .162	800 660 615 560 880 510 400 630
2	Ceramic	not available	AISI 4150 AISI 4150	.160 .160	.016 .016	Dry Dry	.400 .200	2000 620

Sources: 1 - Fundamentals of tool design. ASTME. A.R. Konecny, W.J. Potthoff   2 - Theory of Metal Cutting P.N. Black

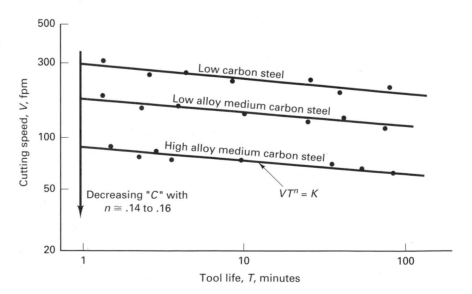

FIGURE 22-17 Log-log tool life plots for three steel work materials cut with HSS tool material.

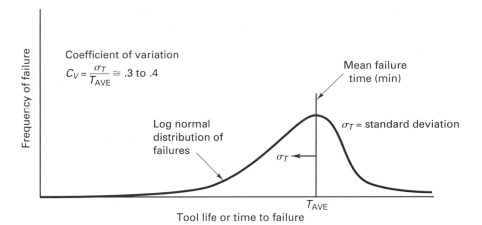

FIGURE 22-18 Tool life viewed as random variable has a log normal distribution with a large coefficient of variation.

tool life distributions are usually log normal and typically have a large standard deviation. As shown in Figure 22-18, tool life distributions have a large coefficients of variation, C_V so tool life is not very predictable.

Other criteria that can be used to define tool death in addition to wear limits are surface finish, failure to conform to size (tolerances), increases in cutting forces and power, time required to drill a hole under a given load, or complete failure of the tool. In automated processes, it is very beneficial to be able to monitor the tool wear on-line, so that the tool can be replaced prior to failure, wherein defective products may also result. The feed force has been shown to be a good, indirect measure of tool wear. That is, as the tool wears and dulls, the feed force increases more than the cutting force increases.

Once criteria for failure have been established, tool life is that time elapsed between start and finish of the cut, in minutes. Other ways to express tool life, other than time, include:

1. Volume of metal removed between regrinds or replacement of tool
2. Number of holes drilled with a given tool. See Figure 22-19
3. Number of pieces machined per tool

FIGURE 22-19 Tool life data for various coated drills. Notice how TiN-coated HSS drills outperform uncoated drills.

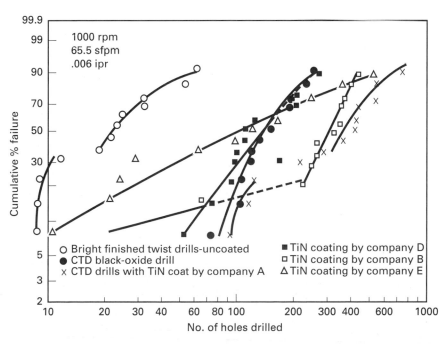

Drill performance based on the number of holes drilled with 1/4 in. diam. drills in T–1 structural steel.

RECONDITIONING CUTTING TOOLS

In the reconditioning of tools by sharpening and recoating, care must be taken in grinding the tool's surfaces. The following guidelines should be observed:

1. Resharpen to original tool geometry specifications. Restoring the original tool geometry will help the tool achieve consistent results on subsequent uses. Computer numerical control (CNC) grinding machines for tool resharpening have made it easier to restore a tool's original geometry.

2. Grind cutting edges and surfaces to a fine finish. Rough finishes left by poor and abusive regrinding hinder the performance of resharpened tools. For coated tools, tops of ridges left by rough grinding will break away in early tool use, leaving uncoated and unprotected surfaces that will cause premature tool failure.

3. Remove all burrs on resharpened cutting edges. If a tool with a burr is coated, premature failure can occur because the burr will break away in the first cut, leaving an uncoated surface exposed to wear.

4. Avoid resharpening practices that overheat and burn or melt (called *glazed over*) the tool surfaces, as this will cause problems in coating adhesion. Polishing or wire brushing of tools causes similar problems.

The cost of each recoating is about one-fifth the cost of purchasing a new tool. By recoating, the tooling cost per workpiece can be cut by between 20 and 30%, depending on the number of parts being machined.

■ 22.7 ECONOMICS OF MACHINING

The cutting speed has such a great influence on the tool life compared to the feed or the depth of cut that it greatly influences the overall economics of the machining process. For a given combination of work material and tool material, a 50% increase in speed results in a 90% decrease in tool life, while a 50% increase in feed results in a 60% decrease in tool life. A 50% increase in depth of cut produces only a 15% decrease in tool life. Therefore, in limited-horsepower situations, depth of cut and then feed should be maximized while speed is held constant and horsepower consumed is maintained within limits. As cutting speed is increased, the machining time decreases but the tools wear out faster and must be changed more often. In terms of costs, the situation is as shown in Figure 22-20, which shows the effect of cutting speed on the cost per piece.

FIGURE 22-20 Cost per unit for a machining process versus cutting speed.

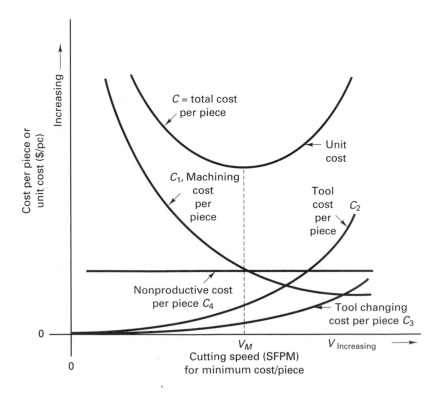

The total cost per operation is comprised of four individual costs: machining costs, tool costs, tool changing costs, and handling costs. The machining cost is observed to decrease with increasing cutting speed because the cutting time decreases. Cutting time is proportional to the machining costs. Both the tool costs and the tool changing costs increase with increases in cutting speeds. The handling costs are independent of cutting speed. Adding up each of the individual costs results in a total unit cost curve which is observed to go through a minimum point. For a turning operation, the total cost per piece, C, equals

$$C = C_1 + C_2 + C_3 + C_4 \tag{22-7}$$

$$= \text{Machining cost} + \text{tooling cost} + \text{tool changing cost} + \text{handling cost per piece.}$$

Expressing each of these cost terms as a function of cutting velocity will permit the summation of all the costs.

$$C_1 = T_m \times C_o \qquad\qquad \text{where } C_o = \text{operating cost (\$/min) and}$$
$$T_m = \text{cutting time (min/piece)}$$

$$C_2 = \left(\frac{T_m}{T}\right) C_t \qquad\qquad \text{where } T = \text{tool life (min/tool)}$$
$$\text{and } C_t = \text{initial cost of tool (\$)}$$

$$C_3 = t_c \times C_o\left(\frac{T_m}{T}\right) \qquad\qquad \text{where } t_c = \text{time to change tool (min)}$$

$$\text{and } \frac{T_m}{T} = \text{number of tool changes per piece}$$

$$C_4 = \qquad\qquad \text{labor, overhead, and machine tool costs consumed while part is being loaded, unloaded, tools advanced, machine broken down, etc.}$$

$$\text{Since } T_m = L/Nf_r \text{ for turning}$$

$$= \pi DL/12Vf_r$$

$$\text{and } T = (K/V)^{1/n} \text{ rewriting equation (22-3)}$$

$$C \text{ can be expressed in terms of } V.$$

$$C = \frac{L\pi DC_o}{12Vf_r} + \frac{C_t V^{1/n}}{K^{1/n}} + \frac{t_c C_o V^{1/n}}{K^{1/n}} + C_4 \tag{22-8}$$

To find the minimum, take $dC/dV = 0$ and solve for V

$$V_m = \left[\frac{1}{n} - 1\right]\left[\frac{C_t + (C_o \times t_c)}{C_o}\right] \tag{22-9}$$

A word of caution here is appropriate. Note that this derivation was totally dependent upon the Taylor tool life equation. Such data may not be available because they are expensive and time-consuming to obtain. Even when the tool life data are available, this procedure assumes that the tool fails only by whichever wear mechanism (flank or crater) was described by this equation and by no other failure mechanism. Recall that tool life had a very large coefficient of variation and was probabilistic in nature. This derivation assumes that for a given V, there is one T and this simply is not the case. The model also assumes that the workpiece material is homogeneous, the tool geometry is preselected, the depth of cut and feed rate are known and remain unchanged during the entire process, sufficient horsepower is available for the cut at the economic cutting conditions, and the cost of operating time is the same whether the machine is cutting or not cutting.

Table 22-7 shows an example of the analysis of tooling economics, where a comparison is being made between four different tools, all used for turning hot rolled 8620 steel with triangular inserts. Operating costs for the machine tool were $60 per hour. The low force groove insert has only three cutting edges available instead of six. It takes

TABLE 22-7.   Cost Comparison of Four Tool Materials, Based on Equal Tool Life of 40 Piece per Cutting Edge

	Uncoated	TiC-coated	Al_2O_3-coated	Al_2O_3LFG
Cutting speed (surface ft/min)	400	640	1100	1320
Feed (in/rev)	0.020	0.022	0.024	0.028
Cutting edges available per insert	6	6	6	3
Cost of an insert ($/insert)	4.80	5.52	6.72	6.72
Tool life (pieces/cutting edge)	192	108	60	40
Tool change time per piece (min)	0.075	0.075	0.075	0.075
Nonproductive cost per piece ($/pc)	0.50	0.50	0.50	0.50
Machining time per piece (min/pc)	4.8	2.7	1.50	1.00
Machining cost per piece ($/unit)	4.8	2.7	1.5	1.00
Tool change cost per piece ($/pc)	0.08	0.08	0.08	0.08
Cutting tool cost per piece ($/pc)	0.02	0.02	0.03	0.06
Total cost per piece ($/pc)	5.40	3.30	2.11	1.64
Production rate (pieces/hr)	11	18	29	38
Improvement in productivity based on pieces/hr (%)	0	64	164	245

Source: Data from T. E. Hale et al., "High Productivity Approaches to Metal Removal," *Materials Technology*, Spring 1980, p. 25.

3 min to change inserts and $\frac{1}{2}$ min to unload a finished part and load in a new 6-in.-diameter bar stock. The length of cut was about 24 in. This table should be carefully analyzed and studied so that each line is understood. Note that the cutting tool cost per piece was three times higher for the low force groove tool over the carbide but really of no consequence since the major cost per piece comes from two sources: the machining cost per piece and the nonproductive cost per piece.

MACHINABILITY

Machinability is a much maligned term which has many different meanings but generally refers to the ease with which a metal can be machined to an acceptable surface finish. The principal definitions of the term are entirely different, the first based on material properties, the second based on tool life, and the third based on cutting speed.

1. Machinability is defined by the ease or difficulty with which the metal can be machined. In this light, specific energy, specific horsepower, and shear stress are used as measures, and in general, the larger the shear stress or specific power values, the more difficult the material is to machine, requiring greater forces and lower speeds. In this definition, the material is the key.

2. Machinability is defined by the relative cutting speed for a given tool life while cutting some material, compared to a standard material cut with the same tool material. As shown in Figure 22-21, tool-life curves are used to develop machinability ratings. In steels, the material chosen for the standard material was B1112 steel, which has a tool life of 60 min at a cutting speed of 100 sfpm. Material X has a 70% rating, which implies that steel X has a cutting speed of 70% of B1112 for equal tool life. Note that this definition assumes that the tool fails when machining X by whatever mechanism

FIGURE 22-21   Machinability ratings defined by deterministic tool life curves.

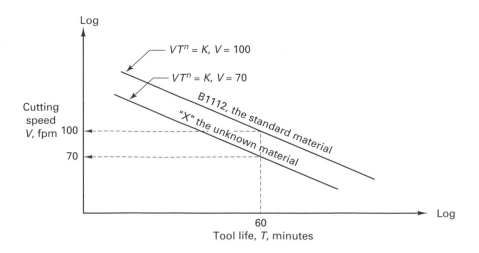

dominated the tool failure when machining the B1112. There is no guarantee that this will be the case. ISO standard 3685 has machinability index numbers based on 30 min of tool life with flank wear of 0.33 mm.

3. Cutting speed is measured by the maximum speed at which a tool can provide satisfactory performance for a specified time under specified conditions. See ASTM standard E 618-81: "Evaluating machining performance of ferrous metals using an automatic screw/bar machine."

4. Other definitions of machinability are based on the ease of removal of the chips (chip disposal), the quality of the surface finish of the part itself, the dimensional stability of the process, or the cost to remove a given volume of metal.

Further definitions are being developed based on the probabilistic nature of the tool failure, in which machinability is defined by a tool reliability index. Using such indexes, various tool replacement strategies can be examined and optimum cutting rates obtained. These approaches account for the tool life variability by developing coefficients of variation for common cutting tool/work material combinations.

The results to date are very promising. One thing is clear, however, from this sort of research: Although many manufacturers of tools have worked at developing materials that have greater tool life at higher speeds, few have worked to develop tools that have less variability in tool life at all speeds. The reduction in variability is fundamental to achieving smaller coefficients of variation, which typically are of the order of 0.3 to 0.4. This means that a tool with a 100-min. average tool life has a standard deviation of 30 to 40 min, so there is a good probability that the tool will fail early. In automated equipment, where early, unpredicted tool failures are extremely costly, reduction of the tool life variability will pay great benefits in improved productivity and reduced costs.

■ 22.8 CUTTING FLUIDS

From the day that Frederick W. Taylor demonstrated that a heavy stream of water flowing directly on the cutting process allowed the cutting speeds to be doubled or tripled, *cutting fluids* have flourished in use and variety and are employed in virtually every machining process. The cutting fluid acts primarily as a coolant and secondly as a lubricant, reducing the friction effects at the tool/chip interface and the work/flank regions. The cutting fluids also carry away the chips and provide friction (and force) reductions in regions where the bodies of the tools rub against the workpiece. Thus in processes such as drilling, sawing, tapping, and reaming, portions of the tool apart from the cutting edges come in contact with the work, and these (sliding friction) contacts greatly increase the power needed to perform the process, unless properly lubricated.

The reduction in temperature greatly aids in retaining the hardness of the tool, thereby extending the tool life or permitting increased cutting speed with equal tool life. In addition, the removal of heat from the cutting zone reduces thermal distortion of the work and permits better dimensional control. Coolant effectiveness is closely related to the thermal capacity and conductivity of the fluid used. Water is very effective in this respect but presents a rust hazard to both the work and tools and also is ineffective as a lubricant. Oils offer less effective coolant capacity but do not cause rust and have some lubricant value. In practice, straight cutting oils or emulsion combinations of oil and water or wax and water are frequently used. Various chemicals can also be added to serve as wetting agents or detergents, rust inhibitors, or polarizing agents to promote formation of a protective oil film on the work. The extent to which the flow of a cutting fluid washes the very hot chips away from the cutting area is an important factor in heat removal. Thus the application of a coolant should be copious and of some velocity.

The possibility of a cutting fluid providing lubrication between the chip and the tool face is an attractive one. An effective lubricant can modify the geometry of chip formation yielding a thinner, less-deformed, and cooler chip. Such action could discourage the formation of a built-up edge on the tool and thus promote improved surface finish. However, the extreme pressure at the tool/chip interface and the rapid movement of the chip away from the cutting edge make it virtually impossible to maintain a conventional

TABLE 22-8.   Cutting Fluid Contaminants

Category	Contaminants	Effects
Solids	Metallic fines, chips	Scratch product's surface
	Grease and sludge	Plug coolant lines
	Debris and trash	Wear on tools and machines
Tramp fluids	Hydraulic oils (coolant)	Decrease cooling efficiency
	Water (oils)	Cause smoking
		Clog paper filters
		Grow bacteria faster
Biologicals	Bacteria	Acidify coolant
(coolants)	Fungi	Break down emulsions
	Mold	Cause rancidity, dermatitis
		Require toxic biocides

hydrodynamic lubricating film at the tool/chip interface. Consequently, any lubrication action is associated primarily with the formation of solid chemical compounds of low shear strength on the freshly cut chip face, thereby reducing chip/tool shear forces or friction. For example, carbon tetrachloride is very effective in reducing friction in machining several different metals and yet would hardly be classified as a good lubricant in the usual sense. Chemically active compounds, such as chlorinated or sulfurized oils, can be added to cutting fluids to achieve such a lubrication effect. Extreme-pressure lubricants are especially valuable in severe operations, such as internal threading (tapping), where the extensive tool/work contact results in high friction with limited access for a fluid. In addition to functional effectiveness as coolant and lubricant, cutting fluids should be stable in use and storage, noncorrosive to work and machines, and nontoxic to operating personnel. The cutting fluid should also be restorable by using a closed recycling system that will purify the used coolant and cutting oils. Cutting fluids become contaminated in three ways (Table 22-8). All these contaminants can be eliminated by filtering, hydrocycloning, pasteurizing and centrifuging. Coolant restoration eliminates 99% of the cost of disposal and 80% or more of new fluid purchases. See Figure 22-22 for a schematic of a coolant recycling system.

FIGURE 22-22   A well-designed recycling system for coolants will return more than 99% of the fluid for reuse.

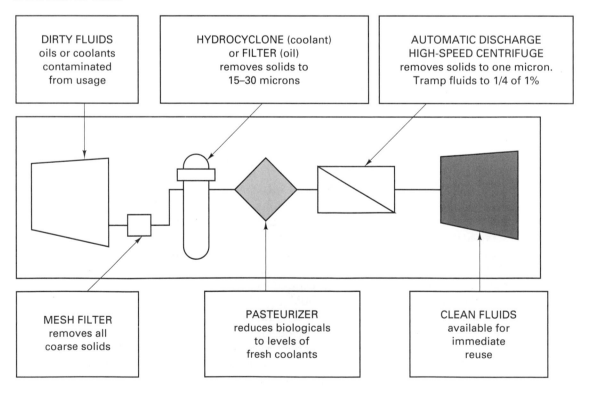

■ KEY WORDS

aluminum oxide	cermets	cutting fluids	HSS high speed steel	sintered
BUE (built-up edge)	chemical vapor deposition	cutting tool materials	machinability	stellite
carbides	coated tools	DCL (depth-of-cut line)	physical vapor deposition	titanium nitride
cast cobalt alloy	crater wear	diamonds	polycrystalline	wear land
ceramics	cubic boron nitride (CBN)	hot hardness	single-crystal	

■ REVIEW QUESTIONS

1. For metal cutting tools, what is the most important material property (ie., the most critical characteristic)? Why?
2. What is hot hardness?
3. What is impact strength and how is it measured?
4. Why is impact strength an important property in cutting tools?
5. What is HIP, and how is it used for tool fabrication?
6. What are the primary considerations in tool selection?
7. What is the general strategy behind coated tools?
8. What is a cermet?
9. How is a CBN tool manufactured?
10. F. W. Taylor was one of the discoverers of high speed steel. What else is he well known for?
11. What casting process do you think was used to fabricate cast cobalt alloys?
12. What does the sintering step do in the manufacture of carbides?
13. What does *cemented* mean in the manufacture of carbides?
14. What advantage do ground carbide inserts have over pressed carbide inserts?
15. What is a chip groove?
16. What is the DCL?
17. Suppose you made four beams out of carbide, HSS, ceramic, and cobalt. The beams are identical in size and shape, differing only in material. Which beam would
    a. Deflect the most, assuming the same load?
    b. Resist penetration the most?
    c. Bend the farthest without breaking?
    d. Support the greatest compressive load?
18. Why are multiple coats or layers put on the carbide base for coated tools?
19. What are the most common surface treatments for
    a. High-speed steels?
    b. Carbides?
    c. Ceramics?
20. What makes the process which makes TiC coatings for tools a problem? [See Equation (22-1).]
21. Why does a TiN-coated tool consume less power than an uncoated HSS under exactly the same cutting conditions?
22. For what work material are CBN tools more commonly used and why?
23. Why is CBN better for machining steel than diamond?
24. What is a coefficient of variation?
25. What is meant by the statement "Tool life is a random variable"?
26. What is the typical value of a coefficient of variation in metal-cutting tool life distributions?
27. How is machinability defined?
28. What are the chief functions of cutting fluids?
29. How are carbide tools manufactured?
30. Why is the PVD process used to coat HSS tools?
31. Why is there no universal cutting tool material?
32. What is an 18-4-1 HSS composed of?
33. Over the years, tool materials have been developed which have allowed significant increases in MRR. Nevertheless, HSS is still widely used. Under what conditions might HSS be the material of choice?
34. Why is the rigidity of the machine tool an important consideration in the selection of the cutting tool material?
35. Explain how it can be that the tool wears when it may be four times as hard as the work material.
36. What is a honed edge on a cutting tool and why is it done?

■ PROBLEMS

1. In Figure 22-A are data for cutting speed and tool life. Determine the constants for the Taylor tool life equation for these data. What do you think the tool material might have been?
2. Suppose you have a turning operation using a tool with a zero back rake and 5° end relief. The insert flank has a wear land on it of .020 in. How much has the diameter of the workpiece grown (increased) due to this flank wear, assuming the tool has not been reset to compensate for the flank wear?
3. In Figure 22-B, a single point tool is shown. Fill in the blanks with names for the tool nomenclature:

    $A =$ _____ $D =$ _____ $G =$ _____

    $B =$ _____ $E =$ _____

    $C =$ _____ $F =$ _____

4. Following data have been obtained for machining AA390 Aluminum, a Si-Al alloy.

Workpiece Material	Tool Material	Cutting Speed (m/min) for Tool Life (m/min) of:		
		20 min	30 min	60 min
Sand casting	Diamond polycrystal	731	642	514
Permanent Mold casting	Diamond polycrystal	591	517	411
PMC with flood cooling	Diamond polycrystal	608	554	472
Sand casting	WC-K-20	175	161	139

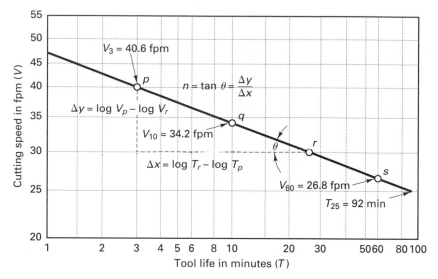

FIGURE 22-A

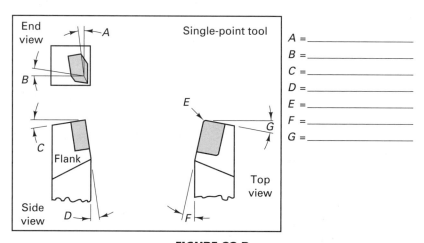

FIGURE 22-B

Compute the K and n values for the Taylor tool life equation. How do these n values compare to the typical values?

5. In Figure 22-C, the insert at the top is set with a zero side cutting edge angle. The insert at the bottom is set so that the edge contact length is increased from 0.250 in. depth-of-cut to 0.289 in. The feed was 0.010 ipr.

   a. Determine the side cutting edge angle for the offset tool.
   b. What is the uncut chip thickness in the offset position?
   c. What effect will this have on the forces and the process?

6. Tool cost is often used as the major criterion for justifying tool selection. Either silicon nitride or PCBN insert tips can be used to machine (bore) a cylinder block on a transfer line at a rate of 312,000 parts/yr (material gray cast iron). The operation requires 12 inserts (two per tool), as six bores are machined simultaneously. The machine was run at 2600 sfpm with a feed of 0.014 in. at 0.005 in. DOC for finishing. Here are some additional data.

	SiN	PCBN
Tips in use per part	12	12
Tool life (parts per tool)	200	4700
Cost per tip	$1.25	$28.50

   a. Which tool material would you recommend?
   b. On what basis?

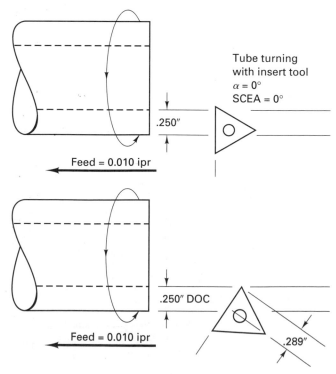

Tube turning
with insert tool
$\alpha = 0°$
SCEA = 0°

.250"

Feed = 0.010 ipr

.250" DOC

.289"

Feed = 0.010 ipr

FIGURE 22-C

7. A 2 in. diameter bar of steel was turned at 284 rpm and tool failure occurred in 10 min. The speed was changed to 132 rpm and the tool failed in 30 min of cutting. Assume that a straight-line relationship exists. What cutting speed should be used to obtain a 60-min tool life of V_{60}?

8. The following measured data were taken from an orthogonal test cutting AISI 1015 steel using a tungsten carbide tool:

Cutting speed	500 ft/min
Feed	0.010 in.
Width of cut	0.100 in.
Chip thickness	0.022 in.
Thrust force	140 lb
Rake angle of tool	10°

   Compute:
   a. Chip thickness ratio
   b. Shear plane angle
   c. Friction angle
   d. Coefficient of friction
   e. Friction force
   f. Shear force
   g. Shear stress on shear plane
   h. Shear velocity
   i. Shear strain
   j. Specific shear energy (in lb-in/in³)

9. Refer to Problem 1. Show the relationship between cutting speed and tool temperature. What does this mean with regard to tool failure?

10. The outside diameter of a roll for a steel (AISI 1015) rolling mill is to be turned. In the final pass, the starting diameter = 26.25 in. and the length 48.0 in. The cutting conditions will be feed = 0.0100 in/rev and depth of cut = 0.125 in. A cemented carbide cutting tool is to be used, and the parameters of the Taylor tool life equation for this setup are $n = 0.25$ and $K = 1300$. It is desirable to operate at a cutting speed such that the tool will not need to be changed during the cut. Determine the cutting speed that will make the tool life equal to the time required to complete this turning operation. (Problem suggested by: Groover, *Fundamentals of Modern Manufacturing: Materials, Processes, and Systems*, 2nd Ed., John Wiley & Sons, 2002.)

11. Using data from Problems 8 and 10, determine the necessary horsepower for the machine tool to make this cut.

12. Figure 22-D shows a sketch of a single point tool and its associated tool signature. Put the signature from the tool in Figure 22-B in the same order as shown in Figure 22-D. Which tool would produce the larger F_c given both are cutting at the same V, f_r, and DOC in the same material?

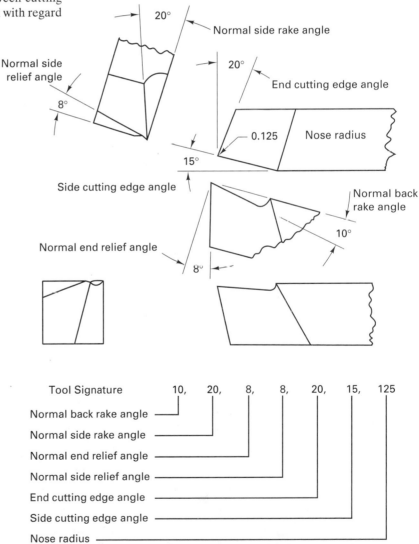

Tool Signature	10,	20,	8,	8,	20,	15,	125
Normal back rake angle							
Normal side rake angle							
Normal end relief angle							
Normal side relief angle							
End cutting edge angle							
Side cutting edge angle							
Nose radius							

FIGURE 22-D

CHAPTER 23

TURNING AND BORING AND RELATED PROCESSES

■ 23.1 INTRODUCTION

Turning is the process of machining external cylindrical and conical surfaces. It is usually performed on a lathe, as shown in Figure 23-1. Details on lathes are shown later in this chapter. As indicated in Figure 23-2, relatively simple work and tool movements are involved in turning a cylindrical surface. The workpiece is rotated into a longitudinally fed, single-point cutting tool. If the tool is fed at an angle to the axis of rotation, an external conical surface results. This is called *taper turning*. If the tool is fed at 90° to the axis of rotation, using a tool that is wider than the width of the cut, the operation is called *facing*, and a flat surface is produced.

FIGURE 23-1 Standard engine lathe performing a turning operation, shown in inset.

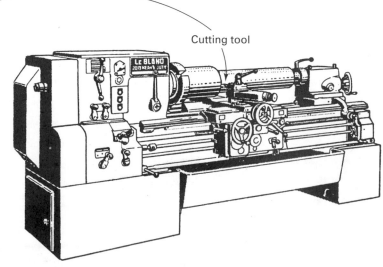

Cutting tool

548

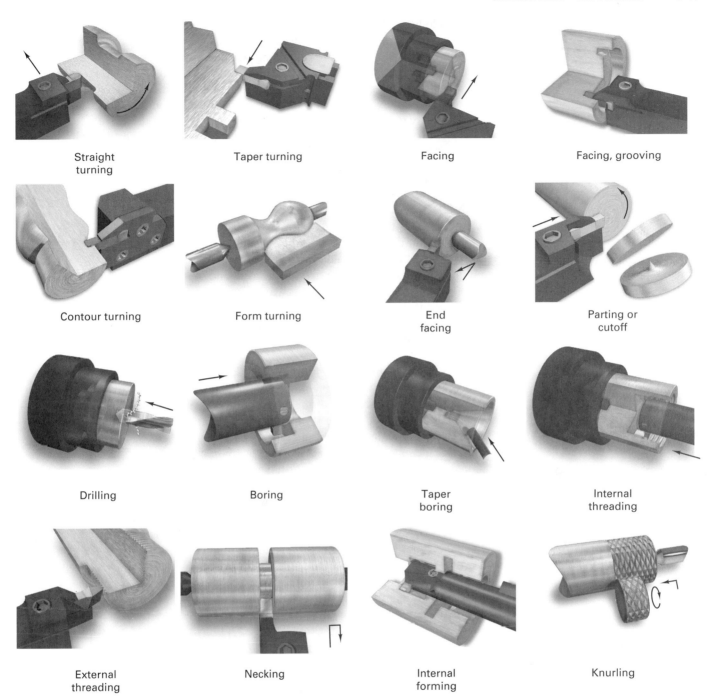

Straight turning	Taper turning	Facing	Facing, grooving
Contour turning	Form turning	End facing	Parting or cutoff
Drilling	Boring	Taper boring	Internal threading
External threading	Necking	Internal forming	Knurling

FIGURE 23-2 Basic turning machines can rotate the work and feed the tool longitudinally for turning and can perform other operations by feeding transversely. Depending on what direction the tool is fed and on what portion of the rotating workpiece is being machined, the operations have different names. The arrows indicate the tool motion relative to the workpiece.

By using a tool having a specific form or shape and feeding it radially or inward against the work, external cylindrical, conical, and irregular surfaces of limited length can also be turned. The shape of the resulting surface is determined by the shape and size of the cutting tool. Such machining is called *form turning*. If the tool is fed all the way to the axis of the workpiece, it will be cut in two. This is called *parting* or *cutoff* and a simple, thin tool is used. A similar tool is used for *necking* or *partial cutoff*.

Boring is a variation of turning. Essentially, boring is internal turning. Boring can use single-point cutting tools to produce internal cylindrical or conical surfaces. It does not create the hole but rather, machines or opens the hole up to a specific size. Boring can be done on most machine tools that can do turning. However, boring also can be done using a rotating tool with the workpiece remaining stationary. Also, specialized machine tools have been developed that will do boring, drilling, and reaming but will not do turning. Other operations, like *threading* and *knurling*, can be done on machines used for turning. In addition, drilling, reaming, and tapping can be done on the rotation axis of the work.

In recent years, turning centers have been developed that use turrets to hold multiple edge rotary tools in powered heads. Some machines feature two spindles with automatic transfer from one to the other and two turrets of tools.

■ 23.2 FUNDAMENTALS OF TURNING, BORING, AND FACING

TURNING

Turning constitutes the majority of lathe work. The cutting forces, resulting from feeding the tool from right to left, should be directed toward the headstock to force the workpiece against the workholder and thus provide better work support.

If good finish and accurate size are desired, one or more roughing cuts usually are followed by one or more finishing cuts. Roughing cuts may be as heavy as proper chip thickness, cutting dynamics, tool life, lathe horsepower, and the workpiece permit. Large *depths of cut* and smaller feeds are preferred to the reverse procedure, because fewer cuts are required and less time is lost in reversing the carriage and resetting the tool for the following cut.

On workpieces that have a hard surface, such as castings or hot-rolled materials containing mill scale, the initial roughing cut should be deep enough to penetrate the hard materials. Otherwise, the entire cutting edge operates in hard, abrasive material throughout the cut, and the tool will dull rapidly. If the surface is unusually hard, the cutting speed on the first roughing cut should be reduced accordingly.

Finishing cuts are light, usually being less than 0.015 in. in depth, with the feed as fine as necessary to give the desired finish. Sometimes a special finishing tool is used, but often the same tool is used for both roughing and finishing cuts. In most cases, one finishing cut is all that is required. However, where exceptional accuracy is required, two finishing cuts may be made. If the diameter is controlled manually, a short finishing cut ($\frac{1}{4}$ in. long) is made and the diameter checked before the cut is completed. Because the previous micrometer measurements were made on a rougher surface, it may be necessary to reset the tool in order to have the final measurement, made on a smoother surface, check exactly.

In turning, the primary cutting motion is rotational with the tool feeding parallel to the axis of rotation (Figure 23-3). The rpm of the rotating workpiece N_s establishes the cutting velocity V at the cutting tool. The feed f_r is given in inches per revolution (ipr). The depth of cut is d, where

$$d = \text{DOC} = (D_1 - D_2)/2 \text{ in inches} \tag{23-1}$$

The length of cut is the distance traveled parallel to the axis L plus some allowance or overrun A to allow the tool to enter and/or exit the cut.

FIGURE 23-3 Basics of the turning process normally done on a lathe. The arrows indicate the motion of the tool relative to the work.

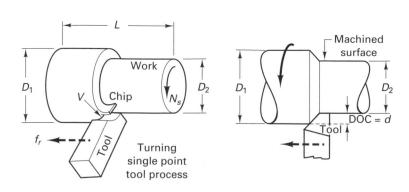

Once the cutting speed, feed, and depth of cut have been selected for a given material being cut with a tool of known cutting-tool material, the rpm value for the machine tool can be determined.

$$N_s = \frac{12V}{\pi D_1} \tag{23-2}$$

(using the larger diameter); where the factor of 12 is used to convert feet to inches. The cutting time is

$$T_M = \frac{L + A}{f_r N_s} \tag{23-3}$$

where A is overrun allowance. The metal removal rate is

$$\text{MRR} = \frac{\text{volume removed}}{\text{time}} = \frac{(\pi D_1^2 - \pi D_2^2)L}{4L/f_r N}$$

(omitting the allowance term). By rearranging and subbing for N_s, an exact expression for MRR is obtained:

$$\text{MRR} = 12V f_r \frac{(D_1^2 - D_2^2)}{4D_1} \tag{23-4}$$

Rewriting the last term

$$\frac{D_1^2 - D_2^2}{4D_1} = \frac{D_1 - D_2}{2} \times \frac{D_1 + D_2}{2D_1}$$

Therefore, since

$$d = \frac{D_1 - D_2}{2} \quad \text{and} \quad \frac{D_1 + D_2}{2D_1} \cong 1 \quad \text{for small } d$$

then,

$$\text{MRR} \cong 12 V f_r d \quad \text{in}^3/\text{min.} \tag{23-5}$$

Note that Equation (23-5) is an approximate equation that assumes that the depth of cut d is small compared to the uncut diameter D_i.

BORING

Boring always involves the enlarging of an existing hole, which may have been made by a drill or may be the result of a core in a casting. An equally important and concurrent purpose of boring may be to make the hole concentric with the axis of rotation of the workpiece and thus correct any eccentricity that may have resulted from the drill drifting off the center line. Concentricity is an important attribute of bored holes.

When boring is done in a lathe, the work usually is held in a chuck or on a faceplate. Holes may be bored straight, tapered, or to irregular contours. Figure 23-4a shows the relationship of the tool and the workpiece for boring. Think of boring as internal turning while

FIGURE 23-4 Basic movement of boring, facing, and cutoff (or parting) processes.

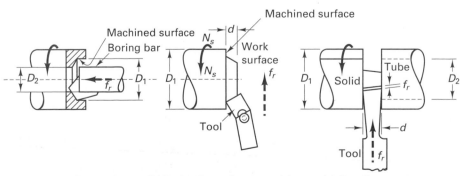

(a) Boring a drilled hole   (b) Facing from the cross side   (c) Cutoff or parting

feeding the tool parallel to the rotation axis of the workpiece, with two important differences. First, the relief and clearance angles on the tool should be larger and the tool overhand (length to diameter) must be considered with regard to stability and deflection problems.

Given V and f_r, for a cut of length L, the cutting time is

$$T_m = \frac{L + A}{f_r N_s} \tag{23-6}$$

where $N_s = 12V/\pi D_1$ for D_1, the diameter of bore, and A, the overrun allowance. The metal removal rate is

$$\text{MRR} = \frac{L(\pi D_1^2 - \pi D_2^2)/4}{L/f_r N}$$

where D_2 is the original hole diameter.

$$\text{MRR} \cong 12 V f_r d \tag{23-7}$$

(omitting allowance term), where d is the depth of cut.

In most respects, the same principles are used for boring as for turning. Again, the tool should be set exactly at the same height as the axis of rotation. Larger end clearance angles help to prevent the heel of the tool from rubbing on the inner surface of the hole. Because the tool overhang will be greater, feeds and depths of cut may be reduced to reduce forces that cause tool vibration and chatter. In some cases, the boring bar may be made of tungsten carbide because of this material's greater stiffness.

FACING

Facing is the producing of a flat surface as the result of the tool being fed across the end of the rotating workpiece, as shown in Figure 23-4b. Unless the work is held on a mandrel, if both ends of the work are to be faced, it must be turned end for end after the first end is completed and the facing operation repeated.

The cutting speed should be determined from the largest diameter of the surface to be faced. Facing may be done either from the outside inward or from the center outward. In either case, the point of the tool must be set exactly at the height of the center of rotation. Because the cutting force tends to push the tool away from the work, it is usually desirable to clamp the carriage to the lathe bed during each facing cut to prevent it from moving slightly and thus producing a surface that is not flat.

In the facing of castings or other materials that have a hard surface, the depth of the first cut should be sufficient to penetrate the hard material to avoid excessive tool wear.

In facing, the tool feeds perpendicular to the axis of the rotating workpiece. Because the rpm is constant, the cutting speed is continually decreasing as the axis is approached. The length of cut L is $D_1/2$ or $(D_1 - D_2)/2$ for a tube.

$$T_m = \text{Cutting time} = \frac{L + A}{f_n N} \text{ minutes}$$

$$= \frac{\dfrac{D_1}{2} + A}{f_r N}$$

$$\text{MRR} = \frac{\text{VOL}}{T_m} = \frac{\pi D_1^2 \, d f_r N}{4L} = 6 V f_r d \quad \text{in}^3/\text{min} \tag{23-8}$$

where d = depth of cut, and $L = D_1/2$ is the length of cut.

PARTING

Parting is the operation by which one section of a workpiece is severed from the remainder by means of a cutoff tool, as shown in Figure 23-4c. Because parting tools are quite thin and must have considerable overhang, this process is more difficult to perform accurately. The tool should be set exactly at the height of the axis of rotation, be kept sharp, have proper clearance angles, and be fed into the workpiece at a proper and uniform feed rate.

In parting or cutoff work, the tool is fed (plunged) perpendicular to the rotational axis, as it was in facing. The length of cut for solid bars is $D_1/2$. For tubes,

$$L = \frac{D_1 - D_2}{2}$$

In cutoff operations, the width of the tool is d in inches, the width of the cutoff operation. The equations for T_m and MRR are then basically the same as for facing.

In boring, facing, and cutoff operations, the speeds and feeds selected are generally less than those recommended for turning because of the large overhang of the tool often needed to complete the cuts. Recall the basic equation for deflection of a cantilever beam, modifying for machining,

$$\delta = \frac{Pl^3}{3EI} = \frac{F_c l^3}{3EI} \qquad (23\text{-}9)$$

In Equation (23-9), l represents the overhang of the tool, which greatly affects the deflection δ, so it should be minimized whenever possible. In Equation (23-9),

$$E = \text{modulus of elasticity (lb}^2/\text{in}^2)$$

$$I = \text{moment of inertia of cross-section}$$

where

$$I = \pi D_1^2/64 \text{ solid round bar}$$

$$I = \pi(D_1^4 - (D_2^4))/64 \text{ bar with bored hole}$$

$$D_1 = \text{diameter of bar}$$

$$D_2 = \text{inside diameter of bar}$$

Deflection is proportional to the fourth power of the boring bar diameter and the third power of the bar overhang. Select the largest diameter bar diameter, minimize the overhang, use carbide shank boring bars ($E \cong 80,000,000$ psi), and select tool geometries that direct cutting forces into the feed direction to minimize chatter. The reduction of the feed or depth of cut reduces the forces operating on the tools. The cutting speed usually controls the occurrence of chatter and vibration.

Any imbalance in the cutting forces will deflect the tool to the side, resulting in loss of accuracy in cutoff lengths. At the outset, the forces will be balanced if there is no side rake on the tool. As the cutoff tool reaches the axis of the rotating part, the tool will be deflected away from the spindle, resulting in a change in the length of the part.

PRECISION BORING

Sometimes bored holes are slightly bell-mouthed because the tool deflects out of the work as it progresses into the hole. This often occurs in castings where the holes have draft angles, which increase the depth of cut as the tool progresses down the bore. This problem usually may be corrected by repeating the cut with the same tool setting, however, the total cutting time for the part is increased. Alternately, a more robust setup can be used. Large holes may be precision-bored using the setup shown in Figure 23-5, where a pilot bushing is placed in the spindle to mate with the hardened ground pilot of the boring bar. This setup eliminates the cantilever problems common to boring.

Because the rotational relationship between the work and the tool is a simple one and is employed on several types of machine tools, such as lathes, drilling machines, and milling machines, boring is very frequently done on such machines. However, several machine tools have been developed primarily for boring, especially in cases involving large workpieces or for large-volume boring of smaller parts. Such machines as these are also capable of performing other operations, such as milling and turning. Because boring frequently follows drilling, many boring machines also can do drilling, permitting both operations to be done with a single setup of the work.

DRILLING

Drilling discussed in Chapter 24 can be done on lathes with the drill mounted in the tailstock quill of engine lathes or the turret on turret lathes and fed against a rotating

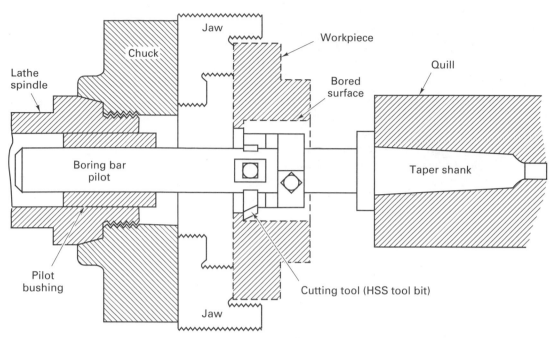

FIGURE 23-5 Pilot boring bar mounted in tailstock of lathe for precision boring large hole in casting. The size of the hole is controlled by the rotation diameter of the cutting tool.

workpiece. Straight-shank drills can be held in Jacobs chucks or drills with taper shanks mounted directly in the quill hole can drill holes online (center of rotation). Drills can also be mounted in the turrets of modern turret centers and fed automatically on the rotational axis of the workpiece or off axis with power heads as shown later in Figures 23-14 and 23-15. It also is possible to drill on a lathe with the drill bit mounted and rotated in the spindle while the work remains stationary, supported on the tailstock or the carriage of the lathe.

Usual speeds are used for drilling in a lathe. Because the feed may be manually controlled, care must be exercised, particularly in drilling small holes. Coolants should be used where required. In drilling deep holes, the drill should be withdrawn occasionally to clear chips from the hole and to aid in getting coolant to the cutting edges. This is called peck drilling. See Chapter 24 for further discussion on drilling.

REAMING

Reaming on a lathe involves no special precautions. Reamers are held in the tailstock quill, taper-shank types being mounted directly and straight-shank types by means of a drill chuck. Rose-chucking reamers usually are used (see Chapter 24). Fluted-chucking reamers also may be used, but these should be held in some type of holder that will permit the reamer to float (i.e., have come compliance) in the hole and conform to the geometry created by the boring process.

KNURLING

Knurling produces a regularly shaped, roughened surface on a workpiece. Although knurling also can be done on other machine tools, even on flat surfaces, in most cases it is done on external cylindrical surfaces using lathes. Knurling is a chipless, cold-forming process, using a tool of the type shown in Figure 23-6. The two hardened rolls are pressed against the rotating workpiece with sufficient force to cause a slight outward and lateral displacement of the metal to form the knurl in a raised, diamond pattern. Another type of knurling tool produces the knurled pattern by cutting chips. Because it involves less pressure and thus does not tend to bend the workpiece, this method is often preferred for workpieces of small diameter and for use on automatic or semiautomatic machines.

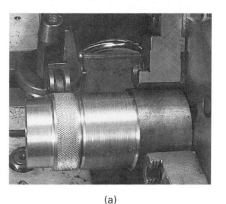
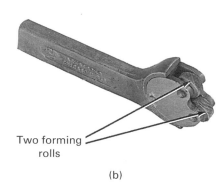

Two forming rolls

(a)　　　　　　　　(b)

FIGURE 23-6 (a) Knurling in a lathe, using a forming-type tool, and showing the resulting pattern on the workpiece; (b) knurling tool with forming rolls. *(Courtesy of Armstrong Brothers Tool Company.)*

SPECIAL ATTACHMENTS

For engine lathes, taper turning and *milling* can be done on a lathe but requires special attachments. The *milling attachment* is a special vise that attaches to the cross slide to hold work. The milling cutter is mounted and rotated by the spindle. The work is fed by means of the cross-slide screw. *Tool-post grinders* are often used to permit grinding to be done on a lathe. Taper turning will be discussed later.

Duplicating attachments are available that, guided by a template, will automatically control the tool movements for turning irregularly shaped parts. In some cases, the first piece, produced in the normal manner, may serve as the template for duplicate parts. To a large extent, duplicating lathes using templates have been replaced by numerically controlled lathes and milling is done with power tools in NC turret lathes.

DIMENSIONAL ACCURACY

Dimensional accuracy in turning operations is controlled by many factors, including the wear at the nose of the tool (see Figure 23-7). Precision is influenced by deflection due to the cutting forces and surface roughness. Tool wear causes the workpiece dimension to change from the initial diameter when the tool is sharp to the diameter obtained after the tool has worn. The cutting forces increase as the tool wears, resulting in increased deflection between the workpiece and the cutting tool. Built-up edge (BUE) may form at the tip of the cutting tool. The BUE has the tendency to change the actual diameter of the workpiece. Thus, to hold close tolerances, the size of the wear land, the magnitude of the radial (thrust) force, and the elimination of the BUE should be taken into account.

Dimensional accuracy will also be influenced by the workpiece shape, the material, the rigidity of all elements, the surface finish, and vibrations. For example, holding the dimensional accuracy of a boring operation on a deep hole is a problem, due to the deflection (rigidity) of the boring bar.

Turned surfaces display characteristic turning grooves that are produced by the feed and the tool tip corner radius, as shown in Figure 23-7. The roughness resulting from feed marks from a round-nosed tool can be approximated by the formula

$$y = CR - \frac{\sqrt{CR^2 - f_r^2}}{4} \cong \frac{f_r^2}{8CR} \qquad (23\text{-}10)$$

where y is the roughness height, CR the corner radius of insert, and f_r the feed rate (in./rev). To improve the surface finish, reduce the feed and increase the corner radius.

Other factors like BUE formations, cutting-edge sharpness, and tool-wear grooves in the flank wear area also affect the surface finish in turning. Flank wear and BUE can combine to affect both surface finish and accuracy as shown in Figure 23-7. Wear on the corner radius may cause grooves and nicks, which produce additional surface roughness on the finish-turned surfaces. Thus, to hold the surface roughness within specified limits, minimize tool wear and use small feeds and large-corner-radius tools. To minimize BUE formation, employ cutting speeds higher than those used in rough turning operations.

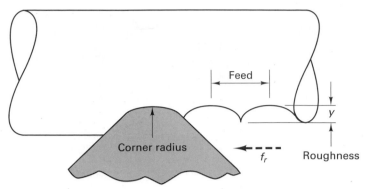

The feed and the corner radius of the cutting tool influence the surface roughness

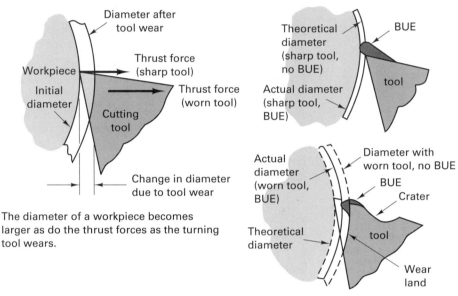

The diameter of a workpiece becomes larger as do the thrust forces as the turning tool wears.

Regardless of whether the tool is dull or sharp, a built-up edge (BUE) causes the diameter of the workpiece to be smaller than desired.

FIGURE 23-7   Accuracy and precision in turning is a function of many factors, including tool wear and BUE.

■ 23.3 LATHE DESIGN AND TERMINOLOGY

Knowing the terminology of a machine tool is fundamental to understanding how it performs the basic processes, how the workholding devices are interchanged, and how the cutting tools are mounted and interfaced to the work. *Lathes* are machine tools designed primarily to do turning, facing, and boring. Very little turning is done on other types of machine tools, and none can do it with equal facility. Lathes also can do facing, drilling, and reaming, and recent designs permit milling and drilling operations using live (also called powered) spindles in multiple tool turrets, so their versatility permits multiple operations to be done with a single setup of the workpiece. Consequently, the lathe is the most common machine tool.

Lathes in various forms have existed for more than 2000 years, but modern lathes date from about 1797, when Henry Maudsley developed one with a leadscrew, providing controlled, mechanical feed of the tool. This ingenious Englishman also developed a change-gear system that could connect the motions of the spindle and leadscrew and thus enable threads to be cut.

LATHE DESIGN

The essential components of an *engine lathe* (Figure 23-8) are the bed, headstock assembly, tailstock assembly, carriage assembly, quick-change gearbox, and the leadscrew and

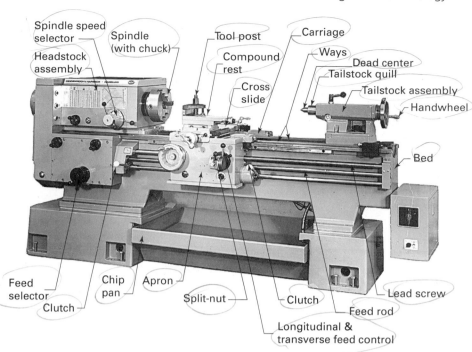

FIGURE 23-8 Modern engine lathe, with the principal parts named. *(Courtesy of Heidenreich & Harbeck.)*

feed rod. The *bed* is the base and backbone of a lathe. The bed usually is made of well-normalized or aged gray or nodular cast iron and provides a heavy, rigid frame on which all the other basic components are mounted. Two sets of parallel, longitudinal *ways*, inner and outer, are contained on the bed. On modern lathes, the ways are surface-hardened and precision-machined, and care should be taken to assure that the ways are not damaged. Any inaccuracy in them usually means that the accuracy of the entire lathe is destroyed.

The *headstock*, mounted in a fixed position on the inner ways, provides powered means to rotate the work at various rpm values. Essentially, it consists of a hollow *spindle*, mounted in accurate bearings, and a set of transmission gears—similar to a truck transmission—through which the spindle can be rotated at a number of speeds. Most lathes provide from 8 to 18 choices of rpm. On modern lathes all the rpm rates can be obtained merely by moving from two to four levers. An increasing trend is to provide a continuously variable spindle rpm through electrical or mechanical drives.

The accuracy of a lathe is greatly dependent on the *spindle*. It carries the workholders and is mounted in heavy bearings, usually preloaded tapered roller or ball types. The spindle has a hole extending through its length, through which long bar stock can be fed. The size of this hole is an important dimension of a lathe because it determines the maximum size of bar stock that can be machined when the materials must be fed through the spindle.

The spindle protrudes from the gearbox and contains means for mounting various types of workholding devices (chucks, face and dog plates, collets). Power is supplied to the spindle from an electric motor through a V-belt or silent-chain drive. Most modern lathes have motors of from 5 to 25 hp to provide adequate power for carbide and ceramic tools at the higher cutting speeds.

For the classic engine lathe, the *tailstock* assembly consists, essentially, of three parts. A lower casting fits on the inner ways of the bed, can slide longitudinally, and can be clamped in any desired location. An upper casting fits on the lower one and can be moved transversely upon it, on some type of keyed ways, to permit aligning the tailstock and headstock spindles (for turning tapers). The third major component of the assembly is the *tailstock quill*. This is a hollow steel cylinder, usually about 2 to 3 in. in diameter, that can be moved longitudinally in and out of the upper casting by means of a handwheel and screw. The open end of the quill hole has a Morse taper. Cutting tools or a *lathe center* are held in the quill. A graduated scale usually is engraved on the outside of the quill to aid in controlling its motion in and out of the upper casting. A locking device permits clamping the quill in any desired position. In recent years, dual-spindle NC turning centers have emerged, where a subspindle replaces the tailstock assembly.

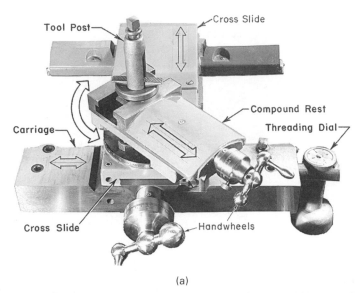

(a)

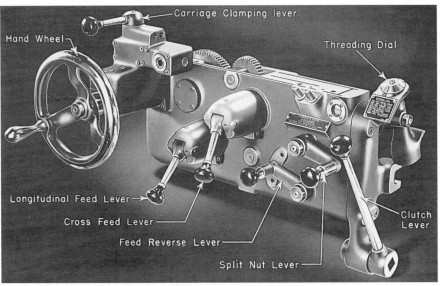

(b)

FIGURE 23-9 The carriage, cross slide, and apron assembly for an engine lathe.

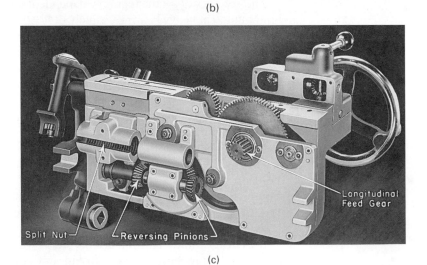

(c)

Parts can be automatically transferred from the spindle to the subspindle for turning the back end of the part. See Chapter 32 for a discussion of NC turning centers.

The *carriage assembly* (Figure 23-9), together with the apron, provides the means for mounting and moving cutting tools. The *carriage*, a relatively flat H-shaped casting, rides on the outer set of ways on the bed. The *cross slide* is mounted on the carriage and can be moved by means of a feed screw that is controlled by a small handwheel and a graduated dial. The cross slide thus provides a means for moving the lathe tool in the facing or cutoff direction.

On most lathes, the tool post is mounted on a *compound rest*. The compound rest can rotate and translate with respect to the cross slide, permitting further positioning of the tool with respect to the work. The *apron*, attached to the front of the carriage, has the controls for providing manual and powered motion for the carriage and powered motion for the cross slide. Figure 23-9 shows front and rear views of a typical apron. The carriage is moved parallel to the ways by turning a handwheel on the front of the apron, which is geared to a pinion on the back side. This pinion engages a rack that is attached beneath the upper front edge of the bed in an inverted position.

Powered movement of the carriage and cross slide is provided by a rotating *feed rod*, shown in Figure 23-8. The feed rod, which contains a keyway, passes through the two reversing bevel pinions (Figure 23-9) and is keyed to them. Either pinion can be activated by means of the feed reverse lever, thus providing "forward" or "reverse" power to the carriage. Suitable clutches connect either the rack pinion or the cross-slide screw to provide longitudinal motion of the carriage or transverse motion of the cross slide.

For cutting threads, a *leadscrew* (Figure 23-8) is used. When a friction clutch is used to drive the carriage, motion through the leadscrew is by a direct, mechanical connection between the apron and the leadscrew. A *split nut* (Figure 23-9) is closed around the leadscrew by means of a lever on the front of the apron directly driving the carriage without any slippage.

Modern lathes have *quick-change gearboxes*, driven by the spindle, that connect the feed rod and leadscrew. The associated gearing, the leadscrew, and feed rod connect the carriage to the spindle, so the cutting tool can be made to move a specific distance, either longitudinally or transversely, for each revolution of the spindle. The calculations for turning rpm and feed in inches per revolution are "mechanically related." Typical lathes may provide as many as 48 feeds, ranging from 0.002 to 0.118 in. (0.05 to 3 mm) per revolution of the spindle, and, through the leadscrew, leads from $1\frac{1}{2}$ to 92 threads per inch.

TURNING AND BORING TAPERS

The turning and boring of uniform tapers are common lathe operations. Such tapers can be specified either in degrees of included angle between the sides or as the change in diameter per unit of length—millimeters per meter or inches per foot.

Four methods are available for turning external tapers on a lathe, and three for boring internal tapers. The simplest method employs the compound rest. (Refer to Figures 23-8 and 23-9 for details and terminology of an engine lathe.) This method is suitable for both external and internal tapers. However, because the length of travel of the compound rest is quite limited—seldom over a few inches—only short tapers can be turned or bored by this method. This method is partially useful for steep tapers. The compound rest is swivelled to the desired angle and locked in position. The compound slide is then fed manually to produce the desired taper. The tool should be set at exactly the height of the axis of rotation of the workpiece in all taper turning and boring operations.

Because the graduated scale on the base of the compound rest usually is calibrated only to 1° divisions, it is difficult to make the angle setting with accuracy. If accuracy is required, tapers made by this method are checked by means of plug or ring gages, and the setting of the compound rest readjusted until the gage fits perfectly. Also, the compound rest cannot be set directly to the correct angle if the taper is dimensioned in millimeters per meter or inches per foot.

Both external and internal tapers can be made on a lathe by using a *taper attachment*, such as shown in Figure 23-10. In this device there is an *extension* bolted to the rear of the carriage. When the carriage is moved, the cross slide is caused to move transversely.

A raised *guide bar* is pivoted to any desired angle (within its limits). A *guide shoe* slides on the guide bar, so that when the carriage is moved longitudinally along the ways of the lathe, the guide follows the guide bar and moves the cross slide and tool post transversely to provide the proper taper angle.

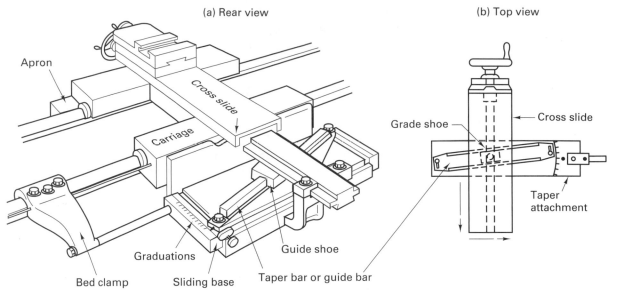

FIGURE 23-10 (a) Rear view of carriage of an engine lathe showing the taper attachment that moves the cross slide transversely when carriage moves, but only if the bed clamp is fastened. (b) The cross slide movements produced by the taper attachment.

Graduations of taper in millimeters per meter or inches per foot are provided at one end (degrees at the other end), so the attachment can be set to the desired taper. While taper attachments provide an excellent and convenient method of cutting tapers, they ordinarily can be used only for tapers of less than 0.5 mm/m or 6 in./ft.

External tapers also can be turned on workpieces that are mounted between centers by *setting over the tailstock*. This method is illustrated in Figure 23-11. The tailstock

FIGURE 23-11 Method of turning tapers by offsetting the tailstock on an engine lathe, with work held between centers.

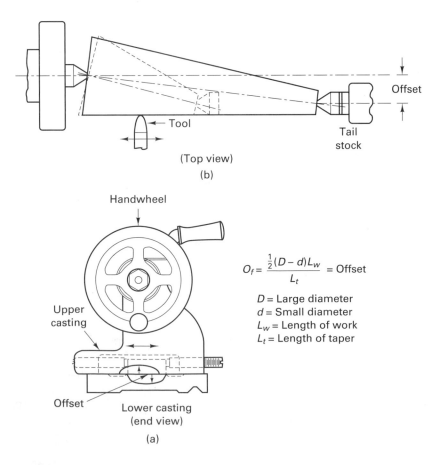

$$O_f = \frac{\frac{1}{2}(D-d)L_w}{L_t} = \text{Offset}$$

D = Large diameter
d = Small diameter
L_w = Length of work
L_t = Length of taper

is moved out of line with the headstock spindle. The set-off distance from the centerline is given by the formula in the figure. This method is limited to small tapers and is seldom used in mass production.

In specifying tapers on drawings, one must remember that it is difficult for the machinist to measure the smaller diameter of a taper accurately if it is the end of a workpiece.

Both internal and external tapers can be programmed on a numerically controlled (NC) lathe. The movement of two perpendicular axes can be programmed, resulting in a tapered surface. See Chapter 32 for a discussion of interpolation using NC equipment.

SIZE DESIGNATION OF LATHES

The size of a lathe is designated by two dimensions. The first is known as the *swing*. This is the maximum diameter of work that can be rotated on a lathe. Swing is approximately twice the distance between the line connecting the lathe centers and the nearest point on the ways. The maximum diameter of a workpiece that can be mounted between centers is somewhat less than the swing diameter because the workpiece must clear the carriage assembly as well as the ways. The second size dimension is the *maximum distance between centers*. The swing thus indicates the maximum workpiece diameter that can be turned in the lathe, while the distance between centers indicates the maximum length of workpiece that can be mounted between centers.

TYPES OF LATHES

Lathes used in manufacturing can be classified as speed, engine, toolroom, turret, automatics, tracer, and numerical control turning centers. Speed lathes usually have only a headstock, a tailstock, and a simple tool post mounted on a light bed. They ordinarily have only three or four speeds and are used primarily for wood turning, polishing, or metal spinning. Spindle speeds up to about 4000 rpm are common.

Engine lathes are the type most frequently used in manufacturing. Figure 23-1 and Figure 23-8 are examples of this type. They are heavy-duty machine tools with all the components described previously and have power drive for all tool movements except on the compound rest. They commonly range in size from 12 to 24 in. swing and from 24 to 48 in. center distances, but swings up to 50 in. and center distances up to 12 ft are not uncommon. Very large engine lathes (36- to 60-ft-long beds) therefore capable of performing roughing cuts in iron and steel at depths of cut of $\frac{1}{2}$ to 2 in., cutting speed 50 to 200 sfpm with WC tools run at 0.010 to 0.100 in./rev. To perform such heavy cuts requires rigidity in the machine tool, the cutting tools, the workholder, and the workpiece (using steady rests and other supports) and large horsepower (50 to 100 hp).

Most engine lathes are equipped with chip pans and a built-in coolant circulating system. Smaller engine lathes, with swings usually not over 13 in., also are available in *bench type*, designed for the bed to be mounted on a bench or table.

Toolroom lathes have somewhat greater accuracy and, usually, a wider range of speeds and feeds than ordinary engine lathes. Designed to have greater versatility to meet the requirements of tool and die work, they often have a continuously variable spindle speed range and shorter beds than ordinary engine lathes of comparable swing, since they are generally used for machining relatively small parts. They may be either bench or pedestal type.

Several types of special-purpose lathes are made to accommodate specific types of work. On a *gap-bed lathe*, for example, a section of the bed, adjacent to the headstock, can be removed to permit work of unusually large diameter to be swung. Another example is the *wheel lathe*, which is designed to permit the turning of railroad-car wheel-and-axle assemblies.

Although engine lathes are versatile and very useful, the time required for changing and setting tools and for making measurements on the workpiece is often a large percentage of the cycle time. Often, the actual chip-production time is less than 30% of the total cycle time. Methods to reduce setup and tool change time are discussed elsewhere. Much of the operator's time is consumed by simple, repetitious adjustments and in watching chips being made. The placement of machines into cells, as discussed in Chapter 42, greatly increases the productivity of the workers because they can run more than one machine. Turret lathes, screw machines, and other types of semiautomatic and automatic lathes have been highly developed and are widely used in manufacturing as another means to improve cutting productivity.

Turret Lathes. The basic components of a *turret lathe* are depicted in Figure 23-12. Basically, a longitudinally feedable, hexagon turret replaces the tailstock. The turret, on which six tools can be mounted, can be rotated about a vertical axis to bring each tool into operating position, and the entire unit can be translated parallel to the ways, either manually or by power, to provide feed for the tools. When the turret assembly is backed away from the spindle by means of a capstan wheel, the turret indexes automatically at the end of its movement, thus bringing each of the six tools into operating position in sequence.

The square turret on the cross slide can be rotated manually about a vertical axis to bring each of the four tools into operating position. On most machines, the turret can be moved transversely, either manually or by power, by means of the cross slide, and

FIGURE 23-12 Block diagrams of ram and saddle turret lathe.

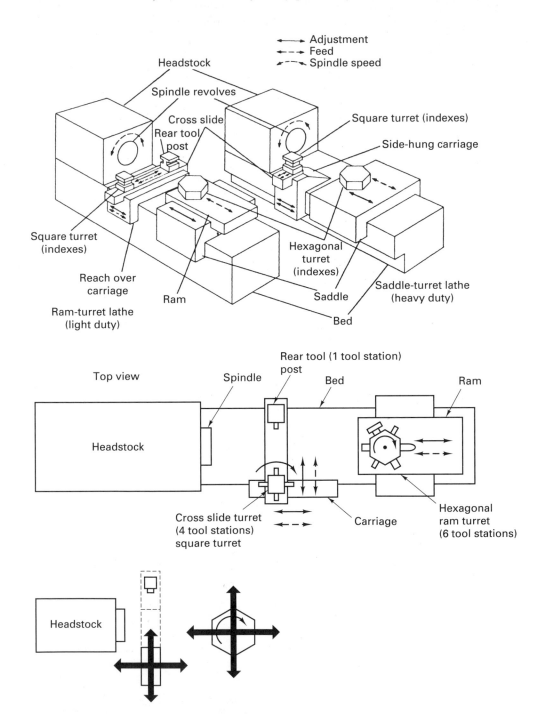

longitudinally through power or manual operation of the carriage. In most cases, a fixed tool holder also is added to the back end of the cross slide; this often carries a parting tool.

Through these basic features of a turret lathe, a number of tools can be set up on the machine and then quickly be brought successively into working position so that a complete part can be machined without the necessity for further adjusting, changing tools, or making measurements.

The two basic types of turret lathes are the ram-type turret lathe and the saddle-type turret lathe. In the *ram-type turret lathe*, the ram and turret are moved up to the cutting position by means of the capstan wheel, and the power feed is then engaged. As the ram is moved toward the headstock, the turret is automatically locked into position so that rigid tool support is obtained. Rotary stopscrews control the forward travel of the ram, one stop being provided for each face on the turret. The proper stop is brought into operating position automatically when the turret is indexed. A similar set of stops usually is provided to limit movement of the cross slide. The *saddle-type turret lathe* provides a more rugged mounting for the hexagon turret than can be obtained by the ram-type mounting. In *saddle-type lathes*, the main turret is mounted directly on the saddle, and the entire saddle and turret assembly reciprocates. Larger turret lathes usually have this type of mounting. However, because the saddle-turret assembly is rather heavy, this type of mounting provides less rapid turret reciprocation. When such lathes are used with heavy tooling for making heavy or multiple cuts, a *pilot arm* attached to the headstock engages a pilot hole attached to one or more faces of the turret to give additional rigidity. Turret lathe headstocks can shift rapidly between spindle speeds and brake rapidly to stop the spindle very quickly. They also have automatic stock-feeding for feeding bar stock through the spindle hole. If the work is to be held in a chuck, some type of air-operated chuck or a special clamping fixture is often employed to reduce the time required for part loading and unloading.

Vertical Turret Lathes. In machining large and/or heavy parts, such as pinions, couplings, and ring gears, vertical turret lathes are used. These are essentially regular turret lathes turned on end. Their rotary work tables commonly range from 12 to 48 in. in diameter and are equipped with both removable chuck jaws and T-slots for clamping the work. The saddle carries 6- to 12-sided turrets. Usually, each motion of successive tools can be controlled by means of stops so that duplicate workpieces can be machined with one tooling setup. Figure 23-13 shows a CNC vertical turning lathe. Vertical lathes are an excellent alternative to large horizontal CNC lathes. Gravity-aided seating of large/heavy workpieces allows a high degree of process repeatability. A smaller footprint, lower initial cost, and increased productivity are all advantages when compared to traditional horizontal lathes.

FIGURE 23-13 CNC vertical turning centers use turrets or automatic tool changers and circular tables.

CNC Turning Centers. Turret lathes have evolved into computer numerical controlled (CNC) turning centers. These machines contain *X*, *Y*, *Z*, and *C* axis control, so turning and milling can be done on the same machine. The turret can hold 8 to 10 tools, with many of them powered (have individual motors), see Figure 23-14. Some machines have automatic tool change (ATC) capability with additional tools in a magazine. The *C* axis control provides 360° location on the spindle, so the spindle can be held in any orientation while the power tools operate. More complex versions of these machines have two turrets and two spindles, and parts can be automatically transferred from the main spindle to the subspindle as shown in Figure 23-15.

Automatic Turret Lathes. After a turret lathe is tooled, the skill required of the operator is very low, and the motions are simple and repetitive. As a result, several types of automatic turret lathes have been developed that require no operator. One type uses buttons and knobs on a control panel to define (program) machine motions. A second type has a turret, the movement of which is controlled by setting trip blocks and pins. Ordinary turret lathes use the 10-station tooling setups for complete machining of a

FIGURE 23-14 CNC turning centers have turrets that hold single-point and powered tools, and *C*-axis control, on the spindle to stop it in any orientation.

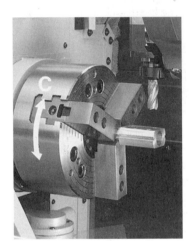

Spindle

X-axis milling

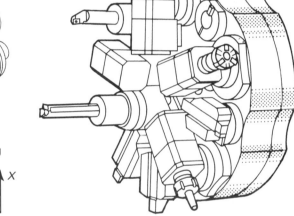

Turret with 10 tool locations

X-axis rotating tool stroke drawing

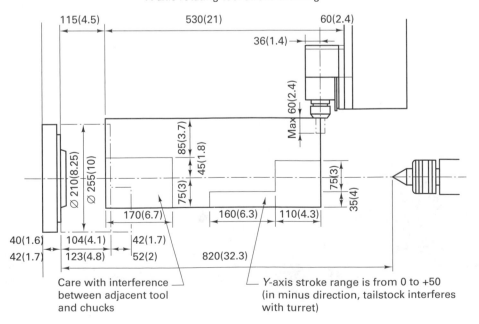

Care with interference between adjacent tool and chucks

Y-axis stroke range is from 0 to +50 (in minus direction, tailstock interferes with turret)

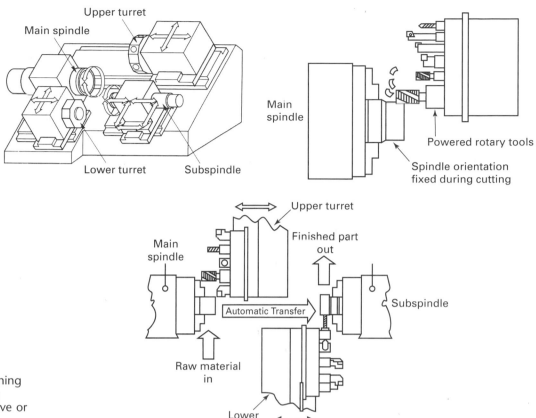

FIGURE 23-15 CNC turning centers now have multiple spindles and turrets with live or powered (rotary) tools.

piece and minimize machine-controlling time. However, an operator is required to control the machine and to feed the work into machining position. Automatic turret lathes can eliminate these last two functions. They usually have provision for manual operation and are not quite as productive as *screw machines*, which are lathes designed for completely automatic operation. Screw machines were originally designed for machining small parts, such as screws, bolts, bushings, and so on, from bar stock—hence the name *screw machines*. Now they are used to produce a wide variety of parts, covering a considerable range of sizes, and are even used for some chucking-type work.

Single-Spindle Automatic Screw Machines. There are two common types of *single-spindle screw machines*. One, an American development and commonly called the turret type (Brown & Sharpe), is shown in Figure 23-16. The other is of Swiss origin and is referred to as the Swiss type. The *Brown & Sharpe screw machine* is essentially a small automatic turret lathe, designed for bar stock, with the main turret mounted in a vertical plane on a ram. Front and rear toolholders can be mounted on the cross slide. All motions of the turret, cross slide, spindle, chuck, and stock-feed mechanism are controlled by cams. The turret cam is essentially a program that defines the movement of the turret during a cycle. These machines are usually equipped with an automatic rod-feeding magazine that feeds a new length of bar stock into the collet (the workholding device) as soon as one rod is completely used.

Often, screw machines of the Brown & Sharpe type are equipped with a transfer or "picking" attachment. This device picks up the workpiece from the spindle as it is cut off and carries it to a position where a secondary operation is performed by a small, auxiliary power head. In this manner screwdriver slots are put in screw heads, small flats are milled parallel with the axis of the workpiece, or holes are drilled normal to the axis.

On the *Swiss-type automatic screw machine*, the cutting tools are held and moved in radial slides (Figure 23-17). Disk cams move the tools into cutting position and provide feed into the work in a radial direction only; they provide any required longitudinal feed by reciprocating the headstock.

Most machining on Swiss-type screw machines is done with single-point cutting tools. Because they are located close to the spindle collet, the workpiece is not subjected

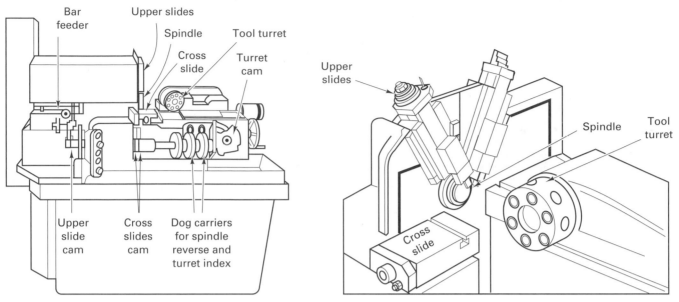

FIGURE 23-16   On the turret-type single-spindle automatic, the tools must take turns to make cuts.

to much deflection. Consequently, these machines are particularly well suited for machining very small parts.

Both types of single-spindle screw machines can produce work to close tolerances, the Swiss-type probably being somewhat superior for very small work. Tolerances of 0.0002 to 0.0005 in. are not uncommon. The time required for setting up the machine is usually an hour or two and can be much less. One person can tend many machines, once they are properly tooled. They have short cycle times, frequently less than 30 seconds per piece.

Multiple-Spindle Automatic Screw Machines.   Single-spindle screw machines utilize only one or two tooling positions at any given time. Thus the total cycle time per workpiece is the sum of the individual machining and tool-positioning times. On *multiple-spindle screw machines*, sufficient spindles, usually four, six, or eight, are provided so that all tools cut simultaneously. Thus the cycle time per piece is equal to the maximum cutting time of a single tool position plus the time required to index the spindles from one position to the next.

The two distinctive features of multiple-spindle screw machines are shown in Figure 23-18. First, the six spindles are carried in a rotatable drum that indexes in order

FIGURE 23-17   Close-up view of a Swiss-type screw machine, showing the tooling and radial tool sides, actuated by rocker arms. *(Courtesy of George Gorton Machine Corporation.)*

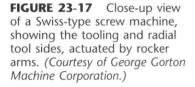

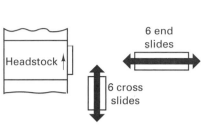

Headstock

6 end slides

6 cross slides

All spindles on multiple-spindle automatic have the same tool path

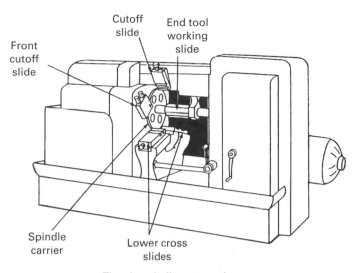

Front cutoff slide

Cutoff slide

End tool working slide

Spindle carrier

Lower cross slides

The six spindle automatic

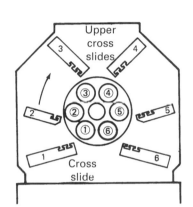

Upper cross slides

Cross slide

Spindle arrangement for 6 spindle automatic. The barstock is usually fed to a stop at position 6. The cutoff position is the one preceding the bar feed position.

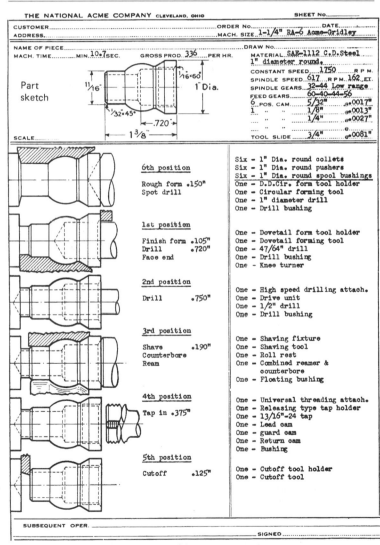

THE NATIONAL ACME COMPANY CLEVELAND, OHIO

SHEET No.

CUSTOMER.. ORDER No............................ DATE...................
ADDRESS... MACH. SIZE 1-1/4" RA-6 Acme-Gridley........

NAME OF PIECE... DRAW No...................................
MACH. TIME........MIN. 10.7 SEC.   GROSS PROD. 336 PER HR.   MATERIAL SAE-1112 C.D. Steel
                                                                    1" diameter round.
Part sketch   CONSTANT SPEED 1750 R.P.M.
                                                                    SPINDLE SPEED 617 R.P.M. 162 ET.
                                                                    SPINDLE GEARS 32-44 Low range
                                                                    FEED GEARS 60-40-44-56
                                                                    6 POS. CAM 5/32" .0017"
                                                                    1 1/8" .0013"
                                                                    1/4" .0027"
SCALE   TOOL SLIDE 3/4" .0081"

6th position

Rough form .150"
Spot drill

Six - 1" Dia. round collets
Six - 1" Dia. round pushers
Six - 1" Dia. round spool bushings
One - D.D. Cir. form tool holder
One - Circular forming tool
One - 1" diameter drill
One - Drill bushing

1st position

Finish form .105"
Drill   .720"
Face end

One - Dovetail form tool holder
One - Dovetail forming tool
One - 47/64" drill
One - Drill bushing
One - Knee turner

2nd position

Drill   .750"

One - High speed drilling attach.
One - Drive unit
One - 1/2" drill
One - Drill bushing

3rd position

Shave   .190"
Counterbore
Ream

One - Shaving fixture
One - Shaving tool
One - Roll rest
One - Combined reamer & counterbore
One - Floating bushing

4th position

Tap in .375"

One - Universal threading attach.
One - Releasing type tap holder
One - 13/16"-24 tap
One - Lead cam
One - guard cam
One - Return cam
One - Bushing

5th position

Cutoff   .125"

One - Cutoff tool holder
One - Cutoff tool

SUBSEQUENT OPER.

SIGNED

Tooling sheet for making a part on a six-spindle.

FIGURE 23-18 The multiple automatic makes all cuts simultaneously and then performs the noncutting functions (tool withdrawal, index, bar feed) at high speed.

to bring each spindle into a different working position. Second, a nonrotating tool slide contains the same number of toolholders as there are spindles and thus provides and positions a cutting tool (or tools) for each spindle. Tools are fed by longitudinal reciprocating motion. Most machines have a cross slide at each spindle position so that an additional tool can be fed from the side for facing, grooving, knurling, beveling, and cutoff operations. These slides also are shown in Figure 23-18. All motions are controlled automatically.

Study the processing steps on the tooling sheet for making a part shown in Figure 23-18. With a tool position available on the end tool slide for each spindle (except for a stock-feed stop at position 6), when the slide moves forward, these tools cut essentially simultaneously. At the same time, the tools in the cross slides move inward and make their cuts. When the forward cutting motion of the end tool slide is completed, it moves away from the work, accompanied by the outward movement of the radial slides. The spindles are indexed one position, by rotation of the spindle carrier, to position each part for the next operation to be performed. At spindle position 5, finished pieces are cut off. Bar stock 1 in. in diameter is fed to correct length for the beginning of the next operation. Thus a piece is completed each time the tool slide moves forward and back.

Multiple-spindle screw machines are made in a considerable range of sizes, determined by the diameter of the stock that can be accommodated in the spindles. There may be four, five, six, or eight spindles. The operating cycle of the end tool slide is determined by the operation that requires the longest time.

Once a multiple-spindle screw machine is set up, it requires only that the bar stock feed rack be supplied and the finished products checked periodically to make sure that they are within desired tolerances. One operator usually services many machines.

Most multiple-spindle screw machines use cams to control the motions. Setting up the cams and the tooling for a given job may require from 2 to 20 hours. However, once such a machine is set up, the processing time per part is very short. Often, a piece may be completed every 10 seconds. Typically, a minimum of 2000 to 5000 parts are required in a lot to justify setting up and tooling a multiple-spindle automatic screw machine. The precision of multiple-spindle screw machines is good, but seldom as good as that of single-spindle machines. However, tolerances from 0.0005 to 0.001 in. on the diameter are typical.

Although screw machines and automatic turret lathes are automatic *types* of lathes, the term *automatic lathe* is generally applied to a lathe that is semiautomatic and makes simultaneous cuts using "massed" tooling (Figure 23-19) but does not use screw-machine or turret principles. The tools are fed and retracted automatically by means of cam-controlled mechanisms. In most cases an operator is required to load and unload the

FIGURE 23-19 Movements of the tool blocks for the machining process in a single-spindle automatic lathe, with the metal to be removed by each tool indicated. *(Courtesy of Gisholt Corporation.)*

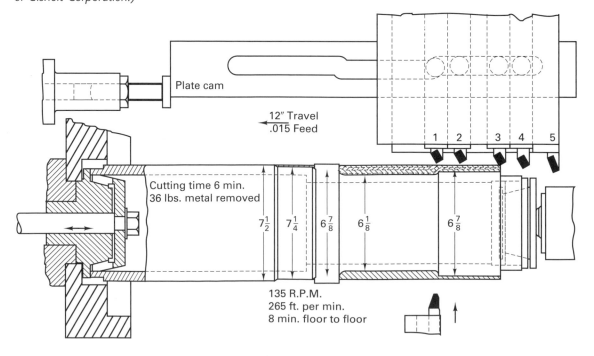

machines, so they do not repeat cycles automatically. The majority of automatic lathes have only a single spindle, but some specialized multispindle machines are also used.

In a typical *single-spindle automatic lathe*, the massed cutting tools are held in *tool blocks*, or *slides*, which are power actuated and controlled to move the tools into position and feed them along the work. Sometimes the front block provides only radial motion for facing, form cutting, and cutoff operations. The rear block has both radial and longitudinal motions, which are controlled by a plate cam, as illustrated in Figure 23-19, which shows the motions of the tool blocks and the portion of the metal removed by each tool. On some machines a third overhead tool block is also provided.

In most cases the work is held between centers, utilizing various types of power-actuated chucks, collets, and tailstocks so that the work-handling time is minimized. In some cases the work is fed into the machine from a hopper-type feeding device and clamped and discharged automatically.

The total machining cycle on an automatic lathe is usually very short, often less than one minute. Sometimes a part is put successively into two to four automatic lathes to complete its machining. Because they are fairly flexible, quite a variety of shapes and sizes can be handled in one lathe by changing the tooling setup. Tracer and NC lathes will be discussed in Chapter 32.

■ 23.4 TYPES OF BORING MACHINES

VERTICAL BORING AND TURNING MACHINES

Figure 23-20 shows the basic elements of a vertical boring and turning machine. These machine tools are structurally similar to double-housing planers except that the table rotates

FIGURE 23-20 Block diagram of a vertical boring and turning machine.

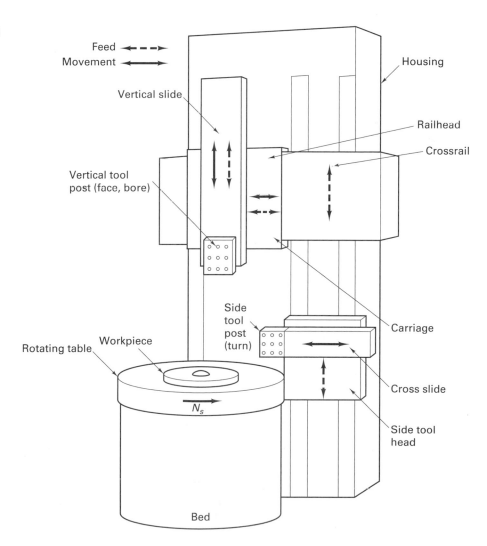

instead of reciprocating. Functionally, a vertical boring machine essentially is the same as a vertical turret lathe, but it usually has two main tool heads instead of a turret. Thus turning, facing, and usually boring (but not milling) are done on vertical boring machines.

Vertical boring machines come with tables ranging from about 3 to 40 ft in diameter. Toolheads have both horizontal and vertical feed and therefore can be used for boring and for facing cuts. Usually, one or both can also be swivelled about a horizontal axis to permit boring at an angle. Most machines also have a side toolhead, sometimes provided with a four-sided turret. This toolhead has vertical and horizontal feed and is used primarily for turning. Single-point tools customarily are used for turning, facing, and boring operations.

Many modern boring machines are numerically controlled (see Chapter 32). This permits the operator to make tool settings and adjustments merely by the input of numerical data. The tool can be moved very quickly to the proper position for the next cut, reducing the amount of machine-controlling time and increasing the productivity of these large and costly machines.

JIG BORERS

Jig borers are very precise vertical-type boring machines designed for use in making *jigs* and *fixtures*. From the viewpoint of boring operations, they contain no unusual features, except that the spindle and spindle bearings are constructed with very high precision. Their unique features are in the design of the worktable controls, which permits very precise movement and control, thus making them especially useful in layout work. Modern NC machining centers are capable of doing similar work, often eliminating the need for jig borers.

HORIZONTAL BORING (DRILLING AND MILLING) MACHINES

Horizontal boring machines are very versatile and thus particularly useful in machining large parts. The basic components of these machines are indicated in Figure 23-21. The essential features are as follows:

1. A rotating spindle that can be fed horizontally (tool rotates).
2. A table that can be moved and fed in two directions in a horizontal plane.
3. A headstock that can be moved vertically.
4. An outboard bearing support for a long boring bar.

The spindle will accept both drills and milling cutters in addition to boring bars. A wide range of rpm values is provided, and heavy bearings are incorporated that will absorb thrust in all directions. The spindle is also provided with longitudinal power feed so that drilling and boring can be done over a considerable distance without moving the table.

Boring on this type of machine is done by means of a rotating single-point tool. The tool can be mounted in either a stub-type bar, held only in the spindle, as shown in Figure 23-22, or in a long line-type bar that has its outer end supported in a bearing on the outboard column, as shown in Figure 23-21. The outboard bearing provides rigid support for the boring bar and permits very accurate work to be done. However, because of the flexibility inherent in a long boring bar and offset tool holder, horizontal boring machines are used primarily for boring holes less than 12 in. in diameter, for long holes, or for a series of in-line holes. Unless they are very long or the shape of the workpiece prevents it, larger holes usually are bored on a vertical boring mill.

MASS-PRODUCTION BORING MACHINES

Special boring machines are built for machining specific parts in mass production. The workpiece usually remains stationary, and boring is done by one or more rotating boring tools, typically carried in a reciprocating powerhead, such as is shown in Figure 23-23. In most cases the operation is automatic once the workpiece is placed in the workholding device. Such machines are usually very accurate and often are equipped with automatic gaging and sizing controls. (See discussion of transfer lines in Chapter 32.)

With rotating workpieces, the size of the hole is controlled by transverse movement of the tool holder. When boring is done with a rotating tool, size is controlled by changing

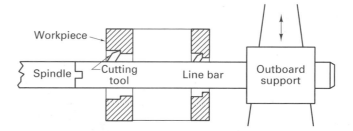

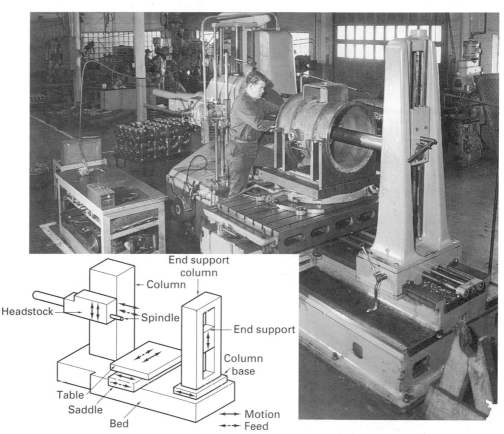

FIGURE 23-21 Boring a weldment on horizontal boring, drilling, and milling machine. A line boring bar extends through the workpiece and is used with an outboard bearing support. *(Courtesy of Lucas Machine Division, The New Britain Machine Company.)*

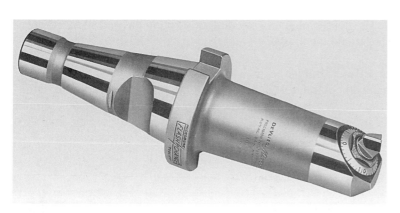

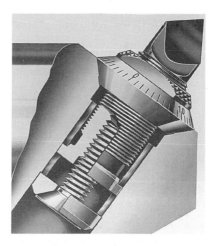

FIGURE 23-22 Adjustable boring tool for stub boring. The single-point tool is adjustable, as shown in sectional view on the right. *(Courtesy of DeVlieg Machine Company.)*

the offset radius of the cutting-tool tip with respect to the axis of rotation. The type of boring tool shown in Figure 23-22 has precise control and is used on larger-scale manufacturing. Two or more adjustable cutting tools can be built into a single bar, thus permitting more than one diameter to be bored simultaneously. For boring relatively long holes, the type of boring bar shown in Figure 23-24 has a special advantage. As shown,

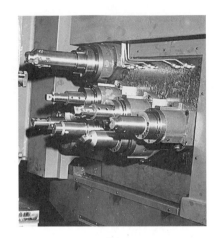

(a)   (b)

FIGURE 23-23 (a) Production-type boring machine, having multiple heads, that completes a part in 51 seconds. (b) Close-up view of one multiple-spindle boring head on a production-type machine. *(Courtesy of Health Machine Company.)*

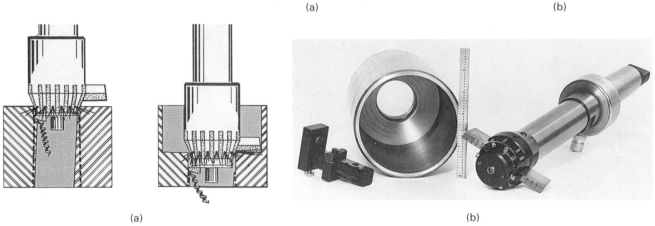

(a)   (b)

FIGURE 23-24   Boring tool employing a centering tool and conical guide, for boring large holes in a single operation. *(Courtesy of Vernon Devices, Incorporated.)*

the smaller, forward bit corrects misalignment of the original hole and provides a guide hole for the nose cone. The nose cone then provides good alignment and support for the rear bit, which bores the final hole to size.

BORING MACHINE PRECISION

As with turning, the precision obtainable in boring depends considerably on the rigidity of the tool support. On specialized, production-type boring machines, tolerances are readily held to within 0.0005 in. on small diameters, whereas on general-purpose machines tolerances of 0.001 in. are typical unless the boring bar overhang becomes excessive.

■ 23.5 CUTTING TOOLS FOR LATHES

LATHE CUTTING TOOLS

Most lathe operations are done using single-point *cutting tools*, such as those illustrated in Figure 23-25. On right-hand (and left-hand turning) and facing tools, the cutting usually takes place on the side of the tool; therefore, the side rake angle is of primary importance, particularly when deep cuts are being made. On the round-nose turning tools, cutoff tools, finishing tools, and some threading tools, cutting takes place on or near the tip of the tool, and the back rake is therefore of importance. Such tools are used with relatively light depths of cut.

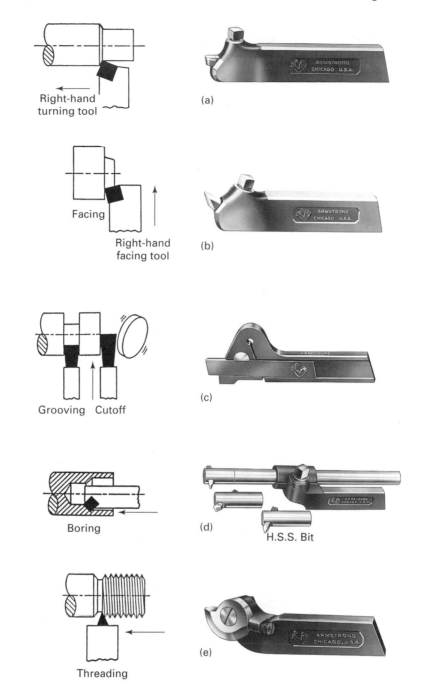

FIGURE 23-25 Common types of forged toolholders (a) right-hand turning, (b) facing, (c) grooving cutoff, (d) boring, (e) threading. *(Courtesy of Armstrong Brothers Tool Company.)*

Because tool materials are expensive, it is desirable to use as little as possible. At the same time, it is essential that the cutting tool be supported in a strong, rigid manner to minimize deflection and possible vibration. Consequently, lathe tools are supported in various types of heavy, forged steel tool holders, as shown in Figure 23-25. The high-speed steel (HSS) tool bit should be clamped in the toolholder with minimum overhang, otherwise tool chatter and a poor surface finish may result.

In the use of carbide, ceramic, or coated carbides for mass production work, throwaway inserts are used that can be purchased in a great variety of shapes, geometrics (nose radius, tool angles, and groove geometry), and sizes (see Figure 23-26 for some examples).

When several different operations on a lathe are performed repeatedly in sequence, the time required for changing and setting tools may constitute as much as 50% of the total cycle time. Quick-change tool holders (Figure 23-27) are used to reduce manual tool-changing time. The individual tools, preset in their holders, can be interchanged in the special tool post in a few seconds. With some systems, a second tool may

Insert shape	Available cutting edges	Typical insert holder
Round	4–10 on a side 8–20 total	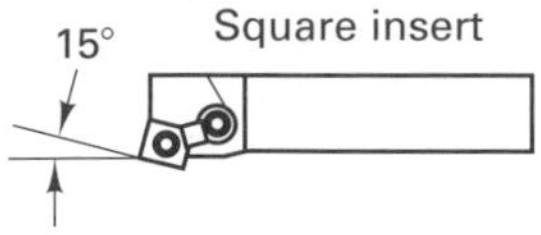 15° Square insert
80°/100° diamond	4 on a side 8 total	
Square	4 on a side 8 total	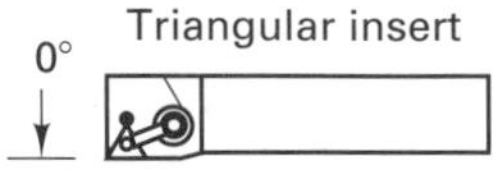 0° Triangular insert
Triangle	3 on a side 6 total	
55° diamond	2 on a side 4 total	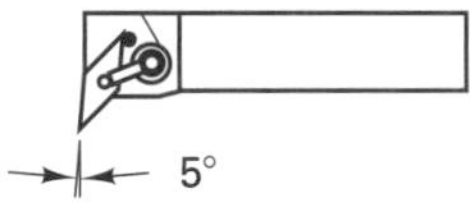 35° diamond
35° diamond	2 on a side 4 total	5°

FIGURE 23-26   Typical insert shapes, available cutting edges per insert and insert holders for throwaway insert cutting tools. *(Adapted from Turning Handbook of High Efficiency Metal Cutting, courtesy of General Electric Company.)*

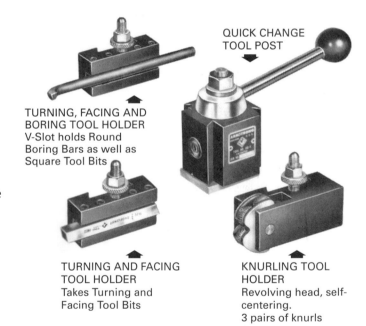

FIGURE 23-27   Quick-change tool post and accompanying toolholders. *(Courtesy of Armstrong Brothers Tool Company.)*

QUICK CHANGE TOOL POST

TURNING, FACING AND BORING TOOL HOLDER
V-Slot holds Round Boring Bars as well as Square Tool Bits

TURNING AND FACING TOOL HOLDER
Takes Turning and Facing Tool Bits

KNURLING TOOL HOLDER
Revolving head, self-centering.
3 pairs of knurls

be set in the tool post while a cut is being made with the first tool and can then be brought into proper position by rotating the post.

In lathe work, the nose of the tool should be set exactly at the same height as the axis of rotation of the work. However, because any setting below the axis causes the work to tend to "climb" up on the tool, most machinists set their tools a few thousandths of an inch above the axis, except for cutoff, threading, and some facing operations.

FORM TOOLS

In Figure 23-18, the use of form tools was shown in automatic lathe work. Form tools are made by grinding the inverse of the desired work contour into a block of HSS or tool steel. A threading tool is often a form tool. Although form tools are relatively expensive to manufacture, it is possible to machine a fairly complex surface with a single inward

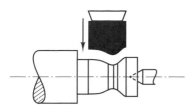

Form turning

FIGURE 23-28 Circular and block types of form tools. *(Courtesy of Speedi Tool Company, Incorporated.)*

feeding of one tool. For mass-production work, adjustable form tools of either flat or rotary types, such as are shown in Figure 23-28, are used. These, of course, are expensive to make initially but can be resharpened by merely grinding a small amount off the face and then raising or rotating the cutting edge to the correct position.

The use of form tools is limited by the difficulty of grinding adequate rake angles for all points along the cutting edge. A rigid setup is needed to resist the large cutting forces that develop with these tools. Light feeds with sharp, coated HSS tools are used on multiple-spindle automatics, turret lathes, and transfer line machines.

TURRET-LATHE TOOLS

In turret lathes, the work is generally held in collets and the correct amount of bar stock is fed into the machine to make one part. The tools are arranged in sequence at the tool stations with depths of cut all preset. The following factors should be considered when setting up a turret lathe.

1. *Setup time:* time required to install and set the tooling and set the stops. Standard toolholders and tools should be used as much as possible to minimize setup time. Setup time can be greatly reduced by eliminating adjustment in the setup. (See Chapter 43.)
2. *Workholding time:* time to load and unload parts and/or stock.
3. *Machine-controlling time:* time required to manipulate the turrets. Can be reduced by combining operations where possible. Dependent on the sequence of operations established by the design of the setup.
4. *Cutting time:* time during which chips are being produced. Should be as short as is economically practical and represent the greatest percentage of the total cycle time possible.
5. *Cost:* cost of the tool, setup labor cost, lathe operator labor cost, and the number of pieces to be made.

There are essentially 11 tooling stations, as shown in Figure 23-29, with six in the turret, four in the indexable tool post, and one in the rear tool post. The tooling is more

FIGURE 23-29 Turret-lathe tooling setup for producing part shown. Numbers in circles indicate the sequence of the operations from 1 to 9. The letters "A" through "F" refer to the surfaces being machined. Operation 3 is a combined operation. The roll turner is turning surface F while tool 3 on the square post is turning surface B. The first operation stops the stock at the right length. The last operation cuts the finished bar off and puts a chamfer on the bar, which will next be advanced to the stock stop.

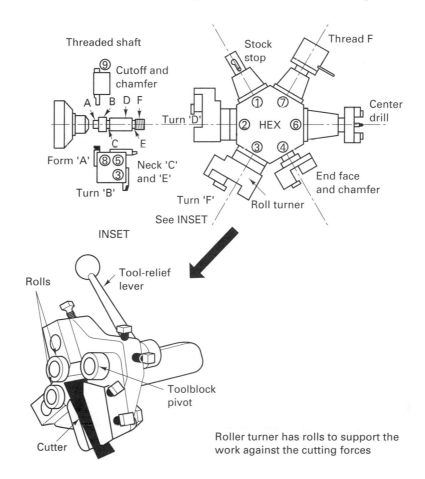

rugged in turret lathes because heavy, simultaneous cuts are often made. Tools mounted in the hex turret that are used for turning are often equipped with pressure rollers set on the opposite side of the rotating workpiece from the tool to counter the cutting forces.

Turret lathes are most economical in producing lots too large for engine lathes but too small for automatic screw machines or automatic lathes. In recent years much of this work has been assumed by numerical control lathes or turning centers. For example, the component (threaded shaft) shown in Figure 23-29 could have also been made on an NC turret lathe with some savings in cycle time.

■ 23.6 WORKHOLDING IN LATHES

WORKHOLDING DEVICES FOR LATHES

Five methods are commonly used for supporting workpieces in lathes:

1. Held between centers
2. Held in a chuck
3. Held in a collet
4. Mounted on a face plate
5. Mounted on the carriage

In the first four of these methods, the workpiece is rotated during machining. In the fifth method, which is not used extensively, the tool rotates while the workpiece is fed into the tool.

LATHE CENTERS

Workpieces that are relatively long with respect to their diameters are usually machined between centers. See Figure 23-30. Two *lathe centers* are used, one in the spindle hole and the other in the hole in the tailstock quill. Two types are used, called *dead* and *live*. Dead centers are *solid*, that is, made of hardened steel with a Morse taper on one end so that it will fit into the spindle hole. The other end is ground to a 60° taper. Sometimes the tip of this taper is made of tungsten carbide to provide better wear resistance. Before a center is placed in position, the spindle hole should be carefully wiped clean. The presence of foreign material will prevent the center from seating properly and it will not be aligned accurately.

Live centers of the type shown in Figure 23-31 are designed so that the end that fits into the workpiece is mounted on ball or roller bearings. It is free to rotate. No lubrication is required. Live centers may not be as accurate as the solid type and therefore are not often used for precision work.

Before a workpiece can be mounted between lathe centers, a 60° center hole must be drilled in each end. This can be done in a drill press or in a lathe by holding the work in a

FIGURE 23-30   Work being turned between centers in a lathe, showing the use of a dog and dog plate. *(Courtesy of South Bend Lathe.)*

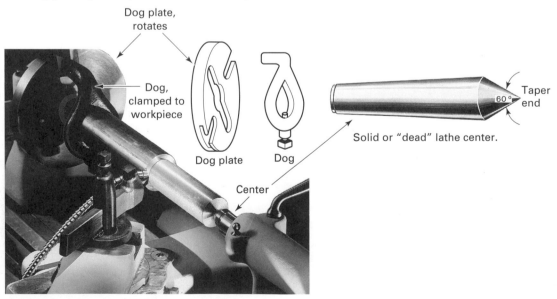

FIGURE 23-31 "Live" type of lathe center. *(Courtesy of Motor Tool Manufacturing Company.)*

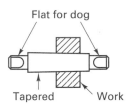

Plain solid mandrel

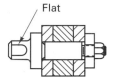

Gang mandrel

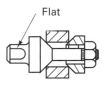

Cone mandrel

FIGURE 23-32 Three types of mandrels, which are mounted between centers for lathe work.

chuck. A combination center drill and countersink ordinarily is used, with care taken that the center hole is deep enough so that it will not be machined away in any facing operation and yet is not drilled to the full depth of the tapered portion of the center drill (see Chapter 24).

Because the work and the center of the headstock end rotate together, no lubricant is needed in the center hole at this end. The center in the tailstock quill does not rotate; adequate lubrication must be provided. A mixture of white lead and oil is often used. Failure to provide proper lubrication at all times will result in scoring of the workpiece center hole and the center, and inaccuracy and serious damage may occur. Live centers are often used in the tailstock to overcome these problems.

The workpiece must rotate freely, yet no looseness should exist. Looseness usually will be manifested in "chattering" of the workpiece during cutting. The setting of the centers should be checked after cutting for a short time. Heating and thermal expansion of the workpiece will reduce the clearances in the setup.

A mechanical connection must be provided between the spindle and the workpiece to provide rotation. This is accomplished by a *lathe dog* and *dog plate*. The dog is clamped to the work. The *tail* of the dog enters a slot in the dog plate, which is attached to the lathe spindle in the same manner as a lathe chuck. For work that has a finished surface, a piece of soft metal, such as copper or aluminum, can be placed between the work and the dog setscrew clamp to avoid marring (Figure 23-30).

MANDRELS

Workpieces that must be machined on both ends or are disk-shaped are often mounted on *mandrels* for turning between centers. Three common types of mandrels are shown in Figure 23-32. *Solid mandrels* usually vary from 4 to 12 in. in length and are accurately ground with a 1 : 2000 taper (0.006 in./ft). After the workpiece is drilled and/or bored, it is pressed on the mandrel. The mandrel should be mounted between centers so that the cutting force tends to tighten the work on the mandrel taper. Solid mandrels permit the work to be machined on both ends as well as on the cylindrical surface. They are available in stock sizes but can be made to any desired size.

Gang (or disk) mandrels are used for production work because the workpieces do not have to be pressed on and thus can be put in position and removed more rapidly. However, only the cylindrical surface of the workpiece can be machined when this type of mandrel is used. *Cone mandrels* have the advantage that they can be used to center workpieces having a range of hole sizes.

LATHE CHUCKS

Lathe chucks are used to support a wider variety of workpiece shapes and to permit more operations to be performed than can be accomplished when the work is held between centers. Two basic types of chucks are used (Figure 23-33).

Three-jaw, self-centering chucks are used for work that has a round or hexagonal cross section. The three jaws are moved inward or outward simultaneously by the rotation of a spiral cam, which is operated by means of a special wrench through a bevel gear. If they are not abused, these chucks will provide automatic centering to within about 0.001 in. However, they can be damaged through use and will then be considerably less accurate.

Each jaw in a *four-jaw independent chuck* can be moved inward and outward independent of the others by means of a chuck wrench. Thus they can be used to support a wide variety of work shapes. A series of concentric circles engraved on the chuck face aid in adjusting the jaws to fit a given workpiece. Four-jaw chucks are heavier and more rugged than the three-jaw type, and because undue pressure on one jaw does not destroy the accuracy of the chuck, they should be used for all heavy work. The jaws on both three- and four-jaw chucks can be reversed to facilitate gripping either the inside or the outside of workpieces.

Combination four-jaw chucks are available in which each jaw can be moved independently or can be moved simultaneously by means of a spiral cam. Two-jaw chucks are also available. For mass-production work, special chucks often are used in which the jaws are actuated by air or hydraulic pressure, permitting very rapid clamping of the work. See Figure 23-34 for schematic. The rapid exchange of tooling is a key manufacturing strategy in manufacturing cells. Chuck jaw sets are dedicated and customized for

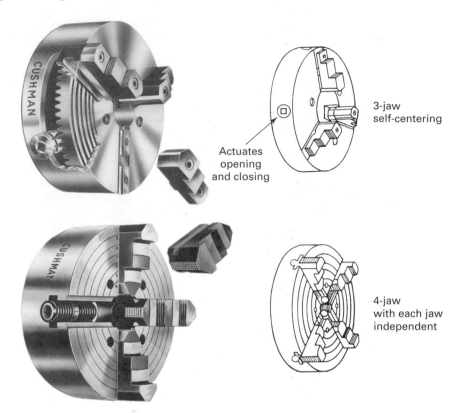

FIGURE 23-33 The jaws on chucks for lathes (four-jaw independent) or (three-jaw, self-centering) can be removed and reversed.

3-jaw self-centering

Actuates opening and closing

4-jaw with each jaw independent

specific parts. The first time a chuck jaw set is used, each jaw is marked with the number of the jaw slot where it was installed and an index mark that corresponds with the alignment of the jaw serrations and the first tooth on the chuck master jaw. The jaws can now be reinstalled on the chuck exactly where they were bored. The adjustability of the chuck body lets the operator dial in part concentrically without resetting the jaw.

COLLETS

Collets are used to hold smooth cold-rolled bar stock or machined workpieces more accurately than with regular chucks. As shown in Figure 23-35, collets are relatively thin tubular steel bushings that are split into three longitudinal segments over about two-thirds of their length. At the split end, the smooth internal surface is shaped to fit the piece of stock

FIGURE 23-34 Hydraulically actuated through-hole 3-jaw power chuck shown in section view to left and in the spindle of the lathe above connected to the actuator.

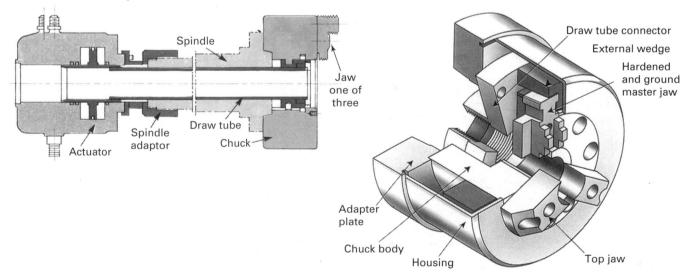

Spindle

Jaw one of three

Actuator

Spindle adaptor

Draw tube

Chuck

Draw tube connector

External wedge

Hardened and ground master jaw

Adapter plate

Chuck body

Housing

Top jaw

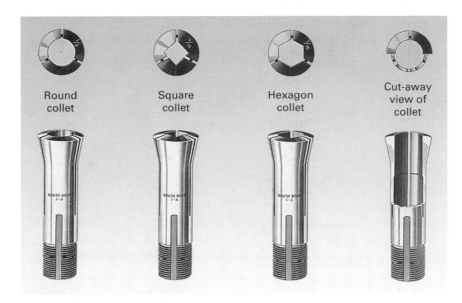

FIGURE 23-35 Several types of lathe collets. *(Courtesy of South Bend Lathe.)*

that is to be held. The external surface of the collet is a taper that mates with an internal taper of a collet sleeve. When the collet is pulled inward into the spindle (by means of the draw bar), the action of the two mating tapers squeezes the collet segments together, causing them to grip the workpiece (see Figure 23-36).

Collets are made to fit a variety of symmetrical shapes. If the stock surface is smooth and accurate, good collets will provide very accurate centering, with runout less than 0.0005 in. However, the work should be no more than 0.002 in. larger or 0.005 in. smaller than the nominal size of the collet. Consequently, collets are used only on drill-rod, cold-rolled, extruded, or previously machined stock.

Collets that can open automatically and feed bar stock forward to a stop mechanism are commonly used on automatic lathes and turret lathes. An example of a collet chuck is shown in Figure 23-37. Another type of collet similar to a Jacobs drill chuck has a greater size range than ordinary collets; therefore, fewer are required.

FACE PLATES

Face plates are used to support irregularly shaped work that cannot be gripped easily in chucks or collets. The work can be bolted or clamped directly on the face plate or can be supported on an auxiliary fixture that is attached to the face plate. The latter procedure is time-saving when identical pieces are to be machined.

MOUNTING WORK ON THE CARRIAGE

When no other means is available, boring occasionally is done on a lathe by mounting the work on the carriage, with the boring bar mounted between centers and driven by means of a dog.

FIGURE 23-36 Method of using a drawn-in collet in a lathe spindle. *(Courtesy of South Bend Lathe.)*

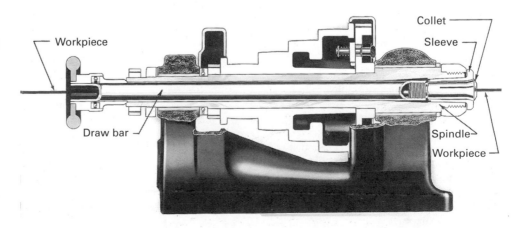

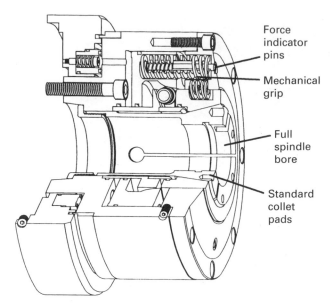

FIGURE 23-37 Schematic of a collet chuck in which the clamping force can be adjusted.

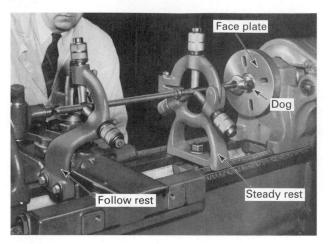

FIGURE 23-38 Cutting a thread on a long, slender workpiece, using a follow rest (*left*) and a steady rest (*right*) on an engine lathe. Note the use of a dog and face plate to drive the workpiece. (*Courtesy of South Bend Lathe.*)

STEADY AND FOLLOW RESTS

If one attempts to turn a long, slender piece between centers, the radial force exerted by the cutting tool, or the weight of the workpiece itself, may cause it to be deflected out of line. Steady rests and follow rests (Figure 23-38) provide means for supporting such work between the headstock and the tailstock. The steady rest is clamped to the lathe ways and has three movable fingers that are adjusted to contact the work and align it. A light cut should be taken before adjusting the fingers to provide a smooth contact-surface area.

A steady rest also can be used in place of the tailstock as a means of supporting the end of long pieces, pieces having too large an internal hole to permit using a regular dead center, or work where the end must be open for boring. In such cases the headstock end of the work must be held in a chuck to prevent longitudinal movement. Tool feed should be toward the headstock.

The follow rest is bolted to the lathe carriage. It has two contact fingers that are adjusted to bear against the workpiece, opposite the cutting tool, in order to prevent the work from being deflected away from the cutting tool by the cutting forces.

■ KEY WORDS

apron	carriage	cutting tools	face plates	headstock	metal removal rate	reaming	taper turning
automatic lathe	chucks	depth of cut	facing	knurling	milling	screw machine	turning
bed	collets	drilling	feed	lathe centers	parting	steady rest	turret lathe
boring	cutoff	engine lathe	follow rest	mandrels	quill	tailstock	workholding

■ REVIEW QUESTIONS

1. How is the tool-work relationship in turning different from facing?
2. What different kinds of surfaces can be produced by turning versus facing?
3. How does form turning differ from ordinary turning?
4. What is the basic difference from facing and a cutoff operation?
5. Which machining operations shown in Figure 23-2 does not form a chip?
6. Why is it difficult to make heavy cuts if a form turning tool is complex in shape?

7. Why is Equation (23-5) an approximate equation?
8. Why is the spindle of the lathe hollow?
9. What function does a lathe carriage have?
10. Why is feed specified for a boring operation different from that specified for turning if the MRR equations are the same?
11. What function is provided by the lead screw on a lathe that is not provided by the feed rod?
12. How can work be held and supported in a lathe?
13. How is a workpiece that is mounted between centers on a lathe driven (rotated)?

14. What will happen to the workpiece when turned, if held between centers, and the centers are not exactly in line?
15. Why is it not advisable to hold hot-rolled steel stock in a collet?
16. How does a steady rest differ from a follow rest?
17. What are the advantages and disadvantages of a four-jaw independent chuck versus a three-jaw chuck?
18. Why should the distance a lathe tool projects from the tool-holder be minimized?
19. What is the difference between a ram and a saddle turret lathe?
20. How can a tapered part be turned on a lathe?
21. Why might it be desirable to use a heavy depth of cut and a light feed at a given speed in turning rather than the opposite?
22. On what diameter is the RPM based for a facing cut, assuming given work and tool materials?
23. Why is it usually necessary to take relatively light feeds and depths of cut when boring on a lathe?
24. How does the corner radius of the tool influence the surface roughness?
25. What effect does a BUE have on the diameter of the workpiece in turning?
26. What important factor must be kept in mind in tooling multiple-spindle screw machines that does not have to be considered in single-spindle machines?
27. What is the reason for having C axis control in a CNC turning center?

28. Why does boring assure concentricity between the hole axis and the axis of rotation of the workpiece (for boring tool), whereas drilling does not?
29. Why are vertical boring mills better suited than a lathe for machining large workpieces?
30. What is the principal advantage of a horizontal boring machine over a vertical boring machine for large workpieces?
31. Where in Chapter 23 is a workpiece being held in a 3-jaw chuck?
32. How is the workpiece in Figure 23-17 being held?
33. In what figures is a dead center shown in Chapter 23?
34. In what figures is a live center shown in Chapter 23?
35. In which figures showing setups do you find the following being used as a workholding device?
    a. Three-jaw chuck   c. Faceplate
    b. Collet   d. Four-jaw chuck?
36. How many form tools are being utilized in the process shown in Figure 23-18 to machine the part?
37. In a production type boring machine with a multiple spindle head, many tools (seven in Figure 23-23) are cutting simultaneously. What about tool wear and tool failure? Do you change all the tools when one fails or do you stop the machine and change each tool as it fails?
38. From the information given in Figure 23-29, start with a piece of round bar stock and show how it progresses, operation by operation, into a finished part—a threaded shaft.

■ PROBLEMS

1. A cutting speed of 200 ft/min has been selected. At what rpm should a 3-in.-diameter bar be rotated?
2. Assume that the workpiece in Problem 23-1 is 8-in. (203.2-mm) long and a feed of 0.020 in. (0.51 mm) per revolution is used. How long will a cut across its entire length require? Don't forget to add an allowance.
3. If the depth of cut in Problem 23-2 is $\frac{1}{8}$ in., what is the metal removal rate (MRR) exactly? What is the MRR approximately?
4. The following data apply for machining a part on a turret lathe and on an engine lathe:

	Engine Lathe	Turret Lathe
Times, in minutes, to machine part	30 min	5 min
Cost of special tooling	0	$300
Time to setup	30 min	3 hr
Labor rates	$8/hr	$8/hr
Machine rates	$10/hr	$12/hr

   a. How many pieces would have to be made for the cost of the engine lathe to just equal the cost of the turret lathe? This is the BEQ.
   b. What is the cost per unit at the BEQ?
5. A hole 89 mm in diameter is to be drilled and bored through a piece of 1340 steel that is 200-mm long, using a horizontal boring, drilling, and milling machine. High-speed tools will be used. The job will be done by center drilling, drilling with an 18-mm drill, followed by a 76-mm drill, then bored to size in one cut, using a feed of 0.50 mm/rev. Drilling feeds will be 0.25 mm/rev for the smaller drill and 0.64 mm/rev for the larg-

er drill. The center drilling operation requires 0.5 min. To set or change any given tool and set the proper machine speed and feed requires 1 min. Select the initial cutting speeds, and compute the total time required for doing the job. (Neglect setup time for the fixture.) This is often referred to as the run time or the cycle time.

6. In Figure 23-A are three plots of unit production cost ($/unit) versus production volume (Q = build quantity). Note that this plot is made on log-log paper. Cost per unit for a particular process decreases with increased volume. For a particular process there is no minimum cost but rather production volumes within which particular processes are most economical.
   a. Each of these curves is a plot of the equation for total cost per unit, which means each is the sum of the fixed cost per unit (setup, tooling, overhead) and the variable costs per unit (direct labor, direct material). From the data on the plots, estimate the fixed costs for the engine lathe, the NC lathe, and the single spindle automatic.
   b. For what build quantities is the NC lathe most economical (approximately)?
   c. What cost per unit does the NC lathe approach as the build quantity becomes very large?
   d. What happens to these plots if you plot them on regular Cartesian coordinates? Try it and comment on what you find.
   e. Many Japanese manufacturers have found innovative ways to eliminate setup time in many of their processes. What is the impact of this on these kinds of plots, on cost per unit economics, and on job shop inventories?
7. The derivation of the approximate Equation (23-5) for the MRR for turning process requires an assumption regarding the diameters of the parts being turned. Determine the error in the equation for Problem 3 in this chapter. What is the assumption?

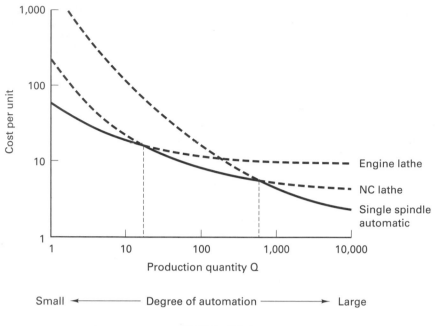

Small ◄——————— Degree of automation ———————► Large

FIGURE 23-A

www.wiley.com/college/degarmo

hapter 23 **CASE STUDY**

Estimating the Machining Time for Turning

As the plant manufacturing engineer at B.R.C., Inc., Jay Langley has been called into the production department to provide an expert opinion on a machining problem. Unfortunately, the only tool or instrument available at the time is a 1-in. micrometer.

Katrin Zachary, the production manager, would like to know the minimum time required to machine a large forging. The 8-ft-long forging is to be turned down from an original diameter of 10 in. to a final diameter of 6 in. The forging has a BHN of 300 to 400. The turning is to be performed on a heavy-duty lathe, which is equipped with a 50-hp motor and a continuously variable-speed drive on the spindle. The work will be held between centers, and the overall efficiency of the lathe has been determined to be 75%.

The forging (or log) is made from medium-carbon, 4345 alloy steel. The steel manufacturer, some basic experimentation, and established knowledge of the product and its manufacture have provided the following information:

1. A tool-life equation developed for the most suitable type of tool material at a feed of 0.020 ipr and a rake angle of $\alpha = 10°$. The equation $VT^n = C$ generally fits the data, with V = cutting speed and T = the time in minutes to tool failure. Two test cuts were run, one at $V = 60$ sfpm, where $T = 100$ min, and another at $V = 85$ sfpm, where $T = 10$ min.

2. According to the vendor, the dynamic shear strength of the material is on the order of 125,000 psi.

3. Jay decides to make two test cuts at the standard feed of 0.020 ipr. He assumes that the chip thickness ratio varies almost linearly between the speeds of 20 and 80 fpm, the values being 0.4 at the speed of 20 fpm and 0.6 at 80 fpm. The chip thickness values were determined by micrometer measurements to determine the value of r_c.

4. The machined forging (log) will be used as a roller in a newspaper press and must be precisely machined. If the log deflects during the cutting more than 0.005 in., the roll will end up barrel-shaped after final grinding and polishing.

How should Jay proceed to estimate the minimum time required to machine this forging, assuming that one finishing pass will be needed when the log has been reduced to 6 in. in diameter? The deflection due to cutting forces must be kept below 0.005 in. at the mid-log location.

Assume that $F_c \times 0.5 = F_f$ and $F_f \times 0.5 = F_R$, and that F_R causes the deflection.

DRILLING AND RELATED HOLE-MAKING PROCESSES

■ 24.1 INTRODUCTION

FIGURE 24-1 Nomenclature and geometry of conventional twist drill. Shank style depends upon the method used to hold the drill. Tangs or notches prevent slippage: (a) Straight shank with tang, (b) tapered shank with tang, (c) straight shank with whistle notch, (d) straight shank with flat notch.

In manufacturing it is probable that more holes are produced than any other shape, and a large proportion of these are made by *drilling*. Of all the machining processes performed, drilling makes up about 25%. Consequently, drilling is a very important process. Although drilling appears to be a relatively simple process, it is really a complex process. Most drilling is done with a tool having two cutting edges or lips as shown in Figure 24-1. This is a twist drill, the most common drill geometry. The cutting edges are at the end of a relatively flexible tool. Cutting action takes place inside the workpiece. The only exit for the chips is the hole that is filled by the drill. Friction results in heat that is additional to that due to chip formation. The counterflow of the chips makes lubrication and cooling difficult. There are four major actions taking place at the point of a drill.

1. A small hole is formed by the web—chips are not cut here in the normal sense.

2. Chips are formed by the rotating lips.

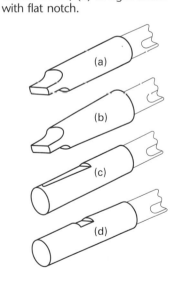

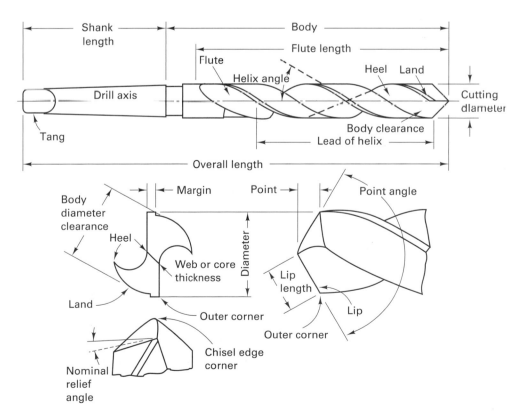

3. Chips are removed from the hole by the screw action of the helical flutes.

4. The drill is guided by lands or margins that rub against the walls of the hole.

In recent years, new drill point geometries and TiN coatings have resulted in improved hole accuracy, longer life, self-centering action, and increased-feed-rate capabilities. However, the great majority of drills manufactured are twist drills. One estimate has U.S. manufacturing companies consuming 250 million twist drills per year.

When HSS drills wear out, the drill is reground. If regrinding is not done properly, the original drill geometry may be lost and so will drill accuracy and precision. Drill performance also depends on the drilling machine tool, the workholding device, the drill holder, and the surface of the workpiece. Poor surface conditions (sand pockets and/or chilled hard spots on castings, or hard oxide scale on hot rolled metal) can accelerate early tool failure and degrade the hole-drilling process.

■ 24.2 FUNDAMENTALS OF THE DRILLING PROCESS

The process of drilling creates two chips. A conventional two-flute drill, with drill of diameter D, has two principal cutting edges rotating at an rpm rate of N and feeding axially. The rpm of the drill is established by the selected cutting velocity or cutting speed

$$N = \frac{12V}{\pi D} \tag{24-1}$$

with V in surface feet per minute (mm/min) and D in inches (mm). This equation assumes that V is the cutting speed at the outer corner of the cutting lip (point X in Figure 24-2).

FIGURE 24-2 Conventional drill geometry viewed from the point showing how the rake angle varies from the chisel edge to the outer corner along the lip. The thrust force increases as the web is approached.

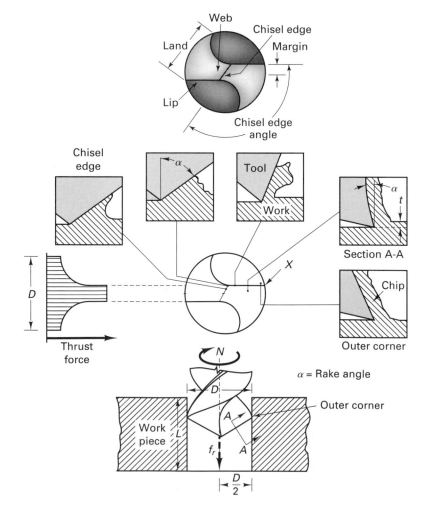

α = Rake angle

The feed, f_r, is given in inches per revolution. The depth of cut in drilling is equal to half the feed rate, or $t = f_r/2$ (see section A-A in Figure 24-2). The feed rate in inches per minute, f_m, is $f_r N$. The length of cut in drilling equals the depth of the hole, L, plus an allowance for approach and for the tip of drill, usually $A = D/2$. As in turning, the speed and feed used in drilling depend upon the material being machined and the cutting tool material. Table 24-1 gives some typical values for V and f_r for carbide indexable-insert drills, a type of drill shown later in the chapter.

After selecting the cutting speed and feed for drilling the hole, the rpm value of the spindle of the machine is determined from Equation (24-1), the maximum velocity

TABLE 24-1.	Recommended Speeds and Feeds for Indexable-Insert Drills		
Material Group	Size Range	Cutting Speed sfm	Feed Rate ipr[1]
Cast iron modular	$\frac{13}{16}-1\frac{1}{8}$	165–300	.004–.008
ductile or malleable	$1-1\frac{3}{8}$	165–300	.005–.010
	$1\frac{1}{4}-1\frac{5}{8}$	165–300	.006–.012
	$1\frac{1}{2}-2\frac{1}{2}$	165–300	.008–.014
	$2\frac{3}{8}-3\frac{1}{2}$	165–300	.010–.015
1000 Series steels	$\frac{13}{16}-1\frac{1}{8}$	300–400	.003–.005
like 1081, 1020, etc.	$1-1\frac{3}{8}$	350–450	.003–.006
	$1\frac{1}{4}-1\frac{5}{8}$	400–550	.004–.007
	$1\frac{1}{2}-2\frac{1}{2}$	450–600	.004–.007
	$2\frac{3}{8}-3\frac{1}{2}$	500–700	.005–.009
Low-carbon	$\frac{13}{16}-1\frac{1}{8}$	200–300	.003–.005
Unalloyed	$1-1\frac{3}{8}$	250–350	.004–.006
case-hardening steels	$1\frac{1}{4}-1\frac{5}{8}$	300–425	.005–.007
	$1\frac{1}{2}-2\frac{1}{2}$	330–490	.005–.008
	$2\frac{3}{8}-3\frac{1}{2}$	350–550	.006–.010
High-carbon alloyed	$\frac{13}{16}-1\frac{1}{8}$	200–300	.003–.005
heat-treated steels	$1-1\frac{3}{8}$	250–325	.004–.006
	$1\frac{1}{4}-1\frac{5}{8}$	300–400	.005–.008
	$1\frac{1}{2}-2\frac{1}{2}$	325–450	.005–.008
	$2\frac{3}{8}-3\frac{1}{2}$	350–500	.006–.010
High tensile steels	$\frac{13}{16}-1\frac{1}{8}$	165–250	.004–.005
	$1-1\frac{3}{8}$	195–300	.004–.006
	$1\frac{1}{4}-1\frac{5}{8}$	230–300	.005–.007
	$1\frac{1}{2}-2\frac{1}{2}$	265–390	.006–.008
	$2\frac{3}{8}-3\frac{1}{2}$	265–425	.006–.009
Stainless steels	$\frac{13}{16}-1\frac{1}{8}$	230–280	.003–.004
	$1-1\frac{3}{8}$	265–300	.004–.005
	$1\frac{1}{4}-1\frac{5}{8}$	280–345	.004–.005
	$1\frac{1}{2}-2\frac{1}{2}$	295–395	.004–.005
	$2\frac{3}{8}-3\frac{1}{2}$	300–400	.004–.006
Titanium steels	$\frac{13}{16}-1\frac{1}{8}$	100–135	.003–.004
	$1-1\frac{3}{8}$	100–150	.004–.007
	$1\frac{1}{4}-1\frac{5}{8}$	115–165	.005–.008
	$1\frac{1}{2}-2\frac{1}{2}$	130–175	.006–.009
	$2\frac{3}{8}-3\frac{1}{2}$	135–190	.006–.010

[1]ipr = Inches per revolution. (*Note:* Ultimate speeds and feeds may differ from recommended speeds and feeds depending upon materials, rigidity of machine and setup, workpiece, and depth of cut.)

TABLE 24-2.   Summary of Drilling Parameters

1. $\text{rpm} = 12V/\pi D$; V = cutting speed; D = drill diameter

2. Penetration rate or feed rate (in./min) = feed (in./rev.) $\times$ rpm; feed in ipr is selected

3. Maximum chip load = Feed (ipr)/2 . . . for two-fluted drill

4. Material removal rate (in.^{3}/min) = $(\pi/4) \times$ (drill diameter)$^2 \times$ feed rate (in./min)

5. Drilling time/hole = $\dfrac{\text{length drilled } + \text{ air cut (allowance)}}{\text{feed rate (in./min)}}$
   $+ \dfrac{\text{rapid traverse length, including withdrawal}}{\text{rapid traverse rate}}$
   $+$ prorated downtime to change drill per hole

6. Cost/hole = (drilling time/hole) $\times$ (labor + machine rate)
   $+$ prorated cost of purchasing and regrinding drill/hole

7. Prorated downtime to change drill/hole = $\dfrac{\text{drill change downtime}}{\text{holes drilled per drill regrind}}$

8. Prorated cost of purchasing and regrinding drill/hole
   $= \dfrac{(\text{purchase cost } + \text{ cost per regrind}) \times (\text{number of regrinds})}{\text{number of holes drilled}}$

occurring at the extreme ends of the drill lips. The velocity is very small near the center of the chisel end of the drill. For drilling, cutting time is

$$T_m = \frac{(L + A)}{f_r N} = \frac{L + A}{f_m} \tag{24-2}$$

The metal removal rate is

$$\text{MRR} = \frac{\text{volume}}{T_m}$$
$$= \frac{\pi D^2 L/4}{L/f_r N} \quad \text{(omitting allowances)} \tag{24-3}$$

which reduces to

$$\text{MRR} = (\pi D^2/4) f_r N \quad \text{in}^3\text{min} \tag{24-4}$$

Substituting for N with Equation (24-1), we obtain an approximate form

$$\text{MRR} \cong 3DVf_r \tag{24-5}$$

Note that the units here are not consistent. A summary of drilling equations and parameters is given in Table 24-2, including some simple cost equations. In line 8, the "number of holes drilled" is the measure of tool life for drills.

■ 24.3   TYPES OF DRILLS

The most common types of drills are *twist drills*. These have three basic parts: the *body*, the *point*, and the *shank*, shown in Figure 24-1. The body contains two or more spiral or helical grooves, called *flutes*, separated by *lands*. To reduce the friction between the drill and the hole, each land is reduced in diameter except at the leading edge, leaving a narrow *margin* of full diameter to aid in supporting and guiding the drill and thus aiding in obtaining an accurate hole. The lands terminate in the point, with the leading edge of each land forming a cutting edge. The flutes serve as channels through which the chips are withdrawn from the hole and coolant gets to the cutting edges. Although most drills have two flutes, some, as shown in Figure 24-3, have three, and some have only one.

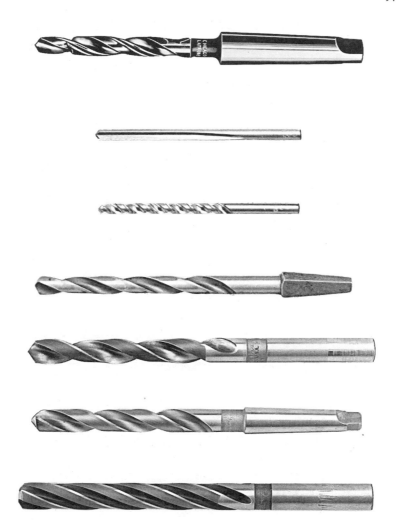

FIGURE 24-3 Types of twist drills and shanks. *Bottom to top:* Straight-shank, three-flute core drill; taper-shank; bit-shank; straight-shank, straight-flute; taper-shank, subland drill.

The principal rake angles behind the cutting edges are formed by the relation of the flute *helix angle* to the work. This means that the rake angle of a drill varies along the cutting edges (or lips), being negative close to the point and equal to the helix angle out at the lip. Because the helix angle is built into the drill, the primary rake angle cannot be changed by normal grinding. The helix angle of most drills is 24°, but drills with larger helix angles—often above 30°—are used for materials that can be drilled very rapidly, resulting in a large volume of chips. Helix angles ranging from 0 to 20° are used for soft materials, such as plastics and copper. Straight-flute drills (zero helix and rake angles) are also used for drilling thin sheets of soft materials. It is possible to change the rake angle adjacent to the cutting edge by a special grinding procedure called *dubbing*.

The cone-shaped point on a drill contains the cutting edges and the various clearance angles. This cone angle affects the direction of flow of the chips across the tool face and into the flute. The 118° cone angle that is used most often has been found to provide good cutting conditions and reasonable tool life when drilling mild steel, thus making it suitable for much general-purpose drilling. Smaller cone angles—from 90 to 118°—are sometimes used for drilling more brittle materials, such as gray cast iron and magnesium alloys. Cone angles from 118 to 135° often are used for the more ductile materials, such as aluminum alloys. Cone angles less than 90° frequently are used for drilling plastics. Many methods of grinding drills have been developed that produce point angles other than 118°.

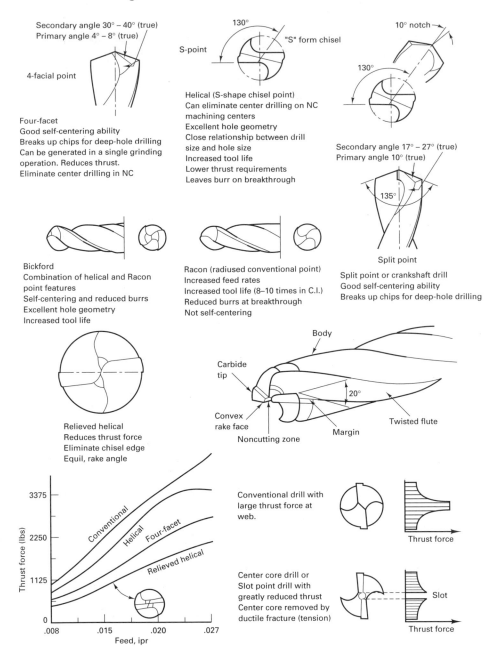

FIGURE 24-4 Variations of the conventional drill geometry are usually aimed at reducing the thrust force at the chisel end of the drill. Thrust force (*lower left*) increases with feed for all drill geometries.

The drill produces a thrust force *T* and a torque *M*. Drill torque increases with feed (in./rev) and drill diameter, while the thrust force is influenced greatly by the web or chisel end design, as shown in Figures 24-2 and 24-4.

The relatively thin *web* between the flutes forms a metal column or backbone. If a plain conical point is ground on the drill, the intersection of the web and the cone produces a straight-line *chisel end*, which can be seen in the end view of Figure 24-2. The chisel point, which also must act as a cutting edge, forms a 56° negative rake angle with the conical surface. Such a large negative rake angle does not cut efficiently, causing excessive deformation of the metal. This results in high thrust forces and excessive heat being developed at the point. In addition, the cutting speed at the drill center is low, approaching zero. As a consequence, drill failure on a standard drill occurs both at the center, where the cutting speed is lowest, and at the outer tips of the cutting edges, where the speed is highest.

When the rotating, straight-line chisel point comes in contact with the workpiece, it has a tendency to slide or "walk" along the surface, thus moving the drill away from the desired location. The conventional point drill, when used on machining centers or high-speed automatics, will require additional supporting operations like center drilling, burr removal, and tool change, all of which increase total production time and reduce productivity.

Many special methods of grinding drill points have been developed to eliminate or minimize the difficulties caused by the chisel point and to obtain better cutting action and tool life (see Figure 24-4 for some examples).

The center core or slot point drill is relatively new to the marketplace. This drill has twin carbide tips brazed on a steel shank and a hole (or slot) in the center. The work material in the slot is not machined but, rather, fractured away. The center core drill has a self-centering action and greatly relieves the thrust force produced by the chisel edge of conventional twist drills. This drill operates at about 30 to 50% less thrust than that of conventional drills. All rake angles of the cutting edge are positive, which further reduces the cutting force.

The conventional point also has a tendency to produce a burr on the exit side of a hole. Some type of chip breaker often is incorporated into drills. One procedure is to grind a small groove in the rake face, parallel with and a short distance back from the cutting edge. Drills with a special chip-breaker rib as an integral part of the flute are available. The rib interrupts the flow of the chip, causing it to break into short lengths.

The split-point drill is a form of *web thinning* to shorten the chisel edge. This design reduces thrust and allows for higher feed rates. Web thinning uses a narrow grinding wheel to remove a portion of the web near the point of the drill. Such methods have had varying degrees of success, and they require special drill-grinding equipment.

Also shown in Figure 24-4 is a four-facet self-centering point that works well in tougher materials. The facets refer to the number of edges on the clearance surfaces exposed to the cutting action. The self-centering drill lasts longer and saves machining time on numerical control centers as they can eliminate the need for center drills. A common aspect in drill-point terminology is total indicator runout (TIR). This is a measure of the cutting lips' relative side-to-side accuracy. The original drill point produced by the manufacturer lasts only until the first regrind; thereafter performance and life depend upon the quality of regrind. Proper regrinding (reconditioning) of a drill is a complex and important operation. If satisfactory cutting and hole size are to be achieved, it is essential that the point angle, lip clearance, lip length, and web thinning be correct. As illustrated in Figure 24-5, incorrect sharpening often results in unbalanced cutting forces

FIGURE 24-5 Typical causes of drilling problems.

(a) Angle unequal   (b) Length unequal

Outer corners break down: Cutting speed too high; Hard spots in material; No cutting compound at drill point; Flutes clogged with chips.

Cutting lips chip: Too much feed; Lip relief too great.

Checks or cracks in cutting lips: Overheated or too quickly cooled while sharpening or drilling.

Chipped margin: Oversize jig bushing.

Drill breaks: Point improperly ground; Feed too heavy; Spring or backlash in drillpress, fixture, or work; Drill is dull; Flutes clogged with chips.

Tang breaks: Imperfect fit between taper shank and socket caused by dirt or chips or by burred or badly worn sockets.

Drill breaks when drilling brass or wood: Wrong type drill; Flutes clogged with chips.

Drill spilts up center: Lip relief too small; Too much feed.

Drill will not enter work: Drill is dull; Web too heavy; Lip relief too small.

Hole rough: Point improperly ground or dull; No cutting compounds at drill point; Improper cutting compound; Feed too great; Fixture not rigid.

Hole oversize: Unequal angle of the cutting edges; Unequal length of the cutting edges; see figure to left.

Chip shape changes while drilling: Dull drill or cutting lips chipped.

Large chip coming from one flute, small chip from the other: Point improperly ground, one lip doing all the cutting.

at the tip, causing misalignment and oversized holes. Drills, even small drills, should always be machine-ground, never hand-ground. Drill grinders, often computer controlled, should be used to assure exact reproduction of the geometry established by the manufacturer of the drill. This is extremely important when drills are used on mass-production or numerically controlled machines. Companies invest huge sums in NC machining centers but overlook the value of a top-quality drill-grinding machine.

Drill shanks are made in several types. The two most common types are the straight and the taper. *Straight-shank* drills are usually used for sizes up to $\frac{1}{2}$ in. and must be held in some type of drill chuck. *Taper shanks* are available on drills from $\frac{1}{8}$ to $\frac{1}{2}$ in. and are common on drills above $\frac{1}{2}$ in. Morse tapers, having a taper of approximately $\frac{5}{8}$ in./ft, are used on taper-shank drills, ranging from a number 1 taper on $\frac{1}{8}$ in. drills to a number 6 on a $3\frac{1}{2}$-in. drill.

Taper-shank drills are held in a female taper in the end of the machine tool spindle. If the taper on the drill is different from the spindle taper, adapter sleeves are available. The taper assures the drill's being accurately centered in the spindle. The *tang* at the end of the taper shank fits loosely in a slot at the end of the tapered hole in the spindle. The drill may be loosened for removal by driving a metal wedge, called a *drift*, through a hole in the side of the spindle and against the end of the tang. It also acts as a safety device to prevent the drill from rotating in the spindle hole under heavy loads. However, if the tapers on the drill and in the spindle are proper, no slipping should occur. The driving force to the drill is carried by the friction between the two tapered members.

Standard drills are available in four size series, the size indicating the diameter of the drill body.

- *Millimeter series:* 0.01- to 0.50-mm increments, according to size, in diameters from 0.015 mm
- *Numerical series:* no. 80 to no. 1 (0.0135 to 0.228 in.)
- *Lettered series:* A to Z (0.234 to 0.413 in.)
- *Fractional series:* $\frac{1}{64}$ to 4 in. (and over) by 64ths.

TiN coating of conventional drills greatly improves drilling performance. The increase in tool life of TiN-coated drills over uncoated drills in machining steel is more than 200–1000%.

DEPTH-TO-DIAMETER RATIO

The depth of the hole to be drilled divided by the diameter of the drill is the depth-to-diameter ratio. Most hole makers consider a ratio of 3 to 1 to be deep hole drilling, after which hole accuracy (location) drilling speed and tool life will be reduced. The bores of rifle barrels were once drilled using conventional drills. Today, *deep-hole drills*, or *gun drills*, are used when deep holes are to be drilled.

The oldest of these deep-hole techniques is *gundrilling*. The original *gundrills* were very likely half-round drills, drilled axially with a coolant hole to deliver cutting fluids to the cutting edge. Modern gundrills typically consist of an alloy-steel-tubing shank with a solid carbide or carbide-edged tip brazed or mechanically fixed to it (see Figure 24-6). Guide pads following the cutting edge by about 90° to 180° are also standard.

The gundrill is a single-lipped tool, and its major feature is the delivery of coolant through the tool at extremely high pressures—typically from 300 psi to 1800 psi, depending on diameter—to force chips back down the flute. Successful application of a gundrill depends almost entirely on the formation of small chips that can be effectively evacuated by the flow of cutting fluids.

Standard gundrills are made in diameters from 0.0078 in. (2 mm) to 2 in. or more. Depth-to-diameter ratios of 100:1 or even 200:1 are possible.

Tolerances in gundrilling for diameters under $\frac{1}{2}$ in. can be held to 0.0005-in. total tolerance, and, over $\frac{1}{2}$ in., the tolerance should not exceed 0.001 in. According to one source, "roundness accuracies of 0.000,08 inch can be attained." Because of the burnishing effect of the guide pads, holes finished to 10–20 μin. can be produced, and as good as 5 μin., under favorable conditions.

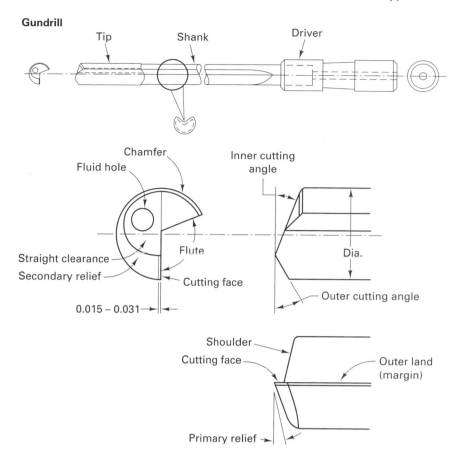

FIGURE 24-6 The gundrill geometry is very different from conventional drills.

Hole straightness is affected by a number of variables, such as diameter, depth, uniformity of workpiece material, condition of the machine, sharpness of the gundrill, feeds and speeds used, and the specific technique used (rotation of the tool, of the work, or both), but deviation should not exceed about 0.002 in. TIR in a 4-in. depth at any diameter, and it can be held to 0.002 in. per foot.

Basic setup for a gundrilling operation, which is generally horizontal, requires a drill bushing very close to the work entry surface and may involve rotating the work or the tool, or both. Best concentricity and straightness are achieved by the work and the tool rotating in opposite directions.

Other deep hole drills are called *BTA* (Boring Trepanning Association) *drills* and *ejector drills*. A deep hole is one in which the length (or depth) of the hole is three or more times the diameter. Coolants can be fed internally through these drills to the cutting edges. See Figure 24-7 for schematic of an ejector drill and the machine tool used for gundrilling. The coolants flush the chips out the flutes. The special design of these drills reduces the tendency of the drill to drift, thus producing a more accurately aligned hole. Figure 24-7 shows some typical BTA deep-hole drilling tools, designed for single-lip end cutting of a hole in a single pass. Solid-deep hole drills have alloy-steel shanks with a carbide-edged tip that is fixed to it mechanically. The cutting edge cuts through the center on one side of the hole, leaving no area of material to be extruded. The cutting is done by the outer and inner cutting angles which meet at a point. Theoretically, the depth of the hole has no limit, but practically, it is restricted by the torsional rigidity of the shank.

Gundrills have a single-lip cutting action. Bearing areas and lifting forces generated by the coolant pressure counteract the radial and tangential loads. The single-lip construction forces the edge to cut in a true circular pattern. The tip thus follows the direction of its own axis. The *trepanning gundrill* leaves a solid core.

Hole straightness is affected by variables such as diameter, depth, uniformity of the workpiece material, condition of the machine, sharpness of cutting edges, feeds and speeds used, and whether the tool or the workpiece is rotated or counterrotated.

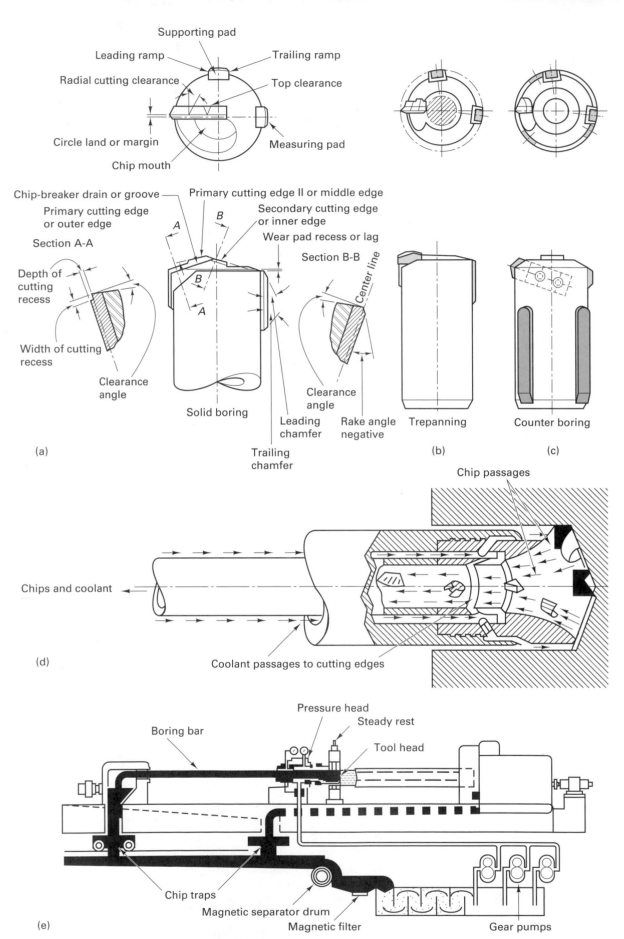

FIGURE 24-7 BTA drills for (a) boring, (b) trepanning, (c) counterboring, (d) deep-hole drilling with ejector drill, (e) horizontal deep-hole-drilling machine. (*S. Azad and S. Chandeashekar, Mechanical Engineering, Sept. 1985, pp. 62, 63.*)

TABLE 24-3. Drilling Processes Compared

	Twist Drill	Pivot (micro)	Spade (Inserted blade)	Indexable-insert Drill	Gundrill	BTA System	Ejector Drill	Trepanning
Diameter, in.	0.020–2	0.001–0.020	1–6	$\frac{5}{8}$–3	0.078–1	$\frac{7}{16}$–8	$\frac{3}{4}$–$2\frac{1}{2}$	$1\frac{3}{4}$–10
Typical range	0.0059	<0.0001	$\frac{5}{8}$ spec, 1	$\frac{5}{8}$	0.039	$\frac{3}{16}$	$\frac{3}{4}$	$1\frac{3}{4}$
Min	$3\frac{1}{2}$ std	$\frac{1}{8}$	std	3	$2\frac{1}{2}$	12	7	>24
Max	6 spec		18					

Depth/diameter Ratio

	Twist Drill	Pivot (micro)	Spade (Inserted blade)	Indexable-insert Drill	Gundrill	BTA System	Ejector Drill	Trepanning
Min practical	no min	no min	no min	<1	1	1	1	10
Common max[1]	5–10	3–10	>40 (horiz)	2–3	100	100	50	100
Ultimate	>50	20	10 (vert) >100 (horiz)	—	200	>100	>50	>100

[1]Maximum depth/diameter ratios in this table are estimates of what can be achieved with special attention and under ideal conditions. Equality of tolerances should not be assumed for the different processes.

FIGURE 24-8 Hole cutter used for thin sheets. *(Courtesy of Armstrong-Blum Manufacturing Company.)*

Two-flute drills are available that have holes extending throughout the length of each land to permit coolant to be supplied, under pressure, to the point adjacent to each cutting edge. These are helpful in providing cooling and also in promoting chip removal from the hole in drilling to moderate depths. They require special fittings through which the coolant can be supplied to the rotating drill, and they are used primarily on automatic and semiautomatic machines. See Table 24-3 for comparison of drilling processes.

Larger holes in thin material may be made with a *hole cutter* (Figure 24-8), whereby the main hole is produced by the thin-walled, multiple-tooth cutter with saw teeth. Hole cutters are often called *hole saws*.

When starting to drill a hole, a drill can deflect rather easily because of the "walking" action of the chisel point. Hole location accuracy is lost. Consequently, to assure that a hole is started accurately, a *center drill* (Figure 24-9) is used prior to a regular chisel-point twist drill. The center drill and countersink tool have a short, straight drill section extending beyond a 60° taper portion. The heavy, short body provides rigidity so that a hole can be started with little possibility of tool deflection. The hole should be drilled only partway up on the tapered section of countersink. The conical portion of the hole serves to guide the drill being used to make the main hole. Combination center drills are made in four sizes to provide a starting hole of proper size for any drill. If the drill is sufficiently large in diameter, or if it is sufficiently short, satisfactory accuracy may often be obtained without center drilling. Special drill holders are available that permit drills to be held with only a very short length protruding.

Because of its flexibility and end-point geometry, a drill may start or drift off centerline during drilling. The use of a center (start) drill will help to assure that a drill will start drilling at the desired location. Nonhomogeneities in the workpiece and imperfect drill

FIGURE 24-9 Sequence of operations required to obtain a hole that is accurate as to size and aligned on center. Inset Combination center drill and countersink. *(Courtesy of Chicago-Latrobe.)*

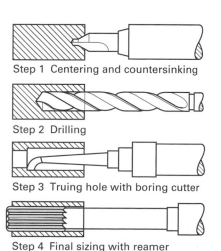

Step 1 Centering and countersinking

Step 2 Drilling

Step 3 Truing hole with boring cutter

Step 4 Final sizing with reamer

Combination center drill and countersink. *(Courtesy of Chicago-Latrobe)*

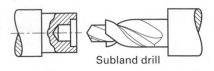

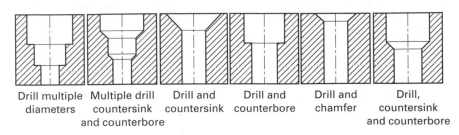

FIGURE 24-10 Special-purpose subland drill (*above*), and some of the operations possible with such drills (*below*).

Subland drill

| Drill multiple diameters | Multiple drill countersink and counterbore | Drill and countersink | Drill and counterbore | Drill and chamfer | Drill, countersink and counterbore |

geometries may also cause the hole to be oversize or off-line. For accuracy, it is necessary to follow center drilling and drilling by boring and reaming. Boring corrects the hole alignment, and reaming brings the hole to accurate size and improves the surface finish.

Special *combination drills* can drill two or more diameters, or drill and countersink and/or counterbore, in a single operation (Figure 24-10). Countersinking and counterboring usually follow drilling. These operations are described in more detail later in this chapter. A *step drill* has a single set of flutes and is ground to two or more diameters. *Subland drills* have a separate set of flutes on a single body for each diameter or operation; they provide better chip flow, and the cutting edges can be ground to give proper cutting conditions for each operation. Combination drills are expensive and may be difficult to regrind but can be economical for production-type operations if they reduce work handling, setups, or separate machines and operations.

Spade drills (Figure 24-11) are widely used for making holes 1 in. or larger in diameter at low speeds or with high feeds (Table 24-4). The workpiece usually has an existing hole, but a spade drill can drill deep holes in solids or stacked materials. Spade drills are less expensive because the long supporting bar can be made of ordinary steel. The drill point can be ground with a minimum chisel point. The main body can be made more rigid because no flutes are required, and it can have a central hole through which a fluid can be circulated to aid in cooling and in chip removal. The cutting blade is easier to sharpen; only the blades need to be TiN-coated.

FIGURE 24-11 (*Top*) Regular spade drill; (*middle*) spade drill with oil holes; (*bottom*) spade drill geometry, nomenclature.

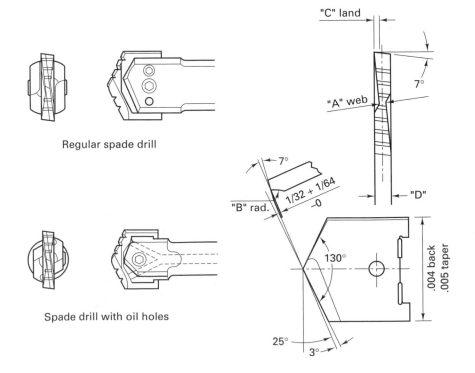

Regular spade drill

Spade drill with oil holes

TABLE 24-4. Recommended Surface Speeds and Feeds for High Speed Steel Spade Drills for Various Materials

Material	Surface Speed (ft per min)
Mild machinery steel 0.2 and 0.3 carbon	65–110
Steel, annealed 0.4 to 0.5 carbon	55–80
Tool steel, 1.2 carbon	45–60
Steel forging	35–50
Alloy steel	45–70
Stainless steel, free machining	50–70
Stainless steel, hard	25–40
Cast iron, soft	80–150
Cast iron, medium hard	55–100
Cast iron, hard, chilled	25–40
Malleable iron	79–90
Brass and bronze, ordinary	200–300
Bronze, high tensile	70–150
Monel metal	35–50
Aluminum and its alloys	200–300
Magnesium and its alloys	250–400

Feed Rates for Spade Drilling (inches per revolution)

Drill Size (inches)	Cast Iron Malleable Iron Brass Bronze	Medium Steel Stainless Steel Monel Metal Drop-Forged Alloys Tool Steel (annealed)	Tough Steel Drop Forging Aluminum
1 to $1\frac{1}{4}$	.010–.020	.008–.014	.006–.012
$1\frac{1}{4}$ to $1\frac{3}{4}$	.010–.024	.008–.018	.008–.017
$1\frac{3}{4}$ to $2\frac{1}{2}$	.010–.030	.010–.024	.010–.017
$2\frac{1}{2}$ to 4	.012–.032	.012–.030	.010–.017
4 to 6	.012–.032	.010–.024	.008–.017

Source: Waukesha Cutting Tools, Inc.

Spade drills are often used to machine a shallow locating cone for a subsequent smaller drill and at the same time to provide a small bevel around the hole to facilitate later tapping or assembly operations. Such a bevel also frequently eliminates the need for deburring. This practice is particularly useful on mass-production and numerically controlled machines.

Carbide-tipped drills and drills with indexable inserts are also available (see Figure 24-12) with one- and two-piece inserts for drilling shallow holes in solid workpieces. *Indexable insert*

FIGURE 24-12 One- and two-lipped insert drills. *(Courtesy of Waukesha.)*

Insert

TABLE 24-5. Indexable Drilling Troubleshooting Guide[1]

Problem	Source	Solution
Insert chipping or breakage[2]	Off-center drill, caused by misalignment	Maintain proper alignment. Concentricity not to exceed ±0.005 TIR.
	Improper seating of tool in toolholder, spindle, or turret	Check tool shank and socket for nicks and dirt. Check parting line between tool shank and socket with feeler gage. Check to see if tool is locked tightly.
	Deflection because of too much overhand and lack of rigidity	With indicator, check if tool can be moved by hand. Check if tool can be held shorter.
	Improper seating of inserts in pockets	Clean pockets whenever indexing or changing inserts. Check pockets for nicks and burrs. Check if inserts rest completely on pocket bottoms.
	Damaged insert screws	Check head and thread for nicks and burns. Do not over-torque screws.
	Improper speeds and feeds	Check recommended guidelines for given materials.
	Insufficient coolant supply	Check coolant flow.
	Improper carbide grade in inboard station	Recommend straight grade for multiple insert drills.
Grooving on back stroke; drill body rubbing hole wall; over or undersize holes	Off-center drill	Maintain proper alignment and concentricity. Check bottom of hole or disk for center stub.
	Deflection	Check setup rigidity. Check speed and feed guidelines.
Poor hole surface finish	Vibrations	Check setup and part rigidity. Check seat in spindle or toolholder. Check speeds and feeds.
	Insufficient coolant pressure and volume	Increase coolant pressure and flow. Is coolant flow constant? Make sure coolant reaches inserts at all times.
	Recutting chips, causing drill to jump	Increase coolant flow. Add coolant grooves.
	Poor chip control; chips trapped in hole	Mostly speed or feed.
	Chatter	Mostly feed rate.
Very short, thick, flat chips	Feed rate too high in relation to cutting speed	Lower feed or increase speed.
Long and stringy chips	Feed rate too low in relation to cutting speed	Increase feed rate or decrease speed. Use dimple inserts.
Unable to loosen insert locking screws	Seized threads, caused by coolant or heat	Apply water and heat resistant lubricant to threads.

[1] *Source: Fundamentals of Indexable Drilling*, K. L. Anderson, *Machining Technology*, vol. 2, No. 3, 1991.
[2] If constant chipping occurs, especially on an inner insert, and conditions are optimum, try an uncoated carbide insert or a grade with higher transverse rapture strength.

drills can produce a hole four times faster than a spade drill because they run at high speeds/low feeds and are really more of a boring operation than a drilling process.

However, to use indexable drills, you must have an extremely rigid machine tool and setup, adequate horsepower and lots of cutting fluid. Indexable drills are roughing tools generating hole tolerances of ±.08 in. and surface finish of 250 rpms or greater. The tool is designed for the inboard insert to cut past the centerline of the tool so the inboard tool is positioned radially below the center. See Table 24-5 for indexable drilling troubleshooting guide.

A high-pressure, pulsating coolant system can generate pressures up to 300 psi and works well with indexable drilling. It can have disadvantages, however. High pressure with pulsating action can decrease chip control and cause drill deflection. A high-pressure coolant stream can flatten chips at the point of forming and forces them into the cut, causing recutting, insert chipping, and poor hole finishes. The pressure can force chips between the drill body and hole diameter wrapping them around the drill. Friction then will weld the chips to the tool body or hole.

The diameter of the hole (D) and the length/diameter ratio usually determine what kind of drill to use. Figure 24-13 explores how drill selection depends on the depth of the hole and the diameter of the drill: Section A shows the drilling areas of relatively shallow holes and small diameters. About half of all the drilling process falls within the category of this section. It is the section for which the majority of the work is done by twist drills and very few cemented carbide drills. Section B is the drilling of deep holes for which cemented carbide gundrills are used. Section C is that of shallow holes having large diameters, for which spade drills are used. Section D is that of deep holes having large diameters, for which BTA tools are used.

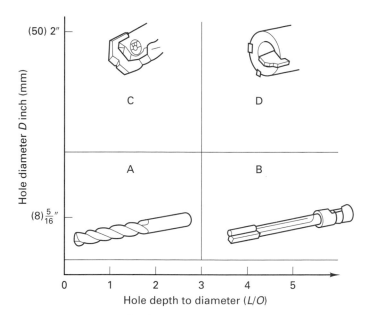

Sector	Typical drill types
A	Twist drill (HSS) Center core drill
B	Twist drill Gundrill BTA Ejector drill
C	Twist drill Indexable insert drill Spade drill Center core drill
D	BTA Ejector drill

FIGURE 24-13   Drill selection depends on hole diameter and hole depth.

MICRODRILLING

As the term suggests, microdrilling is the name for the very special world of miniature-hole machining, involving dimensions at which many workpiece materials no longer exhibit uniformity and homogeneity. Grain borders, inclusions, alloy or carbide segregates, and microscopic voids are problems in microdrilling, where holes of 0.02 to 0.0001 in. have been drilled using pivot drills as shown in Figure 24-14.

Pivot drills are two-lipped (two-fluted), end-cutting tools of relatively simple geometry. Web thickness tapers toward the point, and a generous back-taper is incorporated. For softer workpiece materials, point angles are typically 118°, and lip clearance is 15°. For steels and general use in harder metals, 135° points and 8° clearance are recommended. The chisel edge is similar to that of a twist drill. Pivot drills are made of tungsten-alloy tool steel in standard sizes from 0.0001 in. to 0.125 in. and of sintered tungsten carbide from 0.001 in. to 0.125 in.

Small drills easily deflect and getting accurate and precise holes requires a machine with a high-quality spindle and very sensitive feeding pressure. Speeds and feeds are greatly reduced with frequent pecking to clear the chips. Use a light, lard-based, sulfurized cutting oil.

FIGURE 24-14   Pivot microdrill for drilling very small diameter holes.

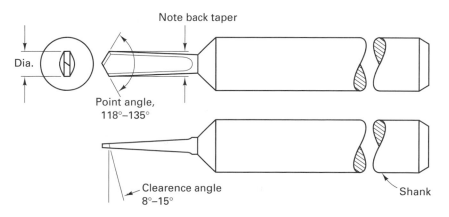

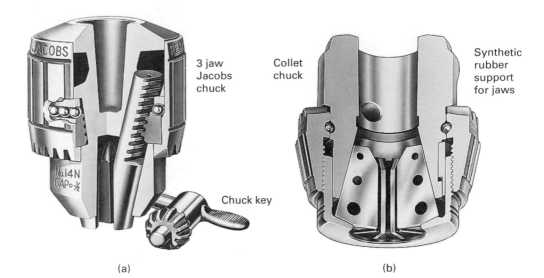

3 jaw
Jacobs
chuck

Chuck key

Collet
chuck

Synthetic
rubber
support
for jaws

FIGURE 24-15 Two of the most commonly used types of drill chucks. *(Courtesy of Jacobs Manufacturing Company.)*

(a)

(b)

■ 24.4 TOOLHOLDERS FOR DRILLS

Straight-shank drills must be held in some type of drill *chuck* (Figure 24-15). Chucks are adjustable over a considerable size range and have radial steel fingers. When the chuck is tightened by means of a chuck key, these fingers are forced inward against the drill. On smaller drill presses, the chuck often is permanently attached to the machine spindle, whereas on larger drilling machines the chucks have a tapered shank that fits into the female Morse taper of the machine spindle. Special types of chucks in semiautomatic or fully automatic machines permit quite a wide range of sizes of drills to be held in a single chuck.

Chucks using chuck keys require that the machine spindle be stopped in order to change a drill. To reduce the downtime when drills must be changed frequently, *quick-change chucks* are used. Each drill is fastened in a simple round collet that can be inserted into the chuck hole while it is turning by merely raising and lowering a ring on the chuck body. With the use of this type of chuck, center drills, drills, counterbores, reamers, and so on can be manually changed in quick succession.

For carbide drills, collet type holders with thrust bearings are recommended (Figure 24-16). For drills using an internal coolant supply, a very rigid chuck with either an inducer or through spindle coolant source is recommended.

Conventional holders such as keyless chucks cannot be used because the gripping strength is limited. Collet holders should be cleaned periodically with oil to remove small chips.

The entire flute length must protrude from the chuck. At maximum hole depth, the length of flute protruding from the hole must be at least 1 to 1.5 times the drill diameter. Radial run out at the drill tip must not exceed 0.001 in.

■ 24.5 WORKHOLDING FOR DRILLING

Work that is to be drilled is ordinarily held in a vise or in specially designed workholders called *jigs*. Workholding devices are the subject of Chapter 29, where the design of workholding devices is discussed. Many examples of drill jigs are shown.

Even in light drilling operations, the work should not be held on the table by hand unless adequate leverage is available. This is a dangerous practice and can lead to serious accidents, because the drill has a tendency to catch on the workpiece and cause it to rotate, especially when the drill exits the workpiece. Work that is too large to be held in a jig can be clamped directly to the machine table using suitable bolts and clamps and the slots or holes in the table. Jigs and workholding devices on indexing machines must be free from play and firmly seated.

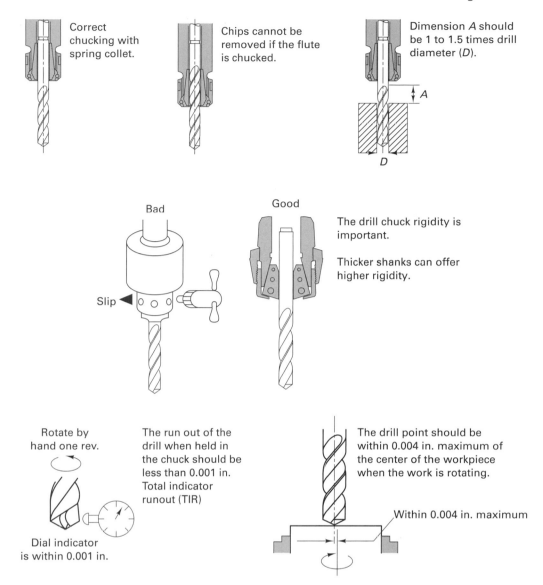

Correct chucking with spring collet.

Chips cannot be removed if the flute is chucked.

Dimension *A* should be 1 to 1.5 times drill diameter (*D*).

Bad

Good

The drill chuck rigidity is important.

Thicker shanks can offer higher rigidity.

Slip

Rotate by hand one rev.

The run out of the drill when held in the chuck should be less than 0.001 in. Total indicator runout (TIR)

Dial indicator is within 0.001 in.

The drill point should be within 0.004 in. maximum of the center of the workpiece when the work is rotating.

Within 0.004 in. maximum

FIGURE 24-16 Correct chucking of carbide drills can prevent many problems.

■ 24.6 MACHINE TOOLS FOR DRILLING

The basic work and tool motions required for drilling—relative rotation between the workpiece and the tool, with relative longitudinal feeding—also occur in a number of other machining operations. Thus drilling can be done on a variety of machine tools such as lathes, vertical milling machines, boring machines, and machining centers. This chapter will focus on those machines that are designed, constructed, and used primarily for drilling.

Machine tools must have sufficient power (torque) and thrust to perform the cut. It is the task of the engineer to select the correct machine or select the cutting parameters (speed and feed) based on the drill diameter, drill material and work material (hardness). The equations given with the plots shown in Figure 24-17 are typical of the empirically tested information provided by cutting tool manufacturers to calculate (estimate) thrust in drilling. Data with K_s (specific cutting energy, in.-lb/in^2) and X and Y (empirical constants) are obtained from cutting tool manufacturers. Much of this kind of data has been developed for high-speed steel tools. When using solid carbide tools, rigid machines such as machining centers or NC turning machines are recommended, whereas a radial drilling machine is not recommended.

Rigidity is especially important in avoiding chatter. A lack of rigidity in the cutting tool, the workpiece, or the machine tool permits the affected members to deflect due to the cutting forces developing the conditions for chatter (see Chapter 21) with the result that the cutting lips have a hammering action against the work. So, use the shortest tool possible.

Axial Thrust Force

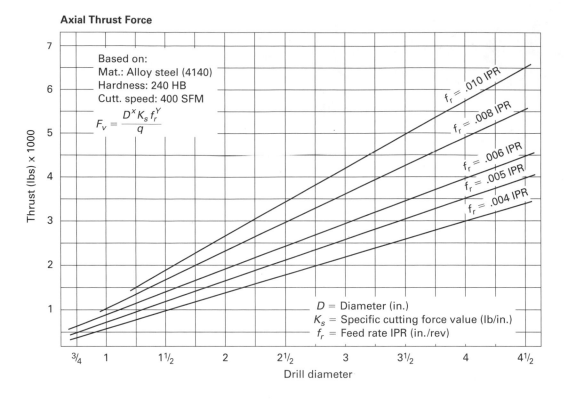

Based on:
Mat.: Alloy steel (4140)
Hardness: 240 HB
Cutt. speed: 400 SFM

$$F_v = \frac{D^x K_s f_r^Y}{q}$$

D = Diameter (in.)
K_s = Specific cutting force value (lb/in.)
f_r = Feed rate IPR (in./rev)

f_r = .010 IPR
f_r = .008 IPR
f_r = .006 IPR
f_r = .005 IPR
f_r = .004 IPR

Thrust (lbs) × 1000

Drill diameter

F_v = Axial thrust
    = $D^{1.15} \times K_s \times f_r^{0.8}$
where
D = Drill diameter (inches)
K_s = Specific cutting energy
    from table (in-lb/in²)
f_r = Feed (in./rev)

MATERIAL	BRINELL HARDNESS		FEED (IPR)							
			.004	.005	.006	.008	.010	.012	.016	.020
		E	0.35	0.39	0.47	0.60	0.70	0.80	0.90	1.08
		C	3.3	3.5	3.8	4.4	4.4	4.4	4.0	3.6
PLAIN CARBON STEEL	140-220		444230	435510	431150	426790	418070	409350	391910	374460
	220-300		493590	483900	479060	474210	464520	454830	435450	416070
FREE MACHINING STEELS	120-180		296150	290340	287440	284530	278710	272900	261270	249640
	180-260		345510	338730	335340	331950	325160	318380	304820	291250
ALLOY STEELS	260-340		493590	483900	479060	474210	464520	454830	435450	416070
STAINLESS	150-200		370190	362930	359300	355660	348390	341120	326590	312050
STEELS	200-300		444230	435510	431150	426790	418070	409350	391910	374460
CAST IRON	180-250		345510	338730	335340	331950	325160	318380	304820	291250
ALUMINUM			148080	145170	143720	142260	139360	136450	130640	124820
TITANIUM			320830	314530	311390	308240	301940	295640	283040	270450
HIGH TEMPERATURE ALLOYS			542950	532290	526970	521630	510970	500310	478990	457680

Values in in.-lb/in²

FIGURE 24-17 Estimating the thrust force in drilling example from Waukesha Cutting Tools.

In addition, backlash in the feed mechanism should be kept at a minimum to reduce strain on the drill when it breaks through the bottom of the hole.

The common name for the machine tool used for drilling is the *drill press*. Drill presses consist of a *base*, a *column* that supports a *powerhead*, a *spindle*, and a *worktable*. On small machines, the base rests on a workbench, whereas on larger machines it rests on the floor (Figure 24-18). The column may be either round or of box-type construction,

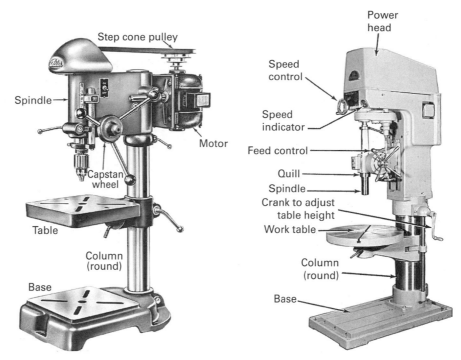

FIGURE 24-18 (*Left*) Fifteen-inch bench-type drill press, usually set on a table. (*Courtesy of Atlas Press Company.*); (*right*) upright drilling machine. (*Courtesy of Buffalo Forge Company.*)

the latter being used on larger, heavy-duty machines, except in radial types. The power-head contains an electric motor and means for driving the spindle in rotation at several speeds. On small drilling machines this may be accomplished by shifting a belt on a step-cone pulley, but on larger machines a geared transmission is used.

The heart of any drilling machine is its spindle. In order to drill satisfactorily, the spindle must rotate accurately and also resist whatever side forces result from the drilling process. In virtually all machines the spindle rotates in preloaded ball or taper-roller bearings. In addition to powered rotation, provision is made so that the spindle can be moved axially to feed the drill into the work. On small machines the spindle is fed by hand, using the handles extending from the capstan wheel; on larger machines power feed is provided. Except for some small bench types, the spindle contains a hole with a Morse taper in its lower end into which taper-shank drills or drill chucks can be inserted.

The worktables on drilling machines may be moved up and down on the column to accommodate work of various sizes. On round-column machines the table can usually be rotated out of the way so that workpieces can be mounted directly on the base. On some box-column machines the table is mounted on a subbase so that it can be moved in two directions in a horizontal plane by means of feed screws.

Figure 24-19 shows some block diagrams and schematics for four common types of drilling machines used in production environments. Drilling machines usually are classified in the following manner:

1. Bench (Figure 24-18)

   a. Plain   b. Sensitive

2. Upright (Figure 24-18)

   a. Single-spindle (Figure 24-19)   b. Turret   c. NC turret (Figure 24-20)

3. Radial (Figure 24-19)

   a. Plain   b. Semiuniversal   c. Universal

4. Gang (Figure 24-19)

5. Multispindle (Figure 24-19)

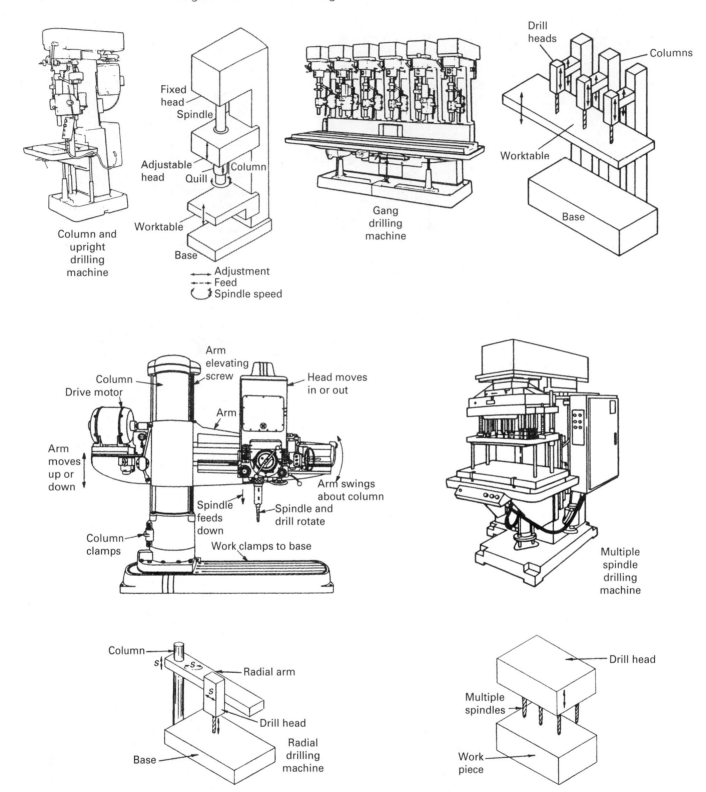

FIGURE 24-19 Four principal types of drilling machines. *(From Manufacturing Producibility Handbook, courtesy of General Electric Company.)*

6. Deep-hole

    a. Vertical   b. Horizontal (Figure 24-6)

7. Transfer

With bench drill presses, holes up to $\frac{1}{2}$ in. in diameter can be drilled. The same type of machine can be obtained with a long column so that it can stand on the floor. The size of bench and upright drilling machines is designated by *twice* the distance from the center line of the spindle to the nearest point on the column, this being an indication of the maximum size of the work that can be drilled in the machines. For example, a 15-in. drill press will permit a hole to be drilled at the center of a workpiece 15 in. in diameter.

Sensitive drilling machines are essentially smaller, plain bench-type machines with more accurate spindles and bearings. They are capable of operating at higher speeds, up to 30,000 rpm. Very sensitive hand-operated feeding mechanisms are provided for use in drilling small holes. Such machines are used for tool and die work and for drilling very small holes, often less than a few thousandths of an inch in diameter, when high spindle speeds are necessary to obtain proper cutting speed and sensitive feel to provide delicate feeding to avoid breakage of the very small drills.

Upright drilling machines usually have spindle speed ranges from 60 to 3500 rpm and power feed rates, from 4 to 12 steps, from about 0.004 to 0.025 in./rev. Most modern machines use a single-speed motor and a geared transmission to provide the range of speeds and feeds. The feed clutch disengages automatically when the spindle reaches a preset depth.

Worktables on most upright drilling machines contain holes and slots for use in clamping work and nearly always have a channel around the edges to collect cutting fluid, when it is used. On box-column machines, the table is mounted on vertical ways on the front of the column and can be raised or lowered by means of a crank-operated elevating screw.

In mass production *gang-drilling machines* are often used when several related operations, such as drilling holes of different sizes, reaming, or counterboring, must be done on a single part. These consist essentially of several independent columns, heads, and spindles mounted on a common base and having a single table. The work can be slid into position for the operation at each spindle. They are available with or without power feed. One or several operators may be used. This machine would be an example of a simple small cell except that the machines are usually not single-cycle automatics.

Turret-type, upright drilling machines, shown in Figure 24-20, are used when a series of holes of different sizes, or a series of operations (such as center drilling, drilling, reaming, and spot facing), must be done repeatedly in succession. The selected tools are mounted in the turret. Each tool can quickly be brought into position merely by rotation

FIGURE 24-20 NC turret drilling machines have replaced the gang-drilling machines in many production facilities.

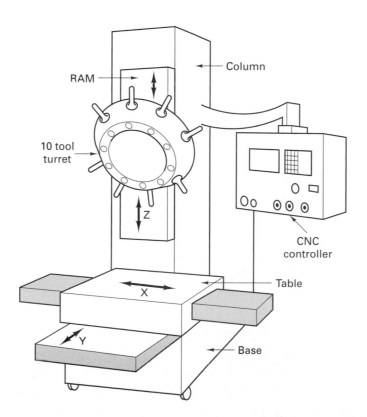

of the turret. These machines automatically provide individual feed rates for each spindle and are often numerically controlled.

Radial drilling machine tools are used on large workpieces that cannot easily be handled manually. As shown in Figure 24-18, these machines have a large, heavy, round, vertical column supported on a large base. The column supports a radial arm that can be raised and lowered by power and rotated over the base. The spindle head, with its speed- and feed-changing mechanism, is mounted on the radial arm. It can be moved horizontally to any desired position on the arm. Thus the spindle can quickly be positioned properly for drilling holes at any point on a large workpiece mounted either on the base of the machine or even sitting on the floor.

Plain radial drilling machines provide only a vertical spindle motion. On *semiuniversal machines*, the spindle head can be pivoted at an angle to a vertical plane. On *universal machines*, the radial arm is rotated about a horizontal axis to permit drilling at any angle.

Radial drilling machines are designated by the radius of the largest disk in which a center hole can be drilled when the spindle head is at its outermost position. Sizes from 3 to 12 ft are available. Radial drilling machines have a wide range of speeds and feeds, can do boring, and include provisions for tapping (internal threading) (see Chapter 30).

Multiple-spindle drilling machines (Figure 24-19) are mass-production machines with as many as 50 spindles driven by a single powerhead and fed simultaneously into the work. Figure 24-21 shows an adjustable multiple-spindle head that can be mounted on a regular single spindle-drill press. Figure 24-21 shows the methods of driving and positioning the spindles, which permit them to be adjusted so that holes can be drilled at any location within

FIGURE 24-21 Three basic types of multiple-spindle drill heads: (*left*) adjustable; (*middle*) geared; (*right*) gearless. (*Courtesy of Zagar Incorporated.*)

Adjustable drill head
Spindle: 6 Production: 50 pieces

Geared drill head
Spindle: 8 Production: 80,000 pieces

Gearless drill head
Spindle: 16 Production: 30,000 pieces

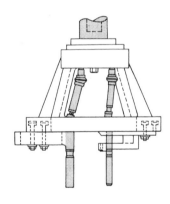

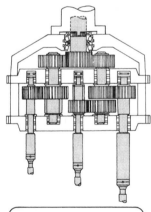

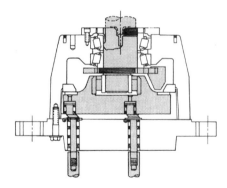

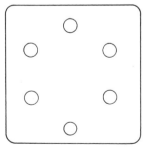

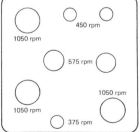

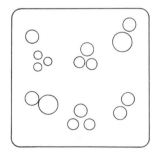

An adjustable drill head should be considered for low production jobs. However, many short-run jobs such as this would be required to justify a multiple-spindle head.

A geared drill head is most appropriate in this situation where there is a large difference in sizes and a high daily production.

Only a gearless head can perform this operation in one pass, due to the close proximity of the spindle centers.

the overall capacity of the head. Special drill jigs are often designed and built for each job to provide accurate guidance for each drill. Although such machines and workholders are quite costly, they can be cost justified when the quantity to be produced will justify the setup cost and the cost of the jig. Reducing setup on these machines is difficult. Numerically controlled drill presses other than turret drill presses are not common because drilling and all its related processes can be done on vertical or horizontal NC machining centers equipped with automatic tool changers (see Chapter 32).

Special machines are used for drilling long (deep) holes, such as are found in rifle barrels, connecting rods, and long spindles. High-cutting speeds, very light feeds, and a copious flow of cutting fluid assure rapid chip removal. Adequate support for the long, slender drills is required. In most cases horizontal machines are used (Figure 24-6). The work is rotated in a chuck with steady rests providing support along its length, as required. The drill does not rotate and is fed into the work. Vertical machines are also available for shorter workpieces. Notice the similarity between this process and boring.

■ 24.7 CUTTING FLUIDS FOR DRILLING

For shallow holes, the general rules relating to cutting fluids, as given in Chapter 22, are applicable. When the depth of the hole exceeds one diameter, it is desirable to increase the lubricating quality of the fluid because of the rubbing between the drill margins and the wall of the hole. The effectiveness of a cutting fluid as a coolant is quite variable in drilling. While the rapid exit of the chips is a primary factor in heat removal, this action also tends to restrict entry of the cutting fluid. This is of particular importance in drilling materials that have poor heat conductivity. Recommendations for cutting fluids for drilling are given in Table 24-6.

If the hole depth exceeds two or three diameters, it is usually advantageous to withdraw the drill each time it has drilled about one diameter of depth, to clear chips from the hole. Some machines are equipped to provide this "pecking" action automatically.

Where cooling is desired, the fluid should be applied copiously. For severe conditions, drills containing coolant holes have a considerable advantage. Not only is the fluid supplied near the cutting edges, but the coolant flow aids in flushing the chips from the hole. Where feasible, drilling horizontally has distinct advantages over drilling vertically downward.

■ 24.8 COUNTERBORING, COUNTERSINKING, AND SPOT FACING

Drilling often is followed by *counterboring*, *countersinking*, or *spot facing*. As shown in Figure 24-22, each provides a bearing surface at one end of a drilled hole. They are usually done with a special tool having from three to six cutting edges.

TABLE 24-6.	Cutting Fluids for Drilling
Work Material	Cutting Fluid
Aluminum and its alloys	Soluble oil, kerosene, and lard-oil compounds; light, nonviscous neutral oil; kerosene and soluble oil mixtures
Brass	Dry or a soluble oil; kerosene and lard-oil compounds; light, nonviscous neutral oil
Copper	Soluble oil, strained lard oil, oleic-acid compounds
Cast iron	Dry or with a jet of compressed air for cooling
Malleable iron	Soluble oil, nonviscous neutral oil
Monel metal	Soluble oil, sulfurized mineral oil
Stainless steel	Soluble oil, sulfurized mineral oil
Steel, ordinary	Soluble oil, sulfurized oil, high extreme-pressure-value mineral oil
Steel, very hard	Soluble oil, sulfurized oil, turpentine
Wrought iron	Soluble oil, sulfurized oil, mineral-animal oil compound

- Neat oil can be used effectively with the solid carbide drills for low speed drilling (up to 130 sfpm).
- If the work surface becomes hard or blue in color, decrease the rpm and use neat oil.
- For heavy duty cutting, emulsion type oil containing some extreme pressure additive is recommended.
- A volume of 3.0 gal/min at a pressure of 37–62 lb/in^2 is recommended.
- A double stream supply of fluid is recommended.

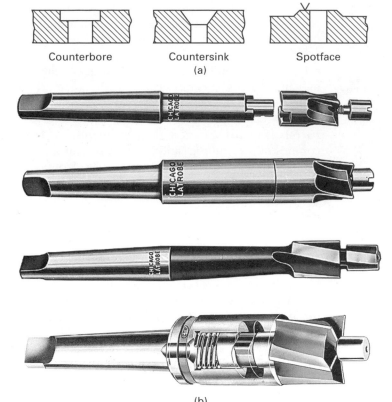

FIGURE 24-22 (a) Surfaces produced by counterboring, countersinking, and spot facing. (b) Counterboring tools: (*bottom to top*) interchangeable counterbore; solid, taper-shank counterbore with integral pilot; replaceable counterbore and pilot; replaceable counterbore, disassembled. *(Courtesy of Ex-Cell-O Corporation and Chicago Latrobe Twist Drill Works.)*

Counterboring provides an enlarged cylindrical hole with a flat bottom so that a bolt head, or a nut, will have a smooth bearing surface that is normal to the axis of the hole; the depth may be sufficient so that the entire bolt head or nut will be below the surface of the part. The pilot on the end of the tool fits into the drilled hole and helps to assure concentricity with the original hole. Two or more diameters may be produced in a single counterboring operation. Counterboring also can be done with a single-point tool, although this method ordinarily is used only on large holes and essentially is a boring operation. Some counterboring tools are shown in Figure 24-22b.

Countersinking makes a beveled section at the end of a drilled hole to provide a proper seat for a flat-head screw or rivet. The most common angles are 60, 82, and 90°. Countersinking tools are similar to counterboring tools except that the cutting edges are elements of a cone, and they usually do not have a pilot because the bevel of the tool causes them to be self-centering.

Spot facing is done to provide a smooth bearing area on an otherwise rough surface at the opening of a hole and normal to its axis. Machining is limited to minimum depth that will provide a smooth, uniform surface. Spot faces thus are somewhat easier and more economical to produce than counterbores. They usually are made with a multiedged end-cutting tool that does not have a pilot, although counterboring tools are frequently used.

■ 24.9 REAMING

Reaming removes a small amount of material from the surface of holes. It is done for two purposes: to bring holes to a more exact size and to improve the finish of an existing hole. Multiedge cutting tools are used as shown in Figure 24-23. No special machines are built for reaming. The same machine that was employed for drilling the hole can be used for reaming by changing the cutting tool.

To obtain proper results, only a minimum amount of materials should be left for removal by reaming. As little as 0.005 in. is desirable, and in no case should the amount exceed 0.015 in. A properly reamed hole will be within 0.001 in. of correct size and have a fine finish.

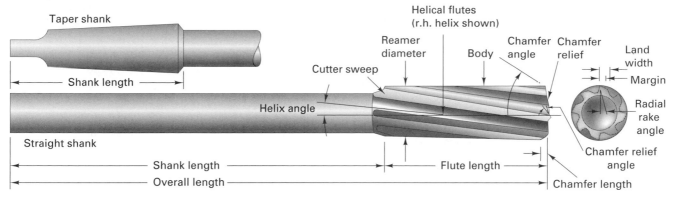

Chucking reamer

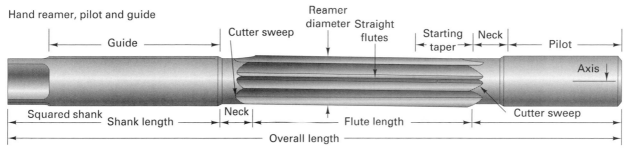

FIGURE 24-23   Standard nomenclature for hand and chucking reamers.

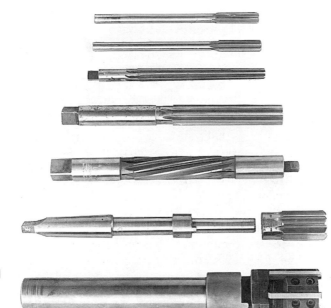

FIGURE 24-24   Types of reamers: (*top to bottom*) Straight-fluted rose reamer, straight-fluted chucking reamer, straight-fluted taper reamer, straight-fluted hand reamer, expansion reamer, shell reamer, adjustable insert-blade reamer.

TYPES OF REAMERS

The principal types of reamers, shown in Figures 24-23 and 24-24, are:

1. Hand reamers
   a. Straight   b. Taper
2. Machine or chucking reamers
   a. Rose   b. Fluted
3. Shell reamers
4. Expansion reamers
5. Adjustable reamers

Hand reamers are intended to be turned and fed by hand and to remove only a few thousandths of an inch of metal. They have a straight shank with a square tang for a wrench. They can have straight or spiral flutes and be solid or expandable. The teeth have relief along their edges and thus may cut along their entire length. However, the reamer is tapered from 0.005 to 0.010 in. in the first third of its length to assist in starting it in the hole, and most of the cutting therefore takes place in this portion.

Machine or *chucking reamers* are for use with various machine tools at slow speeds. The best feed is usually two to three times the drilling feed. Machine reamers have chamfers on the front end of the cutting edges. The chamfer causes the reamer to seat firmly and concentrically in the drilled hole, allowing the reamer to cut at full diameter. The longitudinal cutting edges do little or no cutting. Chamfer angles are usually 45°. Reamers have straight or tapered shanks, and straight or spiral flutes. *Rose chucking reamers* are ground cylindrical and have no relief behind the outer edges of the teeth. All cutting is done on the beveled ends of the teeth. *Fluted chucking reamers*, on the other hand, have relief behind the edges of the teeth as well as beveled ends. They can therefore cut on all portions of the teeth. Their flutes are relatively short, and they are intended for light finishing cuts. For best results they should not be held rigidly but permitted to float and be aligned by the hole.

Shell reamers often are used for sizes over $\frac{3}{4}$ in. in order to save cutting-tool material. The shell, made of tool steel for smaller sizes and with carbide edges for larger sizes or for mass-production work, is held on an arbor that is made of ordinary steel. One arbor may be used with any number of shells. Only the shell is subject to wear and need be replaced when worn. They may be ground as rose or fluted reamers.

Expansion reamers can be adjusted over a few thousandths of an inch to compensate for wear or to permit some variation in hole size to be obtained. They are available in both hand and machine types.

Adjustable reamers have cutting edges in the form of blades that are locked in a body. The blades can be adjusted over a greater range than expansion reamers. This permits adjustment for size and to compensate for regrinding. When the blades become too small from regrinding, they can be replaced. Both tool steel and carbide blades are used.

Taper reamers are used for finishing holes to an exact taper. They may have up to eight straight or spiral flutes. Standard tapers, such as Morse, Jarno, or Brown & Sharpe, come in sets of two. The *roughing reamer* has nicks along the cutting edge to break up the heavy chips that result as a cylindrical hole is cut to a taper. The *finishing reamer* has smooth cutting edges.

REAMING PRACTICE

If the material to be removed is free-cutting, reamers of fairly light construction will give satisfactory results. However, if the material is hard, then tough, solid-type reamers are recommended, even for fairly large holes.

To meet quality requirements, including both finish and accuracy (tolerances on diameter, roundness, straightness, and absence of bell-mouth at ends of holes), reamers must have adequate support for the cutting edges, and reamer deflection must be minimal. Reaming speed is usually two-thirds the speed for drilling the same materials. However, for close tolerances and fine finish, speeds should be slower.

Feeds are usually much higher than those for drilling and depend upon material. A feed of between 0.0015 and 0.004 in. per flute is recommended as a starting point. Use the highest feed that will still produce the required finish and accuracy. Recommended cutting fluids are the same as those for drilling.

Reamers, like drills, should not be allowed to become dull. The chamfer must be reground long before it exhibits excessive wear. Sharpening is usually restricted to the starting taper or chamfer. Each flute must be ground exactly even, or the tool will cut oversize.

Reamers tend to chatter when not held securely, when the work or workholder is loose, or when the reamer is not properly ground. Irregularly spaced teeth may help reduce chatter. Other cures for chatter in reaming are to reduce the speed, vary the feed rate, chamfer the hole opening, use a piloted reamer, reduce the relief angle on the chamfer, or change cutting fluid. Any misalignment between the workpiece and the reamer will cause chatter and improper reaming.

■ KEY WORDS

center core	deep hole drilling	gundrill	multiple-spindle	shell reamer	trepanning
drill	drill press	helix angle	drilling machine	spade drill	turret drilling
chisel end	drilling	indexable insert	radial drilling	spot facing	machine
chuck	flute	drill	machine	subland drill	twist drill
counterboring	gang-drilling	jig	reaming, machine	tang	web
countersinking	machine	lip	reaming, hand	thrust force	

■ REVIEW QUESTIONS

1. What functions are performed by the flutes on a drill?
2. What determines the rake angle of a drill? See Figure 24-2.
3. Basically, what determines what helix angle a drill should have?
4. When a large-diameter hole is to be drilled, why is a small-diameter hole often drilled first?
5. Equation 24-4 for the MRR for drilling can be thought of as _____ times _____ where $f_r N$ is the feed rate of the drill bit.
6. Are the recommended surface speeds for spade drills given in Table 24-4 typically higher or lower than those recommended for twist drills? How about the feeds?
7. What can happen when an improperly ground drill is used to drill a hole?
8. Why are most drilled holes oversize with respect to the nominally specified diameter?
9. What are the two primary functions of a combination center drill?
10. What is the function of the margins on a twist drill?
11. What factors tend to cause a drill to "drift" off the centerline of a hole?
12. The drills shown in Figure 24-12 have coolant passages in the flutes. What is the purpose of these holes?
13. In drilling, the deeper the hole, the greater the torque. Why?
14. Why do cutting fluids for drilling usually have more lubricating qualities than those for most other machining operations?
15. Find the drills shown in this chapter that have oil holes. (*Hint:* there are seven.)
16. How does a gang-drilling machine differ from a multiple-spindle drilling machine?
17. How does the thrust force vary with feed?
18. What may result from holding the workpiece by hand when drilling?
19. What is the rationale behind the operation sequence shown in Figure 24-9?
20. In terms of thrust, what is unusual about the slot-point drill compared to other drills?
21. What is the purpose of spot facing?
22. How does the purpose of counterboring differ from that of spot facing?
23. What are the primary purposes of reaming?
24. What are the advantages of shell reamers?
25. A drill that operated satisfactorily for drilling cast iron gave very short life when used for drilling a plastic. What might be the reason for this?
26. What precautionary procedures should be used when drilling a deep, vertical hole in mild steel when using an ordinary twist drill?
27. What is the advantage of a spade drill? Is it really a drill?
28. What is a "pecking" action in drilling?
29. Why does drill feed increase with drill size?
30. Suppose you specified a feed that was too large. What kinds of problems do you think this might cause? See Table 24-5 for help.

■ PROBLEMS

1. Suppose you wanted to drill a 1.5-in. diameter hole through a piece of 1020 cold-rolled steel that is 2 in. thick, using an indexable-insert drill. Values for feed and cutting speed need to be specified by the student, along with an appropriate allowance. Is this the correct tool? What other drill types could be used?
2. How much time will be required to drill the hole in Problem 1?
3. What is the metal-removal rate when a $1\frac{1}{2}$-inch-diameter hole, 2 in. deep, is drilled in 1020 steel at a cutting speed of 200 fpm with a feed of 0.010 ipr?
4. If the specific horsepower for the steel in Problem 3 is 0.9, what horsepower would be required?
5. If the specific power of AISI steel of 0.7, and 75% of the output of the 1.5-kW motor of a drilling machine is available at the tool, what is the maximum feed that can be used in drilling a 2-in. diameter hole with a carbide drill? (Use the cutting speed suggested in Problem 3.)
6. Show how the approximate equation 24-5 for MRR in drilling was obtained. What assumption was needed?
7. In Table 24-2, is the equation for MRR, line 4, the same as Equation (24-4)? Explain.
8. In Table 24-2, what is the prorated downtime or prorated cost of purchasing? How are these prorated values calculated?
9. An indexable-insert drill can produce a hole four times faster than a spade-drill but may cost (with inserts) 50% to 75% more than the equivalent spade blade and holder. For making only a couple of holes, the extra cost is not justified.

   Manufacturer's charts will help determine the best feed and speed to run the drills. For example, a $1\frac{1}{2}$-in. hole is to be drilled in 4140 steel annealed to Bhn 275. For the spade drill, speed is 80 sfm; feed, 0.009 ipr; and spindle rotation, 204 rpm. For the indexable-insert drill, speed is 358 sfm, feed 0.007 ipr, and spindle rotation 891 rpm. Determine the number of holes needed to justify the extra cost of the indexable-insert drill. Some additional cost data are given below.

   Ignore tool life and assume that the blades and the indexable drills make about the same number of holes. (Why is this a reasonable assumption?) The holes are 3 in. deep, with no allowance needed. Cost of drills:

Spade blade	Indexable-insert drill
$139.00 holder	$273.00 drill
+ 21.90 per table	+ 12.80 per two inserts
$160.90	$285.80

Assume for this example that a $45/hr machine rate includes the cost of labor and machine burden.

10. Let us assume that you are drilling eight holes, equally spaced in a bolt-hole circle. That is, there would be holes at 12, 3, 6, and 9 o'clock and four more holes equally spaced between them. The diameter of the bolt hole circle is 6 in. The designer says that the holes must be $45 \pm 1°$ from each other around the circle.
   a. Compute the tolerance between hole centers.
   b. Do you think a typical multiple-spindle drill setup could be used to make this bolt circle—using eight drills all at once? Why or why not?
   c. Do you think that the use of a jig may help improve the situation?

www.wiley.com/college/degarmo

Chapter 24 CASE STUDY

Bolt-Down Leg on a Casting

Figure CS-24 shows the design of one of four legs on a casting made by the Hunter Company. These legs are used to attach the device to the floor. The section drawing to the right shows the typical loading to which the leg is subjected. The company is currently drilling the bolt hole and then counterboring the land. Manufacturing has experienced some difficulty in machining these legs. They report a lot of drill breakage. Quality control reports that distances between the four holes are frequently too large. Sales has recently reported that a substantial number of in-service failures have occurred with these legs. Kavit Antani, a recently graduated manufacturing engineer, has sent you a sketch showing where the legs typically fail. The casting is manufactured from gray cast iron, using the sand-casting process.

1. What machining difficulties would you expect to have in making the legs?
2. What caused the distance between the holes to be too large?
3. What do you think has caused the failures?
4. What do you recommend to solve this problem in the future in terms of materials, design, and manufacture?
5. What do you recommend to be done with the units in the field to stop the failures?

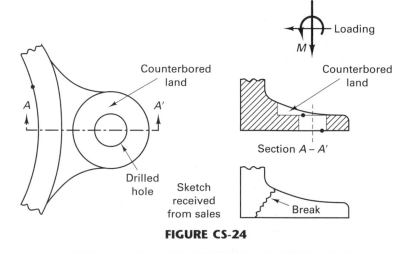

FIGURE CS-24

MILLING

■ 25.1 INTRODUCTION

Milling is a basic machining process by which a surface is generated by progressive chip removal. The workpiece is fed into a rotating cutting tool. Sometimes the workpiece remains stationary, and the cutter is fed to the work. In nearly all cases, a multiple-tooth cutter is used so that the material removal rate is high. Often the desired surface is obtained in a single pass of the cutter or work and, because very good surface finish can be obtained, milling is particularly well suited and widely used for mass-production work. Many types of milling machines are used, ranging from relatively simple and versatile machines that are used for general-purpose machining in job shops and tool-and-die work (these are NC or CNC machines) to highly specialized machines for mass production. Unquestionably, more flat surfaces are produced by milling than by any other machining process.

The cutting tool used in milling is known as a *milling cutter*. Equally spaced peripheral teeth will intermittently engage and machine the workpiece. This is called *interrupted cutting*. The workpieces are typically held in fixtures, as described in Chapter 29.

■ 25.2 FUNDAMENTALS OF MILLING PROCESSES

Milling operations can be classified into two broad categories called peripheral milling and face milling. Each has many variations. In *peripheral milling* the surface is generated by teeth located on the periphery of the cutter body (Figure 25-1). The surface is parallel with the axis of rotation of the cutter. Both flat and formed surfaces can be produced by this method, the cross section of the resulting surface corresponding to the axial contour of the cutter. This process, often called *slab milling*, is usually performed on horizontal spindle milling machines. In slab milling, the tool rotates (mills) at some rpm (N_s) while the work feeds past the tool at a table feed rate f_m in inches per minute, which depends on the feed per tooth, f_t.

As in the other processes, the cutting speed V and feed per tooth are selected by the engineer or the machine tool operator. As before, these variables depend upon the work material, the tool material, and the specific process. The cutting velocity is that which occurs at the cutting edges of the teeth in the milling center. The rpm of the spindle is determined from the surface cutting speed, where D is the cutter of diameter in inches according to

$$N_s = \frac{12V}{\pi D} \qquad (25\text{-}1)$$

The depth of cut, called DOC in Figure 25-1, is simply the distance between the old and new machined surface.

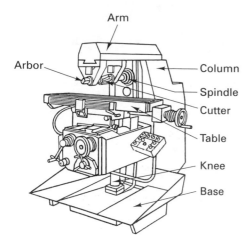

(a) Plain horizontal knee-type milling machine

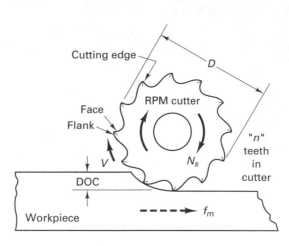

(b) Slab milling—multiple tooth

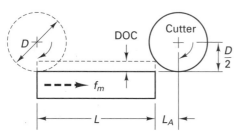

(c) Allowances for cutter approach

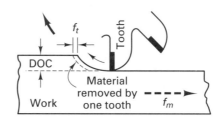

(d) Feed per tooth

FIGURE 25-1   Peripheral milling can be performed on a horizontal spindle milling machine. The cutter rotates at rpm N_s removing metal at cutting speed V. The allowance for starting and finishing the cut depends on the cutter diameter and depth of cut, DOC. The feet per tooth, f_t, is selected by the operator.

The width of cut is the width of the cutter or the work, in inches, and is given the symbol W. The length of the cut L is the length of the work plus some allowance L_A for approach and overtravel. The feed of the table f_m, in inches per minute, is related to the amount of metal each tool removes during a revolution, the feed per tooth f_t, according to

$$f_m = f_t N_s n \tag{25-2}$$

where n is the number of teeth in the cutter (teeth/rev.).

The *cutting time*

$$T_m = \frac{L + L_A}{f_m} \tag{25-3}$$

where

$$\text{the length of approach} = L_A = \sqrt{\frac{D^2}{4} - \left(\frac{D}{2} - \text{DOC}\right)^2} = \sqrt{t(D - \text{DOC})} \tag{25-4}$$

$$\text{MRR} = \frac{\text{volume}}{T_m} = \frac{LW\,\text{DOC}}{T_m} = Wf_m\,\text{DOC}\ \text{in}^3/\text{min} \tag{25-5}$$

ignoring L_A. Values for f_t are given in Table 25-1 along with recommended cutting speeds in feet per minute.

TABLE 25-1. Suggested Starting Feeds and Speeds Using High-Speed Steel and Carbide Cutters

Material	Feed (in./tooth) Speed (fpm)	Carbide Cutters						High-Speed Steel Cutters					
		Face Mills	Slab Mills	End Mills	Full and Half Side Mills	Saws	Form Mills	Face Mills	Slab Mills	End Mills	Full and Half Side Mills	Saws	Form Mills
Malleable iron	Feed per tooth	.005–015	.005–015	.005–010	.005–010	.003–004	.005–010	.005–015	.005–015	.003–015	.006–012	.003–006	.005–010
Soft/Hard	Speed, fpm	200–300	200–300	200–350	200–300	200–350	175–275	60–100	60–90	60–100	60–100	60–100	60–80
Cast steel	Feed per tooth	.008–015	.005–015	.003–010	.005–010	.002–004	.005–010	.010–015	.010–015	.005–010	.005–010	.002–005	.008–012
Soft/Hard	Speed, fpm	150–350	150–350	150–350	150–350	150–300	150–300	40–60	40–60	40–60	40–60	40–60	40–60
100–150	Feed per tooth	.010–015	.008–015	.005–010	.008–012	.003–006	.004–010	.015–030	.008–015	.003–010	.010–020	.003–006	.008–010
BHN Steel	Speed, fpm	450–800	450–600	450–600	450–800	350–600	350–600	80–130	80–130	80–140	80–130	70–100	70–100
150–250	Feed per tooth	.010–015	.008–015	.005–010	.007–012	.003–006	.004–010	.010–020	.008–015	.003–010	.010–015	.003–006	.006–010
BHN Steel	Speed, fpm	300–450	300–450	300–450	300–450	300–450	300–450	50–70	50–70	60–80	50–70	50–70	50–70
250–350	Feed per tooth	.008–015	.007–012	.005–010	.005–012	.002–005	.003–008	.005–010	.005–010	.003–010	.005–010	.002–005	.005–010
BHN Steel	Speed, fpm	180–300	150–300	150–300	160–300	150–300	150–300	35–60	35–50	40–60	35–50	35–50	35–50
350–450	Feed per tooth	.008–015	.007–012	.004–008	.005–012	.001–004	.003–008	.003–008	.005–008	.003–101	.003–008	.001–004	.003–008
BHN Steel	Speed, fpm	125–180	100–150	100–150	125–180	100–150	100–150	20–35	20–35	20–40	20–35	20–35	20–35
Cast Iron Hard	Feed per tooth	.005–010	.005–010	.003–008	.003–010	.002–003	.005–010	.005–012	.005–010	.003–008	.005–010	.002–004	.005–010
BHN 180–225	Speed, fpm	125–200	100–175	125–200	125–200	125–200	100–175	40–60	35–50	40–60	40–60	35–60	35–50
Cast Iron Medium	Feed per tooth	.008–015	.008–015	.005–010	.005–012	.003–004	.006–012	.010–020	.008–015	.003–010	.008–015	.003–005	.008–012
BHN 180–225	Speed, fpm	200–275	175–250	200–275	200–275	200–250	175–250	60–80	50–70	60–90	60–80	60–70	50–60
Cast Iron Soft	Feed per tooth	.015–025	.010–020	.005–012	.008–015	.003–004	.008–015	.015–030	.010–025	.004–010	.010–020	.002–005	.010–015
BHN 150–180	Speed, fpm	275–400	250–350	275–400	275–400	250–350	250–350	80–120	70–110	80–120	80–120	70–110	60–80
Bronze	Feed per tooth	.010–020	.010–020	.005–010	.008–012	.003–004	.008–015	.010–025	.008–020	.003–010	.008–015	.003–005	.008–015
Soft/Hard	Speed, fpm	300–100	300–800	300–1000	300–1000	300–1000	200–800	50–225	50–200	50–250	50–225	50–250	50–200
Brass	Feed per tooth	.010–020	.010–020	.005–010	.008–012	.003–004	.008–015	.010–025	.008–020	.005–015	.008–015	.003–005	.008–015
Soft/Hard	Speed, fpm	500–1500	500–1500	500–1500	500–1500	500–1500	500–1500	150–300	100–300	150–350	150–350	150–300	100–300
Aluminum Alloy	Feed per tooth	.010–040	.010–030	.003–015	.008–025	.003–006	.008–015	.010–040	.015–040	.015–040	.010–030	.004–008	.010–020
Soft/Hard	Speed, fpm	2000 UP	2000 UP	2000 UP	2000 UP	2000 UP	2000 UP	300–1200	300–1200	300–1200	300–1200	300–1000	300–1200

*Generally, lower end of range used for inserted blade cutters, higher end of range for indexable insert cutters.

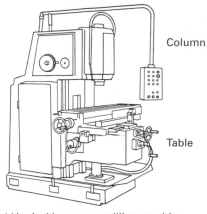

Column

Table

(a) Vertical knee-type milling machine

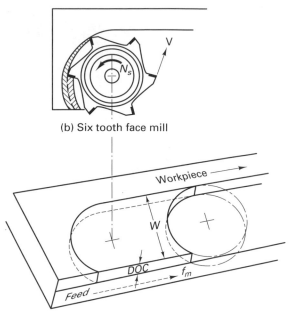

(b) Six tooth face mill

Workpiece

W

DOC

f_m

Feed

(c) Face milling over part of surface

Top views

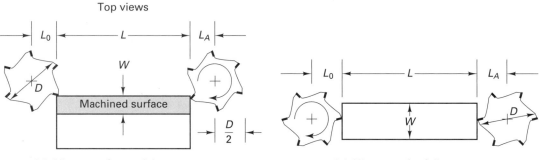

L_0 L L_A

W

D

Machined surface

$\dfrac{D}{2}$

(d) Allowance for partial coverage

L_0 L L_A

W

D

(e) Allowance for full coverage

FIGURE 25-2 Face milling is often performed on a vertical spindle milling machine using a multiple tooth cutter ($n = 6$ teeth) rotating at N_s rpm to produce cutting speed V. The workpiece feeds at rate f_m in inches per minute past the tool. The allowance depends on the tool diameter and the width of cut.

In *face milling* and *end milling*, the generated surface is at right angles to the cutter axis (Figure 25-2). Most of the cutting is done by the peripheral portions of the teeth, with the face portions providing some finishing action. Face milling is done on both horizontal- and vertical-spindle machines.

The tool rotates (face mills) at some rpm (N_s) while the work feeds past the tool. The rpm is related to the surface cutting speed, where the cutting diameter is D, according to Equation (25-1). The depth of cut is DOC, in inches, as shown in Figure 25-2c. The width of cut is W, in inches, and may be width of the workpiece or width of the cutter, depending on the setup. The length of cut is the length of the workpiece L plus an allowance for approach L_A and overtravel L_0, in inches. The feed rate of the table f_m, in inches per minute, is related to the amount of metal each tooth removes during a pass over the work, called the feed per tooth f_t, so $f_m = f_t N_s n$, where the number of teeth in the cutter is n. The cutting time

$$T_m = \frac{L + L_A + L_0}{f_m} \text{ min} \tag{25-6}$$

The *metal removal rate*

$$\text{MRR} = \frac{\text{vol}}{T_m} = \frac{LW\,\text{DOC}}{T_m} = f_m W\,\text{DOC in}^3/\text{min}$$

(ignore L_0 and L_A). The length of approach is usually equal to the length of overtravel which usually equals $D/2$ in. For a setup where the tool does not completely pass over the workpiece,

$$L_0 = L_A = \sqrt{W(D - W)} \quad \text{for} \quad W < \frac{D}{2} \tag{25-7}$$

$$L_0 = L_A = \frac{D}{2} \quad \text{for} \quad W \geq \frac{D}{2} \tag{25-8}$$

End milling is a very common operation performed on both vertical and horizontal spindle milling machines or machining centers. Figure 25-3 shows a vertical spindle milling process, cutting a step or pocket in the workpiece. The cutter can cut on both the sides and ends of the tool. If you were performing this operation on a block of metal (for example, 430F stainless steel), you as the manufacturing engineer would select a specific machine tool. You would have to determine how many passes (rough and finish cuts) were needed to produce the geometry specified in the design. Why? The number of passes is related to the cutting time by the total length of cut.

Vertical spindle milling, using an end mill, can produce a step in the workpiece. In Figure 25-3, an end mill with 6 teeth on a 2-in. diameter is used to cut a step in 430F stainless. The DOC (depth of cut) is .375 in. and the depth of immersion (DOI) is 1.25 in. The tool deflects due to the cutting forces, so the cut needs to be made at full immersion; but there may not be enough power for a full DOC. How many passes are needed to mill the step? The vertical milling machine tool available has a 5 hp motor with an 80% efficiency. The specific horsepower for 430F stainless is 1.3 hp/in^3/min.

The maximum amount of material that can be removed per pass is usually limited by the available power.

$$\text{Total DOC} = .375 \text{ in.}$$

$$\text{hp}_s = \text{hp}/f_m \times \text{DOC} \times \text{DOI} \tag{25-9}$$

FIGURE 25-3   End milling on a vertical spindle machine using a cutter which cuts on the end and sides. Below is a schematic showing DOC and DOI while milling a step in a block of metal.

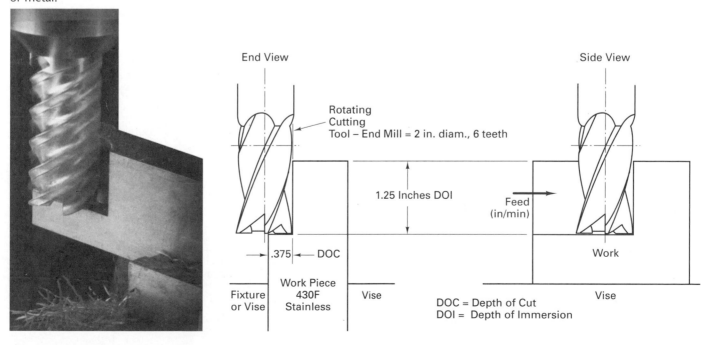

Select $f_t = 0.005$ ipt and $V = 250$ fpm from the table of recommended feeds and speeds. Calculate the spindle rpm:

$$N_s = \frac{12 \times 250}{3.14 \times 2} = 477 \text{ rpm of cutter}$$

calculate the table feed rate;

$$f_m = f_t \times n \times N = 0.005 \times 6 \times 477 = 14.31 \text{ in./min}$$

But the actual table feed rates for the selected machine are 11 in./min or 16 in./min, so, being conservative, select:

$$f_m = \text{table feed rate} = 11.00 \text{ in./min}$$

calculate the dept of cut, from equation 25-9;

$$\text{DOC} \cong \frac{5 \times .8}{1.3 \times 11.00 \times 1.25} \cong 0.225 \text{ in. maximum}$$

Therefore, 2 passes are needed because $(0.375/0.225 = 1.6)$

So, $0.375 - 0.225 = 0.150$ in. second pass DOC

2 passes: DOC $= 0.225$ rough cut

DOC $= \underline{0.150}$ finish cut

0.375 Total DOC

Note for a DOC $= 0.150$, the feed per tooth would be only slightly increased to 0.0051 ipt

$$f_t = \frac{5 \times .8}{1.3 \times 6 \times 477 \times 0.150 \times 1.25} = 0.0051 \text{ in./tooth}$$

You may want to change f_t to improve the surface finish, with smaller f_t, usually resulting in better surface finished; however, there are other factors to consider, like machining time.

UP VERSUS DOWN MILLING

For either slab end or face milling, surfaces can be generated by two distinctly different methods (Figure 25-4). *Up milling* is the traditional way to mill and is called

FIGURE 25-4 Climb cut or down milling versus conventional cut or up milling for slab or face or end milling.

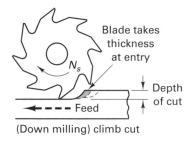

(Down milling) climb cut

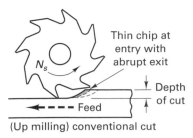

(Up milling) conventional cut

Peripherial or slab milling

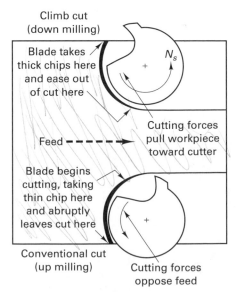

Face or end milling

conventional milling. The cutter rotates against the direction of feed of the workpiece. In *climb* or *down milling*, the cutter rotation is in the same direction as the feed rate. The method of chip formation is completely different in the two cases.

In up milling, the chip is very thin at the beginning where the tooth first contacts the work, then it increases in thickness, becoming a maximum where the tooth leaves the work. The cutter tends to push the work along and lift it upward from the table. This action tends to eliminate any effect of looseness in the feed screw and nut of the milling machine table and results in a smooth cut. However, the action also tends to loosen the work from the clamping device; therefore, greater clamping forces must be employed with the danger of deflecting the part. In addition, the smoothness of the generated surface depends greatly on the sharpness of the cutting edges. In up milling, chips can be carried into the newly machined surface, causing the surface finish to be poorer (rougher) than in down milling.

In down milling, maximum chip thickness occurs close to the point at which the tooth contacts the work. Because the relative motion tends to pull the workpiece into the cutter, any possibility of looseness in the table feed screw must be eliminated if down milling is to be used. It should never be attempted on machines that are not designed for this type of milling. Virtually all modern milling machines are capable of doing down milling and it is a most favorable application for carbide cutting edges. Because the material yields in approximately a tangential direction at the end of the tooth engagement, there is less tendency (than when up milling is used) for the machined surface to show toothmarks, and the cutting process is smoother with less chatter. Another advantage of down milling is that the cutting force tends to hold the work against the machine table, permitting lower clamping forces. However, the fact that the cutter teeth strike against the surface of the work at the beginning of each chip can be a disadvantage if the workpiece has a hard surface, as castings sometimes do. This may cause the teeth to dull rapidly. Metals that readily workharden should be climb milled.

Milling is an interrupted cutting process wherein entering and leaving the cut subjects the tool to impact loading, cyclic heating, and cycle cutting forces. As shown in Figure 25-5, the cutting force, F_c, builds rapidly as the tool enters the work at Ⓐ and progresses to Ⓑ, peaks as the blade crosses the direction of feed at Ⓒ, decreases to Ⓓ, and then drops to zero abruptly upon exit. The diagram does not indicate the impulse loads caused by impacts. The interrupted-cut phenomenon explains in large part why milling cutter teeth are designed to have small positive or negative rakes, particularly when the tool material is carbide or ceramic. These brittle materials tend to be very strong in compression, and negative rake results in the cutting edges being placed in compression by the cutting forces rather than tension. Cutters made from HSS are made with positive rakes, in the main, but must be run at lower speeds. Positive rake tends to lift the workpiece, while negative rakes compress the workpiece and allow heavier cuts to be made. Table 25-2 summarizes some additional milling problems.

FIGURE 25-5 Conventional face milling (*left*) with cutting force diagram for F_c (*right*) showing the interrupted nature of the process. (*From* Metal Cutting Principles, *2nd ed., Ingersoll Cutting Tool Company.*)

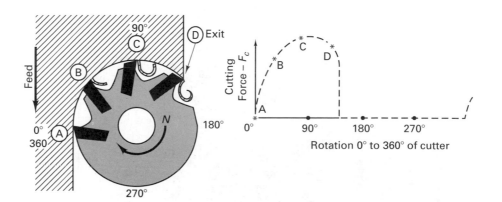

TABLE 25-2. Probable Causes of Milling Problems

Problem	Probable Cause	Cures
Chatter (vibration)	1. Lack of rigidity in machine, fixtures, arbor, or workpiece 2. Cutting load too great 3. Dull cutter 4. Poor lubrication 5. Straight-tooth cutter 6. Radial relief too great 7. Rubbing, insufficient clearance	Use larger arbors. Change rpm (cutting speed). Decrease feed per tooth or number of teeth in contact with work. Sharpen or replace inserts. Flood coolant. Use helical cutter. Check tool angles.
Loss of accuracy (cannot hold size)	1. High cutting load causing deflection 2. Chip packing, between teeth 3. Chips not cleaned away before mounting new piece of work	Decrease number of teeth in contact with work or feed per tooth. Adjust cutting fluid to wash chips out of teeth.
Cutter rapidly dulls	1. Cutting load too great 2. Insufficient coolant	Decrease feed per tooth or number of teeth in contact. Add blending oil to coolant.
Poor surface finish	1. Feed too high 2. Tool dull 3. Speed too low 4. Not enough cutter teeth	Check to see if all teeth are set at same height.
Cutter digs in (hogs into work)	1. Radial relief too great 2. Rake angle too large 3. Improper speed	Check to see that workpiece is not deflecting and is securely clamped.
Work burnishing	1. Cut is too light 2. Tool edge worn 3. Insufficient radial relief 4. Land too wide	Enlarge feed per tooth. Sharpen cutter.
Cutter burns	1. Not enough lubricant 2. Speed too high	Add sulfur-based oil. Reduce cutting speed. Flood coolant.
Teeth breaking	1. Feed too high 2. Depth of cut too large	Decrease feed per tooth. Use cutter with more teeth. Reduce table feed rate.

■ 25.3 MILLING TOOLS AND CUTTERS

Most milling work today is done with face mills and end mills. The face mills use indexable carbide insert tooling while the end mills are either solid HSS or insert tooling (Figure 25-6). Basically, *mills* are shank-type cutters having teeth on the circumferential surface and one

FIGURE 25-6 Solid end mills are often coated. Insert tooling end mills come in a variety of sizes and are mounted on taper shanks.

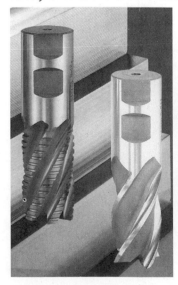

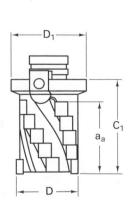
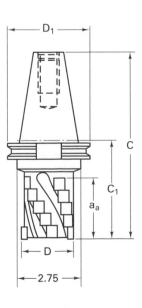

end. They thus can be used for facing, profiling, and end milling. The teeth may be either straight or helical, but the latter is more common. Small end mills have straight shanks, whereas taper shanks are used on larger sizes (Figure 25-6).

Plain end mills have multiple teeth that extend only about halfway toward the center on the end. They are used in milling slots, profiling, and facing narrow surfaces. *Two-lip mills* have two straight or helical teeth that extend to the center. Thus they may be sunk into material, like a drill, and then fed lengthwise to form a groove, a slot, or a pocket.

Shell end mills are solid multiple-tooth cutters, similar to plain end mills but without a shank. The center of the face is recessed to receive a screw head or nut for mounting the cutter on a separate shank or a stub arbor. One shank can hold any of several cutters and thus provides great economy for larger-sized end mills.

Hollow end mills are tubular in cross section, with teeth only on the end but having internal clearance. They are used primarily on automatic screw machines for sizing cylindrical stock, producing a short cylindrical surface of accurate diameter.

Face mills have a center hole so that they can be arbor mounted. Face milling cutters are widely used in both horizontal and vertical spindle machine tools and come in a wide variety of sizes (diameters and heights) and geometries (round, square, triangular, etc.) as shown in Figure 25-7.

The insert can usually be indexed four times and must be well supported. Either the power or the rigidity of the machine tool will be the limiting factor although sometimes setup can be the limiting factor.

FIGURE 25-7 Face mills come in many different designs using many different insert geometries and different mounting arbors.

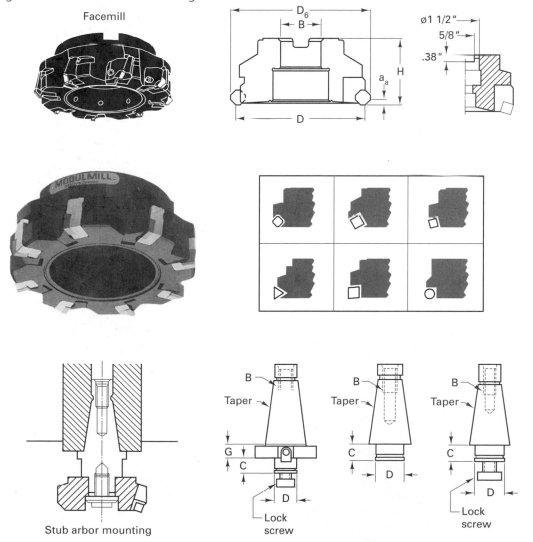

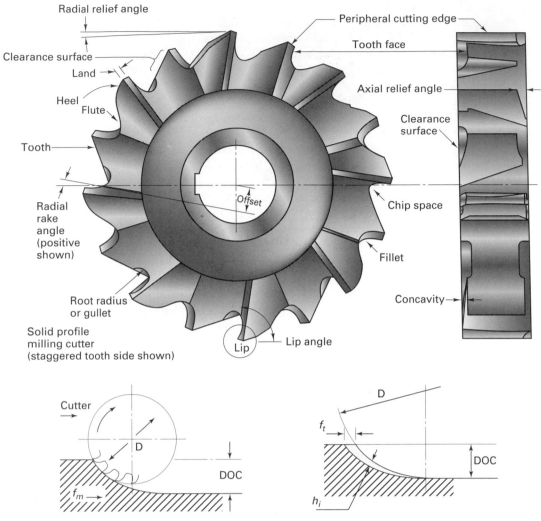

Radial relief angle

Clearance surface

Land

Heel

Flute

Tooth

Radial rake angle (positive shown)

Root radius or gullet

Solid profile milling cutter (staggered tooth side shown)

Lip

Lip angle

Offset

Peripheral cutting edge

Tooth face

Axial relief angle

Clearance surface

Chip space

Fillet

Concavity

Cutter

D

DOC

f_m

D

f_t

DOC

h_i

FIGURE 25-8 The side milling cutter can cut on sides and ends of the teeth so it makes slots or grooves. However, as shown below, only a few teeth are engaged at any one point in time causing heavy torsional vibrations. The average chip thickness, h_i, will be less than the feed per tooth, f_t.

Another common type of arbor mounted milling cutter is called a side mill because it cuts on the ends and sides of the cutters. Figure 25-8 shows the geometry of a staggered tooth side milling cutter.

Staggered-tooth milling cutters are narrow cylindrical cutters having staggered teeth, and with alternate teeth having opposite helix angles. They are ground to cut only on the periphery, but each tooth also has chip clearance ground on the protruding side. These cutters have a free cutting action that makes them particularly effective in milling deep slots. *Staggered-tooth cutters* are really special *side milling cutters* which are similar to plain milling cutters except that the teeth extend radially part way across one or both ends of the cylinder toward the center. The teeth may be either straight or helical. Frequently, these cutters are relatively narrow, being disklike in shape. Two or more side milling cutters often are spaced on an arbor to straddle the workpiece (called *straddle milling*) and two or more parallel surfaces are machined at once.

In Figure 25-9 insert-tooth side mills are arranged in a gang milling setup to cut 3 slots in the workpiece simultaneously. Thus the desired part geometry is repeatedly produced by the setup as the position of the cutters is fixed. However, in side and face milling operations only a few teeth are engaged at any point in time, resulting in heavy torsional vibrations detrimental to the resulting machined product. A flywheel can solve this problem and in many cases be the key to improved productivity.

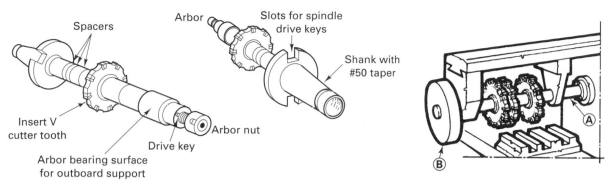

FIGURE 25-9   Arbor (*two views*) used on a horizontal spindle milling machine on left. On right, a gang milling setup showing 3 side milling cutters mounted on an arbor (*A*), with an outboard flywheel (*B*).

For the gang milling as shown in Figure 25-9 the diameter of the flywheel should be as large as possible. (The moment of inertia increases with the square of the radius.) The best position of the flywheel is inboard on the arbor at A, but depending on the setup, this may not be possible so then position B should be chosen. It is important that the distance between the cutters and flywheel is as small as possible.

A flywheel can be built up from a number of carbon steel discs, each having a center hole and keyway to fit the arbor, so the weight can be easily varied.

Interlocking slotting cutters consist of two cutters similar to side mills but made to operate as a unit for milling slots. The two cutters are adjusted to the desired width by inserting shims between them.

Slitting saws are thin, plain milling cutters, usually from $\frac{1}{32}$ to $\frac{3}{16}$ in. thick, which have their sides slightly "dished" to provide clearance and prevent binding. They usually have more teeth per unit of diameter than ordinary plain milling cutters and are used for milling deep narrow slots and cutting-off operations.

In milling, the average chip thickness (h_i) is not the same as the feed per tooth. For example a thickness (h_i) of .004 in. corresponds to 0.012 in. feed per tooth in most side and facemilling operations.

If the radial depth of cut (DOC) is very small compared to the cutter diameter, D, use this formula:

$$\text{Feed per tooth} = f_t = .004 \sqrt{\frac{D}{\text{DOC}}} \, (\text{ipt})$$

Note: For calculating the table feed use half the number of inserts in a full side and face mill to arrive at the effective number of teeth.

Table feed rate (ipm) = rpm × number of effective teeth × feed per tooth.

Another method of classification for face and end mill cutters relates to the direction of rotation. A *right-hand cutter* must rotate counterclockwise when viewed from the front end of the machine spindle. Similarly, a *left-hand cutter* must rotate clockwise. All other cutters can be reversed on the arbor to change them from one hand to the other. Positive rake angles are used on general-purpose HSS milling cutters. Negative rake angles are commonly used on carbide- and ceramic-tipped cutters employed in mass-production milling in order to obtain the greater strength and cooling capacity. TiN coating of these tools is quite common, resulting in significant increases in tool life.

Plain milling cutters used for plain or slab milling have straight or helical teeth on the periphery and are used for milling flat surfaces. *Helical mills* (Figure 25-10) engage the work gradually, and usually more than one tooth cuts at a given time. This reduces shock and chattering tendencies and promotes a smoother surface. Consequently, this type of cutter usually is preferred over one with straight teeth.

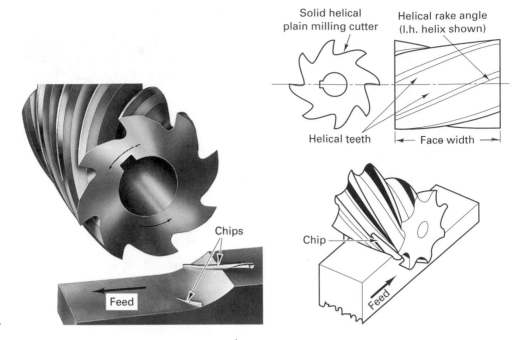

FIGURE 25-10 The chips are formed progressively by the teeth of a plain helical-tooth milling cutter during up milling.

Angle milling cutters are made in two types: single-angle and double-angle. Angle cutters are used for milling slots of various angles or for milling the edges of workpieces to a desired angle. *Single-angle cutters* have teeth on the conical surface, usually at an angle of 45 to 60° to the plane face. *Double-angle cutters* have V-shaped teeth, with both conical surfaces at an angle to the end faces but not necessarily at the same angle. The V-angle usually is 45, 60, or 90°.

Form milling cutters have the teeth ground to a special shape—usually an irregular contour—to produce a surface having a desired transverse contour. They must be sharpened by grinding only the tooth face, thereby retaining the original contour as long as the plane of the face remains unchanged with respect to the axis of rotation. Convex, concave, corner-rounding, and gear-tooth cutters are common examples (Figure 25-11). Solid HSS cutters of simple shape and reasonably small size are usually more economical in initial cost than inserted-blade cutters. However, inserted-blade cutters may be lowest in overall cost on large production jobs.

Form-relieved cutters permit the lowest cost where intricately shaped cuts are involved. Solid or carbide insert tool cutters may be justified by high production requirements.

Most larger-sized milling cutters are of the *inserted-tooth type*. The cutter body is made of steel, with the teeth made of high-speed steel, carbides, or TiN carbides, fastened to the body by various methods. An insert tooth cutter uses indexable carbide or ceramic inserts as shown in Figure 25-7. This type of construction reduces the amount of costly material that is required and can be used for any type of cutter but most often is used with face mills.

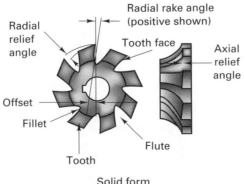

Solid form relieved milling cutter

FIGURE 25-11 Solid form relieved milling cutter.

T-slot cutters are integral-shank cutters with teeth on the periphery and *both* sides. They are used for milling the wide groove of a T-slot. To use them, the vertical groove must first be made with a slotting mill or an end mill to provide clearance for the shank. Because the T-slot cutter cuts on five surfaces simultaneously it must be fed with care.

Woodruff keyseat cutters are made for the single purpose of milling the semi-cylindrical seats required in shafts for Woodruff keys. They come in standard sizes corresponding to Woodruff key sizes. Those below 2 in. in diameter have integral shanks; the larger sizes may be arbor-mounted.

Occasionally, *fly cutters* may be used for face milling or boring. Both operations may be done with a single tool at one setup. A single-point cutting tool is attached to a special shank, usually with provision for adjusting the effective radius of the cutting tool with respect to the axis of rotation. The cutting edge can be made in any desired shape and, because it is a single-point tool, is very easy to grind.

■ 25.4 MACHINES FOR MILLING

The four most common types of manually controlled milling machines are listed below in order of increasing power (and therefore metal removal capability):

1. Ram type milling machines   **3.** Fixed-bed type milling machines

2. Column-and-knee type milling machines   **4.** Planer-type milling machines

   a. Horizontal spindle

   b. Vertical spindle

Milling machines whose motions are electronically controlled are listed in order of increasing production capacity and decreasing flexibility:

1. Manual data input milling machines

2. Programmable CNC milling machines

3. Machining centers (tool and pallet exchange capability)

4. FMC and FMS

5. Transfer lines

BASIC MILLING MACHINE CONSTRUCTION

Most basic milling machines are of *column-and-knee* construction, employing the components and motions shown in Figure 25-12. The column, mounted on the base, is the main supporting frame for all the other parts and contains the spindle with its driving mechanism. This construction provides controlled motion of the worktable in three mutually perpendicular directions: (1) through the *knee* moving vertically on ways on the front of the column, (2) through the *saddle* moving transversely on ways on the knee, and (3) through the *table* moving longitudinally on ways on the saddle. All these motions can be imparted either by manual or powered means. In most cases, a powered rapid traverse is provided in addition to the regular feed rates for use in setting up work and in returning the table at the end of a cut.

The ram type milling machine is one of most versatile and popular milling machines using the knee and column design. Ram-type machines have a head equipped with a motor stopped pulley and belt drive, and spindle. The ram, mounted on horizontal ways at the top of the column, supports the head and permits positioning of the spindle with respect to the table. Ram type milling machines are normally 10 hp or less, light duty milling, drilling, reaming, etc. (Figure 25-13).

Milling machines having only the three mutually perpendicular table motions are called *plain column-and-knee type*. These are available with both horizontal and vertical spindles (Figure 25-12). On the older horizontal spindle type machines, an adjustable over-arm provides an outboard bearing support for the end of the cutter arbor which was shown in Figure 25-9. These machines are well suited for slab, side, or straddle milling.

In some vertical-spindle machines the spindle can be fed up and down, either by power or by hand. Vertical-spindle machines are especially well suited for face- and end-milling operations. They also are very useful for drilling and boring, particularly where holes must be spaced accurately in a horizontal plane, because of the controlled table motion.

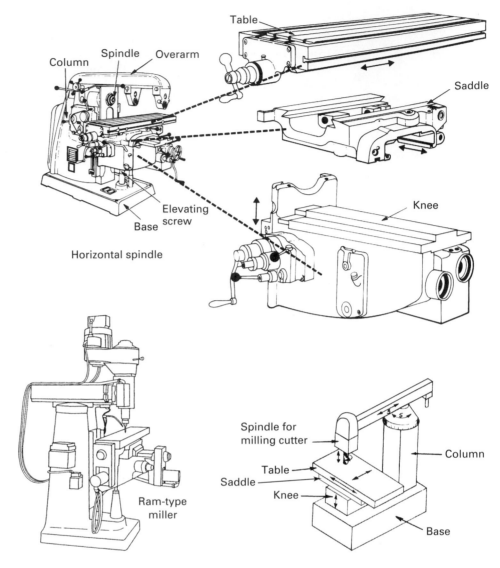

FIGURE 25-12 Major components of a plain column-and-knee-type milling machine, which can have horizontal spindle shown above or a turret machine with a vertical spindle shown below. The table holding the workpiece and workholder can be translated in *X*, *Y*, and *Z* directions with respect to the tool.

Turret-type column-and-knee milling machines have dual heads that can be swivelled about a horizontal axis on the end of a horizontally adjustable ram. This permits milling to be done horizontally, vertically, or at any angle. This added flexibility is advantageous when a variety of work has to be done, as in tool and die or experimental shops. They are available with either plain or universal tables.

Universal column-and-knee milling machines differ from plain column-and-knee machines in that the table is mounted on a housing that can be swivelled in a horizontal plane, thereby increasing its flexibility. Helices, as found in twist drills, milling cutters, and helical gear teeth, can be milled on universal machines.

BED-TYPE MILLING MACHINES

In production manufacturing operations, ruggedness and the capability of making heavy cuts are of more importance than versatility. *Bed-type milling machines* (Figure 25-14), are made for these conditions. The table is mounted directly on the bed and has only longitudinal motion. The spindle head can be moved vertically in order to set up the machine for a given operation. Normally, once the setup is completed, the spindle head is clamped in position and no further motion of it occurs during machining. However, on some machines vertical motion of the spindle occurs during each cycle.

FIGURE 25-13 The ram type knee-and-column milling machine is one of the most versatile and popular milling machine tools ever designed.

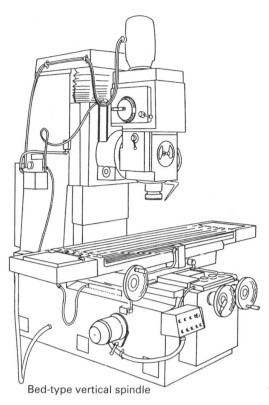

Bed-type vertical spindle

FIGURE 25-14 Bed-type vertical spindle heavy duty production machine tools for milling.

After such milling machines are set up, little skill is required to operate them, permitting faster learning time for the operators. Some machines of this type are equipped with automatic controls so that all the operator has to do is load and unload workpieces into the fixture and set the machine into operation. For stand-alone machines, a fixture can be located at each end of the table so that one workpiece can be loaded while another is being machined.

Bed-type milling machines with single spindles sometimes are called *simplex milling machines*; they are made with both horizontal and vertical spindles. Bed-type machines also are made in *duplex* and *triplex* types, having two or three spindles respectively, permitting the simultaneous milling of two or three surfaces at a single pass.

PLANER-TYPE MILLING MACHINES

Planer-type milling machines (Figure 25-15) utilize several milling heads, which can remove large amounts of metal while permitting the table and workpiece to feed quite slowly. Often only a single pass of the workpiece past the cutters is required. Through the use of different types of milling heads and cutters, a wide variety of surfaces can be machined with a single setup of the workpiece. This is an advantage when heavy workpieces are involved.

ROTARY-TABLE MILLING MACHINES

Some types of face milling in mass-production manufacturing are often done on *rotary-table milling machines*. Roughing and finishing cuts can be made in succession as the workpieces are moved past the several milling cutters while held in fixtures on the rotating table. The operator can load and unload the work without stopping the machine.

PROFILERS AND DUPLICATORS

Milling machines that can duplicate external or internal geometries in two dimensions are called *profilers* or tracer-controlled machines. As shown in Figure 25-16, a tracing probe

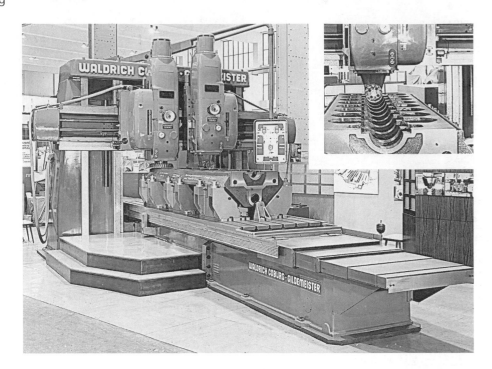

FIGURE 25-15 Large planer-type milling machine. Inset shows 90° head being used. *(Courtesy of Cosa Corporation.)*

follows a two-dimensional pattern or template and, through electronic or air-actuated mechanisms, controls the cutting spindles in two mutually perpendicular directions.

Hydraulic tracer control is not new, but a great variety of control systems for contouring and straight-line milling are now available. These range from simple single-axis control systems for die sinking to control for profiling and three-axis contouring.

Hydraulic tracers have transformed standard and special milling machines into production tools with virtually unlimited capabilities for contouring and straight-line milling. They can produce shapes that are impossible to produce manually, such as free-form dies and molds.

All hydraulic tracers work basically the same way, in that they utilize a stylus connected to a precision servomechanism for each axis of control. Figure 25-14 illustrates a stylus that is linked mechanically to the servos in a two-axis tracer valve. The servos are connected to hydraulic actuators on the machine slides. As the stylus traces a template, the servos control the motion of the slides so that the milling cutter duplicates the template shape onto the workpiece.

As the tracing stylus is tilted toward the pattern, the servovalve is shifted off null, resulting in Y slide motion opposite to that of the stylus direction. When the stylus is in contact with the pattern, the servovalve is at its null position and Y slide motion is stopped.

Duplicators produce forms in three dimensions. A tracing probe follows a three-dimensional master. Often the probe does not actually contact the master, a variation in the length of an electric spark between the probe and the master controlling the drives to the quill and the table, thereby avoiding wear on the master or possible deflection of the probe. On some machines, the ratio between the movements of the probe and cutter can be varied.

Duplicators are widely used to machine molds and dies and sometimes are called *die-sinking machines*. They are used extensively in the aerospace industry to machine parts from wrought plate or bar stock as substitutes for forgings when the small number of parts required would make the cost of forging dies uneconomical.

NC and CNC type machines and their applications are discussed in Chapter 32.

When purchasing or using a milling machine, consider the following issues:

1. Spindle orientation
2. Machine capability and capacity
3. Automatic tool changing

The choice of spindle orientation, horizontal or vertical, depends on the parts to be machined. Relatively flat parts are usually done on vertical machines. Cubic parts are usually done on a horizontal machine, where chips tend to fall free of the part. Operations like slotting

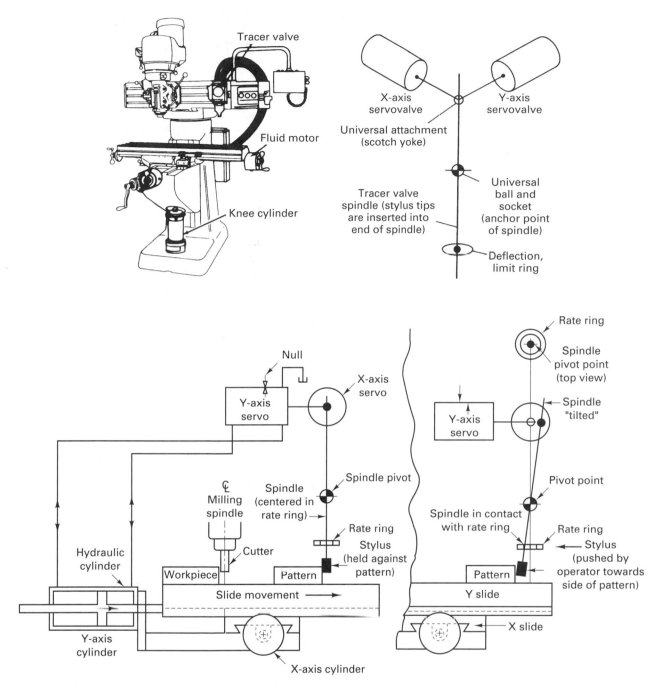

FIGURE 25-16   Small knee-type vertical spindle milling machine modified with tracer control system for *X* and *Y* control.

and side milling are best done on horizontal machines with outboard supports for the arbor. Use the largest diameter arbor possible to reduce twist and deflection due to cutting forces.

Machine capability refers to the tolerances while capacity refers to the size of parts and the power available.

As with all tooling applications, the tolerances that can be maintained in milling are dependent upon the rigidity of the workpiece, the accuracy and rigidity of the machine spindle, the precision and accuracy of the workholding device, and the quality of the cutting tool itself. Milling produces forces that contribute to chatter and vibration because of the intermittent cutting action. Soft materials tend to adhere to the cutter teeth and make it more difficult to hold tolerances. Materials such as cast iron and aluminum are easy to mill.

Within these criteria, properly maintained cutters used in rigid spindles on properly fixtured workpieces can expect to machine within tolerances ±0.0005 in. with surface flatness tolerances of 0.001 in./ft. Such tolerances are also possible on "slotting" operations with milling cutters, but +0.001 in. to +0.002 in. is more probable. Flatness specifications are more difficult to maintain in steel and easier to maintain in some types of aluminum, cast iron, and other nonferrous material.

Part size is the primary factor in selecting the machine size but the length of the tooling as mounted in the spindle must be considered. Horsepower required at the spindle depends on the MRR and the materials (unit horsepower, hp_s). Remember, coated inserts allow the MRR (the cutting speed) to be increased and available power may be exceeded.

Finally, the capacity of the tool changers on machining centers is limited by the number, size, and weight of the tools, especially if large diameter tools are being employed. These often have to be stored in every other space in the storage mechanism.

ACCESSORIES FOR MILLING MACHINES

The usefulness of ordinary milling machines can be greatly extended by employing various accessories or attachments. Here are some examples.

A horizontal milling machine can be equipped with a vertical milling attachment to permit vertical milling to be done. Ordinarily, heavy cuts cannot be made with such an attachment.

The *universal milling attachment* (Figure 25-17) is similar to the vertical attachment but can be swivelled about both the axis of the milling machine spindle and a second, perpendicular axis to permit milling to be done at any angle.

The *universal dividing head* is by far the most widely used milling machine accessory, providing a means for holding and indexing work through any desired arc of rotation. The work may be mounted between centers (Figure 25-17) or held in a chuck that is mounted in the spindle hole of the dividing head. The spindle can be tiled from about 5° below horizontal to beyond the vertical position.

Basically, a dividing head is a rugged, accurate, 40 : 1 worm-gear reduction unit. The spindle of the dividing head is rotated one revolution by turning the input crank 40 turns. An index plate mounted beneath the crank contains a number of holes, arranged in concentric circles and equally spaced, with each circle having a different number of holes. A plunger pin on the crank handle can be adjusted to engage the holes of any circle. This permits the crank to be turned an accurate, fractional part of a complete circle as represented by the increment between any two holes of a given circle on the index plate. Utilizing the 40 : 1 gear ratio and the proper hole circle on the index plate, the spindle can be rotated a precise amount by the application of either of the following rules:

$$\text{number of turns of crank} = \frac{40}{\text{cuts per revolution of work}}$$

$$\text{holes to be indexed} = \frac{40 \times \text{holes in index circle}}{\text{cuts per revolution of work}}$$

FIGURE 25-17   End milling a helical groove on a horizontal spindle milling machine using a universal dividing head and a universal milling attachment. *(Courtesy of Cincinnati Milacron, Inc.)*

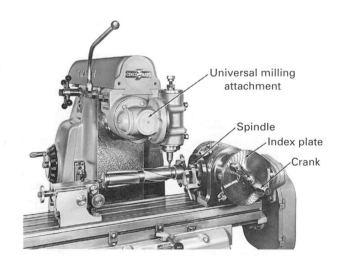

If the first rule is used, an index circle must be selected that has the proper number of holes to be divisible by the denominator of any resulting fractional portion of a turn of the crank. In using the second rule, the number of holes in the index circle must be such that the numerator of the fraction is an even multiple of the denominator. For example, if 24 cuts are to be taken about the circumference of a workpiece, the number of turns of the crank required would be $1\frac{2}{3}$. An index circle having 12 holes could be used with one full turn plus eight additional holes. The second rule would give the same result. Adjustable *sector arms* are provided on the index plate that can be set to a desired number of holes, less than a full turn, so that fractional turns can be made readily without the necessity for counting holes each time. Dividing heads are made having ratios other than 40 : 1. The ratio should be checked before using.

Because each full turn of the crank on a standard dividing head represents 360/40, or 9° of rotation of the spindle, indexing to a fraction of a degree can be obtained. For example, the space between two adjacent holes on a 36-hole circle represents $\frac{1}{4}°$. Indexing can be done in three ways. *Plain indexing* is done solely by the use of the 40 : 1 ratio in the dividing head. In *compound indexing*, the index plate is moved forward or backward a number of hole spaces each time the crank handle is advanced. For *differential indexing* the spindle and the index plate are connected by suitable gearing so that as the spindle is turned by means of the crank, the index plate is rotated a proportional amount.

The dividing head also can be connected to the feed screw of the milling-machine table by means of gearing. This procedure is used to provide a definite rotation of the workpiece with respect to the longitudinal movement of the table, as in cutting helical gears. This procedure is illustrated in Chapter 30.

■ 25.5 WORKHOLDING DEVICES FOR MILLING

T-slots are provided on milling machine tables so that workpieces can be clamped directly to the table. More often various workholding devices, called *vises* or *fixtures*, are utilized. Smaller workpieces usually are held in a vise mounted on the table. A universal vise (Figure 25-18) is particularly useful in tool-and-die work. Fixtures designed to specifically hold a part in the correct location with respect to the tool are used for larger volumes as shown in Figure 25-19. Fixtures reduce the time it takes to put the part in the machine and assure repeatable location with respect to the cutting tools. Fixtures provide clamping forces that counteract the cutting forces. The design of workholding devices is discussed in Chapter 29.

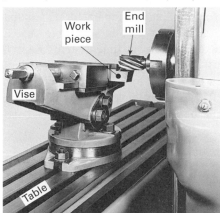

FIGURE 25-18 Universal vise on a horizontal spindle milling machine. *(Courtesy of Cincinnati Milacron, Inc.)*

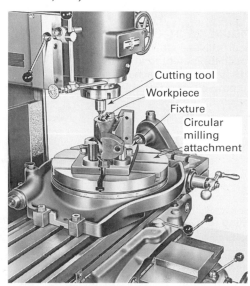

FIGURE 25-19 End milling a circular slot using a fixture mounted on a circular-milling attachment in a vertical spindle milling machine. *(Courtesy of Cincinnati Milacron, Inc.)*

MILLING SURFACE FINISH

The average surface finishes that can be expected on free machining materials range from 60 to 150 μin. Conditions can exist, however, that can produce wide variations on either side of these ranges.

■ 25.6 THE HISTORY OF THE MILLING MACHINE*

FIGURE 25-20 A Lincoln Miller, built in Hartford, CT, at the Phoenix Iron Works around 1860. Features an adjustment for cutter height (at either end of the cone pulley) and an outboard bearing support (on floor).

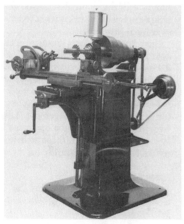

FIGURE 25-21 The Brown & Sharpe No. 1 universal milling machine, patented in 1861, was the first milling machine to employ an adjustable knee.

Traditionally, the metal parts of a musket or pistol were made by a skilled craftsman who finished each piece by hand-filing it. This process was time consuming, and the results were not uniform.

Dissatisfied with this approach, the United States Army Ordnance Department called for uniformity as it organized arms production for the new American Army. The U.S. government funded the development of techniques and machines able to produce interchangeable parts, a revolutionary concept in 1790. This concept was inspired by the work of Major Louis de Tousard of France, who emphasized the need to devise a system of uniformity based on scientific theory.

In 1798, Americans Simeon North and Eli Whitney were awarded contracts to manufacture small arms. These arms-makers followed the lead of French armorer Honoré Blanc, who pioneered the concept of interchangeable parts in Europe. The quest for uniformity led to the development of new machines, such as the milling machine.

Eli Whitney often is credited with being the father of the milling machine, but it probably was invented by Simeon North around 1816. The government advanced large sums of money to arms-makers like North to pay for the development of new tools and facilities. In exchange, the government had the right to use any mechanical innovations developed.

The Army Ordnance Department made information about the advancements available to different public and private armories. This, combined with skilled machinists changing jobs, contributed to the improvements being widely adopted throughout the metalworking industry.

THE MILLING MACHINE

The first milling machine was extremely crude—little more than a hand-cranked lathe headstock adapted for plain milling. This early design was improved and popularized by John Hall at the Harpers Ferry, V.A., armory.

Early milling machines had two major mechanical drawbacks: They didn't support the outboard end of the arbor, and they didn't permit the worktable to be raised and lowered. Innovators solved these problems. The Lincoln Miller machine (Figure 25-20) overcame the problem of cutter-height adjustment with a spindle that could be moved up or down in relation to the worktable. It also had a spindle support—either an overarm or an additional, outboard bearing post.

In 1861, Brown & Sharpe Manufacturing Company developed the universal milling machine (Figure 25-21). Its use of an adjustable knee and simplified gearing arrangement was a major breakthrough in machine design. Joseph Brown designed the machine to cut flutes on twist drills. These drills were used in large numbers to drill holes in the percussion nipples of the new percussion-lock rifles. The universal milling machine's versatility made it very popular, both in job shops and the toolrooms of large manufacturers. The Lincoln Miller, or plain milling machine, continued to be produced and was used mainly for routine jobs.

*Adapted from *Cutting Tool Engineering*, October 1990, p. 90 written by Peter Liebhold, museum specialist, Division of Engineering and Industry, the Smithsonian Institute, Washington, D.C.

■ KEY WORDS

climb (down) milling	duplicator	interrupted cutting	milling	slab milling
column-and-knee milling machine	end milling	machining center	peripheral milling	staggered-tooth milling cutter
	face milling	metal removal rate	profiler	straddle milling
conventional (up) milling	fixtures	milling machines	saddle	Woodruff keyseater
cutting time	insert-tooth milling cutter	milling cutters		

■ REVIEW QUESTIONS

1. Why does milling have a high metal removal rate?
2. For producing flat surfaces in mass-production machining how does face milling differ basically from peripheral milling?
3. Which type of milling (up or down) do you think uses the least power? Explain.
4. Why may down milling dull the cutter more rapidly than up milling in machining sand castings?
5. What are the parameters you need to specify in order to calculate MRR?
6. What kind of milling do you have if the diameter D of the cutter shown in Figure 25-2b is less than the width W of the workpiece?
7. What is the advantage of a helical-tooth cutter over a straight-tooth cutter for slab milling?
8. What would the cutting force diagram for F_c look like if all eight teeth are considered? See right side of Figure 25-5 inset.
9. Could the stub arbor mounted face mill shown in Figure 25-7 be used to machine a T-slot? Why or why not?
10. In a typical solid arbor milling cutter shown in Figure 25-8, why are the teeth staggered?
11. How would you set up a plain column-and-knee milling machine to make it suitable for milling the flutes on a large twist drill?
12. How would you set up a horizontal milling machine to cut both sides of a block of metal simultaneously?
13. What is the distinctive feature of a column-and-knee type milling machine?
14. Explain how controlled movements of the work in three mutually perpendicular directions are obtained in column-and-knee-type milling machines.
15. What is the function of the rate ring in the tracer milling machine?
16. The Bridgeport vertical spindle milling machine is perhaps the single most popular machine tool. Virtually every factory (or shop) that does machining has one or more of these machines. What made it so popular and widely copied?
17. What is the basic principle of a universal dividing head?
18. The input end of a universal dividing head can be connected to the feed screw of the milling machine table. For what purpose?
19. What is the purpose of the hole-circle plate on a universal dividing head?
20. Explain how a standard universal dividing head, having hole circles of 21, 24, 27, 30, and 32 holes, would be operated to cut an 18-tooth gear.
21. Explain how table feed (ipm) and spindle rpm are specified or computed for a milling machine.
22. Why must the number of teeth on the cutter be known when calculating milling machine table feed, in in./min?
23. Why is the question of up or down milling more critical in horizontal slab milling than in vertical spindle (end or face) milling?
24. Suppose you wanted to machine a cast iron with BHN of 275. The process to be used was face milling and a HSS cutter was going to be used. What feed and speed values would you select?

■ PROBLEMS

1. In Question 24, you were asked to select a feed per tooth and a cutting speed for a milling process. Reasonable values for feed and speed are 0.005 to 0.010 in. per tooth and 200 sfpm. Compute the input values for the machine tool. Use the information in Problem 6 as input when needed.
2. How much time will be required to face mill an AISI 1020 steel surface that is 12 in. long and 5 in. wide, using a 6-in. diameter, eight-tooth tungsten carbide inserted-tooth cutter? Select values of feed per tooth and cutting speed from Table 25-1.
3. If the depth of cut is 0.35 in., what is the metal removal rate in Problem 2?
4. Estimate the power required for the operation of Problem 3. Do not forget to consider Figure 25-5.
5. Examine the part shown in Figure 41-22. The slot on the left end must be produced by machining. Provide a process plan (a description of how the part would be machined and the details regarding cutting tools, such as material, sizes, and so on) for specified machine tools, cutting parameters, and any other information needed to make this component.
6. A gray cast iron surface 6 in. wide and 18 in. long may be machined on either a vertical milling machine, using an 8 in.-diameter face mill having 10 inserted HSS teeth, or on a hydraulic shaper with a HSS tool. If milling is used, the feet per tooth is 0.01 in. For

shaping, a feed of 0.015 in. per stroke would be used. Setup time for the milling machine would be 60 min and for the shaper would be 10 min. The lot size is 10 parts. The time required to load and unload either machine is 2 min. Machine-hour charges for the milling machine and shaper would be $24.50 and $16.50, respectively. Labor cost would be $8.75 per hour in each case. Which machine would be more economical for this job?

7. In Problem 6, what percentage of the total time for the job is consumed in nonmetal-removal activities (setup time and part loading/unloading) assuming that 10 parts are to be made?
8. In Figure 21-10, figure on left, the feed is .006 in. per tooth. The cutter is rotating at an rpm which will produce the desired surface cutting speed of 125 sfpm. The cutter diameter is 5 in. The depth of cut is 0.5 in. The block is 2 in. wide. What is the feed, in inches per minute, of the milling machine table?
9. What is the MRR for this situation described in Problem 8? Is up milling or down milling depicted?
10. Suppose you want to do the job described in Problem 6 by slab milling. You have selected a 6-in.-diameter cutter with an eight tooth TiN coated HSS cutter. The cutting speed will be 500 sfpm and the feed per tooth will be 0.010 in./tooth. Determine the input parameters for the machine (rpm of arbor and table feed), then calculate the T_m, and MRR. Compare these answers with what you got for face milling or shaping the block.

www.wiley.com/college/degarmo

CHAPTER 26

SHAPING, PLANING, BROACHING, SAWING, AND FILING

■ **26.1 INTRODUCTION TO SHAPING AND PLANING**

The processes of *shaping* and *planing* are among the oldest single-point machining processes. Shaping has largely been replaced by milling and broaching, as a production process while planing still has applications in producing long flat cuts, like those in the ways of machine tools. From a consideration of the relative motions between the tool and the workpiece, shaping and planing both use a straight-line cutting motion with a single-point cutting tool to generate a flat surface.

In shaping, the workpiece is fed at right angles to the cutting motion between successive strokes of the tool, as shown in Figure 26-1, where f_c is the feed per stroke, V is the cutting speed, and t is the DOC. (In planing, discussed next, the workpiece is reciprocated and the tool is fed at right angles to the cutting motion.) For either shaping or planing, the tool is held in a clapper box which prevents the cutting edge from being damaged on the return stroke of the tool.

In addition to plain flat surfaces, the shapes most commonly produced on the shaper and planer are those illustrated in Figure 26-2. Relatively skilled workers are required to operate shapers and planers, and most of the shapes that can be produced on them also can be made by much more productive processes, such as milling, broaching, or grinding. Consequently, except for certain special types, planers that will do only planing have become obsolete, and shapers are used mainly in tool-and-die work, in very low volume production, or in the manufacture of gear teeth. (See Chapter 30.)

In shaping, the cutting tool is held in the tool post located in the ram, which reciprocates over the work with a forward stroke, cutting at velocity V and a quick return stroke at velocity V_R. The rpm of drive crank (N_s) drives the ram and determines the velocity of the operation. See Figure 26-1d. The stroke ratio

$$R_s = \frac{\text{cutting stroke angle}}{360°} = \frac{200°}{360°} = \frac{5}{9} \qquad (26\text{-}1)$$

The tool is advancing 55% of the time. The number of strokes per minute is N_s, determined by the rpm of the drive crank. Feed, f_c, is in inches per stroke and is at right angles to the cutting direction.

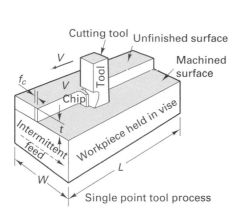

(a) Basic geometry for shaping and planing

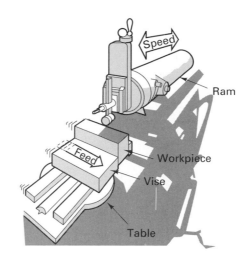

(b) Shaper speed and feed relationship

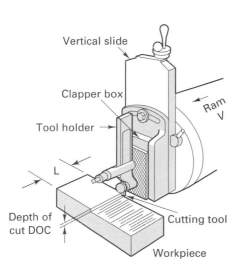

(c) Shaper tool holder, clapper box and workpiece

FIGURE 26-1 Basics of shaping and planing. (a) The cutting speed, V, and feed per stroke f_c. (b) The machine tool. (c) The cutting tool is held in a clapper box so the tool does not damage the workpiece on the return stroke. (d) The ram of the shaper carries the cutting tool at cutting velocity V and reciprocates at velocity V_R by the rotation of a bull wheel turning at rpm N_s.

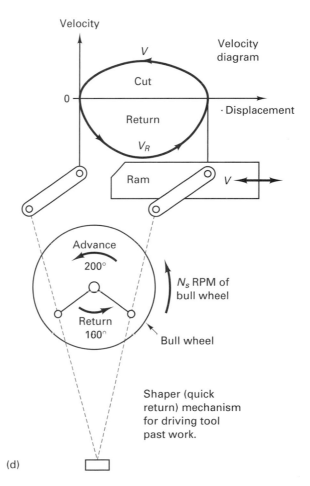

(d)

The length of stroke l must be greater than the length of the workpiece (or length of cut) L, since velocity is position variant. Let l = twice the length of the block being cut or $2L$. The cutting velocity V is assumed to be twice the average forward velocity V, of the ram. The general relationship between cutting speed and rpm is

$$V = \frac{\pi D N_s}{12}\ \text{ft/min} \qquad (26\text{-}2)$$

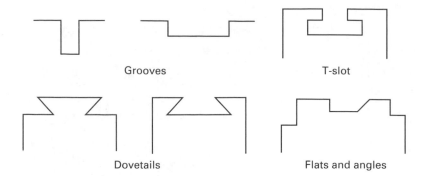

FIGURE 26-2 Types of surfaces commonly machined by shaping and planing.

where D is the diameter (of rotational member) in inches. For shaping, the cutting speed

$$V = \frac{2lN_s}{12R_s} \text{ ft/min} \tag{26-3}$$

Once a cutting speed (V) is selected, the rpm, N_s, of the machine can be calculated. Tables for suggested feed values, f_c, in inches per stroke (or cycle) and recommended depths of cut are also available. The maximum depth of cut is based on the horsepower available to form the chips. This calculation requires that the metal removal rate (MRR) be known. The MRR is the volume of metal removed per unit time:

$$\text{MRR} = \frac{LW\text{DOC}}{T_m} \text{ in}^3/\text{min} \tag{26-4}$$

where W is the width of block being cut and L is the length of block being cut, so volume of cut $= WL\text{DOC}$ where DOC is the depth of cut and T_m is the time in minutes to cut that volume.

In general, T_m is the total length of the cut divided by the feed rate. For shaping, T_m is the width of block divided by the feed rate f_c of the tool moving across the width. Thus, for shaping,

$$T_m = \frac{W}{N_s \times f_c} \text{ min} \tag{26-5}$$

Also,

$$T_m = \frac{S}{N_s} \tag{26-6}$$

where the number of strokes for the job is $S = \dfrac{W}{f_c}$ for a surface of width W.

MACHINE TOOLS FOR SHAPING

Shapers, as machine tools, usually are classified according to their general design features as follows:

1. Horizontal   **2.** Vertical   **3.** Special

   a. Push-cut   a. Regular or slotters

   b. Pull-cut or draw cut shaper   b. Keyseaters

They are also classified as to the type of drive employed: *mechanical drive* or *hydraulic drive*. Most shapers are of the *horizontal push-cut* type (Figure 26-3), where cutting occurs as the ram *pushes* the tool across the work.

On horizontal push-cut shapers, the work is usually held in a heavy vise mounted on the top side of the table. Shaper vises have a very heavy movable jaw, because the vise must often be turned so that the cutting forces are directed against this jaw.

In clamping the workpiece in a shaper vise, care must be exercised to make sure that it rests solidly against the bottom of the vise (on parallel bars) so that it will not be deflected by the cutting force, and that it is held securely yet not distorted by the clamping pressure.

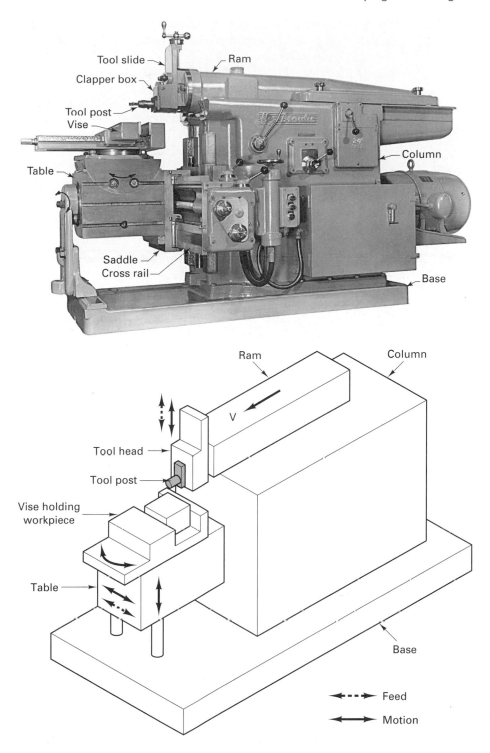

FIGURE 26-3 Details of a horizontal, push cut shaper.

Most shaping is done with simple high-speed steel or carbide-tipped cutting tool bits held in a heavy, forged toolholder. Although shapers are versatile tools, the precision of the work done on them is greatly dependent on the operator. Feed dials on shapers nearly always are graduated in 0.001-in. divisions, and work is seldom done to greater precision than this. A tolerance of 0.002 to 0.003 in. is desirable on parts that are to be machined on a shaper, because this gives some provision for variations due to clamping, possible looseness or deflection of the table, and deflection of the tool and ram during cutting.

PLANING MACHINES

Planing can be used to produce horizontal, vertical, or inclined flat surfaces on large workpieces (too large for shapers). However, planing is much less efficient than other basic machining processes, such as milling, that will produce such surfaces. Consequently, planing and planers have largely been replaced by planer milling machines or machines which can do both milling and planing.

Figure 26-4 shows the basic components and motions of planers. In most planing, the action is opposite to that of shaping. The work is moved past one or more stationary single-point cutting tools. Because a large and heavy workpiece and table must be reciprocated at relatively low speeds, several tool heads are provided, often with multiple tools in each head. In addition, many planers are provided with tool heads arranged so that cuts occur on both directions of the table movement. However, because only single-point cutting tools are used and the cutting speeds are quite low, planers are low in productivity as compared with some other types of machine tools.

Planers are made in four basic types. Figure 26-4 depicts the most common double-housing type. It has a closed housing structure, spanning the reciprocating worktable, with

FIGURE 26-4 Schematic of planers. (a) Double-housing planer; (b) open-sided planer; (c) open-sided planer; (d) interchangeable multiple tool holder for use in planers. *(Photograph courtesy Gebr Boehringer GmbH.)*

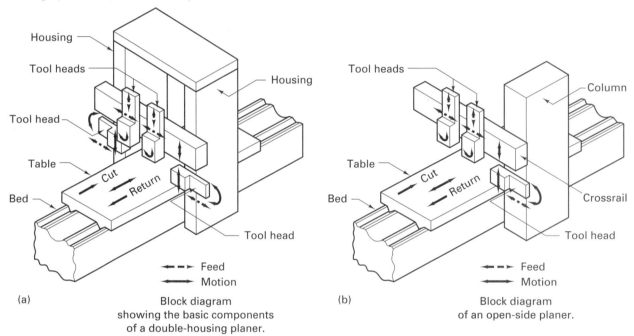

(a)   Block diagram showing the basic components of a double-housing planer.

(b)   Block diagram of an open-side planer.

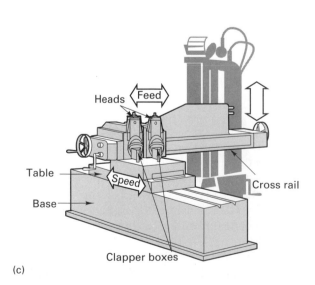

(c)

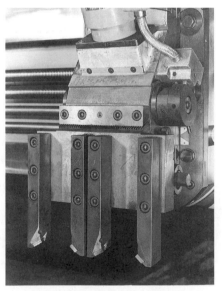

(d)

Four planer tools in toolholder

a cross rail supported at each end on a vertical column and carrying two tool heads. An additional tool head usually is mounted on each column, so that four tools (or four sets) can cut during each stroke of the table. Obviously, the closed-frame structure of this type of planer limits the size of the work that can be machined. Open-side planers have the cross rail supported on a single column. This design provides unrestricted access to one side of the table to permit wider workpieces to be accommodated. Some open-side planers are convertible, in that a second column can be attached to the bed when desired so as to provide added support for the cross rail.

WORKHOLDING AND SETUP ON PLANERS

Workpieces in planers are usually large and heavy. They must be securely clamped to resist large cutting forces and the high-inertia forces that result from the rapid velocity changes at the ends of the strokes. Special stops are provided at each end of the workpiece to prevent it from shifting.

Considerable time usually is required to set up the planer, thus reducing the time the machine is available for producing chips. Sometimes special setup plates are used for quick setup of the workpiece. Another procedure is to use two tables. Work is set up on one table while another workpiece is being machined on the other. The tables can be fastened together for machining long workpieces.

The large workpieces can usually support heavy cutting forces, so large depths of cut are recommended, which decrease the cutting time. Consequently, planer tools usually are quite massive and can sustain the large cutting forces. Usually, the main shank of the tools is made of plain-carbon steel, with tips made from high-speed steel or carbide. Chip breakers should be used to avoid long and dangerous chips in ductile materials.

Theoretically, planers have about the same precision as shapers. The feed and other dimension-controlling dials usually are graduated in 0.001-in. divisions. However, because larger and heavier workpieces are usually involved, and much longer beds and tables, the working tolerances for planer work are somewhat greater than for shaping.

■ 26.2 INTRODUCTION TO BROACHING

The process of *broaching* is one of the most productive of the basic machining processes. The machine tool is called a *broaching machine* or a *broach* and the cutting tool is also called the *broach*. Figure 26-5 shows the basic shape of a conventional pull broach. Broaching competes economically with milling and boring and is capable of producing

FIGURE 26-5 (a) Photo of pull broach. (b) Basic shape and nomenclature for a conventional pull (hole) broach. Section A-A' shows the cross section of a tooth.
P–pitch
n_r–number of roughing teeth
n_s–number of semifinishing teeth
n_f–number of finishing teeth

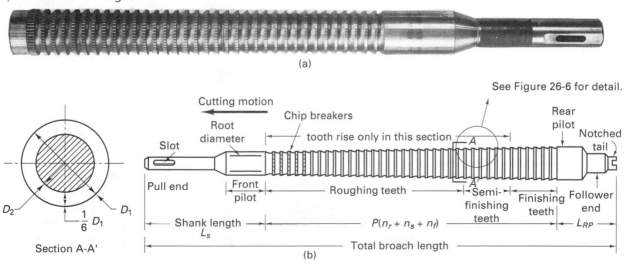

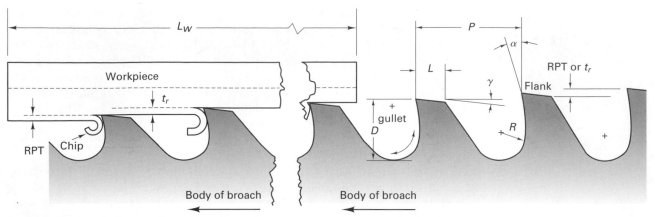

FIGURE 26-6 The feed in broaching depends on the rise per tooth t_r (RPT). The sum of the RPT gives the depth of cut, DOC. P–pitch of teeth, D–depth of teeth (0.4P), L–land behind cutting edge (0.25P), R–radius of gullet (.25P), α–hook angle or rake angle, γ–backoff angle or clearance angle.

precision-machined surfaces. The broach finishes an entire surface in a single pass. Broaches are used in production to finish holes, splines, and flat surfaces. Typical workpieces include small to medium sized castings, forgings, screw machine parts, and stampings.

The feed per tooth in broaching is the change in height of successive teeth. This is called the rise per tooth (RPT or t_r, see Figure 26-6). Broaching looks similar to sawing except that the saw makes many passes through the cut, whereas the broach produces a finished part in one pass. The heart of this process lies in the broaching tool, in which roughing, semifinishing, and finishing teeth are combined into one tool as shown in Figure 26-5. Broaching is unique in that it is the only one of the basic machining processes in which *feed*, which determines the chip thickness, is built into the cutting tool. The machined surface is always the inverse of the profile of the broach, and, in most cases, it is produced with a single linear stroke of the tool across the workpiece (or the workpiece across the broach).

A broach is composed of a series of teeth, each tooth standing slightly higher than the last. This rise per tooth (RPT) also known as *step* or the feed per tooth, determines the amount of material removed. There is no feeding of the broaching tool required. The frontal contour of the teeth determines the shape of the resulting machined surface. As the result of these conditions built into the tool, no complex motion of the tool relative to the workpiece is required and the need for highly skilled machine operators is minimized.

Figure 26-7 shows a *pull broach* in a vertical pull-down broaching machine. The pull end of the broach is passed through the part and a key mates to the slot. The broach is pulled through the part. The broach is retracted (pulled up) out of the part. The part is transferred from the left fixture to the right fixture. One finished part is completed in every time cycle.

■ 26.3 FUNDAMENTALS OF BROACHING

In broaching, the tool (or work) is translated past the work (or tool) with a single stroke of velocity V. The feed is provided by a gradual increase in height of successive teeth. The rise per tooth varies depending on whether the tooth is for roughing (t_r), semifinishing (t_s), or final sizing or finishing (t_f). In a typical broach there are three to five semifinishing and finishing teeth specified. The number of roughing teeth must be determined so that broach length, which is needed to estimate the cutting time, can be calculated. Other lengths needed for a typical pull broach are shown in Figure 26-6. The chip breakers in the first section of roughing teeth may be extended to more teeth if the cut is heavy or material difficult to machine. The distance between the teeth, called the pitch, P, is important because it determines the tooth construction and strengths and the number of teeth actually cutting at a given instant. It is preferable that at least two teeth are in contact with the workpiece at any instant.

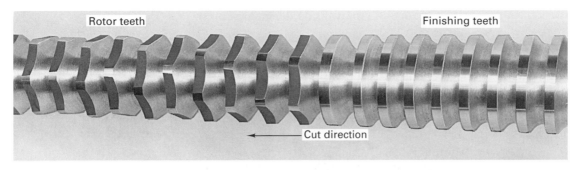

(a) Rotor or jump tooth broach design.

(b) Round, push-type broach with chip-breaking notches on alternate teeth except at the finishing end.

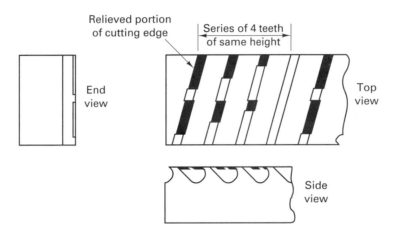

(c) Notched tooth, flat broach

(d) Progressive surface broach. *(Courtesy of Detroit Broach & Machine Company)*

FIGURE 26-7 Methods to decrease force or break up chip rings in broaches. (a) Rotor or jump tooth; (b) notched tooth, round; (c) notched tooth, flat design overlapping teeth permits large RPTs without increasing chip load; (d) progressive tooth design for flat broach.

The pitch or distance between teeth is

$$P \cong .35\sqrt{L_w} \qquad (26\text{-}7)$$

where length of cut usually equals L_w as shown in Figure 26-5.

The number of roughing teeth is

$$n_r = \frac{\text{DOC} - n_f t_f - n_s t_s}{t_r}$$

where DOC is the total amount of metal to be removed and t_r is the rise per tooth.

The overall length of the broach $L_B = (n_r + n_s + n_f)P + L_s + L_{RP}$ for pull broach.

The length of stroke $L = L_B - L_w$, in inches, if the broach moves past work, or $L_w + L_B$ if the work moves past the broach. The cutting time

$$T_m = \frac{L}{12V} \qquad (26\text{-}8)$$

where V is the cutting speed, in surface feet per minute.

The metal removal rate depends upon the number of teeth (roughing) contacting the work.

$$\text{MRR (per tooth)} = 12t_r WV \text{ in}^3/\text{min per roughing tooth}, \quad (26\text{-}9)$$

where W is the width of broach tooth.

The number of roughing teeth in contact with part $n \cong L_w/P$ for a broach longer than the part.

$$\text{MRR (for process)} = 12t_r WVn \text{ in}^3/\text{min} \quad (26\text{-}10)$$

where n is usually rounded off to next largest whole number.

The pull broach must be strong enough so that it will not be pulled apart. The strength of a pull broach is determined by its minimum cross section, which occurs either at the root of the first tooth or at the pull end:

$$\text{allowable pull} = \frac{\text{area of minimum section} \times \text{Y.S. of broach material}}{\text{factor of safety}} \quad (26\text{-}11)$$

The *push broach* must be strong enough so that it will not buckle. If the length-to-diameter ratio, L/D_r, is greater than 25, the broach must be considered a long column which can buckle if overloaded. Let

L = length from push end to first tooth

D_r = root diameter at $0.5L$

S = factor of safety

$$\text{allowance load} = \frac{13.5 \times 10^6 \times D_r^4}{SL^2} \quad (26\text{-}12)$$

For L/D_r less than 25, the normal broach loads are not critical.

Calculation of the total push or pull load depends upon the number of teeth engaged, n estimated from L_w/P; the width of the cut, W; the RPT per tooth engaged, t_r; and the shear strength of the metal being machined.

The force necessary to operate a broach depends upon the material being broached, the conditions of the tool, and the nature of the process. An empirical constant is required in the force calculation to account for the large amount of rubbing (friction) between the tool, the chips captured in the tooth gullet, and the workpiece.

Let F_{CB} be the broach pull force in pounds:

$$F_{CB} \cong 5\tau_s nt_r W \quad (26\text{-}13)$$

τ_s is found in Chapter 21 depending upon the BHN for the metal. This force estimate can be used to estimate the horsepower needed for the broaching machine.

THE ADVANTAGES AND LIMITATIONS OF BROACHING

Because of the features built into a broach, it is a simple and rapid method of machining. There is a close relationship between the contour of the surface to be produced, the amount of material that must be removed, and the design of the broach. For example, the total depth of the material to be removed cannot exceed the total step provided in the broach, and the step of each tooth must be sufficient to provide proper chip thickness for the type of material to be machined. Consequently, either a special broach must be made for each job, or the workpiece must be designed so that a standard broach can be used. Broaching is widely used and particularly well suited for mass production because the volume can easily justify the cost of the broaching tool which can be easily $15,000 to $30,000 per tool. It is also used for certain simple and standardized shapes, such as keyways, where inexpensive standard broaches can be used.

Broaching originally was developed for machining internal keyways. However, its obvious advantages quickly led to its development for mass-production machining of various surfaces, such as flat, interior or exterior, cylindrical and semicylindrical and many irregular surfaces. Because there are few limitations as to the contour form that broach teeth may have, there is almost no limitation in the shape of surfaces that can be produced by broaching. The only physical limitations are that there must be no obstruction to interfere with the passage of the entire tool over the surface to be machined and that the workpiece must be strong enough to withstand the forces involved. In internal broaching, a hole must exist in the workpiece into which the broach may enter. Such a hole can be made by drilling, boring, or coring.

Broaching usually produces better accuracy and finish than can be obtained by drilling, boring, or reaming. Although the relative motion between the broaching tool and the work usually is a single linear one, a rotational motion can be added to permit the broaching of spiral splines or gun-barrel rifling.

BROACH DESIGN

Broaches commonly are classified by design features as shown below:

Purpose	Motion	Construction	Function
Single	Push	Solid	Roughing
Combination	Pull	Built-up	Sizing
	Stationary		Burnishing

Figure 26-5 shows the principal components of a pull broach and the shape and arrangement of the teeth. Each tooth is essentially a single-edge cutting tool, arranged much like the teeth on a saw except for the step, which determines the depth cut by each tooth as shown in Figure 26-6. The rise per tooth, which determines the chip load, varies from about 0.006 in. for roughing teeth in machining free-cutting steel to a minimum 0.001 in. for finishing teeth. Typically the RPT is 0.003 to 0.006 in. in surface broaching and 0.0012 to .0025 in. on the diameter for internal broaching. The exact amount depends on several factors. Too-large cuts impose undue stresses on the teeth and the work; too-small cuts result in rubbing rather than cutting action. The strength and ductility of the metal being cut are the primary factors.

Where it is desirable for each tooth to take a deep cut, as in broaching castings or forgings that have a hard, abrasive surface layer, *rotor-cut* or *jump-cut* tooth design may be used (Figure 26-7). In this design, two or three teeth in succession have the same diameter, or height, but each tooth of the group is notched or cut away so that it cuts only a portion of the circumference or width. This permits deeper but narrower cuts by each tooth without increasing the total load per tooth. This tooth design also reduces the forces and the power requirements. Chip-breaker notches are also used on round broaches to break up the chips (Figure 26-7b).

A similar idea can be used for flat surfaces. Tooth loads and cutting forces also can be reduced by using the *double-cut* construction shown in Figure 26-7c. Four consecutive teeth get progressively wider. The teeth remove metal over only a portion of their width until the fourth tooth completes the cut.

Another technique for reducing tooth loads utilizes the principle illustrated in Figure 26-7d. Employed primarily for broaching wide, flat surfaces, the first few teeth in *progressive* broaches completely machine the center, while succeeding teeth are offset in two groups to complete the remainder of the surface. Rotor, double-cut, and progressive designs require the broach to be made longer than if normal teeth were used, and they therefore can be used only on a machine having adequate stroke length.

The cutting edges of the teeth on surface broaches may be either normal to the direction of motion or at an angle of from 5 to 20°. The latter, *shear-cut*, broaches provide smoother cutting action with less tendency to vibrate. Other shapes that can be broached are shown in Figure 26-8 along with push- or pull-type broaches used for the job.

The pitch of the teeth and the gullet between them must be sufficient to provide ample room for the chips. All chips produced by a given tooth during its passage over the full length of the workpiece must be contained in the space between successive teeth. At the same time, it is desirable to have the pitch sufficiently small so that at least two or three teeth are cutting at all times.

The *hook* determines the primary rake angle and is a function of the material being cut. It is 15 to 20° for steel and 6 to 8° for cast iron. *Back-off* or end clearance angles are from 1 to 3° to prevent rubbing.

Most of the metal removal is done by the *roughing teeth*. *Semifinishing teeth* provide surface smoothness, whereas *finishing teeth* produce exact size. On a new broach all the finishing teeth usually are made the same size. As the first finishing teeth become

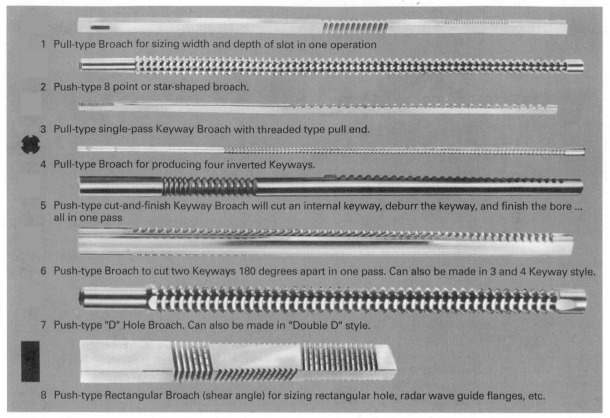

1 Pull-type Broach for sizing width and depth of slot in one operation

2 Push-type 8 point or star-shaped broach.

3 Pull-type single-pass Keyway Broach with threaded type pull end.

4 Pull-type Broach for producing four inverted Keyways.

5 Push-type cut-and-finish Keyway Broach will cut an internal keyway, deburr the keyway, and finish the bore ... all in one pass

6 Push-type Broach to cut two Keyways 180 degrees apart in one pass. Can also be made in 3 and 4 Keyway style.

7 Push-type "D" Hole Broach. Can also be made in "Double D" style.

8 Push-type Rectangular Broach (shear angle) for sizing rectangular hole, radar wave guide flanges, etc.

FIGURE 26-8 Examples of push-and-pull-type broaches. *(Courtesy of DuMont Corporation.)*

worn, those behind continue the sizing function. On some round broaches, *burnishing teeth* are provided for finishing. These have no cutting edges but are rounded disks that are from 0.001 to 0.003 in. larger than the size of the hole. The resulting rubbing action smooths and sizes the hole. They are used primarily on cast iron and nonferrous metals.

The *pull end* of a broach provides a means of quickly attaching the broach to the pulling mechanism. The *front pilot* aligns the broach in the hole before it begins to cut, and the *rear pilot* keeps the tool square with the finished hole as it leaves the workpiece. *Shank length* must be sufficient to permit the broach to pass through the workpiece and be attached to the puller before the first roughing tooth engages the work. If a broach is to be used on a vertical machine that has a tool-handling mechanism, a *tail* is necessary.

A broach should not be used to remove a greater depth of metal than that for which it is designed—the sum of the steps of all the teeth. In designing workpieces, a minimum of 0.020 in. should be provided on surfaces that are to be broached, and about 0.025 in. is the practical maximum.

BROACHING SPEEDS, ACCURACY, FINISH

Depending on the metal being cut, cutting speeds for broaching range from low (25 to 20 sfpm) to high (120 ft/min) while completing the surface in a single stroke, so the productivity is high. A complete cycle usually requires only from 5 to 30 seconds, with most of that time being taken up by the return stroke, broach handling, and workpiece loading and unloading. Such cutting conditions facilitate cooling and lubrication and result in very low tool wear rates, which reduce the necessity for frequent resharpening and prolong the life of the expensive broaching tool.

For a given cutting speed and material, the force required to pull or push a broach is a function of the tooth width, the step, and the number of teeth cutting. Consequently, it is necessary to design or specify a broach within the stroke length and power limitations of the machine on which it is to be used. The average machining precision is typically ±0.001 in. (±0.02 mm) tolerance with surface finish 120 to 60 RMS or better. Burrs are minimal on the exit side of cuts.

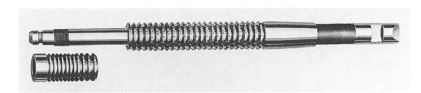

FIGURE 26-9 Shell construction for a pull broach.

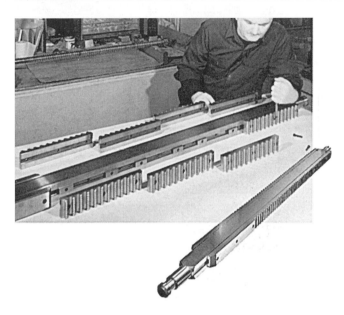

FIGURE 26-10 A modularly constructed broach is cheaper to build and can be sharpened in sections.

BROACHING MATERIALS AND CONSTRUCTION

Because of the low cutting speeds employed, most broaches are made of alloy or high-speed tool steel. Carbide-tipped broaches are seldom used for machining steel parts or forgings, as the cutting edges tend to chip on the first stroke, probably due to a lack of rigidity in the machine tool/cutting tool combination. TiN coating of HSS broaches are becoming more common, greatly prolonging the life of broaches. When they are used in continuous mass-production machines, particularly in surface broaching of cast iron, tungsten carbide teeth may be used, permitting the broach to be used for long periods of time without resharpening.

Internal broaches are of solid construction, or quite often they are made of *shells* mounted on an arbor (Figure 26-9). When the broach (or a section of it) is subject to rapid wear, a single shell can be replaced. This will be much cheaper than replacing an entire solid broach. Shell construction, however, is initially more expensive than a solid broach of comparable size.

Small surface broaches may be of solid construction, but larger ones usually use modular construction (Figure 26-10). Building in sections makes the broach easier and cheaper to construct and sharpen. It also often provides some degree of interchangeability of the sections for different parts, bringing down the tool cost significantly.

SHARPENING BROACHES

Most broaches are resharpened by grinding the hook faces of the teeth. The lands of internal broaches must not be reground because this would change the size of the broach. Lands of flat surface broaches sometimes are ground, in which case all of them must be ground to maintain their proper relationship.

■ 26.4 BROACHING MACHINES

Because all the factors that determine the shape of the machined surface and that determine all cutting conditions except speed are built into the broaching tool, broaching machines are relatively simple. Their primary functions are to impart plain reciprocating motion to the broach and to provide a means for handling the broach automatically.

Most broaching machines are driven hydraulically, although mechanical drive is used in a few special types. The major classification relates to whether the motion of the broach is vertical or horizontal, as given in Table 26-1.

TABLE 26-1.	Broaching Machines
Vertical	
Push-broaching	Arbor press with guided ram 5–50 ton capacity Internal broaching
Pull-down	Double ram design most common Long changeover times
Pull-up	Ram above table pulling broach up Machines with multiple rams common
Surface	No handling of broach Multiple slides
Horizontal	**Short Cycle Times**
Pull	Longer strokes and broaches Basically vertical machines laid on side
Surface	Broaches stationary, work moves on conveyor Work held in fixtures
Continuous	Conveyor chain holds fixtures
Rotary	Broach stationary, work translates beneath tool Work held in fixtures

The choice between vertical and horizontal machines is determined primarily by the length of the stroke required and the available floor space. Vertical machines seldom have strokes greater than 60 in. because of height limitations. Horizontal machines can have almost any length of stroke, but they require greater floor space. The most common machine is the vertical pull-down machine. The worktable, usually having a spherical-seated workholder, sits below the broach elevator with a pulling mechanism below the table. As shown in Figure 26-11, when the elevator raises the broach above the table,

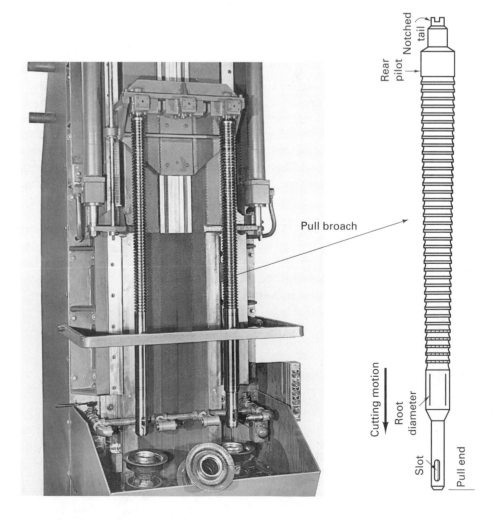

FIGURE 26-11 Vertical pull-down broaching machine shown with parts in position ready for the two broaches to be inserted. An extra part is shown lying at the front of the machine.

the work can be placed into position. The elevator then lowers the pilot end of the broach through the hole in the workpiece, where it is engaged by the puller. The elevator then releases the upper end of the broach, and it is pulled through the workpiece. The workpieces are removed from the table, and the broach is raised upward to be engaged by the elevator mechanism. In some machines with two rams, one broach is being pulled down while the work is being unloaded and the broach raised at the other station. In Figure 26-11, the part is being broached in two passes, first on the left, then on the right.

■ 26.5   INTRODUCTION TO SAWING

Sawing is a basic machining process in which chips are produced by a succession of small cutting edges, or *teeth*, arranged in a narrow line on a saw "blade." As shown in Figure 26-12, each tooth forms a chip progressively as it passes through the workpiece, and the chip is contained within the space between two successive teeth until these teeth pass from the work. Because sections of considerable size can be severed from the workpiece with the removal of only a small amount of the material in the form of chips, sawing is probably the most economical of the basic machining processes with respect to the waste of material and power consumption, and in many cases with respect to labor.

In recent years vast improvements have been made in saw blades and sawing machines, resulting in improved accuracy and precision of the process. Most sawing is done to sever bar stock and shapes into desired lengths for use in other operations. There are many cases in which sawing is used to produce desired shapes. Frequently, and especially for producing only a few parts, contour sawing may be more economical than any other machining process.

■ 26.6   SAW BLADES

Saw blades are made in three basic configurations. The first, commonly called a *hacksaw* blade, is straight, relatively rigid, and of limited length with teeth on one edge. The second is sufficiently flexible so that a long length can be formed into a continuous band with teeth on one edge; these are known as *bandsaw blades*. The third form is a rigid disk having teeth on the periphery; these are called *circular saws* or *cold saws*. Figure 26-13 gives the standard nomenclature for a saw blade.

All saw blades have certain common and basic features: (1) material, (2) tooth form, (3) tooth spacing, (4) tooth set, and (5) blade thickness or gage. Small hacksaw blades usually are made entirely of tungsten or molybdenum high-speed steel. Blades for power-operated hacksaws often are made with teeth cut from a strip of high-speed steel that has been electron-beam-welded to the heavy main portion of the blade, which is made from a

FIGURE 26-12   Formation of chips in sawing.

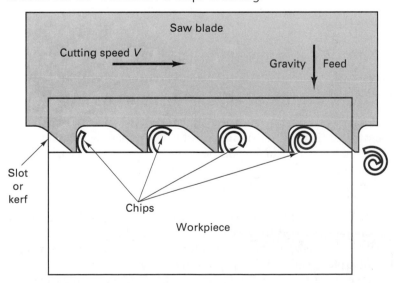
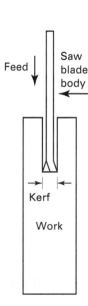

MATRIX MODIFIED MIX-TOOTH	M-42 COBALT WELDED-EDGE	M-2 HIGH-SPEED WELDED-EDGE	HARD BACK CARBON	FLEXIBLE BACK CARBON
The best all-purpose welded-edge blade for sawing varying sizes, shapes and cross-sections. Cobalt-tough for cutting wide range of materials. Welded to length and coil stock.	For high production cutting of solids, super-alloys, tool steels, high temperature alloys. Welded to length and coil stock.	The original and widely used welded-edge band blade for general purpose sawing. Welded to length and coil stock.	Hardened back provides greater beam strength for more accurate sawing. Welded to length and coil stock.	Recommended for Contour Saws running over 3000 SFPM. Welded to length and coil stock.

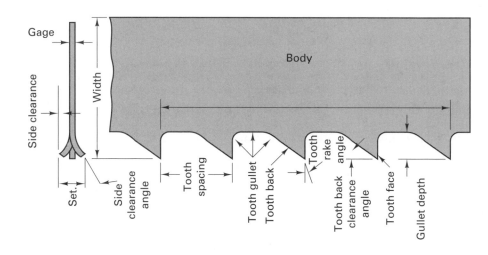

Standard design has zero rake

Skip-tooth blade clears chips, cuts non-ferrous

Hook tooth with 10-deg rake for large sections

Variable pitch can change by section or individually

Variable pitch with 5-deg rake is more aggressive

Variable pitch with 10-deg rake sheds chips better

Tooth Set

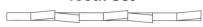

Raker set has a straight tooth between one left and one right

Wavy set for thin sections has progressive set, both directions

Straight set—left, then right—is for better finish

Cluster set has only a few straight teeth

FIGURE 26-13 Bandsaw blade designs and nomenclature (*above*). Tooth set patterns (*left*) and tooth designs (*right*).

tougher and cheaper alloy steel (Figure 26-13). Bandsaw blades frequently are made with this same type of construction but with the main portion of the blade made of relatively thin, high-tensile-strength alloy steel to provide the required flexibility. Bandsaw blades also are available with tungsten carbide teeth and TiN coatings. The three most common *tooth forms* are regular, skip tooth, and hook. *Tooth spacing* is very important in all sawing because it determines three factors. First, it controls the size of the teeth. From the viewpoint of strength, large teeth are desirable. Second, tooth spacing determines the space (*gullet*) available to contain the chip that is formed. The chip cannot drop from this space until it emerges from the slot cut in the workpiece, called the *kerf*. *Tooth set* refers to the manner in which the teeth are offset from the centerline in order to make the kerf wider than the gage (the thickness of the back) of the blade. This permits the saw to move more freely in the kerf and reduces friction and heating. The kerf-gullet space must be such that there is no crowding of the chip and no tendency for chips to become wedged between the teeth and not drop out when the saw emerges from the cut.

Third, tooth spacing determines how many teeth will bear against the work. This is very important in cutting thin material, such as tubing. At least two teeth should be in contact with the work at all times. If the teeth are too coarse, only one tooth rests on the work at a given time, permitting the saw to rock, and the teeth may be stripped from the saw. Hand hacksaw blades have 14 to 32 teeth per inch. In order to make it easier to start a cut, some hand hacksaw blades are made with a short section at the forward end having teeth of a special form with negative rake angles. Tooth spacing for power hacksaw blades ranges from 4 to 18 teeth per inch.

Raker-tooth saws are used in cutting most steel and iron. *Straight-set teeth* are used for sawing brass, copper, and plastics. Saws with *wave-set teeth* are used primarily for cutting thin sheets and thin-walled tubing.

The gage or *blade thickness* of nearly all hand hacksaw blades is 0.025 in. Saw blades for power hacksaws vary in thickness from 0.050 to 0.100 in. Hand hacksaw blades come in two standard lengths, 10 and 12 in. All are $\frac{1}{2}$ in. wide. Blades for power hacksaws vary in length from 12 to 24 in. and in width from 1 to 2 in. Wider and thicker blades are desirable for heavy-duty work. The blade should be at least twice as long as the maximum length of cut that is to be made.

Bandsaw blades are available in straight, raker, wave, or combination sets. In order to reduce the noise from high-speed bandsawing, it is becoming increasingly common to use blades that have more than one pitch, size of teeth, and type of set. Blade width is very important in bandsawing because it determines the minimum radius that can be cut. The most common widths are from $\frac{1}{16}$ to $\frac{1}{2}$ in., although wider blades can be obtained. Because wider blades are stronger, as wide a blade as possible should be used. However, cutting small radii requires a narrower and weaker blade. Bandsaw blades come in tooth spacings from 2 to 32 teeth per inch.

Circular saws for cutting metal are often called *cold saws* to distinguish them from friction-type disk saws, which heat the metal to the melting temperature at the point of metal removal. Cold saws cut rapidly and produce chips like a milling cutter while producing surfaces that are comparable in smoothness and accuracy with surfaces made by slitting saws in a milling machine or by a cutoff tool in a lathe.

Disk or *circular saws* necessarily differ somewhat from straight blade forms. Because they must be relatively large in comparison with the work, only the sizes up to about 18 in. in diameter have teeth that are cut into the disk (Figure 26-14). Larger saws

FIGURE 26-14 Circular sawing a structural shape, using (*left* to *right*) an insert tooth, a segmental tooth, and an integral-tooth circular saw blade.

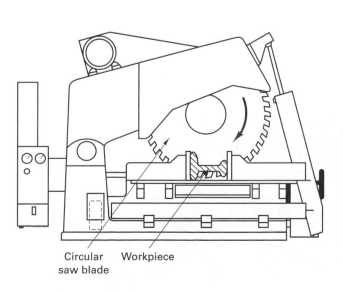

Circular
saw blade   Workpiece

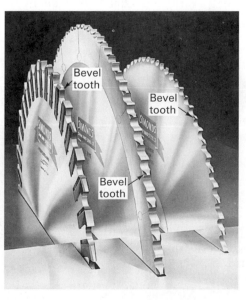

use either *segmented* or *inserted* teeth. The teeth are made of high-speed steel, or tungsten carbide. The remainder of the disk is made of ordinary, less expensive, and tougher steel. *Segmental* blades are composed of segments mounted around the periphery of the disk, usually fitted with a tongue and groove and fastened by means of screws or rivets. Each segment contains several teeth. If a single tooth is broken, only one segment need be replaced to restore the saw to an operating condition.

A common tooth form used in circular saws has every other tooth beveled on both sides. Sometimes the first tooth is beveled on the left side, the second tooth on both sides, the third tooth on the right side, the fourth tooth on the left side, and so forth. Another method is to bevel the opposite sides of successive teeth. Beveling is done to produce a smoother cut. Precision circular saws made from carbide are becoming available which are very thin (0.03 in.) and have high cutting-off accuracy, around ±0.0008 in., with negligible burrs.

■ 26.7 TYPES OF SAWING MACHINES

Metal-sawing machines may be classified as follows:

1. Reciprocating saw
   a. Manual hacksaw
   b. Power hacksaw (Figure 26-15)

2. Bandsaw
   a. Vertical cutoff (Figure 26-16)
   b. Horizontal cutoff (Figure 26-17)
   c. Combination cutoff and contour (Figure 26-18)
   d. Friction

3. Circular saw (Figure 26-14)
   a. Cold saw
   b. Steel friction disc
   c. Abrasive disc

POWER HACKSAWS

As the name implies, power hacksaws are machines that mechanically reciprocate a large hacksaw blade (Figure 26-15). These machines consist of a bed, a workholding frame, a power mechanism for *reciprocating* the saw frame, and some type of feeding mechanism. Because of the inherent inefficiency of cutting in only one stroke direction, they have often been replaced by more efficient, horizontal bandsawing machines.

BANDSAWING MACHINES

The earliest metal-cutting bandsawing machines were direct adaptations from wood-cutting bandsaws. Modern machines of this type are much more sophisticated and versatile and have been developed specifically for metal cutting. To a large degree they were made possible by the development of vastly better and more flexible bandsaw

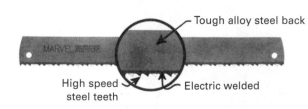

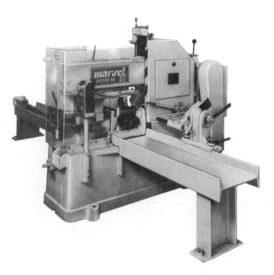

FIGURE 26-15 Power hacksaw blade (*above*) and hacksaw with automatic bar feeding (*right*) cutting two pieces of round stock.

blades and simple flash-welding equipment, which can weld the two ends of a strip of bandsaw blade together to form a band of any desired length. Three basic types of band-sawing machines are in common use.

Upright, cutoff, bandsawing machines (Figure 26-16) are designed primarily for cutoff work on single stationary workpieces that can be held on a table. On many

FIGURE 26-16 Vertical bandsaw cutting a piece of pipe, showing head tilted 45°.

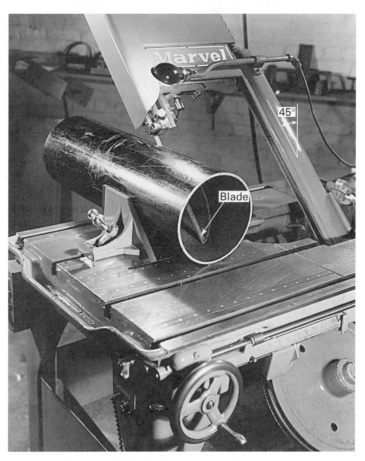

FIGURE 26-17 Front view and rear view of a horizontal bandsawing machine. Inset shows blade changing operation.

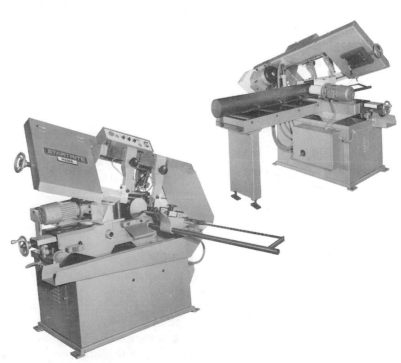

Blade changing

Easy blade loading from the top on all models and quick removal of guards.

FIGURE 26-18 Contour bandsawing on vertical bandsawing machine, shown in inset.

machines the blade mechanism can be tilted to about 45°, as shown, to permit cutting at an angle. They usually have automatic power feed of the blade into the work, automatic stops, and provision for supplying coolant.

Horizontal metal-cutting bandsawing machines were developed to combine the flexibility of reciprocating power hacksaws and the continuous cutting action of vertical bandsaws. These heavy-duty automatic bandsaws feed the saw vertically by a hydraulic mechanism and have automatic stock feed that can be set to feed the stock laterally any desired distance after a cut is completed and automatically clamp it for the next cut. Such machines can be arranged to hold, clamp, and cut several bars of material simultaneously. CNC bandsaws are available with automatic storage and retrieval systems for the bar stock. Smaller and less expensive types have swing-frame construction, with the bandsaw head mounted in a pivot on the rear of the machine. Feed is accomplished by gravity through rotation of the head about the pivot point. Because of their continuous cutting action, horizontal bandsawing machines are very efficient (Figure 26-17).

Combination cutoff and contour bandsawing machines (Figure 26-18) can be used not only for cutoff work but also for contour sawing. They are widely used for cutting irregular shapes in connection with making dies and the production of small numbers of parts and are often equipped with rotary tables. Additional features on these machines include a table that pivots so that it can be tilted to any angle up to 45°. Usually, these machines have a small flash butt welder on the vertical column, so that a straight length of bandsaw blade can be welded quickly into a continuous band. A small grinding wheel is located beneath the welder so that the flash can be ground from the weld to provide a smooth joint that will pass through the saw guides. This welding and grinding unit makes it possible to cut internal openings in a part by first drilling a hole, inserting one end of the saw blade through the hole, and then butt-welding the two ends together. When the cut is finished, the band is again cut apart and removed from the opening. The cutting speed of the saw blade can be varied continuously over a wide range to provide correct operating conditions for any material. A method of power feeding the work is provided, sometimes gravity-actuated.

Contour-sawing machines are made in a wide range of sizes, the principal size dimension being the throat depth. Sizes from 12 to 72 in. are available. The speeds available on most machines range from about 50 to 2000 ft/min. Modern horizontal bandsaws are accurate to ±0.002 in. per vertical inch of cut but have feeding accuracy of only ±0.005 in., subject to the size of the stock and the feed rate. Repeatability from one feed to the next may be ±0.010 to ±0.020 in.

CNC controlled sawing centers with microprocessor controls have opened up new automation aspects for sawing. Such control systems can improve accuracy to within ±0.005 in. over entire cuts by controlling saw speed, blade feed pressure, and feed rate.

Special bandsawing machines are available with very high speed ranges, up to 14,000 ft/min. These are known as *friction bandsawing machines*. Material is not cut by chip formation. Instead, the friction between the rapidly moving saw blade and the work is sufficient to raise the temperature of the material at the end of the kerf to or just below the melting point, where its strength is very low. The saw blade then pulls the molten, or weakened, material out of the kerf. Consequently, the blades do not need to be sharp; they frequently have no teeth, only occasional notches in the blade to aid in removing the metal.

Almost any material, including ceramics, can be cut by friction sawing. Because only a small portion of the blade is in contact with the work for an instant and then is cooled by its passage through the air, it remains cool. Usually, the major portion of the work, away from contact with the saw blade, also remains quite cool. The metal adjacent to the kerf is heat-affected, recast, and sometimes harder than the bulk metal. It is also a very rapid method for trimming the flash from sheet-metal parts and castings.

CUTTING FLUIDS

Cutting fluids should be used for all bandsawing, with the exception that cast iron is always cut dry. Commercially available oils or light cutting oils will give good results in cutting ferrous materials. Beeswax or paraffin are common lubricants for cutting aluminum and aluminum alloys.

FEEDS AND SPEEDS

Because of the many different types of feed involved in bandsawing, it is not practical to provide tabular feed or pressure data. Under general conditions, however, an even pressure, without forcing the work, gives best results. A nicely curled chip usually indicates an ideal feed pressure. Burned or discolored chips indicate excessive pressure, which can cause tooth breakage and premature wear.

Most bandsaws provide recommended cutting speed information right on the machine, depending upon the material being sawed. In general, HSS blades are run at 200 to 300 ft/min when cutting 1-in.-thick, low- and medium-carbon steels. For high-carbon steels, alloy steels, and tool and die steels, the range is from 150 to 225 ft/min, and most stainless steels are cut at 100 to 125 ft/min.

CIRCULAR-BLADE SAWING MACHINES

Machines employing rotating circular or cold saw blades are used exclusively for cutoff work. These range from small, simple types, in which the saw is fed manually, to very large saws having power feed and built-in coolant systems, commonly used for cutting off hot-rolled shapes as they come from a rolling mill. In some cases friction saws are used for this purpose, having disks up to 6 ft in diameter and operating at surface speeds up to 25,000 ft/min. Steel sections up to 24 in. can be cut in less than 1 minute by this technique.

Although technically not a sawing operation, cutoff work up to about 6 in. is often done utilizing thin *abrasive* disks. The equipment used is the same as for sawing. It has the advantage that very hard materials that would be very difficult to saw can be cut readily. A thin rubber- or resinoid-bonded abrasive wheel is used. Usually a somewhat smoother surface is produced.

■ 26.8   INTRODUCTION TO FILING

Basically, the metal-removing action in filing is the same as in sawing, in that chips are removed by cutting teeth that are arranged in succession along the same plane on the surface of a tool, called a *file*. There are two differences: (1) the chips are very small, and therefore the cutting action is slow and easily controlled, and (2) the cutting teeth are much wider. Consequently, fine and accurate work can be done.

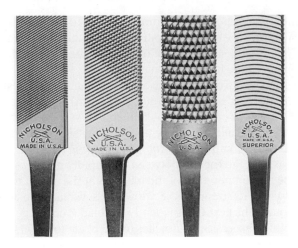

FIGURE 26-19   Four cuts of files. *Left to right:* Single, double, rasp, and curved (Vixen). *(Courtesy of Nicholson File Company.)*

■ 26.9 TYPES OF FILES

Files are classified according to the following:

1. The type, or *cut*, of the teeth
2. The degree of coarseness of the teeth
3. Construction
  a. Single solid units for hand use or in die-filing machines
  b. Band segments, for use in band-filing machines
  c. Disks, for use in disk-filing machines

Four types of *cuts* are available. *Single-cut files* have rows of parallel teeth that extend across the entire width of the file at the angle of from 65 to 85°. *Double-cut files* have two series of parallel teeth that extend across the width of the file. One series is cut at an angle of 40 to 45°. The other series is coarser and is cut at an opposite angle that varies from about 10 to 80°. A *vixen-cut file* has a series of parallel curved teeth, each extending across the file face. On a *rasp-cut* file, each tooth is short and is raised out of the surface by means of a punch. These four types of cuts are shown in Figure 26-19.

The coarseness of files is designated by the following terms, arranged in order of increasing coarseness: *dead smooth, smooth, second cut, bastard, coarse,* and *rough.* There is also a series of finer Swiss pattern files, designated by numbers from 00 to 8.

Files are available in a number of cross-sectional shapes: *flat, round, square, triangular,* and *half-round.* Flat files can be obtained with no teeth on one or both narrow edges, known as *safe edges.* Safe edges prevent material from being removed from a surface that is normal to the one being filed. Most files for hand filing are from 10 to 14 in. in length and have a pointed *tang* at one end on which a wood or metal handle can be fitted for easy grasping.

■ 26.10 FILING MACHINES

Although an experienced machinist can do very accurate work by hand filing, it is a slow and tiresome task. Consequently, three types of filing machines have been developed that permit quite accurate results to be obtained rapidly and with much less effort. *Die-filing machines* hold and reciprocate a file that extends upward through the work-table. The file rides against a roller guide at its upper end, and cutting occurs on the downward stroke; therefore, the cutting force tends to hold the work against the table. The table can be tilted to any desired angle. Such machines operate at from 300 to 500 strokes per minute, and the resulting surface tends to be at a uniform angle with respect to the table. Quite accurate work can be done. Because of the reciprocating action, approximately 50% of the operating time is nonproductive.

Band-filing machines provide continuous cutting action. Most band filing is done on contour bandsawing machines by means of a special band file that is substituted for the usual bandsaw blade. The principle of a band file is shown in Figure 26-20. Rigid, straight

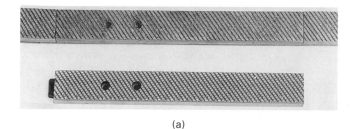

(a)

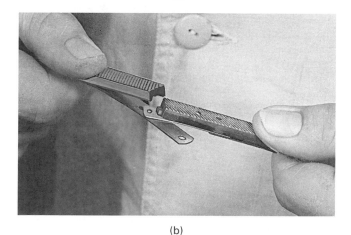

(b)

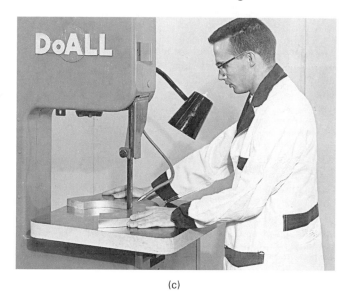

(c)

FIGURE 26-20   Bank file segments (a) are joined together to form a continuous band (b) which runs on a band filing machine (c). *(Courtesy of DoALL Co.)*

file segments, about 3 in. long, are riveted to a flexible steel band near their leading ends. One end of the steel band contains a slot that can be hooked over a pin in the other end to form a continuous band. As the band passes over the drive and idler wheels of the machine, it flexes so that the ends of adjacent file segments move apart. When the band becomes straight, the ends of adjacent segments move together and interlock to form a continuous straight file. Where the file passes through the worktable, it is guided and supported by a grooved guide, which provides the necessary support to resist the pressure of the work against the file. Band files are available in most of the standard cuts and in several widths and shapes. Operating speeds range from about 50 to 250 ft/min.

Although band filing is considerably more rapid than can be done on a die-filing machine, it usually is not quite as accurate. Frequently, band filing may be followed by some finish filing on a die-filing machine.

Some *disk-filing machines* are used, having files in the form of disks (Figure 26-21). These are even simpler than die-filing machines and provide continuous cutting action. However, it is difficult to obtain accurate results by their use.

FIGURE 26-21   Disk-type filing machine and some of the available types of disk files. *(Courtesy of Jersey Manufacturing Company.)*

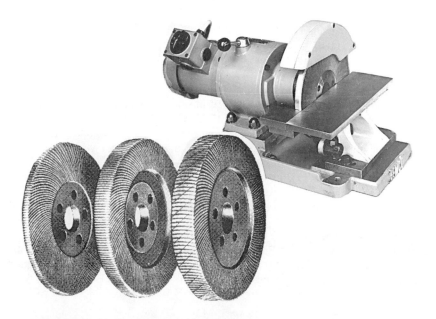

■ KEY WORDS

band-filing machine	burnishing teeth	filing	planing	reciprocating saw	shaping
bandsaw	circular saw	hacksaw	pull broach	rise per tooth	surface broach
broach	cold saw	kerf	push broach	sawing	tooth-set
broaching					

■ REVIEW QUESTIONS

1. What is unique about broaching, as compared with the other basic machining processes?
2. Can a thick saw blade be used as a broach? Why or why not?
3. Broaching machines are simpler in a basic design than most other machine tools. Why is this?
4. Why is broaching particularly well suited for mass production?
5. In designing a broach, what would be the first thing you have to calculate?
6. Why is it necessary to relate the design of a broach to the specific workpiece that is to be machined?
7. What two methods can be utilized to reduce the force and power requirements for a particular broaching cut?
8. For a given job, how would a broach having rotor-tooth design compare in length with one having regular, full-width teeth?
9. Why are the pitch and radius of the gullet between teeth on a broach of importance?
10. Why are broaching speeds usually relatively low, as compared with other machining operations?
11. What are the advantages of shell-type broach construction?
12. Why are most broaches made from alloy or high-speed steel rather than from tungsten carbide?
13. What are the advantages of TiN coated broaching tools?
14. For mass-production operations, which process is preferred, pull-up broaching or pull-down broaching?
15. What is the difference between the roughing teeth and the finishing teeth in a typical pull broach?
16. The sides of a square, blind hole must be machined all the way to the bottom. The hole is drilled to full depth and the bottom end milled flat. Is it possible to machine the hole square by broaching? Why or why not?

17. The interior, flat surfaces of socket wrenches, which have one "closed" end, often are finished to size by broaching. By examining one of these, determine what design modification was incorporated to make broaching possible.
18. Why is sawing one of the most efficient of the chip-forming processes?
19. Explain why tooth spacing is important in sawing.
20. What is the tooth gullet for a saw blade?
21. Explain what is meant by the "set" of the teeth on a saw blade.
22. How is tooth set related to saw kerf?
23. Why can a bandsaw blade not be hardened throughout the entire width of the band?
24. What are the advantages of using circular saws?
25. Why have bandsawing machines largely replaced reciprocating saws?
26. Explain how the hole in Figure 26-18 is made on a contour bandsawing machine.
27. How do you calculate or estimate T_m for a horizontal bandsaw cutting a 3 in. round of 1040 steel?
28. What is the disadvantage of using gravity to feed a saw in cutting round bar stock?
29. To what extent is filing different from sawing?
30. What is a safe edge on a file?
31. Why is an end-filing machine more efficient that a die-filing machine?
32. How does a rasp-cut file differ from other types of files?
33. How does the process of shaping differ from planing?
34. How is feed per stroke in shaping related to feed per tooth in milling?
35. What are some ways to improve the efficiency of a planer? Do any of these apply to the shaper?

■ PROBLEMS

1. A surface 12 in. long is to be machined with a flat, solid broach that has a rise per tooth of 0.0047 in. What is the minimum cross-sectional area that must be provided in the chip gullet between adjacent teeth?
2. The pitch of the teeth on a simple surface broach can be determined by Equation 26-1. If a broach is to remove 0.25 in. of material from a gray iron casting that is 3 in. wide and 17.75 in. long, and if each tooth has a rise per tooth of .004 in., what will be the length of the roughing section of the broach?
3. Estimate the (approximate maximum) horsepower needed to accomplish the operation described in Problem 2 at a cutting speed of 10 m/min. (*Hint:* First find the HP used per tooth and determine the maximum number of teeth engaged at any time. What are those units?)
4. Estimate the approximate force acting in the forward direction during cutting for the conditions stated in Problems 2 and 3.
5. In making a 6-in. cut in a piece of AISI 1020 cold rolled steel that is 1 in. thick, the material is fed to a bandsaw blade with

teeth having a pitch of 1.27 mm (20 pitch) at the rate of 0.0001 in. per tooth. Estimate the cutting time for the cut.
6. The strength of a pull broach is determined by its minimum cross section, which usually occurs either at the root of the first tooth or at the pull end. Suppose the minimum root diameter is D_r, the pull end diameter is D_p and the width of the pull slot is W. Write an equation for the allowable pull, in psi, using 200,000 as the yield strength for the broach material.
7. Suppose you want to shape a block of metal 7 in. wide and 4 in. long ($L = 4$ in.), using a shaper as set up in Figure 26-1. You have determined for this metal that the cutting speed should be 25 sfpm, the depth of cut needed here for roughing is 0.25 in., and the feed will be 0.1 in. per stroke. Determine the approximate crank rpm and then estimate the cutting time and the MRR.
8. Could you have saved any time in Problem 7 by cutting the block in the 7-in. direction? Redo with $L = 7$ and $W = 4$ in.
9. Derive Equation (26-6), the shaping cutting speed.

10. How many strokes per minute would be required to obtain a cutting speed of 36.6 m (120 ft) per minute on a typical mechanical-drive shaper if a 254-mm (10-in.) stroke is used?

11. How much time would be required to shape a flat surface 254 mm (10 in.) wide and 203 mm (8 in.) long on a hydraulic drive shaper, using a cutting speed of 45.7 m (150 ft) per minute, a feed of 0.51 mm (0.020 in.) per stroke, and an over-run of 12.7 mm $\left(\frac{1}{2} \text{ in.}\right)$ at each end of the cut?

12. What is the metal-removal rate in Problem 11 if the depth of cut is 6.35 mm $\left(\frac{1}{4} \text{ in.}\right)$?

13. If the work material in Problem 11 is 6 gray cast iron, what will be the power required?

14. A planer has a 10-hp motor, and 75% of the motor output is available at the cutting tool. The specific power for cutting cast iron certain metal is 0.03 W/mm³ or .67 hp/in³/min. What is the maximum depth of cut that can be taken in shaping a surface in this material if the surface is 305 × 305 mm (12 × 12 in.), the feed is 0.25 in. per stroke, and the cutting speed is 54.9 mm (180 ft) per minute?

15. Calculate the T_m for planing the block of cast iron in Problem 14 and then estimate T_m for milling the same surface. You will have to determine which milling process to use and select speeds and feeds for an HSS cutter.

www.wiley.com/college/degarmo

*C*hapter 26   CASE STUDY

The Socket with the Triangular Hole

The Cochran Company estimated that its annual requirements for the socket component shown in Figure CS-26 will be at least 50,000 units and that this volume will continue for at least 5 years. Consequently, it wants to consider all practical methods for making the component and has assigned you the task of determining these methods and of recommending which should be explored in detail to determine the most effective and economical process. The specifications call for a lightweight metal, such as an aluminum alloy, that must have a tensile strength of at least 138 MPa and an elongation of at least 3%.

1. Briefly outline five practicable methods for producing the part. Reference figures, photos, and sections of the text where the processes are described.

2. Select two processes which appear most likely to be economical and thus could be investigated more fully. Give the reasons for your selection of the two. It is likely that a sequence or set of processes will be needed.

3. Detail one of the two finalists giving costs per unit estimates.

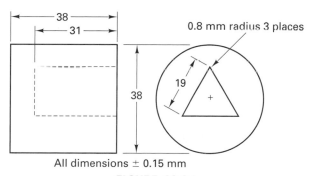

All dimensions ± 0.15 mm

FIGURE CS-26

CHAPTER 27

ABRASIVE MACHINING PROCESSES

■ 27.1 INTRODUCTION

Abrasive machining is a material-removal process that involves the interaction of abrasive grits with the workpiece at high cutting speeds and shallow penetration depths. The chips that are formed resemble those formed by other machining processes. Unquestionably, abrasive machining is the oldest of the basic machining processes. Museums abound with examples of utensils, tools, and weapons that ancient peoples produced by rubbing hard stones against softer materials to abrade away unwanted portions, leaving desired shapes. For centuries, only natural abrasives were available for grinding while other more modern basic machining processes were developed using superior cutting materials. However, the development of manufactured abrasives and a better fundamental understanding of the abrasive machining process have resulted in placing abrasive machining and its variations among the most important of all the basic machining processes.

The results that can be obtained by abrasive machining range from the finest and smoothest surfaces produced by any machining process, in which very little material is removed, to rough, coarse surfaces that accompany high material-removal rates. The abrasive particles may be (1) free (see Chapter 31); (2) mounted in resin on a belt (called *coated product*); or most commonly (3) close packed into wheels or stones, with abrasive grits held together by bonding material (called *bonded product* or a grinding wheel). Figure 27-1 shows

FIGURE 27-1 Schematic of a surface grinding, showing infeed and cross feed motions along with cutting speeds, V_s, and workpiece velocity, V_w.

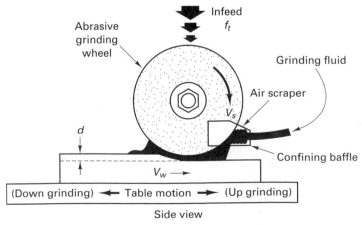

Side view

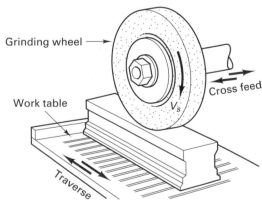

Oblique view

a surface grinding process using a grinding wheel. The depth of cut, d, is determined by the infeed and is usually very small, 0.002 to 0.005 in., so the arc of contact (and the chips) is small. The table reciprocates back and forth beneath the rotating wheel. The work feeds into the wheel in the cross feed direction. After the work is clear of the wheel, the wheel is lowered and another pass is made again removing a couple of thousandths of inches of metal. The metal-removal process is basically the same in all abrasive machining processes but with important differences due to spacing of active grains (grains in contact with work) and the rigidity and degree of fixation of the grains. Table 27-1 summarizes the primary abrasive processes and Table 27-2 the important parameters.

Compared to machining, abrasive machining processes have three unique characteristics. First, each cutting edge is very small, and many of these edges can cut simultaneously. When suitable machine tools are employed, very fine cuts are possible, and fine surfaces and close dimensional control can be obtained. Second, because extremely hard abrasive grits can be produced, including diamonds, very hard materials, such as hardened steel, glass, carbides, and ceramics, can readily be machined. As a result, the abrasive machining processes are not only important as manufacturing processes, they are indeed essential. Many of our modern products, such as modern machine tools, automobiles, space vehicles, and aircraft, could not be manufactured without these processes. Third, in grinding, you have no control over the actual tool geometry (rake angles, cutting edge radius) or all the cutting parameters (depth of cut). As a result of these parameters and variables, grinding is a complex process.

TABLE 27-1.   Abrasive Machining Processes

Process	Particle Mounting	Features
Grinding	Bonded	Uses wheels, accurate sizing, finishing, low MRR. Can be done at high speeds (over 12,000 sfpm)
Creep feed grinding	Bonded open, soft	Uses wheels with long cutting arc, very slow feed rate, and large depth of cut
Abrasive machining[a]	Bonded	High MRR, to obtain desired shapes and approximate sizes
Snagging	Bonded belted	High MRR, rough rapid technique to clean up and deburr castings, forgings
Honing	Bonded	"Stones" containing fine abrasives; primarily a hole-finishing process
Lapping	Free	Fine particles embedded in soft metal or cloth; primarily a surface-finishing process
Abrasive waterjet	Free in jet	Waterjets with velocities up to 3000 ft/sec carry abrasive particles (silica and garnet). See Chapter 31.

[a]The term *abrasive machining* applied to one particular form of the grinding process is unfortunate, because all these processes are machining with abrasives.

TABLE 27-2.   Grinding Parameters*

Independent Parameters/Controllable	Dependent Variables/Resulting Effects
Grinding wheel selection	Forces per unit width of wheel
Abrasive type	Normal
Grain size	Tangential
Hardness grade	
Openness of structure	Surface finish
Bonding media	
	Material removal rate (MRR)
Dressing of wheel	
Type of dressing tool	Wheel wear (G or grinding ratio)
Feed and depth of cut	
Sharpness of dressing tool	Thermal effects
Machine settings	Wheel surface changes
Wheel speed	
Infeed rate	Chemical effects
Cross feed rate	
Workpiece speed	Horsepower
Rigidity of setup	
Type and quality of machine	
Grinding fluid	
Type	
Cleanliness	
Method of application	

*The parameters involved in the grinding process can be grouped by their independence or dependence. Independent variables are those that are controllable (by the machine operator) while the dependent variables are the resultant effects of those inputs. For grinding, the variables are grouped as shown.

■ 27.2 ABRASIVES

An *abrasive* is a hard material that can cut or abrade other substances. Natural abrasives have existed from the earliest times. For example, sandstone was used by ancient peoples to sharpen tools and weapons. Early grinding wheels were cut from slabs of sandstone, but because they were not uniform in structure throughout, they wore unevenly and did not produce consistent results. *Emery*, a mixture of alumina (Al_2O_3) and magnetite (Fe_3O_4), is another natural abrasive still in use today and is used on coated paper and cloth (emery paper). *Corundum* (natural Al_2O_3) and diamonds are other naturally occurring abrasive materials. Today, the only natural abrasives that have commercial importance are quartz, sand, garnets, and diamonds. For example, *quartz* is used primarily in coated abrasives and in air blasting, but artificial abrasives are also making inroads in these applications. The development of artificial abrasives having known uniform properties has permitted abrasive processes to become precision manufacturing processes.

Hardness, the ability to resist penetration, is the key property for an abrasive. Table 27-3 lists the primary abrasives and their approximate Knoop hardness (kg/mm^2). The particles must be able to decompose at elevated temperatures. Two other properties are significant in abrasive grits, attrition, and friability. *Attrition* refers to the abrasive wear action of the grits resulting in dulled edges, grit flattening, and wheel glazing. *Friability* refers to the fracture of the grits and is the opposite of toughness. In grinding, it is important that grits be able to fracture to expose new, sharp edges.

Artificial abrasives date from 1891, when E. G. Acheson, while attempting to produce precious gems, discovered how to make *silicon carbide* (SiC). Silicon carbide is made by charging an electric furnace with silica sand, petroleum coke, salt, and sawdust. By passing large amounts of current through the charge, a temperature of over 4000°F is maintained for several hours, and a solid mass of silicon carbide crystals results. After the furnace has cooled, the mass of crystals is removed, crushed, and graded (sorted) into various desired sizes. As can be seen in Figure 27-2, the resulting grits, or grains, are irregular in shape, with cutting edges having every possible rake angle. Silicon carbide crystals are very hard (Knoop 2480), friable, and rather brittle. This limits their use. Silicon carbide is sold under the trade names Carborundum and Crystolon.

Aluminum oxide (Al_2O_3) is the most widely used artificial abrasive. Also produced in an arc furnace from bauxite, iron filings, and small amounts of coke, it contains aluminum hydroxide, ferric oxide, silica, and other impurities. The mass of aluminum oxide that is formed is crushed, and the particles are graded to size. Common trade names for aluminum oxide abrasives are Alundum and Aloxite. Although aluminum oxide is softer (Knoop 2100) than silicon carbide, it is considerably tougher. Consequently, it is a better general-purpose abrasive.

Diamonds are the hardest of all materials. Those that are used for abrasives are either natural, off-color stones (called *garnets*) that are not suitable for gems, or small, synthetic stones that are produced specifically for abrasive purposes. Manufactured stones appear to be somewhat more friable and thus tend to cut faster and cooler. They do not perform as satisfactorily in metal-bonded wheels. Diamond abrasive wheels are used extensively for sharpening carbide and ceramic cutting tools. Diamonds also are

TABLE 27-3. Knoop Hardness Values for Common Abrasives

Abrasive Material	Year of Discovery	Hardness (Knoop)	Temperature of Decomposition in Oxygen (°C)	Comments and Uses
Quartz (SiO_2)	?	320		Sand blasting
Aluminum oxide	1893	1600–2100	1700–2400	Softer and tougher than silicon carbide; used on steel, iron, brass Silicon
Carbide	1891	2200–2800	1500–2000	Used for brass, bronze, aluminum, and stainless and cast iron
Borazon [cubic boron nitride stainless (CBN)]	1957	4200–5400	1200–1400	For grinding hard, tough tool steels, stainless steel, cobalt and nickel based, superalloys, and hard coatings
Diamond (synthetic)	1955	6000–9000	700–800	Used to grind nonferrous materials, tungsten carbide, and ceramics

FIGURE 27-2 Loose abrasive grains at high magnification, showing their irregular, sharp cutting edges. *(Courtesy of Norton Company.)*

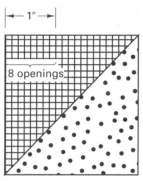

Screen no. 8
Grain size 8

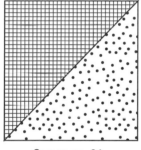

Screen no. 24
Grain size 24

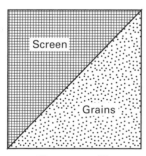

Screen no. 60
Grain size 60

FIGURE 27-3 Typical screens for sifting abrasives into sizes. The larger the screen number (of opening per linear inch), the smaller the grain size. *(Courtesy of Carborundum Company.)*

used for truing and dressing other types of abrasive wheels. Diamonds are usually used only when cheaper abrasives will not produce the desired results. Garnets are used primarily in the form of very finely crushed and graded powders for fine polishing.

Cubic boron nitride (CBN) is not found in nature. It is produced by a combination of intensive heat and pressure in the presence of a catalyst. CBN is extremely hard, registering at 4700 on the Knoop scale. It is the second hardest substance created by nature or manufactured and is often referred to, along with diamonds, as a superabrasive. Hardness, however, is not everything.

CBN far surpasses diamond in the important characteristic of thermal resistance. At temperatures of 650°C, at which diamond may begin to revert to plain carbon dioxide, CBN continues to maintain its hardness and chemical integrity. When the temperature of 1400°C is reached, CBN changes from its cubic form to a hexagonal form and loses hardness. CBN can be used successfully in grinding iron, steel, alloys of iron, Ni-based alloys, and other materials. CBN works very effectively (long wheel life, high G ratio, good surface quality, no burn or chatter, low scrap rate, and overall increase in parts/shift) on hardened materials (R_c50 or higher). It can also be used for soft steel under selected situations. CBN does well at conventional grinding speeds (6000 to 12,000 ft/min), resulting in lower total grinding cost/piece in conventional equipment. CBN can also perform well at high grinding speeds (12,000 ft/min and higher) and will enhance the benefits from future machine tools. CBN can solve difficult-to-grind jobs, but it also generates cost benefits in many production grinding operations despite its higher cost. CBN is manufactured by the General Electric Company under the trade name of Borazon.

ABRASIVE GRAIN SIZE AND GEOMETRY

To enhance the process capability of grinding, abrasive grains are sorted into sizes by mechanical sieving machines. The number of openings per linear inch in a sieve (or screen) through which most of the particles of a particular size can pass determines the grain size (Figure 27-3).

A No. 24 grit would pass through a standard screen having 24 openings per inch but would not pass through one having 30 openings per inch. These numbers have since been specified in terms of millimeters and micrometers (see ANSI B74.12 for details). Commercial practice commonly designates grain sizes from 4 to 24, inclusive, as *coarse*; 30 to 60, inclusive, as *medium*; and 70 to 600, inclusive, as *fine*. Grains smaller than 220 are usually termed *powders*. Silicon carbide is obtainable in grit sizes ranging from 2 to 240 and aluminum oxide in sizes from 4 to 240. Superabrasive grit sizes normally range from 120 grit for CBN to 400 grit for diamond. Sizes from 240 to 600 are designated as *flour* sizes. These are used primarily for lapping, or in fine honing stones for fine finishing tasks.

The grain size is closely related to the surface finish and metal removal rate. In grinding wheels and belts, coarse grains cut faster while fine grains provide better finish, as shown in Figure 27-4.

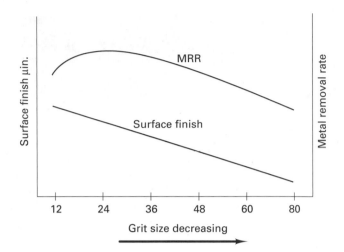

FIGURE 27-4 MRR and surface finish versus grit size.

The grain diameter can be estimated from the screen number (S), which corresponds to the number of openings per inch. The mean diameter of the grain (g) is related to the screen number by $g \cong 0.7/S$.

Regardless of the size of the grain, only a small percentage (2 to 5%) of the surface of the grain is operative at any one time. That is, the depth of cut for an individual grain (the actual feed per grit) with respect to the grain diameter is very small. Thus the chips are small. As the grain diameter decreases, the number of active grains per unit area increases and the cuts become finer because grain size is the controlling factor for surface finish (roughness). Of course, the MRR also decreases.

The grain shape is also important, because it determines the tool geometry—that is, the back rake angle and the clearance angle at the cutting edge of the grit (Figure 27-5). In the figure, γ is the clearance angle, θ is the wedge angle, and α is the rake angle. The cavities between the grits provide space for the chips. The volume of the cavities must be greater than the volume of the chips generated during the cut.

Obviously, there is no specific rake angle but rather a distribution of angles. Thus a grinding wheel can present to the surface rake angles ranging from +45° to −60° or greater. Grits with large negative rake angles or rounded cutting edges do not form chips but will rub or *plow* a groove in the surface (Figure 27-6). Thus abrasive machining is a mixture of *cutting, plowing, and rubbing* with the percentage of each being highly dependent on the geometry of the grit. As the grits are continuously abraded, fractured, or dislodged from the bond, new grits are exposed and the mixture of cutting, plowing, and rubbing is changing continuously. A high percentage of the energy used for rubbing and plowing goes into the workpiece. In cutting, 95 to 98% of the energy (the heat) goes into the chip. Figure 27-7 shows a scanning electron microscope (SEM) micrograph of a ground surface with a plowing track.

FIGURE 27-5 The rake angle of abrasive particles can be positive, zero, or negative. The cavity or voids between the grains must be large enough to hold all the chips during the cut.

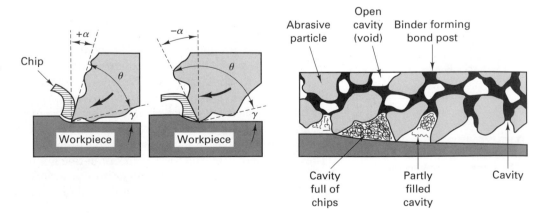

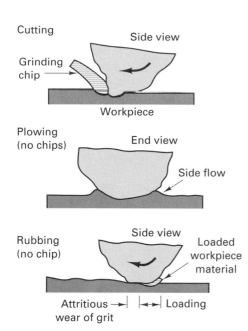

Cutting

Grinding chip

Side view

Workpiece

Plowing (no chips)

End view

Side flow

Rubbing (no chip)

Side view

Loaded workpiece material

Attritious → | ←→ | Loading
wear of grit

FIGURE 27-6 The grits interact with the surface three ways: cutting, plowing, and rubbing.

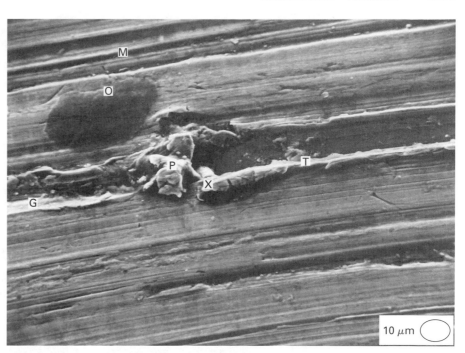

10 μm

FIGURE 27-7 SEM micrograph of a ground steel surface showing a plowed track (T) in the middle and a machined track (M) above. The grit fractured, leaving a portion of the grit in the surface (X), a prow formation (P), and a groove (G) where the fractured portion was pushed farther across the surface. The area marked (O) is an oil deposit.

In grinding, the chips are small but are formed by the same basic mechanism of compression and shear as discussed in Chapter 21 for regular metal cutting. Figure 27-8 shows steel chips from a grinding process at high magnification. They show the same structure as chips from other machining processes. Chips from the grinding process often have sufficient heat energy to burn or melt in the atmosphere. The sparks observed during grinding steel with no cutting fluid are really burning chips as shown in Figure 27-9. The feeds and depths of cut in grinding are small while the cutting speeds are high, resulting in high specific horsepower numbers. Because cutting is obviously more efficient than plowing or rubbing, grain fracture and grain pullout are natural phenomena used to keep the grains sharp. As the grains become dull, cutting forces increase, and there is an increased tendency for the grains to fracture or break free from the bonding material.

FIGURE 27-8 SEM micrograph of stainless steel chips from a grinding process. The tops (T) of the chips have the typical-shear-front-lamella structure while the bottoms (B) are smooth; 4800×.

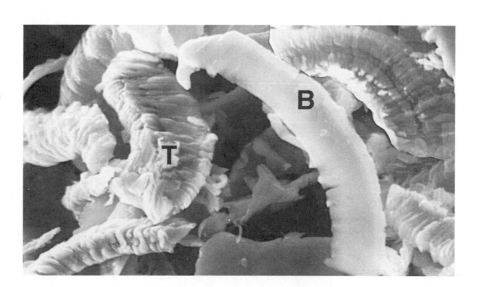

FIGURE 27-9 Plunge grinding a steel shaft produces a shower of sparks, chips burning in atmosphere.

■ 27.3 GRINDING WHEEL STRUCTURE AND GRADE

Grinding, wherein the abrasives are bonded together into a wheel, is the most common abrasive machining process. The performance of grinding wheels is greatly affected by the bonding material and the spatial arrangement of the particles grits.

The spacing of the abrasive particles with respect to each other is called *structure*. Close-packed grains have dense structure; open structure means widely spaced grains. Open-structure wheels have larger chip cavities but fewer cutting edges per unit area (Figure 27-10a).

The fracturing of the grits is controlled by the bond strength, which is known as the *grade*. Thus, grade is a measure of how strongly the grains are held in the wheel. It really is dependent on two factors: the strength of the bonding materials and the amount of the bonding agent connecting the grains. The latter factor is illustrated in Figure 27-10b. Abrasive wheels are really porous. The grains are held together with "posts" of bonding material.

FIGURE 27-10 Meaning of grinding wheel "structure" and grade. (a) The structure of a grinding wheel depends on the spacing of the grits. (b) The grade of a grinding wheel depends on the amount of bonding agent (posts) holding of abrasive grains in the wheel.

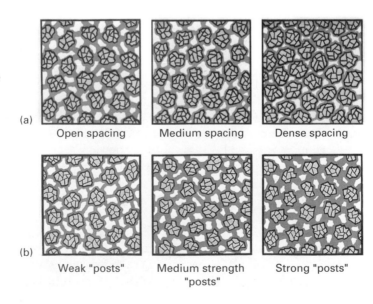

(a) Open spacing   Medium spacing   Dense spacing

(b) Weak "posts"   Medium strength "posts"   Strong "posts"

If these posts are large in cross section, the force required to break a grain free from the wheel is greater than when the posts are small. If a high dislodging force is required, the bond is said to be *hard*. If only a small force is required, the bond is said to be *soft*. Wheels commonly are referred to as hard or soft, referring to the net strength of the bond, resulting from both the strength of the bonding material and its disposition between the grains.

G RATIO

The loss of grains from the wheel means that the wheel is changing size. The grinding ratio or *G* ratio is defined as the cubic inches of stock removed divided by the cubic inches of wheel lost. In conventional grinding, the *G* ratio is in the range 20:1 to 80:1. The *G* ratio is a measure of grinding production and reflects the amount of work a wheel can do during its useful life. As the wheel loses material, it must be reset or repositioned to maintain workpiece size.

A typical vitrified grinding wheel will consist of 50 vol % abrasive particles, 10 vol % bond, and 40 vol % cavities; that is, the wheels have porosity. The manner in which the wheel performs is influenced by the following factors.

1. The mean force required to dislodge a grain from the surface (the grade of the wheel)
2. The cavity size and distribution or the porosity (the structure)
3. The mean spacing of active grains in the wheel surface (grain size and structure)
4. The properties of the grain (hardness, attrition, and friability)
5. The geometry of the cutting edges of the grains (rake angles and cutting edge radius compared to depth of cut)
6. The process parameters (speeds, feeds, cutting fluids) and type of grinding (surface, cylindrical)

It is easy to see why grinding is a complex process, difficult to control.

BONDING MATERIALS FOR GRINDING WHEELS

Bonding material is a very important factor to be considered in selecting a grinding wheel. It determines the strength of the wheel, thus establishing the maximum operating speed. It determines the elastic behavior or deflection of the grits in the wheel during grinding. The wheel can be hard or rigid, or it can be flexible. Finally, the bond determines the force required to dislodge an abrasive particle from the wheel and thus plays a major role in the cutting action. Bond materials are formulated so that the ratio of bond wear matches the rate of wear of the abrasive grits. Bonding materials in common use are

1. *Vitrified bonds.* They are composed of clays and other ceramic substances. The abrasive particles are mixed with the wet clays so that each grain is coated. Wheels are formed from the mix, usually by pressing, and then dried. They are then fired in a kiln, which results in the bonding material's becoming hard and strong, having properties similar to glass. Vitrified wheels are porous, strong, rigid, and unaffected by oils, water, or temperature over the ranges usually encountered in metal cutting. The operating speed range in most cases is 5500 to 6500 ft/min, but some wheels now operate at surface speeds up to 16,000 ft/min.
2. *Resinoid*, or phenolic resins. Because plastics can be compounded to have a wide range of properties, such wheels can be obtained to cover a variety of work conditions. They have, to a considerable extent, replaced shellac and rubber wheels. Composite materials are being used in rubber-bonded or resinoid-bonded wheels that are to have some degree of flexibility or are to receive considerable abuse and side loading. Various natural and synthetic fabrics and fibers, glass fibers, and nonferrous wire mesh are used for this purpose.
3. *Silicate* wheels use silicate of soda (waterglass) as the bond material. The wheels are formed and then baked at about 500°F for a day or more. Because they are more brittle and not so strong as vitrified wheels, the abrasive grains are released more readily. Consequently, they machine at lower surface temperatures than vitrified wheels and are useful in grinding tools when heat must be kept to a minimum.

4. *Shellac-bonded* wheels are made by mixing the abrasive grains with shellac in a heated mixture, pressing or rolling into the desired shapes, and baking for several hours at about 300°F. This type of bond is used primarily for strong, thin wheels having some elasticity. They tend to produce a high polish and thus have been used in grinding such parts as camshafts and mill rolls.

5. *Rubber* bonding is used to produce wheels that can operate at high speeds but must have a considerable degree of flexibility so as to resist side thrust. Rubber, sulfur, and other vulcanizing agents are mixed with the abrasive grains. The mixture then is rolled out into sheets of the desired thickness, and the wheels are cut from these sheets and vulcanized. Rubber-bonded wheels can be operated at speeds up to 16,000 ft/min. They commonly are used for snagging work in foundries and for thin cutoff wheels.

6. Superabrasive wheels are either electroplated (single layer of superabrasive plated to OD of a steel blank) or a thin segmented drum of vitrified CBN surrounds a steel core. The steel core provides dimensional accuracy and the replaceable segments provide durability, homogeneity, and repeatability while increasing wheel life. The later type of wheels can use resin, metal, or vitrified bonding. Selection of bond grade and structure (also called abrasive concentration) is critical.

For the electroplated wheels, nickel is used to attach a single layer of CBN (or diamond) to the OD of an accurately ground or turned steel blank. For the vitrified wheel, superabrasives are mixed with bonding media and molded (or preformed and sintered) into segments or a ring. The ring is mounted on a split steel body. Porosity is varied (to alter structure) by varying preform pressure or by using "pore forming" additives to the bond material which are vaporized during the sintering cycle. The steel-cored segmented design can rotate at 40,000 sfpm (200 m/s) whereas a plain vitrified wheel may burst at 20,000 fpm.

ABRASIVE MACHINING

The condition wherein very rapid metal removal can be achieved by grinding is the one to which some have applied the term *abrasive machining*. The metal-removal rates are compared with, or exceed, those obtainable by milling or turning or broaching, and the size tolerances are comparable. It obviously is just a special type of grinding, using abrasive grains as cutting tools, as do all other types of abrasive machining. Abrasive grinding will produce sufficient localized plastic deformation and heat in the surface so as to develop tensile residual stresses, layers of overtempered martensite (in steels), and even microcracks, because this process is quite abusive (Figure 27-11).

FIGURE 27-11 Typical residual stress distributions produced by surface grinding with different grinding conditions for abusive, conventional, and low stress grinding. Material 4340 steel. *(From M. Field and W. P. Kosher, "Surface integrity in grinding," in* New Developments in Grinding, *Carnegie-Mellon University Press, Pittsburgh, 1972, p. 666.)*

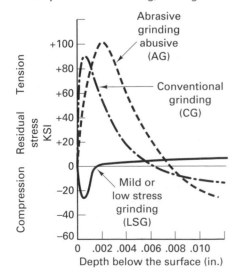

Grinding conditions

	Abusive AG	Conventional CG	Low stress LSG
Wheel	A46MV	A46KV	A46HV or A60IV
Wheel speed ft/min.	6,000–18,000	4,500–6,500	2500–3000
Down feed in./pass	.002–.004	.001–.003	.0002–.005
Cross feed in./pass	.040–.060	.040–.060	.040–.060
Table speed ft/min	40–100	40–100	40–100
Fluid	Dry	Sol oil (1:20)	Sulfurized oil

SNAGGING

Snagging is a type of rough manual grinding that is done to remove fins, gates, risers, and rough spots from castings or flash from forgings, preparatory to further machining. The primary objective is to remove substantial amounts of metal rapidly without much regard for accuracy so this is a form of abrasive machining except that pedestal-type or *swing grinders* ordinarily are used. Portable electric or hand air grinders also are used for this purpose and for miscellaneous grinding in connection with welding.

LOW-STRESS GRINDING

Conventional grinding should be replaced by procedures that develop lower surface stresses in those applications where service failures due to fatigue or stress corrosion are possible (Figure 27-11). This is accomplished by employing softer grades of grinding wheels, reducing the grinding speeds and infeed rates, using chemically active cutting fluids (e.g., highly sulfurized oil or KNO_2 in water), as outlined in the table of grinding conditions in Figure 27-11. These procedures may require the addition of a variable-speed drive to the grinding machine. Generally, only about 0.005 to 0.010 in. of surface stock needs to be finish ground in this way, as the depth of the surface damage due to conventional grinding or abusive grinding is 0.005 to 0.007 in. High-strength steels, high-temperature nickel, and cobalt-based alloys and titanium alloys are particularly sensitive to surface deformation and cracking problems from grinding. Other postprocessing processes, such as polishing, honing, and chemical milling plus peening, can be used to remove the deformed layers in critically stressed parts. It is strongly recommended, however, that testing programs be used along with service experience on critical parts before these procedures are employed in production. See the discussion of surface integrity and residual stresses in Chapter 31.

DRESSING AND TRUING

As the wheel is used, there is a tendency for the wheel to become *loaded* (metal chips become lodged in the cavities between the grains). Also, the grains dull or glaze (grits wear, flatten, and polish). Unless the wheel is cleaned and sharpened (or dressed), the wheel will not cut as well and will tend to plow and rub more. Figure 27-12 shows an arrangement for stick *dressing* a grinding wheel. The dulled grains cause the cutting forces on the grains to increase, ideally resulting in the grains' fracturing or being pulled out of the bond, thus providing a continuous exposure of sharp cutting edges. Such a continuous action ordinarily will not occur for light feeds and depths of cut. For heavier cuts, grinding wheels do become somewhat self-dressing, but the workpiece may become overheated and turn a bluish temper color (this is called *burn*) before the wheel reaches a fully dressed condition. A burned surface, the consequence of an oxide layer formation, results in the scrapping of several workpieces before parts of good quality are ground.

Grinding wheels lose their geometry during use. *Truing* restores the original shape. A single-point diamond tool can be used to *true* the wheel while fracturing abrasive grains to expose new grains and new cutting edges on worn, glazed grains (Figure 27-13). Truing can also be accomplished by grinding the grinding wheel with controlled-path or powered rotary devices using conventional abrasive wheels. The precision in generating a trued wheel surface by these methods is poorer than by the method described earlier.

FIGURE 27-12 Schematic arrangement of stick dressing.

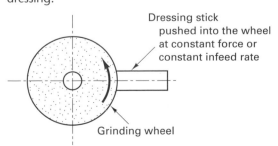

Dressing stick pushed into the wheel at constant force or constant infeed rate

Grinding wheel

FIGURE 27-13 Diamond nibs may be used for truing wheels in batch operations.

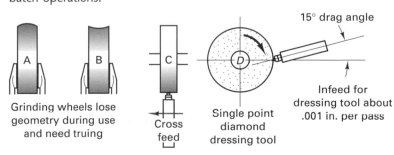

Grinding wheels lose geometry during use and need truing

Cross feed

Single point diamond dressing tool

15° drag angle

Infeed for dressing tool about .001 in. per pass

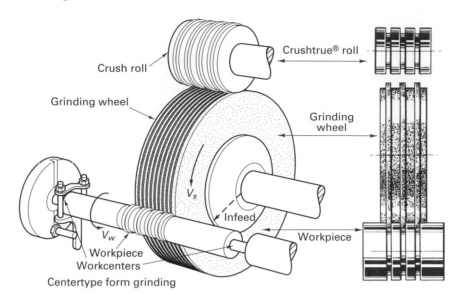

FIGURE 27-14 Crush roll dressing and truing of a grinding wheel doing plunge cut grinding on a cylinder held between centers.

Resin-bonded wheels can be trued by grinding with hard ceramics such as tungsten carbide. The procedure for truing and dressing a CBN wheel in a surface grinder might be as follows: Use 0.0002-in. downfeed per pass and cross feed slightly more than half the wheel thickness at moderate table speeds. The wheel speed is the same as the grinding speed. The grinding power will gradually increase, as the wheel is getting dull, while being trued. When the power exceeds normal power drawn during workpiece grinding, stop the truing operation. Dress the wheel face open using a J-grade stick, with abrasive one grit size smaller than CBN. Continue the truing. Repeat this cycle until the wheel is completely trued.

Modern grinding machines are equipped so that the wheel can be dressed and/or trued continuously or intermittently while grinding continues. A common way to do this is by *crush dressing* (Figure 27-14). Crush dressing consists of forcing a hard roll (WC or HSS) having the same contour as the part to be ground against the grinding wheel while it is revolving—usually quite slowly. A water-based coolant is used to flood the dressing zone at 5 to 10 gal/min. The crushing action fractures and dislodges some of the abrasive grains, exposing fresh sharp edges, allowing free cutting for faster infeed rates. This procedure usually is employed to produce and maintain a special contour to the abrasive wheel. This is also called wheel profiling. Crush dressing is a very rapid method of dressing grinding wheels, and because it fractures abrasive grains, results in free cutting and somewhat cooler grinding. The resulting surfaces may be slightly rougher than when diamond dressing is used.

■ 27.4 GRINDING WHEEL IDENTIFICATION

Most grinding wheels are identified by a standard marking system that has been established by the American National Standards Institute, Inc. This system is illustrated and explained in Figure 27-15. The first and last symbols in the marking are left to the discretion of the manufacturer.

GRINDING WHEEL GEOMETRY

The shape and size of the wheel are critical selection factors. Obviously, the shape must permit proper contact between the wheel and all of the surface that must be ground. Grinding wheel shapes have been standardized and eight of the most commonly used types are shown in Figure 27-16. Types 1, 2, and 5 are used primarily for grinding external or internal cylindrical surfaces and for plain surface grinding. Type 2 can be mounted for grinding either on the periphery or the side of the wheel. Type 4 is used with tapered safety flanges so that if the wheel breaks during rough grinding, such as snagging, these flanges will prevent the pieces of the wheel from flying and causing damage. Type 6, the straight cup, is used primarily for surface grinding but can also be used for certain types of offhand grinding. The flaring-cup type of wheel is used for tool grinding. Dish-type wheels are used for grinding tools and saws.

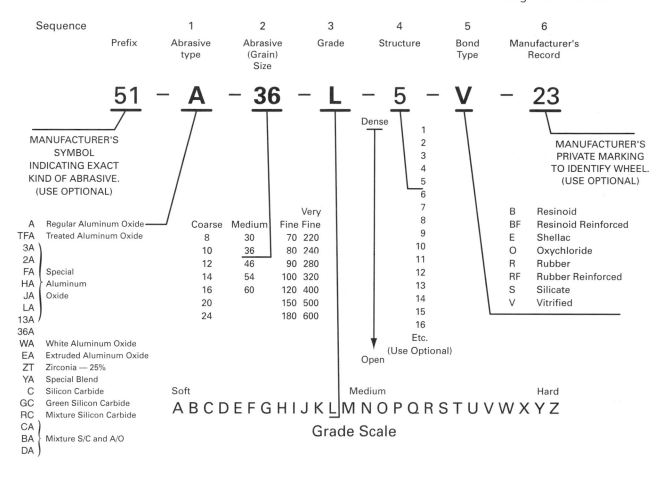

Standard bonded-abrasive wheel-marching system (ANSI *Standard* B74.13-1977)

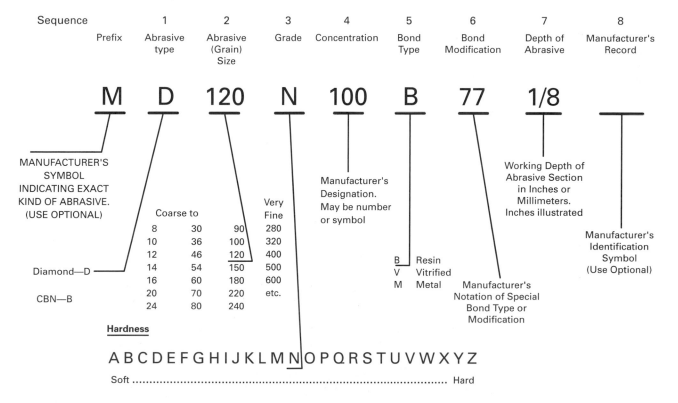

Wheel-marching system for diamond and cubic boron nitride wheels (ANSI *Standard* B74.13-1977).

FIGURE 27-15   Standard marking system for grinding wheels (ANSI standard B74. 13-1970).

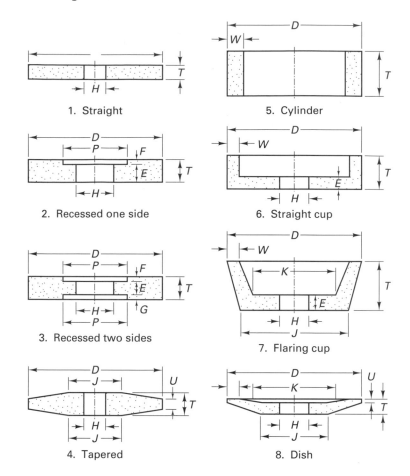

FIGURE 27-16 Standard grinding-wheel shapes commonly used. *(Courtesy of Carborundum Company.)*

Type 1, the straight grinding wheels, can be obtained with a variety of standard faces. Some of these are shown in Figure 27-17.

The size of the wheel to be used is determined, primarily, by the spindle rpm values available on the grinding machine and the proper cutting speed for the wheel, as dictated by the type of bond. For most grinding operations the cutting speed is about 2500 to 6500 ft/min. Different types and grades of bond often justify considerable deviation from these speeds. For certain types of work using special wheels and machines, as in thread grinding and "abrasive machining," much higher speeds are used.

The operation for which the abrasive wheel is intended will also influence the wheel shape and size. The major use categories are:

1. *Cutting off*: for slicing and slotting parts; use thin wheel, organic bond
2. *Cylindrical between centers*: grinding outside diameters of cylindrical workpieces

FIGURE 27-17 Standard face contours for straight grinding wheels. *(Courtesy of Carborundum Company.)*

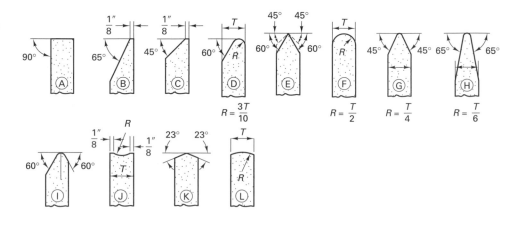

3. *Cylindrical, centerless*: grinding outside diameters with work rotated by regulating wheel

4. *Internal cylindrical*: grinding bores and large holes

5. *Snagging*: removing large amounts of metal without regard to surface finish or tolerances

6. *Surface grinding*: grinding flat workpieces

7. *Tool grinding*: for grinding cutting edges on tools such as drills, milling cutters, taps, reamers, and single-point high-speed-steel tools

8. *Offhand grinding*: work or the grinding tool is hand-held

In many cases, the classification of processes coincides with the classification of machines that do the process. Other factors that will influence the choice of wheel to be selected include the workpiece material, the amount of stock to be removed, the shape of the workpiece, and the accuracy and surface finish desired. Workpiece material has a great impact on choice of the wheel. Hard, high-strength metals (tool steels, alloy steels) are generally ground with aluminum oxide wheels or cubic boron nitride wheels. Silicon carbide and CBN are employed in grinding brittle materials (cast iron and ceramics) as well as softer, low-strength metals such as aluminum, brass, copper, and bronze. Diamonds have taken over the cutting of tungsten carbides, and CBN is used for precision grinding of tool and die steel, alloy steels, stainless, and other very hard materials. There are so many factors that affect the cutting action that there are no hard and fast rules with regard to abrasive selection.

Selection of grain size is determined by whether course or fine cutting and finish are desired. Course grains take larger depths of cut and cut more rapidly. Hard wheels with fine grains leave smaller tracks and therefore usually are selected for finishing cuts. If there is a tendency for the work material to load the wheel, larger grains with more open structure may be used for finishing.

BALANCING GRINDING WHEELS

Because of the high rotation speeds involved, grinding wheels must never be used unless they are in good balance. A slight imbalance will produce vibrations that will cause waviness in the work surface. It may cause a wheel to break, with the probability of serious damage and injury. The wheel should be mounted with proper bushings so that it fits snugly on the spindle of the machine. Rings of blotting paper should be placed between the wheel and the flanges to assure that the clamping pressure is evenly distributed. Most grinding wheels will run in good balance if they are mounted properly and trued. Most machines have provision for compensating for a small amount of wheel imbalance by attaching weights to one mounting flange. Some have provision for semiautomatic balancing with weights that are permanently attached to the machine spindle.

SAFETY IN GRINDING

Because the rotational speeds are quite high, and the strength of grinding wheels usually is much less than the materials being ground, serious accidents occur much too frequently in connection with the use of grinding wheels. Virtually all such accidents could be avoided and are due to one or a combination of four causes. First, grinding wheels occasionally are operated at unsafe and improper speeds. All grinding wheels are clearly marked with the maximum rpm value at which they should be rotated. They are all tested to considerably above the designated rpm and are safe at the specified speed *unless abused. They should never, under any condition, be operated above the rated speed.* Second, a most common form of abuse, frequently accidental, is dropping the wheel or striking it against a hard object. This can cause a crack (which may not be readily visible), resulting in subsequent failure of the wheel while rotating at high speed under load. If a wheel is dropped or struck against a hard object, it should be discarded and never used unless tested at above the rated speed in a properly designed test stand. A third common cause of grinding wheel failure is improper use, such as grinding against the side of a wheel that was designed for grinding only on its periphery. The fourth and most common cause of injury from grinding is the absence of a proper safety guard over the wheel and/or over the eyes or face of the operator. The frequency with which operators will remove safety guards from grinding equipment or fail to use safety goggles or face shields is amazing and inexcusable.

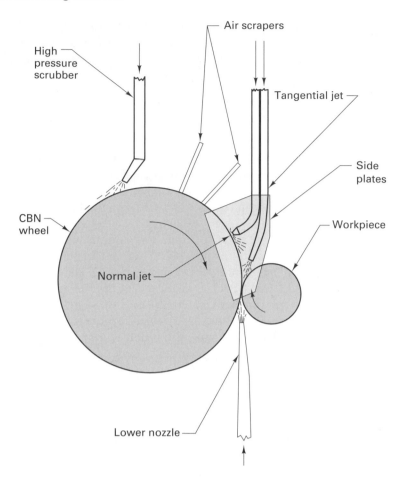

FIGURE 27-18 Coolant delivery system for optimum CBN grinding. (*Source*: Production Grinding with CBN, Dr. M.P. Hitchiner, CBN Grinding Systems Manager, Universal Beck, Romulus, MI, *Machining Technology*, Vol. 2, No. 2, 1991.)

USE OF CUTTING FLUIDS IN GRINDING

Because grinding involves cutting, the selection and use of a cutting fluid is governed by the basic principles discussed in Chapter 22. If a fluid is used, it should be applied in sufficient quantities and in a manner that will assure that the chips are washed away and not trapped between the wheel and the work. This is of particular importance in grinding horizontal surfaces. In hardened steel, the use of a fluid can help to prevent fine microcracks that result from highly localized heating. The air scraper shown in Figure 27-1 permits the cutting fluid (lubricant) to get onto the face of the wheel. Metal air scrapers disrupt the air flow. Upper and lower nozzles cool the grinding zone while a high pressure scrubber helps deter loading of the wheel.

Much snagging and off-hand grinding is done dry. On some types of material, dry grinding produces a better finish than can be obtained by wet grinding.

Grinding fluids strongly influence the performance of CBN wheels. Straight, sulfurized, or sulfochlorinated oils can enhance performance considerably (Figure 27-18) when used with straight oils.

■ 27.5 GRINDING MACHINES

Grinding machines commonly are classified according to the type of surface they produce. Table 27-4 presents such a classification, with further subdivision to indicate characteristic features of different types of machines within each classification. Grinding on all machines is done in three ways. In the first, the depth of cut (d_t) is obtained by *infeed*—moving the wheel down into the work or the work up into the wheel (Figure 27-1). The desired surface is then produced by traversing the wheel across (cross feed) the workpiece, or vice versa (Figure 27-19). In the second method, known as *plunge-cut* grinding, the basic movement is of the wheel being fed radially into the work while the latter revolves on centers. It is similar to form cutting on a lathe; usually a formed grinding wheel is used (Figure 27-14). In the third method, the work is fed very slowly

TABLE 27-4. Grinding Machines

Type of Machine	Type of Surface	Specific Types or Features
Cylindrical external	External surface on rotating, usually cylindrical parts	Work rotated between centers Centerless Chucking Tool post Crankshaft, cam, etc.
Cylindrical internal	Internal diameters of holes	Chucking Planetary (work stationary) Centerless
Surface conventional	Flat surfaces	Reciprocating table or rotating table Horizontal or vertical spindle
Creep feed	Deep slots, profiles in hard steels, carbides, and ceramics using CBN and diamond	Rigid, chatter-free, creep feed rate Continuous dressing Heavy coolant flows NC or CNC control Variable speed wheel
Tool grinders	Tool angles and geometries	Universal Special
Other	Special or any of the above	Disk, contour, thread, flexible shaft, swing frame, snag, pedestal, bench

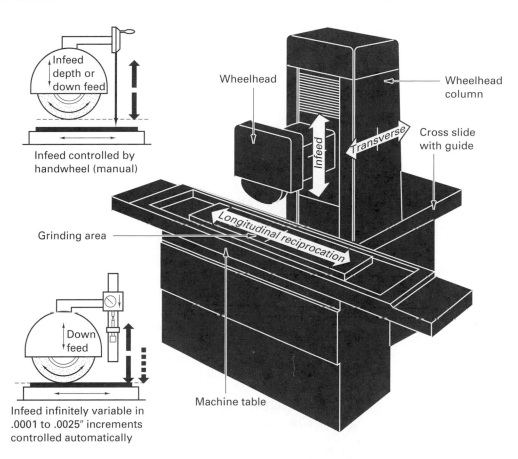

FIGURE 27-19 Horizontal spindle surface grinder, with insets showing movements of wheelhead.

past the wheel and the total downfeed or depth (d) is accomplished in a single pass (Figure 27-20). This is called *creep feed grinding* (CFG). See Table 27-5, which compares CFG to conventional and high-speed grinding for CBN applications.

Grinding machines that are used for precision work have certain important characteristics that permit them to produce parts having close dimensional tolerances. They are constructed very accurately, with heavy, rigid frames to assure permanency of alignment. Rotating parts are accurately balanced to avoid vibration. Spindles are mounted in very accurate bearings, usually of the preloaded ball-bearing type. Controls are provided so that all movements that determine dimensions of the workpiece can be made with accuracy—usually to 0.001 or 0.00001 in.

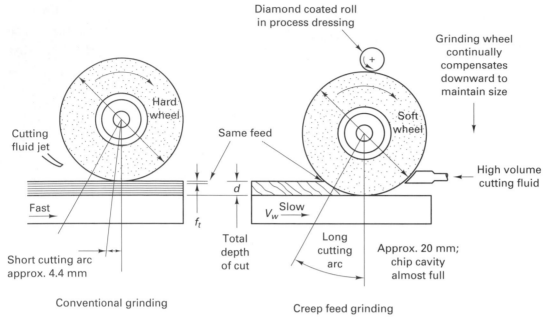

FIGURE 27-20 Conventional grinding contrasted to creep feed grinding. Note that crush roll dressing is indicated.

TABLE 27-5.	Starting Conditions for CBN Grinding		
Grinding Variable	Conventional Grinding	Creep Feed Grinding	High-speed Grinding
Wheel speed (fpm)	5500–9500 versus 4500–6500 vitrified	5000–9000 versus 3000–5000	12000–25000
Table speed V_w (fpm)	80–150	0.5–5	5–20
Feed (f_t) in./pass	0.0005–0.0015	0.100–0.250	250–.500
Grinding fluids	10% heavy duty soluble oil or 3–5% light duty soluble for light feeds	Sulfurized or sulfochlorinated straight grinding oil applied at 80 to 100 gallons per minute at 100 psi or more	

The abrasive dust that results from grinding must be prevented from entering between moving parts. All ways and bearings must be fully covered or protected by seals. If this is not done, the abrasive dust between moving parts becomes embedded in the softer of the two, causing it to act as lap and abrade the harder of the two surfaces, resulting in permanent loss of accuracy.

These special characteristics add considerably to the cost of these machines and require that they be operated by trained personnel. Production-type grinders are more fully automated and have higher metal-removal rates and excellent dimensional accuracy. Fine surface finish can be obtained very economically.

CYLINDRICAL GRINDING

Center-type cylindrical grinding is commonly used for producing external cylindrical surfaces. Figures 27-14 and 27-21 show the basic principles and motions of this process. The grinding wheel revolves at an ordinary cutting speed, and the workpiece rotates on centers at a much slower speed, usually from 75 to 125 ft/min. The grinding wheel and the workpiece move in opposite directions at their point of contact. The depth of cut is determined by infeed of the wheel or workpiece. Because this motion also determines the finished diameter of the workpiece, accurate control of this movement is required. Provision is made to traverse the workpiece with the wheel or the work can be reciprocated past the wheel. In very large grinders, the wheel is reciprocated because of the massiveness of the work. For form or plunge grinding, the detail of the wheel is maintained by periodic crush roll dressing.

A *plain center-type cylindrical grinder* is shown in Figure 27-21. On this type the work is mounted between headstock and tailstock centers. Solid dead centers are always used in the tailstock, and provision usually is made so that the headstock center can be operated either dead or alive. High-precision work usually is ground with a dead

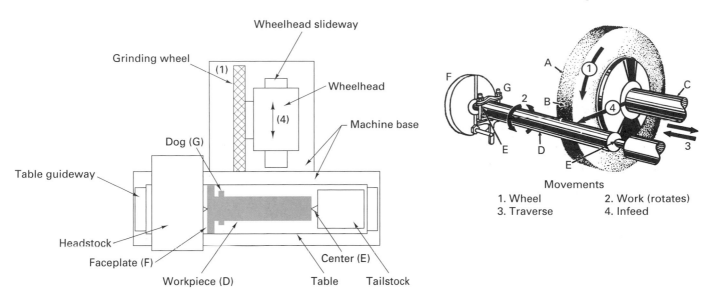

FIGURE 27-21 Cylindrical grinding between centers.

headstock center, because this eliminates any possibility that the workpiece will run out of round due to any eccentricity in the headstock.

The table assembly can be reciprocated, in most cases, by using a hydraulic drive. The speed can be varied, and the length of the movement can be controlled by means of adjustable trip dogs.

Infeed is provided by movement of the wheelhead at right angles to the longitudinal axis of the table. The spindle is driven by an electric motor that is also mounted on the wheelhead. If the infeed movement is controlled manually by some type of vernier drive to provide control to 0.001 in. or less, the machine is usually equipped with digital readout equipment to show the exact size being produced. Most production-type grinders have automatic infeed with retraction when the desired size has been obtained. Such machines are usually equipped with an automatic diamond wheel-truing device that dresses the wheel and resets the measuring element before grinding is started on each piece.

The longitudinal traverse should be about one-fourth to three-fourths of the wheel width for each revolution of the work. For light machines and fine finishes, it should be held to the smaller end of this range. The depth of cut (infeed) varies with the purpose of the grinding operation and the finish desired. When grinding is done to obtain accurate size, infeeds of 0.002 to 0.004 in. commonly are used for roughing cuts. For finishing, the infeed is reduced to 0.00025 to 0.0005 in. The design allowance for grinding should be from 0.005 to 0.010 in. on short parts and on parts that are not to be hardened. On long or large parts and on work that is to be hardened, a grinding allowance of from 0.015 to 0.030 in. is desirable. When grinding is used primarily for metal removal (called abrasive machining), infeeds are much higher, 0.020 to 0.040 in. being common. Continuous downfeed often is used, rates up to 0.100 in./min being common.

Grinding machines are available in which the workpiece is held in a chuck for grinding both external and internal cylindrical surfaces. *Chucking-type external grinders* are production-type machines for use in rapid grinding of relatively short parts, such as ball-bearing races. Both chucks and collets are used for holding the work, the means dictated by the shape of the workpiece and rapid loading and removal.

In chucking-type internal grinding machines, the chuck-held workpiece revolves, and a relatively small, high-speed grinding wheel is rotated on a spindle arranged so that it can be reciprocated in and out of the workpiece. Infeed movement of the wheelhead is normal to the axis of rotation of the work (Figure 27-21).

CENTERLESS GRINDING

Centerless grinding makes it possible to grind both external and internal cylindrical surfaces without requiring the workpiece to be mounted between centers or in a chuck. This eliminates the requirement of center holes in some workpieces and the necessity for mounting the workpiece, thereby reducing the cycle time.

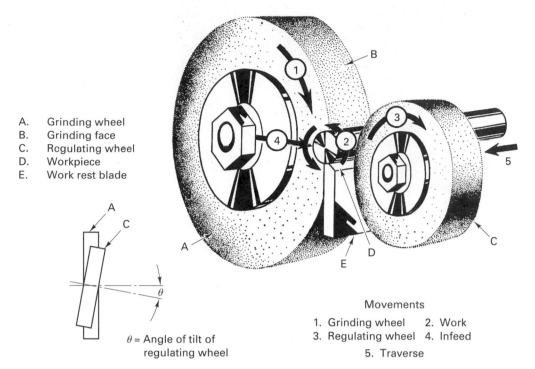

A. Grinding wheel
B. Grinding face
C. Regulating wheel
D. Workpiece
E. Work rest blade

θ = Angle of tilt of regulating wheel

Movements

1. Grinding wheel   2. Work
3. Regulating wheel 4. Infeed
            5. Traverse

FIGURE 27-22 Centerless grinding showing the relationship between the grinding wheel, the regulating wheel, and the workpiece in centerless method. *(Courtesy of Carborundum Company.)*

The principle of *centerless external grinding* is illustrated in Figure 27-22. Two wheels are used. The larger one operates at regular grinding speeds and does the actual grinding. The smaller wheel is the *regulating* wheel. It is mounted at an angle to the plane of the grinding wheel. Revolving at a much slower surface speed—usually 50 to 200 ft/min—the regulating wheel controls the rotation and longitudinal motion of the workpiece and usually is a plastic- or rubber-bonded wheel with a fairly wide face.

The workpiece is held against the work-rest blade by the cutting forces exerted by the grinding wheel and rotates at approximately the same surface speed as that of the regulating wheel. This axial feed is calculated approximately by the equation

$$F = ND \sin \theta \tag{27-1}$$

where

F = feed (mm/min or in./min)
D = diameter of the regulating wheel (mm or in.)
N = revolutions per minute of the regulating wheel
θ = angle of inclination of the regulating wheel

Centerless grinding has several important advantages:

1. It is very rapid; infeed centerless grinding is almost continuous.

2. Very little skill is required of the operator.

3. It can often be made automatic (single-cycle automatic).

4. Where the cutting occurs, the work is fully supported by the work rest and the regulating wheel. This permits heavy cuts to be made.

5. Because there is no distortion of the workpiece, accurate size control is easily achieved.

6. Large grinding wheels can be used, thereby minimizing wheel wear.

Thus centerless grinding is ideally suited to certain types of mass-production operations. The major disadvantages are as follows:

1. Special machines are required that can do no other type of work.

2. The work must be round—no flats, such as keyways, can be present.

3. Its use on work having more than one diameter or on curved parts is limited.

4. In grinding tubes, there is no guarantee that the OD and ID are concentric.

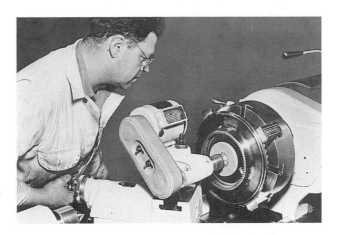

FIGURE 27-23 Grinding a bearing seat by means of a tool-post grinder on a lathe. *(Courtesy of Dunore Company.)*

Special centerless grinding machines are available for grinding balls and tapered workpieces. The centerless grinding principle can also be applied to internal grinding, but the external surface of the cylinder must be finished accurately before the internal operation is started. However, it assures that the internal and external surfaces will be concentric. The operation is easily mechanized for many applications.

TOOL-POST GRINDERS

Tool-post grinders (Figure 27-23) are used on lathes, occasionally, to grind cylindrical parts. The wheelhead is either a high-speed electric or air motor with the grinding wheel often mounted directly on the motor shaft. The entire mechanism is mounted either on the tool post or on the compound rest. The lathe spindle provides rotation for the workpiece, and the lathe carriage is used to reciprocate the wheelhead. Although tool-post grinders are versatile and useful, care should be taken to cover the ways of the lathe with a closely woven cloth to provide protection from the abrasive dust that can become entrapped between the moving parts.

SURFACE GRINDING MACHINES

Surface grinding machines are used primarily to grind flat surfaces. However formed, irregular surfaces can be produced on some types of surface grinders by use of a formed wheel. There are four basic types of surface grinding machines, differing in the movement of their tables and the orientation of the grinding wheel spindles (Figure 27-24):

1. Horizontal spindle and reciprocating table

2. Vertical spindle and reciprocating table

3. Horizontal spindle and rotary table

4. Vertical spindle and rotary table

The most common type of surface grinding machine has a reciprocating table and horizontal spindle (Figures 27-24a, 27-1 and 27-19). The table can be reciprocated longitudinally either by handwheel or by hydraulic power. The wheelhead is given transverse (crossfeed) motion at the end of each table motion, again either by handwheel or by hydraulic power feed. Both the longitudinal and transverse motions can be controlled by limit switches. Infeed or downfeed on such grinders is controlled by handwheels or automatically. The size of such machines is designated by the size of the surface that can be ground.

In using such machines, the wheel should overtravel the work at both ends of the table reciprocation, so as to prevent the wheel from grinding in one spot while the table is being reversed. The transverse or crossfeed motion should be one-fourth to three-fourths of the wheel width between each stroke.

Vertical-spindle reciprocating-table surface grinders differ basically from those with horizontal spindles only in that their spindles are vertical and that the wheel diameter must exceed the width of the surface to be ground. Usually, no traverse motion of either the table or the wheelhead is provided. Such machines can produce very flat surfaces,

Rotary-table surface grinders can have either vertical or horizontal spindles, but those with horizontal spindles are limited in the type of work they will accommodate and therefore are not used to a great extent. *Vertical-spindle rotary-table surface grinders* are primarily production-type machines. They frequently have two or more grinding heads, and therefore

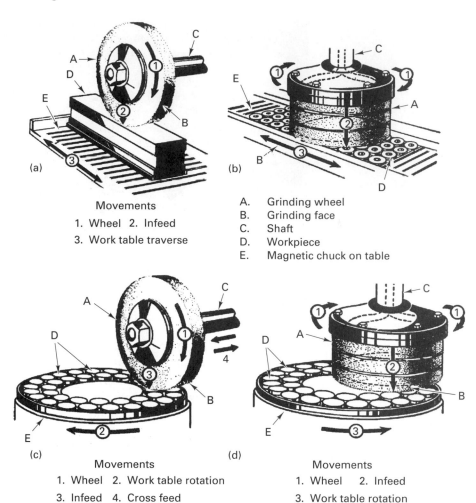

FIGURE 27-24 Surface grinding: (a) horizontal surface grinding and reciprocating table; (b) vertical spindle with reciprocating table; (c) and (d) both horizontal and vertical spindle machines can have rotary tables. *(Courtesy of Carborundum Company.)*

Movements
1. Wheel 2. Infeed
3. Work table traverse

A. Grinding wheel
B. Grinding face
C. Shaft
D. Workpiece
E. Magnetic chuck on table

Movements
1. Wheel 2. Work table rotation
3. Infeed 4. Cross feed

Movements
1. Wheel 2. Infeed
3. Work table rotation

both rough grinding and finish grinding are accomplished in one rotation of the workpiece. The work can be held either on a magnetic chuck or in special fixtures attached to the table.

By using special rotary feeding mechanisms, machines of this type often are made automatic. Parts are dumped on the rotary feeding table and fed automatically onto workholding devices and moved past the grinding wheels. After they pass the last grinding head, they are automatically unloaded.

CREEP FEED GRINDING MACHINES

This is a grinding method, often done in the surface grinding mode, that is markedly different from conventional surface grinding. As shown in Figure 27-20, in contrast to conventional techniques, the depth of cut is increased 1000 to 10,000 times and the work feed is decreased in the same proportion; hence the name *creep feed grinding* (Tables 27-4 and 27-5). The long arc of contact between the wheel and the work increases the cutting forces and the power required. Therefore, the machine tools to perform this type of grinding must be specially designed with high static and dynamic stability, stick-slip free ways, adequate damping, increased horsepower, infinitely variable spindle speed, variable but extremely consistent table feed (especially in the low ranges), high-pressure cooling systems, integrated devices for dressing the grinding wheels, and specially designed (soft with open structure) grinding wheels. The process is mainly being applied to grinding deep slots with straight parallel sides or to grinding complex profiles in difficult-to-grind materials. The process is capable of producing extreme precision at relatively high metal removal rates. Because the process can operate at relatively low surface temperatures, the surface integrity of the metals being ground is good.

However, in CFG, the grinding wheels must maintain their initial profile much longer. This process is greatly enhanced by continuous dressing (form-truing and dressing the grinding wheel throughout the process rather than between cycles). Continuous crush dressing results in higher MRRs, improved dimensional accuracy and form tolerance, reduced grinding forces (and power), and reduced thermal effects while sacrificing wheel wear. Creep feed grinding eliminates preparatory operations such as milling or broaching since profiles are ground into the solid workpiece. This can result in significant savings in unit part costs.

DISK-GRINDING MACHINES

Disk grinders have relatively large side-mounted abrasive disks. The work is held against one side of the disk for grinding. Both single and double disk grinders are used; in the latter type the work is passed between the two disks and is ground on both sides simultaneously. On these machines, the work is always held and fed automatically. On small, single-disk grinders the work can be held and fed by hand while resting on a supporting table. Although manual disk grinding is not very precise, flat surfaces can be obtained quite rapidly with little or no tooling cost. On specialized, production-type machines, excellent accuracy can be obtained very economically.

TOOL AND CUTTER GRINDERS

Simple, single-point tools often are sharpened by hand on bench or pedestal grinders (*offhand grinding*). More complex tools, such as milling cutters, reamers, hobs, and single-point tools for production-type operations, require more sophisticated grinding machines, commonly called *universal tool and cutter grinders*. These machines are similar to small universal cylindrical center-type grinders, but they differ in four important respects:

1. The headstock is not motorized.
2. The headstock can be swivelled about a horizontal as well as a vertical axis.
3. The wheelhead can be raised and lowered and can be swivelled through at 360° rotation about a vertical axis.
4. All table motions are manual. No power feeds being provided.

Specific rake and clearance angles must be created, often repeatedly, on a given tool or on duplicate tools. Tool and cutter grinders have a high degree of flexibility built into them so that the required relationships between the tool and the grinding wheel can be established for almost any type of tool. Although setting up such a grinder is quite complicated and requires a highly skilled worker, after the setup is made for a particular job, the actual grinding is accomplished rather easily. Figure 27-25 shows several typical setups on a tool and cutter grinder.

Hand-ground cutting tools are not accurate enough for automated machining processes. Many NC machine tools have been sold on the premise that they can position work to very close tolerances—within ±0.0001 to 0.0002 in.—only to have the initial workpieces produced by those machines out of tolerance by as much as 0.015 to 0.020 in. In most instances, the culprit was a poorly ground tool. For example, a twist drill with a point ground 0.005 in. off center can "walk" as much as 0.015 in., thus causing poor hole

FIGURE 27-25 Three typical setups for grinding single- and multiple-edge tools on a universal tool and cutter grinder. (a) Single-point tool is held in a device that permits all possible angles to be ground. (b) Edges of a large hand reamer are being ground. (c) Milling cutter is sharpened with a cupped Al_2O_3 grinding wheel.

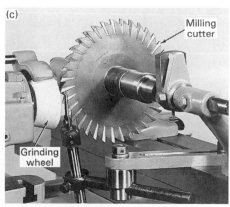

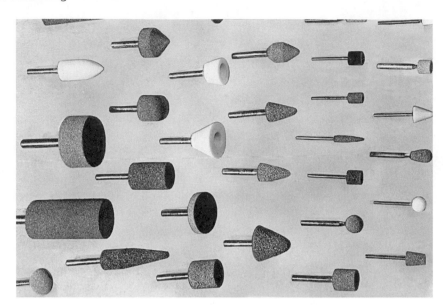

FIGURE 27-26 Examples of mounted abrasive wheels and points. *(Courtesy of Norton Company.)*

location. Many companies are turning to CNC grinders to handle the regrinding of their cutting tools. A six-axis CNC grinder is capable of restoring the proper tool angles (rake and clearance), concentricity, cutting edges, and dimensional size.

MOUNTED WHEELS AND POINTS

Mounted wheels and points are small grinding wheels of various shapes that are permanently attached to metal shanks that can be inserted in the chucks of portable, high-speed electric or air motors. They are operated at speeds up to 100,000 rpm, depending on their diameters, and are used primarily for deburring and finishing in mold and die work. Several types are shown in Figure 27-26.

COATED ABRASIVES

Coated abrasives are being used increasingly in finishing both metal and nonmetal products. These are made by gluing abrasive grains onto a cloth or paper backing (Figure 27-27). Synthetic abrasives—Aluminum Oxide, Silicon Carbide, Aluminum, Zirconia, CBN and Diamond—are used most commonly, but some natural abrasives—sand, flint, garnet, and emery—also are employed. Various types of glues are utilized to attach the abrasive grains to the backing, usually compounded to allow the finished product to have some flexibility.

Coated abrasives are available in sheets, rolls, endless belts, and disks of various sizes. Some of the available forms are shown in Figure 27-27. Although the cutting action of coated abrasives basically is the same as with grinding wheels, there is one major difference: They have little tendency to be self-sharpened when dull grains are pulled from the backing. Consequently, when the abrasive particles become dull or the belt loaded, the belt must be replaced. Finer grades result in finer first cuts but slower MRR. This versatile process is now widely used for rapid stock removal as well as fine surface finishing.

■ 27.6 DESIGN CONSIDERATIONS IN GRINDING

Almost any shape and size of work can be finished on modern grinding equipment, including flat surfaces, straight or tapered cylinders, irregular external and internal surfaces, cams, antifriction bearing races, threads, and gears. For example, the most accurate threads are formed from solid cylindrical blanks on special thread grinding machines. Gears that must operate without play are hardened and then finish ground to close tolerances. Two important design recommendations are to reduce the area to be ground and to keep all surfaces that are to be ground in the same or parallel planes (Figure 27-28). This is an example of *design for manufacturing* (DFM).

Abrasive machining can remove scale as well as parent metal. Large allowances of material, needed to permit conventional metal-cutting tools to cut below hard or abrasive inclusions, are not necessary for abrasive machining. An allowance of 0.015 in.

Belt composition

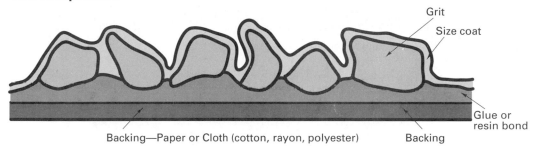

Grit
Size coat
Glue or resin bond
Backing
Backing—Paper or Cloth (cotton, rayon, polyester)

Grit size—Grade		
vs	Approx.	Finish (rms)
24	300	μ in.
36	250	"
50	140	"
80	125	"
120	60-80	"
150	40-60	"

Bonds			
Name	Make coat	Size coat	Backing
Glue bond	Gluc	Glue	Non WP
Modified glue	Mod. glue	Mod. glue	"
Resin over glue	Glue	Resin	"
Resin over resin	Resin	Resin	"
Waterproof	Resin	Resin	WP

WP = waterproof

Platen grinder

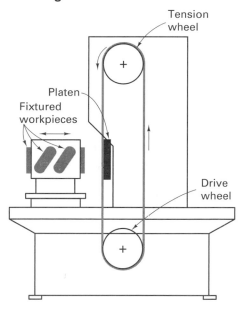

Tension wheel
Platen
Fixtured workpieces
Drive wheel

FIGURE 27-27 Belt composition for coated abrasives (*top*). Platen grinder (*right*) and examples of belts and disks for abrasive machining.

is adequate, assuming, of course, that the part is not warped or out of round. This small allowance requirement results in savings in machining time, in material (often 60% less metal is removed), and in shipping of unfinished parts.

■ 27.7 HONING

Honing is a stock-removal process that uses fine abrasive stones to remove very small amounts of metal. Cutting speed is much lower than that of grinding. The process is used to size and finish bored holes, remove common errors left by boring (taper, waviness, and tool marks) or remove the tool marks left by grinding. The amount of metal removed is typically about 0.005 in. or less. Although honing occasionally is done by hand, as in finishing the face of a cutting tool, it usually is done with special equipment. Most honing is done on internal cylindrical surfaces, such as automobile cylinder walls. The honing stones usually are held in a honing head, with the stones being held against the work with controlled light pressure. The honing head is not guided externally but, instead, *floats* in the hole, being guided by the work surface (Figure 27-29).

Original design of base plate

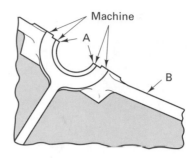

Original design of crankshaft bearing bracket

Redesigned to reduce weight and grinding time

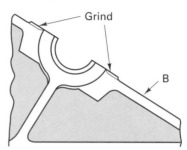

Redesign eliminated shoulders and made part suitable for grinding in single setup

FIGURE 27-28 Reducing area to be ground and keeping all surface to be ground in the same or parallel planes are two important design recommendations. (From *Machine Design,* June 1,1972, p. 87.)

The stones are given a complex motion so as to prevent a single grit from repeating its path over the work surface. Rotation is combined with an oscillatory axial motion. For external and flat surfaces, varying oscillatory motions are used. The length of the motions should be such that the stones extend beyond the work surface at the end. A cutting fluid is used in virtually all honing operations. The critical process parameters are

FIGURE 27-29 Schematic of honing head showing the manner in which the stones are held. The rotary and oscillatory motions combine to produce a crosshatched lay pattern. Typical values for V_c and P_s are given below.

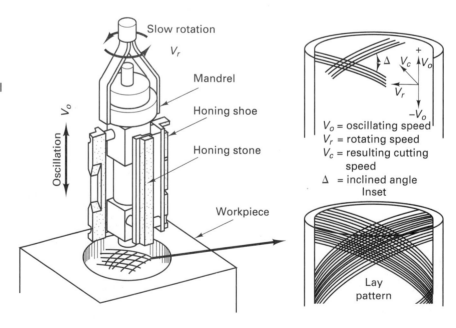

For:	Honing Parameters	Conventional Abrasives	Diamonds	CBN
High MRR	V_c (m/min)	20–30	40–70	35–90
	P_s (N/mm²)	1–2	2–8	2–4
Best quality service	V_c (m/min)	5–30	40–70	20–60
	P_s (N/mm²)	0.5–1.5	1.0–3.0	1.0–2.0

rotational speed, V_r, oscillation speed, V_o, the length and position of stroke, and the honing stick pressure. V_c and the inclination angle are both products of V_o and V_r.

HONING STONES

Virtually all honing is done with stones made by bonding together various fine artificial abrasives. *Honing stones* differ from grinding wheels in that additional materials, such as sulfur, resin, or wax, are often added to the bonding agent to modify the cutting action. The abrasive grains range in size from 80 to 600 grit. The stones are equally spaced about the periphery of the tool. Reference values for V_c and honing stick pressure, P_s, for various abrasives are shown in Figure 27-29.

Single- and multiple-spindle honing machines are available in both horizontal and vertical types. Some are equipped with special, sensitive measuring devices that collapse the honing head when the desired size has been reached.

For honing single, small, internal cylindrical surfaces, a procedure is often used wherein the workpiece is manually held and reciprocated over a rotating hone. If the volume of work is sufficient, honing is a fairly inexpensive process. A complete honing cycle, including loading and unloading the work, is often less than 1 minute. Size control within 0.0003 in. is achieved routinely.

■ 27.8 SUPERFINISHING

Superfinishing is a variation of honing that is typically used on flat surfaces. The process is:

1. Very light, controlled pressure, 10 to 40 psi
2. Rapid (over 400 cycles per minute), short strokes—less than $\frac{1}{4}$ in.
3. Stroke paths controlled so that a single grit never traverses the same path twice
4. Copious amounts of low-viscosity lubricant–coolant flooded over the work surface

This procedure, illustrated in Figure 27-30, results in surfaces of very uniform, repeatable smoothness.

Superfinishing is based on the phenomenon that a lubricant of a given viscosity will establish and maintain a separating, lubricating film between two mating surfaces if their roughness does not exceed a certain value and if a certain critical pressure, holding them apart, is not exceeded. Consequently, as the minute peaks on a surface are cut away by the honing stone, applied with a controlled pressure, a certain degree of smoothness is achieved. The lubricant establishes a continuous film between the stone and the workpiece and separates them so that no further cutting action occurs. Thus, with a given pressure, lubricant, and honing stone, each workpiece is honed to the same degree of smoothness.

Superfinishing is applied to both cylindrical and plane surfaces. The amount of metal removed usually is less than 0.002 in., most of it being the peaks of the surface roughness. Copious amounts of lubricant-coolant maintain the work at a uniform temperature and wash away all abraded metal particles to prevent scratching.

■ 27.9 LAPPING

Lapping is an abrasive surface-finishing process wherein fine abrasive particles are *charged* (caused to become embedded) into a soft material, called a *lap*. The material of the lap may range from cloth to cast iron or copper, but it is always softer than the

FIGURE 27-30 In superfinishing and honing, a film of lubricant is established between the work and the abrasive stone as the work becomes smoother.

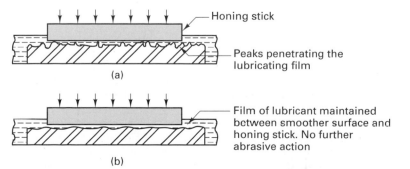

material to be finished, being only a holder for the hard abrasive particles. Lapping is applied to both metals and nonmetals.

As the charged lap is rubbed against a surface, the abrasive particles in the surface of the lap remove small amounts of material from the surface to be machined. Thus the abrasive does the cutting, and the soft lap is not worn away because the abrasive particles become embedded in its surface instead of moving across it. This action always occurs when two materials rub together in the presence of a fine abrasive: The softer one forms a lap, and the harder one is abraded away.

In lapping, the abrasive is usually carried between the lap and the work surface in some sort of a vehicle, such as grease, oil, or water. The abrasive particles are from 120 grit up to the finest powder sizes. As a result, only very small amounts of metal are removed, usually considerably less than 0.001 in. Because it is such a slow metal-removing process, lapping is used only to remove scratch marks left by grinding or honing, or to obtain very flat or smooth surfaces, such as are required on gage blocks or for liquidtight seals where high pressures are involved.

Materials of almost any hardness can be lapped. However, it is difficult to lap soft materials because the abrasive tends to become embedded. The most common lap material is fine-grained cast iron. Copper is used quite often and is the common material for lapping diamonds. For lapping hardened metals for metallographic examination, cloth laps are used.

Lapping can be done either by hand or by special machines. In hand lapping, the lap is flat, similar to a surface plate. Grooves usually are cut across the surface of a lap to collect the excess abrasive and chips. The work is moved across the surface of the lap, using an irregular, rotary motion, and is turned frequently to obtain a uniform cutting action.

In lapping machines for obtaining flat surfaces, workpieces are placed loosely in holders and are held against the rotating lap by means of floating heads. The holders, rotating slowly, move the workpieces in an irregular path. When two parallel surfaces are to be produced, two laps may be employed, one rotating below and the other above the workpieces.

Various types of lapping machines are available for lapping round surfaces. A special type of centerless lapping machine is used for lapping small cylindrical parts, such as piston pins and ball-bearing races.

Because the demand for surfaces having only a few micrometers of roughness on hardened materials has become quite common, the use of lapping has increased greatly. However, it is a very slow method of removing metal, obviously costly compared with other methods, and should not be specified unless such a surface is absolutely necessary.

■ KEY WORDS

abrasive machining	coated abrasive	diamond	G ratio	honing	rubber bond	snagging
aluminum oxide	corundum	dressing	garnet	lapping	shellac bond	surface grinding
attrition	creep feed grinding	emery	grade	quartz	silicate bond	truing
CBN	crush dressing	friability	grinding	resinoid bond	silicon carbide	vitrified bond
centerless grinding	cylindrical grinding					

■ REVIEW QUESTIONS

1. What are machining processes that use abrasive particles for cutting tools called?
2. What is attrition in an abrasive grit?
3. Why is friability an important grit property?
4. Explain the relationship between grit size and surface finish.
5. Why is aluminum oxide used more frequently than silicon carbide as an abrasive?
6. Why is CBN superior to silicon carbide as an abrasive in some applications?
7. What materials commonly are used as bonding agents in grinding wheels?
8. Why is the grade of a bond in a grinding wheel important?
9. How does grade differ from structure in a grinding wheel?
10. What is crush dressing?
11. How does loading differ from glazing?

12. What is meant by the statement that grinding is a mixture of processes?
13. What is accomplished in dressing a grinding wheel?
14. How does abrasive machining differ from ordinary grinding?
15. What is a grinding ratio or G ratio?
16. How is the feed of the workpiece controlled in centerless grinding?
17. Why is grain spacing important in grinding wheels?
18. Why should a cutting fluid be used in copious quantities when doing wet grinding?
19. How does plunge-cut grinding compare to cylindrical grinding?
20. If grinding machines are placed among other machine tools, what precautions must be taken?
21. What is the purpose of low-stress grinding?
22. How is low-stress grinding done?
23. The number of grains per square inch which actively contact

and cut a surface decreases with increasing grain diameter. Why is this so?

24. Why are centerless grinders so popular in industry compared to center-type grinders?

25. Explain how a SEM micrograph is made. Check the Internet or the library to find the answer.

26. Why are vacuum chucks and magnetic chucks widely used in surface grinding and not in milling (see Chapter 29 on workholders for more discussion)?

27. How does creep feed grinding differ from conventional surface grinding?

28. Why does a lap not wear, since it is softer than the material being lapped?

29. How do honing stones differ from grinding wheels?

30. What is meant by "charging" a lap?

31. Why is a honing head permitted to float in a hole that is being honed?

32. In what respect does a coated abrasive differ from an abrasive wheel?

33. Figure out why the bottoms of chips shown in Figure 27-8 are so smooth. The magnification of the micrograph is 4800X. How thick are these chips?

34. What is the inclined angle in honing and what determines it?

35. What are the common causes of grinding accidents?

36. What other machine tool does a surface grinder resemble?

37. Figure 27-11 showed residual stress distributions produced by surface grinding. What is a residual stress?

38. In grinding, what is infeed versus crossfeed?

■ PROBLEMS

1. Perhaps you have observed the following wear phenomena: A set of marble or wooden stairs shows wear on the treads in the regions where people step when they climb (or descend) the stairs. The higher up the stairs, the less the wear on the tread. Given that soles of shoes (leather, rubber) are far softer than marble or granite, explain:
   a. Why and how the stairs wear.
   b. Why the lower stairs are more worn than the upper stairs.

2. Explain why it is that a small particle of a material can be used to abrade a surface made of the same material (i.e., why does the small particle act harder or stronger than the bulk material)?

3. In grinding, both the wheel and workpiece are moving (or rotating). Using the data in Figure 27-11 and assuming that you are doing surface grinding (see Figure 27-1), what are some typical MRR values? How do these compare to MRR values for other machining processes, such as milling? What is the significance of this?

 www.wiley.com/college/degarmo

*C*hapter 27 CASE STUDY

Aluminum Retainer Rings

The Langley Corporation has to make 5000 retainer rings, as shown in Figure CS-27. It is essential that the surfaces be smooth with no sharp corners on the circumferential edges. Determine the most economical method for manufacturing these rings.

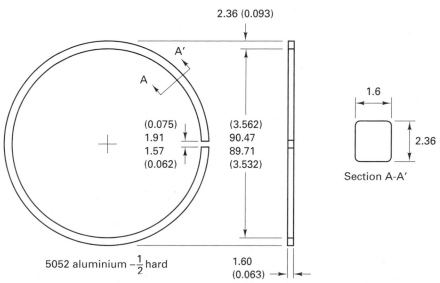

FIGURE CS-27 Aluminum snap-in retainer ring.

CHAPTER 28

NONTRADITIONAL MACHINING

■ 28.1 INTRODUCTION

Machining processes that involve chip formation have a number of inherent limitations which limit their application in industry. Large amounts of energy are expended to produce unwanted chips which must be removed and discarded. Much of the machining energy ends up as undesirable heat that often produces problems of distortion and surface cracking. Cutting forces require that the workpiece be held which can also lead to distortion. Unwanted distortion, residual stress, and burrs caused by the machining process often require further processing. Finally, some geometries are too delicate to machine while others are too complex. Figure 28-1 shows some geometries which are difficult to machine by conventional methods.

In view of these limitations, many *nontraditional machining* (NTM) methods have been developed since World War II to address the growing list of machining requirements which cannot be handled by conventional machining alone. Advantages of NTM methods may include the ability to machine:

- Complex geometries beyond simple planar or cylindrical features
- Parts with extreme surface finish and tolerance requirements
- Delicate components that cannot withstand large cutting forces
- Parts without producing burrs or inducing residual stresses
- Brittle materials or materials with very high hardness

For the purposes of our discussion, NTM processes can be divided into four groups based upon the material removal mechanism:

1. *Chemical*—Chemical reaction between a liquid reagent and the workpiece results in etching.
2. *Electrochemical*—An electrolytic reaction at the workpiece surface is responsible for material removal.
3. *Mechanical*—High-velocity abrasives or liquids remove material.
4. *Thermal*—High temperatures in very localized regions evaporate materials.

Table 28-1 gives a summary of the characteristics for common NTM processes while Table 28-2 shows some characteristics of NTM finishing processes. When examining these tables, recall that conventional end milling has these typical values:

- Feed rate—25 to 5000 mm/min (5 to 200 in./min)
- Surface finish—1.5 to 3.75 μm (60 to 150 μin.) AA
- Dimensional accuracy—0.025 to 0.05 mm (0.001 to 0.002 in.)
- Workpiece/feature size—61 cm × 61 cm (24 in. × 24 in.); 2.5 cm (1 in.) deep

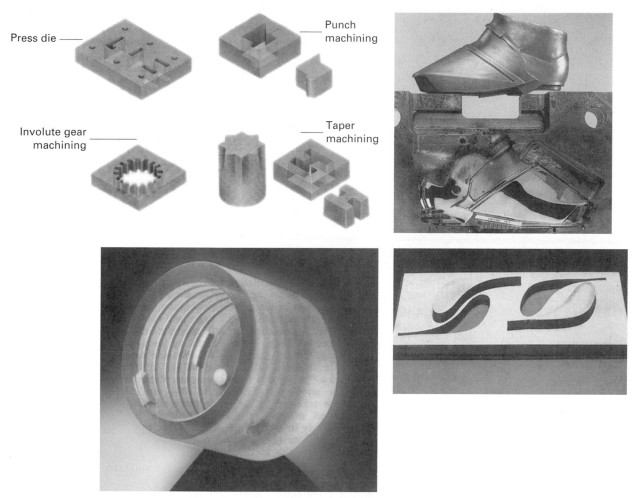

FIGURE 28-1   Nontraditional machining techniques and geometries: (*Top left*) Some difficult-to-machine shapes produced by wire EDM. (*Top right*) Injection mold for a ski boot made in hard tool steel by EDM—note superior surface finish. (*Bottom left*) A quartz housing with two internal grooves cut with ultrasonics. (*Bottom right*) Contour cut turbine blade profile made by CO_2 laser.

In comparison, NTM processes typically have lower feed rates and require more power consumption when compared to machining. However, some processes permit batch processing which increases the overall throughput of these processes and enables them to compete with machining. A major advantage of some NTM processes is that feed rate is independent of the material being processed. As a result, these processes are often used for difficult to machine materials. NTM processes typically have better accuracy and surface finish with the ability of some processes to machine larger feature sizes at lower capital costs. In most applications, NTM requires part specific tooling while general-purpose cutting and workholding tools make machining very flexible. There are numerous hybrid forms of all these processes, developed for special applications but only the main NTM processes are described here due to space limitations.

■ 28.2   Chemical Machining Processes

CHEMICAL MACHINING
Chemical machining (CHM) is the simplest and oldest of the chipless machining processes. The use of CHM dates back 4,500 years to the Egyptians who used it to etch jewelry. In modern practice it is applied to parts ranging from very small microelectronic circuits to very large engravings up to 15 m (50 ft) long. Typically metals are chemically machined although methods do exist for etching ceramics and even glass.

TABLE 28-1. Summary of NTM Processes

Process	Typical Penetration or Feed Rate, mm/m (ipm)	Typical Surface Finish AA, μm (μin.)	Typical Accuracy, mm (in.)	Typical Workpiece or Feature Size, cm (in.)	Comments
Chemical					
Chemical milling (cut-and-peel)	0.013 to 0.076 (.0005 to .003)	1.6 to 6.35; as low as 0.2 (63 to 250; 8)	greater than .127 (.005)	as large as 365 × 1524 (144 × 600); up to 1.27 (.5) thick	No burrs; no surface stresses; tooling cost low
Photochemical machining	as above	as above	.025 to .05 (.001 to .002)	30 × 30 (12 × 12); up to 0.15 (.06) thick	Limited to thin material; burr-free blanking of brittle material; tooling cost low; used in microelectronics
Electrochemical					
Electrochemical machining (ECM)	2.5 to 12.7 (0.1–0.5)	0.4 to 1.6 (16–63)	0.013 to 0.13 (.0005–.005); 0.05 (.002) in cavities	30 × 30 (12 × 12); 5 (2) deep	Stress-free, burr-free metal removal in hard to machine metals; tool design expensive; disposal of wastes a problem; MRR independent of hardness; deep cuts will have tapered walls
Electrostream drilling	1.5 to 3 (.06–0.12)	0.25 to 1.6 (10–63)	0.025 (.001) or 5% of hole dia.	up to 0.5 (.2) thick	Charged high-velocity stream of electrolyte; hole diameters down to 0.127 mm (.005 in.); 40 : 1 hole aspect ratios possible
Shaped tube electrolytic machining (STEM)	as above	0.8 to 3.1 (32–125)	0.025 to 0.125 (.001–.005)	routinely up to 127 (5) thick	Special form of ECM using conductive tube with insulated surface and acidic electrolyte; 300 : 1 hole aspect ratios; hole diameters down to 0.5 mm (.02 in.)
Mechanical					
Abrasive-jet machining	76 (3)	0.25 to 1.27 (10–50)	0.12 (.005)	up to 0.15 (.06) thick	Used for cutting brittle materials; produces tapers; inexpensive to implement; can cut up to 6.3 mm (0.25 in.) thick glass
Abrasive waterjet machining	15 to 450 (.6–18)	2.0 to 6.35 (80–250)	0.13 to 0.38 (.005–.015)	up to 20 (8) thick	Use in glass, titanium, composites, nonmetals and heat-sensitive or brittle materials; produces tapered walls in deep cuts; no burrs
Ultrasonic machining (impact grinding)	0.5 to 3.8 (.02–.15)	0.4 to 1.6; as low as 0.15 (16–63; 6)	0.013 to 0.025 (.0005–.001)	up to 100 cm² (16 in²)	Most effective in hard materials, $R_C > 40$; tool wear and taper limit hole aspect ratio at 2.5 : 1
Waterjet machining	250 to 200,000 (10–7900) soft materials	1.27 to 1.9 (50–100)	0.13 to 0.38 (.005–.015)	up to 2.5 (1) thick	Used on leather, plastics, and other non-metals; pressures of 60,000 psi and jet velocity of up to 3000 ft/sec
Thermal					
Electrical-discharge machining (EDM)	up to 0.5 (.02)	0.8 to 2.7 (32–105)	0.013 to 0.05 (.0005–.002)	up to 200 × 200 (79 × 79); 5 (2) deep	Widely used and disseminated; dies expensive; cuts any conductive material regardless of hardness; forms recast layer
Electron-beam machining (EBM)	30 to 1500 (1.2–60)	0.8 to 6.35 (32–250)	0.005 to 0.025 (.0002–.001)	0.025 to 0.63 (.01–.25) thick	Capable of micromaching thin materials; hole sizes down to 0.05 mm (.002 in.); 100 : 1 hole aspect ratios; requires high vacuum
Laser-beam machining (LBM)	100 to 2500 (4–100)	0.8 to 6.35 (32–250)	0.013 to 0.13 (.0005–.005)	up to 2.5 (1) thick	Capable of drilling holes down to 0.127 mm (.005 in) at 20 : 1 aspect ratio in seconds; has heat-affected zone and recast layers which may require removal
Plasma arc cutting (PAC)	250 to 5000 (10–200)	0.6 to 12.7 (25–500)	0.5 to 3.2 (.02–.125)	up to 15 (6) thick	Clean rapid cuts and profiles in almost all plates; 5 to 10 degree taper; cheaper capital equipment
Precision PAC	as above	as above	0.25 (.01)	up to 1.5 (.625) thick	Special form of PAC limited to thin sheets of material; straighter, smaller kerf
Wire EDM	100 to 250 (4–10)	0.8 to 1.6; as low as 0.38 (32–64; 15)	0.0025 to 0.1 (.0001–.004)	as large as 100 × 160 (40 × 64); up to 45 (18) thick	Special form of EDM using traveling wire; cuts straight narrow kerfs; wire diameters as small as 0.05 mm (.002 in.); CNC machines permit complex geometries

TABLE 28-2. Summary of NTM Finishing Processes

Process	Typical Surface Finish AA, μm (μin)	Typical Accuracy, mm (in.)	Comments
Chemical			
Chemical mechanical polishing (CMP)	0.002 (.08)	flatness on the order of 5×10^{-5} (2×10^{-6})	Polishing with chemical etching assist; used in microelectronics in metal interconnection; polishes metals and dielectrics; selective etch of metals in presence of dielectrics possible
Electrochemical			
Electrochemical grinding	0.2 to 0.8 (8–32)	0.013 to 0.025 (.0005–.001)	Special form of ECM; grinding with ECM assist; good for grinding hard conductive materials like tungsten carbide tool bits; no heat-damage, burrs, or residual stresses
Electropolishing	1.6 to 6.35; as low as 0.02 (4 to 32; 1)	used to obtain finish	High quality, no stress surface; removes residual stresses; makes corrosion-resistant surfaces
Mechanical			
Abrasive flow machining	0.76 to 7.6; as low as 0.05 (30–300; 2)	0.025 to 0.05 (.001–.002)	Typically used to finish inaccessible internal passages; often used to remove recast layer produced in EDM; used for burr removal (not good for finishing blind holes)
Thermal			
Thermal deburring (combustion machining)	burr-free	used for deburring	Vaporizes burrs and fins on cast or machined parts; deburrs steel gears automatically

In CHM, material is removed from a workpiece by selectively exposing it to a chemical reagent or etchant. The mechanism for metal removal is the chemical reaction between the etchant and the workpiece resulting in dissolution of the workpiece. One means for accomplishing CHM is called *gel milling* where the etchant is applied to the workpiece in gel form. However, the most common method of CHM involves covering selected areas of the workpiece with a *maskant* (or etch resist) and imparting the remaining exposed surfaces of the workpiece to the etchant. The general material-removal steps for CHM are:

1. *Cleaning.* Contaminants on the surface of the workpiece are removed to prepare for application of the maskant and permit uniform etching. This may include degreasing, rinsing, and/or pickling.

2. *Masking.* If selective etching is desired, an etch-resistant maskant is applied and selected areas of the workpiece are exposed through the maskant in preparation for etching.

3. *Etching.* The part is either immersed in an etchant or an etchant is continuously sprayed onto the surface of the workpiece. The chemical reaction is halted by rinsing.

4. *Stripping.* The maskant is removed from the workpiece and the surface is cleaned and desmutted as necessary.

Lateral dimensions in CHM are controlled in large part by the patterned maskant. Masking can be performed in one of several ways depending upon the level of precision required in CHM. The simplest method of applying a maskant is the *cut-and-peel* method. In this procedure the maskant material, typically neoprene, polyvinyl chloride, or polyethylene, is applied to the entire surface of the workpiece by dipping or spraying. Once the coating dries, it is then selectively removed in those areas where etching is desired by scribing the maskant with a knife and peeling away the unwanted portions. When volume permits, scribing templates may be used to improve accuracy. Cut-and-peel coatings are thick ranging from 0.025 to 0.13 mm (0.001 to 0.005 in.). Because of this thickness, the maskant can withstand exposure to the etchant for extended periods of time necessary to remove large volumes of material. This technique is generally preferred

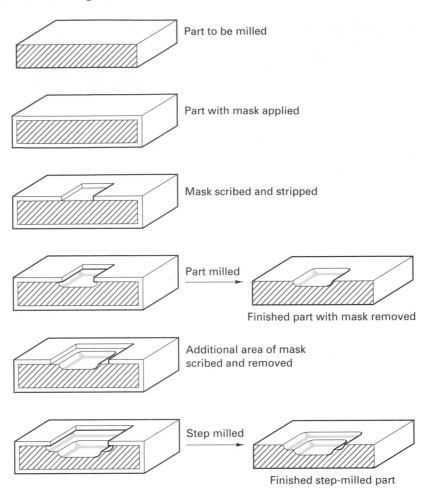

Part to be milled

Part with mask applied

Mask scribed and stripped

Part milled

Finished part with mask removed

Additional area of mask scribed and removed

Step milled

Finished step-milled part

FIGURE 28-2 Steps required to produce a stepped contour by chemical machining.

when the workpiece is not flat or is very large or for low-volume work where the development of screens or phototools necessary for other masking methods is not justified. The scribe-and-peel method is the only method capable of permitting *stepped machining* as shown in Figure 28-2.

Another method used to apply maskants is *screen printing* involving the use of traditional silk-screening technology. The method applies the maskant through a mask made from a fine silk mesh or stainless steel screen. Masks are typically formed by application, exposure, and development of a light-sensitive emulsion on top of the screen. The screen is pressed against the surface of the workpiece and the maskant is rolled on. Screen printing is good for high-volume, low-precision applications with tolerances typically in the 0.05 to 0.18 mm (0.002 to 0.007 in.) range. Etch depth is limited to about 1.5 mm (0.06 in.) by the thickness of the maskant typically on the order of 0.05 mm (0.002 in.).

The most common and most precise method for creating maskants involves the use of UV light-sensitive emulsions, called *photoresists*. In this method, photoresists are applied to the surface of the workpiece and selectively exposed to an intense ray of UV light through a photographic negative of the image to be patterned. The use of photoresists in CHM, called *photochemical machining* (PCM), is described in the following section.

Etch rates in CHM are very slow compared with other nontraditional machining processes. However, etching proceeds on all exposed surfaces simultaneously which significantly increases the overall material removal rate on large parts. The etch rate in CHM is directly proportional to the etchant concentration directly adjacent to the area being machined. For parts machined by immersion, the uniformity of the etchant concentration within the bath can be improved by agitation. If the bath is not agitated properly, several defect conditions can result as shown in Figure 28-3. *Islands*, or isolated high spots, can be the result of improper agitation on large parts. Islands can also be formed due to inadequate cleaning or inhomogeneity with the work material. An *overhang* can result if the etchant is not properly agitated, particularly on deep cuts. Other defects such as *dishing* and *pitting* (due to unequal etch rates) can also be caused by

FIGURE 28-3 Typical chemical milling defects: (a) overhang; deep cuts with improper agitation; (b) islands: isolated high spots from dirt, residual maskant or work material inhomogeneity; (c) dishing: thinning in center due to improper agitation or stacking of parts in tank.

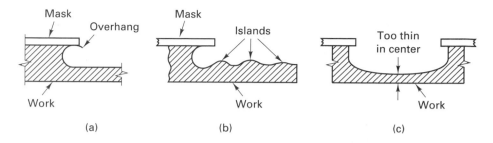

improper agitation. Other causes of unequal etch rates can be due to selective etching at microcracks and grain boundaries in the material.

Single-sided, blind etching of the part is called *chemical milling* or, when the photoresist method of applying maskants is used, *photochemical milling*. Chemical milling is so named because its earliest usage was for replacing mechanical milling on large components. Chemical milling is often used to remove weight on aircraft components as shown in Figure 28-4. Through-etching of the workpiece is called *chemical* (or *photochemical*) *blanking*. The process competes with blanking, laser cutting, and electrical discharge machining (EDM) for through-cutting of thin material sheets. Chemical blanking is typically performed using double-sided etching to increase production rates and minimize taper on the etched walls of the feature. A key requirement for chemical blanking is registration of top-side and bottom-side screens or phototools during masking. Because of the precision required, chemical blanking is not performed with the cut-and-peel method of masking.

Photochemical Machining. Figure 28-5 shows the specific steps that are involved when chemical machining is performed with the use of photoresists. These are as follows:

1. Clean the workpiece.
2. Coat the workpiece with a photoresist usually by hot-roller lamination of dry-film photoresists onto both sides of the workpiece although liquid photoresists may also be applied by dipping, flowing, rolling, or electrophoresis (i.e., migration of charged molecules in the presence of an electric field). For liquid photoresists, the coating is heated in an oven to remove solvents.

FIGURE 28-4 Iconel 718 aircraft engine parts. These sheet metal parts for a jet fighter engine are chemically milled to remove weight. (*Left*) As-formed workpiece; (*middle left*) workpiece coated with liquid rubber, fiberglass scribing template in place; (*middle right*) scribed workpiece; (*right*) finished part. About 0.035 in. of stock is removed from the 0.070-in.-thick workpieces. Tolerances are held to ±0.004 in.

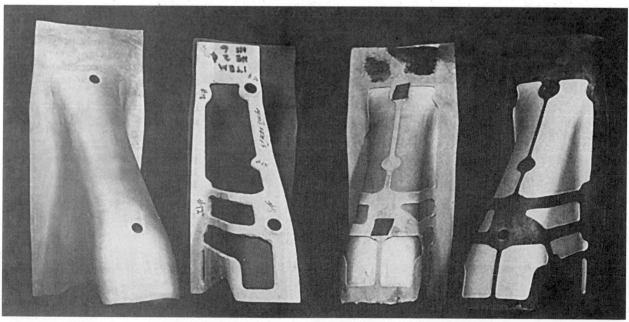

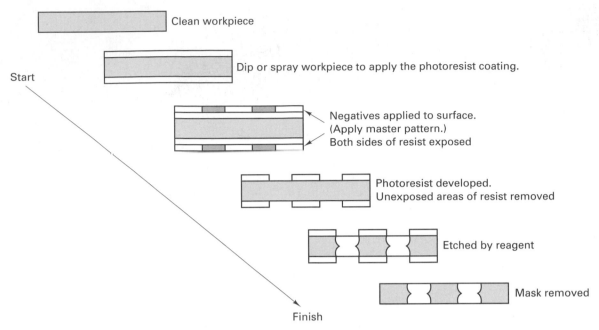

Start

Clean workpiece

Dip or spray workpiece to apply the photoresist coating.

Negatives applied to surface.
(Apply master pattern.)
Both sides of resist exposed

Photoresist developed.
Unexposed areas of resist removed

Etched by reagent

Mask removed

Finish

FIGURE 28-5 Basic steps in photochemical machining (PCM).

3. Prepare the "artwork." A drawing of the workpiece is made on a CAD system.

4. Develop the phototool. The CAD file is used to derive a photographic negative of the workpiece. Several methods may be used. Typically, the CAD drawing is downloaded to a laser imaging system that exposes the desired image directly onto photographic (e.g., silver halide) film. In the past, oversized artwork was used to increase the accuracy of the phototool through photographic reduction of the artwork.

5. Expose the photoresist. Bring the phototool in contact with the workpiece using a vacuum frame to ensure good contact and expose the workpiece to intense UV light.

6. Develop the photoresist. Exposure of the photoresist to intense UV light alters the chemistry of the photoresist making it more resistant to dissolution in certain solvents. By placing the exposed maskant in the proper solvent, the unexposed areas of the resist are removed exposing the underlying material for etching. All residue is rinsed away.

7. Spray the workpiece with (or immerse it in) the reagent.

8. Remove the remaining maskant.

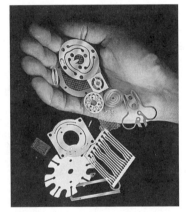

FIGURE 28-6 Typical parts produced by chemical machining or blanking.

PCM has been widely used for the production of small, complex parts such as printed circuit boards and very thin parts that are too small or too thin to be blanked or milled by ordinary sheet metal forming or machining operations, respectively (see Figure 28-6). Refinements to the PCM process are used in the microelectronics industry to etch metal and dielectric thin films (see microelectronics Chapter 34).

Design Factors in Chemical Machining. When designing parts that are to be made by chemical machining, several unique factors related to the process must be kept in mind. First, if artwork is used, dimensional variations can occur through size changes in the artwork or phototool film due to temperature and humidity changes. These can be controlled or eliminated by putting the artwork on thicker polyester films or glass, by controlling the temperature or humidity in the artwork and phototool production areas, or by use of a direct-write laser imaging system.

The second item that must be considered is the *etch factor*, or sometimes referred to as *etch radius*, which describes the undercutting of the maskant. The etchant acts isotropically on whatever surface is exposed. Areas that are exposed longer will have more metal removed from them. Consequently, as the depth of the etch increases, there is a tendency to undercut or etch under the maskant (Figure 28-7). The etch factor, E, in chemical machining is defined as:

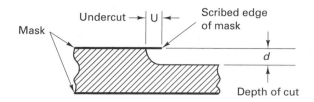

FIGURE 28-7 Undercutting in photochemical machining.

$$E = \frac{U}{d} \tag{28-1}$$

where d is the depth of cut and U is the undercut as defined in Figure 28-7. In photochemical machining, the term *anisotropy* (also sometimes referred to as etch factor) is used to describe the directionality of the cut. Using the same definitions from Figure 28-7, the anisotropy, A, of a material/etchant interaction in photochemical machining is defined as:

$$A = \frac{d}{U} \tag{28-2}$$

which is the inverse of Equation (28-1). In many electrical and electronic products, anisotropies much greater than one are desirable to permit greater densities of electrical and electronic components and wires.

The effect of undercutting is shown for both chemical milling and chemical blanking in Figure 28-8. An allowance for the etch factor must be taken into account in designing the part and the artwork or scribing template. In the case of chemical milling, the width of the opening in the maskant, W_m, must be reduced by an amount sufficient to compensate for the undercut under both sides of the maskant:

$$W_m = W_f - 2(E \cdot d) \tag{28-3}$$

FIGURE 28-8 Effect of the etch factor in chemical machining from one side and both sides of a plate.

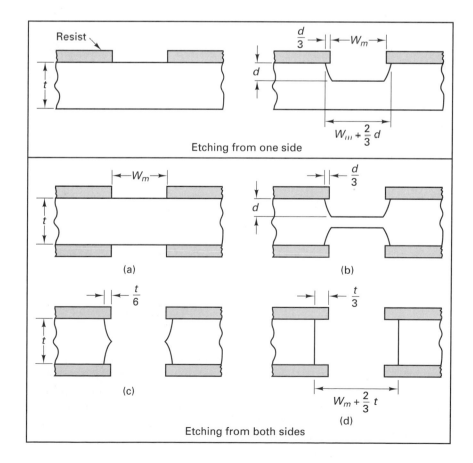

Etching from one side

(a) (b)

(c) (d)

Etching from both sides

TABLE 28-3.	Etch Rates and Etch Factors for Some Common Metal-Etchant Combinations in CHM		
Metal	Preferred Etchant	Penetration Rate (mm/mm)	Etch Factor
Aluminum	FeCl$_3$	0.025	1.7:1
Copper	FeCl$_3$	0.05	2.7:1
Nickel alloys	FeCl$_3$	0.018	2.0:1
Phosphor-bronze	Chromic acid	0.013	2.0:1
Silver	FeNO$_3$	0.02	1.5:1
Titanium	HF	0.25	2.0:1
Tool steel	HNO$_3$	0.018	1.5:1

Source: Benedict, *Nontraditional Manufacturing Processes*, Marcel Deckker, 1987, p. 200.

where W_f is the final desired width of the cut. This allowance is considered a minimum allowance; it has been found that results will vary based on etching conditions, and actual etch allowances will have to be somewhat greater and adapted to specific conditions.

Figure 28-8c shows that in double-sided chemical blanking, a sharp edge remains along the line at which breakthrough occurs. Because such an edge is usually objectionable, etching ordinarily is continued to produce nearly straight sidewalls as shown in Figure 28-8d. In order to achieve nearly straight sidewalls, it is typical to allow the process to continue for the amount of time necessary to do a through-cut from one side.

The anisotropy of the cut defines the maximum limit for aspect ratio (i.e., ratio of depth-to-width) of the cut which is a measure of how deep and narrow of a cut can be made. In chemical blanking, by using a double-sided etch, the maximum aspect ratio of the cut may be effectively doubled if the process is stopped at breakthrough (Figure 28-8c). This is the effective maximum aspect ratio which may be achieved in CHM. Consequently, it is difficult to produce deep, narrow cuts in materials using CHM. Table 28-3 shows some etch rates and etch factors for common metal and etchant combinations in CHM.

Advantages and Disadvantages of Chemical Machining. Chemical machining has a number of distinct advantages when compared with other machining and forming processes. Except for the preparation of the artwork and phototool, screen or scribing template, the process is relatively simple, does not require highly skilled labor, induces no stress or cold working in the metal, and can be applied to almost any metal—aluminum, magnesium, titanium, and steel being the most common. Large areas can be machined; tanks for parts up to 12 by 50 ft and spray lines up to 10 ft wide are available. Machining can be done on parts of virtually any shape. Thin sections, such as honeycomb, can be machined because there are no mechanical forces involved. CHM is very useful and economical for weight reduction. Figure 28-4 shows a typical large, thin part that has been chemically milled to reduce the weight of the part.

The tolerances expected with CHM range from ±0.0005 in. on small etch depths up to ±0.004 in. in depth in routine production involving substantial depths. Tolerances in chemical milling increase with the depth of the cut and with faster etch rates and vary for different materials (Figure 28-9). The surface finish is generally good, as shown in Table 28-1.

In using CHM, some disadvantages and limitations should be kept in mind. CHM requires the handling of dangerous chemicals and the disposal of potentially harmful byproducts although some recycling of chemicals may be possible. The metal removal rate is slow in terms of the unit area exposed, being about 0.2 to 0.04 lb/min per square foot exposed in the case of steel. However, because large areas can be exposed all at once, the overall removal rate may compare favorably with other metal-removal processes, particularly when the work material is not machinable or the workpiece is thin and fragile, unable to sustain large cutting forces.

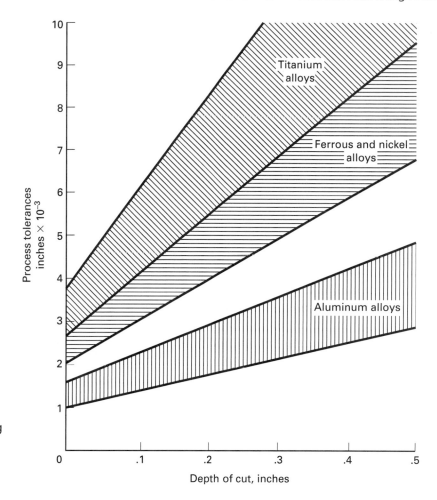

FIGURE 28-9 Chemical milling tolerance bands showing variation with respect to depth of cut and different metals.

The soundness and homogeneity of the metal are very important. Wrought materials should be uniformly heat treated and stress relieved prior to processing. Although chemical machining induces no stresses, it may release existing residual stresses in the metal and thus cause warpage. Castings can be chemically machined provided that they are not porous and have uniform grain size. Lack of the latter can cause nonuniform etching rates producing islands. Because of the different grain structures that exist near welds, weldments usually are not suitable for chemical machining. Preferential etching due to *intergranular attack* can also be an issue in CHM.

CHEMICAL-MECHANICAL POLISHING

Chemical-mechanical polishing (CMP), or chemo-mechanical polishing, uses the synergy of chemistry and mechanical grinding to obtain flatness on the order of 50 nm. CMP is used in the fabrication of integrated circuits (IC) to obtain planar surfaces after dielectric and metal depositions during interconnection of circuit components. As shown in Figure 28-10, the process involves rotation of the wafer as it is pressed against a slurry-filled abrasive pad which rotates counter to the wafer. The slurry contains both a particle abrasive as well as an etchant and is deposited directly onto the abrasive pad by a mechanical arm. The wafer is held in a carrier by a backing pad designed to distribute the mechanical force evenly over the surface of the wafer. Fused silica in a weak KOH solution is a common slurry for oxide polishing. Ferricyanide-phosphate with silica or alumina is used to polish tungsten. Raised features are etched much faster than flat regions since mechanical pressure concentrated on the raised features enhances etching. For dielectric polishing, typical etch rates are on the order of 300 nm/minute.

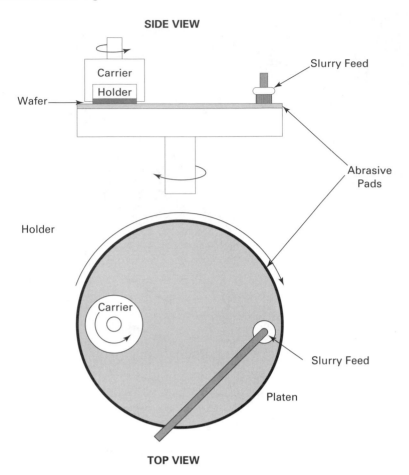

SIDE VIEW

Carrier

Holder

Wafer

Slurry Feed

Abrasive
Pads

Holder

Carrier

Slurry Feed

Platen

TOP VIEW

FIGURE 28-10 Schematic of chemical-mechanical polishing (CMP).

■ 28.3 ELECTROCHEMICAL MACHINING PROCESSES

ELECTROCHEMICAL MACHINING

Electrochemical machining, commonly designated ECM, removes material by anodic dissolution with a rapidly flowing electrolyte. It is basically a deplating process in which the tool is the cathode and the workpiece is the anode; both must be electrically conductive. The electrolyte, which can be pumped rapidly through or around the tool, sweeps away any heat and waste product (sludge) given off during the reaction. The sludge is captured and removed from the electrolyte through filtration. The shape of the cavity is defined by the tool which is advanced by means of a servomechanism that controls the gap between the electrodes (i.e., the interelectrode gap) to a range from 0.003 to 0.03 in. (0.01 in. typical). The tool advances into the work at a constant feed rate, or penetration rate, that matches the deplating rate of the workpiece. The electrolyte is a highly conductive solution of inorganic salt, usually NaCl, KCl, and $NaNO_3$ (or other proprietary mixtures) and is operated at about 75 to 150°F with flow rates ranging from 50 to 200 ft/sec. The temperature of the electrolyte is maintained through appropriate temperature controls. Tools are usually made of copper or brass and sometimes stainless steel. The process is shown schematically in Figure 28-11.

The behavior of the ECM process is governed by the laws of electrolysis (the use of electrical current to bring about chemical change). Faraday's first law of electrolysis states that the amount of chemical change (material removed) during electrolysis is proportional to the charge (number of electrons) passed. This can be expressed mathematically as:

$$n = K \cdot I \cdot t \tag{28-4}$$

where *n* is the theoretical number of moles of material removed, *I* is the applied current which is assumed to stay constant and *t* is the time over which the current is applied. The term *K* is a proportionality constant which is inversely proportional to the valence of the workpiece material.

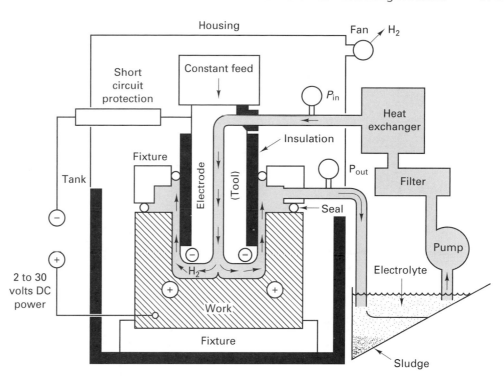

FIGURE 28-11 Schematic diagram of electrochemical machining process (ECM).

To convert Equation (28-4) into a useful expression for estimating the theoretical material removal rate, MRR_i, the number of moles, n, can be converted into a material volume by multiplying through the atomic weight, A_w, and dividing by the density, p, of the workpiece material. Further, by dividing through by the time, t, an expression for the volumetric material removal rate is obtained,

$$MRR_i = \left(\frac{K \cdot A_w}{p} \right) \cdot I = MRR_s \cdot I \qquad (28\text{-}5)$$

where MRR_s is a proportionality constant called the specific material removal rate based on the valence, atomic weight, and density of the workpiece material. In suitable applications, material removal rates on the order of 0.1 in³/min per 1000 A can be expected. Table 28-4 shows the specific material removal rate for some different materials.

TABLE 28-4.	Material Removal Rates for ECM of Alloys Assuming 100% Current Efficiency	
Alloy	Theoretical Removal Rates for 1,000 Amperes per Square Inch	
	in³/min	cm³/min
4340 steel	0.133	2.18
17-4 PH	0.123	2.02
A-286	0.117	1.92
M252	0.110	1.80
Rene4I	0.108	1.77
Udimet 500	0.110	1.80
Udimet 700	0.108	1.77
L605	0.107	1.75

Note: Rates listed were calculated using Faraday's law and valences as follows:

Aluminum	3	Copper	2	Silicon	0
Carbon	0	Iron	4	Titanium	4
Columbium	3	Manganese	3	Tungsten	6
Cobalt	2	Molybdenum	4	Vanadium	5
Chromium	3	Nickel	2		

An estimate for the theoretical feed rate, f_i, or penetration rate, in ECM can be made by dividing Equation (28-5) by the area of workpiece material, A, exposed to the ECM tool at the interelectrode gap. This yields:

$$f_r = \text{MRR}_s \cdot \frac{I}{A} = \text{MRR}_s \cdot J \qquad (28\text{-}6)$$

Equations (28-4) through (28-6) assume that the efficiency, E, for metal removal by electrolysis is 100% though normally it is in the range of 90% to 100% depending on the current, voltage and other operating conditions. Therefore, Equations (28-5) and (28-6) may be modified as follows:

$$\text{MRR} = E \cdot \text{MRR}_i \qquad (28\text{-}7)$$

$$f_r = E \cdot f_i \qquad (28\text{-}8)$$

where MRR and f_r are the actual material removal rate and actual feed rate, respectively.

As shown in Equation (28-6), the penetration rate in ECM is primarily a function of the current density, J, and the physical properties of the material. Current densities from 50 to 1500 A/in^2 are used and typical penetration rates in metals are from 0.02 to 0.75 in./min. Figure 28-12 shows that the penetration rate in ECM is also a function of the interelectrode gap. As the interelectrode gap is reduced, the resistance across the gap is also reduced permitting a larger flow of electrons. One disadvantage of reducing the interelectrode gap is that the likelihood of plating anodic material onto the cathode is increased, thereby requiring higher electrolyte flow rates.

The importance of these implications is that the penetration rate, while a function of current density and interelectrode gap, is not directly affected by the hardness or toughness of the work material. This makes ECM advantageous for the machining of certain high-strength materials with poor machinability. Penetration rates up to 0.1 in./min are obtained routinely in Waspalloy, a very hard metal alloy.

Pulsed-current ECM (PECM) has recently shown the potential to improve accuracies and surface finish in traditional ECM. In PECM, high current densities (>100 A/cm^2) are pulsed on for durations on the order of 1 ms and pulsed off for intervals on the order of 10 ms. The relaxation (pulse off) interval permits reaction by-products to be removed from the interelectrode gap at low electrolyte flow rates without electrolytic deposition on the ECM tool. As a result, high current densities can be used at small interelectrode distances improving both removal rates and precision. Because PECM allows lower electrolyte flow rates, it has been proposed to remove recast layers from the surface of dies produced by electrical-discharge machining (EDM). Some efforts are being made to integrate this technology into EDM platforms.

Pulsed currents have also been used in *electrochemical micromachining* (EMM). In one variation of the process, photolithographic masks are used to concentrate material removal in selective areas of thin films. In addition to providing smoother surfaces

FIGURE 28-12 Relationship of current density, penetration rate, and machining gap in electrochemical machining.

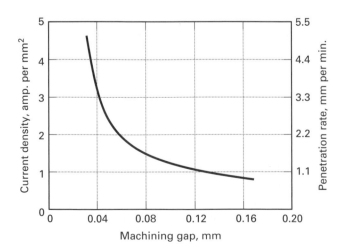

(on the order of electropolished surfaces) than those found in traditional ECM, the high current density ($\sim 100\,\text{A/cm}^2$) results in less taper in etching profiles and more uniform material removal across the workpiece which are important for shrinking feature sizes in micromachining.

Electrochemical polishing is a modification of the ECM process that operates essentially the same as ECM, but with a much slower penetration rates. Current density is lowered which greatly reduces the material removal rate and produces a fine finish on the order of 10 μin. Electrochemical polishing must be differentiated from *electropolishing* which operates without the use of a part-specific hard tool. Electropolishing is discussed in greater detail in Chapter 31.

Electrochemical Hole Machining.

Several electrochemical hole-drilling processes have been developed for drilling small holes with high aspect (depth-to-diameter) ratios in difficult to machine materials. In *electrostream drilling*, large numbers of holes (over 50) can be simultaneously gang drilled in nickel and cobalt alloys with diameters down to 0.005 in. at aspect ratios of 50:1. Machining is performed by a high-velocity stream of charged, acidic electrolyte ejected from the capillary end of a drawn glass tube (the tool). An internal electrode (e.g., a small titanium wire) is fed into the large end of the tube and placed close to the capillary. The electrode is used to charge the electrolyte which is pumped through the tube. An acidic electrolyte is used so that the sludge byproduct goes into solution instead of clogging flow. Voltages used in the process are 10 to 20 times higher than that of typical ECM processes. This technique was originally developed to drill high-incident-angle, cooling holes in turbine blades for jet engines.

A second process, known as the *shaped-tube electrolytic machining* (STEM) process, was also created in response to unique challenges presented in the jet engine industry. Like the electrostream process, STEM is also capable of gang drilling small holes in difficult-to-machine materials. However, the STEM process is generally not capable of drilling holes smaller than about 0.02 in. STEM is capable of make shaped holes with aspect ratios as high as 300:1. Holes up to 24 in. in depth have been drilled. Like the electrostream process, it uses an acidic electrolyte to minimize clogging due to sludge build up. The major differences between the STEM process and the electrostream process are the reduced voltage levels (5 to 10 V dc) and the special electrodes, which are long, straight, metallic tubes coated with an insulator (Figure 28-13). The insulator helps to eliminate taper by constraining the electrolytic action between the bottom of the tool and the workpiece. Titanium is often used for its ability to resist acids. The electrolyte is pressure-fed through the tube and returns through the gap (0.001 to 0.002 in.) between the insulated tube wall and the hole wall. Electrolyte concentrations may include up to 10% sulfuric acid. Lower concentrations may be used to increase tool life.

Electrochemical Grinding.

Electrochemical grinding, commonly designated ECG, is a low-voltage, high-current variant of ECM in which the tool cathode is a rotating, metal-bonded, diamond grit grinding wheel. The setup shown in Figure 28-14 for grinding a cutting tool is typical. As the electric current flows through the electrolyte between the workpiece and the wheel, some surface metal is removed electrochemically and some is changed to a metal oxide, which is ground away by the abrasives. As the oxide film is removed, new surface metal is exposed to electrolytic action. Most of the material removal is electrochemical with only about 10% of the material being removed by grinding. The MRR is dependent on many variables. Table 28-5 gives some typical values for the MRR.

The wheels used in ECG must be electrically conductive abrasive wheels. For most metals, resin-bonded aluminum oxide wheels are recommended. The resin bond is loaded with copper to provide for negligible electrical resistance. The wheels are dressable, using a variety of wheel-dressing measures. Once dressed, the wheels can then be used for precision form-grinding operations.

The abrasive particles are hard, nonconducting materials such as aluminum oxide, diamond, or borazon (cubic boron nitride). In addition to increasing the efficiency of the process, the abrasives act as an insulating spacer maintaining a separation of from 0.0005 to 0.003 in. (0.012 to 0.05 mm) between the electrodes. A dead short would result if the

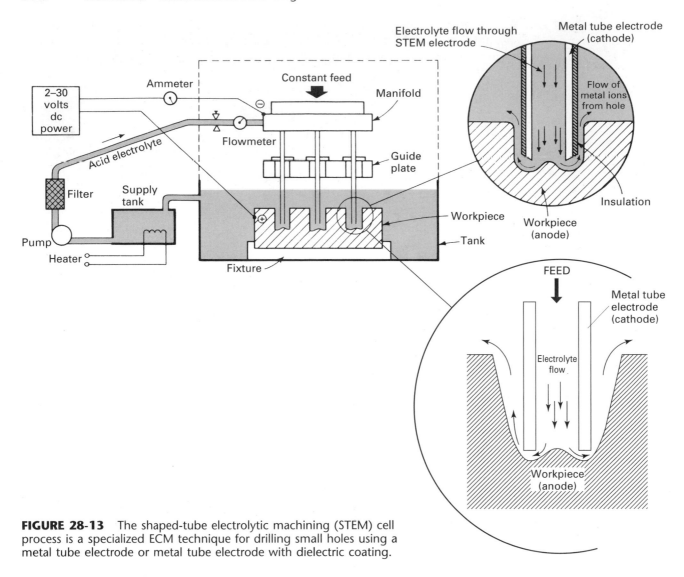

FIGURE 28-13 The shaped-tube electrolytic machining (STEM) cell process is a specialized ECM technique for drilling small holes using a metal tube electrode or metal tube electrode with dielectric coating.

FIGURE 28-14 Equipment setup and electrical circuit for electrochemical grinding.

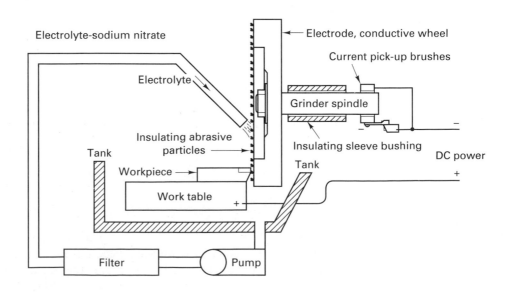

TABLE 28-5.   Metal Removal Rates for ECG for Various Metals

Metal	Valency	Density lb/in³	Density g/cm³	Metal-Removal Rate at 1000 A lb/hr	Metal-Removal Rate at 1000 A in³/min	Metal-Removal Rate at 1000 A cm³/min
Aluminum	3	0.098	2.67	0.74	0.126	2.06
Beryllium	2	0.067	1.85	0.37	0.092	1.50
Chromium	2	0.260	7.19	2.14	0.137	2.25
	3			1.43	0.092	1.51
	6			0.71	0.046	0.75
Cobalt	2	0.322	8.85	2.42	0.125	2.05
	3			1.62	0.084	1.38
Niobium	3	0.310	8.57	2.55	0.132	2.16
(columbium)	4			1.92	0.103	1.69
	5			1.53	0.082	1.34
Copper	1	0.324	8.96	5.22	0.268	4.39
	2			2.61	0.134	2.20
Iron	2	0.284	7.86	2.30	0.135	2.21
	3			1.53	0.090	1.47
Magnesium	2	0.063	1.74	1.00	0.265	4.34
Manganese	2	0.270	7.43	2.26	0.139	2.28
	4			1.13	0.070	1.15
	7			0.65	0.040	0.66
Molybdenum	3	0.369	10.22	2.63	0.119	1.95
	4			1.97	0.090	1.47
	6			1.32	0.060	0.98
Nickel	2	0.322	8.90	2.41	0.129	2.11
	3			1.61	0.083	1.36
Silicon	4	0.084	2.33	0.58	0.114	1.87
Silver	1	0.379	10.49	8.87	0.390	6.39
Tin	2	0.264	7.30	4.88	0.308	5.05
	4			2.44	0.154	2.52
Titanium	3	0.163	4.51	1.31	0.134	2.19
	4			0.99	0.101	1.65
Tungsten	6	0.697	19.3	2.52	0.060	0.98
	8			1.89	0.045	0.74
Uranium	4	0.689	19.1	4.90	0.117	1.92
	6			3.27	0.078	1.29
Vanadium	3	0.220	6.1	1.40	0.106	1.74
	5			0.84	0.064	1.05
Zinc	2	0.258	7.13	2.69	0.174	2.85

Source: 1985 SCTE Conference Proceedings, ASM, Metals Park, Ohio, 1986.

insulating particles were absent. The particles also serve to grind away any passivating oxide that should form during electrolysis. The process is used for shaping and sharpening carbide cutting tools, which cause high wear rates on expensive diamond wheels in normal grinding. ECG greatly reduces wheel wear. Fragile parts (e.g., honeycomb structures), surgical needles, and tips of assembled turbine blades have also been ECG-processed successfully. The lack of heat damage, burrs, and residual stresses is very beneficial, particularly when coupled with MRRs that are competitive with conventional grinding and far less wheel wear.

Electrochemical Deburring.   *Electrochemical deburring* is a deburring process which works on the principle that electrolysis is accelerated in areas with small interelectrode gaps and prevented in areas with insulation between electrodes. The cathodic tool in electrochemical deburring is stationary and generally shaped as a negative of the workpiece to focus the electrolysis to the region of the workpiece where burrs are to be removed. Portions of the tool unused for deburring are coated with insulation to prevent the electrolytic reaction. The process does not require a feed mechanism. A fixture made of insulating material is used to position the workpiece with respect to the cathodic tool. Because of the small amount of material removed, electrochemical deburring generally requires short cycle times under one minute.

Design Factors in Electrochemical Machining.   In general, current densities tend to concentrate at sharp edges or features and, therefore, produce rounded corners. Therefore, ECM processes should not be used to create sharp corners or pockets in the workpiece although sharp corners are possible on through features. Due to the need for an interelectrode gap, the actual feature size will be larger than the tool size. The overcut on the side of the tool is normally on the order of 0.005 in. Taper should be expected in all features. As shown in Figure 28-13a, taper is due to electrolytic reaction between the side of the tool and the workpiece. Depending on the tool design, taper can be held to 0.001 in./in. For micromachining applications, an insulator may be coated on the side of the tool to prevent the taper effects (Figure 28-13b).

Control of the electrolyte flow can be difficult in parts with irregular shapes. Changes in electrolyte concentration due to varying flow patterns can change the local resistance across the interelectrode gap resulting in local variations in removal rates and tolerances. In addition, high electrolytic flow rates can cause erosion of workpiece features. Therefore, complex geometries requiring tortuous interelectrode flow paths are discouraged.

Generally, no detrimental effects to the properties of the workpiece material are expected when using ECM. However, the lack of compressive residual stresses imparted to the surface of the workpiece during operation can have a negative impact on the fatigue resistance of the part when compared with mechanically machined parts. Further, if parameters are set improperly, the process may favor electrolysis at the grain boundaries resulting in intergranular attack which may also have a negative effect on the fatigue resistance of the part. If possible, shot peening or some similar process may be used to impart compressive stresses to the surface of the workpiece, thereby improving its fatigue resistance.

Advantages and Disadvantages of Electrochemical Machining. ECM is well suited for the machining of complex two-dimensional shapes into delicate or difficult-to-machine geometries made from poorly machinable but conductive materials. The principal tooling cost is for the preparation of the tool electrode, which can be time consuming and costly, requiring several cut-and-try efforts for complex shapes since it is difficult to predict the precise final geometry due to variable current densities produced by certain electrode geometries (e.g., corners) or electrolyte flow variations. There is no tool wear during actual cutting which suggests that the process becomes more economical with increasing volume. The process produces a stress-free surface which can be advantageous especially for small, thin parts. The ability to cut a large area simultaneously (as in chemical and electrical discharge machining) makes the production of small parts very productive. However, as in chemical machining, ECM requires the disposal of environmentally harmful by-products.

■ 28.4 MECHANICAL NTM PROCESSES

ULTRASONIC MACHINING

Ultrasonic machining (USM), sometimes called *ultrasonic impact grinding*, employs an ultrasonically vibrating tool to impel the abrasives in a slurry at high velocity against the workpiece. The tool is fed into the part as it vibrates along an axis parallel to the tool feed at an amplitude on the order of several thousandths of an inch and a frequency of 20 kHz. As the tool is fed into the workpiece, a negative of the tool is machined into the workpiece. The cutting action is performed by the abrasives in the slurry which is continuously flooded under the tool. The slurry is loaded up to 60% by weight with abrasive particles. Lighter abrasive loadings are used to facilitate the flow of the slurry for deep drilling (up to 2 in. deep). Boron carbide, aluminum oxide, and silicon carbide are the most commonly used abrasives in grit sizes ranging from 400 to 2000. The amplitude of the vibration should be set approximately to the size of the grit. The process can use shaped tools to cut virtually any material but is most effective on materials with hardnesses greater than R_C 40 including brittle and nonconductive materials such as glass. Figure 28-15 shows a simple schematic of this process.

USM uses piezoelectric or magnetostrictive transducers to impart high-frequency vibrations to the toolholder and tool. Abrasive particles in the slurry are accelerated to great speed by the vibrating tool. The tool materials are usually brass, carbide, mild steel, or tool steel and will vary in tool wear depending on their hardness. Wear ratios

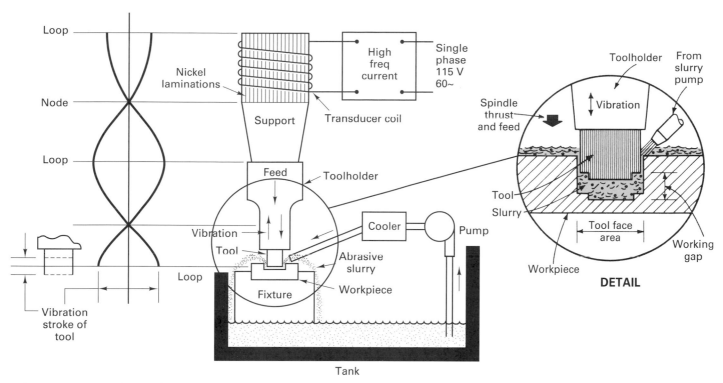

FIGURE 28-15 Sinking a hole in a workpiece by ultrasonic machining.

(workpiece material removed versus tool material lost) from 1:1 (for tool steel) to 100:1 (for glass) are possible. Because of the high number of cyclic loads, the tool must be strong enough to resist fatigue failure.

The cut will be oversize by about twice the size of the abrasive particles being used, and holes will be tapered, usually limiting the hole depth-to-diameter ratio to about 3:1. Surface roughness is controlled by the size of the abrasive particles (finer finish with smaller particles). Holes, slots, or shaped cavities can be readily eroded in any hard material, conductive or nonconductive, metallic, ceramic, or composite. Advantages of the process include that it is one of the few machining methods capable of machining glass. Also, it is the safest machining method. Skin is impervious to the process because of its ductility. High-pitched noise can be a problem due to secondary vibrations. In addition to machining, ultrasonic energy has also been employed for coining, lapping, deburring, and broaching. Plastics can be welded using ultrasonic energy (see Chapter 36).

WATERJET CUTTING

Waterjet cutting (WJC), also known as *waterjet machining* or *hydrodynamic machining*, uses a high-velocity fluid jet impinging on the workpiece to perform a slitting operation (Figure 28-16). Water is ejected from a nozzle orifice at high pressure (up to 60,000 psi). The jet is typically 0.003 to 0.020 in. in diameter and exits the orifice at velocities up to 3000 ft/sec. Key process parameters include water pressure, orifice diameter, water flow rate, and working distance (distance between the workpiece and the nozzle).

Nozzle materials include synthetic sapphire due to its machinability and resistance to wear. Tool life on the order of several hundred hours is typical. Mechanisms for tool failure include chipping from contaminants or constriction due to mineral deposits. This emphasizes the need for high levels of filtration prior to pressure intensification. In the past, long-chain polymers were added to the water to make the jet more coherent (i.e., not come out of the jet dispersed). However, with proper nozzle design, a tight, coherent waterjet may be produced without additives.

The advantages of WJC include the ability to cut materials without burning or crushing the material being cut. Figure 28-17 shows a comparison (end view) of cutting

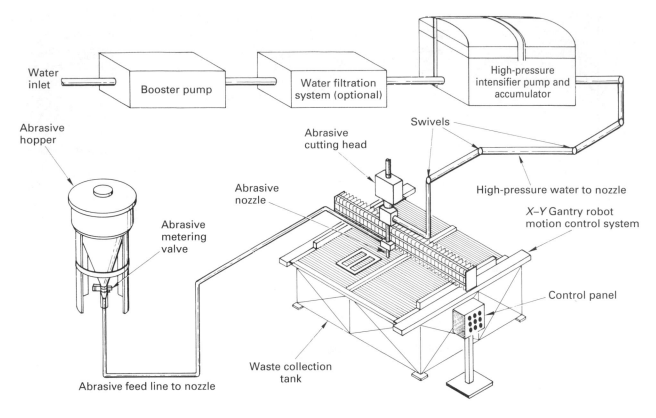

Water inlet

Booster pump

Water filtration system (optional)

High-pressure intensifier pump and accumulator

Abrasive hopper

Abrasive cutting head

Swivels

Abrasive nozzle

High-pressure water to nozzle

Abrasive metering valve

X–Y Gantry robot motion control system

Control panel

Abrasive feed line to nozzle

Waste collection tank

Abrasive cutting head

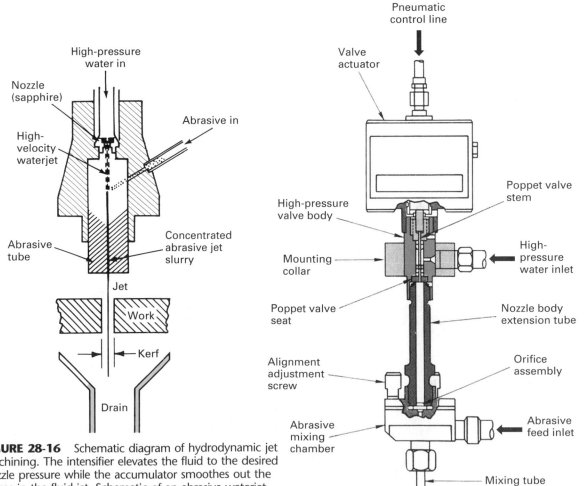

High-pressure water in

Nozzle (sapphire)

High-velocity waterjet

Abrasive in

Abrasive tube

Concentrated abrasive jet slurry

Jet

Work

Kerf

Drain

Pneumatic control line

Valve actuator

High-pressure valve body

Poppet valve stem

Mounting collar

High-pressure water inlet

Poppet valve seat

Nozzle body extension tube

Alignment adjustment screw

Orifice assembly

Abrasive mixing chamber

Abrasive feed inlet

Mixing tube

FIGURE 28-16 Schematic diagram of hydrodynamic jet machining. The intensifier elevates the fluid to the desired nozzle pressure while the accumulator smoothes out the pulses in the fluid jet. Schematic of an abrasive waterjet machining nozzle is shown on the right.

corrugated boxboard with a mechanical knife and with WJC. The mechanism for material removal is simply the impinging pressure of the water exceeding the compressive strength of the material. This limits the materials that can be cut by the process to leather, plastics, and other soft nonmetals which is the major disadvantage of the process. Alternative fluids (alcohol, glycerine, cooking oils) have been used in processing meats, baked goods, and frozen foods. Other disadvantages include that the process is noisy and requires operators to have hearing protection.

ABRASIVE WATERJET CUTTING

The majority of the metalworking applications for waterjet cutting require the addition of abrasives. This process is known as *abrasive waterjet cutting* (AWC). A full range of materials, including metals, plastics, rubber, glass, ceramics, and composites, can be machined by AWC. Cutting feed rates vary from 20 in./min for acoustic tile to 50 in./min for epoxies and 500 in./min for paper products.

Abrasives are added to the waterjet in a mixing chamber on the downstream side of the waterjet orifice. A single, central water jet with side feeding of abrasives into a mixing chamber was shown in Figure 28-16. In the mixing chamber, the momentum of the water is transferred to the abrasive particles and the water and particles are forced out through the AWC nozzle orifice also called the mixing tube. This design can be made quite compact; however, it also experiences rapid wear in the mixing tube. An alternate configuration is to feed the abrasives from the center of the nozzle with a converging set of angled waterjets imparting momentum to the abrasives. This nozzle design produces better mixing of the water and abrasives as well as increased nozzle life. The inside diameter of the mixing tube is normally from 0.04 to 0.125 in. in diameter. These tubes are normally made of carbide.

Generally, the kerf of the cut is about 0.001 in. greater than the nozzle orifice. AWC requires control of additional process parameters over waterjet machining including abrasive material (density, hardness, shape), abrasive size or grit, abrasive flow rate (pounds per minute), abrasive feed mechanism (pressurized or suction), and AWC nozzle (design, orifice diameter, and material). Typical AWC systems operate under the following conditions: water pressures of 30,000 to 50,000 psi; water orifice diameters from 0.01 to 0.022 in.; and working distances of 0.02 to 0.06 in. Working distances are much smaller than in WJC to minimize the dispersion of the abrasive waterjet prior to entering the material. Abrasive materials used include garnet, silica, silicon carbide, or aluminum oxide. Abrasive grit sizes range from 60 to 120 and abrasive flow rates from 0.5 to 3 lb/min. For many applications, the AWC tool is combined with a CNC controlled X–Y table, which permits contouring and surface engraving.

AWC can be used to cut any material through the appropriate choice of the abrasive, waterjet pressure, and feed rate. Figure 28-18 gives cutting speeds for various metals. The ability for the abrasive waterjet to cut through thick materials (up to 8 in.) is attributed to the reentrainment of abrasive particles in the jet by the workpiece material. AWC is particularly suited for composites as the cutting rates are reasonable and they do not delaminate the layered material. In particular, AWC is used in the airplane industry to cut carbon-fiber composite sections of the airplane after autoclaving.

ABRASIVE JET MACHINING

One of the least expensive of the nontraditional processes is *abrasive jet machining* (AJM). AJM removes material by a focused jet of abrasives and is similar in many respects to AWC with the exception that momentum is transferred to the abrasive particles by a jet of inert gas. Abrasive velocities on the order of 1000 ft/sec are possible with AJM. The small mass of the abrasive particles produces a micro-scale chipping action on the workpiece material. This makes AJM ideal for processing hard, brittle materials including glass, silicon, tungsten, and ceramics. It is not compatible with soft, elastic materials. A schematic of the AJM process is shown in Figure 28-19.

Key process parameters include working distance, abrasive flow rate, gas pressure, and abrasive type. Working distance and feed rate are controlled by hand. If necessary,

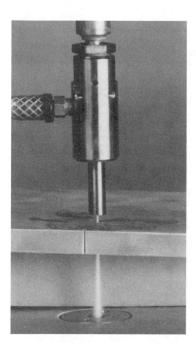

FIGURE 28-17 Machining metal with abrasive waterjet is a routine operation in many factories.

Cutting speeds with abrasive waterjet

Material	Thickness (in.)	Nozzle speed (in./min)	Edge quality (comments)
Aluminum	0.130	20-40	good
Aluminum tube	0.220	50	burred
Aluminum casting	0.400	15	
Aluminum	0.500	6-10	
Aluminum	3.0	0.5-5	
Aluminum	4.0	0.2-2	
Brass	0.125	18-20	good or small burr
Brass	0.500	4-5	
Brass	0.75	0.75 3	striations at 1 +
Bronze	1.100	1.0	good
Copper	0.125	22	good
Copper-nickel	0.125	12-14	fair edge
Copper-nickel	2.0	1.5-4.0	fair edge
Lead	0.25	10-50	good to striated
Lead	2.0	3-8	slower = better
Magnesium	0.375	5-15	good
Armor plate	0.200	1.5-15	good
Carbon steel	0.250	10-12	good
Carbon steel	0.750	4-8	good to bad edge
Carbon steel	3.0	0.4	good w. sm. nozzle
4130 carbon steel	0.5	3.0	
Mild steel	7.5	0.017-0.05	
High-strength steel	3.0	0.38	
Cast iron	1.5	1.0	good edge
Stainless steel	0.1	10-15	good to striated
Stainless steel	0.25	4-12	good to striated
Stainless steel	1.0	1.0	65-150 RMS
15-5 PH stainless	4.0	0.3	striated
Inconel 718	1.25	0.5-1.0	good
Inconel	0.250	8-12	good to striated
Inconel	2-2.5	0.2	good to fair
Titanium	0.025-0.050	5-50	good
Titanium	0.500	1-6	65-150 RMS

Material	Thickness (in.)	Nozzle speed (in./min)	Edge quality (comments)
Titanium	2.0	0.5-1.0	125 RMS
Tool steel	0.250	3-15	125 RMS
Tool steel	1.0	2-5	
Nonmetals			
Acrylic	0.375	15-50	good to fair
C-glass	0.125	100-200	shape dependent
Carbon/carbon comp.	0.125	50-75	good
Carbon/carbon comp.	0.500	10-20	good
Epoxy/glass composite	0.125	100-250	good
Fiberglass	0.100	150-300	good
Fiberglass	0.250	100-150	good
Glass (plate)	0.063	40-150	good
Glass (plate)	0.75	10-20	125 RMS
Graphite/epoxy	0.250	15-70	good to practical
Graphite/epoxy	1.0	3-5	good
Kevlar (steel reinf.)	0.125	30-50	good
Kevlar	0.375/0.580	10-25	good
Kevlar	1.0	3-5	good
Lexan	0.5	10	good
Phenolic	0.25-0.50	10-15	good
Plexiglass	0.175/0.50	25	
Rubber belting	0.300	200	good
Ceramic matrix composites			
Toughened zirconia	0.250	1.5	
SiC fiber in SiC	0.125	1.5	
Al_2O_3/CoCrAly (60%/40%)	0.125	2	
SiC/TiB_2 (15%)	0.250	0.35	
Metal matrix composites			
Mg/B_4C (15%)	0.125	35	fair
Al/SiC (15%)	0.500	8-12	good to fair
Al/Al_2O_3 (15%)	0.250	15-20	good to fair

Table 2. Cutting speeds with simple waterjet

Material	Thickness (in.)	Nozzle speed (in./min)	Edge quality (comments)
ABS plastic	0.087	20-50	100% separation
Aluminum	0.050	2-5	burr
Cardboard	0.055	240-600	slits very well
Delrin	0.500	2-5	good to stringers
Fiberglass	0.100	40-150	good to raggy
Formica	0.040	1450	
Graphite composite	0.060	25	
Kevlar	0.040-0.250	50-3	fair, some furring

Material	Thickness (in.)	Nozzle speed (in./min)	Edge quality (comments)
Lead	0.125	10	good, slight burr
Plexiglass	0.118	30-35	fair
Printed circuit bd.	0.050-0.125	50-5	good
PVC	0.250	10-20	good to fair
Rubber	0.050	2400-3600	good
Vinyl	0.040	2000-2400	good
Wood	0.125	40	fair

Comment on these tables: In trying to provide data on waterjet and abrasive waterjet cutting we have collected material from diverse sources. But we must note that most of the data presented is not from uniform tests. Also, note that in many cases data was largely absent on such parameters as pump horsepower, waterjet pressure, abrasive-particle rate of flow or type or size, and standoff distance. So these cutting rates vary widely in value—from laboratory control to shop floor ballpark estimates. Many of the top speeds cited either represent cuts made to illustrate speed alone, without regard to surface quality, or may reflect data from machines with very high power output. (*American Machinist*, October 1989.)

FIGURE 28-18   Typical values for through-cutting speeds for simple waterjet and abrasive waterjet of machining metals and nonmetals.

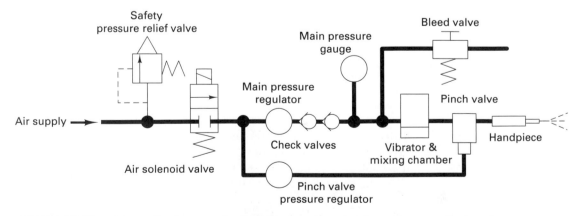

FIGURE 28-19   Schematic diagram of an AJM system showing how gas propellant and abrasive powder are combined and delivered to the handpiece.

a hard mask can be placed on the workpiece to control dimensions. Abrasives are typically smaller than those used in AWC. Abrasives are typically not recycled since the abrasives are cheap and are only used on the order of several hundred grams per hour. To minimize particulate contamination of the work environment, a dust-collection hood should be used in concert with the AJM system.

ABRASIVE FLOW MACHINING

Abrasive flow machining (AFM) involves the use of a semisolid liquid or gel laden with abrasives to flow over or through a workpiece to perform edge finishing, deburring, radiusing, polishing, or minor surface machining. Aluminum oxide, silicon carbide, boron carbide, and diamond are used as abrasives in grit sizes ranging from 8 to 700. The part is mounted in the machine and pistons drive the abrasive back and forth through the part. This process works for most materials and is useful for polishing and deburring features normally inaccessible in internal passageways. An example of how the surface finish of a hole is improved is shown in Figure 28-20.

FIGURE 28-20 An abrasive flow machine produces the pressure necessary to force media through or across a workpiece. Abrasive action takes place wherever flow is restricted as the media is forced back and forth between two opposed media cylinders. The before and after surfaces are remarkably different as the above SEM micrographs show.

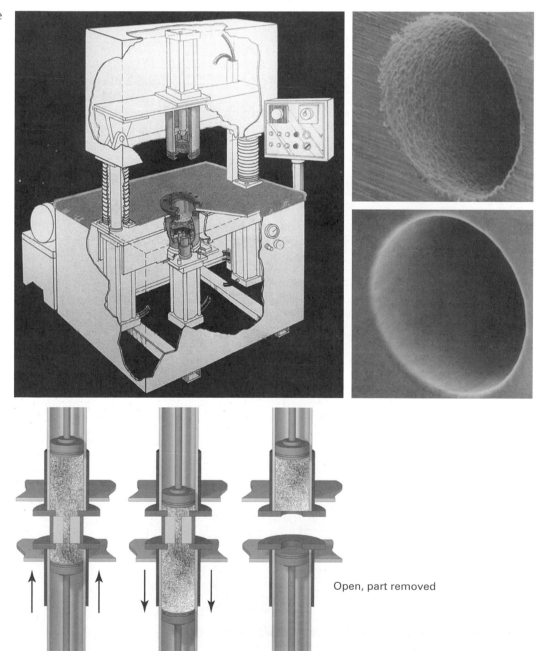

Open, part removed

The AFM process should be considered if a workpiece requires laborious hand finishing of complex shapes or internal edges that are difficult to reach. The process may also be used when bulk methods of deburring (e.g., tumbling or vibrating) are not satisfactory. The process is effective in removing recast layers produced by EDM or laser beam machining discussed in the following sections.

■ 28.5 THERMAL NTM PROCESSES

ELECTRICAL DISCHARGE MACHINING

As early as the 1700s, Benjamin Franklin wrote of witnessing the removal of metal by electrical sparks. However, it was not until the 1940s before development of the *electrical discharge machining* (EDM) process, also known as electric or electro discharge machining, began in earnest. Today, it is one of the most widely used of the nontraditional machining processes.

EDM processes remove metal by discharging electric current from a pulsating DC power supply across a thin interelectrode gap between the tool and the workpiece. See Figure 28-21 for a schematic. The gap is filled by a dielectric fluid which becomes locally ionized at the point where the interelectrode gap is the narrowest, generally, where a high point on the workpiece comes close to a high point on the tool. The ionization of the dielectric fluid creates a conduction path in which a spark is produced. The spark produces a tiny crater in the workpiece by melting and vaporization, and consequently tiny, spherical "chips" are produced by resolidification of the melted quantity of workpiece material. Bubbles from discharge gases are also produced. In addition to machining the workpiece, the high temperatures created by the spark also melt or vaporize the tool creating tool wear. The dielectric fluid is pumped through the interelectrode gap and flushes out the chips and bubbles while confining the sparks. Once the highest point on the workpiece is removed, a subsequent spark is created between the tool and the next highest point and so the process proceeds into the workpiece. Literally hundreds of thousands of sparks may be generated per second. This material removal mechanism is described as *spark erosion*.

Two different types of EDM exist based on the shape of the tool electrode used. In *ram EDM*, also known as *sinker EDM* or simply EDM, the electrode is a die in the shape of the negative of the cavity to be produced in a bulk material. By feeding the die into the workpiece, the shape of the die is machined into the workpiece. In *wire EDM*, also known

FIGURE 28-21   EDM or spark erosion machining of metal, using high-frequency spark discharges in a dielectric, between the shaped tool (cathode) and the work (anode). The table can make *X–Y* movements.

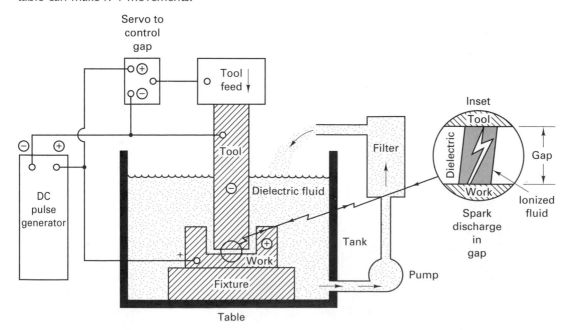

EXAMPLES OF WORKPIECES

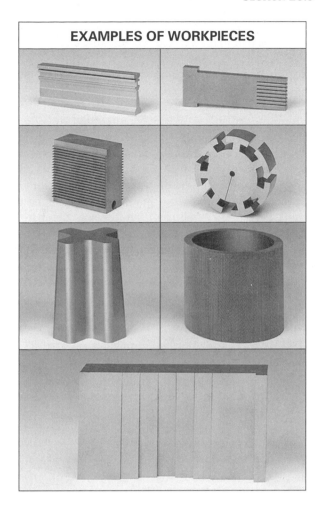

FIGURE 28-22   Examples of wire EDM workpieces made on NC machine (Hatachi).

as *electrical discharge wire cutting*, the electrode is a wire used for cutting through-cut features driving the workpiece with a CNC table. Wire EDM is capable of producing complex two-dimensional patterns (Figure 28-22) in hard-to-machine materials.

EDM processes are slow compared to more conventional methods of machining and they produce a matte surface finish composed of many small craters. While feed rates in EDM are slow, EDM processes can still compete with conventional machining in producing complex geometries particularly in hardened tool materials. As a result, one of the biggest applications of EDM processes is tool and die making. Another drawback of EDM is the formation of a recast or remelt layer on the surface and a heat-affected zone below the surface of the workpiece. Figure 28-23 shows a scanning electron micrograph

FIGURE 28-23   SEM micrograph of EDM surface (*right*) on top of a ground surface in steel. The spherodial nature of debris on the surface is in evidence around the craters (300X).

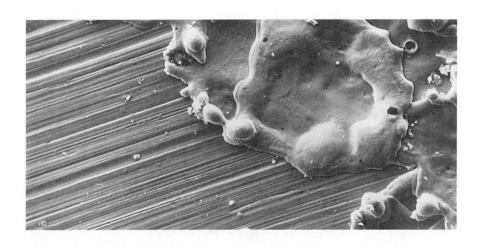

of a recast layer on top of a ground surface. Note the small spheres in the lower right corner attached to the surface representing chips which did not escape the surface. Below the recast layer is a heat-affected zone on the order of 0.001 in. thick. The effect of the recast layer and heat-affected zone is poor surface finish as well as poor surface integrity and poor fatigue strength.

MRR and surface finish are both controlled by the spark energy. In modern EDM equipment, the spark energy is controlled by a dc power supply. The power supply works by pulsing on and off the current at certain frequencies (between 10 and 500 kHz). The on-time as a percentage of the total cycle time (inverse of the frequency) is called the *duty cycle*. EDM power supplies must be able to control the pulse voltage, current, duration, duty cycle, frequency, and electrode polarity. The power supply controls the spark energy mainly by two parameters: current on-time and discharge current. Figure 28-24 shows the effect of current on-time and discharge current on crater size. Larger craters are good for high MRRs. Conversely, small craters are good for finishing operations. Therefore, generally, higher duty cycles and lower frequencies are used to maximize MRR. Further, higher frequencies and lower discharge currents are used to improve surface finish while reducing the MRR. Higher frequencies generally cause increased tool wear.

FIGURE 28-24 The principles of metal removal for EDM.

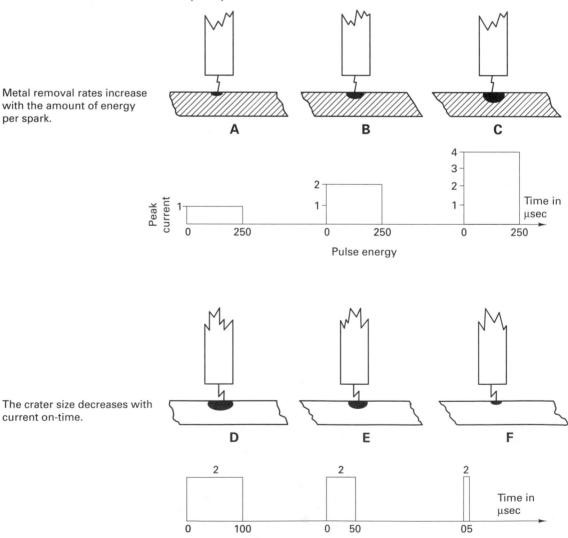

Ram EDM.   An estimate for the MRR in ram EDM for a given workpiece material has been given in Weller:

$$\text{MRR} = \frac{C \cdot I}{T_m^{1.23}} \qquad (28\text{-}9)$$

where MRR is the material removal rate in in³/min (cm³/min), C is a proportionality constant equal to 5.08 in British standard units (39.86 in SI units); I is the discharge current in amps; and T_m is the melting temperature of the workpiece material, °F (°C). Melting points of selected metals are shown in Table 28-6. While this equation shows that the MRR is based mainly on the discharge current, it is recognized that MRR is more a function of current density. For graphite electrodes, a generally accepted rule-of-thumb for the maximum current density is 65 amps per square centimeter of surface area. Given these guidelines, an expression for the maximum MRR, MRR_{max}, can be derived:

$$\text{MRR}_{max} = \frac{C_{max} \cdot A_s}{T_m^{1.23}} \qquad (28\text{-}10)$$

where C_{max} is a proportionality constant equal to 51.2 in British standard units (2591 in SI units) and A_s is the bottom-facing surface area of the electrode in in² (cm²). This formula suggests that the MRR in EDM is also a function of the tooling geometry. The larger and more highly contoured the electrode surface, the slower the MRR at the same discharge current.

The size of the cavity cut by the EDM tool will be larger than the tool. That is, the distance between the surface of the electrode and the surface of the workpiece represents the *overcut* and is constrained by the minimum interelectrode distance necessary for a spark, which is essentially constant over all areas of the electrode, regardless of size or shape. Typical overcut values range from 0.0005 to 0.02 in. Overcut depends on the gap voltage plus the chip size, which varies with the amperage. EDM equipment manufacturers publish overcut charts for the different power supplies for their machines, and these values can be used by tool designers to determine the appropriate dimensions of the electrode. The dimensions of the tool are basically equal to the desired dimensions of the part less the overcut values.

While different materials are used for the tool electrodes, graphite is the most widely used with approximately 85% of electrodes. The choice of electrode material depends on its machinability and cost as well as the desired MRR, surface finish, and tool wear. Equations (28-9) and (28-10), show that the most important material characteristic for MRR is melting temperature. This relationship also applies to tool wear. The higher the melting temperature of the electrode, the less tool wear (i.e., material removed). In addition to its good machinability, graphite has a very high sublimation temperature (3500°C) which is good for minimizing tool wear. In addition to its melting temperature, materials

TABLE 28-6.	Melting Temperatures for Selected EDM Workpiece Materials
Metal	**Melting Temperature, °F (°C)**
Aluminum	1200 (660)
Carbon Steel	2500 (1371)
Cobalt	2696 (1480)
Copper	1980 (1082)
Manganese	2300 (1260)
Molybdenum	4757 (2625)
Nickel	2651 (1455)
Titanium	3308 (1820)
Tungsten	6098 (3370)

with high densities and high specific heats tend toward less tool wear. Cheaper metallic electrode materials with lower melting temperatures can be used in cases where low temperature metals are to be machined. And in some cases requiring good surface finish, a case can be made for metallic electrodes since spark energies are generally lower resulting in reduced temperatures in the interelectrode gap. Copper, brass, copper–tungsten, aluminum, 70/30 zinc–tin, and other alloys have all been employed as electrode (tool) materials for different reasons. In addition to tool material selection, tool wear may be minimized by using the proper polarity across the electrodes. To minimize tool wear in most electrode materials, the tool should be kept positive and the workpiece negative although larger MRRs are possible with reversed polarity.

The dielectric fluid has four main functions: electrical insulation between the tool and workpiece; spark conductor; flushing medium; and coolant. The fluid must ionize to provide a channel for the spark and deionize quickly to become an insulator. Perhaps the most important factor in EDM is flushing of the interelectrode gap to remove residual materials and gas. Filters in the fluidic circuit are used to remove these wastes from the dielectric fluid. In addition, the fluid must carry away the heat produced in the process. A gross temperature change in the dielectric fluid significantly changes the properties of the fluid. Therefore, a heat exchanger is added to the fluidic circuit to remove heat from the dielectric fluid. Common dielectric fluids include paraffin, kerosene, and silicon-based dielectric oil. Polar compounds, such as glycerine water (90:10) with triethylene oil as an additive, have been shown to improve the MRR and decrease the tool wear when compared with traditional dielectric fluids, such as kerosene. The dielectric materials must be safe for inhalation and skin contact as operators are in constant in contact with the fluid.

Wire EDM. *Wire EDM,* shown in Figure 28-25, involves the use of a continuously moving conductive wire as the tool electrode. The tensioned wire of copper, brass, tungsten, or molybdenum is used only once, traveling from a take-off spool to a take-up spool while being "guided" to produce a straight narrow kerf in plates up to 3 in. thick. The wire diameter ranges from 0.002 to 0.01 in. with positioning accuracy up to ±0.0002 in. in machines with NC or tracer control. The dielectric is usually deionized water because of its low viscosity. This process is widely used for the manufacture of punches, dies, and stripper plates, with modern machines capable of routinely cutting die relief, intricate openings, tight radius contours, and corners.

Advantages and Disadvantages of EDM. EDM is applicable to all materials that are fairly good electrical conductors, including metals, alloys and most carbides. The hardness, toughness, or brittleness of the material imposes no limitations. EDM provides a relatively simple method for making holes and pockets of any desired cross section in materials that are too hard or too brittle to be machined by most other methods. The process leaves no burrs on the edges. About 80 to 90% of the EDM work performed in the world is in the manufacture of tool and die sets for injection molding, forging,

FIGURE 28-25 Schematic diagram of equipment for wire EDM using a moving wire electrode.

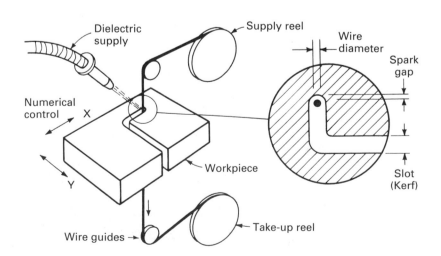

stamping, and extrusions. The absence of almost all mechanical forces makes it possible to EDM fragile or delicate parts without distortion. EDM has been used in micromachining to make feature sizes as small as 0.0004 in. (0.01 mm).

On most materials, the process produces a thin, hard recast surface, which may be an advantage or a disadvantage, depending on the use. When the workpiece material is one that tends to be brittle at room temperature, the surface may contain fine cracks caused by the thermally induced stresses. Consequently, some other finishing process often is used subsequent to EDM to remove the thin recast and heat-affected layers, particularly if the product will be fatigued. Fumes, resulting from the bubbles produced during spark erosion, are given off during the EDM process. Fumes can be toxic when electrical discharge machining boron carbide, titanium boride, and beryllium posing a significant safety issue although machining of these materials is hazardous in many other processes as well.

PARTICLE BEAM MACHINING

As a metals-processing tool, the electron beam is used mainly for welding, to some extent for surface hardening, and occasionally for cutting (mainly drilling). *Electron beam machining* (EBM) is a thermal process that uses a beam of high-energy electrons focused on the workpiece to melt and vaporize metal. This process shown in Figure 28-26 is performed in a vacuum chamber (10^{-5} torr). The electron beam is produced in the electron gun (also under vaccum) by thermionic emission. In its simplest form, a filament (tungsten or lanthanum-hexaboride) is heated to temperatures in excess of 2000°C where a stream (beam) of electrons (more than 1 billion per second) is emitted from the tip of the filament. Electrostatic optics are used to focus and direct the beam. The desired beam path can be programmed with a computer to produce any desired pattern in the workpiece. The diameter of the beam is on the order of 0.0005 to 0.001 in., and holes or narrow slits with depth-to-width ratios of 100:1 can be "machined" with great precision in any material. The interaction of the beam with the surface produces dangerous X-rays; therefore, electromagnetic shielding of the process is necessary. The layer of recast material and the depth of the heat damage are very small. For micromachining applications, MRRs can exceed that of EDM or ECM. Typical tolerances are about 10% of the hole diameter or slot width. These machines require high voltages (50 to 200 kV) to accelerate the electrons to speeds of 0.5 to 0.8 the speed of light and should be operated by fully trained personnel.

· *Ion beam machining* (IBM) is a nano-scale (10^{-9}) machining technology used in the microelectronics industry to cleave defective wafers for characterization and failure analysis. IBM uses a focused ion beam created by thermionic emission similar to EBM to machine features as small as 50 nm. The ion beam may be focused down to 50 nm diameter and is focused and positioned by an electrostatic optics column. Current densities up to 5 A/cm^2 and voltages between 4 and 150 kV provide ion energies up to 300 keV. Target substrates as large as $7'' \times 7'' \times \frac{1}{4}''$ thick can be processed.

FIGURE 28-26   Electron beam machining uses high-energy electron beam (10^9 W/in^2) to melt and vaporize metal in 0.001-in.-diameter spots.

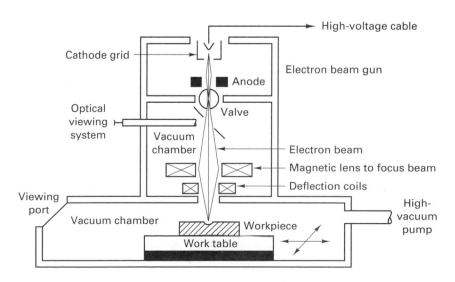

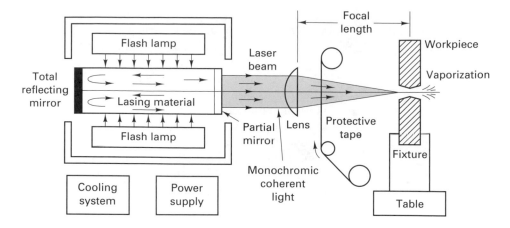

FIGURE 28-27 Schematic diagram of a laser beam machine, a thermal NTM process that can micromachine any material.

LASER BEAM MACHINING

Laser beam machining (LBM) uses an intensely focused, coherent stream of light (a laser) to vaporize or chemically *ablate* materials. A schematic of the LBM process is shown in Figure 28-27. Lasers are also used for joining (welding, crazing, soldering), heat treating materials (see Chapters 5 and 36 for a discussion), and freeform fabrication (see Chapter 33). Power density and interaction time are the basic parameters in laser processing as shown in Figure 28-28. Drilling requires higher power densities and shorter interaction times compared to most other applications.

The material removal mechanism in LBM is dependent upon the wavelength of the laser used. At UV wavelengths (i.e., between about 200 and 400 nm), the material removal mechanism in polymers (for example) is generally thermal evaporation. Below 400 nm, polymeric material is typically removed by chemical ablation. In ablation, the chemical bonds between atoms are broken by the excess amount of laser energy absorbed by the valence electrons in the material. The advantage of chemical ablation is that since it is not a thermal process, it does not result in a heat-affected zone.

Laser light is produced within a laser cavity which is a highly reflective cavity containing a laser rod and a high intensity light source, or laser lamp. The light source is used to "pump up" the laser rod which includes atoms of a lasing media which is capable of absorbing the particular wavelength of light produced by the light source. When an atom of lasing media is struck by a photon of light, it becomes energized. When a second photon strikes the energized atom, the atom gives off two photons of identical wavelength moving in the same direction and with the same phase. This process is called *stimulated emission*. As the

FIGURE 28-28 Power densities and interaction times in laser processing vary with the application.

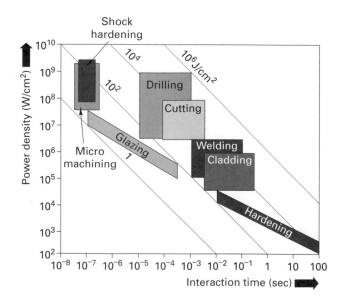

two photons now stimulate further emission from other energized atoms, a cascading of stimulated emission ensues. To increase the number of stimulated emissions, the laser rod has mirrors on both ends that are precisely parallel to one another. Only photons moving perpendicular to these two mirrors stay within the laser rod causing additional stimulated emission. One of the mirrors is partially transmissive and permits some percent of the laser energy to escape the cavity. The energy leaving the laser rod is the laser beam.

Table 28-7 lists some commercially available lasers for material processing. The most common industrial laser is the CO_2 laser. The CO_2 laser is a gas laser which uses a tube of helium and carbon dioxide as the laser rod. Output is in the far infrared range (10.6 μm) and the power can be up to 10 kW. Nd:YAG lasers are called solid state lasers. The laser rod in these lasers is a solid crystal of yttrium, aluminum, and garnet which has been doped with neodymium atoms (the lasing media). The output wavelength is in the near infrared range (1064 nm) and power up to 500 W is common. For micromachining applications, modifications can be made to a 50 W Nd:YAG laser to output at one-half (532 nm), one-third (355 nm), and one-fourth (266 nm) of this wavelength with roughly an order of magnitude reduction in power from 1064 to 532 to 266 nm. More recently, gas lasers called *excimer* (from *ex*cited d*imer*) lasers have been developed with laser rods consisting of excited complex molecules (usually noble gas halides) called dimers. Excimer lasers are pulsed lasers which output in the near and deep UV range at powers up to 100 W. Excimer lasers are significantly more expensive to purchase and operate than CO_2 or Nd:YAG lasers.

Lasers produce highly collimated, coherent (in phase) light which when focused to a small diameter produce high power densities which are good for machining. It is generally accepted that in order to evaporate materials, infrared power densities in excess of 10^5 W/mm^2 are needed. For CO_2 lasers, these levels are directly achievable. However, in Nd:YAG lasers, these high power conditions would significantly decrease the life of the laser lamp. Therefore, Nd:YAG lasers make use of a Q-switch which breaks up the continuous light stream into a series of higher power pulses. The Q-switch is an acousto-optic switch which momentarily stores laser energy prior to releasing a large laser pulse. Pulsed Nd:YAG lasers may have peak powers in excess of 30 kW at 1064 nm and

TABLE 28-7. Commercial Lasers Available for Machining, Welding, and Trimming

Laser Type	Wavelength (μm)	Mode of Operation	Power (W)	Pulse Pulses per Second	Length of Time	Application	Comments
Argon	0.4880 0.5145	Pulse	20 peak; 0.005 average	60	50 μs	Scribing thin films	Power low
Ruby	0.6943	Pulse	2×10^5 peak	Low (5 to 10)	0.2–7 ms	Large material removal in one pulse, drilling diamond dies, spot welding	Often uneconomical
Nd-glass	1.06	Pulse	2×10^6 peak	Low (0.2)	0.5–10 ms	Large material removal in one pulse	Often uneconomical
Nd-YAG[a]	1.06	Continuous	1,000	—	—	Welding	Compact, economical at low powers
Nd-YAG	1.06	Repetitively Q-switched	3×10^5 peak; 30 average	1–24,000	50–250 ns	Resistor trimming, electronic	Compact and economical
Nd-YAG	1.06	Pulse	400	300	0.5–7 ms	Spot weld, drill	
CO_2[b]	10.6	Continuous	15,000	—	—	Cutting organic materials, oxygen-assisted metal	Very bulky at high powers
CO_2	10.6	Repetitively Q-switched	75,000 peak 1.5 average	400	50–200 ns	Resistor trimming	Bulky but economical
CO_2	10.6	Pulse	100 average	100	100 μs and up	Welding, hole production, cutting	Bulky but economical
KrF (excimer)	0.248	Pulse	Up to several kW	100–1,000	15–45 ns	Micromachining, industrial materials processing, and laser annealing	Short wavelength, high energy, and high average power
XeCl (excimer)	0.308	Pulse	200	300	40–50 ns		

Source: Modified from J. F. Ready, Selecting a laser for material working, *Laser Focus* (March 1970), p. 40.
[a] Neodynium-yttrium aluminum garnet.   [b] CO_2 plus He plus N_2 mixture.

peak power densities in excess of 2×10^7 W/mm^2, which are easily enough for metal sublimation. With this magnitude of power density, the thermal effects of LBM are minimal.

Applications of LBM are widely varied. Most CO_2 industrial laser CNC machining centers can focus the beam down to a diameter of about 0.005 in. Applications for these systems range from cigarette paper cutting to drilling microholes in turbine engine blades to cutting steel plate for chain saw blades. For printed circuit board and chip scale packaging applications, Nd:YAG, excimer, and now pulsed CO_2 lasers are being used to drill holes down to 0.001 in. in milliseconds in polyimide or polyester films. However, this drilling is limited to rather thin stock (0.01 in.) as the cutting speed drops off rapidly with penetration into the material. Hole depth-to-diameter ratios of 10:1 are common and hole geometry is irregular. Recast and heat-affected zones exist adjacent to the cuts. Deep UV excimer and Nd:YAG lasers are ideal for micromachining applications. Deep UV lasers use primarily a chemical ablation mechanism for material removal. In addition to providing the potential for eliminating thermal effects in machining, deep UV lasers (having lower wavelengths) may be focused to tighter diameters than lasers with higher wavelengths. While less powerful than infrared lasers, deep UV lasers (because of chemical ablation mechanisms) experience much greater energy efficiencies in cutting. Holes as small as 60 μin. (1.5 μm) in diameter have been machined in thin film materials with excimer lasers.

In LBM, the wavelength used to process the material is mainly determined by the optical characteristics (reflectivity, absorptivity, and transmissivity) of the workpiece. Not all materials can be machined by all lasers. Protective eyeware is necessary when working around laser equipment because of the potential damage to eyesight from either direct or scattered laser light.

PLASMA ARC CUTTING

Plasma arc cutting (PAC) uses a superheated stream of electrically ionized gas to melt and remove material (Figure 28-29). The 20,000 to 50,000°F plasma is created inside a water-cooled nozzle by electrically ionizing a suitable gas such as nitrogen, hydrogen, argon, or mixtures of these gases. The process can be used on almost any conductive metal. The

FIGURE 28-29 Plasma arc machining.

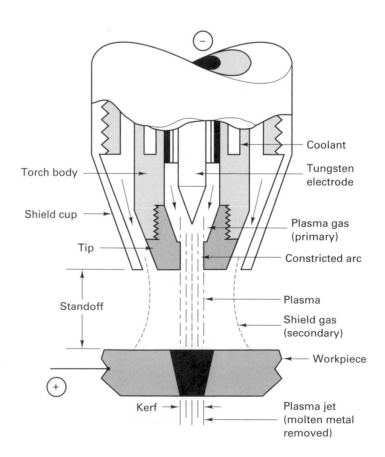

Torch body

Shield cup

Tip

Standoff

Kerf

Coolant

Tungsten electrode

Plasma gas (primary)

Constricted arc

Plasma

Shield gas (secondary)

Workpiece

Plasma jet (molten metal removed)

plasma arc is a mixture of free electrons, positively charged ions, and neutral atoms. The arc is initiated in a confined gas-filled chamber by a high-frequency spark. The high-voltage, DC power sustains the arc, which exits from the nozzle at near sonic velocity. The workpiece is electrically positive. The high-velocity gases melt and blow away the molten metal "chips." Dual-flow torches use a secondary gas or water shield to assist in blowing the molten metal out of the kerf, giving a cleaner cut. The process may be performed under water using a large tank to hold the plates being cut. The water assists in confining the arc and reducing smoke. The main advantage of PAC is speed. Mild steel $\frac{1}{4}$ in. thick can be cut at 125 in./min. Speed decreases with thickness. Greater nozzle life and faster cutting speeds accompany the use of water-injection type torches. Control of nozzle standoff from the workpiece is important. One electrode size can be used to machine a wide variety of materials and thicknesses by suitable adjustments to the power level, gas type, gas flow rate, traverse speed, and flame angle. PAC is sometimes called plasma beam machining.

PAC can machine exotic materials at high rates. Profile cutting of metals, particularly of stainless steel and aluminum, has been the most prominent commercial application. However, mild steel, alloy steel, titanium, bronze, and most metals can be cut cleanly and rapidly. Multiple-torch cuts are possible on programmed or tracer-controlled cutting tables on plates up to 6 in. thick in stainless steel. Smooth cuts free from contaminants are a PAC advantage. Well-attached dross on the underside of the cut can be a problem and there will be a heat-affected zone (HAZ). The depth of the HAZ is a function of the metal, its thickness, and the cutting speed. Surface heat treatment and metal joining are beginning to use the plasma torch. See Chapter 36 for more discussion on plasma arc processes.

Some of the drawbacks of the PAC process include poor tolerances, tapered cuts, and double arcing leading to premature wear on the nozzle. *Precision PAC*, also called *high-definition plasma* and *fine plasma cutting*, uses a special nozzle, where either a high-flow vortex or a magnetic field causes the plasma to spin rapidly and stabilizes the plasma pressure (Figure 28-30). The fast-spinning plasma results in a finely defined beam that cuts a narrow kerf with a perpendicular edge. Precision PAC also reduces the problems of HAZ and dross on the bottom of parts.

THERMAL DEBURRING

Thermal deburring, also known as the *thermal energy method*, has been developed for the removal of burrs and fins by exposing the workpiece to hot corrosive gases for a short period of time typically on the order of a few milliseconds (Figure 28-31). The hot gases are formed by combusting (with a spark plug) explosive mixtures of oxygen and fuel (e.g., natural gas) in a chamber holding the workpieces. A thermal shock wave, moving at Mach 8 (2700 m/s; 6000 mph) with temperatures up to 3300°C (6000°F), vaporizes the burr in about 25 milliseconds. Because of the intense heat, the burrs and fins are unable to dissipate the

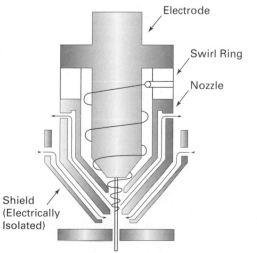

FIGURE 28-30 Precision plasma cutting nozzle.

Arc Current Density: 60,000 amps/sq. in
Kerf Width: ~0.04″ (1 mm) on 1/8″ (3 mm) mild steel

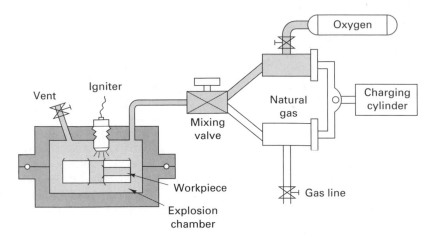

FIGURE 28-31
Thermochemical machining process for the removal of burrs and fins.

heat fast enough to the surrounding workpiece and, consequently, sublimate. The workpiece remains unaffected and relatively cool because of its low surface-to-mass ratio and the short exposure time. Small amounts of metal are removed from all exposed surfaces, and this must be permissible if the process is to be used. Consequently, while large burrs and fins can be removed with this process, the procedure usually can be used only for removing small burrs. Care must be taken with parts with thin cross sections. Maximum burr thickness should be about $\frac{1}{15}$ of the thinnest feature on the workpiece.

Thermal deburring will remove burrs or fins from a wide range of materials, but it is particularly effective with materials of low thermal conductivity which easily oxidize. It will deburr thermosetting plastics but not thermoplastic materials. Any workpiece of modest size requiring manual deburring or flash removal should be considered a candidate for thermal deburring. Die castings, gears, valves, rifle bolts, and similar small parts are deburred readily, including blind, internal and intersecting holes in inaccessible locations. Carburetor parts are processed in automated equipment. The major advantage of thermal deburring is that fine burrs are removed much more quickly and cheaply than if they were removed by hand. Uniformity of results and greater quality assurance over hand deburring are also advantages of thermal deburring. One outcome of thermal deburring is that the part is coated with a fine oxide dust which can be removed easily by solvents. Capital costs can be several hundred thousand dollars and the maximum workpiece dimensions on the order of 250 mm (10 in.) by 690 mm (27 in.). See chapter on surface finishing for additional discussions on deburring.

■ KEY WORDS

ablate
abrasive flow machining (AFM)
abrasive jet machining (AJM)
abrasive waterjet cutting (AWC)
anisotropy
chemical blanking
chemical machining (CHM)
chemical-mechanical polishing
chemical milling
cut-and-peel
dimer
dishing
duty cycle
electrical discharge machining (EDM)
electrical discharge wire cutting
electrochemical deburring

electrochemical grinding (ECG)
electrochemical machining (ECM)
electrochemical micromachining
electrochemical polishing
electron beam machining (EBM)
electropolishing
electrostream drilling
etch factor
etch radius
fine plasma cutting
gel milling
heat-affected zone (HAZ)
high-definition plasma
hydrodynamic machining
intergranular attack
ion beam machining (IBM)

islands
laser beam machining (LBM)
maskant
nontraditional machining (NTM)
overcut
overhang
photochemical blanking
photochemical machining (PCM)
photochemical milling
photoresists
pitting
plasma arc cutting (PAC)
precision PAC
pulsed-current ECM
ram EDM
screen printing

shaped-tube electrolytic machining (STEM)
sinker EDM
spark erosion
stepped machining
stimulated emission
thermal deburring
thermal energy method
ultrasonic impact grinding
ultrasonic machining (USM)
waterjet cutting (WJC)
waterjet machining
wire EDM (WEDM)
wire EDM (see *electrical discharge wire cutting*)
wire cutting

■ Review Questions

1. What are the four basic types of NTM processes?
2. Why might chipless machining processes have greater importance in the future?
3. How do the MRRs for most NTM processes compare to conventional metal cutting?
4. What are the steps in chemical machining using photosensitive resists?
5. Why is it preferable in chemical machining to apply the etchant by spraying instead of immersion?
6. What are the advantages of chemical blanking over regular blanking using punch and die methods?
7. Explain how multiple depths can be produced by chemical machining.
8. Would it be feasible to produce a groove 2 mm wide and 3 mm deep by chemical machining?
9. A drawing calls for making a groove 23 mm wide and 3 mm deep by chemical machining. What should be the width of the opening in the maskant?
10. Could an ordinary steel weldment be chemically machined? Why or why not?
11. How would you produce a tapered section by chemical machining?
12. What is the principle application of thermochemical machining?
13. Is ECM related to chemical machining?
14. What effect does work-material hardness have on the metal-removal rate in ECM?
15. Explain the basic principle involved in electrochemical deburring.
16. What is the principal cause of tool wear in ECM?
17. How is STEM related to ECM or the electrostream process?
18. Would electrochemical grinding be a suitable process for sharpening ceramic tools? Why or why not? What about using ultrasonics?
19. Upon what factors does the metal-removal rate depend in ECM?
20. Why is the tool insulated in the ECM schematic?
21. Is ultrasonic machining really a chipless process?
22. Where, in the ultrasonic stroke, is the acceleration and deceleration of the tool the greatest? Given that $F = ma$, what does this mean about the force given to the grits in the slurry?
23. What is the nature of the surface obtained by electrodischarge machining?
24. What is the principal advantage of using a moving-wire electrode in electrodischarge machining?
25. What effect would increasing the voltage have on the metal-removal rate in electrodischarge machining? Why?
26. If the metal from which a part is to be made is quite brittle and the part will be subjected to repeated tensile loads, would you select ECM or electrodischarge machining for making it? Why?
27. If you had to make several holes in a large number of delicate parts, would you prefer ECM, EDM, EBM, or LBM? Why?
28. What process would you recommend to make many small holes in a very hard alloy where the holes will be used for cooling and venting?
29. Why are the specific power values so large for LBM?
30. Explain (using a little physics and metallurgy) why the "chips" in a thermal process like EDM are often hollow spheres?
31. One of the problems with waterjet cutting is that the process is very noisy. Why?
32. In AWC, what keeps the abrasive jet from machining the orifice?
33. Would AFM be useful for finishing the insides of an engine block? Why or why not?
34. What is thermal deburring and why does it not work on thermoplastics?

CHAPTER 29

WORKHOLDING DEVICES

■ 29.1 INTRODUCTION

In the chapters on machining processes, the manner in which workpieces are mounted and held in the various machine tools was discussed. Workholding devices—that is, jigs and fixtures—are critical components in the manufacturing of interchangeable parts. *Workholders* hold and locate the work in the machine tool with respect to the cutting tool. This book has many examples of fixtures holding workpieces in machine tools providing location with respect to the cutting tools. With workholders, process accuracy and precision (repeatability) can be achieved which otherwise would be impossible with a given combination of cutting tools and machine tools. In this chapter, workholding devices (jigs and fixtures) will be considered as important production tools or adjuncts, with primary attention being directed toward their functional characteristics, their relationship to the machine tools, and the manufacturing processes.

In recent years, workholding devices have become more flexible: that is

- Able to hold more than one part
- Able to be quickly exchanged

Flexible workholders are critical elements in lean manufacturing cells where components are made in families of parts (groups of parts of similar design). Further, being able to change from one device to another quickly to accommodate different parts means smaller lot sizes can be run, which reduces inventory levels in the plants. These flexibility requirements add significantly to the complexity of conventional jig and fixture design. Let's begin with a discussion of the basics of jig and fixture design.

■ 29.2 CONVENTIONAL FIXTURE DESIGN

In the conventional method of fixture design, tool designers rely on their experience and intuition to design simple, single-purpose fixtures for specific machining operations, often using a trial-and-error method until the workholders perform satisfactorily. Of course, these designers should calculate the clamping forces or stress distributions in the fixturing elements to make sure that the loads will not deform the fixtures or the workpieces elastically or plastically. In the design of the workholding devices, two primary functions must be considered: *locating and clamping*. *Locating* refers to orienting and positioning the part in the machine tool with respect to the cutting tools to achieve the required specifications. *Clamping* refers to holding or maintaining the part in that location during the cutting operations (resisting the cutting forces).

Jigs and *fixtures* are specially designed and built workholding devices that hold the work during machining or assembly operations. In addition, a jig determines a location dimension that is produced by machining or fastening. For example, location dimensions determine the position of a hole on a plate (Figure 29-1). Consider the subject of dimensioning as used in drafting practice. Dimensions are of two types: *size* and

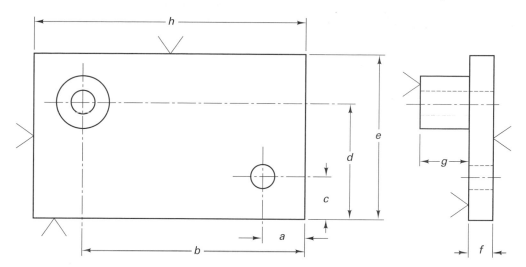

FIGURE 29-1   Drawing of a plate showing locating dimension (*a*, *b*, *c*, *d*) versus sizing dimension (*e*, *f*, *g*, *h*).

location. Size dimensions denote the size of geometrical shapes—holes, cubes, parallelepipeds, and so on—of which objects are composed. *Location* dimensions, on the other hand, determine the position or location of these geometrical shapes *with respect to each other*. Thus *a* and *c* in Figure 29-1 are location dimensions, whereas *e* and *g* are size dimensions. With location dimensions in mind, one can precisely define a jig as follows: *A jig is a special workholding device that, through built-in features, determines location dimensions that are produced by machining or fastening operations.* The key requirement of a jig is that it determines a location dimension. Thus, jigs accomplish layout by means of their design.

In establishing location dimensions, jigs may do a number of other things. They frequently guide tools, as in drill jigs, and thus determine the location of a component geometrical shape. However, they do not always guide tools. In the case of welding jigs, component parts are held (located) in a desired relationship with respect to each other while an unguided tool accomplishes the fastening. The guiding of a tool is not a necessary requirement of a jig.

Similarly, jigs usually hold the work that is to be machined, fastened, or assembled. However, in certain cases, the work actually supports the jig. Thus, although a jig *may* incidentally perform other functions, the basic requirement is that, through qualities that are built into it, certain critical dimensions of the workpiece are determined.

A *fixture* is *a special workholding device that holds work during machining or assembly operations and establishes size dimensions.* The key characteristic is that it is a *special* workholding device, designed and constructed for a particular part or shape. A general-purpose device, such as a vise, chuck or clamp, is usually not considered to be a fixture. Thus a fixture has as its specific objective the facilitating of *setup*, or making the part holding easier. Because many jigs hold the work while determining critical location dimensions, they usually meet all the requirements of a fixture. Alternatively, many fixtures are used in NC machines holding parts where holes are located and drilled according to a program. (See the discussion of machining centers in Chapter 30.) So the strict definition of jigs and fixtures has been blurred by the changes in technology.

In designing workholders, the designer must consider whether the part is a casting, forging, or bar stock. With castings and forgings, variations in shape and size must be accommodated in the design, and usually a machining operation is required to establish a reference surface (called the datum surface) to aid initial fixturing.

In parts cut from bar stock, allowances must be made for inaccuracies and irregularities produced by the cutoff operations. Table 29-1 provides a summary of design criteria for workholders.

To meet all the design criteria for workholders is impossible. Compromise is inevitable. Still, it is useful to know the optimal design objectives to illustrate the positioning, holding, and supporting functions that fixtures must fulfill.

TABLE 29-1. Design Criteria for Workholders

Positive location A fixture must, above all else, hold the workpiece precisely in space to prevent each of 12 kinds of degrees of freedom-linear movement in either direction along the X, Y, and Z axes and rotational movement in either direction about each axis.

Repeatability Identical workpieces should be located by the workholder in precisely the same space on repeated loading and unloading cycles. It should be impossible to load the workpiece incorrectly. This is called "fool proofing" the jig or fixture.

Adequate clamping forces The workholder must hold the workpiece immobile against the forces of gravity, centrifugal forces, inertial forces, and cutting forces but not distort the part. Milling and broaching operations, in particular, tend to pull the workpiece out of the fixture, and the designer must calculate these machining forces against the fixture's holding capacity. The device must be rigid.

Reliability The clamping forces must be maintained during machine operation every time the device is used. The mechanism must be easy to maintain and lubricate.

Ruggedness Workholders usually receive more punishment during the loading and unloading cycle than during the machining operation. The device must endure impact and abrasion for at least the life of the job. Elements of a device that are subject to damage and wear should be easily replaceable.

Design and construction ease Workholders should use standard elements as much as possible to allow the engineer to concentrate on function rather than on construction details. Modular fixtures epitomize this design rule as the entire workholder is made from standard elements, permitting a bolt-together approach for substantial time and cost savings over custom workholders.

Low profile Workholder elements must be clear of the cutting-tool path. Designing lugs on the part for clamping can simplify the fixture and allow proper tool clearance.

Workpiece accommodation Surface contours of castings or forgings vary from one part to the next. The device should tolerate these variations without sacrificing positive location or other design objectives.

Ergonomics and safety Clamps should be selected and positioned to eliminate pinch points and facilitate ease of operation. The workholder elements should not obstruct the loading or unloading of workpieces. In manual operations, the operator should not have to reach past the tool to load or unload parts. A rule sometimes used is that the operator can repeatedly exert a force of 30 to 40 lb to open or close a clamp but greater forces than this can cause ergonomic problems.

Freedom from part distortion Parts being machined can be distorted by gravity, the machining forces, or the clamping forces. Once clamped into the device, the part must be unstressed or, at least, undistorted. Otherwise, the newly machined surfaces take on any distortions caused by the clamping forces.

Flexibility The workholding device can locate and restrain more than one type (design) of part. Many different schemes are being proposed to provide workholder flexibility. Modular vise fixturing, programmable clamps using air-activated plungers, part encapsulation with a low-melting point alloy, and NC-controlled clamping machines are some of the more recently developed systems. Despite their flexibilities, these clamping systems have some significant drawbacks. They are expensive, and the individual systems may not integrate well into individual machine tools. [See the discussions of intermediate jig concept and group jigs for additional thoughts on flexibility.]

■ 29.3 DESIGN STEPS

The classical design of a workholder (e.g., a drill jig) involves the following steps:

1. Analyze the drawing of the workpiece and determine (visualize) the machining operations required to machine it. Note the critical (size and location) dimensions and tolerances.

2. Determine the orientations of the workpiece in relation to the cutting tools and their movements.

3. Perform an analysis to estimate the magnitude and direction of the cutting forces (see Chapter 21).

4. Study the standard devices available for workholders and for the clamping functions. Can an off-the-shelf device be modified? What standard elements can be used?

5. Form a mental picture of the workpiece in position in the workholder in the machine tool with the cutting tools performing the required operation(s). See the figures in chapters on machining for examples.

6. Make a three-dimensional sketch of the workpiece in the workholder in its required position to determine the location of all the elements: clamps, locator buttons, bushings, and so on.

7. Make a sketch of the workholder and workpiece in the machine tool to show the orientation of these elements with respect to the cutting tool in the machine tool.

3-2-1 LOCATION PRINCIPLE

After determining the orientation of the workpiece in the workholder, the next step is to locate it in that position. This location is also used for all similar workpieces. The designer must select or design locating devices (supports) *which ensure that every workpiece placed in the device occupies the same position with respect to the cutting tools.* Thus, when the machining operation is performed, the workpieces are processed identically. This is, of course, the key to making interchangeable parts. In locating the workpiece, the basic *3-2-1 principle*

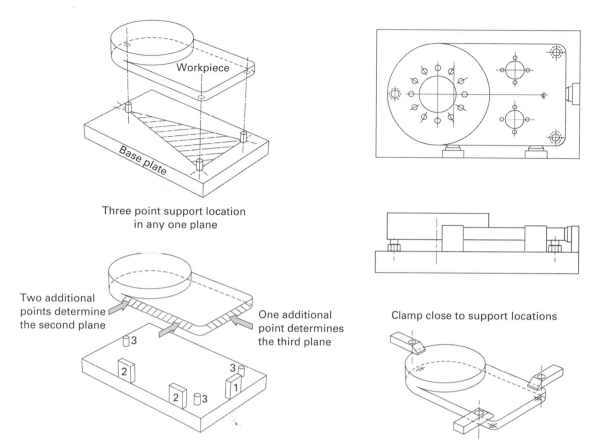

FIGURE 29-2   Workpiece location is based on the 3-2-1 principle. Three points will define a base surface, two points in a vertical plane will establish an end reference, and one point in a third plane will positively locate a part.

of location is used (Figure 29-2). For positive location, the fixture must position the workpiece in each of three perpendicular planes. Positioning processes can vary greatly, but workholder design always begins by defining the first plane of reference with three points. Once the object is defined in a single plane, supported at three points (like a three-legged stool on a floor), a second plane can be assigned that is perpendicular to the first. To do this, the object is brought up against any two points in the second plane. To continue the example, the stool is slid along the floor until two legs touch a wall.

A third plane, perpendicular to each of the other two, is then defined by designating one point on it. As long as an object is in contact with three points on the first plane, two points on the second, and a single point on the third, it is positively located in space. The location points within each plane should be selected as far apart as possible for maximum stability.

In practice, it is often necessary to support a workpiece on more points than this 3-2-1 formula dictates. The machining of a large rectangular plate, for example, typically requires support at four or more points. However, any extra points must be established carefully to support the workpiece in a plane defined by three—and only three—points.

Appropriate clamping devices are selected so that the clamping forces hold the workpiece in the proper location and resist the effects of the cutting forces, centrifugal forces, and vibrations. If possible, the machining forces should act into the location points, not into the clamps, so that smaller clamps can be used. In reality, the worker often determines clamping force when loading the part into the workholder.

Fixtures are usually fastened to the table of the machine tool. Although used primarily on milling and broaching machines, fixtures are also designed and used to hold workpieces for various operations on most of the standard machine tools and machining centers. Some additional design rules for fixture design are given in Table 29-2.

TABLE 29-2.	Twenty Principles for Workholder Design

1. Determine the critical surfaces or points for the part.
2. Decide on locating points and clamping arrangements.
3. For mating parts, use corresponding locating points or surfaces to ensure proper alignment when assembled.
4. Try to use 3-2-1 location, with 3 assigned to largest surface. Additional points should be adjustable.
5. Locating points should be visible so that the operator can see if they are clean. Can they be replaced if worn?
6. Provide clamps that are as quick acting and easy to use as is economically justifiable for rapid loading and unloading.
7. Clamps should not require undue effort by operator to close or to open, nor should they harm hands and fingers during use.
8. Clamps should be integral parts of device. Avoid loose parts that can get lost.
9. Avoid complicated clamping arrangements or combinations that can wear out or malfunction. Keep it simple.
10. Locate clamps opposite locaters (if possible) to avoid deflection/distortion during machining and springback afterward.
11. Take the thrust of the cutting forces on the locaters (if possible) and not on the clamps. See Figure 29-4.
12. Arrange the workholder so that the workpiece can easily be loaded and unloaded from the device and so that it can be loaded only in the correct manner (mistake-proof) and in such a way that the location can be found quickly (visually).
13. Consistent with strength and rigidity, make the workholder as light as possible.
14. Provide ample room for chip clearance and removal. See Figures 29-5 and 29-6.
15. Provide accessibility for cleaning.
16. Provide for entrance and exit of cutting fluid (which may carry off chips) if one is to be used.
17. Provide four feet on all movable workholders.
18. Provide hold-down lugs on all fixed workholders.
19. Provide keys to align fixtures on machine tables so fixtures can be replaced in exactly the same position.
20. Do not sacrifice safety for production.

■ 29.4 CLAMPING CONSIDERATIONS

Clamping of the work is closely related to support of the work. Any clamping, of course, induces some stresses into the part that can cause some distortion of the workpiece, usually elastic. If this distortion is measurable, it will cause some inaccuracy in final dimensions, as illustrated in an exaggerated manner in Figure 29-3. The obvious solution is to spread the clamping forces over a sufficient area to reduce the stresses to a level that will not produce appreciable distortion.

The clamping forces should direct the work against the points of location and work support. Clamped surfaces often have some irregularities that may produce force components in an undesired direction. Consequently, clamping forces should be applied in directions that will assure that the work will remain in the desired position.

Whenever possible, jigs and fixtures should be designed so that the forces induced by the cutting process act to hold the workpiece in position against the supports. These forces are predictable, and proper utilization of them can materially aid in reducing the magnitude of the clamping stresses required. In addition to locating the work properly, the stops or work-supporting areas must be arranged so as to provide adequate support against the cutting forces. As shown in Figure 29-4a, having the cutting force act against a fixed portion of the jig or fixture and not against a movable section permits lower clamping forces to be used. Figure 29-4b illustrates the principle of keeping the points of clamping as nearly as possible in line with the action forces of the cutting tool so as to reduce their tendency to pull the work from the clamping jaws. Compliance with this principle results both in lower clamping stresses and less massive clamping devices. Don't forget that downmilling produces different forces than upmilling. The location points should be as far apart as possible but positioned so as not to allow the cutting force to distort the work, as shown in Figure 29-4c. The cutting forces may distort the work, with resulting inaccuracy or broken tools. These design suggestions materially reduce vibration and chatter during the cutting process.

As many operations as possible and practical should be performed with each clamping of the workpiece. This principle has both physical and economic aspects. Because some stresses result from each clamping, with the possibility of accompanying distortion, greater

FIGURE 29-3 Exaggerated illustration of the manner in which excessive clamping forces can affect the final dimensions of a workpiece.

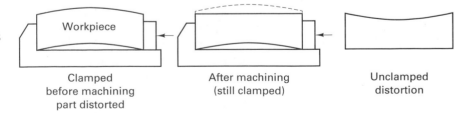

Clamped before machining part distorted

After machining (still clamped)

Unclamped distortion

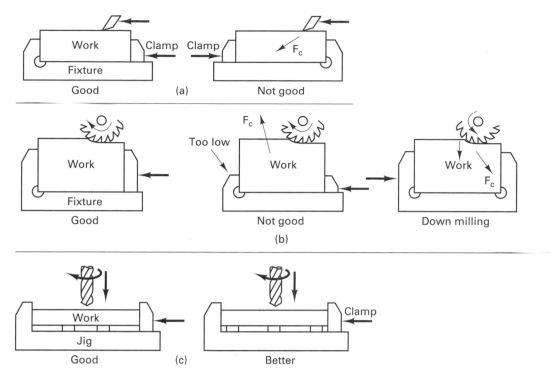

FIGURE 29-4   Proper work support to resist the forces imposed by cutting tools. In (c), three buttons form triangle for work to rest on.

accuracy is achieved if multiple operations are performed with each clamping. From the economic viewpoint, if the number of jigs or fixtures is reduced, less capital will be required and less time will be spent handling the workpiece loading and unloading.

■ 29.5 CHIP DISPOSAL

When jigs or fixtures are used in connection with chip-making operations, adequate provision must be made for the easy removal of the chips. This is essential for several reasons. First, if chips become packed around the tool, heat will not be carried away and tool life can be decreased. Figure 29-5 illustrates how insufficient clearance between the end of a drill bushing and the workpiece can prevent the chips from escaping, whereas too much clearance may not provide accurate drill guidance and can result in broken drills.

A second reason why chips must be removed is so that they do not interfere with the proper seating of the work in the jig or fixture (Figure 29-6). Even though chips and dirt always have to be cleaned from the locating and supporting surfaces by a worker or by automatic means, such as an air blast, the design details should be such that chips and other debris will not readily adhere to, or be caught in or on, the locating surfaces, corners, or overhanging elements and thereby prevent the work from seating properly. Such a condition results in distortion, high clamping stresses, and incorrect workpiece dimensions.

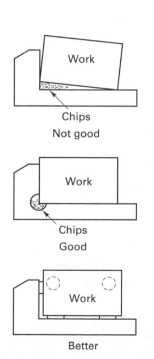

FIGURE 29-6   Methods of providing chip clearance to assure proper seating in the work.

FIGURE 29-5   Proper clearance between drill bushing and top of workpiece is important.

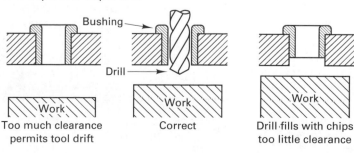

■ 29.6 UNLOADING AND LOADING TIME

The cost of the workholders must be justified by the quantities of production involved, and their primary purpose is to increase productivity and quality. While work is being put into or being taken out of jigs and fixtures, the machines with which they are used are not making chips. The loading and unloading time plus the machining time (also called the run time) plus any delay times equals the cycle time for a part. The loading and unloading time is greatly influenced by the choice of clamps.

There are many ways in which jigs and fixtures can be made easier to load and unload. Some clamping methods can be operated more readily than others. For example, in the drill jig shown in Figure 29-7, a *knurled clamping screw* is used to hold the block against the buttons at the end of the jig. To clamp or unclamp the block in this direction requires several

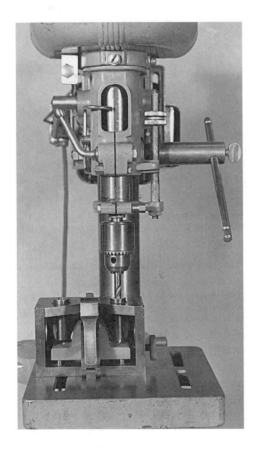

FIGURE 29-7 (*Lower left*) Part to be drilled, (*lower right*) Box drill jig for drilling two holes. (*Upper left*) Jig in drill press. (*Upper right*) Drill being guided by drill bushing.

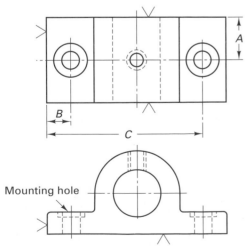

motions. On the other hand, a *cam latch* is used to close the jig and hold the workpiece against the rear locating buttons. This type of latch can be operated with a single motion.

Certainly, the device should be designed so that the part cannot be loaded incorrectly. Defect prevention is often accomplished by the clamping device so that a part loaded improperly cannot be clamped. Ease of operation of workholders not only directly increases the productivity of such equipment but also results indirectly in better quality and fewer lost-time accidents.

The workholder is as critical as the machine tool and the cutting tool to the final quality of the part. The use of the workholder eliminates manual layout of the desired features of the part on the raw material. Manual layout requires a highly skilled worker and is very time consuming. The workholder permits a lower-skilled person to achieve quality and repeatable production with far greater efficiency.

■ 29.7 EXAMPLE OF JIG DESIGN

Some of these principles of work location and tool guidance are illustrated in Figure 29-7. The two mounting holes in the base of the bearing block are to be located and drilled. The dimensions A, B, and C are determined by the jig. While it is not specified on the drawing, there is one other location dimension that must be controlled. The axes of the mounting holes must be at right angles to the bottom surface of the block, so the bottom must be machined (milled) prior to this drilling step.

The way in which the part dimensions are obtained in the finished workpiece is as follows. The surfaces marked with a V are reference (or location) surfaces and are finished (machined) prior to insertion of the part into the drill jig. The part rests in the jig on four buttons marked X in Figure 29-7. These buttons, made of hardened steel, are set into the bottom plate of the jig and are accurately ground so that their surfaces are in a single plane. The left-hand end of the part is held against another button Y. This locating button is built into the jig so that its surface is at right angles to the plane of the X buttons. When the block is placed in the jig, its rear surface rests against three more buttons marked Z. These buttons are located and ground so that their surfaces lie in a plane that is at right angles to the planes of both the X and Y buttons. The part is held in its located position by the two clamps marked C.

The use of four buttons on the bottom of this jig (X buttons) appears not to adhere to the 3-2-1 principle stated previously. However, although only two X buttons would have been required for complete location, the use of only two buttons would not have provided adequate support during drilling. The thrust from the drills would have dislodged the part from the locators. Thus the 3-2-1 principle is a *minimum* concept and often must be exceeded.

To assure that the mounting holes are drilled in their proper locations, the drill must be located and then guided during the drilling process. This is accomplished by the two drill bushings marked K. Such drill bushings are accurately made of hardened steel with their inner and outer cylindrical surfaces concentric. The inner diameter is made slightly larger than the drill—usually 0.0005 to 0.002 in.—so that the drill can turn freely but not shift appreciably. The bushings are accurately mounted in the upper plate of the jig and positioned so that their axes are exactly perpendicular to the plane of the X buttons, at a distance A from the Z buttons and at distances B and C, respectively, from the plane of the Y button. Note that the bushings are sufficiently long that the drill is guided close to the surface where it will start drilling. Consequently, when the workpiece is properly placed and clamped in the jig, the drill will be located and guided by the bushings so that the critical dimensions on the workpiece will be correct. The right hole will be drilled in a vertical spindle drill press (not running) and then the jig will be shifted (manually by the operator) to the right and the left hole will be drilled. The box construction is rigid but open for chip removal.

■ 29.8 TYPES OF JIGS

Jigs are made in several basic forms and carry names that are descriptive of their general configurations or predominant features. Several of these are illustrated in Figure 29-8.

A *plate jig* is one of the simplest types, consisting only of a plate that contains the drill bushings and a simple means of clamping the work in the jig or the jig to the work. In the

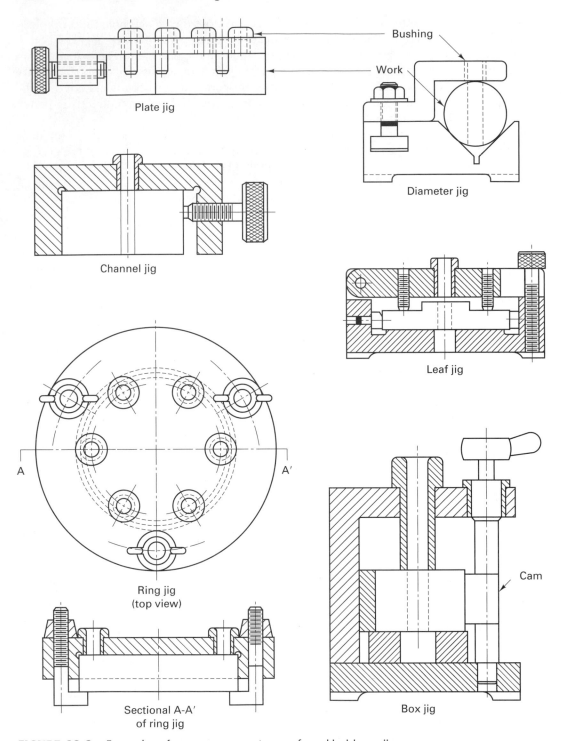

FIGURE 29-8 Examples of some common types of workholders—jigs.

latter case, wherein the jig is clamped to the work, the device is sometimes called a *clamp-on jig*. Such jigs frequently are used on large parts, where it is necessary to drill one or more holes that must be spaced accurately with respect to each other, or to a corner of the part, but that need not have an exact relationship with other portions of the work.

Channel jigs also are simple and derive their name from the cross-sectional shape of the main member. They can be used only with parts having fairly simple shapes.

Ring jigs are used only for drilling round parts, such as pipe flanges. The clamping force must be sufficient to prevent the part from rotating in the jig.

Diameter jigs provide a means of locating a drilled hole exactly on a diameter of a cylindrical or spherical piece.

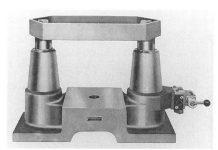
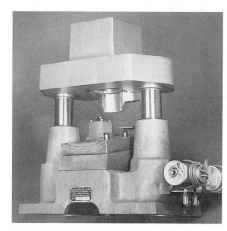

FIGURE 29-9 Two types of universal jigs: Manual (*left*) and power-actuated (*center*). (*right*) Completed jig, made from unit shown at center. *(Courtesy of Cleveland Universal Jig Division, The Industrial Machine Company.)*

Leaf jigs derive their name from the hinged leaf or cover that can be swung open to permit the workpiece to be inserted and then closed to clamp the work in position. Drill bushings may be located in the leaf as well as in the body of the jig to permit locating and drilling holes on more than one side of the workpiece. Such jigs are called *rollover jigs* or *tumble jigs* when they require turning to permit drilling from more than one side.

Box jigs are very common, deriving their name from their boxlike construction. They have five fixed sides and a hinged cover or leaf, which opens to permit loading the workpiece, and a cam that locks the workpiece in place. Usually, the drill bushings are located in the fixed sides to assure retention of their accuracy. The fixed sides of the box usually are fastened by means of dowel pins and screws so that they can be taken apart and reassembled without loss of accuracy. Because of their more complex construction, box jigs are costly, but their inherent accuracy and strength can be justified when there is sufficient volume of production. They have two obvious disadvantages: (1) It usually is more difficult to put work into them than into simpler types, and (2) there is a greater tendency for chips to accumulate within them. Figure 29-7 shows a box-type jig.

Because jigs must be constructed very accurately and be made sufficiently rugged so as to maintain their accuracy despite the use (and abuse) to which they inevitably are subjected, they are expensive. Consequently, several methods have been devised to aid in lowering the cost of manufacturing jigs. One way to reduce this cost is to use simple, standardized plate and clamping mechanisms called *universal jigs* (Figure 29-9). These can easily be equipped with suitable locating buttons and drill bushings to construct a jig for a particular job. Such universal jigs are available in a variety of configurations and sizes, and because they can be produced in quantities, their cost is relatively low. However, the variety of work that can be accommodated by such jigs obviously is limited.

While the drill bushing should be spaced far enough from the work to allow chips to escape without entering the bushing, when drilling into an angled surface, the bushing should be very close. Once the drill has penetrated to at least one half of the drill diameter, the bushing should be retracted to provide chip clearance. Design of the drill jig must not obstruct coolant flow to where it is needed. Bushing length should be 1.75 to 2.5 times the drill diameter.

■ 29.9 CONVENTIONAL FIXTURES

Many examples of conventional fixtures have appeared in the text. Production milling, broaching, and boring processes as performed on NC machines, conventional equipment, or machining centers routinely use fixtures to locate and hold the part properly with respect to the cutting tools on the machine tool. Like cutting tools, fixtures are sold separately and are not usually supplied by the machine tool builder. Traditionally, beginning with Eli Whitney, manufacturers have designed and built custom-made, dedicated fixtures. Because of the pressure of shorter production runs and smaller lot sizes, many companies are turning to modified fixturing approaches. The greatest advantage of these systems is that the fixture can be constructed quickly.

Perhaps the most common fixture uses the vise as its base element. Figure 29-10 shows a schematic and photo of a typical commercially available vise that can be adapted for use as a fixture. As shown, the vise jaws are readily modified to conform to the 3-2-1 location principle and provide adequate clamping forces for most every machining operation. Four vises (shown in Figure 29-10) can be mounted on a subplate for rapid insertion and location in the machine or four vices can be mounted on a tombstone for milling parts in a CNC machine.

The chucks used in lathes are really general-purpose fixtures for rotational parts. Newer chuck designs have greatly improved their flexibility (the range of diameters the chuck can accommodate in a given setup and speed of setup). Figure 29-11 shows a complete change of top jaws for a three-jaw chuck being done in less than 5 minutes. The

FIGURE 29-10 The conventional or standard vise (*top left* and *right*) can be modified with removable jaw plates to adapt to different part geometries. These vises can be integrated into milling fixtures (*right middle* and *bottom*).

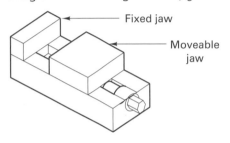

- Fixed jaw
- Moveable jaw

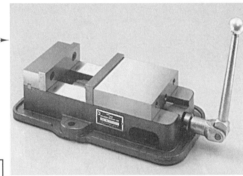

Conventional or standard vise

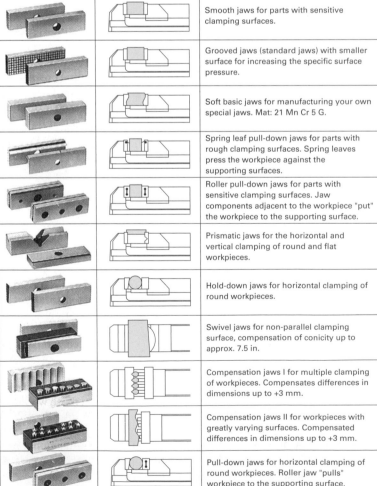

		Smooth jaws for parts with sensitive clamping surfaces.
		Grooved jaws (standard jaws) with smaller surface for increasing the specific surface pressure.
		Soft basic jaws for manufacturing your own special jaws. Mat: 21 Mn Cr 5 G.
		Spring leaf pull-down jaws for parts with rough clamping surfaces. Spring leaves press the workpiece against the supporting surfaces.
		Roller pull-down jaws for parts with sensitive clamping surfaces. Jaw components adjacent to the workpiece "put" the workpiece to the supporting surface.
		Prismatic jaws for the horizontal and vertical clamping of round and flat workpieces.
		Hold-down jaws for horizontal clamping of round workpieces.
		Swivel jaws for non-parallel clamping surface, compensation of conicity up to approx. 7.5 in.
		Compensation jaws I for multiple clamping of workpieces. Compensates differences in dimensions up to +3 mm.
		Compensation jaws II for workpieces with greatly varying surfaces. Compensated differences in dimensions up to +3 mm.
		Pull-down jaws for horizontal clamping of round workpieces. Roller jaw "pulls" workpiece to the supporting surface.

1. Pre-assembled Mini-System and top jaws.
2. Assembly being inserted into the Master Jaw.
3. Quickly retighten cap screw.
4. All 3 jaws changed in 5 minutes or less.

FIGURE 29-11 Quick-changing of the top jaws on a three-jaw chuck. *(Courtesy of Huron Machine Products.)*

normal time for this part of the setup might exceed 15 minutes. New quick-change insert top jaws may even snap in by hand with no jaw nuts, keys, screws, or tools. Jaws that can be exchanged by robots can also be designed.

Most producers of chucks use some variation of Equation (29-1) to compute the maximum rpm rate at which the chuck can run:

$$S_m^2 = \frac{F_m}{3 \times (2.84 \times 10^{-5}) \times W \times D} \tag{29-1}$$

where

S_m = maximum rpm value at which gripping force equals $\frac{1}{3} F_m$

F_m = maximum rated gripping force, at rest (lb)

W = combined weight of jaws (lb)

D = distance from spindle centerline to center of jaw mass (in.)

Thus, with this equation, a 10-in. power chuck with a published rating, F_m, of 13,200 lb would retain one-third of its initial gripping force at 2507 rpm. (Check this calculation using $W = 8$ lb, $D = 3.1$ in.) The higher the rpm value, the greater the centrifugal force factor. This is an important factor in high-speed machining operations in which the part is rotating.

■ 29.10 MODULAR FIXTURING

Modular fixtures have all the same design criteria as those of conventional fixtures, plus one more, *versatility*. Modular fixture elements must be useful for a variety of machining applications and easily adaptable to different workpiece geometries. Individual fixture designs can be photographed or entered into a CAD library for future reference. After the job is done, the fixture itself can be dismantled and the elements returned to the toolroom.

The erector-set approach uses either T-slot or dowel-pin designs. Figures 29-12 and 29-13 show two examples of modular fixturing. The designs begin with base plates. Elements for locating and clamping are added to the subplate. Rectangular, square, and round are the typical patterns for the subplates. Also shown are the typical components for modular fixturing systems used for mounting points, locators, attachments, and so on. The standard elements needed to construct the fixture include riser blocks, vee blocks, angle plates, cubes, box parallels, and the like. Smaller elements such as locator pins, supports, pads, and clamps are added to the subplate on the larger structural elements. Mechanical clamping devices are shown, but power-assisted clamps are available. The base and fixturing elements are made to tolerances of ±0.0002 to 0.0004 in. in flatness, parallelism, and size. Figure 29-13 shows a part in a dedicated fixture compared to a modular fixture. The dedicated fixture represents a capital investment that must be absorbed by the job and must be maintained after the job is complete. The modular fixture is disassembled and the elements reused later in fixtures

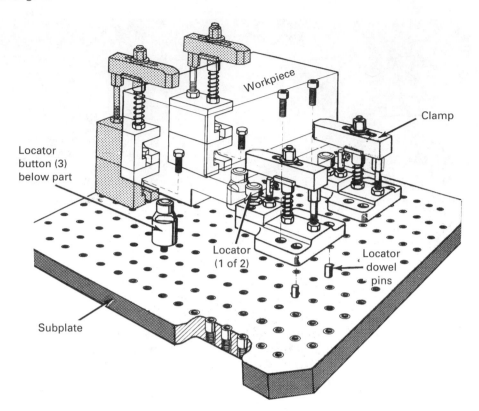

FIGURE 29-12 Modular fixturing begins with a subplate (grid base) and adds locators and clamps.

FIGURE 29-13 Dedicated fixture on left versus modular fixture on the right. (From *Manufacturing Engineering*, Jan. 1984.)

for other parts. Modular fixtures are commonly used for prototype tooling and small batch production runs. They are being incorporated more frequently into regular production as users gain confidence in this approach.

■ 29.11 SETUP AND CHANGEOVER

Every part coming out of the workholder should be the same, resulting in interchangeable parts. But what about the first part? What about the initial setup of the workholder into the machine? In many cases this setup operation takes hours and the machine is not producing anything during this time. Rapid exchanges of workholding devices is a key technique in modern manufacturing systems. (See Chapter 43 for a discussion of the elimination of setup and SMED.) Reducing setup times permits shorter production runs (smaller lot sizes). Do not confuse initial setup (of workholders) with part loading and unloading or tool changing. The trick with initial setup is to do it quickly and to get the first part out of the process as a good part, with no adjustment of the machine, the tooling, or the workholder. Quick tool-and-die exchange is a critical component in the strategy for the factory with a future.

One approach is shown in Figure 29-14 where instead of six different jigs, a master jig is made (also called a group jig) for a family of similar components and then a set of adapters is made which customizes the jig for each part. This concept of group jigs and fixtures originated from the technology (GT) concept. GT is discussed in Chapter 41 as a method to form cellular manufacturing systems by determining a family of parts (see Figure 29-15) from which an imaginary part, called the composite part, is designed which has all the key factors of all the parts in the family. In other words, the composite part is an envelope, the shape of which encompasses the shapes of all the parts in the family. The theory is that if the tooling is designed for the composite part, any part that fits

FIGURE 29-14 Master jig designed for part family: (a) part family of round plates (six parts); (b) group jig for drilling, showing adapter and part A.

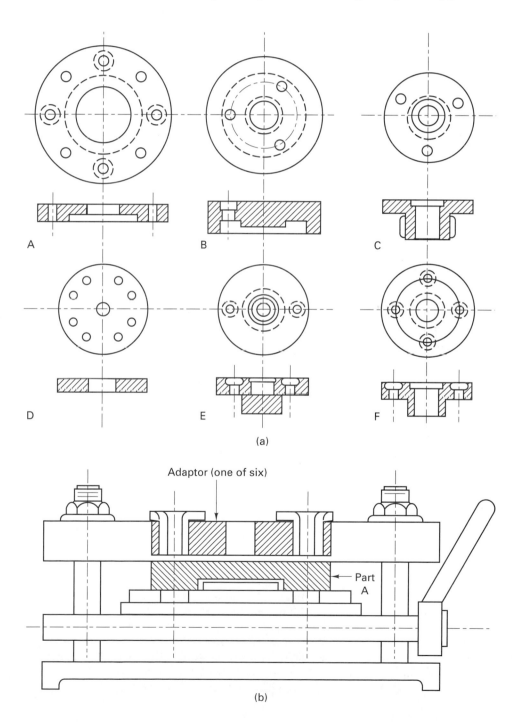

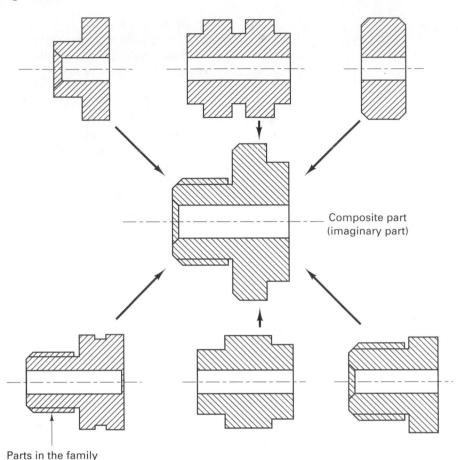

FIGURE 29-15 A part family of six parts is used to design a composite part which is used to design the group jig or fixture.

within the envelope could be machined without any tooling changes. This part is used for designing the workholder. The workholding devices should be able to accommodate all the parts within the parts family. For manufacturing cells, the workholders will also have to compensate for variation in cutting forces, centrifugal forces, and so on. Group workholders are designed to accept every part family member, with or without adapters that accommodate minor part variations.

INTERMEDIATE JIG CONCEPT

One way to achieve rapid fixture exchange is to employ the intermediate jig concept. This means that the workholding devices are designed so that they all appear the same to the machine tool but different to the parts. This usually requires one to construct intermediate jig or fixture plates to which the jig or fixture is attached. The jigs or fixtures are all different but the plates are all identical.

The cassette tape for your VCR is an example of an intermediate workholder. To the VCR, every cassette appears to be the same and can be quickly loaded and unloaded with one handling—that is, one touch. From the outside, every tape appears to be the same, but on the inside, every tape is different. If you think about the workholding devices in terms of the intermediate jig concept, you can quickly achieve one touch setups.

Figure 29-16 shows an example of the *intermediate jig concept*, also discussed in Chapter 43, applied to lathes and chucks. An adapter or intermediate fixture is bolted to the machine spindle and is a permanent part of the machine tool. The intermediate fixture will accept mating chucks that have been preset for the workpiece prior to insertion. Different chuck designs mount interchangeably on the common actuator. This method greatly reduces setup time and permits the operator to perform chuck maintenance and retooling (setup) while the machine is running. These chucks can be exchanged automatically.

Many companies are selling quick change fixtures for CNC milling machines and machining centers using the intermediate jig concept. The fixture shown in Figure 29-17 combines vises with standard bases onto a modular base.

FIGURE 29-16 Example of the intermediate jig concept applied to lathe chucks. *(Courtesy of Sheffer Collet Company.)*

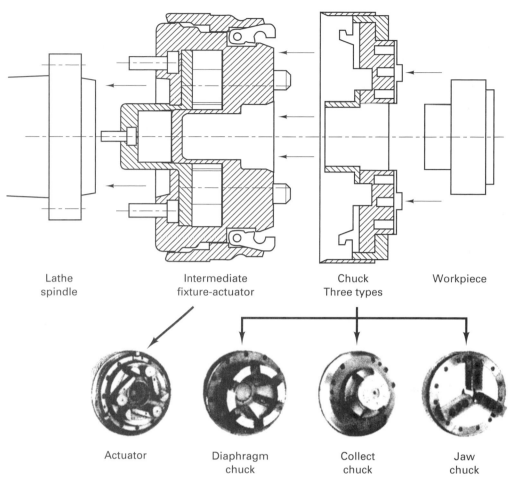

Lathe spindle | Intermediate fixture-actuator | Chuck Three types | Workpiece

Actuator | Diaphragm chuck | Collect chuck | Jaw chuck

FIGURE 29-17 Example of a fixture where workholders can be quickly exchanged. (Chick Multi-Lok Vise Fixturing System with snap-on QuickChange™)

■ 29.12 CLAMPS

Manual clamps include screw, strap, swing, edge, cam, toggle, and C-clamps, each with certain strengths and weaknesses.

In Figure 29-18 typical types of clamps that are used in fixtures are shown. The *strap clamp* comes in many forms and sizes and is simple, low cost, and flexible. The force can be applied by a hand knob, a cam, or a wrench turning down a nut. A conventional *toggle clamp* accommodates only small thickness variation from part to part, yet provides an excellent, consistent clamping force.

Figure 29-19 shows some examples of power-actuated clamps. Power clamping provides more consistent clamping forces than do manual clamps, especially in applications that promote operator fatigue. The higher cost must be weighed against the capability for consistent and repeatable operation, automatic adjustment of holding forces, remote actuations, and automating sequencing of clamping actions. *Extending clamps* operate in a manner similar to that of a manual clamp-strap assembly. They extend forward horizontally, then clamp down. *Edge clamps* have a very low profile. They clamp down and forward simultaneously.

■ 29.13 OTHER WORKHOLDING DEVICES

ASSEMBLY JIGS

Because *assembly jigs* usually must provide for the introduction of several component parts and the use of some type of fastening equipment, such as welding or riveting, they commonly are of the open-frame type. Such jigs are widely used in the automobile and aircraft industries. Large jigs of this type are shown in Figure 29-20 for the assembly of aircraft wings. This jig is constructed mainly of reinforced concrete.

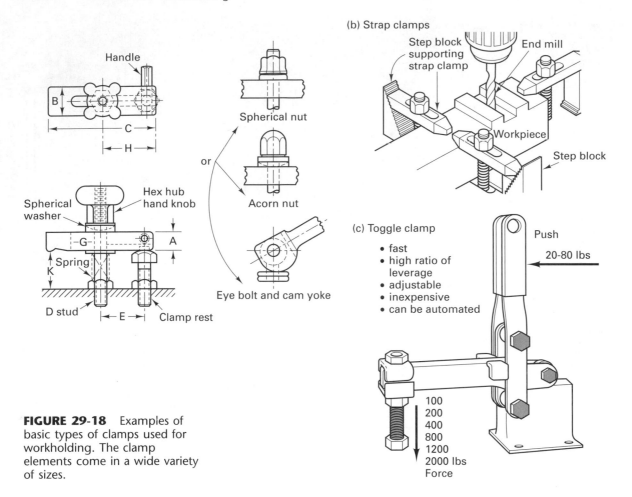

FIGURE 29-18 Examples of basic types of clamps used for workholding. The clamp elements come in a wide variety of sizes.

FIGURE 29-19 Examples of power clamping devices: (a) extending clamp; (b) edge clamp.

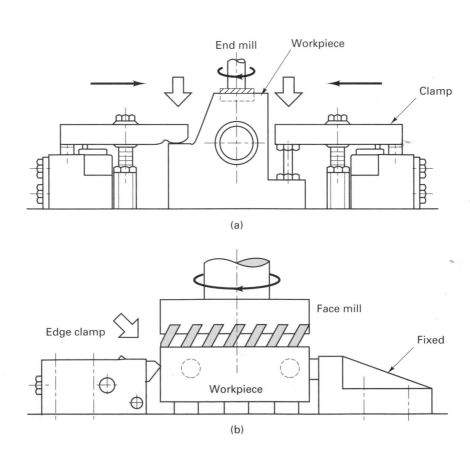

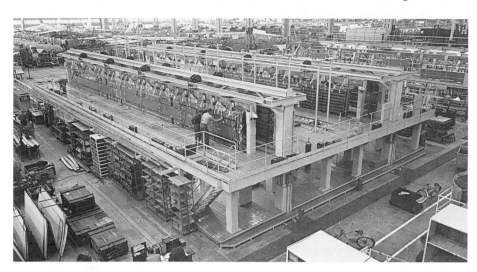

FIGURE 29-20 Example of large assembly jig for an airplane wing.

MAGNETIC WORKHOLDERS

Because of the light cuts and low cutting forces, workpieces can be held in a different manner on surface grinders than on other machine tools. *Magnetic chucks* are used for ferromagnetic materials. To obtain high accuracy, it is desirable to reduce clamping forces and distribute them over the entire area of the workpiece. Also, grinding is frequently done on quite thin or relatively delicate workpieces, which would be difficult to clamp by normal methods. In addition, there often is the problem of grinding a number of small, duplicate workpieces. Magnetic chucks solve all these problems very satisfactorily. Magnetic chucks are available in disk or rectangular shapes. Dry-disk rectifiers are used to provide the necessary direct-current power. Some magnetic chucks utilize permanent magnets as shown in Figure 29-21 and can be tilted so that angles can be ground.

Magnetic chucks provide an excellent means of holding workpieces provided that the cutting or inertial forces are not too great. The holding force is distributed over the entire contact surface of the work, the clamping stresses are low, and therefore there is little tendency for the work to be distorted. Consequently, pieces can be held and ground accurately. Also, a number of small pieces can be mounted on a chuck and ground at the same time. Magnetic chucks provide great part-to-part repeatability because the holding power from one part to the next is the same. Initial setup is usually fast, simple, and relatively inexpensive. Parts loading and unloading is also relatively easy.

It often is necessary to demagnetize work that has been held on a magnetic chuck. Some electrically powered chucks provide satisfactory demagnetization by reversing the direct current briefly when the power is shut off.

FIGURE 29-21 Example of magnetic chuck. *(Courtesy of O. S. Walker.)*

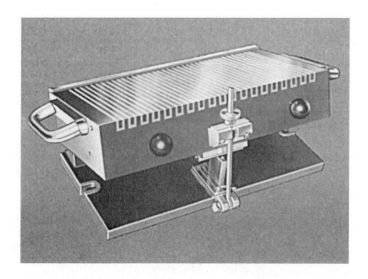

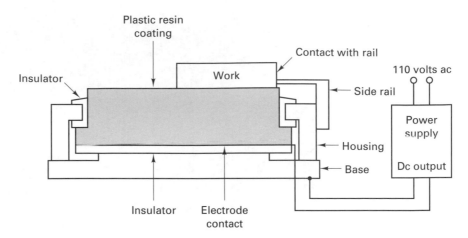

FIGURE 29-22 Principle of electrostatic chuck.

ELECTROSTATIC WORKHOLDERS

Magnetic chucks can be used only with ferromagnetic materials. Electrostatic chucks can be used with any electrically conductive material. This principle (Figure 29-22) directs that work be held by mutually attracting electrostatic fields in the chuck and the workpiece. These provide a holding force of up to 20 psi (21,000 Pa). Nonmetal parts usually can be held if they are flashed (i.e., coated) with a thin layer of metal. These chucks have the added advantage of not inducing residual magnetism in the work.

VACUUM CHUCKS

Vacuum chucks are also available. In one type, illustrated in Figure 29-23, the holes in the work plate are connected to a vacuum pump and can be opened or closed by means of valve screws. The valves are opened in the area on which the work is to rest. The other type has a porous plate on which the work rests. The workpiece and plate are covered with a polyethylene sheet. When the vacuum is turned on, the film forms around the workpiece, covering and sealing the holes not covered by the workpiece and thus producing a seal. The film covering the workpiece is removed or the first cut removes the film covering the workpiece. Vacuum chucks have the advantage that they can be used on both nonmetals and metals and can provide an easily variable force. Magnetic, electrostatic, and vacuum chucks are used for some light milling and turning operations.

FIGURE 29-23 Cutaway view of a vacuum chuck. *(Courtesy of Dunham Tool Company, Inc.)*

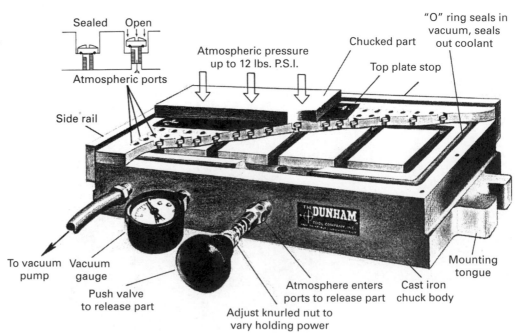

■ 29.14 ECONOMIC JUSTIFICATION OF JIGS AND FIXTURES

As discussed previously, workholders are expensive, even when designed and constructed by using standard components. Obviously, their cost is a part of the total cost of production, and one must determine whether they can be justified economically by the savings in labor and machine cost and improvements in quality that will result from their use. Often it is only through the use of such devices that the design specifications can be met and sustained from part to part.

To determine the economic justification of any special tooling, the following factors must be considered:

1. The cost of the tooling

2. Interest or profit charges on the tooling cost

3. The savings resulting from the use of the tooling; can result from reduced cycle times or improved quality or lower-cost labor

4. The savings in machine cost due to increased productivity

5. The number of units that will be produced using the tooling

The economic relationship between these factors can be expressed in the following manner:

savings per piece (exclusive of tooling costs) $\geq$ additional cost per piece

$$\left\{\begin{array}{c}\text{total cost per piece}\\\text{without tooling}\end{array}\right\} - \left\{\begin{array}{c}\text{total cost per piece}\\\text{using tooling}\\\text{(exclusive of tooling cost)}\end{array}\right\} \geq \text{tooling cost per piece}$$

$$\underbrace{\begin{array}{c}\text{labor cost}\\\text{per piece}\\\text{without}\\\text{tooling}\end{array} + \begin{array}{c}\text{machine and}\\\text{overhead cost}\\\text{per piece}\\\text{without tooling}\end{array}} - \underbrace{\begin{array}{c}\text{labor cost}\\\text{per piece}\\\text{with}\\\text{tooling}\end{array} + \begin{array}{c}\text{machine and}\\\text{overhead cost}\\\text{per piece}\\\text{with tooling}\end{array}} \geq \underbrace{\begin{array}{c}\text{cost}\\\text{of}\\\text{tooling}\end{array} + \begin{array}{c}\text{interest on}\\\text{tooling cost}\end{array}}$$

$$[(R)(t) + (R_m)(t)] - [(R_t)(t_t) + (R_m)(t_t)] \geq \frac{C_t + (C_t/2)(n)(i)}{N} \qquad (29\text{-}2)$$

where

$$R = \text{labor rate per hour, without tooling}$$
$$R_t = \text{labor rate per hour, using tooling}$$
$$t = \text{hours per piece, without tooling}$$
$$t_t = \text{hours per piece, using tooling}$$
$$R_m = \text{machine cost per hour, including all overhead}$$
$$C_t = \text{cost of the special tooling}$$
$$n = \text{number of years tooling will be used}$$
$$i = \text{interest rate (or what invested capital is worth)}$$
$$N = \text{number of pieces that will be produced with the tooling}$$

Equation (29-2) can be expressed in a simpler form:

$$(R + R_m)t - (R_t + R_m)t_t \geq \frac{C_t}{N}\left(1 + \frac{n \times i}{2}\right) \qquad (29\text{-}3)$$

This equation assumes straight-line depreciation and computes interest on the average amount of capital invested throughout the life of the tooling.[1] When the time over which the tooling is to be used is less than one year, companies often do not include an interest cost. If this factor is neglected, the right-hand term of Equation (29-3) reduces to C_t/N.

The equations assume that the material cost will be the same regardless of whether or not special tooling is used. This is not always true. Although these equations are not

[1]For the use of more sophisticated economic analysis, see C. S. Park, *Contemporary Engineering Economics*, 2nd ed., Wiley, 1999.

completely accurate for all cases, they are satisfactory for determining tooling justification in most cases, because the life of such tooling seldom exceeds five years and more frequently does not exceed two years. The equation does not include the cost of poor quality. This can be included by estimating the decrease in the number of defective parts when the workholder is used versus when it is not used.[2]

The following example illustrates the use of Equation (29-3) to determine tooling justification for a dedicated jig. In drilling a series of holes on a radial drill, the use of a drill jig will reduce the time from $\frac{1}{2}$ hour per piece to 15 minutes per piece. If a jig is not used, a machinist, whose hourly rate is $18.00/hr must be used. If the jig is used, the job can be done by a machinist whose rate is $12.00 per hour. The hourly rate for the radial drill is $32.00 per hour.

The cost of making the jig would include $350 for design, $150 for material, and 50 hours of toolmaker's labor, which is charged at the rate of $22.00 per hour to include all machine and overhead costs in the toolmaking department. Investment capital is worth 16% to the company. It is estimated that the jig would last three years and that it would be used for the production of 300 parts over this period. Is the jig justified?

The cost of the jig, C_t, is estimated to be $1,600.00.

$$C_t = \$350 + \$150 + \$150 + \$22.00 \times 50\,\text{hr} = \$1,600.00$$

Substituting the values given in Equation (29-3), we find:

$$(18.00 + 32.00)\,0.5 - (12.00 + 32.00)\,0.25 \geq \$\frac{1600}{300}\left(1 + \frac{3 \times 0.16}{2}\right)$$

$$\text{or } 14.00 > 13.76$$

So this jig is not justified based on cost savings.

One could also ask how many parts would have to be produced with the jig to break even (i.e., increased costs just equal savings.) By omitting the value 300 in the solution above and solving for N, it is found that at least 1627 pieces would have to be produced annually with the jig for it to break even.

It should be noted that Equations (29-2) and (29-3) assume that the time (of the people and machines) saved by the use of the special tooling can be used for other operations. If this is not the case, the cost analysis should be altered to take this important fact into account. Otherwise, the tooling justification may be substantially in error.

The application of group technology and NC machines may eliminate the need for designing and building a new jig or fixture every time a new part is designed. New measures of manufacturing productivity that include terms for quality and flexibility are being developed.

---

[2]For a discussion of new measures and methods on cost accounting, see C. S. Park, "Counting the Costs," Jan. 1987, *Mechanical Engineering*, p. 66.

■ KEY WORDS

3-2-1 principle	channel jigs	fixture	leaf jig	plate jig	toggle clamp
assembly jig	clamping	intermediate jig concept	location	ring jig	vacuum chuck
box jig	electrostatic workholder	jig	magnetic chuck	strap clamp	workholder

■ REVIEW QUESTIONS

1. What are the two primary functions of a workholding device?
2. What distinguishes a jig from a fixture?
3. An early treatise defined a jig as "a device that holds the work and guides a tool." Why was this definition incorrect?
4. Why would an ordinary vise not be considered to be a fixture?
5. What basic criteria should be considered in designing jigs and fixtures?
6. What are the critical surfaces of a part (i.e., what makes a part surface critical)? (This question requires an understanding of basic part drawings.)
7. What difficulties can result from not keeping clamping stresses low in designing jigs and fixtures?
8. Explain the 3-2-1 concept for workpiece location.
9. Which of the basic design principles relating to jigs and fixtures would most likely be in conflict with the 3-2-1 location concept?

10. What are two reasons for not having drill bushings actually touching the workpiece? Notice in Figure 29-8 how many of the designs violate this rule. It is not uncommon to have conflicts and trade-offs in fixture design situations.

11. Why does the use of down milling often make it easier to design a milling fixture than if up milling were used?

12. What do we do to workholding devices to make them more flexible?

13. A large assembly jig for an airplane-wing component gave difficulty when it rested on four-point support but was satisfactory when only three supporting points were used. The assembled wing components were not consistent in shape. Why?

14. Explain why the use of a given fixture may not be economical when used with one machine tool but may be economical when used in conjunction with another machine tool.

15. What are roll-over jigs, and what advantages do they offer?

16. In the clamps shown in Figure 29-18, what is the purpose of the spherical washer?

17. What are other common types of clamps?

18. What is the purpose of dimensioning the strap clamp assembly in Figure 29-18 with letters?

19. Figure 29-2 showed locator buttons. Why not have the part rest on the flat plate?

20. In Figure 29-7, where aren't there three points put on the X plane, two points on the Z plane and one point on the Y plane?

21. Which set of locators in Figure 29-7 establishes the A dimension?

22. To prepare the workpiece shown in Figure 29-7, which surface would you have milled first, the bottom, the back, or the front?

23. For the part shown in Figure 29-7, why don't you drill the holes first, then mill? Why mill at all?

24. The holes are countersunk after they are drilled. How must the jig be designed to put the countersinks on the mounting holes while the part is in the jig or would this operation be done afterward?

■ PROBLEMS

1. Using the following values, determine the number of pieces that would have to be made to justify the use of a jig costing $3,000.

$$R = \$5.75$$
$$R_t = \$4.50$$
$$t_t = 1\tfrac{1}{4}$$
$$t = 2\tfrac{1}{2}$$
$$i = 10\%$$
$$R_m = \$4.50$$
$$n = 3$$

2. Suppose in the sample problem in Section 29.14 that modular fixturing is used, which reduces the toolmaker's labor to 4 hours and the design cost to $100 (4 hours at $25 per hour), that the material cost (modular elements) for the subplate structural elements, clamps, and so on, was $600. What is the break-even quantity for a modular fixture? (*Note:* The modular fixture is used for the job, then disassembled and returned to the tool room. The parts are reused in other workholders.)

3. Here is a typical problem in clamping force analysis (Figure 29-A). A milling cutter is face milling the top surface of a 1500-lb casting. The cutting force is 1800 lb, and the thrust force is 900 lb. The feed force is assumed to be zero for this analysis. The unknown forces are as follows:

F_R = total force from all clamps on right side
F_L = total force from all clamps on left side
R_1 = horizontal reaction force from the fixed stop
R_2 = vertical reaction force from the fixed stop
R_3 = vertical reaction force from the right side
N = normal force ($F_L + F_R + 1500\,\text{lb} + 900\,\text{lb}$)
μ = coefficient of friction (0.19)

What are the required clamping forces, F_R and F_L?
Hint: For a static condition, the sum of the X direction (horizontal) is zero, the sum of the Y direction (vertical) is zero, and the sum of moments around any point is zero. In order to solve the equations, you will have to assume that some of the forces are zero.

4. (Problem in loading and unloading time)
Suppose the jig in Figure 29-7 could be improved by replacing the screw clamps with toggle clamps. Many things

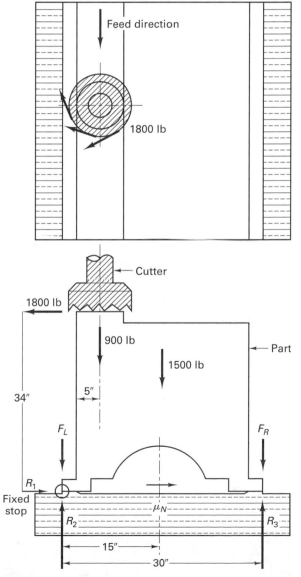

FIGURE 29-A Forces on a 1500-lb work piece produced by face milling.

are needed to be able to cost justify the improvement in the jig. This problem requires that the engineer estimate or determine the following:

1. How much time is saved with a toggle clamp (in the loading and unloading) cycle?

2. How much does a toggle clamp cost?

3. How much will it cost to modify the existing jig? Currently, for this job, the machining cycle time to drill the two holes is 1 min and the unload/load time is 30 sec, the operator is getting $12 per hour and the machine cost is $30 per hour.

www.wiley.com/college/degarmo

*C*hapter 29   CASE STUDY

Overhead Crane Installation

To install an overhead crane (Figure CS-29) in one bay of the Strom Manufacturing Company, an assembly plant, brackets for the rails of the crane are to be mounted on eight columns, four on each side of the bay area facing each other. The rails for the crane will span four columns. Each bracket on each column will need six holes in a circular pattern. The holes must be accurately spaced within 5 minutes of arc of each other. The axis of the holes must be parallel and normal to the face of the columns. The center of the bolt-hole circle must be at a height of at least 20 ft from the floor, but the centers of all eight bolt-hole circles must be on the same parallel plane, so that the rails for the crane are level and parallel with each other. Four of the columns along the wall have their faces flush with the wall surface so that mechanical clamping or attachments cannot be used. The building code will permit no welding of anything to these columns.

1. How would you proceed to get the bolt holes located in the right position on the beams?

2. How would you get the hole patterns located properly with respect to each other on all eight beams?

3. List the equipment you will need.

4. Make a sketch of any special tool you recommend.

(*Hint:* Check Chapter 29 for drill jig designs and ask your favorite civil engineer for suggestions regarding surveying.)

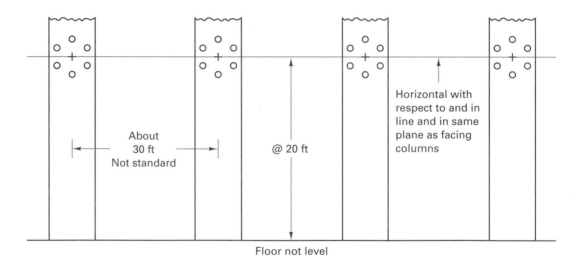

FIGURE CS-29   Overhead crane installation at the Strom Manufacturing Company.

CHAPTER 30

THREAD AND GEAR MANUFACTURING

■ 30.1 INTRODUCTION

Screw threads and gears are important machine elements. *Threading*, *thread cutting*, or *thread rolling* refers to the manufacture of threads on external diameters. *Tapping* refers to machining threads in (drilled) holes. Without these processes our present technological society would come to a grinding halt. More screw threads are made each year than any other machined element. They range in size from those used in small watches to threaded shafts 10 in. in diameter. They are made in quantities ranging from one to several million duplicate threads. Their precision varies from that of inexpensive hardware screws to that of lead screws for the most precise machine tools. Consequently, it is not surprising that many very different procedures have been developed for making screw threads and that the production cost by the various methods varies greatly. Fortunately, some of the most economical methods can provide very accurate results. However, as in the design of most products, the designer can greatly affect the ease and cost of producing specified screw threads. Thus, understanding thread-making processes permits the designer to specify and incorporate screw threads into designs while avoiding needless and excessive cost.

   Gears transmit power or motion mechanically between parallel, intersecting, or nonintersecting shafts. Although usually hidden from sight, gears are one of the most important mechanical elements in our civilization, possibly even surpassing the wheel, since most wheels would not be turning were power not being applied to them through gears. They operate at almost unlimited speeds under a wide variety of conditions. Millions are produced each year in sizes from a few millimeters up to more than 30 ft in diameter. Often the requirements that must be, and are routinely, met in their manufacture are amazingly precise. Consequently many special machines and processes have been developed for producing gears. Let us begin by discussing threads.

SCREW-THREAD STANDARDIZATION AND NOMENCLATURE

A screw thread is a ridge of uniform section in the form of a helix on the external or internal surface of a cylinder, or in the form of a conical spiral on the external or internal surface of a frustrum of a cone. These are called *straight* or *tapered* threads, respectively. Tapered threads are used on pipe joints or other applications where liquid-tight joints are required. Straight threads, on the other hand, are used in a wide variety of applications, most commonly on fastening devices, such as bolts, screws, and nuts, and as

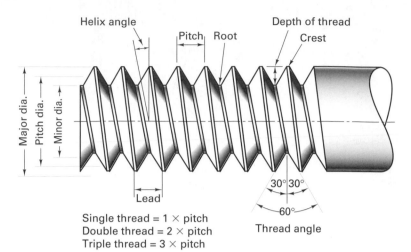

FIGURE 30-1   Standard screw-thread nomenclature.

integral elements on parts that are to be fastened together. But, as mentioned previously, they find very important applications in transmitting controlled motion, as in lead screws and precision measuring equipment.

The standard nomenclature for screw-thread components is illustrated in Figure 30-1. The symbol P, the pitch, refers to the distance from a point on one screw thread to the corresponding point on the next thread, measured parallel to the length axis of the part. In 1948, representatives of the United States, Canada, and Great Britain adopted the Unified and American Screw Thread Standards, based on the form shown in Figure 30-2. In 1968 the International Organization for Standards (ISO) recommended the adoption of a set of metric standards, based on the basic thread profile. It appears likely that both types of threads will continue to be used for some time to come.

In both the *Unified* and *ISO systems*, the crests of external threads may be flat or rounded. The root usually is made rounded to minimize stress concentrations at this critical area. The internal thread has a flat crest in order to mate with either a rounded or V-root of the external thread. A small round is used at the root to provide clearance for the flat crest of the external thread.

In the metric system, the *pitch* always is expressed in millimeters, whereas in the American (Unified) system, it is a fraction having as the numerator 1 and as the denominator the number of threads per inch. Thus, a $\frac{1}{16}$ pitch is $\frac{1}{16}$ of an inch. Consequently, in the Unified system, threads more commonly are described in terms of threads per inch rather than by the pitch.

While all elements of the thread form are based on the *pitch diameter*, screw-thread sizes are expressed in terms of the *outside*, or *major* diameter and the *pitch* or *number of threads per inch*. In threaded elements, *lead* refers to the axial advance of the element during one revolution; therefore lead equals pitch on a single-thread screw.

TYPES OF SCREW THREADS

Eleven types, or series, of threads are of commercial importance, several having equivalent series in the metric system and Unified systems. See Figure 30-3. As has been indicated, the Unified threads are available in a coarse (UNC and NC), fine (UNF and NF), extra-fine (UNEF and NEF), and three-"pitch" (8, 12, and 16) series, the number of threads per inch being according to an arbitrary determination based on the major diameter.

Many nations have now adopted ISO threads into their national standards. Besides metric ISO threads, there are also inch-based ISO threads, namely the UN series with which people in the United States, Canada, and Great Britain are familiar. ISO offers a wide range of metric sizes. Individual countries have the choice of accepting all or a selection of the ISO offerings.

The size listings of metric threads start with "M" and continue with the outside diameter in millimeters. Most ISO metric thread sizes come in coarse, medium, and fine pitches. When a coarse thread is designated, it is not necessary to spell out the pitch. For

(a) External thread

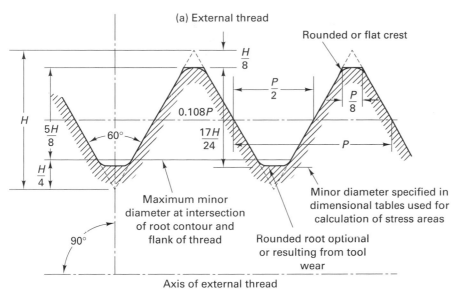

(b) Internal thread

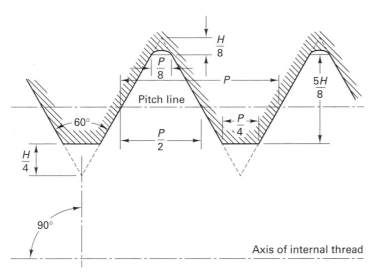

(c) Metric ISO/R68-1969

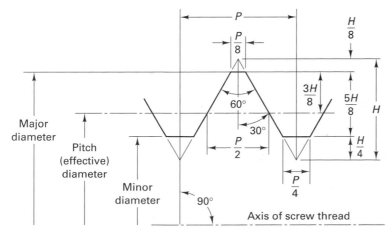

D = Major diameter of thread of nut ⎱ Normal
d = Major diameter of screw ⎰ diameter
D_1 = Minor diameter of thread of nut
d_1 = Minor diameter of screw
D_2 = Pitch diameter of thread of nut
d_2 = Pitch diameter of screw
H = Height of the complete theoretical
    thread profile
H_1 = Engagement
P = Pitch

$$H = \frac{\sqrt{3}}{2} P$$

FIGURE 30-2   Basic profiles of Unified and American screw-threads (a) external, (b) internal, and (c) metric.

Types of Screw Threads

1. *Coarse-thread series* (UNC and NC). For general use where not subjected to vibration.

2. *Fine-thread series* (UNF and NF). For most automotive and aircraft work.

3. *Extra-fine-thread series* (UNEF and NEF). For use with thin-walled material or where a maximum number of threads are required in a given length.

4. *Eight-thread series* (8UN and 8N). Eight threads per inch for all diameters from 1 to 6 in. Used primarily for bolts on pipe flanges and cylinder-head studs where an initial tension must be set up to resist steam or air pressures.

5. *Twelve-thread series* (12UN and 12N). Twelve threads per inch for diameters from $\frac{1}{2}$ through 6 in. Not used extensively.

6. *Sixteen-thread series* (16UN and 16N). Sixteen threads per inch for diameters from $\frac{3}{4}$ through 6 in. Used for a wide variety of applications that require a fine thread.

7. *American Acme thread.* See Figure 30-4. This thread and the following three are used primarily in transmitting power and motion.

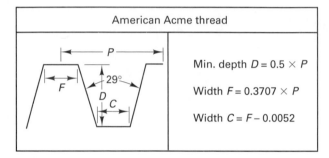

8. *Buttress thread.*

9. *Square thread.*

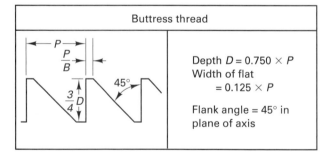

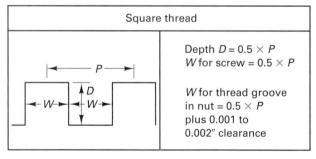

10. *29° Worm thread.*

11. *American, standard pipe thread.* This thread is the standard tapered thread used on pipe joints in this country. The taper on all pipe threads is $\frac{3}{4}$ in./ft.

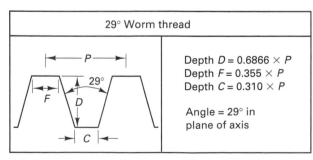

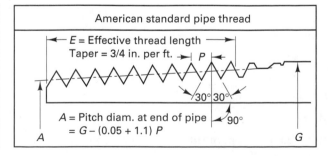

FIGURE 30-3   Types of screw threads.

example, a coarse 10 mm OD thread is called out as "M 12." This thread has a pitch of 1.75 mm, but the pitch may be omitted from the call-out. A fine 12 mm OD thread is available. It has a 1.25 mm pitch and must be designated "M 12 × 1.25." An extra-fine 12 mm OD thread having 0.75 mm pitch would receive the designation "M 12 × 0.75."

The sign "×" is not employed as a multiplication symbol in metric practice, but is used to relate these two attributes of the threads. The full description of a thread fastener obviously includes information beyond the thread specification. Hcad type, length, length of thread, design of end, thread runout, heat treatment, applied finishes, and other data may be needed to fully specify a bolt besides the designation of the thread. The "×" sign should not be used to separate any of the other characteristics.

Here is another example of an ISO thread designation.

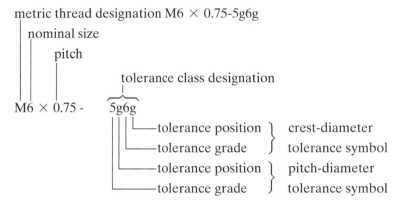

metric thread designation M6 × 0.75-5g6g

nominal size

pitch

tolerance class designation

M6 × 0.75 - 5g6g

tolerance position } crest-diameter
tolerance grade } tolerance symbol
tolerance position } pitch-diameter
tolerance grade } tolerance symbol

In the ISO system, tolerances are applied to "positions" and "grades." Tolerance positions denote the limits of pitch and crest diameters, using "e" (large), "g" (small), and "H" (no allowance) for internal threads. The grade is expressed by numerals 3 through 9. Grade 6 is roughly equivalent to U.S. grades 2A and B, medium quality, general purpose threads. Below 6 is fine quality and/or short engagement. Above 6 is coarse quality and/or long length of engagement.

In the Unified system, screw threads are designated by symbols as follows:

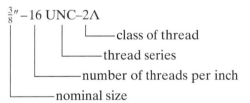

$\frac{3}{8}'' - 16$ UNC–2A

class of thread

thread series

number of threads per inch

nominal size

This type of designation applies to right-hand threads. For left-hand threads, the letters LH are added after the class of thread symbol.

In the Unified system, manufacturing tolerances are specified by three classes. Class 1 is for ordnance and other special applications. Class 2 threads are the normal production grade, and Class 3 threads have minimum tolerances where tight fits are required. The letters A and B are added after the class numerals to indicate external and internal threads, respectively.

The availability of fasteners, particularly nuts, containing plastic inserts to make them self-locking and thus able to resist loosening due to vibration, and the use of special coatings that serve the same purpose, have resulted in less use of finer-thread-series fasteners in mass production. Coarser-thread fasteners are easier to assemble and less subject to cross-threading (binding).

■ 30.2 THREAD MAKING

Three basic methods are used to produce threads: *cutting*, *rolling*, and *casting*. Although both external and internal threads can be cast, relatively few are made in this manner, primarily in connection with die casting, investment casting, or thc molding of plastics. Today, by far, the largest number of threads are made by rolling. Both external and internal threads can be made by rolling, but the material must be ductile. Because rolling is a less flexible process than thread cutting, it is restricted essentially to standardized and simple parts. Consequently, large numbers of external and internal threads still are made by cutting processes including grinding and tapping.

External Thread Cutting Methods	Internal Thread Cutting Methods
Threading on an engine lathe	Threading (on an engine lathe or NC lathe)
Threading on a NC lathe	
With a die held in a stock (manual)	With a tap and holder (manual NC, machine, semiautomatic, or automatic)
With an automatic die (turret lathe or screw machine) or NC lathe	With a collapsible tap (turret lathe, screw machine, or special threading machine)
By milling	By milling
By grinding	

CUTTING THREADS ON A LATHE

Lathes provided the first method for cutting threads by machine. Although most threads are now produced by other methods, lathes still provide the most versatile and fundamentally simple method. Consequently, they often are used for cutting threads on special workpieces where the configuration or nonstandard size does not permit them to be made by less costly methods.

There are two basic requirements for thread cutting on a lathe. First, an accurately shaped and properly mounted tool is needed because thread cutting is a form-cutting operation. The resulting thread profile is determined by the shape of the tool and its position relative to the workpiece. Second, the tool must move longitudinally in a specific relationship to the rotation of the workpiece, because this determines the *lead* of the thread. This requirement is met through the use of the *lead screw* and the *split nut*, which provide positive motion of the carriage relative to the rotation of the spindle.

To cut a thread, it also is essential that a constant positional relationship be maintained between the workpiece, the cutting tool, and the lead screw. If this is not done, the tool will not be positioned correctly in the thread space on successive cuts. Correct relationship is obtained by means of a *threading dial* (Figure 30-4), which is driven directly by the lead screw through a worm gear. Because the workpiece and the lead screw are directly connected, the threading dial provides a means for establishing the desired positional relationship between the workpiece and the cutting tool. The threading dial is graduated into an even number of major and half divisions. If the feed mechanism is engaged in accordance with the following rules, correct positioning of the tool will result:

1. *For even-number threads:* at any line on the dial
2. *For odd-number threads:* at any numbered line on the dial
3. *For threads involving $\frac{1}{2}$ numbers:* at any odd-numbered line on the dial
4. *For $\frac{1}{4}$ or $\frac{1}{8}$ threads:* return to the original starting line on the dial

To start cutting a thread, the tool usually is fed inward until it just scratches the work, and the cross-slide dial reading is then noted or set at zero. The split nut is engaged and the tool permitted to run over the desired thread length. When the tool reaches the end of the thread, it is quickly withdrawn by means of the cross-slide control. The split nut is then disengaged and the carriage returned to the starting position, where the tool is clear of the workpiece. At this point the future thread will be indicated by a fine scratch line. This permits the operator to check the thread lead by means of a scale or thread gage to assure that all settings have been made correctly.

Next, the tool is returned to its initial zero depth position by returning the cross slide to the zero setting. By using the compound rest, the tool can be moved inward the proper depth for the first cut. A depth of 0.010 to 0.025 in. usually is used for the first cut and smaller amounts on each successive cut, until the final cut is made with a depth of only 0.001 to 0.003 in. to produce a good finish. When the thread has been cut nearly to its full depth, it is checked for size by means of a mating nut or thread gage. Cutting is continued until a proper fit is obtained.

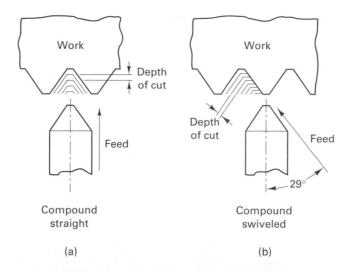

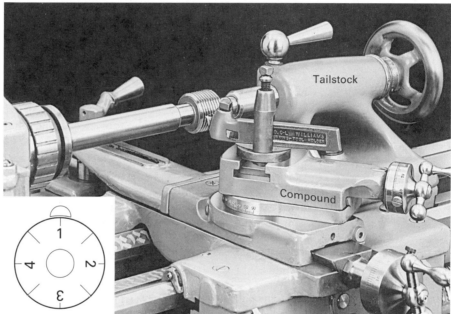

FIGURE 30-4 Cutting a screw thread on a lathe, showing the method of supporting the work and the relationship of the tool to the work with the compound swivelled. Inset shows face of threading dial.

Figure 30-4 illustrates two methods of feeding the tool into the work. If the tool is fed radially, cutting takes place simultaneously on both sides of the tool. With this true form-cutting procedure, no rake should be ground on the tool, and the top of the tool must be horizontal and be set exactly in line with the axis of rotation of the work. Otherwise, the resulting thread profile will not be correct. An obvious disadvantage of this method is that the absence of side and back rake results in poor cutting (except on cast iron or brass). The surface finish on steel usually will be poor. Consequently, the second method commonly is used, with the compound swivelled 20°. The cutting then occurs primarily on the left-hand edge of the tool, and some side rake can be provided.

Proper speed ratio between the spindle and the lead screw is set by means of the gear-change box. Modern industrial lathes have ranges of ratios available so that nearly all standard threads can be cut merely by setting the proper levers on the quick-change gear box.

Cutting screw threads on a lathe is a slow, repetitious process that requires considerable operator skill. The cutting speeds usually employed are from one third to one half of regular speeds to enable the operator to have time to manipulate the controls and to ensure better cutting. The cost per part can be high, which explains why other methods are used whenever possible.

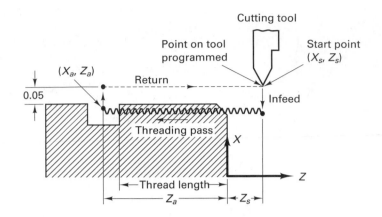

FIGURE 30-5 Canned subroutines called G codes are used on CNC lathes to produce threads.

X_a	Specifies the absolute X coordinate of the tool after axial infeed.
$G32$	Initiates the single-pass threading cycle.
Z_a	Specifies the absolute Z coordinate of the tool after the threading pass.
Fn	n specifies the feed rate
$X_s Z_s$	Specifies the absolute X and Z coordinates of the start point.

CUTTING THREADS ON A CNC LATHE

CNC lathes and turning centers can be programmed to machine straight, tapered, or scroll threads. Threads machined using the same type of special tool have the thread shape shown in Figure 30-4. The tool is positioned at a specific starting distance from the end of the work (Figure 30-5). This distance will vary from machine to machine. Its value can be found in the machine's programming manual. The CNC software will have a set of preprogrammed machining routines (called G codes) specifically for threading. Beginning at the start point, the tool accelerates to the feed rate required to cut the threads. The tool creates the thread shape by repeatedly following the same path as axial infeed is applied. For standard Vee threads, the infeed can be applied along a 0° or 29° angle. The depth of cut for the first pass is the largest. The cutting depth is then decreased for each successive pass until the required thread depth is achieved. A final finishing pass is then made with the tool set at the thread depth.

CUTTING THREADS WITH DIES

Straight and tapered external threads up to about $1\frac{1}{2}$ in. in diameter can be manually cut quickly by means of threading dies (Figure 30-6a). Basically, these dies are similar to hardened, threaded nuts with multiple cutting edges. The cutting edges at the starting end are beveled to aid in starting the dies on the workpiece. As a consequence, a few threads at the inner end of the workpiece are not cut to full depth. Such threading dies are made of carbon or high-speed tool steel.

FIGURE 30-6 (a) Solid threading die; (b) solid-adjustable threading die; (c) threading-die stock for round die (die removed). *(Courtesy of TRW-Greenfield Tap & Die.)*

(a) (b)

(c)

Solid-type dies are seldom used in manufacturing because they have no provision for compensating for wear. The solid-adjustable type (Figure 30-6(b)) is split and can be adjusted over a small range by means of a screw to compensate for wear or to provide a variation in the fit of the resulting screw thread. These types of threading dies usually are held in a *stock* for hand rotation. A suitable lubricant is desirable to produce a smoother thread and to prolong the life of the die, since there is extensive friction during the cutting process.

SELF-OPENING DIE HEADS

A major disadvantage of solid-type threading dies is that they must be unscrewed from the workpiece to remove them. They therefore are not suitable for use on high-speed, production-type machines and *self-opening die heads* are used instead on turret lathes, screw machines, NC lathes, and special threading machines for cutting external threads.

There are three types of self-opening die heads, all having four sets of adjustable, multiple-point cutters that can be removed for sharpening or for interchanging for different thread sizes. This permits one head to be used for a range of thread sizes (see Figure 30-7). The cutters can be positioned radially or tangentially, resulting in less tool

FIGURE 30-7   Self-opening die heads, with (a) radial cutter, (b) tangential cutters, (c) circular cutters, and (d) terminology of circular chasers and their relation to the work. *(Courtesy of Geometric Tool Company, Warner & Sawsey Company, National Acme Company, and TRW-Greenfield Tap & Die, respectively.)*

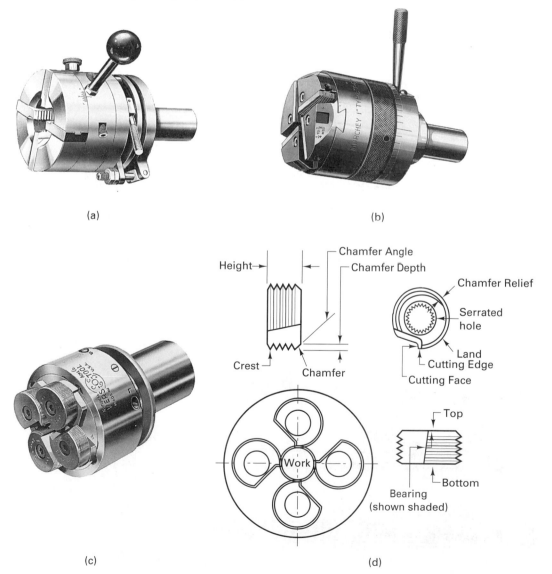

(a)

(b)

(c)

(d)

flank contact and friction rubbing. In some self-opening die heads, the cutters are circular, with an interruption in the circular form to provide an easily sharpened cutting face. The cutters are mounted on the holder at an angle equal to the helix angle of the thread.

As the name implies, the cutters in self-opening die heads are arranged to open automatically when the thread has been cut to the desired length, thereby permitting the die head to be quickly withdrawn from the workpiece. On die heads used on turret lathes, the operator usually must reset the cutters in the closed position before making the next thread. The die heads used on screw machines and automatic threading machines are provided with a mechanism that automatically closes the cutters after the heads are withdrawn.

Cutting threads by means of self-opening die heads frequently is called *thread chasing*. However, some people apply this term to other methods of thread cutting, even to cutting a thread in a lathe.

■ 30.3 INTERNAL THREAD CUTTING–TAPPING

The cutting of an internal thread by means of a multiple-point tool is called *thread tapping*, and the tool is called a *tap*. A hole of diameter slightly larger than the minor diameter of the thread must already exist, made by drilling/reaming, boring, or die casting. For small holes, solid *hand taps* (Figure 30-8) are usually used. The flutes create cutting edges on the thread profile and provide space for the chips and the passage of cutting fluid. Such

FIGURE 30-8 Terminology for a plug tap with photographs of taper (t), plug (p), and bottoming (b) taps which are used serially in threading holes. *(Courtesy of TRW-Greenfield Tap & Die.)*

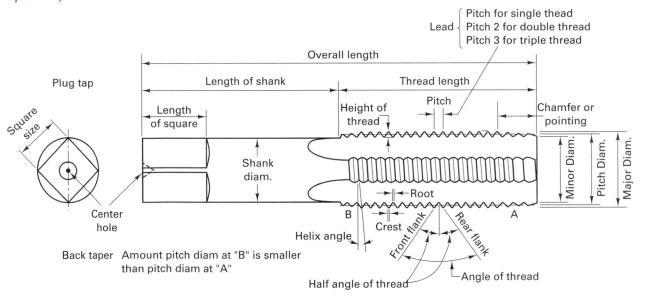

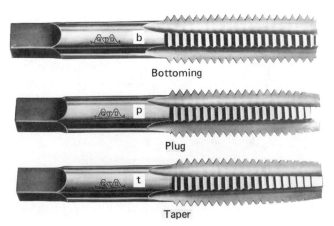

FIGURE 30-9 *(Left to right)* Spiral fluted tap; spiral point tap; spiral point tap cutting chips; fluteless bottoming tap and fluteless plug tap for cold-forming internal threads; cross section of fluteless forming tap. *(Courtesy of TRW-Greenfield Tap & Die.)*

taps are made of either carbon or high-speed steel and are now routinely coated with TiN. The flutes can be either straight, helical, or spiral.

Hand taps (Figure 30-8) have square shanks and are usually made in sets of three. The *taper tap* has a tapered end that will enter the hole a sufficient distance to help align the tap. In addition, the threads increase gradually to full depth, and therefore this type of tap requires less torque to use. However, only a through hole can be threaded completely with a taper tap because it cuts to full depth only behind the tapered portion. A blind hole can be threaded to the bottom using three types of taps in succession. After the taper tap has the thread started in proper alignment, a *plug tap*, which has only a few tapered threads to provide gradual cutting of the threads to depth, is used to cut the threads as deep into the hole as its shape will permit. A *bottoming tap*, having no tapered threads, is used to finish the few remaining threads at the bottom of the hole to full depth. Obviously, producing threads to the full depth of a blind hole is time-consuming, and it also frequently results in broken taps and defective workpieces. Such configurations usually can be avoided if designers will give reasonable thought to the matter.

Taps operate under very severe conditions because of the heavy friction (high torque) involved and the difficulty of chip removal. Also, taps are relatively fragile. *Spiral-fluted taps* (Figure 30-9) provide better removal of chips from a hole, particularly in tapping materials that produce long, curling chips. They also are helpful in tapping holes where the cutting action is interrupted by slots or keyways. The *spiral point* cuts the thread with a shearing action that pushes the chips ahead of the tap so that they do not interfere with the cutting action and the flow of cutting fluid into the hole.

COLLAPSING TAPS

Collapsing taps are similar to self-opening die heads in that the cutting elements collapse inward automatically when the thread is completed. This permits withdrawing the tap from the workpiece without the necessity of unscrewing it from the thread. They can either be self-setting, for use on automatic machines, or require manual setting for each cycle. Figure 30-10 shows some of the types available.

HOLE PREPARATION

Drilling is the most common method of preparing holes for tapping, and when close control over hole size is required, reaming may also be necessary. The drill size determines the final thread contour and the drilling torque. Unless otherwise specified, the tap drill size for most materials should produce approximately 75% thread, that is, 75% of full thread depth.

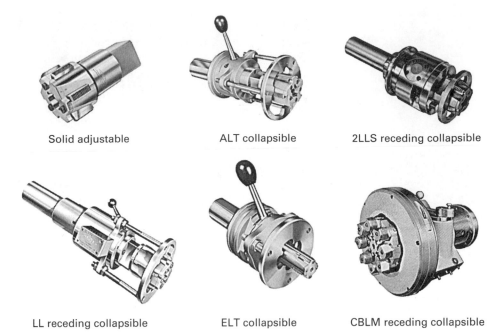

Solid adjustable · ALT collapsible · 2LLS receding collapsible

LL receding collapsible · ELT collapsible · CBLM receding collapsible

FIGURE 30-10 Solid adjustable and collapsible taps. *(Courtesy of Teledyne Landis Machine.)*

TAPPING IN MACHINE TOOLS

Solid taps also are used in tapping operations on machine tools, such as lathes, drill presses, and special tapping machines. In tapping on a drill press, a tapping attachment often is used. These devices rotate the tap slowly when the drill press spindle is fed downward against the work. When the tapping is completed and the spindle raised, the tap is automatically driven in the reverse direction at a higher speed to reduce the time required to back the tap out of the hole. Some modern machine tools provide for extremely fast spindle reversal for backing taps out of holes.

When solid taps are used on a screw machine or turret, the tap is prevented from turning while it is being fed into the work. As the tap reaches the end of the hole, the tap is free to rotate with the work. The work is then reversed and the tap, again prevented from rotating, is backed out of the hole.

The machine should have adequate power, rigidity, speed and feed ranges, cutting fluid supply, and positive drive action. Chucks, tap holders, and collets should be checked regularly for signs of wear or damage. Accurate alignment of the tap holder, machine spindle, and workpiece is vital to avoid broken taps or bell-mouthed, tapered, or oversized holes.

TAPPING CUTTING TIME

The equation to calculate the cutting time for tapping is (approximately):

$$T_m = \frac{Ln}{N} = \frac{\pi D L n}{12V} + A_L + A_R \tag{30-1}$$

where

$$N = \text{spindle rpm}$$

$$T_m = \text{cutting time (min)}$$

$$D = \text{tap diameter (in.)}$$

$$L = \text{depth of tapped hole or length of cut (in.)}$$

$$n = \text{number of threads per inch (tpi)(feed rate)}$$

$$V = \text{cutting speed (sfpm)}$$

$$A_L = \text{allowance to start tap (min)}$$

$$A_R = \text{allowance to withdraw the tap (min)}$$

SPECIAL THREADING AND TAPPING MACHINES

Special machines are available for production threading and tapping. Threading machines usually have one or more spindles on which a self-opening die head is mounted, with suitable means for clamping and feeding the workpiece. Special tapping machines using self-collapsing taps substituted for the threading dies are also available. More commonly, tapping machines resemble drill presses, modified to provide spindle feeds both upward and downward, with the speed and feed more rapid on the upward motion.

COMMON TAPPING PROBLEMS

Tap overloading is often caused by poor lubrication, lands that are too wide, chips packed in the flutes, or tap wear. Surface roughness in the threads has many causes. A negative grind on the heel will prevent the tap from tearing the threads when backing out.

When a tap loses speed or needs more power, it generally indicates that the tap is dull (or improperly ground), or the chips are packed in flutes (loaded). The flutes may be too shallow or the lands too deep. When tapping soft ductile metals, loading can usually be overcome by polishing the tap before usage.

Improper hole size due to drill wear increases the percentage of threads being cut. Dull tools can also produce a rough finish or work-harden the hole surface and cause the tap to dull more quickly. Check to see that the axis of the hole and tap are aligned. If the tap cuts when backing out, check to see if the hole is oversize.

TAPPING HIGH STRENGTH MATERIALS

High-strength, thermal-resistant materials, sometimes called "exotics," cause special problems in tapping. A variety of materials are classified as exotics: stainless steel, precipitation-hardened stainless steel, high-alloy steels, iron-based superalloys, titanium, Inconcel, Hastelloy, Monel, and Waspalloy. Their most important attribute is their high strength-to-weight ratio.

Each material presents different problems to efficient tapping, but they all share certain similarities. Toughness and general abrasiveness top the list. It is also difficult to impart a good surface finish to exotics; heat tends to localize in the shear zone and exotics tend to workharden and grab the tool.

A tap's chamfer and first full thread do virtually all the cutting. The remaining ground threads serve merely as chasers. Because of this, taps to thread exotics are increasingly manufactured with short threads and reduced necks. They diminish problems caused by material closure, and provide more space for coolant and chip ejection.

When tapping exotics, the largest tap core diameter possible should be applied. Cutting 75 or 65% threads places less stress on the taps, lengthening tool life and reducing breakage. To cut threads in exotic alloys successfully, taps must combine geometries specifically tailored for those materials and be made of premium tool steels subjected to precisely controlled heat-treatment processes.

Stainless steel is known to workharden and to have slow heat-dissipation characteristics; Stainless steel requires a tap geometry with a positive 6° to 9° rake, preventing workhardening and reducing torque. Grinding an appropriate eccentric thread and back-taper relief onto the tap will reduce friction. A surface treatment promotes lubricity. That is, the tool should be made from high-vanadium, high-cobalt tool steel and have a surface treatment so coolant adheres to it. Stainless steel generates long chips and requires a tap with a 38° helix angle and adequate flute depth to promote chip evacuation. Proper hook and radial relief guarantee accurate thread-hole size and long tool life.

When tapping a titanium alloy, the material's tendency to concentrate heat in a small contact area must be considered. Concentrated heat often leads to excessive cutting-edge wear. Titanium generates average-to-short chips, is abrasive, and is prone to chip welding and high friction. These characteristics degrade tap performance and shorten tool life.

Taps for threading titanium are constructed of premium tool steel, nitrided for hardness. Titanium nitride (TiN) coatings cannot be used because they react chemically with the workpiece material, causing rapid tap failure. For tapping through-holes, a 3° to 5° rake and short thread design with high eccentric relief will promote efficient chip evacuation.

Nickel-based Inconel, Monel, Waspalloy, and Hastelloy present severe tapping problems. Among their machining characteristics are toughness, workhardening, heat retention, and built-up-edge (BUE). Taps designed to thread these materials need tremendous

TABLE 30-1.	Cutting Fluids for Tapping (HSS Tools)
Work Material	Cutting Fluid
Aluminum	Kerosene and lard oil; kerosene and light-base oil
Brass	Soluble oil or light-base oil
Naval brass	Mineral oil with lard or light-base oil
Manganese bronze	Mineral oil with lard or light-base oil
Phosphor bronze	Mineral oil with lard or light-base oil
Copper	Mineral oil with lard or light-base oil
Iron, cast malleable	Dry or soluble oil
	Soluble oil or sulfur-base oil
Magnesium	Light-base oil diluted with kerosene
Monel metal	Sulfur-base oil
Steels:	
Up to 0.25 carbon	Sulfur-base or soluble oil
Free machining	Sulfur-base or soluble oil
0.30–0.60 carbon, annealed	Sulfur-base oil
0.30–0.60 carbon, heat treated	Chlorinated sulfur base oil
Tool, high-carbon, HSS	Chlorinated sulfur base oil
Stainless	Chlorinated sulfur base oil
Titanium	Chlorinated sulfur base oil
Zinc die castings	Kerosene and lard oil

stability and a strong cross-sectional construction. The most popular tap materials for nickel alloys are high-vanadium, high-cobalt tool steel or powdered metal (PM) tool steel. Tapping blind holes in these alloys requires a 3° to 5° rake to shear and deflect the cutting forces downward toward the tap's root, its strongest area. A 26° helix angle promotes chip evacuation, and a nitride or TiN coating reduces friction and tool wear.

Because of nickel alloys' toughness, taps to cut them should have the longest taper possible. This allows the cutting edges to progressively gain thread height before the first full thread begins its cut, distributing the load over a wider area. Spiral-pointed, straight-flute taps have a four- to five-thread taper. For tapping blind holes, the first two or three threads—more if possible—should be tapered.

CUTTING FLUID FOR TAPPING

Cutting fluids should be kept as clean as possible and should be supplied in copious quantities to reduce heat and friction and to aid in chip removal. Long tap life has been reported to result from routing high pressure coolants through the tap to flush out the chips and cool the cutting edges. Recommended cutting fluids are listed in Table 30-1.

■ 30.4 THREAD MILLING

Highly accurate threads, particularly in larger sizes, are often form-milled. Either a single- or a multiple-form cutter may be used. A single-form cutter having a single annular row of teeth is tilted at an angle equal to the helix angle of the thread and is fed inward radially to full depth while the work is stationary. The workpiece then is rotated slowly, and the cutter simultaneously is moved longitudinally, parallel with the axis of the work (or vice versa), by means of a lead screw, until the thread is completed. The thread can be completed in a single cut, or roughing and finish cuts can be used. This process is used primarily for large-lead or multiple-lead threads.

Some threads can be milled more quickly by using a multiple-form cutter having multiple rows of teeth set perpendicular to the cutter axis (the rows having no lead). The cutter must be slightly longer than the thread to be cut. It is set parallel with the axis of the workpiece and fed inward to full-thread depth while the work is stationary. The work then is rotated slowly for a little over one revolution, and the rotating cutter is simultaneously moved longitudinally with respect to the workpiece (or vice versa) according to the thread lead. When the work has revolved one revolution, the thread is complete. This process cannot be used on threads having a helix angle greater than about 3°, because clearance between the sides of the threads and the cutter depends on the cutter diameter's being substantially less than that of the workpiece. Thus, although the process is rapid, its use is restricted to threads of substantial diameter and not more than about 2 in. long.

As shown in Figure 30-11, advances in CNC computer controls have led to thread milling on three-axis machines. Today's CNC can helically interpolate the axial feed controlling the thread pitch with circular feed controlling the circumference of the thread. The cutter has teeth shaped like the desired thread form. The cutter rotates at high speeds while its axis slowly moves around the part in a planetary arc just over 360°. The cutter advances axially a distance equal to one pitch to generate the helical path. Thread milling has advantages in diameters over 1.5 in. including better surface finish and concentricity and the ability to produce right- or left-hand threads with the same tool.

FIGURE 30-11 Thread milling on a three-axis NC machine can produce a complete thread in a single feed revolution. (*American Machinist*, November, 1988.)

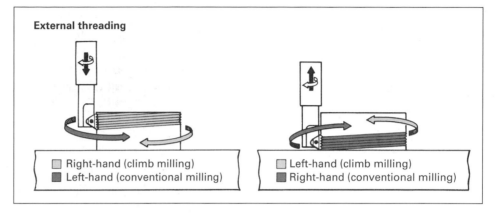

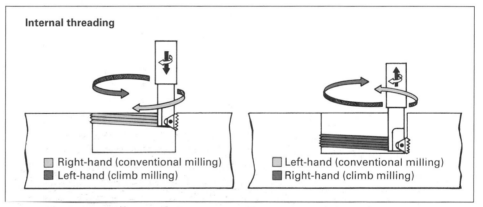

Thread milling on machining centers with multitooth indexable-carbide-insert cutters was introduced about fifteen years ago. The cutter can produce a finished thread in one helical pass.

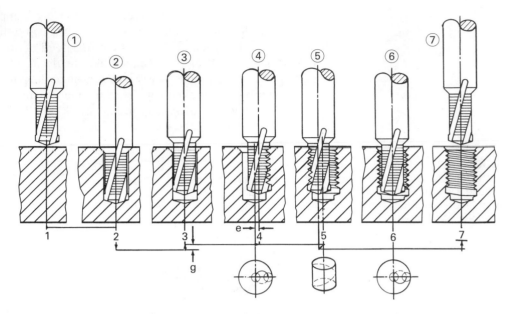

FIGURE 30-12 The process of high-speed thrilling (drilling plus threading) a hole includes (1) approach, (2) drill plus chamfer, (3) retract one thread pitch, (4) radially ramp to the major thread diameter, (5) thread-mill with helical interpolation, (6) return the tool to the centerline of the hole, and (7) retract from the finished hole. At 20,000 rpm, a hole can be thrilled in Aluminum in less than 2 seconds. (Fred Mason, *American Machinist*, November, 1988.)

Thrilling produces threaded holes by combining short hole drilling with thread milling using a combination tool with a drill point and a thread mill body. The details of the process are shown in Figure 30-12 and can be done on any CNC machining center. Compared to tapping, the process eliminates two tools, two tool holders, and two tool change cycles as the single tool combines the drill ream and tap functions into one tool. Threaded-hole depths are limited to about three hole diameters.

■ 30.5 THREAD GRINDING

Grinding can produce very accurate threads, and it also permits threads to be produced in hardened materials. Three basic methods are used. *Center-type grinding with axial feed* is the most common method, being similar to cutting a thread on a lathe. A shaped grinding wheel replaces the single-point tool. Usually, a single-ribbed grinding wheel is employed, but multiple-ribbed wheels are used occasionally. The grinding wheels are shaped by special diamond dressers or by crush dressing and must be inclined to the helix angle of the thread. Wheel speeds are in the high range. Several passes usually are required to complete the thread.

Center-type infeed thread grinding is similar to multiple-form milling in that a multiple-ribbed wheel, as wide as the length of the desired thread, is used. The wheel is fed inward radially to full thread depth, and the thread blank is then turned through about $1\frac{1}{2}$ turns as the grinding wheel is fed axially a little more than the width of one thread. *Centerless thread grinding* is used for making headless set screws. The blanks are hopper-fed to the regulating wheel which causes them to traverse the grinding wheel face from which they emerge in completed form. Production rates of 60 to 70 screws of $\frac{1}{2}$-inch length per minute are possible.

■ 30.6 THREAD ROLLING

Thread rolling is used to produce threads in substantial quantities. This is a cold-forming process operation in which the threads are formed by rolling a thread blank between hardened dies that cause the metal to flow radially into the desired shape. Because no metal is removed in the form of chips, less material is required, resulting in substantial savings. In addition, because of cold working, the threads have greater strength than cut threads, and a smoother, harder, and more wear-resistant surface is obtained. In addition, the process is fast, with production rates of one per second being common. The quality of cold-rolled products is consistently good. Chipless operations are cleaner and there is a savings in material (15% to 20% savings in blank stock weight is typical).

Thread rolling is done by four basic methods. The simplest of these employs one fixed and one movable flat rolling die (Figure 30-13). After the blank is placed in position

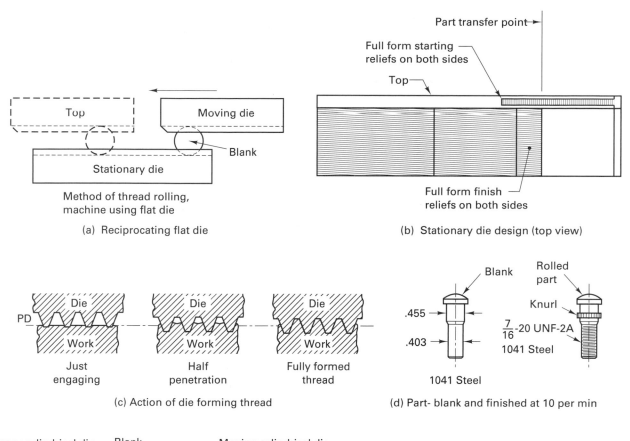

(a) Reciprocating flat die

(b) Stationary die design (top view)

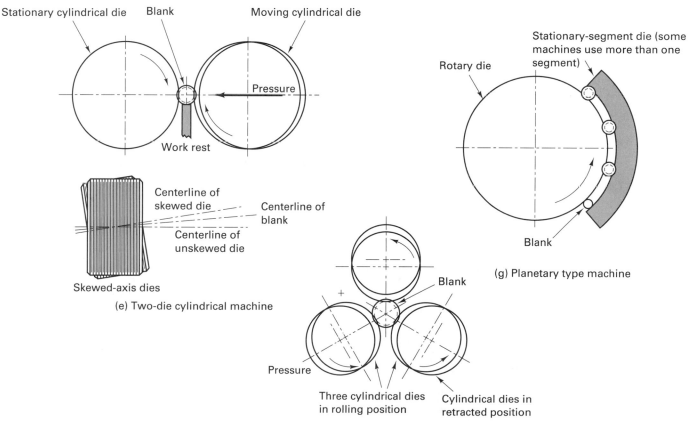

(c) Action of die forming thread

(d) Part- blank and finished at 10 per min

(e) Two-die cylindrical machine

(f) Three-die cylindrical machine

(g) Planetary type machine

FIGURE 30-13   Roll forming threads using flat die thread rolling process shown in (a) and (b). The threads forming action is shown in (c) and the product in (d). Three variations of cylindrical rolling are shown in (e), (f), and (g).

on the stationary die, movement of the moving die causes the blank to be rolled between the two dies and the metal in the blank is displaced to form the threads. As the blank rolls, it moves across the die parallel with its longitudinal axis. Prior to the end of the stroke of the moving die, the blank rolls off the end of the stationary die, its thread being completed.

One obvious characteristic of a rolled thread is that its major diameter always is greater than the diameter of the blank. When an accurate class of fit is desired, the diameter of the blank is made about 0.002 in. larger than the thread-pitch diameter. If it is desired to have the body of a bolt larger than the outside diameter of the rolled thread, the blank for the thread is made smaller than the body.

Thread rolling can be done with cylindrical dies. Figure 30-13 illustrates the three-roll method commonly employed on turret lathes and screw machines. Two variations are used. In one, the rolls are retracted while the blank is placed in position. The rolls then move inward radially, while rotating, to form the thread. More commonly the three rolls are contained in a self-opening die head similar to the conventional type used for cutting external threads. The die head is fed onto the blank longitudinally and forms the thread progressively as the blank rotates. With this procedure, as in the case of cut threads, the innermost $1\frac{1}{2}$ to 2 threads are not formed to full depth because of the progressive action of the rollers.

The two-roll method is commonly employed for automatically producing large quantities of externally threaded parts up to 6 in. in diameter and 20 in. in length. The planetary type machine is for mass production of rolled threads on diameters up to one inch.

Not only is thread rolling very economical, the threads are excellent as to form and strength. The cold working contributes to increased strength, particularly at the critical root areas. There is less likelihood of surface defects (produced by machining) which can act as stress raisers.

Large numbers of threads are rolled on thin, tubular products. In this case external and internal rolls are used. The threads on electric lamp bases and sockets are examples of this type of thread.

CHIPLESS TAPPING

Unfortunately, most internal threads cannot be made by rolling; there is insufficient space within the hole to permit the required rolls to be arranged and supported, and the required forces are too high. However, many internal threads, up to about $\frac{1}{2}$ inch in diameter, are cold-formed in holes in ductile metals by means of *fluteless taps*. Such a tap and its special cross section are shown in Figure 30-9. The forming action is essentially the same as in rolling external threads. Because of the forming involved and the high friction, the torque required is about double that for cutting taps. Also, the hole diameter must be controlled carefully to obtain full thread depth without excessive torque. However, fluteless taps produce somewhat better accuracy than cutting taps and tap life is often greater than that of HSS machine taps. A lubricating fluid should be used, water-soluble oils being quite effective. Fluteless taps are especially suitable for forming threads in dead-end holes because no chips are produced. They come in both plug and bottoming types.

MACHINING VERSUS ROLLING THREADS

Threads are machined or cut when full thread depth is needed (more than one pass necessary), for short production runs; when the blanks are not very accurate; when proximity to the shoulder in end threading is needed; for tapered threads; or when the workpiece material is not adaptable for rolling.

■ 30.7 GEAR MAKING

As with threads, we need to have an introductory understanding of the product before we can understand the process.

GEAR THEORY AND TERMINOLOGY

Basically, gears are modifications of wheels, with *gear teeth* added to prevent slipping and to assure that their relative motions are constant. However, it should be noted that the relative surface velocities of the wheels (and shafts) are determined by the diameters of the wheels.

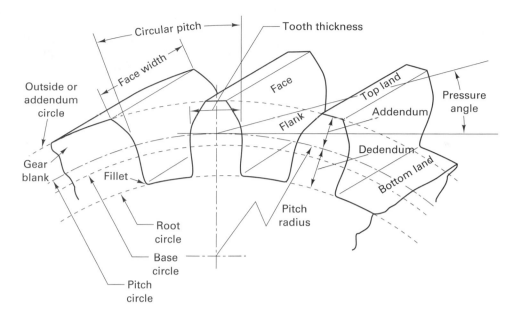

FIGURE 30-14   Gear-tooth nomenclature.

Although wooden teeth or pegs were attached to disks to make gears in ancient times, the teeth of modern gears are produced by machining or forming teeth on the outer portion of the wheel. The *pitch circle* (Figure 30-14 and Figure 30-15) corresponds to the diameter of the wheel. Thus the angular velocity of a gear is determined by the diameter of this imaginary pitch circle. All design calculations relating to gear performance are based on the pitch-circle diameter or, more simply, the *pitch diameter* (PD).

For two gears to operate properly, their pitch circles must be tangential to each other. The point at which the two pitch circles are tangent, at which they intersect the centerline connecting their centers of rotation, is called the *pitch point*. The common normal at the point of contact of mating teeth must pass through the pitch point. This condition is illustrated in Figure 30-15.

To provide uniform pressure and motion and to minimize friction and wear, gears are designed to have rolling motion between mating teeth rather than sliding motion. To achieve this condition, most gears utilize a tooth form that is based on an *involute*

FIGURE 30-15   Tangent pitch circles between two gears produce a pitch point.

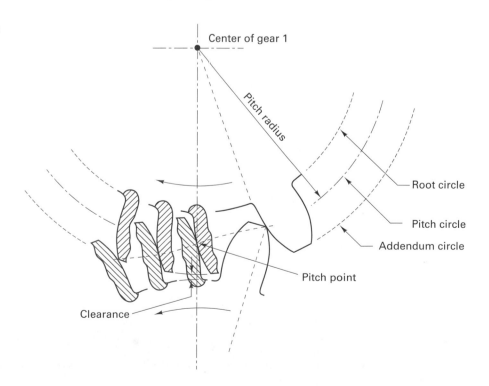

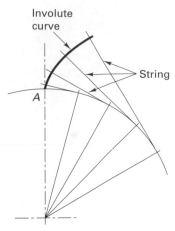

Involute curve

String

A

FIGURE 30-16 Method of generating an involute curve by unwinding a string from a cylinder.

curve. This is the curve that is generated by a *point* on a straight line when the line rolls around a *base circle.* A somewhat simpler method of developing an involute curve is that shown in Figure 30-16. By unwinding a tautly held string from a point on the base circle, point A, an involute curve is generated.

There are three other reasons for using the involute form for gear teeth. First, such a tooth form provides the desired pure rolling action. Second, even if a pair of involute gears is operated with the distance between the centers slightly too large or too small, the common normal at the point of contact between mating teeth will always pass through the pitch point. Obviously, the theoretical pitch circles in such cases will be increased or decreased slightly. Third, the *line of action,* or *path of contact,* that is, the locus of the points of contact of mating teeth, is a straight line that passes through the pitch point and is tangent to the base circles of the two gears.

Cutting an involute shape in gear blanks can be done by simple form cutting, (i.e., milling the shape into the workpiece) or by generating. Generating involves relative motion between the workpiece and the cutting tool. True involute tooth form can be produced by a cutting tool that has straight sided teeth. This permits a very accurate involute tooth profile to be obtained through the use of a simple and easily made cutting tool. The straight sided teeth are given a rolling motion relative to the workpiece to create the curved gear-tooth face, that is, the involute shape.

The basic size of gear teeth may be expressed in two ways. The common practice, especially in the United States and England, is to express the dimensions as a function of the *diametral pitch* (DP). DP *is the number of teeth (N) per unit of pitch diameter* (PD); thus DP = $N/$PD. Dimensionally, DP involves inches in the English system and millimeters in the SI system, and it is a measure of tooth size. Metric gears use the module system (*M*), defined as *the pitch diameter divided by the number of teeth,* or $M = $ PD$/N$. It thus is the reciprocal of diametral pitch and is expressed in millimeters. Any two gears having the same diametral pitch or module will mesh properly if they are mounted so as to have the correct distances and relationship.

The important tooth elements can be specified in terms of the diametral pitch or the module and are as follows:

1. *Addendum:* the radial distance from the pitch circle to the outside diameter.

$$\text{Addendum} = \frac{1}{\text{DP}} \text{ inches}$$

2. *Dedendum:* the radial distance from the pitch circle to the root circle. It is equal to the addendum plus the *clearance,* which is provided to prevent the outer corner of a tooth from touching against the bottom of the tooth space.

3. *Circular pitch:* the distance between corresponding points of adjacent teeth, measured along the pitch circle:

$$\pi/\text{diametral pitch}$$

4. *Tooth thickness:* the thickness of a tooth, measured along the pitch circle. When tooth thickness and the corresponding *tooth space* are equal, no *backlash* exists in a pair of mating gears.

5. *Face width:* the length of the gear teeth in an axial plane.

6. *Tooth face:* the mating surface between the pitch circle and the addendum circle.

7. *Tooth flank:* the mating surface between the pitch circle and the root circle.

8. *Pressure angle:* the angle between a tangent to the tooth profile and a line perpendicular to the pitch surface.

Four shapes of involute gear teeth are used in the United States:

1. $14\frac{1}{2}°$ pressure angle, full depth (used most frequently)

2. $14\frac{1}{2}°$ pressure angle, composite (seldom used)

3. $20°$ pressure angle, full depth (seldom used)

4. $20°$ pressure angle, stub tooth (second most common)

TABLE 30-2.	Formula for Calculating the Standard Dimensions for Involute Gear Teeth	
	$14\frac{1}{2}°$, Full Depth	$20°$, Stub Tooth
Pitch diameter (PD)	$\dfrac{N}{DP}$	$\dfrac{N}{DP}$
Addendum	$\dfrac{1}{DP}$	$\dfrac{0.8}{DP}$
Dedendum	$\dfrac{1.157}{DP}$	$\dfrac{1}{DP}$
Outside diameter	$\dfrac{N+2}{DP}$	$\dfrac{N+1.6}{DP}$
Clearance	$\dfrac{0.157}{DP}$	$\dfrac{0.2}{DP}$
Tooth thickness	$\dfrac{1.508}{DP}$	$\dfrac{1.508}{DP}$

DP = Number of teeth (N) per unit of pitch diameter (PD)

In the $14\frac{1}{2}°$ full-depth system, the tooth profile outside the base circle is an involute curve. Inward from the base circle the profile is a straight radial line that is joined with the bottom land by a small fillet. With this system, the teeth of the basic rack have straight sides.

The $14\frac{1}{2}°$ composite system and the $20°$ full-depth system provide somewhat stronger teeth. However, with the $20°$ full-depth system considerable undercutting occurs in the dedendum area; therefore, stub teeth often are used. The addendum is shortened by 20%, thus permitting the dedendum to be shortened a similar amount. This results in very strong teeth without undercutting. Table 30-2 gives the formulas for computing the dimensions of gear teeth in the $14\frac{1}{2}°$ full-depth and $20°$ stub-tooth systems.

PHYSICAL REQUIREMENTS OF GEARS

A consideration of gear theory leads to five requirements that must be met in order for gears to operate satisfactorily:

1. The actual tooth profile must be the same as the theoretical profile.

2. Tooth spacing must be uniform and correct.

3. The *actual* and theoretical pitch circles must be coincident and be concentric with the axis of rotation of the gear.

4. The face and flank surfaces must be smooth and sufficiently hard to resist wear and prevent noisy operation.

5. Adequate shafts and bearings must be provided so that desired center-to-center distances are retained under operational loads.

The first four of these requirements are determined by the material selection and manufacturing process. The various methods of manufacture that are used represent attempts to meet these requirements to varying degrees with minimum cost, and their effectiveness must be measured in terms of the extent to which the resulting gears embody these requirements. Before looking at the ways to manufacture gears, let's look at some examples of gears.

■ 30.8 GEAR TYPES

The more common types of gears are shown in Figure 30-17. *Spur gears* have straight teeth and are used to connect parallel shafts. They are the most easily made and the cheapest of all types.

The teeth on *helical gears* lie along a helix, the angle of the helix being the angle between the helix and a pitch cylinder element parallel with the gear shaft. Helical gears can connect either parallel or nonparallel nonintersecting shafts. Such gears are stronger and quieter than spur gears because the contact between mating teeth increases more gradually and more teeth are in contact at a given time. Although they usually are slightly

Different types of gears

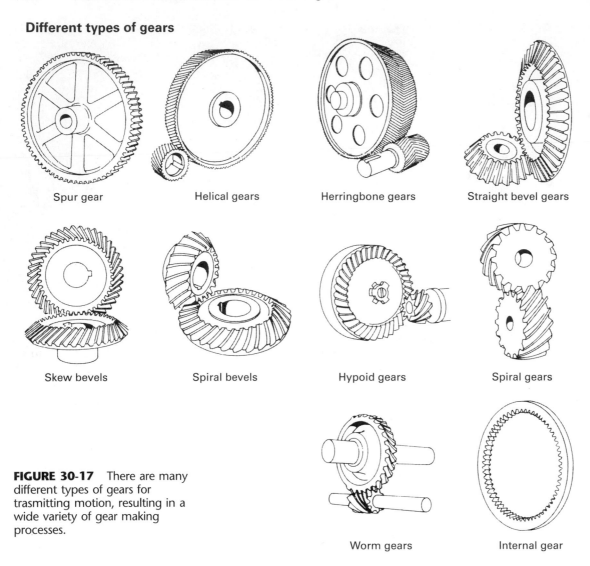

Spur gear   Helical gears   Herringbone gears   Straight bevel gears

Skew bevels   Spiral bevels   Hypoid gears   Spiral gears

Worm gears   Internal gear

FIGURE 30-17   There are many different types of gears for trasmitting motion, resulting in a wide variety of gear making processes.

more expensive to make than spur gears, they can be manufactured in several ways and are produced in large numbers.

Helical gears have one disadvantage. When they are in use, a side thrust is created that must be absorbed in the bearings. *Herringbone gears* neutralize this side thrust by having, in effect, two helical-gear halves, one having a right-hand and the other a left-hand helix. The *continuous* herringbone type is rather difficult to machine but is very strong. A modified herringbone type is made by machining a groove, or gap, around the gear blank where the two sets of teeth would come together. This provides a runout space for the cutting tool in making each set of teeth.

A *rack* is a gear with infinite radius, having teeth that lie on a straight line on a plane. The teeth may be normal to the axis of the rack or helical so as to mate with spur or helical gears, respectively.

A *worm* is similar to a screw. It may have one or more threads, the multiple-thread type being very common. Worms usually are used in conjunction with a *worm gear*. High gear ratios are easily obtainable with this combination. The axes of the worm and worm gear are nonintersecting and usually are at right angles. If the worm has a small helix angle, it cannot be driven by the mating worm gear. This principle frequently is employed to obtain nonreversible drives. Worm gears usually are made with the top land concave to permit greater area of contact between the worm and the gear. A similar effect can be achieved by using a *conical worm*, in which the helical teeth are cut on a double-conical blank, thus producing a worm that has an hour-glass shape.

Bevel gears, teeth on a cone, are used to transmit motion between intersecting shafts. The teeth are cut on the surface of a truncated cone. Several types of bevel gears are made, the types varying as to whether the teeth are straight or curved and whether the axes of the mating gears intersect. On *straight-tooth* bevel gears the teeth are straight, and if extended all would pass through a common apex. *Spiral-tooth* bevel gears have teeth that are segments of spirals. Like helical gears, this design provides tooth overlap so that more teeth are engaged at a given time and the engagement is progressive. *Hypoid* bevel gears also have a curved-tooth shape but are designed to operate with nonintersecting axes. Rear-drive automobiles used hypoid gears in the rear axle so that the drive shaft axis can be below the axis of the axle and thus permit a lower floor height. *Zerol* bevel gears have teeth that are circular arcs, providing somewhat stronger teeth than can be obtained in a comparable straight-tooth gear. They are not used extensively. When a pair of bevel gears are the same size and have their shafts at right angles, they are termed *miter gears*.

A *crown gear* is a special form of bevel gear having a 180° cone apex angle. In effect, it is a disk with the teeth on the side of the disk. It also may be thought of as a rack that has been bent into a circle so that its teeth lie in a plane. The teeth may be straight or curved. On straight-tooth crown gears the teeth are radial. Crown gears seldom are used, but they have the important quality that they will mesh properly with a bevel gear of any cone angle, provided that the bevel gear has the same tooth form and diametral pitch. This important principle is incorporated in the design and operation of two very important types of gear-generating machines that will be discussed later.

Most gears are of the external type, the teeth forming the outer periphery of the gear. Internal gears have the teeth on the inside of a solid ring, pointing toward the center of the gear.

■ 30.9   GEAR MANUFACTURING

Whether produced in large or small quantities, in cells, or job shop batches, the sequence of processes for gear manufacturing requires four sets of operations, see Figure 30-18.

FIGURE 30-18   The typical gear making process (for very accurate gears) involves both hobbing and shaving followed by grinding after heat treating.

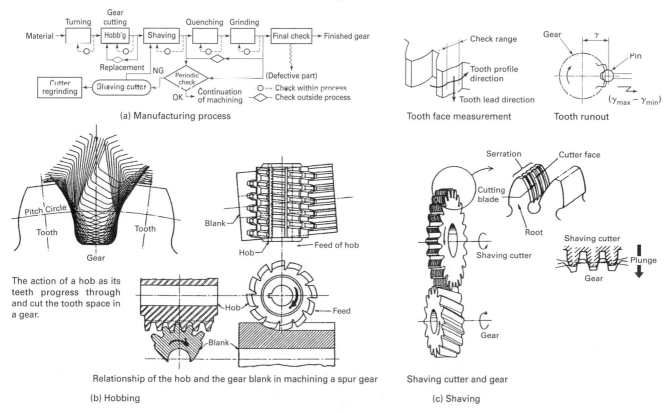

(a) Manufacturing process

(b) Hobbing

(c) Shaving

Gear Manufacturing Process

1. Blanking (turning)   3. Heat treatment
2. Gear cutting (hobbing and shaving)   4. Grinding

Blanking refers to the initial forming or machining operations that produce a semi-finished part ready for gear cutting, starting from a piece of raw material. Turning on chuckers or lathes, face and centering of shafts, milling, and sometimes grinding fall into this category of operations. Good quality blanks are essential in precision gear manufacturing.

Hobbing, shaping, and shaving machines are the most frequently used machines for gear cutting, producing gears for automotive, truck, agricultural, and construction equipment. Other processes used in industrial gear production include broaching, rolling, grinding, milling, and shaving. The process selected depends on finding a cost-effective application based on quality specification, production volumes, and economic conditions.

The gear cutting or machining operations can be divided into operations executed prior to heat treatment, when the material is still soft and easily machinable, and after heat treatment, performed on parts that have acquired high hardness and strength.

Heat treatment gives the material the strength and durability to withstand high loads and wear but results in a reduction in dimensional and geometrical accuracy. The metallurgical transformations that occur during hardening, quenching, and tempering cause a general quality deterioration in the gears. Therefore, precision grinding operations are used on external and internal bearing diameters, critical length dimensions, and fine surface finishes after heat treatment. Cylindrical grinders, angle-head grinders, internal grinders, and surface grinders are commonly used.

Gears are made in very large numbers by cold-roll forming, and in addition, significant quantities are made by extrusion, by blanking, by casting, and some by powder metallurgy and by a forging process. However, it is only by machining that all types of gears can be made in all sizes, and although roll-formed gears can be made with accuracy sufficient for most applications, even for automobile transmissions, machining still is unsurpassed for gears that must have very high accuracy. Also, roll forming can be used only on ductile metals.

■ 30.10 MACHINING OF GEARS

FORM MILLING

Form cutting or *form milling* on a horizontal milling machine is illustrated in Figure 30-19. The multiple tooth form cutter has the same form as the *space* between adjacent teeth. The tool

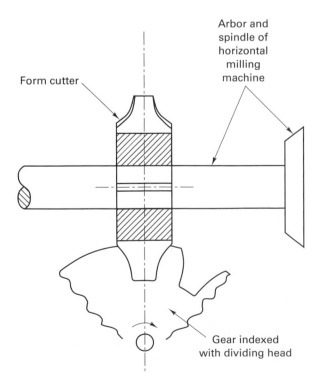

FIGURE 30-19 The basic method of machining a gear using a form cutter (*shown left*) to mill out the space between the teeth using the form cutter (*middle*) or the stocking cutter (*right*) to machine the gear. (*Courtesy of Brown and Sharpe Manufacturing Company.*)

Form cutter

Arbor and spindle of horizontal milling machine

Gear indexed with dividing head

is fed radially toward the center of the gear blank to the desired tooth depth, then across the tooth face to obtain the required tooth width. When one tooth space has been completed, the tool is withdrawn, the gear blank is indexed using a dividing head, and the next tooth space is cut. In machining gears by the form-cutting process, the form cutter is mounted on the machine arbor, and the gear blank is mounted on a mandrel held between the centers of some type of indexing device. Basically, form cutting is a simple and flexible method of machining gears. The equipment and cutters required are relatively simple, and standard machine tools (milling machines) often are used. However, in most cases the procedure is quite slow, and considerable care is required on the part of the operator. Therefore, this process usually is employed where only one or a few gears are to be made. When a helical gear is to be cut, the table must be set at an angle equal to the helix angle, and the dividing head is geared to the longitudinal feed screw of the table so that the gear blank will rotate as it moves longitudinally.

Standard cutters usually are employed in form-cutting gears. In the United States, these come in eight sizes for each diametral pitch and will cut gears having any number of teeth. A single cutter will not produce a theoretically perfect tooth profile for all sizes of gears in the range for which it is intended. However, the change in tooth profile over the range covered by each cutter is very slight, and most of the time, satisfactory results can be achieved. When greater accuracy is required, half-number cutters can be obtained. Cutters are available for all common diametral pitches and $14\frac{1}{2}°$ and $20°$ pressure angles. If the amount of metal that must be removed to form a tooth space is large, roughing cuts may be taken with a *stocking cutter*. The stepped sides of the stocking cutter remove most of the metal and leave only a small amount to be removed subsequently by the regular form cutter in a finish cut.

Straight-tooth bevel gears can be form cut on a milling machine, but this is seldom done. Because the tooth profile in bevel gears varies from one end of the tooth to the other, after one cut is taken to form the correct tooth profile at the smaller end, the relationship between the cutter and the blank must be altered. Shaving cuts then are taken on the side of each tooth to form the correct profile throughout the entire tooth length.

Although the form cutting of gears on a milling machine is a flexible process and is suitable for gears that are not to be operated at high speeds or that need not operate with extreme quietness, the process is slow and requires skilled labor. Semiautomatic machines are available for making gears by the form-cutting process. The procedure utilized is essentially the same as on a milling machine, except that, after setup, the various operations are completed automatically. Gears made on such machines are no more accurate than those produced on a milling machine, but the possibility of error is less, and they are much cheaper because of reduced labor requirements. For large quantities, however, form cutting is not used.

BROACHING

Broaching is another way to form cut teeth. All the tooth spaces are cut simultaneously and the tooth is formed progressively. The circular table in Figure 30-20 holds 10 sets of progressive

FIGURE 30-20 Blind gear spline broaching machine for producing internal gears. The machine has automated pick-and-place arms for load/unload. *(Courtesy of Apex Broach and Machine Company.)*

tooling. The table rotates moving one set of tooling at a time under two workpieces. The arms load and unload a set of parts every 15 seconds so the cycle time is very quick. Excellent gears can be made by *broaching*. However, a separate broach must be provided for each size of gear. The tooling tends to be expensive, restricting this method to large volumes.

GEAR GENERATING

Most high-quality gears that are made by machining are made by the *generating process*. This process is based on the principle that any two involute gears, or any gear and a rack, of the same diametral pitch will mesh together properly. Utilizing this principle, one of the gears (or the rack) is made into a cutter by proper sharpening. It can be used to cut into a mating gear blank and thus generate teeth on the blank. The two principal methods for gear generating are *shaping* and *hobbing*.

SHAPING

To carry out the shaping process, the cutter and the gear blank must be attached rigidly to their respective shafts, and the two shafts must be interconnected by suitable gearing so that the cutter and the blank rotate positively with respect to each other and have the same pitch-line velocities. To start cutting the gear, the cutter is reciprocated vertically and is fed radially into the blank between successive strokes. When the desired tooth depth has been obtained, the cutter and blank are then slightly indexed after each cutting stroke. The resulting generating action is indicated schematically in the upper diagram of Figure 30-21a and shown in the cutting of an actual gear tooth in Figure 30-21b.

Figure 30-22 shows a machine called a gear shaper. *Gear shapers* generate gears by a reciprocating tool motion. The gear blank is mounted on the rotating table (or vertical spindle) and the cutter on the end of a vertical, reciprocating spindle. The spindle

FIGURE 30-21 (*Top*) Generating action of a Fellows gear-shaper cutter; (*bottom*) series of photographs showing various stages in generating one tooth in a gear by means of a gear shaper, action taking place from right to left, corresponding to a diagram above. One tooth of the cutter was painted white. (*Top: Courtesy of Fellows Gear Shaper Company.*)

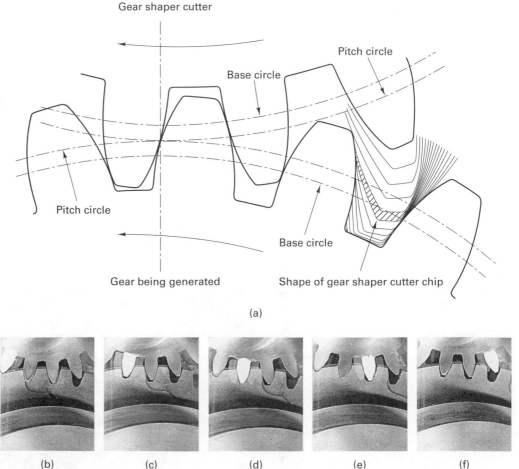

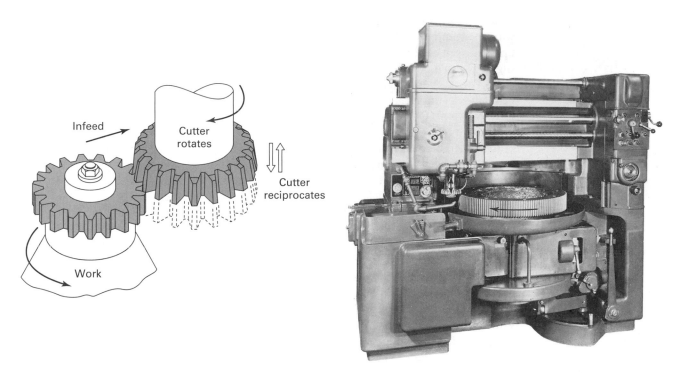

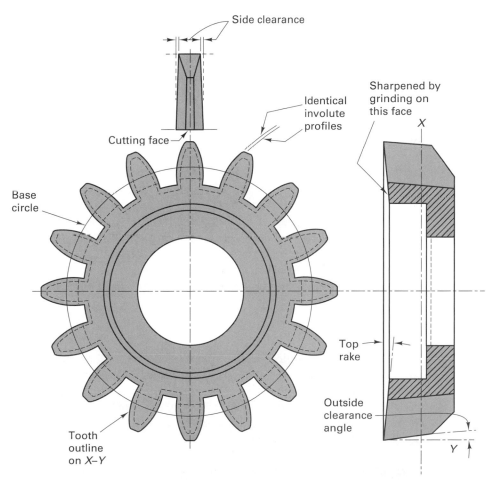

FIGURE 30-22   This machine tool is a gear shaper. The blank is rotating while the cutter is reciprocating vertically, as shown in the inset. The tool is very complex and is shown in detail below. *(Courtesy Fellows Gear Shaper Company.)*

and the table are connected by means of gears so that the cutter and gear blank revolve with the same pitch-line velocity. Cutting occurs on the downstroke (sometimes on the upstroke). At the end of each cutting stroke, the spindle carrying the blank retracts slightly to provide clearance between work and tool on the return stroke.

The conventional cutter for gear shaping is made from high-speed steel which can be coated to improve wear life with a superhard layer of titanium nitride (TiN) using the physical vapor deposition (PVD) process. Recently, new throwaway disk-shape insert tools have been developed for gear shaping, eliminating regrinding and recoating operations on the conventional tools. Regrinding the conventional cutter required two adjustments (resetting operations) on the machine tool. Because the throwaway blades are all sized the same, machine resetting after cutting tool changeover is eliminated.

Either straight- or helical-tooth gears can be cut on gear shapers. To cut helical teeth, both the cutter and the blank are given an oscillating rotational motion during each stroke of the cutter, turning in one direction during the cutting stroke and in the opposite direction during the return stroke. Because the cutting stroke can be adjusted to end at any desired point, gear shapers are particularly useful for cutting cluster gears. Some machines can be equipped with two cutters simultaneously to cut two gears, often of different diameters. Gear shapers also can be adapted for cutting internal gears.

Special types of gear shapers have been developed for mass-production purposes. The *rotary gear shaper* essentially is 10 shaper units mounted on a rotating base and having a single drive mechanism. Nine gears are cut simultaneously while a finished gear is removed and a new blank is put in place on the tenth unit. *Planetary gear shapers* holding six gear blanks move in planetary motion about a large, central gear cutter. The cutter has no teeth in one portion to provide a space where the gear can be removed and a new blank placed on the empty spindle.

CNC gear shapers are now available with hydromechanical stroking systems which produce a uniform cutting velocity during the cutting portion of the downstroke. These machines can operate at 500 to 1700 rpm and use TiN-coated cutters to enhance tool life.

Vertical shapers for gear generating have a vertical ram and a round table that can be rotated in a horizontal plane by either manual or power feed. These machine tools are sometimes called *slotters*. Usually, the ram is pivoted near the top so that it can be swung outward from the column through an arc of about 10°.

Because one circular and two straight-line motions and feeds are available, vertical shapers are very versatile tools and thus find considerable use in one-of-a-kind manufacturing. Not only can vertical and inclined flat surfaces be machined, but external and internal cylindrical surfaces can be generated by circular feeding of the table between strokes. This may be cheaper than turning or boring for very small lot sizes. A vertical shaper can be used for generating gears or machining curved surfaces, interior surfaces, and arcs by using a stationary tool and rotating the workpiece. A *keyseater* is a special type of vertical shaper designed and used exclusively for machining keyways on the inside of wheel and gear hubs. For machining continuous herringbone gears, a *Sykes gear-generating machine* is used.

HOBBING

Involute gear teeth could be generated by a cutter that has the form of a rack. Such a cutter would be simple to make but has two major disadvantages. First, the cutter (or the blank) would have to reciprocate, with cutting occurring only during one stroke direction. Second, because the rack would have to move longitudinally as the blank rotated, the rack would need to be very long (or the gear very small) or the two would not be in mesh after a few teeth were cut. A *hob* overcomes the preceding two difficulties. As shown in Figure 30-23, a hob can be thought of, basically, as one long rack tooth that has been wrapped around a cylinder in the form of a helix and fluted at intervals to provide a number of cutting edges. Relief is provided behind each of the teeth. The cross section of each tooth, normal to the helix, is the same as that of a rack tooth. (A hob can also be thought of as a gashed worm gear.)

The action of a *hobbing* machine cutting a spur gear is illustrated in Figure 30-23. To cut a spur gear, the axis of the hob must be set off from the normal to the rotational axis of

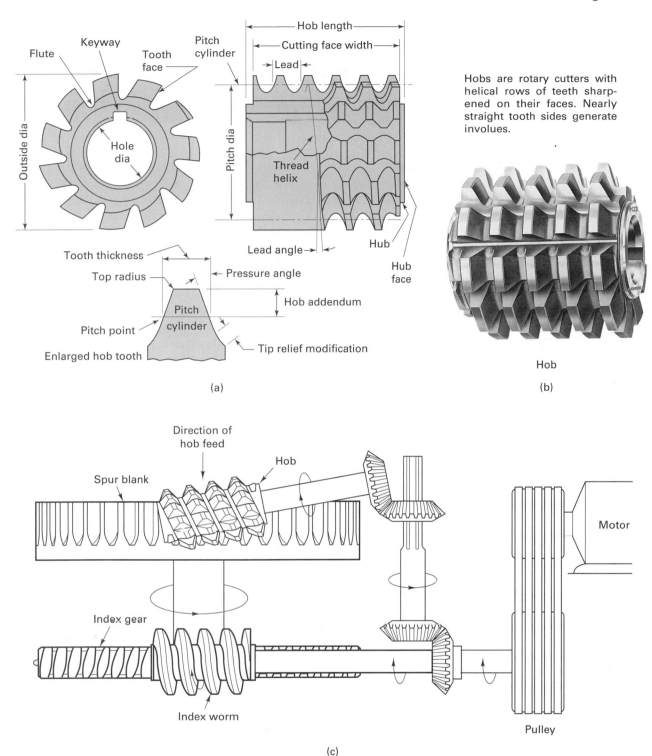

Hobs are rotary cutters with helical rows of teeth sharpened on their faces. Nearly straight tooth sides generate involues.

FIGURE 30-23 (a) Three views of hob and gear blanks (small spur gear), (b) Hob and (c) schematic of mechanism of a hobbing machine, shown hobbing a large spur gear.

the blank by the helix angle of the hob. In cutting helical gears, the hob must be set over an additional amount equal to the helix angle of the gear. The cutting of a gear by means of a hob is a continuous action. The hob and the blank are connected by proper gearing so that they rotate in mesh. To start cutting a gear, the rotating hob is fed inward until the proper setting for tooth depth is obtained. The hob is then fed in a direction parallel with the axis of rotation of the blank. As the gear blank rotates, the teeth are generated and the feed of the hob across the face of the blank extends the teeth to the desired tooth face width.

Hobbing is rapid and economical. More gears are cut by this process than by any other. The process produces excellent gears and can also be used for splines and sprockets. Single-, double-, and triple-thread hobs are used. Multiple-thread types increase the production rate but do not produce accuracy as high as single-thread hobs.

Gear-hobbing machines are made in a wide range of sizes. Machines for cutting accurate large gears frequently are housed in temperature-controlled rooms, and the temperature of the cutting fluid is controlled to avoid dimensional change due to variations in temperature.

COLD-ROLL FORMING

The manufacture of gears by *cold-roll forming* has been highly developed and widely adopted in recent years. Currently, millions of high-quality gears are produced annually by this process; many of the gears in automobile transmissions are made this way. As indicated in Figure 30-24, the process is basically the same as that by which screw threads are roll-formed, except that in most cases the teeth cannot be formed in a single rotation of the forming rolls; the rolls are fed inward gradually during several revolutions.

Because of the metal flow that occurs, the top lands of roll-formed teeth are not smooth and perfect in shape; a depressed line between two slight protrusions can often be seen, as shown encircled in Figure 30-24. However, because the top land plays no part in gear-tooth action, if there is sufficient clearance in the mating gear, this causes no difficulty. Where desired, a light turning cut is used to provide a smooth top land and correct addendum diameter.

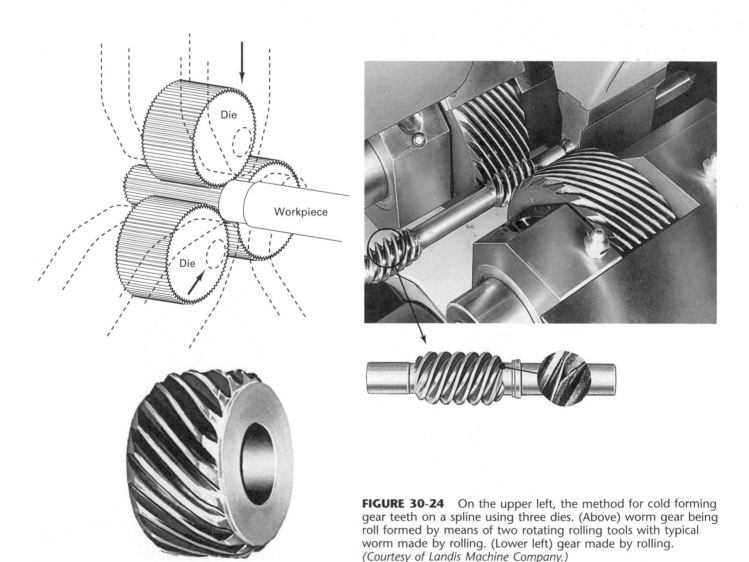

FIGURE 30-24 On the upper left, the method for cold forming gear teeth on a spline using three dies. (Above) worm gear being roll formed by means of two rotating rolling tools with typical worm made by rolling. (Lower left) gear made by rolling. *(Courtesy of Landis Machine Company.)*

The hardened forming rolls are very accurately made, and the roll-formed gear teeth usually have excellent accuracy. In addition, because the severe cold working produces tooth faces that are much smoother and harder than those on ordinary machined gears, they seldom require hardening or further finishing, and they have excellent wear characteristics.

The process is rapid (up to 50 times faster than gear machining) and easily mechanized. No chips are made and thus less material is needed. Less skilled labor is required. Small gears often are made by rolling a length of shaft and then slicing off the individual gear blanks. Usually, soft steel is required and 4 to 5 in. diameter is about the limit, with fewer than six teeth, coarser than 12 diametral pitch, and no pressure angle less than 20°.

OTHER GEAR-MAKING PROCESSES

Gears can be made by the various casting processes. *Sand-cast gears* have rough surfaces and are not accurate dimensionally. They are used only for services where the gear moves slowly and where noise and inaccuracy of motion can be tolerated. Gears made by *die casting* are more accurate and have fair surface finish. They can be used to transmit light loads at moderate speeds. Gears made by *investment casting* may be accurate and have good surface characteristics. They can be made of strong materials to permit their use in transmitting heavy loads. In many instances, gears that are to be finished by machining are made from cast blanks, and in some larger gears the teeth can be cast to approximate shape to reduce the amount of machining.

Large quantities of gears are produced by *blanking* in a punch press. The thickness of such gears usually does not exceed about $\frac{1}{16}$ in. By shaving the gears after they are blanked, excellent accuracy can be achieved. Such gears are used in clocks, watches, meters, and calculating machines. *Fine blanking* is also used to produce thin, flat gears of good quality.

High-quality gears, both as to dimensional accuracy and surface quality, can be made by the *powder metallurgy process*. Usually, this process is employed only for small sizes, ordinarily less than 1 in. in diameter. However, larger and excellent gears are made by forging powder metallurgy preforms. This results in a product of much greater density and strength than usually can be obtained by ordinary powder metallurgy methods, and the resulting gears give excellent service at reduced cost. Gears made by this process often require little or no finishing.

Large quantities of plastic gears are made by *plastic molding*. The quality of such gears is only fair, and they are suitable only for light loads. Accurate gears suitable for heavy loads frequently are machined out of laminated plastic materials. When such gears are mated with metal gears, they have the quality of reducing noise.

Quite accurate small gears can be made by the *extrusion* process. Typically, long lengths of rod, having the cross section of the desired gear, are extruded. The individual gears are then sawed from this rod. Materials suitable for this process are brass, bronze, aluminum alloys, magnesium alloys, and occasionally, steel.

Flame machining (oxyacetylene cutting) can be used to produce gears that are to be used for slow-moving applications wherein accuracy is not required.

A few gears are made by the hot roll-forming process. In this process a cold master gear is pressed into a hot blank as the two are rolled together.

■ 30.11 GEAR FINISHING

To operate efficiently and have satisfactory life, gears must have accurate tooth profiles and the faces of the teeth must be smooth and hard. These qualities are particularly important when gears must operate quietly at high speeds. When they are produced rapidly and economically by most processes except cold-roll forming, the tooth profiles may not be as accurate as desired, and the surfaces are somewhat rough and subject to rapid wear. Also, it is difficult to cut gear teeth in a hardened gear blank, and therefore, economy dictates that the gear be cut in a relatively soft blank and subsequently be heat-treated to obtain greater hardness. Such heat treatment usually results in some slight distortion and surface roughness. Although most roll-formed gears have sufficiently accurate profiles, and the tooth faces are adequately smooth and frequently have sufficient hardness, this process is feasible only for relatively small gears. Consequently, a large proportion of high-quality gears are given some

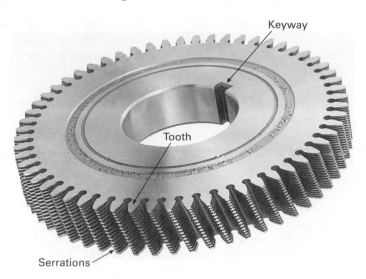

FIGURE 30-25 Rotary gear shaving cutter (see Figure 30-18 and Figure 30-26).

type of finishing operation after they have received primary machining or after heat treatment. Most of these finishing operations can be done quite economically because only minute amounts of metal are removed and they are fast and often automatic.

Gear shaving is the most commonly used method for finishing spur and helical gear teeth prior to hardening. The gear is run, at high speed, in contact with a shaving tool, usually of the type shown in Figure 30-25. Such a tool is a very accurate, hardened, and ground gear that contains a number of peripheral serrations, thus forming a series of sharp cutting edges on each tooth. The gear and shaving cutter are run in mesh with their axes crossed at a small angle, usually about 10° (Figure 30-26). As they rotate, the gear is reciprocated longitudinally across the shaving tool (or vice versa). During this action, which usually requires less than 1 min, very fine chips are *shaved* from the gear-tooth faces, thus eliminating any high spots and producing a very accurate tooth profile.

Rack shaving cutters sometimes are used for shaving small gears, the cutter reciprocating lengthwise, causing the gear to roll along it, as it is moved sideways across the cutter and fed inward.

Although shaving cutters are costly, they have a relatively long life because only a very small amount of metal is removed, usually 0.001 to 0.004 in. Some gear-shaving machines produce a slight crown on the gear teeth during shaving. Some gears are not hardened prior to shaving, although it is possible to remove very small amounts of metal from hardened gears if they are not too hard. However, modern heat-treating equipment makes it possible to harden gears after shaving without harmful effects, and therefore this practice is followed, if possible.

Roll finishing is a cold-forming process that is used to finish helical gears. The unhardened gear is rolled with two hardened, accurately formed rolling dies. The center distance between the dies is reduced to cold work the surfaces and produce highly accurate tooth forms. High points on the unhardened gear are plastically deformed so that a smoother surface and more accurate tooth form are achieved. Because the operation is one of localized cold working, some undesirable effects may accrue, such as localized residual stresses and nonuniform surface characteristics. Surface finishes of 6 to 8 μin. have been achieved. If roll finishing is to be used, attention must be paid to the prerolled geometry. Designers should consult the manufacturers of gear-rolling machines for specific recommendations.

Grinding is used to obtain very accurate teeth on hardened gears. Two methods are used. One employs a formed grinding wheel that is trued to the exact form of a tooth by means of diamonds mounted on a special holder and guided by a large template. The other method is involute-generation grinding which uses straight-sided grinding wheels which simulate one side of a rack tooth. The surface of the gear tooth is ground as the gear rolls (and reciprocates) past the grinding wheels. Grinding produces very accurate gears, but because it is slow and expensive, it is used only on the highest-quality, hardened gears.

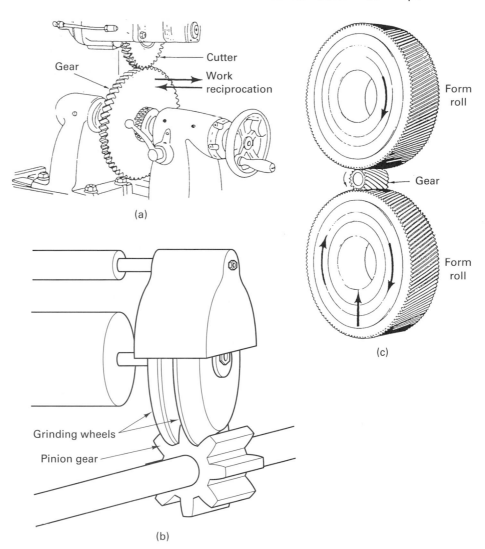

(a)

(c)

(b)

FIGURE 30-26 Three methods for gear finishing are (a) shaving using a special cutter—see Figure 30-25, (b) grinding using form grinding wheels, and (c) using form rolls, i.e., roll forming/finishing.

Lapping can also be used for finishing hardened gears. The gear to be finished is run in contact with one or more cast-iron lapping gears under a flow of very fine abrasive in oil. Because lapping removes only a very small amount of metal, it usually is employed on gears that previously have been shaved and hardened. This combination of processes produces gears that are nearly equal to ground gears in quality but at considerably lower cost.

■ 30.12 GEAR INSPECTION

FIGURE 30-27 Using gear-tooth vernier calipers to check the tooth thickness at the pitch circle.

As with all manufactured products, gears must be checked to determine whether the resulting product meets the design specifications and requirements. Because of their irregular shape and the number of factors that must be measured, inspection of gears is somewhat difficult. Among the factors to be checked are the linear tooth dimensions such as thickness, spacing, depth, and so on; tooth profile, surface roughness; and noise. Several special devices, most of them automatic or semiautomatic, are used for such inspection.

Gear-tooth vernier calipers can be used to measure the thickness of gear teeth on the pitch circle (Figure 30-27). CNC gear inspection machines (Figure 30-28) can quickly check several factors, including variations in circular pitch, involute profile, lead, tooth spacing, and variations in pressure angle. The gear usually is mounted between centers. The probe is moved to the gear through X, Y, and Z translations. The gear is rotated between measurements. The inset in Figure 30-28 shows a typical display for involute profile.

Because noise level is important in many applications, not only from the viewpoint of noise pollution but also as an indicator of probable gear life, special equipment for noise measurement is quite widely used, sometimes integrated into mass-production assembly lines.

FIGURE 30-28 A CNC gear inspection machine has *X, Y, Z* motions plus a rotary table.

■ KEY WORDS

addendum	cold-roll forming	form milling	gear shaping	pitch	self-opening	thread tapping
bevel gear	collapsible tap	gear finishing	gear teeth	pitch circle	die head	threading
blanking	crown gear	gear grinding	herringbone gear	pitch diameter	spur gear	words
bottoming tap	dedendum	gear hobbing	hob	planetary gear	taper tap	worm gear
broaching	diametral pitch	gear roll	hobbing	plus tap	thread cutting	
circular pitch	fluteless tap	forming	involute curve	rack	thread rolling	

■ REVIEW QUESTIONS

1. How does the pitch diameter differ from the major diameter for a standard screw thread?
2. For what types of threads are the pitch and the lead the same?
3. What is the helix angle of a screw thread?
4. Why are pipe threads tapered?
5. What are three basic methods by which external threads be produced?
6. Explain the meaning of $\frac{1}{4}''$-20 UNC-3A.
7. What is meant by the designation M20 × 2.5-6g6g? (What does the "×" mean?)
8. What are two reasons fine-series threads are being used less now than in former years?
9. In cutting a thread on a lathe, how is the pitch controlled?
10. What is the function of a threading dial on a lathe?
11. In Chapter 23, what figure(s) showed a threading dial?

12. What controls the lead of a thread when it is cut by a threading die?
13. What is the basic purpose of a self-opening die head?
14. What is the reason for using a taper tap before a plug tap in tapping a hole?
15. What difficulties are encountered if full threads are specified to the bottom of the dead-end hole?
16. Can a fluteless tap be used for threading a hole in gray cast iron? Why or why not?
17. What provisions should a designer make so that a dead-end hole can be threaded?
18. What is the major advantage of a spiral-point tap?
19. Can a fluteless tap be used for threading to the bottom of a dead-end hole? Why or why not?
20. Is it desirable for a tapping fluid to have lubricating qualities? Why?

21. How does thread milling differ from thread turning?
22. What are the advantages of making threads by grinding?
23. Why has thread rolling become the most commonly used method for making threads?
24. How may you determine whether a thread has been produced by rolling rather than by cutting?
25. Why is the involute form used for gear teeth?
26. What is the diametral pitch of a gear?
27. What is the relationship between the diametral pitch and the module of a gear?
28. On a sketch of a gear, indicate the pitch circle, addendum circle, dedendum circle, and the circular pitch.
29. What five requirements must be met for gears to operate satisfactorily? Which of these are determined by the manufacturing process?
30. What are the advantages of helical gears compared with spur gears?
31. What is the principal disadvantage of helical gears?
32. What difficulty would be encountered in hobbing a herringbone gear?
33. What is the only type of machine on which full-herringbone gears can be cut?
34. What modification in design is made to herringbone gears to permit them to be cut by hobbing?
35. Why are not more gears made by broaching?
36. What is the most important property of a crown gear?
37. What are three basic processes for machining gears?
38. Which basic gear-machining process is utilized in a Fellows gear shaper?
39. When a helical gear is machined on a milling machine, the table lead screw and the universal dividing head have to be connected by a gear train. Why?
40. Why is a gear-hobbing machine much more productive than a gear shaper?
41. In gear shaping and gear hobbing, the tooth profiles are generated. What does that mean?
42. What are the advantages of cold-roll forming for making gears?
43. Why is cold-roll forming not suitable for making gray cast iron gears?
44. Under what conditions can shaving not be used for finishing gears?
45. What inherent property accrues from cold-roll forming of gears that may result in improved gear life?
46. Can lapping be used to finish cast iron gears?
47. What factors usually are checked in inspecting gears?
48. What are the basic methods for gear finishing?

■ PROBLEMS

1. Calculate the cutting time needed to tap a $\frac{3}{4} \times 2$ in. hole using a cutting speed of 30 sfpm. The tap has 10 tpi.
2. The new manufacturing manager has recommended to you that chipless tapping be adopted for tapping holes in the deep, dead-end holes on the 2 cylinder engine blocks that the company makes. Chipless tapping can run at twice the speed of the conventional tapping process, works well on deep holes, and provides better quality and finish with longer tool life. The tapping process is the bottleneck process in machining all the cast iron engine blocks, plus the fact about 10% of the blocks have to be scrapped due to broken taps. What do you recommend?
3. A hob that has a pitch diameter of 76.2 mm is used to cut a gear having 36 teeth. If a cutting speed of 27.4 m/min is used, what will be the rpm value of the gear blank?
4. If the gear in Problem 3 has a face width of 76.2 mm, a feed of 1.9 mm/rev of the workpiece is used, and the approach and overtravel distances of the hob are 38 mm, how much time will be required to hob the gear?
5. In Figure 30-19 a form-milling operation and cutters are shown. The gear is to be made from 4340 steel, R_C50 prior to heat treat and final grind. Select the proper speeds and feeds for the job (the cutter is 4 in. in diameter) and compute the cutting time to mill this gear. Up milling versus down milling?
6. A gear broaching machine of the type shown in Figure 30-20 can do the gear in 15 seconds (about 240 parts per hour). How many additional gears per year are needed to cover the broaching tooling cost if each broach on the machine cost $250?

Do you think the broach tool life is sufficient to handle that number of parts? What about TiN coating the broaches (cost $10.00 per broach)?

7. Assume that 10,000 spur gears, $1\frac{1}{8}$ in. in diameter and 1 in. thick are to be made of 70-30 brass. What manufacturing method would you consider?
8. If only three gears described were to be made, what process would you select?
9. KC Stern, a design engineer for Boeing Commercial Airplane Group, is doing some design work and has a question regarding the hole size that is drilled before tapping internal threads. The book states, "A hole diameter slightly larger than the minor diameter of the thread must already exist, made by drilling/reaming, boring or die casting."

    Conventional wisdom would lead one to believe that a hole slightly smaller should exist before tapping threads. The *Machinery's Handbook* seems to agree with this. Most tap drill sizes listed are smaller than the minimum minor diameter listed for a particular thread size in MIL-S-008879B. KC has found at least one exception to this but the difference was only a couple of thousandths of an inch.

    KC consulted local sources on this matter and no one will commit to the correctness or incorrectness of the book. A number of people are scratching their heads on this one. Your assistance in this matter is greatly appreciated. Should the hole be slightly smaller or slightly larger than the minor diameter of the thread and why?

www.wiley.com/college/degarmo

Chapter 30 CASE STUDY

Vented Cap Screws

The machine shop at the Pogi Space Laboratories received an order from their engineering department for 100 of the vented cap screws shown in Figure CS-30. The machine shop foreman returned the order, stating that there was no practical way to make these other than by hand. The design engineer insisted that the vent slot, as designed, was essential to assure no pressure buildup around the threads of the screw body in the intended application, and that he knew they could be made because he had seen such cap screws.

To accommodate the shop, however, the design engineer redesigned the part, now specifying that a vent hole be drilled parallel to the axis of the cap at location D. The small hole, with cross-sectional area equal to the previous 0.062 in. × 0.031 in. slot, would be drilled $\frac{1}{2}$ in. deep through the cap to connect to the longitudinal slot. This would eliminate the right-angle slot geometry.

1. What do you think of this solution? What problems will be incurred in the manufacture of the vented cap screw?
2. As the manufacturing engineer, working with the design engineer, can you suggest an alternative redesign that would make a slot practical to manufacture? Show a sketch of your redesigned cap screw.

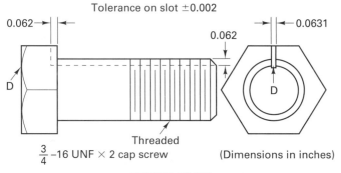

Tolerance on slot ±0.002

0.062

0.062

0.0631

D

D

Threaded

$\frac{3}{4}$ –16 UNF × 2 cap screw

(Dimensions in inches)

FIGURE CS-30

SURFACE TREATMENTS, FINISHING, AND INTEGRITY

■ 31.1 INTRODUCTION

Many manufacturing processes influence surface properties which in turn may significantly affect the way the component functions in service. The demands for greater strength and longer life in components often depend on changes in the surface properties rather than the bulk properties. These changes may be mechanical, thermal, chemical, and/or physical and therefore are difficult to describe in general terms. For example, Inconel 718 can have a fatigue limit as high as 540 MPa after gentle grinding and as low as 150 MPa after EDMing (Figure 31-1).

FIGURE 31-1 Fatigue strength of Inconel 718 components after surface finishing by grinding or EDM, (Field and Kahles, 1971).

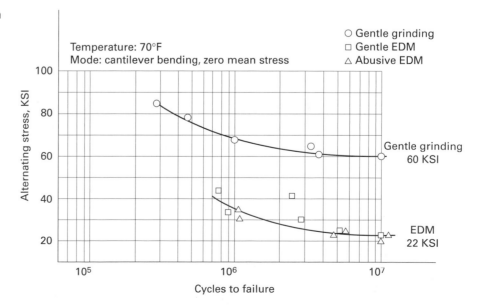

777

Many metal-cutting processes specified by the manufacturing engineer to produce a specific geometry can often have the effect of producing alterations in the surface material of the component, which in turn, produces changes in performance.

The term "surface integrity" was coined by Field and Kahles in 1964 in reference to the nature of the surface condition that is produced by the manufacturing process. If we view the process to have five main components (workpiece, tool, machine tool, environment, and process variables) we see that surface properties can be altered by all of these parameters (see Table 31-1) by producing:

- High temperatures involved in the machining process
- Plastic deformation of the work material (residual stress)
- Surface geometry (roughness, cracks, distortion)
- Chemical reactions, particularly between the tool and the workpiece

More specifically, surface integrity refers to the unimpaired or enhanced surface condition of a component or specimen which influences its performance. Change in surface can lead to failure in base metal.

Surface integrity has two aspects: topography characteristics and surface layer characteristics. Topography is made up of surface roughness, waviness, errors of form, and flaws (Figure 31-2).

A typical roughness profile includes the peaks and valleys that are considered separately from waviness. Flaws also add to texture, but should be measured independent of it. Changes in the surface layer, as a result of processing, include plastic deformation, residual stresses, cracks, and other metallurgical changes (hardness, overaging, phase

---

TABLE 31-1. Characteristics of Manufacturing Processes That Affect Surface Integrity

Workpiece–Tool–Machine–Environment–Process Variables

Workpiece characteristics	**Tool characteristics**
Geometry	Tool body
Shape	Type of tool
Dimensions	Size
Material	Shape
Type	Number of cutting edges
Route of manufacture	Cutting edge
Mechanical properties	Shape (angles)
Elastic constants	Nose geometry/topography
Plastic constants	Microgeometry
Physical properties	Wear
Melting point	Material
Thermal diffusivity, conductivity,	Type
capacity	Coating
Coefficient of thermal expansion	Type
Phase transformations	Thickness
Chemical properties	Number and kind of layers
Chemical composition	Mechanical properties
Chemical affinity to tool material	Elastic constants
and environment	Plastic properties
Metallurgical properties	Physical properties
Structure	Thermal diffusivity, conductivity,
Grain size	capacity
Hardness	Coefficient of thermal expansion
	Chemical properties
Environment characteristics	Chemical composition
Type of medium (gas, fluid, mist)	Chemical affinity to tool material
Lubricity	Metallurgical properties
Cooling ability	Structure
Flow rate	Grain size
Temperature	
Chemical composition	**Process variables**
	Speed
Machine tool characteristics	Feed
Error motions	Depth of cut

Advanced Manufacturing Engineering, Vol. 1, July, 1989.

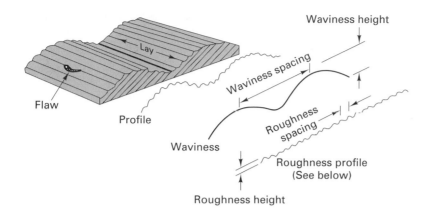

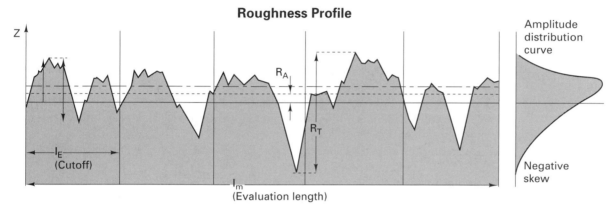

Roughness Profile

R_T = Maximum roughness depth (peak to valley) along l_m

R_A = Arithmetic roughness average

FIGURE 31-2   The surface roughness produced by the process can influence the performance of the component.

changes, recrystallization, intergranular attack, and hydrogen embrittlement). See Figure 31-2. The surface layer will always sustain local surface deformation due to a machining pass, a traditional process.

Surface integrity has become the subject of intense interest because the traditional, nontraditional, and post treatment methods used to manufacture hardware can change the material's properties. Although the consequence of these changes become a design problem, the preservation of properties is a manufacturing consideration. Designs that require a high degree of surface integrity are the ones that:

- are highly stressed
- employ low safety factors
- operate in severe environments
- must have prime reliability
- have a high surface areas to volume ratio
- are made with alloys that are sensitive to processing

Surface integrity should be a joint concern of manufacturing and engineering. Manufacturing must balance cost and producibility with design requirements. It bears repeating to say that engineering must design components with knowledge of manufacturing processes. A reduction in fatigue life resulting from processing can be reversed with a post treatment. This is another example of design for manufacturing.

All of the processes used to manufacture components are important if their effects are present in the finished part. It is convenient to divide processes that are used to manufacture parts into three categories: traditional, nontraditional, and finishing treatments. In *traditional processes* the tool contacts the workpiece. Examples are grinding, milling, and turning. These material removal processes will inflict damage to the surface if improper parameters are used. Examples of improper parameters are dull tools, excessive infeed, inadequate coolant, and improper grinding wheel hardness. The *nontraditional processes* have intrinsic characteristics that, even if well controlled, will change the surface. In these processes the workpiece does not touch the tool. Electrochemical machining (ECM), electrical discharge machining (EDM), and chemical milling are examples of nontraditional methods. Such methods can leave stress-free surfaces, remelted layers, and excessive surface roughness. *Finishing treatments* can be used to negate or remove the impact of both traditional and nontraditional processes as well as providing good surface finish. For example, residual tensile stresses can be removed by shot peening or roller burnishing. Chemical milling can remove the recast layer left by EDM.

The objectives of the surface modification processes can be quite varied. Some are designed to clean surfaces and remove the kinds of defects that occur during processing or handling (such as scratches, pores, burrs, fins, and blemishes). Others further improve or modify the products' appearance, providing features such as smoothness, texture, or color. Numerous techniques are available to improve resistance to wear or corrosion, or to reduce friction or adhesion to other materials. Scarce or costly materials can be conserved by making the interior of a product from a cheaper, more common material and then coating or plating the product surface.

As with all other processes, surface treatment requires time, labor, equipment, and material handling, and all of these have an associated cost. Efficiencies can be realized through process optimization and the integration of surface treatment into the entire manufacturing system. Design modifications can often facilitate automated or bulk finishing, eliminating the need for labor-intensive or single-part operations. Process selection should further consider the size of the part, the shape of the part, the quantity to be processed, the temperatures required for processing, the temperatures encountered during subsequent use and any dimensional changes that might occur due to the surface treatment. Through knowledge of the available processes and their relative advantages and limitations, finishing costs can often be reduced or eliminated while maintaining or improving the quality of the product.

In addition to the above, the field of surface finishing has recently undergone another significant change. Many chemicals that were once "standard" to the field, such as cyanide, cadmium, chromium, and chlorinated solvents, have now come under strict government regulation. Wastewater treatment and waste disposal have also become significant concerns. As a result, processes may have to be modified, or replacement processes may have to be used.

Because of their similarity to other processes, many surface finishing techniques have been presented elsewhere in the book. The *case hardening* techniques, both selective heating (flame, induction, and laser hardening) and altered surface chemistry (diffusion methods such as carburizing, nitriding, and carbonitriding), are presented in Chapter 5 as variations of heat treating. *Shot peening* and *roller burnishing* are presented in Chapter 19 as cold-working processes. *Roll bonding* and explosive bonding are discussed in Chapter 20 as means of producing laminar composites. *Hard facing* and metal spraying are included in Chapter 36 as adaptations of welding techniques. Chemical vapor deposition and physical vapor deposition are discussed in Chapters 22 and 34. Sputtering and ion implantation are also discussed in Chapter 34 as processes needed in electronics manufacturing. In the present chapter we focus on techniques for cleaning and surface preparation as well as the remaining methods of surface finishing or surface modification.

■ 31.2 MECHANICAL CLEANING AND FINISHING

ABRASIVE CLEANING

It is not uncommon for the various manufacturing processes to produce certain types of surface contamination. Sand from the molds and cores used in casting often adheres to product surfaces. Scale (metal oxide) can be produced whenever metal is processed at elevated temperatures. Oxides such as rust can form if material is stored between operations. These and other contaminants must be removed before decorative or protective surfaces can be produced. While vibratory shaking can be useful, some form of *abrasive cleaning* is usually required to remove the foreign material. In the most common method, sand, steel grit, metal shot, fine glass shot, or other form of abrasive is mechanically impelled against the surface to be cleaned. When sand is used, it should be clean, sharp-edged silica sand. Steel grit tends to clean more rapidly and generates much less dust, but is more expensive and less flexible.

When the parts are large, it may be easier to bring the cleaner to the part rather than the part to the cleaner. A common technique for such applications is *sand blasting* or *shot blasting,* where the abrasive particles are carried by a high-velocity blast of air emerging from a nozzle with about a $\frac{3}{8}$ in. opening. Air pressures between 60 and 100 psi are common when cleaning ferrous metals, and 10 to 60 psi is common for non-ferrous. The abrasive may be sand or shot, or materials such as walnut shells, dry-ice pellets, or even baking soda. Pressurized water can also be used as a carrier medium.

When production quantities are large or the parts are small, the operation can be conducted in an enclosed hood, with the parts traveling past stationary nozzles. For large parts or small quantities, the blast may be delivered manually. Protective clothing and breathing apparatus must be provided and precautions taken to control the spread of the resulting dust. The process may even require a dedicated room or booth that is equipped with integrated air pollution control devices.

From a manufacturing perspective, these processes are limited to surfaces that can be reached by the moving abrasive and cannot be used when sharp edges or corners must be maintained (since the abrasive tends to round the edges).

BARREL FINISHING OR TUMBLING

Barrel finishing or *tumbling* is an effective means of finishing large numbers of small parts. In the Middle Ages, wooden casks were filled with abrasive stones and metal parts and were rolled about until the desired finish was obtained. Today, modifications of this technique can be used to deburr, radius, descale, remove rust, polish, brighten, surface-harden, or prepare parts for further finishing or assembly. The amount of stock removal can vary from as little as 0.0001 to as much as 0.005 in.

In the typical operation, the parts are loaded into a special barrel or drum until a predetermined level is reached. Occasionally, no other additions are made, and the parts are simply tumbled against one another. In most cases, however, an additional media of metal slugs or abrasives (such as sand, granite chips, slag, or ceramic pellets) is added. Rotation of the barrel causes the material to rise until gravity causes the uppermost layer to cascade downward in a "landslide" movement, as depicted in Figure 31-3. The

FIGURE 31-3   Schematic of the flow of material in tumbling or barrel finishing. The parts and media mass typically account for 50 to 60% of capacity.

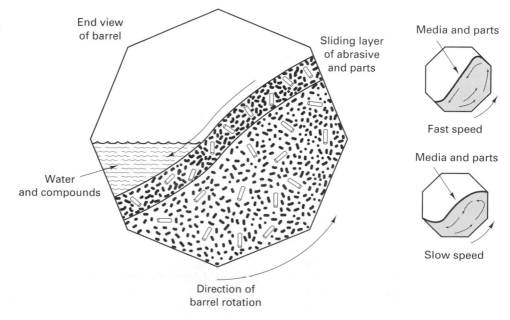

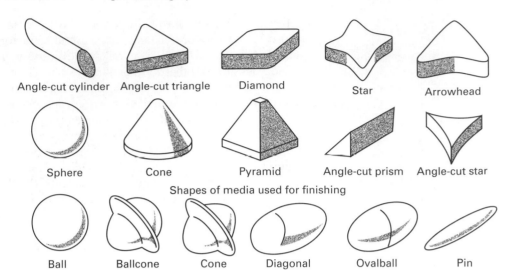

Shapes of media used for finishing

Steel media shapes used for burnishing

FIGURE 31-4 Synthetic abrasive media is available in a wide variety of sizes and shapes. Through proper selection, the media can be tailored to the product being cleaned.

sliding produces abrasive cutting that can effectively remove fins, flash, scale, and adhered sand. Since only a small portion of the load is exposed to the abrasive action, long times may be required to process the entire contents.

Increasing the speed of rotation adds centrifugal forces that cause the material to rise higher in the barrel. The enhanced action can often accelerate the process, provided that the speed is not so great as to destroy the cascading action and that the additional action does not damage the workpiece. By a suitable selection of abrasives, filler, barrel size, ratio of workpieces to abrasive, fill level and speed, a wide range of parts can be tumbled successfully. Delicate parts may have to be attached to racks within the barrel to reduce their movement while permitting the media to flow around them.

Natural and synthetic abrasives are available in a wide range of sizes and shapes, including those depicted in Figure 31-4, that enable the finishing of complex parts with irregular openings. The various media are often mixed in a given load, so that some will reach into all sections and corners to be cleaned.

Tumbling is usually done dry, but it can also be performed with an aqueous solution in the barrel. Chemical compounds can be added to the media to assist in cleaning, or descaling, or to provide features such as rust inhibition. Support equipment usually assists with loading and unloading the barrels, and with the separation of the workpieces from the abrasive media. The latter operation often uses mesh screens with selected size openings.

Barrel tumbling can be a very inexpensive way to finish large quantities of small parts and produce rounded edges and corners. Unfortunately, the abrasive action occurs on all surfaces and cannot be limited to selected areas. The cycle time is often long, and the process can be quite noisy.

In the *barrel burnishing* process, no cutting action is desired. Instead, the parts are tumbled against themselves, or with media such as steel balls, shot, rounded-end pins, or ballcones. If the original material is free of visible scratches and pits, the combination of peening and rubbing will reduce minute irregularities and produce a smooth, uniform surface.

Barrel burnishing is normally done wet, using a solution of water and lubricating or cleaning agents, such as soap or cream of tartar. Because the rubbing action between the work and the media is very important, the barrel should not be loaded more than half full, and the volume ratio of media to work should be about 2 : 1, so the workpieces rub against the media and not each other. The speed of rotation should be set to maintain the cascading action and not fling the workpieces free of the tumbling mass.

Centrifugal barrel tumbling places the tumbling barrel at the end of a rotating arm. This adds centrifugal force to the weight of the parts in the barrel and can accelerate the process by as much as 25 to 50 times.

In *spindle finishing* the workpieces are attached to rotating shafts, and the assembly is immersed in media that is moving in a direction opposite to part rotation. This

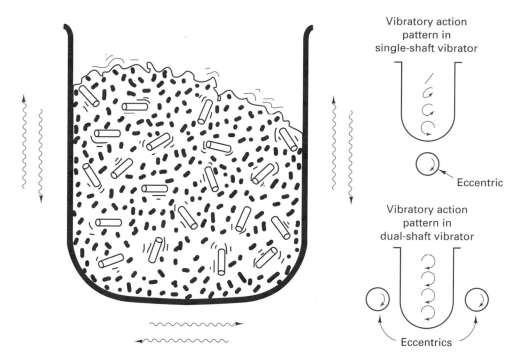

FIGURE 31-5 Schematic diagram of a vibratory finishing tub loaded with parts and media. The single eccentric shaft drive provides maximum motion at the bottom, which decreases as one moves upward. The dual shaft design produces more uniform motion of the tub, and reduces processing time.

process is commonly applied to cylindrical parts and avoids the impingement of workpieces on one another. The abrasive action is accelerated, but time is required for fixturing and removal of the parts.

VIBRATORY FINISHING

Vibratory finishing is a versatile process widely used for deburring, radiusing, descaling, burnishing, cleaning, brightening, and fine finishing. In contrast to the barrel processing, *vibratory finishing* is performed in open containers. As illustrated in Figure 31-5 tubs or bowls are loaded with workpieces and media and are vibrated at frequencies between 900 and 3600 cycles per minute. The specific frequency and amplitude are determined by the size, shape, weight, and material of the pan, as well as the media and compound. Because the entire load is under constant agitation, cycle times are less than with barrel operations. The process is less noisy and is easily controlled and automated. In addition, the open tubs allow for direct observation during the process, which can also deburr or smooth internal recesses or holes.

MEDIA

The success of any of the mass-finishing processes depends greatly on *media* selection and the ratio of media to parts, as presented in Table 31-2. As developed in this table, a significant objective of the media may be to prevent the parts from impinging upon one another as it simultaneously cleans and finishes. Fillers, such as scrap punchings, minerals, leather scraps, and sawdust, are often added to provide additional bulk and cushioning.

TABLE 31-2. Typical Media-to-Part Ratios for Mass Finishing

Media/Part Ratio by Volume	Typical Application
0:1	Part-on-part processing or burr removal without media
1:1	Produces very rough surfaces and is suitable for parts in which part-on-part damage is not a problem
2:1	Somewhat less severe part-on-part damage, but more action from less media
3:1	May be acceptable for very small parts and very small media. Part-on-part contact is likely on larger and heavier parts
4:1	In general, a good average ratio for many parts; a good ratio for evaluating a new deburring process
5:1	Better for nonferrous parts subject to part-on-part damage
6:1	Suitable for nonferrous parts, especially preplate surfaces on zinc parts with resin-bonded media
8:1	For improved preplate surfaces with resin-bonded media
10:1	Produces very fine finishes

Source: American Machinist, August 1983.

Natural abrasives include slag, cinders, sand, corundum, granite chips, limestone, and hardwood shapes, such as pegs, cylinders, and cubes. Synthetic media typically contain 50 to 70 wt% of abrasives, such as alumina (Al_2O_3), emery, flint, and silicon carbide. This material is embedded in a matrix of ceramic, polyester, or resin plastic, which is softer than the abrasive, and erodes, allowing the exposed abrasive to perform the work. The synthetics are generally produced by some form of casting operation, so their sizes and shapes are consistent and reproducible (as opposed to the random sizes and shapes of the natural media). Steel media with no added abrasive is frequently specified for burnishing and light deburring.

Media selection should also be correlated with part geometry, since the abrasives should be able to contact all critical surfaces without becoming lodged in recesses or holes. This requirement has resulted in a wide variety of sizes and shapes, including those presented in Figure 31-4. The different abrasives, sizes, and shapes can be selected or combined to perform tasks ranging from light deburring with a very fine finish to heavy cutting with a rough surface.

COMPOUNDS

A variety of functions are performed by the *compounds* that are added in addition to the media and workpieces. These compounds can be liquid or dry, abrasive or nonabrasive, and acid, neutral, or alkaline. They are often designed to assist in deburring, burnishing, and abrasive cutting, as well as to provide cleaning, descaling, or corrosion inhibition.

In deburring and finishing, many small particles are abraded from both the media and the workpieces, and these must be suspended in the compound solution to prevent them from adhering to the parts. Deburring compounds also act to keep the parts and media clean and to inhibit corrosion. Burnishing compounds are often selected for their ability to develop desired colors and enhance brightness.

Cleaning compounds such as dilute acids and soaps are designed to remove excessive soils from both the parts and media and are often specified when the incoming materials contain heavy oil or grease. Corrosion inhibitors can be selected for both ferrous and nonferrous metals and are particularly important when steel media is being used.

Another function of the compounds may be to condition the water when aqueous solutions are being used. Consistent water quality, in terms of "hardness" and metal ion content, is important to ensure uniform and repeatable finishing results. Liquid compounds may also provide cooling to both the workpieces and the media.

SUMMARY OF MASS-FINISHING METHODS

The barrel and vibratory finishing processes are really quite simple and economical and can process large numbers of parts in a batch procedure. Soft, nonferrous parts can be finished in as little as 10 minutes, while the harder steels may require 2 hours or more. Sometimes the operations are sequenced, using progressively finer abrasives. Figure 31-6 shows a variety of parts before and after the mass-finishing operation, using the triangular abrasive shown with each component.

Despite the high volume and apparent success, these processes may still be as much art as science. The key factors of workpiece, equipment, media, and compound are all interrelated, and the effect of changes can be quite complex. Media, equipment, and compounds are often selected by trial and error, with various approaches being tested until the desired result is achieved. Even then, maintenance of consistent results may still be difficult.

FIGURE 31-6 A variety of parts before and after barrel finishing with triangular-shaped media. *(Courtesy of Norton Company.)*

BELT SANDING

In the *belt sanding* operation, the workpieces are held against a moving abrasive belt until the desired degree of finish is obtained. Because of the movement of the belt, the resulting surface contains a series of parallel scratches with a texture set by the grit of the belt. When smooth surfaces are desired, a series of belts may be employed, with progressively finer grits.

The ideal geometry for belt sanding is a flat surface, for the belt can be passed over a flat table where the workpiece can be held firmly against it. Belt sanding is frequently a hand operation and is therefore quite labor intensive. Furthermore, it is difficult to apply when the geometry includes recesses or interior corners. As a result, belt sanding is usually employed when the number of parts is small and the geometry is relatively simple. See Chapter 27.

WIRE BRUSHING

High-speed rotary *wire brushing* is sometimes used to clean surfaces and can also impart some small degree of material removal or smoothing. The resulting surface consists of a series of uniform curved scratches. For many applications, this *may* be an acceptable final finish. If not, the scratches can easily be removed by barrel finishing or buffing.

Wire brushing is often performed by hand application of a small workpiece to the brush, or the brush to a larger workpiece. Automatic machines can also be used where the parts are moved past a series of rotating brushes. In another modification the brushes are replaced with plastic or fiber wheels that are loaded with abrasive.

BUFFING

Buffing is a polishing operation in which the workpiece is brought into contact with a revolving cloth wheel that has been charged with a fine abrasive, such as polishing rouge. The "wheels" are made of disks of linen, cotton, broadcloth, or canvas, and achieve the desired degree of firmness through the amount of stitching used to fasten the layers of cloth together. When the operation calls for very soft polishing or polishing into interior corners, the stitching may be totally omitted, the centrifugal force of the wheel rotation being sufficient to keep the layers in the proper position. Various types of polishing compounds are also available, with many consisting of ferric oxide particles in some form of binder or carrier.

The buffing operation is very similar to the lapping process that was discussed in Chapter 27. In buffing, however, the abrasive removes only minute amounts of metal from the workpiece. Fine scratch marks can be eliminated and oxide tarnish can be removed. A smooth, reflective surface is produced. When soft metals are buffed, a small amount of metal flow may occur, which further helps to reduce high spots and produce a high polish.

In manual buffing, the workpiece is held against the rotating wheel and manipulated to provide contact with all critical surfaces. Once again, the labor costs can be quite extensive. If the workpieces are not too complex, semiautomatic machines can be used, where the workpieces are held in fixtures and move past a series of individual buffing wheels. By designing the part with buffing in mind, good results can be obtained quite economically.

ELECTROPOLISHING

Electropolishing is the reverse of electroplating (discussed later in this chapter) since material is removed from the surface rather than being deposited. A dc electrolytic circuit is constructed with the workpiece as the anode. As current is applied, material is stripped from the surface, with material removal occurring preferentially from any raised location. Unfortunately, it is not economical to remove more than about 0.001 in. of material from any surface. However, if the initial surface is sufficiently smooth (less than 8 μin. rms), and the grain size is small, the result will be a smooth polish with irregularities of less than 2 μin.—a mirrorlike finish.

Electropolishing was originally used to prepare metallurgical specimens for examination under the microscope. It was later adopted as a means of polishing stainless steel sheets and other stainless products. It is particularly useful for polishing irregular shapes that would be difficult to buff.

■ 31.3 CHEMICAL CLEANING

Chemical cleaning operations are effective means of removing oil, dirt, scale, or other foreign material that may adhere to the surface of a product, as a preparation for subsequent painting or plating. Because of environmental, health, and safety concerns, however, many processes that were once the industrial standard have now been eliminated or substantially modified. While the major concern with the mechanical methods has usually been airborne particles, the chemical methods often require the disposal of spent or contaminated solutions, and occasionally use hazardous, toxic, or environmentally unfriendly materials. Chlorofluorocarbons (CFCs) and carbon tetrachloride, for example, have been identified as ozone-depleting chemicals and have been phased out of commercial use. Process changes to comply with added regulations can significantly shift process economics. Manufacturers must now ask themselves if a part really has to be cleaned, what soils have to be removed, how clean the surfaces have to be, and how much they are willing to pay to accomplish that goal. Selection of the specific method will depend on the quantity of parts to be processed, part configuration, part material, desired surface finish, temperature of the process, and other variables.

ALKALINE CLEANING

Alkaline cleaning is basically the "soap and water" approach to parts cleaning and is a commonly used method for removing a wide variety of soils (including oils, grease, wax, fine particles of metal, and dirt) from the surfaces of metals. The cleaners are usually complex solutions of alkaline salts, additives to enhance cleaning or surface modification, and surfactants or soaps that are selected to reduce surface tension and displace, emulsify, and disperse the insoluble soils. The actual cleaning occurs as a result of one or more of the following mechanisms: (1) saponification, the chemical reaction of fats and other organic compounds with the alkaline salts; (2) displacement, where soil particles are lifted from the surface; (3) dispersion or emulsification of insoluble liquids; and (4) dissolution of metal oxides.

Alkaline cleaners can be applied by immersion or spraying, and are usually heated to accelerate the cleaning action. The cleaning is then followed by a water rinse to remove all residue of the cleaning solution, as well as flush away some small amounts of remaining soil. A drying operation may also be required since the aqueous cleaners do not evaporate quickly, and some form of corrosion inhibitor (or rust preventer) may be required, depending on subsequent use.

Environmental issues relating to alkaline cleaning include (1) reducing or eliminating phosphate effluent, (2) reducing toxicity and increasing biodegradability, and (3) and recycling the cleaners to extend their life and reduce the volume of discard.

SOLVENT CLEANING

In *solvent cleaning,* oils, grease, fats, and other surface contaminants are removed by dissolving them in organic solvents derived from coal or petroleum, usually at room temperature. The common solvents include petroleum distillates (such as kerosene, naphtha, and mineral spirits), chlorinated hydrocarbons (such as methylene chloride and trichloroethylene), and liquids such as acetone, benzene, toluene, and the various alcohols. Small parts are generally cleaned by immersion, with or without assisting agitation, or by spraying. Products that are too large to immerse can be cleaned by spraying or wiping. The process is quite simple, and capital equipment costs are rather low. Drying is usually accomplished by simple evaporation.

Solvent cleaning is an attractive means of cleaning large parts, heat-sensitive products, materials that might react with alkaline solutions (such as aluminum, lead, and zinc), and products with organic contaminants (such as soldering flux or marking crayon). Virtually all common industrial metals can be cleaned, and the size and shape of the workpiece are rarely a limitation. Insoluble contaminants, such as metal oxides, sand, scale, and the inorganic fluxes used in welding, brazing, and soldering, cannot be removed by solvents. In addition, resoiling can occur as the solvent becomes contaminated. As a result, solvent cleaning is often used for preliminary cleaning.

Many of the common solvents have been restricted because of health, safety, and environmental concerns. Fire and excessive exposure are common hazards. Adequate ventilation is critical. Workers should use respiratory devices to prevent inhalation of vapors and wear protective clothing to minimize direct contact with skin. In addition, solvent wastes are often considered to be "hazardous materials" and may be subject to high disposal cost.

VAPOR DEGREASING

In *vapor degreasing,* the vapors of a chlorinated or fluorinated solvent are used to remove oil, grease, and wax from metal products. A nonflammable solvent, such as trichloroethylene, is heated to its boiling point, and the parts to be cleaned are suspended in its vapors. The vapor condenses on the work and washes the soluble contaminants back into the liquid solvent. Although the bath becomes dirty, the contaminants rarely volatilize at the boiling temperature of the solvent. Therefore, vapor degreasing tends to be more effective than cold solvent cleaning, since the surfaces always come into contact with clean solvent. Since the surfaces become heated by the condensing solvent, they dry almost instantly when they are withdrawn from the vapor.

Vapor degreasing is a rapid process that has almost no visible effect on the surface being cleaned. It can be applied to all common industrial metals, but the solvents may attack rubber, plastics, and organic dyes that might be present in product assemblies. A major limitation is the inability to remove insoluble soils, forcing the process to be coupled with another technique, such as mechanical or alkaline cleaning. Since hot solvent is present in the system, the process is often accelerated by coupling the vapor cleaning with an immersion or spray using the hot liquid.

Unfortunately, environmental issues have forced the almost complete demise of the process. While the vapor degreasing solvents are chemically stable, have low toxicity, are nonflammable, evaporate quickly, and can be recovered for reuse, they also have been identified as ozone-depleting compounds and have essentially been banned from use. It would be nice if there were a replacement solvent that could be used in the same process, or a replacement process that offered all of the qualities of a vapor degreaser, but these do not exist. Most manufacturers have been forced to convert to some form of water-based process using alkaline, neutral, or acid cleaners, or a process using chlorine-free, hydrocarbon-based solvents.

ULTRASONIC CLEANING

When high-quality cleaning is required for small parts, *ultrasonic cleaning* may be preferred. Here the parts are suspended or placed in wire mesh baskets that are then immersed in a liquid cleaning bath, often a water-based detergent. The bath contains an ultrasonic transducer that operates at a frequency that causes cavitation in the liquid. The bubbles that form and implode provide the majority of the cleaning action, and if gross dirt, grease, and oil are removed prior to the immersion, excellent results can usually be obtained in 60 to 200 seconds. Because of the ability to use water-based solutions, ultrasonic cleaning has replaced many of the environmentally unfriendly solvent processes.

ACID PICKLING

In the *acid pickling* process, metal parts are first cleaned to remove oils and other contaminants and then dipped into dilute acid solutions to remove oxides and dirt that are left on the surface by the previous processing operations. The most common solution is a 10% sulfuric acid bath at an elevated temperature between 150 and 185°F. Muriatic acid is also used, either cold or hot. As the temperature increases, the solutions can become more dilute.

After the parts are removed from the pickling bath, they should be rinsed to flush the acid residue from the surface and then dipped in an alkaline bath to prevent rusting. When it will not interfere with further processing, an immersion in a cold *milk of lime* solution is often used. Caution should be used to avoid overpickling, since the acid attack can result in a roughened surface.

■ 31.4 COATINGS

Each of the surface finishing methods previously presented has been a material removal process, designed to clean, smooth, and otherwise reduce the size of the part. Many other techniques have been developed to add material to the surface of a part. If the material is deposited as a liquid or organic gas (*or* from a liquid or gas medium), the process is called *coating*. If the added material is a solid during deposition, the process is known as *cladding*.

PAINTING, WET OR LIQUID

Paints and enamels are by far the most widely used finish on manufactured products, and a great variety are available to meet the wide range of product requirements. Most of today's commercial paints are synthetic organic compounds that contain pigments and dry by polymerization or by a combination of polymerization and adsorption of oxygen. Water is the most common carrying vehicle for the pigments. Heat can be used to accelerate the drying, but many of the synthetic paints and enamels will dry in less than an hour without the use of additional heat. The older oil-based materials have a long drying time and require excessive environmental protection measures. For these reasons they are seldom used in manufacturing applications.

Paints are used for a variety of reasons, usually to provide protection and decoration but also to fill or conceal surface irregularities, change the surface friction, or modify the light or heat absorption or radiation characteristics. Table 31-3 provides a list of some of the more commonly used organic finishes, along with their significant characteristics. *Nitracellulose lacquers* consist of thermoplastic polymers dissolved in organic solvent. Although fast drying (by the evaporation of the solvent) and capable of producing very beautiful finishes, they are not sufficiently durable for most commercial applications. The *alkyds* are a general-purpose paint but are not adequate for hard-service applications. *Acrylic enamels* are widely used for automotive finishes and may require catalytic or oven curing. *Asphaltic paints*, solutions of asphalt in a solvent, are used extensively in the electrical industry, where resistance to corrosion is required and appearance is not of prime importance.

When considering a painted finish, the temptation is to focus on the outermost coat, to the exclusion of the underlayers. In reality, painting is a complex system that includes the substrate material, cleaning, and other pretreatments (such as anodizing, phosphating, and various conversion coatings), priming, and possible intermediate layers. The method of application is another integral feature to be considered.

PAINT APPLICATION METHODS

In manufacturing, almost all painting is done by one of four methods: *dipping, hand spraying, automatic spraying,* or *electrostatic spray finishing.* In most cases, at least two coats are required. The first (or *prime*) coat serves to (1) assure adhesion, (2) provide a leveling effect

TABLE 31-3.	Commonly Used Organic Finishes and Their Qualities		
Material	Durability (Scale of 1–10)	Relative Cost (Scale of 1–10)	Characteristics
Nitrocellulose lacquers	1	2	Fast drying; low durability
Epoxy esters	1	2	Good chemical resistance
Alkyd-amine	2	1	Versatile; low adhesion
Acrylic lacquers	4	1.7	Good color retention; low adhesion
Acrylic enamels	4	1.3	Good color retention; tough; high baking temperature
Vinyl solutions	4	2	Flexible; good chemical resistance; low solids
Silicones	4–7	5	Good gloss retention; low flexibility
Flouropolymers	10	10	Excellent durability; difficult to apply

by filling in minor porosity and other surface blemishes, and (3) improve corrosion resistance, and thus prevent later coatings from being dislodged in service. These properties are less easily attainable in the more highly pigmented paints that are used in the final coats to promote color and appearance. When using multiple coats, however, it is important that the carrying vehicles for the final coats do not unduly soften the underlayers.

Dipping is a simple and economical means of paint application when all surfaces of the part are to be coated. The products can be manually immersed into a paint bath or passed through the bath while on or attached to a conveyor. Dipping is attractive for applying prime coats and for painting small parts where spray painting would result in a significant waste due to overspray. Conversely, the process is unattractive if only some of the surfaces require painting, or where a very thin, uniform coating would be adequate, as on automobile bodies. Other difficulties are associated with the tendency of paint to run, producing both a wavy surface and a final drop of paint attached to the lowest drip point. Good-quality dipping requires that the paint be stirred at all times and be of uniform viscosity.

Spray painting is probably the most widely used paint application process because of its versatility and the economy in the use of paint. In the conventional technique, the paint is atomized and transported by the flow of compressed air. In a variation known as *airless spraying,* mechanical pressure forces the paint through an orifice at pressures between 500 and 4500 psi. This provides sufficient velocity to produce atomization and also propel the particles to the workpiece. Because no air pressure is used for atomization, there is less spray loss (paint efficiency may be as high as 99%) and less generation of gaseous fumes.

Hand spraying is probably the most versatile means of application but can be quite costly in terms of labor and production time. When air or mechanical means provide the atomization, the workers must exercise considerable skill to obtain the proper coverage without allowing the paint to "run" or "drape." Only a very thin film can be deposited at one time, usually less than 0.001 in. As a result, several coats may be required with intervening time for drying.

One means of applying thicker layers in a single application is known as *hot spraying*. Special solvents are used which reduce the viscosity of the material when heated. Upon atomization, the faster evaporating solvents are removed, and the drop in temperature produces a more viscous, run-resistant material that can be deposited in thicker layers.

When producing large quantities of similar or identical parts, some form of automatic system is usually employed. The simplest automatic equipment consists of some form of parts conveyor that transports the parts past a series of stationary spray heads. While the concept is simple, the results may be unsatisfactory. A large amount of paint is wasted and it is difficult to get uniform coverage.

Industrial robots can be used to move the spray heads in a manner that mimics the movements of a human painter, maintaining uniform separation distance and minimizing waste. This is an excellent application for the robot, since a monotonous and repetitious process can be performed with consistent results. In addition, use of a robot removes the human from an unpleasant, and possibly unhealthy, environment. See Chapter 32, robotic applications.

Both manual and automatic spray painting can benefit from the use of *electrostatic deposition*. A dc electrostatic potential is applied between the atomizer and the workpiece. The atomized paint particles assume the same charge as the atomizer and are therefore repelled. The oppositely charged workpiece then attracts the particles, with the actual path of the particle being a combination of the kinetic trajectory and the electrostatic attraction. The higher the dc voltage, the greater the electrostatic attraction. Overspraying can be reduced by as much as 60 to 80%, as well as the generation of airborne particles and other emissions. Unfortunately, part edges and holes receive a heavier coating than flat surfaces due to the concentration of electrostatic lines of force on any sharp edge. Recessed areas will receive a reduced amount of paint, and a manual touch-up may be required using conventional spray techniques. Despite these limitations, electrostatic spraying is an extremely attractive means of painting complex-shaped products where the geometry would tend to create large amounts of overspray.

In an electrostatic variation of airless spraying, the paint is fed onto the surface of a rapidly rotating cone or disk that is also one electrode of the electrostatic circuit.

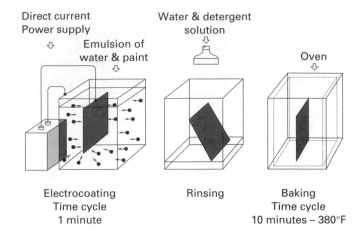

FIGURE 31-7 Basic steps in the electrocoating process.

Centrifugal force causes the thin film of paint to flow toward the edge, where charged particles are spun off without the need for air assist. The particles are then attracted to the workpiece, which serves as the other electrode of the electrostatic circuit. Because of the effectiveness of the centrifugal force, paints can be used with high solids content, reducing the amount of volatile emissions and enabling a thicker layer to be deposited in a single application.

Electrocoating or *electrodeposition* applies paint in a manner similar to the electroplating of metals. As shown schematically in Figure 31-7, the paint particles are suspended in an aqueous solution and are given an electrostatic charge by applying a dc voltage between the tank (cathode) and the workpiece (anode). As the electrically conductive workpiece enters and passes through the tank, the paint particles are attracted to it and deposit on the surface, creating a uniform, thin coating that is more than 90% resin and pigment. When the coating reaches a desired thickness, determined by the bath conditions, no more paint is deposited. The workpiece is then removed from the tank, rinsed in a water spray, and baked at a time and temperature that depends on the particular type of paint. Baking of 10 to 20 minutes at 375°F is somewhat typical.

Electrocoating combines the economy of ordinary dip painting with the ability to produce thinner, more uniform coatings. The process is particularly attractive for applying the prime coat to complex structures, such as automobile bodies, where good corrosion resistance is a requirement. Hard-to-reach areas and recesses can be effectively coated. Since the solvent is water, no fire hazard exists (as with the use of many solvents), and air and water pollution is reduced significantly. In addition, the process can be readily adapted to conveyor line production.

DRYING

Most paints and enamels used in manufacturing require from 2 to 24 hours to dry at normal room temperature. This time can be reduced to between 10 minutes and 1 hour if the temperature can be raised to between 275 and 450°F. As a result, elevated temperature drying is often preferred. Parts can be batch processed in ovens, or continuously passed through heated tunnels or under panels of infrared heat lamps.

Elevated-temperature drying is rarely a problem with metal parts, but other materials can be damaged by exposure to the moderate temperatures. For example, when wood is heated, the gases, moisture, and residual sap are expanded and driven to the surface beneath the hardening paint. Small bubbles tend to form that roughen the surface, or break, producing small holes in the paint.

PAINTING, DRY OR POWDER

Powder coating is yet another variation of electrostatic spraying, but here the particles are solid rather than liquid. Several coats, such as primer and finish, can be applied and then followed by a single baking, in contrast to the baking after each coat that is required in the conventional spray processes. In addition, the overspray powder can often

TABLE 31-4. Thermosetting Powders for Powder Coating (Dry Painting) Have a Wide Variety of Properties and Applications. (*Manufacturing Engineering*, June 1989.)

Generic Powder	Typical Properties	Typical Applications	Relative Cost
Hybrid (combination of polyester and epoxy)	Very good chemical resistance Very good mechanical properties Fair exterior color/gloss retention	Water heaters, radiators, transformer covers, office furniture, power tools, and shelving	1 (Least expensive)
Epoxy	Excellent chemical resistance Excellent mechanical properties Very good exterior color/gloss retention	Metal furniture, underbody automotive parts, major appliances, microwave ovens, shelving, and fire extinguishers	2
TGIC polyester	Good chemical resistance Very good mechanical properties Very good exterior color/gloss retention	Architectural aluminum, outdoor furniture, farm equipment, fence poles, and air conditioning units	3
Polyurethane	Good chemical resistance Good mechanical properties Very good exterior color/gloss retention	Automotive wheels and trim, playground equipment, and fluorescent light fixtures	4
Acrylic	Very good chemical resistance Good mechanical properties Very good exterior color/gloss retention	Washing machines, oven parts, refrigerators, microwave ovens, and aluminum extrusions	5 (Most expensive)

be collected and reused. While volitized solvents are no longer a concern, operators must now address the possibility of powder explosion, as well as the health hazards of airborne particles.

Modern powder technology can produce a good high quality finish with superior surface properties and usually at a lower cost than liquid painting. Powder painting is more efficient in the use of materials (the overspray can be captured and reused) and lower energy requirements. The economic advantages must be weighed against the limitations of powder coating. Dry systems have a longer color change time than wet systems. The process is not good for large objects (massive tanks) or heat sensitive objects. It is not easy to produce film thickness less than 1 mil (0.03 mm).

Table 31-4 provides details on powders that are used in powder coatings. Thermoplastics can also be used but thermosetting powders are most common.

The elements of a powder coating system are shown in Figure 31-8. The following aspects of the process must be considered.

- Types of guns—corona charged or tribo charged
- Number of guns—depends on many factors like parts per hour, size of parts, line speed, and powder types.
- Color change time/frequency
- Safety
- Curing oven—coated parts are put in ovens to melt, flow, and cure the powder

HOT-DIP COATINGS

Large quantities of metal products are given corrosion-resistant coatings by direct immersion into a bath of molten metal. The most common coating materials are zinc, tin, aluminum, and tene (an alloy of lead and tin).

Hot-dip galvanizing is the most widely used method of imparting corrosion resistance to steel. (The zinc acts as a sacrificial anode, protecting the underlying iron.) After the products, or sheets, have been cleaned to remove oil, grease, scale, and rust, they are fluxed by dipping into a solution of zinc ammonium chloride and dried. Next, the article is completely immersed in a bath of molten zinc. The zinc and iron react metallurgically to produce a coating that consists of a series of zinc–iron compounds and a surface layer of nearly pure zinc.

The coating thickness is usually specified in terms of weight per unit area. Values between 0.5 and 3.0 oz/ft^2 are typical, with the specific value depending on the time of

Powder application equipment

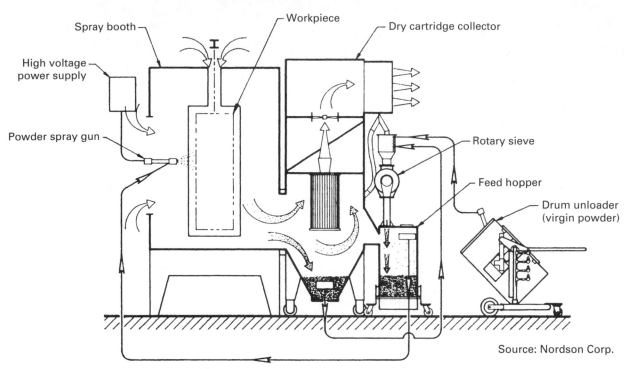

FIGURE 31-8 A schematic of a powder coating system. The wheels on the color modules permit it to be exchanged with a spare module to obtain the next color.

immersion and speed of withdrawal. Thinner layers can be produced by incorporating some form of air jet or mechanical wiping as the product is withdrawn. Since the corrosion resistance is provided through the sacrificial action of the zinc, the thin layers do not provide long-lasting protection. Extremely heavy coatings, on the other hand, may tend to crack and peel. The appearance of the coating can be varied through both the process conditions, as well as alloy additions of tin, antimony, lead, and aluminum. When the coatings are properly applied, bending or forming can often follow galvanizing without damage to the integrity of the coating.

The primary limitations to hot-dip galvanizing are the size of the product (which is limited to the size of the tank holding the molten zinc), and the "damage" that might occur when a metal is exposed to the temperatures of the molten material (approximately 850°F).

Tin coatings can also be applied by immersing in a bath of molten tin with a covering of flux material. Because of the high cost of tin and the relatively thick coatings applied by hot dipping, most tin coatings are now applied by electroplating. *Terne coating* utilizes an alloy of 15 to 20% tin and the remainder lead. This material is cheaper than tin and can provide satisfactory corrosion resistance for many applications.

CHEMICAL CONVERSION COATINGS

In *chemical conversion coating,* the surface of the metal is chemically treated to produce a nonmetallic, nonconductive surface that can impart a range of desirable properties. The most popular types of conversion coatings are chromate and phosphate.

Aluminum, magnesium, zinc, and copper (as well as cadmium and silver) can all be treated by a *chromate* conversion process that usually involves immersion in a chemical bath. The surface of the metal is convened into a layer of complex chromium compounds that can impart colors ranging from clear, through blue-bright, yellow, brown, olive drab, and black. Most of the films are soft and gelatinous when they are formed, but harden upon drying. They can be used to (1) impart exceptionally good corrosion resistance; (2) act as an intermediate bonding layer for paint, lacquer, or other organic finishes; or (3) provide specific colors by adding dyes to the coating when it is in its soft condition.

Phosphate coatings are formed by immersing metals (usually steel or zinc) in baths where metal phosphates (iron, zinc, and manganese phosphates are all common) have been dissolved in solutions of phosphoric acid. The resultant coatings can be used to precondition surfaces to receive and retain paint or enhance the subsequent bonding with rubber or plastic. In addition, phosphate coatings are usually rough and can provide an excellent surface for holding oils and lubricants. This feature can be used during manufacturing, where the coating holds the lubricants that assist in forming, or in the finished product, as with the black-color bolts and fasteners, whose corrosion resistance is provided by a phosphate layer impregnated with wax or oil.

BLACKENING OR COLORING METALS

Many steel parts are treated to produce a black, iron oxide coating (Fe_3O_4), a lustrous surface that is resistant to rusting when handled. Since this type of oxide forms at elevated temperatures, the parts are usually heated in some form of special environment, such as spent carburizing compound or special blackening salts.

Chemical solutions can also be used to blacken, blue, and even "brown" steels. Brown, black, and blue colors can also be imparted to tin, zinc, cadmium, and aluminum through chemical bath immersions or wipes. The surfaces of copper and brass can be made to be black, blue, green, or brown, with a full range of tints in between.

ELECTROPLATING

Large quantities of metal *and plastic* parts are electroplated to produce a metal coating that imparts corrosion or wear resistance, improves appearance (through color or luster), or increases the overall dimensions. Virtually all commercial metals can be plated, including aluminum, copper, brass, steel, and zinc-based die castings. Plastics can be electroplated, provided that they are first coated with an electrically conductive material.

The most common platings are zinc, chromium, nickel, copper, tin, gold, platinum, and silver. The electrogalvanized zinc platings are thinner than the hot-dip coatings and can be produced without subjecting the base metal to the elevated temperatures of molten zinc. Nickel plating provides good corrosion resistance but is rather expensive and does not retain its lustrous appearance. Consequently, when lustrous appearance is desired, a chromium plate is usually specified. Chromium is seldom used alone, however. An initial layer of copper produces a leveling effect and makes it possible to reduce the thickness of the nickel layer that typically follows to less than 0.0006 in. The final layer of chromium then provides the attractive appearance. Gold, silver, and platinum platings are used in both the jewelry and electronics industries, where the thin layers impart the desired properties while conserving the precious metals.

Hard chromium plate, with Rockwell hardnesses between 66 and 70, can be used to build up worn parts to larger dimensions and to coat tools and other products that need reduced surface friction and good resistance to both wear and corrosion. Hard chrome coatings are always applied directly to the base material and are usually much thicker than the decorative treatments, typically ranging from .003 to .010 in. thick. Even thicker layers are used in applications such as diesel cylinder liners. Since hard chrome plate does not have a leveling effect, defects or roughness in the base surface will be amplified. If smooth surfaces are desired, subsequent grinding and polishing may be necessary.

Figure 31-9 depicts the typical electroplating process. A dc voltage is applied between the parts to be plated (which is made the cathode) and an anode material that is either the metal to be plated or an inert electrode. Both of these components are immersed in a conductive electrolyte, which may also contain dissolved salts of the metal to be plated, as well as additions to increase or control conductivity. In response to the applied voltage, metal ions migrate to the cathode, lose their charge, and deposit on the surface. While the process is simple in its basic concept, the production of a high-quality plating requires selection and control of a number of variables, including the electrolyte and the concentrations of the various dissolved components, the temperature of the bath, and the electrical voltage and current. The interrelation of these features adds to the complexity and makes process control an extremely challenging problem.

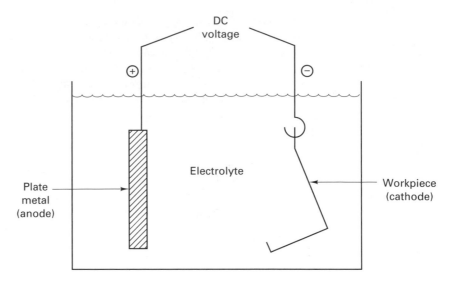

FIGURE 31-9 Basic circuit for an electroplating operation, showing the anode, cathode (workpiece), and electrolyte (conductive solution).

The surfaces to be plated must also be prepared properly if satisfactory results are to he obtained. Pinholes, scratches, and other surface defects must be removed if a smooth, lustrous finish is desired. Combinations of degreasing, cleaning, and pickling are used to assure a chemically clean surface, one to which the plating material can adhere.

As shown in Figure 31-10 the plated metal tends to be preferentially attracted to corners and protrusions. This makes it particularly difficult to apply a uniform plating to irregular shapes, especially ones containing recesses, corners, and edges. Design features can be incorporated to promote plating uniformity, and improved results can often be obtained through the use of multiple spaced anodes, or anodes whose shape resembles that of the workpiece.

Electroplating is frequently performed as a continuous process, where the individual parts to be plated are hung from conveyors. As they pass through the process, they are lowered into successive plating, washing, and fixing tanks. Ordinarily, only one type of workpiece is plated at a time, because the details of solutions, immersion times, and current densities are usually changed with changes in workpiece size and shape.

FIGURE 31-10 Design recommendations for electroplating operations.

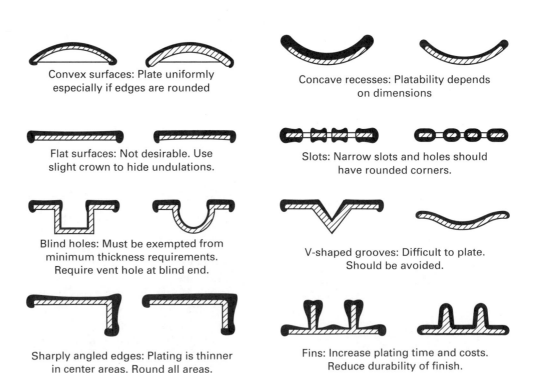

Convex surfaces: Plate uniformly especially if edges are rounded

Concave recesses: Platability depends on dimensions

Flat surfaces: Not desirable. Use slight crown to hide undulations.

Slots: Narrow slots and holes should have rounded corners.

Blind holes: Must be exempted from minimum thickness requirements. Require vent hole at blind end.

V-shaped grooves: Difficult to plate. Should be avoided.

Sharply angled edges: Plating is thinner in center areas. Round all areas.

Fins: Increase plating time and costs. Reduce durability of finish.

In the *electroforming* process, the coating becomes the final product. Metal is electroplated onto a mandrel (or mold) to a desired thickness and is then stripped free to produce small quantities of molds or other intricate-shaped sheet-type products.

ANODIZING

Anodizing is an electrochemical process, which is somewhat the reverse of electroplating, which produces a conversion type coating on aluminum that can improve corrosion and wear resistance and impart a variety of decorative effects. If the workpiece is made the anode of an electrolytic cell, instead of a plating layer being deposited on the surface, a reaction progresses inward, increasing the thickness of the hard hexagonal aluminum oxide crystals on the surface. The hardness depends on thickness, density, and porosity of the coating, which are controlled by the cycle time and applied currents along with the chemistry, concentration and temperature of the electrolyte. The surface texture very nearly duplicates the prefinishing texture, so a buffing prefinish produces a smooth lustrous coating while sand blasting produces a grainy or satiny coating.

The flow diagram in Figure 31-11 shows the anodizing process. Coating thicknesses range from 0.1 mils to 0.25 mils. Note that the product dimensions will increase, however, because the aluminum oxide coating occupies about twice the volume of the metal from which it formed.

The nature of the developed coating is controlled by the electrolyte. If the oxide coating is not soluble in the anodizing solution, it will grow until the resistance of the oxide prevents current from flowing. The resultant coating is thin, nonporous, and nonconducting, and finds use in a variety of electrical applications.

If the oxide coating is slightly soluble in the anodizing solution, dissolution competes with oxide growth, and a porous coating will be produced, where the pores provide for continued current flow to the metal surface. As the coating thickens, the growth rate decreases until it achieves steady state, where the growth rate is equal to the rate of dissolution. This condition is determined by the specific conditions of the process, including voltage, current density, electrolyte concentration, and electrolyte temperature. Sulfuric, chromic, oxalic, and phosphoric acids all produce electrolytes that dissolve oxide, with a sulfuric acid solution being the most common.

In a process variation known as *color anodizing,* a sulfuric acid bath is used to produce a layer of microscopically porous oxide that is transparent on pure aluminum and somewhat opaque on alloys. When this material is immersed in a dye solution, capillary

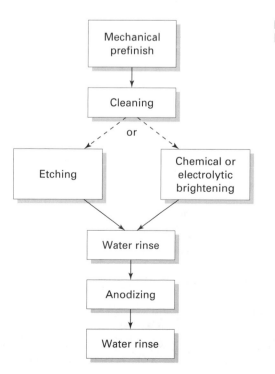

FIGURE 31-11   The anodizing process has many steps.

action pulls the dye into the pores. The dye is then trapped in place by a sealing operation, usually performed simply by immersing the anodized metal in a bath of hot water. The aluminum oxide coating is converted to a monohydrate with accompanying increase in volume. The pores close and become resistant to further staining or the leaching out of the dye.

While most people are familiar with the variety of colors in aluminum athletic goods, such as softball hats, the actual applications range from giftware, through automotive trim, to architectural use. Aluminum can be made to look like gold, copper, or brass, or take on a variety of colors with a combined metallic luster that cannot be duplicated by other methods.

If PTFE (Teflon) is introduced into the pores, coatings can be produced that couple high hardness and low friction. The porous oxide layer can also be used to enhance the adhesion of an additional layer of material, such as paint, or carry lubricant during a subsequent forming operation. Since the coating is integral to the part, subsequent operations can often be performed without destroying its integrity or reducing its protective qualities.

Anodizing can also be performed on other metals, such as magnesium, and the process is similar to the passivation of stainless steel.

ELECTROLESS PLATING

When using electroplating, it is almost impossible to obtain a uniform plating thickness on even moderately complex shapes, the platings cannot be applied to nonconductors, and a large amount of energy is required. For these reasons, a substantial effort has been directed toward the development of plating techniques that do not require an external source of electricity. These methods are known as *electroless,* or *autocatalytic plating.* Considerable success has been achieved with nickel, but copper and cobalt, as well as some of the precious metals, can also be deposited.

In the electroless process, complex plating solutions (containing metal salts, reducing agents, complexing agents, pH adjusters, and stabilizers) are brought into contact with a substrate surface that acts as a catalyst or has been pretreated with catalytic material. The metallic ion in the plating solution is reduced to metal and deposits on the surface. Since the deposition is purely a chemical process, the coatings are uniform in thickness, independent of part geometry. Unfortunately, the rate of deposition is considerably slower than with electroplating.

Probably the most popular of the electroless coatings is electroless nickel, and various methods exist for its deposition using both acid and alkaline solutions. The coatings offer good corrosion resistance, as well as hardnesses between Rockwell C 49 and 55. In addition, the hardness can be increased further to as high as Rockwell C 80 by subsequent heat treatment.

ELECTROLESS COMPOSITE PLATING

A very useful adaptation of the electroless process has been developed wherein minute particles are codeposited along with the electroless metal to produce composite material coatings. Finely divided, solid particles, with diameters between 1 and 10 μin, are added to the plating bath and deposit up to 50 vol% with the matrix. While it may appear that a large variety of materials could be codeposited, commercial applications have largely been limited to diamond, silicon carbide, aluminum oxide, and Teflon (PTFE).

Figure 31-12 shows a deposit of silicon carbide particles in a nickel alloy matrix, where the particles constitute about 25% by volume. The coating offers the same corrosion resistance as nickel, but the high hardness of the silicon carbide particles (about 4500 on the Vickers scale, where tungsten carbide is 1300 and hardened steel is about 900) contributes outstanding resistance to wear and abrasion. Since the deposition is electroless, the thickness of the coating is not affected by the shape of the part. Applications include the coating of plastic-molding dies, for use where the polymer resin contains significant amounts of abrasive filler.

MECHANICAL PLATING

Mechanical plating, also known as peen plating or impact plating, is an adaptation of barrel finishing in which coatings are produced by cold-welding soft, malleable metal powder onto the substrate. Numerous small products are first cleaned and may be given

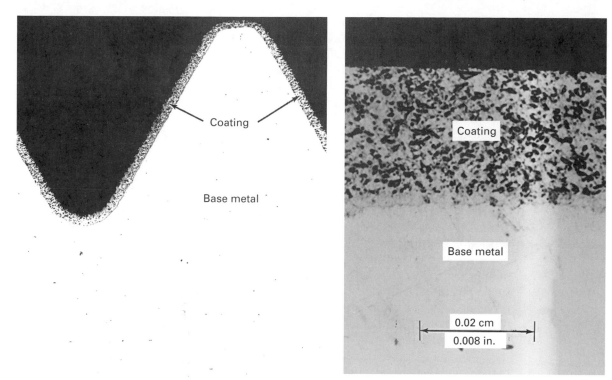

FIGURE 31-12   (*Left*) Photomicrograph of nickel carbide plating produced by electroless deposition. Notice the uniform thickness coating on the irregularly shaped product. (*Right*) High magnification cross section through the coating. *(Courtesy of Electro-Coatings Inc.)*

a thin galvanic coating of either copper or tin. They are then placed in a tumbling barrel, along with a water slurry of the metal powder to be plated, glass or ceramic tumbling media, and chemical promoters or accelerators. The media particles peen the metal powder onto the surface, producing uniform thickness deposits (possibly a bit thinner on edges and thicker in recesses—the opposite of electroplating!). Any metal that can be made into fine powder can be deposited, but the best results are obtained for soft materials, such as cadmium, tin, and zinc. Since the material is deposited mechanically, the coatings can be layered or involve mixtures with bulk chemistries that would be chemically impossible due to solubility limits. The fact that the coatings are deposited at room temperature, and in an environment that does not induce hydrogen embrittlement, makes mechanical plating an attractive means of coating hardened steels.

PORCELAIN ENAMELING

Metals can also be coated with a variety of glassy, inorganic materials that impart resistance to corrosion and abrasion, decorative color, electrical insulation, or the ability to function in high-temperature environments. Multiple coats may be used, with the first or ground coat being selected to provide adhesion to the substrate and the cover coat to provide the surface characteristics. The material is usually applied in the form of a multicomponent suspension or slurry (by dipping or spraying), which is then dried and fired. An alternative dry process uses electrostatic spraying of powder and subsequent firing. During the firing operation, which may require temperatures in the range of 800 to 8000°F, the coating materials melt, flow, and resolidify. Porcelain enamel is often found on the inner, perforated tubs of many washing machines and may be used to impart the decorative exterior on cookpots and frying pans.

■ 31.5 VAPORIZED METAL COATINGS

Vapor deposition processes can be classified into two main categories: *physical vapor deposition* (PVD) and *chemical vapor deposition* (CVD). While sometimes used as though it were a specific process, the term *PVD* applies to a group of processes in which the material to be deposited is carried physically to the surface of the workpiece. Vacuum metallizing and sputtering are key PVD processes, as are complex variations, such as ion plating. All are carried out in some form of vacuum, and most are line-of-sight processes in which the target surfaces must be positioned relative to the source. In contrast, the CVD processes deposit material through chemical reactions, and generally require significantly higher temperatures. Tool steels treated by CVD may have to be heat-treated again, while most PVD processes can be conducted below normal tempering temperatures. See Chapter 22 and Chapter 34 for additional discussions on PVD and CVD processes.

■ 31.6 CLAD MATERIALS

Clad materials are actually a form of composite in which the components are joined as solids, using techniques such as roll bonding, explosive welding, and extrusion. The most common form is a laminate, where the surface layer provides properties such as corrosion resistance, wear resistance, electrical conductivity, thermal conductivity, or improved appearance, while the substrate layer provides strength or reduces overall cost. Alclad aluminum is a typical example. Here surface layers of weaker but more-corrosion-resistant single-phase aluminum alloys are applied to a base of high-strength but less-corrosion-resistant, age-hardenable material. Aluminum-clad steel meets the same objective but with a heavier substrate, and stainless steel can be used to clad steels, reducing the need for nickel and chromium alloy additions throughout.

Wires and rods can also be made as *claddings*. Here the surface layer often imparts conductivity, while the core provides strength or rigidity. Copper-clad steel rods that can be driven into the ground to provide electrical grounding for lightning rod systems are one example.

■ 31.7 TEXTURED SURFACES

While technically not the result of a surface finishing process or operation, *textured surfaces* can be used to impart a number of desirable properties or characteristics. The types of textures that are often rolled onto the sheets used for refrigerator panels serve to conceal dirt, smudges, and fingerprints. Embossed or coined protrusions can enhance the grip of metal stair treads and walkways. Corrugations provide enhanced strength and rigidity. Still other textures can be used to modify the optical or acoustical characteristics of a material.

■ 31.8 COIL-COATED SHEETS

Traditionally, sheet metal components, such as appliance panels, have been fabricated from bare metal sheets. Pans are blanked and shaped by the traditional metal-forming operations, and the shaped panels are then finished on an individual basis. This requires individual handling and the painting or plating of geometries that contain holes, bends, and contours. In addition, there is the time required to harden, dry, or cure the applied surface finish.

An alternative approach is to apply the finish to the sheet material after rolling but before coiling. *Coatings* can be applied continuously to one or both sides of the material while it is in the form of a flat sheet. Thus the coiled material is effectively prefinished, and efforts need to be taken to protect the surface during the blanking and forming operations used to produce the final shape. Various paints have been applied successfully, as well as a full spectrum of metal coatings and platings. The sheared edges will not be coated, but if this feature can be tolerated, the additional measures to protect the surface may be an attractive alternative to the finishing of individual components.

■ 31.9 EDGE FINISHING AND BURRS

Burrs are the small, sometimes flexible projections of material that adhere to the edges of workpieces that are formed by cutting, punching, or grinding, like the exit-side burrs formed in the milled slot of Figure 31-13. Dimensionally, they are typically only 0.003 in. thick and 0.001 to 0.005 in. in height, but if not removed, they can lead to assembly failures, short circuits, injuries to workers, or even fatigue failures.

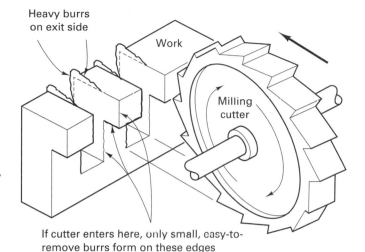

Heavy burrs
on exit side

Work

Milling
cutter

If cutter enters here, only small, easy-to-remove burrs form on these edges

FIGURE 31-13 Schematic showing the formation of heavy burrs on the exit side of a milled slot. *(From L. X. Gillespie, American Machinist, November 1985.)*

A number of different processes have been used for *burr removal,* including some discussed previously in this chapter and others presented as special types of machining. These include grinding, chamfering, barrel tumbling, vibratory finishing, centrifugal and spindle finishing, abrasive jet machining, water jet cutting, wire brushing, belt sanding, chemical machining, electropolishing, buffing, electrochemical machining, filing, ultrasonic machining, and abrasive flow machining (see Chapter 27).

Other methods may be quite specialized to the field of burr removal, such as thermal energy deburring. Here the parts are loaded into a chamber, which is then filled with a combustible gas mixture. When the gas is ignited, the short-duration wavefront heats the small burrs to as much as 6000°F, while the remainder of the workpiece rarely exceeds 300°F. The burrs are vaporized in less than 20 ms, including those in inaccessible or difficult-to-reach locations. Since the process does not use abrasive media, there is no change to any of the product dimensions. The product surfaces are rarely affected by the generated heat, and the cycle (including loading and unloading) can be repeated as many as 100 times an hour. Unfortunately, there is a thin recast layer and heat-affected zone that forms where the burrs were removed. This region is usually less than 0.001 in. thick but may be objectionable in hardened steels and highly stressed parts.

Of all of the burr-removal methods, tumbling and vibratory finishing are usually the most economical, typically costing in the neighborhood of a few cents per part. Since most of the common methods also remove metal from exposed surfaces and produce a radius on all edges, it is important that the parts be designed for deburring. Table 31-5 provides a listing of the various deburring processes, as well as the edge radius, stock loss, and surface finish that would result from removal of a "typical burr" of 0.003 in. thickness.

DESIGN TO FACILITATE OR ELIMINATE BURR REMOVAL

By knowing how and where burrs are likely to *form,* the engineer may be able to design parts to make the burrs easy to remove or even eliminate them. As shown in Figure 31-14 extra recesses or grooves can eliminate the need for deburring, since the burr produced by a cutoff tool or slot milling cutter will now lie below the surface. In this approach, one must determine whether it is cheaper to perform another machining operation (*undercutting* or grooving) or to remove the resulting burr.

Chamfers on sharp corners can also eliminate the need to deburr. The chamfering tool removes the large burrs formed by facing, turning, or boring and produces a relief for mating parts. The small burr formed during chamfering may be allowable or can easily be removed. Often, it may be preferable to give the manufacturer the freedom to use either a chamfer (produced by machining) or an edge radius (formed during the deburring operation) on all exposed corners or edges.

TABLE 31-5. Recommended Allowances for Deburring Processes[a]

Process	Edge Radius, mm (in.)	Stock Loss, mm (in.)	Surface Finish, μm AA (μin. AA[b])
Barrel tumbling	0.08–0.5 (0.003–0.020)	0–0.0025 (0–0.001)	1.5–0.5 (60–20)
Vibratory deburring	0.08–0.5 (0.003–0.020)	0–0.025 (0–0.001)	1.8–0.9 (70–35)
Centrifugal barrel tumbling	0.08–0.5 (0.003–0.020)	0–0.025 (0–0.001)	1.8–0.5 (70–20)
Spindle finishing	0.08–0.5 (0.003–0.020)	0–0.025 (0–0.001)	1.8–0.5 (70–20)
Abrasive-jet deburring	0.08–0.25 (0.003–0.010)	0–0.05 (0–0.002)[c]	0.8–1.3 (30–50)
Water-jet deburring	0–0.13 (0–0.005)(p)	0(p)	
Liquid hone deburring	0–0.13 (0–0.005)	0–0.013 (0–0.0005)	
Abrasive-flow deburring	0.025–0.5 (0.001–0.020)	0.025–0.13 (0.001–0.005)[d]	1.8–0.5 (70–20)
Chemical deburring	0–0.5 (0–0.002)	0–0.025 (0–0.001)	1.3–0.5 (50–20)
Ultrasonic deburring	0–0.05 (0–0.002)	0–0.025 (0–0.001)	0.5–0.4 (20–15)
Electrochemical deburring	0.05–0.25 (0.002–0.010)	0.025–0.08 (0.001–0.003)[e]	
Electropolish deburring	0–0.25 (0–0.010)	0.025–0.08 (0.001–0.003)[e]	0.8–0.4 (30–15)
Thermal-energy deburring	0.05–0.5 (0.002–0.020)	0	1.5–1.3(p) (60–50)
Power brushing	0.08–0.5 (0.003–0.020)	0–0.013 (0–0.0005)	
Power sanding	0.08–0.8 (0.003–0.030)[f]	0.013–0.08 (0.0005–0.003)	1.0–08 (40–30)
Mechanical deburring	0.08–1.5 (0.003–0.060)		
Manual deburring	0.05–0.4 (0.002–0.015)[g]		

[a]Based on a burr 0.08 mm (0.003 in.) thick and 0.13 mm (0.005 in.) high in steel. Thinner burrs can generally be removed much more rapidly. Values shown are typical. Stock-loss values are for overall thickness or diameter. Location A implies that loss occurs over external surfaces, B that loss occurs over all surfaces, and C that loss occurs only near edge. (p) indicates best estimate.
[b]Values shown indicate typical before and after measurements in a deburring cycle.
[c]Abrasive is assumed to contact all surfaces.
[d]Stock loss occurs only at surfaces over which medium flows.
[e]Some additional stray etching occurs on some surfaces.
[f]Flat sanding produces a small burr and no radius.
[g]Chamfer is generally produced with a small burr.
Source: L. X. Gillespie, *American Machinist,* November 1985.

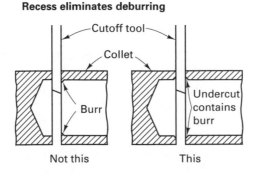

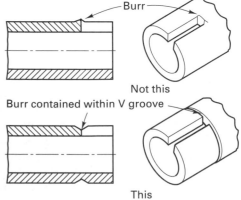

FIGURE 31-14 Designing extra recesses and grooves into a part may eliminate the need to deburr. *(From L. X. Gillespie, American Machinist, November 1985.)*

SURFACE INTEGRITY

It is important to understand that the various manufacturing and surface finishing processes each impart distinct properties to the materials which will influence the performance of the product. The achievement of satisfactory product performance obviously depends on a good design, high-quality manufacturing (including surface treatment), and proper assembly. The failure of parts in service, however, is usually the result of a combination of factors. A brief survey of features associated with surfaces and surface processing follows.

Each of the various machining processes produce characteristic surface textures (roughness, waviness, and lay) on the workpieces. See Chapter 10 and Table 31-6 for summary of surface integrity effects produced by various processes. In addition, the various processes tend to produce changes in the chemical, physical, mechanical, and metallurgical properties on or near the surfaces that are created. For the most part, these changes are limited to a depth of 0.005 to 0.050 in. below the surface. The effects can be beneficial or detrimental, depending on the process, material, and function of the product.

Machining processes (both chip forming and chipless) induce plastic deformation into the surface layer, as shown in Figure 31-15. The cut surfaces are generally left with tensile residual stresses, microcracks, and a hardness that is different from the bulk material. Processes such as EDM and laser machining leave a layer of hard, recast metal on the surface that usually contains microcracks. Ground surfaces can have either residual tension or residual compression, depending on the mix between chip formation and plowing or rubbing during the grinding operation. If sufficient heat is generated, phase transformations can occur in the surface and subsurface regions.

FIGURE 31-15   Plastic deformation in the surface layer after cutting. *(Kruszynski, B. W. and C. W. Cuttervelt*, Advanced Manufacturing Engineering, *Vol. 1, 1989.)*

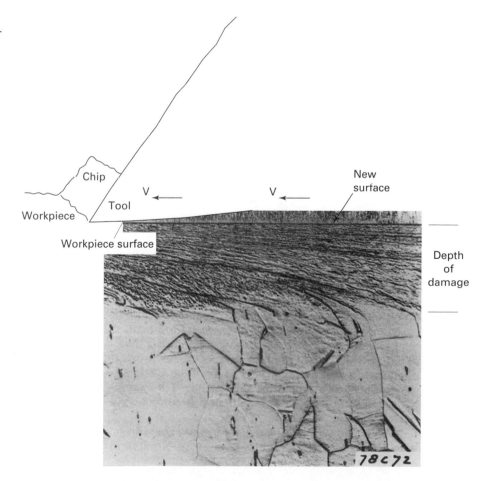

TABLE 31-6. Surface Integrity Effect Observed in Various Metal Removal Processes (Bellows and Koster 1972, cited in Drozda and Wick 1983, and Field and Kahles 1971).

Process \ Effects	Depth of Plastic Deformation	Plastically Deformed Debris, BUE	Hardness Alteration	Cracks	Residual Stresses	Intergranular Attack	Metallurgical Transformation	Heat Affected Zone
Turning, Milling, Drilling	✓	✓	✓	✓	✓ typically tensile, ✓ compressive for hard materials (Matsumoto 1986) & high-speed finishing with 0° rake angle (Sadat 1990)	○	✓	✓
Grinding	✓	✓	✓	✓	✓ compressive to tensile	○	✓	✓
EDM	○	○	✓	✓	✓	○	✓	✓
ECM,	○	○		✓ ECM		✓		
Chemical Milling	○	○	✓ (softening)	○ Chem. Mill	○		✓	○
Laser Beam Machining	○	○	no data	✓	✓	○	✓	✓

✓ = observed   ○ = not observed

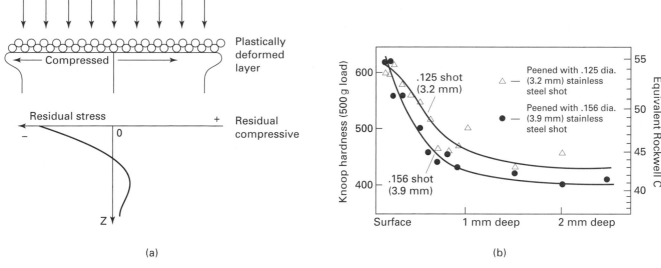

(a)   (b)

FIGURE 31-16 (a) Mechanism for formation of residual compressive stresses in surface by cold plastic deformation. (Shot peening). (b) Hardness increased in surface due to shot peening.

Processes such as *roller burnishing* (described in Chapter 19) produce a smooth surface with compressive residual stresses. Shot peening (and tumbling) can increase the hardness in the surface and introduce a residual compressive stress as shown in Figures 31-16a and 31-16b. Welding processes produce tensile residual stresses as the deposited material shrinks upon cooling. Similar shrinkage occurs in castings, but the resulting stresses may be complex due to the variation of shrinkage or the lack of restraint. Tensile stresses on the surface can often be offset by a subsequent exposure to shot peening or tumbling.

In summary, the surface and subsurface regions of a material can be significantly altered due to (1) plastic strain or plastic deformation, (2) high temperatures, (3) differential expansions or contractions due to temperature changes or variations, and (4) chemical reactions.

To illustrate the complex nature of surface effects, consider Figure 31-17 which shows the depth of "surface damage" due to machining as a function of the rake angle of the tool. To increase the cutting speed (and thereby increase the rate of production), an engineer might change from a high-speed tool steel cutter with a large rake angle (such as 30°) to a carbide tool with a zero rake. While the resulting surface finish may be similar, the depth of "surface damage" is doubled. Failures may occur in service, whereas previous parts had performed quite admirably.

FIGURE 31-17 The depth of damage to the surface of a machined part increases with decreasing rake angle of the cutting tool.

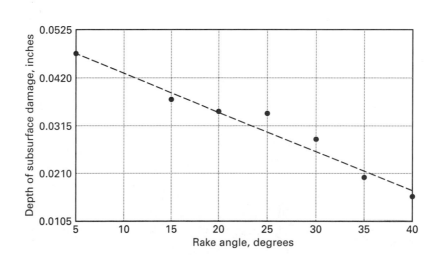

■ 31.10 FATIGUE FAILURES ASSOCIATED WITH SURFACE INTEGRITY

Fatigue failure occurs as the result of repeated loading at some point typically below the yield strength of the material. Fatigue failures have been shown to almost always "nucleate on or near the surface of a component" (Field and Kahles 1971). Fine surface cracks begin at discontinuities (such as microcracks, grooves, ridges, cavities, machining marks, imbedded particles, etc.) at the surface, and the cracks propagate with repeated cyclic loads.

Tensile residual stresses in the altered surface layer have an additive effect on the applied stresses in the component. This means that tensile residual stresses in the material add to external stresses to the component, reducing its fatigue strength. Alternatively, as shown in Figure 31-18, compressive residual stresses subtract from tensile external stresses, and since tensile stresses are those ultimately responsible for fatigue failure, the fatigue strength of the material is increased.

Figure 31-18 shows how residual stresses couple with applied stresses to affect product performance. Suppose that a round beam has a load applied to it so that it is bent while rotating. At the top of the rotation, the surface is in tension, and at the bottom, it is in compression. The result is a condition of cyclic fatigue and the likelihood of a service life limited by fatigue failure. If the part is roller burnished or shot peened, the compressive residual stress pattern of the middle figure is added to the applied stresses, producing the net pattern shown at the bottom. The net effect is a lowering of the peak tensile stress experienced by the surface and a related extension in fatigue life. The specific results will depend on the details of the process. For *shot peening*, the key variables include shot size, shot velocity, exposure time, distance between the nozzle and the surface, and the angle of impact.

Figure 31-19 presents the results of a study in which specimens were prepared by milling and turning and then either polished, shot-peened, or roller-burnished. If an applied stress between 41,000 and 42,000 psi is experienced in a fatigue application, the difference in fatigue life between a milled specimen and one that has been milled and roller burnished is 610,000 cycles (90,000 cycles as opposed to 700,000 cycles). In essence, roller burnishing serves to induce a sevenfold extension to the fatigue life of the product. Similar results have been observed in the resistance to stress-corrosion cracking.

FIGURE 31-18 (*Top*) A cantilever-loaded (bent) rotating beam, showing the normal distribution of surface stresses (i.e., tension at the top and compression at the bottom). (*Center*) The residual stresses induced by roller burnishing or shot peening. (*Bottom*) Net stress pattern obtained when loading a surface-treated beam. The reduced magnitude of the tensile stresses contributes to increased fatigue life.

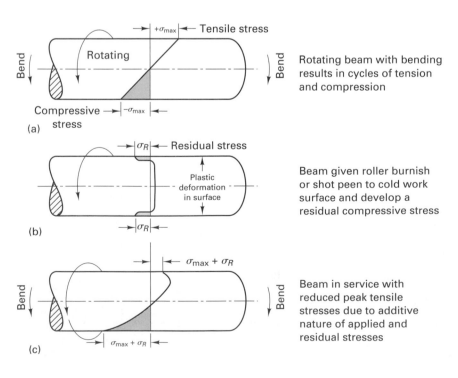

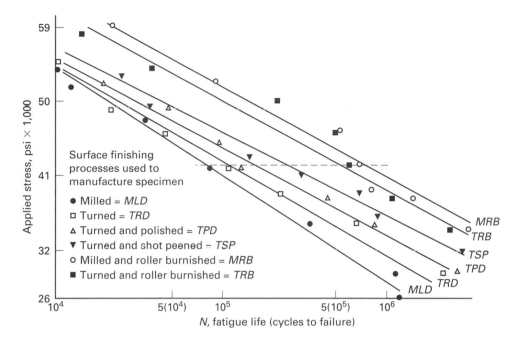

FIGURE 31-19 Fatigue life of rotating beam 2024-T4 aluminum specimens with a variety of surface finishing operations. Note the enhanced performance that can be achieved by shot peening and roller burnishing.

As the data above shows, both the designer and the manufacturer need to be aware of the effects that manufacturing processes can have on the performance of a product. Maintaining the proper sequence of operations may be as important to the surface properties as the selection of the processes and control of the operating parameters.

■ KEY WORDS

abrasive cleaning	chemical conversion	electroforming	hot-dip coating	sand blasting
acid pickling	coating	electroless composite	mechanical cleaning	shot peening
alkaline cleaning	chemical vapor	plating	mechanical plating	solvent cleaning
anodizing	deposition	electroless plating	media	spray painting
barrel finishing	chemical cleaning	electroplating	paint	surface integrity
barrel burnishing	chromate	electropolishing	phosphate	textured surfaces
belt sanding	cladding	electrostatic	porcelain enameling	tumbling
buffing	coating	deposition	powder coating	ultrasonic cleaning
burr removal	coil-coated sheets	fatigue	prime coat	vacuum metallizing
case hardening	color anodizing	finishing compounds	residual stresses	vapor degreasing
centrifugal barrel	dipping	hard facing	roller burnishing	vibratory finishing
tumbling	electrocoating	hard chromium plate	roughness	wire brushing

■ REVIEW QUESTIONS

1. What are some possible objectives of surface modification processes?
2. What are some of the factors that should be considered when selecting a surface modification process?
3. How is the surface and its integrity altered by the process of metal cutting?
4. What are some of the sources of foreign material on the surface of manufactured products?
5. What are some the common abrasive media used in blasting or abrasive cleaning operations?
6. What types of quantities and part sizes are most attractive for barrel finishing operations?
7. Describe why there might be an optimum fill level in barrel finishing; an optimum rotational speed.
8. Describe the primary differences between barrel finishing and vibratory finishing.

9. What are some of the attractive features of synthetic abrasive media?
10. What are some of the possible functions of the compounds that are used in abrasive finishing operations?
11. What is the ideal geometry for belt sanding? What types of geometric features would be difficult for belt sanding?
12. How is electropolishing different from electroplating?
13. What are some of the mechanisms of alkaline cleaning, and what types of soils can be removed?
14. What types of surface contaminants cannot be removed by solvent cleaning?
15. In view of its many attractive features, why has vapor degreasing become an unattractive process?
16. What is the primary type of surface contaminant removed by acid pickling?
17. How are burrs made by the milling process? Figure 31-13.

18. Describe the process of thermal energy deburring.
19. What is the difference between coating and cladding operations?
20. What are some of the reasons that paints may be specified for manufactured items?
21. What are some of the functions of a prime coat in a painting operation? What features are desired in the final coat?
22. What produces atomization and propulsion in airless spraying?
23. What features make industrial robots attractive for spray painting? See chapters 32 and 43.
24. What are some of the attractive features of electrostatic spraying?
25. Why would it be difficult to apply electrostatic spray painting to products made from wood or plastic?
26. What are some of the metal coatings that can be applied by the hot-dip process?
27. What are the two most common types of chemical conversion coatings?
28. How can nonconductive materials such as plastic be coated by electroplating?
29. What are the attractive properties of hard chrome plate?
30. What are some of the common process variables in an electroplating cell?
31. Why is it difficult to mix parts of differing size and shape in an automated electroplating system?
32. How is electroforming different from electroplating?
33. When anodizing aluminum, what features determine the thickness of the resulting oxide when the oxide is not soluble in the electrolyte? When it is partially soluble?
34. What produces the various colors in the color anodizing process?
35. What are some of the attractive features of electroless plating?
36. What types of particulate composites can be deposited by electroless plating?
37. What is mechanical plating?
38. What are some of the attractive properties of a porcelain enamel coating?
39. What deburring processes are available that were not described in this chapter?
40. Do all machining processes leave a residual stress?
41. Why would the depth of damage (i.e., plastic deformation) increase with negative rake angle cutting tools?
42. Why do fatigue failures almost always occur at the surface?
43. What is a "stress raiser"?
44. What types of surface features or surface modifications result from machining-type processes?
45. Why might processes that produce residual compressive stresses on product surfaces be attractive for mechanically loaded products?

■ PROBLEMS

1. Fishermen are among the most superstitious people in the world, and their superstitions affect the type of equipment that they use. As a result, hook manufacturers generally offer their products in a wide range of colors and finishes. Your company manufactures a range of hooks from AISI 1080 carbon steel wire, forming them to precision shape (eye, bends, barbs, and point) and then heat treating them by a quench and temper treatment.

   Consider the size and shape of the product and the various properties that are required. The hooks must be strong enough to resist bending, but not so brittle that they might break. They must be corrosion-resistant to both fresh and salt water, and the desired appearance must be provided without fouling the point or the barbs. If the surface is applied before heat treatment, it must endure that process and maintain its appearance. If it is applied afterward, it cannot weaken or embrittle the hook.
   a. Of the various surface modification processes, which ones might be attractive for such an application? (*Note:* Make sure that the process is appropriate! For example, barrel plating would probably produce a hopelessly snarled mass of wires!)
   b. For each of the possible processes, describe the advantages, limitations, possible colors or finishes, and relative cost.

2. Select one or more of the following products (as directed) and recommend a surface treatment or coating. Consider the appropriateness of the technique to the size, shape, quantity, and material. Cite the specific features that make the recommended treatment the most attractive. What, if any, are the primary limitations or production concerns?
   a. The exterior housing for the motor and drive unit of a chain saw that has been made as a magnesium alloy die casting.
   b. Large quantities of steel bolts that are intended for use in outdoor construction. They have been fabricated from 4140 steel, have a shank diameter of $\frac{1}{2}$ in., and have been quenched and tempered to a final hardness of Rockwell C 45.
   c. The handle of a household utility knife (retractable-blade cutter) has been made as a two-part zinc die casting.
   d. The scoop portion of an inexpensive ice cream scoop that has been made as a zinc die casting.
   e. A decorative handle for a kitchen cabinet that is made as a zinc die casting.
   f. The exterior of an office filing cabinet that has been made from low-carbon steel sheet.
   g. A high-quality combination wrench (open-end and box-end) that has been forged from 4147 steel bar stock.
   h. Case of a moderately priced wristwatch that has been fabricated from yellow brass (to have a gold appearance).
   i. Tubular frame of a lightweight bicycle that has been made from age-hardened aluminum.
   j. The basket section of a grocery-store shopping cart that has been fabricated from welded steel mesh.
   k. The exterior of an automobile muffler to be fabricated from steel sheet. Describe how the coating treatment might best be integrated into the fabrication sequence.
   l. An inexpensive interior door knob that has been fabricated from deep-drawn cartridge brass sheet.
   m. High-quality steel sockets for a socket-wrench set. These have been forged from AISI 4145 bar stock and subsequently heat treated by a quench and temper process.
   n. Refrigerator door panels that have been fabricated from textured AISI 1010 steel sheet.
   o. The interior and exterior surfaces of a 1000-gallon water storage tank that is fabricated by welding 5000 series aluminum plates. The water will be held at room temperature and is intended for human consumption.
   p. A standard office paper clip.
   q. A flashlight case that has been fabricated from deep-drawn yellow brass sheet.

r. High-speed drill bits that have been fabricated from Ml tool steel.

s. Injection-molded ABS plastic wheel covers for cars that are intended to look like chrome-plated metal.

t. Inexpensive household scissors that have been cast from gray cast iron.

u. The blade of a high-quality screwdriver that has been forged from AISI 1053 steel and quenched and tempered to Rockwell C 55.

v. The exterior of high-quality, thick-walled cast aluminum cookware.

w. A bathroom sink basin made from deep-drawn 1008 steel sheet.

x. The body section of a child's toy wagon that has been deep-drawn from 1008 steel sheet.

*C*hapter 31 **CASE STUDY**

Burrs on Tonto's Collar

The Tonto Company requires 25,000 collars as shown in Figure CS-31. They are to be made from AISI 304 (18-8) stainless steel by machining on a screw machine using transfer and slotting attachments. The slotting process is creating large, thick burrs on the surface where the cutter exits the workpiece, as shown in sections A-A' (*left*). These burrs are causing major problems in an assembly process and you have been assigned the problem. The part designer has recommended a re-designed part to eliminate the undercut and make the burrs more accessible; see view A-A' (*right*). The burrs

on the internal surface will still remain, however, and are still inaccessible.

1. What are the methods that are commonly used in industry to remove burrs?

2. Outline the sequence of steps needed to produce this part on a screw machine from $\frac{5}{8}$-in. stainless steel bar that would serve to eliminate the burrs. It is helpful to show the part progression.

3. Determine two other practical ways of making these collars. Estimate the cost of these alternative methods versus the screw machine method.

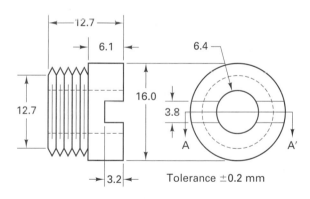

Tolerance ±0.2 mm

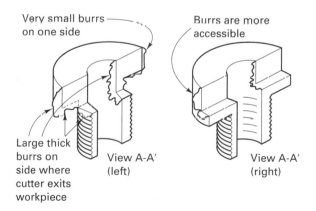

CHAPTER 32

MANUFACTURING AUTOMATION

■ 32.1 INTRODUCTION

The term *automation* has many definitions. Apparently, it was first used in the early 1950s to mean automatic handling of materials, particularly equipment used to unload and load stamping equipment. It has now become a general term referring to services performed, products manufactured and inspected, information handling, materials handling, and assembly, all done automatically (i.e., as an automatic operation without human involvement).

In 1962, Amber & Amber presented their *Yardstick for Automation*, which is based on the concept that all work requires energy and information, and certain functions must be provided by workers or machines. Whenever a machine replaces a human function or attribute, it is considered to have taken an "order" of automaticity. The chart that they developed has been updated (see Table 32-1). Notice that each order of automation is tied to the human attribute that is being replaced (mechanized or automated) by the machine. Therefore, the A(0) level of automation, in which no human attribute was mechanized, covers the Stone

TABLE 32-1. Yardstick for Automation

Order of Automation	Human Attribute Replaced	Examples
A(0)	*None:* lever, screw, pulley, wedge	Hand tools, manual machine
A(1)	*Energy:* muscles replaced with power	Powered machines and tools, Whitney's milling machine
A(2)	*Dexterity:* self-feeding	Single-cycle automatics
A(3)	*Diligence:* no feedback but repeats cycle automatically	Repeats cycle; open-loop numerical control or automatic screw; transfer lines
A(4)	*Judgment:* positional feedback	Closed loop; numerical control; self-measuring and adjusting
A(5)	*Evaluation:* adaptive control; deductive analysis; feedback from the process	Computer control; model of process required for analysis and optimization
A(6)	*Learning:* from experience	Limited self-programming; some artificial intelligence (AI); expert systems
A(7)	*Reasoning:* exhibits intuition; relates causes and effects	Inductive reasoning; advanced AI in control software
A(8)	*Creativeness:* performs design unaided	Originality
A(9)	*Dominance:* supermachine, commands others	Machine is master (Hal in *2001, A Space Odyssey*)

Source: G. Amber & P. Amber, *Anatomy of Automation*, Prentice Hall, Englewood Cliffs, N.J., 1962. Used by permission, modified by Black.

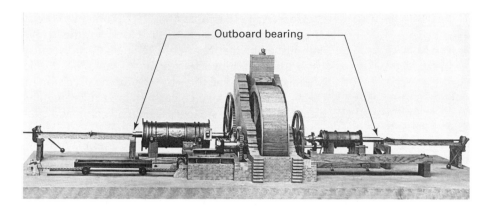

Outboard bearing

FIGURE 32-1 Model of Wilkinson's horizontal boring machine. *(British Crown Copyright, Science Museum, London.)*

Age through the Iron Age. Two of the earliest machine tools were the crude lathes the Etruscans used for making wooden bowls around 700 B.C. and the windlass-powered broach for machining of grooves into rifle barrels used over 300 years ago.

If the first industrial revolution is tied to the machines that made cotton, the second industrial revolution can be tied to the development of powered machine tools, dating from 1775 when the energetic, "iron-mad" John Wilkinson constructed a horizontal boring machine in England for machining internal cylindrical surfaces, such as piston-type pumps. In Wilkinson's machine, a model of which is shown in Figure 32-1, the boring bar extended through the casting to be machined and was supported at its outer end by a bearing. Modern boring machines still employ this basic design. Wilkinson reported that his machine could bore a 57-in.-diameter cylinder to such accuracy that nothing greater than an English shilling (about $\frac{1}{16}$ in. or 1.59 mm) could be inserted between the piston and the cylinder. This machine tool made Watt's steam engine a reality. At the time of his death, Wilkinson's industrial complex was the largest in the world.

The next machine tool was developed in 1794 by Henry Maudsley. It was an engine lathe with a practical slide tool rest. This machine tool, shown in Figure 32-2, was the forerunner of the modern engine lathe. The lead screw and change gear mechanism, which enabled threads to be cut, were added about 1800. The first planer was developed in 1817 by Richard Roberts in Manchester, England. Roberts was a student of Maudsley, who also had a hand in the career of Joseph Whitworth, the designer of screw threads. Roberts also added back gears and other improvements to the lathe. The first horizontal milling machine is credited to Eli Whitney in 1818 in New Haven, Connecticut. The development of machine tools that not only could make specific products but could also produce other machines to make other products was fundamental to the second industrial revolution.

While early work in machine tools and precision measurement was done in England, the earliest attempts at interchangeable manufacturing apparently occurred almost simultaneously in Europe and the United States with the development of *filing jigs*, with which duplicate parts could be hand-filed to substantially identical dimensions. In 1798, Eli Whitney, using this technique, was able to obtain and eventually fulfill a

FIGURE 32-2 Maudsley's screw-cutting lathe *(British Crown Copyright, Science Museum, London.)*

contract from the United States government to produce 10,000 army muskets, the parts of each being interchangeable. However, this truly remarkable achievement was accomplished primarily by painstaking handwork and not by specified machines.

Joseph Whitworth, starting about 1830, accelerated the use of Wilkinson's and Maudsley's machine tools by developing precision measuring methods. Later he developed a measuring machine using a large micrometer screw. Still later, he worked toward establishing thread standards and made plug-and-ring gauges. His work was valuable because precise methods of measurement were the prerequisite for developing interchangeable parts, a requirement for later mass production.

The next significant machine tool was the drill press with automatic feed, developed by John Nasmyth, another student of Maudsley, in 1840 in Manchester, England. Surface-grinding machines came along about 1880, and the era was completed with the development of the bandsaw blades that could cut metal. In total, there were eight basic machine tools in the first industrial revolution for machining: lathe, milling machine, drill press, broach, boring mill, planer (shaper), grinder, and saw. The first factories were developed so that power could be added to drive the machines. This is the A(1) level of automation. See Chapters 41–43 for further discussion on the industrial revolution.

A(2) LEVEL IS SINGLE CYCLE AUTOMATICS

The A(2) level of automation was clearly delineated when machine tools became single-cycle, self-feeding machines displaying *dexterity*. Many examples of this level of machine are given in this text. They exist in great numbers in many factories today as milling, drilling, and turning machines. The A(2) level of machine can be loaded with a part and the cycle initiated by the worker. The machine completes the processing cycle and stops automatically. *Mechanization* refers to the first and second orders of automaticity, which includes semiautomatic machines. Virtually all of the machine tools described in previous chapters are A(2) machines. The machines used in manned manufacturing cells described in Chapters 41–43 are basically A(2) or A(3) machines. Some examples of A(2) machines are shown in Figure 32-3.

A(3) IS REPEAT CYCLE AUTOMATICS

The A(3) level of automation requires that the machine be *diligent*, or repeat cycles automatically. These machines are open loop, meaning that they do not have *feedback* and are controlled by either an internal fixed program such as a cam, or are externally programmed with a tape, programmable logic controller (PLC) and more recently, a computer. Figure 32-4 provides some examples of A(3) machines that include transfer lines.

A(3), A(4), and A(5) levels are basically superimposed on A(2)-level machines, which must be A(1) by definition. The A(3) level also includes robots, numerical control (NC) machines that have no feedback, and many special-purpose machine tools (see Figure 32-4).

Automation as we know it today begins with the A(3) level. In recent years, this level has taken on two forms: *hard (or fixed-position) automation* and *soft (or flexible or programmable) automation*. Instructions to the machine, telling it what to do, how to do it, and when to do it, are called the *program*. In hard-automation transfer lines, the programming consists of cams, stops, slides, and hard-wired electronic circuits using relay logic. A widely used example of the A(3) level of automation is the automatic screw machine. If the machine is programmed with a tape, a programmable controller (PC), a handheld control box, microprocessor, or computers, control instructions are easily changed, with the software making the system or device much more adaptable.

Transfer Machines. A *transfer machine* is an automated flow line. Workpieces are automatically transferred from station to station, from one machine to another. Operations are performed sequentially. Ideally, work stations perform the operation(s) simultaneously on separate workpieces, with the number of parts equal to the number of stations. Each time the machine cycles, a part is completed. Figure 32-5 shows an example of a transfer line.

Transferring is usually accomplished by one or more of four methods. Frequently, the work is pulled along supporting rails by means of an endless chain that moves intermittently as required. Alternately, the work is pushed along on continuous rails by air or hydraulic

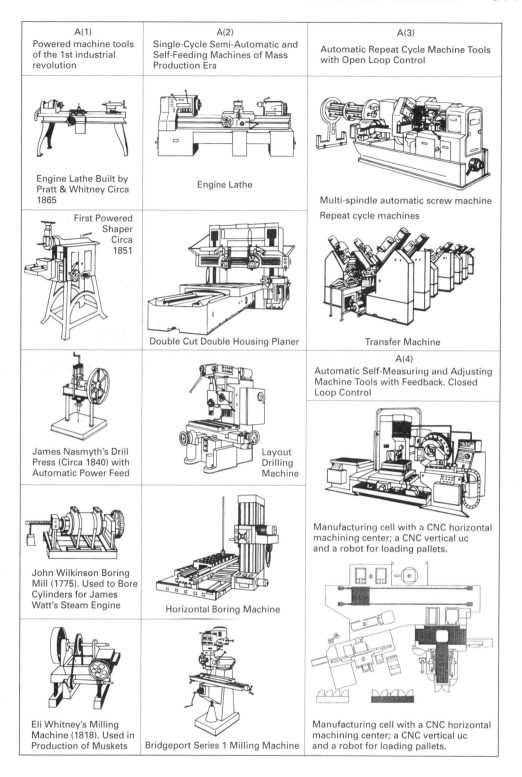

A(1)	A(2)	A(3)
Powered machine tools of the 1st industrial revolution	Single-Cycle Semi-Automatic and Self-Feeding Machines of Mass Production Era	Automatic Repeat Cycle Machine Tools with Open Loop Control

Engine Lathe Built by Pratt & Whitney Circa 1865

Engine Lathe

Multi-spindle automatic screw machine

Repeat cycle machines

First Powered Shaper Circa 1851

Double Cut Double Housing Planer

Transfer Machine

James Nasmyth's Drill Press (Circa 1840) with Automatic Power Feed

Layout Drilling Machine

A(4)
Automatic Self-Measuring and Adjusting Machine Tools with Feedback. Closed Loop Control

John Wilkinson Boring Mill (1775). Used to Bore Cylinders for James Watt's Steam Engine

Horizontal Boring Machine

Manufacturing cell with a CNC horizontal machining center; a CNC vertical uc and a robot for loading pallets.

Eli Whitney's Milling Machine (1818). Used in Production of Muskets

Bridgeport Series 1 Milling Machine

Manufacturing cell with a CNC horizontal machining center; a CNC vertical uc and a robot for loading pallets.

FIGURE 32-3 Machine tools of the first industrial revolution A(1) and the mass production era A(2) with examples of A(3) open loop level and A(4) closed loop level.

pistons. A third method, restricted to lighter workpieces, is to move them by an overhead chain conveyor, which may lift and deposit the work at the machining stations.

A fourth method often is employed when a relatively small number of operations, usually three to ten, are to be performed. The machining heads are arranged radially around a rotary indexing table, which contains fixtures in which the workpieces are mounted (Figure 32-6). The table movement may be continuous or intermittent. Face-milling operations sometimes are performed by moving the workpieces past one or more vertical-axis heads at the required feed rate. Such circular configurations have the advantage of being compact and of permitting the two pieces to be loaded and unloaded at a single station without having to interrupt the machining.

ORDER OF AUTOMATICITY	HUMAN ATTRIBUTE MECHANIZED	DISCUSSION	EXAMPLES
A (3) Automatic repeat cycle or Open loop control	DILIGENCE Carries out routine instructions without aid of man Open end or nonfeedback	All automatic machines Loads, processes, unloads, repeats– System assumed to be doing okay. Probability of malfunctions negligible Obeys fixed internal commands or external program	• Record player with changer • Automatic screw machine • Bottling machines • Clock works • Donut maker • Spot welder • Engine production lines • Casting lines • Newspaper printing machines • Transfer machines
A (4) Self-measuring and adjusting feedback or closed loop systems	JUDGMENT Measures and compares result (output) to desired size or position (input) and adjusts to minimize any error	Self-adjusting devices Feedback from product position, size, velocity, etc. 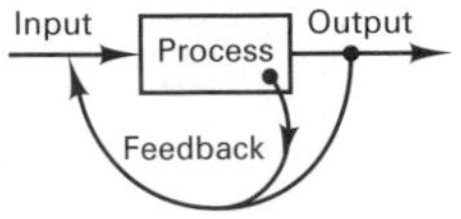 Input → Process → Output; Feedback Multiple loops are possible	• Product control • Can filling • N/C machine tool with position control • Self-adjusting grinders • Windmills • Thermostats • Waterclock • Fly ball governor on steam engine
A (5) Adaptive or computer control: Automatic cognition; Fuzzy logic control	EVALUATION Senses multiple factors on process performance, evaluates and reconciles them Use mathematical algorithms or fuzzy logic	Process performance must be expressed as equation 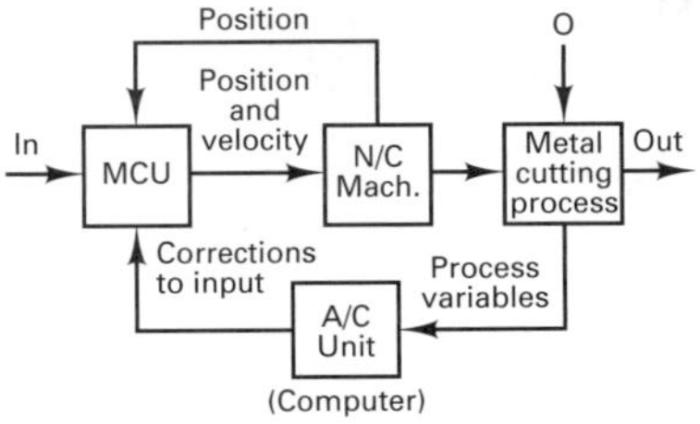 In → MCU → N/C Mach. → Metal cutting process → Out; Position; Position and velocity; Corrections to input; A/C Unit (Computer); Process variables	CN/C machine with A/C capability Maintaining pH level Turbine fuel control Maintaining constant cutting force
A (6) Expert systems or neural networks or limited self-programming	LEARNING BY EXPERIENCE	Subroutines are a form of limited self-programming Trial and error sequencing Develops history of usage	Phone circuits Elevator dispatching

FIGURE 32-4   Yardstick for Automation: Levels A(3) to A(6).

Means must be provided for positioning the workpieces correctly as they are transferred to the various stations. One method is to attach the work to carrier pallets or fixtures that contain locating holes or points that mate with retracting pins or fingers at each work station. Excellent station-to-station precision is obtained as the fixtures thus are located and then clamped in the proper positions. However, carrier fixtures are costly and must be identical. When it is possible, they are eliminated, and the workpiece is transported between the machines on rails, which locate the parts by self-contained holes or surfaces. This procedure eliminates the labor required for fastening the workpieces to the carrier pallets as well as the pallets.

In the design of large transfer machines, the matter of the geometric arrangement of the various production units must be considered. Whether transfer fixtures or pallets must be used is an important factor. These fixtures and pallets usually are quite heavy. Consequently, when they are used, a closed rectangular arrangement is often employed so that the fixtures are returned to the loading point automatically. If no fixtures or pallets are required, straight-line configurations can be employed. Whether pallets or fixtures must be used depends primarily on the degree of precision needed as well as on the size, rigidity, and design of the workpieces. If no transfer pallet or fixture is to be used, locating bosses or points must be designed or machined into the workpiece.

The matter of tool wear and replacement is of great importance when a large number of operations are incorporated in a single production unit. Tools must be replaced before they

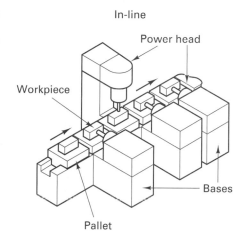

FIGURE 32-5 A classic example of an inline transfer machine with inset showing two workpieces mounted on pallets for transfer from machine to machine.

FIGURE 32-6 Rotary transfer lines also called dial index machines typically come in 6–10 stations. The power heads can be vertical or horizontal and can hold multiple tools. The workpiece on the right is mounted in a specialty designed fixture (one of 10) which must be identical.

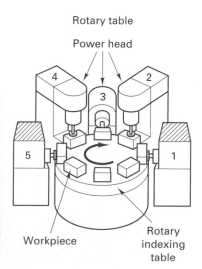

become worn and produce defective parts. Transfer machines often have more than 100 cutting tools. If the entire complex machine had to be shut down each time a single tool became dull and had to be replaced, overall productivity would be very low. This is avoided by designing the tooling so that certain groups have similar tool lives, monitoring tool thrust and torque and shutting down the machine before the tooling has deteriorated. All the tools in the affected group can be changed so that repeated shutdowns are not necessary. If the transfer machine is equipped with ac drives, diagnostic feedback information of individual processes is easily accomplished. Programmable drives replace feedboxes, limit switches, and hydraulic cylinders and eliminate changing belts, pulleys, or gears to change feed rates and depths of cut, thus making the system much more flexible.

Methods have been developed for accurately presetting tools and for changing them rapidly. Tools can be changed in a few minutes, thereby reducing machine downtime. Increasingly, tools are preset in standard quick-change holders with excellent accuracy, often to within 0.0002 in. (0.005 mm).

In large transfer lines, to prevent entire machines from being shut down when one or two stations become inoperative, the individual machines are grouped in sections with 10 to 12 stations per section. A small amount of buffer storage (*banking* of workpieces) is provided between the sections. This permits production to continue on all remaining sections for a short time while one section is shut down for tool changing or repair.

The most significant problem deals with designing the line itself so that it operates efficiently as a whole. Transfer machines or, for that matter, any system wherein a number of processes are connected sequentially, will require that the line be *balanced*. *Line balancing* means that the process time at each station must be the same, with the total nonproductive time for all other stations minimized. Theoretically, there will be one station that will have the longest time, and this station will control the cycle time for all the stations. Computer algorithms have been developed to deal with line balancing, which, when the line is very large, becomes a rather complex problem. Ford Motor has a single transfer line comprising 15 transfer machines, two assembly machines, part washers, gages, and inspection equipment. One of the transfer machines has 30 stations just for performing rough milling on cylinder heads. The machines perform a variety of drilling, milling, rough and finish boring, reaming, and end milling operations. This line produces over 200,000 engine blocks and cylinder heads per year (100 cylinder heads per hour).

Transfer machines are usually A(3)-level machines but can be A(4)-, or A(5)-level machines, depending on whether they have the built-in capacity for sensing when corrective action is required and how such corrections are made. Sensing and feedback control systems are essential requirements for the fourth level and all higher levels of automation.

Many machines have built-in *feedforward* devices. This means that the system takes information from the input side of the process rather than the output side of the process and uses that information to alter the process. For example, the temperature of the billet as it enters a hot-rolling or hot-extrusion process can be sensed and used as feedforward information to alter the process parameters. This would be an A(5) or adaptive control example. Sensors can be located in three positions, ahead of the process, in the process, or on the output side of the process. The feedforward concept can also be applied at the A(4) level. Suppose that you have a transfer line that processes flywheel housings from castings that are similar but will have slight differences in size, which will alter depth of cut during different machining operations. When the housings are fed into the machine, a sensing device contacts the casting and determines the variation from the nominal size. The sensing and feedforward system then selects the proper tooling for the housing and adjusts cutting parameters accordingly.

It is common practice to equip transfer machines with automated inspection stations or probing heads that determine whether the operation was performed correctly and detect whether any tool breakage has occurred that might cause damage in subsequent operations. For example, after drilling, a hole is checked to make certain that it is clear prior to tapping.

Automation and transfer principles are also used very successfully for assembly operations. In addition to saving labor, automatic testing and inspection can be incorporated into such machines at as many points as desired. Such in-process inspection should be used

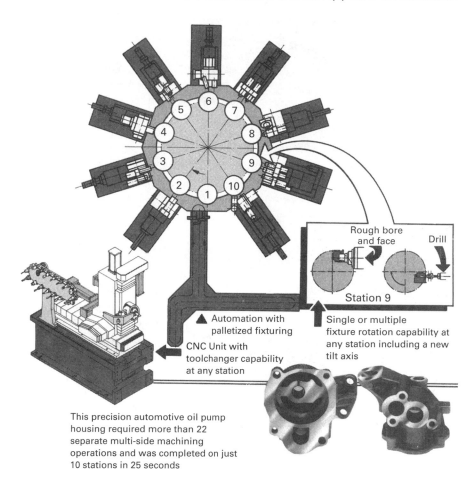

FIGURE 32-7   Horizontal, rotary transfer machine with CNC units in place of conventional slide units. *(Courtesy of Jestadt.)*

Rough bore and face
Drill
Station 9

▲ Automation with palletized fixturing

CNC Unit with toolchanger capability at any station

Single or multiple fixture rotation capability at any station including a new tilt axis

This precision automotive oil pump housing required more than 22 separate multi-side machining operations and was completed on just 10 stations in 25 seconds

to prevent defects from being made rather than finding defects after they are made. This assures superior quality. When defective assemblies are simply discovered and removed for rework or scrapped, the cause of the problem is not necessarily corrected.

In many cases, some manual operations are combined with some automatic operations. For example, one transfer machine for assembling steering knuckle, front-wheel hub, and disk-brake assemblies has 16 automatic and five manual work stations. As with manufacturing operations, automatic assembly often can be greatly improved through proper part design.

Figure 32-7 shows a dial-indexing transfer machine that has CNC machines replacing slide units. In addition to CNC units with tool changers, this system features multiaxis fixture positioning, palletized automation, in-process gaging, size control, fault diagnostics, and excellent process capability, so this machine has A(4) capability.

■ 32.2 NUMERICAL CONTROL AND THE A(4) LEVEL OF AUTOMATION

A(4) LEVEL HAS FEEDBACK

A(4) level of automation required that *human judgment* be replaced by a capability in the machine to measure and compare results with desired position or size and adjustments to minimize errors. This is *feedback* or closed-loop control. The first numerically controlled machine tool was developed in 1952. It had positional feedback control and is generally recognized as the first A(4) machine tool. By 1958, the first NC *machining center* was being marketed by Kearney and Trecker. A machining center was a compilation of many machine tools capable of performing many processes: in this case, milling, drilling, tapping, and boring (see Figure 32-8). It can automatically change tools to give it greater flexibility. Almost from the start, computers were needed to help program these machines. Within 10 years, NC machine tools became computer numerical control (CNC) machine tools with an onboard microprocessor and could be programmed directly.

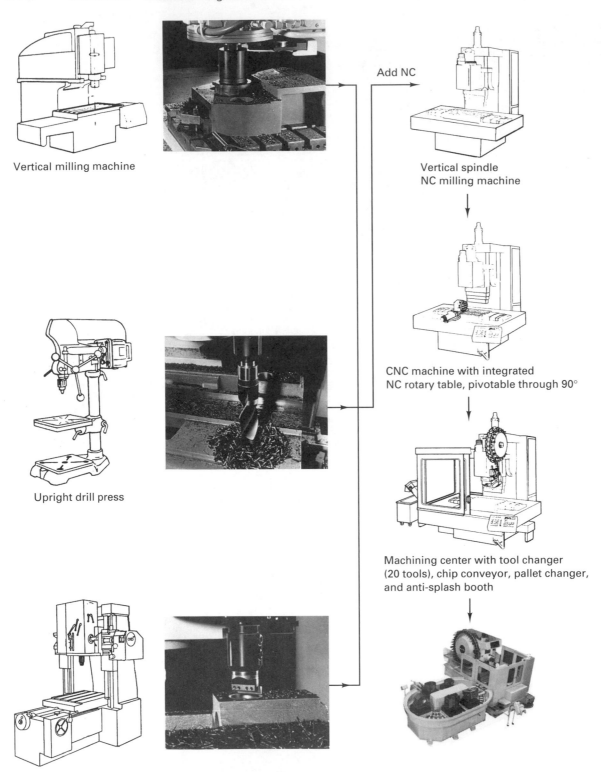

Vertical milling machine

Add NC

Vertical spindle
NC milling machine

Upright drill press

CNC machine with integrated
NC rotary table, pivotable through 90°

Machining center with tool changer
(20 tools), chip conveyor, pallet changer,
and anti-splash booth

FIGURE 32-8 Evolution of the machining center from A(3) to NC to CNC to machining center occurred very quickly.

With the advent of the NC type of machine and more recently programmable robots two types of automation were defined. *Hard* or *fixed automation* is exemplified by transfer machines or automatic screw machines, and *flexible* or *programmable automation* is typified by CNC machines or robots that can be taught or programmed externally by means of computers. The control was in computer software rather than mechanical hardware.

If you are not familiar with the concept of feedback control, a simple A(4) level is one in which some aspect (usually position) of the process is measured using a detection

Open loop A(3)

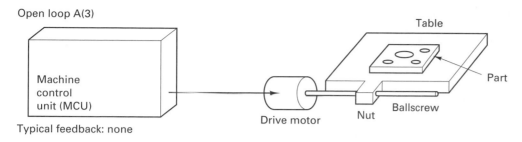

Typical feedback: none

Closed loop – motor feedback A(4)

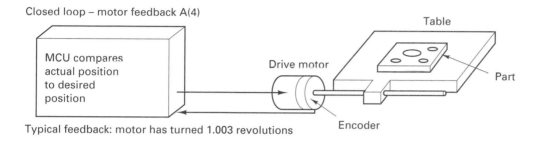

Typical feedback: motor has turned 1.003 revolutions

Closed loop-ballscrew feedback A(4)

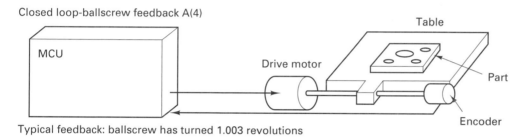

Typical feedback: ballscrew has turned 1.003 revolutions

Closed loop – worktable position feedback A(4)

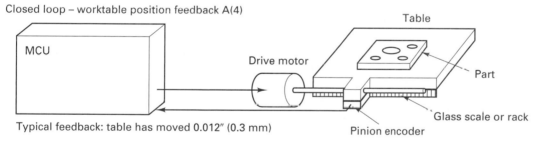

Typical feedback: table has moved 0.012″ (0.3 mm)

FIGURE 32-9   Open loop NC versus three position control schemes for NC and CNC machine tools.

device (sensor). This information is fed back to an electronic comparator, housed in the machine control unit (MCU), which makes comparisons with the desired level of operation. If the output and input are not equal, an error signal is generated and the table is adjusted to reduce the error.

Figure 32-9 shows the difference between an open loop A(3) machine and a closed loop A(4) machine, with feedback provided on the location of the table and the part with respect to the axis of the spindle of the cutting tool. Three position control schemes are shown. (See discussion on machining centers for additional NC basics.)

In CNC turning machines, the feedback is on the tool tip with respect to the rotating part creating tool paths. Figure 32-10 shows how a part can be turned (machined) from a round bar in a CNC lathe. A program is written which directs the machine to execute the necessary roughing and finishing passes.

$$\text{Number of rough passes} = \frac{\text{stock diameter} - \text{minimum diameter} + \text{finish}}{\text{depth of cut} \times 2}$$

In this case, eight roughing passes and one finishing cut were specified.

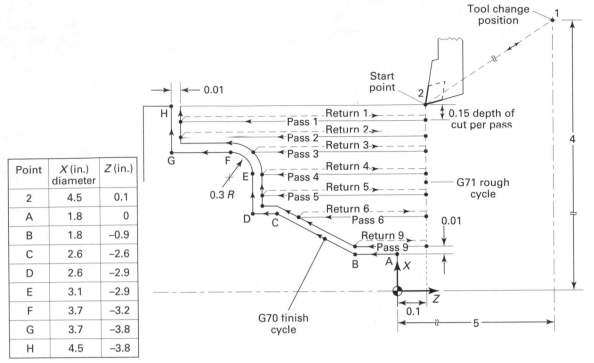

Point	X (in.) diameter	Z (in.)
2	4.5	0.1
A	1.8	0
B	1.8	-0.9
C	2.6	-2.6
D	2.6	-2.9
E	3.1	-2.9
F	3.7	-3.2
G	3.7	-3.8
H	4.5	-3.8

FIGURE 32-10 The tool paths necessary to rough and finish turn a part in a CNC lathe are computer generated. See also Figure 32-32.

BRIEF HISTORY OF NUMERICAL CONTROL

The advent and wide-scale adoption of numerically (tape- and computer-) controlled machine tools has been the most significant development in machine tools in the past 50 years. These machines raised automation to a new level, A(4), by providing positional feedback as well as programmable flexibility. Numerical control of machine tools created entirely new concepts in manufacturing. Certain operations are now routine that previously were very difficult if not impossible to accomplish.

In earlier years, highly trained NC programmers were required. The development of low-cost, solid-state microprocessing chips has resulted in machines that can be quickly programmed, by skilled machinists after only a few hours of training, using only simple machine shop language. As a consequence, there are few manufacturing facilities today, from the largest down to the smallest job shops, that do not have one or more numerically controlled machine tools in routine use.

NC came into being to fill a need. The U.S. Air Force (USAF) and the airframe industry were seeking a means to manufacture complex contoured aircraft components to close tolerances on a highly repeatable basis. John Parsons of the Parsons Corporation of Traverse, Michigan, had been working on a project for developing equipment that would machine templates to be used for inspecting helicopter blades. He conceived of a machine controlled by numerical data to make these templates and took his proposal to the USAF. Parsons convinced the USAF to fund the development of a machine. The Massachusetts Institute of Technology (MIT) was subcontracted to build a prototype machine. The prototype was a conventional two-axis tracer mill retrofitted with servomechanisms. As luck would have it, the servomechanism lab was located next to a lab where one of the very first digital computers (Whirlwind) was being developed. This computer generated the digital numerical data for the servomechanisms.

By 1962, NC machines accounted for about 10% of total dollar shipments in machine tools. Today, about three-fourths of the $35 billion spent for machine tools (drill presses, milling machines, lathes, and machining centers) goes for NC equipment.

Early on, NC machines were continuous-path or contouring machines where the entire path of the tool was controlled with close accuracy in regard to position and velocity. Today, milling machines, machining centers, and lathes are the most popular applications of continuous-path control requiring feedback control. Next, point-to-point machines were produced in which the path taken between operations is relatively unimportant and therefore not monitored continuously. Point-to-point machines are used primarily for drilling, milling straight cuts, cutoff, and punching. Automatic tool changers, which require that the tools be precisely set to a given length prior to installation in the machines, permitted the merging of many processes into one machine (Figure 32-11). Machines are often equipped with two pallets so that one can be set up while the other is working. The two- (or four-sided) "tombstone" fixture shown has multiple mounting

FIGURE 32-11   Horizontal machining center with four axes of control (X, Y, Z, R table) receives inputs to the control panel from many sources.

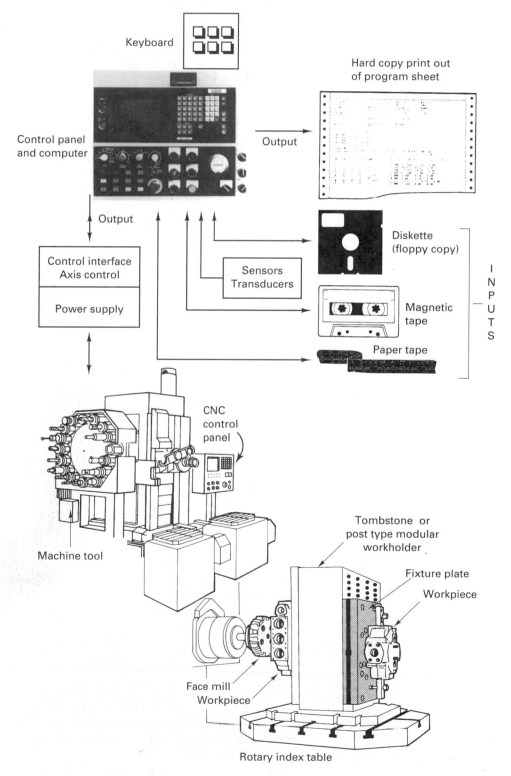

and locating holes for attaching part-dedicated fixture plates, which greatly extends the utility of a horizontal machining center.

During the early days of NC, the machine tools had to be programmed using a machine control language, a difficult, time-consuming task prone to error. Many companies developed computer languages to aid the NC programming task, each developing a different language to describe tool geometry, tool movements, and machining instructions. Confusion reigned.

The USAF sponsored the placement of many NC machines in the major aerospace companies. The companies soon concluded that a universally accepted NC programming language was needed. Under the auspices of the Aerospace Industries Association (AIA), these companies, with the USAF, sponsored additional research at MIT to develop a computer language that would use simple English-like statements to produce an output that would control the NC machines. This language, called *APT* (automatically programmed tools), was announced by MIT in 1959 when computer technology was in its adolescence. Hundreds of thousands of parts have been programmed using APT, initially on large mainframe computers. The output from the APT program had to be converted in the language of a particular machine. This was called *postprocessing*. Traditional postprocessing yields NC workpiece programs that are not exchangeable. To machine an identical workpiece on another machine, the program must be postprocessed again unless the machine and control are exactly the same.

With the arrival of high-resolution *computer-aided design* (CAD) graphics, many people thought that APT would be phased out. But for parts where complex tool control is required, APT is still used. The chief problems in NC programming were tool radius compensation and tool path interpolation (discussed later). Computer software with the capability to perform linear, circular, parabolic, and other kinds of interpolations had to be developed. The capabilities were included in APT and each software program is now routinely available on CNC machines.

In the 1960s, it was envisioned that a large computer could be used directly to control, in real time, a number of machines. A limited number of *direct numerical control* (DNC) systems were developed, with the idea that the programs were to be sent directly to the machines (eliminating paper tape handling). The mainframe computer would be shared on a real-time basis by many machine tools. The machine operator would have access to the main computer through a remote terminal at the machine, while management would have up-to-the-minute data on production status and machine utilization. This version of DNC had very few takers. Instead, NC machines became *computer numerical control* (CNC) machines through the development of small inexpensive computer microprocessors with large memories. These onboard computers have every sort of input imaginable (see Figure 32-11). Functions such as program storage, tool offset and tool compensation, program-editing capability, various degrees of computation, and the ability to send and receive data from a variety of sources including remote locations are routinely available through the onboard computer. The computer can store multiple-part programs, recalling them as needed for different parts. Immediately, it was found that the machine tool operator could readily learn how to program these machines (manually) for many component parts, often eliminating the need for a part programmer.

In recent years the DNC concept has been revived with the small computers at the machines being networked to a larger computer to provide enhanced memory and computer computational capacity. Minicomputers, supervising and controlling a *flexible manufacturing system* (FMS) or manufacturing cell, are networked to a large central computer. Therefore, DNC now means *distributed numerical control*, with the distribution of NC programs by a central computer to individual CNC units. The FMS is a special type of manufacturing system. Historically speaking, the first examples of FMS systems appeared in the late 1960s, but few companies adopted them because of their high initial cost.

FLEXIBLE MANUFACTURING SYSTEMS

The most publicized type of modern manufacturing systems is known as the *flexible manufacturing system* (FMS). The development of FMSs began in England and the United States in the 1960s. By combining the repeatability and productivity of the transfer line with the programmable flexibility of the NC machine, a variety of parts could be produced on the same set of machines. In the United States the first systems were called variable mission or flexible manufacturing systems. In the late 1960s Sundstrand installed a system for machining aircraft speed drive housings that was used over 30 years. Over all, however, very few of these systems were sold until the late 1970s and early 1980s, when a worldwide FMS movement began. But even today international trade in FSM is not significant, and there are fewer than 2000 systems in the world (less than 0.1% of the machine tool population). There is also some evidence that the market for these large expensive systems became saturated around the mid-1980s.

The FMS permits (schedules) the products to take random paths through the machines. This system is fundamentally an automated, conveyorized, computerized job shop. The system is complex to schedule. Because the machining time for different parts varies greatly, the FMS is difficult to link to an integrated system and often remains an island of very expensive automation.

About 60–70% of FMS implementations are for components consisting of non-rotational (prismatic) parts such as crankcases and transmission housings as shown in Figure 32-12, an FMS with two machine tools serviced by a pallet system and an

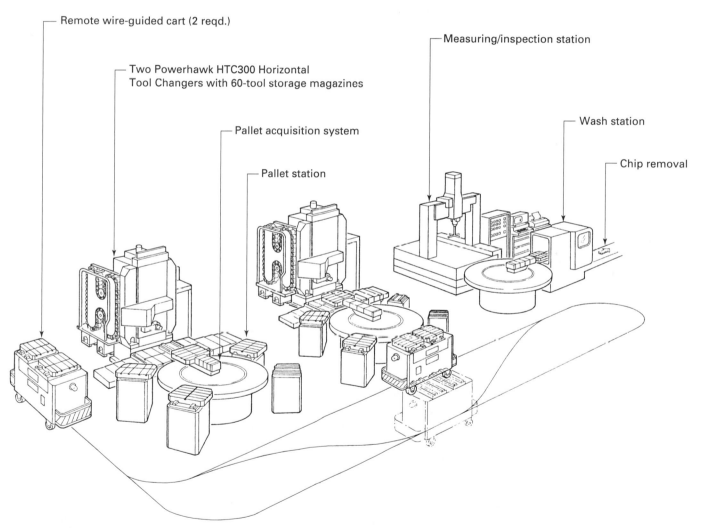

Application:

An aircraft parts manufacturer needed parts transfer mobility, in/out parts queue, cutting tool library and quality control management for production of high-technology parts.

Wire-guided vehicles offer interdepartment transfer capability as well as in-cell transport. The Q.C. center manages the machining accuracy for continuous flow of acceptable parts. Parts are scheduled in batch and/or random, controlled by a management computer.

The machines are equipped with telemetry probes, adaptive control, bulk tool storage and complete tool management.

FIGURE 32-12   An example of an FMS for aircraft parts with a wire guided vehicle for part transfer.

TABLE 32-2. Common Features of Flexible Manufacturing Systems
Pallet changers
Multiple machine tools: NC or CNC
Automated material handling system (to deliver parts to machines)
Computer control for system: DNC
Multiple parts: Medium sized lots (200–10,000) with families of parts
Random sequencing of parts to machines (optional)
Automatic tool changing
Inprocess inspection
Parts washing (optional)
Automated storage and retrieval (optional, to deliver parts to system)

automated guided vehicle (AGV). The balance of FMS installations are for rotational parts or a mixture of both types of parts.

The number of machine tools in an FMS varies from 2 to 10, with 3 or 4 being typical. Annual production volumes for the systems are usually in the range of 3000 to 10,000 parts, the number of different parts ranging from 2 to 20, with 8 being typical. The lot sizes are typically 20 to 100 parts and the typical part has a machining time of about 30 minutes with a range of 6 to 90 minutes per part. Each part typically needs two or three chucking or locating positions and 30 or 40 machining operations. NC machining centers were used on older FMSs. In recent years, CNC machine tools have been favored, leading to a considerable number of systems being operated under direct numerical control (DNC). The machining centers always have tool changers. To overcome the limitation of a single spindle, some systems are being built with head changers. These are sometimes referred to as modular machining centers.

Common features of FMSs (see Table 32-2) are *pallet changers,* underfloor conveyor systems for the collection of chips (not shown), and a conveyor system that delivers parts to the machine. This is also an expensive part of the system, as the conveyor systems are either powered rollers, mechanical pallet transfer conveyors, *automated guided vehicles* (AGVs) operating on underground towlines or buried guidance cables (see Figures 32-12 and 32-13). The carts are more flexible than the conveyors. The AGVs also serve to connect the islands of automation, operating between FMSs, replacing human guided vehicles (forklift trucks).

Pallets are a significant cost item for the FMS because the part must be accurately located on the pallet and the pallet accurately located in the machine. Since many pallets are required for each different component, a lot of pallets are needed and they typically represent anywhere from 15 to 20% of the total system cost. FMSs cost about $1 million per machine tool. Thus the seven-machine FMS costs $6 million for hardware and software, with the transporter costing over $1 million.

Computer Control in FMS. The CNC machines receive programs as needed from a host minicomputer, which acts as a supervisory computer for the system, tracking the status of any particular machine in the system. In recent years, in-process inspection and automatic tool position correction for tool wear have been added features along with diagnostic routines to computer-monitor the condition of the machines. However, a common problem with these systems is the monitoring of the tool condition and performance. Most installations also incur problems in the performance and reliability of the software and the control systems. It takes just as long to debug software as it does to debug hardware, and delays of 2 to 6 months in start-up are not uncommon.

Figure 32-14 shows the levels of computer control in the FMS. (Note that operators are typically needed to load workpieces, unload finished parts, change worn tools, and perform equipment maintenance and repairs.) CNC and DNC functions can be incorporated into a single FMS. The system can usually monitor piece-part counts, tool changes, and machine utilization, with the computer also providing supervisory control of the production. The workpieces are launched randomly into the system, which identifies each part in the family and routes it to the proper machines. The systems generally

Types of AGVs

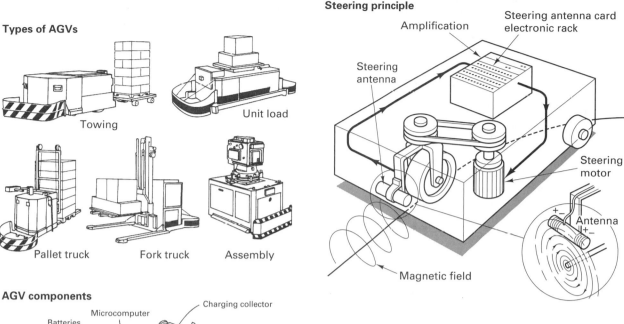

Towing

Unit load

Pallet truck

Fork truck

Assembly

Steering principle

Amplification

Steering antenna card
electronic rack

Steering
antenna

Steering
motor

Antenna

Magnetic field

AGV components

Charging collector

Microcomputer

Rear of carrier

Batteries

Steering
assembly

Emergency
stop button

Safety
bumper

Audible signal
beeper

Hour meter

Battery indicator
meter

Tape switch bumper

Fuse compartment

Main switch

Release button

Battery disconnect

Front of carrier

Spacing
transceiver

Guidepath frequency
indicator light

Emergency
stop button

Emergency
stop light

Emergency
stop light

FIGURE 32-13   Automated guided vehicles can
be used as a material handling system for
automated assembly as well as in FMSs, delivering
workpieces and tooling to the machines. One
means of guidance is a buried-wire guidepath
tracked by an antenna on the vehicle. The need
for guidepath flexibility has inspired new
guidance systems (infrared and inertial), adding
further to the cost.

display reduced manufacturing lead time, low in-process inventory, and high machine tool
utilization, with reduced indirect and direct labor. The materials-handling system must
be able to route any part to any machine in any order and provide each machine with a
small queue of "banked parts" waiting to be processed so as to maximize machine utilization. Convenient access for loading and unloading parts, compatibility with the control system, and accessibility to the machine tools are other necessary design features for
the material handling system.

The computer control for an FMS system has three levels. The master control monitors the entire system for tool failures or machine breakdowns, schedules the work, and
routes the parts to the appropriate machine. The DNC computer distributes programs
to the CNC machines and supervises their operations, selecting the required programs and transmitting them at the appropriate time. It also keeps track of the
completion of the cutting programs and sends this information to the master computer.
The bottom level of computer control is at the machines themselves.

It is difficult to design an FMS because it is, in fact, a very complex assembly
of elements that must work together. Designing the FMS to be flexible is difficult.
Many companies have found that between the time they ordered their system and
had it installed and operational, design changes had eliminated a number of parts
from the FMS. That is, the system was not as flexible as they thought. Figure 32-15
shows some more typical FMS designs. If the system has only one or two machines,
it is often called a flexible manufacturing cell (FMC). However, the typical FMS installation may not be as flexible (in terms of different parts it can make) as it was

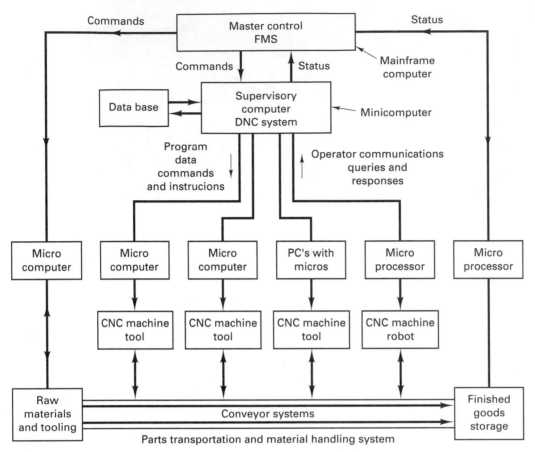

FIGURE 32-14 Three levels of computer control in an FMS system.

thought to be when it was designed. FMSs are, in fact, classic examples of supermachines. Such large, expensive systems must be examined with careful and complete planning. It is important to remember that even though they are often marketed and sold as a *turnkey* installation (the buyer pays a lump sum and receives a system that can be turned on and run), this is only rarely possible with a system that has so many elements that must work together reliably. Taken in the context of integrated manufacturing systems, large FMSs may be difficult to synchronize to the rest of the system. The flexibility of the FMS requires variable speeds and cycles, numerical control, and a supervisory computer to coordinate cell operation. In the long run, smaller manned or unmanned cells may well be the better solution, in terms of system flexibility. Perhaps a better name for these systems is *variable mission or random path manufacturing systems*.

The use of computer coding and classification systems, a group technology technique, to identify the initial family of parts around which the FMS is designed greatly improves the FMS design. One might say that FMSs were developed before their time, since they are being more readily accepted since GT has been used (at least conceptually) to identify families of parts for the system to produce.

As an FMS generally needs about three or four workers per shift to load and unload parts, change tools, and perform general maintenance, it cannot really be said to be self-operating. FMS systems are rarely left untended, as in third-shift operations. Other than the personnel doing the loading and unloading, the workers in the FMS are usually highly skilled and trained in NC and CNC. Most installations run fairly reliably (once they are debugged) over three shifts, with uptime ranging from 70 to 80%, and many are able to run one shift untended.

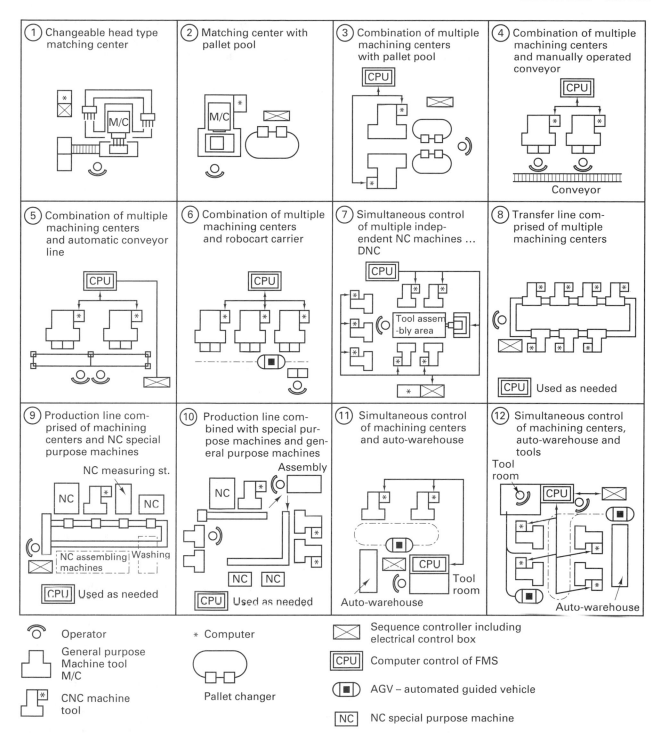

FIGURE 32-15   Examples of machining centers and FMS designs.

■ 32.3 ROBOTICS

Robots, or steel-collared workers, are typically A(3), A(4), or A(5) level machines. As defined by the Robot Institute of America, "A *robot* is a reprogrammable, multifunctional manipulator designed to handle material, parts, tools, or specialized devices through variable programmed motions for the performance of a variety of tasks." The word *robot* was coined in 1921 by Karel Capek in his play *R.U.R.* (Rossum's Universal Robots). The term is derived from the Czech word for "worker." The principal inventor of the underlying control technology for robots was developed by George Devol, who worked for

Remington-Rand in the 1950s. His patents were purchased by CONDEC, Inc., which developed the first commercial robot, called *Unimate*. Much of the work for the next 20 years was spearheaded by Joseph Engelberger, who came to be called the father of robotics. Another famous author, Isaac Asimov, depicted robots in many of his stories and gave three laws that hold quite well for industrial robotic applications. Asimov's three laws of robotics were:

1. A robot may not injure a human being or, through inaction, allow a human being to be harmed. (Safety first.)
2. A robot must obey orders given by human beings except when that conflicts with the First Law. (A robot must be programmable.)
3. A robot must protect its own existence unless that conflicts with the First or Second Law. (Reliability.)

In considering the use of a robot, the following points should be considered. Anything that makes a job or task easy for the robot to do makes the job easy for a human being to do as well. Why? The robot is a severely handicapped worker and is several orders of magnitude less flexible than a human being. Robots cannot think or solve problems on the plant floor. The big advantage of the robot is that it will do a job in an exact cycle time, and do it every time, whereas a human being often cannot. This is an important feature in robotic manufacturing and assembly cells.

For our purposes, if a machine is programmable, capable of automatic repeat cycles, and can perform manipulations in an industrial environment, it is an industrial robot.

All robots have the following basic components:

1. *Manipulators:* the mechanical unit, often called the *arm*, that does the actual work of the robot. It is composed of mechanical linkages and joints with actuators to drive the mechanism directly or indirectly through gears, chains, or ball screws.
2. *Feedback devices:* transducers that sense the positions of various linkages and joints and transmit this information to the controllers in either digital or analog form [A(4)-level robots].
3. *End effector:* the *hand* or *gripper* portion of the robot, which attaches the end of the arm and performs the operations of the robot.
4. *Controller:* the brains of the system that direct the movements of the manipulator. In higher-level robots, computers are used for controllers. The functions of the controller are to initiate and terminate motion, store data for position and motion sequence, and interface with the outside world, meaning other machines and human beings.
5. *Power supply*: electric, pneumatic, and hydraulic power supplies used to provide and regulate the energy needed for their manipulator's actuators.

The early robots were programmed using analog rather than digital control technology. The next generation of robots was more accurate and precise, with reliable electric (digital) controls. Many robot manufacturers made it hard to integrate their robots with other CNC equipment by retaining proprietary operating systems. Today, Japan leads the world in the implementation of robots into the factory. Many of the commercially available robots have one of the mechanical configurations shown in Figure 32-16. Cylindrical coordinate robots have a work envelope (shaded region) that is a portion of a sphere. Jointed-spherical coordinate robots have a jointed arm and a work envelope that approximates a portion of a sphere. Rectangular coordinate robots with a rectangular work envelope have been developed for high precision in assembly applications. Another design called SCARA (selective compliance assembly robot arm) also has good positioning accuracy, high speeds, and lost cost for a robot with three or four controllable axes, which is usually adequate for assembly tasks.

Figure 32-17 shows the six axes or motion typical for a jointed-arm robot. Robots may have two or three additional minor axes of motion at the end of the arm (commonly called the *wrist*). These three movements are *pitch* (vertical movement), *yaw*

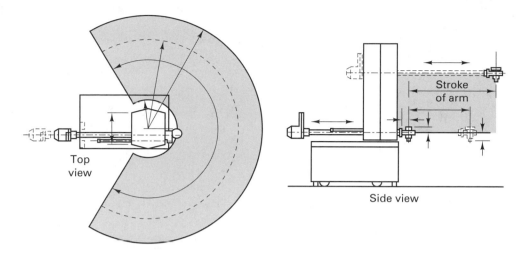

(a) Cylindrical coordinate

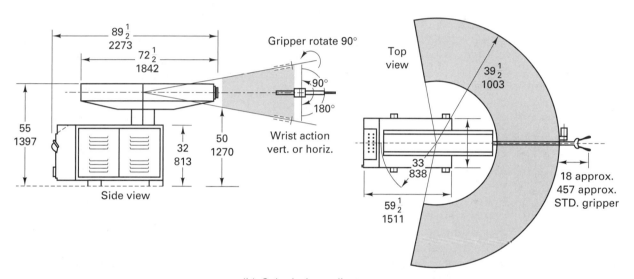

(b) Spherical coordinate

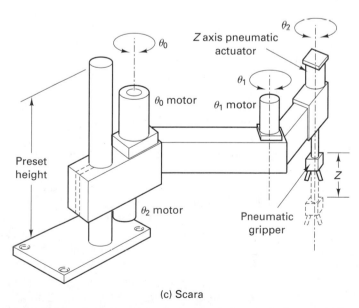

(c) Scara

FIGURE 32-16   Three basic designs for industrial robots. Shaded areas indicate work envelopes—region in space where robot can reach.

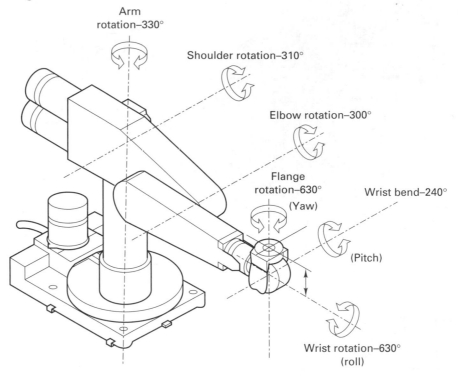

Arm
rotation–330°

Shoulder rotation–310°

Elbow rotation–300°

Flange
rotation–630°
(Yaw)

Wrist bend–240°

(Pitch)

Wrist rotation–630°
(roll)

FIGURE 32-17 The six axes of motion of a robot include roll, pitch, and yaw movement of the wrist.

(horizontal motion), and *roll* (wrist rotation). The *hand* (or *gripper* or *end effector)*, which is usually custom-made by the user, attaches to the wrist.

Industrial robots used in industry today fill three main functions: material handling, assembly, and material processing. They have, for the most part, very primitive motor and intelligence capabilities, because most robots are A(3)-level machines. The sensory-interactive control, decision-making, strength-to-size, and artificial intelligence capabilities of robots are far inferior to those of human beings at this time. With regard to shortcomings in performance, the robot's major stumbling blocks are its accuracy and repeatability (i.e., process capability or dexterity), but robot capabilities are progressing steadily, with improvements in controls, ease of programming, operating speeds, and precision.

Robots are having a strong impact in certain industrial environments, often doing those jobs that are unhealthy, hazardous, extremely tedious, and unpleasant. Robots perform well doing paint spraying, loading and unloading small forgings or die-casting machines, spot welding, and so on.

The A(3)-level robot, usually called a *pick-and-place machine,* is capable of performing only the simplest repeat-cycle movements, on a point-to-point basis, being controlled by an electronic or pneumatic control with manipulatory movement controlled by end stops.

All A(3) robots are usually small robots with relatively high-speed movements, good repeatability (0.010 in.), and low cost. They are simple to program, operate, and maintain but have limited flexibility in terms of program capacity and positioning capability.

To raise the robot to an A(4)-level machine, sensory devices must be installed in the joints of the arm(s) to provide positional feedback and error signals to the servomechanisms, just as was the case for NC machines. The addition of an electronic memory and digital control circuitry allows this level of robot to be programmed by a human being guiding the robot through the desired operations and movements using a hand controller (Figure 32-18). The handheld control box has rate-control buttons for each axis of motion of the robot arm. When the arm is in the desired position, the record or program button is pushed to enter that position or operation into the memory. This is similar to point-to-point NC machines, as the path of the robot arm movement is defined by selected endpoints when the program is played back. The electronic memory can usually store multiple programs and randomly access the required one, depending on the job to be done. This allows for a product mix to be handled without stopping to reprogram the machine. The addition of a computer, usually a minicomputer, makes it possible to program the robot to move its hand or gripper in straight lines or other geometric paths between given points, but the robot is still essentially

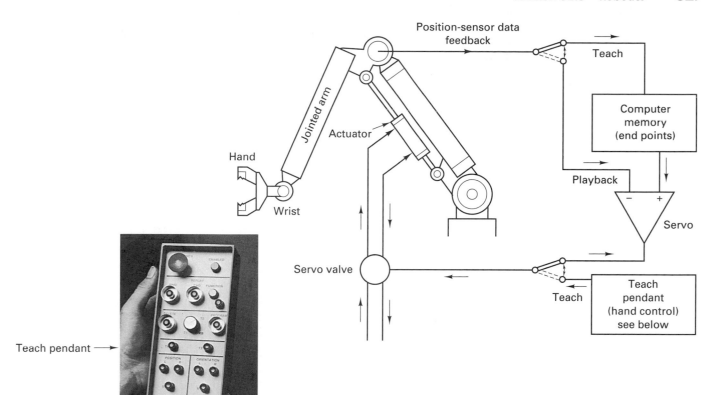

FIGURE 32-18   Manual programming of an industrial robot is accomplished by using a handheld "teach pendant." The robot is guided through a sequence of operations. The successive positions of all the robotic joints are stored in an electronic memory. This establishes the endpoints of the motions of the arm. By switching from "teach" to "playback," the stored endpoints are replayed. *(Courtesy of Cincinnati Milacron, Inc.)*

a point-to-point machine. Using computer simulations, robots can be programmed offline to do assembly and processing tasks, but they must usually be taught the final location of the points by the operator.

There are three ways of controlling point-to-point motion independently: (1) sequential joint control, (2) noncoordinate joint control, and (3) terminal coordinate joint control. Point-to-point servo-controlled robots have the following common characteristics: high load capacity, large working range, and relatively easy programming, but the path followed by the manipulator during operation may not be the path followed during teaching.

To make an A(4) robot continuous-path, position and velocity data must be sampled on a time basis rather than as discretely determined points in space. Due to the high rate of sampling, many spatial positions must be stored in the computer memory, thus requiring a mass storage system.

Tactile and Visual Sensing in Robots.   Many of the industrial robots in use at this time are A(4) point-to-point machines. Most robots operate in systems wherein the items to be handled or processed are placed in precise locations with respect to the robots. Even robots with computer control that can follow a moving auto conveyor line while performing spot-welding operations have point-to-point feedback information, but this is satisfactory for most industrial applications.

To expand the capability of this handicapped worker made of steel, sensors are used to obtain information regarding position and component status. Tactile sensors provide information about force distributions in the joints and in the hand of the robot during manipulations. This information is then used to control movement rates. Visual sensors collect data on spatial dimensions by means of image recording and analysis. Visual sensors are used to identify workpieces; determine their position and orientation; check position, orientation, geometry, or speed of parts; determine the correct welding path or point; and so on.

TABLE 32-3. Sensors Used on Robots with Some Typical Applications

Sensor Type	Design	Application
Visual	Video pickup tubes (TV camera) Semiconductor sensors (lasers) Fiber optics	Position detection, part inspection Parts detection, identification, sorting Consistency testing (e.g., in manipulation, welding, and assembly) Guidance and control
Tactile	Feelers Pin matrix Load cells (piezo, capacitance) Conductive elastometers Silicon	Position detection Tool monitoring (e.g., in casting, cleaning, grinding, manipulation, and assembly) Force sensing, part identification Object recognition, pressure
Electrical (inductive capacitative)	Shunt (current determination) Capacitor Coil	Position detection Status determination (e.g., in manipulation and welding)

To provide a robot with tactile or visual capability, powerful computers and sophisticated software are required, but this appears to be the most logical manner to raise the robot to the A(5) level of automation, at which it can adapt to variations in its environment. Vision systems can locate parts moving past a robot on a conveyor, identify those parts that should be removed from the conveyor, and communicate this information to the robot. The robot tracks the moving part, orients its gripper, picks up the part, and moves it to the desired work station. See Table 32-3.

ROBOTIC APPLICATIONS

The current generation of robots is finding applications in the areas described next.

1. *Die casting.* In single- or multishift operations, custom or captive shops, robots unload machines, quench parts, operate trim presses, load inserts, ladle metal, and perform die lubrication. Die life is increased because die casting machines can be operated without breaks or shutdowns. Die temperatures remain stable and better controlled with uniform cycle times.

2. *Press transfer.* Robots in sheet metal press transfer lines guarantee consistent throughput shift after shift. Large and unwieldy parts can be handled at piece rates as high as 400 per hour, with no change in cycle time due to fatigue. Robots are adaptable for long- or short-run operations. Programming for new part sizes can be accomplished in minutes.

3. *Material handling.* Strength, dexterity, and a versatile memory allow robots to pack goods in complex palletized arrays or to transfer workpieces (to and from) moving or indexing conveyors from machines. These are boring, labor-intensive operations. Operating costs are reduced when robots feed forge presses and upsetters. They work continuously without fatigue or the need for relief in the hot, hostile environments commonly found in forging. Robots can easily manipulate the hot parts in the presses (Figure 32-19).

4. *Investment casting.* Scrap rates as high as 85% have been reduced to less than 5% when molds are produced by robots. The smooth, controlled motions of the robot provide consistent mold quality impossible to achieve manually.

5. *Material processing.* Product quality is improved and sustained with point-to-point, continuous-path robots in jobs such as routing, flame cutting, mold drying, polishing, and grinding. Once programmed, the robot will process each part with the same high quality. Figure 32-19 shows a robot cell with a vision system for drilling, tapping, buffing, filing, and deburring a family of plastic workpieces. The deburring program and the robot's hand are changed automatically when the robot retrieves a different workpiece. The input conveyor holds about a 3-hour supply of parts for the cell. The camera equipment is placed about 8 ft over the belt. When the controller receives a request

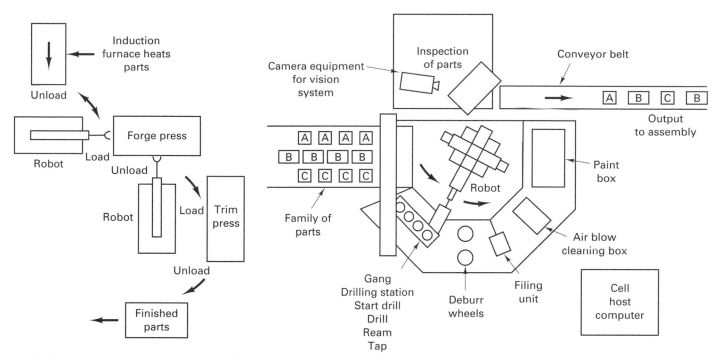

FIGURE 32-19 Robots on the factory floor can do material handling as well as processing.

for certain parts from assembly, the vision system identifies the raw material on the input side and the robot retrieves it and starts it through the cell.

6. *Welding.* Robots spot-weld cars and trucks for almost every major manufacturer in the world, with uniformity of spot location and weld integrity. With the addition of seam-tracking capability, robots can be used for arc welding, increasing arc time and freeing operators from hazardous environments, reducing the cost of worker protection, and improving the consistency of weld quality.

7. *Assembly.* The replacement of pneumatic and hydraulic systems with electric motors has improved the accuracy and precision of robots, permitting them to be more widely used in assembly. To perform assembly tasks using robots requires consideration of the entire system, from part presentation through joining, test, and inspection. Most important, the parts must be designed for robotic assembly. Flexible assembly, as this is now referred to, usually addresses mid-volume products. As opposed to hard automation that uses special-purpose fixtures, part feeders, and work heads, flexible (or robotic) assembly uses general-purpose and programmable equipment and combinations of visual and tactile sensing so that a variety of parts can be assembled using the same equipment. This requires multiple degrees of freedom and general-purpose grippers, which usually decreases the accuracy and precision of the robot, so a vision system is needed to compensate.

8. *Painting.* The automobile painting process uses robots to apply paint to thicknesses of about 0.1 mm using repeated painting and drying cycles. The painting process is the highest energy user in the entire automobile manufacturing and assembly process. Each car gets about 100 liters of paint. The system, shown in Figure 32-20, can paint about 100,000 cars per year at a rate of 30 cars per hour, using an air volume of 860,000 in^3. This is a huge volume of air recirculated every 30 minutes. The figure shows how the typical paint booth was redesigned to reduce energy consumption while improving quality, using robots to apply the materials.

The integration of robots into cellular arrangements with machine tools to process families of component parts where the robot performs tasks right along with one or more human beings is very efficient. The robot can perform part loading and unloading, and material processing (like joining) when machines are grouped properly in a machining cell, see Chapter 43.

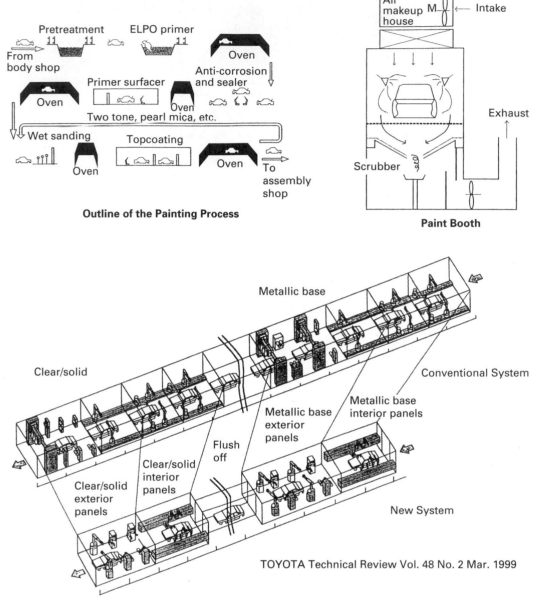

Outline of the Painting Process

Paint Booth

TOYOTA Technical Review Vol. 48 No. 2 Mar. 1999

FIGURE 32-20   The painting process for automobiles uses many robots to consistently apply the primer and topcoating to the car bodies.

Totally unmanned cells can facilitate maximum automation and productivity while maintaining programmable flexibility in producing small to medium-sized production lots of parts from compatible parts families. The robots can also change tools in the machines and even the workholding devices, thereby adding more flexibility to the cell. Unmanned manufacturing cells help to achieve maximum machine tool utilization by greatly increasing the percentage of time the machines spend cutting, which in turn increases the output for the same investment.

■ 32.4 A(4) VERSUS A(5) LEVEL OF AUTOMATION

In the yardstick for automation, the NC or CNC machine represents the A(4) level. The next level of automation, A(5), requires that the control system perform an evaluation function of the process. Figure 32-21 compares block diagrams for A(4) closed-loop control (with encoder) to A(5) adaptive control. A(5) requires a feedback loop from the process or the product and seeks to either constrain or optimize the process. Note how this loop lies inside the position feedback loop. The multiple loops can cause conflicts

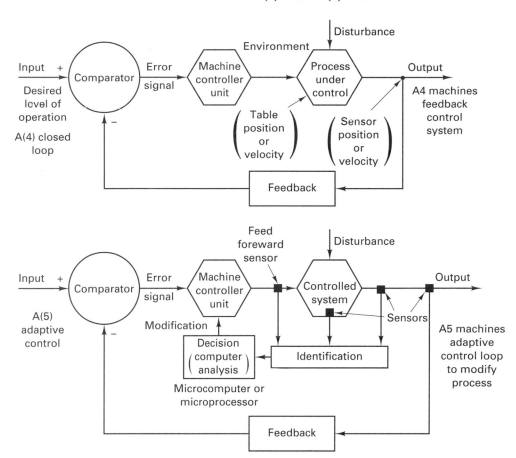

FIGURE 32-21 Block diagrams of A(4) and A(5) level of control: A(4), closed loop NC; A(5) closed loop NC with adaptive control loop.

in the controller's decision-making software. CNC machines, with their onboard computers, are potentially capable of the A(5) level of automation. In the standard versions of NC and CNC machines in use today, speed and feed are fixed in the program unless the operator overrides them at the control panel of the machine. If either speed or feed is too high, the result can be rapid tool failure, poor surface quality, or damaged parts. If the speed and feed are too low, production time is greater than desired for best productivity. An *adaptive control* (A/C) system that can sense deflection, force, heat, geometry, torque, and the like will use these measurements to make decisions about how the input parameters might be altered to *constrain* or *optimize* the behavior of the process. This means that the computer must have *mathematical models* in its software that describe how this process behaves and mathematical functions that state what is to be constrained or optimized (surface quality, cutting force, MRR, power consumed, and so on).

A(5) LEVEL REQUIRES EVALUATION

The A(5) level requires that machines perform *evaluation* of the process itself. Thus the machine must be cognizant of the multiple factors on which the process performance is predicted, evaluate the current setting of the input parameters versus the outputs from the process, and then determine how to alter the inputs to optimize the process.

A(5) level, typified by *adaptive control* (A/C), emulates human evaluation. Basically, A(5) machines are capable of adapting the process itself so as to optimize it in some way. This level of automation requires that the system have a computer programmed with models (mathematical equations) that describe how this process behaves, how this behavior is bounded, and what aspect of the process or system is to be constrained or optimized. *This modeling obviously requires that the process be sufficiently well understood theoretically so that equations (models) can be written that describe how the real process works.* This level of automation has been achieved for continuous processes (oil refineries, for example) where the theory (of heat transfer and fluid dynamics) of the process is well understood and parameters are easier to measure. Unfortunately, the theory of metal forming and metal removal

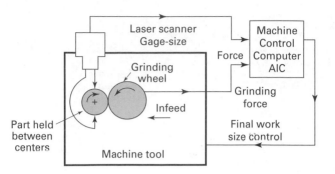

FIGURE 32-22 Developing adaptive control systems for metal cutting machines usually requires that the part is measured as well as the cutting forces.

is less well understood and parameters are usually difficult to measure. These processes have resisted adequate theoretical modeling, and as a consequence there are very few A(5) machines on the shop floor. A/C applications are important where the components vary in size (like castings), which alters the depth of cut and so changes the cutting forces, perhaps producing chatter. Similarly changes in the hardness of the workpiece, perhaps produced by improper heat treating, prior nonuniform rolling, or surface processing, can produce significant variations in the cutting forces, again leading to dynamic instability or quality problems.

The basic elements of the adaptive control loop are:

1. *Identification:* measurements from the process itself, its output, or process inputs using sensors to determine operating conditions
2. *Decision analysis:* optimization of the process in the computer
3. *Modification:* signal to the controller to alter the inputs
4. *Monitor:* continuous control over process

To raise the automatic CNC grinder to the A(5) level, assume that it is a cylindrical center-type grinder on which part deflection as well as part size are measured (see Figure 32-22). In this process, the cutting forces tend to deflect the part more as the grinding wheel gets farther from the centers. The adaptive control software program would have equations that relate deflection to grinding forces and infeed rates. The infeed would be decreased to reduce the force and minimize deflection. Notice that the overall system would still have to compensate for grit dulling and grit attrition that accompanies wheel wear and which also alters the grinding forces. The A(5) level reflects deductive reasoning whereby particular outcomes are predicted from general theory.

A/C systems that place a constraint on a process variable (such as forces, torque, or temperature) are called *adaptive control constraint* (ACC) systems. Thus, if the thrust force and the cutting force (and hence the torque) increase excessively (for example, because of the presence of a hard region in a casting), the adaptive control system changes the speed or the feed to lower the cutting force to an acceptable level. Note that altering the feed may cause a change in the surface finish that may be unacceptable.

Without adaptive control (or without the direct intervention of the operator) excessive cutting forces may cause the tools to break or cause the workpiece to deflect resulting in a loss of size in the part, so the system tries to maintain a constant force.

A/C systems that optimize an operation are called *adaptive control optimization* (ACO) systems. Optimization may involve maximizing material-removal rate between tool changes (or resharpening) or improving surface finish. Currently, most systems are based on ACC, because the development and the proper implementation of ACO is complex.

A(6) LEVEL OF AUTOMATION AND BEYOND

There are very few examples of A(5) machines on the factory floor and fewer at the A(6) level, wherein the machine control has expert systems capability. *Expert systems* try to infuse the software with the deductive decision-making capability of the human brain by having the system get smarter *through experience.* The software is designed to emulate human learning by configuring neural networks where the computer relates inputs to outputs. *Artificial intelligence* (AI) carries this step higher by *infecting* the control software with programs that exhibit the ability to reason inductively.

The A(6) level tries to *relate cause to effect*. Suppose that the effect was tool failure. At the A(5) level, the system would detect the increase in deflection due to increased forces (due to the tool's dulling) and reduce the feed to reduce the force. However, it might simultaneously try to increase the speed to maintain the MRR constant, which would increase the rate of tool wear. This was not the desired result. At the A(6) level, multiple factors are evaluated so that the system can recognize the need to change the tool rather than reduce the cutting feed. That is, the system learned by experience that feed reduction or speed increase was the wrong decision, and that it was tool wear that caused force to increase. Such systems are often called *expert systems* when the software contains the collective experience of human experts or prior replications of the same process. The use of neural networks will allow the process to learn (i.e., the expert system will become more *intelligent* about the process).

An alternative approach is being evaluated using *fuzzy logic* control. The behavior of the process is described with linguistic terms. Fuzzy logic control statements replace closed-form mathematical models. The decision analysis element depends on a set of if–then statements rather than specific equations.

The A(6) level reflects the beginnings of artificial intelligence (AI), in which the control software is infected with elements (subroutines) that permit some thinking on the part of the software. Few, if any, such systems exist on the factory floor.

The A(7) level reflects the next level of AI, whereby inductive reasoning is used. The system software can determine a general principle (the theory) based on the particular facts (the database) collected. Levels A(6) and A(7) are the subjects of intensive worldwide research efforts. Levels A(8) and A(9) are left to the whims of the science fiction writer.

Automation involves machines, or integrated groups of machines, that automatically perform required machining, forming, assembly, handling, and inspection operations. Through sensing and feedback devices, these systems automatically make necessary corrective adjustments. That is, *human thinking must be automated*. There are relatively few completely automated systems, but there are numerous examples of highly mechanized machines. While the potential advantage of a completely automated plant are tremendous, in practice, step-by-step automation of individual operations is required. Therefore, it is important to have a piecewise plan to convert from the classical job shops to the simplified, integrated factory of the future that can be automated, an approach discussed in more detail in Chapters 41–43. However, the most serious limitation in automation is available capital, as the initial investments for automation equipment and installations are large. Because proper engineering economics analysis must be employed to evaluate these investments, students who anticipate a career in manufacturing should consider a course in this area a firm requirement.

■ 32.5 COMPUTER-INTEGRATED MANUFACTURING

A number of definitions have been developed for *computer-integrated manufacturing* (CIM). However, a CIM system is commonly thought of as an integrated system that encompasses all the activities in the production system from the planning and design of a product through the manufacturing system, including control. CIM is an attempt to combine existing computer technologies to manage and control the entire business. CIM is an approach that very few companies have adopted at this time, since surveys show that only 1 or 2% of U.S. manufacturing companies have approached full-scale use of FMS and CAD/CAM, let alone CIM systems.

As with traditional manufacturing approaches, the purpose of CIM is to transform product designs and materials into salable goods at a minimum cost in the shortest possible time. CIM begins with the design of a product (CAD) and ends with the manufacture of that product (CAM). With CIM, the customary split between the design and manufacturing functions is (supposed to be) eliminated.

CIM differs from the traditional job shop manufacturing system in the role the computer plays in the manufacturing process. Computer-integrated manufacturing

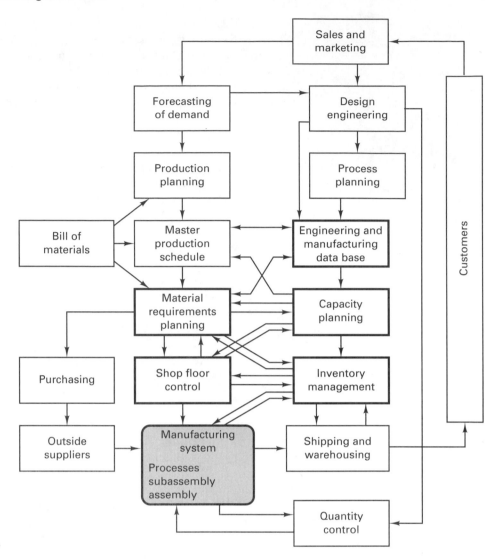

FIGURE 32-23 Cycles of activities in a computer-integrated manufacturing system.

systems are basically a network of computer systems tied together by a single integrated database. Using the information in the database, a CIM system can direct manufacturing activities, record results, and maintain accurate data. CIM is the computerization of design, manufacturing, distribution, and financial functions into one coherent system.

Figure 32-23 presents a block diagram illustrating the functions and their relationship in CIM. These functions are identical to those found in a traditional production (planning and control) system for a job shop-flow shop manufacturing system. With the introduction of computers, changes have occurred in the organization and execution of production planning and control through the implementation of such systems as material requirements planning (mrp), capacity planning, inventory management, shop floor control, and cost planning and control.

Engineering and manufacturing databases contain all the information needed to fabricate the components and assemble the products. The design engineering and process planning functions provide the inputs for the engineering and manufacturing database. This database includes all the data on the product generated during design, such as part geometric data, parts lists, and material specifications. The bill of materials is shown separately, but it is a key part of the database. Figure 32-24 shows how the CAD/CAM database is related to the design and manufacturing activities. Included in the CAM is a CAPP (*Computer-aided process planning*) module, which acts as the interface between CAD and CAM.

Capacity planning is concerned with determining what labor and equipment capacity is required to meet the current master production schedule as well as the long-term future production needs of the firm. Capacity planning is typically performed in terms of labor and/or machine hours available. The master schedule is transformed into

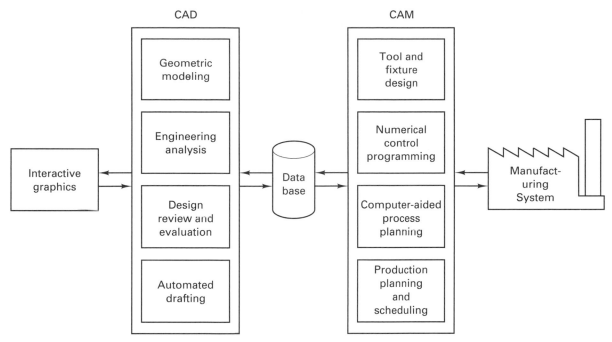

FIGURE 32-24   Database relationship to design and manufacturing.

material and component requirements using mrp. These requirements are then compared with available plant capacity over the planning horizon. If the schedule is incompatible with capacity, adjustments must be made either in the master schedule or in plant capacity. The possibility of adjustments in the master schedule is indicated by the arrow in Figure 32-23 leading from capacity planning to the master schedule.

The term *shop floor control* in Figure 32-23 refers to a system for monitoring the status of manufacturing activities on the plant floor and reporting the status to management so that effective control can be exercised. The cost planning and control system consists of the database to determine expected costs to manufacture each of the firm's products. It also consists of the cost collection and analysis software to determine what the actual costs of manufacturing are and how these costs compare with the expected costs.

■ 32.6 COMPUTER-AIDED DESIGN

A major element of a CIM system is a *computer-aided design* (CAD) system. CAD involves any type of design activity that makes use of the computer to develop, analyze, or modify an engineering design. The design-related tasks performed by a CAD system are:

- Geometric modeling
- Engineering analysis
- Design review and evaluation
- Automated drafting

Geometric modeling corresponds to the synthesis phase of the design process. To use geometric modeling the designer constructs the graphical image of the object on the CRT. The object can be represented using several different methods. The simplest systems use wire frames to represent the object. The wire frame geometric modeling is classified as:

- Two-dimensional representation used for flat objects
- Three-dimensional modeling of more complex geometrics

Other enhancements to wire-frame geometric modeling are:

- Color graphics
- Dashed lines to portray rear edges that would be invisible from the front
- Removal of hidden lines
- Surface representation that makes the object appear solid to the viewer

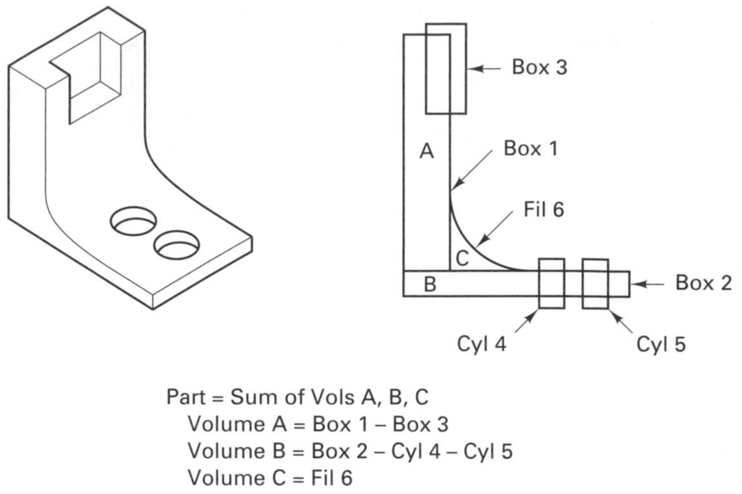

FIGURE 32-25 The CAD (computer-aided design) part design is constructed by summing basic solid geometry shapes.

Part = Sum of Vols A, B, C
  Volume A = Box 1 – Box 3
  Volume B = Box 2 – Cyl 4 – Cyl 5
  Volume C = Fil 6

The more advanced method of geometric modeling is solid modeling in three dimensions. This method uses solid geometry shapes called *primitives* to construct the object as shown in Figure 32-25. That is, the object is a collection of geometric shapes which can be understood by both the CAD and the CAM computers.

Engineering analysis is required in the formulation of any engineering design project. The analysis may involve stress calculations, finite element analysis for heat transfer computations, or differential equations to describe the dynamic behavior of the system. Generally, commercially available general-purpose programs are used to perform these analyses.

Design review and evaluation techniques check the accuracy of the CAD design. Semiautomatic dimensioning and tolerancing routines that assign size specification to surfaces help to reduce the possibility of dimensioning errors. The designer can zoom in on the part design details for close scrutiny. Many systems have a layering feature that involves overlaying the geometric image of the final shape of a machined part on top of the image of a rough part. Other features are interference checking and animation, which enhance the designer's visualization of the operation of the mechanism and help to ensure against interferences.

Automated drafting involves the creating of hard-copy engineering drawings from the CAD database. Typical features of CAD systems include automatic dimensioning, generation of cross-hatched areas, scaling of drawings, ability to develop sectional views, enlarged views of particular part details, and the ability to rotate the part and perform transformations such as oblique, isometric, and perspective views.

Using the CAD database, a group technology *coding and classification* system can also be developed. Parts coding involves assigning letters or numbers to parts to define their geometry or manufacturing process sequence. The codes are used to classify or group similar parts into families for manufacturing using groups of machines. Designers can use the classification and coding system to retrieve existing part designs rather than design new parts.

COMPUTER-AIDED PROCESS PLANNING

Computer-Aided Process Planning (CAPP) uses computer software to determine how a part is to be made. If group technology is used, parts are grouped into part families according to how they are to be manufactured. For each part family, a standard process plan is established. That is, each part in the family is a variation of the same theme. The standard process plan is stored in computer files and then retrieved for new parts that belong to that family.

Figure 32-26 explains the CAPP process. The user initiates the procedure by entering the part code. The CAPP program then searches the part family file to determine whether a match exists. If the file contains an identical code number, the standard machine routing and operation sequence are retrieved for display to the user. The standard operation sequence is examined by the user to permit any necessary editing of the plan to make it compatible with the new part design. This is *variant* CAPP. After editing, the process plan formatter prepares the paper documents (the route sheet and operation sheets for the job shop).

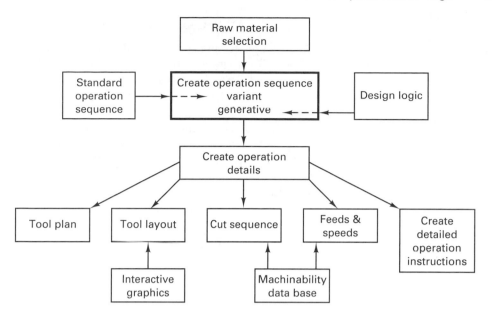

FIGURE 32-26 CAPP (computer-aided process planning) can be variant or generative.

If an exact match cannot be found, a new plan has to be *generated*, based on design logic and the examination of geometry and tolerance information and then comparing these requirements to the machine capability. The software determines what processes can produce what surfaces and establishes the machining constraints. Once the process plan for a new part code number has been entered and verified, it becomes the standard process for future parts of the same classification. The process plan formatter may include software to compute machining conditions, layouts, and other detailed operation information for the job shop.

COMPUTER-AIDED MANUFACTURING

Another major element of CIM is *computer-aided manufacturing* (CAM). An important reason for using a CAD system is that it provides a database for manufacturing the product. However, not all CAD databases are compatible with manufacturing software. The tasks performed by a CAM system are:

- Numerical control or CNC programming
- Production planning and scheduling
- Tool and fixture design

Numerical control can use special computer languages. APT and COMPACT II are two of the early language-based computer-assisted programming systems. These systems take the CAD data and adapt them to the particular machine control unit/machine tool combination used to make the part. The development of computer-aided design in the 1980s led to an attractive alternative method for NC programming. However, virtually all CAD graphics systems were developed as part (or product) design tools, with NC capability added as an afterthought. CAD programs depict the final part and do not usually deal with how the part is processed from raw material to finished goods. The major advantage of NC graphics is visualization of a part. You can see how to machine it. Even so, it takes considerable training (months) to achieve proficiency in the use of CAD-NC, or, as it is commonly called, CAD-CAM.

APT was designed in exactly the opposite way to CAD-NC. APT starts with the cutter and proceeds to "machine" the raw material into the final part shape. However, APT users must be able to visualize or imagine the part and "see" it being machined in their mind's eye, so APT also requires considerable training to become expert. For modern CAM applications for complex parts, the software simulates the machining steps necessary to generate the part. However, the CAD capability to design a complex part may still exceed the CAM capability of the CAD system. It is estimated that 5 to 10% of the parts in the aerospace and defense industries still require APT programming.

■ 32.7 Basic Principles of Numerical Control

NC uses a processing language to control the movement of the cutting tool or workpiece or both. The programs contain information about the machine tool and cutting tool geometry, the part dimensions (from rough to finish size), and the machining parameters (speeds and feeds). Thus NC machines can duplicate consecutive parts and a part made at a later date will be the same as one made today. Repeatability and quality are improved. Workholding devices can be made more universal, and setup time can be reduced, along with tool-change time, thus making programmable machines economical for producing small lots or even a single piece. When combined with the managerial and organizational strategies of group technology (GT) and cellular manufacturing, programmable machines lead to tremendous improvements in quality and productivity. GT basically leads to the creation of families of parts made in machining cells containing flexible, programmable machines. The compatibility of the components (similarly in process and sequences of processes) greatly enhances the productivity (utility) of the programmable equipment.

A side result has been the decrease in the non-chip-producing time of machine tools. The operator was relieved of the jobs of changing speeds and feeds and locating the tool relative to the work. Even simple forms of NC and digital readout equipment have provided both greater productivity and increased accuracy. Most early NC machine tools were developed for special types of work where accuracies of as much as 0.00005 in. were required, and many NC machines are built to provide accuracies of at least 0.0001 in., regardless of whether it was needed. This forced many machine tool builders to redesign their machines and improve their quality because the operator was not available to compensate for the machine (positioning) error. While most NC machine tools today will provide greater accuracy than is required for most jobs, the tendency is toward greater accuracy and precision (i.e., better quality) but at no increase in cost. Therefore, NC machines will continue to be the very backbone of the machine tool business. Hopefully, all builders will one day arrive at a common language, so that if one can program one machine, one can program them all.

As the name implies, *numerical control* is a method of controlling the motion of machine components by means of numbers or coded instructions.

Assume that three 1-in. holes in the part shown in Figure 32-27 are to be drilled and bored on a vertical spindle machine. The centers of these holes must be located relative to each other and with respect to the left-hand edge of the workpiece (*X* direction) and the bottom edge (*Y* direction). The depth of the hole will be controlled by the *Z* (or *W*) axis. For this part, this is the zero reference point.

The holes will be produced by center drilling, hole drilling, boring, reaming, and counterboring (five tool changes). If this were done conventionally or in manned cells, three or four different machines might be required. On the NC machining center, it only requires changing the tool. The movements of the table are controlled by coordinate systems. The accurate positioning of the cutting tool with respect to the work is established by zero points. The center of the table or a point along the edge of the traverse range is commonly used. The workpiece is positioned on the table with respect to this zero point on the table. For our example, the lower left-hand corner of the part is placed on the table 12 in. in the *X*+direction, 4 in. in the *Y*+direction and 2 in. in the *Z*+ direction with respect to the machine zero point.

A software program is written that instructs the table to move with respect to the axis of the spindle to bring the holes to the correct location for machining. The machine shown in Figure 32-27 is called a five-axis machine because it has five movements (shown by the dark arrows) under numerical control. No fixture was shown in this example.

HOW A(4) WORKS ON CNC MACHINES

NC and CNC machines can be subdivided into three types shown in Figure 32-28. In *point-to-point machines*, the tool path is not controlled. *Straightline* controls permit axially parallel traverses at desired table feed rates, but only one axis drive is operated at a time. *Contouring* permits two or three axes to be controlled simultaneously, permitting two- or three-dimensional geometries to be generated. Another term for contouring is *continuous path*. Most machining centers have contouring capability and are closed loop. The point-to-point machines are often open loop rather than closed loop.

FIGURE 32-27 The part to be machined on the NC machine has a zero reference point. The machine also has a zero point.

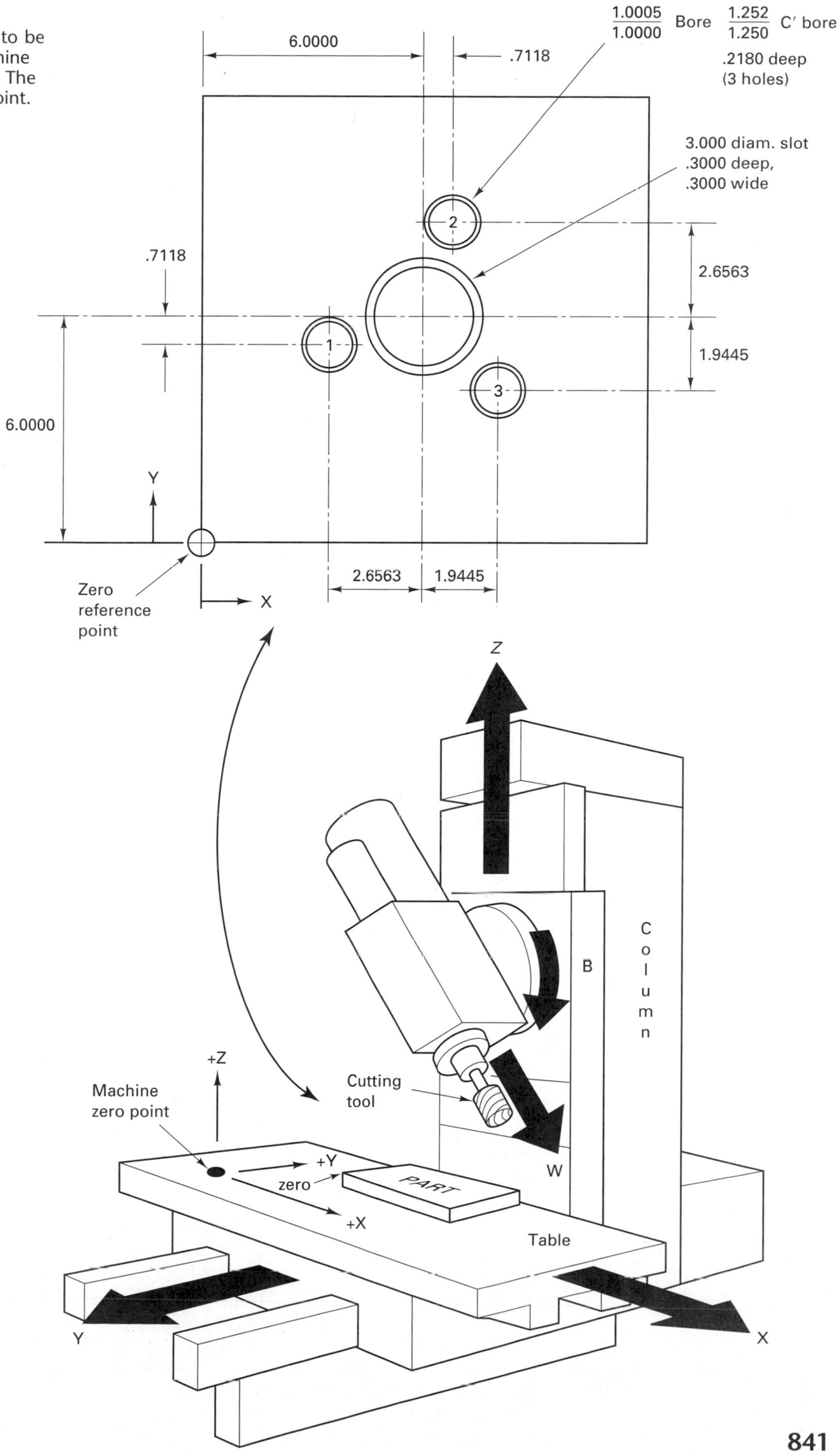

6.0000

.7118

$\frac{1.0005}{1.0000}$ Bore $\frac{1.252}{1.250}$ C' bore

.2180 deep
(3 holes)

3.000 diam. slot
.3000 deep,
.3000 wide

.7118

2.6563

1.9445

6.0000

Y

Zero
reference
point

X

2.6563

1.9445

Z

Column

B

W

+Z

Cutting
tool

Machine
zero point

+Y

zero

PART

+X

Table

Y

X

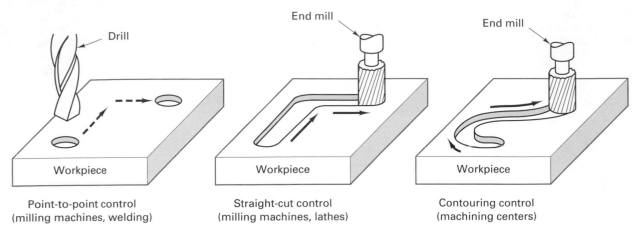

Point-to-point control
(milling machines, welding)

Straight-cut control
(milling machines, lathes)

Contouring control
(machining centers)

FIGURE 32-28 NC and CNC systems are subdivided into three basic categories: point-to-point controls, straight-cut controls, and contouring controls.

The X axis of the three-axis vertical spindle CNC machine tool shown in Figure 32-27 will be used to explain how the closed-loop positional control works. The CNC control system, shown schematically in Figure 32-29, uses a resolver or encoder to provide axis-position feedback to the *machine control unit* (MCU). A closed-loop control requires a transducer or sensing device to detect machine table position (and velocity for contouring) and transmit that information back to the MCU to compare the current status with the desired state. If they are different, the control unit produces a signal to the drive motors to move the table, reducing the error signal and ultimately moving the table to the desired position at the desired velocity. At this point, the command counter reaches zero, meaning that the correct number of pulses has been sent to move the table to the desired position. In a closed-loop system, a comparator is used to compare feedback pulses with the original value, generating an error signal. Thus, when the machine control unit receives a signal to execute this command, the table is moved to the specified location, with the actual position being monitored by the feedback transducer. Table motion ceases when the error signal has been reduced to zero and the function (drilling a hole) takes place. Closed-loop systems tend to have greater accuracy and respond faster to input signals but may exhibit stability problems (oscillating about a desired value instead of achieving it) not found in open-loop machines.

For most NC controls, the feedback signals are supplied by transducers actuated either by the feed screw or by the actual movement of the component. The transducers may provide either *digital* or *analog* information (signals). The resolver in Figure 32-29 measures (indirectly) the movement of the table by the rotation of the ball screw. A pulse disc on the end of the ball screw converts analog movement back into digital pulses, which are used to calculate table movement. The tachometer on the drive motor measures the table velocity.

Two basic types of digital transducers are used. One supplies *incremental* information and tells how much motion of the input shaft or table has occurred since the last time. The information supplied is similar to telling newspaper carriers that papers are to be delivered to the first, fourth, and eight houses from a given corner on one side of the block. To follow the instructions, the carriers would need a means of counting the houses (pulses) as they passed them. They would deliver papers as they counted 1, 4, and 8. The second type of digital information is *absolute* in character, with each pulse corresponding to a specific location of the machine components. To continue the carrier analogy, this would correspond to telling the carriers to deliver papers to the houses having house numbers 2400, 2406, and 2414. In this case it would only be necessary for the carriers (machine component) to be able to read the house numbers (addresses) and stop to deliver a paper when arriving at a proper address. This *address* system is a common one in numerical-control systems, because it provides absolute location information relative to a machine zero point.

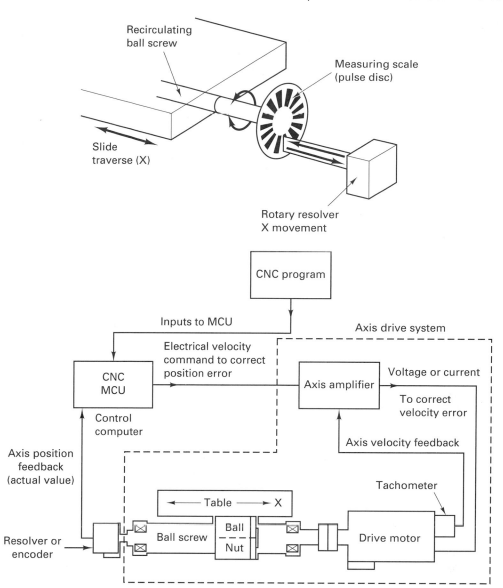

FIGURE 32-29 A CNC control system includes a velocity loop within an axis drive system and a position loop where table positions are measured with a resolver.

When analog information is used, the signal is usually in the form of an electric voltage that varies as the input shaft is rotated or the machine component is moved, the variable output being a function of movement. The movement is evaluated by measuring, or matching, the voltage, or by measuring the ratios between the applied and feedback voltages; this eliminates the effect of supply-voltage variations.

The input information (i.e., the location of the holes) is given in binary form to the machine control unit in the form of a punched tape, magnetic tape in a cassette, or a floppy disk, as shown in Figure 32-11. The data can be input directly via the control panel. This command signal is converted into pulses by the machine control unit (MCU), which in turn drives the servomotor or stepper motor.

Alternating-current servomotors are rapidly replacing direct-current motors on new CNC machine tools. The reasons for change are better reliability, better performance-to-weight ratio, and lower power consumption.

If the system is point-to-point (or positioning), the control system disregards the paths between points. Some positioning systems provide for control of straight cuts along the machine axes and produce diagonal paths at 45° to the axes by maintaining one-to-one relationships between the motions of perpendicular axes. Contouring systems require directional changes at controlled velocity. Any lost motion in the system can distort the part. In contouring, it is usually necessary to control multiple paths between points by interpolating intermediate coordinate positions. As many of these systems as desired can be combined to provide control in several axes—two- and three-axis controls are most common, but some machines have as many as seven. In many, conversion to either English or metric measurement is available merely by throwing a switch.

The components required for such a numerical control system now are standardized items of hardware. In most cases the drive motor is electric, but hydraulic systems are also used. They are usually capable of moving the machine elements such as tables at high rates of speed, up to 200 in./min being common. Thus exact positioning can be achieved more rapidly than by manual means. The transducer can be placed on the drive motor or connected directly to the lead screw, with special precautions being taken, such as the use of extra-large screws and ball nuts, to avoid backlash and to assure accuracy. In other systems, the encoder is attached to the machine table, providing direct measurement of the table position. Various degrees of accuracy are obtainable. Guaranteed positioning accuracies of 0.001 or 0.0001 in. are common, but greater accuracies can be obtained at higher cost. Most NC systems are built into the machines, but they can be retrofitted to some machine tools.

Initially, NC machines provided tool change; tool setting; and speed, feed, and depth-of-cut settings for positioning the work relative to the tool, with the remaining functions controlled by the operator. Gradually, the functions were incorporated into the control system so that the machines could change the tools automatically, change the speeds and feeds as needed for different operations, position the work relative to the tools, control the cutter path and velocity, reposition the tool rapidly between operations, and start and stop the sequence as needed.

Recirculating ball screw drives, of the type shown in Figure 32-30, greatly reduce the backlash in the drive systems, helping to eliminate problems of servoloop oscillation and machine instability. Using such hardware, NC machines are manufactured with greater accuracy and repeatability and more rapid table movements than is possible for conventional machines.

Some of the functions in programmable machines require feedforward or preset loops. The machine must know in advance the rough dimensions of a casting or a forging so that it can determine how many roughing cuts are needed prior to the finishing cut. However, for most CNC work, the operator (or part programmer) still plans the sequence of operations; selects the cutting tools and workholding devices; and selects the speeds, feeds, and depths of cut. Common machining routines such as pocket milling or peck drilling have been programmed into many CNC machines. These are called *canned cycles*. As shown in Figure 32-31, the operator merely supplies the information requested by the control menu, and the machine fills in the necessary data for the previously written program to perform the desired machining routines.

To ensure accurate machining of a workpiece on a CNC machine, the control system has to *know* certain dimensions of the tools. These *tool dimensions* are referenced to a fixed *setting point* on the toolholder. For the milling cutter, the dimensions are length L and cutter radius. For the turning tool, the dimensions are length L and transverse overhang. These dimensions are part of the information given to the operator on the setting sheet (see Figure 32-32). The program assumes that the tools will have the specified dimensions.

FIGURE 32-30 The ball lead screw provides great accuracy and position to NC and CNC machine tools.

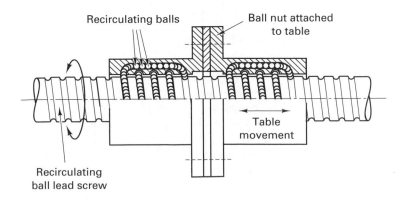

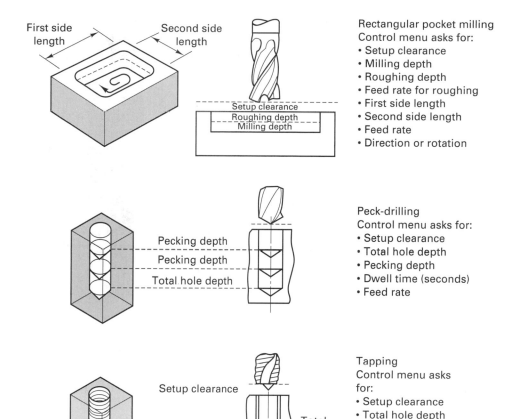

Rectangular pocket milling
Control menu asks for:
• Setup clearance
• Milling depth
• Roughing depth
• Feed rate for roughing
• First side length
• Second side length
• Feed rate
• Direction or rotation

Peck-drilling
Control menu asks for:
• Setup clearance
• Total hole depth
• Pecking depth
• Dwell time (seconds)
• Feed rate

Tapping
Control menu asks for:
• Setup clearance
• Total hole depth
• Dwell time (seconds)
• Feed rate

Source: Heidenhain Corp. (Elk Grove Village, Ill.)

FIGURE 32-31 Canned or preprogrammed machining routines greatly simplify programming CNC machines. *(Courtesy of Heidenkain Corporation, Elk Grove Village, Illinois.)*

PART PROGRAMMING

Obviously, the preparation of the control program for use in NC is a critical step. Standard languages and programs have been developed. For older machines, punched tapes (see Figure 32-33) are used.

The basic steps can be illustrated by reference to the part shown in Figure 32-27. The first step is to modify the drawing to establish the zero reference axes: the X, Y, and Z directions. The zero reference point is the lower left-hand corner of the part. The hole labeled 1 is located $+3.3437$ in the X direction and ± 5.2882 in the Y direction, with

FIGURE 32-32 The location of the corner of the end mill (*left*) or the tip of a single point tool (*right*) must be known with respect to the tool setting points so tool dimensions are accurately set.

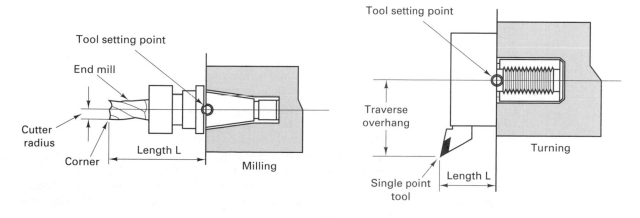

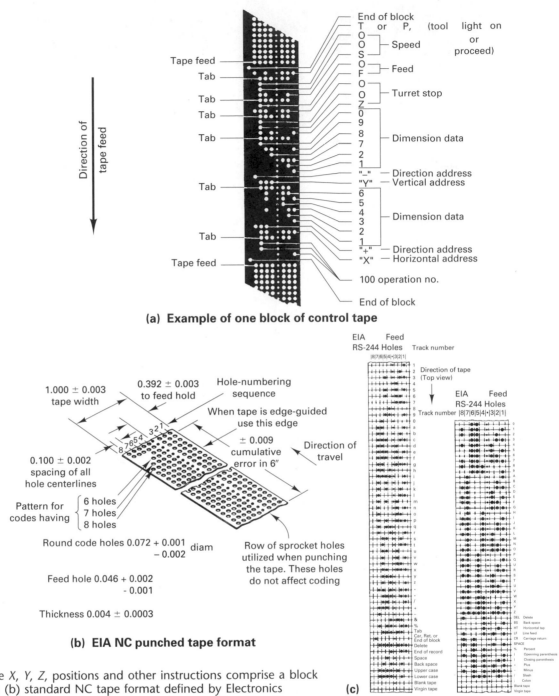

(a) Example of one block of control tape

(b) EIA NC punched tape format

(c)

FIGURE 32-33 (a) The X, Y, Z, positions and other instructions comprise a block of machine information; (b) standard NC tape format defined by Electronics Industries Association; (c) original standard for tape coding as defined by EIA was RS-244. The ASCII subset is now EIA standard RS-358. Both codes are used.

respect to its zero point. The part should be dimensioned with respect to point reference zero. This may require redrawing and redimensioning the part. Obviously, this step can be avoided if the original drawing is made in the desired form. The setup instructions given to the operator establish the position of the workpiece properly on the machine table with respect to the axis of the spindle or the machine zero reference point.

The second step is to make a *part program*. The program (1) defines the sequence of operations required to fabricate the part; (2) gives the X, Y, and Z coordinate positions of the operations; and (3) specifies the spindle traverse that determines the depth of the cut, the spindle speed, and feed, and also determines whether the same tool can continue the next operation or whether a tool change is required. The last four items are specified by code symbols, or NC words (see Table 32-4 and Figure 32-33). The NC words are put together in a specified order to define a *block* of information needed to execute

TABLE 32-4. Definitions of Common NC Words

NC Word	Use
N	*Sequence number:* identifies the block of information
G	*Preparatory function:* requests different control functions, including preprogramed machining routines
X, Y, Z, B	*Dimensional coordinate data:* linear and angular motion commands for the axis of the machine
F	*Feed function:* sets feed rate for this operation
S	*Speed function:* sets cutting speed for this operation
T	*Tool function:* tells the machine the location of the tool in the toolholder or tool turret
M	*Miscellaneous function:* turns coolant on or off, opens spindle, reverses spindle, tool change, etc.
EOB	*End of block:* indicates to the MCU that a full block of information has been transmitted and the block can be executed

an operation. By convention, the data are usually arranged in blocks in the sequential order shown in the table.

After the program has been written, the tape is prepared or the program directly entered into the control panel of a CNC machine. Usually, longer programs are entered by tape or disk; and short programs, entered manually. The inputs may be in binary code to convert the numerical and instructional information into commands the machine understands. Binary numbers facilitate the rapid and accurate operation of control systems and computers. The conversion from Arabic to binary is done by the control system or computer, not by the programmer.

Punched tape is not a recent idea. The old player-piano roll was a form of tape control, and punched cards had been used for many years for controlling complicated weaving and business machines. Thus tape control of machine tools is an extension of an existing basic concept in which holes, representing information that has been punched into the tape, are read by sensing devices and used to actuate the devices that control various electrical or mechanical mechanisms.

Historically, four basic types of tape format are used for NC input to communicate dimensional and nondimensional information: *fixed-sequential format, work-address format, tab-sequential format,* and *word-address* format. Most new NC or CNC systems use the word-address format, which allows the words to be presented in any order and is the most flexible.

After the program is converted to tape or disk, but before it is used, it is *verified,* or checked, to make sure that it is correct. The verification step can use the computer monitor, which simulates the part being made by tracing all the toolwork paths as they would occur on the machine tool. Sometimes a sample part is machined in plastic or wax for checking the part specifications from the drawing against the real part.

Today, CNC machines permit the user to program the interface between the machine tool and the control, greatly reducing the number of machining system components and interconnections. Current CNCs have extensive self-diagnostics and performance-monitoring systems. The greatest advances in CNC technology, however, are in part programming, where easy-to-use, menu-driven software makes programming almost as simple as setting up the machine manually. With older NC machines, the operator may override the tape when necessary but cannot reprogram the machine unless a new tape is prepared. The CNC machine has the capability of reading a program into its computer memory, and the program can be modified at the machine like any other computer program.

On CNC machines, the machine tool operator may perform all the programming steps right at the console of the machine, programming the processing steps for the part directly into the computer memory. The program can be saved by having the machine print out a copy of the program, which can be used later for reorders of the same part. Features such as program edit, canned routines, program storage, diagnostics, constant surface speed, and tape punch are common on today's CNC machines.

As more and more design work is done on the computer (CAD) using databases and software that are compatible to the machine tools, there will be less dependency on tape for program storage and more utilization of floppy disks, hard disks, and other typical computer storage means. For example, a machining cell composed of NC machine tools designed for a family of 10 component parts may be able to make the 10 different parts without needing retooling or refixturing, but it will still have 10 different programs for these parts for each machine. If the programs are in the stored computer, they can be readily accessed, but if they are stored on tape, delays will occur in dumping the different programs in to or out of the control computer.

Cutter Offset and Interpolation in Numerical Control. Although the majority of NC and tape-controlled machine tools do not provide for machining contoured surfaces, most CNC and DNC machines do. The required curves and contours are generated approximately by a series of very short, straight lines or segments of some type of regular curves, such as hyperbolas. This is called interpolation. The program fed to the machine is arranged to approximate the required curve within the desired accuracy. Figure 32-34 illustrates how a desired straight line or curved surface can be approximated by means of short segments. Interpolation refers to the fact that curved surfaces as generated by machine tools must be approximated by a series of very short, straight-line movements in the X, Y, and Z directions. The length of the segment must be varied in accordance with the deviation permitted. Most machine tools with contouring capability will produce a surface that is within 0.001 in. of the one desired, and many will provide considerably better performance. Most contouring machines have either two- or three-axis capability, but a good many have up to five-axis capability.

In milling machines, the centerline of the cutter is offset from the desired surface by the radius of the cutter. The path that the cutter needs to take to generate the desired geometry is not simply the perimeter or profile of the part. Thus cutter offset programs must be included in the software. Obviously, contour machining (with cutter offset and interpolation) requires that complex information be entered into the MCU because the number of straight-line or curved segments may be quite large. Manual programming of the tape can be quite laborious. Computer programming can translate simple commands into the complex information required by the machine.

COMPUTER LANGUAGES FOR NUMERICAL CONTROL

The control mechanisms on NC machines do not understand ordinary shop language that people use to describe what machining, or machine-work movements, must take place. In addition, the effects of various sizes and types of cutting tools, each requiring different

FIGURE 32-34 Two classic problems in NC programming are the determination of cutter offset and interpolation of cutter paths.

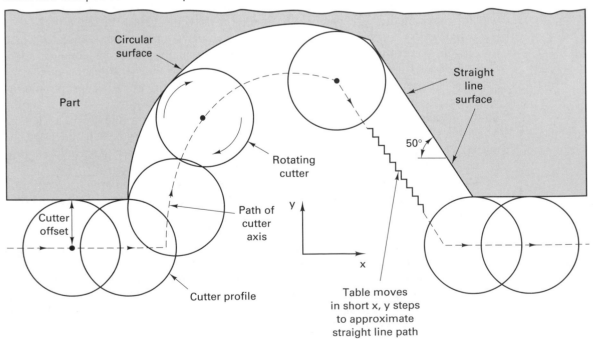

offsets, must be accounted for. Tool radius offset accounts for the fact that the center of tool rotation (as held in the spindle) must be offset from the workpiece surface to be generated. The path of the cutter centerline will have dimensions different from those of the surface. The capabilities of the machine tool regarding available power, speeds, feeds, table travel, and so on, must be taken into account. This barrier has been diminished by the use of CAD-based NC programming, which has reduced the use of APT, a special programming language that the computer can understand and convert into the commands required by NC machine controls. APT was the first widely used language in the United States and is used for both positioning and continuous path programming. APT was designed to run on mainframe computers with large memories. ADAPT was developed by IBM, had many of the features of APT but is designed for smaller computers. Another language, AUTOSPOT, was developed by IBM for point-to-point positioning, but today's version can be used for contouring. Today it is possible to implement APT on personal computers, but unlike many of the newer CAD/CAM programs, the APT system does not have visual aids or colorful graphics. See Figure 32-35 for a sample of APT programming for a bolt-hole circle with eight holes.

The part programmer prepares a manuscript that specifies the part geometry, the tool path, and the operations needed for their sequence, using English-like statements. Auxiliary machine functions, such as "coolant on," are also programmed. The program is entered into the computer. The computer performs an input translation, followed by the required arithmetic calculations including such things as cutter offset computations, to obtain the coordinate points the cutter must follow. The individualities of the specific machine tools, relative to the available speeds, feeds, accelerations, and so on, are provided for by a postprocessor program. The output from the postprocessor is the required NC tape (or disk or direct program), which is fed to the machine tool.

The workpiece, no matter how complex, can be conceived of as a compilation of points, straight lines, planes, circles, cylinders, and other mathematically defined geometries. Historically, the part programmer's job has been to translate the component parts into basic geometric elements, defining each geometric element in terms of workpiece dimensions. In recent years, much of this work has become routine in CAD systems. However, NC programming languages and CAD software are often incompatible, which has hindered the CAD-to-CAM step.

FIGURE 32-35   English and metric APT programs expedite the writing of NC program code.

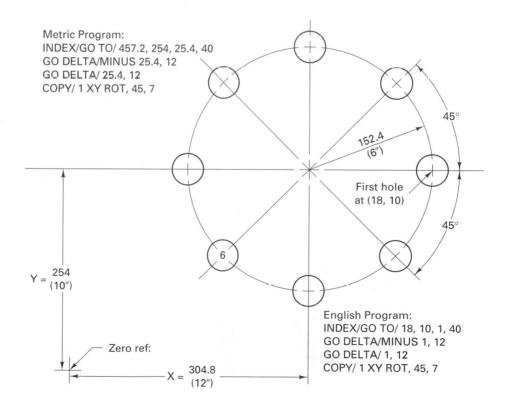

Metric Program:
INDEX/GO TO/ 457.2, 254, 25.4, 40
GO DELTA/MINUS 25.4, 12
GO DELTA/ 25.4, 12
COPY/ 1 XY ROT, 45, 7

152.4
(6")

45°

First hole
at (18, 10)

45°

English Program:
INDEX/GO TO/ 18, 10, 1, 40
GO DELTA/MINUS 1, 12
GO DELTA/ 1, 12
COPY/ 1 XY ROT, 45, 7

Y = 254
(10")

Zero ref:

X = 304.8
(12")

■ 32.8 MACHINING CENTER FEATURES AND TRENDS

Computer and numerical control is used on a wide variety of machine tools. These range from single-spindle drilling machines, which often have only two-axis control and can be obtained for about $10,000, to machining centers, such as shown in Figure 32-36. The machining center can do drilling, boring, milling, tapping, and so on, with four-axis control. It can automatically select and change 40–180 preset tools. The table can move left/right and in/out, and the spindle can move up/down and in/out, with positioning accuracy in the range 0.00012 in. with repeatability to ±.00004 in. over 40 in. of travel. The machine has automatic tool change and automatic work transfer capability, so that workpieces can be loaded and unloaded while machining is in process. Such a machine can cost over $200,000. Between these extremes are numerous machine tools that do less varied work than the highly sophisticated machining centers but which combine high output and minimum setup time with remarkable flexibility (large number of tool motions provided).

CNC TURNING CENTERS

The modern lathe has CNC control and tools mounted in turrets on slanted beds. The tailstock has been replaced by a live, powered spindle and chuck. On some lathes the concept of automatic tool changing has been implemented. The tools are held on a rotating tool magazine and a gantry-type tool changer is used to change the tools. Each magazine holds one type of cutting tool. This is an example of the trend of providing greater versatility along with high productivity in lathes. The versatility is being further increased by combining both rotary-work and rotary-tool operations—turning and milling in a single machine. Live, powered, or driven tools replace regular tools in the turrets and perform milling and drilling operations when the spindle is stopped. There

FIGURE 32-36 Modern machining centers will typically have horizontal spindles with rpms up to 15,000, dual pallets and cutting tool magazines holding 40 to 100 tools.

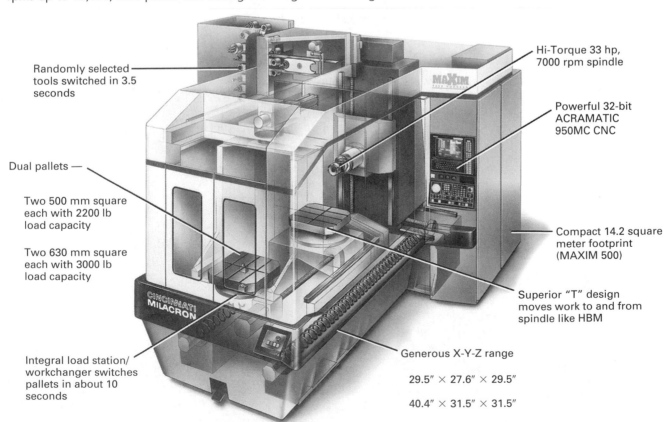

Randomly selected tools switched in 3.5 seconds

Dual pallets —

Two 500 mm square each with 2200 lb load capacity

Two 630 mm square each with 3000 lb load capacity

Integral load station/ workchanger switches pallets in about 10 seconds

Hi-Torque 33 hp, 7000 rpm spindle

Powerful 32-bit ACRAMATIC 950MC CNC

Compact 14.2 square meter footprint (MAXIM 500)

Superior "T" design moves work to and from spindle like HBM

Generous X-Y-Z range

29.5″ × 27.6″ × 29.5″

40.4″ × 31.5″ × 31.5″

are numerous tape-controlled machines that provide four- and five-axis contouring capability. Tools are changed in 6 seconds or less. It is also common to provide two or more worktables, permitting work to be set up while machining is done on the workpiece in the machine, with the tables being interchanged automatically. Consequently, the productivity of such machines can be very high, with the chip-producing time often approaching 50% of the total.

Numerical control has been applied to a wide variety of other production processes. NC turret punches with *X-Y* control on the table, CNC wire EDM machines, laser and machining, flame cutters, and many other machines are readily available.

Some new trends are being observed in the development of machining centers such as smaller, compact machining centers with higher spindle speeds. Machines with four- and five-axis capability are readily available. Modern machining centers have contributed significantly to improved productivity in many companies. They have eliminated the time lost in moving workpieces from machine to machine and the time needed for workpiece loading and unloading for separate operations. In addition, they have minimized the time lost in changing tools, carrying out gaging operations, and aligning workpieces on the machine.

The latest generation of machining centers is aimed at further improving utilization by reducing the time when machines are stopped, either during pauses in a shift or between shifts. Delays are caused by tool breakage, unforeseen tool wear, limited number of tools, or an inadequate number of available workpieces. Machines are fitted with tool breakage monitors, tool wear compensating devices, and means for increasing the number of tools and workpieces available.

Probes on NC machines can greatly improve the process capability of the machine tool. There is a big difference between the claimed program resolution for an NC machine and the accuracy and precision (the process capability) of the actual parts. As shown in Figure 32-37, true positioning accuracy and precision are affected by machine alignment, machine and fixture setup, variations in the workholding device, raw material variations, workpiece location in the fixture variations, and cutting tool tolerances. Thus the finished workpiece may be unacceptable even though the machine is more than capable of producing the part to the design specifications. The part program has no assurance that the part is properly located in the fixture or that the fixture is properly located on the table of the machine. However, a probe, carried in the tool storage magazine and mounted when needed in the spindle like a cutting tool, can establish the location of the surface features relative to each other and to the spindle axis within 0.0005 in. (Figure 32-38). The machine controller, using the probe data, will then shift the program reference data accordingly. The probe can be used to determine the amount of material on a rough casting, locate a corner of a part, define the center of a hole, or

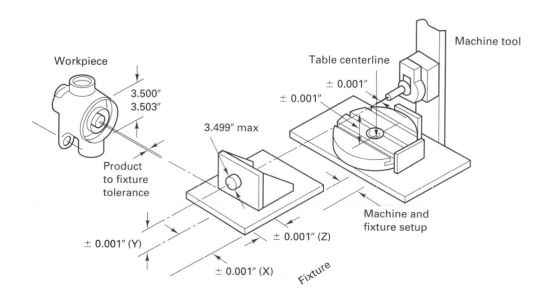

FIGURE 32-37 Process capability in NC machines is affected by many factors.

FIGURE 32-38 (a) Probe carried in the tool changer can be mounted in the spindle (b) for checking the location of part features accurately.

check for the presence or absence of a feature. All of the variability described in Figure 32-37 can be compensated for except for variations in the cutting-tool geometry or tool wear. A probe mounted on the machine tool can be used to automatically update tool-offset data in the control computer. Thus the machine tool can function as a coordinate measuring machine. By comparing the actual touched location with the programmed location, the measuring routine determines appropriate compensation.

Many of the new machining centers are equipped with novel automatic pallet changers and workpiece loading and unloading devices. Robots are increasingly being applied for workpiece handling in machine groups, including manufacturing cells. In some cases the robot is also used for tool-changing functions.

■ 32.9 ULTRA HIGH-SPEED MACHINING CENTERS (UHSMCs)

Anyone involved in the product development process knows that the longest lead time path from a new product (like a car) design to a finish product coming off the final assembly line always includes the die making process. For example, in the automotive industry, each new car design may have many die sets for forged and sheet metal parts. Machining of the dies is a key process in the conventional die making process (see Figure 32-39), where the major steps are milling, EDM, polishing, die assembly, tryout, and modification.

Ultra high speed machining centers (UHSMCs) are available to shorten the die making process lead time. The new machining centers have highly accurate high-speed spindles capable of rpms of 30,000 to 50,000, cutting feed rates of 60 m/min and cutting feed accelerations of 9.8 m/sec². The machine requires a spindle with high stiffness utilizing ceramic ball bearings with a constant-pressure preload mechanism and jet lubrication for superior performance.

Robust machining centers of this sort are capable of very high metal removal rates particularly in materials like aluminum. However, before investing in a high speed machining center (HSMC), these issues must be addressed. The new machine will usually require the purchase of new cutting tools and toolholders. At the higher rpms and feed rates, smaller diameter tools are easier to balance. Tungsten carbide is the primary tool material. Make sure the insert retention mechanism is adequate in case of a failure. Other major problems will include chatter and removal of the large volume of chips.

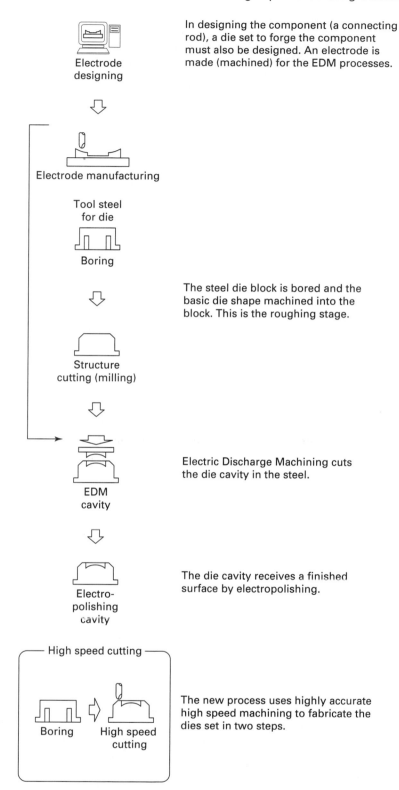

In designing the component (a connecting rod), a die set to forge the component must also be designed. An electrode is made (machined) for the EDM processes.

Electrode designing

Electrode manufacturing

Tool steel for die

Boring

The steel die block is bored and the basic die shape machined into the block. This is the roughing stage.

Structure cutting (milling)

Electric Discharge Machining cuts the die cavity in the steel.

EDM cavity

The die cavity receives a finished surface by electropolishing.

Electro-polishing cavity

High speed cutting

Boring   High speed cutting

The new process uses highly accurate high speed machining to fabricate the dies set in two steps.

FIGURE 32-39 Process for making forging dies—old versus new.

UHSMCs require exceptionally accurate, stiff spindles using ceramic ball bearings, raising the possible rpm as compared to regular ball bearings, air bearings, or magnetic bearings. As shown schematically in Figure 32-40, synchronized sets of ball screws are used to feed the tool in the X and Y directions to reduce the errors caused by distortion in the frame during rapid acceleration of the tool. These machines have usually four or five axes under control and represent the pinnacle of machine tool development at this time.

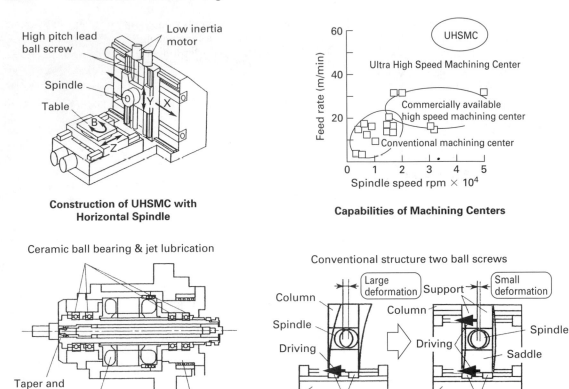

High pitch lead ball screw

Low inertia motor

Spindle

Table

Construction of UHSMC with Horizontal Spindle

Capabilities of Machining Centers

Ceramic ball bearing & jet lubrication

Taper and face contact

Built-in motor

Constant pressure preload mechanism

Construction of Spindle

Conventional structure two ball screws

Large deformation

Column

Spindle

Driving

Bed

Support

Support

Small deformation

Column

Driving

Bed

Spindle

Saddle

Support

Feed Mechanism of Conventional (X-axis) MC vs UHSMC

FIGURE 32-40 Ultra high speed machining centers (UHSMCs) are being developed with ceramic ball bearings in the spindles, synchronized ball screws on the X axis to reduce distortion (due to inertia) in the moving components. *(Development of Ultra High Speed Machining Center, Toyota Technical Review Vol. 49 No. 1 September, 1999.)*

■ 32.10 SUMMARY

The yardstick for automation developed in 1962 still serves us well in 2002 with some minor additions. Clearly, Numerical Control machines, robots, FMSs, and computers are seen as critical elements of the advanced manufacturing technologies available for the next decade, which is being touted as the time when computer-integrated manufacturing (CIM) will become a widespread reality. Computer technology abounds: computer-aided design; computer-aided manufacturing with NC, CNC, or A/C and DNC; computer-aided process planning; computer-aided testing and inspection (CATI); artificial intelligence; smart robots; and much more. But a word of caution: Any company can buy computers, robots, and other pieces of automation hardware and software. The secret to manufacturing success lies in the design of a unique manufacturing system so that it can achieve superior quality at low cost with on-time delivery and still be flexible. Flexibility means the system can readily adapt to changes in the customer demand (both volume and mix) while quickly implementing design changes. This requires a visionary management team and a change in culture on the factor floor (i.e., an empowered and involved workforce). Changing how people work in a manufacturing system means you have to redesign the system. No better explanation of this can be found than at the Toyota Motor Company. Led by their vice president for manufacturing, Taiichi Ohno, who conceived, developed, and implemented Toyota's unique manufacturing system, this company has emerged as the world leader in car production. The implementation of this system has saved many companies (Harley Davidson, for example) and carried many others to the top position in their industry. The Toyota system is unique and as revolutionary today as were the American Armory System (job shops) and the Ford system for mass production in their day. It is significant that virtually every manufacturing system or technology cited in this chapter is practiced at Toyota. This new system is now being called *lean production* (to contrast it to mass production). Toyota does not use the word *CIM* because the computer is only a tool used in their system, a system that recognizes people as the most flexible element. This new system is discussed in Chapters 41–43. Students of industrial, mechanical, and manufacturing engineering are well advised to be knowledgeable about this unique system.

■ KEY WORDS

A/C (adaptive control)
ACC (adaptive control constraint)
ACO (adaptive control optimization)
AGV (automated guided vehicle)
automation
CAD (computer-aided design)
CAM (computer-aided manufacturing)
canned cycles
CAPP (computer-aided process planning)

CIM (computer-integrated manufacturing)
closed-loop control
CNC (computer numerical control)
contouring
controller
DNC (distributed numerical control)
DNC (direct numerical control)
end effector
feedback device
feedforward

FMS (flexible manufacturing system)
HSMC (high speed machining center)
machining center
manipulators
marketing
material requirements planning
MCU (machine control unit)
mechanization
MPS (master production schedule)

MRP (manufacturing resource planning)
numerical control
open-loop control
pallet changer
part program
PLC (programmable logic controller)
point-to-point machines
robotic cell
transfer machine
UHSMCs (ultra high speed machining centers)

■ REVIEW QUESTIONS

1. What human attribute is replaced by a machine capability for the first five levels of automation?
2. Give an everyday example of a household device or appliance that exhibits automation levels A(1), A(2), A(3), and A(4).
3. Explain how a windmill is an A(4) device or machine (i.e., self-adjusting).
4. What device in the typical modern bathroom has a feedback control device in it?
5. In terms of a transfer line, explain what is meant by *line balancing*. Discuss using the rotary transfer line shown in Figure 32-6.
6. How is a machining center different from a milling machine?
7. What role did John Parsons play in the development of NC?
8. What was DNC as first practiced, and what does DNC usually mean today?

9. How does an adaptive control system differ from a numerical control system?
10. How does feedforward differ from feedback in a process control system?
11. Why was it necessary for machine tool builders to improve the lead screws on their machines when they made them into NC machines?
12. Explain what is meant by *interpolation* in NC programming.
13. Explain the problem of cutter offset in NC programming, using Figure 32-A with a $\frac{3}{4}$-in. diameter cutter end milling on the bottom and right side.
14. Some of the functions performed by the operator in piece-part manufacturing are very difficult to automate completely. Name the functions and explain why.

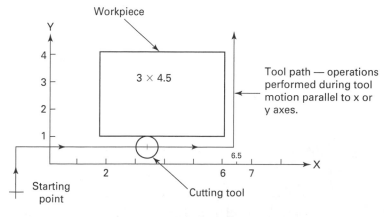

Straight-cut NC operation

FIGURE 32-A

15. The first NC machines were closed-loop control. Later some machines were open loop. What change did this require on the part of machine tool builders?
16. Can a continuous-path NC machine be open loop? Why or why not?
17. Why do you think there are no NC shapers or broaches?
18. Why isn't manual programming used for continuous-path NC?
19. There are three basic closed loop feedback A(4) schemes used on CNC machines. Which is the most accurate?
20. What is the difference between the zero reference point and the machine zero point?
21. What is an encoder?
22. Why does contouring require both velocity and positional feedback?
23. What is a peck-drilling subroutine for a CNC machine?
24. What is pocket milling, and what kind of milling cutter is usually used to perform it?
25. How are probes used in CNC machines to improve process capability?
26. What are G words used for in NC?
27. Flexible manufacturing systems use CNC machines for processing and AGVs, robots, or conveyors to transport parts. What differentiates the FMS from a transfer line?

28. Why do you think that transfer lines are using more programmable machines and PLCs?
29. A standard piece of CAD analysis is called FEA. What does a finite element analysis do?
30. What is the difference between the two types of CAPP?
31. The A(6) level tries to relate causes to effect using AI. What is AI?
32. What are the main areas in which robots are being utilized in industry today?
33. What are the basic components of all robots?
34. What are work envelopes for industrial robots?
35. How is positional feedback obtained in robots?
36. Compare a rotary transfer machine with an unmanned robotic cell.
37. Compare a human worker in a manned cell to the robotic worker in the unmanned cell. What are the advantages of one over the other?
38. What is tactile sensing? Name some examples of tactile sensing in the home.
39. The vast majority of robots are used in automobile fabrication for welding and painting. Why?
40. What are the major changes that are being introduced into UHSMCs?

■ PROBLEMS

1. What are the X and Y dimensions for the center position of holes 2 and 3 in the part shown in Figure 32-27?
2. Configurations obtained from continuous-path machining are the result of a series of straight-line, parabolic-span or higher-order curves. The degree to which curved surfaces correspond to their design depends on how many lines or spans are used. Four equal chords in a circle describe a square (Figure 32-C). Six make a hexagon, and, as the number increases, the lines themselves come closer to a perfect circle. The number of lines needed is determined by a maximum tolerance allowed between the design of the curved section and the actual chord programmed. This is the dimension T. The program for a parabolic span control unit requires enough spans for any deviation to stay within an acceptable tolerance. For a tolerance of $T = 0.001$ in., how long should the span be for a curve with a 5-in. radius.? Assume that the arc is part of a circle. What is the span angle here, in degrees?

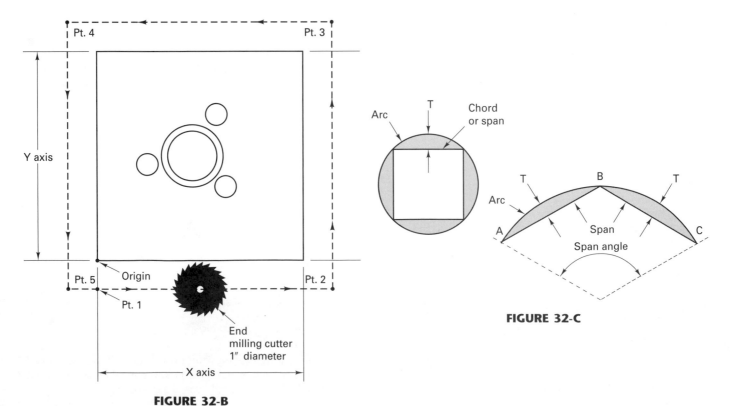

FIGURE 32-B

FIGURE 32-C

3. In Problem 2, suppose that the acceptable tolerance was 0.0001 in. Determine the span angle.

4. Suppose that the plate shown in Figure 32-B was to be profile milled around the periphery with a 1-in.-diameter end-milling cutter. The dashed line is the cutter path. The programmer must calculate an offset path to allow for cutter diameter. Since the programmed points are followed by the cutter centerline and the profile is made at the tool's periphery, the programmer called for a $\frac{1}{2}$-in. cutter offset. Working with computer assistance, the programmer would describe the part profile to be machined and specify the cutter. The computer would generate the cutter path. Complete the table below to specify the cutter path, starting with the origin at the zero reference point. Move the tool around the plate counterclockwise.

Programmed Point Locations

PT	X	Y
1		
2		
3		
4		
5		
1		

5. Suppose that surface finish is very important for the profile milling job described in Problem 4. Thus downmilling is going to be used. Rewrite the NC program points to accommodate this requirement. Show the new path on a sketch such as Figure 32-B.

6. An example of how the APT language would be used on a part is shown in Figure 32-35. Essentially, it is necessary to tell the computer the location of a center of the bolt hole circle or one of the holes in the circle with respect to a zero point, the number of holes in the circle, and the radius of the circle. Numerals 18, 10, and 1 (457.2, 254, 25.4) are X, Y, and Z coordinates for table and tool. Numeral 40 is table movement rate. Numeral 12 is feed rate for drill. Numeral 45 is 45° of rotation. Numeral 7 is the instruction for seven duplicate holes to be drilled. Calculate the location of hole number 6 in X and Y English numbers.

7. Table 32-A gives the Arabic and comparative binary numbers. The number 182 is expressed in Arabic thus as:

$$1 \times 100^2 = 100$$
$$8 \times 10^1 = 80$$
$$2 \times 10^0 = 2$$
$$\overline{182}$$

In binary, the number 182 is 10110110. Show how this is obtained.

8. You have received the part drawing for a typical lathe part that will require turning, facing, grooving, boring, and threading as it is machined from a casting. See Figure 32-D. Unfortunately, you do not yet know how many of these parts will be ordered this year, so you do not know what the build quantity will be. To be prepared, you have developed some cost data for the manufacture of the part by four different lathe processes (see Table 32-B). Complete the table by determining the run cost per batch, the cost per unit at the various quantities, and the total cost per batch.

Answer the following questions regarding this situation:

1. Of the four costs listed for each process, which costs are fixed and which are variable?

TABLE 32-A.   Comparison of Arabic and Binary Numbers. Note: Any number raised to the zero power equals 1.

Arabic	Binary	Power of 2
0	0	
1	1	2^0
2	10	2^1
3	11	
4	100	2^2
5	101	
6	110	
7	111	
8	1000	2^3
9	1001	
10	1010	
11	1011	
12	1100	
13	1101	
14	1110	
15	1111	
16	10000	2^4
17	10001	
18	10010	
19	10011	
20	10100	
21	10101	
22	10110	
23	10111	
24	11000	
25	11001	
26	11010	
27	11011	
28	11100	
29	11101	
30	11110	
31	11111	
32	1000000	2^5
64	1000000	2^6
128	10000000	2^7

examples:

Value of 182 expressed in Arabic numbers

$$1 \times 10^2 = 100$$
$$8 \times 10^1 = 80$$
$$2 \times 10^0 = 2$$
$$182 \qquad \overline{182}$$

Value of 182 expressed in binary numbers

$$1 \times 2^7 = 128$$
$$0 \times 2^6 = 0$$
$$1 \times 2^5 = 32$$
$$1 \times 2^4 = 6$$
$$0 \times 2^3 = 0$$
$$1 \times 2^2 = 4$$
$$1 \times 2^1 = 2$$
$$0 \times 2^0 = 0$$
$$10110110 \qquad \overline{182}$$

2. What is included in the cost per unit? Do you need to estimate the machining time per piece and the cycle time per piece, including the time to change parts and setups in order to compute run cost?

3. How would you go about estimating this time, and what time elements might be included in the cycle time in addition to the machining time?

4. How would you use this estimate of time in the cost table? Show the calculation.

5. Make a plot of cost in dollars versus quantity, with all four methods on one plot.

6. Make a plot of cost per unit versus quantity, again with all four methods on one plot. Find the breakeven quantities. (*Hint:* Did you plot the data on log paper?)

TABLE 32-B.   Cost Data for Lathe Processes

	Make Quantity				
	10,000 Units	1,000 Units	100 Units	10 Units	1 Units
Cost to produce on six-spindle automatic					
Total cost of batch	—	—	—	—	—
Engineering 2.5 hr at $40/hr	50.00	50.00	50.00	50.00	50.00
Tooling (cutting tools and workholders)	600.00	600.00	600.00	600.00	600.00
Sctup 8 hr at $15/hr	120.00	120.00	120.00	120.00	120.00
Run cost per batch: 50¢ per piece	5000.00	500.00	50.00	5.00	0.50
Cost each	—	—	—	—	—
Cost to produce on turret lathe					
Total cost of batch	—	—	—	—	—
Engineering 2 hr at $20/hr	40.00	40.00	40.00	40.00	40.00
Tooling	150.00	150.00	150.00	150.00	150.00
Setup 4 hr at $20/hr	48.00	48.00	48.00	48.00	48.00
Run cost per batch: $8 per piece	—	—	—	—	—
Cost each	—	—	—	—	—
Cost to produce on engine lathe					
Total cost of batch	—	—	—	—	—
Engineering 1 hr at $20/hr	20.00	20.00	20.00	20.00	20.00
Tooling (no cost)	—	—	—	—	—
Setup 2 hr at $12.00/hr	24.00	24.00	24.00	24.00	24.00
Run cost per batch: $12	120000.00	12000.00	1200.00	120.00	12.00
Cost each	12.00		—	—	—
Cost to produce on NC lathe					
Total cost of batch	—	—	—	—	—
Engineering and programming	150.00	150.00	150.00	150.00	150.00
Tooling	100.00	100.00	100.00	100.00	100.00
Setup 1 hr at $20/hr	20.00	20.00	20.00	20.00	20.00
Run cost per batch: $2 per piece	—	—	—	—	—
Cost each	—	—	—	—	—

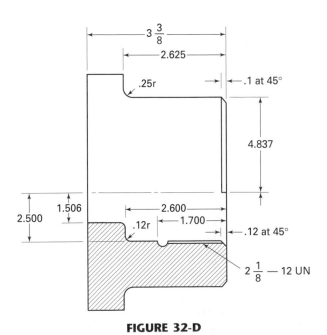

FIGURE 32-D

7. Discuss the breakeven quantities that you found in part 5 versus part 6. (They should be the same.)

8. When would you use the NC lathe? The turret lathe? When would you use the turret lathe if you have no NC lathe?

9. As shown in Figure 32-E, a keyway and two holes are to be cut using an end mill on a vertical milling machine. The end mill is a versatile milling tool that can be employed to cut metal using either its cylindrical sides, its ends, or both simultaneously. In the following NC word address program, feeds are expressed in inches per minute and spindle speed is expressed in revolutions per minute. Some word address definitions are also provided next. The end mill is 0.500 in. in diameter.

1. Sketch the programmed path of the tool, labeling the endpoints. A simple x-y-plane view of the part is recommended, but you may use the 3-D figure provided if you can make your sketch clear and incorporate z-axis position.

2. Briefly state in English what has been commanded in each program block (line). You may use your sketched tool path for reference. For example, for line 1, you might state, "Rapid transverse in x-y plane to point #1".

3. How deep are the holes?

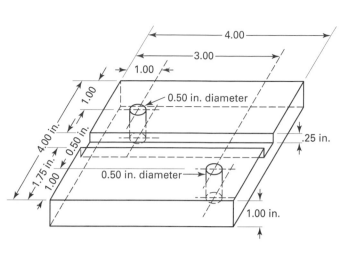

FIGURE 32-E

Dimensions shown on figure: 4.00, 3.00, 1.00, 1.00, 0.50 in. diameter, 4.00 in., 0.50 in., 1.75 in., 1.00, .25 in., 0.50 in. diameter, 1.00 in., 1.00 in.

Word Address (Absolute Programming)

N01 GO G17 X1.00 Y3.00 EOB v
N02 G20 Z-0.75 F2.87 S573 M3 EOB
N03 G20 20.00 EOB
N04 GO G17 X-0.50 Y2.00 EOB
N05 G20 Z-0.50 EOB
N06 G17 X5.25 F2.67 S382 EOB
N07 G20 20.00 EOB
N08 GO G17 X3.00 Y1.00 EOB
N09 G20 Z-0.75 F2.87 S573 EOB
N10 G20 20.00 EOB
N11 M5 EOB
N12 M2 EOB

Zero plane for the z axis is 0.25 inches above the workpiece.

Word definitions

GO	rapid transverse	MO	no M code
G1	linear interpolation	M2	end of program
G7	go home	M3	start spindle rotation
G17	axis selection x-y	M5	stop spindle
G20	axis selection z		
G90	absolute programming		
G91	incremental programming		

*C*hapter 32 CASE STUDY

Steam Line Holes

An underground steam line for a military base in Alaska, approximately 1.6 km (1 mile) in length, used the units depicted in Figure CS-32, each unit being 9.14 m (30 ft) in length. As shown, the steam line for each unit was enclosed in a cylindrical conduit made of sheet steel and weighed approximately 356 kg (785 lb). The conduit was to serve as the return drain line and to provide insulation. Each length of steam line was supported within the conduit by means of three U-shaped legs, made by cold-bending hot-rolled steel bar stock. See section A–A' for a cross section. These legs were welded to the pipe about 2 m from each end.

The units were fabricated in California and were transported to Alaska by a sequence of truck, barge, and railroad. Because of an early winter, only one-half of the line was installed the first summer. Before work was resumed the fol-

lowing spring, someone decided to test the line and found that there were numerous holes in the conduit. The holes (slits in casing) were located where the edges of the U-shaped supporting legs contacted the outside conduit (casing) on the top of the casing. As the result of these holes, the project was delayed and a costly litigation ensued.

1. Why do you think these holes occurred?
2. Was it a design error? A fabrication error? A service error? A service environment error?
3. What error(s) caused the failures? When did the defects actually develop?
4. What action would you recommend to avoid this problem in the future?
5. What should be done about the part of the line already installed?

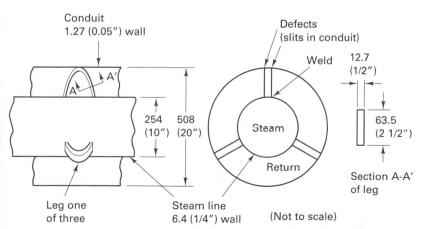

FIGURE CS-32 Schematic diagram showing method of supporting steam line in drainage conduit for an underground steam line.

CHAPTER 33

RAPID PROTOTYPING AND FREEFORM FABRICATION

■ 33.1 INTRODUCTION

Most manufacturing companies are involved in product development, a process wherein the product concept must be transitioned from engineering drawings to a working physical product. Producing the first physical model, or *prototype*, of the product is called *prototyping*. Prototyping is important because it is the ultimate means for verifying the form, fit, and function of a product. Prototypes are made in low volume at high unit cost because all the tooling costs associated with making the prototypes are spread over a small number of parts. The fabrication of specialized tooling (e.g., patterns or molds for casting; dies for forming; fixtures for machining) results in long lead times for getting the prototypes made and tested. *Rapid prototyping* (RP) methods assist in making the product development process cheaper and faster, which can ultimately impact customer satisfaction and corporate profitability by helping the company get the product to market first.

Over the past couple of decades, many industry groups have used the term *rapid prototyping* strictly in reference to mechanical product development (e.g., prototyping of mechanical components for products like copy machines, computers, cellular phones, automobiles, aircraft subassemblies, and medical diagnostic equipment, among many others). However, software developers, printed-circuit-board manufacturers, and microelectronic companies also use the term rapid prototyping to describe methods and technologies used to accelerate product development in those industries. The information presented in this chapter focuses on RP for mechanical components.

Other terms used to describe RP are *desktop manufacturing*, which refers to the size and convenience of many new RP technologies and is analogous to the term desktop publishing used in the publishing industry; *3D printing*, which draws a parallel with the laser and inkjet printers used to produce two-dimensional images; freeform fabrication, or layered manufacturing; and tool-less manufacturing, described in more detail later.

Most methods of mechanical component RP are some form of *freeform fabrication* (FF), referring to the freedom to produce largely unconstrained mechanical part geometries. Some processes are even capable of producing a ship in a bottle without any support posts. FF processes use layered manufacturing techniques to build up parts in small material increments called *voxels* (i.e., volume elements). As shown in Figure 33-1, the first step in all FF processing is the *preprocessing* step used to convert the part geometry into a sequence of *tool paths* used to program the FF machine tools. The computer-aided design (CAD) part shown on the computer screen is redrawn in layers starting from the bottom and moving up. During this step, *bases* and *supports* may be added to the part to allow the part to be separated from the machine tool after processing and to support suspended

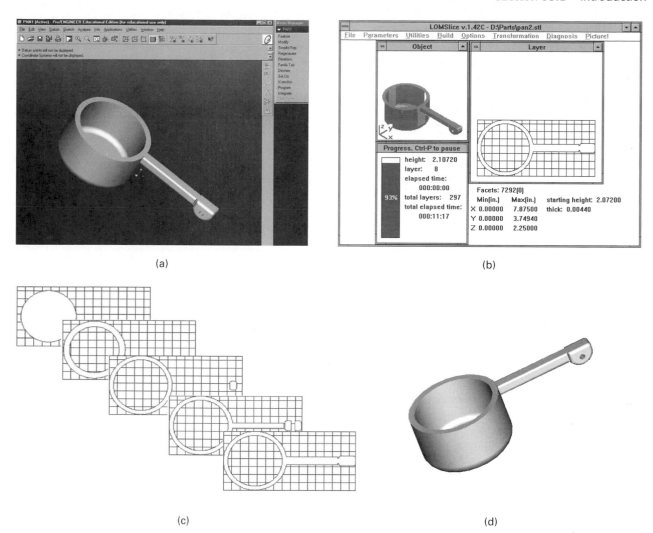

(a)

(b)

(c)

(d)

FIGURE 33-1 Conceptual framework for producing rapid prototypes: (a) development of a virtual model in CAD; (b) model is converted to STL file format and loaded into CAM software (for Laminated Object Manufacturing in this case); (c) CAM software slices model to generate tool paths; and (d) final prototype is produced layer-by-layer in FF machine tool.

geometries such as cantilevers. Computer-aided manufacturing (CAM) software is used to virtually slice (CAD) files of the part geometry into a series of layers. Next, each layer is reduced to a set of tool paths used to guide the selective deposition of energy or material during fabrication. Software in support of FF preprocessing is similar in concept to the CAM software needed for reducing CAD geometries to a series of tool paths in support of computer numerical control (CNC) machining. Once preprocessing has been completed, the prototype can be automatically fabricated one layer at a time within the FF machine tool. Figure 33-2 shows a part processed by stereolithography. The preprocessed data are used to guide the delivery of energy or material to pattern each layer in the prototype, bonding new layers to previous layers. In stereolithography, a laser beam selectively polymerizes the liquid photopolymer by scanning the laser beam over the surface using a X-Y raster scanning pattern. By processing each layer one on top of another, a 3D physical model of the CAD part is produced. After processing, the physical prototype will typically require some *postprocessing*, which may include finishing depending upon the process and the application. In the case of stereolithography, removal of excess resin from the exterior of the prototype as well as postcuring of the prototype in an ultraviolet (UV) oven is necessary prior to finishing the prototype.

FF processes are *additive*, meaning they build up the prototype in thin layers. This is in contrast to machining processes, which are *subtractive*, meaning they remove material from a bulk workpiece. While metal removal processes are used in prototyping, certain geometries cannot be produced on CNC machining centers. Further, machining

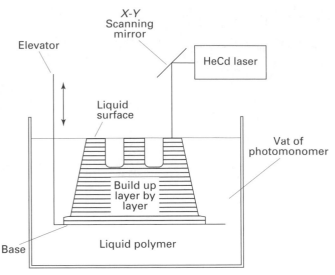

FIGURE 33-2 The stereolithography apparatus (SLA) (*on right*) can build a prototype part (*on left*) in plastic, layer by layer, using a laser to photopolymerize the liquid photopolymer. *(Source: Cutting Tool Engineereing, Dic 1989.)*

is hampered by special tooling and fixturing requirements, which can significantly increase development times and costs.

The allure of FF is that it results in *tool-less manufacturing*, that is, the ability to make any product geometry, no matter how complicated, without requiring costly part-specific tooling like dies or fixtures. FF processes have the unique ability to prototype complex, freeform surfaces and contours, as quickly as simple planar and cylindrical features. For simple geometries, subtractive machining processes using general-purpose tooling provide a much quicker and cheaper means for prototyping. For the complex geometries produced by CAD solid modeling, FF processes are required. In fact, the time required to produce prototypes has actually decreased despite the increased complexity of CAD geometries largerly due to the introduction of FF processes. In addition to rapid prototyping, FF processes are also used for test marketing, rapid manufacturing (e.g., small batches of plastic parts), custom manufacturing (e.g., one-of-a-kind prostheses), and rapid tooling. Rapid tooling involves the application of FF processes to the rapid fabrication of prototype and production tooling used to make products. FF processes have been applied in fields as diverse as architecture, medicine, archaeology, and the arts.

Many original equipment manufacturers of FF machine tools are beginning to take advantage of this differentiation in applications by building equipment that addresses specific markets. Several companies are now offering desktop-sized *concept modelers* as three dimensional printers to be used for design verification applications. This is in contrast to the larger *functional modelers* that build larger, more accurate, and more durable prototypes. Concept modelers are designed for application in office environments, whereas functional modelers are more for use in shop floor or laboratory environments. It is expected that the lower-cost concept modelers will become more commonplace in industry for the purpose of quickly verifying product concepts, whereas functional modelers will continue to serve testing and tooling markets.

■ 33.2 BASIC CONCEPTS IN FREEFORM FABRICATION

Over 40 different types of FF processes have been identified in various stages of development. These FF processes can be classified into four basic groups:

1. *Photopolymer-based.* Photopolymerization is the mode of material addition.
2. *Deposition-based.* Material is physically deposited.
3. *Powder-based.* Powder is selectively bound together.
4. *Lamination-based.* Layers are bonded together and patterned.

Table 33-1 gives a summary of the processing characteristics for various FF machine tools. Not included in this table are subtractive methods of prototyping such as CNC machining. When examining this table, recall that conventional turning has these typical values:

MRR	10 to 200 in^3/min. (160 to 3300 cm^3/min.)
Surface finish	32 to 250 μin. (0.8 to 6.4 μm) AA
Dimensional accuracy	0.0005 to 0.002 in. (0.0125 to 0.05 mm)

Material properties (metals like aluminum and steel)
- Tensile strength   15 to 150 ksi (100 to 1000 MPa)
- Elastic Modulus   10 to 200 GPa (1,500 to 30,000 ksi)
- Cost   0.05 to 1 \$/in^3 (3 to 61 \$/dm^3)

Typical work envelope   24 in. dia. $\times$ 48 in. (61 cm dia. $\times$ 122 cm)

FF processes typically have low material addition rates compared to the material removal rates in machining. They typically have poorer accuracy and surface finish with smaller work envelopes. Because FF processes are limited to a few specialized materials, material properties tend to be worse and material costs higher. The ultimate advantage of FF processes is the production of complex mechanical geometries with minimal lead time and no part-specific tooling. Each FF process has its own unique set of advantages and limitations. Before proceeding with a description of the major FF processes, let us review some basic concepts and terminology associated with freeform fabrication.

BUILD TIME

A major motivating factor in the development of FF processes has been the reduction in product development time. Therefore, the build time of FF processes is a major concern. Build time in FF processes consists of three major components: preprocessing, fabrication, and postprocessing. Preprocessing involves the conversion of CAD solid models into the control data needed to operate FF machine tools and make the part. Postprocessing involves any manual finishing of the part after the automated fabrication cycle.

Because FF processes involve a high degree of automation, computer planning and control software are essential to their operation. Planning software is needed to generate control data from CAD geometric data during preprocessing. This function is generally handled by specialized CAM software unique to each FF process. Typically, information regarding the geometry of the part is fed to the planning software in the form of STL files. STL files are faceted solid representations of the part generated by *tessellation*, that is, representation of the part surface as an interconnected network of triangles for the purpose of reducing the electronic file size. The STL format is the default standard for the FF industry having been developed for the first FF process, *stereolithography*, in the late 1980s. All major CAD solid modeling packages allow solid models to be exported in STL format.

The planning software essentially slices the tessellated solid representations into a series of cross-sectional layers on the order of 75 to 250 μm thick. These slices are reduced to a series of tool paths used to guide the selective deposition of energy or materials used to pattern each layer in the process. The process control software then uses these slices to operate the FF machine tool. Process control is handled differently for different processes and equipment. In some processes, the layer thickness is difficult to control, so in-process measurement of the build height is needed. Using these data, the control software can modify the slice data, if necessary, to compensate for inaccuracies in thickness.

While variations in preprocessing and postprocessing times exist among FF processes and machine tools, preprocessing times are becoming less important with the development of faster computers. The largest component of build time, and consequently cost, is the actual time required to fabricate the model. Currently, pre- and postprocessing costs added together range from 10% to 50% of fabrication costs with most processes averaging 20%. This can vary depending upon the geometric complexity of the part as well as the number of parts produced at one time (i.e., the batch size). All FF processes have the ability to nest multiple workpieces within their respective work envelopes (i.e., their maximum work space), which can save time and money. When batch sizes exceed about 10 to 20 parts, postprocessing times can become quite significant

TABLE 33-1. Summary of Freeform Fabrication Process Capability

Equipment	Typical Material Addition Rate, in³/min. (cm³/min.)	Typical Surface Finish AA, μin. (μm)	Typical Accuracy, in. (mm)	Typical Tensile Strength, ksi (MPa)	Typical Elastic Modulus, ksi (Goa)	Typical Material Cost, US$/in³ (US$/dm³)	Maximum Work Envelope, in. (cm) (W × L × H)
Concept Modelers							
Sander's Modelmaker	—	32 to 63 (.8 to 1.6)	.001 to .012 (.025 to .3)	—	—	—	12 × 6 × 9 (30 × 15 × 23)
Z Corp 3D Printer (monochrome)	1.4 to 4.7 (23.7 to 77.7)	—	—	4 to 10 MPa	—	1 to 2 (61 to 122)	8 × 10 × 8 (20.3 × 25.4 × 20.3)
Functional Modelers							
Stereolithography (SL)	.07 to 4.2 (1.2 to 69)	16 to 100 (.4 to 2.5)	.003 to .01 (.075 to .25)	5 to 10 (35 to 70)	150 to 500 (1 to 3.5)	2.25 to 2.8 (135 to 170)	20 × 20 × 24 (50 × 50 × 60)
Solid ground curing (SGC)	.12 to .3 (2 to 4.8)	—	.004 to .02 (.1 to .5) or 0.1%	2 to 5 (13 to 35)	30 to 135 (.2 to .9)	1.2 (71)	20 × 14 × 20 (50 × 35 × 60)
Fused deposition modeling (FDM)	.0003 to .15 (.006 to 2.4)	—	.005 to .03 (.125 to .75)	.5 to 5 (3.5 to 35)	37.5 to 90 (.25 to .6)	2.8 to 4.2 (170 to 250)	24 × 20 × 24 (60 × 50 × 60)
Selective laser sintering (SLS)	.07 to .46 (1.2 to 7.5)	200 to 500 (5 to 12.5)	.003 to .016 (.075 to .4)	3.5 to 5 (23 to 33)	180 to 210 (1.2 to 1.4)	2.25 to 2.5 (135 to 150)	13 dia. × 15 (33 dia. × 38)
Soligen (3DP)	1.5 (24.5)	472 (12)	.005 (.125)	—	—	9 (551)	16 × 16 × 16 (41 × 41 × 41)
Laminated object manufacturing (LOM)	1.8 to 2.9 (30 to 48)	100 to 400 (2.5 to 10)	.005 to .01 (.127 to .25)	8.5 (57)	1000 (6.7)	.07 (4)	22 × 32 × 20 (56 × 81 × 51)

depending upon the process used. The postprocessing time for lamination-based process-es can be quite long compared to other processes.

In general, for FF equipment, the fabrication time is made up of two components: lay-ering and patterning. Layering involves the bulk deposition of the raw material to be pat-terned. (Deposition-based processes do not require a separate layering step.) Of these two components, the patterning step is usually the longest. As a result, the material addition rates (MARs) associated with the patterning step are most representative of the total fabrication time for models. The MAR can be defined as the volume of material added per unit time.

VOXEL GEOMETRY

Figure 33-3 shows a schematic of a generic laser-based FF process involving the depo-sition of laser light energy into an existing layer of material for purposes of selective, lo-calized material consolidation or material removal. The *xy*-plane is oriented parallel to the material layers and the material surface. The *z*-axis is perpendicular to the material layers. This coordinate system is consistent among all FF processes. For processes in-volving scanning methods during patterning, the *x*-axis is congruent with the direction of the scan. This coordinate system is important for indicating the anisotropy of various material properties, including yield strength and surface texture. In general, material properties tend to be better in the *xy*-plane than along the *z*-axis.

Because most FF processes involve some type of scanning technique for depositing energy or material, the MAR can be best understood by the shape of the voxel (i.e., the *voxel geometry*). The term *voxel* is derived from the phrase "volume element." Voxel is to 3D geometries what the term pixel (derived from picture element) is to 2D images. The voxel can be thought of as the fundamental building block of an FF process. Understanding the nature of the voxel geometry is useful in that it determines the thickness of layers, distance between adjacent scans, and ultimately the number of scans needed to fabricate a layer. The voxel geometry can also impact the surface finish or texture of the prototype part. For scanning-type FF processes, it is important to understand how the material and process pa-rameters of a process affect the voxel geometry. A picture of the voxel geometry produced in laser-based FF processes is shown in Figure 33-4.

FIGURE 33-3 Schematic of a generic laser-based FF process.

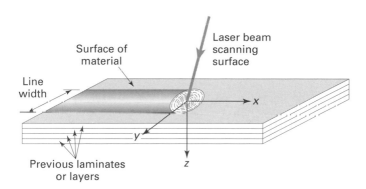

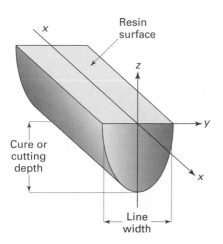

FIGURE 33-4 The laser produces a single cure line of photopolymer, called a voxel (30 pixels) in the laser-based FF process.

STAIRSTEPPING AND SURFACE FINISH

Surface finish and surface texture are great concerns in many RP applications such as those involving the use of prototypes as investment casting patterns or as aerodynamic test models. Most FF processes require additional secondary processes, such as grinding or polishing, to produce prototypes with good surface finishes. These additional process steps can add additional time and cost to prototyping efforts and can also limit the degree of geometric accuracy.

Currently very little data exist concerning the surface finish produced by FF processes. In general, it is expected that powder-based FF processes will produce poorer surface finishes than photopolymer-based FF processes. Further, as mentioned, it is expected that surface geometry will be rougher in the z-axis than along the xy-plane. By and large, the most dominant surface feature in most RP applications is the stair-stepping effect caused by orienting a flat or contoured surface not orthogonal to the xy-plane (or slice plane). This stair-stepping effect, as illustrated in Figure 33-5, is common to all current FF processes. In general, higher resolution of contoured surfaces can be obtained by orienting the surface orthogonal to the slice plane. Higher resolutions can also be obtained by reducing the layer thickness during the build cycle. However, as shown in Figure 33-6, a trade-off typically exists between the build speed of the machine and the thickness of a layer.

FIGURE 33-5 One of the problems with SL methods is stair stepping of curved surfaces and rounded corners.

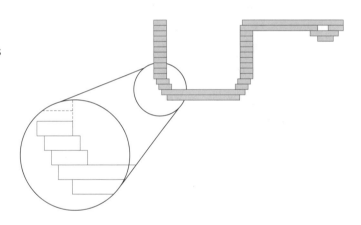

FIGURE 33-6 Build time versus layer thickness for various laser power levels. *(Reference: Jacobs, P.F.; Rapid Prototyping and Manufacturing: Fundamentals of Stereolithography, Dearborn, MI: SME, 1992.)*

Part: Slab 6 × 6 × 0.25   Hatch: 10 mils
Resin: XB 5081-1 $D_p = 7.1$ $E_c = 5.6$
$W_o = 4.5$ mils $T_{oh} = 35$ sec.   O.C = −1 mil

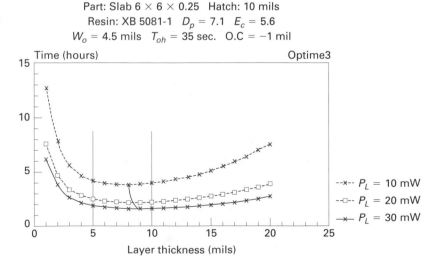

DIMENSIONAL ACCURACY

Several points must be considered when evaluating the accuracy of prototypes made on FF processes. First, and most important, is that operating conditions greatly affect the dimensional accuracy of prototypes. That is, a prototype fabricated under one set of processing conditions may have a different overall accuracy than a part built under another set of conditions. Build rate, part geometry, preprocessing, postprocessing, type of material, and, to some degree, even ambient temperature and humidity can all affect the accuracy of FF prototypes. Therefore, it is difficult to specify a precise dimensional tolerance, which will be met under all operating conditions.

Another point to consider is the size of the parts to be fabricated. For most processes, the part accuracy greatly improves as the measured dimension decreases. This is largely due to phase changes in the material as a result of processing. Specifically, a material is transformed from a liquid to a solid or, in some cases, from a solid to a liquid and then back to a solid again. In each case, the phase change from liquid to solid involves an increase in density and a resulting shrinkage. The total volumetric shrinkage varies from process to process and from material to material. However, all FF processes experiencing a phase change involve some level of volumetric shrinkage.

Phase change volumetric shrinkage is a stochastic process. The term *random noise shrinkage* has been introduced to describe the variation from deterministic behavior. For many different phase change processes, the resulting random noise shrinkage, σ_s, is directly proportional to the mean process shrinkage, $\bar{S}$, or:

$$\sigma_s = K \cdot \bar{S} \qquad (33\text{-}1)$$

where K is a proportionality constant referred to as the *random shrinkage coefficient* for a given process. The value of K can be determined by collecting a statistical significant set of dimensional measurements for a given material, geometry, and process combination and dividing the standard deviation by the mean. For example, K has been found to be approximately 0.096 for the shrinkage associated with powder processing of tooling surfaces from a mandrel.

For the case of perfect isotropic shrinkage, the linear shrinkage in a process may be approximated as one-third the volumetric shrinkage. Consequently, a material exhibiting 3% volumetric shrinkage would measure 1% undersize in all directions, provided the shrinkage was perfectly isotropic. Assuming that the mean process shrinkage is isotropic and substituting the percent linear shrinkage S for the mean process shrinkage in Equation (33-1), we have:

$$\sigma_s = K \cdot (S_v/3) \qquad (33\text{-}2)$$

where σ_s is a percentage estimator for the dimensional accuracy in a process and S_v is the percent volumetric shrinkage for a material. Table 33-2 shows some values for S_v for various materials. The implication of Equation (33-2) is that the larger the volumetric shrinkage of the material, the greater its tolerance. Further, since the random noise shrinkage cannot be compensated for in advance, the key to accuracy and repeatability in phase change FF processes is the reduction of the volumetric shrinkage of the material.

The random noise shrinkage is a best-case estimator of the process tolerances as other factors also affect dimensional accuracy. Equation (33-2) assumes that the shrinkage is isotropic. However, shrinkage within all FF processes happens anisotropically since the phase change happens layer by layer, voxel by voxel. Moreover, for FF processes involving thermal phase change, interior part temperatures during the build will always lag surface temperatures creating a temperature gradient and subsequently residual stresses within the material as it cools. These residual stresses can lead to part warpage, which also nullifies the isotropic condition described earlier. Additionally, constrained shrinkage associated with real part geometries will also result in numerous small shrinkage variations due to residual stresses (e.g., a thin wall section cools faster than an adjoining thick wall section).

TABLE 33-2.   Volumetric Shrinkage of Some Freeform Fabrication Materials

Vendor	Material	Volumetric Shrinkage* (% increase)	Final Density (g/cc)
Helium-cadmium-tuned photopolymer resins for SLA-190, -250:			
Ciba-Geigy/Cibatool SL	XB 5081-1 XB 5134-1 XB 5149	7.0	1.2
Du Pont/SOMOS	2110	3.0	1.2
	3110	6.0	1.2
Allied-Signal	2201	2.5	1.16
Loctite	8100	6.5	1.19
	8101	4.8	1.14
Asahi Denka Kogyo	HSS-660	—	1.153
	HSS-661	—	1.14
	HS-663	—	1.13
Argon ion-tuned photopolymer resins for SLA-500, Stereos:			
Ciba-Geigy/Cibatool SL	XB 5131 XB 5154	7.0	1.2
Du Pont/SOMOS	2100	3.0	1.2
	3100	6.0	1.16
Asahi Denka Kogyo	HS-671	—	1.16
	HS-672	—	1.16
JFC	SCR-100	—	1.08
Broad-Band photopolymer resins for Solider:			
Coates/Solimer	Type G Type F	—	1.1
DSM/DeSolite	4112-143-1	—	1.12
Heat-sensitive adhesive sheet (roll) for FOM:			
Helisys	Paper	—	0.83
Thermoplastic powder for Sinterstation:			
BF Goodrich / Laserite	LN-4000	67.0	1.02
	LPC-3000	46.0	0.82
	LWX-2010	56.0	0.98
Thermoplastic filaments for 3D Modeler:			
Stratasys	Casting Wax	—	0.92
	Plastic200	—	0.9
	Plastic300	—	1.1

*A dash indicates the shrinkage is unknown

■ 33.3 PHOTOPOLYMER-BASED FF PROCESSES

STEREOLITHOGRAPHY AND SCANNED LASER PHOTOPOLYMERIZATION

Stereolithography (SL) employs ultraviolet (UV) radiation in the form of a computer-controlled HeCd laser to selectively cure a photopolymer. A diagram of the process built by 3D Systems is shown in Figure 33-2. Specifically, like all FF processes, the SL process starts by converting a 3D CAD solid model into a series of very thin cross-sectional layers, or slices, as though the object were cut into multiple layers. Next, using the tool path data from the CAM software, the SL process "draws" the cross section onto the surface of liquid resin with a UV laser. The small but intense spot of UV light causes the resin to locally solidify (polymerize) wherever it is scanned. Once a layer is complete, an elevator within the vat of liquid photopolymer descends into the vat, which allows the resin to wash across the surface of the part, forming a new layer. A mechanical carriage sweeps across the surface to level and remove excess material. This leaves a uniform layer of liquid resin ready to be patterned with the laser. These steps are repeated over and over until the desired 3D geometry is complete. Once completed, excess partially cured resin is removed from the outside of the prototype with the use of a solvent. Also, the prototype is typically postcured in an ultraviolet oven to achieve more thorough polymerization and, consequently, better material properties.

SL is classified as a scanned laser polymerization (SLP) process. Since its inception, at least six additional SLP technologies have become commercially available worldwide. Figure 33-7 shows some of the different layering methods adopted by various SLP processes. Each of these methods has its own set of technological advantages and issues. Most SLP processes use the *descending platform* method which benefits from lower viscosity resins, faster layering and better flatness of liquid layers during build with faster drainage of resin from objects after build. However, low-viscosity resins are more sensitive to vibration, and therefore, vibration suppression must be addressed within these machines. Also, the low molecular weights of most low-viscosity resins result in weaker solid materials due to the fewer number of cross-links produced among the short oligomers used to reduce viscosity.

Alternatives to this approach that have been tried or are in use include the *ascending suspension* method and the *masked-lamp descending platform* method. In the ascending suspension method, the part is built from the top cross section down. As the elevator is raised to scan each successive cross section, resin flows between the transparent window and the suspension substrate. In contrast, the masked-lamp descending platform method simply creates photomasks for each layer that rest upon a

FIGURE 33-7   Schematics of four different SL methods: (*top left*) descending platform; (*top right*) ascending suspension; (*bottom left*) ascending surface; (*bottom right*) masked-lamp descending platform where entire layer is exposed all at once. *(Source: Paul, B.K., and C. Ruud. "Rapid Prototyping and Freeform Fabrication," in Integrated Product, Process and Enterprise Design, B. Wang (Ed.), London Chapman & Hall, 1997 pp. 193–242.)*

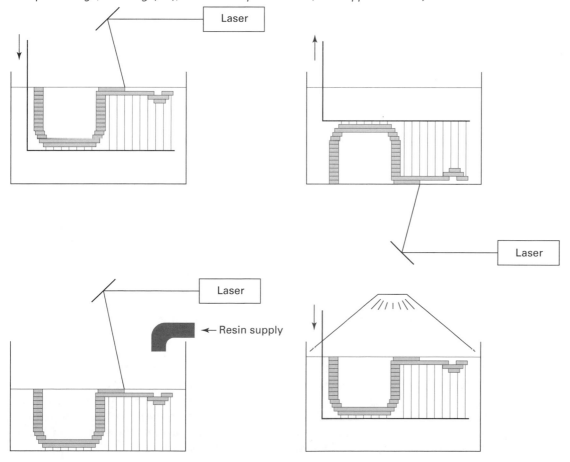

transparent material that is in contact with the resin surface. The advantage of this method lies in potentially higher throughputs because each layer is produced by a single photoexposure rather than a serial laser scan. Both of these methods have the potential for shorter layering setups between layers and, therefore, decreased build time. Also, these methods are much less sensitive to vibration since the polymer surface is produced by contact with a solid surface, therefore, making lower viscosity photopolymers desirable. However, separation of the cured photopolymer from the photomask must be addressed in these methods. The final layering method, that is, the *ascending surface* method, has been used to decrease layering setup times between layers.

In SLP processes, the focused laser spot is typically positioned with the use of orthogonal, servocontrol-led, galvanometer-driven mirrors. First, the laser beam traces the boundaries of the particular slice cross section being drawn. Once the boundaries have been completed, the laser beam then scans the remainder of the cross section to solidify any remaining portions. This filling in of the remainder of the cross section is known as *hatching*.

Hatching would be straightforward if the photopolymer did not undergo shrinkage during phase change. Due to the phase change, various styles of hatching have been developed to minimize the effects of this shrinkage on material properties and dimensional accuracy. One method to speed up hatching involves spacing scan lines such that small pockets of uncured resin remain that can be photocured during postprocessing. Special hatch styles also exist for producing investment-casting patterns by making drainable honeycomb structures allowing unneeded, uncured resin to be removed during postprocessing. This technique permits the user to take advantage of the fact that the value-added nature of additive processes is at the surface of the part, not in the bulk.

In developing hatching schemes, layer thickness and distance between adjacent scans must be considered. These parameters, and others like surface finish, are dependent upon the voxel geometry. The voxel geometry shown in Figure 33-4 represents the volume of material that hardens within the liquid resin as the laser beam is scanned over its surface. The voxel geometry is produced as the photons in the laser beam penetrate the liquid resin and are absorbed by resin molecules. If the density of photon absorption becomes high enough, the resin material photopolymerizes to the point of gelation corresponding to the transition from liquid to solid. The major assumptions leading to the parabolic shape of the scanned SLP voxel geometry are:

1. The photon density (photons per unit area), or exposure, is Gaussian for the laser beam. Therefore there is no true beam diameter. The beam diameter is approximated as two times the radial distance at which the exposure is $1/e^2 (\sim 13.5\%)$ of the exposure in the center of the beam.

2. Photon absorption within the resin follows the Beer-Lambert Law, which states that the absorption within the resin follows an exponential decay from the surface to a point, D_p, the penetration depth, at which point the absorption is $1/e$ (36.8%) of the surface absorption.

Linewidth, defined by y, and layer thickness, defined by z, are intimately related. The maximum z-value will occur at $y = 0$, which constrains the layer thickness. The maximum y-value occurs at $z = 0$, which constrains the linewidth. Changing power or scanning speed will impact both the layer thickness and the linewidth.

Advantages of SLP technology include high accuracy and good surface finish. SLP processes tend to offer the best dimensional accuracies of all FF processes with repeatable accuracies down to 0.03% over a 60-mm dimension, and surface finishes below 16 μin. CLA on surfaces perpendicular to the beam axis are typical. A major disadvantage of all SLP processes involves the use of expensive photopolymers, which are often highly toxic. Prices for a typical photopolymer resin may run in the hundreds of dollars per pound. Further, many of the liquid resins used are acrylics, which can cause skin irritation or other toxic effects if handled improperly. Some resins also contain suspected carcinogens. Thus, safety precautions must be taken when handling these raw materials, which generally eliminates the accessibility of SLP technology within typical engineering office environments. Because of their toxicity, it is expected that future environmental requirements may pose further difficulties for photopolymer-based freeform fabrication.

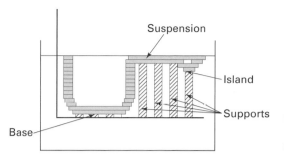

FIGURE 33-8 Many of FF processes require the addition of a base and supports.

Another disadvantage of these processes involves the need for additional preprocessing and postprocessing due to the requirement of *supports*. As shown in Figure 33-8, supports are extra material used to support material layers that do not have a bulk solid substrate underneath. Supports act like fixtures in conventional machining by holding the part in place during processing and are required only in the photopolymer and deposition-based FF processes. Supports are needed when fabricating a part with a suspended feature, which is cantilevered out away from the bulk part and which might distort due to phase change shrinkage as the resin is cured (e.g., pan handle in Figure 33-8). Supports can also provide a means of securing isolated segments or islands that would otherwise float away during processing (e.g., island in Figure 33-8). In addition to wasting material, supports must be designed into the part before processing and removed after processing, negating some of the advantages of tool-less production. Manual support generation can be tedious and time consuming, especially for complex parts. Automatic support generation software is now available. Another additional postprocessing step involved with most SLP processes involves a post cure of the part in an ultraviolet oven for purposes of achieving full resin strength.

In the past, many SLP resins had other less than desirable properties. Early in their development, many acrylate polymers were prone to a great deal of volumetric shrinkage resulting in curling and poor dimensional accuracy, and poor impact resistance due to absorbed water, causing swelling and dimensional instability. SLP resins are now available for prototyping rigid, high-accuracy parts with little shrinkage or warpage or with more impact resistance or elasticity. Throughout the world more RP applications and process research have been carried out on SLP equipment than any other FF process. Over all, SLP is a maturing technology, and significant improvements in build time, surface finish, dimensional accuracy, and material properties are expected to continue.

SOLID GROUND CURING AND UV LAMP PHOTOPOLYMERIZATION

Solid ground curing (SGC) cures layer upon layer of photopolymer with the use of a photomask and a high-intensity UV lamp similar to the masked-lamp descending platform method mentioned in the previous section. However, the implementation is much different. The SGC process has been capable of some of the highest production rates among FF processes largely because of its unique method of layer support enabling a higher density of parts during batch production. As such, the SGC process warrants an explanation of its own.

A diagram of the SGC process is shown in Figure 33-9. Like SLP technology, the SGC process receives its control data from CAM software used to slice a solid representation of the part (STL file) into thin cross sections. Fabrication of a layer begins with the development of a photomask through a process known as *ionography*. This process, shown in Figure 33-9 as the *mask plotter cycle*, is similar to the xerography process used in photocopiers. A pattern of static charge is put down on the glass based on control data from the slice file. Black powder, or toner, is electrostatically attracted to the charged areas of the glass. The resulting photomask is used to selectively expose a layer of photopolymer. Subsequently, the photomask is erased and recycled by removing the charge and powder from the glass plate.

Once a layer of photopolymer has been exposed under the photomask, it is then further processed within a *model grower cycle*. First, any unexposed excess resin is removed from the layer. Then, a layer of liquid wax is applied and solidified to fill any voids left by removing the unexposed resin. Finally, the entire layer of wax and photopolymer is milled down to a specific thickness by a face milling operation. The process is repeated for the next slice, each layer adhering to the previous one, until the object is finished.

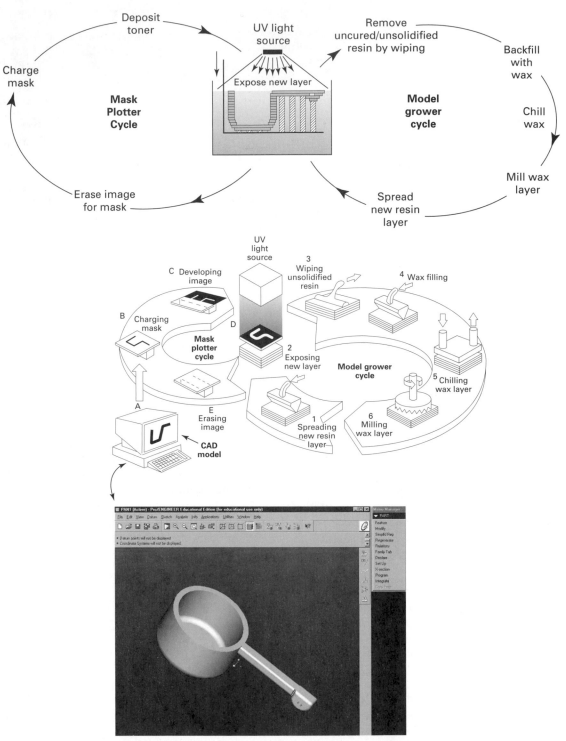

FIGURE 33-9   The solid ground curing process operates the mask plotter cycle (A, B, ... E) and the model grower cycle (1, 2, ... 6) synchronously.

When the object is removed from the machine, it is essentially in the form of a large block of wax. The wax can be removed by melting or rinsing with hot water until the finished prototype is revealed. Alternately, if the processing is done at a supplier, the wax can be left on the part for shipping or security purposes and simply removed by the end user.

Dimensional accuracy with the SGC process tends to be good. Claimed accuracy for the process is 0.1% up to a maximum of 0.02 in. The z-axis accuracy of the SGC process tends to be better than for other FF processes. This improvement can be attributed to the milling operation. However, the overall level of accuracy of the SGC process tends to be less than that of SLP processes. This might be due in part to the hatching schemes available to SLP processes, which cannot be used in the SGC process. In addition, the basic SGC image mask is generated at 300 dots per inch, corresponding to a current system resolution limit of 0.0033 in. (0.084 mm) per spot.

In addition to producing a precise, uniform layer thickness, the milling operation is also performed to promote adhesion between layers by roughening the surface. Adhesion between layers is a more difficult problem within masked-lamp FF processes than SLP processes, the difference being that the photopolymer is not exposed to oxygen inhibition within masked-lamp processes. Within open-air, descending-platform, SLP processes, oxygen from the air inhibits initiation of the polymerization process within a thin surface layer of the exposed photopolymer. This thin liquid layer aids in the adhesion between model layers. In an enclosed, masked-lamp process, the layers of photopolymer are not exposed to air and thus do not retain this thin layer of unpolymerized resin. As a result, a method such as milling is needed for improving the adhesion between layers.

An advantage of the SGC process is that the time to build every layer is the same, independent of either part geometry or the number of parts being built. Therefore, build time can be predicted quite accurately by multiplying the time per layer by the number of layers. Further, the use of photomasking technology provides the SGC process with a potential for high MARs and fast build times. Once a mask is developed, layers can be photopolymerized within three seconds. However, because of the number of steps involved, the process cycle time is about one minute per layer.

Though the SGC process does not have the fastest MAR of FF processes, it does have one of the largest production rates. This is because the process can fabricate parts in much larger batches than most other processes. In the SGC process, all of the work envelope can be dedicated to the fabrication of parts. In SLP and other processes, much of the work envelope is consumed by supports needed to reinforce parts during fabrication. This allows the SGC process to have a very high part density during hatch fabrication relative to SLP processes.

One extraordinary feature of the SGC process is its ability to fabricate preassembled structures, which are multiple-piece assemblies, fabricated already assembled. As a result, no assembly is required, and since the part can be fabricated as one piece, dimensional tolerances do not inhibit the assembly of the part. This ability to fabricate preassembled structures is unique among existing FF processes. Figure 33-10 shows some example parts made in the Cubital Solider machine.

A disadvantages of the process is the number of process steps, which leads to larger, heavier, and more expensive equipment. Current implementations of the process

FIGURE 33-10 Solid Ground Curing (SGC), using the Cubital Solider machine, (*above*) is the process of exposing an entire layer of photopolymer using a strong UV light passing through a mask and fully curing all exposed areas. Unexposed resin is vacuumed off and replaced with a water soluble wax. The resin/wax layer is then machined to precise thickness and process begins again. Once complete, the wax must be removed. Some typical parts are shown (*below*). (*Source: Modern Casting, March, 1996.*)

Build Envelope:
20 × 14 × 20 in.
Layer Thickness:
0.004–0.006 in.
Materials: UV liquid polymer.
Software Required:
Cubital's proprietary.

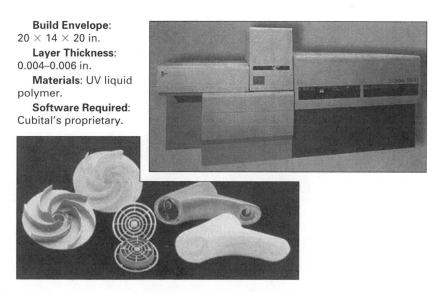

weigh over four tons and are more than 10 feet in length. Further, because of the complexity of the process, reliability can be a problem, requiring a full-time operator. In addition, many users have experienced difficulties with the removal of wax, especially from smaller features such as those found in preassembled items.

■ 33.4 DEPOSITION-BASED FF PROCESSES

FUSED DEPOSITION MODELING AND EXTRUDED DEPOSITION

Extruded deposition-based (ED) FF processes, also known as fused deposition modeling, produce laminated three-dimensional objects through robotically guided extrusion of a thermoplastic. A schematic of the process is shown in Figure 33-11. A spool of thermoplastic filament is unwound and fed through a robotic extruding head. The extruding head melts the thermoplastic, which bonds with the layer below. As a whole, the process resembles a pen plotter except that the plots are three dimensional.

The build cycle for ED processes is much simpler than other FF processes. As a result, ED processes have the advantage of being compact and low cost, ideal for application in a design engineering office environment. ED machines are among the lowest capital cost per unit work volume available. The footprint for the smallest machine is about 0.67 m and weighs less than 210 lb. In addition, the process uses nontoxic thermoplastic materials, a requirement in office environments. Further, it does not require venting or postprocessing, which would require more office space. As a desktop unit, ED processes are ideal for concept-modeling applications early in product development cycles.

Disadvantages of the process include the limited selection of engineering materials. Some of the materials available in the process include machinable wax, investment casting wax, nylon, and ABS. Polyester compounds have recently been introduced. Though ED materials are nontoxic, most of them have poor mechanical properties (with the exception of ABS).

While the parts fabricated by the process do not require postprocessing, the build time of the process is the longest among FF processes. Further, the surface finish of ED parts tend to be the poorest among FF processes. Finally, the material cost of ED filaments is among the highest per unit volume.

INKJET DEPOSITION

Inkjet deposition-based (ID) FF processes use an inkjet mechanism to selectively deposit thermoplastic and wax materials. A diagram of the process is shown in Figure 33-12. Build materials are selectively deposited onto a substrate by a scanning inkjet printing head as a series of uniformly spaced micro-droplets. The droplets adhere to each other and solidify to produce a uniform mass.

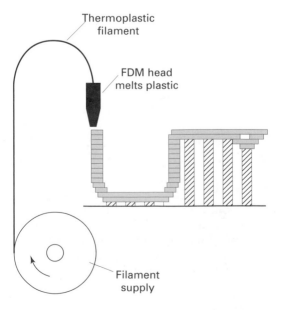

Thermoplastic filament

FDM head melts plastic

Filament supply

FIGURE 33-11 Some fused deposition processes are "office friendly" due to their small size and use of nontoxic materials.

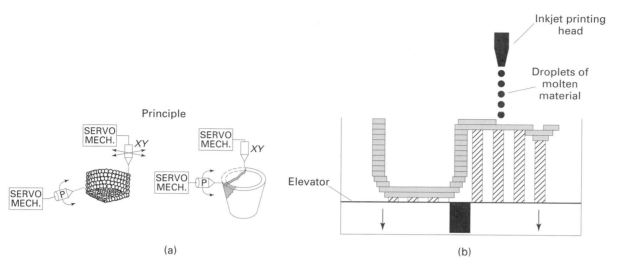

FIGURE 33-12 The inkjet deposition-based FF process uses Ballistic particle manufacturing concept to make solid objects.

Like SLP and ED technology, ID systems require the fabrication of support structures for unsupported features. The supports are deposited in a perforated pattern to facilitate removal. Some systems separately print support material as well as build material. Alternatively, the inkjet can be mounted on a 5-axis positioning mechanism so that overhanging features can be deposited without support, as the figure shows.

Dimensional accuracy and surface finish are very good with ID technology. Dimensional accuracy for the placement of droplets may be to within 250 μin. in the xy-plane. Vertical free-standing walls as narrow as 0.004 in. thick with reasonably smooth side walls have been produced. In some versions of the process, milling of each layer is performed immediately following the deposition cycle to improve dimensional accuracy. As in other FF processes, surface quality can be enhanced by pretracing the surface of each cross section prior to filling the interior with build material.

Build times tend to be very long with inkjet deposition processes. To address build speed, some vendors have incorporated multiple jets on a single printing head. Different hatching patterns may also be used to improve accuracy and build time.

Material selection is currently very limited in ID processes. Typical materials include low melting temperature thermoplastic and wax. Other issues with the technology include cleanliness in office environments. Ongoing research efforts in inkjet deposition include a process that injects streams of UV-curable binder resins under a UV light and jetting systems that can directly deposit metal alloys.

■ 33.5 POWDER-BASED FF PROCESSES

SELECTIVE LASER SINTERING AND SCANNED LASER FUSION AND SINTERING

FF processes based on scanned laser fusion and sintering (SLFS) rely on CO_2 lasers to fuse or sinter selected areas of loose powder. An illustration of the selective laser sintering (SLS) process is given in Figure 33-13. An elevator within the process vertically positions the powder bed. As in other processes, CAM data are prepared by virtual slicing of an STL file.

The process begins by depositing a layer of heat-fusible powder across the part build chamber. A CO_2 laser is then scanned over the layer to selectively fuse together those areas defined by the geometry of the cross section. The interaction of the laser beam with the powder elevates the powder temperature to the point of melting, fusing the powder particles to one another and the layer below them. The unfused material remains in place as the support structure. Afterward, an additional layer of powdered material is spread out and leveled over the top surface of the powder bed. Successive layers of powder are deposited and selectively sintered until the part is complete. After processing, the part is removed from the build chamber, and the loose powder is removed for reuse. Postprocessing, such as sanding, may be required depending upon the intended application.

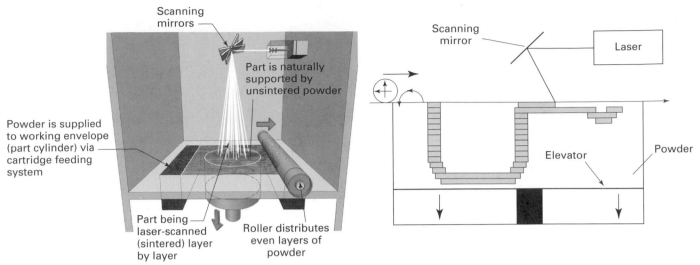

FIGURE 33-13   Selective laser sintering process uses a laser to scan and sinter powder into solid shapes layer by layer. (Kochran, 1993).

SLFS processes usually do not require the design, fabrication, and removal of special layer support structures because the excess powder acts as a *natural* support. Thus, time and materials are not wasted in building and removing support structures. Further, excess powder from one build can be reused in another build.

SLFS processes can use a wide range of build materials, including plastics, waxes, metals, and ceramics. A variety of polymeric materials are available for the SLS process including nylons, polyamides, polycarbonates, elastomers, and acrylic styrene. Because of the short durations of localized heating during SLFS processing, the primary mechanisms for binding together and densifying particles are *fusion*, based on melting and resolidification, and *sintering*, based on viscous flow transport mechanisms. Sintering is a process in which solid particles bond together at temperatures just below the melting point of a material based on the accelerated atomic transport in solid matter. The driving force for the consolidation of powder particles via sintering is the high surface energy associated with unsintered powder. Sintering is the preferred method for binding together thermoplastic powders due to their low activation energies for viscous flow. For amorphous materials, such as polyvinyl chloride (PVC) or polycarbonate, the absorbed laser energy causes the powder particles to soften and bind to one another at their points of contact, producing a porous material. The porosity of these materials may be as much as 40% of the part volume. Crystalline materials, such as nylon or wax, are locally melted in SLFS processes and can essentially produce fully dense parts.

Sintering of metal and ceramic powders requires a much longer period of time than is available during the SLFS processing. Therefore, SLFS processes are used to bind together polymer-coated metal and ceramic powders into a net shape, which are fired and densified within a sintering furnace. The metal and ceramic powders are coated with a very thin ($\sim 1\ \mu$m) layer of polymer binder. The laser heats the polymer coating, fusing it with adjacent coatings and resulting in powder binding. When finished, the "green" part is removed from the powder work volume and placed into a sintering furnace for densification. Once in the furnace, the polymer binder is evaporated away, leaving a porous metal or ceramic preform, which is eventually densified upon heating to a higher temperature. Because of this ability to fabricate in metals and ceramics, SLFS processes have been among the first FF processes to offer direct fabrication of metal tooling for processes such as injection molding and die casting. Early metals available for rapid tooling fabrication in SLS machine tools have included tool steel infiltrated with copper for full densification.

Because of the need to significantly raise the temperature of materials, SLFS processes require much more powerful lasers than SLP processes. SLFS processes use a CO_2 laser, which is about 1000 times more powerful than the HeCd laser used in some SLP technology. The CO_2 laser is also about 1000 times more efficient at creating laser energy, and subsequently any difference in power consumption is minimal. Because of the mechanisms involved, SLFS processes tend to have significantly slower scan speeds and build times than SLP processes.

To minimize the required laser power and increase throughputs, the powder is maintained at temperatures just below the fusing point. This also helps to decrease the material distortion that occurs during processing due to thermal residual stresses. This has led to improved dimensional stability within the process, but requires more complexity in machine design. Since the process is carried out at elevated temperature, the chamber is typically filled with nitrogen to avoid oxygen contamination of the bonding surfaces, as well as the potential combustion or explosion hazard associated with the high surface energy of powder particles. Cooling the chamber gas is also necessary.

Dimensional accuracy and surface roughness tend to be poorer in SLFS processes than in SLP processes. Like SLP processes, dimensional stability in SLFS processes is a function of the volumetric shrinkage of materials. However, the volumetric shrinkage associated with the melting and fusing of nylon and wax powders to fully dense materials can be quite large in comparison with other processes. Therefore, trade-offs exist among the density, the mechanical properties and shrinkage, and the dimensional accuracy of these materials.

In addition to affecting the density of materials, material porosity is the dominant source of surface roughness. Particles in the SLS process are approximately spherical and range in diameter from 50 to 125 μm. For amorphous materials that only fuse at their respective contact points, the resulting porosity can lead to significant surface roughness. Further, older SLFS systems may scan the surfaces of powder beds using a rastering method. *Rastering* methods involve sequential, line-by-line scanning similar to the way a printer creates a two-dimensional graphics object. While all FF processes are subject to stair-stepping errors in the *z*-axis, raster scanning also produces similar errors in the *xy*-plane (also known as *horizontal aliasing*).

A major advancement in SLFS processing has been the direct fusing of uncoated metal powders. In this process, high-power Nd:YAG lasers are used to fuse high-temperature metal powder directly into a solid. Powder delivery nozzles shoot streams of metal powder at the focal point of the laser. The powder streams melt and fuse with the underlying layer. The laser remains stationary, while the part is scanned underneath the laser on an *xy*-positioning stage. After each layer, the laser and powder delivery nozzle move upward and begin patterning the next layer, saving layering time. By varying the powder mix, it is possible to make parts out of a variety of metals, alloys, and nonmetallic materials. This technology is initially being targeted for the rapid manufacture and rapid repair of metal molds for injection molding of plastic parts. As expected, long build times, poor surface finish, and poor dimensional accuracy due to thermal residual stresses are issues that must be addressed.

SOLIGEN AND SELECTIVE INKJET BINDING

Several FF processes have taken advantage of inkjet printing technology to selectively bind together powder layers. A schematic of a selective inkjet binding (SIB) process is shown in Figure 33-14. This technology is sometimes referred to as three-dimensional printing (3DP) technology. Like the SLFS processes described earlier, the SIB processes begin by loosely depositing a thin layer of powder for processing. An inkjet printing head scans the powder surface and selectively injects a binder (such as colloidal silica) into the powder. The binder joins the powder together into those areas defined by the geometry of the cross section. As in SLFS processes, the unbound powder can be used to support the next layer. When the shape is completely built up, the part is removed from the unbound powder.

SIB processes are similar to SLFS processes in that metal and ceramic structures require postprocessing in a sintering furnace. The green structures produced by SIB processes consist of metal or ceramic powder particles bound together by either organic or inorganic compounds. Like the SLFS process, these green structures require sintering to achieve higher material densities and better mechanical properties. For metal powders, such as stainless steel, postprocessing after sintering can also include infiltration of the metal matrix by a lower-temperature metal to increase densification.

The main advantages of SIB processes are the variety of engineering materials that can be used and the fact that SIB processes do not involve thermal binding mechanisms.

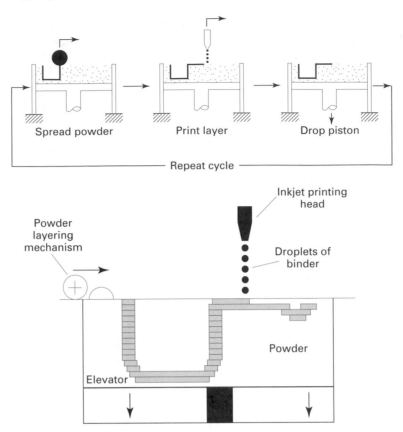

FIGURE 33-14 Schematics of the selective inkjet binding processes which begin with the material in a powder form.

Therefore, SIB processes do not expose parts to thermal residual stresses and their related problems. Further, SIB processes have been used to inject inorganic binders such as colloidal silica into ceramic powders. Rather than being evaporated away like organic binders, these inorganic binders are fused together with the ceramic powder and, therefore, reduce the volumetric shrinkage associated with organic binders and SIB technology can selectively modify the microstructure of materials by selectively injecting different compositions and amounts of inorganic binder into the powder bed. Applications of SIB processes include the fabrication of investment casting shells, metal tooling, and concept modeling.

■ 33.6 LAMINATION-BASED FF PROCESSES

Lamination-based FF processes involve the sequential bonding and patterning of material sheets or laminae. Typically, laminae are made of paper layers bonded to one another with the use of a thermally activated organic compound and patterned by a laser or some other type of cutter. A diagram of the laminated object manufacturing (LOM) process is shown in Figure 33-15.

The process begins by delivering a sheet of paper onto the working surface either by roller or some other means. Next, a heated roller is rolled over the surface of the paper to bond the paper sheet to the layer below it. Once the paper sheet has been bonded, a laser or cutter is guided in a cookie-cutter pattern to cut the 2D cross section assigned to that layer. These steps are repeated over and over until part completion.

One unique characteristic of lamination processes is the method for extracting the design after fabrication. Measures must be taken to remove the excess paper mass, which accumulates around the periphery of the part. Removal of the excess material is handled by drawing a consistent set of cross-hatches in the excess material regions during patterning. Over many layers, these cross-hatches form the boundaries of small blocks, which must be removed from around the finished part. Unfortunately, this can result in a great deal of wasted time and material.

The lamination processes do not require the fabrication of any specialized support structures. The excess solid material provides a natural support of the part while it is being fabricated. Thus, complex cantilevered geometries can be fabricated as easily as

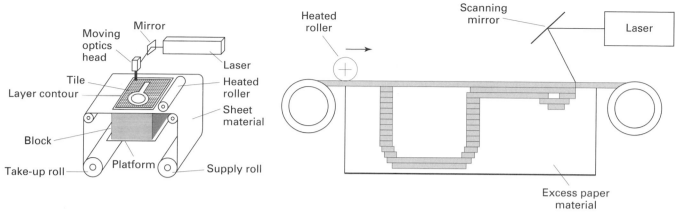

FIGURE 33-15 Schematic view of the laminated object manufacturing system which delivers material in sheets.

any other structure. Equipment and materials tend to be cheaper due to the simplicity of the process and reduced maintenance requirements.

Attempts have been made to increase the variety of materials available to lamination processes. Using paper laminates, the properties of the finished part are similar to those of birch wood, so pattern makers and model makers in woodworking shops use this process. Other sheet materials have been introduced with limited success including polyester and metal and ceramic tape. Material properties for metal and ceramic from laminated processes are among the highest achieved by FF processes. A comparison of material properties of ceramic bars fabricated by conventional powder pressing and the LOM process is shown in Table 33-3. The fully dense ceramic bars were made by replacing the paper rolls in the process with ceramic tape. Once the part was removed from the excess material, it was fired in a sintering oven to achieve full densification. The excellent material properties achieved by this process are accounted for by the high green density of the ceramic tape. However, the high cost of metal and ceramic tape combined with the potential for large amounts of material waste make laminated processing of metals and ceramics costly.

Lamination-based FF processes should have a faster build time, but benchmarking studies reveal the lamination processes consistently finish in the back of the pack. Why? Many FF processes are able to increase production rate and thus reduce build time, by building in batch sizes larger than one. Batching parts together in lamination processes is difficult due to the difficulty in removing excess material, which does not allow parts to be spaced closely together.

Since the material does not go through a phase change (e.g., liquid to solid), residual stresses are reduced resulting in less warpage and better dimensional stability. The process cannot currently achieve tolerances much better than about 0.01 in. (250 μm), leaving considerable room for improvement. However, since the paper does not experience a significant phase change, laminated processes can hold this tolerance over the entire work envelope, providing possibly the best dimensional accuracies at larger scales.

TABLE 33-3. Material Properties of Pressed and LOM-Fabricated Ceramic Bars

Method	Direction of Test	Fixture Strength MPa	Vickers Hardness Gpa	Fracture Toughness MPa* m$^{1/2}$	Green Density g/cm^3	Sintered Density g/cm^3	Shrinkage %
LOM	Parallel	314	20.2	4.3	2.55	3.88	14.1
	Perpendicular	311	20.1	3.9			
Pressed	Parallel	336	21.8	4.0	2.34	3.89	15.8
	Perpendicular	325	19.8	3.7			
Commercial Grade		379	14.1	4–5			

Source: C. Griffin, J. Daufenbach, and S. McMillin. (1994) "Desktop manufacturing: LOM vs. Pressing," *American Ceramic Society Bulletin*, 73(8): pp. 109–113.

Since the material is hygroscopic in nature, laminated parts must be surface treated with water-resistant resins to prevent swelling.

Layer thickness is limited to the lamina thickness, which decreases the flexibility in build time and surface roughness. To compensate, some lamination processes offer the capability of laminating two layers at once. In addition to these constraints, z-axis dimensions tend to be worse in laminated processes since there is less control over layer thickness.

■ 33.7   RAPID TOOLING

Many attempts have been made in the past to use RP models as patterns or molds for basic manufacturing processes. The use of FF technology in the rapid development of such tooling has become known as *rapid tooling*. Rapid tooling is an excellent application for FF technology for several reasons:

1. Mechanical tooling is critical to most manufacturing technology, and therefore, rapid tooling can be applied universally across many industries.
2. Mechanical tooling is a low-production-volume, one-of-a-kind commodity.
3. Mechanical tooling typically has an associated long lead time and high cost and, therefore, requires a high production volume of product to be justified.

As a rapid source of cheap yet effective tooling, FF technology has the potential to redefine economic order quantities. Future companies may be able to justify die casting, stamping, or injection molding runs of a few hundred or a few thousand parts.

Several levels of rapid tooling have begun to appear in industry. The most basic level of rapid tooling is called *soft tooling*. Soft tools are those required for small production quantities. Typically, they are low in cost and made of "softer" materials than conventional tooling. Soft tools may be used when a manufacturer wants to prototype a functional polymeric or metallic part with mechanical properties similar to those produced in production processes.

One type of soft tooling involves the fabrication of rubber or epoxy molds for vacuum casting of polymers or spin casting of low-temperature metal alloys. In particular, silicone room temperature vulcanizing (RTV) rubber can readily be molded around RP patterns to produce mold cavities compatible with the vacuum casting of polyurethanes and other polymers. Soft tooling for spin casting low-temperature metals, including zinc alloys, has been created using vulcanized rubber molds. In one process, a room-temperature vulcanizing mold is produced using an RP model as a pattern. This low-temperature RTV mold is used to spin cast a pewter pattern, which is used as a permanent pattern to develop a heat-curable silicone rubber mold. The heat-cured mold can withstand temperatures on the order of 550°C at a rate of 50 to 60 cycles per hour for hundreds of cycles. Pattern materials other than pewter can be used as long as they can withstand a temperature of 190°C under several thousands of pounds of pressure without distorting, melting, breaking, or outgassing.

RP models can be used as permanent or temporary patterns for sand casting or plaster mold casting. In addition, many attempts have been made to use RP models as expendable patterns using the investment casting process. Table 33-4 provides an overview of the advantages and disadvantages of using various RP models as investment casting patterns.

Prototype tooling is low-production-volume tooling that may be used in production processes that normally have high production volumes. Prototype tooling may be used for low-volume production runs or to produce functional prototypes that have material properties that depend upon processing conditions. One approach to prototype tooling involves the fabrication of metal shells that can be used to produce engineering tooling assemblies (ETAs). ETAs are typically made up of three parts: the metal shell (i.e., the contoured mold surface), a metal frame, and a backing material. The metal shell, on the order of 2 mm in thickness, is fabricated by depositing a metal onto a mandrel in the shape of the desired mold surface. The mandrel can be fabricated using any one of the FF technologies. The resulting metal shell is then separated from the mandrel and attached to a hollow frame which is reinforced with a backing material such as a metal-reinforced epoxy or chemically bonded ceramic. Production runs in excess of

TABLE 33-4. Comparison of Rapid Prototyping (RP) Processes for Producing Investment Casting Patterns

	SLA (solid acrylic)	SLA (Quick Cast)	SLS (wax)	SLS (poltcarb)	LOM (paper)	FDM (wax)	SGC (solid acrylic)
General compatibility with investment casting process	low	moderate/ good	excellent	good	moderate/ good	good/ excellent	low
Casting method	flask mold only	flask or shell mold	flask or shell mold	flask or shell mold	flask or shell mold	flask or shell mold	flask mold only
Pattern accuracy	excellent	excellent	fair	good	good	—	—
Thermal expansion before melt out/burn out	high	high	negligible	moderate/ low	low	negligible	high
Residue after melt out/burn out	moderate/ high	low	none	low	high	none	moderate/ high
Surface finish	good	good	poor	fair	fair	poor	good

Source: B. Sarkis and S. Kennerknecht. (1994) "Rapid Prototype Casting: The Fundamentals of Producing Functional Metal Parts from Rapid Prototyping Models," *Proc. Fifth International Conf on Rapid Prototyping*, Dayton, OH, pp. 291–300.

5,000 have been made on a plastics compression molding machine using an ETA with a pure Ni shell backed with chemically bonded ceramics. The fabrication process for an ETA is shown in Figure 33-16.

Both thermal spraying and electroforming have been used to fabricate metal casting shells for ETAs. Low-melting-point alloys (tin/lead-based) can be sprayed similar to a paint sprayer. Zinc and aluminum-based alloys have been deposited with a system that feeds two metal wires into a gun where an electric arc between them causes melting and an exhaust jet sprays the melt onto the mandrel. Higher-temperature metals have been sprayed by first coating the mandrel with a thin metallic layer via electroless plating or physical vapor deposition. The metallic layer helps to transmit the heat more readily across the mandrel surface, preventing softening and material property degradation in the mandrel. Problems associated with metal spraying that must be compensated for include high internal tensile residual stresses and poor shape definition for narrow slots and small diameter holes.

FIGURE 33-16 Fabrication steps to make tooling using RP to reduce delivery time. *(Source: Net Shape Nickel Ceramic Composite Tooling from RP Models, S. Wise, Rapid Prototyping, Vol. 3, No. 1, SME-RPA.)*

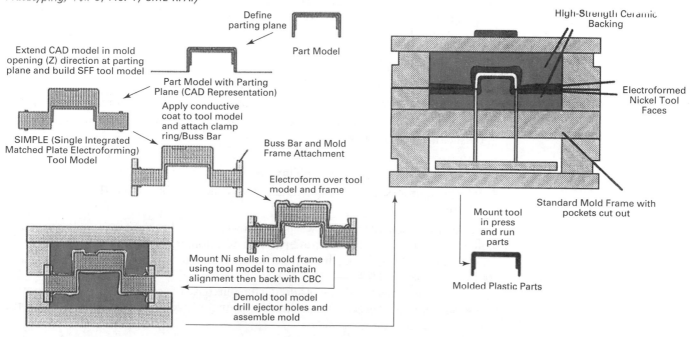

Electroforming, involving electroless deposition or electroplating, may also be used to make metal shells. Issues with electroforming include poor shape definition in narrow slots and nonuniform deposition around internal and external corners.

■ 33.8 ECONOMIC CONSIDERATIONS

When deciding to use FF processes, several questions must be answered including whether to buy FF equipment or to outsource work to vendors. Many FF equipment vendors exist along with many job shops and service bureaus that offer their products. If purchasing, a company must decide which FF machine to buy. If outsourcing, the company must determine which vendor to choose. In addition, the company must recognize and prepare for the effect that FF technology will have on the organization.

Answering these questions begins by establishing a purpose for using the FF technology. FF technology can be used for many different applications including concept modeling, form/fit verification, marketing demonstrations, functional testing, prototype tooling, and production tooling. Comparisons must be made between prototype requirements and equipment capabilities. For example, full-scale models of an engine block can only be fabricated on three or four of the FF processes mentioned because of constraints in the size of the work envelope. Other prototype requirements are important as well including material properties, dimensional tolerances, surface finish, build time, and geometric complexity. In general, material removal processes will produce better dimensional accuracy, surface finish, and material properties, whereas FF processes provide better geometric complexity and shorter lead time. Lead times depend upon the amount of specialized fixturing and cutting tools required for machining which depends upon the geometric complexity of the part. Generally, lead times are better on machining processes for simple geometries and are better on additive processes for complex geometries.

Finally, a comparison of costs and benefits can help make the final decision regarding whether to buy or outsource FF technology. To buy an FF machine tool, several costs must be factored into its procurement, including: equipment costs, support technology costs (including CAD systems, cleaning equipment, postcuring devices, etc.), freight and installation costs (including any building modifications), and personnel costs (including any new hires or training). Beyond procurement are the costs required for its operation including: material costs, maintenance costs (including service contracts, preventive maintenance, equipment downtime etc.), facility costs (including floor space, utilities, etc.), and labor costs.

Experienced personnel are very important when operating some FF processes. For example, SL technology has over 30 input variables, which require fine adjustment in order to make good parts, and typically one person per FF machine is needed for simple part finishing. Site preparation can also be a factor. Certain FF systems use toxic resins that require high-capacity venting or cleaning solvents, that must be changed frequently. Many FF systems require additional room for postprocessing equipment including cleaning and postcuring. In all, the personnel, maintenance, training, and site preparation costs must be anticipated in order for the FF system to be profitable.

Benefits resulting from the use of FF equipment may include:

1. Reduction of current operating costs (including labor, quality, purchasing, etc.).
2. Improved sales and marketing due to the ability to respond to customer requests for bids with actual models.
3. Improved product development including improved customer satisfaction and improved product manufacturability.
4. Improved process development as lead times and costs are reduced, leading to more iterations.

However, these benefits must be weighed against the time required to learn how to make high-quality, dimensionally accurate parts. In addition, if a company has not had experience with CAD solid modeling prior to its involvement with FF technology, the learning time will be greater.

Of utmost importance in considering the purchase of an FF system is consideration of the batch size of prototypes to be fabricated on the machine. Nesting parts within the build cycle can make the difference between economical and expensive FF processing of prototypes. For example, the laminated processes tend to not permit large batch sizes; however, they are the fastest and cheapest when building large, individual parts.

Prior to procurement of FF equipment, it is generally a good idea to test a particular FF technology by using FF services at a job shop or service bureau. In addition to trying FF technology through a service bureau, many small and medium-sized companies prefer to use service bureaus over the costs of purchasing an FF machine.

■ KEY WORDS

3D printing
additive
ascending surface
ascending suspension
(CAD) computer-aided design
(CAM) computer-aided manufacturing
(CNC) computer numerical control
concept modelers

deposition-based
descending platform
desktop manufacturing
(ED) extruded deposition-based
(FF) freeform fabrication
functional modelers
fusion
hatching
(ID) inkjet deposition-based

ionography
lamination-based
layered manufacturing
mask plotter cycle
masked-lamp descending platform
niodel grower cycle
photopolymer-based
postprocessing
powder-based
preprocessing

production tooling
prototype
prototype tooling
prototyping
random noise shrinkage
random shrinkage coefficient
(RP) rapid prototyping
rapid tooling
(SLFS) scanned laser fusion and sintering

(SLP) scanned laser polymerization
sintering
soft tooling
(SL) stereolithography
subtractive
tessellation
tool paths
tool-less manufacturing
voxel
voxel geometry

■ REVIEW QUESTIONS

1. What is a prototype?
2. Why is the activity of prototyping so important?
3. Why is the activity of rapid prototyping so expensive?
4. In addition to mechanical component industries, what are other industries that have developed rapid prototyping methods not addressed in this chapter?
5. List five other names that are often used synonymously to describe rapid prototyping technology. For each name, provide a rationale or draw a parallel for why the name is used.
6. What three steps used in the general sequence are common to all FF processes?
7. What is meant by the statement that FF processes are additive?
8. What is the distinction between a concept modeler and a functional modeler with respect to FF hardware?
9. What are the four classifications of FF processes?
10. How do the capabilities of FF processes in general compare with those of material removal processes such as turning?
11. What does the term *tessellation* mean, and what is it used for?
12. What does preprocessing involve in FF processes? What does postprocessing involve?
13. Of the three components of build time in FF processes, which tends to be the longest? Why?
14. What is a voxel? Why is it significant in FF processing?
15. What is the most dominant surface effect caused by FF processes? How can this effect be minimized?
16. What is random noise shrinkage, and why is it a factor in determining the dimensions of prototypes produced in FF processes? What is the implication of the random noise shrinkage?
17. According to Table 33-2, which material would be most likely to have the worst dimensional accuracy?
18. In addition to the random noise shrinkage, what other general factors affect dimensional accuracy in FF processes?
19. What are the patterning and layering mechanisms for stereolithography?

20. What postprocessing is required for SLA prototypes?
21. What alternative layering mechanisms exist for scanned laser polymerization processes?
22. What is hatching? What advantage is provided by investment casting hatch styles?
23. What are the two factors that lead to the parabolic shape of the voxel geometry in scanned laser photopolymerization processes?
24. Assuming everything else remained constant, what would happen to the line width and line depth of a voxel if the power were decreased? If the scanning speed were decreased?
25. What are the advantages of SLP processes? Disadvantages?
26. What are the patterning and layering mechanisms for solid ground curing?
27. What type of postprocessing is required for lamp photopolymerization processes?
28. What are the advantages of lamp photopolymerization processes? Disadvantages?
29. Why can the packing density of prototypes in the SGC process be significantly higher than in the SLP processes?
30. What are the patterning and layering mechanisms for fused deposition modeling?
31. What type of postprocessing is required for ED processes?
32. What are the advantages of ED processes? Disadvantages?
33. What are the patterning and layering mechanisms for inkjet deposition processes?
34. What type of postprocessing is required for ID processes?
35. What are the advantages of ID processes? Disadvantages?
36. What are the patterning and layering mechanisms for selective laser sintering?
37. What type of postprocessing is required for SLFS processes?
38. What are the advantages of SLFS processes? Disadvantages?
39. What is the difference between fusing and sintering?
40. How is the processing of metal and ceramic powder different from the processing of thermoplastic powder in SLFS processes?

41. What are the patterning and layering mechanisms for selective inkjet binding processes?
42. What type of postprocessing is required for SIB processes?
43. What are the advantages of SIB processes? Disadvantages?
44. What are the patterning and layering mechanisms for lamination-based processes?
45. What type of postprocessing is required for lamination-based processes?
46. What are the advantages of lamination-based processes? Disadvantages?
47. What is rapid tooling? Why is it such a good match for FF technology?
48. Name three different types of rapid tooling. Briefly describe each.
49. What are engineering tooling assemblies (ETAs)? Why are ETAs significant?
50. List several applications that FF technology can be used for.

*C*hapter 33 CASE STUDY

Flywheel for a High-Speed Computer Printer

Figure CS-33 shows a $5\frac{3}{4}$ in. diameter flywheel for use in a large, commercial-grade high-speed computer printer. The part is $3\frac{3}{4}$ in. thick to provide a total weight of approximately 12 lb, assuming fabrication from a ferrous-based metal. The only mechanical requirement is a Rockwell B hardness equal to or greater than 50. The most restrictive requirement appears to be dimensional precision, since tooth, bore, hub and diameter dimensions must all be within 0.001 in. of specifications.

1. Based on the size and shape of the product, what are some possible means of producing the part using a rapid prototyping process? Would you make the prototype out of metal or plastic? Would you need to machine the prototype?

2. What are some possible ways to produce this part, assuming that a ferrous material will be used? Select an appropriate material for each of the alternatives discussed. (*Note:* Casting alloys should be matched with casting processes, high machinability alloys would be favored for cutting applications, etc.)

   Which of the above alternatives do you feel would be the "best" solution to the problem? Why? For this system, outline the specific steps that would be necessary to produce the part from reasonable starting material.

   For your proposed solution in part 2, would any additional heat treatment or surface treatment be required? If so, what would you recommend?

FIGURE CS-33

CHAPTER 34

MICROELECTRONIC MANUFACTURING AND ELECTRONIC ASSEMBLY

■ 34.1 INTRODUCTION

Miniaturized microelectronic circuits are in common use today in wristwatches, portable CD players, cellular phones, home entertainment systems, fax machines, artificial hearts, military satellites, automotive fuel injection systems, and cardiac defibrillators, among others. Over the past three decades, the number of components per integrated circuit has increased from 2,300 in 1971 to 42 million in 2001, while the number of calculations per second has increased 100,000 times, from 10,000 to over a billion. The key to this progress has been the development of large-batch-size, semiconductor-processing methods coupled with miniaturization of electrical components and connectors. Unlike most other manufacturing processes in this book, semiconductor processes and other electrical and electronic manufacturing processes are concerned mainly with the manipulation of electrical properties rather than mechanical properties.

■ 34.2 HOW ELECTRONIC PRODUCTS ARE MADE

The goal of all electronics is the processing and manipulation of electrical signals represented most fundamentally by the flow of electrons. A hierarchy for producing electronic products is illustrated in Figure 34-1. At the lowest level, microelectronic fabrication methods produce entire *integrated circuits* (ICs) of solid-state (no moving parts) components, complete with wiring and connections, on a single piece of semiconductor material. Arrays of ICs are produced on thin, round disks of semiconductor material called *wafers*. Once the semiconductor wafer has been processed, the finished wafer is sectioned into individual ICs, or *chips*. Next, these chips are individually housed within various types of IC packages for connection to other electronic components and protection from environmental elements. These IC packages, along with other discrete components (e.g., resistors, capacitors, etc.), are then combined together into even larger circuits on *printed circuit boards* (PCBs). This is sometimes referred to as *electronic assembly*. Electronic packages at this level are called *cards* or *printed wiring assemblies* (PWAs). Next, series of cards are combined on a *backpanel* PCB, also known as a *motherboard* or simply a *board*. This level of packaging is sometimes referred to as *card-on-board* packaging. Ultimately, card-on-board assemblies are put into housings and integrated with power supplies and other electronic peripherals through the use of cables to produce final commercial products.

In general, the lower the level of integration (i.e., the physically smaller the circuit and its components) within this hierarchy, the less expensive it is to produce in terms of cost per functional element. This is because, to some extent, the manufacturing cost per IC is about the same regardless of how many components are packaged onto the chip.

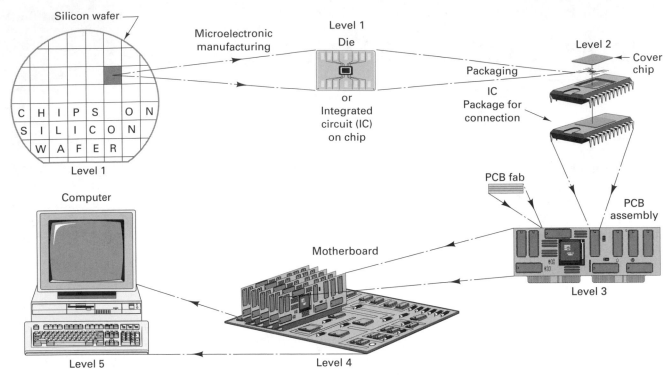

FIGURE 34-1 The hierarchy for producing electronic products has many levels. *(Source: Minges, M. L., Electronic Materials Handbook, Volume 1. Packaging, Materials Park, OH: ASM International, 1989.)*

At the same time, the lower the level of integration is, the less flexibility in configuring the electronic system for different commercial applications. This balance between cost and flexibility is primarily what drives designers to implement circuits at various hierarchical levels.

■ 34.3 SEMICONDUCTORS

Semiconductors, such as silicon, gallium arsenide, and germanium, are materials that can be made to be either electrically conducting or electrically insulating by changing the type and concentration of impurity atoms found within the material. Like metals, all semiconductors have crystalline microstructures exhibiting long-range order in the form of a lattice. However, unlike metals, semiconductor atoms are characterized as having half-filled valence shells, and so, when placed into a lattice, the semiconductor atoms form covalent bonds. Figure 34-2a shows a schematic of a lattice of covalently bonded silicon atoms.

At room temperature (25°C), silicon permits a small amount of electrical conductivity that is too small for most electronic applications. The electrical conductivity of semiconductors can be altered by inserting impurity atoms into the semiconductor lattice. The process of modifying the electrical properties of semiconductors by introducing impurity atoms is commonly referred to as *doping*. Figure 34-2b shows the same lattice as before with the middle silicon atom having been replaced by a phosphorous atom. Because phosphorous is a Column V element on the periodic chart, the phosphorous atom has one more valence electron than the surrounding silicon atoms. As such, the phosphorous atom is considered a donor of electrons to the silicon lattice and the phosphorus-doped semiconductor is now called an *n-type* (negatively charged type) *semiconductor*. N-type semiconductors have extra valence electrons, which are free to move about, providing increased electrical conductivity. Similarly, Figure 34-2c shows a third silicon lattice, this time with the middle silicon atom replaced by a boron atom (a Column III element). This lattice has a shortage of electrons represented as electron *holes* and is therefore termed a *p-type* (positively charged type) *semiconductor*.

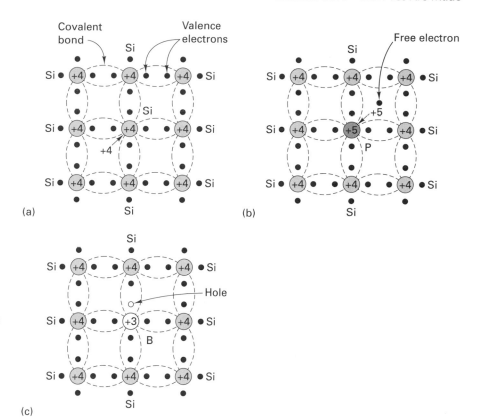

FIGURE 34-2 (a) Schematic of a lattice of silicon atoms; (b) doping with impurity atoms changes conductivity to n-type semiconductor or (c) to p-type, positively charged semiconductor. *(Source: Millman, J., Microelectronics: Digital and Analog Circuits and Systems, New York: McGraw-Hill, 1979.)*

Silicon is the most widely used semiconductor. It is plentiful and can be readily produced in single crystal form. Also, the native oxide, silicon dioxide, can be used both as a dieletric layer and a diffusion mask during processing.

■ 34.4 How ICs Are Made

The ability to selectively modify the electrical properties of semiconductors is the backbone of microelectronic manufacturing as shown in the following examples for producing IC components. A simple bipolar diode (allows current flow in only one direction) may be fabricated by forming two adjacent regions of n-type and p-type semiconductors whose electrical properties have been modified through the placement of impure, secondary atoms into the semiconductor lattice. At the interface of the regions, a so-called *p-n junction* is formed (Figure 34-3a). In the p-n junction, the excess electrons (from the n-type semiconductor) and holes (from p-type) recombine to form a *depletion region* where all charge mobility (i.e., via electrons and holes) is effectively eliminated. While the recombination of holes and electrons fills out the valence shells in the lattice, an imbalance in charge exists, creating an electrostatic potential called the *barrier potential*. The barrier potential for a p-n junction in silicon is approximately 0.7 volt. Application of a negative potential to the cathode and a positive potential to the anode (forward bias) at a level greater than the barrier potential of the p-n junction results in a flow of electrons (or holes) as shown in Figure 34-3b. In this state, the diode acts as a closed switch with very little electrical resistance. Application of a reverse bias (Figure 34-3c) causes the diode to act as an open switch with very high electrical resistance.

As shown in Figure 34-4, the manufacturing fabrication sequence for making a simple bipolar diode has many steps, beginning with the production of a silicon wafer from a predoped, single crystal ingot (*boule*) which is cut into wafers, lapped, and polished to produce silicon wafers. The wafers are placed in vacuum chambers, where an oxide layer is grown on the surface of the wafer to act as a mask during subsequent doping of the substrate. The oxide layer is patterned using photolithography in combination with etching. Photolithography is used to produce a polymeric mask over the oxide layer, which will allow only select areas of the oxide layer to be etched. After etching, the

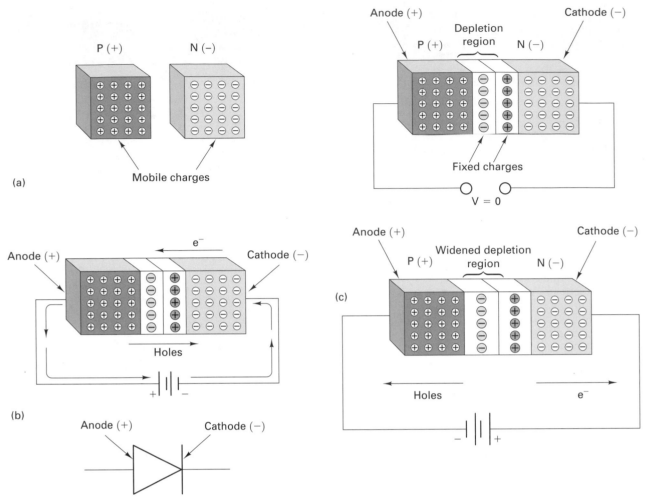

FIGURE 34-3 The diode is produced with a p-n junction or interface at (a) which allow electrons to flow at (b) or have high electrical resistance at (c). (Source: Texas Engineering Extension Service, Semiconductor Processing Overview, College Station, Texas: The Texas A&M University System, 1996.)

polymeric mask is removed from the silicon dioxide layer, and the n-type silicon is doped (by diffusion) with boron to produce a p-type region. After doping, the silicon dioxide mask is removed, and a second silicon dioxide layer is grown and patterned to establish openings in the silicon dioxide layer above the n-type and p-type regions. Next, a thin metal film is deposited on top of the silicon dioxide to provide an electrical pathway allowing the p-type and n-type regions of the diode to be connected to an external power supply. Photolithography and etching are used once again to pattern the thin film into leads and contact pads large enough for biasing the device. To protect the final integrated device from mechanical damage and moisture, a final passivation coating is added.

This example shows the production of a single IC component. Typically, multiple components and, further, multiple circuits are produced in parallel during IC fabrication. As shown in Table 34-1, IC fabrication has evolved from the original small scale integration (SSI) architecture of the 1960s with 2–50 electronic components per circuit to the ultra-large-scale integration (ULSI) architectures of today with tens of millions of components per circuit. The classification of ICs by scale of integration represents the successive advancement of semiconductor processing technologies to provide lower-cost, higher-performance ICs. Each increase in the number of components represented a breakthrough in miniaturization technology (e.g., photolithography and clean rooms) that permitted the fabrication of smaller IC components with improved performance. To achieve lower cost, manufacturing processing technology breakthroughs were needed to make miniaturization technologies possible and economical. Today this trend of seeking higher performance at lower cost continues.

1. Material preparation (P⁺ wafer)

2. Epitaxial growth (P⁻)

3. Mask oxide and photolithography

4. Etch and diffusion and oxide removal

5. Mask removal and fresh oxidation (gate oxide)

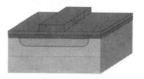

6. Deposited polysilicon

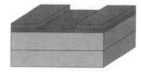

7. Photolithography

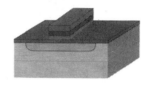

8. Etch

9. Photolithography

10. Ion implantation

11. Oxide deposition

12. Photolithography

13. Etch

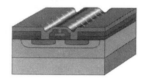

14. Metallization

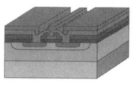

15. Photolithography

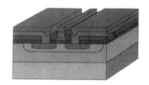

16. Etch

17. Final overcoat

FIGURE 34-4   The manufacture of a simple metal-oxide-semiconductor (MOS) field effect transistor device requires many steps as shown here. *(Source: Texas Engineering Extension Service, Semiconductor Processing Overview, College Station, Texas: The Texas A&M University System, 1996.)*

TABLE 34-1.   Classification in the Development of IC Architectures

Class	Number of Electrical Components per IC	Applications
SSI	2–50	Basic logic
MSI	50–5,000	Encoders, multiplexers, etc.
LSI	5,000–100,000	First generation microprocessors, memory ICs, early calculators, and electronic watches
VLSI	100,000–1,000,000	Integration of microprocessor, memory and I/O on single chip, digital signal processors, computer workstations and microcomputers
ULSI	Greater than 1,000,000	4–64 Mb memory ICs, latest microprocessors, advanced workstations and microcomputers

■ 34.5 HOW THE SILICON WAFER IS MADE

One of the key reasons that single crystal silicon is the most widely used semiconductor is that it can be refined and grown economically in single crystal form. Under equilibrium conditions, molten silicon when cooled produces a polycrystalline structure. However, under controlled conditions, silicon can be grown from a single seed in a large single crystal ingot called a *boule*. The technique used most often for growing single crystal silicon is called the *Czochralski method*. In the Czochralski method, a small *seed crystal* is lowered into molten silicon and raised slowly, allowing the crystal to grow from the seed (Figure 34-5). The size of the seed crystal is about 0.5 cm diameter by 10 cm long. Its crystallographic orientation is critical because it defines the crystallographic orientation and, therefore, the electrical properties, within the boule. The melt consists of electronic-grade (99.999999999% pure) polycrystalline silicon (polysilicon). If desirable, dopant may be added to the melt, although alloying complicates the crystal growth process. The silicon is melted in a fused silica crucible within a furnace chamber backfilled with an inert gas such as argon. The crucible is heated to approximately 1500°C and maintained at slightly above the melting point with a graphite resistance heater.

Once grown, the boule is characterized for resistivity and crystallographic defects. Table 34-2 provides a list of specifications for a silicon wafer. If acceptable, the unusable portions of the boule are cut off, and the outside of the body is ground into a cylindrical ingot. For diameters below 300 mm, *flats* are ground along the length of the ingot to

TABLE 34-2.   Typical Specifications for State-of-the-Art Silicon Wafer

Cleanliness (particle/cm^2)	<0.03
Oxygen concentration (cm^{-3})	Specified ±3%
Carbon concentration (cm^{-3})	<1.5 × 10^{17}
Metal contaminants bulk (ppb)	<0.001
Grown in dislocation (cm^{-2})	<0.1
Oxidation induced stacking faults (cm^{-3})	<3
Diameter (mm)	≥150
Thickness (μm)	625 or 675
Bow (μm)	10
Global flatness (μm)	3
Cost ($/cm^2)	0.2

Source: Campbell, S. A., *The Science and Engineering of Microelectronics*, Oxford: Oxford University Press, 2001.

FIGURE 34-5   In the Czochralski method, a small seed crystal is used to grow large single crystals of silicon. *(Source: MEMC Electronics International Website, www.memc.com.)*

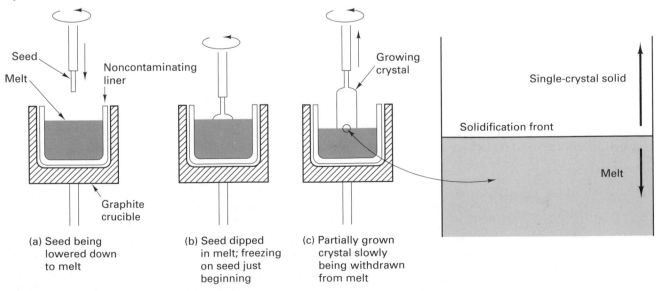

(a) Seed being lowered down to melt

(b) Seed dipped in melt; freezing on seed just beginning

(c) Partially grown crystal slowly being withdrawn from melt

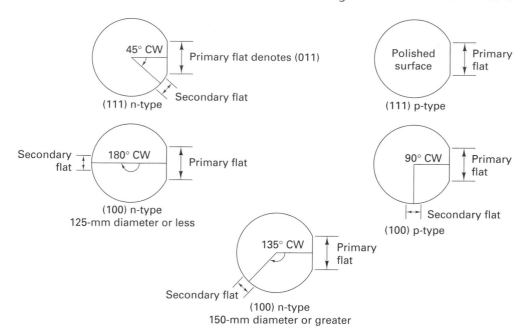

FIGURE 34-6 Flats are ground on the boule to denote the (011), (111), (100) planes, n-type and p-type materials. (*Source: El-Kareh, B., Fundamentals of Semiconductor Processing Technology, Boston: Kluwer Academic Publishers, 1995.*)

denote crystal orientation and dopant type (Figure 34-6). The largest flat, called the primary flat, denotes the (011) plane. Flats are used to properly orient wafers during IC processing. In larger diameter boules, notches are cut along the length of the boule to increase the surface area available for IC processing. Afterward, the boule is chemically etched to remove any damage imparted while grinding the flat.

Next, the boule is sliced into wafers using a wire or diamond saw. Geometric concerns resulting from wafer slicing include flatness and bowing of the wafer. The wafers are typically ground on the edge because edge-rounded wafers handle better and have less mechanical damage during IC processing, and the pile up of photoresists on the edge of the wafer during photolithography is minimized. Finally, a series of processing steps are needed to remove any sawing damage, including lapping, chemical etching, and polishing.

Single-crystal silicon, with few lattice imperfections, is necessary to produce the high yields required in IC processing. Several sources of crystalline defects exist during processing of the wafer, including contamination, improper pull rates, temperature gradients during pulling, and residual stress during wafer machining. Some methods exist for controlling crystalline defects during processing, such as by rotating the solidified boule and the melt in opposite directions during growth to minimize unbalanced growth caused by temperature gradients. However, not all defects can be avoided. To keep unwanted impurities and defects from diffusing into active regions of the wafer (i.e., where IC components are made), a strategy known as *gettering* is used. Gettering involves the use of hard-to-move crystalline defects in inactive regions of the wafer (i.e., away from where components will be made) to trap other impurities and defects that may otherwise diffuse into active regions thereby impairing device performance.

■ 34.6 FABRICATING IC ON SILICON WAFERS

The first level of electronic manufacturing involves the manufacture of the ICs or chips. This is a complex process involving many steps, the sequence of which depends on the particular electrical device. The initial steps of doping by diffusion or ion implantation and oxidation are performed in large machines that manipulate the wafers in and out of various vacuum chambers in the correct sequence and duration.

Doping can be accomplished in bulk by alloying at the time of crystal formation. However, selective doping is required for IC production. Selective doping in most early IC devices involved thermal diffusion; more recently, as device dimensions have continued to shrink, ion implantation has become more suitable to better control the depth and concentration of dopant atoms in the wafer. The doped lateral geometry is primarily defined by the use of a low diffusivity mask (e.g., oxide mask) patterned by lithography methods (covered later in this chapter). The depth and concentration are controlled by the method of doping and its process parameters.

One method for doping semiconductor materials involves diffusion. *Diffusion* can be defined as the random migration of particles from regions of high concentration to regions of lower concentration. Any solid solution that contains a concentration gradient will experience a redistribution of solute (dopant) concentration over time. The source of this migration is the random motion characteristic of atoms above 0° K. Single atom movements can cause atoms to swap lattice locations with adjacent atoms, move to adjacent vacancies, or move interstitially.

Diffusion doping of silicon substrates is usually carried out in two steps. First, a *predeposition* step is used to deposit a fixed quantity (dose) of dopant atoms through an oxide mask and into the substrate. This may be done by placing the wafer in a furnace having a gaseous atmosphere containing the required source concentration of dopant. Predeposition can also happen by solid-source and liquid-source doping. In solid-source doping a solid disk of the dopant material is placed in a furnace *boat* adjacent to the wafer where it is heated and evaporated onto the wafer. In liquid-source doping, an inert gas is bubbled through an isotropic solution (bath) containing compounds with the desired dopant. The partial pressure of dopant in the furnace is controlled by the temperature of the bath, the pressure of the gas above the liquid, and the flow of other inert gases into the furnace. Liquid-source doping has gained significant acceptance because of improved purity levels. Disadvantages include high corrosivity and sensitivity to temperature changes in the bath.

Once the dose is deposited in the predeposition step, a *drive-in* step is used to redistribute the dose to achieve the proper depth and concentration. The drive-in step is performed in a vacuum oven without the presence of the dopant source. The advantage of thermal diffusion is that it is fast relative to other doping processes. The disadvantage is less control over the depth and concentration of dopant profiles.

As the overall size of IC devices has decreased, the required thickness of doped regions has also decreased, requiring greater control and precision of doped dimensions. Therefore, doping by thermal diffusion has been replaced by ion implantation within the current generation of IC devices. Ion implantation involves electrostatically accelerating a beam of ionized atoms or molecules toward the wafer surface, allowing the resultant kinetic energy to drive the particles into the substrate. Ion implantation has been found to control the amount of impurity and the depth of impurity penetration much better than thermal diffusion.

One disadvantage of ion implantation is that the kinetic energy of the ion particles damages the silicon substrate. The resulting lattice damage can significantly affect the electrical and chemical properties of the single crystal substrate. This damage can be minimized by annealing the substrates at temperatures up to 1000°C after ion implantation. However, annealing at these temperatures can create problems of its own, causing redistribution of dopant profiles within other previously processed regions of the device. To compensate, *rapid thermal processing* technologies have been developed to reduce the time the wafer is exposed to high temperature. In rapid thermal annealing (RTA), the wafer rests on quartz pins and is heated using a bank of high-intensity filament lamps. Problems with RTA include temperature measurement and thermal uniformity across the wafer. Excessive temperature gradients across the wafer can lead to plastic deformation in the wafer such as warpage and/or slip. RT technologies have been extended to include rapid thermal oxidation, chemical vapor deposition, and epitaxial growth, among others.

Under exposure to oxygen, a silicon surface oxidizes to form silicon dioxide (SiO_2), the same underlying chemical makeup of window glass. Silicon dioxide is an excellent dielectric material and so can be used as the "gate" dielectric in a MOSFET device or as an isolation layer between layers of metal wires which interconnect IC components. Thick oxides formed by thermal oxidation are generally used as masks during doping. The major objective in thermal oxidation is to create an oxide layer of uniform thickness. While silicon readily oxidizes at room temperature, deep penetration of the oxide into the single crystal is accelerated at high temperatures by thermal diffusion. From this standpoint, thermal oxidation is similar to diffusion doping. Other methods for producing thin oxide layers (e.g., for device isolation) do exist and are briefly discussed in the section on deposition processes.

Though similar in some ways, atomic diffusion in oxidation is different than in doping. Compared with the number density of silicon (on the order of 5×10^{22} atoms/cm^3), final dopant concentrations for active device regions within semiconductor substrates are small (on the order of 10^{17} atoms/cm^3), whereas oxygen concentrations are of the same magnitude (10^{22} atoms/cm^3). Due to the large oxygen concentrations and the stoichiomety of the reactions, about 46% of the silicon surface is "consumed" during oxidation. That is, for every 1 μm of SiO$_2$ grown, about 0.46 μm of silicon is used.

The thermal oxidation of silicon is achieved by heating the substrate to temperatures typically in the range of 900–1200°C within an oxygen atmosphere. The atmosphere in the furnace can either contain pure oxygen (dry oxidation) or part oxygen and water vapor (wet oxidation). Initially, the growth of silicon dioxide is a surface reaction, and the growth rate depends on the reaction rate at the silicon surface. The chemical reaction for dry oxidation at the wafer surface is:

$$\text{Si} + \text{O}_2 \longrightarrow \text{SiO}_2 \tag{34-1}$$

Thin gate oxides can be prepared with a very high uniformity over the wafer and from wafer to wafer using dry oxidation.

Because growth rates in wet oxidation are higher than in dry oxidation, wet oxidation is often used to grow thicker oxides. For thicker oxides, the arriving oxidant molecules must diffuse through the growing SiO$_2$ layer to get to the silicon surface in order to react. The primary reason for the faster growth rate in wet oxidation is because water vapor molecules are smaller than molecular oxygen and, therefore, diffuse more easily within silicon. One disadvantage of wet oxidation is that the oxide layer is not as dense. Therefore, wet oxidation is used in applications that are not subjected to electrical stress such as for diffusion masks.

Techniques like diffusion and oxidation are used to modify the electrical properties of the silicon wafer. Additional techniques are needed to transfer the shape of the integrated circuit from the designer's workstation to the semiconductor wafer. In particular, lithography and etching are two intermediate steps necessary to pattern the silicon dioxide films formed as diffusion masks and as electrically insulating layers in components. In addition, these two steps are also needed to pattern the various conductive and insulating thin films necessary to fabricate and interconnect IC components.

Lithography is the process of transferring the geometric patterns of the IC design to a thin layer of polymer, called a *resist*, producing a resist mask on the surface of the silicon wafer. The purpose of the resist mask is to serve as a temporary barrier to etching or implantation, allowing for the selective patterning of *thin films* (e.g., thin films of deposited polysilicon, oxide, or metal for component fabrication, insulation, or interconnection) or the selective doping of semiconductor substrates underneath the resist in various steps of IC processing. Lithography is the most complicated, expensive, and critical process in mainstream microelectronic fabrication. A typical silicon IC device technology may involve 15–20 different lithography patterns, each with feature sizes, or *linewidths*, as small as 0.18 μm. Needless to say, the technologies needed to meet these requirements are expensive. In the early 1990s, using dynamic random access memory (DRAM) ICs as an example, lithography accounted for roughly one-third of the total fabrication cost.

Several different lithography methods exist including: (1) photolithography; (2) X-ray lithography; (3) electron-beam (e-beam) lithography; and (4) ion-beam lithography. The difference in the techniques is the source of ionizing radiation used to expose the resist. The first two methods involve the use of electromagnetic radiation, whereas the latter two involve particle radiation (i.e., an electron or ion beam). Lithography techniques based on electromagnetic radiation are *through-mask* lithographic techniques, requiring the use of a lithography mask to selectively pattern the resist, whereas techniques based on particle radiation are *direct-write* techniques indicating that the particle beams scan the pattern onto the resist directly without the use of a mask. Photolithography is the most common method and will be discussed here.

In photolithography, UV sources of radiation are used to expose UV-sensitive materials called *photoresists*, or simply *resists*. The photomasks, sometimes called *reticles*,

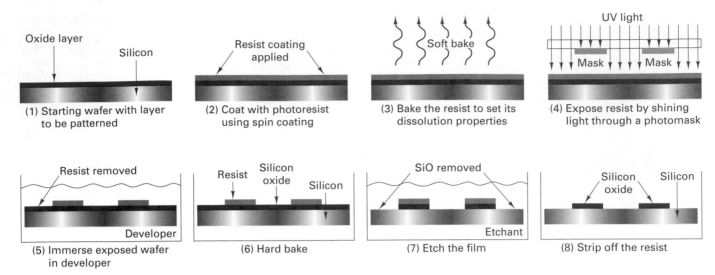

FIGURE 34-7 The process of making an IC using photolithography has many steps. (*Source: Campbell, S. A., The Science and Engineering of Microelectronic, Oxford: Oxford University Press, 2001.*)

are used to mask or screen parts of the surface from etching or doping processes. The photomask is a thin, high-optical-purity quartz plate onto which a thin film of opaque material (such as chromium) has been deposited and patterned (selectively etched). The photomask is patterned with the aid of computer-aided manufacturing (CAM) techniques using the original IC design data from computer-aided design (CAD) systems.

The photolithography process involves the sequence of steps shown in Figure 34-7. First, a liquid photoresist is applied to the surface of the silicon oxide layer over the silicon wafer. Typically, this is done with a process known as *spin coating*. In spin coating, centrifugal forces are used to produce a photoresist layer of uniform thickness. Next, the coated wafer is *soft baked* on a hot plate or in an oven. In this step, solvents used to reduce the viscosity of the photoresist during spin coating are evaporated, and adhesion between the wafer and the photoresist is improved. After soft bake, the photoresist is *exposed* using a photomask to transmit a pattern of electromagnetic radiation onto the surface of the photoresist. This step is performed using a machine called a *stepper*, because the lithographic pattern of the device is indexed or stepped across the wafer, subjecting it to repeated exposures—one for each chip you are making. Once the resist has been exposed, the wafer is *developed* in a chemical solvent. Development removes the unwanted resist materials, exposing the underlying material to be etched. Next, the resist is *hard baked* to remove any remaining solvents after development and to further toughen the remaining resist against downstream etching or implantation processes. Hard bakes generally take longer and are at slightly higher temperatures than soft bakes. Once the downstream etching or implantation has made use of the resist, a photoresist *stripping* step is necessary for removal of the resist.

During exposure, the UV radiation that is transmitted through the photomask selectively modifies the molecular weight of the polymer in desired regions. In *positive* photoresists, the UV radiation is responsible for decreasing the molecular weight in these regions by breaking molecular bonds. The lower molecular weight of these regions makes them more soluble in the chemical developer. In *negative* photoresists, the UV radiation increases the molecular weight of the exposed resist through cross-linking, making the exposed region more insoluble in the developer. These two types of photoresists are contrasted in Figure 34-8.

Obviously, the most important requirement of the photoresist is that it resists the downstream etching or implantation process. Other requirements important to the function of resists are their *resolution* and *sensitivity*. Resolution refers to the smallest linewidth that can be reproduced repeatably by the resist. The resolution of the resist is strongly a function of the source of ionizing radiation or the exposure machine tool used. Sensitivity refers to the amount of ionizing energy required to sufficiently modify

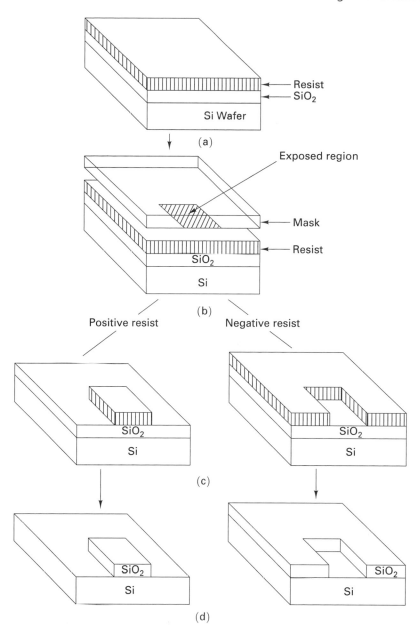

FIGURE 34-8 Photoresist material can be made soluble (positive) or insoluble (negative) to the developer. *(Source: Jaeger, R. C., Introduction to Microelectronic Fabrication (Modular Series on Solid State Device Volume 5), New York: Addison-Wesley, 1990.)*

the solubility of the resist. The more sensitive a resist is, the shorter the exposure cycle time and the greater the throughput. A final requirement of the resist is that it adheres to the substrate.

In the past, the most commonly used light source in photolithography was the mercury arc lamp. Its most useful wavelengths for photolithography occur at 436 and 365 nm (blue and UV light, respectively)—the so-called mercury g-line and i-line. As a result, g-line and i-line photoresists have long been used as standards within the IC industry. Negative photoresists were popular in the early history of IC processing because of their low cost and good adhesion, but positive photoresists are now most widely used since they offer better process control for small geometric features.

Schematic of the exposure step in photolithography are shown in Figure 34-9. Exposure begins with photomask alignment. The photomask is aligned with the wafer so that the pattern can be transferred onto the wafer surface. Each pattern after the first one requires photomask alignment to the previous pattern. For linewidths on the order of 0.18 μm (current IC resolutions), misregistration errors as small as 6 nm (10^{-9} m) can have detrimental effects on device performance. This registration requirement contributes to the high cost of lithography equipment.

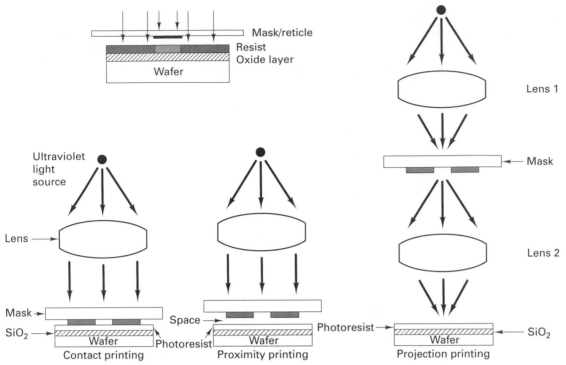

FIGURE 34-9   The exposure step in photolithography is shown in upper left with three primary exposure methods. *(Source: Jaeger, R. C., Introduction to Microelectronic Fabrication (Modular Series on Solid State Device Volume 5), Addison-Wesley Publishing Company, New York 1990 and Zant, P. V., Microchip Fabrication: A Practical Guide to Semiconductor Processing, New York: McGraw-Hill, 2000).*

Once the photomask has been accurately aligned with the pattern on the wafer's surface, the photoresist is exposed through the photomask with a high intensity ultraviolet light. Three primary exposure methods exist: contact printing, proximity printing, and projection printing, as shown in Figure 34-9.

In *contact printing*, the resist-coated silicon wafer is brought into physical contact with the photomask. The wafer is held on a vacuum chuck, and the whole assembly rises until the wafer and photomask contact each other. The photoresist is exposed with UV light, while the wafer is in contact position with the photomask. Because of the contact between the resist and photomask, very high resolution is possible in contact printing (e.g., 1-micron features in 0.5 micron of positive resist). The problem with contact printing is that debris trapped between the resist and the photomask can damage the photomask and cause defects in the resist mask.

Proximity printing is similar to contact printing except that a small gap, 10 to 25 microns wide, is maintained between the wafer and the photomask during exposure. This gap minimizes but may not eliminate resist mask damage entirely due to particles between the photomask and the wafer. Proximity printing offers higher throughput than the other methods but is limited in resolution. Approximately 2- to 4-micron resolution is possible with proximity printing.

Projection printing avoids photomask and resist-mask damage entirely. An image of the photomask is projected onto the resist-coated wafer, which can be many centimeters away. To achieve high resolution, only a small portion of the resist layer can be imaged. Thus the need to scan or *step* the small image over the surface of the wafer. Projection printers that step the photomask image over the wafer surface are called step-and-repeat systems, or *steppers*. Step-and-repeat projection printers are capable of submicron resolution.

After photolithography, the next step is the permanent removal of an underlying film or substrate by *etching*, by chemical or physical means or both. Typical materials etched during semiconductor processing include silicon dioxide to make diffusion masks, dielectric layers, and thin metal films for device fabrication and interconnection. Typical etch rates in semiconductor processing are on the order of several hundred to several thousand Å/min.

The objective in etching is to produce the proper lateral dimensions of the IC in the target material while minimizing the removal of the mask and substrate (underlying the target) materials. Lateral dimensions in etching are controlled in large part by a resist (or perhaps an oxide) mask patterned as described in the lithography section. Deviations from the lateral dimensions in the resist mask are called *etch bias*.

Etching through the resist mask is typically accomplished by either wet chemical etching or dry plasma etching. *Wet etching* involves the immersion of the lithographically patterned wafer in a liquid etchant. The etchant removes material exposed through the resist mask, creating soluble by-products. A rinsing procedure is used to terminate the etching process. Critical parameters in wet etching processes include immersion time, etchant concentration, and etchant temperature. Poor control of process parameters can cause underetching or overetching. Underetching of oxide masks may cause electrical opens in doped regions. In thin films, underetching can cause electrical shorts. Overetching results in etch bias due to undercutting of the mask. Undercutting is the lateral extent of the etch beneath the mask. Overetching can also cause damage to the properties of substrate materials or to the geometry of the resist mask, resulting in further bias.

One way to minimize damage in mask and substrate materials is to use an etchant with high *selectivity*. Selectivity of an etchant refers to the ratio of its etch rate in the target material to its etch rate in the mask or substrate material. As an example, hydrofluoric (HF) acid has a nearly infinite selectivity for SiO_2 over silicon in the making of a diffusion mask. However, one disadvantage of using HF to etch SiO_2 is that the etch process is *isotropic*, meaning that it proceeds equally in all directions. As shown in Figure 34-10, isotropic etching results in an etch bias caused by undercutting.

Dry etching refers to those plasma-assisted etching techniques sometimes called gas-phase chemical etching. There are three main types of plasma-assisted etching, with the main difference being the gas pressure (vacuum) inside the plasma and, consequently, the kinetic energy generated by ions formed within the plasma. *Plasma etching* involves the use of a partially ionized gas (plasma) to chemically react with the target material surface, producing gaseous by-products.

At the opposite end of the dry-etching spectrum is *sputter etching* or ion milling. Sputter etching involves no chemical reaction with the target. Etching of target materials simply involves the physical removal of target atoms as electrostatically accelerated plasma ions slam into the target substrate. As such, sputter etching is the micromechanical equivalent of sandblasting. High etch anisotropy is possible with sputter etching, meaning the etch is very directional with very little undercutting.

A cross between plasma etching and sputter etching is ion-assisted etching, better known as *reactive ion etching* (RIE). In RIE, plasma ions bombard the target material, creating physical damage, which increases the rate of chemical etching.

Table 34-3 compares the anisotropies, resist selectivities, and etch rates of the dry etching processes.

FIGURE 34-10   Deviations from the lateral dimensions in the resist mask are called etch bias, produced here by isotropic behavior of the etchant. *(Source: Campbell, S. A., The Science and Engineering of Microelectronic, Oxford: Oxford University Press, 2001.)*

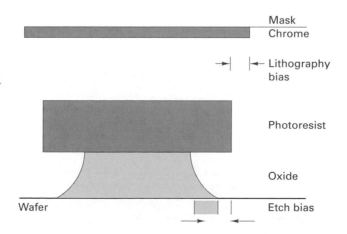

TABLE 34-3.   Types of Dry Etching

	Plasma Etching	Reactive Ion Etching	Sputter Etching
Relative excitation energy	Low	Medium	High
Relative chance of radiation damage	Low	Medium	High
Relative selectivity	High	Medium	Low
Undercut, directionality	Isotropic	Directional from quasi-isotropic (slope) to anisotropic (vertical profile)	Highly anisotropic
Pressure (vacuum)	Greater than 100 mtorr	Approximately 100 mtorr	Less than 100 mtorr
Etch rate	High	Medium	Low

■ 34.7 THIN FILM DEPOSITION

In semiconductor processing, many thin layers of material must be deposited on top of the semiconductor substrate to build IC component features such as transistor gates and to interconnect IC components to form electrical circuits. These layers of material are often well below 1 μm in thickness, and so the term *thin films* is used to describe them. Table 34-4 illustrates some typical ways that thin films are used in semiconductor processing. Various thin film deposition processes are necessary to accomplish these objectives. In general, deposition processes can be broken down into physical vapor deposition (PVD) and chemical vapor deposition (CVD) processes. PVD processes include both evaporation and sputtering.

The simplest form of PVD is *evaporation*. In evaporation, the substrate is coated by condensation of a metal vapor. The vapor is formed from a source material called the *charge*, which is heated within a crucible in moderate vacuum (below 1 millitorr) at a high temperature (greater than 1000°C). Heat energy is provided by either electrical resistance or an electron beam. Electron-beam heating has the advantage of reducing contamination of the deposited thin film since it does not require crucible heating and, consequently, outgassing of the crucible.

While early semiconductor technologies utilized evaporation, it is not used in mainstream processes today. The major disadvantage of evaporation is poor *step coverage*.

TABLE 34-4.   Some Common Applications of Deposited Thin Films and the Processes Used to Make Them

Function	Process						
	VPE	MBE	APCVD	LPCVD	PECVD	Sputtering	Evaporation
Component Fabrication							
Growth of higher purity semiconductor for increased device performance	Single-crystal Si	Single-crystal GaAs					
Masking layer during oxidation			Si_3N_4				
Dopant source for diffusion				Doped polysilicon			
Metal and Dielectric Layers							
Dielectric layer for component fabrication			BPSG	SiO_2, SiO_3N_4			
Conduction path for component fabrication				Doped polysilicon			
Component Interconnection							
Dielectric layer for component interconnection			SiO_2, PSG		SiO_2, PSG	SiO_2, PSG	
Conduction path for component interconnection				W, TiN	TiN	Al, Cu	Al, Cu
IC Packaging							
Passivation of the IC after processing				SiO_2, PSG	SiO_2, Si_3N_4, PSG		

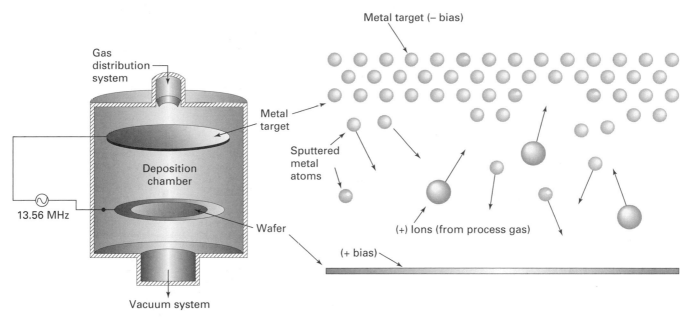

FIGURE 34-11   Sputtering is a PVD method for depositing thin films on microelectronic devices. *(Source: Texas Engineering Extension Service, Semiconductor Processing Overview, College Station, Texas: The Texas A&M University System, 1996.)*

Step coverage is important to avoid opens in wires that connect components during late-stage processing where the surface of the wafer can have severe topology as a result of the many deposition and etching steps. Step coverage has become even more important because the lateral dimensions of circuits continue to decrease with little change in vertical dimensions, resulting in device features with higher aspect ratios. Another disadvantage of evaporation is that it is limited primarily to the deposition of pure metals, although deposition of alloys can be accomplished with difficulty. Dielectrics and polysilicon cannot be deposited by evaporation. Finally, the uniformity of film thickness is hard to control over a large substrate.

An alternative PVD method for metal deposition is *sputtering*. The physics of sputtering are much the same as reactive ion etching. As shown in Figure 34-11, two electrodes are placed several centimeters apart in a low-pressure gas (typically, argon at about 100 millitorr). A potential is placed across the electrodes forming a plasma. Plasma ions are accelerated toward the cathode on which is placed the charge material. At moderate ion energies, atoms and clusters of atoms are ejected from the charge material surface and accelerated toward the wafer.

One advantage of sputtering over evaporation is better step coverage due largely to greater transport energies leading to enhanced surface mobility of the atoms on the wafer. Further, sputtering can be performed on a wide range of materials including elemental metals, alloys, and dielectrics. For metals, a simple dc power source can be used to generate the plasma. For dielectrics, an RF plasma is required. Due to its advantages, sputtering has replaced evaporation for most silicon-based technologies though deposition rates for sputtering are lower than those for evaporation and require more expensive equipment.

CHEMICAL VAPOR DEPOSITION

Chemical vapor deposition (CVD) processes involve the growth of a thin film on a heated substrate by chemical reactions between the substrate and a gaseous compound containing reacting species. In general, CVD techniques provide the advantage of uniform step coverage and, therefore, have become the preferred deposition method for many materials. Figure 34-12 shows a simple configuration for an *atmospheric pressure CVD* (APCVD) system. The reactor consists of a tube with a heated susceptor on which the wafer rests. An inlet and outlet permits the flow of gasses over the surface of the wafer.

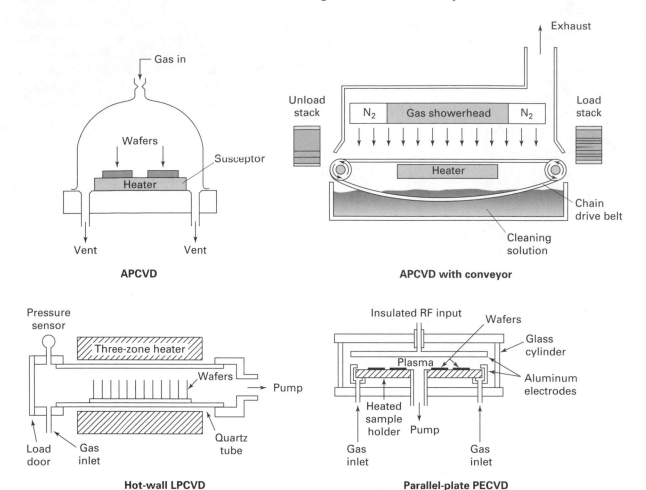

FIGURE 34-12 Chemical vapor deposition (CVD) processes include atmospheric pressure CVD (*upper left*); APCVD with conveyor (*upper right*); hot-wall, low-pressure CVD (*lower left*); cold-wall plasma enhanced CVD. (*Source: Campbell, S. A., The Science and Engineering of Microelectronic, Oxford University press, Oxford, 2001; Jaeger, R. C., Introduction to Microelectronic Fabrication (Modular Series on Solid State Device Volume 5), Addison-Wesley Publishing Company, New York, 1990; and Madou, M., Fundamentals of Microfabrication, New York: CRC Press, 1997.*)

A common thin film deposited in APCVD reactors is silicon dioxide used for passivating circuits. The chemical reaction for the deposition of silicon dioxide is:

$$SiH_4(g) + O_2(g) - (h) \longrightarrow SiO_2(s) + 2H_2(g)$$

where the parenthetical entities (g), (s), and (h) represent gas, solid, and heat, respectively. In this reaction, silane (SiH_4) and molecular oxygen enter the reactor at the inlet, silicon dioxide is deposited onto the wafer, and molecular hydrogen leaves the reactor at the outlet (along with any unused silane). Under proper conditions, this reaction takes place on the surface of the wafer at around 425°C.

APCVD can be performed at temperatures much lower than thermal oxidation, which has advantages in midstage processing of dielectrics. APCVD processes are also attractive because of high deposition rates and simple equipment design. To increase production, wafers can be conveyed through the APCVD reactor on a heated chain conveyor, fed one wafer at a time by multiwafer cassettes. In addition, APCVD can be used to deposit phosphorous-doped SiO_2, or phosphosilicate glass (PSG) otherwise known as *p-glass*, by adding phosphine (PH_3) to the reaction. P-glass can be used to smooth the wafer topology and getter wafer impurities, also during midstage processing. Problems with APCVD include impurities and poor control over film thickness.

Because CVD processes involve chemical reactions, one distinction from other processes involves the location of those reactions. Gas-phase (homogeneous) reactions resulting in solid particulates are generally undesirable because the ensuing particulate deposition produces thin films with poor morphology, increased contamination, and inconsistent properties. Chemical reactions on the wafer surface (heterogeneous reactions) result in the deposition of a solid, thin film with more uniform properties. In APCVD, homogeneous reactions are reduced by the introduction of diluent gases. However, greater control over gas-phase reactions can be obtained with the use of low gas pressures on the order of several hundred millitorr. This process is commonly called *low-pressure CVD* (LPCVD).

Most polycrystalline silicon, or *polysilicon*, is deposited through LPCVD. The chemical reaction for the deposition of polysilicon is carried out at 600°C on the wafer surface. Polysilicon is often used for making gate electrodes during component fabrication. While polysilicon can be deposited within an APCVD reactor, the uniformity of the film thickness is hard to control, which is problematic for gate electrodes. Therefore, polysilicon is normally deposited in an LPCVD reactor where uniformity is easier to control. Another benefit of LPCVD is that it consumes much less carrier gas, which reduces gas expense and handling.

To understand the reason for the improved process control in LPCVD reactors, it is important to differentiate CVD processes that are *reaction-rate limited* from those that are *mass-transport limited*. In reaction-rate limited processes, the deposition rate is controlled by the chemical reaction rate at the surface of the wafer. In contrast, mass-transport limited processes are controlled by the concentration of gases at the surface of the wafer. These two limits on process kinetics account for the major differences in CVD reactor designs.

Most LPCVD processes are reaction-rate limited. Since chemical reaction rates are heavily temperature dependent, thermal uniformity tends to be a design requirement for LPCVD reactors. However, at the low gas pressures in LPCVD reactors, there is more distance between molecules than in APCVD reactors, and consequently, there are fewer interactions between molecules. Therefore, it is more difficult to transfer energy between molecules and attain thermal equilibrium within LPCVD reactors. Because thermal equilibrium is hard to achieve within LPCVD reactors, most LPCVD reactors keep all surfaces within the reactor at the same temperature to minimize thermal gradients within the reactor. Because the walls of these reactors are heated, they are called hot-wall reactors. The ability to maintain thermal stability within hot-wall reactors is the reason for the improved process control of LPCVD reactors.

One disadvantage of hot-wall reactors is that the thin film is deposited along the walls of the reactor as well as on the wafer surface. Over time, these deposited films can flake off and contaminate the wafer surface. As a result, hot-wall LPCVD reactors must be dedicated to the growth of only one material, which reduces their flexibility. In some cases, cold-wall reactors can be used to reduce deposition on the walls. Cold-wall reactors have been used successfully to deposit tungsten for component interconnection, which has the advantage of reducing the size of metal contacts. However, cold-wall reactors do not permit the same level of temperature control and, therefore, do not permit the same level of deposition uniformity as hot-wall reactors.

In general, LPCVD reactors require higher capital expense due to their vacuum requirements and permit lower deposition rates than APCVD reactors. However, since LPCVD reactors typically are not mass-transport limited, wafers may be processed in higher densities within the reactor. Batch sizes in hot-wall LPCVD reactors may be as high as several hundred wafers, which more than makes up for the loss of deposition rate. With such large batch sizes, depletion of reacting species can cause variation in deposition rates from the front to the back of the reactor. This can be accommodated by setting up a temperature gradient from the front to the back of the reactor, resulting in higher chemical reaction rates in the back of the reactor.

In other CVD processes where the deposition rate is mass-transport limited, the major design requirement of reactors is to permit uniform transport of reactant gasses to all parts of the wafers. As a result, these reactors have geometries optimized for gas flow and have excellent gas flow controls. One example of this was shown in the conveyorized APCVD reactor in Figure 34-12. In such reactors, the natural convection of the reactant gas between the wafer surface and the cold walls of the reactor can cause circulation of the gas and make it difficult to control the concentration of the gas at the wafer surface. Consequently, this can make the deposition uniformity between wafers hard to control. These problems can be addressed with proper flow design of the reactor and appropriate parametric design.

Many integrated electronic features require the deposition accuracy of LPCVD. However, many LPCVD processes require high temperatures that are not compatible with late-stage processing, because high temperatures will cause diffusion of previously deposited material layers. Plasma-enhanced CVD (PECVD) has been used effectively to lower the processing temperatures needed to sustain the necessary chemical reactions. One manifestation of a cold-wall PECVD reactor is shown in Figure 34-12. In this case, the wafer rests between electrodes through which an AC potential is applied at radio frequency.

One application of PECVD is in passivating the IC after processing. This is done by depositing silicon nitride as a durable, inert coating to protect the circuit from moisture and scratches. A chemical reaction for sustaining silicon nitride passivation is:

$$3\,SiH_4(g) + 4\,NH_3(g) - (h) \longrightarrow Si_3N_4(s) + 12\,H_2(g)$$

An LPCVD reactor would drive this process at 900°C, but by substituting dichlorosilane ($SiCl_2H_2$), the temperature can be driven lower (700 to 900°C). PECVD can drive this reaction at 300 to 400°C. Both hot-wall and cold-wall PECVD reactors have been used for silicon nitride passivation.

Issues with PECVD include low deposition rates, poor throughput, poorer step coverage, and more impurities than LPCVD. Deposition uniformity can be a concern in hot-wall reactors due to gas depletion. Some effort has been made to improve the throughput of cold-wall reactors by permitting multiple process steps to be performed in a single vacuum chamber.

EPITAXIAL GROWTH

In some cases, it is desirable to deposit a thin film of single-crystal semiconductor material onto the silicon wafer prior to semiconductor processing. *Epitaxy* is the growth of a single crystal, thin film of identical crystallographic orientation as the surface on which it is grown. Epitaxy is used for the purpose of improving semiconductor properties or fabricating abrupt transitions between doped layers that would otherwise be hard to form by diffusion or ion implantation. Epitaxial layers, or epi-layers, are grown under tighter specifications than bulk single crystals, resulting in fewer crystal defects, higher purity, more uniform dopant distributions, and sharper transitions between doped layers. In early semiconductor processing, n-type epi-layers were grown on top of p-type substrates for standard buried-collector bipolar processing. Due to the high temperatures involved (leading to the solid-state diffusion of dopants), current mainstream silicon manufacturing uses epitaxial deposition mainly for thick layers (1 to 10 microns) from which devices may be fabricated.

The mainstream method for silicon epitaxial deposition is vapor-phase epitaxy (VPE). VPE is an extension of LPCVD. One difference is that the VPE of single crystal silicon is performed at higher temperatures (1000°C) than the LPCVD of polysilicon. Since surface reaction rates are much faster at higher temperatures, VPE reactions tend to be mass-transport limited, and as a result, the reactor is designed to optimize gas flow to the wafer. Many other CVD techniques are able to produce semiconductor epitaxial growth including ultra-high vacuum and laser, optical, and x-ray assisted CVD. However, VPE is the mainstream epitaxial method used in silicon fabs. Non-CVD methods (liquid-phase, molecular beam, ion beam, and clustered ion beam epitaxy) have been found more important for depositing compound semiconductors such as GaAs. CVD techniques such as metallorganic CVD are beginning to show promise for GaAs as well.

IC COMPONENT INTERCONNECTION

Up to this point, most of the discussion has focused on individual IC components. The components must be *interconnected* (connected together) with the use of thin film metal wires between thin film dielectric layers. The deposition of thin film metals and dielectrics for component interconnection is called *metallization*.

As shown in Figure 34-13, the geometry of metallization involves metal wires, or *lines*, in between layers of dielectric. In metallization, metal deposition has conventionally been accomplished by PVD processes, whereas dielectric deposition has been

FIGURE 34-13 Schematic of a two-level metal interconnect structure typical of metallization process. *(Source: Jaeger, R. C., Introduction to Microelectronic Fabrication (Modular Series on Solid State Device Volume 5), New York: Addison-Wesley, 1990.)*

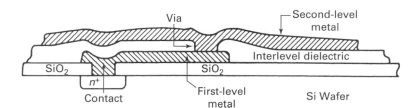

accomplished by either PECVD or sputtering. In order to interconnect the IC components, access must be made between the first level of lines and the semiconductor. These access points are called *contacts*. To permit lines to cross over one another, access must be made between each layer of lines. These access points are called *vias*. Holes for contacts and vias are produced by pattern transfer with photolithography and etching steps.

Modern ICs typically have between one and six metallization layers. During metallization, as each layer of metal and dielectric is deposited and etched on top of another, the topology of the wafer can become quite severe. If the topology of the wafer becomes too pronounced, the result can be shorts in metal wires due to poor step coverage during metal deposition. Further, as the feature resolution of photolithography exposure systems continues to improve, the depth of focus of these systems decreases, making photolithography on uneven topology difficult. To avoid these problems, interlayer dielectric layers must be *planarized* (made flat). As mentioned earlier, one method of reducing topology is to use p-glass as a dielectric interlayer, since it flows at a relatively low temperature and smooths out peaks and valleys. Another method is to perform an etch on the dielectric, which will preferentially attack the high points on the surface. For devices with less than 0.5 μm (transistor) gate thicknesses involving a large number of metallization layers, *chemical mechanical polishing* (CMP) has become the standard in planarization. CMP is similar to conventional mechanical polishing with a wet abrasive slurry, except that the slurry contains a chemical etchant as well. This process is discussed in more detail in Chapter 27.

Metals used for interconnection are those that exhibit low electrical resistance and good adhesion to dielectric insulating layers. Aluminum is a popular metal for interconnecting IC components. Small amounts of copper may be added to reduce the potential for *electromigration* effects in which the applied current to the device can induce the undesirable mass transport of metal atoms over time. As the device and wire sizes continue to decrease, electromigration, which can result in short or open circuits, is becoming a bigger problem. In addition to aluminum alloys, other alloys and pure metals such as tungsten, titanium, and copper are being considered for metallization for their ability to resist electromigration. Other key characteristics of deposited metal films include low film reflectivity (to reduce interference with optical alignment during photolithography) and low residual stress.

To finish component interconnection, a series of connectivity and functional tests called *wafer testing* are performed on each separate IC on the wafer, and the wafer is cut into individual ICs called *dies* or *chips*. The purpose of wafer testing is to eliminate any unnecessary packaging of defective ICs. At this stage, the wafer has an array of ICs that has been produced on it. Wafer testing is performed by computer-controlled probing equipment that introduces electrical signals into each IC by contacting each set of bonding pads on the wafer with needle-like probes. The multiprobe procedure involves indexing each IC under a probe head, which has a probe for each bonding pad. ICs that fail the test are marked with an ink dot and discarded. After testing, the wafer is diced into individual dies or chips. Wafer dicing is typically performed by *diamond sawing*, to give clean edges with minimal damage.

IC YIELD AND ECONOMICS

The larger the IC, the greater the chance for a defect to appear and render the IC inoperative. At first it might appear more economical to build very simple, and therefore very small, circuits on the grounds that more of them would likely be functional. However, in the late 1990s die area was increasing at a rate of about 12% per year, so by the turn of the century, ICs with over ten million transistors had been produced. While it is true that small circuits are inexpensive, the cost of packaging, testing, and assembling the completed circuits into an electronic system must be taken into account. Once the ICs are separated into individual chips, each chip must be handled individually. From that point on, the cost of any processing is not spread over hundreds or thousands. Thus packaging and testing costs often dominate the other production costs in the fabrication of ICs.

One way to improve the economics of microelectronic manufacturing is to increase wafer sizes. The key benefit from processing larger wafers is an increase in the percentage of useable area. Larger wafers have a smaller proportion of the area being affected by edge losses and wafer dicing. Since the mid-1980s, wafer diameters have increased threefold from 100 mm to 300 mm, which required the development of new equipment throughout the semiconductor manufacturing process. A second strategy for improving semiconductor economics involved increasing the number of chips per wafer by decreasing IC dimensions. IC dimensions have decreased more than 50-fold in the past 30 years. The smallest feature size in 1971 was 10 microns. By 2001, transistors with gate features as small as 0.18 micron were made. Again, the catalyst for this improvement was an investment in the process technology, in particular, photolithography.

Perhaps the most effective method of improving IC economics has been the improvement in die yield. Die yield improvement is much more desirable because considerable improvements in economics can be had without making large capital investments. The die yield depends on the wafer yield (the fraction of silicon wafers that started versus those that finished the process) which involves the processing yield (the fraction of good die per wafer), the assembly yield (the fraction of die that are packaged), and the burn-in yield (the fraction of packaged die that survives wafer testing). The largest contributors to lower yields are generally the wafer and processing yields. Wafer yields are driven by large-area defects, which might be the result of poor process control in deposition or etching that would eliminate the usefulness of the entire wafer. Processing yields are generally driven by point defects such as particle contamination, although large area defects can also affect processing yields.

A single, submicron dust particle trapped between the photoresist and reticle in a photolithographic step can cause a point defect that will result in the malfunction of an entire IC. As a result, all microelectronic manufacturing is conducted in *clean rooms*, where special clothing must be worn to prevent dust particle contamination of wafers being processed. The air is continuously filtered and recirculated using high-efficiency particulate-arresting (HEPA) and ultra HEPA (ULPA) filters to keep the dust level at a minimum. Clean rooms are specified by their class of cleanliness with respect to federal Standard 209D. Class 100,000 indicates that the filtration in the clean room limits the number of 0.5-micron diameter particles to 100,000 per cubic foot of volume. Wafers are commonly processed in Class 100 clean rooms.

■ 34.8 IC PACKAGING

Several levels of packaging and assembly are necessary to integrate the IC chip with other electronic devices to make it part of a fully functional commercial or military product. IC packaging serves to distribute electronic signals and power as well as provide mechanical interfacing to test equipment and printed circuit boards (PCBs). In addition to this interconnection role, IC packages protect the delicate circuitry from mechanical stresses and electrostatic discharge during handling and corrosive environments during its operational life. Finally, because of the high density of the integrated circuits, dissipation of heat generated in the circuits has become more critical.

PACKAGE TYPES

ICs come in a variety of packages made from a variety of materials. Figure 34-14 shows a cutaway view of the most well-known IC chip package; the dual in-line package (DIP) refers to the two sets of in-line pins that go into holes in the PCB. The DIP, like all other IC packages, is made up of a leadframe and a package body. Typically composed of a copper alloy (sometimes with an aluminum coating), the leadframe provides electrical interface between the IC and the PCB. The DIP body is made from a low-cost epoxy, which facilitates mass production. In high-reliability applications (e.g., military), where hermetic (air-tight) sealing of the package is important, ceramic package bodies are used.

Generally, IC packages are grouped mainly based on the arrangement, shape, and quantity of leads. Lead pitch refers to the center-to-center distance between leads on an IC package. In conformance to standard-setting bodies, such as the Electronics

Plastic DIP

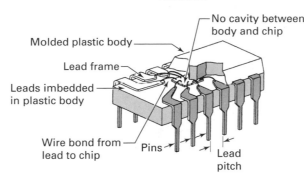

No cavity between body and chip

Molded plastic body

Lead frame

Leads imbedded in plastic body

Wire bond from lead to chip

Pins

Lead pitch

FIGURE 34-14   The dual-in-line package (DIP) has a leadframe and package body. The leads on the chips are connected to the pins. *(Source: Seraphim, D. P., Lasky, R. C., Li, C.-Y., Principles of Electronic Packaging, New York: McGraw-Hill, 1989.)*

Industries Association (EIA) in the United States and EIA Japan, lead pitches above 20 mils (0.02 in.) are measured in inches. Below 20 mils, lead pitches are measured in millimeters.

There are two methods by which components are connected to the circuit on the PCB. The DIP is the leading example of *through-hole* (TH) technology, also known as *pin-in-hole* (PIH) technology, where IC packages and discrete components are inserted into metal-plated holes in the PCB and soldered from the underside of the PCB. In *surface mount* (SM) technology, electronic components are placed onto solder paste pads that have been dispensed onto the surface of the PCB. Figure 34-15 shows the cross section of solder joints for typical SM and TH-packaged components on a PCB.

SM packages are more cost-effective in electronic assembly, and this SM technology has replaced a lot of the TH technology, but not entirely, since not all electronic components can be purchased in an SM package. SM packages are designed for automated production and allow for higher circuit board density than TH components. The manufacturing challenges associated with SM technology include weaker joint strength and solderabilty issues relating to lower in-process lead temperatures. Also, TH components have only one lead geometry, whereas SM components have many different designs. The key packaging families for TH technology are dual in-line packages (DIPs) and *pin grid arrays* (PGAs).

In SM technology, IC packages cannot be discussed separately from lead geometry. Lead geometries affect the electrical performance, size constraints on the PCB, and ease of assembly of the IC package. The most basic form of SM lead is the *butt lead*, or I lead. See Figure 34-16. Butt leads are normally formed by clipping the leads on TH component. This technique is sometimes used to convert an existing TH component to an SM component. Consequently, butt-leaded components do not typically save any space on the PCB. However, they can reduce costs by eliminating the need to perform TH soldering of the PCB after SM soldering. Butt-lead components tend to result in the lowest solder joint strengths, and therefore reliability is an issue.

Gull wing leads bend down and out, whereas *J-leads* bend down and in. Gull wing leads allow for thinner package sizes and smaller leads, which is important for compact applications such as laptop computers. In addition, packages with gull wing leads are compatible with most reflow soldering processes and have the ability to self-align during reflow if they are slightly misoriented. Gull wing leads are compatible with fine pitch packages, but inspection of solder joints is difficult in its final soldered configuration. Gull wing leads are also susceptibile to lead damage and deviation from lead coplanarity. J-leads are sturdier than gull wings and stand up better in handling. The solder joint of J-leads face out, making inspection easier. J-leads have a higher profile than gull wings, which can be a disadvantage for compact applications. At the same time, this higher standoff makes postsolder cleaning easier. J-leads can be used for packages with between 20 and 84 leads.

Solder balls are increasingly being used to provide SM interconnection through *ball grid arrays* (BGAs). Figure 34-17 shows a BGA package. BGAs provide high lead density because the solder balls are arrayed across the entire bottom surface of the package. Lead counts on BGAs can go as high as 2,400, with most in the 200 to 500 lead range. Because of their arrayed nature, BGAs do not need as fine of a pitch (40 to 50 mils) as *quad flat packages* (QFPs), which can help in electronic assembly yields. To further boost yield, the solder balls on BGAs have excellent self-aligning capability during reflow and require

FIGURE 34-15 Here is a summary of the various types of packaging used for ICs. *(Source: Manzione L. T., Plastic Packaging of Microelectronic Devices, New York: Van Nostrand Reinhold, 1990.)*

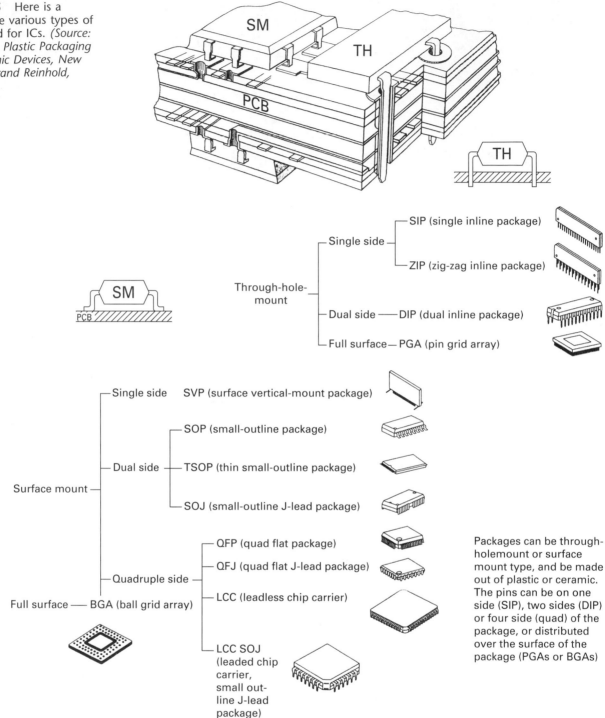

Packages can be through-holemount or surface mount type, and be made out of plastic or ceramic. The pins can be on one side (SIP), two sides (DIP) or four side (quad) of the package, or distributed over the surface of the package (PGAs or BGAs)

less coplanarity (6 to 8 mils) than other leads. The downside of BGAs is the difficulty associated with cleaning, inspection, and rework of solder joints and the lack of compatibility with some reflow methods since joints are out of sight beneath the package.

PACKAGING PROCESSES

The first step in IC packaging is to attach the die to the package. Die attachment techniques include *wire bonding, tape-automated bonding* (TAB), and *flip-chip* technology. In wire bonding, also known as *chip-and-wire attachment*, the chip is attached to the package with an adhesive, and a wire is attached to bonding pads on the chip and on the package. Gold wire as thin as 25 microns and aluminum wire as thin as 50 microns can be attached in wirebonding. As shown in Figure 34-18 for gold wire, the ball bond at the die pad is formed by melting the wire tip and compressing it against the die pad. After die pad bonding,

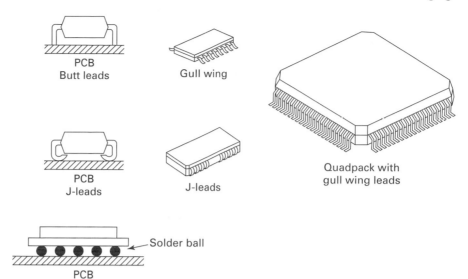

FIGURE 34-16   Basic lead geometries for surface mounted packages include butt leads, gull wing leads, J-leads, solder balls, and plastic quad flat pack.

FIGURE 34-17   Ball grid arrays (BGAs) provide high numbers of connections for leads using solder balls arranged across the entire bottom of the package. *(Source: Prasad, R., Surface Mount Technology. Principles and Practice, New York: Chapman & Hall, 1997, p. 493.)*

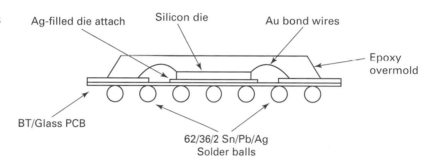

the wire is then looped out and ultrasonically or thermosonically welded to the leadframe of the package. In ultrasonic welding, frictional energy, caused by placing the vibrating wire in contact with the leadframe, causes heating, melting, and coalescence of the two materials. Thermosonic welding is ultrasonic welding with the addition of heat.

In TAB attachment, a thin polymer tape carrying the lead circuitry (see Figure 34-19) is aligned with the die, and the leads are bonded under temperature and pressure to the IC

FIGURE 34-18   Thermosonic ball-wedge bonding of a gold wire. (a) Gold wire in a capillary; (b) ball formation accomplished by passing a hydrogen torch over the end of the gold wire or by capacitance discharge; (c) bonding accomplished by simultaneously applying a vertical load on the ball while ultrasonically exciting the wire (the chip and substrate are heated to about 150°C); (d) a wire loop and a wedge bond ready to be formed; (e) the wire is broken at the wedge bond; (f) the geometry of the ball-wedge bond that allows high-speed bonding. Because the wedge can be on an arc from the ball, the bond head or package table does not have to rotate to form the wedge bond. *(Source: Semiconductor International magazine, May, Des Plaines, Il: Cahners Publishing Co., 1982.)*

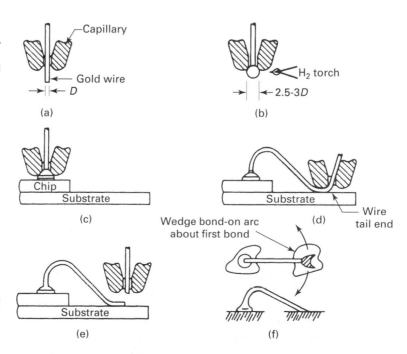

FIGURE 34-19 Tape automated bonding (TAB) uses a polymer type to carry the leads to the chip for bonding. *(Source: Jaeger, R. C., Introduction to Microelectronic Fabrication (Modular Series on Solid State Device Volume 5), New York: Addison-Wesley, 1990.)*

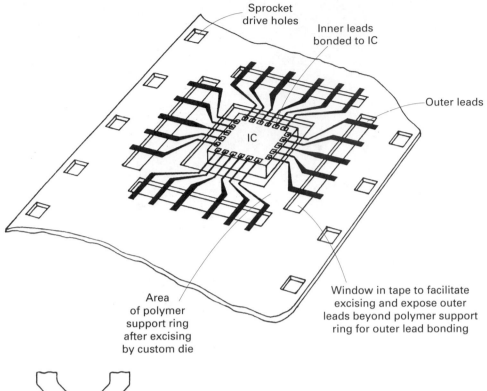

Sprocket drive holes

Inner leads bonded to IC

Outer leads

IC

Area of polymer support ring after excising by custom die

Window in tape to facilitate excising and expose outer leads beyond polymer support ring for outer lead bonding

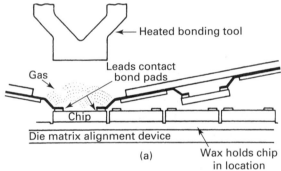

Heated bonding tool

Leads contact bond pads

Gas

Chip

Die matrix alignment device

Wax holds chip in location

(a)

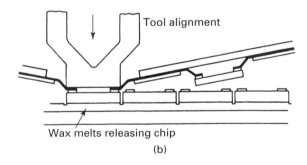

Tool alignment

Wax melts releasing chip

(b)

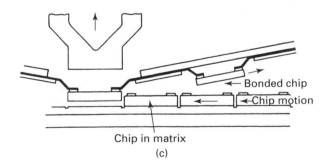

Bonded chip

Chip motion

Chip in matrix

(c)

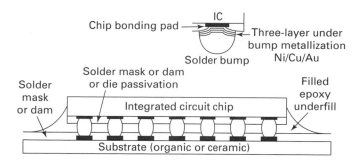

FIGURE 34-20   Flip-chips have the chip turned over so that the bonding pads on the chip and the package face each other. *(Source: Harper, C. A., Electronic Packaging and Interconnection Handbook, New York: McGraw-Hill, 2000.)*

chip. In flip-chip attachment, the chip is turned over so that the bonding pads on the chip and on the package face each other. Flip-chip technology is more common for direct chip attachment to the PCB but is becoming more important for chip-scale packages, as explained later. As shown in Figure 34-20, flip chips are normally attached to the package with a solder bump.

After die attachment, the package is sealed. Plastic packages are either premolded or postmolded. Premolded packages are sealed adhesively with a lid (Figure 34-21). Postmolded packages are sealed via a transfer molding or injection molding process (Figure 34-21). Prior to molding, the die is adhered and wire bonded to the lead frame, which is automatically inserted into the mold. The postmolding process is relatively harsh on the die and wire bonds and can cause major yield and reliability problems. To keep out environmental contaminates, ceramic packages are hermetically sealed by glass using either eutectic AuSi or silver-loaded glass adhesive technologies.

Once the package is sealed, leads are typically formed and may require a solder dip. BGA packages differ from the other packages in that the BGA is interconnected through the use of laminated substrates (plastic or ceramic) similar to PCB processing instead of through leadframes. In BGAs, the outermost layer of interconnection is covered with a solder mask, and openings in the solder mask allow for solder ball attachment during solder dipping.

FIGURE 34-21   Premolded packages (*on left*) are sealed adhesively with a lid while postmolded packages are sealed via a transfer molding or injecting molding process. *(Source: Manzione L. T., Plastic Packaging of Microelectronic Devices, New York: Van Nostrand Reinhold, 1990.)*

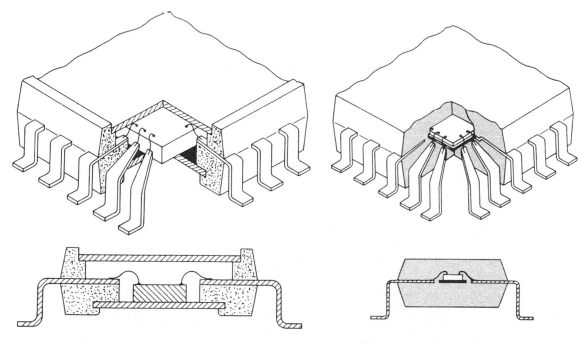

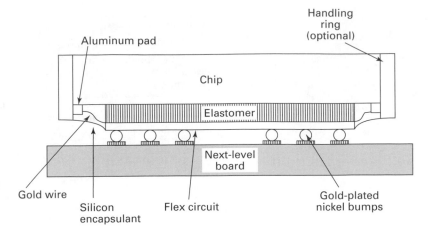

FIGURE 34-22   The Tessera micro BGA package is an example of chip scale packaging. *(Source: Prasad, R., Surface Mount Technology. Principles and Practice, New York: Chapman & Hall, 1997, p. 493.)*

A more advanced option for interconnecting ICs with PCBs is called *direct chip attachment* (DCA), also known as *chip-on-board* or direct die mounting. As suggested, DCA directly attaches the chip to the board using any of three die-attachment technologies mentioned previously. On paper, flip-chip technology has the greatest potential for DCA. However, one challenge associated with flip-chip technology is the coefficient of thermal expansion (CTE) mismatch between the chip and the PCB substrate, particularly as chip sizes increase and solder joint sizes decrease. As a result, the development of underfill encapsulants has become increasingly important for the reinforcement of the mechanical and thermal properties of flip-chip solder joints.

Disadvantages of DCA technologies include shipping and handling of the bare chip and the need for electronic assembly manufacturers to purchase die-attachment equipment. As a compromise, *chip scale packaging* (CSP) has been developed to help downstream processes take advantage of DCA technology. CSP is defined as any packaging that adds no more than 20% of additional board area to the chip. The micro BGA (MBGA) package shown in Figure 34-22 is one example of CSP.

Another alternative to single-chip carriers is multichip carriers or *multichip modules* (MCMs). MCMs are chip carriers that package more than one chip through direct chip attachment to fine line, thin film conductors within a ceramic carrier. MCMs are an extension of hybrid circuits that use refractory substrates and thick and thin film metallization processes for interconnection. The MCM is essentially a mini-PCB with DCA interconnection between the chips and the board. Usually the die is mounted in the MCM with flip-chip technology. The major advantage of MCMs is the reduction in electronic single path distance between ICs. The MCM replaces the typical die-wirebond-pin-board-pin-wirebond-die path with a much shorter die-bump-wire-bump-die path.

Selection of the final chip package for an IC device is dependent on several factors including size, weight, cost, number of leads, power handling, signal delay, electrical noise, and cooling requirements, among others.

■ 34.9 PRINTED CIRCUIT BOARDS

The *printed circuit board* (PCB), or printed wiring board, connects the IC with other components to produce a functional circuit. Specifically, a PCB is a laminated set of dielectric layers or laminates of bulk sheet materials that have metallic circuits that are used to interconnect the various packaged components. As shown in Figure 34-23, each PCB laminate comprises a base, tracks, and pads. The base material must be electrically insulating to provide support to all components making up the circuit. Pads on the laminate are connected by conductive tracks or traces (usually copper) that have been deposited onto the surface of the base. Screen printing was the first technology used to make circuits, hence the term *printed* circuits. Today, metal for traces and pads is deposited by electroless plating and electroplating. Surface mount (SM) components are connected to the PCB at pads (lands) or, in the case of through-hole (TH) technology, at insertion holes.

Typical base materials used may be epoxy-impregnated fiberglass, polyimide, or ceramic. Criteria used for substrate material selection are shown in Table 34-5. Epoxy-impregnated

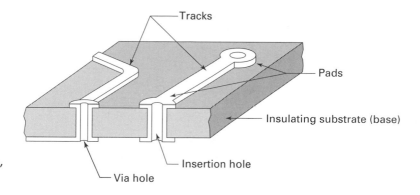

FIGURE 34-23   Double-sided PCB laminate has a base, tracks, and pads.

TABLE 34-5.   Substrate Selection Criteria

Design Parameters	Material Properties								
	Transition Temperature	Coefficient of Thermal Expansion	Thermal Conductivity	Tensile Modulus	Flexural Modulus	Dielectric Constant	Volume Resistivity	Surface Resistivity	Moisture Absorption
Temperature & Power Cycling	×	×	×	×					
Vibration				×	×				
Mechanical Shock				×	×				
Temperature & Humidity	×	×				×	×	×	×
Power Density	×		×						
Chip Carrier Size		×		×					
Circuit Density						×	×	×	
Circuit Speed						×	×	×	

fiberglass is the cheapest substrate for interconnecting leaded packages. Fiberglass is used to increase the mechanical stiffness of the device for handling, while epoxy resin imparts better ductility. Prior to impregnation, the uncured epoxy resin is referred to as *A-stage*. The fiberglass is impregnated on a continuous line where A-stage resin infiltrates the fiberglass mat in a dip basin, and the soaked fabric passes through a set of rollers to control thickness and an oven where the resin is partially cured (Figure 34-24). The resulting glass-resin sheet is called *B-stage* or *prepreg*. Multiple prepregs are then pressed together between electro-formed copper foil under precise heat and pressure conditions to form a copper-clad laminate. The fully cured glass-epoxy core is called *C-stage* material.

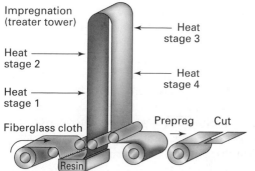

Impregnation (treater tower)

Heat stage 3

Heat stage 2

Heat stage 4

Heat stage 1

Fiberglass cloth   Prepreg   Cut

Resin

FIGURE 34-24   The PCB is made up of laminates of epoxy-impregnated fiberglass manufactured on a machine like this. *(Source: Seraphim, D. P., Lasky, R. C., Li, C.-Y., Principles of Electronic Packaging, New York: McGraw-Hill, 1989.)*

TABLE 34-6.   Laminate Materials Used in Printed Circuit Boards

Common Designation	Resin System	Base Material	Description
XXXP	Phenolic	Paper	Punchable at room temperature.
XXXPC	Phenolic	Paper	Punchable at or above room temperature. XXXP and XXXPC are widely used in high volume, single-sided consumer products.
C-10[a]	Epoxy	Glass fibers	General-purpose material system.
G-11	Epoxy	Glass fibers	Same as G-10, but can be used at higher temperatures.
FR-2	Phenolic	Paper	Same as XXXPC, but has a flame retardant (FR) system that renders it self-extinguishing.
FR-3	Epoxy	Paper	Punchable at room temperature and flame retardant.
FR-4[a]	Epoxy	Glass fibers	Same as G-10, but is flame retardant.
FR-5	Epoxy	Glass fibers	Same as FR-4, but has better strength and electrical properties at higher temperatures.
FR-6	Polyester	Glass fibers	Designed for low capacitance or high impact resistance and is flame retardant.
Polyimide[a]	Polymide	Glass fibers	Better strength and demonstrated stability to a higher temperature than FR-4.

Many different types of epoxy-impregnated fiberglass exist, as shown in Table 34-6. FR-4 (flame retardant) and G-10 are the most popular PCB substrates in use today. Polyimide (without reinforcing fiberglass) is also now widely used in consumer products. PCBs using polyimide substrates are known as flexible printed circuits, or simply *flex* circuits, emphasizing the lack of rigidity. Flex circuits offer the advantages of reduced size and weight as well as the ability to route printed circuits around corners or other nonplanar geometry. Polyester is also used as a flex circuit substrate, although polyimide is the most popular due to its high temperature stability.

Ceramic substrates (typically alumina) are used primarily to minimize thermal stresses on joints in military applications that use leadless ceramic packages. In addition, ceramic substrates are used for hybrid circuits involving both semiconductor and thick-film components. By thick-film, it is meant components that are screen printed as opposed to deposited by thin film technology (i.e., evaporation, sputtering or electroplating).

PCBs can be single-sided, double-sided, or multilayer. Single-sided PCBs simply have metallic circuits on one side of the laminate. Through-hole (TH), single-sided PCBs have insertion holes that extend through the board to the other side where TH components may be inserted into the board (Figure 34-25a). Surface mount (SM) components are simply mounted onto the pads on the same side as the circuit and do not require through-holes. Double-sided PCBs are used in cases where circuits must "jump," or cross over, one another. In this case, *via holes* (or simply vias) are needed to route the circuits over one another (Figure 34-25b). Vias are essentially metal-filled holes through the laminate material that connect a circuit on one side to the other. The metal inside of the via is electroplated. Vias that are also used as insertion holes are called plated through-holes (PTH). As the number of packaged components on the board increases, the complexity of the circuits increases, giving rise to the need for multilayer PCBs in which multiple single and double-sided boards are laminated together using prepreg. Vias that pass from an outermost track on one side of the board to the outermost track on the other side are called *through vias* (Figure 34-25c). Vias within a laminate core on the inside of a multilayer PCB are called *buried vias*. Vias that come out on only one side of a multilayer PCB are called *blind or partially buried vias*. Multilayer PCBs can have as many as 20 layers, although four to eight are more common.

Production of a multilayer PCB from a C-stage laminate begins with a process known as *inner layer circuitization*. Inner layer circuitization may be either subtractive or additive. Subtractive circuitization (Figure 34-26) for glass-epoxy PCBs begins with a double-sided, copper-clad, C-stage laminate known as a *panel*, which has been sheared to size. A film of dry photoresist is applied by hot roller to the copper surface. Next, the circuit pattern (traces and pads) is transferred to the photoresist by exposure through a reticle and chemical development of the photoresist in a photolithographic process similar to that used in semiconductor processing. The copper is then selectively etched through the resulting etch mask, and the resist is subsequently stripped from the laminate. Afterward, registration holes are drilled relative to locator marks, called *fiducials*, produced in the copper layer during the lithography and etching processes.

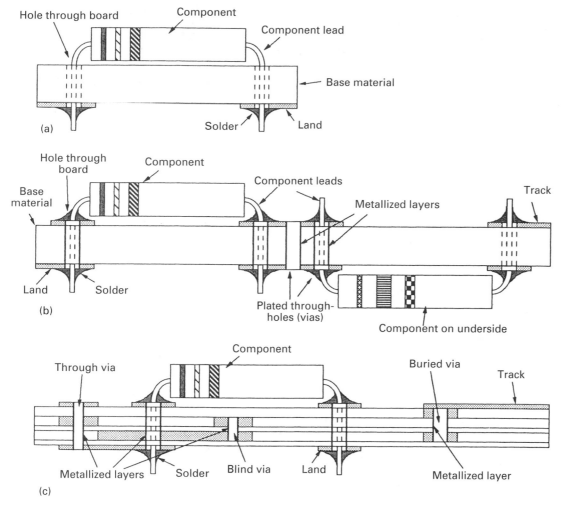

FIGURE 34-25   PCBs can be single-sided, double-sided, or multilayer. *(Source: Judd, M. and Brindley, K., Soldering in Electronics Assembly, Boston: Reed International Books, 1992.)*

FIGURE 34-26   PCBs can have inner layer circuits made by either a subtractive or an additive process. *(Source: Seraphim, D. P., Lasky, R. C., Li, C.-Y., Principles of Electronic Packaging, New York: McGraw-Hill, 1989.)*

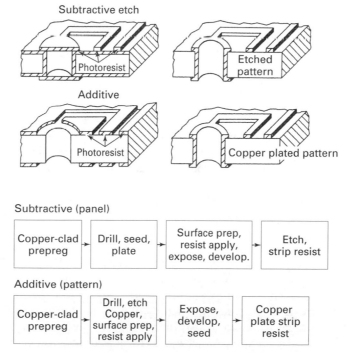

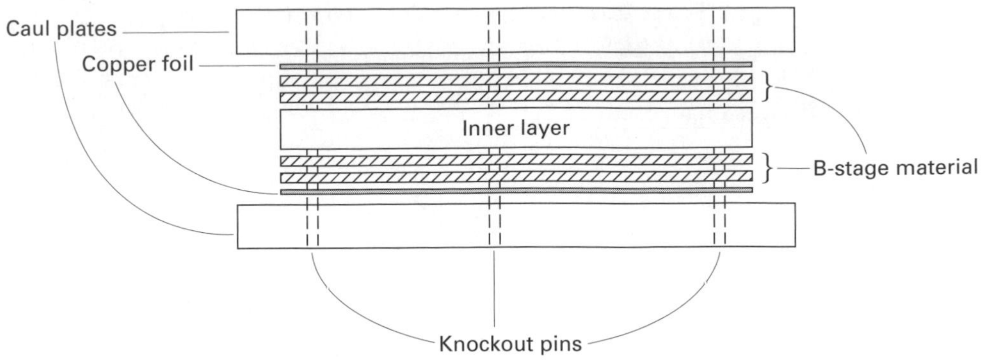

FIGURE 34-27 PCBs are often multilayers of laminations as shown here with a four-layer foil construction. *(Source: Minges, M. L., Electronic Materials Handbook, Volume 1. Packaging, Materials Park, OH: ASM International, 1989.)*

In additive circuitization (Figure 34-26), copper is selectively deposited instead of etched away. The process begins with a bare glass-epoxy laminate cut to size with registration holes. If necessary, via holes may be drilled. An etch mask is exposed and developed, exposing the underlying dielectric including all via holes. The exposed dielectric surfaces are *buttercoated*, or *seeded*, to permit electroless deposition by adsorbing a catalyst (usually palladium) from solution onto the surface of the dielectric. A thin layer of electroless copper is deposited on the seeded dielectric, followed by electroplating of thicker copper layers. Afterward, the resist is stripped. The additive process has the advantage of providing higher resolution for circuitry with finer lines and higher density but tends to be less economical.

The final multilayer board is produced by lamination of inner layers, drilling and preparation of via holes, and circuitization of outer layers. In lamination, as shown in Figure 34-27, a stack of inner layers is bonded together between B-stage prepreg of the appropriate shape and size under time, temperature, and pressure. Usually the inner layers are stacked up between copper layers on the top and bottom of the stack. Prior to lamination, the copper surfaces of the inner layers are oxidized to improve adhesion between layers. Alignment between layers is critical during lamination and is controlled with pins and registration holes. After lamination, excess resin that oozed out during bonding is sheared off. Next, via holes are drilled, deburred, and plated. After drilling, some epoxy smear may exist on the copper surface within the via, which will prevent electrical connection or weaken the mechanical integrity of the via. Therefore, an acidic solution is used to remove excess epoxy and actually to slightly etch back epoxy within the hole so that the copper layers protrude slightly, which improves the connection between copper layers within the via. Subsequently, the epoxy surface within the via is seeded, and a thin layer of electroless copper is deposited. Outer layer circuitization involves the same method of dry resist patterning to produce a lithographic mask. One difference between inner and outer layer circuitization is that copper is electrolytically deposited onto exposed copper after masking to ensure good electrical contact between the inner and outer copper layers. To finish the PCB, the copper layer is etched, a photosensitive solder mask/encapsulant is applied, and the remaining exposed pads are "pre-tinned" with solder.

The manufacturing process for producing flex circuits is similar to the process just described. Flex circuits are typically not as complicated as glass-epoxy PCBs and so have fewer layers. The typical process makes use of a two-sided, copper-clad, polyimide film usually bonded with an epoxy resin. Encapsulation is performed with the use of flexible cover layers, which are either photosensitive or precut and bonded. After encapsulation, the final shape of the flex circuit is cut to size either by shearing (high volume) or by lasers and water jets (prototyping).

With the advent of CSP and MCM packaging technologies, the requirements for track and pad densities have increased. The practical limit of mechanical drilling is a diameter of about 200 microns. *Microvias* are via holes made by photoimaging, laser ablation, and plasma etching that extend well below 200 microns. Microvias as small as 25 microns can be made that increase the density of pads and tracks eightfold over conventional mechanical drilling technologies. Advantages of such small vias include the elimination of bonding pads for direct trace-to-trace connections. Microvias can be used to produce *built-up multilayers*

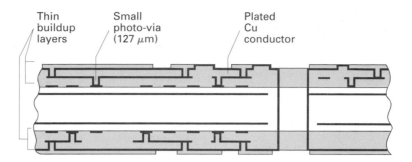

Thin buildup layers

Small photo-via (127 μm)

Plated Cu conductor

FIGURE 34-28 Microvias can be used to produce built-up microlayers. *(Source: Harper, C. A., Electronic Packaging and Interconnection Handbook, New York: McGraw-Hill, 2000.)*

(Figure 34-28). Built-up multilayers are made by deposition and processing of one dielectric layer at a time, similar to IC interconnection. Connections between dielectric layers are made by drilling and electroplating into microvias. Circuit routing can be made extremely efficient, which results in optimally short signal lengths for high-performance applications.

■ 34.10 ELECTRONIC ASSEMBLY

The term *electronic assembly* is generally reserved for the third level of electronics manufacturing involving the soldering of packaged ICs and other discrete components onto PCBs using either through-hole (TH) and/or surface mount (SM). As explained in the IC packaging section, TH technology refers to the insertion of packaged leads into plated through-holes (PTH) in the PCB and soldering of the terminals from the backside. SM technology involves temporary attachment of components to the surface of the PCB via a flux-containing solder paste, which is reflowed within an oven. SM components are much smaller and have much different leads. Passive (non-IC) SM components have terminations rather than leads that permit better shock- and vibration-resistance as well as reduced inductance and capacitance losses.

The sequence of operations for SM and TH assembly is shown in Figure 34-29. Insertion can be performed either manually or with automatic insertion machines. After insertion, leads are generally clinched and trimmed if necessary to avoid *bridging* between joints during soldering, which can cause electrical shorting of the circuit. Generally, soldering of TH components is performed automatically through a process known as *wave soldering*. Wave soldering involves the conveyance of a preheated and prefluxed PCB over a standing wave of solder created by pumping action. The combination of capillary action and pumping action permits flow of the solder from the underside of the board into the joint. Cleanliness of the PCB is critical for wetting of the lead and PTH. A high-pressure air jet is used to blow off excess solder from the underside of the board to prevent solder bridging. Postsolder cleaning of the board includes degreasing and defluxing.

One key consideration for TH solder joints is joint strength. The trade-off is the clearance between the insertion lead and insertion hole. As the clearance decreases, joint strength increases. However, with smaller clearances it is more difficult to insert pins in holes. Clearances on the order of 0.25 mm are typical. Another factor affecting joint strength involves the *clinching* of leads. Clinched lead joints are much stronger than unclinched joints. Because the mechanical strength of TH joints is generally superior to SM joints, large, heavy components are generally attached with TH technology.

SM assembly involves application of solder paste to the lands on the surface of the PCB, placement of SM components on top of this paste, and reflow of the solder paste within an oven. Solder paste consists of small spherical particles of solder less than a tenth of a millimeter in diameter together with flux and solvents used to dissolve the flux (imparting tackiness) and thicken the paste. At the time of application, the paste has the consistency of peanut butter and is applied by screening, stenciling, or dispensing. In screening and stenciling, a solder paste printer is used to apply solder paste through a mask (screen or stencil) by running a squeegee over the surface of the mask. The mask is typically held off of the surface by a distance on the order of 0.5 mm, known as the *snap off distance* (see Figure 34-30). As the squeegee passes over the mask surface, the mask is pressed against the PCB, allowing contact between the paste and the lands on the board. After the squeegee has passed,

FIGURE 34-29 The assembly process steps for making a TH PCB (*above*). The steps for SM assembly are given below along with the steps for mixed technologies—both TH and SM. (*Source: Haskard, M. R., Electronic Circuit Cards and Surface Mount Technology: A Guide to Their Design, Assembly, and Application, New York: Prentice Hall, 1992.*)

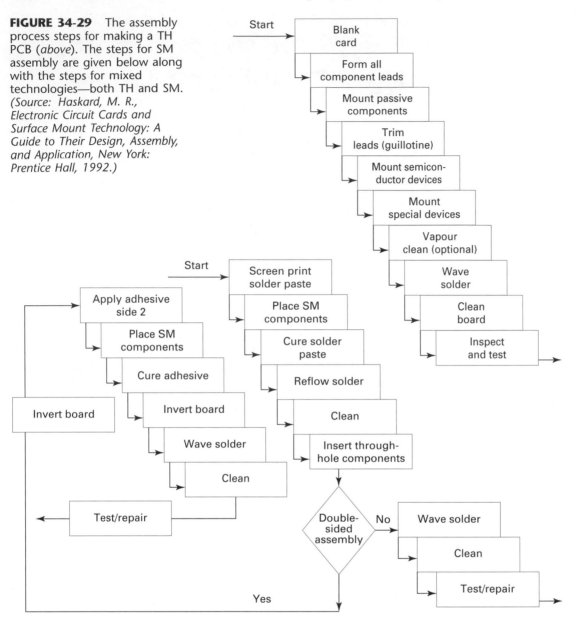

FIGURE 34-30 Schematic for applying solder paste on a substrate by squeegee in a screen printing process in SM technology. (*Source: Prasad, R., Surface Mount Technology. Principles and Practice, New York: Chapman & Hall, 1997, p. 493.*)

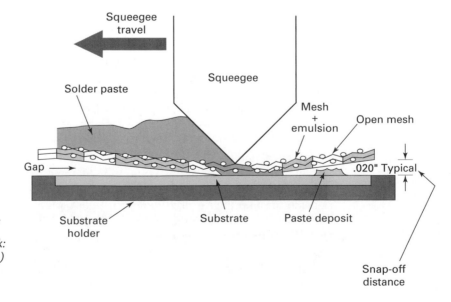

the mask snaps back from the surface, leaving an island of solder paste on the PCB lands. Stencils are typically metal sheets or wire mesh that have been chemically etched using a lithographic process. Screens are typically formed by application, exposure, and development of a photosensitive emulsion on top of a wire mesh. Advantages of metal sheet stencils include longevity and multilevel (pads of varying thicknesses) printing, and the screens are cheaper to make. To decrease tooling costs during product development, pastes can also be dispensed without a mask through a syringe needle. Dispensing generally requires pastes with lower viscosity, which can lead to other problems including solder paste *slump* (spreading out of the solder pastes after application).

Once the solder paste is positioned on the board, a component placement machine, also known as a *pick-and-place* machine, is used to place the components onto the solder paste pads. The flux in the solder paste is tacky and holds positioned components in place until oven soldering. Components are fed to a robotic manipulator that has a vacuum chuck or a mechanical chuck, or both. Component feeders deliver components to the manipulator. Several types of feeders exist including tape (or reel), bulk, tube (or stick), and waffle pack. The feeder system must be carefully selected based on the desired quantity per feeder, availability, part identification, component cost, inventory turns, and potential for damage during shipping and handling. Tape feeders are widely utilized and are most desirable for high-volume placement. Tube feeders are useful for smaller-volume assemblers, even though costs per component are higher. Waffle packs are flat-machined plates with inset pockets to hold various chips. In general, waffle packs increase the cost of assembly. However, some IC packages, like the bumperless, fine-pitch QFPs, require a high level of protection during handling to minimize lead damage, so this component requires the tape feeding mechanism. Bulk feeding of IC components, through the use of a vibratory bowl, may be useful for prototyping environments.

The economics of SM technology are driven by component placement equipment, which determines the throughput of the SM line and is the source (at least partially) of most defects requiring rework. Further, placement equipment strongly influences start-up costs because it may involve as much as 50% of the capital equipment cost in setting up a line. Key criteria in the selection of placement equipment include placement accuracy, placement rate, maximum PCB size, types and sizes of components, and maximum number of feeders, among others. In general, placement equipment has been classified as four discrete types: (1) high throughput, (2) high flexibility, (3) high flexibility and high throughput, and (4) low cost and low throughput with high flexibility. High-throughput placement machines are called *chip shooters*. Chip shooters are typically dedicated to the placement of passive (resistors, capacitors, etc.) and small active (IC) components and can place components at rates up to 60,000 components per hour with linear repeatability around 0.05 to 0.1 mm and rotational accuracy of 0.2 to 0.5 degrees over a 350 by 450 mm area.

After components are placed, the PCB is placed in a *reflow* oven where the solder paste melts causing a fluxing action, which permits the melted solder to wet the leads and the PCB lands. To achieve this, the PCB must be exposed to an appropriate *thermal profile*, or time-temperature curve, as it passes through the oven. Figure 34-31 shows a common thermal profile for SM reflow. At a minimum, the thermal profile must include at least four zones. The first zone, called preheating, is used to drive off any nonflux volatiles within the paste. The second zone, the soak zone, is used to bring the entire assembly up to just below the reflow temperature of the paste. The third (reflow) zone quickly raises the temperature of the solder paste above the reflow temperature allowing for fluxing and wetting of solder joints. The fourth zone cools the assembly permitting solidification. Reflow soldering is generally done in infrared (IR) reflow ovens, and heating involves both IR radiation as well as gas-forced convection. The minimum number of heating zones for a reflow oven must be three (the fourth is a cooling zone) but can contain as many as 20 to provide better control over the thermal profile.

An alternative to IR reflow soldering is *vapor phase soldering*, or *condensation soldering*, involving the condensation of a hot perfluorocarbon vapor onto the assembly surface, releasing the latent heat of vaporization into the solder joints and substrates.

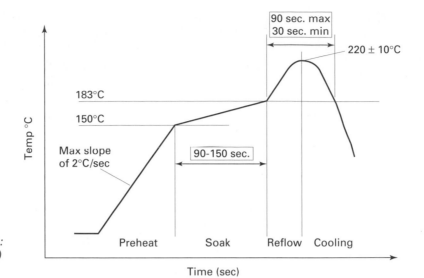

FIGURE 34-31   The typical thermal profile used for SM reflow. *(Source: Prasad, R., Surface Mount Technology. Principles and Practice, New York: Chapman & Hall, 1997, p. 493.)*

By and large, this process has been replaced by IR reflow soldering due to improved process reliability and control. Other alternatives to IR reflow soldering include laser, hot bar, and hot belt reflow soldering.

For a variety of reasons, SM technology and through-hole insertion technology are mixed on the same PCB. Some components are not available in SM packages. Some components are large and require the added strength provided by TH solder joints. Some components require more heat dissipations than SM can accommodate. Thus both methods will continue to be used in the future.

■ KEY WORDS

back panel	clinching	gull wing	pin-in-hole	sputter etching
ball grid arrays	contact printing	integrated circuit	plasma etching	sputtering
bipolar junction	Czochralski method	J-leads	polishing	stripping
transistor	depletion region	linewidths	polysilicon	surface mount
blind vias	diamond sawing	metal-oxide-semi-	prepreg	tape-automated
board	die yield	conductor (MOS)	printed circuit board	bonding
boat	donor	metallization	programmable logic	thin films
boule	dopants	microvias	arrays	through vias
bridging	doping	motherboard	projection printing	through-hole
buried via	drain	n-type semiconductors	proximity printing	vapor phase
butt lead	electronic assembly	hole	quad flat packages	soldering
buttercoated	epitaxial growth	p-channel	reactive ion etching	vias
card-on-board	etching	p-n junction	seed crystal	wafer testing
chemical vapor	evaporation	p-type semiconductors	semiconductor	wafers
deposition (CVD)	feeders	photomask	snap off distance	wave soldering
chip carriers	field-effect transistors	photoresists	soft baked	wet etching
chip shooter	flip-chip	pick-and-place	solder balls	wire bonding
chip	gettering	pin grid arrays	spin coating	

■ REVIEW QUESTIONS

1. What is the goal of the field of electronics?
2. What are the major advantages of integrated circuits over electronic assemblies? What is the advantage of electronic assemblies?
3. How many levels of electronic manufacturing exist? Name each level.
4. What is a semiconductor?
5. Name three common semiconductor materials.
6. What is meant by the term *doping*?
7. What is the difference between n-type and p-type semiconductors?
8. What are electron holes?

9. Give three reasons why silicon is the most popular semiconductor used today.
10. What is a p-n junction? What can it be used for?
11. What is the barrier potential of a p-n junction? Why does it exist?
12. List the sequence of steps necessary to produce a bipolar diode.
13. What is meant by the term ULSI? How is it different from VLSI?
14. In general, what technological breakthroughs were necessary to advance to each successive level of integration?
15. What is a silicon *boule*?
16. Why is it advantageous to add impurities during the formation of the single crystal?
17. Why are notches or flats cut or ground along the periphery of the single-crystal ingot?
18. Why are single-crystal ingots with diameters 300 mm and larger marked with a notch instead of a flat?
19. What are some geometric concerns involved with wafer production?
20. What does *gettering* mean in relation to wafer production?
21. Name three methods for doping a silicon wafer.
22. What are some advantages of ion implantation over thermal diffusion?
23. Why must ion-implanted substrates be annealed?
24. Why are rapid thermal processing technologies generally advantageous?
25. What are two ways in which silicon dioxide is commonly used in microelectronic manufacturing?
26. Give two reasons why wet oxidation is better suited to making thicker oxides (e.g., diffusion masks) than dry oxidation.
27. How is the lateral geometry of the IC and its components patterned into microelectronic materials?
28. What is the most complicated, expensive, and critical step in microelectronics manufacturing?
29. In addition to photolithography, name three other lithographic methods that may be used for pattern transfer.
30. List the photolithographic steps necessary to produce a resist mask on a silicon substrate.
31. Of the two major classifications of photoresists, which type cross-links under exposure to electromagnetic energy of the proper wavelength, resulting in longer polymer chains?
32. List four requirements of a photoresist.
33. What were the two most important wavelengths for photolithography provided by the mercury arc lamp?
34. List the three types of exposure methods used in photolithography. Give an advantage of each.
35. What is the difference between *wet* and *dry* etching?
36. What is undercutting?
37. What are some possible defects that can result from underetching? From overetching?
38. List and describe two properties of etchants.
39. Compare the three dry etch processes including a description of their etch mechanisms and their advantages.
40. What are thin films? Why are they important to microelectronic manufacturing?
41. List two different types of physical vapor deposition including the physical mechanisms and advantages of each.
42. List three different forms of chemical vapor deposition and indicate in what application each might be used.
43. How are undesirable gas-phase reactions controlled within APCVD and LPCVD reactors?

44. What is the key difference in the reactor designs of APCVD and LPCVD processes?
45. What are the two types of LPCVD reactor designs? List advantages and disadvantages of each.
46. What is the advantage of using PECVD processes?
47. What is epitaxy, and why is it important for microelectronic manufacturing?
48. In metallization, what is the difference between a contact and a via?
49. What is planarization, and why is it needed?
50. Give three methods for planarizing interconnect layers.
51. What is electromigration, and why is it a concern in IC processing?
52. What is the purpose of wafer testing?
53. What is meant by the term *chip*?
54. What drives the increase in component density and die area within microelectronic manufacturing?
55. Why are clean rooms so important to microelectronic processing?
56. What two subcomponents make up an IC package?
57. What are the advantages of surface mount technology and through-hole (or pin-in-hole) technology for attachment of IC packages and discrete electrical components to boards?
58. Name the two key classes of TH packages.
59. Name the four different types of SM lead geometries, and discuss the advantages of each.
60. List the key steps involved in conventional IC packaging.
61. List three techniques for attaching and electrically connecting dies to IC packages.
62. What is meant by direct chip attachment, and how does it differ from more conventional IC packaging? What are some disadvantages to direct chip attachment?
63. What are chip-scale packages, and how do they differ from direct chip attachment methods?
64. What are multichip modules, and why are they advantageous for IC packaging?
65. What is a printed circuit board (PCB)? What three elements does a PCB consist of?
66. Name three alternative materials used as dielectrics within PCBs, and give the major reason why each is used.
67. What is the difference between plated through-holes and via holes?
68. What is the difference between blind, buried, and through vias?
69. List the four major process steps for producing a multilayer PCB.
70. List the two methods for inner layer circuitization, and discuss the physical process and advantages of each.
71. What are built-up mulitlayers and microvias? How are they different from laminates and vias?
72. List the four major steps for assembling a TH printed wiring assembly (only TH technology).
73. Why are leads trimmed and clinched after through-hole insertion?
74. List the four major steps for assembling an SM printed wiring assembly (only SM technology).
75. Name four methods for feed components to robotic manipulators in pick-and-place robots. Discuss the application of each.
76. What step in the SM assembly process most strongly affects startup and operations costs for an SM assembly line? Why?
77. What four zones are necessary within an SM solder reflow thermal profile?

CHAPTER 35

FUNDAMENTALS OF JOINING

■ 35.1 INTRODUCTION TO WELDING PROCESSES

Because of a large size, high degree of shape complexity, or wide variation in required properties, it is not uncommon for manufactured products to contain components that are actually joined assemblies of two or more smaller pieces. These pieces may be smaller and therefore easier to handle, simpler shapes that are easier to manufacture, or segments that have been made from different materials. A wide variety of *consolidation processes* have been developed to facilitate the joining or assembly. The metallurgical processes of welding, brazing, and soldering are all quite familiar, as well as the use of discrete fasteners, such as nuts, bolts, screws, and rivets. Adhesive bonding has grown with new developments in polymeric materials and the need for low-temperature joining of composite materials. Lesser-known techniques include shrink fits, slots and tabs, and a wide variety of other mechanical methods. From a technical viewpoint, powder metallurgy is actually another consolidation process, since the end product is built up by the joining of a multitude of individual particles.

We will begin our survey of consolidation processes with a spectrum of techniques known by the generic term of *welding*. Welding is the permanent joining of two materials, usually metals, by *coalescence*, which is induced by a combination of temperature, pressure, and metallurgical conditions. The particular combination of these variables can range from high temperature with no pressure to high pressure with no increase in temperature. Because welding can be accomplished under a wide variety of conditions, a number of different processes have been developed. As a result, welding has become the dominant method of joining in manufacturing, and a large fraction of metal products would have to be drastically modified, or would be far more costly, if welding were not available.

Coalescence between two metals requires sufficient proximity and activity between the atoms of the pieces being joined to cause the formation of common crystals. The ideal metallurgical bond, for which there would be no noticeable or detectable joint, would require (1) perfectly smooth, flat, or matching surfaces; (2) clean surfaces free from oxides, absorbed gases, grease, and other contaminants; (3) metals with no internal impurities; and (4) two metals that are both single crystals with identical crystallographic structure and orientation. These conditions would be difficult to obtain under laboratory conditions, and are virtually impossible to achieve in normal production. Consequently, the various joining methods have been designed to overcome or compensate for the various deficiencies.

Surface roughness can be overcome either by force, causing plastic deformation and flattening of the high points, or by melting the two surfaces so that fusion occurs. The various processes also entail different approaches to cleaning the metal surfaces prior to welding and preventing further oxidation or contamination during the joining process. In solid-state welding, contaminated layers are generally removed by mechanical or chemical cleaning prior to welding, or by causing sufficient metal flow along the interface so that the impurities are squeezed out of the joint. In *fusion welding*, where molten material is produced and high temperatures accelerate the reactions between the metal and its surroundings, contaminants are often removed from the pool of molten metal through the use of fluxing agents. When welding is performed in a vacuum, the contaminants are removed much more easily, and coalescence is easier to achieve. In the vacuum of outer space, mating parts may weld under extremely light loads, even when welding is not intended.

If the process heat is sufficient to induce melting, the structure of the metal may be significantly altered. Even in the absence of melting, the heating and cooling of the welding process can affect the metallurgical structure and quality of both the weld and the adjacent material. The possible consequences of heating and cooling should be a major consideration when selecting a joining process.

The production of a high-quality weld, therefore, requires (1) a source of satisfactory heat and/or pressure, (2) a means of protecting or cleaning the metals to be joined, and (3) caution to avoid, or compensate for, harmful metallurgical effects.

■ 35.2 CLASSIFICATION OF WELDING AND THERMAL CUTTING PROCESSES

Wherever possible, this book will utilize the nomenclature of the American Welding Society (AWS). This organization has classified the various welding processes in the manner presented in Figure 35-1 and has assigned short letter symbols to facilitate their designation. These processes provide a variety of ways of achieving coalescence and make it possible to produce effective and economical welds in nearly all metals and combinations of metals. Chapter 36 will present the oxyfuel gas processes; Chapter 37 will discuss arc welding; and Chapter 38 will cover resistance, solid state, and other processes.

For many years, welding equipment (such as oxyfuel torches and electric arc units) has been used to cut metal sheets and plates. Developed originally for salvage and repair work, then used for preparing plates for welding, this type of equipment is now widely used to cut sheets and plates into desired shapes for a variety of uses and operations. Laser and electron-beam equipment can now cut both metals and nonmetals at speeds up to 25 m/min (1000 in./min), and accuracies of up to 0.25 mm (0.01 in.) are readily attainable. Figure 35-2 summarizes the commonly used *thermal cutting* processes and provides their AWS designations. Since cutting is usually an adaptation of welding, the individual processes will be presented with discussion of both their welding and cutting capabilities.

FIGURE 35-1 Classification of common welding processes along with their AWS (American Welding Society) designations.

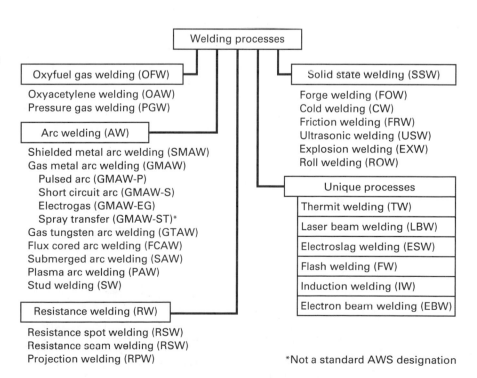

*Not a standard AWS designation

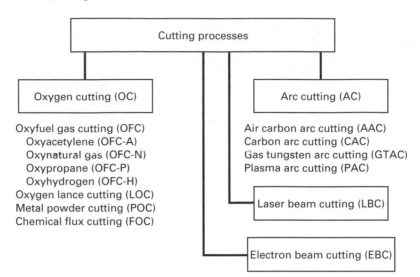

FIGURE 35-2 Classification of thermal cutting processes along with their AWS (American Welding Society) designations.

■ 35.3 WELDING BASICS AND COMMON CONCERNS

Many of the problems that are inherent to welding and joining can be avoided by properly considering the characteristics and requirements of the particular process. Proper design of the joint is extremely critical. Selection of a specific process requires an understanding of the large number of available options, the variety of possible joint configurations, and many other variables that must be specified. Heating, melting, and resolidification can cause problems, including drastic changes in the properties of base and filler materials. Weld metal properties can also be changed by dilution of the filler by melted base metal, vaporization of various alloy elements, and gas-metal reactions.

Various types of weld defects can also be produced. These include cracks in various forms, cavities (both gas and shrinkage), inclusions (slag, flux, and oxides), *incomplete fusion* between the weld and base metals, *incomplete penetration* (insufficient weld depth), unacceptable weld shape or contour, arc strikes, spatter, undesirable metallurgical changes (aging, grain growth, or transformations), and excessive distortion.

■ 35.4 TYPES OF FUSION WELDS AND TYPES OF JOINTS

Figure 35-3 illustrates four basic types of fusion welds. *Bead welds*, or surfacing welds, are made directly onto a flat surface and therefore require no edge preparation. Since the penetration depth is limited, bead welds are used primarily for joining thin sheets of metal, building up surfaces, and depositing hard-facing (wear-resistant) materials.

Groove welds are used when full-thickness strength is desired on thicker material. Some sort of edge preparation is required to form a groove between the abutting edges. V, double-V, U, and J (one-sided V) configurations are most common and are often produced by oxyacetylene flame cutting. The specific type of groove usually depends upon the thickness of the joint, the welding process to be employed, and the position of the work. The objective is to obtain a sound weld throughout the full thickness with a minimum amount of additional weld metal. If possible, single-pass welding is preferred, but multiple passes may be required, depending upon the thickness of the

FIGURE 35-3 Four basic types of fusion welds.

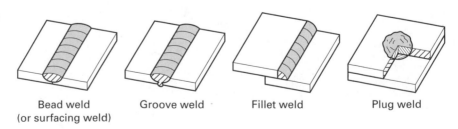

Bead weld
(or surfacing weld)   Groove weld   Fillet weld   Plug weld

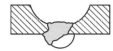

Insert in place

Insert tack-welded

Insert consumed

Completed weld

FIGURE 35-4 The use of a consumable backup insert in making a fusion weld. *(Courtesy Arcos Corporation.)*

material and the welding process being used. As shown in Figure 35-4, special consumable inserts can be used to ensure proper spacing between the mating edges and good quality in the root pass. These inserts are particularly useful in pipeline welding and other applications where welding must be performed from only one side of the work.

Fillet welds are used for tee, lap, and corner joints, and require no special edge preparation. The size of the fillet is measured by the leg of the largest 45° right triangle that can be inscribed within the contour of the weld cross section. This is shown in Figure 35-5, which also depicts the proper shape for fillet welds to avoid excess metal deposition and reduce stress concentration.

Plug welds attach one part on top of another and are often used to replace rivets or bolts. A hole is made in the top plate and welding is started at the bottom of this hole.

Figure 35-6 shows five basic types of joints (*joint configurations*) that can be made with the use of bead, groove, and fillet welds, and Figure 35-7 shows some of the methods

FIGURE 35-5 Preferred shape and the method of measuring the size of fillet welds.

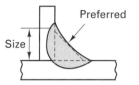
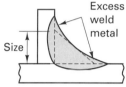
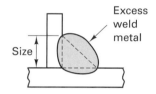

FIGURE 35-6 Five basic joint designs for fusion welding.

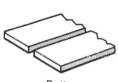 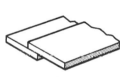 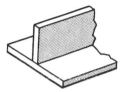 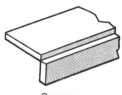 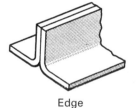

Butt   Lap   Tee   Corner   Edge

FIGURE 35-7 Various weld procedures used to produce welded joints. *(Courtesy Republic Steel Corporation.)*

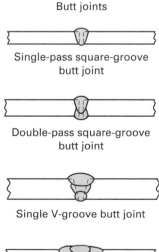

Butt joints

Single-pass square-groove butt joint

Double-pass square-groove butt joint

Single V-groove butt joint

Double V-groove butt joint

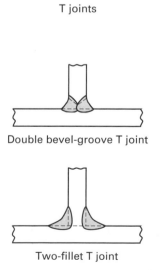

T joints

Double bevel-groove T joint

Two-fillet T joint

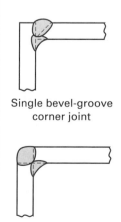

Corner joints

Single bevel-groove corner joint

Two fillet corner joint

to construct these joints. In selecting the type of weld joint to be used, a primary consideration should be the type of loading that will be applied. A large portion of what are erroneously called "welding failures" can more accurately be attributed to inadequate loading considerations. Cost and accessibility for welding are other important factors when specifying joint design, but should be viewed as secondary to loading. Cost is affected by the amount of required edge preparation, the amount of weld metal that must be deposited, the type of equipment that must be used, and the speed and ease with which the welding can be accomplished.

■ 35.5 DESIGN CONSIDERATIONS

Welding is a unique process that cannot be directly substituted for other methods of joining without proper consideration to its particular characteristics and requirements. Unfortunately, welding is also easy and convenient, and the considerations of proper design and implementation are often overlooked.

One very important fact is that welding produces *monolithic*, or one-piece, structures. When two pieces are welded together, they become one continuous piece. This can cause significant complications. For example, a crack in one piece of a multipiece structure may not be catastrophic, because it will seldom progress beyond the single piece in which it occurs. However, when a large structure, such as a ship hull, pipeline, storage tank, or pressure vessel, consists of many pieces welded together, a crack that starts in a single plate or weld can propagate for a great distance and cause complete failure. Obviously, this kind of failure is not the fault of the welding process itself, but is simply a reflection of the monolithic nature of the welded product.

It is also important to note that a given material in small pieces may not behave as it does in a larger size. This feature is clearly illustrated in Figure 35-8, which shows the relationship between energy absorption and temperature for the same steel tested as a small Charpy impact specimen (see Chapter 2 and Figure 2-20) and as a large, welded structure. In the form of a Charpy bar, the material exhibits ductile behavior and good energy absorption at temperatures down to 25°F (−4°C). When welded into a large structure, however, brittle behavior is observed at temperatures as high as 110°F (43°C). More than one welded structure has failed because the designer failed to consider the effect of size on the notch-ductility of metals used in large welded structures.

Another common error is to make welded structures too rigid, thereby restricting their ability to redistribute high stresses and avoid failure. Considerable thought may be required to design structures and joints that provide sufficient flexibility, but the multitude of successful welded structures attests to the fact that such designs are indeed possible.

Accessibility, welding position, component match-up, and the specific nature of a joint are other important considerations in welding design.

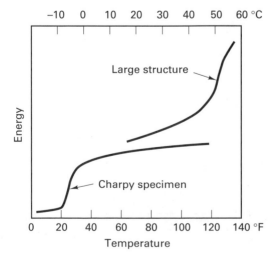

FIGURE 35-8 Effect of size on the transition temperature and energy-absorbing ability of a certain steel. While the larger structure absorbs more energy because of its size, it becomes brittle at a much higher temperature.

■ 35.6 HEAT EFFECTS

WELDING METALLURGY

Heating and cooling are essential and integral components of almost all welding processes and tend to produce metallurgical changes that are often undesirable. In *fusion welding*, the heat is sufficient to melt some of the *base metal*, and this is often followed by a rapid cooling. Thermal effects tend to be most pronounced for this type of welding, but also exist to a lesser degree in processes where the heating–cooling cycle is less severe. If the thermal effects are properly considered, adverse results can usually be avoided or minimized, and excellent service performance can be obtained. If they are overlooked, however, the results can be disastrous.

Because such a wide range of metals are welded and such a variety of processes are used, welding metallurgy is an extensive subject, and the material presented here serves only as an introduction. In fusion welding, a pool of molten metal is created, with the molten metal coming from either the parent plate alone (*autogenous welding*) or a mixture of parent and filler material. Figure 35-9 shows a butt weld between plates of material A and material B. A backing strip of material C is used with filler metal of material D. In this situation, the molten pool is actually a complex alloy of all four materials. The molten material is held in place by a metal "mold" formed by the surrounding solids. Since the molten pool is usually small compared to the surrounding metal, fusion welding can often be viewed as *casting a small amount of molten metal into a metal mold*. The resultant structure and its properties can be best understood by first analyzing the casting and then considering the effects of the associated heat treatment on the adjacent base material.

Figure 35-10 shows a typical microstructure produced by a fusion weld. In the center of the weld is a region composed of metal that has solidified from the molten state.

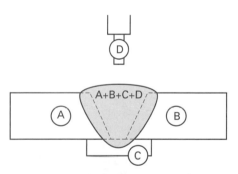

FIGURE 35-9 Schematic of a butt weld between a plate of metal A and a plate of metal B, with a backing plate of metal C and filler of metal D. The resulting weld nugget becomes a complex alloy of all four metals.

FIGURE 35-10 Grain structure and various zones in a fusion weld.

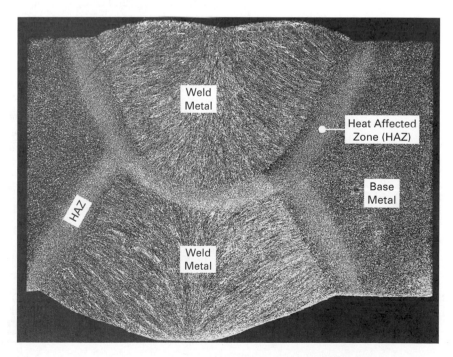

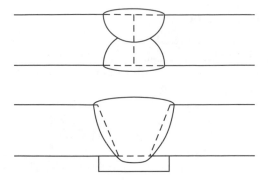

FIGURE 35-11 Comparison of two butt-weld designs. In the top weld, a large percentage of the weld pool is base metal. In the bottom weld, most of the weld pool is filler metal.

The material in this *weld pool*, or *fusion zone*, is actually a mixture of parent metal and electrode or filler metal, with the ratio depending upon the process used, the type of joint, and the edge preparation. Figure 35-11 compares two butt weld designs where the weld pool in the upper design would contain a large percentage of base metal and the weld pool in the lower design would be largely filler material. The metal in the fusion zone is cast material with a microstructure reflecting the cooling rate of the weld. This region cannot be expected to have the same properties and characteristics as the wrought material being welded since their processing histories and resulting structures may be different. Adequate mechanical properties, therefore, can only be achieved by selecting filler rods or electrodes, which have properties *in their as-deposited condition* that equal or exceed those of the wrought parent metal. It is not uncommon, therefore, for the filler metal to have a different chemistry than the metal being welded. The matching or exceeding of base metal strength, in the as-solidified condition, is the basis for several AWS specifications for electrodes and filler rods. The grain structure in the fusion zone may be fine or coarse, equiaxed or dendritic, depending upon the type and volume of weld metal and the rate of cooling, but most electrode and filler rod compositions tend to produce fine, equiaxed grains.

The pools of molten metal created by fusion welding are prone to all of the problems and defects associated with metal casting, such as gas porosity, inclusions, blowholes, cracks, and shrinkage. Since the amount of molten metal is usually small compared to the total mass of the workpiece, rapid solidification and rapid cooling of the solidified metal are quite common. Associated with these conditions may be the entrapment of dissolved gases, chemical segregation, grain-size variation, grain shape problems, and orientation effects.

Adjacent to the fusion zone, and wholly within the base material, is the ever-present and generally undesirable *heat-affected zone (HAZ)*. In this region, the parent metal is not melted, but is subjected to elevated temperatures for a brief period of time. Since the temperature and its duration vary widely with location, fusion welding might be more appropriately described as "a casting coupled with an abnormal, widely varying, heat treatment." The adjacent metal may well experience sufficient heat to bring about structure and property changes, such as phase transformations, recrystallization, grain growth, precipitation or precipitate coarsening, embrittlement, or even cracking. The variation in thermal history can produce a variety of microstructures and a range of properties. In steels, the structures can range from hard, brittle martensite all the way through coarse pearlite and ferrite.

Because of the altered structure, the heat-affected zone may no longer possess the desirable properties of the parent material, and since it was not molten, it cannot assume the properties of the solidified weld metal. Consequently, this is often the weakest area in the as-welded joint. Except where there are obvious defects in the weld deposit, most welding failures originate in the heat-affected zone. This region extends outward from the weld to the location where the base metal has experienced too little heat to be affected or altered by the welding process. Figure 35-12 presents a schematic of a fusion weld in a steel where part of the heat-affected zone has been heated above the transformation temperature. Standard terminology is presented for the various regions and interfaces.

Because of the melting, solidification, and exposure to a range of high temperatures, the structure and properties of welds can be extremely complex and varied. Through

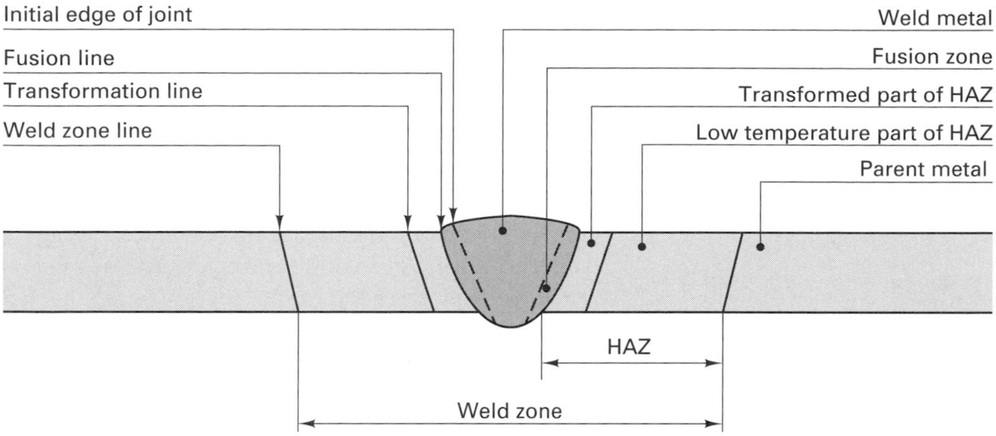

FIGURE 35-12 Schematic of a fusion weld in steel, presenting proper terminology for the various regions and interfaces. Part of the heat-affected zone has been heated above the transformation temperature and will form a new structure upon cooling. The remaining segment of the heat-affected zone experiences heat alteration of the initial structure. *(Courtesy Sandvik AB.)*

proper concern, however, associated problems can often be reduced or totally eliminated. Consideration should first be given to the thermal characteristics of the various processes. Table 35-1 classifies some of the more common welding processes with regard to their *rate of heat input*. Processes with low rates of heat input (slow heating) tend to produce high total heat content within the metal, slow cooling rates, large heat-affected zones, and resultant structures with lower strength and hardness, but higher ductility. High-heat-input processes, on the other hand, have low total heats, fast cooling rates, and small heat-affected zones. The size of the heat-affected zone will also increase with increased starting temperature, decreased welding speed, increased thermal conductivity of the base metal, and a decrease in base metal thickness. Weld geometry is also important, with fillet welds producing smaller heat-affected zones than butt welds.

If the as-welded results are unacceptable, the entire welded assembly might be heat treated after welding. Structure variations can be reduced or eliminated, but the results are restricted to those that can be produced by heat treatment. The structures and properties associated with cold working, for example, could not be achieved. In addition, problems may be encountered in trying to achieve controlled heating and cooling (heat treatment) within the large, complex-shaped structures commonly produced by welding. Moreover, furnaces, quench tanks, and related equipment may not be available to handle the full size of welded assemblies.

An alternative technique to reduce the variation in microstructure, or at least the sharpness of the variation, is to *preheat* either the entire base metal or the segments adjacent to the joint just prior to welding. This heating serves to reduce the cooling rate of both the weld deposit and the immediately adjacent metal in the heat-affected zone. The slower cooling rate results in the production of a softer, more ductile structure that is more resistant to cracking. Preheating is more common with alloy steels and thicker

TABLE 35-1. Classification of Common Welding Processes by Rate of Heat Input	
Low Rate of Heat Input	High Rate of Heat Input
Oxyfuel welding (OFW)	Plasma arc welding (PAW)
Electroslag welding (ESW)	Electron-beam welding (EBW)
Flash welding (FW)	Laser welding (LBW)
	Spot and seam resistance welding (RW)
	Percussion welding
Moderate Rate of Heat Input	
Shielded metal arc welding (SMAW)	
Flux cored arc welding (FCAW)	
Gas metal arc welding (GMAW)	
Submerged arc welding (SAW)	
Gas tungsten arc welding (GTAW)	

sections, and is particularly important with the high-thermal-conductivity metals, such as copper and aluminum, where the cooling rate would otherwise be extremely rapid.

If the carbon content of plain carbon steels is greater than about 0.3%, the cooling rates encountered in normal welding may be sufficient to produce hard, untempered martensite, with the accompanying loss of ductility. Alloy steels possess higher hardenability, so the likelihood of martensite formation will be even greater with these materials. Therefore, special pre- and postwelding heat cycles (preheat and *postheat*) may be required when welding higher carbon and alloy steels. For plain carbon steels, a preheat temperature of 200 to 400°F (100 to 200°C) is usually adequate. Because they can be welded without the need for preheating or postheating, low-carbon, low-alloy steels are extremely attractive for welding.

In joining processes where little or no melting occurs, there is often considerable pressure applied to the heated metal (as in forge or resistance welding). The weld region experiences deformation, and the resultant structure is similar to that of wrought material.

Our discussion of metallurgical effects has largely focused on steels. It should be noted, however, that other metals also exhibit heat-related changes in their structure and properties. The exact effects of the heating and cooling associated with welding will depend upon the specific transformations and structural changes that can occur within the materials being joined.

THERMAL EFFECTS IN BRAZING AND SOLDERING

In brazing and soldering, there is no melting of the base metal, but the joint still contains a solidified region and heat-affected sections will be produced within the base material. For these processes, however, another thermal effect may be quite significant. The base and filler metals are usually of radically different chemistries, and the elevated temperatures of joining also promote interdiffusion. Intermetallic phases can form at the interface and alter the properties of the joint. If present in small amounts, they can enhance bonding and provide strength reinforcement. Most *intermetallic compounds*, however, are quite brittle. Too much intermetallic material can result in significant loss of both strength and ductility.

THERMAL-INDUCED RESIDUAL STRESSES

Another effect of heating and cooling is the introduction of *residual stresses*. In welding, these may be of two types and are most pronounced in fusion welding, where maximum heating occurs. Their effects can be observed in the form of dimensional changes, distortion, or cracking.

Residual welding stresses are the result of restraint to thermal expansion and contraction offered by the pieces being welded. Consider a rectangular bar of metal that is uniformly heated and cooled. When heated, the material expands and becomes larger in length, width, and thickness. Upon cooling, the material contracts, and each dimension returns to its original value. Now clamp the ends of the bar in a vise so that lengthwise expansion cannot take place and repeat the thermal cycle. Upon heating, all of the expansion is restricted to the width and thickness, but the contraction upon cooling will occur uniformly. The resulting rectangle will be shorter, thicker, and wider than the original specimen.

Now apply these principles to a weld between two plates, like that illustrated in Figure 35-13. As the weld is made, the liquid region conforms to the shape of the "mold," and the adjacent plate material becomes hot and expands. The molten pool can absorb expansion of the plate material perpendicular to the line of weldment, but expansion parallel to the weld line tends to be restrained by the adjacent plate material that has remained cooler and is therefore stronger. This resistance or restraining force is often sufficient to induce plastic deformation of the hot, weak, heat-affected zone, which responds to its thermal expansion by becoming thicker instead of longer.

After the weld pool solidifies, both the weld metal and adjacent heat-affected region cool and contract. The cooler surrounding metal resists this contraction. The weld region wants to contract, but is restrained and forced to remain in a "stretched" condition, known as residual tension (region T). Likewise, this contracting region exerts forces that try to squeeze the adjacent material, producing regions of residual compression (regions C). While the net force must be zero, in keeping with the equilibrium laws of

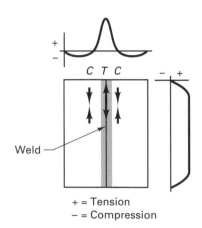

+ = Tension
− = Compression

FIGURE 35-13   Schematic of the longitudinal residual stresses in a fusion-welded butt joint.

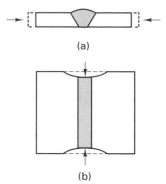

FIGURE 35-14 Shrinkage of a typical butt weld in the transverse (a) and longitudinal (b) directions as the material responds to the induced stresses. Note that restricting transverse motion will place the entire weld in transverse tension.

physics and mechanics, the localized variations can be substantial. As Figure 35-13 depicts, a high residual tension is observed in the weld metal, which becomes compressive and then returns to zero as one moves away from the weld centerline. The magnitude of the residual tension will be relatively uniform along the weld line, except at the ends where the stresses can be relieved by a pulling in of the edges.

Thermal contractions occur both parallel (longitudinal) and perpendicular (transverse) to the weld line. Lateral movement of the material being welded can often compensate for the transverse contractions. The width of the welded assembly simply becomes less than that of the positioned components at the time of welding. Figure 35-14(a) depicts this reduction in width, while Figure 35-14(b) illustrates the longitudinal contractions that generate the complex stresses of Figure 35-13.

Components being joined during fabrication typically have considerable freedom of movement, but welds made on nearly completed structures or repair welds often join components that are somewhat restrained. If the welded plates in Figure 35-14(a) are restrained from horizontal movement, *additional stresses will be induced.* These residual stresses are known as *reaction stresses*, and they can contribute to cracking or failure during use. Their magnitude is an inverse function of the length between the weld joint and the point of restraint. They can have magnitudes up to the yield strength of the parent metal (yielding would occur to relieve higher stresses), but are generally less than the stresses that occur parallel to the weld.

EFFECTS OF THERMAL STRESSES

A common result of the thermal stresses induced by welding is a *distortion* or warping of the assembly. Figure 35-15 depicts the distortions that can occur during various welding configurations. No fixed rules can be provided to avoid warping because the specific conditions that cause it can vary widely. The following suggestions, however, can help to reduce distortion problems.

Total heat input to the weld should be minimized. Welds should be made with the least amount of weld metal necessary to form the joint. Faster welding speeds reduce the welding time and also reduce the volume of metal that is heated. Welding sequences should be designed to use as few passes as possible, and the base material should be permitted to have a high freedom of movement. When constructing a multiweld assembly, it is beneficial to weld toward the point of greatest freedom, such as from the center to the edge.

The initial components can also be oriented out of position, so that the subsequent distortion will move them to the desired final shape. Another common procedure is to completely restrain the components during welding, thereby forcing some plastic flow in the joint and surrounding material. This procedure is used most effectively on small weldments where the high reaction stresses are not likely to cause cracking. Still another procedure is to balance the resulting thermal stresses by depositing the weld metal in a specified pattern, such as short lengths along a joint or on alternating sides of a plate. Warping can also be reduced by the use of *peening*. As the weld bead surface is hammered with the peening tool or material, the metal is flattened and tries to spread. Being held back by the underlying material, the surface becomes compressed or squeezed. Surface rolling of the weld bead area can have the same effect. In both processes, the compressive stresses induced by the surface deformation serve to offset the tensile stresses induced by welding.

Residual stresses should not have a harmful effect on the strength performance of weldments, except in the presence of notches or in very rigid structures where no plastic flow can occur. These two conditions should not exist if the welds have been properly designed and proper workmanship has been employed. Unfortunately, it is easy to inadvertently join heavy sections and produce rigid configurations that will not permit the small amounts of elastic or plastic movement required to reduce highly concentrated stresses. In addition, geometric notches, such as sharp interior corners, are often incorporated into welded structures. Other harmful "notches," such as gas pockets, rough beads, porosity, and arc "strikes" can serve as initiation sites for weld failures, but these can be avoided by proper welding procedures, good workmanship, and adequate supervision and inspection.

The residual stresses of welding can cause additional warpage when subsequent machining removes metal and upsets the stress equilibrium balance. Consequently,

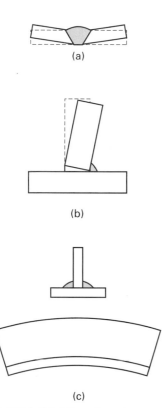

FIGURE 35-15 Distortions or warpage that may occur as a result of welding operations: (a) V-groove butt weld where the top of the joint contracts more than the bottom; (b) one-side fillet weld in a T-joint; (c) two-fillet weld T-joint with a high vertical web.

welded assemblies that are to undergo subsequent machining are frequently given a *stress relief* heat treatment prior to that operation.

While the reaction stresses can also contribute to distortion, they are most often associated with cracking during or immediately following the welding operation (as the weld is cooling). This cracking is most likely to occur when there is great restraint to the shrinkage that occurs transverse to the direction of welding. When a multipass weld is being made, cracking tends to occur in the early beads where there is insufficient weld metal to withstand the shrinkage stresses. This can be quite serious if the resulting crack goes undetected and is not chipped out and repaired, or melted and resolidified during subsequent passes.

To minimize the possibility of fracture in weldments, the joints should be designed to keep restraint to a minimum. The metals and alloys of the structure should be selected with welding in mind (more problems exist with higher-carbon steels, higher-alloy steels, and high-strength materials), and special consideration should be given when welding thicker materials. Crack-prevention efforts can also be directed at maintaining the proper size and shape of the weld bead. While a concave fillet is desirable when machining, a concave weld profile has a greater tendency to crack upon cooling since contraction actually increases the length of the surface. With a convex profile, the length of the surface will contract simultaneously with the volume, reducing the possibility of surface tension and cracking. Weld beads with high penetration (high depth/width ratio) are also more prone to cracking.

Still other methods to suppress cracking focus on reducing the stresses by making the cooling more uniform or relaxing them by promoting plasticity in the metals being welded. The metals to be welded may be preheated and additional heat may be applied between the welding passes to retard cooling. Some welding codes also require weldments to be subjected to a thermal stress relief after welding but prior to use. Hydrogen dissolved in the molten weld metal can also induce cracking. Slower welding and cooling will allow any hydrogen to escape and the use of low-hydrogen electrodes and low-moisture fluxes will reduce the likelihood of hydrogen being present.

■ 35.7 WELDABILITY OR JOINABILITY

It is important to note that not all joining processes are compatible with all engineering materials. While they imply a distinct measure of a material's ability to be welded or joined, the terms *weldability* or *joinability* are actually quite nebulous. One process might produce excellent results when applied to a given material, whereas another may produce a dismal failure. Within a given process, the quality of results may vary greatly with variations in the process parameters, such as electrode material, shielding gases, welding speed, and cooling rate.

Table 35-2 shows the compatibility of the various joining processes with some of the major classes of engineering materials. In each case, the process is classified as recommended (R), commonly performed (C), performed with some difficulty (D), seldom used (S), and not used (N). It should be noted, however, that the classifications are generalizations, and exceptions often exist within both the family of materials (such as the various types of stainless steels) and the types of processes (arc welding here encompasses a large variety of specific processes). Nevertheless, the table can serve as a guideline to assist in process selection.

■ 35.8 SUMMARY

If the potential benefits of welding are to be obtained and harmful side effects are to be avoided, proper consideration should be given to the selection of the process, the design of the joint, and the effects of heating and cooling on both the weld and parent material. Parallel considerations apply to brazing and soldering operations, with additional attention to the effects of interdiffusion of the filler and base metals. Flame and arc-cutting operations involve localized heating and cooling, and they also create altered structures in the heat-affected zone. Since many cut products undergo further welding or machining, however, these regions of undesirable structure may not be retained in the final product.

TABLE 35-2. Weldability or Joinability of Various Engineering Materials[a]

Material	Arc Welding	Oxyacetylene Welding	Electron Beam Welding	Resistance Welding	Brazing	Soldering	Adhesive Bonding
Cast iron	C	R	N	S	D	N	C
Carbon and low-alloy steel	R	R	C	R	R	D	C
Stainless steel	R	C	C	R	R	C	C
Aluminum and magnesium	C	C	C	C	C	S	R
Copper and copper alloys	C	C	C	C	R	R	C
Nickel and nickel alloys	R	C	C	R	R	C	C
Titanium	C	N	C	C	D	S	C
Lead and zinc	C	C	N	D	N	R	R
Thermoplastics	Heated tool R	Hot gas R	N	Induction C	N	N	C
Thermosets	N	N	N	N	N	N	C
Elastomers	N	N	N	N	N	N	R
Ceramics	N	S	C	N	N	N	R
Dissimilar metals	D	D	C	D	D/C	R	R

[a]C, commonly performed; R, recommended (easily performed with excellent results); D, difficult; N, not used; S, seldom used.

■ KEY WORDS

autogenous weld
base metal
bead weld
  (or surfacing weld)
coalescence
consolidation processes

distortion
fillet weld
fusion weld
fusion zone
groove weld

heat-affected zone (HAZ)
incomplete fusion
incomplete penetration
intermetallic
  compound

joint configuration
monolithic
peening
plug weld
postheat

preheat
rate of heat input
reaction stresses
residual stresses
stress relief

thermal cutting
welding
weldability
weld metal
  (or weld pool)

■ REVIEW QUESTIONS

1. What types of design features favor manufacture as a joined assembly as opposed to single piece?
2. What types of manufacturing processes fall under the classification of *consolidation processes*?
3. Define *welding*.
4. What four conditions are required to produce an ideal metallurgical bond?
5. What are some of the ways in which welding processes compensate for the inability to meet the conditions of an ideal bond?
6. What are some possible problems associated with the high temperatures that are commonly used in welding?
7. What is thermal cutting?
8. What are some of the common types of weld defects?
9. What are the four basic types of fusion welds?
10. What types of weld joints commonly employ fillet welds?
11. What are some of the factors that influence the cost of making a weldment?
12. Why is it important to consider welded products as monolithic structures?
13. How does the fracture resistance and temperature sensitivity of a steel vary with changes in material thickness?
14. How might excessive rigidity actually be a liability in a welded structure?
15. In what way is the weld pool segment of a fusion weld like a small metal casting?

16. Why is it possible for the fusion zone to have a chemistry that is different from the filler metal?
17. Why is it not uncommon for the selected filler metal to have a chemical composition that is different from the material being welded?
18. What are some of the defects or problems that can occur in the molten metal region of a fusion weld?
19. Why can the resulting material properties vary widely in welding heat-affected zones?
20. What are some of the structure and property modifications that can occur in welding heat-affected zones?
21. Why do most welding failures occur in the heat-affected zone?
22. What are some of the characteristics and consequences of welding with processes that have low rates of heat input?
23. What are some of the difficulties or limitations encountered in heat treating large, complex welded structures?
24. What is the purpose of pre- and postheating in welding operations?
25. What heat-related metallurgical effects may produce adverse results when brazing or soldering?
26. What are some of the undesirable consequences of residual stresses?
27. What is the cause of reaction-type residual stresses?
28. How are reaction stresses affected by the distance between the weld and the point of fixed constraint?

29. What are some of the techniques that can reduce the amount of distortion in a welded structure?

30. Residual stresses should not have a harmful effect on load-bearing abilities, except under what conditions?

31. Why might a welded structure warp if the structure is machined after welding?

32. What are some of the techniques that can be employed to reduce the likelihood of cracking in a welded structure?

33. Why are the terms *weldability* and *joinability* somewhat nebulous?

■ PROBLEMS

1. Through the 1940s the hulls of ocean-going freighters were constructed by riveting plates of steel together. When the defense efforts of World War II demanded accelerated production of freighters to supply U.S. troops overseas, construction of the hulls was converted to welding. The resulting Liberty Ships proved quite successful but also drew considerable attention when minor or moderate impacts (usually under low-temperature conditions) produced cracks of lengths sufficient to scuttle the ship, often up to 50 feet or more. Since the material was essentially the same and the only significant process change had been the conversion from riveting to welding, the welding process was blamed for the failures. Is this a fair assessment? What do you think may have contributed to the problem? What evidence might you want to gather to support your beliefs?

2. Two pieces of AISI 1025 steel are being shielded-metal-arc welded with E6012 electrodes. Some difficulty is being experienced with cracking in the weld beads and in the heat-affected zones. What possible corrective measures might you suggest?

3. Figure P35-1 schematically depicts the design of a go-cart frame with cross bars and seat support. The assembly is to be constructed from hot-rolled, low-carbon, box-channel material with miter, butt, and fillet welds at the 12 numbered joints. Due to the solidification shrinkage and subsequent thermal contraction of the joint material, the welds are best made when one or more of the sections are unrestrained. If the structure is too rigid at the time of welding, the associated dimensional changes are restricted, causing the generation of residual stresses that can lead to distortion, cracking, or tears.

   a. Consider the 12 welds in the proposed structure and recommend a welding sequence that would minimize the possibility of hot tears and cracks due to the welding of a restrained joint.

   b. Your company is developing a computer-assisted design program. Suggest one or more rules that may be programmed to aid in the selection of an acceptable weld sequence.

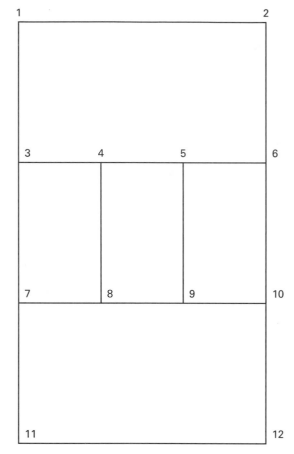

FIGURE P35-1

www.wiley.com/college/degarmo

CHAPTER 36

GAS FLAME PROCESSES

■ 36.1 OXYFUEL GAS WELDING

OXYFUEL GAS WELDING PROCESSES

Oxyfuel gas welding (OFW) refers to a group of welding processes that use as their heat source the flame produced by the combustion of a fuel gas and oxygen. It was the development of a practical *torch* to burn acetylene and oxygen, shortly after 1900, that brought welding out of the blacksmith's shop, demonstrated its potential, and started its development as a manufacturing process. Other processes have largely replaced gas-flame welding in large-scale manufacturing, but the process is still popular for small scale and repair operations because of its portability, versatility (most ferrous and nonferrous metals can be welded), and the low capital investment required. Acetylene is still the principal fuel gas.

The combustion of oxygen and *acetylene* (C_2H_2) by means of a welding torch of the type shown in Figure 36-1 produces a temperature of about 3250°C (5850°F) in a two-stage reaction. In the first stage, the supplied oxygen and acetylene react to produce carbon monoxide and hydrogen:

$$C_2H_2 + O_2 \longrightarrow 2CO + H_2 + \text{heat}$$

This reaction occurs near the tip of the torch and generates intense heat. The second stage of the reaction involves the combustion of the CO and H_2 and occurs just beyond the first combustion zone. The specific reactions of the second stage are:

$$2CO + O_2 \longrightarrow 2CO_2 + \text{heat}$$

$$H_2 + \tfrac{1}{2}O_2 \longrightarrow H_2O + \text{heat}$$

FIGURE 36-1 Typical oxyacetylene welding torch and cross-sectional schematic. *(Courtesy of Victor Equipment Company.)*

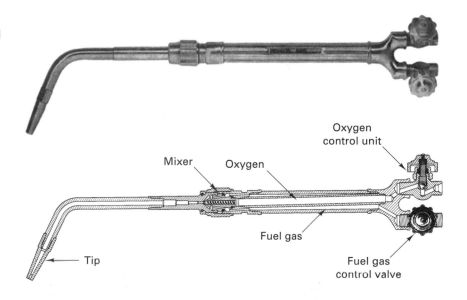

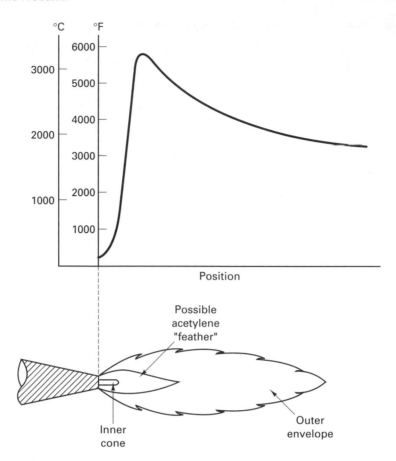

FIGURE 36-2 Typical oxyacetylene flame and the associated temperature distribution.

The oxygen for these secondary reactions is generally obtained from the surrounding atmosphere.

The two-stage combustion process produces a flame having two distinct regions. As shown in Figure 36-2, the maximum temperature occurs near the end of the inner cone, where the first stage of combustion is complete. Most welding should be performed with the torch positioned so that this point of maximum temperature is just above the metal being welded. The outer envelope of the flame serves to preheat the metal and, at the same time, provides shielding from oxidation, since oxygen from the surrounding air is consumed in the secondary combustion.

Three different types of flames can be obtained by varying the oxygen/acetylene (or oxygen/fuel gas) ratio. If the ratio is about 1:1 to 1.15:1, all reactions are carried to completion and a *neutral flame* is produced. Most welding is done with a neutral flame, since it will have the least chemical effect on the heated metal.

A higher ratio, such as 1.5:1, produces an *oxidizing flame*, which is hotter than the neutral flame (about 3600°C or 6000°F) but similar in appearance. Such flames are used when welding copper and copper alloys but are generally considered harmful when welding steel because the excess oxygen reacts with the carbon in the steel, lowering the carbon in the region around the weld.

Excess fuel, on the other hand, produces a *carburizing flame*. The excess fuel decomposes to carbon and hydrogen, and the flame temperature is not as great (about 3050°C or 5500°F). Flames with a slight excess of fuel are reducing flames. No carburization occurs, but the metal is well protected from oxidation. Flames of this type are used in welding Monel (a nickel–copper alloy), high-carbon steels, and some alloy steels, and for applying some types of hard-facing material.

For welding purposes, oxygen is usually supplied in relatively pure form from pressurized tanks, but, in rare cases, air can also be used. The acetylene is usually obtained in portable storage tanks that hold up to 8.5 m³ (300 ft³) at 1.7 MPa (250 psi) pressure. Because acetylene is not safe when stored as a gas at pressures above 0.1 MPa (15 psi), it is usually dissolved in acetone. The storage cylinders are filled with a porous filler. Acetone is absorbed into the voids in the filler material and serves as a medium for dissolving the acetylene. Alternative fuel gases include propane, propylene, and stabilized methyl acetylene propadiene, best known by the trade name of *MAPP* gas. While flame temperature is slightly lower, these gases can be safely stored in ordinary pressure tanks. Three to four times as much gas can be stored in a given volume, and cost per cubic foot can be less than acetylene.

The pressures used in gas-flame welding range from 0.006 to 0.1 MPa (1 to 15 psi) and are controlled by pressure regulators on each tank. Because mixtures of acetylene and oxygen or air are highly explosive, precautions must be taken to avoid mixing the gases improperly or by accident. All acetylene fittings have left-hand threads, while those for oxygen are equipped with right-hand threads. This prevents improper connections.

The tip size (or orifice diameter) of the torch can be varied to control the shape of the inner cone and the flow rate of the gases. Larger tips permit greater flow of gases, resulting in greater heat input without the higher gas velocities that might blow the molten metal from the weld puddle. Larger torch tips are used for the welding of thicker metal.

USES, ADVANTAGES, AND LIMITATIONS

Almost all oxyfuel gas welding is *fusion welding*. The metals to be joined are simply melted where a weld is desired and no pressure is required. Because a slight gap often exists between the pieces being joined, *filler metal* can be added in the form of a solid metal wire or rod. Welding rods come in standard sizes, with diameters from 1.5 to 9.5 mm ($\frac{1}{16}$ to $\frac{3}{8}$ inch) and lengths from 0.6 to 0.9 m (24 to 36 in.). They are available in standard grades that provide specified minimum tensile strengths or in compositions that match the base metal. Figure 36-3 shows a schematic of oxyfuel gas welding using a consumable welding rod.

To promote the formation of a better bond, *fluxes* may be used to clean the surfaces and remove contaminating oxide. In addition, the gaseous shield produced by vaporizing flux can prevent further oxidation during the welding process, and the slag produced by solidifying flux can protect the weld pool as it cools. Flux can be added as a powder, the welding rod can be dipped in a flux paste, or the rods can be precoated.

The oxyfuel gas welding (OFW) processes can produce good-quality welds if proper caution is exercised. Welding can be performed in all positions, the temperature of the work can be easily controlled, and the puddle is visible to the welder. However, exposure of the heated and molten metal to the various gases in the flame and atmosphere makes it difficult to prevent contamination. Since the heat source is not concentrated, heating is rather slow. A large area of metal is heated, and distortion is likely to occur. Thus, in production applications, the flame-welding processes have largely been replaced by arc welding. Nevertheless, flame welding is still quite common in field work, in maintenance and repairs, and in fabricating small quantities of specialized products.

FIGURE 36-3 Schematic of oxyfuel gas welding with a consumable welding rod.

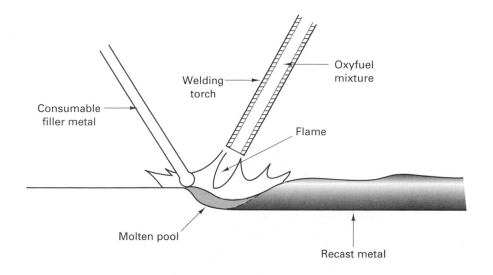

TABLE 36-1.	Process Summary: Oxyfuel Gas Welding (OFW)
Heat Source:	Fuel gas—oxygen combustion
Protection:	Gases produced by combustion
Electrode:	None
Material Joined:	Best for steel and other ferrous metals
Rate of Heat Input:	Low
Weld Profile (D/W):	1/3
Max. Penetration:	3 mm
Assets:	Cheap, simple equipment, portable, versatile
Limitations:	Large HAZ, slow

TABLE 36-2.	Engineering Materials and Their Compatibility with Oxyfuel Welding
Material	Oxyfuel Welding Recommendation
Cast iron	Recommended with cast iron filler rods; braze welding recommended if there are no corrosion objections
Carbon and low-alloy steels	Recommended for low-carbon and low-alloy steels, using rods of the same material; more difficult for higher carbon
Stainless steel	Common for thinner material; more difficult for thicker
Aluminum and magnesium	Common for aluminum thinner than 1 in.; difficult for magnesium alloys
Copper and copper alloys	Common for most alloys; more difficult for some types of bronzes
Nickel and nickel alloys	Common for nickel, Monels, and Inconels
Titanium	Not recommended
Lead and zinc	Recommended
Thermoplastics, thermosets, and elastomers	Hot-gas welding used for thermoplastics, not used with thermosets and elastomers
Ceramics and glass	Seldom used with ceramics, but common with glass
Dissimilar metals	Difficult; best if melting points are within 50°F; concern for galvanic corrosion
Metals to nonmetals	Not recommended
Dissimilar nonmetals	Difficult

Oxyfuel equipment is quite portable, relatively inexpensive, and extremely versatile. This single process can be used for welding, brazing, and soldering, and as a heat source for bending, forming, straightening, and hardening. With the modifications to be discussed shortly, it can also perform flame cutting. Table 36-1 summarizes some of the key features of oxyfuel gas welding. Table 36-2 shows its compatibility with some common engineering materials. The thickness of the material being joined, however, is usually less than 6.5 mm ($\frac{1}{4}$ inch).

PRESSURE GAS WELDING

Pressure gas welding (PGW) is a process that uses equipment similar to the oxyfuel gas process to produce butt joints between the ends of objects such as pipe and railroad rail. The ends are heated with a gas flame to a temperature below the melting point, and the soft metal is then forced together under pressure. Pressure gas welding, therefore, is actually a form of solid-state welding where the gas flame simply softens the metal.

■ 36.2 OXYGEN TORCH CUTTING

PROCESSES

The most common *thermal cutting* process is *oxyfuel gas cutting* (OFC), commonly called flame cutting. In some cases the metal is merely melted by the flame of the oxyfuel gas torch and blown away to form a gap, or *kerf*, as illustrated in Figure 36-4. When ferrous metal is cut, however, the process becomes one where the iron actually burns (or oxidizes) at high temperatures according to one or more of the following reactions:

$$Fe + O \longrightarrow FeO + heat$$

$$3Fe + 2O_2 \longrightarrow Fe_3O_4 + heat$$

$$4Fe + 3O_2 \longrightarrow 2Fe_2O_3 + heat$$

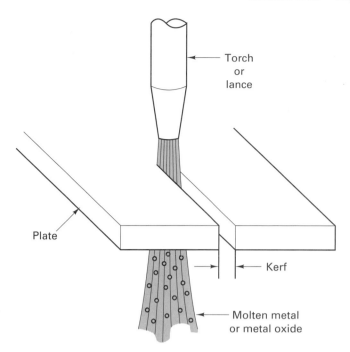

FIGURE 36-4   Flame cutting of a metal plate.

Because these reactions do not occur until the metal is above 815°C (1500°F), the oxyfuel flame is first used to raise the metal to the temperature where burning can be initiated. Then a stream of pure oxygen is added to the torch (or the oxygen content of the oxyfuel mixture is increased) to oxidize the iron. The liquid iron oxide and unoxidized molten iron are then expelled from the joint by the kinetic energy of the oxygen-gas stream. Because of the low rate of heat input and the need for preheating ahead of the cut, oxyfuel cutting produces a relatively large heat-affected zone and associated distortion compared to competing techniques. Therefore, the process is best used where the edge finish or tolerance is not critical, and the edge material will either be subsequently welded or removed by machining.

Theoretically, the heat supplied by the oxidation will be sufficient to keep the cut progressing, but additional heat is often necessary to compensate for losses to the atmosphere and the surrounding metal. If the workpiece is already hot from other processing, such as solidification or hot working, no supplemental heating is required, and a supply of oxygen through a small pipe is all that is needed to initiate and continue a cut. This is known as *oxygen lance cutting* (LOC). A workpiece temperature of about 1200°C (2200°F) is required to sustain continuous cutting.

Oxyfuel gas cutting works best on metals that oxidize readily but do not have high thermal conductivities. Carbon and low-alloy steels can be readily cut in thicknesses from 5 mm to in excess of 75 cm (30 inches). Stainless steels contain oxidation-resistant ingredients and are difficult to cut, as are aluminum and copper alloys. Cutting speeds are relatively slow, but the low cost of both the required equipment and its operation make the process attractive for many applications.

FUEL GASES FOR OXYFUEL GAS CUTTING

Acetylene is by far the most common fuel used in oxyfuel gas cutting, and the process is often referred to as *oxyacetylene cutting* (OFC-A). Figure 36-5 shows a typical cutting torch. The tip contains a circular array of small holes through which the oxygen–acetylene mixture is supplied to form the heating flame. A larger hole in the center supplies a stream of oxygen and is controlled by a lever valve. The rapid flow of the cutting oxygen not only oxidizes the hot metal, but also blows the formed oxides from the cut.

If the torch is adjusted and manipulated properly, it is possible to produce a relatively smooth cut. Cut quality, however, depends upon careful selection of the process variables, including preheat conditions, oxygen flow rate, and cutting speed. Oxygen purities over 99.5% are required for the most efficient cutting. If the purity drops to

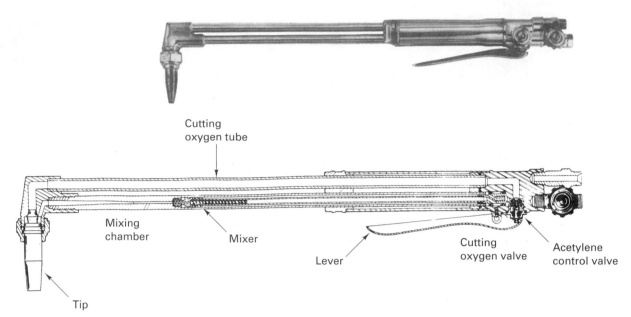

FIGURE 36-5   Typical oxyacetylene cutting torch and cross-sectional schematic. *(Courtesy of Victor Equipment Company.)*

98.5%, cutting speed will be reduced by 15%, oxygen consumption will increase by 25%, and the quality of the cut will diminish.

Cutting torches can be manipulated manually. However, when the process is applied to manufacturing, the desired path is usually controlled by mechanical or programmable means. Specialized equipment has been designed to produce straight cuts in flat stock and square-cut ends on pipe. The marriage of computer numerically controlled (CNC) machines and cutting torches has also proven to be quite popular. This approach, along with the use of robot-mounted torches, provides great flexibility along with good precision and control.

Fuel gases other than acetylene can be used for oxyfuel gas cutting, the most common being natural gas (OFC-N) and propane (OFC-P). While their flame temperatures are lower than acetylene, their use is generally a matter of economics and gas availability. For certain special work, hydrogen can also be used (OFC-H).

STACK CUTTING

When a modest number of duplicate parts are to be cut from thin sheet, but not enough to justify the cost of a blanking die, stack cutting may be the answer. The sheets should be flat, smooth and free of scale, and they should be clamped together tightly so that there are no intervening gaps that could interrupt uniform oxidation or permit slag or molten metal to be entrapped. Obviously, stack cutting will produce a less accurate cut than could be achieved by using a blanking die.

METAL POWDER CUTTING, CHEMICAL FLUX CUTTING, AND OTHER THERMAL METHODS

Modified torch techniques may be required when thermal cutting hard-to-cut materials. Metal powder cutting (POC) injects iron or aluminum powder into the flame to raise its cutting temperature. Chemical flux cutting (FOC) adds a fine stream of special flux to the cutting oxygen to increase the fluidity of the high-melting-point oxides. Both of these methods, however, have largely been replaced by plasma arc cutting (PAC), which is discussed as an extension of plasma arc welding in Chapter 37. Laser and electron-beam cutting are presented with their welding parallels in Chapter 38.

UNDERWATER TORCH CUTTING

The thermal cutting of materials underwater presents a special challenge. A specially designed torch, like the one shown in Figure 36-6, is used to cut steel. An auxiliary skirt surrounds the main tip, and an additional set of gas passages conducts a flow of compressed

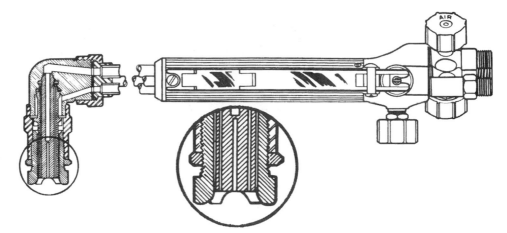

FIGURE 36-6 Underwater cutting torch. Note the extra set of gas openings in the nozzle to permit the flow of compressed air and the extra control valve. *(Courtesy of Bastian-Blessing Company.)*

air that provides secondary oxygen for the oxyacetylene flame and expels water from the zone where the burning of metal occurs. The torch is either ignited in the usual manner before descent or by an electric spark device after being submerged. Acetylene gas is used for depths up to about 7.5 m (25 ft). For greater depths, hydrogen is used, since the environmental pressure is too great for the safe use of acetylene.

■ 36.3 FLAME STRAIGHTENING

Flame straightening is basically the use of controlled, localized *upsetting* as a means of straightening warped or buckled plates. Figure 36-7 illustrates the theory of the process. If a straight piece of metal is heated in a localized area, such as the shaded area of the upper diagram, the metal on side *b* will be upset (i.e., plastically deformed) as it softens and tries to expand against the cooler restraining metal. When the upset portion cools, it will contract, and the resulting piece will be shorter on side *b*, forcing it to bend to the shape in the lower diagram.

If the starting material is bent or warped, as in the lower segment of Figure 36-7, the upper surface can be heated. Upsetting and subsequent thermal contraction will shorten the upper surface at *a'*, bringing the plate back to a straight or flat configuration. Such a procedure can be used to restore structures that have been bent in an accident, such as automobile frames.

A similar process can be used to flatten metal plates that have become dished. Localized spots about 50 mm (2 in.) in diameter are quickly heated to the upsetting temperature while the surrounding metal remains cool. Cool water is then sprayed on the plate, and the contraction of the upset spot brings the buckle into an improved degree of flatness. To remove large buckles, the process may have to be repeated at several spots within the buckled area.

Several cautions should be noted. When straightening steel, consideration should be given to the possible phase transformations that could occur during the heating and cooling. Since rapid cooling is used and martensite may form, a subsequent tempering operation may be required. In addition, one should also consider the residual stresses

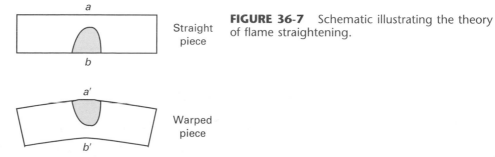

FIGURE 36-7 Schematic illustrating the theory of flame straightening.

that are induced and their effect on subsequent cracking, stress-corrosion cracking, and other modes of failure. The effects of phase transformations and residual stresses have been discussed more fully in Chapter 35.

Also, flame straightening should not be attempted with thin material. For the process to work, the metal adjacent to the heated area must have sufficient rigidity to induce upsetting. If the material is too thin, localized heating and cooling will simply transfer the buckle from one area to another.

■ KEY WORDS

acetylene	flame straightening	kerf	neutral flame	oxyfuel gas cutting	oxygen lance cutting	torch
carburizing flame	flux	MAPP	oxidizing flame	oxyfuel gas welding	thermal cutting	upsetting
filler metal	fusion welding					

■ REVIEW QUESTIONS

1. Why does an oxyfuel gas welding torch usually have a flame with two distinct regions?
2. What is the location of the maximum temperature in an oxy-acetylene flame?
3. What function or functions are served by the outer zone of the welding flame?
4. What three types of flames can be produced by varying the oxygen/fuel ratio?
5. What are some of the attractive features of MAPP gas?
6. Why might a welder want to change the tip size (or orifice diameter) in an oxyacetylene torch?
7. What is filler metal, and why might it be needed to produce a joint?
8. What is the role of a welding flux?
9. Oxyfuel gas welding has a low rate of heat input. What are some of the adverse features that result from the slow rate of heating?
10. What are some of the more attractive features of the oxyfuel gas process?

11. In what way does the torch cutting of ferrous metals differ from cutting nonoxidizing metals?
12. Why might it be possible to cut cast hot steel strands with only an oxygen lance as they emerge from the continuous casting operation? (*Note:* The metal is still red hot, having just solidified from the liquid state.)
13. How does an oxyacetylene cutting torch differ from an oxyacetylene welding torch?
14. What are some of the ways in which cutting torches can be mechanically manipulated?
15. What modification must be incorporated into a cutting torch to permit it to cut metal underwater?
16. If a curved plate is to be straightened by flame straightening, should the heat be applied to the longer or shorter surface of the arc? Why?
17. Why does the flame-straightening process not work for thin sheets of metal?

CHAPTER 37

ARC PROCESSES

■ 37.1 ARC WELDING

With the development of commercial electricity in the late nineteenth century, it was soon recognized that an *arc* between two electrodes was a concentrated heat source that could approach 4000°C (7000°F) in temperature. As early as 1881, various attempts were made to use such an arc as the heat source for fusion welding. Carbon was initially selected as one *electrode* and the metal workpiece became the other. Figure 37-1 depicts the basic electrical circuit. *Filler metal* was provided by a metallic wire or rod that was independently fed into the arc. As the process developed, the filler metal replaced carbon as the upper electrode. The metal wire not only carried the welding current, but as it melted in the arc, it also supplied the necessary filler.

The results of these early efforts were extremely uncertain. Because of the instability of the arc, a great amount of skill was required to maintain it, and contamination of the weld resulted from the exposure of hot metal to the atmosphere. There was little or no understanding of the metallurgical effects and requirements of arc welding. Consequently, while the great potential was recognized, very little use was made of the process until after World War I. Shielded metal electrodes were developed around 1920. These electrodes enhanced the stability of the arc by shielding it from the atmosphere and provided a fluxing action to the molten pool. The major problems of arc welding were overcome, and the process began to expand rapidly.

All *arc-welding* processes employ the basic circuit depicted in Figure 37-1. Welding currents vary from 1 to 4000 amps, with the range from 100 to 1000 being most typical. Voltages are generally in the range of 20 to 50 volts. If direct current is used and the electrode is made negative, the condition is known as straight polarity (SPDC) or *DCEN*, for direct-current electrode-negative. Electrons are attracted to the positive workpiece, while ionized atoms in the arc column are accelerated toward the negative electrode. Since the ions are far more massive than the electrons, the heat of the arc is more concentrated at the electrode. DCEN processes are characterized by fast melting of the

FIGURE 37-1 Basic circuit for arc welding.

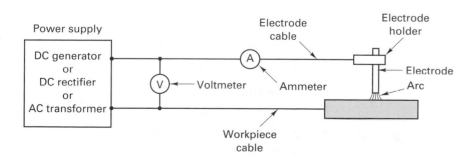

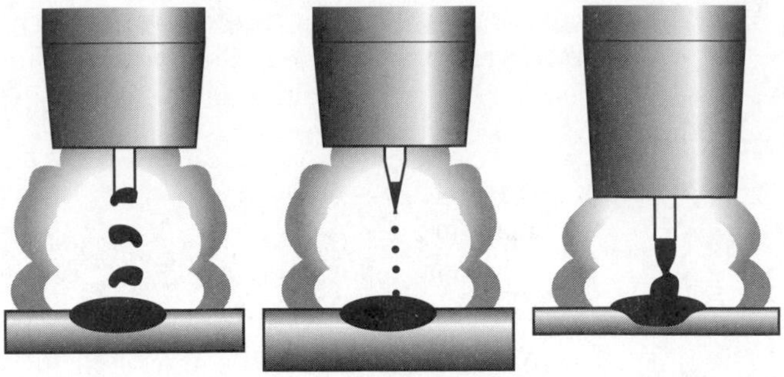

FIGURE 37-2 Three modes of metal transfer during arc welding. *(Courtesy of Republic Steel Corporation.)*

Globular   Spray   Short circuit

electrode (high metal deposition rates) and a shallow molten pool on the workpiece (weld penetration). If the work is made negative and the electrode positive, the condition is known as reverse polarity (RPDC) or *DCEP*, for direct-current electrode-positive. The positive ions impinge on the workpiece, breaking up any oxide films and giving deeper penetration. Metal deposition rate is lower, however. Sinusoidal *alternating current* provides a 50–50 average of the above two modes and is a popular alternative to the dc conditions. The new *variable polarity* power supplies also alternate between DCEP and DCEN conditions, but use rectangular waveforms to vary the fraction of time in each mode, as well as the frequency of switching. Weld characteristics can now be varied over a continuous range between DCEN and DCEP conditions.

In one group of arc welding processes, the electrode is consumed (*consumable electrode processes*) and thus supplies the metal needed to fill the joint. Consumable electrodes have a melting temperature below the temperature of the arc. Small droplets are melted from the end of the electrode and pass to the workpiece. The size of these droplets varies greatly and the transfer mechanism depends on the type of electrode, welding current, and other process parameters. Figure 37-2 depicts metal transfer by the globular, spray, and short-circuit transfer modes. As the electrode melts, the arc length and the electrical resistance of the arc path will vary. To maintain a stable arc and satisfactory welding conditions, the electrode must be moved toward the work at a controlled rate. Manual arc welding is almost always performed with shielded (covered) electrodes. Continuous bare-metal wire can be used as the electrode in automatic or semiautomatic arc welding, but this is always in conjunction with some form of shielding and arc-stabilizing medium and automatic feed-control devices that maintain the proper arc length.

The second group of arc welding processes employs a tungsten electrode, which is not consumed by the arc, except by relatively slow vaporization. In these *nonconsumable electrode processes*, a separate metal wire is required to supply the filler metal.

Because of the wide variety of processes available, arc welding has become a widely used means of joining material. Each process and application, however, requires the selection or specification of the welding voltage, welding current, arc polarity (straight polarity, reversed polarity, or alternating), arc length, welding speed (how fast the electrode is moved across the workpiece), arc atmosphere, electrode or filler material, and flux. Filler materials must be selected to match the base metal with respect to properties and/or alloy content (chemistry). For many of the processes, the quality of the weld also depends on the skill of the operator. Automation and robotics are reducing this dependence, but the selection and training of welding personnel are still of great importance.

■ 37.2 CONSUMABLE ELECTRODE ARC WELDING

Four processes make up the bulk of consumable electrode arc welding:

1. Shielded metal arc welding (SMAW)
2. Flux-cored arc welding (FCAW)
3. Gas metal arc welding (GMAW)
4. Submerged arc welding (SAW)

These processes all have a medium rate of heat input and produce a fusion zone whose depth is approximately equal to its width. Because the fusion zone is composed of metal from both pieces, plus melted filler, the electrode must be of the same material as that being welded, and the processes cannot be used to join dissimilar metals or ceramics.

SHIELDED METAL ARC WELDING

Shielded metal arc welding (SMAW), also called *stick welding*, is the most common of the arc welding processes because of its wide versatility and because it requires only low-cost equipment. The key to the process is a finite-length electrode that consists of metal wire, usually from 1.5 to 6.5 mm in diameter and 20 to 45 cm in length. Surrounding the wire is a bonded coating containing chemical components that add a number of desirable characteristics, including all or a number of the following:

1. Vaporize to provide a protective atmosphere (a gas shield around the arc and pool of molten metal).

2. Provide ionizing elements to help stabilize the arc, reduce weld metal spatter, and increase efficiency of deposition.

3. Act as a *flux* to deoxidize and remove impurities from the molten metal.

4. Provide a protective *slag* coating to accumulate impurities, prevent oxidation, and slow the cooling of the weld metal.

5. Add alloying elements.

6. Add additional filler metal.

7. Affect arc *penetration* (the depth of melting in the workpiece).

8. Influence the shape of the weld bead.

Coated electrodes are classified by the tensile strength of the deposited weld metal, the welding position in which they may be used, the preferred type of current and polarity (if direct current), and the type of coating. A four- or five-digit system of designation has been adopted by the American Welding Society (AWS) and is presented in Figure 37-3. As an example, type E7016 is a low-alloy steel electrode that will provide a deposit with a minimum tensile strength of 70,000 psi (485 MPa) in the non-stress-relieved condition; it can be used in all positions, with either alternating or reverse-polarity direct current; and it has a low-hydrogen plus potassium coating. To assist in identification, all electrodes are marked with colors in accordance with a standard established by the National Electrical Manufacturers Association. Electrode selection consists of determining the electrode coating, coating thickness, electrode composition, and electrode diameter.

FIGURE 37-3   Designation system for arc-welding electrodes.

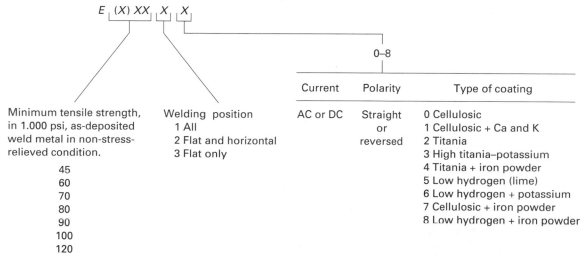

	Current	Polarity	Type of coating
Minimum tensile strength, in 1.000 psi, as-deposited weld metal in non-stress-relieved condition. 45 60 70 80 90 100 120	Welding position 1 All 2 Flat and horizontal 3 Flat only	AC or DC	Straight or reversed	0 Cellulosic 1 Cellulosic + Ca and K 2 Titania 3 High titania–potassium 4 Titania + iron powder 5 Low hydrogen (lime) 6 Low hydrogen + potassium 7 Cellulosic + iron powder 8 Low hydrogen + iron powder

A variety of electrode coatings have been developed. The cellulose and titania (rutile) coatings contain SiO_2; TiO_2; small amounts of FeO, MgO, and Na_2O; and volatile matter. Upon decomposition, the volatile matter may release hydrogen, which can dissolve in the weld metal and lead to embrittlement or cracking in the joint. Low-hydrogen electrodes are available with compositions designed to provide shielding without the emission of hydrogen. Since many of the electrode coatings can absorb moisture, and this is another source of undesirable hydrogen, the coated electrodes are often baked just prior to use.

To initiate a weld, the operator briefly touches the tip of the electrode to the workpiece and quickly raises it to a distance that will maintain a stable arc. The intense heat quickly melts the tip of the electrode wire, the coating, and portions of the adjacent base metal. As part of the electrode coating melts and vaporizes, it forms a protective atmosphere of CO, CO_2, and other gases that stabilizes the arc and protects the molten and hot metal from contamination. Other coating components surround the metal droplets with a layer of liquid flux and slag. The fluxing constituents unite with any impurities in the molten metal and float them to the surface to be entrapped in the slag coating that forms over the weld. The slag coating protects the cooling metal from oxidation and slows down the cooling rate to prevent the formation of hard, brittle structures. The glassy slag is easily chipped from the weld when it has cooled. Figure 37-4 illustrates the shielded metal arc welding process, and Figure 37-5 provides a schematic of metal deposition from a shielded electrode.

Iron powder can be added to the electrode coating to significantly increase the amount of weld metal that can be deposited with a given size electrode wire and current. Alloy elements can also be incorporated into the coating to adjust the chemistry of the weld. Special contact or drag electrodes utilize coatings that are designed to melt more slowly than the filler wire. If these electrodes are tracked along the surface of the work, the faster-melting center wire will be recessed by the proper length to maintain a stable arc.

Since electrical contact must be maintained with the center wire, SMAW electrodes are finite-length "sticks." Length is limited since the current must be supplied near the arc, or the electrode will tend to overheat and ruin the coating. Overheating also restricts the weld currents to values below 300 amps (generally about 40 amps per millimeter of

FIGURE 37-4 Shielded metal arc welding (SMAW).

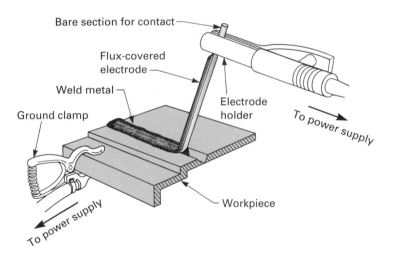

FIGURE 37-5 Schematic diagram of shielded metal arc welding (SMAW). *(Courtesy of American Iron and Steel Institute, Washington, D.C.)*

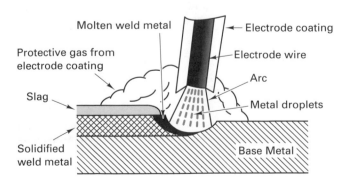

TABLE 37-1. Process Summary: Shielded Metal Arc Welding (SMAW)	
Heat Source:	Electric arc
Protection:	Slag from flux and gas from vaporized coating material
Electrode:	Discontinuous, consumable
Material Joined:	Best for steel
Rate of Heat Input:	Medium
Weld Profile (D/W):	1
Current:	<300 amps
Max. Penetration:	3–6 mm
Assets:	Cheap, simple equipment
Limitations:	Discontinuous, shallow welds, requires slag removal

electrode diameter). As a result, the arc temperatures are somewhat low, and penetration is generally less than 5 mm ($\frac{3}{16}$ in.). Welding of material thicker than 5 mm will require multiple passes, and the slag coating must be removed between each pass.

The shielded metal arc process is best used for welding steels; carbon steels, alloy steels, stainless steels, and cast irons can all be welded. Welds can be made in all positions. DCEP conditions are used to obtain the deepest possible penetration, with alternate modes being employed when welding thin sheet. The mode of metal transfer is either globular or short circuit.

Shielded metal arc welding is a simple, inexpensive, and versatile process, requiring only a power supply, power cables, electrode holder, and a small variety of electrodes. The equipment is portable and can even be powered by gasoline or diesel generators. Therefore, it is a popular process in job shops and is used extensively in repair operations. Unfortunately, the process is discontinuous, produces shallow welds, and requires slag removal after each welding pass. Table 37-1 presents a process summary for shielded metal arc welding.

FLUX-CORED ARC WELDING

Flux-cored arc welding (FCAW) overcomes some of the shielded metal arc limitations by moving the powdered flux to the interior of a continuous tubular electrode (Figure 37-6). When the arc is established, the vaporizing flux again produces a protective atmosphere and also forms a slag layer over the weld pool that will require subsequent removal. Alloy additions (metal powders) can be blended into the flux to create a wide variety of filler

FIGURE 37-6   Schematic representation of the flux-cored arc welding (FCAW). *(Courtesy of The American Welding Society, New York.)*

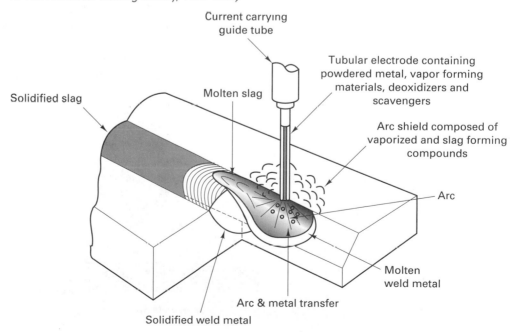

TABLE 37-2.	Process Summary: Flux-Cored Arc Welding (FCAW)
Heat Source:	Electric arc
Protection:	Slag and gas from flux (optional secondary gas shield)
Electrode:	Continuous, consumable
Material Joined:	Best for steel
Rate of Heat Input:	Medium
Weld Profile (D/W):	1
Current:	<500 amps
Max. Penetration:	6–10 mm
Assets:	Continuous electrode
Limitations:	Requires slag removal

metal chemistries. Compared to the stick electrodes of the shielded metal arc process, the flux-cored electrode is both continuous and less bulky, since binders are no longer required to hold the flux in place.

The continuous electrode is fed automatically through a welding gun, with electrical contact being maintained through the bare metal exterior of the wire at a position near the exit of the gun. Overheating of the electrode is no longer a problem, and welding currents can be increased to about 500 A. The higher heat input increases penetration depth to about 1 cm ($\frac{3}{8}$-in.). The process is best used for welding steels, and welds can be made in all positions. Direct current electrode positive (DCEP) conditions are almost always used for the enhanced penetration. High deposition rates are possible, but the slag layer must still be removed from the final weld, and the equipment cost is greater than SMAW because of the need for a controlled wire feeder and more costly power supply.

The basic flux-cored arc welding process is a self-shielding process in which the shielding gas is provided by the vaporization of flux components. In a modification of the process, enhanced weld properties can be achieved by integrating a flow of externally supplied shielding gas, such as CO_2, into the torch. The better protection results in cleaner welds.

Table 37-2 presents a process summary of flux-cored arc welding.

GAS METAL ARC WELDING

If the supplemental shielding gas flowing through the torch (described above) becomes the primary protection for the arc and molten metal, there is no longer a need for the volatilizing flux, and the electrode can now become a continuous, solid, uncoated metal wire (or a continuous hollow tube with powdered alloy additions in the center, known as a metal-cored electrode). The resulting process, shown in Figure 37-7, is known as *gas metal arc welding (GMAW)* and was formerly referred to as metal inert-gas welding (MIG). The arc is still maintained between the workpiece and the automatically fed bare-wire electrode, which continues to provide the necessary filler metal. The welding current, penetration depth, and process cost all remain similar to the flux-cored process. Because shielding is provided by the flow of gas, and fluxing and slag-forming agents are no longer required, the gas metal arc process can be applied to all metals. Argon, helium, and mixtures of the two can be used for welding virtually any metal but are used

FIGURE 37-7 Schematic diagram of gas metal arc welding (GMAW). *(Courtesy of American Iron and Steel Institute, Washington, D.C.)*

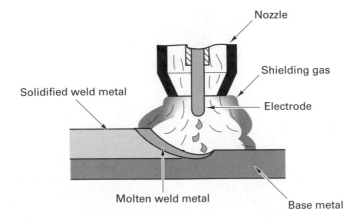

primarily with the nonferrous metals. When welding steel, some O_2 or CO_2 is usually added to improve the arc stability and reduce weld spatter. The cheaper CO_2 can be used alone when welding steel, provided that a deoxidizing electrode wire is employed. Since these shielding gases only provide protection and do not remove existing contamination, starting cleanliness is critical to production of a good weld.

The specific shielding gases can have considerable effect on the nature of metal transfer from the electrode to the work and also influence the heat transfer behavior, penetration, and tendency for undercutting (weld pool extending laterally beneath the surface of the base metal). Helium produces the hottest arc and deepest penetration. Argon is intermediate, and CO_2 yields the lowest arc temperatures and shallowest penetrations.

Electronic controls can be used to alter the welding current, enabling further control of the mechanism of metal transfer, shown previously in Figure 37-2. The lowest currents and voltages and the use of CO_2 shield gas promote *short-circuit transfer* (GMAW-S). The advancing electrode makes direct contact with the weld pool, and the short circuit causes a rapid rise in current. Big molten globs form on the tip of the electrode and then separate, forming a gap between the electrode and workpiece. This gap reinitiates a brief period of arcing, but the rate of electrode advancement exceeds the rate of melting in the arc, and another short circuit occurs. The power conditions, therefore, oscillate between arcing and short circuiting at a rate of 20 to 200 cycles per second. This mode is preferred when joining very thin materials and can be used in all welding positions, but suffers from a high degree of spatter. If the voltage and amperage are increased, the mode becomes one of *globular transfer*. The electrode melts from the heat of the arc, and metal drops form with a diameter approximately equal to that of the electrode wire. Gravity and electromagnetic forces then transfer the drops to the workpiece at a rate of several per second. Since gravity plays a role in metal transfer, there is a companion limitation on the positions of welding. With higher currents and voltages, argon gas shielding, and DCEP conditions, *spray transfer* (GMAW-ST) occurs. Small droplets emerge from a pointed electrode at a rate of hundreds per second. Because of their small size and the greater electromagnetic effects, the droplets are easily propelled across the arc in any direction, irrespective of the effects of gravity. Spray transfer is accompanied by deep penetration and low spatter. The biggest problem with out-of-position welding may be keeping the rather large molten weld pool in place until it solidifies.

Pulsed spray transfer (GMAW-P) was invented in the 1960s to overcome some of the limitations of conventional spray transfer. In this mode, a low welding current is first used to create a molten globule on the end of the filler wire. A burst of high current then "explodes" the globule and transfers the metal across the arc in the form of a spray. By alternating low and high currents at a rate of 60 to 120 times per second, the filler metal is transferred in a succession of rapid bursts, similar to the emissions of a rapidly squeezed aerosol atomizer. With the pulsed form of deposition, there is less heat input to the weld, and the weld temperatures are reduced. Thinner material can be welded; distortion is reduced; workpiece discoloration is minimized; heat-sensitive parts can be welded; high-conductivity metals can be joined; electrode life is extended; electrode cooling techniques may not be required; and fine microstructures are produced in the weld pool. Welds can be made in all positions, and the use of pulsed power lowers spattering and improves the safety of the process. The high speed of the process is attractive for productivity, and the energy or power required to produce a weld is lower than with other methods (reduced cost). Controls can be adjusted to alter the shape of the weld pool and vary the penetration.

In general, the gas metal arc process is fast and economical and currently accounts for nearly half of all weld metal deposition. There is no frequent change of electrodes as with the shielded metal arc process. No flux is required, and no slag forms over the weld. Thus multiple-pass welds can be made without the need for intermediate cleaning. The process can be readily automated, and the lightweight, compact welding unit lends itself to robotic manipulation. A direct current electrode positive (DCEP) arc is generally used because of its deep penetration, spray transfer, and the ability to produce smooth welds with good profile. Process variables include type of current, current magnitude, shielding gas, electrode diameter, electrode composition, electrode stickout (extension beyond the gun), welding speed, welding voltage, and arc length. Table 37-3 provides a process summary for gas metal arc welding.

TABLE 37-3.	Process Summary: Gas-Metal Arc Welding (GMAW)
Heat Source:	Electric arc
Protection:	Externally supplied shielding gas
Electrode:	Continuous, consumable
Material Joined:	All common metals
Rate of Heat Input:	Medium
Weld Profile (D/W):	1
Current:	<500 amps
Max. Penetration:	6–10 mm
Assets:	No slag to remove
Limitations:	More costly equipment than SMAW or FCAW

In a process modification known as advanced gas metal arc welding (AGMAW), a second power source is used to preheat the filler wire before it emerges from the welding torch. Less arc heating is needed to produce a weld, so less base metal is melted, producing less dilution of the filler metal and less penetration.

SUBMERGED ARC WELDING

No shielding gas is used in the *submerged arc welding (SAW)* process. Instead, a thick layer of granular flux is deposited just ahead of a solid bare-wire consumable electrode, and an arc is maintained beneath the blanket of flux with only a few small flames being visible. A portion of the flux melts and acts to remove impurities from the rather large pool of molten metal, while the unmelted excess provides additional shielding. The molten flux solidifies into a glasslike covering over the weld. This layer, along with the flux that is not melted, provides good thermal insulation. The slow cooling of the weld metal helps to produce soft, ductile welds. Upon further cooling, the solidified flux cracks loose from the weld (due to the differential thermal contraction) and is easily removed. The unmelted granular flux is recovered by a vacuum system and reused. Figure 37-8 provides a schematic of the process.

Submerged arc welding is most suitable for making flat butt or fillet welds in low-carbon steels (<0.3% carbon). With some preheat and postheat precautions, medium-carbon and alloy steels and some cast irons, stainless steels, copper alloys, and nickel alloys can also be welded. The process is not recommended for high-carbon steels, tool steels, aluminum, magnesium, titanium, lead, or zinc. The reasons for this incompatibility are somewhat varied, including the unavailability of suitable fluxes, reactivity at high temperatures, and low sublimation temperatures.

Because the arc is totally submerged, high welding currents can be used (600 to 2000 A). High welding speeds, high deposition rates, deep penetration, and high cleanliness (due to the flux action) are all characteristic of submerged arc welding. A welding speed of 0.75 m/min in 2.5-cm thick steel plate is typical. Single-pass welds can be made with penetrations up to 2.5 cm (1.0 in.), and greater thicknesses can be joined by multiple passes. Because the metal is deposited in fewer passes than with alternative processes, there is less possibility of entrapped slag or voids, and weld quality is further enhanced. For even higher deposition rates, multiple electrode wires can be employed.

Limitations to the process include the need for extensive flux handling, possible contamination of the flux by moisture (leading to porosity in the weld), the large volume of slag that must be removed, the high heat inputs that promote large-grain-size structures, and the slow cooling rate (which permits segregation and possible hot cracking). Welding is restricted to the horizontal position, since the flux and slag are held in place by gravity. In addition, chemical control is quite important, since the electrode material often composes over 70% of the molten weld region.

The electrodes are generally classified by composition and are available in diameters ranging from 1 to 10 mm (0.045 to $\frac{3}{8}$ in.). The larger electrodes can carry higher currents and enable more rapid deposition, but penetration is shallower. Electrodes for welding alloy steels can take several forms: solid wire of the desired alloy, plain carbon steel with the alloy additions being incorporated into the flux, and tubular metal with the alloy additions in the hollow core. Various fluxes are also available and are selected for compatibility with the weld metal. All are designed to have low melting temperatures, good fluidity, and brittleness after cooling.

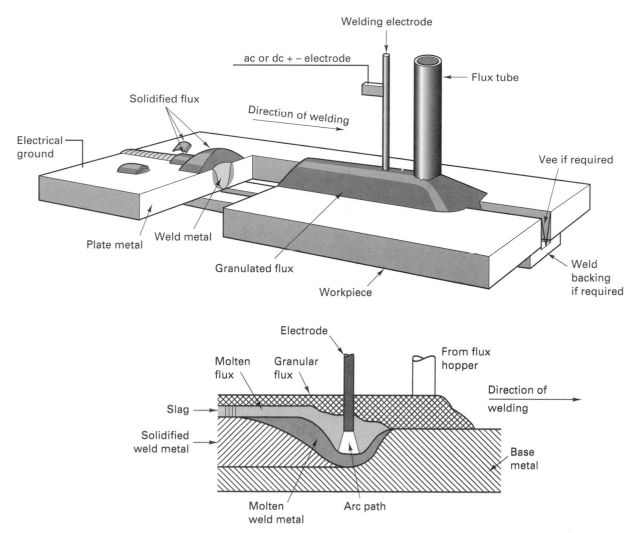

FIGURE 37-8   (*Top*) Basic features of submerged arc welding (SAW). (*Courtesy of Linde Division, Union Carbide Corporation.*) (*Bottom*) Cutaway schematic of submerged arc welding. (*Courtesy of American Iron and Steel Institute, Washington, D.C.*)

In a modification of the submerged arc process known as *bulk welding*, iron powder is first deposited into the joint (ahead of the flux) as a means of increasing deposition rate. A single weld pass can then produce enough filler metal to be equivalent to seven or eight conventional submerged arc passes.

Table 37-4 provides a summary of the submerged arc welding process.

STUD WELDING

Stud welding (SW) is an arc-welding process used to attach studs, screws, pins, or other fasteners to a metal surface. A special gun is used, such as the one shown in Figure 37-9.

TABLE 37-4.   Process Summary: Submerged Arc Welding (SAW)

Heat Source:	Electric arc
Protection:	Granular flux provides slag and an isolation blanket
Electrode:	Continuous, consumable
Material Joined:	Best for steel
Rate of Heat Input:	Medium
Weld Profile (D/W):	1
Current:	<1000 amps
Max. Penetration:	25 mm
Assets:	High-quality welds, high deposition rates
Limitations:	Requires slag removal, difficult for overhead and out-of-position welding, joints often require backing plates

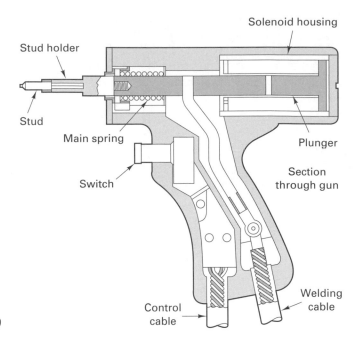

FIGURE 37-9 Schematic diagram of a stud welding gun. *(Courtesy of American Machinist.)*

The inserted stud acts as an electrode and a dc arc is established between the end of the stud and the workpiece. After a small amount of metal is melted, the two pieces are brought together under light pressure and allowed to solidify. Automatic equipment controls the arc, its duration, and the application of pressure to the stud.

Figure 37-10 shows some of the wide variety of studs that are specially made for this process, and often contain a recessed end that is filled with flux. A ceramic ferrule, such as the one shown in the center photo of Figure 37-10, is usually placed over the end of the stud before it is positioned in the gun. During the arc, the ferrule serves to concentrate the heat and isolate the hot metal from the atmosphere. It also confines the molten or softened metal and shapes it around the base of the stud, as shown in the photo on the right of Figure 37-10. After the weld has cooled, the brittle ceramic can be broken free and removed. Since burn-off or melting reduces the length of the stud, the original dimensions should be selected to compensate.

FIGURE 37-10 (*Left*) Types of studs used for stud welding. (*Center*) Stud and ceramic ferrule. (*Right*) Stud after welding and a section through a welded stud. *(Courtesy of Nelson Stud Welding Co.)*

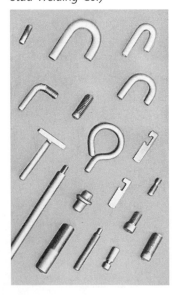
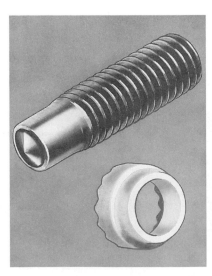
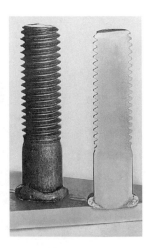

Stud welding requires almost no skill on the part of the operator. Once the stud and ferrule are placed in the gun and the gun positioned on the work, all the operator has to do is pull the trigger. The cycle is executed automatically and takes less than 1 second. Thus the process is well suited to manufacturing and can be used to eliminate the drilling and tapping of many special holes. Production-type stud welders can produce over 1000 welds per hour.

■ 37.3 NONCONSUMABLE ELECTRODE ARC WELDING

GAS TUNGSTEN ARC WELDING

Gas tungsten arc welding (GTAW) was formerly known as tungsten inert-gas (TIG) welding, or Heliarc welding when helium was the shielding gas. A nonconsumable tungsten electrode provides the arc but not the filler metal. The pointed electrode is positioned in a special holder through which inert gas (argon, helium, or a mixture of them) flows to provide a protective shield around the electrode, the arc, the pool of molten metal, and the adjacent heated areas. (*Note:* CO_2 cannot be used in this process since it provides inadequate protection for the hot tungsten electrode.) Operating in an inert environment, the tungsten electrode, which is often alloyed with thorium oxide, zirconium oxide, cerium oxide, or lanthanum oxide to provide better current-carrying and electron-emission characteristics and longer electrode life, is not consumed at the temperatures of the arc. The arc length remains constant, and the arc is very stable and easy to maintain. Figure 37-11 shows a typical GTAW torch with cables and passages for gas flow, power, and cooling water.

In applications where there is a close fit between the pieces being joined, no filler metal may be needed. When filler metal is required, it must be supplied as a separate wire, as illustrated in Figure 37-12. The filler metal is generally selected to match the chemistry and/or tensile strength of the metal being welded. Where high deposition rates are desired, a separate resistance heating circuit can be provided to preheat the filler wire. As shown in Figure 37-13, the deposition rate of heated wire can be several times that of a cold wire. By oscillating the filler wire from side to side while making a weld pass, the deposition rate can be further increased. The hot-wire process is not practical when welding copper or aluminum, however, because it is difficult to preheat the low-resistivity filler wire.

With skilled operators, gas tungsten arc welding can produce high-quality welds that are very clean and scarcely visible. Since no flux is employed, no special cleaning or slag removal is required. However, the surfaces to be welded must be clean and

FIGURE 37-11   Welding torch used in nonconsumable electrode, gas tungsten arc welding (GTAW), showing feed lines for power, cooling water, and inert gas flow. *(Courtesy of Linde Division, Union Carbide Corporation.)*

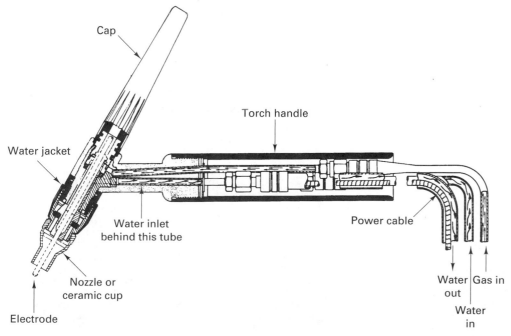

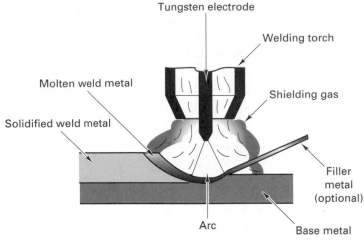

FIGURE 37-12 Schematic diagram of gas tungsten arc welding (GTAW). *(Courtesy of American Iron and Steel Institute, Washington, D.C.)*

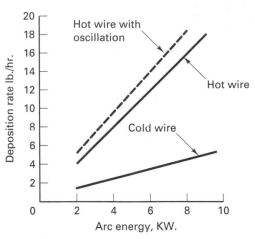

FIGURE 37-13 Comparison of the metal deposition rates in GTAW with cold, hot, and oscillating hot filler wire. *(Courtesy of Welding Journal.)*

free of oil, grease, paint, and rust, because the inert gas does not provide any cleaning or fluxing action. It is also important to control the arc length throughout the process. Since the arc is somewhat bell-shaped, decreasing the stand-off distance will decrease the melt and heat-affected widths on the workpiece. However, if the hot tungsten electrode comes into contact with the workpiece or molten pool, it will contaminate the electrode.

All metals and alloys can be welded by this process, and the use of inert gas makes it particularly attractive for the reactive metals, such as aluminum, magnesium, and titanium, as well as the high-temperature refractory metals. Maximum penetration is obtained with direct current electrode negative (DCEN) conditions, although alternating current may be specified to break up surface oxides (as when welding aluminum). DCEP or reverse-polarity conditions are used only when welding thin pieces where shallow penetrations are desired. Weld currents should be kept low since this mode tends to melt the tungsten electrode. Weld voltage is typically 20 to 40 V, and weld current varies from less than 125 A for DCEP to 1000 A for DCEN. A high-frequency, high-voltage, alternating current is often superimposed on the regular ac or dc welding current to make it easier to start and maintain the arc. The pulsed arc gas tungsten arc welding (GTAW-P) modification offers all of the advantages previously cited for pulsed gas metal arc, including the ability to weld thinner materials due to the lower heat input and lower temperatures.

The GTAW process can be used in all positions, has a medium rate of heat input, and produces welds where the depth is approximately equal to the width. The materials being welded are generally thinner than 6.5 mm ($\frac{1}{4}$ in.). Table 37-5 summarizes the process.

TABLE 37-5.	Process Summary: Gas Tungsten Arc Welding (GTAW)
Heat Source:	Electric arc
Protection:	Externally supplied shielding gas
Electrode:	Nonconsumable
Material Joined:	All common metals
Rate of Heat Input:	Medium
Weld Profile (D/W):	1
Current:	<500 amps
Max. Penetration:	3 mm
Assets:	High-quality welds, no slag to be removed
Limitations:	Slower than consumable electrode GMAW

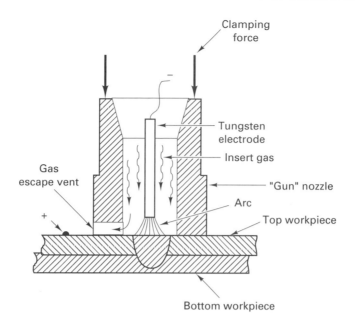

FIGURE 37-14 Process schematic of spot welding by the inert-gas-shielded tungsten arc process.

FIGURE 37-15 Making a spot weld by the inert-gas-shielded tungsten arc process. *(Courtesy of Air Reduction Company Inc.)*

GAS TUNGSTEN ARC SPOT WELDING

A variation of gas tungsten arc welding can be used to produce *spot welds* between two pieces of metal where access is limited to one side of the joint or where thin sheet is being attached to heavier material. The basic procedure is illustrated in Figures 37-14 and 37-15. A modified inert-gas tungsten arc gun is used with a vented nozzle on the end. The nozzle is pressed firmly against the material, holding the pieces in reasonably good contact. (The workpieces must be sufficiently rigid to sustain the contact pressure.) Inert gas, usually argon or helium, flows through the nozzle to provide a shielding atmosphere. Automatic controls advance the electrode to initiate the arc and then retract it to the correct distance for stabilized arcing. The duration of arcing is timed automatically to produce an acceptable spot weld. The depth and size of the weld nugget are controlled by the amperage, time, and type of shielding gas.

In arc spot welding, the weld nugget begins to form at the surface where the gun makes contact. With the more standard resistance spot-welding methods, the weld nugget forms at the interface between the two members. Each technique has distinct advantages and disadvantages.

PLASMA ARC WELDING

In *plasma arc welding (PAW)* the arc is maintained between a nonconsumable electrode and either the welding gun (*nontransferred arc*) or the workpiece (*transferred arc*) as illustrated in Figure 37-16. The nonconsumable tungsten electrode is set back within the "torch" in such a way as to force the arc to pass through or be contained within a small-diameter nozzle. An inert gas (usually argon) is forced through this constricted arc, where it is heated to a high temperature and forms a hot, fast-moving *plasma*. The emerging gas then transfers its heat to the workpiece and melts the metal. This flow is called the *orifice gas*. When needed, filler metal can be fed into the plasma column, which is usually surrounded by a second flow of inert gas used to provide shielding for the weld pool.

Plasma arc welding is characterized by a high rate of heat input and temperatures on the order of 16,500°C (30,000°F). This in turn offers fast welding speeds, narrow welds with deep penetration (a depth-to-width ratio of about 3), a narrow heat-affected zone, reduced distortion, and a process that is insensitive to variations in arc length since the plasma column is cylindrical. Figure 37-17 presents a comparison of the nonconstricted GTAW arc and the constricted arc of the plasma arc process.

With a low-pressure plasma and currents between 20 and 100 A, the metal simply melts and flows into the joint. At higher pressures and currents in excess of 100 A, a "keyhole" effect occurs in which the plasma gas creates a hole completely through the sheet (up to 20 mm thick) that is surrounded by molten metal. As the torch is moved,

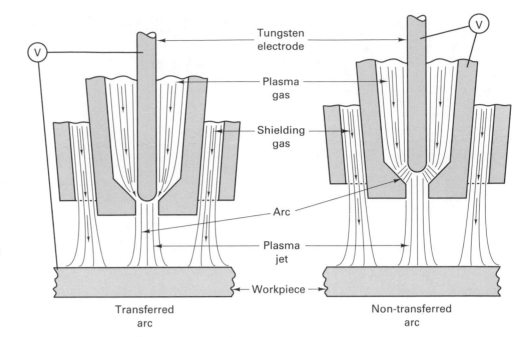

FIGURE 37-16 Two types of plasma arc torches.
(*Left*) transferred arc;
(*right*) nontransferred arc.

Tungsten electrode

Plasma gas

Shielding gas

Arc

Plasma jet

Workpiece

Transferred arc

Non-transferred arc

liquid metal flows to fill the keyhole. If the gas pressure is increased even further, the molten metal is expelled from the region, and the process becomes one of plasma cutting, which is discussed later in this chapter. Welds can be made in all positions, and nearly all metals and alloys can be welded.

Many plasma torches employ a small nontransferred arc within the torch to heat the orifice gas and ionize it. The ionized gas then forms a good conductive path for the main transferred arc. This dual-arc technique permits instant ignition of a low-current arc, which is more stable than that of an ordinary plasma torch. Separate dc power supplies are used for the pilot and main arcs. Microplasma, or needle arc, torches can operate with very low currents (0.1 to 20 A) and still produce stable arcs. They are quite useful for welding very thin sheet.

Table 37-6 provides a process summary for plasma arc welding.

FIGURE 37-17 Comparison of the nonconstricted arc of gas tungsten arc welding and the constricted arc of the plasma arc process. (*Courtesy ASM International, Materials Park, Ohio.*)

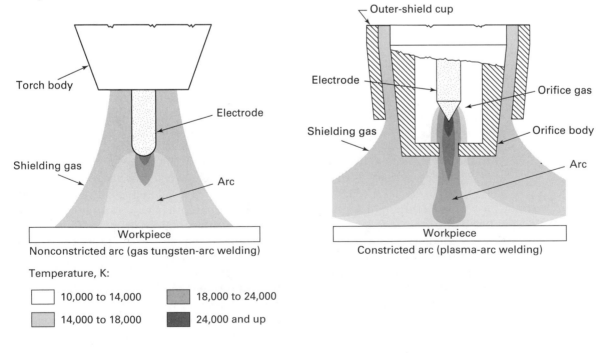

Torch body

Electrode

Shielding gas

Arc

Outer-shield cup

Electrode

Orifice gas

Shielding gas

Orifice body

Arc

Workpiece

Workpiece

Nonconstricted arc (gas tungsten-arc welding)

Constricted arc (plasma-arc welding)

Temperature, K:

10,000 to 14,000	18,000 to 24,000
14,000 to 18,000	24,000 and up

TABLE 37-6.	Process Summary: Plasma Arc Welding (PAW)
Heat Source:	Plasma arc
Protection:	Externally supplied shielding gas
Electrode:	Nonconsumable
Material Joined:	All common metals
Rate of Heat Input:	High
Weld Profile (D/W):	3
Current:	<500 amps
Max. Penetration:	12–18 mm
Assets:	Can have long arc length
Limitations:	High initial equipment cost, large torches may limit accessibility

■ 37.4 WELDING EQUIPMENT

POWER SOURCES FOR ARC WELDING

Arc welding requires a large electrical current, often in the range of 100 to 1000 A. The load voltage is usually between 20 and 50 V, with the actual voltage across the arc having a lower value depending on the arc length. Both dc and ac *power supplies* are available and generally employ the "drooping voltage" characteristics shown in Figure 37-18. These characteristics are designed to minimize changes in welding current as the welding voltage fluctuates within anticipated limits.

In the past, most direct current for welding was provided by gasoline- or diesel-powered motor-generator sets, and these are still used when welding is to be performed in remote locations. Most welding today, however, uses solid-state transformer-rectifier machines, such as the one shown in Figure 37-19. Operating on a three-phase electrical line, these machines can usually provide both ac and dc output.

If only ac welding is to be performed, relatively simple transformer-type power supplies can be used. These are usually single-phase devices with low power factors. When multiple machines are to be operated, as in a production shop, they are often connected to the various phases of a three-phase supply to help balance the load.

Inverter-based power supplies, introduced in the 1980s, provide great flexibility. Through solid-state electronics, these machines can quickly modify the nature and shape of the pulse waveform or momentarily change the power supply output. Through feedback and control logic, the power supply can now adjust to compensate for changes in a number of process variables.

JIGS AND POSITIONERS

Jigs or fixtures (also called positioners) are frequently used to hold the work in production welding. By positioning and manipulating the workpiece, the welding operations

FIGURE 37-18 Drooping-voltage characteristics of typical arc-welding power supplies. (*Left*) Direct current; (*right*) alternating current.

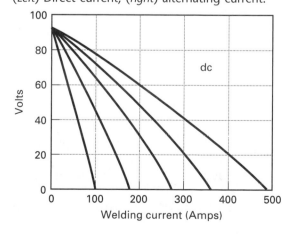
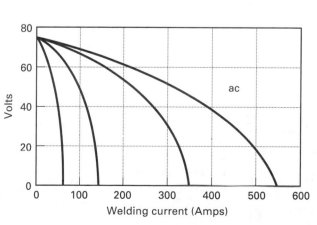

FIGURE 37-19   Rectifier-type ac and dc welding power supply. *(Courtesy of Lincoln Electric Company.)*

can often be performed in a more favorable orientation. Parts can also be mounted on numerically controlled (NC) tables that manipulate and position them with respect to the welding tool.

■ 37.5   ARC CUTTING

Virtually all metals can be cut by electric arc methods. In these processes the material is melted by the intense heat of the arc and then permitted, or forced, to flow away from the region of the slit or notch (*kerf*). Most of the techniques are simply adaptations of the arc-welding procedures discussed in this chapter. Each has its inherent characteristics and capabilities, including tolerance, thickness capability, kerf width, edge squareness, size of the heat-affected zone, and cost. Selection depends upon factors such as tolerance requirements, the subsequent processes that will be performed on the cut part, and the end use of the product.

CARBON ARC AND SHIELDED METAL ARC CUTTING

The carbon arc cutting (CAC) and shielded metal arc cutting (SMAC) methods use the arc from a carbon or shielded metal arc electrode to melt the metal, which is then removed from the cut by gravity or the force of the arc itself. Use of these processes is generally limited to small shops, garages, and homes, where the equipment investment is small.

AIR CARBON ARC CUTTING

In air carbon arc cutting, the arc is again maintained between a carbon electrode and the workpiece, but high-velocity jets of air are directed at the molten metal from holes in the electrode holder (Figure 37-20). While there is some oxidation, the primary function of the air is to blow the molten material from the cut. Air carbon arc cutting is particularly effective for cutting cast iron and preparing steel plates for welding. Speeds up to 0.6 m/min are possible, but the process is quite noisy, and hot metal particles tend to be blown over a substantial area.

FIGURE 37-20   Gun used in the arc-air cutting process. Note the air holes surrounding the electrode in the holder. *(Courtesy of Jackson Products.)*

OXYGEN ARC CUTTING

In oxygen arc cutting (AOC), an electric arc and a stream of oxygen are combined to make the cut. The electrode is a coated ferrous-metal tube. The coated metal serves to establish a stable arc, while oxygen flows through the bore and is directed on the area of incandescence. With easily oxidized metals, such as steel, the arc preheats the base metal, which then reacts with oxygen, becomes liquefied, and is expelled by the oxygen stream.

GAS METAL ARC CUTTING

If the wire feed rate and other variables of gas metal arc welding (GMAW) are adjusted so that the electrode penetrates completely through the workpiece, cutting rather than welding will occur, and the process becomes gas metal arc cutting (GMAC). The wire feed rate controls the quality of the cut, and the voltage determines the width of the slit or kerf.

GAS TUNGSTEN ARC CUTTING

Gas tungsten arc cutting (GTAC) employs the same basic circuit and shielding gas as used in gas tungsten arc welding, with a high-velocity jet of gas added to expel the molten metal.

PLASMA ARC CUTTING

The torches used in plasma arc cutting (PAC) produce the highest temperatures available from any practical source. With the nontransferred-type of torch, the arc column is completely within the nozzle, and a temperature of about 16,500°C (30,000°F) is obtained. With the transfer-type torch, the arc is maintained between the electrode and the workpiece, and temperatures can be as high as 33,000°C (60,000°F). Ionized gases flowing at these temperatures and near supersonic speeds are capable of cutting virtually any electrically conductive material simply by melting it and blowing it away from the cut.

Early efforts to employ this technique showed that the speed, versatility, and operating cost were far superior to those of the oxyfuel cutting methods. However, the early systems could not constrict the arc sufficiently to produce the quality of cut needed to meet the demands of manufacturing. Therefore, plasma arc cutting was generally limited to those materials that could not be cut by the oxidation-type of cutting techniques. In the 1970s radial impingement of water on the arc was found to produce the desired constriction. It provides an intense, highly focused source of heat, and water-injected torches can now cut virtually any metal in any position. Magnetic fields have also been used to constrict the arc and can produce high-quality cuts without the need for water impingement.

Compared to oxyfuel cutting, plasma cutting is more economical (cost per cut is a fraction of oxyfuel), more versatile (can cut all metals as easily as mild steel), and much faster (typically, five to eight times faster than oxyfuel). Cutting speeds up to 7.5 m/min have been obtained in 6-mm thick aluminum, and up to 2.5 m/min in 12.5-mm thick steel. The combination of the extremely high temperatures and jetlike action of the plasma produces narrow kerfs and remarkably smooth surfaces, nearly as smooth as can be obtained by sawing. Plasma-cut surfaces are often within 2° of vertical, and surface oxidation is nearly eliminated by the cooling effect of the water spray. In addition, the heat-affected zone in the metal is only one-third to one-fourth as large as that produced by oxyfuel cutting, and a preheat cycle is not required in the cutting of steel. Heat-related distortion is extremely small.

Transferred-type torches are usually used for cutting metals, while the nontransferred type are employed with the low-conductivity nonmetals. Ordinary air or inexpensive nitrogen can be used as the plasma gas for the cutting of all types of metal. Oxygen plasma systems were introduced in the 1980s and are used on carbon and low-alloy steel products with thicknesses ranging from 2 to 32 mm (up to $1\frac{1}{4}$ in.). When cutting thick sections (greater than 12 mm), an argon–hydrogen mixture may be preferred to provide a deeper-penetrating arc. A secondary flow of shielding gas (nitrogen, air, or carbon dioxide) may be used to help cool the torch, blow the molten metal away, shield the arc, and prevent oxidation of the cut surface. The arc-constricting water flow can also serve as a shielding medium.

FIGURE 37-21   Cutting sheet metal with a plasma torch. *(Courtesy of GTE Sylvania.)*

During the 1990s *high-density*, or *precision plasma*, systems began to appear. Various designs are used to restrict the orifice (i.e., superconstrict the plasma), producing vertical edges (less than 1° taper), close tolerances ($\frac{1}{3}$ that of conventional), and dross-free plasma cutting of thin materials. The lower-amperage torches (10 to 100 A) are limited to cutting carbon and low-alloy steels less than 16-mm ($\frac{5}{8}$-in.) thick and higher-performance metals (such as stainless and high-strength steels, nickel alloys, titanium, and aluminum) less than 12-mm ($\frac{1}{4}$-in.) thick. In addition, the cutting speeds are slower than conventional plasma cutting, but there is no reduction in the size of the heat-affected zone. *Pulsed plasma arc cutting*, another recent development, can reduce heat input to the workpiece while producing kerfs that are 50% narrower and cleaner edges on the cuts.

Combining a plasma torch with robotic or CNC manipulation can provide fast, clean, and accurate cutting, like that shown in Figure 37-21. It is not uncommon for the cutting table to be placed under water as a means of reducing noise, air pollution, dust, and arc glare (dyes are placed in the water). Plasma arc torches can also be incorporated into punch presses to provide a manufacturing machine with outstanding flexibility in producing cut and punched products from a variety of materials.

◼ 37.6 METALLURGICAL AND HEAT EFFECTS IN THERMAL CUTTING

When used for cutting, the flame and arc processes expose materials to high localized temperatures and can produce harmful metallurgical effects. If the cut edges will be subsequently welded, or if they will be removed by machining, there is little cause for concern. When the edges are retained in the finished product, consideration should be given to the effects of cutting heat and their interaction with the applied loads. In some cases, additional steps may be required to avoid or overcome harmful consequences.

For carbon steels with less than 0.25% carbon, thermal cutting does not produce serious metallurgical effects. However, in steels of higher carbon content, the metallurgical changes can be quite significant, and preheating and/or postheating may be required. For alloy steels, additional consideration should be given to the effects of the various alloy elements.

Because of the low rate of heat input, oxyacetylene cutting will produce a rather large *heat-affected zone*. The arc-cutting methods produce intermediate effects that are quite similar to those of arc welding. Plasma arc cutting is so rapid, and the heat is so localized, that the original properties of a metal are only modified within 1.5 mm of the cut.

All of the thermal cutting processes produce some *residual stresses*, with the cut surface generally in tension. Except in the case of thin sheet, warping should not occur. However, if subsequent machining removes only a portion of the cut surface, or does not penetrate to a sufficient depth, the resulting imbalance in residual stresses can induce distortion. It may be necessary to remove all cut surfaces to a substantial depth to ensure good dimensional stability.

Thermal cutting can also introduce geometrical features into the edge. All flame- or arc-cut edges are rough to varying degrees and thus contain notches that can act as stress raisers and reduce the endurance or fracture strength. If cut edges are to be subjected to high or repeated tensile stressing, the cut surface and the heat-affected zone should be removed by machining or at least subjected to a stress-relief heat treatment.

■ KEY WORDS

arc	DCEP	heat-affected zone	plasma arc welding	slag
arc cutting	electrode	kerf	power supply	spot weld
arc welding	filler metal	nonconsumable	pulsed arc	spray transfer
alternating current	flux	electrode process	residual stresses	straight polarity
bulk welding	flux-cored arc welding	nontransferred arc	reverse polarity	stud welding
consumable electrode	gas metal arc welding	orifice gas	shielded metal arc	submerged arc welding
process	gas tungsten arc welding	penetration	welding	transferred arc
DCEN	globular transfer	plasma	short-circuit transfer	variable polarity

■ REVIEW QUESTIONS

1. What sorts of problems plagued early attempts to develop arc welding?
2. What are the three basic types of current and polarity that are used in arc welding?
3. What is the difference between a consumable and nonconsumable electrode? For which processes does a filler metal have to be added by a separate mechanism?
4. What are the three types of metal transfer that can occur during arc welding?
5. What are some of the process variables that must be specified when setting up an arc-welding process?
6. What are the four primary consumable arc-welding processes?
7. What are some of the functions of the electrode coatings used in shielded metal arc welding?
8. How are welding electrodes commonly classified, and what information does the designation usually provide?
9. Why are shielded metal arc electrodes often baked just prior to welding?
10. What are some of the functions of the gases that form when the electrode coating melts and vaporizes?
11. What is the function of the slag coating that forms over a shielded metal arc weld?
12. What benefit can be obtained by placing iron powder in the coating of shielded metal arc electrodes that will be used to weld ferrous metals?
13. Why are shielded metal arc electrodes generally limited in length, forcing the process to be one of intermittent operation?
14. What is the advantage of placing the flux in the center of an electrode (flux-cored arc welding) as opposed to a coating on the outside (shielded metal arc welding)?
15. What feature enables the welding current in FCAW to be higher than in SMAW?
16. What are some of the advantages of gas metal arc welding compared to the shielded metal arc process?
17. Describe the roles of argon, helium, and carbon dioxide gases in creating a high-temperature arc and promoting weld penetration.
18. Describe the metal transfer that occurs during pulsed arc gas metal arc welding.
19. What are some of the benefits that can be obtained by the reduced heating of the pulsed arc process?
20. What are some of the primary process variables in the gas metal arc welding process?
21. What are some of the functions of the flux in submerged arc welding?
22. From a production viewpoint, what are some of the attractive features of submerged arc welding? Major limitations?
23. What is the primary goal or objective in bulk welding?
24. What is the special-use application of stud welding?
25. What is the function of the ceramic ferrule placed over the end of the stud in stud welding?
26. Why is carbon dioxide not an acceptable shielding gas for the gas tungsten arc process?
27. What can be done to increase the rate of filler metal deposition during gas tungsten arc welding?
28. What are some of the attractive features of gas tungsten arc welding?
29. How are the spot welds produced by gas tungsten arc spot welding different from those made by conventional resistance spot welding?
30. How is the heating of the workpiece during plasma arc welding different from the heating in other arc welding techniques?
31. Describe the two different gas flows in plasma arc welding.
32. What are some of the attractive features of plasma arc welding?
33. What is the primary difference between plasma arc welding and plasma arc cutting?
34. What is the *kerf* in thermal cutting operations?
35. What is the purpose of the oxygen in oxygen arc cutting?
36. Why is plasma arc cutting an attractive way of cutting high-melting-point materials?
37. What techniques can be used to constrict the arc in plasma arc cutting, to produce a narrower, more controlled cut?
38. Compared to oxyfuel cutting, what are some of the attractive features of the plasma technique?
39. Describe the relative size of the heat-affected zone for the various cutting processes: oxyfuel, arc, and plasma.
40. Why might the residual stresses induced during cutting operations be objectionable?
41. Why might it be wise to machine away the thermally cut edge and heat-affected zone of metal that will be used as a stressed machine part?

*C*hapter 37 CASE STUDY

Bicycle Frame Repair

As a new employee in a bicycle shop, you learn that customers will frequently seek advice regarding various types of bicycle repair. Bent or broken frames appear to be a common problem, and you learn that inexpensive bicycles generally have frames made from welded low-carbon steel tubing. This material can be heated to straighten and is easily repair-welded using a wide variety of welding techniques. As you move to the more costly, lightweight, or high-performance models, you learn that repair is generally not quite as simple.

a. In one case, the frame of a high-quality, lightweight bicycle had been fabricated from aluminum alloy tubing, with joints made by adhesive bonding of the tubes to connectors that incorporate either internal lugs or external sleeves. When the frame fractured near one of the joints, the owner contacted a local auto body shop and requested that a repair be made using conventional gas tungsten arc welding. The welder was familiar with the welding of aluminum, and the repair seemed to be of good quality. Shortly thereafter, however, the frame broke again. This time the fracture was adjacent to the repair weld and the characteristics of the break were different. While the first fracture was somewhat brittle in nature, the second appeared to be more ductile, with evidence of metal flow prior to fracture. Since the second fracture occurred in the tube material and not the weld itself, the welder felt that the failure was not relat-

ed to the attempted repair and the material in the tubing must be defective.

   1. If the material had been cold-drawn aluminum tubing (i.e., strain hardened), explain what may have occurred during the repair. What is the probable cause of the second fracture? Was the weld in any way defective? Was the second failure related to the welding repair?
   2. If the tubing had been strengthened by age hardening, could the same results have occurred? Explain.
   3. Is there a better means of repairing the original fracture? What would have been your recommendation?

b. You learn that titanium offers the strength of heat-treated steel at approximately half of the weight. Magnesium, while not as strong as steel or aluminum, is the lightest-weight engineering metal. Could bicycle frames be constructed from these materials, and if so, how would they be assembled?

c. If the bicycle frame deflects, motion of the cyclist and related energy are wasted. Therefore, a rigid frame is desirable. Beryllium are an extremely rigid, lightweight metal. Can it be used as a material for bicycle frames?

d. Composite materials can be used to produce tailored sets of properties. Fiber-reinforced composites can have extremely high rigidity in the direction of fiber orientation, coupled with extremely light weight. If you were to assemble a composite frame using fiber-reinforced tubing, how would you join the assembly?

CHAPTER 38

RESISTANCE WELDING AND OTHER WELDING PROCESSES

■ 38.1 INTRODUCTION

As indicated in the lists of Figures 35-1 and 35-2, there are a number of welding and cutting processes that utilize heat sources other than oxyfuel flames and electric arcs. We will begin this chapter by considering a group of processes that use electrical resistance heating to form the joint. A second series of processes create joints without any melting of the workpiece or filler material, known as solid-state welding processes. The remaining processes include some that are quite old and others that are among the newest in manufacturing. A special section will be devoted to the joining of plastics, and the chapter will close with a section on the welding-related processes of surfacing and thermal spray coating.

■ 38.2 THEORY OF RESISTANCE WELDING

In *resistance welding*, both heat and pressure are used to induce coalescence. Electrodes are placed in contact with the material, and electrical resistance heating is used to raise the temperature of the workpieces and the interface between them. The same electrodes that supply the current also apply the pressure, which is usually varied throughout the weld cycle. A certain amount of pressure is applied initially to hold the workpieces in contact and thereby control the electrical resistance at the interface. When the proper temperature has been attained, the pressure is increased to induce coalescence. Because pressure is utilized, coalescence occurs at a lower temperature than that required for oxyfuel gas or arc welding. In fact, melting of the base metal does not occur in many resistance welding operations. Resistance welding processes might well be considered as a form of solid-state welding, although they are not officially classified as such by the American Welding Society.

In some resistance welding processes, additional pressure is applied immediately after coalescence to provide a certain amount of forging action. Accompanying the deformation is a certain amount of grain refinement. Additional heating can also be employed after welding to provide tempering and/or stress relief.

The required temperature can often be attained, and coalescence can be achieved in a few seconds or less. Consequently, resistance welding is a very rapid and economical process, extremely well suited to automated manufacturing. No filler metal is required, and the tight contact maintained between the workpieces excludes air and eliminates the need for fluxes or shielding gases.

HEATING

The heat for resistance welding is obtained by passing a large electrical current through the workpieces for a short period of time. The amount of heat input can be determined by the basic relationship

$$H = I^2 Rt$$

where

H = total heat input in joules

I = current in amperes

R = electrical resistance of the circuit in ohms

t = length of time during which current is flowing in seconds

It is important to note that the workpieces actually form part of the electrical circuit, as illustrated in Figure 38-1 and that the total resistance between the electrodes consists of three distinct components:

1. The resistance of the workpieces themselves

2. The contact resistance between the electrodes and the workpieces

3. The resistance between the surfaces to be joined, known as the *faying surfaces*

Since the maximum amount of heat is generated at the point of maximum resistance, it is desirable to have this be the location where the weld is to be made. Therefore, it is essential to keep resistances 1 and 2 as low as possible with respect to resistance 3. Resistance 1 is determined by the type and thickness of the metal being joined. This is usually much less than the other two resistances because of the large areas involved and the relatively high electrical conductivity of most metals. Resistance 2 (the resistance between the electrodes and workpieces) can be minimized by using electrode materials that are excellent electrical conductors, by controlling the size and shape of the electrodes, and by applying proper pressure. Any change in the pressure between the electrodes and workpieces, however, also affects the contact between the faying surfaces. Therefore, only limited control of the electrode-to-work resistance can be obtained by variation in pressure.

The final resistance, that between the faying surfaces, is a function of the quality (surface finish or roughness) of the surfaces; the presence of nonconductive scale, dirt, or other contaminants; the pressure; and the contact area. These factors must all be controlled if uniform resistance welds are to be produced.

As indicated in Figure 38-2, the objective of resistance welding is to bring both of the faying surfaces to the proper temperature, while simultaneously keeping the remaining material and the electrodes relatively cool. The electrodes are usually water cooled to keep their temperature low and thereby extend their useful life.

FIGURE 38-1 The basic resistance welding circuit.

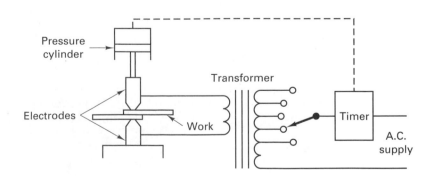

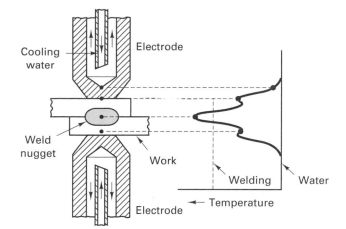

FIGURE 38-2 The desired temperature distribution across the electrodes and workpieces in resistance welding.

PRESSURE

Because the pressure in resistance welding promotes a forging action, resistance welds can be produced at lower temperatures than welds made by other processes. However, the control of both the magnitude and timing of the pressure is very important. If too little pressure is used, the contact resistance will be high and surface burning or pitting of the electrodes may result. If excessive pressure is applied, molten or softened metal may be expelled from between the faying surfaces or the softened workpiece may be indented by the electrodes. Ideally, moderate pressure should be applied to hold the workpieces in place and establish proper resistance at the interface prior to and during the passage of the welding current. The pressure should then be increased considerably just as the proper welding temperature is attained. This completes the coalescence and forges the weld to produce a fine-grained structure.

On small, foot-operated machines, only a single spring-controlled pressure is used. On larger, production-type welders, the pressure is generally applied through controllable air or hydraulic cylinders.

CURRENT AND CURRENT CONTROL

While surface conditions and pressure are important variables, the temperature achieved during resistance welding is primarily determined by the magnitude and duration of the welding current. High currents and short time intervals enable the weld location to attain the desired temperature while minimizing the amount of heat dissipated to the adjacent material.

In production-type welders, the magnitude, duration, and timing of both current and pressure can be programmed to follow specified cycles. Figure 38-3 shows a relatively simple cycle for a resistance weld that includes both forging and postheating operations. The quality of the final weld, therefore, depends more on the development of a proper schedule and the subsequent setup, adjustment, and maintenance of equipment than it does on operator skill.

FIGURE 38-3 Typical current and pressure cycle for resistance welding. The cycle includes forging and postheating operations.

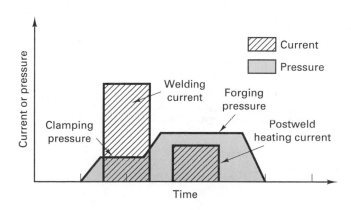

POWER SUPPLY

Since the overall resistance in the welding circuits can be quite low, high currents are generally required to produce a resistance weld. Power transformers convert the high-voltage, low-current line power to the high-current (up to 100,000 A), low-voltage (0.5 to 10 V) power required for welding. Smaller machines may utilize single-phase circuitry, but the larger units generally operate on three-phase power. Many resistance welders use dc welding current, obtained through solid-state rectification of the three-phase power. These machines reduce the current demand per phase, give a balanced load, and produce excellent welds.

■ 38.3 RESISTANCE WELDING PROCESSES

RESISTANCE SPOT WELDING

Resistance spot welding (RSW) is the simplest and most widely used form of resistance welding, providing a fast, economical means of joining overlapped materials that will not require subsequent disassembly. Even with all of the advances in technology, resistance spot welding is still the dominant method for joining sheet material, and the average steel-bodied automobile still contains between 2000 and 5000 spot welds. Figure 38-4 presents a schematic of the process. Overlapped metal sheets are positioned between water-cooled electrodes, which have reduced areas at the tips to produce welds that are usually from 1.5 to 13 mm ($\frac{1}{16}$ to $\frac{1}{2}$ in.) in diameter. The electrodes close on the work, and the controlled cycle of pressure and current is applied to produce a weld at the metal interface. The electrodes are then opened, and the work is removed.

A satisfactory spot weld, like the one shown in Figure 38-5, consists of a *nugget* of coalesced metal formed between the faying surfaces. There should be little indentation

FIGURE 38-4 The arrangement of the electrodes and the work in resistance spot welding.

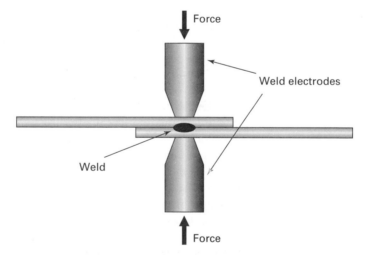

FIGURE 38-5 A spot-weld nugget between two sheets of 1.3-mm (0.05-in.) aluminum alloy. The nugget is not symmetrical because the radius of the upper electrode was greater than that of the lower electrode. *(Courtesy Lockheed Aircraft Corporation.)*

FIGURE 38-6 Tear test of a satisfactory spot weld, showing how failure occurs outside of the weld.

of the metal under the electrodes. As shown in Figure 38-6, the strength of the weld should be such that in a tensile or tear test, the weld will remain intact while failure occurs in the heat-affected zone surrounding the nugget. If proper current density and timing, electrode shape, electrode pressure, and surface conditions are maintained, sound spot welds can be obtained with excellent consistency.

Spot-Welding Equipment. A variety of spot-welding equipment is available to meet the various needs of production operations. For light-production work where complex current-pressure cycles are not required, a simple *rocker-arm machine* is often used. The lower electrode arm is stationary, while the upper electrode, mounted on a pivot arm, is brought down into contact with the work by means of a spring-loaded foot pedal. Rocker-arm machines are available with throat depths up to about 1.2 m (48 in.) and transformer capacities up to 50 kVa. They are used primarily on steel.

Larger spot welders, and those used at high production rates, are generally of the press type (Figure 38-7). On these machines, the movable electrode is controlled by an air or hydraulic cylinder, and complex pressure cycles can be programmed. Capacities up to 500 kVa with a 1.5-m (60-in.) throat depth are quite common. Special-purpose press-type welders can employ multiple welding heads to make up to 200 simultaneous spot welds in under 60 seconds.

Portable spot-welding guns have been instrumental in extending the application of the spot-welding process. These guns are connected to a stationary power supply and control unit by flexible air hoses, electrical cables, and water-cooling lines. By bringing the process to the work, spot welding can be performed on products that are too large to be positioned on a welding machine. The portable units can also be installed on industrial robots, which can be programmed to position the gun at specified locations and produce quality spot welds in a highly automated fashion. Robotic spot welding is the most common means of joining sheet metal components in the automotive industry.

Electronic advances in the late 1980s enabled the welding transformer to be integrated into the welding gun. By transforming the power immediately adjacent to the area of use, the small integral transformer guns, or *transguns*, offer reduced power losses and enhanced process efficiency. However, if accurate positioning is required in an articulated system like an industrial robot, the added weight of the integral transformer may become a disadvantage. Servomotors have also been added to spot-welding equipment to provide control of the electrode positioning, speed of closure, the level of applied torque or pressure, and the rate at which the load is applied.

Electrodes. Resistance spot-welding electrodes must conduct the welding current to the work, set the current density at the weld location, apply force, and help dissipate heat during the noncurrent portions of the welding cycle. Electrical and thermal

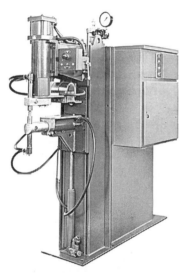

FIGURE 38-7 Single-phase, air-operated, press-type resistance welder with microprocessor control. *(Courtesy Sciaky Bros., Inc.)*

TABLE 38-1.   Metal Combinations That Can Be Spot Welded

Metal	Aluminum	Brass	Copper	Galvanized Iron	Iron (Wrought)	Monel	Nichrome	Nickel	Nickel Silver	Steel	Tin Plate	Zinc
Aluminum	×										×	×
Brass		×	×	×	×	×	×	×	×	×	×	×
Copper		×	×	×	×	×	×	×	×	×	×	×
Galvanized iron		×	×	×	×	×	×	×	×	×	×	
Iron (wrought)		×	×	×	×	×	×	×	×	×	×	
Monel		×	×	×	×	×	×	×	×	×	×	
Nichrome		×	×	×	×	×	×	×	×	×	×	
Nickel		×	×	×	×	×	×	×	×	×	×	
Nickel silver		×	×	×	×	×	×	×	×	×	×	
Steel		×	×	×	×	×	×	×	×	×	×	
Tin plate	×	×	×	×	×	×	×	×	×	×	×	
Zinc	×	×	×									×

conductivity properties are important considerations for electrode selection. Hot compressive strength must be sufficient to resist electrode deformation during pressure applications. In addition, the electrode should not melt under welding conditions and should be of a composition that does not alloy with the material being welded—a phenomenon that promotes sticking or galling and electrode wear. The Resistance Welder Manufacturers Association (RWMA) has standardized various electrode geometries and has approved a variety of electrode materials, including copper-base alloys, refractory metals, and refractory–metal composites.

Spot-Weldable Metals.   While steel is clearly the most common metal that is spot welded, one of the greatest advantages of the process is that virtually all of the commercial metals can be joined, and most of them can be joined to each other. In only a few cases do the welds tend to be brittle. Table 38-1 shows some of the many combinations of metals that can be successfully spot welded.

While spot welding is primarily used to join wrought sheet material, other forms of metal can also be welded. Sheets can be spot welded to rolled shapes and steel castings, as well as some types of nonferrous die castings. Most metals require no special preparation, except to be sure that the surface is free of corrosion and is not badly pitted. For best results, aluminum and magnesium should be cleaned immediately prior to welding by some form of mechanical or chemical technique. Metals that have high electrical conductivity require clean surfaces to ensure that the electrode-to-metal resistance is low enough for adequate temperature to be developed within the metal itself. Silver and copper are especially difficult to weld because of their high thermal conductivity. Higher welding currents coupled with water cooling of the surrounding material may be required if adequate welding temperatures are to be obtained.

When the two pieces being joined are of the same thickness, the practical limit for spot welding is about 3 mm ($\frac{1}{8}$ in.) for each sheet. Sheets of differing thickness can also be joined, and thin pieces can be attached to material that is considerably thicker than 3 mm. When metals of different thickness or different conductivity are to be welded, however, a larger electrode or one with higher conductivity is often used against the thicker or higher-resistance material to ensure that both workpieces will be brought to the desired temperature in a simultaneous fashion.

RESISTANCE SEAM WELDING

Resistance seam welds (RSEW) can be made by two distinctly different processes. In the first process, sheet metal segments are joined to produce gas- or liquid-tight vessels, such as gas tanks, mufflers, and simple heat exchangers. The weld is made between overlapping sheets of metal, and the seam is simply a series of overlapping spot welds,

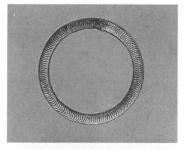

FIGURE 38-8 Seam welds made with overlapping spots of varied spacing. *(Courtesy Taylor-Winfield Corporation.)*

like those shown in Figure 38-8. The basic equipment is the same as for spot welding, except that the electrodes now assume the form of rotating disks, like those shown schematically in Figure 38-9. As the metal passes between the electrodes, timed pulses of current form the overlapping welds. The timing of the welds and the movement of the work are controlled to ensure that the welds overlap, and the workpieces do not get too hot. The welding current is usually a bit higher than in conventional spot welding (to compensate for the short circuit of the adjacent weld), and the workpiece is often cooled by a flow of air or water. In a variation of the process, a continuous current is passed through the rotating electrodes to produce a continuous seam. This form of seam welding is best suited for thin materials, but metals up to 6 mm ($\frac{1}{4}$ in.) can be joined. A typical welding speed is about 2 m/min for thin sheet.

The second type of resistance seam welding is used to produce butt welds between thicker metal plates. The electrical resistance of the abutting metals is still used to generate heat, but high-frequency current (up to 450 kHz) is now employed to restrict the flow of current to the surfaces to be joined and their immediate surroundings. (*Note:* This is similar to the results obtained in the parallel process of high-frequency induction heating.) When the abutting surfaces attain the desired temperature, they are pressed together to form a weld.

Resistance butt welding sees its greatest use in the manufacture of pipes and tubes, as illustrated in Figure 38-10, but the process is also used to fabricate simple structural shapes from sections of plate. Material from 0.1 mm (0.004 in.) to more than 20 mm ($\frac{3}{4}$ in.) in thickness can be welded at speeds up to 80 m/min (250 fpm). The combination of high-frequency current and high welding speed produces a very narrow heat-affected zone. Almost any type or combination of metal can be welded, including difficult dissimilar metals and the high-conductivity metals, such as aluminum and copper.

PROJECTION WELDING

In a mass production operation, conventional spot welding is plagued by two significant limitations. Because the small electrodes provide both the high currents and the required pressure, the electrodes generally require frequent attention to maintain their geometry.

FIGURE 38-9 Schematic representation of seam welding.

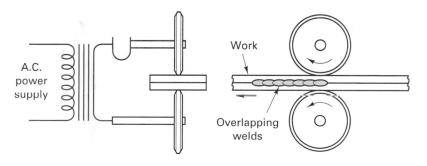

FIGURE 38-10 Using high-frequency ac current to produce a resistance seam weld in butt-welded tubing. Arrows from the contacts indicate the path of the high-frequency current.

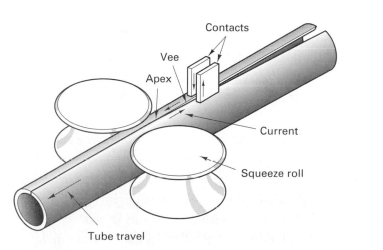

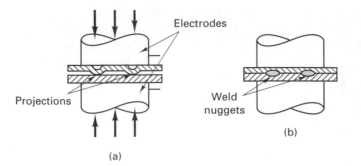

FIGURE 38-11 Principle of projection welding (a) prior to application of current and pressure and (b) after formation of the welds.

In addition, the process is designed to produce only one spot weld at a time. If increased strength is required, multiple welds will be needed, and this means multiple operations. *Projection welding* (RPW) provides a means of overcoming both of these limitations.

Figure 38-11 illustrates the principle of projection welding. A dimple is *embossed* into one of the workpieces at the location where a weld is desired. The two workpieces are then placed between large-area electrodes in a press machine, and pressure and current are applied as in spot welding. Since the current must flow through the points of contact (i.e., the dimples), the heating is concentrated where the weld is desired. As the metal heats and becomes plastic, the pressure causes the dimple to flatten and form a weld.

Because the projections are press-formed into the sheet, they can often be produced during other blanking and forming operations with virtually no additional cost. Moreover, the dimples or projections can be made in almost any shape—for example, round, oval, or circular ring-shaped—to produce welds of shapes that optimize a given design. It is important, however, that the shape be such that the weld will form outward from the center of the projection. Since the heat tends to develop in the part bearing the projections, it is best to incorporate them into the heavier of the two pieces to be joined or the metal with the higher conductivity.

Multiple dimples can be incorporated into a sheet, and therefore multiple welds can be made at one time. The number of projections is limited only by the ability of the machine to provide the required current and pressure and the need to uniformly distribute both. If more than three projections are to be made at one time, however, the height of all projections should be uniform to assure uniform contact and heating.

The attractiveness of the projection welding process is further enhanced by the fact that conventional spot-welding machines can be converted to projection welding simply by changing the size and shape of the electrodes. In addition, projection welding leaves no indentation mark on the exterior surfaces, a definite advantage over spot welding when good surface appearance is required.

In a variation of the process, bolts and nuts can be attached to other metal parts by projection welding. Contact is made at a projection that has been machined or forged onto the bolt or nut, current is applied, and the pieces are pressed together to form a weld.

■ 38.4 ADVANTAGES AND LIMITATIONS OF RESISTANCE WELDING

The resistance welding processes have a number of distinct advantages that account for their wide use, particularly in mass production operations:

1. They are very rapid.
2. The equipment can often be fully automated.
3. They conserve material, since no filler metal, shielding gases, or flux is required.
4. There is minimal distortion of the parts being joined.
5. Skilled operators are not required.
6. Dissimilar metals can be easily joined.
7. A high degree of reliability and reproducibility can be achieved.

TABLE 38-2.	Process Summary: Resistance Welding (RW)
Heat Source:	Electrical resistance heating with high current
Protection:	None; isolation of weld site is adequate
Material Joined:	All common metals (steel, aluminum and copper)
Rate of Heat Input:	High
Weld Profile (D/W):	Does not apply
Maximum Penetration:	Does not apply
Assets:	High speed; Small HAZ; no flux, filler metal, or shielding gas required; adaptable to mass production
Limitations:	Equipment is more expensive than arc welding; welds are weaker than arc welds; requires access to both sides of a joint

The primary limitations of resistance welding include

1. The equipment has a high initial cost.

2. There are limitations to the thickness of material that can be joined (generally less than 6 mm or $\frac{1}{4}$ in.), and the type of joints that can be made (mostly *lap joints*). Access to both sides of the joint is usually required to apply the proper electrode force or pressure.

3. Skilled maintenance personnel are required to service the control equipment.

4. For some materials, the surfaces must receive special preparation prior to welding.

The resistance welding processes are among the most common techniques for high-volume joining. The rapid heat inputs, short welding times, and rapid quenching by both the base metal and the electrodes can produce extremely high cooling rates in and around the weld. While these conditions can be quite attractive for most nonferrous metals, untempered martensite can form in steels containing more than 0.15% carbon. For these materials, some form of postweld heating is generally required to eliminate possible brittleness.

Table 38-2 provides a process summary for resistance welding.

■ 38.5 SOLID-STATE WELDING PROCESSES

FORGE WELDING

Being the most ancient of the welding processes, *forge welding* (FOW) has both historical and practical value, as it helps us to understand how and why the modern welding practices were developed. The armor makers of ancient times occupied positions of prominence in their society, largely because of their ability to join pieces of metal into a single, strong product. More recently, the village blacksmith was another master at the art of forge welding. With his hammer and anvil, coupled with skill and training, he could create a wide variety of useful shapes from metal.

Using a charcoal forge, the blacksmith heated the pieces to be welded to a practical forging temperature and then prepared the ends by hammering so that they could be fitted together without excessive thickness. The ends were then reheated and dipped into a borax flux. Heating was continued until the blacksmith judged (by color) that the workpieces were at the proper temperature for welding. They were then withdrawn from the heat and either struck on the anvil or hit by the hammer to knock off any loose scale or impurities. The ends to be joined were then overlapped on the anvil and hammered to the degree necessary to produce an acceptable weld.

By the correct combination of heat and deformation, a competent blacksmith could produce joints that might be every bit as strong as the original metal. However, because of the crudeness of the heat source, the uncertainty of temperature, and the difficulty in maintaining metal cleanliness, a great amount of skill was required and the results were highly variable.

FORGE-SEAM WELDING

Although forge welding has largely been replaced by other joining methods, a large amount of *forge-seam welding* is still used in the manufacture of pipe. As discussed in Chapter 17, a heated strip of steel is first formed into a cylinder, and the edges are pressed together in

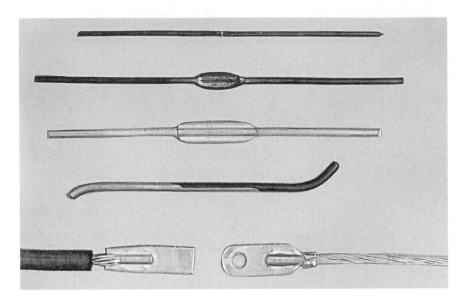

FIGURE 38-12 Small parts joined by cold welding. *(Courtesy of Koldweld Corporation.)*

either a lap or a butt configuration. Welding is the result of pressure and deformation when the metal is pulled through a conical welding bell or passed between welding rolls.

COLD WELDING

Cold welding is a unique variation of forge welding that uses no heating but produces metallurgical bonds by means of cold plastic deformation. The surfaces to be joined are first cleaned and placed in contact. They are then subjected to localized pressures sufficient to cause about 30 to 50% localized cold working. While some heating will occur due to the severe deformation, the primary factor in producing coalescence is the high localized pressure acting on newly formed surface material. The cold-welding process is generally confined to the joining of small parts made from soft, ductile metal, such as the electrical connections shown in Figure 38-12.

ROLL WELDING OR ROLL BONDING

In the *roll welding* or *roll bonding* (ROW) process, two or more sheets or plates of metal can be joined by passing them simultaneously through a rolling mill. As the materials are reduced in thickness, the length and/or width must increase to compensate. The newly created uncontaminated interfaces are pressed together and coalescence is produced. The process can be performed either hot or cold, and can be used to join either similar or dissimilar metals (such as the Alclad aluminums or steel with a stainless steel cladding). The resulting bond can be quite strong, as evidenced by the roll-bonded "sandwich" material used in the production of U.S. coinage.

By coating portions of one interface surface with a material that prevents bonding, the roll-bonding process can be used to produce sheets that are bonded, or not bonded, in selected regions. Upon subsequent heating, the coating material volatilizes, and the no-bond regions expand to produce flow paths for gases or liquids. A common example of this technique is in the manufacture of refrigerator freezer panels, like those shown in Figure 38-13, where inexpensive sheet metal is used to produce structural panels that also serve to conduct the coolant.

FRICTION WELDING AND INERTIA WELDING

In *friction welding* (FRW) the heat required to produce the joint is generated by friction heating at the interface. The components to be joined are first prepared to have smooth, square-cut surfaces. As shown in Figure 38-14, one piece is then held stationary while the other is mounted in a motor-driven chuck or collet and rotated against it at high speed. A low contact pressure may be applied initially to permit cleaning of the surfaces by a burnishing action. The pressure is then increased, and contact friction quickly generates enough heat to raise the abutting surfaces to the welding temperature. As soon as this temperature is reached, rotation is stopped and the pressure is maintained or increased to complete the weld. The

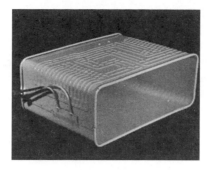

FIGURE 38-13   Examples of roll-bonded refrigerator freezer evaporators. Note the raised channels that have been formed between the bonded sheets. *(Courtesy Olin Brass, East Alton, IL.)*

(a)

(b)

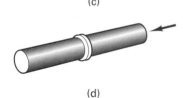

(c)

(d)

FIGURE 38-14   Sequence for making a friction weld.
(a) Components with square surfaces are inserted into a machine where one part is rotated and the other is held stationary.
(b) The components are pushed together with a low axial pressure to clean and prepare the surfaces.
(c) The pressure is increased, causing an increase in temperature, softening, and possibly some melting.
(d) Rotation is stopped and the pressure is increased rapidly, creating a forged joint with external flash.

softened metal is squeezed out to form a *flash* and a "forged" structure is formed in the joint. If desired, the flash can be removed by subsequent machining. Friction welding has been used to join steel bars up to 20 cm (8 in.) in diameter and tubes of even larger diameter. The process is also ideal for welding dissimilar metals with very different melting temperatures and physical properties, such as copper to aluminum, titanium to copper, and nickel alloys to steel. Figure 38-15 shows a schematic of the equipment required for friction welding.

Inertia welding is a modification of friction welding where the moving piece is attached to a rotating flywheel (Figure 38-16). The flywheel is brought to a specified rotational speed and is then separated from the driving motor. The rotating and stationary components are then pressed together, and the kinetic energy of the flywheel is converted into frictional heat at the interface between the two pieces. The weld is formed when the flywheel stops its motion and the pieces remain pressed together. Since the conditions of inertia welding are easily duplicated, welds of consistent quality can be produced, and the process can be readily automated.

FIGURE 38-15   Schematic diagram of the equipment used for friction welding. *(Courtesy of Materials Engineering.)*

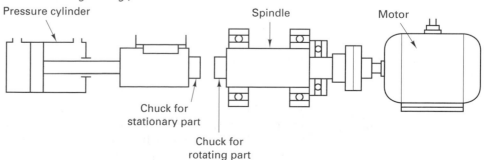

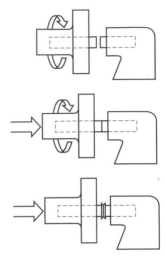

FIGURE 38-16 Schematic representation of the various steps in inertia welding.

FIGURE 38-17 Some typical friction-welded parts. (*Top*) Impeller made by joining a chrome–moly steel shaft to a nickel–steel casting. (*Center*) Stud plate with two mild steel studs joined to a square plate. (*Bottom*) Tube component where a turned segment is joined to medium-carbon steel tubing. (*Courtesy of Newcor Bay City, Division of Newcor, Inc.*)

With inertia welding, the time required to form a weld can be very short, often on the order of several seconds. Because of the high rate of heat input and the limited time for heat to flow away from the joint, both the weld and heat-affected zones are usually very narrow. Oxides and other surface impurities tend to be displaced radially into the upset *flash*, which can be removed after welding if desired. Because virtually all of the energy is converted to heat, the process is very efficient. No material is melted, so joints can be formed with a wide variety of metals or combinations of metals, including some not normally considered compatible, such as aluminum to steel. (*Note:* Cast irons, free-machining metals, and some bearing materials are the exception, since the graphite or lead smears across the surface, reducing the friction heating and preventing good solid-state bonding.) Grain size tends to be refined during the hot deformation, so the strength of the weld is about the same as the base metal. In addition, the friction processes are environmentally attractive since no smoke, fumes, or gases are generated, and no fluxes are required.

Because of the rotational motion, both of the above processes require that one of the components have rotational symmetry. They are used primarily to join round bars or tubes of the same size or connect bars or tubes to flat surfaces. Linear, orbital, and angular reciprocating motions can extend the friction welding processes to noncircular shapes, such as square and rectangular bar. One, or preferably both, of the components must be sufficiently ductile (when hot) to permit deformation during the forging stage. Figure 38-17 shows some typical friction welded parts.

FRICTION STIR WELDING

Butt welds have been made between plates of the lower-melting-point metals (both wrought and cast alloys), as well as thermoplastic polymers, using a friction process known as *friction stir welding*. This relatively new process, first performed in 1991, is illustrated in Figure 38-18. The frictional heating is generated by a nonconsumable probe that is rotated at high speed between the abutting edges of rigidly clamped plates. As the probe traverses the interface, a plasticized region is continually created. The softened material flows to the back of the advancing probe and coalesces to form a solid-state bond. Because of the refined grain structure, the strength, ductility, fatigue life, and toughness of the resulting weld are all quite good. (Welds in aircraft aluminum are 30 to 50% stronger than those formed by arc welding.) No filler material or shielding gas is required, and there are no hot cracking or porosity problems. Total heat input and associated distortion are both low. Joint preparation is minimal and surface oxides need not be removed. Welding can be performed in any position and requires access to only one side of the plate, but weld speed is slower than most fusion processes.

As shown in Figure 38-18, the key process variables include probe geometry (diameter, depth, and profile), shoulder diameter, rotation speed, downward force, and travel speed. Plates to be butt welded require little edge preparation, and the low heat distortion and absence of weld spatter eliminate many postweld operations.

Aluminum plates up to 50-mm (2-in.) thick have been successfully welded from a single side, and aluminum up to 75-mm (3-in.) thick has been welded by a two-sided process. Both wrought and cast alloys can be joined and they can also be joined to each other. Copper, lead, magnesium, titanium, and zinc have all been successfully welded, along with thin steel plate. The process can even be used for purposes other than creating a joint. Key surfaces of large aluminum castings can be enhanced by the refined microstructure produced by the passage of the probe, and reinforcement particles can be stirred into a material to create a particle-reinforced composite surface on a standard alloy substrate.

ULTRASONIC WELDING

Ultrasonic welding (USW) is a solid-state process in which coalescence is produced by the localized application of high-frequency (10,000 to 200,000 Hz) shear vibrations to surfaces that are held together under rather light normal pressure. Although there is some increase in temperature at the faying surfaces, they generally do not exceed one-half of the melting point of the material (on an absolute temperature scale). Instead, it appears that the rapid reversals of stress along the contact interface break up and disperse the oxide films and surface contaminants, allowing the clean metal surfaces to coalesce into a high-strength bond.

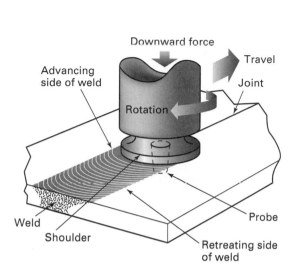

FIGURE 38-18 Schematic of the friction stir welding process. The rotating probe generates frictional heat, while the shoulder provides additional friction heating and prevents expulsion of the softened material from the joint.

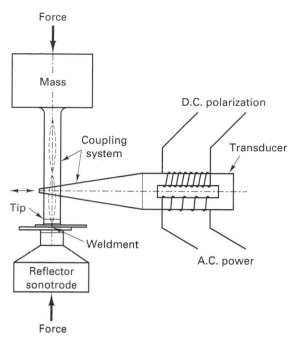

FIGURE 38-19 Schematic diagram of the equipment used in ultrasonic welding.

Figure 38-19 depicts the basic components of the ultrasonic welding process. The ultrasonic transducer is essentially the same as that employed in ultrasonic machining. It is coupled to a force-sensitive system that contains a welding tip on one end, either stationary for spot welds, or rotating for seams. The pieces to be welded are placed between this tip and a reflecting anvil, thereby concentrating the vibratory energy.

Ultrasonic welding is restricted to the lap joint welding of thin materials—sheet, foil, and wire—or the attaching of thin sheets to heavier structural members. The maximum thickness is about 2.5 mm (0.1 in.) for aluminum and 1.0 mm (0.04 in.) for harder metals. As indicated in Table 38-3, the process is particularly valuable because of the number of metals and dissimilar metal combinations that can be joined. It is even possible to bond metals to nonmetals, such as aluminum to ceramics or glass. Because the temperatures are low and no arcing or current flow is involved, the process is also attractive for heat-sensitive electronic components. Intermetallic compounds

TABLE 38-3.	Metal Combinations Weldable by Ultrasonic Welding								
Metal	Aluminum	Copper	Germanium	Gold	Molybdenum	Nickel	Platinum	Silicon	Steel	Zirconium
Aluminum	X	X	X	X	X	X	X	X	X	X
Copper		X		X		X	X		X	X
Germanium			X	X		X	X	X		
Gold				X		X	X	X		
Molybdenum					X	X			X	X
Nickel						X	X		X	X
Platinum							X		X	
Silicon										
Steel									X	X
Zirconium										X

seldom form, and there is no contamination of the weld or surrounding area. The equipment is simple and reliable, and only moderate skill is required of the operator. The required surface preparation is less than for most competing processes (such as resistance welding), and less energy is needed to produce a weld. Typical applications include joining the dissimilar metals in bimetallics, making microcircuit electrical contacts, welding refractory or reactive metals, bonding ultrathin metal, and encapsulating explosives or chemicals.

DIFFUSION WELDING

Diffusion welding (DFW) or *diffusion bonding* occurs when properly prepared surfaces are maintained in contact under sufficient pressure and time at elevated temperature. In contrast to the deformation welding methods, plastic flow is limited and the principal bonding mechanism is atomic diffusion. A well-prepared interface can be viewed as a planar grain boundary with intervening voids and impurities. Under low pressure and elevated temperature, atomic diffusion will provide the necessary void shrinkage and grain boundary migration to form a metallurgical bond.

The quality of a diffusion weld depends on the surface condition of the materials, temperature, time at temperature, pressure, and the possible use of intermediate material layers, which can either promote diffusion or prevent the formation of undesirable intermetallic compounds. Some intermediate layers melt to form a temporary liquid that significantly accelerates the rate of atom movement.

Diffusion bonding is frequently used to join dissimilar metals and composite materials. Furnaces with inert or protective atmospheres can be used to produce high-quality joints with the reactive metals, such as titanium, beryllium, and zirconium, and the high-temperature refractory metals. Since the bonding process is quite slow, multiple parts are generally loaded into a furnace, or application is restricted to low-volume production.

EXPLOSIVE WELDING

Explosive welding (EXW) is used primarily to bond sheets of corrosion-resistant metal to heavier plates of base metal (a cladding operation), particularly when large areas are involved. As shown in Figure 38-20, the bottom sheet or plate is positioned on a rigid base or anvil, and the top sheet is inclined to it with a small open angle between the surfaces to be joined. An explosive material, usually in the form of a sheet, is placed on top of the two layers of metal and detonated in a progressive fashion, beginning where the surfaces touch. A compressive stress wave, on the order of thousands of megapascals (hundreds of thousands of pounds per square inch), sweeps across the surface of the plates. Surface films are liquefied or scarfed off the metals and are jetted out of the interface. The clean metal surfaces are then thrust together under high contact pressure. The result is a low-temperature weld with an interface configuration consisting of a series of interlocking ripples. Since the bond strength is quite high, explosively clad plates can be subjected to a wide variety of subsequent processing, including further reduction in thickness by rolling. Because it is a solid-state welding process, numerous combinations of dissimilar metals can be joined.

FIGURE 38-20 (*Left*) Schematic of the explosive welding process. (*Right*) Explosive weld between mild steel (*left*) and stainless steel (*right*), showing the characteristic wavy interface.

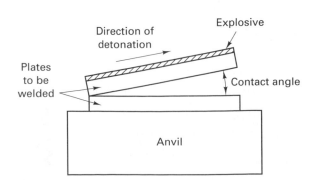
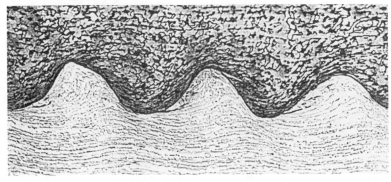

■ 38.6 OTHER WELDING AND CUTTING PROCESSES

THERMIT WELDING

In *thermit welding* (TW) superheated molten metal and slag are produced from an exothermic chemical reaction between a metal oxide and a metallic reducing agent. The name *thermit* usually refers to a mechanical mixture of about one part (by weight) finely divided aluminum and three parts iron oxide, plus possible alloy additions. When this mixture is ignited by a magnesium fuse (the ignition temperature is about 1150°C or 2100°F), it reacts according to the following chemical equation:

$$8\,Al + 3\,Fe_3O_4 \longrightarrow 9\,Fe + 4\,Al_2O_3 + heat$$

A temperature of over 2750°C (5000°F) is produced in about 30 seconds, superheating the molten iron, which then flows by gravity into a prepared joint, providing both heat and filler metal. Runners and risers must be provided, as in a casting, to channel the molten metal and compensate for solidification shrinkage.

Steels and cast irons can be welded using the process described above. Copper, brass, and bronze can be joined using a starting mixture of copper oxide and aluminum. Nickel, chromium, and manganese oxides have also been used in the thermit welding of the more exotic metals.

The thermit welding process is extremely old and has been replaced to a large degree by alternative methods. Nevertheless, it is still effective and can be used to produce economical, high-quality welds in thick sections of material, particularly in remote locations or where more sophisticated welding equipment is not available. One such application is the field repair of large steel castings that have broken or cracked.

ELECTROSLAG WELDING

Electroslag welding (ESW), depicted in Figure 38-21, is a very effective process for welding thick sections of steel plate. There is no arc involved (except to start the weld), so the process is entirely different from submerged arc welding, and the electrical resistance of the metal being welded plays no part in producing the heat. Instead, heat is derived from the passage of electrical current through a liquid slag. Resistance heating

FIGURE 38-21   (*Left*) Arrangement of equipment and workpieces for making a vertical weld by the electroslag process. (*Right*) Cross section of an electroslag weld, looking through the water-cooled copper slide.

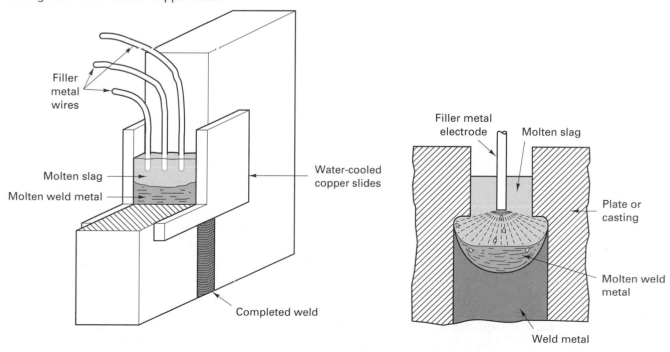

(a)   (b)

raises the temperature of the slag to around 1750°C (3200°F). The molten slag then melts the edges of the pieces that are being joined, as well as continuously fed solid or flux-cored electrodes, which supply the filler metal. Multiple electrodes are often used to provide an adequate supply of filler and maintain the molten pool. Under normal operating conditions, there is a 65-mm (2.5-in.) deep layer of molten slag, which serves to protect and cleanse the underlying 12- to 20-mm ($\frac{1}{2}$- to $\frac{3}{4}$-in.) deep pool of molten metal. These liquids are confined to the gap by means of sliding, water-cooled *molding plates* that are usually made of copper. As the weld metal solidifies at the bottom of the pool, the molding plates move upward at a rate that is typically between 12 and 40 mm/min.

Since a vertical joint provides the easiest conditions for maintaining a deep slag bath, the process is used most frequently in this configuration. Circumferential joints can also be produced in large pipe by using special curved slag-holder plates and rotating the pipe to maintain the welding area in a vertical position.

Because large amounts of weld metal and heat can be supplied, electroslag welding is the best of all the welding processes for making welds in thick plates. The thickness of the plates can vary from 12 to 900 mm ($\frac{1}{2}$ to 38 in.), and the length of the weld (amount of vertical travel) is almost unlimited. Edge preparation is minimal, requiring only squared edges separated by 25 to 35 mm. Applications have included building construction, shipbuilding, machine manufacture, heavy pressure vessels, and the joining of large castings and forgings.

Solidification control is vitally important to obtaining a good electroslag weld, since slow cooling tends to produce a coarse grain structure. Cracking tendencies can be suppressed by adjusting the current, voltage, slag depth, number of electrodes, and electrode extension to produce a wide shallow pool of molten metal. A large heat-affected zone and extensive grain growth are common features of the process. However, the long thermal cycle does serve to minimize residual stresses, distortion, and cracking in the heat-affected zone. If good fracture resistance is required, subsequent heat treatment of the welded structure may be necessary.

ELECTRON-BEAM WELDING

Electron-beam welding (EBW) is a fusion welding process in which heating results from the impingement of a beam of high-velocity electrons on the metal to be welded. Originally developed for obtaining ultrahigh-purity welds in reactive and refractory metals, its unique qualities have led to use in a number of applications.

The electron optical system for the process is shown in Figure 38-22. A current heats a tungsten filament to about 2200°C (4000°F) causing it to emit a stream of electrons by thermal emission. By means of a control grid, accelerating voltage, and focusing coils, these electrons are collected into a concentrated beam, accelerated, and directed to a focused spot between 0.8 and 3.2 mm. ($\frac{1}{32}$ to $\frac{1}{8}$ in.) in diameter. Since electrons accelerated at 150 kV achieve speeds nearly $\frac{2}{3}$ the speed of light, the electron beam becomes an extremely intense heat source, capable of producing temperatures in excess of one million degrees Celsius when its kinetic energy is converted to heat. Since the beam is composed of charged particles, it can be positioned and moved by electromagnetic lenses. Unfortunately, the electrons cannot travel well through air. To be effective as a welding heat source, the beam must be generated and focused in a very high vacuum, typically at pressures of 0.01 Pa (1 × 10^{-4} mm Hg) or less.

In many operations, the workpiece must also be enclosed in the high-vacuum chamber and positioned and manipulated in this vacuum. When welding under these conditions, the vacuum ensures degasification and decontamination of the molten weld metal, and welds of very high quality are obtained. However, the size of the vacuum chamber tends to impose serious limitations on the size of the workpiece that can be accommodated, and the need to break and reestablish the high vacuum as pieces are inserted and removed places a considerable restriction on productivity. As a consequence, electron-beam welding machines have been developed that operate at pressures considerably higher than those required for beam generation. Some permit the workpiece to remain outside the vacuum chamber, with the beam emerging through a small orifice in the vacuum chamber to strike the adjacent workpiece. High-capacity vacuum pumps are required to compensate for the leakage through the orifice. Although these machines offer more production freedom, they produce shallower, wider welds since the beam loses energy and diffuses as the pressure increases.

In general, two distinct ranges of voltage are employed in electron beam welding. High-voltage equipment employs 60 to 150 kV and produces a smaller spot size and greater penetration than does the lower-voltage type, which uses from 10 to 50 kV. Because of their high electron velocities, the high-voltage units emit considerable quantities of harmful X-rays and thus require expensive shielding and indirect viewing systems for observing the work. The X-rays produced by the low-voltage machines are sufficiently soft that they are absorbed by the walls of the vacuum chamber, and the parts can be viewed directly through viewing ports.

Almost any metal can be welded by the electron-beam process, including those that are difficult to weld by other methods, such as zirconium, beryllium, and tungsten. Dissimilar metals, including those with extremely different melting points, can also be readily welded, since the intense beam will melt both metals simultaneously. Electron beam welds typically exhibit a narrow profile and remarkable penetrations like those shown in Figure 38-23. The high power and heat concentrations can produce fusion zones with depth-to-width ratios of 25:1. This is coupled with low total heat input, low distortion, and a very narrow heat-affected zone. Heat-sensitive materials can be welded without damage to the base metal. High welding speeds are common; no shielding gas, flux, or filler metal is required; the process can be performed in all positions; and preheat or postheat is generally unnecessary.

FIGURE 38-22 Schematic diagram of the electron-beam welding process.

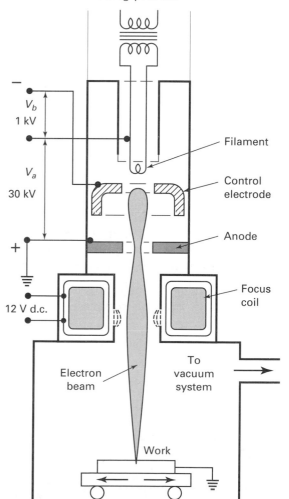

FIGURE 38-23 (*Left to right*) Electron-beam welds in 19-mm thick 7079 aluminum, and 102-mm thick stainless steel. (*Courtesy of Hamilton Standard Division of United Aircraft Corp.*)

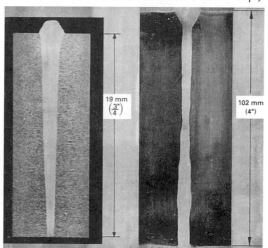

TABLE 38-4. Process Summary: Electron-Beam Welding (EBW)

Heat Source:	High-energy electron beam
Protection:	Vacuum
Electrode:	None
Material Joined:	All common metals
Rate of Heat Input:	High
Weld Profile (D/W):	20
Maximum Penetration:	175 mm (7 in.)
Assets:	High precision; high quality; deep and narrow welds; small HAZ, low distortion; fast welding speed; beam is easily positioned and deflected; no filler metal, flux, or shielding gas required
Limitations:	Beam and target must be within a vacuum; very expensive equipment; workpiece size is limited by the vacuum chamber; significant edge preparation and alignment required; requires safety protection from X-ray and visible radiation

On the negative side, the equipment is quite expensive, and extensive joint preparation is required. Because of the deep and narrow weld profile, joints must be straight and precisely aligned over the entire length of the weld. Machining and fixturing tolerances are often quite demanding. The vacuum requirements tend to limit production rate, and the size of the vacuum chamber may restrict the size of workpiece that can be welded.

The electron-beam process is best employed where welds of extremely high quality are required or where other processes will not produce the desired results. Its unique capabilities have resulted in its routine use in a number of applications, particularly in the automotive and aerospace industries. Table 38-4 provides a process summary for electron-beam welding.

LASER-BEAM WELDING

As a heat source, laser beams can be used for welding, hole making, cutting, cladding, and heat treating of a wide variety of engineering materials. When used for *laser-beam welding* (LBW), the beam can be focused to a diameter of 0.1 to 1.0 mm and provide power density in excess of 10^6 W/mm^2. The high-intensity beam can be used to simply melt the material at the joint, but more often, it produces a very thin column of vaporized metal with a surrounding liquid pool. As the beam traverses, the liquid flows into the joint region to produce a weld with depth-to-width ratio generally greater than 5 : 1. Because of the narrow weld pool geometry, high travel speed of the beam (typically several meters per minute), and low total heat input, the molten metal solidifies quickly, producing a very thin heat-affected zone and little thermal distortion. Laser-beam welding is most effective for simple fusion welds without filler metal (*autogenous welds*), but careful joint preparation is required to produce the narrow gap and necessary level of cleanliness. Filler metal can be added if the gap is excessive, and a low-velocity flow of inert gas (generally helium or argon) is commonly used to protect the weld pool from oxidation.

As shown in Figure 38-24, laser-beam welding and electron-beam welding have some of the highest power densities of the welding processes. In laser welding, the well-collimated beam of intense energy produces deep penetration welds that are similar to electron-beam welds, but the laser-beam technique offers several distinct advantages:

1. The beam can be transmitted through air. A vacuum environment is not required, and the laser can operate at a considerable distance from the workpiece.

FIGURE 38-24 Relative comparison of the power densities of various welding processes. The high power densities of the electron-beam and laser-beam welding processes enable the production of deep, narrow welds with small heat-affected zones. Welds can be made quickly and at high travel speeds.

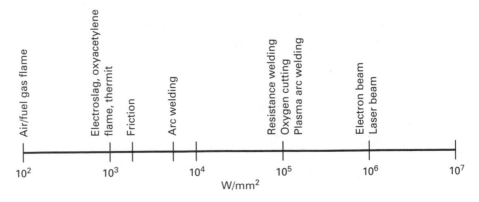

FIGURE 38-25 Laser butt weld of 3-mm (0.125-in.) stainless steel, made at 1.5 m/min with a 1250-W laser. *(Courtesy of Coherent, Inc.)*

2. No X-rays are generated.

3. The laser beam can be easily shaped, directed, and focused with both transmission and reflective optics (lenses and mirrors) and can be transmitted through fiber-optic cables.

4. The only requirement for a weld is optical accessibility. Welds can be made in difficult-to-reach locations and on materials that are encapsulated within transparent containers, such as components in a vacuum tube.

With a sharply focused beam, often less than 0.025 mm (0.01 in.) in width or diameter and having a short exposure time, laser welds can be very small, and the total heat input extremely low, on the order of 0.1 to 10 J. These conditions are ideal for use in the electronics industry, and laser welding is frequently used to connect lead wires to small electronic components. Lap, butt, tee, and cross-wire configurations can all be used. It is even possible to weld wires without removing the polyurethane insulation. The laser simply evaporates the insulation and completes the weld with the internal wire. Figure 38-25 shows a typical laser beam weld and Table 38-5 provides a process summary for laser-beam welding.

A laser welding system consists of an industrial laser, a means of guiding and focusing the beam, and a means of positioning and moving the parts to be welded. The traditional equipment for laser welding is a CO_2 laser with a power output range of 1.5 to 6 kW (much higher outputs are now available). Because of their far-infrared wavelength, the CO_2 lasers require mirror systems or special optical materials to focus and position the beam. They can be used to weld steel up to 25 mm thick at speeds ranging from 1 to 20 m/min. Nd:YAG (neodymium: yttrium–aluminum–garnet) lasers, while more limited in power and capability, operate in a near-infrared wavelength and can utilize conventional glass lenses or delivery by flexible fiber-optic cable (as much as 3000 W of energy can be transmitted up to 150 m through 0.6-mm diameter fiber!).

The industrial lasers lend themselves to automation and robotic manipulation. A Nd:YAG laser with fiber-optic system can distribute its beam to multiple workstations in either a simultaneous or time-sharing fashion, and the individual stations can be as much as 300 ft from the laser. Through use of fiber-optic cables, the laser energy can be piped directly to the end of a robot arm. This eliminates the need to mount and maneuver a heavy, bulky laser and enhances the speed and accuracy of both positioning and manipulation. Cutting, drilling, welding, and heat treating can all be performed with the same unit, and multiple axes of motion can be programmed for specific parts.

The equipment cost for a CO_2 or Nd:YAG laser-beam welding system is quite high, but this cost can be somewhat offset by the faster welding speeds, the ability to weld without filler metal, and low distortion, which enables a reduction in postweld straightening and machining. Caution should be used with such equipment, however, since reflected or scattered laser beams can be quite dangerous, even at great distances from the welding site. Eye protection is a must.

LASER-BEAM CUTTING

Cutting small holes, narrow slots, and closely spaced patterns in a variety of materials is another widely used application of industrial lasers. *Laser-beam cutting* (LBC) begins by "drilling" a hole through the material and then moving the beam in a programmed path. The intense heat from the laser is used to melt and/or evaporate the material being cut.

TABLE 38-5.	Process Summary: Laser-Beam Welding (LBW)
Heat Source:	Laser light
Protection:	None, or externally supplied gas
Electrode:	None
Material Joined:	All common metals
Rate of Heat Input:	High
Weld Profile (D/W):	5
Maximum Penetration:	25 mm (1 in.)
Assets:	High heat-transfer efficiency; can weld any location that is light-accessible; small HAZ; low distortion; can accurately focus the beam with light optics; high welding speed
Limitations:	Possible problems with reflectivity of some metals; good positioning and fit-up required

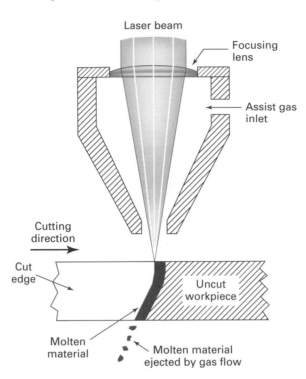

Laser beam

Focusing lens

Assist gas inlet

Cutting direction

Cut edge

Uncut workpiece

Molten material

Molten material ejected by gas flow

FIGURE 38-26 Schematic of laser-beam cutting. The laser provides the heat, and the flow of assist gas propels the molten droplets from the cut.

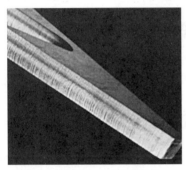

FIGURE 38-27 Surface of 6-mm thick carbon steel cut with a 1250-W laser at 1.8 m/min. *(Courtesy of Coherent, Inc.)*

As shown in Figure 38-26, a stream of *assist gas* is used to blow the molten metal through the cut, cool the workpiece, minimize the heat-affected zone, and possibly assist in combustion of the material. Oxygen is the usual gas for cutting mild steel, since it assists with the exothermic reaction and speeds the cut. Nitrogen or oxygen is used with stainless steel, nitrogen for aluminum, and, because of its high reactivity, titanium requires an inert gas, such as argon. Inert gases or air are used when cutting nonmetallics. Clean, accurate, square-edged cuts are characteristic of the process, and the *kerf* (typically as small as 0.25 mm) and heat-affected zone are narrower than with any other thermal cutting process. No postcut finishing is required in many applications, even though the process does produce a thin recast surface. Figure 38-27 shows a laser cut of 6-mm (0.25-in.) thick carbon steel, cut at 1.8 m/min with a 1250-W laser.

CO_2 and YAG lasers have been used in both continuous and pulsed modes. While cutting speed depends on the material being cut and its thickness, it is greatest in the continuous mode which is preferred for straight and mildly contoured cuts. The pulsed mode is preferred for thin materials and enables tight corners and intricate details to be cut without excessive burning. Metal plates as well as a variety of nonmetals can be cut in thicknesses up to 25 mm (1 in.). Cutting temperatures can be in excess of 11,000°C (20,000°F), and cutting speeds as high as 25 m/min have been observed with some nonmetals.

In addition to the robotic applications described above, lasers have also been mounted on machine tools, such as CNC-type machines, or combined with traditional tools, such as punch presses, to produce extremely flexible hybrid equipment. Because no dedicated dies or tooling are required to produce a cut, and there is no setup time, the laser is an economical alternative to blanking or nibbling for prototype or short-run products or for materials that are difficult to cut by conventional methods, such as plastics, wood, and composites.

Since lasers can cut a wide variety of both metals and nonmetals, laser cutting is becoming a dominant process in the cutting of composite materials. The more uniform the thermal characteristics of the components, the better the cut and the less thermal damage to the material. Kevlar-reinforced epoxy cuts easiest and gives a narrow heat-affected zone. Glass-reinforced epoxy is more difficult because of the greater thermal differences, and graphite-reinforced epoxy is even worse because of the high dissociation temperature and thermal conductivity of the graphite. By the time the graphite has absorbed sufficient cutting heat, the epoxy matrix has been decomposed to a significant depth. The use of lasers for machining is discussed further in Chapter 28.

LASER SPOT WELDING

Lasers have also been used to produce spot welds in a manner that offers unique advantages over the conventional resistance methods. A small clamping force is applied to assure contact of the workpieces, and a fine-focused beam scans the area of the weld. Welding is performed in the keyhole mode, wherein the laser produces a small hole through the molten puddle. As the beam is moved, molten metal flows into the hole and solidifies, forming a fusion-type nugget.

Laser spot welding can be performed with access to only one side of the joint. It is a noncontact process and produces no indentations. No electrodes are involved, so electrode wear is no longer a production problem. Weld quality is independent of material resistance, surface resistance, and electrode condition, and no water cooling is required. The heat input is low, so the size of heat-affected zones is reduced. Speed of welding and strength of the resulting joint are comparable to resistance spot welds.

FLASH WELDING

In *flash welding* (FW), two pieces of metal are first secured in current-carrying grips and lightly touched together. An electric current may be passed through the joint to provide preheat (optional), after which the pieces are withdrawn slightly. An intense flashing arc is thereby created across the gap, which melts the material at both surfaces. The pieces are then forced together under high pressure (on the order of 70 MPa, or 10,000 psi), expelling the liquid and oxides, and upsetting the softened metal. The electric current is turned off, and the force is maintained until solidification is complete. After the product is removed from the machine, the upset portion can be removed by machining. Figure 38-28 shows a schematic of the flash welding process, including both the equipment setup and the completed weld.

To produce a high-quality weld, it is important that the initial surfaces be sufficiently flat and parallel so that the flashing is even across the entire area. The flashing action must be continued long enough to melt the interface and also soften the adjacent metal. Sufficient plastic deformation must occur during the upsetting to enable the impurities and contaminants to be squeezed out into the flash.

Flash welding is usually employed to produce butt welds of similar or dissimilar metals in solid or tubular shape. The equipment required is generally large and expensive, but excellent welds can be made at high production rates.

Percussion welding (PEW) is a similar process, wherein a rapid discharge of stored energy produces a brief period of arcing, which is then followed by the rapid application of force to expel the molten metal and produce the joint. In percussion welding, the arc duration is only 1 to 10 ms. The heat is intense but highly concentrated. Only a small amount of molten metal is produced, little or no upsetting occurs at the joint, and the heat-affected zone is quite small. Application is generally restricted to the butt welding of bar or tubing where heat damage is a major concern.

FIGURE 38-28 Schematic diagram of the flash welding process: (a) equipment and setup and (b) completed weld.

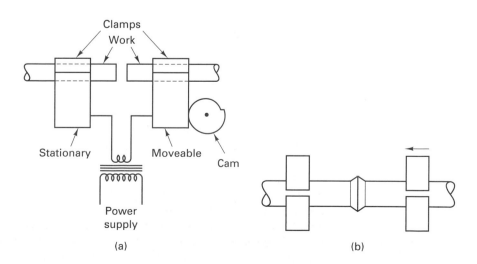

Upset welding (UW) is also similar to flash welding, but there is no period of arcing. While the equipment and geometries are similar, the heating is achieved through electrical resistance. The parts are clamped in the machine, pressure is applied, and high current is passed through the joint. When the abutting surfaces have been heated to a suitable forging temperature, the current is stopped and an upsetting force is applied to produce coalescence. The initial conditions of interface flatness, finish, and alignment must create uniform contact if a good quality weld is to be produced.

■ 38.7 WELDING OF PLASTICS

Mechanical fasteners, adhesives, and welding processes can all be employed to form joints between engineering plastics. Fasteners are quick and are suitable for most materials, but they may be expensive to use, they generally do not provide leaktight joints, and the localized stresses may cause them to pull free of the polymeric material. Threaded metal inserts may have to be incorporated into the plastic components to receive the fasteners, further increasing the product cost. Adhesives can provide excellent properties and fully sound joints, but they are often difficult to handle and relatively slow to cure. In addition, considerable attention is required in the areas of joint preparation and surface cleanliness. In a modification of adhesive bonding, solvents may be used to soften surfaces, which are then pressed together to form a bond. Welding can be used to produce bonded joints with mechanical properties that approach those of the parent material. Unfortunately, only the *thermoplastic polymers* can be welded, since these materials can be melted or softened by heat without degradation. The *thermosetting polymers* do not soften with heat, tending only to char or burn, and must be joined with fasteners or adhesives.

Because the thermoplastics soften at such low temperatures, the heat required to weld these materials is significantly less than that required in the welding of metals. The processes used to weld plastics can be divided into two groups: (1) those that utilize mechanical movement and friction to generate heat, such as ultrasonic welding, spin welding, and vibration welding, and (2) those that involve external heat sources, such as hot-plate welding, hot-gas welding, and resistive and inductive implant welding.

Ultrasonic welding of plastics uses high-frequency mechanical vibrations to create the bond. Parts are held together and are subjected to ultrasonic vibrations (20 to 40 kH frequency and 10 to 100 micron amplitude) perpendicular to the area of contact. The high-frequency stresses generate heat at the joint interface sufficient to produce a high-quality weld in a period of $\frac{1}{2}$ to $1\frac{1}{2}$ seconds. The process can be readily automated, but the tools are expensive and large production runs are generally required. Ultrasonic welding is usually restricted to small components where relative movement is restricted and weld lengths do not to exceed a few centimeters.

In *vibration welding*, relative movement between the two parts is again used to generate the heat, but the direction of movement is now parallel to the interface and aligned with the longest dimension of the joint. The vibration amplitudes are 10 times larger than in ultrasonic welding, and the frequencies are considerably less (on the order of 100 to 240 Hz). When molten material is produced, the vibration is stopped, parts are aligned, and the weld region is allowed to cool and solidify. The entire process takes about 1 to 5 seconds. Long-length, complex joints can be produced at rather high production rates. Nearly all thermoplastics can be joined, independent of whether their prior processing was by injection molding, extrusion, blow molding, thermoforming, foaming, or stamping.

The *friction welding* of plastics (also called *spin welding*) is similar to vibration welding, except the relative motion is now continuous and rotational. The process is essentially the same as the friction welding of metals, but melting now occurs at the joint interface. High-quality welds are produced with good reproducibility, and little end preparation is required. The major limitation is that at least one of the components must exhibit circular symmetry, and the axis of rotation must be perpendicular to the mating surface. Weld strengths vary from 50 to 95% of the parent material in bonds of the same plastic. Joints between dissimilar materials generally have poorer strengths. Butt welds can be made between plates of thermoplastic using the *friction stir welding* process described earlier in this chapter.

Hot-plate welding uses an external heat source and is probably the simplest of the mass production techniques used to join plastics. The parts to be joined are held in fixtures and pressed against the opposite sides of an electrically heated tool. Contact is maintained until the surface has melted and the adjacent material has softened to a specified distance from the interface. The parts separate, the tool is removed, and the two prepared surfaces are pressed together and allowed to cool. Contaminated surface material is usually displaced into a flash region. Weld times are comparatively slow, ranging from 10 seconds to several minutes. The joint strength can be equal to that of the parent material, but the joint design is usually limited to a square-butt configuration, like that encountered when joining sections of plastic pipe. If the bond interface has a nonflat profile, shaped heating tools can be employed. Heated-tool welding can also be used to produce lap seams between flexible plastic sheets. Rollers apply pressure after the material has passed over a heater.

The *hot-gas welding* of plastics is similar to the oxyacetylene welding of metals, and V-groove or fillet welds are the most common joint configuration. A gas (compressed air, nitrogen, hydrogen, oxygen, or carbon dioxide) is heated by an electric coil as it passes through a welding gun, like the one shown in Figure 38-29. The hot-gas stream emerges from the gun at 200 to 300°C (400 to 570°F) and impinges on the joint area. Thin rods of plastic material are heated with the workpiece and are then forced into the softened joint area, providing both the filler material and the pressure needed to produce coalescence. Because this process is usually slow and the results are generally dependent on operator skill, it is seldom used in production applications. It is, however, a popular process for the repair of thermoplastic materials.

Extrusion welding is similar to the hot-gas process, except that the external filler material rod is replaced by a stream of fully molten polymer that emerges from the weld tool as it moves along the joint. This is similar to the wire feed in the metal welding processes.

In the *implant welding* of plastics, metal inserts are placed between the parts to be joined and are then heated by means of induction (radio frequency) or resistance heating. The resistance method requires that a current-carrying path exist along the joint to conduct the current to the implants; this is not required for induction heating. The thermoplastic material melts around the heated implants and flows to form a joint. Since a weld forms only in the vicinity of the implants, the process resembles spot welding and produces joints that are considerably weaker than those formed by processes that bond the entire contact area. When bonding is desired over larger areas, tapes, rods, or gaskets of thermoplastic material laced with iron oxide or metal particles can be used to concentrate the heat at the interface and provide additional filler material. In all cases, the implant material remains as an integral part of the final assembly.

Still other processes to weld thermoplastics are based on *infrared radiation* or *microwave heating*. Laser welding has also been performed on plastics.

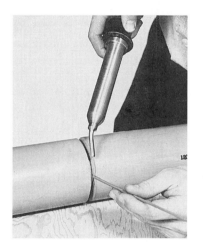

FIGURE 38-29 Using a hot-gas torch to make a weld in plastic pipe.

▪ 38.8 WELDING-RELATED PROCESSES

SURFACING

Surfacing or overlaying is the process of depositing a layer of weld metal on the surface or edge of a base material of different composition. The usual objectives are to obtain improved resistance to wear, abrasion, heat, or chemical attack without having to make the entire piece from an expensive material, one that is difficult to fabricate, or one that would not possess the desired bulk properties. This process is often called *hardfacing*, since the deposited surfaces are generally harder than the base metal. This does not have to be true, however, for in some cases a softer metal (such as bronze) is applied to a harder base material.

Surfacing Materials. The materials most commonly used for surfacing include (1) carbon and low-alloy steels; (2) high-alloy steels and irons; (3) cobalt-based alloys; (4) nickel-based alloys, such as Monel, Nichrome, and Hastelloy; (5) copper-based alloys; (6) stainless steels; and (7) ceramic and refractory carbides, oxides, borides, silicides, and similar compounds.

Surfacing Methods and Applications. Since some of the base metal melts during the deposition, surfacing is actually a variation of fusion welding and can be performed by nearly all of the gas-flame or arc-welding techniques, including oxyfuel gas, shielded metal arc, gas metal arc, gas tungsten arc, submerged arc, and plasma arc. Arc welding is frequently used for the deposition of high-melting-point alloys. Submerged arc welding is used when large areas are to be surfaced or a large amount of surfacing material is to be applied. The plasma arc process further extends the process capabilities because of its extreme temperatures. To obtain true fusion of the surfacing material, a transferred arc is used and the surfacing material is injected in the form of a powder. If a nontransferred arc is used, only a mechanical bond is produced, and the process becomes a form of *metallizing*. Lasers can also be used to perform hardfacing.

THERMAL SPRAY COATING OR METALLIZING

The thermal spray processes offer a means of applying a coating of high-performance material (metals, alloys, ceramics, intermetallics, cermets, carbides, or even plastics) to more economical and more easily fabricated base metals. A wire or rod of the coating material is fed into a gas flame or arc, where it melts and becomes atomized by a stream of gas, such as argon, nitrogen, combustion gases, or compressed air. The gas stream propels the 0.01- to 0.05-mm (0.0004- to 0.002-in.) diameter particles toward the target surface, where they impact ("splat"), cool, and bond. Very little heat is transferred to the substrate, whose peak temperatures generally range from 100 to 250°C (200 to 500°F). As a result, thermal spraying does not induce undesirable metallurgical changes or excessive distortion, and coatings can be applied to thin or delicate targets or to heat-sensitive materials such as plastics. The applied coating can range in thickness from 0.1 to 12 mm (0.004 to 0.5 in.).

Several of the thermal spray processes are adaptations of oxyfuel welding equipment. Figure 38-30 shows a schematic of an oxyacetylene metal spraying gun designed to utilize wire feed. The flame melts the wire and compressed air disintegrates the molten material and propels it to the workpiece. An alternative type of gun uses material in the form of powder, which is gravity or pressure fed into the flame, where it is melted and carried by the flame gas onto the target. The powder feed permits the deposition of material that would be difficult to fabricate into wire, such as cermets, oxides, and carbides. In addition, the droplet size is controlled by the powder, not by the factors that control atomization. The lower temperatures and lower particle velocities of the oxyfuel deposition methods result in coatings with high porosity and low cohesive strength. An adaptation of the process known as high-velocity oxyfuel (HVOF) spraying propels the droplets with a supersonic stream of hot gas. Because the particles impact with high kinetic energies, the resulting coating is dense and well bonded.

The simplest of the electric arc methods is probably wire arc or electric arc spraying. Two oppositely charged electrode wires are fed through the gun, meeting at the tip, where they form an arc. A stream of atomizing gas flows through the gun, stripping off the molten metal to produce a high-velocity spray. Since all of the input energy is used to melt the metal, this process is extremely energy efficient.

Plasma spray metalizing, illustrated in Figure 38-31, is a more sophisticated technique. A plasma-forming gas serves as both the heat source and propelling agent for the coating

FIGURE 38-30   Schematic diagram of an oxyacetylene metal-spraying gun. *(Courtesy of METCO, Inc.)*

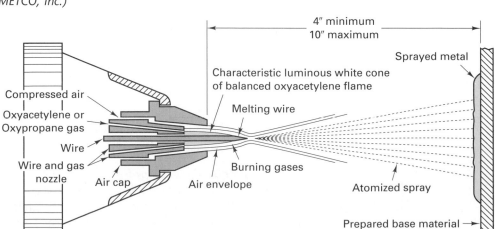

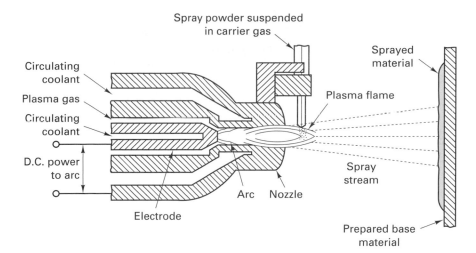

FIGURE 38-31 Schematic diagram of a plasma-arc spray gun. *(Courtesy of METCO, Inc.)*

material. The particles attain high velocity, and therefore produce a dense, strongly bonded coating. Since temperatures can reach 16,500°C (30,000°F), plasma spraying can be used to deposit materials with extremely high melting points. Metals, alloys, ceramics, carbides, cermets, intermetallics, and plastic-based powders have all been successfully deposited.

While thermal spraying or metallizing is similar to surfacing and is often applied for the same reasons, the coatings are usually thinner and the process is more suitable for irregular surfaces or heat-sensitive substrates. The deposition guns can be either handheld or machine driven. A stand-off distance of 0.15 to 0.25 m (6 to 10 in.) is usually maintained between the spray nozzle and the workpiece. Table 38-6 compares the five basic methods of thermal spray deposition.

Surface Preparation for Metallizing. Unlike surfacing, metallizing does not melt the base metal. Since adhesion is entirely mechanical, it is essential that the base metal be prepared in a way that promotes good mechanical interlocking. This material must first be clean and free of dirt, moisture, oil, and other contaminates. The surface is then roughened by one of a variety of methods to create minute crevices that can anchor the solidifying particles. Grit blasting with a sharp, abrasive grit is the most common technique. A surface roughness of 2.5 to 7.5 microns is adequate for most applications.

Characteristics and Applications of Sprayed Metals. During deposition, the atomized, molten, or semimolten particles mix with air and then cool rapidly upon impact with the base metal. The deposited coatings, therefore, consist of bonded particles that span a range of size, shape, and degree of melting. Some particles become oxidized and voids can become entrapped. As a result, the coatings are harder, more porous (0.1 to

| | Heat | | | Particle Impact | | Maximum Spray |
Method	Source	Temperature (°C)	Deposited Materials	Velocity (m/sec)	Adhesion Strength	Rate (kg/hr)
Flame spray						
Wire	Oxyfuel	3,000	Metals	180	Medium	9
Powder	Oxyfuel	3,000	Metals, ceramics, plastics	30	Low	7
High-velocity oxyfuel (HVOF)	Oxyfuel	3,100	Metals, carbides	600–1000	Very high	14
Wire arc	DC arc	5,500	Metals only	250	High	16
Plasma spray	DC arc	5,500 to 16,500	All	250–1200	High to Very high	5–25

TABLE 38-6. Comparison of the Five Basic Thermal Spray Deposition Techniques

15% porosity), and more brittle than the same material in its conventional wrought state. Thermal spray coatings add little, if any, additional strength to a part, since the strength of the porous coating is usually between one-third and one-half of its normal wrought strength. Applications, therefore, generally focus on the ability of the coating to provide resistance to heat, wear, erosion, and/or corrosion, or to restore worn parts to original dimensions and specifications. Some typical applications include:

1. *Protective coatings.* Zinc and aluminum are sprayed on iron and steel to provide corrosion resistance, a process that may well extend the lifetime of bridges, buildings, and other infrastructure items. The interior surfaces of power boilers can be coated with high-chromium alloys to extend wall life through resisting both heat and corrosion.

2. *Building up worn surfaces.* Worn parts may be salvaged by adding new metal to the depleted regions. The repair and restoration of aircraft engine components is probably the largest single use of thermal spraying.

3. *Hard surfacing.* Although metal spraying should not be compared to hard-facing deposits that are applied by welding techniques, it can be useful when thin coatings are considered to be adequate. Typical applications might include automobile cylinder liners and piston rings, thread guides in textile plants, and critical parts within pumps, bearings, and seals.

4. *Applying coatings of expensive metals.* Metal spraying provides a simple method for applying thin coatings of noble metals to surfaces where conventional plating would not be economical.

5. *Electrical properties.* Because metal can be sprayed on almost any surface, metal or nonmetal, it can be used to apply a conductive surface to an otherwise poor conductor or nonconductor. Copper, aluminum, or silver is frequently sprayed on glass or plastics for this purpose. Conversely, alumina (Al_2O_3) can be used to impart insulating or dielectric properties.

6. *Reflecting surfaces.* Aluminum, sprayed on the back of glass by a special fusion process, makes an excellent mirror.

7. *Decorative effects.* One of the earliest and still important uses of metal spraying was to obtain decorative effects. Because sprayed metal can be treated in a variety of ways, such as buffed, wire brushed, or left in the as-sprayed condition, it is frequently specified for manufactured products and architectural materials.

8. *Tailored surface characteristics.* Porous coatings of cobalt or titanium alloys, or certain ceramic materials, have been applied to medical implants to help promote adhesion and ingrowth of bone and tissue.

■ KEY WORDS

assist gas	flash	inertia welding	nugget	surfacing
autogenous weld	flash welding	infrared radiation	percussion welding	thermit
butt welding	forge welding	kerf	projection welding	thermit welding
cold welding	forge-seam welding	lap joint	resistance welding	thermoplastic polymer
diffusion bonding	friction welding	laser-beam cutting	resistance seam welding	thermosetting polymer
electron-beam welding	friction stir welding	laser-beam welding	resistance spot welding	transgun
electroslag welding	hard facing	laser spot welding	rocker arm machine	ultrasonic welding
embossed	hot-gas welding	metallizing	roll bonding	upset welding
explosive welding	hot-plate welding	microwave heating	solid-state welding	vibration welding
faying surfaces	implant welding	molding plates	spot-welding gun	

■ REVIEW QUESTIONS

1. What are the two major roles of applied pressure in resistance welding?

2. Why might resistance welding be considered as a form of solid-state welding?

3. What is the benefit of applying additional pressure and forging the hot metal in the joint?

4. Why is there no need for fluxes or shielding gases in resistance welding?

5. What are the three components that contribute to the total resistance between the electrodes?

6. What measures can be taken to reduce the resistance between the electrodes and the workpieces?

7. What are the possible consequences of too little pressure during the resistance welding cycle? Too much pressure?

8. What is the ideal sequence for pressure application during resistance welding?

9. What magnitudes of current may be used to produce resistance welds?

10. What is the simplest and most widely used form of resistance welding?

11. What is the typical size of a spot-weld nugget?

12. What are the two basic types of stationary spot-welding machines?

13. What is the major advantage of spot-welding guns?

14. What is a transgun? What is a disadvantage of a transgun when accurate positioning is required?

15. What are some of the properties that must be possessed by resistance welding electrodes?

16. What is the most common metal that is spot welded?

17. What is the practical limit of the thicknesses of material that can be readily spot welded?

18. What design features can be altered to permit the joining of different thicknesses or different conductivity metals?

19. What is the difference between the seams produced by roll-spot welding and continuous seam welding?

20. What two limitations of spot welding can be overcome by using the projection approach?

21. What limits the number of projection welds that can be formed in a single operation?

22. What are some of the attractive features of resistance welding when viewed from a manufacturing standpoint?

23. What are some of the primary limitations to the use of resistance welding?

24. What type of metallurgical problem might be encountered when spot welding medium- or high-carbon steels?

25. What were some of the limitations that made the forge welds of a blacksmith somewhat variable in terms of quality?

26. What is the primary feature that promotes coalescence in cold welding?

27. Describe how the roll bonding process can be used to fabricate products that contain pressure-tight, fluid-flow channels that once required the use of metal tubing.

28. How does inertia welding differ from friction welding?

29. How are surface impurities removed in the friction and inertia welding processes?

30. What are some of the geometric limitations of friction and inertia welding?

31. How does friction stir welding differ from friction welding?

32. What are some of the geometric limitations of ultrasonic welding?

33. What are some of the attractive features of ultrasonic welding?

34. What are the conditions necessary to produce high-quality diffusion welds?

35. If the interface of a weld is viewed in cross section, what is the distinctive geometric feature of an explosive weld?

36. In what ways is a thermit weld similar to the production of a casting?

37. What is the source of the welding heat in thermit welding?

38. For what types of applications might thermit welding be attractive?

39. What is the source of the welding heat in electroslag welding?

40. What are some of the various functions of the slag in electroslag welding?

41. Electroslag welding would be most attractive for the joining of what types of geometries and thicknesses?

42. Why is a high vacuum required in the electron beam chamber of an electron-beam welding machine?

43. What types of production limitations are imposed by the high-vacuum requirements of electron-beam welding? What compromises are made when welding is performed on pieces outside the vacuum chamber?

44. What are the major drawbacks of high-voltage electron-beam welding equipment?

45. What are some of the attractive features of electron-beam welding?

46. What are some of the ways in which laser-beam welding is more attractive than electron-beam welding?

47. Why is laser-beam welding an attractive process for use on small electronic components?

48. What is the function of the *assist gas* in laser-beam cutting?

49. What features enable industrial lasers to be integrated easily with robots and CNC machines?

50. What are some of the attractive features of laser spot welding?

51. In the flash welding process, why is it important to have a minimum duration of arcing and minimum amount of upsetting?

52. Why is it difficult to produce a weld between thermosetting polymers but not difficult with the thermoplastics?

53. What are some of the more common techniques for welding thermoplastics? Identify the primary source of heat for each of these processes.

54. What types of materials are applied by surfacing methods?

55. What are some of the primary methods by which surfacing materials can be deposited onto a metal substrate?

56. What are some of the techniques that can be used to apply a thermal spray coating?

57. How is thermal spraying similar to surfacing? How is it different?

58. Why is surface preparation such a critical feature of metallizing?

59. What are some of the more common applications of sprayed coatings?

■ PROBLEMS

1. Many advanced engineering products, as well as composite materials, require the joining of dissimilar materials. Select several of the processes discussed in this chapter and investigate the capability of the process to join dissimilar materials and the associated limitations.

2. The processes described for the joining of plastics focused almost exclusively on the thermoplastic polymers. What types of joining techniques could be applied to thermosetting polymers? To elastomeric polymers? To ceramic materials?

www.wiley.com/college/degarmo

Chapter 38 CASE STUDY

Field Repair to a Power Transformer Case

Electric power transformers do not operate at 100% efficiency, and generally incorporate some means of cooling in their design. Large transformers are often submerged in oil-filled reservoirs, where the volume of oil provides a noncorrosive heat sink. In addition, it is not uncommon for additional features, such as horizontal cooling fins, to be added to the design to aid in dissipating heat from the reservoir.

Figure CS-38 shows the exterior of a large transformer that has been installed in a rural, somewhat-remote location. The reservoir housing has been constructed by welding 9- and 12-mm ($\frac{3}{8}$- and $\frac{1}{2}$-in.) thick, low-carbon steel plates. While the transformer was in use, a service vehicle accidentally backed into the cooling-fin assembly,

producing cracks in several of the fillet areas and a resulting loss of oil. Overheating occurred, and a repair is now necessary. Because of the size of the transformer, some form of on-site repair is preferred. It is your job to determine the procedure and make the necessary arrangements.

1. Consider the full spectrum of welding processes and identify candidates that might be appropriate for this task.
2. For each of the candidate processes, identify its primary advantages and limitations.
3. Which of the candidate processes would you recommend? Why?
4. Describe the procedure that you would outline for such a repair. Are there any special concerns or precautions?

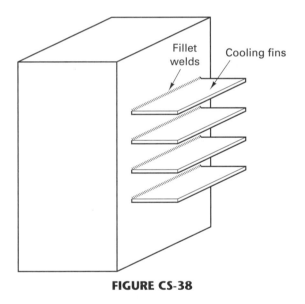

Fillet welds   Cooling fins

FIGURE CS-38

CHAPTER 39

BRAZING AND SOLDERING

■ 39.1 INTRODUCTION

There are many joining or assembly operations where welding may not be the best choice. Perhaps the heat of welding is objectionable, the materials possess poor weldability, welding is too expensive, or the joint involves thin or dissimilar materials. In such cases low-temperature joining methods may be preferred. These include brazing, soldering, adhesive joining, and the use of mechanical fasteners. In brazing and soldering, the metal surfaces are cleaned, the components assembled or fixtured, and a low-melting-point nonferrous metal is then melted, drawn into the space between the two solid surfaces by capillary action, and allowed to solidify. Adhesive bonding utilizes a nonmetallic filler material (often a polymerizable resin) to fill the space between the surfaces to be joined. Variations in surface finish and fit are more tolerable, since capillary action is not required, and oxides may not be a problem since the adhesive may actually bond better to a tight oxide layer. Mechanical fasteners span a wide spectrum, including rivets, bolts, screws, staples, nails, and others. While some forms may be permanent, others offer the advantage of easy disassembly for service or replacement of components.

Brazing and soldering will be explored in this chapter; adhesive bonding and mechanical fasteners are presented in Chapter 40.

■ 39.2 BRAZING

Brazing is the permanent joining of similar or dissimilar metals or ceramics (or composites based on those two materials) through the use of heat and a filler metal whose melting temperature (actually liquidus temperature) is above 450°C (840°F)[1] but below the melting point (or solidus temperature) of the materials being joined. The brazing process is different from welding in a number of ways:

1. The *composition* (or chemistry) of the brazing alloy is significantly different from that of the base metal.
2. The *strength* of the brazing alloy is usually lower than that of the base metal.
3. The *melting point* of the brazing alloy is lower than that of the base metal, so none of the base metal is melted.
4. Bonding requires *capillary action* to distribute the filler metal between the closely fitting surfaces of the joint. The specific flow is dependent upon the viscosity of the liquid, the geometry of the joint, and surface wetting characteristics.

Because of these differences, the brazing process has several distinct advantages:

1. A wide range of metallic and nonmetallic materials can be brazed. The process is ideally suited for joining dissimilar materials, such as ferrous metal to nonferrous metal, metals with widely different melting points, or even metal to ceramic.

[1]This temperature is an arbitrary one, selected to distinguish brazing from soldering.

989

2. Since less heating is required than for welding, the process can be performed quickly and economically.

3. The lower temperatures reduce problems associated with heat-affected zones (or other material property alteration), warping, and distortion. Thinner and more complex assemblies can be joined successfully. Thin sections can be joined to thick.

4. Assembly tolerances are closer than for most welding processes, and joint appearance is usually quite neat.

5. Brazing is highly adaptable to automation and performs well when mass-producing complex or delicate assemblies.

6. A strong permanent joint is formed.

Successful brazing or soldering requires that the parts have relatively good fit-up (i.e., small joint clearances) to promote capillary flow of the filler metal. The parts must be thoroughly cleaned prior to joining, and many parts will require flux removal after joining. In addition, remember that any subsequent heating of the assembly can cause inadvertent melting of the braze metal, thereby weakening or destroying the joint.

Another concern with brazed joints is their enhanced susceptibility to corrosion. Since the *filler metal* is of different composition from the materials being joined, the brazed joint is actually a localized galvanic corrosion cell. Corrosion problems can often be minimized, however, by proper selection of the filler metal.

NATURE AND STRENGTH OF BRAZED JOINTS

Just as in welding, brazing forms a strong metallurgical bond at the interfaces. Clean surfaces, proper clearance, good wetting, and good fluidity will all enhance the bonding. The strength of the resulting bond can be quite high, certainly higher than the strength of the brazing alloy and often greater than the strength of the metal being brazed. Attainment of high bond strength, however, requires optimum processing and design.

Of all of the factors contributing to joint strength, *joint clearance* is the most important. If the joint is too tight, it may be difficult for the braze metal to flow into the gap (leaving unfilled voids), and flux may be unable to escape (remaining in locations that should be filled with braze material). There must be sufficient clearance so that the braze metal will wet the joint and flow into it under the force of capillary action. As the gap is increased beyond this optimum value, however, the joint strength decreases rapidly, dropping off to that of the braze metal itself. If the gap becomes too great, capillary forces may be insufficient to draw the material into the joint or hold it in place during solidification. Figure 39-1 shows the tensile strength of a butt joint braze as a function of joint clearance.

Proper clearance can vary considerably, depending primarily on the type of braze metal being used. The ideal clearance is usually between 0.01 and 0.04 mm (0.0005 and 0.0015 in.), an "easy-slip" fit. A press fit can even be acceptable if fluxes are not used and

FIGURE 39-1 Typical variation of tensile strength with different clearances in a butt-joint braze. *(Courtesy of Handy & Harman.)*

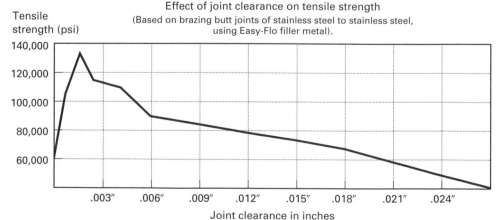

Effect of joint clearance on tensile strength
(Based on brazing butt joints of stainless steel to stainless steel, using Easy-Flo filler metal).

Tensile strength (psi)

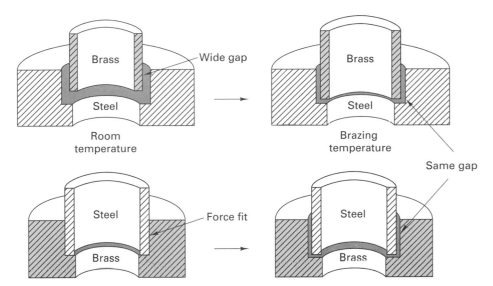

FIGURE 39-2 When brazing dissimilar metals, the initial joint clearance should be adjusted for the different thermal expansions. Proper brazing clearances should exist at the temperature where the filler metal flows.

surface roughness is sufficient to assure adequate flow of the filler metal into the joint. Clearances up to 0.075 mm (0.003 in.) can be accommodated with a more sluggish filler metal, such as nickel. However, when clearances range between 0.075 and 0.13 mm (0.003 and 0.005 in.), acceptable brazing becomes somewhat difficult, and joints with gaps in excess of 0.13 mm (0.005 in.) are almost impossible to braze. Finally, it should be noted that the surfaces to be brazed should be parallel, maintaining the specified gap over the entire area of contact.

It is also important to note that the dimensions cited above are the clearances that should exist *at the temperature of the brazing process*. Any effects of thermal expansion should be compensated when specifying the dimensions of the starting components. This is particularly significant when dissimilar materials are to be joined, for here the joint clearance will change as one material expands at a faster rate than the other. Consider a joint between brass and steel, like the one depicted in Figure 39-2. Brass expands more than steel when temperature is increased. Therefore, if the insert tube is the brass component, the initial fit should be somewhat loose. The brass will expand more than the steel as the temperature is increased, and at the brazing temperature, the gap will assume the desired dimensions. Conversely, if a steel tube is to be inserted into a brass receiver, an initial force fit may be required since the interface will widen as the brass expands more than the steel. It should also be noted that reverse dimensional changes will occur during cooldown, and these can result in significant residual stresses or even cracking of the joint.

Wettability is a strong function of the surface tensions between the braze metal and the base alloy. Generally, the wettability is good when the surfaces are clean and the two metals can form intermediate diffused alloys. Sometimes the wettability can be improved, as is done when steel is tin plated to accept a lead-tin solder, or plated with nickel or copper to enhance brazing. *Fluidity* is a measure of the flow characteristics of the molten braze metal and is a function of the metal, its temperature, surface cleanliness, and clearance.

DESIGN OF BRAZED JOINTS

Because the strength of the filler braze metal is generally less than that of the metals being joined, a good joint design is required if one is to obtain adequate mechanical strength. The desired load-carrying ability is usually obtained by (1) assuring proper joint clearance and (2) providing sufficient area for the bond. Figure 39-3 depicts the two most common types of brazed joints: *butt* and *lap*. Butt joints do not require additional thickness in the vicinity of the joint, but are most often used where the strength requirements are not that critical. The butt geometry restricts the bonding area to the cross-sectional area of the thinner or smaller member. In contrast, lap joints can provide bonding areas that are considerably larger than the butt configuration, and are often preferred when maximum strength

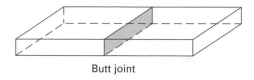

Butt joint

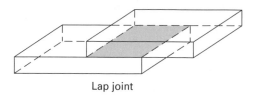

Lap joint

FIGURE 39-3 The two most common types of braze joints are butt and lap. Butt offers uniform thickness across the joint, whereas lap offers greater bonding area and higher strength.

is desired. If the joints are made very carefully, a lap of 1 to $1\frac{1}{4}$ times the material thickness can develop strength as great as that of the parent metal. For joints that are made by routine production, it is best to use a lap equal to three to six times the material thickness to assure that failure will occur in the base metal and not in the brazed joint.

Variations of the two basic joint designs include the *butt-lap* and *scarf* configurations. The butt-lap design is an attempt to combine the advantage of a uniform thickness with a large bonding area and companion high strength. Unfortunately, it also requires a higher degree of joint preparation. The scarf joint maintains uniform thickness and increases bonding area by tilting the butt joint interface. Careful joint preparation and component alignment is required to maintain the desired dimensions of joint clearance. Figure 39-4 shows relatively simple butt, lap, butt-lap, and scarf joints for both flat and tubular parts. Figure 39-5 shows good brazing designs for several types of joint configurations.

FIGURE 39-4 Variations of the butt and lap configurations include the butt-lap and scarf. The four types are shown for both flat and tubular parts.

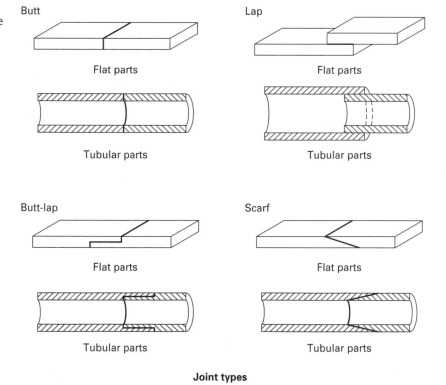

Butt

Flat parts

Tubular parts

Lap

Flat parts

Tubular parts

Butt-lap

Flat parts

Tubular parts

Scarf

Flat parts

Tubular parts

Joint types

FIGURE 39-5 Some common joint designs for assembling parts by brazing.

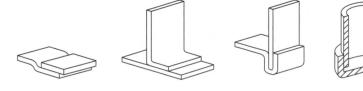

TABLE 39-1. Engineering Materials and Their Compatibility with Brazing

Material	Brazing Recommendation
Cast iron	Somewhat difficult
Carbon and low-alloy steels	Recommended for low- and medium-carbon materials; difficult for high-carbon materials; seldom used for heat treated alloy steels
Stainless steel	Recommended; Silver and nickel brazing alloys are preferred
Aluminum and magnesium	Common for aluminum alloys and some alloys of magnesium
Copper and copper alloys	Recommended for copper and high-copper brasses; somewhat variable with bronzes
Nickel and nickel alloys	Recommended
Titanium	Difficult, not recommended
Lead and zinc	Not recommended
Thermoplastics, thermosets, and elastomers	Not recommended
Ceramics and glass	Not recommended
Dissimilar metals	Recommended, but may be difficult, depending on degree of dissimilarity
Metals to nonmetals	Not recommended
Dissimilar nonmetals	Not recommended

There are also material effects to be considered when designing a brazed joint. Table 39-1 summarizes the compatibility of various engineering materials with the brazing process.

BRAZING METALS

Brazing filler metal can be any metal that melts between 450°C (840°F) and the melting point of the material being joined. Actual selection, however, considers a variety of criteria, including compatibility with the base materials, brazing temperature restrictions, restrictions due to service or subsequent processing temperatures, the brazing process to be used, the joint design, anticipated service environment, desired appearance, desired mechanical properties (such as strength, ductility, and toughness), desired physical properties (such as electrical, magnetic, or thermal), and cost. In addition, the material must be capable of flowing through small capillaries, "wetting" the joint surfaces, and partially alloying with the base metals. The most commonly used brazing metals are copper and copper alloys, silver and silver alloys, and aluminum alloys. Many of the brazing alloys are based on eutectic reactions where the material melts at a single temperature that is lower than the melting points of the individual metals in the alloy. Table 39-2 presents some common braze metal families, the metals they are used to join, and the typical brazing temperatures.

Copper and copper alloys are the most commonly used brazing alloys. Unalloyed copper is used primarily for brazing steel and other high-melting-point materials, such as high-speed steel and tungsten carbide. Its melting point is rather high (about 1100°C), and tight fitting joints are required (gaps less than 0.075 mm). Copper–zinc alloys offer lower melting points and are used extensively for brazing steel, cast irons, and copper. Copper–phosphorus alloys are used for the fluxless-brazing of copper since the phosphorus can reduce the copper oxide film. (These alloys should not be used with ferrous or nickel-based materials, however, since they form brittle compounds with phosphorus and the resulting joints may be brittle.) Manganese bronzes can also be used as filler metal in brazing operations.

TABLE 39-2. Some Common Braze Metal Families, Metals They Are Used to Join, and Typical Brazing Temperatures

Braze Metal Family	Materials Commonly Joined	Typical Brazing Temperature (°C)
Alumimum–silicon	Aluminum alloys	565–620
Copper and copper alloys	Various ferrous metals as well as copper and nickel alloys and stainless steel	925–1150
Copper–phosphorus	Copper and copper alloys	700–925
Silver alloys	Ferrous and nonferrous metals, except aluminum and magnesium	620–980
Precious metals (gold-based)	Iron, nickel, and cobalt alloys	900–1100
Magnesium	Magnesium alloys	595–620
Nickel alloys	Stainless steel, nickel, and cobalt alloys	925–1200

Pure silver can be used for brazing titanium. *Silver solders* (alloys based on silver and copper) have brazing temperatures significantly below that of pure copper and are used in joining steels, copper, brass, and nickel. While these brazing alloys are expensive, only a small amount is required, so the cost per joint is kept low.

Aluminum–silicon alloys, containing about 6 to 12% silicon, are used for brazing aluminum and other aluminum alloys. By using a braze metal that is similar to the base metal, the possibility of galvanic corrosion is reduced. However, these brazing alloys have melting points of about 610°C (1130°F), and the melting temperature of commonly brazed aluminum alloys, such as 3003, is around 670°C (1290°F). Therefore, control of the brazing temperature is critical. In brazing aluminum, proper fluxing action, surface cleaning, and/or the use of a controlled atmosphere or vacuum environment is required to assure adequate flow of the braze metal.

Nickel- and cobalt-based alloys are attractive for joining assemblies that will be subjected to elevated-temperature service conditions and/or extremely corrosive environments. Gold and palladium alloys offer outstanding oxidation and corrosion resistance, as well as good electrical and thermal conductivity. Magnesium alloys can be used to braze other types of magnesium.

A variety of brazing alloys are currently available in the form of amorphous foils, formed by cooling metal at rates in excess of 1 million degrees Centigrade per second. The resulting foils are extremely thin (0.04 mm or 0.0015 in. being typical) and exhibit excellent ductility and flexibility, even when they are made from alloys whose crystalline form is quite brittle. Shaped inserts can be cut or stamped from the foil and inserted into the joint region. Since the braze material is fully dense, no shrinkage or movement occurs during the brazing operation. One amorphous alloy, composed of nickel, chromium, iron, and boron, is used to produce assemblies that can withstand high temperatures. During the brazing operation, the boron diffuses into the base metal, raising the melting point of the remaining filler. Subsequent service temperatures can then be above the melting point of the original braze alloy, and the brazed joint will not melt.

FLUXES

In a normal atmosphere, the heat required to melt the brazing alloy would also cause the formation of surface oxides that oppose the wetting of the surface and subsequent bonding. *Brazing fluxes,* therefore, play an important part in the process by (1) dissolving oxides that may have formed on the surfaces prior to heating, (2) preventing the formation of new oxides during heating, and (3) lowering the surface tension between the molten brazing metal and the surfaces to be joined, thereby promoting the flow of the molten material into the joint. Ideally, the flux will melt and become active at a temperature below the solidus of the filler metal, yet remain active throughout the entire range of temperatures encountered while making the braze.

Surface cleanliness is one of the most significant factors affecting the quality and uniformity of brazed joints. Although fluxes can dissolve modest amounts of oxides, *they are not cleaners.* Before a flux is applied, dirt, grease, oil, rust, and heat-treat scale should be removed from the surfaces that are to be brazed. Cleaning operations can involve water- or solvent-based techniques; high-temperature burn-off of oils, greases, and fuel residues; acid pickling; grit blasting with selected media; other mechanical methods; or exposure to high-temperature reducing atmospheres. The less cleaning the flux has to do prior to heating, the more effective it will be during the brazing operation. Because the presence of surface graphite impairs wetting, cast iron materials often require special treatment. Graphite removal by chemical etching may be required before cast iron can be brazed.

Brazing fluxes usually take the form of chemical compounds in which the most common ingredients are borates, fused borax, fluoroborates, fluorides, chlorides, acids, alkalies, wetting agents, and water. The particular flux should be selected for compatibility with the base metal being brazed and the particular process being used. Paste fluxes are utilized for furnace, induction, and dip brazing, and they are usually applied by brushing. Either paste or powdered fluxes can be used with the torch-brazing process, where application is usually achieved by dipping the heated end of the filler wire into the flux material.

APPLYING THE BRAZING METAL

The brazing filler metal can be applied to joints in several ways. The oldest method (and still a common technique when torch brazing) uses brazing metal in the form of a rod or wire. The joint area is first heated to a temperature high enough to melt the braze alloy and assure that it remains molten while flowing into the joint. The torch is then used to melt the braze metal and capillary action draws it into the prepared gap.

The above method of braze metal application requires considerable labor, and care must be taken to assure that the filler metal has flowed into the inner portions of the joint. To avoid these difficulties, the braze metal is often inserted into the joint prior to heating, usually in the form of wires, shims, powders, or preformed rings, washers, disks, or slugs. In cases where it can be done, rings or shims are fitted into internal grooves in the joint before the parts are assembled. The least expensive approach to brazing is to design assemblies where all of the components maintain a fixed position throughout the brazing cycle. When this is not possible, alignment and clearances can be maintained by tack welding, riveting, staking, expanding or flaring, swaging, knurling, and dimpling. Shims, wires, ribbons, and screens can also be employed to assist in locating pieces or maintaining fit. When the components become more complex, special brazing *jigs and fixtures* are often used to hold the components during the heating. When these are used, however, it is necessary to provide springs that will compensate for thermal expansion, particularly when two or more dissimilar metals are being joined.

When using preloaded joints, care must be exercised to assure that the filler metal is not drawn away from the intended surface by the capillary action of another surface of contact. Capillary action will always pull the molten braze metal into the smallest clearance, whether or not that was the intended location. In addition, the flow of filler metal must not be cut off by inadequate clearances, or the presence of entrapped or escaping air. Fillets and grooves within the joint can also act as reservoirs and trap the filler metal.

An alternative approach to the brazing process is to precoat one or both of the surfaces to be joined with the brazing alloy. Simply placing the materials in contact and heating forms the desired bond. By having the braze material already in place over the full area of contact, the joining operation does not have to rely on capillary action and metal flow. More complex assemblies can be produced than with conventional methods, and the thickness of the braze material is precisely controlled to provide maximum strength to the joint.

HEATING METHODS USED IN BRAZING

Since molten metal tends to flow toward the location of highest temperature, it is important that the heat sources used in brazing provide for control of both the temperature and the uniformity of that temperature throughout the joint. In specifying the heating method, consider a number of factors, including the size and shape of the parts being brazed, the type of material being joined, and the desired quantity and rate of production.

A common source of heat for brazing is a gas-flame torch. In the *torch-brazing* procedure, oxyacetylene, oxyhydrogen, or other gas-flame combinations can be used. Most repair brazing is done in this manner because of its flexibility and simplicity, but the process is also widely used in production applications. Local heating permits the retention of most of the original material strength and permits large components to be joined with little or no distortion. The major drawbacks are the difficulty in controlling the temperature, maintaining uniformity of heating, and meeting the cost of skilled labor. A protective flux is required, and the flux residue must be removed after brazing. In production-type torch brazing, specially shaped torches are often used to speed the heating and aid in reducing the amount of skill required.

If the flux and filler metal can be preloaded into the joints, a number of assemblies can be heated simultaneously in controlled-atmosphere or vacuum furnaces, a process known as *furnace brazing*. If the components are not likely to maintain their alignment, brazing jigs or fixtures must be used. Fortunately, most assemblies that are to be furnace brazed can be designed so that jigs and fixtures are not needed. A light press fit is often sufficient. Figure 39-6 shows several typical furnace-brazed assemblies.

Because excellent control of the furnace temperature is possible and no skilled labor is required, furnace brazing is particularly well suited for mass production operations. Either batch- or continuous-type furnaces can be used, with the latter being more

FIGURE 39-6 Typical furnace-brazed assemblies. *(Courtesy of Pacific Metals Company.)*

suitable for mass production work. Furnace brazing heats the entire assembly in a uniform manner and therefore produces less warpage and distortion than processes that employ localized heating. Extremely complex assemblies can be produced in which multiple joints are formed in a single heating.

A variety of furnace atmospheres can be utilized to reduce oxide films and prevent both the base and filler metals from oxidizing during the brazing operation. A chemical flux may no longer be needed, and the parts will emerge clean and free of contaminants. When reactive materials are to be joined or the joint must meet the highest of standards, a vacuum furnace may be preferred.

A third type of heating is *salt-bath brazing*, where the parts are preheated and then dipped in a bath of molten salt that is maintained at a temperature slightly above the melting point of the brazing metal. This process offers three distinct advantages: (1) The salt bath acts as the brazing flux, preventing oxidation and enhancing wettability. (2) The work heats very rapidly because it is in complete contact with the heating medium. (3) Temperature can be accurately controlled so thin pieces can be attached to thicker pieces without danger of overheating. This last feature makes the process well suited for brazing aluminum, where precise temperature control is often required.

In salt-bath brazing, the parts again must be held in jigs or fixtures (or be prefastened in some manner), and the brazing metal must be preloaded into the joints. To assure that the bath remains at the desired temperature during the immersion process, its volume must be substantially larger than that of the assemblies to be brazed.

In *dip brazing*, the assemblies are immersed in a bath of molten brazing metal. The bath thus provides both the heat and the metal for the joint. Since the braze metal will usually coat the entire workpiece, it is a somewhat wasteful process and is usually employed only for small products.

Induction brazing utilizes high-frequency induction currents as the source of heat, and is therefore limited to the joining of electrically conductive materials. A variety of high-frequency ac power supplies is available in large and small capacities. These are coupled to a simple heating coil designed to fit around the joint. The heating coils are generally formed from copper tubing and typically carry a supply of cooling water. Although the filler metal can be added to the joint manually after it is heated, the usual practice is to use preloaded joints to speed the operation and produce more uniform bonds.

Induction brazing offers the following advantages, which account for its extensive use.

1. The complete heating cycle is very rapid, usually only a few seconds in duration.
2. The operation can be made semiautomatic so that only semiskilled labor is required.
3. Heating can be confined to the specific area of the joint through use of specially designed coils, frequency control, and short heating times. This minimizes softening and distortion and reduces problems associated with scale and discoloration.
4. Uniform results are easily obtained.
5. By making new, and relatively simple heating coils, a wide variety of work can be performed with a single power supply.

Resistance brazing can be used to produce relatively simple joints in metals with high electrical conductivity. The parts to be joined are pressed between two electrodes and a current is passed through. Unlike resistance welding, most of the resistance here is provided by the carbon or graphite electrodes, and the heating of the joint is primarily by conduction from the hot electrodes.

Infrared heat lamps, lasers, and electron beams can also be used to provide the heat required for brazing.

FLUX REMOVAL AND OTHER POSTBRAZE OPERATIONS

Since many of the brazing fluxes become corrosive when in contact with moisture, the flux residue should be removed from the work as soon as brazing is completed. Rapid and complete flux removal is particularly important in the case of aluminum, where chlorides can be particularly detrimental. Fortunately, many brazing fluxes are water soluble, and an immersion in a hot-water tank for a few minutes will often provide satisfactory results. Blasting with grit or sand is another effective method of flux removal, but this procedure may not be attractive if a good surface finish is to be maintained. Fortunately, such drastic treatment is seldom necessary.

Other postbraze operations may include heat treating, cleaning, and inspection. A visual examination is probably the simplest of the inspection techniques and is most effective when both sides of a brazed joint are accessible for examination. A proof test can be performed by subjecting the joint to loads in excess of those expected during service. Leak tests or pressure tests can assure gas- or liquid-tightness. Cracks and other flaws can be detected by dye-penetrant, magnetic particle, ultrasonic, or radiographic examination, as described in Chapter 11. Destructive forms of evaluation include peel tests, tension or shear tests, and metallographic examination.

FLUXLESS BRAZING

Both the application and removal of brazing flux involves significant costs, particularly where complex joints and assemblies are involved. Consequently, a large amount of work has been devoted to the development of procedures where a flux is not required. Controlled furnace atmospheres can make a flux unnecessary by reducing existing oxides and preventing the formation of new ones. Vacuum furnaces can also be used to create and preserve clean brazing surfaces. Special brazing metals have been developed with alloy additions, such as phosphorus, that can also fulfill the role of a flux.

BRAZE WELDING

Braze welding differs from straight brazing in that capillary action is not required to distribute the filler metal. Here the molten filler is simply deposited by gravity, as in oxyacetylene gas welding. Because relatively low temperatures are required and warping is minimized, braze welding is very effective for the repair of steel products and ferrous castings. It is also attractive for joining cast irons since the low heat does not alter the graphite shape, and the process does not require good wetting characteristics. Strength is determined by the braze metal being used and the amount applied. Considerable buildup may be required if full strength is to be restored to the repaired part.

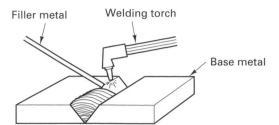

FIGURE 39-7   Schematic of the braze welding process.

Braze welding is almost always done with an oxyacetylene torch. The surfaces are first "tinned" with a thin coating of the brazing metal, and the remainder of the filler metal is then added. Figure 39-7 shows a schematic of braze welding.

■ 39.3 SOLDERING

By definition, *soldering* is a brazing type of operation where the filler metal has a melting temperature (or liquidus temperature if the alloy has a freezing range) below 450°C (840°F). It is typically used for connecting thin metals, connecting electronic components, joining metals while avoiding exposure to high elevated temperatures, and filling surface flaws and defects. Effective soldering generally involves six important steps: (1) design of an acceptable joint; (2) selection of the correct solder for the job; (3) selection of the proper type of flux; (4) cleaning the surfaces to be joined; (5) application of flux, solder, and sufficient heat to allow the molten solder to fill the joint by capillary action and solidify; and (6) removal of the flux residue.

DESIGN AND STRENGTH OF SOLDERED JOINTS

Soldering can be used to join a wide variety of sizes, shapes, and thicknesses, and is used extensively to provide electrical coupling or gas- or airtight seals. While the low joining temperatures are attractive for heat-sensitive materials, soldered joints seldom develop shear strengths in excess of 1.75 MPa (250 psi). Consequently, if appreciable strength is required, soldered joints should be avoided, or, if specified, the contact area should be large or some form of mechanical joint, such as a rolled-seam lock, should be made prior to soldering. Butt joints should never be used, and designs where peeling action is possible should be avoided. Figure 39-8 shows some of the more common solder joint designs: lap, flanged butt, and interlock.

As with brazing, there is an optimal clearance for best performance. For typical solder joints, a clearance of 0.025 to 0.13 mm. (0.001 to 0.005 in.) provides for capillary flow of the solder, expulsion of the flux, and reasonable joint strength. In addition, the parts should be held firmly so that no movement can occur until the solder has cooled to well below the solidification temperature. Otherwise, the resulting joint may contain cracks and have very little strength.

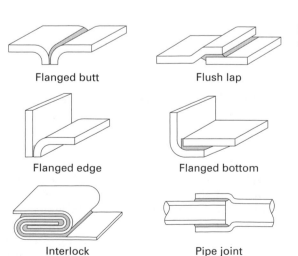

FIGURE 39-8   Some common designs for soldered joints.

Flanged butt

Flush lap

Flanged edge

Flanged bottom

Interlock

Pipe joint

TABLE 39-3.   Engineering Materials and Their Compatibility with Soldering

Material	Soldering Recommendation
Cast iron	Seldom used since graphite and silicon inhibit bonding
Carbon and low-alloy steels	Difficult for low-carbon materials; seldom used for high-carbon materials
Stainless steel	Common for 300 series; difficult for 400 series
Aluminum and magnesium	Seldom used; however, special solders are available
Copper and copper alloys	Recommended for copper, brass, and bronze
Nickel and nickel alloys	Commonly performed using high-tin solders
Titanium	Seldom used
Lead and zinc	Recommended, but must use low-melting-temperature solders
Thermoplastics, thermosets, and elastomers	Not recommended
Ceramics and glass	Not recommended
Dissimilar metals	Recommended, but with consideration for galvanic corrosion
Metals to nonmetals	Not recommended
Dissimilar nonmetals	Not recommended

METALS TO BE JOINED

Table 39-3 summarizes the compatibility of soldering processes with a variety of engineering materials. Copper, silver, gold, and tin-plated steels are all easily soldered. Aluminum has a strong, adherent oxide film that makes soldering difficult. Special fluxes and modified techniques may be required, but adequate joints are indeed possible, as shown by the large number of aluminum radiators currently in automotive use.

SOLDER METALS

Soldering alloys are generally combinations of low-melting-temperature metals, such as lead, tin, bismuth, indium, cadmium, silver, gold, and germanium. Because of their low cost, acceptable mechanical and physical properties, and well-established knowledge base, the most common solders are alloys of lead and tin with the addition of small amounts of antimony, usually less than 0.5%. The three most commonly used alloys contain 60, 50, and 40% tin and all melt below 240°C (465°F). Because tin is expensive, those alloys having higher proportions of tin are used only where their higher fluidity, higher strength, and lower melting temperature are desired. For wiped joints and for filling dents and seams, where the primary desire is appearance and little strength is required, solders containing only 10 to 20% tin are preferred.

Other soldering alloys may be specified for special purposes or where environmental or health concerns dictate the use of leadfree joints. Lead and lead compounds can be quite toxic. Since 1988, the use of lead-containing solders in drinking water lines has been prohibited in the United States, and this ban may well be extended to include other applications and industries. Japan and the European Union have already banned lead in electronic equipment. If substitute solders are to be acceptable, however, they should not only be harmless to the environment, but should also exhibit desirable characteristics in the areas of melting temperature, wettability, electrical and thermal conductivity, thermal-expansion coefficient, mechanical strength, ductility, creep resistance, thermal fatigue resistance, corrosion resistance, manufacturability, and cost. At present, none of the *leadfree solders* meet all of these requirements, and most are deficient in more than one area. In addition, compatible fluxes must be identified, and assembly methods may need to be modified.

Most of the alternative solders have been proposed from other eutectic alloy systems. Tin–antimony and tin–copper alloys are useful in electrical applications and have good strength and creep resistance but high melting points. Bismuth alloys have very low melting points and good fluidity, but suffer from poor wettability. Tin–indium alloys have been used to join metal to glass. They have very low melting points and good wettability, but they are expensive and can be somewhat brittle. Aluminum is often soldered with tin–zinc, cadmium–zinc, or aluminum–zinc alloys. Tin–silver and tin–gold offer possibilities when a somewhat higher melting point is desired (typically above 205°C, or 400°F) coupled with good mechanical strength and creep resistance, but both systems are limited by the high cost of their components. Lead–silver and cadmium–silver alloys can also be used for high-temperature service. The three-component tin–silver–copper system has received considerable attention for electronics applications.

TABLE 39-4. Some Common Solders and Their Properties

Composition (wt %)	Freezing Temperature (°C)			Applications
	Liquidus	Solidus	Range	
Lead–tin solders				
98 Pb–2 Sn	322	316	6	Side seams in three-piece can
90 Pb–10 Sn	302	268	34	Coating and joining metals
80 Pb–20 Sn	277	183	94	Filling and seaming auto bodies
70 Pb–30 Sn	255	183	72	Torch soldering
60 Pb–40 Sn	238	183	55	Wiping solder, radiator cores, heater units
50 Pb–50 Sn	216	183	33	General purpose
40 Pb–60 Sn	190	183	7	Electronic (low temperature)
Silver solders				
97.5 Pb–1 Sn–1.5 Ag	308	308	0	Higher-temperature service
36 Pb–62 Sn–2 Ag	189	179	10	Electrical
96 Sn–4 Ag	221	221	0	Electrical
Other alloys				
45 Pb–55 Bi	124	124	0	Low temperature
43 Sn–57 Bi	138	138	0	Low temperature
95 Sn–5 Sb	240	234	6	Electrical
50 Sn–50 In	125	117	8	Metal-to-glass
37.5 Pb–25 In–37.5 Sn	138	138	0	Low temperature
95.5 Sn–3.9 Ag–0.6 CO	217	217	0	Electrical

Like the filler metal used in brazing and braze welding, solders are available as wire and paste, as well as a variety of standard and special preshaped forms. Table 39-4 presents some of the more common solder alloys with their melting properties and typical applications.

SOLDERING FLUXES

As in brazing, soldering requires that the metal surfaces be clean and free of oxide so that the solder can wet the surfaces and be drawn into the joint to produce an effective bond. Soldering fluxes are used to remove surface oxides and prevent oxide formation during the soldering process, but it is essential that dirt, oil, and grease be removed before the flux is applied. This precleaning or surface preparation can be performed by a variety of chemical or mechanical means, including solvent or alkaline degreasers, acid immersion (pickling), grit blasting, sanding, wire brushing, and other mechanical abrasion techniques.

The soldering fluxes further clean and strip away surface oxides to expose the base metal. They are generally classified as *corrosive* or *noncorrosive*. The most common noncorrosive flux is *rosin* (the residue after distilling turpentine) dissolved in alcohol. Rosin fluxes are suitable for making joints to copper and brass, and to tin-, cadmium-, and silver-plated surfaces, provided that the surfaces have been adequately cleaned prior to soldering. Aniline phosphate is a more active noncorrosive flux, but it has limited use because it emits toxic gases when heated. A wide variety of corrosive fluxes can provide enhanced cleaning action but require complete removal after the soldering operation to prevent corrosion problems during service.

HEATING FOR SOLDERING

The primary requirements for soldering are a source of sufficient heat and a means of transferring it to the metals being joined. Any method of heating that is suitable for brazing can be used for soldering, but furnace and salt-bath heating are seldom used. Wave soldering is used extensively for joining wire ends (particularly in electronics work–see Chapter 34), while dip soldering is used for automobile radiators, and tinning. Induction heating is used when large numbers of identical parts are to be soldered. Most hand soldering, is still done with soldering irons or small oxy-fuel or air-fuel (acetylene, propane, butane or MAPP) torches. For low-melting-point solders, infrared heat sources can also be employed.

In the various processes, the joints can be preloaded with solder, or the filler metal can be supplied from a wire. The particular method of heating usually dictates which procedure is used.

FLUX REMOVAL

After soldering, the flux residues should be removed from the finished joints, either to prevent corrosion or for the sake of appearance. Flux removal is rarely difficult, provided that the type of solvent in the flux is known. Water-soluble fluxes can be removed with hot water and a brush. Alcohol will remove most rosin fluxes. However, when the flux contains some form of grease, as in most paste fluxes, a grease solvent must be used, followed by a hot water rinse. In the past, solvents containing chlorofluorocarbons (CFCs) were the cleaners of choice, but since they have been implicated in the depletion of atmospheric ozone, an alternative means of flux removal should be employed or the process converted to fluxless soldering.

FLUXLESS SOLDERING

Several *fluxless soldering* techniques have been developed using controlled atmospheres (such as hydrogen plasma), thermomechanical surface activation (such as plasma gas impingement), or protective coatings that prevent oxide formation and enhance wetting. Additional successes have been reported with both laser and ultrasonic soldering.

■ KEY WORDS

braze welding	dip brazing	flux	furnace brazing	lap joint	soldering
brazing	filler metal	fluxless brazing	jigs and fixtures	leadfree solder	torch brazing
butt joint	fluidity	fluxless soldering	joint clearance	silver solder	wettability
capillary action					

■ REVIEW QUESTIONS

1. What are some of the alternative processes in making a low-temperature joint?
2. What features are important in a reasonable definition of brazing?
3. What are some key differences between brazing and fusion welding?
4. Why is brazing an attractive process for joining dissimilar materials?
5. Why do brazed joints have an enhanced susceptibility to corrosion?
6. What is the most important factor contributing to the strength of a brazed joint, and therefore what is capillary action?
7. Why is it necessary to adjust the initial room-temperature clearance of a joint between two significantly different metals?
8. What are the two most common types of brazed joints and the attractive features of each?
9. What are some important considerations when selecting a brazing alloy?
10. What are some of the most commonly used brazing metals?
11. Is *silver solder* a solder or a braze filler metal?

12. What are the three primary functions of a brazing flux?
13. Why is it important to preclean brazing surfaces before applying the flux?
14. What is the purpose of brazing jigs and fixtures?
15. What is the primary attraction of furnace brazing operations?
16. Why might reducing atmospheres or a vacuum be employed during furnace brazing operations?
17. Why is dip brazing usually restricted to use with small parts?
18. What are some of the attractive features of induction brazing?
19. Why is flux removal a necessary part of many brazing operations?
20. How does braze welding differ from traditional brazing?
21. What is the primary difference between brazing and soldering?
22. Why is soldering unattractive if a high-strength joint is desired?
23. The most common solders are alloys of what two base metals?
24. What is driving the conversion to leadfree solders?
25. What are some of the difficulties encountered when attempting a conversion to leadfree solder?
26. What are some of the more common heat sources for producing a soldered joint?

■ PROBLEMS

1. A common problem with brazed or soldered joints is galvanic corrosion, since the joint usually involves dissimilar metals in direct metal-to-metal electrical contact.
   a. For each of the various solder or braze joints described below, determine which material will act as the corroding anode.
      i. Two pieces of low-carbon steel being brazed with a copper-base brazing alloy
      ii. A copper wire being soldered to a steel sheet using lead-tin solder

   iii. Pieces of tungsten carbide being brazed into recesses in a carbon-steel plate
   b. How do the various leadfree solders compare to the conventional lead–tin solders with regard to their potential for galvanic corrosion?
   c. If galvanic corrosion becomes a significant and chronic problem in a brazed assembly, what changes might you suggest that could possibly reduce or eliminate the problem?

CHAPTER 40

ADHESIVE BONDING AND MECHANICAL FASTENING

■ 40.1 ADHESIVE BONDING

The *ideal adhesive* bonds to any material, needs no surface preparation, cures rapidly, and maintains a high bond strength under all operating conditions. It also doesn't exist. However, tremendous advances have been made in the development of adhesives that are stronger, easier to use, less costly, and more reliable than many of the alternative methods of joining. From early applications, such as plywood, the use of structural adhesives (where the adhesive is a load-transmitting part of the product) has grown rapidly. Adhesives are everywhere—in construction, packaging, furniture, appliances, electronics, bookbinding, product assembly, and even medical and dental applications. They are used to bond metals, ceramics, glass, plastics, composite materials, woods, and even a variety of roofing materials. Even such quality- and durability-conscious fields as the automotive and aircraft industries make extensive use of adhesive bonding. Adhesives in the automotive industry have advanced from the attaching of interior and exterior trim to the joining of major components, such as door, hood, and trunk assemblies, and the installation of the nonmoving front and rear windows. Adhesive bonding has become the preferred means of assembly for polymeric body panels made from sheet-molding compounds and reaction-injection-molded (RIM) materials. Moreover, since adhesive bonding has the ability to bond such a wide variety of materials, its use has grown significantly with the ever-expanding applications of plastics and composites.

ADHESIVE MATERIALS AND THEIR PROPERTIES

In *adhesive bonding*, a nonmetallic material (the *adhesive*) is used to create a joint between two surfaces. The actual adhesives span a wide range of material types and forms, including *thermoplastic* resins, *thermosetting* resins, artificial *elastomers*, and even some ceramics. They can be applied as drops, beads, pellets, tapes, or coatings (films) and are available in the form of liquids, pastes, gels, and solids. *Curing* can be induced by the use of heat, radiation or light (photoinitiation), moisture, activators, catalysts, multiple-component reactions, or combinations thereof. Intended applications range from full load bearing (structural adhesives), to light-duty holding or fixturing, to basic sealing (the forming of liquid- or gas-tight joints). With such a wide range of possibilities, the selection of the best adhesive for the task at hand can often be quite challenging.

The *structural adhesives* are selected for their ability to effectively transmit load across the joint and include epoxies, cyanoacrylates, anaerobics, acrylics, urethanes, silicones, high-temperature adhesives, and hot melts. Both strength and rigidity may be important, and the bond must be able to be stressed to a high percentage of its maximum load for extended periods of time without failure.

1. *Epoxies*. The thermosetting epoxies are the oldest, most common, and most diverse of the adhesive systems and can be used to join most engineering materials, including metal, glass, and ceramic. They are strong, versatile adhesives that can be designed to offer high adhesion, good tensile and shear strength, toughness, high rigidity, creep resistance, easy curing with little shrinkage, and tolerance to elevated temperatures. Various epoxies can be used over a temperature range from -50 to $+250°C$ (-60 to $500°F$). After curing at room temperature, shear strengths can be as high as 35 to 70 MPa (5,000 to 10,000 psi).

   Single-component epoxies use heat as the curing agent. Most, however, are two-component blends involving a resin and a curing agent, plus possible additives such as accelerators, plasticizers, and fillers that serve to enhance cure rate, flexibility, peel-resistance, impact resistance, or other characteristics. Heat may again be required to drive or accelerate the cure.

   Low peel strength and flexibility limit epoxy adhesives, and the bond strength can be sensitive to moisture and surface contamination. Epoxies are often brittle at low temperatures, and the rate of curing is comparatively slow. Sufficient strength for structural applications is generally achieved in 8 to 12 hours, with full strength often requiring two to seven days.

2. *Cyanoacrylates*. These are liquid monomers that polymerize when spread into a thin film between two surfaces. Trace amounts of moisture on the surfaces promote curing at amazing speeds, often in as little as two seconds. Thus, the cyanoacrylates offer a one-component adhesive system that cures at room temperature with no external impetus. Commonly known as *superglues*, this family of adhesives is now available in the form of liquids, gels, toughened versions designed to overcome brittleness, and even nonfrosting varieties.

   The cyanoacrylates provide excellent tensile strength, fast curing, and good shelf life and adhere well to most commercial plastics. They are limited by their high cost, poor peel strength, and brittleness. Bond properties are poor at elevated temperatures and effective curing requires good component fit (no or very small gaps).

3. *Anaerobics*. These one-component, thermosetting, polyester acrylics remain liquid when exposed to air. When confined to small spaces and shut off from oxygen, as in a joint to be bonded, the polymer becomes unstable. In the presence of iron or copper, it polymerizes into a bonding-type resin, without the need for elevated temperature. Additives can reduce odor, flammability, and toxicity and can speed the curing operation. Slow-curing anaerobics require 6 to 24 hours to attain useful strength. With selected additives and heat, however, curing can be reduced to as little as five minutes.

   The anaerobics are extremely versatile and can bond almost anything, including oily surfaces. The joints resist vibrations and offer good sealing to moisture and other environmental influences. Unfortunately, they are somewhat brittle and are limited to service temperatures below $150°C$ ($300°F$).

4. *Acrylics*. The acrylic-based adhesives offer good strength, toughness, and versatility, and they are able to bond a variety of materials, including plastics, metals, ceramics, and composites, even oily or dirty surfaces. Most involve application systems where a catalyst primer (curing agent) is applied to one of the surfaces to be joined and the adhesive is applied to the other. The pretreated parts can be stored separately for weeks without damage. Upon assembly, the components react to produce a strong bond at room temperature. Heat can often accelerate the curing, and at least one variety cures with ultraviolet light. In comparison to other varieties of adhesives, the acrylics offer strengths comparable to the epoxies, good resistance to water and humidity, and the added advantages of room temperature curing and a no-mix application system. Major limitations include poor strength at high temperatures, flammability, and an unpleasant odor when uncured.

5. *Urethanes*. Urethane adhesives are a large and diverse family of polymers that are generally targeted for applications that involve temperatures below $65°C$ ($150°F$) and components that may experience great elongation. Both one-part thermoplastic and two-part thermosetting systems are available. In general, they cure quickly to handling strength but are slow to reach the full-cure condition. Two minutes to handling with 24 hours to complete cure is common at room temperature.

Compared to other structural adhesives, the urethanes offer low-temperature adhesion coupled with flexibility and toughness. They are somewhat sensitive to moisture, degrade in many chemical environments, and can involve toxic components or curing products.

6. *Silicones.* The silicone thermosets cure from the moisture in the air or adsorbed moisture from the surfaces being joined. They form low-strength structural joints and are usually selected when considerable expansion and contraction are expected in the joint, flexibility is required (as in sheet metal parts), or good gasket or sealing properties are necessary. Metals, glass, paper, plastics and rubbers can all be joined. The adhesives are relatively expensive, and curing is slow, but the bonds that are produced can resist moisture, hot water, oxidation, and weathering, and they retain their flexibility at low temperature.

7. *High-temperature adhesives.* When strength must be retained at temperatures in excess of 300°C (500°F), high-temperature structural adhesives should be specified. These include epoxy phenolics, modified silicones or phenolics, polyamides, and some ceramics. High cost and long cure times are the major limitations for these adhesives, which see primary application in the aerospace industry.

8. *Hot melts.* Hot-melt adhesives can be used to bond dissimilar substrates, such as plastics, rubber, metals, ceramics, glass, wood, and fibrous materials like paper, fabric, and leather. They can produce permanent or temporary bonds, seal gaps, and plug holes. While generally not considered to be true structural adhesives, the hot melts are being used increasingly to transmit loads, especially in composite material assemblies. The joints can withstand exposure to vibration, shock, humidity, and numerous chemicals and offer the added features of sound deadening and vibration damping.

Most hot-melt adhesives are thermoplastic resins that are solid at room temperature, but melt abruptly when heated into the range of 100 to 150°C (200 to 300°F). They are usually applied as heated liquids (between 160 and 180°C) and form a bond as the molten adhesive cools. Another method of application is to position the adhesive in the joint prior to operations such as the paint bake process in automobile manufacture. During the baking, the adhesive melts, flows into seams and crevices, and seals against the entry of corrosive moisture. These adhesives contain no solvents and do not need time to cure or dry. They provide reasonable strength within minutes, but do soften and creep when subsequently exposed to elevated temperatures and can become brittle when cold.

Figure 40-1 provides the distribution of adhesive and sealant products for a recent year, and Figure 40-2 classifies adhesives by end-use markets. Table 40-1 presents a listing of some popular structural adhesives along with their service and curing temperatures and expected strengths. Table 40-2 presents the advantages and disadvantages of various curing processes.

FIGURE 40-1 Distribution among the common types of adhesives and sealants. *(Courtesy of Impact Marketing Consultants, Inc., Manchester Center, VT.)*

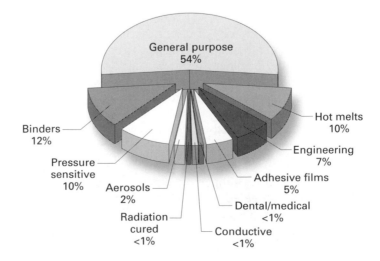

Leading adhesives and sealants products 1998

General purpose 54%

Hot melts 10%

Engineering 7%

Adhesive films 5%

Dental/medical <1%

Conductive <1%

Radiation cured <1%

Aerosols 2%

Pressure sensitive 10%

Binders 12%

Leading adhesives and sealants end-use markets 1998

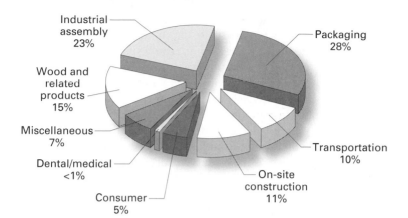

FIGURE 40-2 Distribution of adhesives and sealants by end-use areas. *(Courtesy of Impact Marketing Consultants, Inc., Manchester Center, VT.)*

TABLE 40-1.	Some Common Structural Adhesives, Their Cure Temperatures, Maximum Service Temperatures, and Strengths Under Various Types of Loadings

Adhesive Type	Cure Temperature (°F)	Service Temperature (°F)	Lap Shear Strength (psi at °F)[a]	Peel Strength at Room Temperature (lb/in.)
Butyral–phenolic	275 to 350	−60 to 175	1000 at 175 2500 at RT	10
Epoxy				
Room-temperature cure	60 to 90	−60 to 180	1500 at 180 2500 at RT	4
Elevated-temperature cure	200 to 350	−60 to 350	1500 at 350 2500 at RT	5
Epoxy–nylon	250 to 350	−420 to 180	2000 at 180 6000 at RT	70
Epoxy–phenolic	250 to 350	−420 to 500	1000 at 175 2500 at RT	10
Neoprene–phenolic	275 to 350	−60 to 180	1000 at 180 2000 at RT	15
Nitrile–phenolic	275 to 350	−60 to 250	2000 at 250 4000 at RT	60
Polyimide	550 to 650	−420 to 1000	1000 at 1000 2500 at RT	3
Urethane	75 to 250	−420 to 175	1000 at 175 2500 at RT	50

[a]RT, room temperature.

TABLE 40-2.	Advantages and Disadvantages of Various Structural Adhesive Curing Processes

Curing Process	Advantages	Disadvantages
Mixing reactive components	Good shelf life, unlimited depth of cure, accelerated with heat	High processing costs, mix ratio critical to performance
Anaerobic cure	Single-component adhesive, good shelf life	Poor depth of cure, require primer on many surfaces, sensitive to surface contaminants
Heat cure	Unlimited depth of cure, heat can aid adhesion	Expenses for oven energy cost, heat can adversely affect some substrates
Moisture cure	Room-temperature process, one component, no curing equipment required	Long cure cycles (12–72 hr), minimum % humidity required, limited depth of cure
Light cure	Rapid cure, cure on demand	Expenses for UV light source, limited depth of cure, most allow light to reach bond
Surface-initiated cure	Rapid cure	Poor depth of cure

NONSTRUCTURAL AND SPECIAL ADHESIVES

There are a number of other types of adhesives whose limited load-bearing capabilities place them in a nonstructural classification. Nevertheless, they still play roles in manufacturing through a variety of uses, such as labeling and packaging. The hot-melt adhesives are often placed in this category but can be used for applications in both classifications. *Evaporative adhesives* use an organic solvent or water base, coupled with vinyls, acrylics, phenolics, polyurethanes, or various types of rubbers. Some common evaporative adhesives are rubber cements and floor waxes. *Pressure-sensitive adhesives* are usually based on various rubbers, compounded with additives to bond at room temperature with a brief application of pressure. No cure is involved, and the tacky adhesive-coated surfaces require no activation by water, solvents, or heat. Peel-and-stick labels, cellophane tape, and Post-it notes are examples of this group of adhesives. *Delayed-tack adhesives* are similar to the pressure-sensitive systems, but are nontacky until activated by exposure to heat. They then remain tacky for several minutes to a few days to permit use or assembly.

While most adhesives are electrical and thermal insulators, *conductive adhesives* can be produced by incorporating selected fillers, such as silver, copper, or aluminum, in the form of flakes or powder. Certain ceramic oxide fillers can be used to provide thermal conductivity coupled with electrical insulation.

Still another group of commercial adhesives are those designed to cure by exposure to radiation, such as visible, infrared, or ultraviolet light; microwaves; or electron beams. These *radiation-curing adhesives* offer rapid conversion from liquid to solid at room temperature and a curing mechanism that occurs throughout, rather than progressing from exposed surfaces (as with the competing low-temperature air or moisture cures). Current applications include a wide variety of dental amalgams that can fill cavities or seal surfaces while matching the color of the remaining tooth. In the manufacturing realm, heat-sensitive materials can be effectively bonded, and the rapid cure time significantly reduces the need for fixturing.

DESIGN CONSIDERATIONS

The structural adhesives have been used for a wide range of applications in fields as diverse as automotive, aerospace, appliances, biomedical, electronics, construction, machinery, and sporting goods. Proper selection and use, however, require consideration of a number of factors, including

1. What materials are being joined? What are their porosities, hardnesses, and surface finishes? Will the thermal expansions or contractions be different?

2. How will the joined assembly be used? What type of joint is proposed, what will be the bond area, and what will be the applied stresses? How much strength is required? Will there be mechanical vibration, acoustical vibration, or impacts?

3. What temperatures might be required to affect the cure, and what temperatures might be encountered during service? Consideration should be given to highest temperature, lowest temperature, rates of temperature change, frequency of change, duration of exposure to extremes, the properties required at the various conditions, and differential expansions or contractions.

4. Will there be subsequent exposure to solvents, water or humidity, fuels or oils, light, ultraviolet radiation, acid solutions, or general weathering?

5. What is the desired level of flexibility or stiffness? How much toughness is required?

6. Over what length of time is stability desired? What portion of this time will be under load?

7. Is appearance important?

8. How will the adhesive be applied? What equipment, labor, and skill are required?

9. What will it cost?

Because there is such a large difference in bonding area between the two types, adhesive-bonded joints are often classified as either continuous surface or core-to-face. In

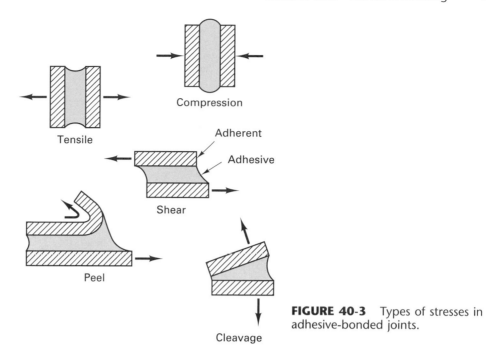

FIGURE 40-3   Types of stresses in adhesive-bonded joints.

continuous-surface bonds, both of the adhering surfaces are relatively large and are of the same size and shape. *Core-to-face bonds* have one *adherend* area that is very small compared to the other, as in bonding the edges of lightweight honeycomb core structures to face sheets.

A major design consideration for both types is the nature of the stresses that the joint will experience. As shown in Figure 40-3, applied stresses can subject the joint to tension (or compression), *shear*, *cleavage*, and *peel*. Most of the structural adhesives are significantly weaker in peel and cleavage than they are in shear or tension. Therefore, adhesively bonded joints should be designed so as much of the stress as possible is in shear or tension where all of the bonded area shares equally in bearing the load. The shear strengths of common structural adhesives range from 14 to 40 MPa (2000 to 6000 psi) at room temperature, while the tensile strengths are only 4 to 8 MPa (600 to 1200 psi). Therefore, the best adhesive-bonded joints will be those that are designed to utilize the superior shear strengths. Creep, vibration and associated fatigue, thermal shock, and mechanical shocks can all induce additional stresses. When vibration or shock loading is expected, the elastomeric adhesives can provide valuable damping and become quite attractive.

Figure 40-4 shows some of the commonly used joint designs and indicates their relative effectiveness. The butt joint is unsatisfactory because it offers only a minimum of bond surface area and little resistance to cleavage. Useful strength is generally obtained by increasing

FIGURE 40-4   Possible designs of adhesive-bonded joints and a rating of their performance in service.

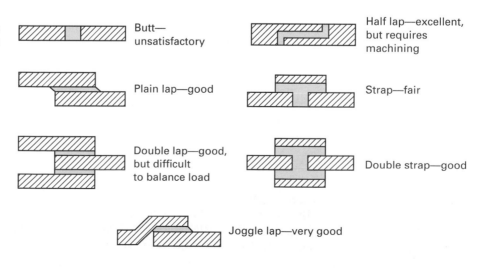

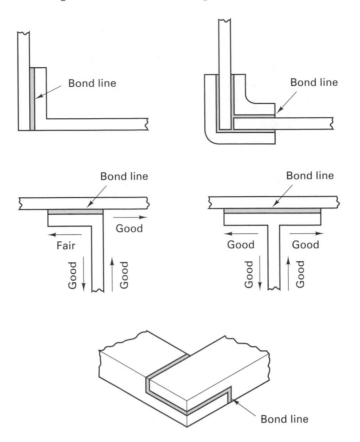

FIGURE 40-5 Adhesively bonded corner and angle joint designs.

the bond area through the addition of straps or the conversion to some form of lap design. Figure 40-5 shows some recommended designs for corner and angle joints. Adhesives can also be used in combination with welding, brazing, or mechanical fasteners. Spot welds or rivets can provide additional strength or simply prevent movement of the components when the adhesive is not fully cured or is softened by exposure to elevated temperature.

To obtain satisfactory and consistent quality in adhesive-bonded joints, it is essential that the surfaces be properly prepared. Procedures vary widely, but frequently require cleaning of the surfaces to be joined. Contaminants and grease usually must be removed to assure adequate wetting of the surfaces by the adhesive. Chemical etching, steam cleaning, or abrasive techniques may be employed to further enhance wetting and bonding. While thick or loose oxide films are detrimental to adhesive bonding, a thin porous oxide can often provide surface roughness and enhance adhesion.

The destructive testing of adhesive joints, or the examination of joint failures, can reveal much about the effectiveness of an adhesive system. If failure occurs by separation at the adhesive-substrate interface, as shown in Figure 40-6(a), it is indicative of a bonding or adhesion problem. If the failure lies entirely within the adhesive as in Figure 40-6(b), then the bonding with the substrate is adequate, but the strength of the adhesive may need to be enhanced. Finally, if failure occurs within the substrate materials, as in Figure 40-6(c), the joint is good, and failure is unrelated to the adhesive bonding operation.

ADVANTAGES AND LIMITATIONS

Adhesive bonding has many obvious advantages. Almost any material or combination of materials can be joined in a wide variety of sizes, shapes, and thicknesses. For most adhesives, the curing temperatures are low, seldom exceeding 180°C (350°F). A substantial number cure at room temperature or slightly above and can provide adequate strength for many applications. As a result, very thin or delicate materials, such as foils, can be joined to each other or to heavier sections. Heat-sensitive materials can be joined without damage, and heat-affected zones are not present in the product. When joining dissimilar materials, the adhesive provides a bond that can tolerate the stresses of differential expansion and contraction.

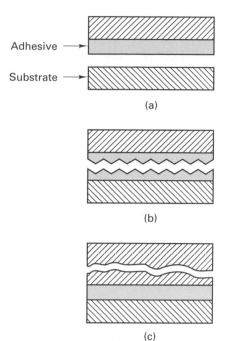

Adhesive ⟶

Substrate ⟶

(a)

(b)

FIGURE 40-6 Failure modes of adhesive joints. (a) Adhesive failure, (b) cohesive failure within the adhesive, and (c) cohesive failure within the substrate.

(c)

Because adhesives bond the entire joint area, good load distribution and fatigue resistance are obtained, and stress concentrations (such as those observed with screws, rivets and spot welds) are avoided. Similarly, because of the large amount of contact area that can usually be obtained, the total joint strength compares favorably with that produced by alternative methods of joining or attachment. (Shear strengths of industrial adhesives can exceed 20 MPa, or 3000 psi.) Additives can be incorporated to enhance strength, increase flexibility, or provide resistance to various environments.

Adhesives are generally inexpensive and frequently weigh less than the fasteners needed to produce a comparable-strength joint. In addition, an adhesive can also provide thermal and electrical insulation; act as a damper to noise, shock, and vibration; stop a propagating crack; and provide protection against galvanic corrosion when dissimilar metals are joined. By providing both a joint and a seal against moisture, gases, and fluids, adhesive-bonded assemblies often offer improved corrosion resistance throughout their useful lifetime. When used to bond polymers or polymer–matrix composites, the adhesive can be selected from the same family of materials to assure good compatibility.

From a manufacturing viewpoint, the formation of a joint does not require the capillary induced flow of material, as in brazing and soldering. The bonding adhesive is applied directly to the surfaces, and the joint is then formed by the application of heat and/or pressure. Most adhesives can be applied quickly, and useful strengths are achieved in a short period of time. Some curing mechanisms take as little as 2 to 3 seconds! Surface preparation may be reduced since bonding can occur with an oxide film in place, and rough surfaces are actually beneficial because of the increased contact area. Tolerances are less critical since the adhesives are more forgiving than alternative methods of bonding. Smooth contours are obtainable, and no holes have to be made, as with rivets or bolts. These factors contribute to reduced manufacturing costs, which can be further reduced through the elimination of the mechanical fasteners and the absence of highly skilled labor. Bonding can often be achieved at locations that would prevent the access of many types of welding apparatus. Robotic dispensing systems can often be utilized.

The major disadvantages of adhesive bonding are

1. There is no universal adhesive. Selection of the proper adhesive is often complicated by the wide variety of available options.

2. Most industrial adhesives are not stable above 180°C (350°F). Oxidation reactions are accelerated, thermoplastics can soften and melt, and thermosets decompose. While some adhesives can be used up to 260°C (500°F), elevated temperatures are usually a cause for concern.

3. Some adhesives shrink significantly during curing.

4. High-strength adhesives are often brittle (poor impact properties). Resilient ones often creep. Some become brittle when exposed to low temperatures.

5. Long-term durability and life expectancy are difficult to predict.

6. Surface preparation and cleanliness, adhesive preparation, and curing can be critical if good and consistent results are to be obtained. Some adhesives are quite sensitive to the presence of grease, oil, or moisture on the surfaces to be joined. Surface roughness and wetting characteristics must be controlled.

7. Assembly times may be greater than for alternative methods depending upon the curing mechanism. Elevated temperatures may be required as well as specialized fixtures.

8. It is difficult to determine the quality of an adhesive-bonded joint by traditional nondestructive techniques, although some inspection methods have been developed that give good results for certain types of joints.

9. Some adhesives contain objectionable chemicals or solvents or produce them upon curing.

10. Many structural adhesives deteriorate under certain operating conditions. Environments that may be particularly hostile include: ultraviolet light, ozone, acid rain (low pH), moisture, and salt. Thus, long-term durability and reliability may be questioned.

11. Adhesively bonded joints cannot be readily disassembled.

Nevertheless, the extensive and successful use of adhesive bonding provides ample evidence that these limitations can be overcome if adequate quality-control procedures are adopted and followed.

It should be emphasized that while the unit area strengths of adhesives are relatively low, this is not a major disadvantage or limitation, since sufficient area can generally be provided in a properly designed joint.

■ 40.2 MECHANICAL FASTENING

INTRODUCTION AND METHODS

Mechanical fastening is a classification that includes a wide variety of techniques and fasteners designed to suit the individual requirements of a multitude of joints and assemblies. Included within this family are: integral fasteners, threaded discrete fasteners (which includes screws, bolts, studs, and inserts), nonthreaded discrete fasteners (such as rivets, pins, retaining rings, staples, and wire stitches), special-purpose fasteners (such as the quick-release and tamper-resistant types), shrink and expansion fits, press fits, and others. Selection of the specific fastener or fastening method depends primarily on the materials to be joined, the function of the joint, strength and reliability requirements, weight limitations, dimensions of the components, and environmental conditions. Other considerations include cost, installation equipment and accessibility, appearance, and the need or desire for disassembly. When disassembly and reassembly are desired (as for parts replacement, maintenance, or repair), threaded fasteners, snap-fits, or other fasteners that can be removed quickly and easily should be specified. Such fasteners should not have a tendency to loosen after installation, however. If disassembly is not necessary, permanent fasteners are often preferred, or threaded fasteners can be coupled with anaerobic adhesives that cure to full strength at room temperature.

A mechanical joint acquires its strength through either mechanical interlocking or interference of the surfaces brought about by the clamping force. No fusion or adhesion of the surfaces is required. The fasteners and fastening processes should be selected to provide the required strength and properties in view of the nature and magnitude of subsequent loading. Consider the possibility of vibrations and/or cyclic stresses that might promote loosening over a period of time. Weight considerations may be significant in certain applications, such as aerospace and automotive. The need to withstand corrosive environments, operate at high or low temperatures, or face other severe conditions may provide additional constraints to the selection of fasteners or fastening processes.

The effectiveness of a mechanical fastener often depends upon (1) the material of the fastener, (2) the fastener design (including the load-bearing area of the head), (3) hole preparation, and (4) the installation procedure. The general desire is to achieve a uniform load transfer, a minimum of stress concentration, and uniformity of installation torque or interference fit. Various means are available for achieving these goals, as described in the following paragraphs.

Integral fasteners are formed areas of a component that interfere or interlock with other components of the assembly and are most commonly found in sheet metal products. Examples include lanced or shear-formed tabs, extruded hole flanges, embossed protrusions, edge seams, and crimps. Figure 40-7 shows some of these techniques, each of which involve some form of metal shearing, and/or forming. The common beverage can include several of these joints—an edge seam to join the top of the can to the body (as in Figure 40-7e) and an embossed protrusion that is subsequently flattened to attach the opener-tab (Figure 40-7d).

Discrete fasteners, like those illustrated in Figure 40-8, are separate pieces whose function is to join the primary components. These include bolts and nuts (with accessory washers, etc.), screws, nails, rivets, quick-release fasteners, staples, and wire stitches. Over 150 billion discrete fasteners are consumed annually in the United States, with a variety so immense that the major challenge is usually selection of an appropriate fastener for the task at hand (and, if possible, an optimum fastener). This task is further complicated by inconsistent nomenclature and identification schemes. Some fasteners are identified by their specific product or application, whereas others are classed by the material from which they are made, their size, their shape, or primary operational features. (Discussion of the primary terms used in identifying discrete fasteners can be found in ANSI Standard B18.12.) A more positive feature is the commercial availability of a wide range of standard and special

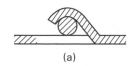

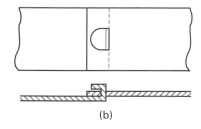

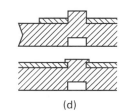

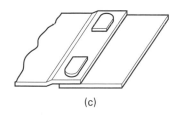

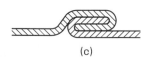

FIGURE 40-7 Several types of integral fasteners: (a) lanced tab to fasten wires or cables to sheet or plate; (b) and (c) assembly through folded-tabs and slots for different types of loading; (d) use of a flattened embossed protrusion; and (e) single-lock seam.

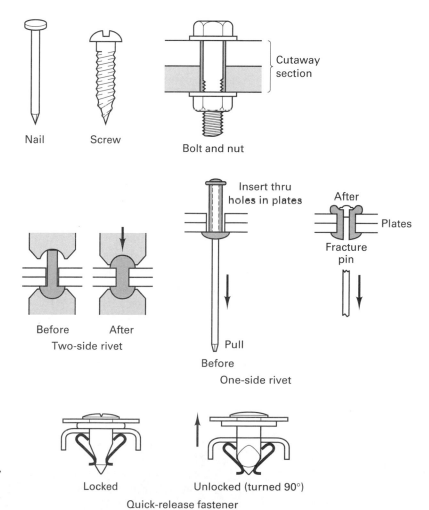

FIGURE 40-8 Various types of discrete fasteners, including a nail, screw, nut-and-bolt, two-side access supported rivet, one-side access blind rivet, and quick-release fastener.

types, sizes, materials, strengths, and finishes such that an appropriate fastener can generally be found for most all joining needs. Discrete fasteners are easy to install, remove, and replace. In addition, most standard varieties are interchangeable. Various finishes and coatings can be applied to withstand a multitude of service conditions.

Shrink and expansion fits form another major class of mechanical joining. Here, a dimensional change is introduced to one or both of the components by heating or cooling (heating one part only, heating one and cooling the other, or cooling one). Assembly is then performed, and a strong interference fit is established when the temperatures return to uniformity. Joint strength can be exceptionally high. By prestressing, a weak, low-cost material can often replace a more costly, stronger one. Similarly, a corrosion-resistant cladding or lining can be applied to a less costly bulk material.

Press fits are similar to shrink and expansion fits, but the results are obtained through mechanical force instead of differential temperatures.

REASONS FOR SELECTION

Mechanical fastening offers a number of attractive features, among them being:

1. Ease of disassembly and reassembly. The threaded fasteners are noteworthy for this feature, and semipermanent fasteners (such as rivets) can be drilled out for a major disassembly.

2. The ability to join similar or different materials in a wide variety of sizes, shapes, and joint designs. Some joint designs, such as hinges and slides, permit limited motion between the components.

3. Low manufacturing cost. The fasteners are usually small formed components which cost little compared to the components being joined. They are readily available in a variety of mass-produced sizes.

4. Installation does not adversely affect the base materials, as is often the case with techniques involving the application of heat and/or pressure.

5. Little or no surface preparation or cleaning is required.

MANUFACTURING CONCERNS

Many mechanical fasteners require that the components contain aligned holes. Castings, forgings, extrusions, and powder metallurgy components can be designed to include integral holes. In addition, holes can be produced by such techniques as punching, drilling, and electrical, chemical, or laser-beam machining. Each technique produces holes with characteristic surface finish, dimensional features, and properties. Holemaking and the proper positioning and alignment of the holes are major considerations in mechanical fastening. See chapter 24 on drilling.

Bolts, coupled with nuts, require access to both sides of an assembly during joining. By inserting the bolt into a threaded (tapped) hole, the nut can be eliminated, and only one-side access is required. Screws also offer one-side joining. If a bolt or screw is sufficiently hard, the fastener can cut its own threads, thereby eliminating the need for a threaded receptacle. Stapling is a fast way of joining thin materials and does not require prior holemaking. Rivets offer good strength, but produce permanent or semipermanent joints. Snap-fits utilize the elasticity of one of the components, and the necessary elastic deformation must be possible without fracture.

DESIGN AND SELECTION

The design and selection of a fastening method requires numerous considerations, including the possible means of joint failure. When a product is assembled with fastened joints, the fasteners are extremely vulnerable sites. Mechanical joints generally fail because of oversight or lack of control in one of four areas: (1) the design of the fastener itself and the manufacturing techniques used to make it, (2) the material from which the fastener is made, (3) joint design, or (4) the means and details of installation. Fasteners may have insufficient strength or corrosion resistance or may be subject to stress corrosion cracking or hydrogen embrittlement. Metal fasteners provide electrical conductivity between the components, and an inappropriate choice of fastener or component material can cause severe galvanic corrosion. Nonmetallic fasteners (such as threaded nylon, fiberglass, or graphite) can be used for low-strength applications where corrosion is a concern, but creep under load is a concern for these materials. Since mechanical fasteners only join at discrete points, gases or liquids can easily penetrate the joint area and further aggravate conditions.

Many failures are the result of poor joint preparation or improper fastener installation. A high percentage of the cracks in aircraft structures originate at fastener holes, and fatigue of fasteners is the largest single cause of fastener failure. Installation frequently imparts too much or too little preload (too tight or too loose). The joint surfaces may not be flat or parallel, and the area under the fastener head may be insufficient to bear the load. Vibrational loosening enhances fastener fatigue. The details of joint design should further consider stress distribution, since much of the load will be concentrated on the fasteners (in contrast to the previously discussed adhesive joints that distribute the load uniformly over the entire joint area).

Nearly all fastener failures can be avoided by proper design and fastener selection. Consideration should be given to the operating environment, required strength, and magnitude and frequency of vibration. Fastener design should incorporate a shank-to-head fillet whenever possible. Rolled threads can be specified for their superior strength and fracture resistance. Corrosion-resistant coatings can be employed for enhanced performance. Joint design should seek to avoid such features as offset or oversized holes. Proper installation and tightening are critical to good performance. Standard sizes, shapes, and grades should be used whenever possible with as little variety as is absolutely necessary.

■ KEY WORDS

adherend	cleavage	curing	hot-melt	peel	structural
adhesive	continuous-surface	discrete fasteners	adhesives	press fit	adhesive
adhesive bonding	bonds	elastomer	integral fasteners	shear	thermoplastic
anaerobic	core-to-face bonds	epoxies	mechanical fastening	shrink fit	thermosetting

■ REVIEW QUESTIONS

1. What would be some of the characteristics of an ideal adhesive?
2. What is a structural adhesive?
3. What are some of the newer applications that have helped promote increased use of adhesive bonding?
4. What are some of the types of materials that have been used as industrial adhesives?
5. What are some of the ways in which adhesives can be cured?
6. Characterize the temperature range over which epoxies might be used, typical values of shear strength, and commonly observed curing times.
7. What promotes the curing of cyanoacrylates? Of anaerobics?
8. What features or characteristics might favor the selection of a silicone adhesive?
9. Describe some types of nonstructural or special adhesives.
10. How can polymeric adhesives be made electrically or thermally conductive?
11. What are some of the temperature considerations that should be made when selecting an adhesive?
12. What are some of the environmental conditions that might reduce the performance or lifetime of a structural adhesive?
13. Why is it desirable for adhesive joints to be designed so the adhesive is loaded in shear or compression?
14. Why are butt joints unattractive for adhesive bonding?
15. What are some common techniques by which surfaces are prepared for adhesive bonding?
16. Why are the structural adhesives an attractive means of joining dissimilar metals or materials? Different sizes or thicknesses?
17. What are some of the other attractive properties of structural adhesives?
18. In what ways might a structural adhesive offer manufacturing ease or reduced manufacturing cost?
19. Why are adhesive joints unattractive for applications that involve exposure to elevated temperature? Are there any low-temperature concerns?
20. In view of the relatively low strengths of the structural adhesives, how can adhesively bonded joints attain strengths comparable to other methods of joining?
21. What factors would influence the selection of a specific type of mechanical fastener or fastening method?
22. What types of fasteners are attractive if the application requires the ability to disassemble and reassemble the product?
23. What is an integral fastener? Provide an example.
24. How are press fits similar to shrink or expansion fits? How do they differ?
25. What are some of the common causes for failure of mechanically fastened joints?
26. From a manufacturing viewpoint, why is it desirable to use standard fasteners and minimize the variety of fasteners within a given product?

■ PROBLEMS

1. Some automakers are using adhesives and sealants that cure under the same conditions used for the paint-bake operation. Determine the conditions used for paint-bake, and identify some adhesives and sealants that could be used. What are some of the pros and cons of such an integration?

2. A contractor has installed aluminum siding on a house with steel nails. Use the galvanic series to evaluate the corrosion properties of this assembly. (*Note:* The aluminum is exposed to air, so it should be considered to be in its passive condition.)

What do you expect will be the outcome of this fastener selection? Can you recommend a better alternative?

3. Mechanical fasteners are an attractive means of joining composite materials because they avoid exposing the composite to heat and/or high pressure. Assume that the composite is a polymer-based fiber-reinforced material with either uniaxial or woven fibers. For this particular system, what are some possible fastener-related problems? Consider joint preparation, assembly, and possible service failure.

www.wiley.com/college/degarmo

*C*hapter 40 CASE STUDY

Golf Club Heads with Insert

You are employed by a small manufacturer of sporting goods equipment, and your design team has recently proposed a new line of high-performance golf clubs. The club head is to be a "standard" AISI 431 martensitic stainless steel investment casting with a metal insert incorporated into the striking face, as shown in Figure CS-40. The insert will be produced by powder metallurgy and will consist of a copper-based alloy laced with particles of tungsten carbide. After a mild acid etching of the copper matrix, the carbide particles will protrude sufficiently to better grip the surface of the ball, emparting an enhanced amount of backspin to better control the "bite" of the ball upon landing. Since its purpose is to modify the striking face, the insert is rather thin, about 1.5 to 3 mm ($\frac{1}{16}$ to $\frac{1}{8}$ in.) in thickness. It must be incorporated into the club face in a manner that does not dampen the impact or compromise the "feel" of the club.

1. You must devise a means of incorporating the insert into the face of the club. What are some possible means of joining or bonding dissimilar metals? What

are the advantages and limitations of each of your alternatives?

2. An additional joint occurs where the club head is attached to the shaft. If a graphite fiber reinforced epoxy is being considered for the shaft, how might it be attached to the stainless steel head? If the shafts were metal, what other methods might be possible?

3. For production simplicity, it might be preferable to use the same joining procedure at all locations. In view of your answers to Questions 1 and 2 above, does this appear to be a possibility for this product? If the composite shaft were selected, which joining method would you recommend?

4. Are there other ways that might be considered to produce a raised-carbide surface on a stainless steel golf club face? What might they be, and what do you see as advantages and limitations? Consider both manufacturing and performance. Would you expect them to be cheaper or more expensive than the proposed insert?

FIGURE CS-40

CHAPTER 41

MANUFACTURING AND PRODUCTION SYSTEMS

■ 41.1 INTRODUCTION

In a factory, *manufacturing processes* are assembled together to form a *manufacturing system* (MS) to produce a desired set of goods. The manufacturing system takes specific inputs and materials, adds value through processes, and transforms the inputs into products for the customer. It is important to distinguish between the manufacturing system and the production system, which is also known as the enterprise system or the whole company. The production system includes the manufacturing system.

As shown in Figure 41-1, the production system services the manufacturing system, using all the other functional areas of the plant for information, design, analysis, and control. These subsystems are connected to each other to produce goods or services, or both.

FIGURE 41-1 Schematic definition of a production system.

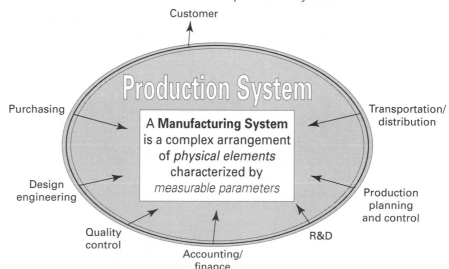

Customer

Production System

Purchasing

Transportation/ distribution

A **Manufacturing System** is a complex arrangement of *physical elements* characterized by *measurable parameters*

Design engineering

Quality control

Accounting/ finance

R&D

Production planning and control

■ 41.2 PRODUCTION SYSTEM (THE ENTERPRISE)

A production system includes all aspects of the business, including design engineering, manufacturing engineering, sales, advertising, production and inventory control (scheduling and distribution), and, most important, the manufacturing system.

■ 41.3 MANUFACTURING SYSTEMS

A sequence of processes and people that actually produce the desired product(s) is called the *manufacturing system.* In Figure 41-2, the manufacturing system is defined as the *complex arrangement of the manufacturing elements* (physical elements) *characterized (and controlled) by measurable parameters* (Black, 1991). The relationship among the elements determines how well the system can run or be controlled. The control of a system refers to the entire manufacturing system (which means the control of the operators in a harmonious way relative to the system's objectives), not merely the individual processes or equipment. The entire manufacturing system must be under daily control to enable the management of material movement, people and processes (scheduling), inventory levels, product quality, production rates, throughput, and, of course, cost.

As shown in Figure 41-2, inputs to the manufacturing system include materials, information, and energy. The system is a complex set of elements that includes machines (or machine tools), people, materials-handling equipment, and tooling. Workers are the internal customers. They process materials within the system, which gain value as the material progresses from process to machine. Manufacturing system outputs may be finished or semifinished goods. Semifinished goods serve as inputs to some other process at other locations. Manufacturing systems are dynamic, meaning that they must be designed to adapt constantly to change. Many of the inputs cannot be fully controlled by management, and the effect of disturbances must be counteracted by manipulating the controllable inputs or the system itself. Controlling the input material availability and/or predicting demand fluctuations may be difficult. A national economic decline or recession can cause shifts in the business environment that can seriously change any of these inputs. In manufacturing

FIGURE 41-2 Definition of a manufacturing system with its inputs and outputs *(From Design of the Factory with a Future, 1991, McGrow-Hill by J.T. Black).*

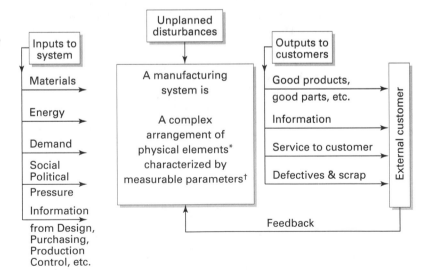

* Physical elements:
- Machine tools for processing
- Tooling (fixtures, dies, cutting tools)
- Material handling equipment (which includes all transportation and storage)
- People (internal customers) operators, workers, associates

† Measurable system parameters:
- Throughput time (TPT)
- Production rate (PR)
- Work-in-process inventory
- % on-time delivery
- % defective
- Daily/weekly/monthly volume
- Cycle time or takt time
- Total cost or unit cost

systems, not all inputs are fully controllable. To understand how manufacturing systems work and be able to design manufacturing systems, computer modeling (simulation) and analysis are used. However, modeling and analysis are difficult because

1. In the absence of a system design, the manufacturing systems can be very complex, difficult to define, and have conflicting goals.

2. The data or information may be difficult to secure, inaccurate, conflicting, missing, or even too abundant to digest and analyze.

3. Relationships may be awkward to express in analytical terms, and interactions may be nonlinear; thus many analytical tools cannot be applied with accuracy. System size may inhibit analysis.

4. Systems are always dynamic and change during analysis. The environment can change the system, and vice versa.

5. All systems analyses are subject to errors of omission (missing information) and commission (extra information). Some of these are related to breakdowns or delays in feedback elements.

Because of these difficulties, digital simulation has become an important technique for manufacturing systems modeling and analysis as well as for manufacturing system design.

CONTROL OF THE MANUFACTURING SYSTEM

In general, a manufacturing system should be an integrated whole, composed of integrated subsystems, each of which interacts with the entire system. The critical control functions are production rate and mix control, inventory control, quality control, and machine tool control (reliability). While the system may have a number of objectives or goals, the users of the system may seek to optimize the whole. Optimizing bits and pieces does not optimize the entire system. System control functions require information gathering, communication capabilities, and decision-making processes that are integral parts of the manufacturing system.

◼ 41.4 CLASSIFICATION OF MANUFACTURING SYSTEMS

Tables in Chapter 1 listed manufacturing industries that vary by the products that they make or assemble. While almost all factories are different, there are five basic manufacturing system designs (MSDs), four classic (or hybrid combinations thereof), and one new manufacturing system that is rapidly gaining acceptance in almost all of these industries. Table 41-1 lists examples of five types of manufacturing systems. These lists are

Type of	Examples	
Manufacturing System	Service[a]	Product[b]
Job shop	Auto repair	Machine shop
	Hospital	Metal fabrication
	Restaurant	Custom jewelry
	University	FMS
Flow shop or flow line	X-ray	TV factory
	Cafeteria	Auto assembly line
	College registration	
	Car wash	
Lean shop or	Fast-food restaurant	Product families
linked-cell	(KFC, Wendy's)	Family of turned parts
	Food court at mall	Composite part families
	10-minute oil change	Design families
Project shop	Producing a movie	Locomotive assembly
	Broadway play	Bridge construction
	TV show	House construction
Continuous process	Telephone company	Oil refinery
	Phone company	Chemical plant

TABLE 41-1. Types and Examples of Manufacturing Systems

[a] Customer receives a service or perishable product.
[b] Products can be for customers or for other companies.

not meant to be complete. The classical systems here are the *job shop*, the *flow shop*, the *project shop*, and the *continuous process*. The *lean shop* is a new kind of system. The *assembly line* is a form of the flow shop, and a system that has *batch flow* might also be added as another manufacturing system (MS). Figure 41-3 shows schematics of the four classical systems along with the linked-cell manufacturing system, or the lean shop.

JOB SHOPS

The job shop's distinguishing feature is its functional design. In the job shop, a variety of products is manufactured, which results in small manufacturing lot sizes, often one of a kind. Job shop manufacturing is commonly done to specific customer order, but in truth, many job shops produce to fill finished-goods inventories. Because the plant must perform a wide variety of manufacturing processes, general-purpose production equipment is required. Workers must have relatively high skill levels to perform a range of different work assignments, often due to a lack of work standardization, defects, and variety in goods produced. Job shop products include space vehicles, aircraft, machine tools, special tools, and equipment. Figure 41-4 depicts the functionally arranged job shop. Production machines are grouped according to the general type of manufacturing process. The lathes are in one department, drill presses in another, plastic molding in still another, and so on. The advantage of this layout is its ability to make a wide variety of products. Each different part requiring its own unique sequence of operations can be routed through the respective departments in the proper order. *Route sheets* are used as the production control device to define the path of the material through the manufacturing system; see Figure 41-18 for example. Forklifts and handcarts are used to move materials from one machine to the next. As the company grows, the job shop evolves into a production job shop making products in large lots or batches.

The production job shop becomes extremely difficult to manage as it grows, resulting in long product throughput times and very large in-process inventory levels. In the job shop, parts spend 95% of the time waiting (delay) or being transported and only 5% of the time on the machine, as shown in Figure 41-5. Thus the time spent actually adding value may only be 2 or 3% of the total time available. The job shop typically builds large volumes of products but still builds lots or batches, usually medium-sized lots of 50 to 200 units. The lots may be produced only once, or they may be produced at regular intervals. The purpose of batch production is often to satisfy continuous customer demand for an item. This system usually operates in the following manner. Because the production rate can exceed the customer demand rate, the shop builds an inventory of the item, then changes over the machines to produce other products to fill other orders. This involves tearing down the setups on many machines and resetting them for new products. When the stock of the first item becomes depleted, production is repeated to build the inventory again.

Some machine tools are designed for higher production rates. For example, automatic lathes capable of holding many cutting tools can have shorter processing times than engine lathes. The machine tools are often equipped with specially designed workholding devices, jigs, and fixtures, which increase process output rate, precision, accuracy, and repeatability.

Industrial equipment, furniture, textbooks, and components for many assembled consumer products (household appliances, lawn mowers, and so on) are made in production job shops. Such systems are called machine shops, foundries, plastic-molding factories, and pressworking shops.

As much as 75% of all piece-part manufacturing is estimated to be in lot sizes of 50 to 100 pieces. Hence job shop production constitutes an important portion of total manufacturing activity. Most service industries have the characteristics of a job shop, as do many of the production systems that exist to serve job shop manufacturing systems.

FLOW SHOP

The flow shop has a product-oriented layout composed mainly of flow lines. When the volume gets very large, especially in an assembly line, this is called *mass production*. This kind of system can have (very) high production rates. Specialized equipment,

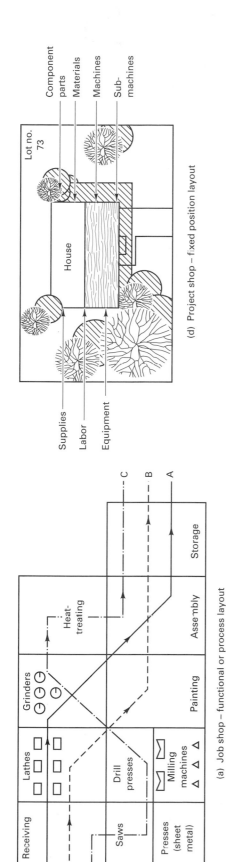

(a) Job shop – functional or process layout

(b) Flow shop – line or product layout

(c) Lean shop (U-shaped cells)

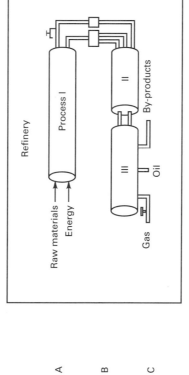

(d) Project shop – fixed position layout

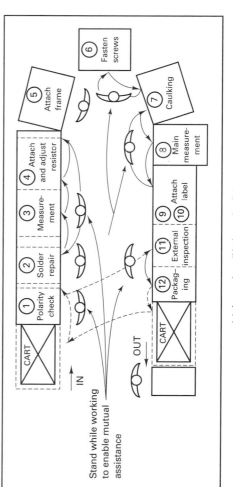

(e) Continuous-process layout

FIGURE 41-3 Schematic layouts of five manufacturing systems: (a) job shop (functional or process layout), (b) flow shop (line or product layout), (c) lean shop (linked-cell layout), (d) project shop (fixed-position layout), and (e) continuous process.

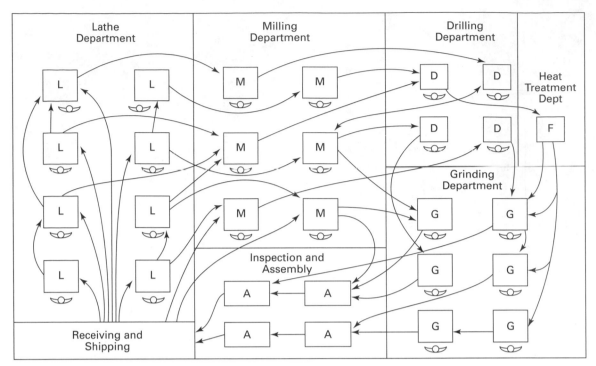

Key: ——→ Movement of parts in containers
⌣ Machine tool operators
▭ Machine tool

FIGURE 41-4 Schematic layout of a job shop where processes are gathered functionally into areas or departments. Each square block represents a manufacturing process. Sometimes called the "spaghetti design."

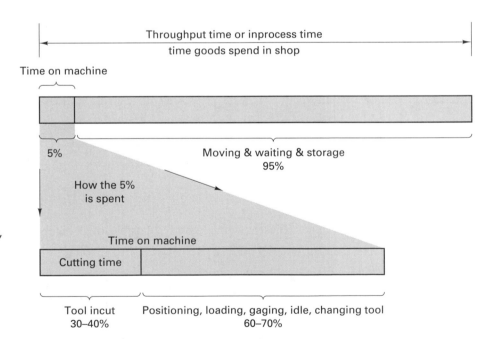

FIGURE 41-5 In the job shop, the parts spend most of the time waiting or moving and only 5% on the machine. *(Courtesy of C. F. Carter, Cincinnati Milacron.)*

dedicated to the manufacture of a particular product, is used. The entire plant may be designed exclusively to produce the particular product or family of products, using special-purpose rather than general-purpose equipment. The investment in specialized machines and specialized tooling is high. Many production skills are transferred from the operator to the machines so that the manual labor skill level in a flow shop tends to be lower than in a production job shop. Items are made to "flow" through a

FLOW SHOP

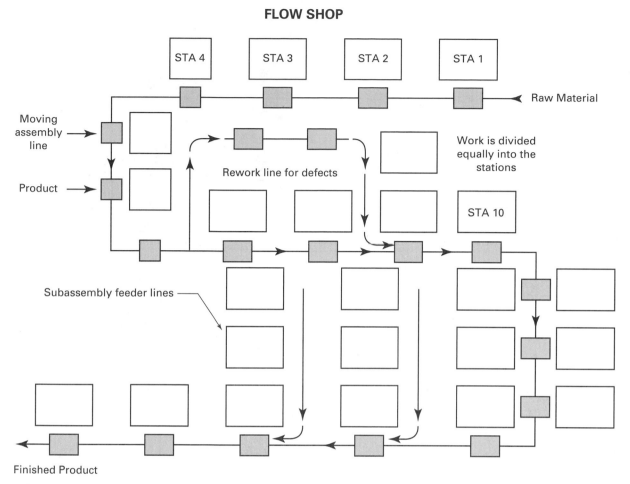

FIGURE 41-6 Schematic of a flow shop manufacturing system.

sequence of operations by materials-handling devices (conveyors, moving belts, and transfer devices). The items move through the operations one at a time. The time the item spends at each station or location is fixed or equal (balanced). A schematic of the flow line is shown in Figure 41-6.

On a much larger scale, Figure 41-7 shows an example of an automated transfer machine for the assembly of engine blocks at the rate of 100 per hour. All the machines are specially designed and built to perform specific tasks and are not capable of making any other products. Consequently, to be economical, such machines must be operated for considerable periods of time to spread the cost of the initial investment over many units. These machines and systems, although highly efficient, can be utilized only to make products in very large volume, hence the term *mass production*. Changes in product design are often avoided or delayed because it would be too costly to change the process or scrap the machines.

However, as we have already noted, products manufactured to meet the demands of free-economy, mass-consumption markets need to incorporate changes in design for improved product performance as well as style changes. Therefore, hard automation systems need to be as flexible as possible while retaining the ability to mass-produce. This has led to the following developments in transfer lines.

First, the machines are constructed from basic building blocks or modular units that accomplish a function rather than produce a specific part. The production machine modules are combined to produce the desired system for making the product. Self-contained power-head production machines are an example of such machines. Automatic transfer devices connect the machines to handle the material (i.e., the thing being produced), moving it automatically from one process to the next.

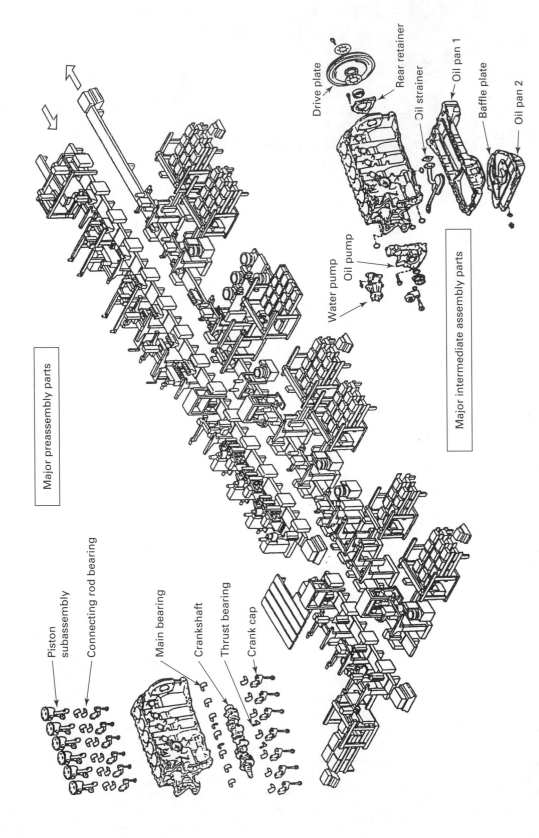

Major preassembly parts

Piston subassembly

Connecting rod bearing

Main bearing

Crankshaft

Thrust bearing

Crank cap

Major intermediate assembly parts

Drive plate

Rear retainer

Oil strainer

Oil pan 1

Baffle plate

Oil pan 2

Water pump

Oil pump

FIGURE 41-7 The transfer line is an example of a very automated flow line.

Second, the incorporation of *programmable logic controllers* (PLCs) and feedback control devices have made these machines more flexible. Modern PLCs have the functional sophistication to perform virtually any control task. These devices are rugged, reliable, easy to program, and economically competitive with alternative control devices. PLCs have replaced conventional hard-wired relay panels in many applications because they are easy to reprogram. Relay panels have the advantage of being well understood by maintenance people and are invulnerable to electronic noise, but construction time is long and tedious. PLCs allow for mathematical algorithms to be included in the closed-loop control system and are being widely used for single-axis, point-to-point control as typically required in straight-line machining, robot handling, and robot-assembly applications. They do not at this time challenge the computer numerical controls (CNCs) used on multiaxis contouring machines. However, PLCs are used for monitoring temperature, pressure, and voltage on such machines. PLCs are used on transfer lines to handle complex material movement problems, gaging, automatic tool setting, on-line tool wear compensation, and automatic inspection, giving these systems flexibility that they never had before.

Third, the transfer line has been combined with CNC machines to form *flexible manufacturing systems* (FMSs). These systems were discussed in Chapter 32.

In the flow line manufacturing system, the processing and assembly facilities are arranged in accordance with the product's sequence of operations. Work stations or machines are arranged in line with only one work station of a type, except where duplicates are needed for balancing the time products take at each station. The line is organized by the processing sequence needed to make a single product or a regular mix of products. A hybrid form of the flow line produces batches of products moving through clusters of work stations or processes organized by product flow. In most cases, the setup times to change from one product to another are long and often complicated.

Various approaches and techniques have been used to develop machine tools that would be highly effective in large-scale manufacturing. Their effectiveness was closely related to the degree to which the design of the products was standardized and the time over which no changes in the design were permitted. If a part or product is highly standardized and will be manufactured in large quantities, a machine that will produce the parts with a minimum of skilled labor can be developed. A completely tooled automatic screw machine is a good example for small parts.

Most factories are mixtures of the job shop with flow lines. Obviously, the demand for products can precipitate a shift from batch to high-volume production, and much of the production from these plants is consumed by that steady demand. Subassembly lines and final assembly lines are further extensions of the flow line, but the latter are usually much more labor intensive.

PROJECT SHOP

In the typical project manufacturing system, a product must remain in a fixed position or location during manufacturing because of its size and/or weight. The materials, machines, and people used in fabrication are brought to the site. Prior to the development of the flow shop, cars were assembled in this way. Today, large products like locomotives, large machine tools, large aircraft, and large ships use fixed-position *layout*. Obviously, fixed-position *fabrication* is also used in construction jobs (buildings, bridges, and dams). See Figure 41-8. As with the fixed-position layout, the product is large, and the construction equipment and workers must be moved to it. When the job is completed, the equipment is removed from the construction site. The project shop invariably has job shop/flow shop elements manufacturing all the subassemblies and components for the large complex project and thus has a functionalized production system.

The project shop often produces one-of-a-kind products with very low production rates—from one per day to one per year. The work is scheduled using project management techniques like the *critical path method* (CPM) or *program evaluation and review technique* (PERT). These methods use precedence diagrams that show the sequence of manufacturing and assembly events or steps and the relationship (precedence) between

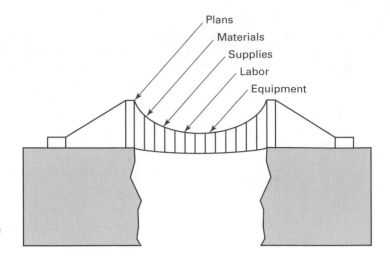

FIGURE 41-8   The project shop involves large assemblies (projects) where components are usually fabricated elsewhere and transported to the site.

the steps. There is always a path through the diagram that consumes the most time, called the *critical path*. If any of the tasks on the longest path are delayed, it is likely the whole project will be delayed. The process engineer must know the relationships (the order) of the processes and be able to estimate the times needed to perform each task. All this (and more) is the responsibility of project management. Project shop manufacturing is labor intensive, with projects typically costing in the millions of dollars.

CONTINUOUS PROCESSES

In the *continuous process,* the project physically flows. Oil refineries, chemical processing plants, and food-processing operations are examples. This system is sometimes called flow *production*, when the manufacture of either complex single parts (such as a canning operation or automotive engine blocks) or assembled products (such as television sets) is described. However, these are not continuous processes but rather high-volume flow lines. In continuous processes, the products really do flow because they are liquids, gases, or powders. Continuous processes are the most efficient but least flexible kinds of manufacturing systems. They usually have the leanest, simplest production systems because these manufacturing systems designs are the easiest to control, having the least work-in-process (WIP). However, these manufacturing systems usually involve complex chemical reactions, and thus a special kind of *manufacturing engineer* (MfE) called the *chemical engineer* (ChE) is usually assigned the task of designing and running the manufacturing system.

LEAN MANUFACTURING SYSTEM

The lean manufacturing employs U-shaped cells or parallel rows to manufacture components. The entire factory is reconfigured. The final assembly lines are converted to mixed-model, final assembly so that the demand for subassemblies and components is *leveled*, making the daily demand for components the same every day. Subassembly lines are also reconfigured into volume-flexible, single-piece flow cells.

The restructuring of the job shop is difficult; the concept is shown in Figure 41-9. The lean shop employs a *linked-cell system* with standing, walking workers. System design requirements are used to design the cells. In a linked-cell system, the key proprietary aspects are the U-shaped manufacturing and assembly cells. In the cells the axiomatic design concept of decoupling is employed to separate (decouple) the processing times for individual machines from the cycle time for the cell, enabling the lead time for a batch of parts to be independent of the processing times for individual machines. This takes all the variation out of the supply chain lead times, so scheduling of the supply system becomes so simple that the supply chain can be operated by a pull system of production control (kanban). Using *kanban*, the inventory levels can be dropped, which decreases the throughput time for the manufacturing system.

Workers in the cells are multifunctional; each worker can operate more than one kind of process and also perform inspection and machine maintenance duties according to a standard work pattern. Cells eliminate the job shop concept of one person–one

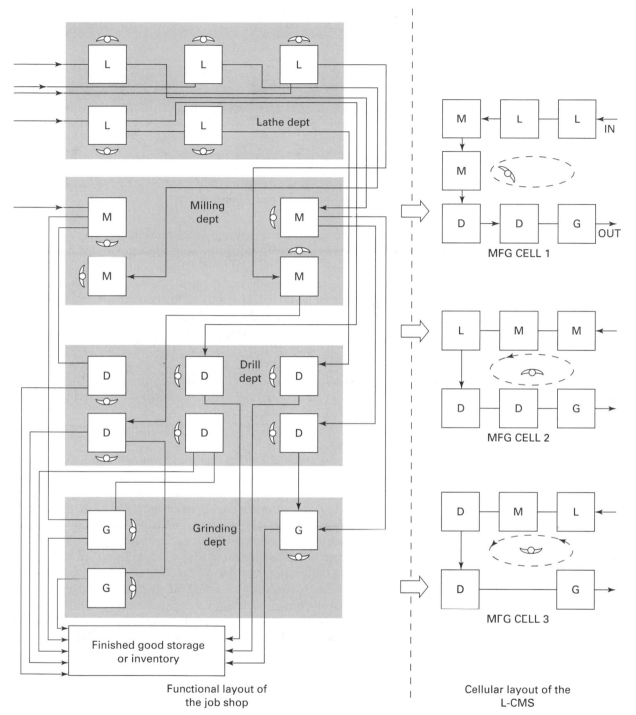

FIGURE 41-9   The job shop portion of the plant requires a systems level conversion to reconfigure it into manufacturing cells, operated by standing and walking workers.

machine and thereby greatly increase worker productivity and utilization. The restriction of the cell to a family of parts makes reduction of setup in the cell possible. The general approach to setup reduction is discussed in Chapter 43 as part of the cell design strategy.

In some cells *decouplers* are placed between the processes, operations, or machines to connect the movement of parts between operators. A decoupler physically holds one unit, decouples the variability in processing time between the machines and enables the separation of the worker from the machine. Decouplers can provide flexibility, part transportation, inspection for defect prevention (pokayoke) and quality control, and process delay for the manufacturing cell. Here is how an inspection decoupler might work. The part is removed from a process and placed in a decoupler. The decoupler inspects the

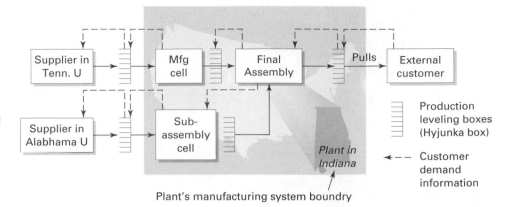

FIGURE 41-10 The linked-cell manufacturing pulls the goods from the suppliers through the assembly to the external customer.

part for a critical dimension and feeds back adjustments to the machine to prevent the machine from making oversize parts as the milling cutter wears down. A process delay decoupler delays the part movement to allow the part to cool down, heat up, cure, or whatever is necessary, for a period of time greater than the cycle time for the cell. Decouplers are vital parts of manned and unmanned cells and will be discussed further in Chapters 42 and 43.

The system design ensures that the right mix and quantity of parts are made according to an averaged customer demand. This system is robust in that it ensures that the right quantity and mix are made even though there is variation or disturbance in the manufacturing process or other operations in the manufacturing system. If process variation or disturbances to the system occur from outside the manufacturing system's boundary (i.e., incoming parts are defective), the system design is robust enough to handle these disturbances due to its feedback control design.

A typical linked-cell system is shown in Figure 41-10 with the customer demand information illustrated by dashed lines.

A specific level of inventory is established after each producing cell. This set-point level is sometimes called the *standard work-in-process* (SWIP) inventory. The SWIP defines the minimum inventory necessary for the system to produce the right quantity and mix to the external customer and to be able to compensate for disturbances and variation to the system.

Let us summarize the characteristics of manufacturing cells. The cell makes parts one at a time in a flexible design. Cell capacity (the cycle time) can be altered quickly to respond to changes in customer demand by increasing or decreasing operators. The cycle time does not depend on the machining time.

Families of parts with similar designs, flexible workholding devices, and tool changers in programmable machines allow rapid changeover from one component to another. Rapid change are over means that quick or one-touch setup is employed, often like flipping a light switch. This leads to small lot production. Significant inventory reductions between the cells are possible and the inventory level can be directly controlled by the user. Quality is controlled within the cell, and the equipment within the cell is maintained routinely by the workers. The utilization of individual machines is not optimized, but the cell as a whole operates to exactly meet the pace of the customer's demand.

The product designer can easily see how parts are made in the cell since all the processes are together. Because quality-control techniques are also integrated into the cells, the designer knows exactly the cell's process capability. The designer can easily configure the future designs to be made in the cell. This is truly designing for manufacturing.

For robotic (unmanned) cells, the robot typically loads and unloads parts for one to five CNC machine tools, but this number may be increased if the robot can become mobile. A machining center can do the same sequence of steps but is not as flexible as a cell composed of multiple simple machines. Cellular layouts facilitate the integration of critical production functions while maintaining flexibility in producing superior-quality products.

The cells facilitate *single-piece-flow* (SPF) and volume flexibility. SPF is the movement of one part at a time between machines. In the main, each machine executes a step in the sequence of processes or operations. The outcome of that step is checked before the part is advanced to the next step. Volume flexibility is achieved by worker–machine separation and the rapid reallocation of operations to workers.

TABLE 41-2.	Characteristics of Basic Manufacturing Systems				
Characteristics	Job Shop	Flow Shop	Project Shop	Continuous Process	Lean Shop
Types of machines	Flexible, general purpose	Single purpose, single function	General purpose, mobile, manual automation	Specialized, high technology	Simple, customized single cycle automatic
Design of processes	Functional or process process	Product flow layout	Project or fixed-position layout	Product	Linked U-shaped cells
Setup time	Long, variable, frequent	Long and complex	Variable, every job different	Rare and expensive	Short, frequent, one touch
Workers	Single function, highly skilled: one worker—one machine	One function, lower-skilled: one worker—one machine	Specialized, highly skilled	Sill level varies	Multifunctional, respected standing
Inventories (WIP)	Large inventory to provide for large variety	Large to provide buffer storage	Variable, usually large	Very small	Small
Lot sizes	Small to medium	Large lot	Small lot	Very large	Small
Manufacturing lead time	Long, variable	Short, constant time	Long, variable lead	Very fast, constant	Short, constant

Table 41-2 provides a comparison of the chief characteristics of the five basic manufacturing systems.

■ 41.5 THE EVOLUTION OF THE FIRST FACTORY

There was a time when there were no *factories*. From 1700 to 1850, most manufacturing was performed by skilled craftsmen in home-based workshops, what historians call *cottage industry*. These artisans were gunsmiths, blacksmiths, toolmakers, silversmiths, and so on. The machines they had for forming and cutting materials were manually powered.

Amber & Amber (1962) in their yardstick for automation pointed out that the first step in mechanizing and automating the factory was to *power the machines*. The need to power the machine tools efficiently created the need to gather the machine tools into one location where the power source was concentrated. By this time, the steel and cast iron to build machine tools were becoming economically available. Railroads were built to transport goods. These were the technologies that influenced the factory design and are now called *enabling technologies*; see Table 41-3. A functional design evolved because of the *method* needed to drive or power the machines.

TABLE 41-3.	The Evolution of Manufacturing System Designs[a]			
	1st Industrial Revolution	2nd Industrial Revolution	3rd Industrial Revolution	4th Industrial Revolution
Time Period (ERA)	1840–1910	1910–1970	1960–2010	2000–20??
Manufacturing System Design	Job shop	Flow shop	Lean shop	IM,C [c]
Layout	Functional layout	Product layout	One-piece flow via L-CMS	Linked assembly of large modules or subassemblies
Enabling Technologies	Power for machines, steel production, and railroad for transportation	Moving final assembly line, standardization leading to true interchangeability, and automatic material handling	U-shaped cells, kanban, and rapid die exchange	Virtual reality/simulation, 3D design using low cost, and very high performance computers[d]
Historical Company Name	Whitney, Colt, and Remington	Singer, Ford	Toyota Motor Company,[b] HP, Omark, Harley Davidson	Boeing, Lockheed, Electric Boat, and Mercedes
Economics	Economy of collected technology	Economy of scale, high volume → low unit cost	Economy of scope, wide variety at low unit cost	Economy of modules, even leaner

[a] MSD (manufacturing system design), including machine tools, material handling equipment, tooling and people
[b] System developed by Taiichi Ohno, who called it the Toyota Production System (TPS)
[c] IM,C (integrated manufacturing, computerized), term coined by J T. Black
[d] Single-source digital product definition and 3D design, a manufacturing system simulated using virtual reality

The first factories were built next to rivers, and the machines were powered by flowing water driving waterwheels, which drove shafts that ran into the factory. See Figure 41-11. Machines of like type were set, all in a line, underneath the appropriate power shaft, i.e., the shaft that turned at the speed needed to drive this type of machine. So all the lathes were collected under their own power shaft, likewise all the milling machines and all the drill presses. Huge leather belts were used to take power off the shaft to the machines. Later, the waterwheel was replaced by a steam engine, which allowed the factory to be built somewhere other than next to a river. Eventually, large electric motors were used, and later individual electric motors for each machine replaced the large steam and electric engines. Nevertheless, the job shop design was replicated, and the functional design held. Many early factories were in the northeastern part of the United States and were manufacturing guns. Filing jigs were used by laborers to create interchangeable parts. In time, this design became known as the *American Armory System*. People from all over the world came to America to observe the American Armory System, and this functionally designed system was duplicated around the industrial world as a key driver in the first industrial revolution.

EVOLUTIONS AND REVOLUTIONS

Revolutions do not happen overnight but require many simultaneous events to occur. An evolution in thought, philosophy, and the mindset of people in the factory, as well as the redesign of the factory are needed. Even though manufacturing is the leading wealth producer in the world, historians have often neglected the technical aspects of history.

FIGURE 41-11   Early factories used water power to drive the machines.

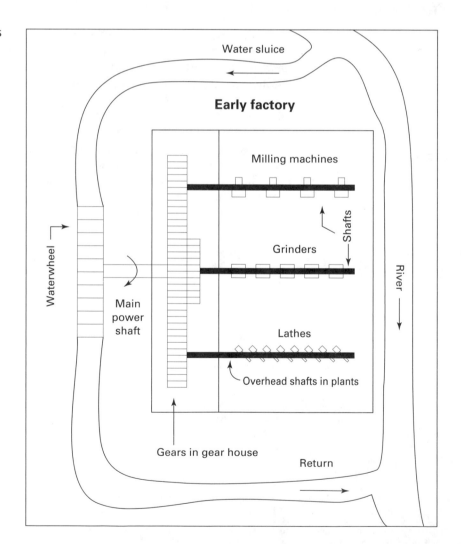

Evolution of the First Industrial Revolution

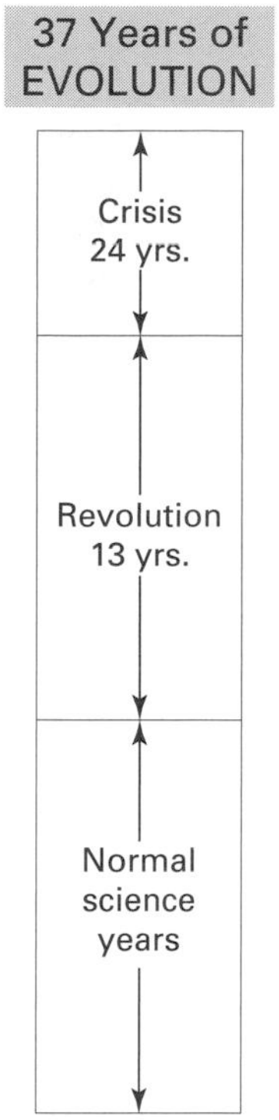 37 Years of EVOLUTION	**Revolutions Don't Happen Overnight**	

<table>
<tr><td>1785</td><td>Thomas Jefferson proposes that Congress mandate interchangeable parts for all musket contracts.</td></tr>
<tr><td>1792</td><td>Eli Whitney invents cotton gin.</td></tr>
<tr><td>*1798*</td><td>*Eli Whitney contract for 4000 muskets in 1.5 years.*</td></tr>
<tr><td>1801</td><td>Eli Whitney demonstrates interchangeability to Congress.</td></tr>
<tr><td>1809</td><td>Eli Whitney delivers, 8.5 years late, non-interchangeable parts.</td></tr>
<tr><td>1811</td><td>John Hall patents breech-loading rifle.</td></tr>
<tr><td>1812</td><td>Roswell Lee becomes superintendent of Springfield Armory.</td></tr>
<tr><td>1815</td><td>Congress orders Ordnance Dept. to require interchangeable parts.</td></tr>
<tr><td>1818</td><td>Blanchard invents trip hammer for making gun barrels.</td></tr>
<tr><td>1819</td><td>Blanchard invents lathe for making gunstocks.</td></tr>
<tr><td>1819</td><td>Lee introduces inspection gages; Springfield Armory.</td></tr>
<tr><td>*1822*</td><td>*John Hall announces success at Harpers Ferry using system of gages to measure parts.* Government report praises effort.</td></tr>
<tr><td>1825</td><td>Eli Whitney dies.</td></tr>
<tr><td>*1834*</td><td>*Simeon North at Middletown, CT, adopts Hall's gages, delivers rifles (parts) interchangeable with Harpers Ferry production.*</td></tr>
<tr><td>1839</td><td>Samuel Colt and Eli Whitney Jr. revolver contract.</td></tr>
<tr><td>1845</td><td>The *Armory Practice* spreads to private contractors.</td></tr>
<tr><td>1860</td><td>Pocket micrometer invented.</td></tr>
</table>

The vertical diagram on the left is divided into: Crisis 24 yrs.; Revolution 13 yrs.; Normal science years.

FIGURE 41-12 Evolution of the first industrial revolution. *Source: David Cochran.*

The model used here is based on Thomas Kuhn's book, *The Structure of Scientific Revolutions*. He said there are three phases in any revolution: the crisis stage, the revolution stage, and the normal science stage, as shown in Figure 41-12. Some people have portrayed this last stage as the diffusion of science. The first industrial revolution can be considered a crisis due to the problems associated with craft production such as producing products in which filing was used to fit products together. This crisis was highlighted in the United States by a congressional mandate to have interchangeable parts for muskets in the field in the late 18th century.

Eli Whitney received a contract from Congress to produce four thousand muskets in one and a half years with interchangeable parts. In wartime there was a need for part interchangeability so that particular musket parts could be replaced in the field, instead of replacing the entire musket. The goal was to reduce replacement time and cost. At that time, muskets consisted of parts that were all craft-made, filed-to-fit products.

Part interchangeability was a manufacturing challenge for the armory system from the beginning, and the ultimate success in achieving part interchangeability was the result of John Hall's work in 1822. His innovation was the ability to measure the product that was made. Instead of attempting to produce parts to a master part, as was the practice at the time, a system of gages was introduced to measure the part. As a result, the engineers and designers had to be able to draw the parts with specifications in terms of their geometric dimensions and tolerances. The entire need for drawings, tolerances, and specifications was the result of part measurement to achieve part interchangeability.

Many people viewed part interchangeability as adding additional manufacturing cost and concluded that it was unnecessary. However, standard measurements and part interchangeability were key enabling technologies for the development of the moving assembly line during the second industrial revolution.

After this evolution occurred, the practice of using gages to achieve part interchangeability was then moved to Middletown, Connecticut, from Harper's Ferry, which had adopted Hall's system of gages. A gun could then be assembled from parts made from the Harper's Ferry and Middletown locations. By 1845, this armory practice spread to private contractors. Soon, the ability to do gaging became ubiquitous throughout the country. In 1860, the pocket micrometer was invented, and the science of measurement spread throughout the world.

As a system design, the job shop still exists. Modern-day versions of the job shop produce large volumes of goods in batches or lots of 50 to 200 pieces. Processes are still grouped functionally but the walls are removed. Between the machines, filling the aisles, are tote boxes filled with components in various stages of completion. While it may be difficult to actually count the inventory, these designs have thousands of parts on the floor at any one time. However, early on the production system for the early job shop was minimal.

■ 41.6 EVOLUTION OF THE SECOND MSD—THE FLOW SHOP

The second industrial revolution was the development of mass production, shown in Figure 41-13. Henry Ford refined the concept of economies of scale with mass production. His primary "mass production" tools were the extreme division of labor, the moving assembly line, the shortage chaser to control minimum and maximum stock levels, and the overarching reliance of assembly cycle time predictability with interchangeable parts.

Ford is credited with many product design innovations, as well as manufacturing innovations. He developed the process for engine blocks. One reason for the Model T's success was a single cast engine block instead of four cylinders bolted together. The use of this manufacturing process innovation decreased the car's weight and increased power.

However, the enabling technology for mass production was interchangeable parts. Ford insisted that every product meet specifications. He was a stickler about keeping all gages calibrated in the factory. In other words, the second industrial revolution was founded on the first industrial revolution concept of part interchangeability.

Flow line manufacturing began in the 1900s for small items and culminated with the moving assembly line at the Ford Motor Company around 1913. This methodology was developed by Ford production engineers led by Charles Sorenson. Today's moving assembly line for automobile production has hundreds of stations where the car is assembled. This requires the work at each station to be balanced where tasks at each station take about the same amount of time. This is called *line balancing*. The moving assembly line makes cars one at a time, in what is now called single piece flow.

FIGURE 41-13   Evolution of the second industrial revolution.

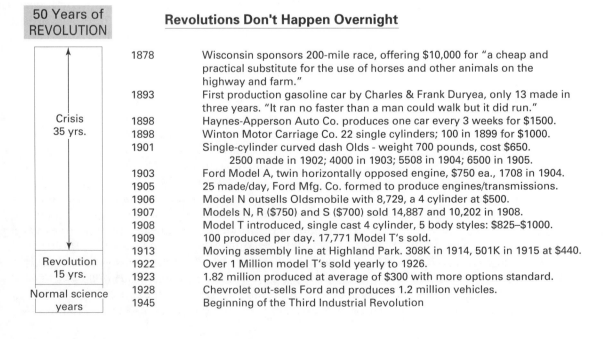

Evolution of the Second Industrial Revolution

50 Years of REVOLUTION		**Revolutions Don't Happen Overnight**
	1878	Wisconsin sponsors 200-mile race, offering $10,000 for "a cheap and practical substitute for the use of horses and other animals on the highway and farm."
	1893	First production gasoline car by Charles & Frank Duryea, only 13 made in three years. "It ran no faster than a man could walk but it did run."
Crisis 35 yrs.	1898	Haynes-Apperson Auto Co. produces one car every 3 weeks for $1500.
	1898	Winton Motor Carriage Co. 22 single cylinders; 100 in 1899 for $1000.
	1901	Single-cylinder curved dash Olds - weight 700 pounds, cost $650. 2500 made in 1902; 4000 in 1903; 5508 in 1904; 6500 in 1905.
	1903	Ford Model A, twin horizontally opposed engine, $750 ea., 1708 in 1904.
	1905	25 made/day, Ford Mfg. Co. formed to produce engines/transmissions.
	1906	Model N outsells Oldsmobile with 8,729, a 4 cylinder at $500.
	1907	Models N, R ($750) and S ($700) sold 14,887 and 10,202 in 1908.
	1908	Model T introduced, single cast 4 cylinder, 5 body styles: $825–$1000.
	1909	100 produced per day. 17,771 Model T's sold.
Revolution 15 yrs.	1913	Moving assembly line at Highland Park. 308K in 1914, 501K in 1915 at $440.
	1922	Over 1 Million model T's sold yearly to 1926.
	1923	1.82 million produced at average of $300 with more options standard.
Normal science years	1928	Chevrolet out-sells Ford and produces 1.2 million vehicles.
	1945	Beginning of the Third Industrial Revolution

Just as in the 1800s, people throughout the world came to observe how this system worked, and the new design methodology was again spread around the world. For many companies, a hybrid system evolved, which included a mixture of job shop and flow shop, with the components made in the job shop feeding the assembly line. This design permitted companies to manufacture large volumes of identical products at low unit cost.

Mass production (or the second industrial revolution) relied on the first industrial revolution for part interchangeability, while producing products in a fixed cycle time with moving assembly lines. To produce at a fixed cycle time, division of labor was used, and unskilled workers replaced the craftsmen in the factory (see Adam Smith's *Wealth of Nations*). With the division of labor, instead of assembling an entire transmission, the workers performed the same small set of tasks on each transmission. As a result, turnover in the factory increased dramatically, so Ford introduced the "five-dollar day," an exorbitant amount of money in those days. In fact, the five-dollar day created the economic system that enabled the emergence of the middle class. The workers were able to buy the products they produced in high-volume mass-production factories.

Between 1928 and 1945, the Ford factory system diffused to other products. During WWII, airplanes were produced at Ford's Willow Run Factory using fixed cycle time, moving assembly lines, and interchangeable parts. Planes were moved down an assembly line at the pace of demand. All kinds of tooling and jigs were designed and built so that interchangeable parts could be assembled onto the aircraft in a fixed cycle time.

The transfer line is a very automated form of the flow line for machining or assembling complex products like engines for cars in large volumes (200,000 to 400,000 per year). The enabling technology here is repeat cycle automatic machines. This system also required interchangeable parts based on precise standards of measurement (gage blocks). The transfer line is designed for large volumes of identical goods. These systems are very expensive and not very flexible.

The parts that feed the flow line were made in the job shop in large lots, held in inventory for long periods of time, then brought to the line where a particular product was being assembled. The line would produce one product over a long run and then switch to another product, which could be made for days or weeks. This mass-production system was in place during WWII and was clearly responsible for producing the military equipment and weapons that enabled the allies to win the war.

This massive production machine thrived after WWII, which enabled automobile producers to make cars in large volumes using the economy of scale. Just when it appeared that nothing could stop this machine, a new player, The Toyota Motor Company, evolved a new manufacturing system that changed the manufacturing world. This system is based on a different system design, which ushered in the third industrial revolution, characterized by global companies producing for worldwide markets.

■ 41.7 THE THIRD MSD—THE LEAN SHOP

The latest MSD is sometimes called *lean production*, enhancing the theory that industrial revolutions are spawned by new manufacturing system designs. A new manufacturing system design brings one or more new companies to the forefront of the industrial world (this time, Toyota). See Figure 41-14 for details of this revolution. Each new design employs some unique enabling technologies, in this case, manufacturing and assembly cells that produce defect-free grads. The new system is being adopted by many companies and is slowly being disseminated around the industrial world.

Black's Theory of Industrial Revolutions (IRs) as outlined in Table 41-4, proposes that we are now 40-plus years into the third industrial revolution. Again, this industrial revolution is not based on computers, hardware, or a particular process, but once again on the design of the manufacturing system—the complex arrangement of physical elements characterized by measurable parameters. Again, people throughout the world went to observe the new design, but this time they went to Japan to try to understand how this tiny nation became a giant in the global manufacturing arena.

After WWII the Japanese were confronted with different requirements of manufacturing than that the Americans leading to the third industrial revolution as shown in

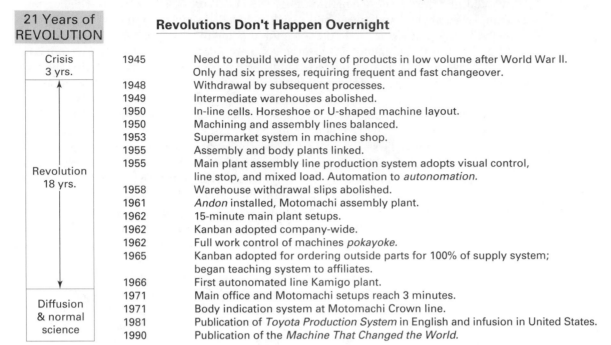

Third Industrial Revolution at Toyota in Japan (Ohno, 1988)

Revolutions Don't Happen Overnight

21 Years of REVOLUTION		
Crisis 3 yrs.	1945	Need to rebuild wide variety of products in low volume after World War II. Only had six presses, requiring frequent and fast changeover.
	1948	Withdrawal by subsequent processes.
	1949	Intermediate warehouses abolished.
	1950	In-line cells. Horseshoe or U-shaped machine layout.
	1950	Machining and assembly lines balanced.
	1953	Supermarket system in machine shop.
	1955	Assembly and body plants linked.
Revolution 18 yrs.	1955	Main plant assembly line production system adopts visual control, line stop, and mixed load. Automation to *autonomation*.
	1958	Warehouse withdrawal slips abolished.
	1961	*Andon* installed, Motomachi assembly plant.
	1962	15-minute main plant setups.
	1962	Kanban adopted company-wide.
	1962	Full work control of machines *pokayoke*.
	1965	Kanban adopted for ordering outside parts for 100% of supply system; began teaching system to affiliates.
	1966	First autonomated line Kamigo plant.
	1971	Main office and Motomachi setups reach 3 minutes.
Diffusion & normal science	1971	Body indication system at Motomachi Crown line.
	1981	Publication of *Toyota Production System* in English and infusion in United States.
	1990	Publication of the *Machine That Changed the World*.

FIGURE 41-14   Third industrial revolution at Toyota in Japan *(Ohno, 1988)*.

TABLE 41-4.	Industrial Revolutions (IRs) Are Driven by New Manufacturing Systems Designs (MSD)	
	1700–1850 Craft/Cottage Productions	No MSD
First IR	1840–1910 First industrial revolution (American armory system) • Creation of factories with powered machines • Mechanization/interchangeable parts	Job shop (functional layout)
Second IR	1910–1970 Second industrial revolution (the Ford system) • Assembly line–flow shop product layout • Economy of scale yields mass production era • Automation (automatic material handling)	Flow shop (product layout)
Third IR	1960–2010 (estimated) Third industrial revolution (lean production) • U-shaped manufacturing cells • One-piece flow • Flexibility • Integrated control functions	Lean shop (Linked-cell layout)

Figure 41-14. The United States had just about infinite capacity to produce. There were rows of stamping presses in the factories, a surplus of resources, pent-up demand, and many people with money, so all the United States had to do was to produce. In Japan, there were very few presses and very little money. One of the first concepts of the Toyota Production System was that all parts had to be good, because there was no excess capacity and no dealers in the United States to fix the defects. The right mix of cars of perfect quality had to be made with limited resources, and they had to be exactly right the first time in spite of any variations or disturbances to the system.

The result was a new manufacturing system design and just as a new MSD pushed Colt and Remington to the forefront in the first industrial revolution, and Ford and Singer in the second industrial revolution, the development of this unique manufacturing system vaulted Toyota into world leadership. Black defines the new physical system design as linked-cell (Black, 1991) or L-CMS for linked-cell manufacturing system. Toyota called it the *Toyota Production System* (TPS). Schonberger called it the JIT/TQC system or *World Class Manufacturing* (WCM) system. In 1990 it was finally given a name that would become universal: lean production. This term was coined by John Krafcik, an engineer in the International Motor Vehicle program at MIT (Womack et al., 1991).

What was different about this system design was the development of manufacturing and assembly cells linked to final assembly by a unique material control system, producing a functionally integrated system for inventory and production control. In cells, processes are grouped according to the sequence and operations needed to make a product. This arrangement is much like that of the flow shop, but is designed for flexibility.

The cell is designed in a *U-shape* or in parallel rows so that the workers can readily rebalance the line and change the output rate while moving from machine to machine loading and unloading parts. Figure 41-15 shows how the job shop in Figure 41-4 can be rearranged into manned cells. Cell 3 has one worker who can make a walking loop around the cell in 60 seconds. The machines in the cell are single cycle automatics, so they can complete the desired processing untended, turning themselves off when done with a machining cycle. The operator comes to a machine, unloads a part, checks the part, loads a new part into the machine, and starts the machining cycle again. The cell usually includes all the processing needed for a complete part or subassembly and may include assembly steps.

To form a linked-cell manufacturing system, the first step is to restructure portions of the job shop, converting it in stages into manned cells. At the same time, the linear flow lines in subassembly are also reconfigured into U-shaped cells, which operate much like the manufacturing cells. The long setup times typical in flow lines must be vigorously attacked and reduced so that the flow lines can be changed quickly from making one product to another. The need to perform line-balancing tasks is eliminated through design. The standing, walking workers are capable of performing multiples of operations. More details are given in Chapter 42.

FIGURE 41-15   The functionally designed job shop can be restructured into manufacturing cells to process families of components at production rates that match part consumption.

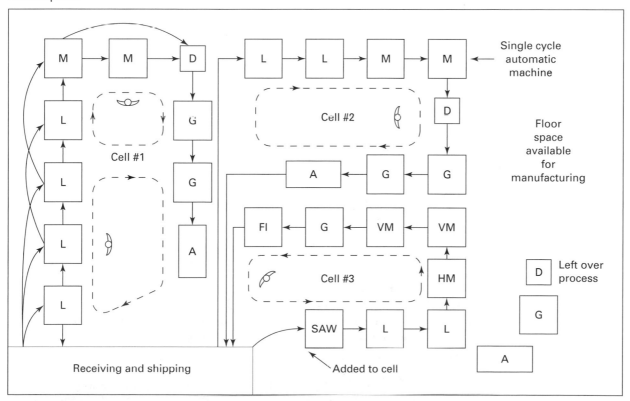

Walking loop of operator........

■ 41.8 THE ENTERPRISE

In earlier chapters the basics of manufacturing processes and manufacturing systems were presented. The manufacturing processes make or assemble products, whereas the systems integrate the processes (with their operations) into selected sequences to produce entire products. We can detail and name the basic processes (casting, forming, machining, etc.) and the operations (transportation, inspection, storage). We can categorize the manufacturing systems by their physical characteristics, (e.g., job shop, flow shop, lean shop) but we cannot yet clearly define the enterprise or the existing productions systems, even though many such systems exist. Production systems are mostly collections of subsystems devised to service the manufacturing system. Their objective is to aid the manufacturing systems in producing products to meet delivery schedules at a target cost that maximizes profit.

Let's compare the university football program to manufacturing. The athletic department can be equated to the production system. The athletic department does many things for the football team (and other teams as well) to help it get ready to play (or produce) but they never carry the ball. So it is that the production system serves the manufacturing system and the individual processes but does not actually make products. The people in the production system may design the product (even choose the color scheme), purchase materials and supplies, plan the implementation of the product into the manufacturing system, sell the products, forecast demand, maintain inventory, hire and fire workers, pay employees, and so on. The functions typically performed in a production system are examined next.

■ 41.9 TYPICAL FUNCTIONAL AREAS

Production systems should serve and support the manufacturing processes and manufacturing systems by providing and transmitting information, energy, knowledge, skills, and services to the plant areas, the company's suppliers, and its customers. Traditionally, the enterprise includes the following departments or functional areas (Figure 41-16):

- Marketing and sales department
- Finance and Accounting (not shown)
- Manufacturing system
- Personnel
- Research and development (not shown)
- Design engineering (product design)
- Purchasing
- Production planning and control (scheduling)
- Inventory control
- Quality control (see Chapter 12) and inspection
- Plant engineering or maintenance

In most companies, the people working in these areas are called *staff* or indirect labor to distinguish them from *line* personnel of the manufacturing system. Although production systems have no standard design, they are usually arranged functionally. To connect the functional areas, informal lines of communication (information flows) are developed. For example, for the job shop, production planning and control is responsible for scheduling the workers, determining what jobs will be done when, in what sequence, and so forth. Figure 41-17 shows the typical communication links for a job shop just in the areas of production control. This network for communication can become complex. Most managerial workers are in the PS, except for foremen, line supervisors, and manufacturing managers.

Often, production workers view the services of quality and inventory control with distrust. Production workers may not understand how control charts monitor their processes or how the materials requirement planning system controls the work-in-process (WIP). An adversarial relationship often develops between the people in the MS and the people in the PS. Computerizing this function merely complicates the problem. This is perhaps the most important difference between the lean production system (discussed

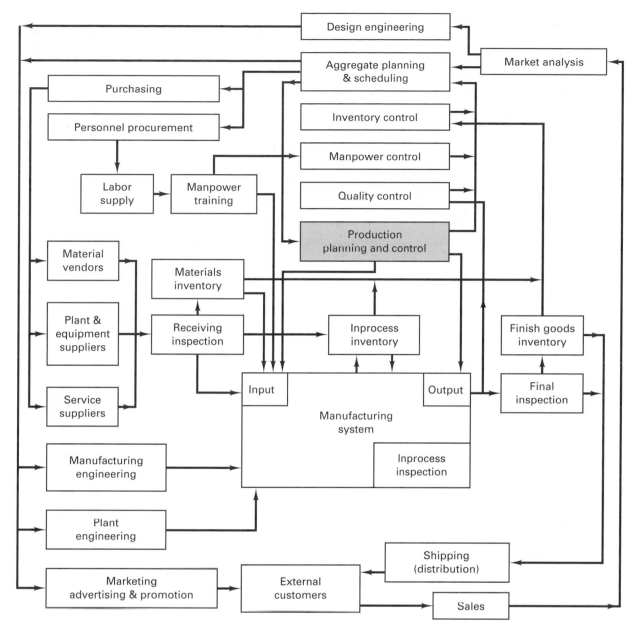

FIGURE 41-16 Classical production system showing inclusion of manufacturing system and major functional elements. See Figure 41-17 for additional details on production planning/control.

in detail in Chapter 42) and the mass production system. That is, in the former, the key functions of the production system are infused into the manufacturing system. The critical control functions not only serve the manufacturing system but also become an integral operational part in the making of the product. Also briefly discussed in Chapter 42 is the ongoing movement to restructure the production system into cross-functional teams, usually organized around product lines or value streams. In this chapter the more traditionally organized production system is presented.

MARKETING

The chief activities of *marketing* are forecasting sales, advertising, and estimating future demand for existing products. Selling the product is the primary interest of marketing. Promotional work, a highly specialized activity, involves advertising and customer relations. Customer service is a critical function for any manufacturing company. Thus marketing provides information and services concerning

1. Sales forecast of future demand for existing products

2. Sales order data

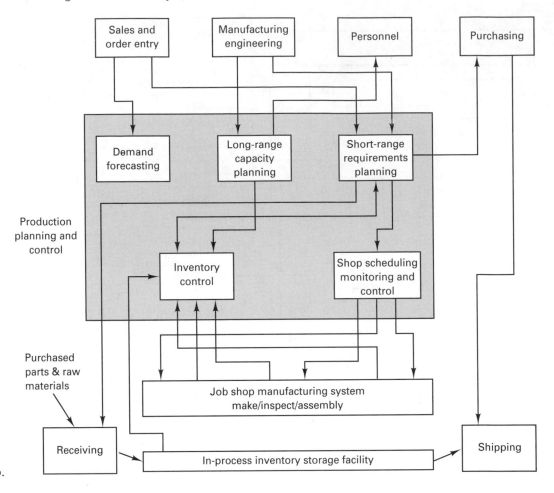

FIGURE 41-17 Flow of information in the planning production planning and control area of the job shop.

3. Customer quality requirements
4. Customer reliability requirements
5. New products or modifications for existing products
6. Customer feedback on products
7. Customer service (repair or replacement of defective products)

There is no piece of information that is more vital to a company, or harder to come by accurately, than future demand. This information is required to plan effectively how much should be produced and to schedule production when changes in demand are predicted. The faster the MS can respond to changes in product demand (for both existing and new products), the better, because the quick response reduces the need to develop accurate long-range forecasts. Short-range forecasts are more accurate than long-range forecasts.

Sales order information is central to production planning and control. Products are either made to stock (finished-goods inventory) or made to fill customer orders. Therefore, the orders determine how much, when, and what kinds of products or services must be produced.

Marketing develops information on new products or new uses for old products. This information usually goes to research and development or to product design engineers. Marketing also gathers customer feedback on existing products. The marketing department, which is in direct customer contact, gathers complaints about product performance and communicates them to design and/or manufacturing. Often, long-time users identify product characteristics that create problems in its use. Clearly, customers want superior-quality products that give them reliable service. In this sense, quality and reliability are related, but they are functionally different. Quality is concerned with the prevention of defects and the conformance to specifications at the time the product is made or sold. Reliability is concerned with the performance of the product over time, while in service with the customer. In general, a superior-quality product is more reliable if the design is good. Failures of products that are well made but perform badly because of faulty design occur infrequently but are usually spectacular and newsworthy.

FINANCE

Financial functions involve management of the company's assets. For the production system, finance provides information and services concerning the following:

1. Internal capital financing
2. Budgeting
3. Investment analysis

Internal financing includes the review of budgets for operating sections, evaluation of proposed capital investments for production facilities, and preparation of financial statements such as balance sheets or profit-and-loss (or income-and-expense) statements.

Periodically, the manufacturing manager, as well as other managers, must submit budgets of expected financial requirements and expenditures to the finance department. The decisions made during budget preparation and the discussions on budget adjustments have a significant impact on the manufacturing system's operation. One of the strongest criticisms of the American system is that decision makers know little about manufacturing processes or systems and therefore make poor investment decisions. Few MBA programs have courses in manufacturing processes. Very few undergraduate business students take courses in manufacturing processes. Managers may not really understand what the company does. However, American managers usually do have problem-solving and decision-making skills for handling investment alternatives that require knowledge of such concepts as rate of return, depreciation, sinking funds, payback periods, and compound interest. Managers must have the financial expertise to understand the complex and constantly changing tax structure, tax regulations, and tax court decisions that affect the company's capital investment decisions.

Therefore, the issue is that an enterprise run by a financial system that does not understand manufacturing as a system may result in the long-term demise of a business. On the other hand manufacturing must take a systems viewpoint before costs can be reduced. MBA programs and management accounting practices in industry are based on the premise that total cost can be reduced by minimizing the cost of each operation. That is, the traditional cost reduction approach tries to optimize bits and pieces of the system to achieve unit cost reduction. In contrast, lean manufacturing is a system design that meets the requirements of internal and external customers. Cost is reduced once the requirements of these two customers are consistently met. The relationships within the system to meet the requirements of the system are emphasized. Wasted work and process delay are systematically removed resulting in reduced cost.

ACCOUNTING

The *accounting* department maintains the company's financial records. Money is used to keep score, so to speak. Accounting also provides the data needed for decision making. For the PS, accounting provides information and services on the following:

1. Cost accounting
2. Special reports
3. Data processing

Cost-accounting information indicates the level of efficiency of various departments and the cost of the products being manufactured. Unit-cost data (cost of materials, direct labor, and overhead) help the company to establish prices. Most American companies view this classical equation as follows:

$$\text{unit cost} + \text{profit} = \text{sales price}$$

If the unit cost goes up, then to maintain profit, the sales price goes up. How many times have you heard this statement: "You'd better buy it now; next year it will cost you more"? Such a statement is an open invitation to those who view the equation in this way.

However, the marketplace and the customer dictate the sales price. The only way to maintain or improve profits is to reduce the cost (unit cost)! Therefore, the best way to view the relationship is

$$\text{sales price} - \text{unit cost} = \text{profit}$$

The purchasing (procurement) department uses manufacturing cost data in analyzing whether a product should be manufactured by the company (in-house) or purchased from a vendor (the classic *make-or-buy* decision). Accounting also produces special reports that monitor the status of the scrap and rework levels, raw-materials inventories, work-in-process inventories, finished-goods inventories, direct-labor hours and overtime, and so on. These reports provide quantitative measures of performance (measurable parameters), which can be compared with the original plans (estimates). These functions are largely useless unless the system design viewpoint is employed wherein systems level performance measures are used.

In large companies, the accounting department often controls the data-processing equipment. In companies that use computers for problem solving instead of for record keeping, data processing is a separate function.

PERSONNEL (HUMAN RESOURCES (HR) DEPARTMENT)

The *personnel* department typically represents workers, one of the key physical elements of the manufacturing system, and provides information and services concerning

1. Recruitment   3. Labor relations
2. Training   4. Safety

Although the personnel department may not hire people directly, it assists the company managers by recruiting, screening, and testing potential employees for jobs in both the manufacturing and the production systems. It also handles the details of terminations and department transfers.

The personnel department can assist in training people. For example, in the area of safety, industrial accidents are both costly and disruptive to the workforce and production schedules. By working closely with the personnel department, management develops and institutes programs that can minimize safety problems. If the company has a union, the personnel department will handle labor relations, grievances, collective bargaining, and problems with the shop stewards and union officials.

RESEARCH AND DEVELOPMENT

Research and development (R&D) involves invention or discovery and innovation and their development in terms of achievable ends, such as new materials, products, processes, tools, and techniques. Many industries show the impact of R&D on their manufacturing systems. For example, for many years, the wood-products industry's manufacturing system produced only lumber products. In recent years R&D efforts have produced new products and processes for making plywood, particleboard (from wood chips), gardening mulch (from bark), laminated beams and panels, and chemicals (from wood).

The folks in the manufacturing system will work with R&D on ideas on the manufacture of new products and processes and on the implementation of new process technology. Modern companies understand that unique process technology can provide a significant competitive edge in the marketplace. Often, R&D also provides ideas for product improvement and may answer questions on economical uses of by-products and waste products from manufacturing operations.

ENGINEERING

Engineering functions are usually staff functions in the production system, providing information and services on the following:

1. Product design engineering or design   4. Plant engineering
2. Manufacturing engineering   5. Quality engineering
3. Industrial engineering

Product Design Engineering.   In discussing design, the relationship of design to manufacturing must be reviewed. Through recognition of the systems approach, the design stage can save many dollars and much time later in manufacturing, inspection, assembly, packaging, and even distribution and marketing. Let us examine some of the simpler aspects of designing for manufacture or assembly, sometimes referred to as *producibility.*

Traditionally, a design will often develop in three phases. In the *conceptual* or *idea* phase, the designer conceives of an idea for a device that will accomplish some function. This stage establishes the functional requirements that must be met by the device. The functional requirements should be independent and cost must be considered at this phase.

In the second, *functional-design* stage, the product is designed so that it will achieve the functional requirements established in the conceptual stage. Often, more than one prototype will be made, suggesting alternative ways in which the functions can be met. At this stage, the designer is usually more concerned with materials than with processes and may ignore the fact that the designed configuration cannot be produced economically utilizing the material being considered.

The third phase of design is called *production design.* Although attention should also be given to the appearance of the product at this stage, particularly if sales appeal is important, the major emphasis is on providing a design that can be manufactured and assembled economically. The *design engineer* must, of course, know that certain manufacturing processes and operations exist that can manufacture the desired product. However, merely knowing that feasible processes exist is not sufficient. The designer must also know their limitations, relative costs, and process capabilities (accuracy, tolerance requirements, etc.) to *design for manufacture* (DFM). If maximum economy is to be achieved, the designer should be aware of the intimate relationship between design details and production operations.

It is extremely important that the relationship between manufacturing (including assembly) and design be given careful consideration throughout the design phase. Changes can be made for pennies in the design room that might cost hundreds or thousands of dollars later in the factory. This type of consideration should be an integral and routine part of planning for manufacturing. Having the design engineers make a working prototype of each new model before final production drawings are made is one approach. If the model performs in accordance with the conceptual requirements, a second model is made using, insofar as possible, the same manufacturing and assembly methods that will be used for actual production. Any design changes that will permit easier and more economical production are incorporated into this second model. If the second model meets the functional requirements of the engineering design group, it is sent to the drafting room, and production drawings are made. This practice eliminates details from products that are costly to produce and the need for a lot of design changes after a part has gone into production.

The designer plays a key role in determining what processes and equipment must be used to manufacture the product, although often indirectly. Clearly, one of the ways in which the designer can indirectly determine the process is through the selection of material. For example, the chassis for record players (stereos) are usually made from zinc and are die cast. Suppose, however, that the designer specifies a composite chassis to be made from fiber-reinforced plastic; then a form-molding process is needed instead of a die-casting machine. The designer specifies a particular joining process when he or she calls for a welded joint. These kinds of direct relationships are pretty obvious. However, other equipment and processes may be specified just as certainly in not so obvious ways.

One of the most common ways in which equipment and processes may be specified indirectly is through dimensional tolerances placed on a drawing. If a tolerance of 0.0002 in. is shown, a grinding operation may be specified just as definitely as if the word *grind* were placed on the drawing. Designers often fail to realize this fact and specify unnecessarily close tolerances; expensive and unnecessary operations result. However, important part requirements may be ignored in manufacturing if dimensions and tolerances are poorly defined. Designers should realize that the dimensions and tolerances they place on a drawing may have implications and results far beyond what they anticipated. Design details are directly related to the processing, making the processing easy, difficult, or impossible and affecting the cost and/or quality.

Through this kind of design for manufacturing thinking, *design engineering* prepares the product design for the customer. If the product is proprietary, the manufacturer is responsible for developing and designing the product. The product design is documented with component drawings, specifications, and a bill of materials (Figure 41-18) that defines how many of each component go into the product. Initial designs may be based on information from R&D. Prototypes are often built for testing and demonstrating product capability. Manufacturing engineering should be consulted on matters of producibility. In a further step sometimes referred to as *value engineering*, engineers look for design changes that could reduce production costs and maintain quality or function. Manufacturing cost estimates are prepared at this point to help determine the market situation for the product.

Upon completion of the design and fabrication of the prototype, company management reviews the design and decides whether to manufacture the item. Engineering management must review and approve the product's design. Many companies call this an *engineering release*.

Corporate management must review and approve the product's general suitability. This second decision represents an authorization to produce the item. The design process material selection factors, manufacturing considerations, materials substitutions, and product liability were also discussed in Chapter 9.

Manufacturing Engineering.   Manufacturing engineers address the planning and management of all manufacturing processes and systems. Using the specifications, *manufacturing engineers* plan the manufacture of the product, determining which machines, operations, workers, tools, and other manufacturing system components should be used

FIGURE 41-18   The process flow chart and the bill of materials (BOM) for a personal computer. See chapter 34.

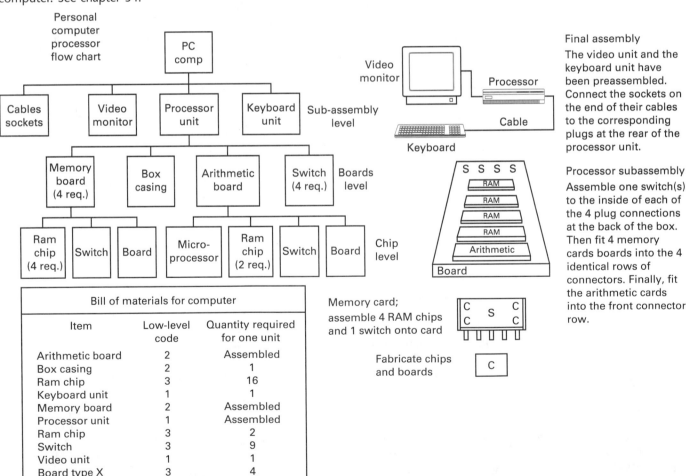

Bill of materials for computer		
Item	Low-level code	Quantity required for one unit
Arithmetic board	2	Assembled
Box casing	2	1
Ram chip	3	16
Keyboard unit	1	1
Memory board	2	Assembled
Processor unit	1	Assembled
Ram chip	3	2
Switch	3	9
Video unit	1	1
Board type X	3	4
Board type Y	3	1
Microprocessor	3	1

to meet quality and functional requirements. In the classic sense, manufacturing engineering includes responsibility for working with and giving advice to product designers on product producibility. Once production has started, engineering design changes are expensive. Manufacturing engineering may also design individual processes for machine tools (or modification), design tooling and specifications (workholding devices and cutting tools and dies), specify the sequence of production processes and operations (process planning), and solve processing problems on the plant floor as well.

In the job shop, *process planning* consists of determining the *sequence* of individual manufacturing processes and operations needed to produce the parts. The *route sheet* and the *operations sheet* are the documents that specify the process sequence through the job shop. The route sheet lists manufacturing operations and associated machine tools for each workpiece (see Figure 41-19, for example). The route sheet travels with the parts, which move in batches (or lots) between the processes. When a cart of parts has to be moved from one point in the job shop to another, the route sheet provides routing (travel) information, telling the material handler which machine in which department the parts must go to next.

The operations sheet describes what machining or assembly operations are done to the parts at particular machines. The operations performed on an engine lathe to make the part shown in Figure 41-20 are shown. Note that the details of speed, feed, and depth of cut are specified (actually, recommended, since the machinist may change them). Over time, many different people may plan the same part; therefore, there can be many different process sequences and many different routes for the same or similar parts.

BASIC REQUIREMENTS FOR PROCESS PLANNING   The first step in planning is to determine the basic job requirements that must be satisfied. These usually are determined by analysis of the drawings and the job orders. They involve consideration and determination of the following:

1. Size and shape of the geometric components of the workpiece
2. Tolerances, as applied by the designer
3. Material from which the part is to be made
4. Properties of material being machined
5. Number of pieces to be produced (see the section in this chapter on quantity versus process and case studies on economic analysis)
6. Machine tools available for this workpiece

Such an analysis for the threaded shaft shown in Figure 41-20 would be as follows:

1. a. Two concentric and adjacent cylinders having diameters of 0.877/0.873 and 0.501/0.499, respectively, and lengths of 2 in. and $1\frac{1}{2}$ in.

    b. Three parallel plain surfaces forming the ends of the cylinders.

    c. A $45° \times \frac{1}{8}$-in. bevel on the outer end of the $\frac{7}{8}$-in. cylinder.

    d. A $\frac{7}{8}$-in. NF-2 thread cut the entire length of the $\frac{7}{8}$-in. cylinder.

2. The tightest tolerance is 0.002 in., and the angular tolerance on the bevel is $\pm 1°$.
3. The material is AISI 1340 cold-rolled steel, BHN 200.
4. The job order calls for 25 parts.

A number of conclusions regarding the processing can be drawn from this analysis. First, because concentric, external, cylindrical surfaces are involved, turning operations are required and the piece should be made on some type of lathe. Second, because 25 pieces are to be made, the use of an engine lathe or a CNC lathe would be preferred over an automatic screw machine. As the company's NC lathe is in use, an engine lathe will be used. Third, because the maximum required diameter is approximately $\frac{7}{8}$-in., 1-in.-diameter cold-rolled stock will be satisfactory; it will provide about $\frac{1}{16}$ in. of material for rough and finish turning of the large diameter. From this information, the operations sheet(s) for a particular engine lathe is prepared.

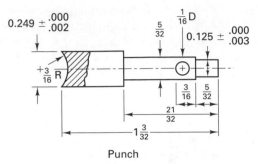

Punch

Matl. – 0.250 dia. AISI 1040
H.T. to 50 R.C. on 0.249 dia.

DARIC INDUSTRIES

ROUTING SHEET

NAME OF PART ___Punch___		PART NO. ___2___	
QUANTITY ___1,000___		MATERIAL ___SAE 1040___	

OPERATION NUMBER	DESCRIPTION OF OPERATION	EQUIPMENT OR MACHINES See Fig 41-4	TOOLING
1	Turn, $\frac{5}{32}$, 0.125, and 0.249 diameters	J & L turret lathe	#642 box tool
2	Cut off to $1\frac{3}{32}$ length	"	#6 cutoff in cross turret
3	Mill $\frac{3}{16}$ radius	#1 Milwaukee	Special jaws in vise $\frac{3}{16}$ form cutter × 4″ D
4	Drill $\frac{1}{16}$ hole	Turret Drill	
5	Heat treat. 1,700° F for 30 minutes, oil quench	Atmosphere furnace	
6	Grind (cutting edges)	Surface grinder	$\frac{3}{16}$ end radius on wheel
7	Check hardness	Rockwell tester	

FIGURE 41-19 Part drawing for a punch (*above*) with route sheet (traveler) for making the punch shown here. The route sheet is used in the job shop (Figure 41-4).

OPERATIONS SHEETS The *operations sheet* shown in Figure 41-20 lists, in sequence, the operations required for machining the threaded shaft shown in the part drawing. A single operation sheet lists the operations that are done in sequence on a single machine. This sheet is for an engine lathe (see Chapter 23). In a manned cell the sheets may cover all the operations for a given part for a group of machines. See Chapter 42 for additional details.

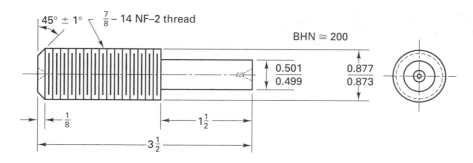

Matl. AISI 1340 Medium carbon steel
BHN $\cong$ 200
Threaded shaft made in quantities of 25 units.

DEBLAKOS INDUSTRIES

OPERATION SHEET

PART NAME: _____ Threaded Shaft 1340 Cold Rolled Steel _____

Part No. 7358-267-10

OPER. NO.	NAME OF OPERATION	MACH. TOOL	CUTTING TOOL	CUTTING SPEED		FEED ipr	DEPTH OF CUT Inches	REMARKS
				ft/min	rpm			
10	Face end of bar	Engine Lathe		120	458	Hand		Use 3-jaw Universal chuck
20	Center Drill End	"	Combination center drill		750	Hand		
30	Cut off to $3\frac{9}{16}$ length	"	Parting tool	120	458	Hand		To prevent chattering, keep overhang of work and tool at a minimum and feed steadily. Use lubricant.
40	Face to length	"	RH facing tool (small radius point)	120	458	Hand	(R) $\frac{1}{8}$ max. (F) .005	Before replacing part in 3-jaw chuck, scribe a line marking the $3\frac{1}{2}$ inch length.
50	Center Drill End	"	Combination center drill		750	Hand		
60	Place between centers, turn $\frac{.501}{.499}$ diameter, and face shoulder	"	RH turning tool (small radius point)	120 160	(R) 458 (F) 611	(R) .0089 (F) .0029	(R) .081 (3) (F) .007	
70	Remove and replace end for end and turn $\frac{.877}{.873}$ diameter	"	RH tools (R) (small radius point) (F) Round nose tool	120 160	(R) 458 (F) 611	(R) .0089 (F) .0029	(R) .057 (F) .005	
80	Produce 45°-chamfer	"	RH round nose tool	120	458	Hand	(R) $\frac{1}{8}$ max. (F) .005	
90	Cut $\frac{7}{8}$ - 14 NF-2 thread	"	Threading tool	60	208		(R) .004 (F) .001	(1) Swivel compound rest to 30 degrees. (2) Set tool with thread gage. (3) When tool touches outside diamter of work set cross slide to zero. (4) Depth of cut for roughing = .004. (5) Engage thread dial indicator on any line. (6) Depth of cut for finishing = .001 Use compound rest.
	Remove burrs and sharp edges	"	Hand file					

Date _____   _____

FIGURE 41-20 Part drawing (*above*) and operations sheet, which provides details of the recommended manufacturing process steps.

Operations sheets vary greatly as to details. The simpler types often list only the required operations and the machines to be used. Speeds and feeds may be left to the discretion of the operator, particularly when skilled workers and small quantities are involved. However, it is common practice for complete details to be given regarding tools, speeds, and often the time allowed for completing each operation. Such data are necessary if the work

Turning, Single Point and Box Tools   MEDIUM-CARBON ALLOY STEELS, WROUGHT

MATERIAL	HARD-NESS	CONDITION	DEPTH OF CUT*	HIGH SPEED STEEL TOOL			CARBIDE TOOL						
							UNCOATED				COATED		
				SPEED	FEED	TOOL MATERIAL AISI	SPEED		FEED	TOOL MATERIAL GRADE C	SPEED	FEED	TOOL MATERIAL GRADE C
							BRAZED	INDEX-ABLE					
			in	fpm	ipr		fpm	fpm	ipr		fpm	ipr	
	Bhn		mm	m/min	mm/r	ISO	mm	m/min	mm/r	ISO	m/min	mm/r	ISO
5. ALLOY STEELS, WROUGHT (cont.)	175 to 225	Hot, Rolled, Annealed or Cold Drawn	.040	135	.007	M2, M3	375	500	.007	C 7	650	.007	CC 7
			.150	105	.015	M2, M3	300	400	.020	C 6	525	.015	CC 6
Medium Carbon			.300	80	.020	M2, M3	240	315	.030	C 6	400	.020	CC 6
			.625	65	.030	M2, M3	190	250	.040	–	–	–	–
1340 4340 81B45			1	41	.18	S4, S5	115	150	.18	P10	200	.18	CP10
1345 50B40 8640			4	32	.40	S4, S5	90	120	.50	P20	150	.40	CP20
4042 50B44 8642			8	24	.50	S4, S5	73	95	.75	P30	120	.50	CP30
4047 5046 8645			16	20	.75	S4, S5	58	76	1.0	–	–	–	–
4140 50B46 86B45	225 to 275	Annealed, Normalized, Cold Drawn or Quenched and Tempered	.040	115	.007	M2, M3	350	465	.007	C 7	600	.007	CC 7
4142 5140 8740			.150	90	.015	M2, M3	280	365	.020	C 6	475	.015	CC 6
4145 5145 8742			.300	70	.020	M2, M3	220	285	.030	C 6	375	.020	CC 6
4147 5147			.625	55	.030	M2, M3	170	225	.040	–	–	–	–
			1	35	.18	S4, S5	105	140	.18	P10	185	.18	CP10
			4	27	.40	S4, S5	85	110	.50	P20	145	.40	CP20
			8	21	.50	S4, S5	67	87	.75	P30	115	.50	CP30
			16	17	.75	S4, S5	52	69	1.0	–	–	–	–
			.040	90	.007	T15, M42⁺	330	440	.007	C 7	575	.007	CC 7
			.150	70	.015	T15, M42⁺	260	340	.015	C 6	450	.015	CC 6
			.300	55	.020	T15, M42⁺	200	270	.020	C 6	350	.020	CC 6
			.625	–	–		–	–	–	–	–	–	–
				27	.18	S9, S11⁺	100	135	.18	P10	175	.18	CP10
				21	.40	S9, S11⁺	79					.40	CP20
					.50								CP30

FIGURE 41-21 Portion of table from the *Machining Data Handbook,* 3rd ed., which gives recommended speeds and feeds for a given material; broken down by operation, material hardness, and condition according to what tool material is being used at a given depth of cut. Such data are considered to be reliable, but their accuracy is not guaranteed. *(Courtesy of Metcut Research Associates, Inc.)*

is to be done on NC machines, and experience has shown that these preplanning steps are advantageous when ordinary machine tools are used.

The selection of speeds and feeds required to manufacture the part may require referring to handbooks in which tables of the type shown in Figure 41-21 are given. For the threaded shaft, this table will give suggested values for the turning and facing operations for either high-speed-steel or carbide tools. Note that the tables are segregated by workpiece material (medium carbon-alloy steels, wrought or cold worked) and then by process, in this case, turning. For these materials, additional tables for drilling and threading would have to be referenced. Notice that the depth of cut dictates the speed and feed selection. The depth of cut for a roughing pass for operation 60 was 0.081 inches, and the finish pass was 0.007 inches. Therefore, for operation 60 three roughing cuts and one finishing cut are used. Why? Could you estimate or determine the cutting time for one of these roughing cuts?

While the operation sheet does not specifically say so, high-speed-steel tools are being used throughout. If the job were being done on an NC lathe, it is likely that all the cutting tools would be carbides which would change all the cutting parameters. Notice also that the job as described for the engine lathe required two setups, a three-jaw chuck to get the part to length and produce the centers and a between-centers setup to complete the part. It was necessary to stop the lathe to invert the part. Do you think that this part could be manufactured efficiently in an NC lathe? How would you orient the part in the NC lathe?

Data of the type shown in Figure 41-21 are often computerized to be compatible with computer-aided manufacturing and computer-aided process planning systems but the manufacturing engineer is still responsible for the final process plan.

As noted earlier, the cutting time, as computed from the machining parameters, represents only 30 to 40% to of the total time needed to complete the part. Referring again to Figure 41-20, the operator will need to pick up the part after it is cut off at operation 30, stop the lathe, open up the chuck, take out the piece of metal in the chuck, scribe a line on the part marking the desired $3\frac{1}{2}$ in. length, place the part back in the chuck, change the cutoff tool to a facing tool, adjust the facing tool to the right height, and move the tool to the proper position for the desired facing cut in operation 40. All these operations take time, and someone must estimate how much time is required for such noncutting operations if an accurate estimate of total time to make the part is to be obtained.

When an operation is machine-controlled, as in making a lathe cut of a certain length with power feed, the required time can be determined by simple mathematics. For example, if a cut is 10 in. in length and the turning speed is 200 rpm, with a feed of 0.005 in. per revolution, the cutting time required will be 10 minutes. See Chapter 23 for details of this calculation. Procedures are available for determining the time required for people-controlled machining elements, such as moving the carriage of a lathe by hand, back from the end of one cut to the starting point of a following cut. Such determinations of time estimates are generally considered the job of the industrial engineer, and space does not permit us to cover this material in this book, but the techniques are well established. Actual time studies, accumulated data from past operations, or some type of motion-time data, such as MTM (methods-time-measurement), can be employed for estimating such times. Each can provide accurate results that can be used for establishing standard times for use in planning. Various handbooks and books on machine-shop estimating contain tables of average times for a wide variety of elemental operations for use in estimating and setting standards. However, such data should be used with great caution, even for planning purposes, and they should never be used as a basis of wage payment. The conditions under which they were obtained may have been very different from those for which a standard is being set.

ROUTE SHEETS *Route sheets* are very useful for general planning in the job shop even though they do not provide detailed information for the operation. The route sheet lists the processes that must be performed to produce the part, in their sequential order, and the machines or work stations and tooling that will be required for each operation. For example, Figure 41-19 shows the drawing of a small, round "punch," and the routing sheet for making this part. Once the routing of a part has been determined, the detailed planning of each operation in the processing can then be done and the operations sheet prepared. The route sheet is also needed by *production planning,* another functional group in the PS that schedules the products (parts) through the machines and assigns the personnel to run the machines. Notice that the route sheet does not provide information on when (i.e., what day and time) the parts are to be made.

QUANTITY VERSUS PROCESS AND MATERIAL ALTERNATIVES Most processes are not equally suitable and economical for producing a wide range of quantities for a given product. Consequently, the quantity to be produced should be considered, and the design should be adjusted to the process that actually is to be used before the design is finalized. As an example, consider the part shown in Figure 41-22. Assume that, functionally, a brass alloy, a heat-treated aluminum alloy, or stainless steel would be suitable materials. What material and process would be most economical if 10, 100, or 1000 parts were to be made?

If only 10 parts were to be made, lathe turning, milling the flat, milling the slot, and drilling and tapping the holes would be very economical. The part could be machined out of bar stock. Casting would require the making of a pattern, which would be about as costly to produce as the part itself. It is likely that a suitable piece of stainless steel, brass, or heat-treated aluminum alloy would be available in the correct diameter

FIGURE 41-22 Pinion coupling to be analyzed for production.

and finish so that the largest diameter would not need any additional machining. Brass may be slightly more expensive than the other materials in low volumes. However, brass material with sawing, turning, milling, and drilling most likely would be the best combination. For 10 parts, the excess cost of brass over stainless steel would not be great, and this combination would require no special consideration on the part of the designer.

For a quantity of 100 parts, an effort to minimize machining costs may be worthwhile. Forging (to a near net shape) might be the most economical process, followed by machining. Stainless steel would be cheaper than any of the other permissible materials. Although the design requirements for forging this simple shape would be minimal, the designer would want to consider them, particularly as to whether the part should be forged, as this will reduce machining time. (Can stainless steel be forged?)

For 1000 parts, entirely different solutions become feasible. The use of an aluminum extrusion, with the individual units being sawed off, might be the most economical solution assuming that the time required to obtain a special extrusion die was not a detriment. Aluminum can be cut at much higher speeds than stainless steel. What material would you recommend for the part now? How many feet of extrusion would be required, including sawing allowance? How much should be spent on the die so that the per-piece cost would not be great and probably would be more than offset by the savings in machining costs? If extrusion is used, the designer should make sure that any tolerances specified are well within commercial extrusion tolerances.

This simple example clearly illustrates how quantity can interact with material and process selection, and the selection of the process may require special considerations and design revisions on the part of the designer. Obviously, if the dimensional tolerances were changed, entirely different solutions might result. When more complex products are involved, these relations become more complicated, but they also are usually more important and require detailed consideration by the designer.

Other duties of the manufacturing engineer may include the responsibility for the design of tools, jigs, and fixtures to produce the product. Just as the engineer who is concerned with manufacturing must understand the operation, functionality, and capability of machine tools but almost never designs them, similarly he or she should have a thorough understanding of the basic principles of jigs and fixtures so as to utilize them effectively. In large companies, the design of the tooling is left to the tool design specialists. In most companies, the manufacturing engineer makes recommendations on new machine tools, cutting tools, workholding devices, and material handling equipment.

After the product is in production, manufacturing problems invariably arise, and because of the functional design and operation of the existing system (the job shop), the manufacturing engineer has the responsibility for solving them. A typical scenario might entail poor-quality materials being received from a vendor and accepted (in error). Use of these poor-quality materials causes fixtures to work improperly, producing defective parts and resulting in components that cannot be assembled. Assembly workers, on incentive pay, will sacrifice quality for the sake of the piece rate, and the company will ship defective products. Finding the cause of such problems is part of the manufacturing engineer's job and is sometimes called *troubleshooting* or *firefighting*. The actual causes of problems may not be eliminated because of the pressure to keep on schedule. Since the causes are not eliminated, the defects keep coming back (in greater numbers) when material shortages occur in the system. Manufacturing engineers can find themselves responsible for a system that cannot possibly be controlled, because of its functional design.

Industrial Engineering. *Industrial engineers* are usually responsible for determining the level of workers, machines, and materials needed on the plant floor to turn the ideas developed in R&D, marketing, and procurement into real products. Industrial engineers look for the "better way" to produce products and services under uncertain conditions and constraints, such as the nature of the plant, materials, machines on hand, personnel, and available capital. The industrial engineering department is responsible for many elements of the MS and PS, including some that overlap with those of the manufacturing engineer. These include

1. Production methods analysis (motion study)
2. Work measurement (time study and time standards)

3. Safety and ergonomics

4. Factory design (layout) and material handling

5. Quality engineering (may be a separate functional group)

6. Plant maintenance information

The industrial engineering determines, standardizes, and analyzes the methods used to produce particular products and services. Motion-study principles, videotapes and movies, SMED (see Chapter 43), and other techniques are used to determine how a product or service can best be produced and to develop efficient work methods. In other words, the sequence of activities as well as the machines, tools, and materials to be used must all be specified in the job shop environment.

After standardization of the job's content, information is obtained on how much time is required to do the job. This information is based on the time required for an average person to produce a given product or service, using average effort under normal working conditions. Time standards and studies are used to develop standard times for the standard methods. The method for a particular job can be analyzed using an *operations analysis sheet.*

Designing the plant layout and the associated materials-handling equipment falls to industrial engineering. Layouts that reduce manufacturing costs with minimized materials handling and inventory are fundamental to integrated manufacturing systems.

Plant Engineering.   Plant engineering, another functional engineering group within the classical production system, is responsible for in-plant construction and *maintenance,* machine tool and equipment repair, heating and air-conditioning system maintenance, and for any other mechanical, hydraulic, or electric problems not necessarily related to the manufacturing system. For example, suppose that a new machine tool were being purchased. Plant engineers would be responsible for seeing to the installation of the machine.

Quality Engineering.   *Quality engineering* is responsible for assuring that the quality of the product and its components meets the standards specified by the designer before, during, and after manufacturing (Figure 41-23). Inprocess quality control inspections are performed at various points throughout the manufacturing system. In the mass production system, materials and parts purchased from outside suppliers are inspected when they are received. The acceptable quality level (AQL) of this incoming material is traditionally around 2%, which means that the company is willing to accept 2% defectives from their vendors. Historically, many companies have believed that it costs too much to reduce defectives below this level. In recent years, goals of perfection, zero defects, and defect rates of parts per million have been the new way of life in the world of quality. Quality is designed into the product and the processes.

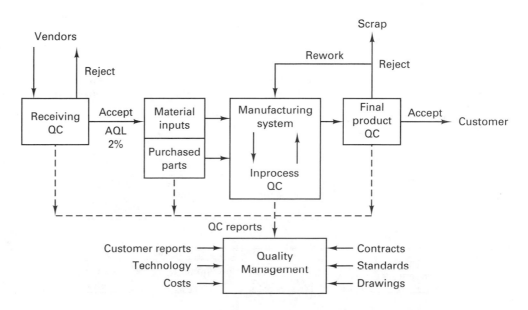

FIGURE 41-23   Classical quality control system, typically used in the job shop.

Parts fabricated inside the company may be inspected many times during processing. Final inspection and testing of the finished product is performed to determine overall functional performance and appearance quality. See Chapters 12 and 43 for additional discussions on quality.

PROCUREMENT AND PURCHASING

The *procurement and purchasing* functions in a company involve primarily the acquisition of specified materials; equipment, services; and supplies of the proper quality, in the correct quantities, at the best prices, and at the correct time. Many departments are involved in procurement, purchasing, manufacturing, marketing, finance, accounting, research and development, and engineering. For production systems, procurement provides information on vendors, prices, new products, and materials and determines the delivery schedule for purchased items.

PRODUCTION PLANNING AND CONTROL

Production planning translates sales into forecasts by part number. The authority to manufacture the product is translated into a *master production schedule,* a key planning document specifying the products to be manufactured, the quantity to be produced, and the delivery date to the customer (Figure 41-24). The master schedule is converted into purchase orders for raw materials, orders for components from outside vendors, and production schedules for parts made in the manufacturing system. *Production control* develops the timing and coordination to ensure that delivery of the final product meets customer demand. Because of the complexity of the job shop MS, production is not controlled very well; therefore, many other control functions are needed.

The scheduling periods used in the master schedule are usually months. The master schedule must take into account the production capacity of the plant (how much can be built in a given period of time). The capacity of a job shop is tremendously variable

FIGURE 41-24 Master production schedule indicates what products to produce and when the products are needed.

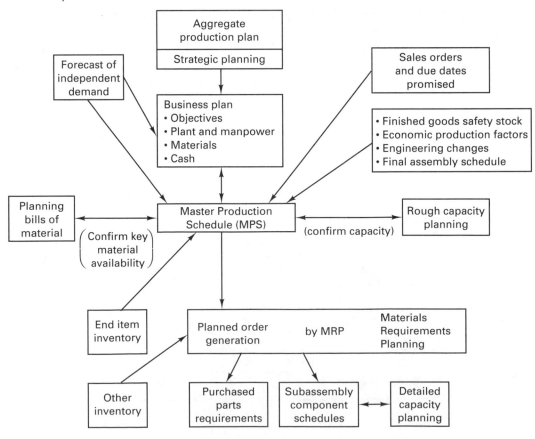

and flexible and is not well controlled. Because of this characteristic, larger quantities of products are often requested in violation of the master schedule.

Based on the master schedule, individual components and subassemblies that make up each product are planned. Raw materials are ordered to make the various components. Purchased parts are ordered from vendors. Planning is a must if the components and assemblies are to arrive when needed, not months before or, even worse, days or weeks late. Many companies use a computer software technique called *material requirements planning* (MRP), which is discussed later in this chapter.

The next task is *production scheduling,* in which start dates and due dates are assigned for the various components to be processed through the factory. Many factors make the scheduling job complex. The number of individual parts can be in the thousands. Each part seems to have its own individual process route through the plant. Parts are often routed through dozens of separate machines in many different departments. The number of machines in the shop is limited, and the machines are different, perform different operations, and have different features, capacities, and capabilities. In effect, the orders compete with each other for machines. In addition to these factors, parts become defective during the processing, cycle times vary, machines break down, and operators expand job times to fit the time available. All these factors destroy the validity of the planning schedule and require (in the classical system) huge amounts of resources (lots of people and paperwork) to manage all the exceptions. Chaos reigns supreme in the large job shop.

In Chapters 42 and 43, we will discuss the linked-cell manufacturing system design which compensates for the above kinds of variations. The WIP in the links between the cells is maintained at a level to cover the expected and unexpected problems. Only final assembly is scheduled. Upstream processes produce parts based on consumption from upstream cells or suppliers.

Dispatching is a production planning and control function requiring voluminous paperwork wherein individual orders comprised of order tickets, route sheets, part drawings, and job instructions are sent to the machine operators or foremen.

Expediters find lost or late materials by tracking the progress of the order against the production schedule. To speed up late orders, the expediter may rearrange the order-processing sequence for a certain machine, coax the foreman to tear down one setup so that another order can be run, or hand-carry parts from one department to the next just to keep production going. Obviously, the master schedule is disrupted. The size of the production control department and the number of expediters in a company is an informal measure of the level of chaos and inefficiency in the production control system.

INVENTORY CONTROL

The life blood of a manufacturing system is its inventory. Inventory represents a major portion of a manufacturing facility's valuable assets (i.e., capital). For most companies, however, excess inventory represents idle investment dollars and wasted storage space. Even though the cost of carrying large inventories is substantial, many reasons are given for having them:

1. Fluctuation in demand and/or supply
2. Protection against process breakdowns
3. Replacement parts for lost batches or defective lots
4. Overproduction in anticipation of future demand
5. Protection from defective parts
6. Goods in transport
7. Just in case they are needed
8. Quantity purchasing

Inventory control governs finished goods, raw materials, purchased components, and work-in-process within the factory. The idea is to achieve a balance between too little inventory (with possible stockouts of raw materials) and too much inventory (with investments and storage space tied up).

The product, when finished, is either shipped directly to the customer or stocked in inventory. *Inventory control is* used to ensure that enough products of each type are available to satisfy customer demand. However, competing with this objective is the company's desire to minimize its financial investment in inventory. Inventory control interfaces with marketing and production control since coordination must exist among the various products' sales, production, and inventory levels. Although none of these three functions can operate effectively without information about what the others are doing, this information is often missing or out of date. Inventory control is often done by the production planning and control department.

The requirements for a total inventory control system are that it

1. Analyze and plan inventory requirements.
2. Purchase raw materials and component parts in the amounts needed according to scheduled usage.
3. Receive and record the receipt of purchased materials.
4. Provide adequate facilities to store raw material, work-in-process, and finished goods inventory.
5. Maintain accurate records of inventories on hand and on order.
6. Install realistic controls for materials in stores and for the issuance of materials, parts, and supplies.

A good inventory control system provides the correct quantity and quality of material at the correct time. This system also maintains accurate records/control of these materials.

Inventory Models. In 1915, Ford Harris and R. H. Wilson derived, independently, the *simple lot-size formula.* This model states (see Figure 41-25) that

$$\text{total annual cost} = \text{purchase cost} + \text{order cost} + \text{holding cost}$$

$$TC = RP + \frac{RC}{Q} + \frac{QiP}{2} \tag{41-1}$$

where

R = annual demand (units)

P = unit cost of an item

C = ordering cost per order

$H = iP$ = holding cost per unit per year

Q = lot size or order quantity (units) or conveyance quantity between departments in a job shop

i = annual holding cost as a fraction of unit cost

FIGURE 41-25 The economic order quantity (EOQ) model minimizes the total (annual) cost of inventory.

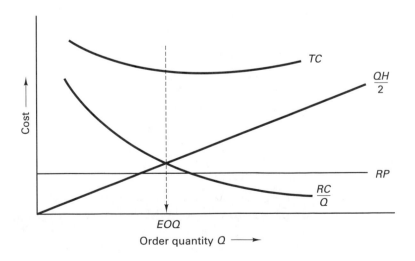

To find the value of Q that minimizes the total cost, use

$$\frac{d(TC)}{dQ} = 0 - \frac{RC}{Q^2} + \frac{iP}{2} = 0$$

$$\frac{RC}{Q^2} = \frac{iP}{2}$$

$$Q = \sqrt{\frac{2RC}{iP}}$$

The model, called the *economic order quantity model,* has also been extended to cover economic production quantity by letting C equal the setup cost. The model assumes that the setup cost is fixed when, in fact, the setup cost can often be significantly reduced. Reduction in the setup cost results in a reduction in lot size. The best lot size is then the smallest lot size that permits a smoothly running manufacturing system.

Many other models have been developed for the inventory process, and many models and systems have since evolved for stock or inventory replenishment, including the *reorder point model.* Figure 41-26 illustrates the relationship between lead time *(L),* order size *(Q),* safety stock (SS), expected demand rate *(D),* and the reorder point (ROP). The reorder point is based on how long it takes to obtain parts (lead time) and how many parts will be used up during this (lead) time.

One of the problems with the reorder point model is the dependence of the model on usage. The model shows that parts are used linearly. In fact, in batch processing, utilization of parts is usually uneven, with large spikes or peaks. That is, 200 of an item may be used in week 2, zero in weeks 3 through 9, and 200 again in week 10. The reorder point model would cause the system to run out of parts in week 2, precipitating a rush order for parts that would remain on the shelf unnecessarily until week 10. The model also breaks down if the average utilization changes. To use the model effectively, demand must be constantly recalculated.

The utilization of certain parts is dependent on the demand for other parts. For example, in the building of cars, the demand for tires depends on how many cars are built. While demand for cars must still be estimated, the demand for tires is known, based on the demand for cars.

FIGURE 41-26   Classical inventory model for ROP (reorder point) methods.

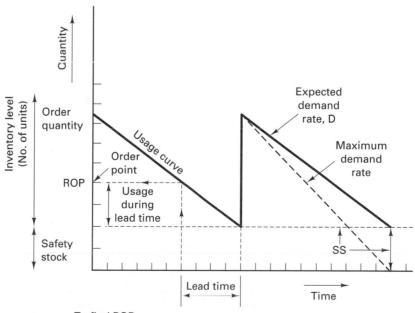

To find ROP
1. Establish lead time.
2. Back lead time off from order receipt time.
3. Go to usage curve.
4. Find order pont on vertical scale.

These two types of demand are categorized as *independent* and *dependent*. The first must be "guessed" (forecasted). The second can be computed for scheduling purposes but not for plant production control. The basis for these computations is a record of the relationship between the independent demand item (cars) and its dependent components, (tires, horn, windows.) This record is the *bill of materials* (see Figure 41-18).

Manufacturing Resource Planning.   To control inventory within the job shop, a computerized system called *material requirements planning* (mrp) was developed (Figure 41-27). Given a schedule showing the expected demand of independent demand items (a master production schedule) and given the relationship between independent and dependent demand items (bills of materials), mrp will calculate the quantities of dependent demand items needed and when they will be needed.

The potential value of this concept is significant. Suppose that a company has 100 finished goods items, 400 assemblies and subassemblies, and 1000 raw material items. Using statistical stock replenishment, it will need to forecast the average demand for 1500 items. Many of these items will have "lumpy" demand, which will cause the stockout/expedite/overstock cycle discussed earlier. With mrp, a *master production schedule* (MPS) for 100 items must be maintained; the other 1400 items will have their exact demand computed for every period.

Almost at the same time that mrp was developed, practitioners began to expand the concept's scope. Just as bills of materials could establish the usage of dependent

FIGURE 41-27   Material requirements planning is a computerized inventory system for the job shop.

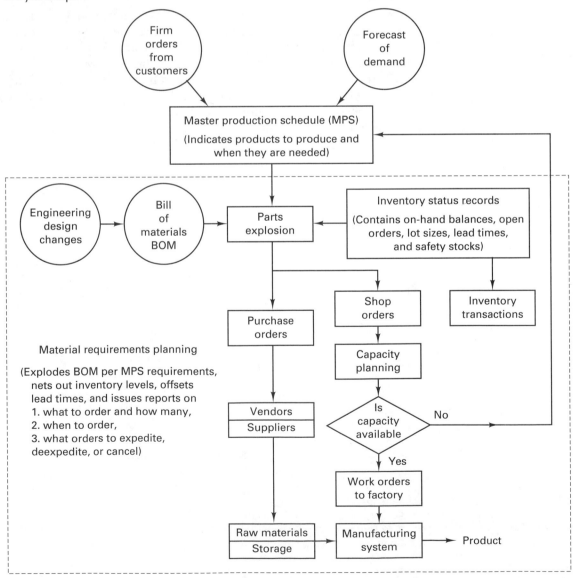

materials, other records could be developed to tell the dependent requirements of labor hours, machine hours, capital, shipping containers—in fact any of the resources required to support the job shop MPS. Material requirements planning has become *manufacturing resource planning* (MRP). More recently, management consultants have been peddling enterprise resource management (ERP) software, which is responsible for final scheduling of production, dispatching, and releasing purchase orders. It will also maintain stock status, monitor output and scrap levels, and compare performance against the plan.

MRP and mrp are good *planning* techniques. There is no inherent capability to *control* and replan. Subsequent sophistication of MRP by adding feedback of actual results has led to *closed-loop manufacturing resource planning* (MRP II). Shop floor control and vendor control systems have been added to the existing software so that revisions of dates and quantities can be taken into account in the next planning cycle.

The job shop is a complex dynamic manufacturing system. Inventory is its life blood. MRP is an attempt to control inventory using computers. Unfortunately, the MRP systems that have evolved are so complex that very few people in the companies that use them really understand them. Would you trust a system you did not understand? What is the value of an inventory control system that the majority of users do not comprehend?

What makes a good production/inventory control system? First, everyone who uses it must understand how it works. It must have accurate information and make accurate predictions or forecasts. The users of the system must act on the information that the system produces. In Chapter 43 a simple, remarkable system is described that has become known as the pull system (or the JIT kanban system). But a word of caution: This inventory control system is not usable in the job shop. It is meant for a new MS called the lean shop with its linked-cell MS, which can replace the job shop.

■ KEY WORDS

accounting
bill of materials
continuous process
customers (internal and external)
decouplers
design engineering
economic order quantity
enterprise

flow shop
Ford Production System
industrial engineering
industrial revolutions
inventory control
job shop
lean production

lean shop
linked-cell
machine tools
manufacturing engineering
manufacturing resource planning
manufacturing system

marketing
mass production
master production schedule
operation sheet
production planning and control
production system

project shop
purchasing
quality control/engineering
route sheet
standard work-in-process (SWIP)
Tooling
Toyota Production System (TPS)

■ REVIEW QUESTIONS

1. What are the major functional elements or departmental areas of the production system?
2. a. What is a route sheet?
   b. What is the function of a route sheet?
   c. Find an example of a route sheet other than the one in the book.
   d. What are other names for a route sheet?
3. What is a process flow chart? How is it related to the bill of material?
4. What is an operations sheet? How is it related to the route sheet?
5. What is the master production schedule (MPS)?
6. What is MRP and how is it related (tied to) the MPS?
7. How is the EOQ calculation used in MRP?
8. What is the functional objectives of production control?
9. a. What is the function of inventory control?
   b. How are inventory levels controlled?
10. What is the difference between production control and inventory control?
11. How does the design of the product influence the design of the manufacturing system, including assembly and the production system?
12. Discuss this statement: "Software can be as costly to design and develop as hardware and will require long production runs to recover, even though these costs may be hidden in the overhead costs."
13. Suppose that the part in Figure 41-20 was to be made in quantities of 250 rather than 25. What processes and machines would you have selected?
14. Explain the difference between a route sheet and an operations sheet (or a process planning sheet).
15. What does the study of ergonomics entail? (This was not discussed in Chapter 41.)
16. How does the mrp use the BOM and the MPS to determine how many of an item should be ordered?
17. Define CIM in your own words.
18. Table 41-1 lists some examples of service job shops. Compare your college to a manufacturing/production system, using the definition of a manufacturing system given in the chapter. Who

is the internal customer in the academic job shop? What or who are the products in the academic job shop?

19. In project shop manufacturing, what is the *critical path*?

20. Explain how function dictates design with respect to the design of footwear. Use examples of different kinds of footwear (shoes, sandals, high heels, boots, etc.) to emphasize your points. For example, cowboy boots have pointed toes so that they slip into the stirrups easily and high heels to keep the foot in the stirrup.

21. Most companies, when computing or estimating costs for a job, will add in an overhead cost, often tying that cost to some direct cost, such as direct labor. What costs are usually included in this overhead cost?

22. What are some things you could do to improve the stability of a manufacturing system?

23. Is there a critical path through the academic job shop.

24. How does the job shop, the flow shop, and the lean shop manufacturing systems perform in terms of quality, cost, delivery, and flexibility?

25. What are the possibilities of incorporating lean manufacturing concepts into a high-volume transfer line for machining?

26. How have the so-called mass plants evolved into lean plants?

27. Quantify the management accounting thinking that drives mass plants. Provide an equation for discussion.

28. How did the Toyota thinking build on Ford's system design?

29. What is the impact of minimizing the unit cost of each operation:
    a. On machine design?
    b. On the workers?
    c. On the factory as *a system*?

30. How did the Ford thinking build on the first industrial revolution?

*C*hapter 41   CASE STUDY

Fire Extinguisher Pressure Gage

As a materials engineer for the War Eagle Extinguisher Company, Carlos has recently been made aware of several potentially hazardous failures that have occurred in his company's products. Bourdon tube pressure gages (Figure CS-41) are used to monitor the internal pressure of the sodium bicarbonate (dry chemical) fire extinguishers. In several extinguishers, a longitudinal crack has formed along the axis of the bourdon tube, a curved tube of elliptical cross section that has been fabricated from phosphor bronze tubing. These cracks are particularly disturbing for several reasons. First, they allow the fire extinguisher to lose pressure and become inoperable. More significantly, however, the cracks allow the tube to deflect elastically in such a manner that the gage still indicates high internal pressure. Thus, while the extinguisher has actually lost all internal pressure and is useless, the reading on the

gage would cause the owner to believe that he still had an operable firefighting device.

Carlos is trying to determine the cause of these failures and suggest appropriate corrective measures. Can you help him?

1. What additional information would you like to have regarding the failed components, their fabrication history, and their service history? Why?

2. What might be some of the possible causes of these failures? What type of evidence would support each possibility? What types of additional tests or investigations might you propose to Carlos.

3. Could these failures have occurred in "normal" use, or is it likely that some form of negligence, abuse, or misuse was involved?

4. What possible corrective or preventative measures might you suggest to Carlos to prevent a recurrence?

FIGURE CS-41

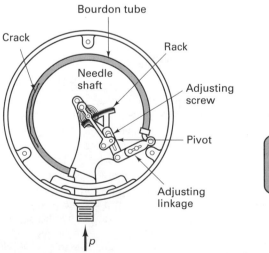

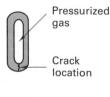

(a) Schematic of a bourdon-tube pressure gage   (b) Cross section of bourden tube

SYSTEM AND CELL DESIGN

■ 42.1 INTRODUCTION

This chapter examines the subject of system and cell design to provide a framework for thinking about manufacturing and enterprise (or production) system design and its associated implementation. Then subassembly and manufacturing cell design concepts will be introduced.

Understanding manufacturing system and production (enterprise) system design requires the understanding of certain system design principles and the system implementation process. System design principles convey the thinking or logic that exists for a given system design. A system design implementation process outlines the steps necessary to put the system design principles into practice. The implementation process assumes that the logic or thought process of a system design can be determined in advance of the actual implementation, which will be discussed in Chapter 43. Therefore, the logic of the system design must be determined prior to any implementation.

■ 42.2 ENTERPRISE (PRODUCTION) SYSTEM DESIGN

An enterprise or production system supports the manufacturing system's value-adding work. A manufacturing system consists of the elements necessary to convert a product from one state to another, that is, to a more value-added state, see Figure 42-1. Elements in a manufacturing system consist of machines, people, material handling devices, buildings, utilities, and the like. The manufacturing system is where the work on the product occurs to convert raw materials and subassemblies into finished products to directly add value to a part.

Defining an enterprise system this way is the result of understanding that the external customer (i.e., marketplace) that actually buys a company's superior-quality products generates income and ultimately profit. This perspective with respect to the interface between manufacturing and the enterprise also recognizes that the manufacturing work within an enterprise represents the beating heart of that enterprise. A football team running its plays is analogous to a manufacturing system producing its goods. In football, the enterprise system comprises coaches, managers, ticket agents, vendors, announcers, and so on, who support the team but do not play on the field.

Manufacturing systems can be designed. However, most scholars and many engineers do not take this view. They tend to look at systems mostly as optimization problems. The systems are assumed to preexist. Axiomatic design teaches us that most optimization problems are the result of coupled designs, wherein the key functional requirements interact. When a design is uncoupled, optimization is not necessary.

Some may view systems in terms of the physical manifestation of the system only. Many textbooks in production and operations management define certain physical system entities to be the solution for certain volume or mix problems. However, a manufacturing system's physical design is not limited to only two factors: mix and volume.

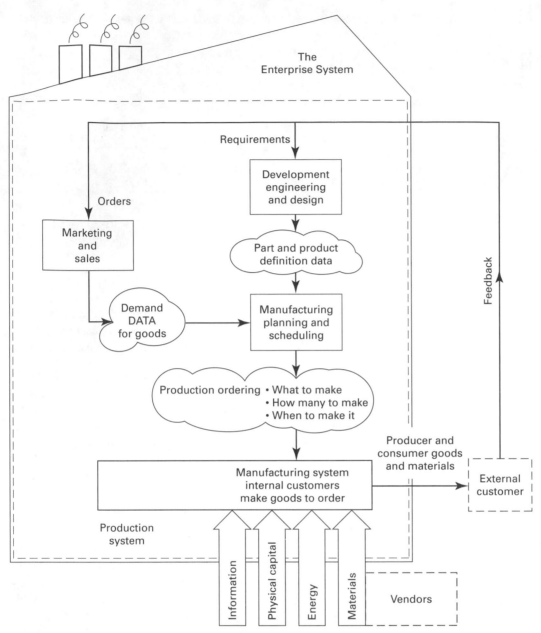

FIGURE 42-1 The enterprise system takes product demand information and product definition data and uses them to plan the activities in the manufacturing system.

There are many factors that affect a manufacturing system's design, and in recent years some significant changes have taken place in the design of manufacturing systems. These changes are fueled by the following trends:

1. The customer's demand for "cheaper, better, faster," meaning reduced cost, better quality, and on-time delivery, has forced many companies to implement lean manufacturing.

2. Manufacturing has become a global operation as auto producers build plants worldwide. The new manufacturing system design must be functional in every country, with any language and culture.

3. The proliferation of variety in products has resulted in a decrease in quantities (lot size). Increases in variety require the design of more flexible manufacturing systems to meet changes in mix and volume rapidly.

4. Increased variety in *materials*, including growth in the use of composite materials with widely diverse properties, has created a need for the proliferation of new manufacturing processes. Manufacturing systems are designed so that new materials and

processes can be readily introduced. In fact, management at some companies believe they have a competitive edge because they have developed proprietary processes that their competition does not have access to.

5. The continued decrease in the number of direct laborers has resulted in the cost of materials (including materials handling and energy) being a major part of the total product cost. Direct labor has dropped to 5 to 10% of the total manufacturing cost in almost every factory.

6. Product reliability has increased in response to the increasing cost of product liability (lawsuits) for use and abuse of the products.

7. The decrease in the time between design of a product and its daily manufacture (shorter time-to-market for new products) has resulted in a more adaptive manufacturing system design.

8. The continued growth of computer use has led to integrated product and system design, including the use of 3D simulation of the manufacturing system.

9. The increased design and use of manufacturing and assembly cells that are able to match the production rates of final assembly has resulted in increased flexibility—the ability to quickly change volume and mix.

COUPLED VERSUS UNCOUPLED DESIGNS

In light of these trends, systems must be designed to achieve certain functional requirements (FRs). The physical manifestation of a system is achievemed when the functional requirements are uncoupled, which simply means there is no interaction between the FRs; that is, the FRs are independent.

However, many traditional notions of business strategy are contrary to the concept of uncoupling to the detriment of the system design. Most business-strategy literature views systems in terms of optimization, not design, problems. Optimization advocates that a business should choose between low cost and high quality. For example, in the mass production system, if a company tried to reduce defect levels below around 2% (the acceptable quality level), the reaction was that cost increased. In this design, cost and quality are coupled.

The system design approach, however, might define low unit cost and superior quality as the functional requirements of a system design. A system design would then determine how to achieve these functional requirements (FRs) through the selection of design parameters (DPs). DPs define a means to achieve the FRs. The choice of the DPs renders uncoupled FRs. In many cases, the trade-offs that business strategists have identified can be explained in terms of path dependency in the design. *Path dependency* means that one solution can affect the achievement of multiple FRs. For example, a company with a faulty manufacturing system design may be the lowest cost producer but may not be achieving the FRs related to safety or quality. If the functional requirements of safety (or quality) are not consistently achieved, it makes no difference that the company is a low cost producer.

An example of a coupled design is shown by the following 2 FRs.

FR-1: Reduce cost   ⟶   DP-1: Reduction in direct labor by automation

FR-2: Rapid delivery   ⟶   DP-2: Inventory reduction

The above is a simple example of a coupled design. The relationship of the FRs relative to the DPs is given by

$$
\begin{matrix} FR\text{-}1 \\ FR\text{-}2 \end{matrix} = \begin{bmatrix} x & x \\ x & x \end{bmatrix} \begin{matrix} DP\text{-}1 \\ DP\text{-}2 \end{matrix}
$$

The design is coupled. As DP-1 is changed, its effect on FR-1 and FR-2 is unpredictable. Likewise, DP-2 affects FR-1 and FR-2 in an unpredictable way. This situation immediately leads one to attempt to define the coefficients of the relationship matrix. Once the coefficients are known, one can begin to solve for the DP-1 and DP-2.

Optimization deals with coupled designs. In some cases, coupled designs lead to unwanted, negative, or unpredictable results. The best design is uncoupled, one DP affects only one FR. The relationship of the single DP to the single FR is one to one, and the DP satisfies the FR in a known and unambiguous way. No optimization is required when a design is uncoupled and the FRs are independent through the selection of the DPs. A new design could be

FR-1:   Reduce cost $\longrightarrow$ DP-1:   Waste & motion reduction

FR-2:   Rapid delivery $\longrightarrow$ DP-2:   System design to eliminate five delays

$$\begin{matrix} \text{FR-2} \\ \\ \text{FR-1} \end{matrix} = \begin{bmatrix} x & o \\ x & x \end{bmatrix} \begin{matrix} \text{DP-2} \\ \\ \text{DP-1} \end{matrix}$$

In this case, the design is partially coupled. The independence of the FRs can be achieved, but only if the DPs are implemented in a proper sequence or context. The path's dependent design indicates that it may be useless to eliminate wasted motion in a system where the five throughput time delays are not reduced. Reducing the five delays that compose throughput time most certainly will reduce cost.

Naturally, the best path of action would be an uncoupled design with independent FRs, but this is difficult to achieve.

ENTERPRISE SYSTEM DESIGN (ESD) PRINCIPLES

A *principle* can be defined as an accepted or professed rule of action or conduct; or an axiom; or even a doctrine.

ESD-P1: One's thinking creates a system design.   System designs are all around us. Every service business, organization, manufacturing company, industry, government entity, and school is the result of a system design. Most of the time, these organizations operate based on implicit knowledge and tacit assumptions and have not been really designed by any one person. The design that exists may have evolved over years and may represent the thinking or mental model of many people, based on conclusions assumed to be true about a system's behavior and performance.

Many systems evolve or change behavior based on the way they are measured, but many behaviors within systems are the result of using measures of the wrong parameters or placing the wrong emphasis of using measures of parameters to manage and operate a business.

The measured parameters are often based on tradition and result in coupled system designs. These measures may not be aligned to achieve the objectives or FRs of the system design. Stressing the use of measures to drive a business often leads to poor performance by the business enterprise. Instead, the need is to design and to emphasize through measures the relationships (FRs and DPs) necessary to achieve the desired business objectives.

Organizations that evolve based on tacit assumptions and coupled designs often work at cross-purposes to achieving their true objectives. Many people in an organization may understand the right thing to do even though the existing measures and the management-accounting approach dictate otherwise. The FR-DP relationships within the enterprise should be designed first, and then the organizational structure aligned to focus on successfully achieving the FRs.

ESD-P2: A system design expresses the relationship(s) between the FRs and DPs.
Functional requirements state what a system must achieve, that is, its objective. However, an objective must not be confused with means. DPs are the means by which FRs are achieved. For example, a common functional requirement for a system may be superior quality (zero defects). The design parameters may be control charts (to track process performance), histograms (to measure process capability) or pokayokes (to prevent defects).

It is a mistake to view DPs as goals, that is, as FRs. A classic example of a DP being thought of as an FR is the use of information technology in manufacturing.

A company invests heavily in computers, CAD, and CIM. In many cases, computerization of the enterprise becomes the FR. Instead, computers should be a DP to solve a business FR.

ESD-P3: Effective system designs are uncoupled. An uncoupled design expresses how a set of DPs should achieve a set of FRs. An uncoupled design is the result of achieving design independence. Path-dependent designs are partially coupled. These designs are acceptable and may be unavoidable. Lean production is the name given to a partially coupled (or path-dependent) system design developed by Toyota (called the Toyota Production System) and its vice president for manufacturing, Taiichi Ohno.

Using the term *lean* to express a system design is a superficial expression of enterprise system design accomplishments. Furthermore, in trying to become lean, a company may cut inventory or people. Many imitators of lean have often missed this point (it is a design problem) and instead have focused on implementing tools such as the 5Ss (described later in this chapter), pokayokes, and cells without first defining the FRs of a stable manufacturing system design. The physical manifestation of a system design changes as the FRs and DPs change. The development of uncoupled system designs is a difficult task, but an organization that recognizes and nurtures an uncoupled design will have superior long-term performance.

ESD-P4: True cost reduction is a result of a stable system design. The first step in the transformation to lean is to recognize what it means to become lean and to recognize why so many companies are not lean. The reason that most companies are not lean begins with how cost is viewed. There are two key points regarding the view of cost:

1. Focusing only on the end results of a system (low cost) does not reduce cost. The key is to focus on what a system must achieve and on the relationships and activities that are necessary to achieve the desired results.

2. Reducing the unit cost for an operation does not reduce total cost (Cochran, 1999). The unit-cost-reduction equation focuses on the reduction of cost of each operation (i.e., turning, grinding, washing, assembly) individually. The approach incorrectly focuses on reducing each operation's unit cost. It incorrectly assumes that total cost is reduced by minimizing the sum of each operation's unit cost, as if each operation is stand-alone and unrelated to other operations. This thinking is shown by

$$TC = \sum_{i=1}^{n} (OP_i)$$

$$\text{Unit Cost } (OP_i) = \frac{DL + OVHD}{n} + MTL$$

where

OP_i	= Unit Cost (\$ /piece)
DL	= direct labor (\$ /shift)
MTL	= material (\$ /piece)
$OVHD$	= overhead based on DL hours
n	= number of parts/shift

This approach does not reduce the complexity of the flow of the parts through a manufacturing system. It also does not focus on whether the manufacturing system can produce products at the pace of customer demand. This costing approach leads to departmental, mass-production-oriented plant designs. It also creates flow complexity between plants because certain operations might be outsourced to other plants that appear to have a lower unit cost for a particular operation or run at "optimal speeds" that produce excess inventory (i.e., overproduce).

Furthermore, the manufacturing engineers are forced to design either high speed (increase n) or highly automated (decrease DL) machines (or lines) to reduce the unit cost ($ /piece) of each operation by the unit-cost (OP_i) equation. The unit-cost equation always points to paying lower labor rates to reduce the unit cost of each manufacturing operation.

Building the part complete (either fabricated or assembled) in a single-piece flow cell reduces flow complexity and total cost. The lean manufacturing system design approach puts capacity in place in accordance with the system design principles of building a part complete, rapid problem-detection time (ability to distinguish abnormal condition from normal condition to facilitate problem identification and resolution), and system robustness to produce the right quantity and right mix. These principles guide the definition of the 6 FRs for stable system design. Manufacturing cost cannot be reduced until the manufacturing system is designed to be stable. A stable manufacturing system design must meet the following six FRs:

Six FRs for a Stable System Design (D. Cochran, MIT)

FR - 1	Right Quantity	Produce the right quantity every shift.
FR - 2	Right Mix	Produce the right mix of products to meet the variety in demand.
FR - 3	Right Quality	Produce and ship perfect quality everyday (zero defects).
FR - 4	Robust	Do FR-1, FR-2, FR-3 in the face of variation in the system or disturbances outside the system.
FR - 5	Rapid Problem Solving	Identify problems quickly and resolve problems in a standardized way. Continuously improve the system.
FR - 6	Safe, Ergonomically Sound	Manufacture in a safe, humane (bright, clean, quiet) environment with ergonomically sound processes.

These six FRs define the system design requirements. This system design can tolerate disturbances and problems (FR-4) to a certain degree. When that is exceeded, the people who are part of the system design know exactly what to do (FR-5) and respond in a standard way.

A system design defines how these requirements are achieved. System design connects the means of achievement (the activities or work) with the functional requirements or needs that a system must meet. These connections include the people within the system.

A stable system is robust (FR-4). The system is able to achieve its FRs even if variation is present. This variation may be due to incoming defects from the supplier, process fallout within the company or machine downtime. These sources of variation can be modeled as disturbances that affect the systems operation. A robust system design compensates for variation up to some preestablished limit.

So, FR-5 means that people must immediately react to a problem condition with respect to achieving FRs 1 through 4. An electronics supplier for Toyota, for example, has the system design parameter of reacting to a problem condition in just 10 seconds or less.

FR-6 requires that all work must be done in a safe, clean, quiet, bright, properly ventilated, and heated or cooled environment that is ergonomically sound. The environment should be one that the system designers themselves would work in day after day.

From an operating perspective, cost cannot be reduced until a system at least achieves the six FRs. What are the implications of these FRs on the operational system design?

For example, in a lean manufacturing system, a cell not producing the right quantity (or at the right pace) would immediately be identified and the problem condition for not producing at the right pace isolated. The supervisory/leadership team would immediately invoke a countermeasure to correct this problem. Management would not simply look at reports a day later to determine that a problem condition had occurred.

A shipment to the customer at the end of shift would not be delayed due to the problem on the line that, in this case, could not be fixed. The right quantity would be shipped to the customer, even though the right quantity had not been made during the shift. Yet, the team would know immediately during the day about the problem condition and would make every effort to catch up and rectify the problem. The manufacturing system was robust enough to meet the customer's FRs in the presence of internal variation. This robust design is accomplished through the design of the manufacturing

Designing the system for the internal customer recognizes these factors:

Factor	Requirements for manufacturing system design - the manufacturing engineer
Safety	Design ergonomically safe equipment to meet all safety standards.
Reliability in equipment	Consistent and durable.
Quality of job and plant environment	Easy to operate; no dirty, unpleasant, labor-intensive work; fail-safe designs.
Maintainability	Easy to maintain and simple.
Robustness to variation	Pull, SWIP, standard work.
Responsibility	Feedback from customers/users involved in decision making during implementation.
Service	Technical support and training manuals.
Continuously improving	Determine normal from abnormal.

Designing the system for the external customer recognizes these factors:

Factor	Requirement for manufacturing system design - the manufacturing engineer
Attractiveness or style	Fit and finish appearance, new technology and features, improvements and innovation.
Quality	High accuracy and precision, reliable, durable, and maintainable
Cost/price	Low initial cost, good operating cost, and long warranty
Delivery/predictable output	Standardized work; mixed-model, small-lot manufacturing; and quick startup for new models.
Flexibility	Model changes easy to do.

FIGURE 42-2 Designing the plant requires a systems viewpoint that considers the user of the plant and the buyer of the products made in the plant.

and subassembly cells. In the next section, an example of a cell will be detailed. It will be shown that a linked-cell system design is the physical manifestation of the six FRs.

System design must start with the recognition that the internal and external customers of a manufacturing system are of primary importance to a system design. Here are some factors that Honda considered in designing its manufacturing systems. See Figure 42-2.

Figure 42-2 shows the factors that a major lean automaker must take into account when dealing with its two customers. Clearly there will be conflicts among the factors, which must be resolved through compromise by the leadership. Leadership must understand how the system works to be able to manage it and teach others how it works. Systems comprise relationships. System designers define the relationships with a system. The key is to recognize whether the relationships are incomplete, redundant, coupled, uncoupled, or path dependent. Coupled relationships can never achieve the desired results.

PHYSICAL SIMULATION AND MANUFACTURING SYSTEM DESIGN

Physical simulation may be used to promote true learning and to teach people how manufacturing and production systems work. The key to promoting true learning is to first challenge people to understand the objectives of a manufacturing system's design and then to associate the physical implementation of the system design with the achievement of the manufacturing system's objective. Physical simulation is a proven way to create and implement successful change within a manufacturing organization. Physical simulation uses toy models of products and processes along with production control tools to simulate the operation of the factory. Figure 42-3 provides a flowchart of the change process to get people to think about redesigning their existing manufacturing system using physical simulation. A manufacturing system can be defined as an arrangement and operation of machine tools, tooling, materials-handling equipment, and people to produce a value-added physical, informational, or service product whose quality, value, and cost are characterized by measurable parameters.

1. *Current state value stream map (VSM).* The first step is to establish the value stream map of the current state manufacturing system and depict the processes and activities in both the material and information flows in the system. Value stream mapping creates a flowchart of the current process from start to finish, determining how the system works. This may be quite a task for large, complex systems.

2. *Current state physical simulation design and current state performance measures (PMs).* These PMs are used to develop a current state physical simulation based upon the

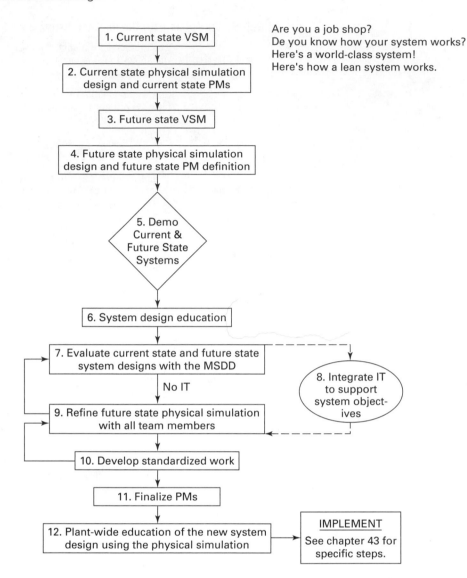

Are you a job shop?
De you know how your system works?
Here's a world-class system!
Here's how a lean system works.

FIGURE 42-3 Flowchart for the MSD change process using physical simulation.

existing manufacturing system's design and operating practices. The physical simulation should be a simplified, toy-scale model of an existing value stream within the system. This step should also capture how the existing system is measured and model the behavior of the people within the system resulting from the existing PMs. The current state physical simulation model provides the basis for people to learn, observe, and think about how to change the system.

3. *Future state VSM.* The material and information flow for the future state value stream should be designed (on paper) so that the objectives of a *stable* manufacturing system will be achieved. A stable manufacturing system design achieves the six FRs.

4. *Future state physical simulation design and future state PM definition.* The future state physical simulation must be designed in alignment with the future state VSM that schematically illustrates the design to achieve system stability. Concurrently, new performance measures should be established for the new manufacturing system. The performance measures should reward the inclusion of the six FRs in the system design. Management must not measure results alone; they must lead the system to re-design and implement FRs and DPs.

5. *Physical simulation to demonstrate the current state versus the future state.* This demonstration is the "gotcha" event, the learning milestone. The demonstration contrasts the operation of the current state system in a stable and an unstable manufacturing system. There is a sharp contrast in the roles of people in operating the stable versus the unstable system. In an unstable system, people's best efforts barely keep the

system operational. The focus is on trying to ship parts, sometimes any part. In contrast, a stable system design enables focused problem identification and improvement processes. People work on improving the work itself and not on merely deadlines with red tags. The simulation illustrates the cost of people not working on improving the system. The model of the stable system captures the hearts and minds of the people in the system, since they are able to see (sometimes for the first time) that their manufacturing system can be truly successful, but only if it is *designed* to achieve the objectives of a stable manufacturing system.

6. *System design education*. Following the "gotcha" milestone, educational workshops are needed to cement the learning. During these workshops the participants learn how to design a manufacturing system to achieve the system objectives (FRs) and means (DPs). These workshops also present case-study research results and the specifics for designing manufacturing systems.

7. *Evaluate current state and future state system designs*. This step ensures that the weaknesses of the existing and new system designs are identified clearly. A questionnaire can be developed to evaluate the current state and future state manufacturing system designs. The questionnaire should show how well the FR-DP pairs are actually achieved by the system design. Every member of the system design team should complete the questionnaire. This evaluation uses a technique called the *manufacturing system design decomposition* (MSDD). The MSDD defines the FRs and DPs that a system design must achieve holistically. It decomposes the objectives and means within a system design to improve quality, decrease delivery response time and to improve delivery reliability. It also states the objectives and means to reduce the root causes of operational cost. VSM identifies the material and information flow necessary to achieve the objectives and means stated by the MSDD.

8. *Integrate the information technology infrastructure to support system objectives*. A benefit of constructing the physical simulation model is that the use of the supporting computer information technology can also be physically simulated. This step requires the live and concurrent operation of the information technology with manufacturing. The physical simulation establishes a laboratory environment in which all of the important interfaces are exposed and tested. An important aspect of this integration is that it to enables all participants to agree on one model for the design and operation of the manufacturing system. The participants learn how their business practices affect manufacturing and, significantly, are able to redesign their business processes to meet the objectives of the manufacturing system. In fact, the participants will realize that they are indeed part of the manufacturing system and learn the impact of their decisions on the design and operation of the manufacturing system.

9. *Refine future state physical simulation with all team members*. Physical simulation requires multiple interactions of refinement to ensure success. All participants are asked to run the physical simulation. Each team member tries to make the simulation system fail by asking questions. The purpose of the step is to make the work methods in the simulation as realistic as possible and to improve the system design's robustness in addressing problem conditions.

10. *Develop standardized work*. Standardized work defines the work methods necessary to operate the manufacturing system. Standardized work affects the work of both salaried and hourly team members. In fact, standardized work defines how management will react to specific problem conditions. Developing standardized work is crucial to the successful launch of the new manufacturing system. The people who operate the new system must know *what* to do, *how* to do it, and *why* they are doing it. More important, they will have to be involved in the cell design process. Understanding standardized work helps the operating personnel understand their role in lean manufacturing. The physical simulation enables the participants to test standardized work methods. The standardized work methods must be written down. Significant changes to written standardized work instructions will be made as a result of testing the standardized work methods with the physical simulation.

11. *Finalize performance measures (PMs).* This step ensures that the PMs that are used to evaluate the new system's performance are aligned with the objectives of the new system design. It can be disastrous to operate a new system design and yet measure its behavior based on an inappropriate set of performance measures. Many systems evolve into physical designs based upon the way they are measured.

12. *Plantwide education of the new system design using the physical simulation.* Once the simulation has been designed and tested, and the standardized work developed, the physical simulation may be used as a powerful teaching and educational tool. This step captures the idea that everyone in the redefined manufacturing system may be taught the new system design using the physical simulation.

■ 42.3 LINKED-CELL MANUFACTURING SYSTEM (L-CMS)

Lean production and lean manufacturing systems evolve from meeting the FRs of a stable system design using the linked-cell manufacturing system (L-CMS). L-CMS was invented at the Toyota Motor Company by Taiichi Ohno, Vice President for Manufacturing, but he never gave it a name. He simply referred to it as the Toyota Production System (TPS). The strategy was brought to the United States by Toyota and many other companies and has been implemented in various forms. The term *lean* to describe the result of TPS was coined by John Krafcik in *The Machine That Changed the World*.

The lean production company uses a system of linked manufacturing and assembly cells to produce low-cost products in a timely fashion. The manufacturing and assembly cells are designed for flexibility—they can handle changes in product design as well as external customer demand. Chapter 43 presents a proven strategy to convert a mass-production system (job shop/flow shop) into a lean shop. Many companies begin the conversion by designing interim cells, using machine tools designed for stand-alone applications in the job shop. The equipment and tooling utilized in the job shop must be modified when it is utilized in cells. In Chapter 43, 10 steps are outlined reflecting the amalgamated experience of many companies that have Americanized and implemented some version of the Toyota Production System. In this chapter, the details of cell design will be presented. The flow of processes defines the machine layout wherein products having common or similar processes are grouped together and quick movement between the processes is provided, along with the means to reduce setup time.

LEAN PRODUCTION SYSTEM DESIGN

The lean manufacturing system shown schematically in Figure 42-4 links many manufacturing and assembly cells to final assembly by means of kanban inventory links. Within the system, a serial set of processes is created that is balanced to takt time (TT). (*Takt* is the German word for a conductor's baton, which is used to keep, for example, orchestra

FIGURE 42-4 The linked-cell manufacturing system has many manufacturing and subassembly cells connected by kanban links to final assembly. The links hold the inprocess inventory.

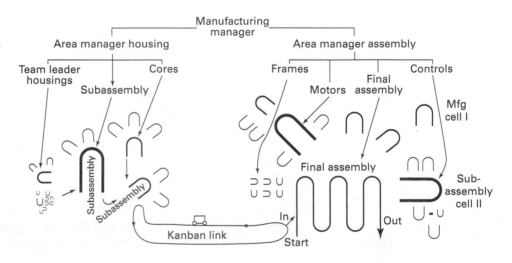

members, all in time.) *Takt time* is the available production time divided by daily demand. Producing to takt time (a DP) meets the FR of right quantity. At the same time, this manufacturing system produces superior quality products at the lowest possible cost in a flexible way. The lean manufacturing system groups the processes into U-shaped (or parallel lines) manufacturing and assembly cells. The processes can be changed over rapidly, so products can be turned out in greater variety in an almost customized fashion with no cost penalty for small production runs (small lots). The manufacturing system is designed such that the throughput time (TPT) is as short as possible. This requires a continuous redesigning of the manufacturing system (which is just another way to say "continuous improvement") to shorten the TPT and increase the speed with which products move through the plant. Therefore, the manufacturing system design (MSD) must be as simple and flexible as possible.

The design of the manufacturing system is an important but neglected aspect of concurrent engineering that has focused on the relationships between product design and process design but has often ignored the system design. If lean MSD and implementation come before concurrent engineering efforts (the latter involving people from design engineering, manufacturing, logistics, marketing, and customer services for the development of new products), the ability of a company to compete in the global world of manufacturing will be enhanced. The manufacturing system is the beating heart of any manufacturing company.

The manufacturing system is defined as a complex arrangement of physical elements characterized by measurable parameters. The physical elements are machine tools, tooling (workholders), materials-handling equipment, and (most important) people. The people who work in the manufacturing system are the internal customers, and the system must be designed to satisfy their needs. At the same time, the manufacturing system must produce products that satisfy the needs of the external customers. In terms of MSD, this is a key concept; that is, the manufacturing system is designed to satisfy the needs of both the internal and the external customers. This complex arrangement is the design of the manufacturing cells and the manufacturing system that satisfies the needs of these two customers.

Different system designs will result in different levels of measurable parameters. Time is probably one of the most critical measurable parameters for evaluating manufacturing performance. By systematically reducing TPT and its variation, companies can achieve world-class status. In the lean manufacturing system, the manufacturing TPT is reduced by the systematic and gradual removal of inventory. The inventory is held in links (between the cells) in the system and controlled by the internal customers (the users of the inventory) using kanban. The linked-cell manufacturing system (L-CMS) shown in Figure 42-4 shows many cells linked to final assembly by inventory links. This is a surprise to many who think that the lean shop has no inventory. This is not correct. The inventory is between the cells in the links and is standardized, controlled, and minimized by the internal customers. The links also provide production control information to the upstream processes and suppliers, telling them what to make, when to make it, and how many to make on a daily basis. See Figure 42-5.

Within the manufacturing cells, the material is referred to as *stock on hand* (SOH) and is not considered to be inventory. SOH material is carried in the machines and/or

FIGURE 42-5 In the L-CMS, the kanban links provide production control information from final assembly back upstream through the supply chain.

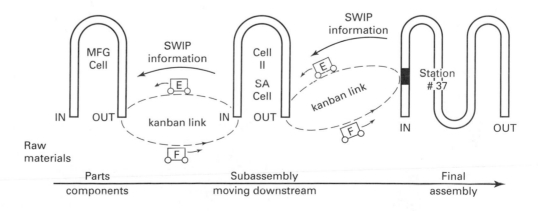

the decoupler elements. Quantity is strictly controlled. The decoupler elements in the cells result in a decoupled design.

The design of the factory and its physical elements must precede the design and manufacture of the product. Because the design includes people, ergonomic issues must be considered. A discussion of the application of design axioms to system design can be found in Cochran (1994 and 2001, in the additional reading section).

SUBASSEMBLY CELLS

The L-CMS is the physical design that meets the FRs of stability. In the lean manufacturing system, there are manufacturing cells and subassembly cells. Cells are designed to match takt time (TT) in an L-CMS. In the manufacturing cells, the machines are capable of completing an operation untended, once initiated by the operator, and when the operation is completed, turning itself off and waiting for the operator. This is called a single-cycle automatic machine tool (SCA). Your toaster is a good example of an SCA.

Assembly cells usually require the operator to remain at the station or machine and complete the process steps before moving on to the next machine. The manufacturing or assembly cells are typically manned by multifunctional operators trained to be able to perform several tasks and operate many machines. The machines are placed next to each other by a U-shaped or parallel row design to achieve one-piece part movement within the cell and volume flexibility.

Figure 42-6 shows an example of a linear flow line with a conveyor. This design is common to the mass production system design. In this design, an attempt is made to make

FIGURE 42-6   A subassembly flow line, with conveyor, can be redesigned into two cells, using walking operators, with stations as part of parallel rows.

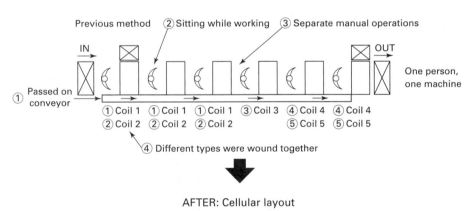

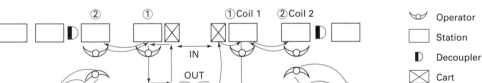

	Before:	After: Per Cell
Output	1105 units/shift	540 units/shift
In-process inventory	180	5 units/cell
Personnel	6	3 operators/cell
Daily output per person	145	180 units
Cycle time	0.44 minute	0.83 minute

Measurable Parameters

the amount of work at each station equal, to minimize the idle time at the stations. This is called *line balancing*. Redesigning the tasks into a pair of cells using standing, walking workers improves the output and makes the area more flexible—easier to change the output as needed—by changing the number of workers in the cell, not by rebalancing the work at individual stations. The cells can be staffed in different ways. The three most common methods are the chase method, the Toyota Sewing System, and the subcell system.

Decouplers are placed between the machines in the cell to connect the flow of parts between the operators. Decouplers are elements of manufacturing cells and are so named because they physically decouple (separate) one station or machine from the other. The term *decoupler* is derived from an axiom design developed by Nam Suh and applied to manufacturing system design (see additional reading section).

The decoupler's major role is to provide flexibility so that one or more operators can perform the operations, changing the cell's output while integrating production control. Decouplers enable the cell to work on a single-piece flow (SPF) basis by decoupling the processes within the cell so that they are independent of one another for time and function. The simplest decouplers hold one unit of stock on hand (SOH) and have specific input and output locations.

Black and Schroer did the first research on decouplers as part of an L-CMS approach. The first integrated pull manufacturing cells with decouplers were simulated, using physical and digital models, to examine the influence of machining time and quality levels on the cells. Based on these studies, the function of decouplers is summarized as follows:

1. Decouplers, which are elements of manufacturing systems that are placed between processes or workstations, enhance the flexibility in the cell by reducing the functional dependency of the side-by-side processes.

2. Decouplers permit flexibility in worker (or robot) assignments and movement. Workers can travel with or against the part flow.

3. Decouplers can perform 100 percent inspection of the parts after the part comes out of the process. The decoupler is usually equipped with a red light if a defect is found and will pass only good parts on to the next process.

4. The decoupler may be designed to feed back quality information to the upstream process so that corrections to the process are made automatically (defect prevention).

5. Decouplers can also perform piece-part manipulations, part transportation, part reorientation, and part reregistration for both manned and robotic cells.

6. Decouplers can branch or combine part flows within the cell, so one machine can feed two or more machines, or two machines can feed one.

7. Decouplers can be designed for handling the family of parts that the cell is producing, so between each machine is one part that has been completely processed by all the machines up to that point in time.

8. Decouplers control the level of SOH within the cell, allowing it to be raised or lowered whenever needed. The capacity of a decoupler never exceeds two unless it is performing a process delay function.

9. Decouplers can be used for process delay, to change the state of the product (heat, cool, cure, degrease, etc.) before the next process. These decouplers may hold parts for multiples of the cycle time.

Obviously, decouplers are uniquely designed elements that help the cells produce superior quality parts. Here is an example of an apparel assembly cell with walking workers and decouplers. In this example, a pair of pants is being assembled manually, so there is variation in the processing times. Normally, decouplers hold one part to connect the flow when multiple workers are in the cell. Increasing decoupler capacity improves cell output as the processing time variation increases, but little improvement occurs after increasing the decoupler capacity beyond two.

Apparel is a labor-intensive industry, in which the operator has incentive to outperform the piece rate, allowing earning and productivity to exceed a predicted level. Most sewing machines are not automatic repeat-cycle processes.

Many companies in the apparel industry have adopted the *Toyota Sewing System* (TSS), which is also called the *modular production system*. TSS was developed by Toyota for making seat covers. TSS was known in the West in 1985 as Toyota's "standing up system" because workers work standing up. The features of TSS (or so-called modules) are similar to manned assembly cells: U-shaped with workers working in teams, making garments on an SPF basis.

TSS has advantages compared with the traditional batch or bundle system, including less floor space, less creasing in garments, and a better working environment.

CELL DESCRIPTION

With its U-shaped layout (Figure 42-7) and ergonomically identical sewing machines, the cell can be operated by a variable number of workers, each cross-trained in all the different processes. Each of the cell's 13 sewing stations had a different mean processing time (Table 42-1). Thus, one of the unique aspects of the design is that processing times at the individual stations need not be balanced. Between each workstation, a decoupler holds a certain number of garments, with one being the minimum number. The garments in the cell are the stock on hand (SOH).

If there is one worker in the cell, the worker moves from machine 1 to machine 13 traveling with the garment. Two workers could follow each other around in the cell, doing all the processes in sequence. This method can be used in an assembly cell in which the workers carry the parts from workstation to workstation. The need for precise line balancing for the entire cell or the partial loops is eliminated, and the apparel assembly cell requires no decouplers because the garment is always with the worker. After completing a garment at station 13, the operator returns to station 1 directly without moving backward and starts making another garment.

The disadvantage of this method is that the slowest worker or the variability in processing time dictates the cell's output. For example, workers may be blocked by station 5, which has the longest processing time in the apparel assembly model. Blocking increases the idle time and decreases the output rate of the cell. Figure 42-8 shows how the output of an assembly cell increases with the number of workers. There is usually almost no blocking with three or fewer workers in the cell. As the variation about the processing time increases, blocking will occur more frequently. The chase method requires every worker to operate all the machines or processes, which usually increases the cycle time, the possibility of making errors, and the training time for new workers.

FIGURE 42-7   The cell is U-shaped. The workers follow each other with garments from station 1 to station 13 in the rabbit chase method.

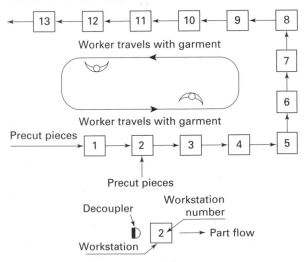

FIGURE 42-8   The output for the apparel assembly cell increases linearly with the number of workers and decreases slightly with variation in processing time.

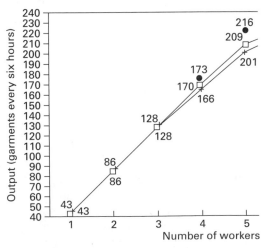

Key:

Standard deviation (%)
● = no variation in processing time
□ = 10% in processing time
+ = 20% in processing time

TABLE 42-1.	Mean Processing Time	
Workstation	Operation	Time (seconds)
1	Sew pockets	30
2	Attach pockets of legs	70
3	Connect legs	60
4	Sew legs	40
5	Sew leg opening	95
6	Turn inside out	5
7	Hem bottom legs	40
8	Attach elastic band	30
9	Stay	5
10	Buttonhole	10
11	Topstitch elastic	15
12	Inspect	20
13	Package garment	80
Total mean processing time per garment		500

TOYOTA SEWING SYSTEM (TSS)

TSS uses multifunctional workers in a cell, with typically three to five workers operating 10 to 15 machines. TSS permits workers to share processes and pass work to one another, just as runners in a relay race pass the baton to one another at 10-meter sections of the track. Thus, the TSS design has processes that are called *relay zones*.

Workers travel counterclockwise with the garments, assembling them as they go from workstations 1 to 13. As long as a worker has a garment to assemble in a succeeding workstation, that worker travels counterclockwise. When the worker is blocked, that worker puts the garment in the decoupler between the workstations and travels clockwise until finding another unfinished garment to assemble. This unfinished garment may be either in a decoupler or at another workstation.

For example, let us look at a five-worker cell (workers A, B, C, D, and E in Figure 42-9). Worker E completes a garment at workstation 13. With no more garments to assemble, worker E then walks clockwise. If worker E finds a decoupler that contains an unfinished garment, worker E travels counterclockwise and assembles the garment. If worker E finds a workstation where a garment is being assembled, worker E takes over the job from that workstation's worker (D) and assembles the garment. Worker D then travels clockwise. If worker D finds a decoupler that contains an unfinished garment, worker D travels counterclockwise and assembles the garment. If worker D finds another workstation where a garment is being assembled, worker D then takes over the job from that workstation's worker C and assembles the garment. Worker C travels clockwise back toward worker B, and worker B loops between worker A and C. Normally, worker D would do operations 7 through 11, but sometimes would do 6 or 12.

If one worker goes on break, the cell keeps going in the same manner with four workers. Over time, the workers develop work patterns, with certain workers taking the entire responsibility for certain workstations, and other workers sharing the responsibility for some workstations.

Note that for this level, the total processing times were perfectly balanced—each worker worked on each garment for 100 seconds. The output now has reached about 210 garments per six hours, or 42 garments per six hours per worker, one less than the output when the cell was run by a single worker.

THE SUBCELL DESIGN

The subcell design allows workers to cross the aisle, essentially separating the cell into sections or subcells with approximately equal processing times. The number of sections depends on the number of workers. Each section can be considered as a subcell linked by decouplers, thus forming a *pull* manufacturing system. The garment is pulled through these subcells one at a time. The subcells start making a garment only when the garment in the decoupler has been removed by the next worker or withdrawn from this cell.

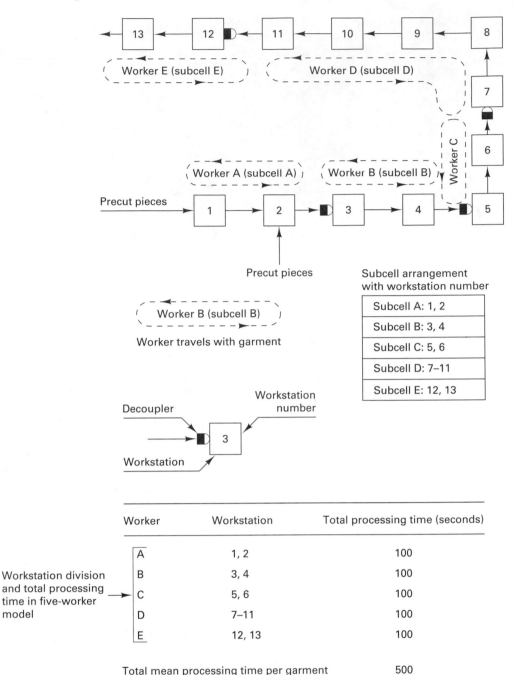

FIGURE 42-9 Processing arrangement for a five-worker model.

Worker	Workstation	Total processing time (seconds)
A	1, 2	100
B	3, 4	100
C	5, 6	100
D	7–11	100
E	12, 13	100
Total mean processing time per garment		500

Workstation division and total processing time in five-worker model

Figure 42-10 shows the worker arrangement for three and four operators. The total processing times are not precisely balanced. Each worker is in charge of a subcell's input and output, thus controlling the SOH within the cell.

Regardless of the worker arrangement the processing time variation increased, the model's output decreased, but increasing decoupler capacity improved the output, reducing the effect of processing-time variation. Increasing the decoupler capacity from one to two has the greatest effect on eliminating the degrading effect of considerable processing time variation.

Comparing the TSS staffing method with the chase method, the TSS was found to be more flexible because workers can help one another instead of staying idle.

The worker balance method also has output approximately the same as the other two methods, showing the flexibility in the U-shaped design. The decouplers separate or decouple each process or subcell from others and link these subcells together to allow the system to have a single-piece flow capability. Simulation studies show that processing time variation

FIGURE 42-10 Processing arrangement for three and four workers in the cell to meet different output requirements.

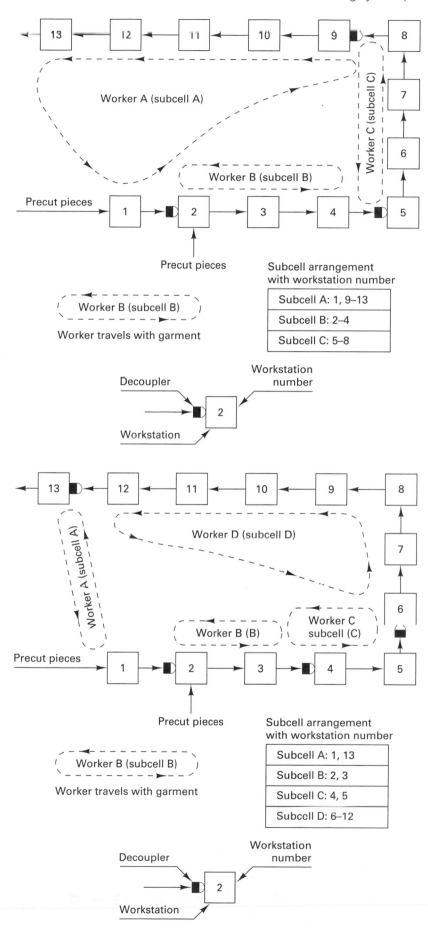

decreases the output and that increasing decoupler capacity improves cell output. But improvements are very marginal when the decoupler capacity is more than two, even for large amounts of processing time variation. In true lean manufacturing and assembly cells, the operations are both automatic and manual. See examples of a lean cell in Chapter 43.

■ 42.4 MANUFACTURING CELL DESIGN

Most companies begin their transition to a new manufacturing system design by implementing a manufacturing cell using existing equipment (machine tools) that were originally purchased for the job shop. Rearranging these machines into a manufacturing cell is the precursor to the design and implementation of true lean cells discussed in Chapter 43. For the interim step, where the company learns how cells work, how setup is eliminated, and how machines can produce perfect parts every time and not break down, the use of existing machines is a low-cost, risk-free approach.

Cells are designed around product families or components. For example, the drive shaft shown in Figure 42-11 comes in four different overall lengths, as shown. The manufacturing engineer has determined that this component is being made using the processes (sequence of operations) shown in Figure 42-12 by dark lines (lathe, lathe, mill, mill, drill and grind). Each machine has an operator who tends the machine and performs the tasks outlined in Figure 42-13, an example of a process planning sheet for the drive shaft. The part begins as a 12-foot bar of cold finished stainless steel, 1.780 + .003 in. in diameter. This is the diameter of the large end of the pinion, which connects to the output of a motor via the four holes on the left end. The 0.50-in. slot carries a drive key. The step on the right end mates with an element in the transmission being driven by the motor. If something in the driven element stops, the four bolts holding the pinion to the motor output shear, the drive key falls out, and damage to the motor or transmission is prevented. The pinions come in different lengths to accommodate different locations of the transmission with respect to the motor.

The job requires two lathes because the part is long and relatively thin, so the turning cuts are likely to cause severe chatter and vibration if both ends are not supported. So the first lathe puts a center in the large end and holds the other end in a collet chuck while operation 10, rough turn 1.45 in., takes place. The 1.45 means the 1.780-in. diameter is rough-turned down to 1.45 in. Therefore, the depth of cut is 1.780 − 1.450

FIGURE 42-11 Part drawing for a family of drive pinions, A, C, C. D.

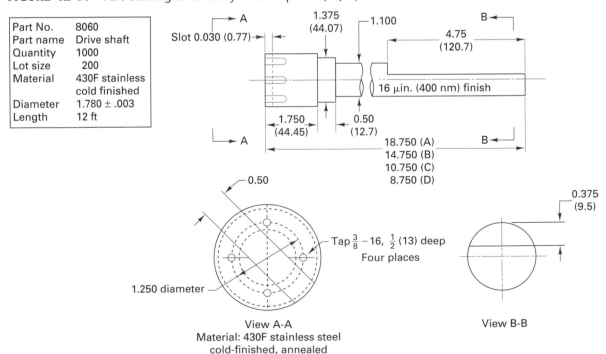

Part No.	8060
Part name	Drive shaft
Quantity	1000
Lot size	200
Material	430F stainless cold finished
Diameter	1.780 ± .003
Length	12 ft

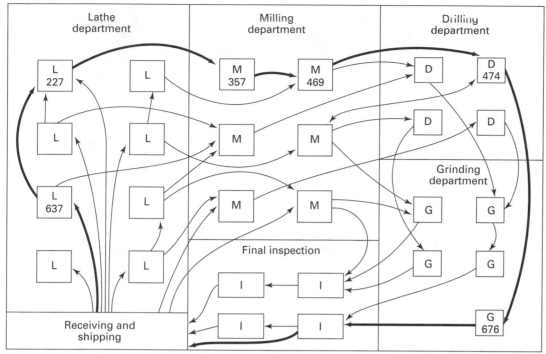

FIGURE 42-12   The drive shaft 8060 is made in a job shop with parts moving from machine to machine in tote boxes that hold 200 parts.

$= .33\,in/2 = 0.165$. Naturally, the MfE would estimate the cutting force and the horse-power required here (see Chapter 23 for details). The cutoff operation produces a bar of the correct length for the second lathe.

In the second lathe, the part is flipped end over end and held on the large diameter with a collect chuck and the right end with a center. After the second lathe, the part moves to vertical mill #357 and the slot 0.50 in the bar is milled. Next, the step is milled and then the holes drilled and tapped. Finally, the bar is ground to the $16\text{-}\mu in.$ finish. For each process operation or step, the cost is estimated. This is simply the estimated cycle time multiplied by the direct labor plus overhead rate. The cycle time is the machining time plus handling time or manual time (locating the workpiece in the machine, loading, unloading, etc.). Here is how the times for operation 10 are computed.

$$3.2\,\text{hrs for setup} \div 200\,\text{units} = .96\,\text{min/part}$$

$$.10067\,\text{hrs/part cycle time} \times 60\,\text{min} = 6.04\,\text{min/part}$$

$$.96\,\text{min/part} + 6.04\,\text{min/part} = 7\,\text{min/part}$$

$$7/60 = 0.117 \times 18.35 \times 1.70 = \$\,3.65/\text{part}$$

The four machining operations plus tool-handling time took 6.04 minutes per part.
The actual machining times were

facing end of bar	=	.3
center drilling end	=	.6
rough turning	=	1.2
cutoff to length	=	.3
Total	=	2.4 min/part

or machining (making chips) actually took 40% of the cycle time.
Here is how rough turning would be calculated

$$T_m = \frac{L + All}{f_r \times N_s}$$

Part no. _____8060_____   **Ordering quantity** _1000_   **Material** ___430F Stainless___
Part name _Drive Pinion_   **Lot requirement** ___200___   steel, 1780 ± 0.003 in.
     cold-finished   12-ft
     bars = 1000 pieces
     Unit material cost _$22.47_

Workstation	Operation no.	Description of operations (list tools and gauges)	Setup hour	Cycle hour/ 100 units	Unit estimate	Labor rate	Labor + overhead rate	Cost for labor + overhead
Engine lathe # 137	10	Face A-A end 0.015, center drill A-A end, rough turn 1.45, cut off to length 18.750.	3.2	10.067	0.117	18.35	1.70	3.65
Engine lathe # 227	11	Center drill BB end, finish turn 1.100, turn 1.735 diam.	3.2	8.067	0.095	18.35	1.70	2.96
Vertical mill # 357	20	End mill 0.50 slot with 1/2 H.S.S. end mill (collet fixture).	1.8	7.850	0.088	19.65	1.85	3.20
Horizontal mill # 469	30	Slab mill 4.75 × 3/8 (nesting vise H.S.S. tool).	1.3	1.500	0.022	19.65	1.80	0.78
NC turret drill press # 474	40	Drill 3/8 holes—4×. tap 3/8–16 (collet fixture).	0.66	5.245	0.056	17.40	2.15	2.10
Cylindrical grinder # 676	50	Grind shaft to 16 μin. finish –1.10.	1.0	10.067	0.110	19.65	1.80	3.89

                                                       $16.58
                           Estimated MFG   22.47
                           cost per unit ⟶ 39.05

FIGURE 42-13 Process planning sheet for drive pinion based on manufacturing the pinion in a job shop.

where $L = 16$ in. for 8060A parts, and feed rates and cutting speeds are selected for high speed steel tools cutting 430F stainless steel.

This is the detail required for manufacturing one component in the job shop area of a mass production manufacturing system, as shown in Figure 42-14. Now let's see what happens when we convert this manufacturing into lean manufacturing using linked cells.

The machine tools are rearranged into a U-shaped cell as shown in Figure 42-15. A horizontal bandsaw has been added to the six machines specified in the process-planning sheet. The saw is used for cutting the 12 bars into the correct length, depending upon which pinion, A, B, C, or D, is being made in the cell. This machine eliminates the need to face and the cutoff operations in the first lathe. All the machine tools are modified to perform the machining cycles untended (single-cycle automatics) and have been equipped with walk-away start switches and other part-checking, safety, and loading devices to reduce the non-machining times in the cycle time. Between each machine, decoupler elements have been added, which help maintain the single-piece flow of parts through the cell. Decouplers can perform other nonprocessing tasks like quality control (inspection) and process delay. See the section on assembly cells and Chapter 43 for additional discussions on decouplers.

The cell produces a family of four parts and can be manned by one, two, three, or four operators. These operators are multifunctional (can perform tasks related to setup, quality, maintenance, problem solving, continuous improvement) and multiprocess

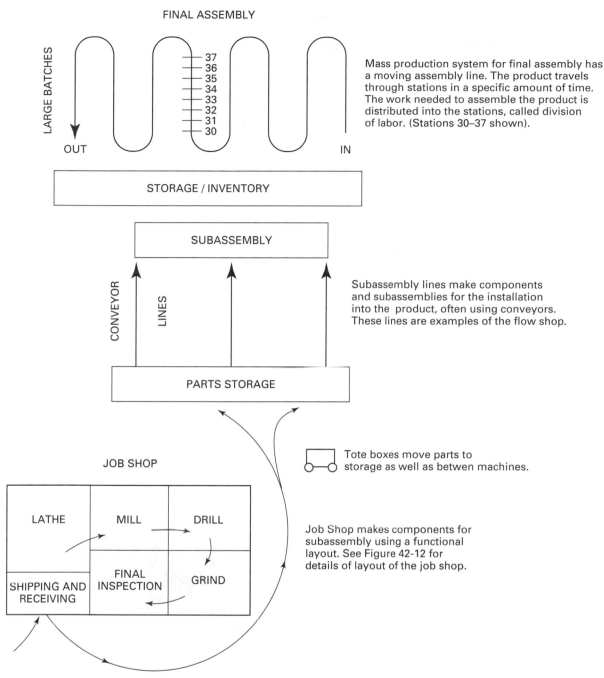

FINAL ASSEMBLY

LARGE BATCHES

— 37
— 36
— 35
— 34
— 33
— 32
— 31
— 30

OUT   IN

Mass production system for final assembly has a moving assembly line. The product travels through stations in a specific amount of time. The work needed to assemble the product is distributed into the stations, called division of labor. (Stations 30–37 shown).

STORAGE / INVENTORY

SUBASSEMBLY

CONVEYOR LINES

Subassembly lines make components and subassemblies for the installation into the product, often using conveyors. These lines are examples of the flow shop.

PARTS STORAGE

Tote boxes move parts to storage as well as betwen machines.

JOB SHOP

| LATHE | MILL | DRILL |
| SHIPPING AND RECEIVING | FINAL INSPECTION | GRIND |

Job Shop makes components for subassembly using a functional layout. See Figure 42-12 for details of layout of the job shop.

FIGURE 42-14 The mass production manufacturing system has large storage areas between the manufacturing areas.

(can operate various processes and do assembly tasks). Each of the machines are at least single cycle automatics; that is, they can complete the machining cycle automatically once it has been initiated by the operator (using a walkaway switch). The machining times for processes are, in minutes: .30, .40, .40, .45, .45, .30, and .45, beginning with a saw and going clockwise (CW) around the cell to the grinder. These processing times (MT) have been shortened by changing to carbide tools and having each machine perform one step in the sequence of operations. The design rule for the cell is

$$MT_{ij} < CT$$

i = number of machines

j = number of components

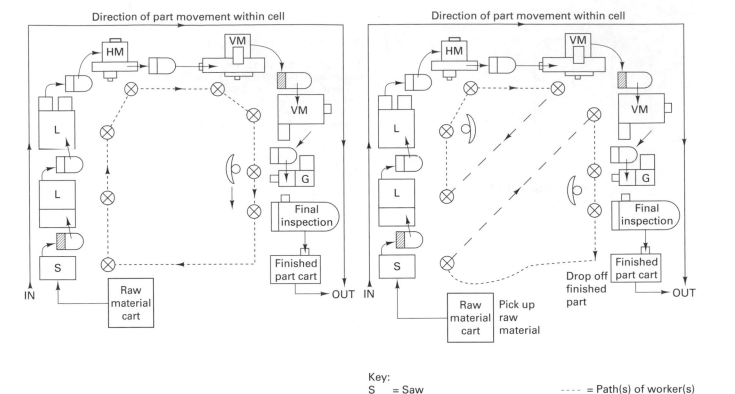

Direction of part movement within cell

Key:
S = Saw
L = Lathe
HM = Horizontal milling machine
VM = Vertical milling machine
G = Grinder
⊗ = Worker positions

---- = Path(s) of worker(s) moving within cell
—— = Material movement paths within cell
▨ = Active (decoupler)
⌣ = Operator

FIGURE 42-15 Manned manufacturing cell with seven machines can be operated by one or two workers (Black, 1991).

The machining time for any machine in the cell for any component in the family is less than the necessary cycle time (CT), which is usually slightly less than the takt time for final assembly. Thus $MT_{ij} \leq CT \cong TT(1 - \text{allowance})$.

The final inspection machine is also automatic and takes about .25 minute to manually check the parts then load the parts in the machine, depending on which component from a part family is being inspected. The time for the inspection process is .30 minute.

The machining times given above are for part A from the part family. The machining times for parts B, C, and D are shorter because the length of some cuts are shorter for these parts. The operator(s) take about .25 minute at each machine to perform various manual operations (standardized work) like unloading the machined part, checking the part, perhaps deburring the part, and loading in the next part to be machined. The operators spend about 3 seconds walking from machine to machine. The aisle between the machines is about 4-ft wide. Table 42-2 summarizes the data for one operator working in the cells. The approximate cycle time (CT) for the cell manned by one

TABLE 42-2. List of Operations and Respective Times

Machine Tool	Operation or Process	Machining Times (MT) in min	Operator Manual Time (min)	Walk Time (min)
Saw	Cut bar to length	0.3	0.25	0.05
Lathe 1	Rough turn	0.4	0.25	0.05
Lathe 2	Finish turn	0.4	0.25	0.05
Horizontal mill	Mill step	0.45	0.25	0.05
Vertical mill	Mill slot	0.45	0.25	0.05
Vertical mill	Drill and tap holes	0.3	0.25	0.05
Surface grinder	Grind slot	0.45	0.25	0.05
Final inspect	Final inspection	0.25	0.25	0.05

operator is 15 sec $\times$ 8 + 8 $\times$ 3 sec = 120 + 24 = 144 sec = 2.4 minutes per part. The cell is producing 25 parts per hour with one operator. The cycle time is the reciprocal of the production rate for the cell. Cycle time is in minutes per part.

Here is the unique aspect to the cell design. Notice that the output cell cycle time does not depend upon the machining times for the processes. The machining times are decoupled from the cycle times (i.e., the schedule) by the cell design. This result is achieved since the machines' processing times are designed to be less than CT. That is, the schedule has been made independent of processing times on individual machines.

The approximate throughput time (TPT) for a part moving through the cell using one operator is 8 $\times$ 2.40 = 19.20 minutes. Notice how the TPT does not depend on the processing times, so machining times can be changed without affecting lead times on the schedule. Therefore, cutting speeds could be reduced (improving tool life) or feed rates could be reduced (improving surface finish) as long as machine time remains less than CT, improving tool life and perhaps quality.

When a cell's cycle time is dictated by the operator, the TPT in these cases depends on the number of cycles necessary to move a part through all the processes multiplied by the cell cycle time, which does not depend on the machining processing time. In review, for $MT_{ij} < CT$

$$\text{Operator cycle time } (CT) = \sum_{i=1}^{n} (\text{Manual Time} + \text{Walk Time})$$

The number of cycles the part spends in the cell depends on the stock on hand in the cell. For one operator, working in a *"make one, check one, move one on"* (MO-CO-MOO) mode with eight machines, there are eight units in the cell in various stages of processing. If the six decouplers are activated, then there are 14 SOH units in the cell.

Suppose we want to increase the output from the cell to meet a new takt time. The cycle time must be reduced. An operator is added, as shown in Figure 42-15.

The addition of the second operator in the cell cuts the cycle time in half, providing the processing time on any machine is less than the cycle time. So for the cell, the $CT = 4 \times 15$ sec. + 4 $\times$ 3 sec = 60 + 23 = 72 sec, or half of what it was previously. Now the cell is capable of producing 50 parts per hour. Since the longest machine processing time was 45 seconds, CT is still greater than MT.

The throughput time for the cell manned by two operators is reduced but not cut in half because now the decouplers are activated to connect the flow of parts. The TPT is the number of cycles needed to advance the part through all the processes at a CT of 72 seconds. Let us assume that two cycles are added to account for the active decouplers which connect the two operators. So, TPT = 10 $\times$ 72 = 720 seconds. The addition of the second worker requires the addition of decouplers in the cell to connect the flow of parts from one worker to the next. The decouplers are shown in the cell by the capital D symbol. The actual loops the workers take in the cell vary from time to time, as shown in Figure 42-16 where two workers are taking different loops than shown in Figure 42-15. Three workers shown in Figure 42-17 would divide up the tasks differently and may even share responsibility for a machine (HM).

For three operators the $CT \approx (15 \times 3) + (3 \times 3) \approx$ or about 51 sec/part, producing 70 to 71 parts/hour. The output is virtually linear with respect to the number of operators. Can you determine the CT, TPT, and hourly output for this cell if four workers are used? How do you think the workers would allocate themselves? Did you keep one worker in control of the input/output points of the cell? In summary because of its unique design features, the cell can be operated by one, two, three, or four workers with the output ramping up with the addition of each worker.

STANDARD OPERATIONS ROUTINE SHEET (SORS)

In an L-CMS, standard work is defined as specifying the content, sequence, and timing of all work. In designing cells and developing standard work, standard operations routine sheets are used. Figure 42-18 shows an example for a manufacturing cell that is producing steering wheels for cars. The cell has a die-casting machine with a 36-second cycle time and produces a casting for the first milling operation, VM-1571. The decoupler

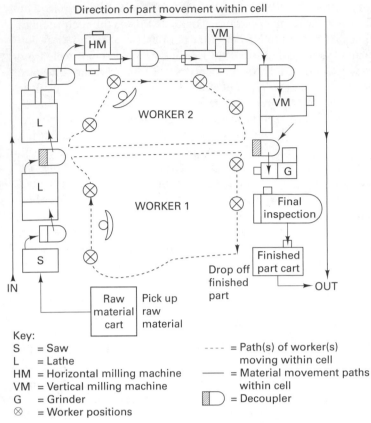

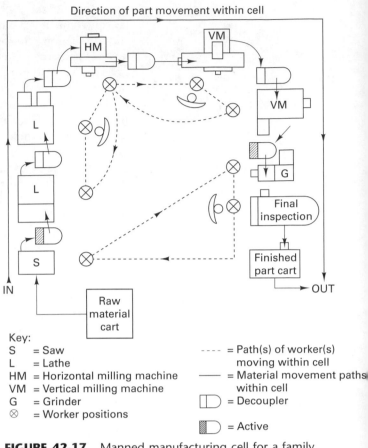

FIGURE 42-16 Manned manufacturing cell with two workers cuts the cycle time in half (Black, 1991).

FIGURE 42-17 Manned manufacturing cell for a family of parts with seven machines operated by three workers.

Standard operations routine sheet

Item no. name of items		Wheel die casting	Standard operations routine sheet		Date of manufacturing	Feb. 5, 1997	Required daily quantity		524	Manual operation $\vdash\!\!\!+\!\!\!+\!\!\!\dashv$
Process		ID rough milling ID finish milling	Takt time 60 sec.		Worker's group	Cell 143	480 minutes/ required quantity (cycle time)		55 sec	Machine processing ‐ ‐ ‐ Worker walking

Work sequence	Name of operation	Time (sec)		Operations time (seconds)
		Manual	Machine	6 12 18 24 30 36 42 48 54 60 66 72 78 84 90 96 102 108 114 120
1/5	Remove materials from die casting	3		
2/6	Rough mill outer end ID	13	27	
3/7	Rough mill inner end ID	12	21	
4	Finish mill outer & inner ID	14	70	
8	Finish mill outer & inner ID	14	70	
				1 min. 50 sec. to complete two parts

Layout Cell 143

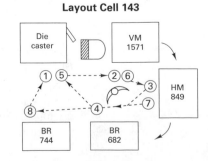

FIGURE 42-18 Example of standard operations routine sheet for a steering wheel cell. The manual times are for unloading (U), loading (L), and checking the components (I). The machines times are for untended machines.

between the die caster and the mill is for water cooling the casting, so the operator can handle it without getting burned. The necessary cycle time for the cell is 55 seconds, because this cell feeds a final assembly line making cars at a rate of 1 car/minute or takt time = 1 minute. The finish milling operations require 70 seconds of machining time, so $MT > CT$. This engineer has chosen to duplicate the finish milling operation. The operator alternates between the BR finish milling machines.

The SORS shows the working relationships between the operator and the machines. The manual operations are

U = unload a part from the machine

L = load a part into the machine

D = deburr a part

I = inspect the surface, size, etc. of the operation just completed

The manufacturing engineer wants to redesign the cell so that the operator is moving counterclockwise (CCW) rather than clockwise (CW). Why does the MfE want to do this? Because most people are right-handed and loading (with the right hand) is more difficult than unloading (with the left hand), and most machine tools designed for the job shop are designed for right-handed people.

Now there is another issue in this cell design—quality. The two machines (BR 744 and BR 682) cannot produce identical components, so variation has been introduced to the MO-CO-MOO methodology. But suppose we could reduce the machining time for finish milling to 35 seconds by separating the finish outer (ID milling) process from the inner ID milling process. Now BR 682 process is in series with BR 744, the variation is eliminated, the product improved through improved manufacturing cell design. Did this alter the CT? Can you redo the layout of the cell and modify the SORS to reflect this change?

Figure 42-19 shows a layout of a cell where the operator is moving upstream, or CCW, while the flow of the parts is CW through five machines and five decouplers.

FIGURE 42-19 (a) Layout and (b) time bar diagram of a five-machine, U-shaped cell with decouplers, operator upstreaming, and variable machine times (MT) (Black, 1991).

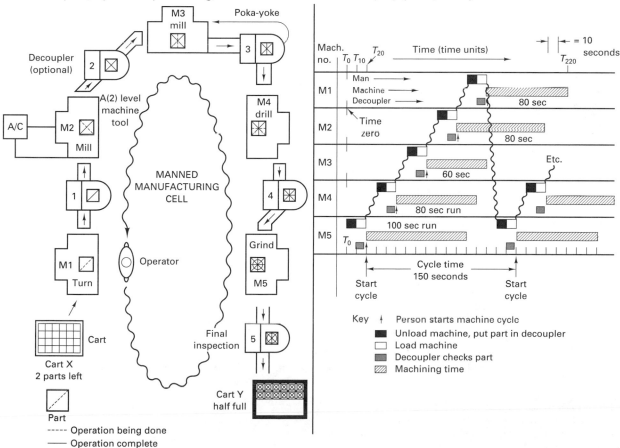

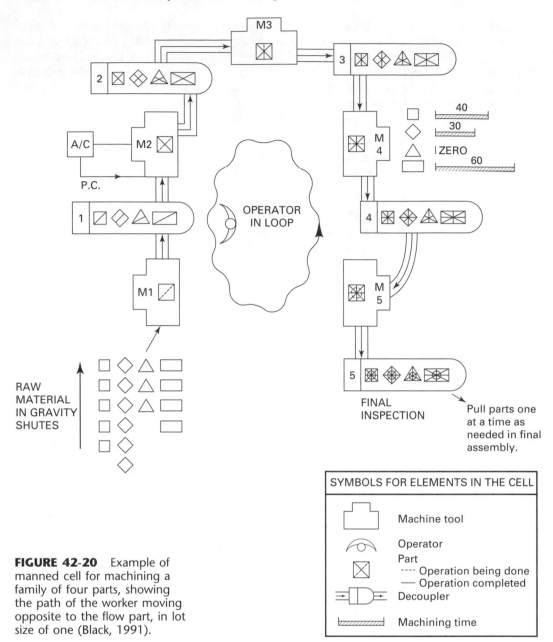

FIGURE 42-20 Example of manned cell for machining a family of four parts, showing the path of the worker moving opposite to the flow part, in lot size of one (Black, 1991).

It may be necessary for the operator to move counter to the part flow in cells that make components in left-hand and right-hand pairs or where the operator is manning a part in a cell. The decouplers transport the parts CW to the downstream machine. The lot size is 24 parts, of which 12 are finished, 10 are still in the cell, and 2 are in the cart on the left. The cycle time is 150 seconds for this one-operator cell. Now suppose we want this cell to produce parts one at a time in lot size of 1 instead of 24. The cell is shown in Figure 42-20.

This cell is designed to fabricate a family of four parts with five machines. The cell operates on a single-piece flow and produces in lot sizes of one as well. That is, any of the machines can quickly switch from one part to another. Between each machine is a decoupler that holds one part of each family member (i.e., four parts) processed through that stage of the cell. The four parts are all different in size and shape. The triangular part is not machined in machine 4 at all, so it has a machining time of zero. The worker is walking opposite to the movement of the parts. The parts are pulled through the cell by the decouplers. An empty decoupler is the signal to make more.

Parts removed from the final inspection decoupler dictate what the cell needs to make and when the cell needs to make it. If a square part is pulled out of the final decoupler, the worker unloads machine 5, removes a square part from decoupler 4, and loads it into machine 5. This creates a need to replace the part in decoupler 4. The worker walks to machine 4, unloads a part, places the part in decoupler 4, removes a square part from decoupler 3, then loads it into machine 4. This creates a demand in decoupler 3 for a part. Thus, the parts are pulled through the system. Obviously, the cell must be fully loaded with parts, processed in stages and waiting in the decouplers.

The cycle time for the manufacture of any of the four parts is the same. Remember, CT is determined by the time it takes for the worker to make one complete walking loop around the cell. This kind of cell requires that setup is virtually eliminated, the processes make no defective parts, and the machines rarely, if ever, break down.

Machine 2 has been equipped with an adaptive control (A/C) feedback loop to control the behavior of the process, optimizing some aspect of it. A/C feedback is not the same as a pokayoke device, which tries to prevent defects from occurring. (See Chapter 32 for discussion on A/C.)

■ 42.5 SUMMARY

The lean shop with its linked-cell (L-CMS) design is an outgrowth of the previous two manufacturing system designs, the job shop and the flow shop. What was different about this system compared to mass production? The job shop as a manufacturing system design evolved during the 1800s. These early factories replaced craft or cottage manufacturing when it became necessary to have powered machines. A functional design evolved because of the *method* needed to drive or power the machines. The job shop design became known to the historians as the American Armory System. The world came to New England to see the American Armory System and it was duplicated around the industrial world.

In the early 1900s, the first vestiges of the flow shop began to emerge. Flow line manufacturing began for small items and culminated with the moving assembly line at the Ford Motor Company. Just as in the 1800s, the world again came to see how this system worked, and this new design methodology was spread around the world. Over time a hybrid system—a mixture of job shop and flow shop evolved. This physical design permitted companies to manufacture large volumes of identical products at low unit costs, and the mass production system evolved. The job shop produced components in large lot sizes according to the economic order quantity strategies. The subassembly lines fed components to storage systems and to final assembly. This system produced the goods that helped America win WWII and become a world leader in manufacturing. The physical design was the result of achieving certain functional requirements like low cost.

After WWII, the Toyota Motor Company developed a new MS design known initially as the TPS. In 1990, it was finally given a name that would become universal, *lean production*. In lean production, final assembly is converted to mixed model so that the demand for subassemblies and components is leveled. The subassembly lines are converted to assembly cells, often eliminating conveyor lines. Third, the job shop is converted to manufacturing cells. The *subassembly* and *manufacturing cells are linked* to final assembly by kanban to form an integrated inventory and production control system. The result is low-cost (high-efficiency), superior quality (no defects), and on-time delivery of unique products from a flexible system. The new design operates on a single-piece flow basis just like final assembly lines. The subassembly lines are designed (flexibly) to handle mixes of models, so that the quantity of all the components pulled into the final product was the same day after day.

This new design is *flexible, controllable, efficient and unique*. Flexibility means the system can readily adapt to changes in customer demand (volume and mix) and changes in product design (new products or changes to the existing product).

Controllability means the design permits the system level control functions (of quality control, inventory control, production control, and machine tool reliability) to be designed into (integrated) the manufacturing system.

For example, Toyota believed in company-wide quality control and taught it to everyone, from the company president on down to every production worker. They were able to change from a country that made junk to a nation that could give customers products of high reliability. How was this accomplished? Through a redesign of the manufacturing system into a single-piece flow methodology throughout where the outcome from each processing step and each assembly step is checked before the part moves to the next step.

The system is designed for efficiency. The efficient use of the space, the people, the machine tools, and the material handling and storage elements is what makes this system lean.

Uniqueness and creativity on the part of the engineers and workers just means they develop unique new processes and work to improve existing processes, reduce equipment failures, reduce changeover/setup times and continuously improve the design of the workplace (methods improvement).

More significantly, the placement of the processes in the manufacturing and assembly cells requires unique processing solutions to bring the processing times well under the takt time. Many of these processes are well-guarded company secrets that only suppliers to Toyota come to learn.

Many believe that the only way by which manufacturing companies can compete is to automate. This approach was known as computer integrated manufacturing (CIM), recently renamed agile manufacturing. The concept to achieve integration through computerization and automation often results in trying to computerize, robotize, or automate very complex manufacturing and assembly processes. This works only where there is little or no variety in the products. Manufacturing system design takes a different approach. Redesign the system into a linked-cell arrangement. The critical control functions are integrated into the manufacturing system, then computerize and automate (IM, C). Experts on CIM now agree that lean manufacturing, especially development of manufacturing and assembly cells, must come before efforts to computerize the system. While costs of these systems are difficult to obtain, the early evidence suggests that the lean shop approach is significantly cheaper than the CIM approach.

The lean factory is based on a different design for the manufacturing system on which the sources of variation in time are minimized and delays in the system are systematically removed. In the linked-cell manufacturing system, in which manufacturing and assembly cells are liked together with a pull system for material and information control, downstream processes dictate upstream production rates. The linked-cell strategy simplifies the manufactoring system, integrates the critical control functions before applying technology (automation, robotization, and computerization), avoids risks, and makes automation easier to implement. This is the strategy that is dominating the new generation of factories (auto plants) both in the United States (and abroad).

■ KEY WORDS

assembly cell	level balancing	physical simulation	throughput time (TPT)
chase method	linked-cell manufacturing	pokayoke	Toyota Sewing System (TSS)
decouplers	system (L-CMS)	single-cycle automatic (SCA)	Toyota Production System
design parameters (DPs)	manufacturing cell	single-piece flow (SPF)	(TPS)
enterprise system design (ESD)	manufacturing system design	stock on hand (SOH)	value stream mapping (VSM)
functional requirements (FRs)	decomposition (MSDD)	Subcell system	
kanban links	performance measures (PMs)	takt time (TT)	

■ REVIEW QUESTIONS

1. Define what an enterprise or production system is.
2. What is a manufacturing system?
3. What is meant by manufacturing system design?
4. What should be the relationship between a manufacturing system and the enterprise system?
5. What are the six FRs of a stable manufacturing system?
   a. W. Edwards Deming said that to know when a problem exists we must distinguish normal from abnormal. How and why does standardized work make this idea possible?
   b. To achieve FRs, what must be added to make a system robust to variation?
   c. How is cost reduced once a system has been designed to be and is stable?
6. Go to several fast-food restaurants and develop a value stream map of the operation. Determine whether each restaurant system design achieves the six FRs of system stability.
7. Repeat Question 6 for your college program. (You are the product receiving value adding knowledge via your courses.)
8. Why is it important for a system design to be robust?
9. What is standardized work? Why is standardized work necessary?
10. What is takt time?
11. If a machine's processing time exceeds cycle time in a manufacturing cell, what alternatives might the manufacturing engineer consider to ensure that a cell meets the necessary cycle time?
12. Why is single-piece flow necessary?

13. Identify the machine design requirements for cells (for example, a walkaway switch).
14. What is the key role of the worker in the cell?
15. a. Assume that the takt time is 60 seconds for a cell, but a finishing process has a cycle time of 65 seconds, and there is no way known to reduce that time. What would you do, ideally, and why?
   b. If a new finishing process machine costs $5,000, what would you do?
   c. If a new machine costs $50,000, what would you do?
16. a. Are parts pushed or pulled through a cell?
   b. How would you know if you observed an operational cell in a plant?
17. a. What is a mistake-proofing device?
   b. Why is mistake proofing important in a cell's operation?
18. a. Assume that you are designing a machine cell for a product that has an expected life of one year. What would you automate? What would you not automate?
   b. If the product life is five years, what would you automate? What would you not automate?
   c. If the product life is 50 years, what would you automate? What would you not automate?
19. Name one product that has a life of 50 years that is produced in volume over 10,000 per year. How is it made?
20. Why do cells use single-cycle automatic machines?

■ PROBLEMS

1. Assume that an assembly line produces three products, A, B, and C (see Figure 42-A). The available operating time per shift is 480 minutes. The cell operates two shifts per day. The average daily demand is 160 of A, 120 of B, 200 of C. What is the takt time?
   The assembly cell has the following design:
   M designates a single-cycle automatic machine
   H designates a manual operation
   Assume the walk time between all stations is three seconds.
   a. The cell has a yield of 97%. What is the necessary cell cycle time?
   b. Develop the standardized work combination time graph for *one* required cycle time period for each operator as shown.
   c. Can the takt time be met? If not, then why not?
   d. What would you try to fix/improve before adding workers?
   e. How would you facilitate this improvement?
   f. Assume certain improvements and provide a workable cell design.
2. The Toyota Sewing System shown in Figure 42-9 does not have a worker covering the input and output stations.
   a. Is this a problem with the TSS?
   b. Suggest a fix for this problem.

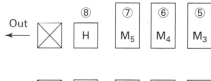

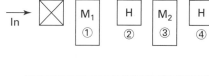

Operation	Load time (sec)	Proc time (sec)	Manual time (sec)
1	3	42	
2			17
3	4	53	
4			4
5	4	18	
6	6	50	
7	9	47	
9			5

FIGURE 42-A

www.wiley.com/college/degarmo

Chapter 42 CASE STUDY

Snowmobile Accident

Katrin B is suing the Pogidog Snowmobile Company and LINUS Components for $750,000 over her friend's death. He was killed while racing his snowmobile through the woods in the upper peninsula of Michigan. Her lawyer claims that he was killed because a tie-rod broke, causing him to lose control and crash into a tree, breaking his neck. While it was impossible to determine whether the tie-rod broke before the crash or as a result of the crash, the following evidence has been put forth. The tie-rod was originally designed and made entirely out of low-carbon steel (heat treated by case hardening) in three pieces, as shown in Figure CS-42. These tie-rods were subcontracted by Pogidog to LINUS Components. LINUS Components changed the material of the tie-rod bolts from steel to a heat-treated aluminum having the same UTS value as the steel. They did this because aluminum rods were easier to thread roll rather than to form by a thread-cutting operation. It was further found that threads on one of the tie-rod bolts were not as completely formed as they should have been. The sleeve of the tie-rod in question was split open and one of the tie-rod bolts was bent.

Katrin's lawyer claimed that the tie-rod was not assembled properly. He claimed that one rod was screwed into the sleeve too far and the other not far enough, thereby giving it insufficient thread engagement. LINUS testified that these tie-rods are hand assembled and checked only for overall length and that such a misassembly was possible. Katrin's lawyer stated that the failure was due to a combination of material change, manufacturing error, and bad assembly, all combining to result in a failure of the tie-rod.

A design engineer for Pogidog testified that the tie-rods were "way overdesigned" and would not fail even with slightly small threads or misassembly. Pogidog's lawyer then claimed that the accident was caused by driver failure and that the tie-rod broke upon impact of the snowmobile with the tree. One of the men racing with Katrin's friend claimed that her friend's snowmobile had veered sharply just before he crashed, but under cross examination he admitted that they had all been drinking that night because it was so cold (he guessed 20° below). Since this accident had taken place over five years ago, he could not remember how much they had had to drink.

You are a member of the jury and have now been sequestered to decide if Pogidog is guilty of negligence resulting in death. The rest of the jury, knowing you are an engineer, has asked for your opinion. What do you think? Who is really to blame for this accident? What actually caused the accident?

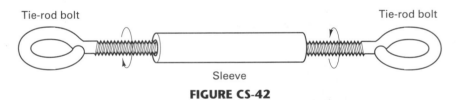

Tie-rod bolt   Tie-rod bolt

Sleeve

FIGURE CS-42

IMPLEMENTATION OF LEAN MANUFACTURING SYSTEMS AND CELLS

■ 43.1 A NEW INTEGRATED MANUFACTURING SYSTEM DESIGN

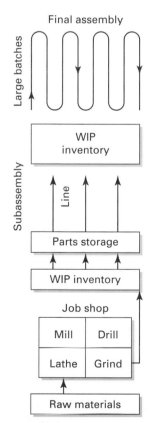

FIGURE 43-1 The MSD called mass production produced large volumes at low unit cost.

The linked-cell manufacturing system (L-CMS) design is an outgrowth of the previous two manufacturing system designs, the job shop and the flow shop. The job shop as a manufacturing system design evolved during the 1800s. These early factories replaced craft or cottage manufacturing when it became necessary to have powered machines. A functional design evolved because of the *method* needed to drive or power the machines. The job shop design became known to the historians as the American Armory System. People from all over the world came to New England to see the this system, and it was duplicated around the industrial world.

In the early 1900s, the first vestiges of the flow shop began to emerge. Flow line manufacturing began for small items and culminated with the moving assembly line at the Ford Motor Company. Just as in the 1800s, the world again came to see how this system worked, and this new design methodology was spread around the world. Over time a hybrid system—a mixture of job shop and flow shop—evolved. This physical design permitted companies to manufacture large volumes of identical products at low unit costs, and the mass production system evolved, see Figure 43-1. The job shop produced components in large lot sizes according to the economic order quantity strategies. The subassemblies lines feed components to storage systems and to final assembly. This system produced the goods that helped America win WWII and become a world leader in manufacturing. The physical design was the result of achieving certain function requirements like low cost.

After WWII, the Toyota Motor Company, led by the genius of a manufacturing engineer named Taiichi Ohno, developed a new manufacturing system design known initially as the Toyota Production System (TPS) and later the just-in-time/total quality control (JIT/TQC) system or world class manufacturing (WCM) system. In 1990, it was finally given a name that would become universal, *lean production*. What was different about this system compared to mass production?

First, in lean production final assembly is converted to mixed model so that the demand for subassemblies and components is leveled. Second, the subassembly lines are converted to cells, often eliminating conveyor lines. Third, the job shop is converted to cells. The *subassembly* and *manufacturing cells are linked* to final assembly by kanban to form an integrated inventory and production control system. The result is low-cost (high-efficiency), superior quality (no defects), and on-time delivery of unique products from a flexible system (Figure 43-2). The new design operates in ways the old "mass production" system could not. The cells within the system could operate on a single-piece flow basis just like final assembly lines. The final assembly lines were designed (flexibly) to handle mixes of models so that quantity of all the components pulled into the final product was the same day after day.

The lean shop is flexible, controllable, efficient and unique.

Flexibility means the system can readily adapt to changes in customer demand (volume and mix) and changes in product design (new products or changes to the existing product).

Controllability means that the system level control functions (for quality control, inventory control, production control, and machine tool reliability) are designed into (integrated) the manufacturing system. This functional integration is quite different from getting the computer-aided design (CAD) computer to talk to the computer-aided manufacturing (CAM) computer. Because the new method is based on a different manufacturing system design rather than expensive computer technology, it requires less capital investment.

The lean shop is designed to produce superior quality products. Toyota believed in total quality control and taught it to everyone, from the company president down to every production worker. They were able to change from a country that made junk to a nation that could give customers products of high reliability. This was accomplished through a redesign of the manufacturing system into a single-piece flow methodology where every part and every assembly is checked after each processing step.

The reduction of variation is the key to continuous improvement. The kinds of variation we need to consider are:

- Variation in quality (defects/million)
- Variation in output (parts/day)
- Variation in process time (hrs/part)
- Variation in cost ($ /part)

FIGURE 43-2 The mass production system is restructured into a lean manufacturing system design to achieve single-piece flow, one of the requirements of a lean production system design.

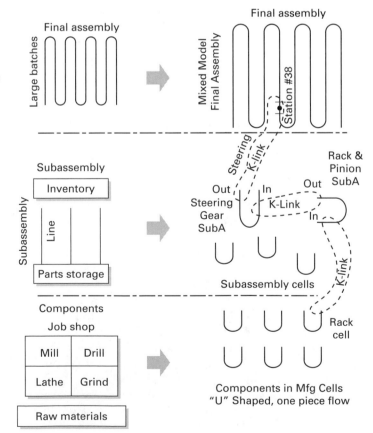

These kinds of variation are reduced through the implementation of lean manufacturing and the continuous changing (redesign) of the MSD.

The system is designed for efficiency. The efficient use of the space, the people, the machine tools, and the material handling and storage elements are what make people call this system lean.

The system is designed for uniqueness and creativity on the part of engineers and workers to develop and improve processes, reduce equipment failures, reduce changeover/setup times and continuously improve the design of the workplace (methods improvement).

More significantly, the placement of the processes in the manufacturing and assembly cells requires unique processing solutions to bring the processing times well under the takt time. While many of these processes are well guarded company secrets, some examples will be given in the section on lean cell design.

■ 43.2 Management Philosophy Must Change

Redesigning the manufacturing and production system requires a change of philosophy within the company. Employee involvement and teamwork are rooted in the idea that no one employee is better than another. Everyone is called an associate. There are no private offices. There is no executive lunch room. There are no preferred parking places (except for the associate of the month). The system in which management tells workers what to do and how to do it must be changed. However, this change requires that the CEO have the courage to shift some of the decision-making power from management to the people on the factory floor. Restructuring the manufacturing system and the production system helps this most difficult transition. The biggest change is psychological—convincing the workers on the factory floor that nothing is more important than what they think and how they feel about the manufacturing system. For this change to work, the operators must have input to the new manufacturing system design. The way the work is done must be clearly defined by the workers. It must be their "cell design." Their achievements must be tracked closely and recognized. People are not going to "bust their guts" unless there is a strong reward system.

Toyota developed a new and different manufacturing system that was flexible, delivered quality products on time at the lowest possible cost, and did so day after day. They educated their work force and placed their best engineering talent on the production floor rather than in the design room. Those who say that the Japanese are not inventive or ingenious have simply looked in the wrong places for evidence. However, much of the unique processing technology was hidden in the manufacturing cells held by the sole-source vendors and suppliers to the main plant (the final assembly plant). This is one of the reasons why lean manufacturers opt for sole or single source suppliers. The proprietary process technology is captured and held in one place. It is difficult for those outside the company to get access to the proprietary process technology.

PRELIMINARY STEPS TO LEAN PRODUCTION

Integration of the production system functions into the manufacturing system requires commitment from top-level management and communication with everyone, particularly manufacturing. Total employee (and union) involvement is absolutely necessary, but it is not usually the union leadership or the production workers who raise barriers to lean production. Those in middle management have the most to lose in systems-level changes as their job functions get integrated into the manufacturing system.

The preliminary steps are as follows:

1. All levels in the plant, from the production workers (the internal customer) to the president, must be educated in lean production philosophy and concepts and understand how lean is different from mass. The physical simulation methodology described in Chapter 42 is very helpful in accomplishing this step.

2. Top down commitment and involvement is critical. The entire company must be involved. The top people must be totally committed to the change, set an example, and be active leaders and, in fact, system designers.

3. Top management must understand that lean design will lead to financial decisions that are opposite to current management accounting practices that have led to mass production in their company.

4. The selection of measurable parameters that will track the change is critical. Everyone in the plant must understand that cost, not price, determines profits. Everyone must be committed to the elimination of waste as these steps are put in place.

5. Encourage your internal customer to set tough goals, realistic measures, and even timelines. Let them know who is the best in the class by using benchmarking.

6. Education and training of the operators is vital. Operators must understand why change is necessary and how to change. The operators must be empowered to design the cell, and then to implement the quality control, machine maintenance, production control, inventory control, process improvements, and setup time reduction and then to make process/cell improvements later.

7. The company must spread the success and reward the teams. The company must share the gains with those who contributed. Many companies feel that bonus payments are the way to go to reward people.

8. The reward structure of middle management must be changed to support the system design.

■ 43.3 LEAN PRODUCTION METHODOLOGY FOR IMPLEMENTATION

The basic implementation strategy outlined here is the amalgamation of the methods used by many companies to successfully implement lean manufacturing. These companies were not suppliers to Toyota and not privy to the Toyota Supplier System training and philosophy.

- Design the lean manufacturing system for existing products to meet the six FRs.
- Integrate the critical control functions.
  - Quality control (zero defects).
  - Production control (right time, right quantity).
  - Inventory control (minimum).
  - Machine tool/equipment control/reliability.
- Design new products that can be made in this system, using concurrent engineering.

This design strategy is broken down into critical steps as outlined in Table 43-1. Notice that *autonomation*, the autonomous control of quality and quantity, comes very late in the methodology (step 9), after functional integration has been achieved. That is, the system must be redesigned, simplified, and integrated before computers and automation are applied. Concurrent engineering becomes the last step as part of an effort to restructure the rest of the enterprise, while improving the cell designs, taking them from interim cells (using single cycle automatics) to lean cells (using machine tools and processes designed for single-piece flow).

This approach implements a manufacturing system design that permits rapid deployment of new products and rapid production of existing products, as well as accommodating changes in the product demand, so it is the winner when it comes to time-based competition. Companies that employ this strategy become the factories with a future because the manufacturing system is simplified before computer technology is applied. This approach avoids large financial risks, and makes the implementation of automation easier.

The order of the design steps outlined in Table 43-1 is important. Many companies have implemented steps 6, 7, and 8 before steps 1–5. Such implementations often result in failure. Even worse, many companies have tried to implement costly CIM strategies without first implementing lean production resulting in computerization of an inefficient manufacturing system.

If you become a supplier for Toyota, they will teach you steps 1 through 7 and your company will become the sole source for this component or subassembly. However, you are free to use this methodology to supply similar components to other companies.

TABLE 43 1. Lean Production Methodology for Implementation	
1. *Design or reconfigure the manufacturing system*: Design and implement manufacturing and assembly cells. The design of the manufacturing system must consider the design of the product and the need of the internal and external customers while meeting the 6 FRs of system stability. FRs = functional requrements	• Standard work for operators • Develop single piece flow in subassembly • Design/implement manufacturing cells
2. *Setup reduction, changing methods and designs to reduce setup time* (Shingo 1985): Setup time is delay time. Affects lot size. Optimum lot size is one. Use SMED because it involves everyone on the factory floor (SMED = single minute exchange of dies). Permits small lots and creates flexibility.	• Teach everyone SMED • Develop one touch setups in the cells
3. *Integrate quality control into the manufacturing system* (Shingo 1986): Does the manufacturing process satisfy the design specifications every time? Inspection to prevent the defect from occurring (pokayokes). System will breakdown if machines and support equipment fail.	• Inspect to prevent defects • Use the 7 tools for quality control and line stop • Teach everyone quality
4. *Integrate preventive maintenance* (Nakajima 1988): Do the machines and people behave reliably? Design equipment to be reliable. Design methods to check people and methods for people to check machines, identify and solve problems.	• Machines designed for reliability using TPM • Operators solve problems • Operators perform daily maintenance
5. *Level and balance the manufacturing system, smoothing the material flow* (Monden 1983): Leveling involves the development of mixed model final assembly. Level or smooth the demand on the cells or the suppliers. Balancing is getting the output from the cells to match the needs of final assembly. Synchronize subassemblies with final assembly.	• Mixed model final assembly; takt time • Balance the output from the cells • Sequence subassemblies with order of assembly
6. *Integrate production control, link the cells, pull material to final assembly*: Control the where, when, and how much material. The design of the manufacturing system defines flow and the kanban operates within the structure. This is integrated production control or kanban (Black 1991).	• Link the cells • Pull material to final assembly • Kanban drives the production
7. *Integrated inventory control*: Reduce the WIP in the links that connect the cells. This is control of the quantity of material in the links. Minimized and optimized and controlled by the internal customers, the users of the materials (Black 1991).	• Gradually remove inventory from links • Expose problems • Solve problem, improve system TPT
8. *Integrate the suppliers: make them JIT manufacturers just like you*: Suppliers become remote cells. Suppliers become partners. Relationship built on trust. This is how real technology transfer takes place.	• Suppliers are sole source • Teach suppliers steps 1 thru 7
9. *Autonomation: autonomous control of quality and quantity within the manufacturing system*: Automate the integrated pull manufacturing system.	• Computerize the integrated system
10. *Design new products concurrently with customers in mind* (Whitney 1992). Design/implement lean manufacturing with lean machine tools.	• Design products in families • Design processes to support SPF

TPM = Total Productive Maintenance   TPT = throughput time   SPF = Single piece flow

STEP 1: RESTRUCTURE THE FACTORY FLOOR

In the lean shop, cells replace the job shop. The first task is to restructure and reorganize the basic manufacturing system into manufacturing cells that fabricate families of parts. This prepares the way for systematically creating a linked-cell system for one-piece movement of parts within cells and for small-lot movement between cells. Creating cells that can operate at the takt time is the first step in designing a manufacturing system in which production control, inventory control, and machine tool maintenance are functionally integrated.

Conversion of the job shop system into a flexible, linked-cell system is a design task. Just as design engineers now try to design products which are simpler and easier to manufacture, manufacturing engineers try to design a factory which is simpler to operate. See Figure 43-3 for an example.

Most companies "design" their first cell by one of the trial-and-error techniques for expediency in gaining experience in cells.

By grouping similar components into families of parts, a group or set of processes can be collected together to make a family. This is a manufacturing cell. The arrangement of the machines in the cell is defined by the sequence of manufacturing processes. The operations in the cell include all sorts of metal cutting operations, heat treating, inspection, assembly, and even grinding and superfinishing. All are done as steps in one cell that makes all the variations of a component, say for a model of a car.

Many companies begin with a pilot cell so that everyone can see how cells function. It will require time and effort to train the operators, and they will need time to adjust to standing and walking. Simply select a product or group of products that seems most logical. The operators must be involved in designing the cell or they will not take

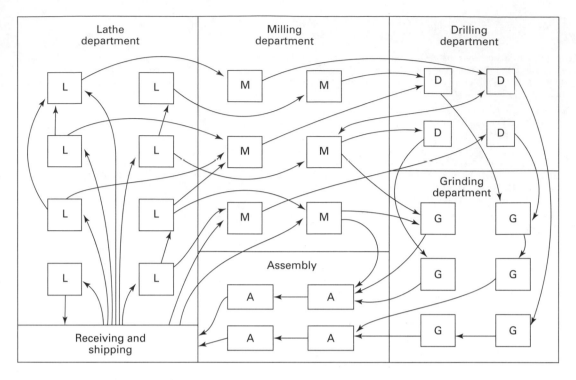

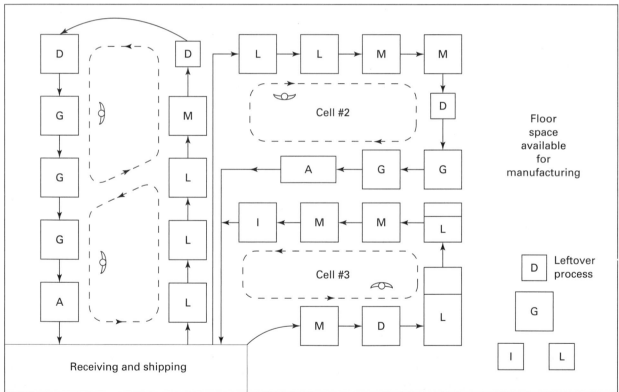

FIGURE 43-3   The job shop is restructured into manufacturing cells which build families of components. See Figure 43-6 for detail on cell #3.

ownership. The pilot cell will show everyone how cells operate and how to reduce setup time on each machine. Machines will not be utilized 100%. Machine utilization rate usually improves but may not be what it was in the functional system where *overproduction* is allowed. The objective in manned linked-cell manufacturing systems design is to utilize the people fully, enlarging and enriching jobs, allowing operators to become multifunctional. The operators learn to operate many different kinds of machines and perform

Interim cells are made up of single cycle automatic
machine tools designed for job shop but modified
for use in the single piece flow cells.

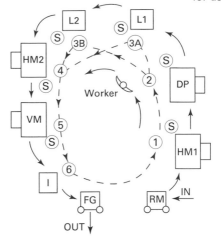

Work Sequence	Name of Operation	Time		
		Manual	Walking	Machine
①	Mills ends HM1	12"	5"	30"
②	Drill hole DP	15"	5"	20"
③A or ③B	Turn, bore L1 or L2	13"	3" 5"	180"
④	Mill flats HM2	12"	5" 3"	20"
⑤	Mill steps VM	13"	7"	30"
⑥	Final inspect-I	15"	5"	—
		80"	30"	280"

Takt Time = 2 minutes
Cycle Time = 80 sec + 30 sec = 110 sec
Longest Machining Time = 180 sec
Total Machining Time = 280 sec per part

All machines in the cell are single cycle automatics,
processing unattended while the operator is doing
manual operations (unload, load, inspect, and deburr)
at other machines or walking from machine to machine.
The time to change tools and workholders (perform
setup) is not shown.

Key:
DP = Drill Press
L = Lathes
FG = Finished goods
RM = Raw material
HM = Horizontal milling machine
VM = Vertical milling machine
——— Material flow
– – –➤ Operator's path
Ⓧ Operation sequence
Ⓢ Walk-away switch

FIGURE 43-4 Example of an interim manned cell part of an L-CMS. This cell operates
with six machines and one multifunctional worker. The lathe operation, #3, is duplicated.

tasks that include quality control, machine tool maintenance, setup reduction, and
continuous improvement. In unmanned cells and systems, the utilization of the equipment is more important because the most flexible and smartest element in the cell, the
operator, has been removed and has been "replaced" by a robot.

The cells are designed in U-shapes or parallel rows so that the workers can move from
machine to machine, unloading, checking, and loading parts. Figure 43-4 shows an example
of a simple manned cell. The cell has one worker who can make a walking loop around the
cell in 110 seconds. (See the breakdown of time in the table in Figure 43-4.) Therefore, the cycle
time for the cell is 110 seconds. Cycle time equals the inverse of the production rate. So

$$CT = \frac{1}{PR}.$$

The machines in the cell are single-cycle automatics, so that they can complete the desired
processing untended, turning themselves off when finished with a machining cycle. The operator comes to a machine, unloads a part, checks the part, loads a new part into the machine,
and starts the machining cycle by hitting a walkaway switch as he moves to the next machine. The cell usually includes all the processing needed for a complete part or subassembly.
The table shows the typical average times for the operator time, human or manual time, and
walking time. The times for the machining cycle are given as machining times. Here is the
axiomatic design rule for cells. The cell is designed such that the machining time for any
part in the family for any machine in the cell is less than the necessary cycle time, CT. That
is, $MT_{ij} < CT$. Thus, machining times are uncoupled from the cycle time.

However, note that the machining time for the third operation is greater than the
necessary cycle time. That is, 180 seconds is greater than 110 seconds. One solution to this
problem is to duplicate the operation. The operator alternates between the two lathes,

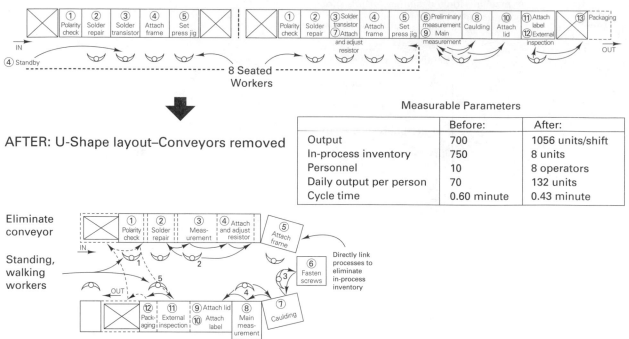

FIGURE 43-5   The traditional assembly line using conveyors can be redesigned into a U-shaped cell. The cell uses walking operators.

visiting each lathe every other trip he or she makes around the cell. This makes the average MT $180/2 = 90$ seconds and less than the cycle time.

At the same time that the manufacturing cells are being designed, the subassembly (flow) lines that use conveyors are reconfigured into U-shaped cells as well (see Figure 43-5) to make these systems operate on a single-piece flow basis. The operations in an assembly cell are usually all manual so the operator must stay at the station until the task is completed. This example shows the marked improvement in productivity that cells can achieve. As with the manufacturing cell, the long setup times typical in flow lines must be vigorously attacked and reduced so that the flow lines can be changed over quickly from making one product to making another. This makes them flexible and compatible with the cells designed to make piece parts and with the other subassembly lines and final assembly lines. Again the cells are designed to manufacture specific groups or families of parts. In the U-shaped layout in Figure 43-5 workers 2, 3, and 4 cover multiple operations. Notice that workers 3 and 4 share operation 7. Workers 1 and 5, are covering operations 1, 11, and 12 using a rabbit chase (workers follow each other around the loop). The need to line balance (make task times equal) the flow line has been eliminated. This is accomplished by using standing, walking workers who are capable of performing multiple operations. For the cell, the cycle time is 0.43 minute (to produce 1056 units per shift) which is 26 seconds per part. The cell must produce four parts every 2 minutes to match the demand at final assembly (each product needs four of these motors).

Standardized work is the basis for maintaining productivity, quality, and safety at high levels. It provides a consistent structure for performing the tasks in the designated takt time while uncovering opportunities for making improvements in work procedures.

There are three aspects in structuring standardized work in a cell

- Takt time   • Working sequence   • Stock-on-hand

Takt time reflects the pace of sales in the marketplace or what may be called the daily demand. The working sequence is the sequence of operations that is the best way to perform a task. The stock-on-hand is the minimum number of workpieces needed to have on hand in a cell to maintain a smooth flow of work.

Standardized work provides detailed, step-by-step guidelines for every job in the lean shop. Team leaders and operators determine the most efficient working sequence and make continuing improvements—kaizen—in that sequence. Kaizen thus begets new patterns of standardized work.

Cells have many features that make them unique and different from other manufacturing systems. Parts move from machine to machine *one at a time* within the cell. For material processing, the *machines are typically* capable of completing a machining cycle initiated by a worker. The U-shape puts the start and finish points of the cell next to each other. Every time the operator completes a walking trip around the cell, a part is completed. The cell is designed so that this cycle time (CT) is equal to or slightly less than the takt time (cycle time) for final assembly. This is referred to as the necessary cycle time. The machining time (MT) for each machine needs only to be less than the time it takes for the operator to complete the walking trip around the cell. Thus, the machining time can be altered without changing the production schedule. Conversely, the production schedule can be altered without changing the machining time.

The cell is designed to make parts *as needed* by downstream processes and operations. There is no overproduction. Overproduction will result in the need to store parts, transport parts to storage, retrieve the parts when needed, keep track of the parts (paperwork), and so on. All this requires people and costs money but adds no value. In assembly cells, the cycle time is readily changed by adding operators.

In manufacturing cells, there is no need to balance the MTs for the machines. This can be a very difficult task in flow lines (transfer lines). In cells, it is necessary only that no MT be greater than the required CT. The machining speeds and feeds can be relaxed to extend the tool life of the cutting tools and reduce the wear and tear on the machines as long as the MT does not equal or exceed the CT.

The fixtures in the machines are designed to hold the family of parts so rapid changeover from one part to another is possible. The fixtures are designed so parts can be easily loaded/unloaded, cannot be loaded incorrectly, and defective parts cannot be loaded.

Between the process steps *decouplers* elements are placed to provide flexibility, part transportation, inspection for defect prevention (pokayoke) and quality control, and process delay for the manufacturing cell. The decoupler element can inspect the part for a critical dimension and feed back adjustments to the machine to prevent the machine from making oversize parts (as the milling cutter wears). A process delay decoupler would delay the part movement to allow the part to cool down, heat up, cure, or whatever is necessary for a period of time greater than the cycle time for the cell. Decouplers and flexible fixtures are vital parts of both manned and unmanned cells.

STEP 2: RAPID EXCHANGE OF TOOLING AND DIES

When cells are formed to make a family of parts, then the problem of changing over the machines from one part in the family to another must be immediately addressed. Therefore, everyone on the plant floor must be taught how to reduce setup time using SMED (single-minute exchange of die). SMED is a four stage methodology developed by Shigeo Shingo to reduce tooling and die exchange times, that is, reduce setup times. A setup reduction team trains the production workers and foremen in the SMED process and demonstrates the methodology on a project, usually the plant's worst setup problem. Reducing setup time is critical to reducing lot size.

The lean production approach to manufacturing demands that small lots be run. This is impossible to do if machine setups take hours to accomplish. The *economic order quantity* (EOQ) formula has been widely used to determine what quantity should run to cost-justify a long and costly setup time. The EOQ was a faulty suboptimal approach that accepted long setup times as a given. Setup times can be reduced, resulting in reduced lot sizes and throughput times while improving the flexibility of the system.

Successful setup reduction is easily achieved when approached from a methods engineering perspective. Much of the initial work in this area uses videotapes with time and motion studies while applying Shingo's SMED rules for rapid exchange of tooling. Setup time reduction occurs in four stages. The initial stage is to determine what currently is being done in the setup operation. The setup operation is usually videotaped and everyone

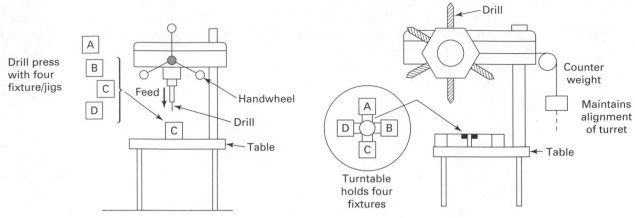

In the job shop, a machine tool with four different fixtures or jigs would need four different setups, each consisting of changing fixtures or jigs and alignment of each.

In the manufacturing cell the four fixtures are mounted on a turntable and are permanently aligned to the spindle when locked in position. A multiple tool turret replaces single spindle. Automatic down feed replaces hand wheel, so the machine is a single cycle automatic.

FIGURE 43-6   The machines in the interim cell are modified to process the family of parts, reducing setup time to 2 or 3 seconds.

concerned gets together and reviews the tape to determine the elemental steps in the setup. The next stage is to separate all setup activities into two categories, *internal* and *external*. Internal elements can be done only when the machine is not running; external elements can be done while the machine is running. This elemental division will usually shorten the lead time considerably. Stages 3 and 4 focus on reducing the internal time. The key here is that the operators perform all the setups and learn how to reduce setup times by applying simple principles and techniques. If a company must wait for the setup reduction people to examine every process, a lean manufacturing system will never be achieved.

The similarity in shape and processes needed in the family of parts allows setup time to be reduced or even eliminated. Initially setup should be less than 10 minutes (SMED). As the cell matures, the setup times are continually reduced. Reducing the setup time until it is equal to or less than the cycle time (1–2 minutes) for the cell is usually quite easily accomplished. This will permit a significant initial reduction in lot size. The next goal is to get setup times down to around 10–15 seconds, what is commonly called one-touch exchange of dies (OTED), see Figure 43-6.

In the last stages of SMED, it may be necessary to invest capital to drive the setup times below 1 minute. Automatic positioning of workholders, intermediate jigs and fixtures, and multiple duplicate workholders represent the typical kinds of hardware needed. The result is that long setup times can be reduced to under 15 seconds in relatively short order.

When the setup time is down to less than the time needed to load, unload, inspect, deburr, and so on at the machine, the operators can quickly change the machines over from one component to the next. Figure 43-7 shows how the setup operation flows through the cell one cycle at a time as the cell is changed over from part A to part B. After each setup, at each machine, defect-free products should be made right from the start. The first part will be good. Ultimately, the ideal condition would be to eliminate setup between parts. This is called no touch exchange of dies (NOTED).

In summary, the savings in setup times are used to decrease the lot size and increase the frequency at which the lot is produced. The smaller the lot the lower the inventory, creating shorter throughput time and improving quality.

STEP 3: INTEGRATE QUALITY CONTROL

A *multiprocess* worker can run more than one kind of manufacturing process. A *multifunctional* worker can do more than operate machines. This team member is also an inspector who understands process capability, quality control, and process improvement. In the lean shop, every worker has the responsibility and the authority to make

Cycles Through the Cell	Four Processes in the Cell			
	Machine 1	Machine 2	Machine 3	Machine 4
1	A	A	A	A
2	A (last A part)	A	A	A
3	Setup change	A (last A part)	A	A
4	B (first B part)	Setup change	A (last A part)	A
5	B	B (first B part)	Setup change	A (last A part)
6	B	B	B (first B part)	Setup change
7	B	B	B	B (first B part)
8	B	B	B	B

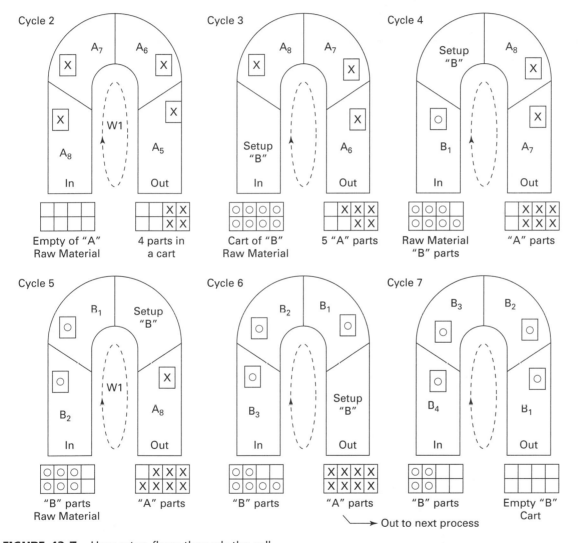

FIGURE 43-7 How setup flows through the cell.

the product right the first time and every time and the authority to stop the process when something is wrong. This integration of quality control into the manufacturing system markedly reduces defects while eliminating inspectors. Cells provide the natural environment for the integration of quality control. The fundamental idea is to inspect to prevent the defect from occurring and to never let a defective product leave the cell.

When every worker is responsible for quality and able to use the seven basic tools of quality control, shown in Figure 43-8, the number of inspectors on the plant floor is markedly reduced. Products that fail to conform to specification are immediately uncovered because they are checked or used immediately.

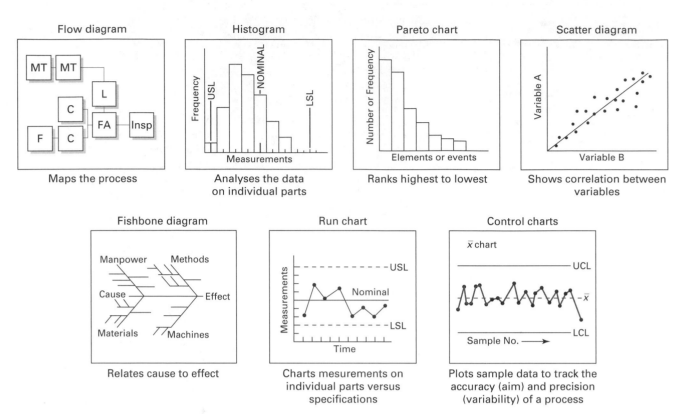

FIGURE 43-8 The seven tools of quality control.

In lean manufacturing every worker is an inspector responsible for quality. What happens when a worker on the assembly line stops the line if they find something wrong? Everyone's attention is focused on the problem holding up production. Problems get solved fast and permanently so the line is not stopped again. *Inspection to prevent the defect from occurring* rather than *inspecting to find* the defect after it has occurred becomes the mode of operation in the lean shop. Ultimately, the concept of *autonomation* evolves (see step 9) which means to automate to prevent the occurrence of defects based on lessons learned from the operation of the cell. Literally, the autonomation means automation with a human touch.

Now here is the fundamental difference between lean production systems and other systems. Through redesign, the cells produce parts one at a time, just like assembly lines. This is called one-piece flow. Within all the production entities, the internal customers operate a *make one, check one, move one on* basis or MO-CO-MOO. The operator is checking what the previous process produced to assure it is right 100% of the time. Pull cords are installed on the assembly lines to stop the lines if anything goes wrong. If workers find defective parts, if they cannot keep up with production, if production is going too fast according to the quantity needed for the day, or if a safety hazard is found, they are obligated to stop the line. The problem is fixed immediately. Meanwhile, the other workers maintain their equipment, change tools, sweep the floor, or practice setups; but the line does not move until the problem is solved.

For manual work on assembly lines, the system for tracking defective work is called *andon*. Andon is actually an electric light board that hangs high above the assembly lines so that everyone can see it. When everything is going okay, the lights are green. But, when a worker on the line needs help, the yellow light can be turned on. Nearby (multifunctional) workers who have finished their jobs within the allotted cycle time move to assist workers having problems (called mutual assistance). If the problem cannot be solved within the cycle time, a red light comes on and the line stops automatically until the problem is solved. Music usually plays to let everyone know there is a problem on the line.

In most cases the red lights go off within 10 seconds and the next cycle begins, a green light comes on, with all the processes beginning together. The name for this system is Yo-i-don, which literally means "ready, set, go." The stages on the assembly line are synchronized. Such systems are built on teamwork and a cooperative spirit among the workers, fostered by a management philosophy based on harmony and trust.

Of course there are more ways to control quality than outlined in Figure 43-8 but these "tools" are fundamental to the integrated process because these tools are used by the operators. Mapping and analyzing the process to understand its behavior is key to finding out what went wrong. By designing the system and all the processes into a single-piece flow methodology, the causes of the defects (the effects) are quickly isolated and identified, thus cures can be readily implemented!

The lean system design strives for continuous improvement in the processing to reduce the variability (or spread in measurements about the process mean). Thus reducing the process variability improves its process capability as shown in Figure 43-9. Process capability, C_p and C_{pk}, were defined in Chapter 12. In lean manufacturing, the continuous

FIGURE 43-9 To achieve six sigma capability from three sigma capability, the precision of the process must be greatly improved, reducing σ, the process standard deviation which measures variability.

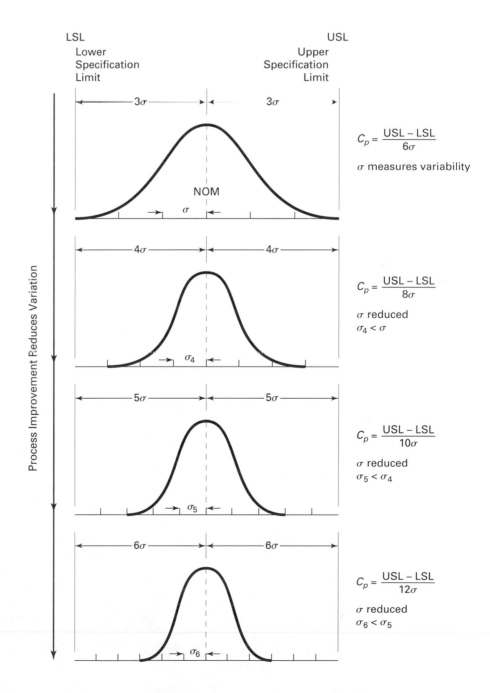

$$C_p = \frac{USL - LSL}{6\sigma}$$

σ measures variability

$$C_p = \frac{USL - LSL}{8\sigma}$$

σ reduced
$\sigma_4 < \sigma$

$$C_p = \frac{USL - LSL}{10\sigma}$$

σ reduced
$\sigma_5 < \sigma_4$

$$C_p = \frac{USL - LSL}{12\sigma}$$

σ reduced
$\sigma_6 < \sigma_5$

improvement aspects are driven by the systematic removal of inventory—see step 7 for further discussion.

So here we understand the key to achieving superior quality in lean manufacturing lies in isolating process steps in a serial fashion in the cells. There is one process for each step. Multiple processes for the same step will increase the process variability. The outcome from each step is checked before the component is moved to the next step. The operators have complete feedback of each step in the processing. This feedback involves all of the operators senses, hearing, sight, smell, touch, and even taste, making them superior to machines. They are on the front line for problem identification and resolution.

STEP 4: MAKE OUTPUT PREDICTABLE: INTEGRATE PREVENTIVE MAINTENANCE

Making machines operate reliably begins with the installation of an integrated preventive maintenance program, giving workers the training and tools to maintain their equipment properly. The excess processing capacity obtained by reducing setup time allows operators to reduce the equipment speeds or feeds and to run processes at less than full capacity. Reducing pressure on workers and processes to produce at full speed or maximum capacity fosters a drive to produce perfect quality.

Producing to takt time, at the pace of customer demand, in a balanced system (step 1) defines the appropriate pace of production for the workers and the processes. Prior to lean, companies thought they could lower cost by running machines as fast as possible, all of the time. Lean recognizes that such strategies result in overproduction and poor quality.

The multifunctional operators are trained to perform routine machine tool maintenance. Just adding lubricants (oiling the machine), checking for wear and tear, replacing damaged nuts and bolts, routinely changing and tightening belts and bolts, and listening for telltale whines and noises that signify impending failures can do wonders for machine tool reliability. The maintenance department must instruct the workers on how to do these things and help them prepare the routine checklists for machine maintenance. The workers are also responsible for keeping their areas of the plant clean and neat. Thus another function that is integrated into the manufacturing system is housekeeping.

The following housekeeping rules are implemented:

1. A place for everything and everything in its place. Everything should be put away so it is ready to use the next time (the next cycle).
2. Each worker is responsible for cleanliness of workplace and equipment.

Naturally, the machines still need attention from the experts in the maintenance department, just as the airplane is taken out of service periodically for engine overhaul and maintenance. One alternative here is to implement two 9-hour shifts (8 work hours plus 30 minutes for lunch plus two 15-minute breaks) separated by two 3-hour time blocks for machine maintenance, tooling changes, restocking, long setups, overtime, earlytime, and so on. This is called the 8-4-8-4 scheme. The main advantage that equipment has over people is that it can decrease variability, but it must be reliable and dependable. Smaller machines are simpler and easier to maintain and therefore are more reliable. Small machines in multiple copies add to the flexibility of the system as well. The linked-cell system permits certain machines in the cells to be slowed down and therefore, like the long-distance runner, to run farther and easier without breakdown. Many observers of the lean shop come away with the feeling that the machines are "babied." In reality, they are being run at the pace needed to meet the daily demand.

True lean producers build and modify much of their manufacturing process technology. This is where they become unique and proprietary. Machine tools are equipped with vibration sensors, temperature monitors, coolant/lubricant sensors, and of course, process controls for single cycle automatic operation of the machines and walkaway switches to start the machining cycles. The sensors provide early warnings of problems or can shut off the machine if a problem is detected. In addition, multiple copies of machines are made for making the similar products. Suppose you are the lean supplier for steering gears. Your plant has a cell for making racks for a rack and pinion steering gear. The Honda manufacturing cell for racks makes 6 different racks for the Accord. In the same plant, the Toyota cell makes racks for Camry and Avalon and can make 10 different kinds of racks. But the two cells are very similar in their design because the racks require a similar set of processes. In the event of a machine failure in the Accord cell, a machine from the Toyota cell can be borrowed, modified, and used during second shift in the Accord cell. Processing capacity and capability is replicated in proven increments. Because the increment (the cell) has an optimal design, this is an economic choice as well as having the security of dealing with a proven manufacturing process technology. Modifying existing equipment shortens the time needed to bring new technology on stream. Manufacturing in multiple versions of small-capacity machines retains the expertise and permits the company to keep improving and mistake-proofing the process. In contrast to this approach is the typical job shop, where a new supermachine (large, expensive, multiple operation) would be purchased and installed when product demand increases. That is, many companies try to increase capacity by buying new, untried manufacturing technology that may take months, even years, to debug and make reliable.

STEP 5: LEVELING, BALANCING, SEQUENCING, AND SYNCHRONIZING THE MANUFACTURING SYSTEM

The lean production system depends upon *smoothing the manufacturing system.* In order to eliminate variation or fluctuation in quantities in feeder processes, it is necessary to eliminate fluctuation in final assembly. This is also called *leveling* the final assembly. It means the demand for the subassemblies and components from the suppliers is leveled. Here is a single example to show the basic idea. First, we need to calculate the production rate and its reciprocal or cycle time which is based on the daily demand.

$$DD = \text{average daily demand for parts} = \frac{\text{monthly demand (forecast plus customer orders)}}{\text{number of days in each month}}$$

$$TT = \frac{\text{available hours/shift}}{\text{average } DD/\text{shift}} = \text{takt time for final assembly}$$

$$TT = 1/\text{production rate for final assembly (FA)}$$

$$PR = \frac{\text{average daily demand (parts)/shift}}{\text{available hours /shift}} = \frac{DD}{\text{hrs/day}} = \frac{\text{parts}}{\text{minute}}$$

But in reality $CT = TT(1 - \text{Allowance})$.

This incredibly simple approach highlights the way in which lean companies calculate takt time for FA and cycle time for cells and processes. The operator's life is simpler when the mass production system has been eliminated and a linked cell system has been installed.

Here is an example of how cycle time is determined for a mix of cars at final assembly. Suppose that the forecast is for the assembly of 240 cars per day and 480 production minutes are available (60 minutes × 8 hours/day). Thus, every cycle = 2 minutes, which is the takt time in this case. Every 2 minutes a car rolls off the line. Suppose that the mix is as given in Table 43-2. The sequencing of the four types (models) of cars going

TABLE 43-2. Example of Mixed Model Final Assembly Line That Determines the Cycle Time for Model

Q	Car Mix for Line Model	Cycle Time by Model (min)	Production Minutes by Model	Sequence (24 cars)
50	Two-door coupe	9.6	100	TDC, TDF, TDF, FDS, FDW
100	Two-door fastback	4.8	200	TDC, TDF, TDF, FDW, FDW
25	Four-door sedan	18.2	50	TDC, TDF, TDF, TDS, TDW
65	Four-door wagon	7.7	130	TDC, TDF, TDF, FDW
	240 cars/8 hr			Takt time = 480 min/240 = 2 min per car

down the assembly is shown in the sequence. The subprocesses that feed the two-door fastback are controlled by the cycle time for this model. Every 4.8 minutes, the rear deck line will produce a rear hatch for the fastback version. Every 4.8 minutes, two doors for the fastback are assembled in synch with the final assembly line. The doors are mounted on the body, initially painted with the body, then removed when the car exits painting, placed on an overhead conveyor which transports the doors to the door line which has the same mix of product and cycle time as the main assembly line. The doors are trimmed (i.e., filled with windows, door handles, etc.) in sequence and in synch with final assembly. If the final assembly line stops, so does the door line, the seat assembly line and other main subassembly lines for parts which are special to the car model. This door is then returned to the line and goes back on the same car body from which it was earlier removed. This is an example of *synchronizing* or *producing in sequence*.

Every car, regardless of model type, has an engine. Engines for this assembly line are produced at a rate of one every 2 minutes. Each engine needs four pistons. Therefore every 2.0 minutes, four pistons are produced. Parts and assemblies are produced in their minimum lot sizes and delivered to the next process, under the control of kanban. Engines are often made elsewhere, shipped to the assembly plant and put in the right *sequence* for the cars coming down the line.

Balancing is making the output from the cells equal to necessary demand for the parts downstream. The parts or components are not made in synch with final assembly, only the daily quantity is the same. In summary, small lot sizes, made possible by setup reduction within the cells, single-unit conveyance within the cells, and standardized cycle times are the keys to having a leveled manufacturing system. One strives to make the cycle time in the cells equal to the takt time for final assembly but at the outset, matching the daily demand is sufficient. Ultimately, every part, sequence of assembly operations, or subassembly has the same number of specified minutes as the final assembly line.

For example, when the car body exits from painting, an electronic order for the seats is issued to the seat suppliers. The seats are made in the same order as the cars on the assembly line. They are made and delivered in the same amount of time as it takes the car to get from paint to the station on the line where seats are installed. That is, seat manufacturing is synchronized. The minimum number of workers needed to produce one unit of output in the necessary cycle time is used. The lean shop is a *balanced and synchronized* manufacturing system. Actually, a synchronized system is the ultimate leveled system. The run size is one unit in final assembly. Many subassemblies are shipped to final assembly in lots or batches and put on the line at the same takt time. These items are put in the correct sequence so they get to the right car at the right time. This is called *sequencing*. The dashboard subassembly may be done in this way or it may be done in a synchronized fashion if paint matching is a problem.

STEP 6: INTEGRATING PRODUCTION CONTROL (PC)

The function of the people who work in production control is to schedule the manufacturing system which means they determine *where* the raw materials, purchased parts, and subassemblies are to go, *when* they should go there, and *how many* should go at any point in time. In short, production control determines where, when and how many. In the traditional mass production system, the PC function is very labor intensive, and many people have tried to computerize it using software called manufacturing or enterprise resources planning. Many companies' experience with these computerized control systems has been that of great expense, wasted time, disappointment, and frustration. Why? Because these software packages were designed for planning, not control. The lean manufacturing approach redesigns the manufacturing system to integrate the production control functions into the system design. Steps 1 through 5 provide the physical stage to integrate production control functions. This is typically achieved by the use of kanban.

Integration of production control is achieved by linking the cells, subassemblies, and final assembly elements utilizing kanban. The layout of the manufacturing system will define paths that parts can take through the plant. By connecting the elements with kanban links, the need for route sheets is eliminated because the job shop is eliminated.

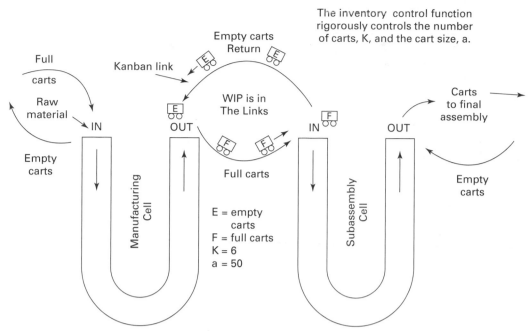

FIGURE 43-10   Cells are linked by a kanban system. The arrival of an empty container at the manufacturing cell is the signal to produce more parts.

The subassemblies and component parts (i.e., the in-process inventory) move within the structure on clearly defined paths. All the cells, processes, subassemblies, and final assemblies are connected by the kanban links, pulling material to final assembly. This is the integration of production control into the manufacturing system, forming an L-CMS.

Kanban is a visual control system that is only good for lean production with its linked cells and its namesakes; it is not good for the job shop. Linking the cells together provides control over the route that the parts must take (while doing away with the route sheet), controls the amount of material flowing between any two points, and provides information about when the parts will be needed.

There are two kinds of kanban: *withdrawal* (or conveyance) *kanban* (WLK) and *production-ordering kanban* (POK). Think of kanban as a link connecting the output side of one cell with the input point of the next cell (see Figure 43-10). This link is filled with carts or containers that hold parts in specific numbers. Every cart holds the same number of parts and has one WLK and one POK. If there are k carts, then

Maximum inventory = k (number of carts) $\times$ a (number of parts in each cart).

The arrival of an empty cart at the manufacturing cell initiates the order (the POK) to make more parts to fill the cart. The kanban cards tell the material handler where to take the parts. Suppose the link has six carts and each cart holds 50 parts, the maximum inventory (WIP) is 300 parts.

The same kind of link connects the subassembly cells to final assembly. All the other cells in an L-CMS are similarly connected by the pull system for production control.

STEP 7: INTEGRATING INVENTORY CONTROL

All of the material in a manufacturing system is considered inventory. There are three basic types—raw material, in-process or work-in-process (WIP), and finished goods. Step 7 involves the integration of the control of the in-process inventory. In the lean system, the work-in-process (WIP) inventory in the system is held in the links. The WIP inventory has been analogized to the water in a river, as shown in Figure 43-11. A high river level is equivalent to a high level of inventory in the system. The high river level covers the rocks in the riverbed. Rocks are equivalent to problems. Lower the level of the river (inventory) and the rocks (problems) are exposed. This analogy, developed by Taiichi Ohno and published by Shigeo Shingo, is quite accurate. In lean manufacturing, the problems receive immediate

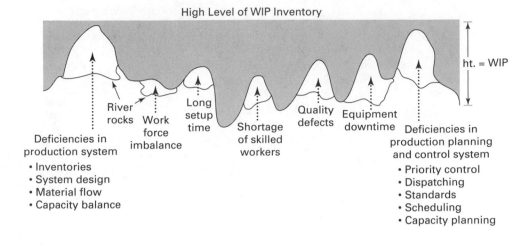

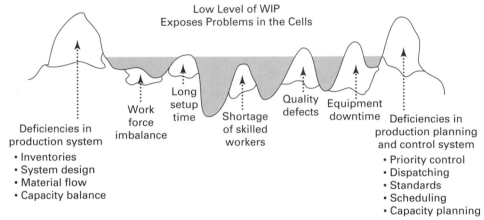

FIGURE 43-11   Lowering the level of inventory in the links uncovers the problems.

attention when exposed. When all the rocks are removed, the river can run very smoothly with very little water. However, if there is no water, the river has dried up. The notion of zero inventory is incorrect. While zero defects is a proper objective, zero inventory is not possible. (Within the cell, parts are already handled one at a time, just as they are in assembly lines. The material in the cell is called the stock-on-hand [SOH] so technically there is no inventory in the cells [hence zero inventory].)

The idea is to minimize the necessary WIP between the cells. This is how the inventory in a lean manufacturing system is controlled. The level of WIP between cells, subassembly, and final assembly is actually controlled by teamleaders in the various departments. The control is integrated and performed at the point of use. Here is what teamleaders do. Suppose that there are six carts in the link and that each cart holds 50 parts. The maximum inventory in this area is therefore 300 parts. The foreman goes to the stock area outside the cell and picks up the kanban cards (one WLK, one POK), which puts one full cart of parts out of commission. The (maximum) inventory level is now 5×50, or 250 parts. The foreman waits until a problem appears. When it appears, the foreman immediately restores the kanban, which restores the inventory to its previous level. The cause of the problem may or may not be identified by the restoration of the inventory, but the condition is relaxed until a solution can be enacted. Once the cause of the problem is identified and solved, the foreman repeats this procedure. If no other problems occur, the foreman then tries to drop the inventory to $4 \times 50 = 200$ parts. This procedure is repeated daily in the links all over the plant. After a few months, the foreman may be down to the three carts of 50 parts. Over the weekend the system will be restored to six carts between the two points, but this time each cart will hold only 25 parts. If everything works smoothly, with the reduced cart quantity, the foreman will soon remove a cart to see what happens. More than likely, some setup times will need to be reduced. In this way, the inventory in the linked-cell system is continually reduced, exposing problems. The problems are solved one by one. The teams work on solving the exposed problems including long setup times. The effect on the system of removing inventory is to continuously improve the TPT. This is how continuous improvement works in the lean production factory.

The minimum level of inventory that can be achieved is a function of many factors: the quality level, the probability of a machine breakdown, the length of the setups, the variability in the manual operations, the number of workers in the cell, parts shortages, the transportation distance, and so on. It appears that the minimum number of carts is three, and, of course, the minimum lot size is one. The significant point here is that inventory becomes a controllable independent variable rather than an uncontrollable variable dependent on cravings of the users of the manufacturing system for more inventory.

STEP 8: INTEGRATING THE SUPPLIERS

In lean system design one tries to reduce the number of suppliers and have a single source for each component or subassembly. Suppliers are educated and encouraged to develop their own lean production system for superior quality, low cost, and rapid on-time delivery. They must be able to deliver perfect parts downstream to their customers when needed and where needed without incoming inspection. The linked-cell network ultimately should include every supplier. Suppliers become remote cells in the L-CMS.

In the traditional job shop environment, the purchasing department permits its suppliers to make weekly/monthly/semiannual deliveries with long lead times—weeks/months are not uncommon. A large safety stock is kept just in case something goes wrong. Quantity variances are large and late and early deliveries are the norm. The situation leads to expediting, where people try to find the parts that are delaying assembly and get them moving (this is a total waste).

In mass production, as a hedge against supplier problems, multiple sources are developed. The suppliers' profit margins are cut thin and the plant is unreliable and may be in jeopardy of bankruptcy. Meanwhile, the purchasing department may claim that pitting one supplier against another gives the company a competitive advantage and lower cost parts.

The lean manufacturing approach to suppliers is quite different. Just-in-time purchasing is a program of continual long-term improvements. The buyer and supplier work together to reduce lead times, lot sizes, inventory levels, and unit cost while improving quality. Both the supplier and the customer become more competitive in the world marketplace.

In this environment, longer-term (18–24 months) flexible contracts are drawn up with three or four weeks lead time at the outset. The buyer supplies updated forecasts every month that are good for 12 months, commits to long-term quality, and perhaps even promises to buy out any excess materials. Exact delivery is specified by midmonth for the next month. Frequent communication between the buyer and the supplier is typical. Kanban controls the material movement between the supplier and the buyer. The supplier is a remote cell. Long-range forecasting for 6 months to 1 year is utilized. As soon as the buyer sees a change, the supplier is informed; this knowledge gives the supplier better visibility instead of a limited lead time view. The supplier has *build-schedule stability* (not "jerking" up and down of the build-schedule) an outcome of leveling final assembly.

The buyer moves toward fewer suppliers, often going to local, sole sourcing. Frequent visits are made to the supplier by the buyer, who may also supply engineering aid (quality, automation, setup reduction, packaging, and the like) to help the supplier become more knowledgeable on how to deliver, on time, the right quantity of parts that require no incoming inspection. This is truly technology transfer. The suppliers learn about lean manufacturing from the customer. The buyer and the seller work together to solve problems.

The advantages of single sourcing are that resources can be focused on selecting, developing, and monitoring one source instead of many. When tooling dollars are concentrated in one source, there is a savings in tooling dollars. The higher volume should lead to lower costs. The supplier is more inclined to do special things for the buyer. The buyer and the supplier learn to trust one another. Quality is more consistent and easier to monitor when there is a single source as there is less variability in the components.

Finally there is the aspect of proprietary processes. Toyota and other users of lean manufacturing have published very little about the manufacturing/assembly cells. Why?

TABLE 43-3.	Managing the Lean Production System (LPS)*

The Lean Production System is the basic philosophy and concepts used to guide production processes and environment. The LPS includes the linked-cell manufacturing system (cells linked by a kanban pull system), the five S's, standard operation, the seven tools of QC, and other key organizational elements.

Kanban pull system (see step 7)	The production processes that use a card system, standard container sizes, and pull versus push production to accomplish just-in-time production.
Five S's (*Seiri, Seiton, Seiketsu, Seisō, Shitsuke*) (see step 4)	The five S's are proper arrangement, orderliness, cleanliness, cleanup, and discipline.
Standard operation in manufacturing cells (see step 1)	The production process used by technicians that combines people and process. The components of standard operation include cycle time, work sequence, and standard stock-on-hand in the cells
Morning meeting	A daily meeting held for the purpose of sharing production and safety information, quite often by a quality circle.
Key points: Process sheets	The process sheets, which are visually posted at each workstation, detail the work sequence and most critical points for performing the tasks.
Tooling parts: Changeover and setup (see step 2)	The machine setup that takes place when an assembly line changes products.
Seven tools of quality (see step 3)	The seven tools to quality are Pareto's diagram, check sheets, histograms, cause-and-effect diagram, run charts for individuals, control charts for samples, and scatter diagrams.
Production behavior	Rules that include information on personal safety, safety equipment, clothing, restricted areas, vehicle safety, equipment safety, and housekeeping.
Visual management	Each line in the plant has a complete set of charts, graphs, or other devices, like andons, for reporting the status and progress of the area.

*From a plant manager of a first tier supplier to Toyota.

This is where the unique process technology exists which give companies like Toyota and Honda the edge in manufacturing. They develop the machine tools and the processes in-house rather than buy them from a machine tool supplier. Machine tool suppliers cannot keep secrets. Lean manufacturers gain their competitive edge by developing unique manufacturing process technology and keeping it "locked up" in the cells.

Table 43-3 summarizes the key points in managing a supplier plant. This list was obtained from the plant manager at a first tier supplier to Toyota Camry.

STEP 9: AUTONOMATION (NOT JUST AUTOMATION)

Autonomation means the autonomous control of quality and quantity. Stop everything immediately when something goes wrong; control the quality at the source instead of using inspectors to find the problem that someone else may have created. The workers in the lean factory inspect each other's work, called successive checking. Taiichi Ohno, former vice president of manufacturing for Toyota, was convinced that Toyota had to raise its quality to superior levels in order to penetrate the world automotive market. He wanted every worker to be personally responsible for the quality of the piece part or product that was produced (called source or self-checking).

The need for automation simply reflects the gradual transition of the factory from manual to automated functions. Some people think of this as computer integrated manufacturing (CIM). Others recognize that people are the most important (and flexible) asset in the company and see the computer as just another tool in the process but not the heart of the system. These companies are moving toward human integrated manufacturing (HIM) where a creative, motivated workforce is seen as the key to lean manufacturing.

Quite often, inspection devices are placed in the machines (source inspection) or in devices (decouplers) between the machines, so the inspection is performed automatically. This generally reduces the cycle time in the cell. Remember, the idea is to prevent the defect from occurring rather than to inspect to find the defect after the part is made. Inspection by a machine instead of by a person can be faster, easier, and more repeatable. This is called *in-process control inspection*.

Autonomation means inspection becomes part of the production process and does not involve a separate location or person to perform it. Parts are 100% inspected by devices that either stop the process if a defect is found or correct the process before the defect can occur (requiring feedback to the controller). The machine may shut off automatically when a problem arises to prevent mass production of defective parts. See

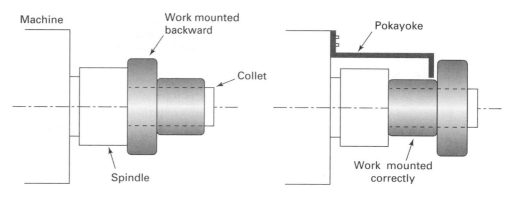

Pokayokes for preventing human and mechanical error are valuable in building quality into the production processes.

The pokayoke device prevents the operator from accidently loading the workpiece on the collet backwards.

(a)

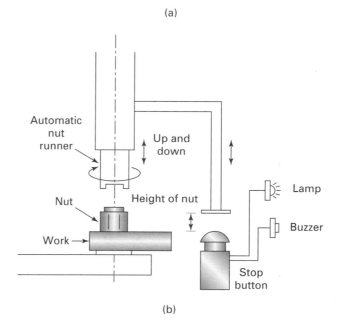

FIGURE 43-12 Pokayoke devices prevent defects. (a) Pokayoke prevents operator from mounting workpiece backward on collet. (b) Pokayoke stops machine if nut is missing during assembly.

(b)

Figure 43-12 for examples of pokayoke devices (which prevent defects). The machine may also shut off automatically when the necessary parts have been made to prevent overproduction. This is part of inventory control.

STEP 10: RESTRUCTURE THE PRODUCTION SYSTEM

Once the factory (the manufacturing system) has been restructured into a lean manufacturing system and the critical control functions well integrated, the company will find it expedient to restructure the rest of the company. This will require removing the functionality of the various departments and forming teams, often along product lines. It will require the implementation of concurrent engineering teams to decrease the time needed to bring new products to market. This movement is gaining strength in many companies and was called *business process reengineering*, where the idea was to restructure the production system to be as waste free as the manufacturing system.

Shifting from one type of manufacturing system design to another will affect product design, tool design and engineering, production planning (scheduling) and control, inventories and their control, purchasing, quality control and inspection, and, of course, the production worker, the foreman, the supervisors, the middle manager, and so on, right up to top management. Such a conversion cannot take place overnight and must be viewed as a *long-term transformation* from one type of *production system* to another. This kind of downsizing can be very traumatic for the business part of the company and usually has a negative impact on the morale of the company. This is why lean manufacturing is difficult to implement.

Therefore, step 10 recognizes the need for the rest of the company to reorganize (get lean). This effort often begins with building *product realization teams* designed to bring new products to the marketplace faster. In the automotive industry, these are called *platform teams* and are an example of concurrent engineering. Platform teams are composed of people from design engineering, manufacturing, marketing, sales, finance, and so on. As the notion of team building spreads and the lean manufacturing system gets implemented, it is only natural that the production system will follow suit. Unfortunately, many companies are restructuring the business part of the company without having done the necessary stcps 1 through 8 to get the manufacturing system lean and efficient. Downsizing the enterprise without simplifying and redesigning the manufacturing system can lead to difficult times for the enterprise. Other companies have tried to automate their way to productivity without first simplifying the system and integrating the function. More difficult times are the outcome.

The design of the manufacturing process technology must be done early in the product development process. The manufacturing process technology within the manufacturing cells must be part of (i.e., elements within) a well-designed, integrated manufacturing system. Flexibility in the design of an integrated manufacturing system means it can readily accept new product designs. Flexibility in the process technology means the process can readily adapt to product design changes and be engineered to accept new products.

■ 43.4 INTERIM CELLS USING NEW MACHINE TOOLS

An interim cell uses equipment the company already owns, rearranged into U-shaped cells for one-piece flow. New equipment that was built for the job shop and stand-alone applications can also be used. The cell shown in Figure 43-13 replaced a transfer line and used all new CNN equipment. This cell is designed to machine 10 different cylinder heads for small gas engines. The company brought together two machine tool companies and

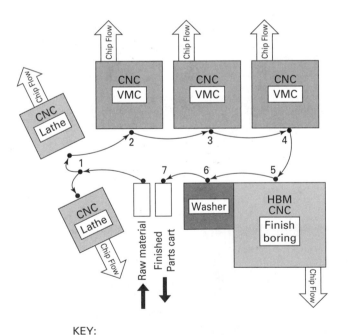

FIGURE 43-13 Interim cell for the manufacturing of a family of ten cylinders, using six CNC machines and one, two, or three operators. VCM = Vertical Milling Machine; HBM = Horizontal Boring Machine.

KEY:
MACHINES & OPERATIONS

1. Lathe
   Face, rough bore,
   Turn register

2. VMC
   Drill, tap

3. VMC
   Drill tap

4. VMC
   Drill, tap

5. HMC
   Finish bore

6. Washer
   Clean, deflash
   cylinder

7. Parts to plating

a workholding supplier as members of the design team. The part volumes for the various cylinder types ranged from 1000 to 600,000 parts per year. The machines were modified to allow the operators to quickly and easily load and unload the machines. Fixtures were designed for quick load and unload and rapid changeover from one part in the family to another. Some of the fixtures have part checking sensors (checking the operation from the previous machine) and pokayoke capability (defect prevention). Each machine can have a different fixture setup performed in less than one minute. The cell outputs about 2500 units per day. A more automated cell was developed to provide additional capacity (replacing other transfer lines). The layout (design) shown in Figure 43-13 is one of many that were evaluated using digital systems simulation.

In the best lean production system facilities, the manufacturing cells use machines for processing that were specifically designed and built for the system. The cells are called lean manufacturing cells.

■ 43.5 LEAN MANUFACTURING CELL DESIGN

Product design (for manufacture and assembly) and MSD must go hand-in-hand, striving for customer satisfaction. The functional requirements of a good MSD are flexibility, controllability, efficiency, and uniqueness. The factors needed to achieve satisfactory usage of the manufacturing system by the internal customers (the workers) are safety, equipment reliability, ergonomically sound equipment, and good service from engineering. Ergonomically sound equipment means that the processes are designed to be easy to operate and easily maintained, are fail-safe, and are not dirty, noisy, labor-intensive, or hazardous.

In the lean manufacturing system, manufacturing engineers are responsible for designing, building, testing, and implementing the manufacturing equipment that will be used by the internal customers in the manufacturing cells (machine tools and processes, tooling [workholders, cutting tools] and material handling devices [decouplers]). The equipment is simple and reliable and can be easily maintained. In general, this flexible, dedicated equipment is built in-house or purchased from contractors and modified for the needs of the cell.

Many companies understand that it is not good strategy simply to buy the manufacturing process technology from another company and then expect to make an exceptional product using the same technology as the competitor. When the process technology is purchased from outside suppliers, any uniqueness aspects will be quickly lost. The lean company must perform research and development (R&D) on manufacturing technologies as well as manufacturing systems in order to produce effective and cost-efficient products. However, an effective, cost-efficient manufacturing system makes R&D in manufacturing process technology pay off.

The proprietary part of lean manufacturing is designing and building machines and material handling equipment (decouplers) and tooling to meet the needs of the cell and the system. Machines are typically single-cycle automatics but may have capacity for process delay. An example of a process delay decoupler is shown in Figure 43-14. This is a process delay decoupler for paint drying that has a capacity for 6 fixtures so each unit gets six minutes of drying time.

EXAMPLE OF A LEAN CELL DESIGN

The lean manufacturing system has some unique characteristics that are embodied in the manufacturing and assembly cells. The cells are manned by multifunctional workers that perform tasks other than handling the material and operating the equipment. These tasks include quality control and inspection (to prevent defects from occurring), machine tool maintenance, setup reduction, and problem solving. The cells are usually U-shaped or rectangular. The cells are designed so the worker can step across the aisle and work on machines on the opposite side. The concept is called "separation of machine's work and man's work."

Understanding lean manufacturing is to understand how a lean manufacturing cell works as part of the lean system. Manufacturing cells, the proprietary element in lean

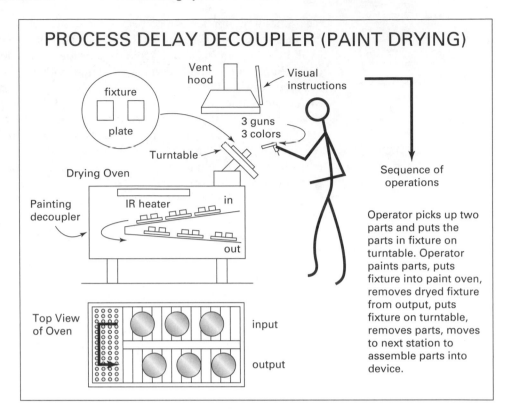

PROCESS DELAY DECOUPLER (PAINT DRYING)

Operator picks up two parts and puts the parts in fixture on turntable. Operator paints parts, puts fixture into paint oven, removes dried fixture from output, puts fixture on turntable, removes parts, moves to next station to assemble parts into device.

FIGURE 43-14 Example of a process delay decoupler element for paint drying. (*Source:* Toyota supplier)

production, are designed with the system requirements in mind. So we begin with an auto assembly line, Figure 43-15. The steering gear shown in Figure 43-16 is installed at station 38. The rack for this gear is made in the manufacturing cell shown in Figure 43-17. The finished rack goes to the rack and pinion assembly cell and then into the steering gear assembly cell. This is a fairly large cell capable of high throughput rates. The cell only produces a part called the rack bar for one model of car but the part comes in 10 different designs. The part requires heat treating, inspection, and mechanical straightening, in addition to numerous machining operations like drilling and tapping, gear teeth milling, deep hole drilling, grinding, and broaching. The point is that all the processing required to produce a finished bar, ready for subassembly, is in the cell.

In total, the rack moves through 26 steps or operations, most of which are machining, performed by single-cycle automatic machines (that turn off after the machining process is complete). However, there are some manual operations in the cell other than loading/unloading. These include steps 13 and 20 to manually straighten the bar (which can warp after heat treating) or the assembly of parts onto the bar.

This cell can make 10 different types of racks for the same model of product. The changeover at any individual machine occurs with "one touch" at the time the operator unloads the previous part for the machine. Many machines are equipped with poka-yoke devices that can prevent the machines or the operator from making mistakes.

This cell is designed a bit differently in that the work arrives and departs from the middle of the cell, but the cell still has a U-shape. Operations 2 through 10 at the start (the right end of the cell) are the same for all the bars in the family. Therefore, this area employs a transfer line using small robots and mechanical arms and levers to move the part from machine to machine.

In the transfer line portion of the cell, steps 2 through 10, the processing time for the deep hole drilling is longer than the cycle time for the cell. The cell puts out one finished rack bar per minute. Since one rack is started through the cell every minute, step 2 is divided into four steps so that the MT for each step is less than one minute. This approach eliminated the deep hole drilling problem and extended the life of the cutting tools. In this area, the machines have automatic repeat cycle capability with automatic transfer devices moving the parts from step 2 to 10. It is very common in the manufacturing cells to have

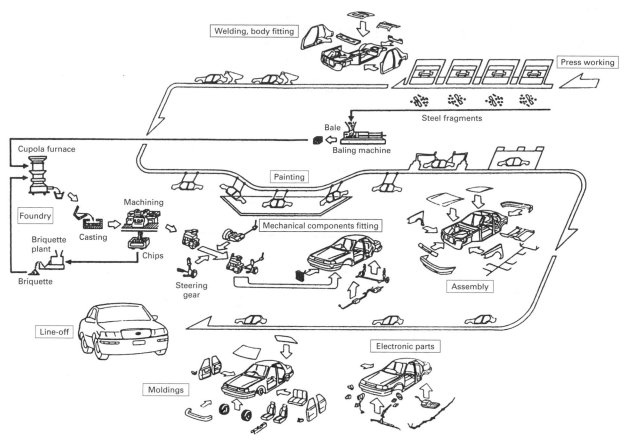

FIGURE 43-15 Automobile assembly plant. The steering gear is installed at station #38 after it has been received from the supplier and brought to the assembly line.

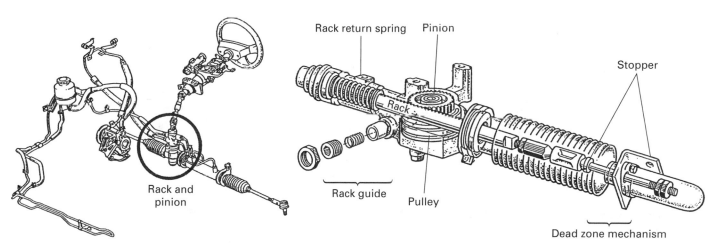

FIGURE 43-16 The steering gear shown above has two major components, the rack bar and the pinion.

a transfer line subsystem or subset of processes where the processing steps are the same for all components in the cell.

Figure 43-18 shows the standard work sheet for the manufacturing cell shown in Figure 43-17. Two operators are shown in the cell. These workers are standing, walking workers who move from machine to machine in loops as shown in the figure. The open circles indicate the positions at the machines when they perform some tasks. Each operator makes the loop in about one minute. Operator 1 addresses 10 stations and operator 2 addresses 11 stations. For the single cycle automatic processes the operators are typically unloading a machine, loading another part into the machine, checking the part they have unloaded, and

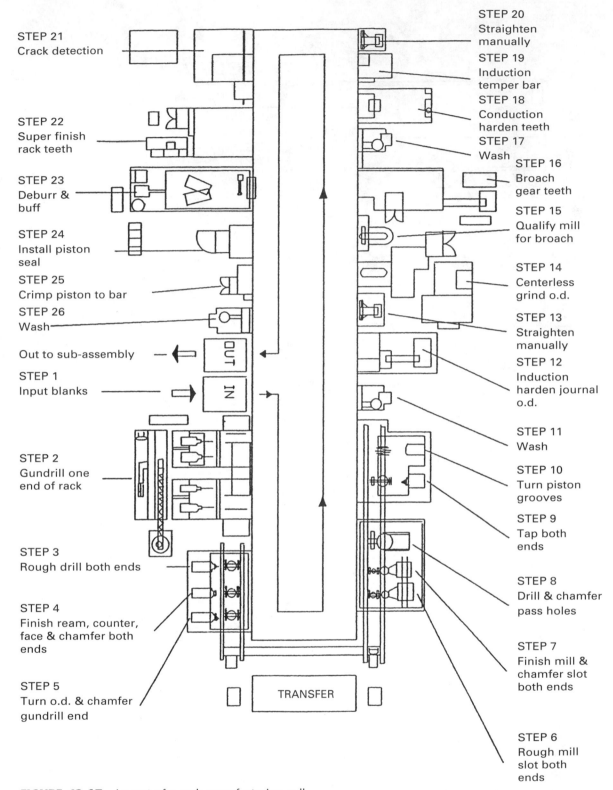

STEP 21
Crack detection

STEP 22
Super finish
rack teeth

STEP 23
Deburr &
buff

STEP 24
Install piston
seal

STEP 25
Crimp piston to bar

STEP 26
Wash

Out to sub-assembly

STEP 1
Input blanks

STEP 2
Gundrill one
end of rack

STEP 3
Rough drill both ends

STEP 4
Finish ream, counter,
face & chamfer both
ends

STEP 5
Turn o.d. & chamfer
gundrill end

STEP 20
Straighten
manually

STEP 19
Induction
temper bar

STEP 18
Conduction
harden teeth

STEP 17
Wash

STEP 16
Broach
gear teeth

STEP 15
Qualify mill
for broach

STEP 14
Centerless
grind o.d.

STEP 13
Straighten
manually

STEP 12
Induction
harden journal
o.d.

STEP 11
Wash

STEP 10
Turn piston
grooves

STEP 9
Tap both
ends

STEP 8
Drill & chamfer
pass holes

STEP 7
Finish mill &
chamfer slot
both ends

STEP 6
Rough mill
slot both
ends

OUT

IN

TRANSFER

FIGURE 43-17 Layout of a rack manufacturing cell.

dropping the part into the decoupler elements between the machines. The stock-on-hand (SOH) in the decouplers and the machines help to maintain the smooth flow of the parts through the machines. The decoupler allows the operator to leave the sequence on one side of the line and pick up the process on the other side. The decouplers also can be designed to perform inspections for part quality or necessary process delays while the parts heat up, cool down, cure, and so forth. The SOH is kept as small as possible.

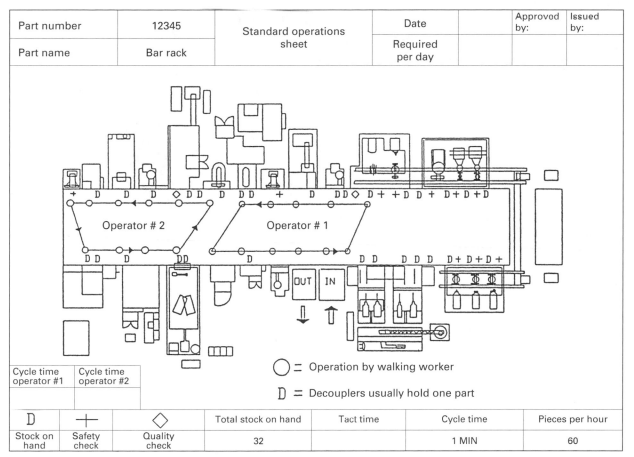

Part number	12345	Standard operations sheet	Date		Approvod by:	Issued by:
Part name	Bar rack		Required per day			

Operator # 2

Operator # 1

OUT   IN

◯ = Operation by walking worker

D = Decouplers usually hold one part

Cycle time operator #1	Cycle time operator #2							
D	+	◇	Total stock on hand	Tact time	Cycle time	Pieces per hour		
Stock on hand	Safety check	Quality check	32		1 MIN	60		

FIGURE 43-18   Standard work sheet for manufacturing cell.

After they have completed all the tasks at a machine, the operators walk to the next machine, hitting a start switch for the machine as they leave (called a walk-away switch).

Sometimes the decoupler elements perform the inspection or checking of the part, but mostly they serve to transport parts from one process to the next. Sometimes the decoupler performs a secondary operation like deburring or degaussing the bar to remove residual magnetic fields. The bar is made from steel, which can become magnetized, causing small chips to adhere to the bar and perhaps cause the bar to be mislocated in the subsequent process.

Notice operator 1 controls both the input and output of the cell and this is by design. One operator always controls the volume of material going through the cell. This design also keeps the SOH quantity constant and keeps the cell working in balance with the final or subassembly lines it is feeding. Operator 1 loads the centerless grinder then moves across the aisle to unload the operation called deburr and buff. Operator 2 unloads the centerless grinder and loads the part in the next process, moving in a counterclockwise loop in the cell from right to left.

At the interface between the two operators, either one can perform the necessary operations depending on when they arrive and when the processes in the machines are finished. That is, the region where the two operators typically meet is really not fixed but changes or shifts depending upon the way parts are moving about the cell. This is called the relay zone. This flexibility requires the workers be cross-trained on all the processes in the cell. The cell is designed so that it can be operated by one, two, three, or even four workers. Changing the number of workers changes the output rate. This is a key to flexibility.

MACHINE TOOL DESIGN FOR LEAN MANUFACTURING CELLS

In Figures 43-19 and 43-20, two examples of machine tools are given, one designed to perform the broaching process in the job shop (see Chapter 26) and the other designed for the lean cell shown in Figure 43-17, step 16.

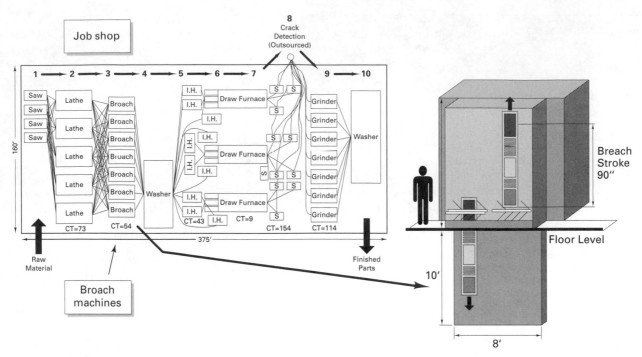

FIGURE 43-19 The broach as designed for the job shop is not a good machine tool for a lean manufacturing cell.

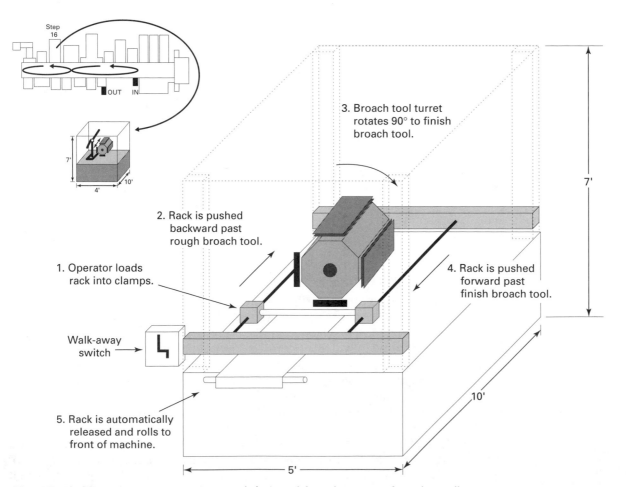

FIGURE 43-20 A broaching machine tool designed for a lean manufacturing cell.

Here is how the cell broach works. While the standard surface broach pushes the broach past the part, the cell broach moves the part (the rack bar) past the broach tool. The broach tool is divided into two segments—a roughing broach and a finishing broach. This allows the total size of the machine to be minimized. Because the cell makes racks for right and left hand steering gears, two sets of broach tools (for two distinct products) are installed in the machine on the broach turret. This turret indexes in 90° increments either between roughing and finishing broaches or between the broaches for left hand and right hand gear teeth.

The operator loads the rack bar into the clamps at the front of the machine. The machine then pushes the rack toward the rear of the machine, past the roughing broach tool. When the part reaches the rear of the machine, the broach turret indexes 90° to the finishing broach tool. The part is pushed back to the front of the machine, past the finishing broach. The clamps then release the rack and it rolls to the front of the machine where it can easily be picked up by the operator. Here are characteristics of the lean cell machine tool design.

- *Flexibility* (process and tooling adaptable to many types of products): Flexibility requires rapid changeover of jigs, fixtures, and tooling for existing products and rapid modification for new designs. The processes have excess capacity—they can run faster if they need to, but they are designed for less-than-full capacity operation.

- *Build exactly what you need*: There are three aspects to this. First, you are not paying for unused capability or options. Second, the machine can have unique capabilities that your competitors do not have and cannot get access to through equipment suppliers. When you purchase equipment from suppliers, you may be paying for capability your competition can get for free (from the supplier). Third, the equipment should allow the operator to stand and walk. Equipment should be the appropriate height to allow the operators to easily perform tasks standing up and then move to the next machine in a step or two (i.e., have a narrow footprint).

- *Maintainability/reliability/durability are all built-in features*: Equipment should be easy to maintain (oil, clean, changeover, replace worn parts, have standardized screws). Many of the cells at the lean suppliers plants are near-clones of each other. The supplier company, being sole-source, has the volume and the expertise to get business from many companies making essentially the same components or subassemblies for many *original equipment manufacturers* (OEMs). The suppliers build a manufacturing cell for each OEM. Some of the equipment can be interchanged from one cell to another in emergencies. The most skilled maintenance personnel must be given this task so that breakdowns in the L-CMS are eventually eliminated.

- Equipment designed to prevent accidents (safety first).

- Equipment designed to be easy to operate, load, and unload (ergonomics).

- Equipment designed to process single units, not batches. Small footprint, low-cost equipment is the best.

- $T_m \leq CT$. The T_m (machining or processing time) should be modified so that it is less than the cycle time (the time in which one unit must be produced). Equipment processing speed should be set in view of the CT, such that $T_m < CT$. The T_m is related to the machining parameters selected (speed, feed, depth of cut). For example, for the turning lathe operation, suppose the part is of length L, then

$$T_m = \frac{L + \text{allowance}}{\text{feed} \times N_s} \quad \text{and} \quad N_s = \frac{12V}{\pi D}$$

where N_s is the rpm of the spindle of the machine, V is the cutting speed (selected), and D is the diameter of the shaft. The tool failure relationship here is (most simply) $VT^n = C$, where T is tool life, and n and C are empirical constants. So, decreasing V increases the tool life and reduces cutting tool replacements and machine tool maintenance.

For example, suppose the T_m is 30 seconds for the $CT = 1$ minute. The cutting speed can be reduced, thereby increasing the tool life and reducing downtime for tool changes. This approach also reduces equipment stoppages, lengthens the life of the equipment, and may improve quality.

- Equipment can have inspection devices (such as sensors, pokayokes, counters) to promote autonomation. Autonomation is the autonomous control of quantity (do not overproduce) and quality (no defects). Often the machine is equipped to count the number of items produced and the number of defects.

- Equipment should be movable. Machines are equipped with casters or wheels, flexible pipes, and flexible wiring. There are no fixed conveyor lines.

- Equipment should be self-cleaning. Equipment disposes of its own chips and trash in the rear.

- Equipment should be profitable at the production volume given to it. Equipment that needs millions of units to be profitable (supermachines) should be avoided because once production volume even slightly exceeds the capacity of the first supermachine, the purchase of another supermachine is necessary, and the new supermachine will not be profitable until it approaches full utilization.

- Each machine or process or operation in the cell is designed for a standing/walking worker coming to the machine from the right or straight on, and leaving to the left. The material is usually moved from right to left since most people are right handed, and it is easier to unload (with the left) than it is to load (with the left).

- The machines are designed to have walk-away switches that the worker hits when leaving the machine. The doors close and the machine begins the processing cycle untended. The machine completes the processing cycle in time T_m. The machine is at least a single-cycle automatic.

- The machines are arranged in the sequence of operations needed to process the part.

- All the processes needed to make the part are in the cell and have an T_m less than the CT or the average CT. This often requires some rather unique process technology.

- The width of the aisle in the cell is about 4 feet so that workers can pass each other in the cell but can also easily step across the aisle. Decouplers located between the processes holds one or two parts. The decouplers are designed to maintain the flow of the parts while decoupling the dependency of the operator on the machine and the dependency of one step or operation on the previous. The decoupler may also transport and reorient the part, as shown in Figure 43-21, and inspect the feature M produced by the previous operation. Decoupler elements like this can deburr the part, degauss the part, hold the part

FIGURE 43-21   This simple decoupler reorients the part, transports the part, inspects the part (surface M), and detects the presence or absence of the part.

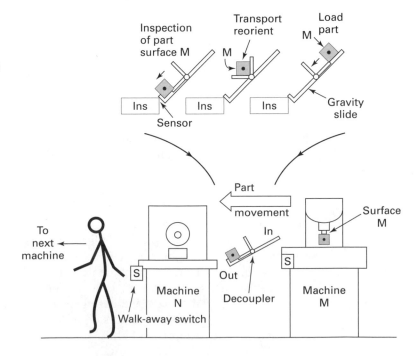

for heating or cooling or curing or drying (this is called process delay). Decouplers are custom designed to hold all the parts in the family with equal facility.

- The cell is equipped with many pokayokes for defect prevention that also perform self-inspections or successive inspections. The devices can be in the decouplers or in the workholding device of the next machine and should be simple.

■ 43.6 ROBOTIC MANUFACTURING CELLS

Unmanned robotic cells will have CNC machine tools, robot for material handling, and decouplers for flexibility and capability.

In robotic manufacturing cells, the robot moves the part from process to process. The cell shown in Figure 43-22 shows a robot with 2 processes and three decouplers that

FIGURE 43-22 Examples of decouplers in robotic cells. (22a) Model of unmanned, robotic manufacturing cell with decouplers, using a pull cell control algorithm. (22b) Part A for manufacturing cell. (22c) Decoupler checks for presence of burrs and part diameter. (22d) U-shaped manned cell with a human and a robot with decouplers integrating the process.

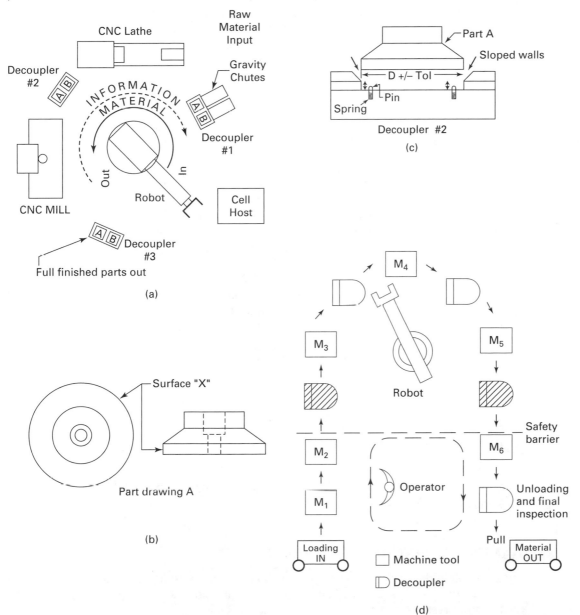

operate on a pull basis. The removal of a part from D3 produces the signal (an empty place that needs to be filled) which launches the cell into action to replace the withdrawn part. Here is how the decoupler works.

Figure 43-22c shows the decoupler that follows the process that produces surface X (see part drawing in Figure 43-22b). This decoupler automatically indicates when a part has been removed, thus producing a signal for the control system that another part is needed.

This decoupler shows the presence of a part when both pins are depressed closing an electric circuit. Springs lift the pins and open the circuit when the part is removed. If the part is slightly tilted (not correctly located) the pins are not fully depressed, the circuit is open, indicating that the part is not correctly located. The walls of the decoupler are sloped so the round part is autonomously centered, repositioned, and located for the robot's next move.

The cutting process (turning) may produce a burr on the lower edge of surface X that will also result in the part not seating correctly and stop the process. Such a burr would result in a misalignment of the part in the next process and create a defect. So this decoupler is a pokayoke, a defect preventer.

The robot does not have to place the part in the decoupler precisely. The decoupler is designed to locate the part automatically. The part must locate precisely in the decoupler so that it can be picked up by the robot and loaded into the fixture of the next process.

Also, the opening in the ring in the decoupler is slightly larger than the largest part diameter (D + tolerance). Oversize parts do not depress the pins indicating the absence of a part and the process is stopped. Tool wear is a common cause of oversize parts.

Figure 43-22d shows a cell with a robot and a human operator working together. The decoupler elements carry the part to and from the robot while protecting the operator from injury. Human/robot cells are in common use in many lean manufacturing facilities but are not widely published.

■ 43.7 ERGONOMICS OF LEAN MANUFACTURING CELLS

Ergonomics deals with the mental, physical, and social requirements of the work and how the work is designed (or modified) to accommodate the human limitation. For example, are the machines in the cell designed to a common height to minimize lifting of parts? Are transfer devices designed for easy slide on/slide off? Are automatic steps equipped with interrupt signaling devices to help the worker monitor the process? When the job is defined as primarily loading/unloading, ergonomic concerns with lifting and placing parts in machines and operating workholding devices must be addressed. In these systems human performance in detecting and correcting cell malfunctions will establish utilization and thus production efficiency. The design of machines for maintainability and diagnostics is critical. The original designer of the cell must incorporate ergonomic issues initially rather than trying to come back later to implement fixes. However, the cell design with its great variety in sequential tasks is ergonomically sound and few repetitive injuries are produced.

Ergonomic machine design considerations include:

- Uniform loading heights
- Location of walkaway start switches
- Minimum reach (bending) into machine
- Eliminating machine obstructions in aisle
- Providing access to machine from the back
- Position of load/unload decoupler (when part requires a two-hand load)
- Part weight, size, shape, burrs, finish, etc.

■ 43.8 WHY LEAN MANUFACTURING PROGRAMS FAIL

There should always be a champion or take-charge person who gets behind the program and sees it through. Lean manufacturing implementations will fail if management loses interest.

Problems in quality will prevent any real reduction in inventory and lead times. If the company cannot get to zero defects and zero breakdowns, they will not get to lean manufacturing.

Watch out for management (managers) with hidden agendas. If the upper management already knows in advance what solution they want and uses the lean process to direct the employees to their (management) desired ends, then the employees will tire of game playing and lose interest.

Middle management opposition will stop lean manufacturing. The engineers, quality or MRP folks come to believe that the effort is a reflection on their expertise or competence and try to quietly sabotage the program. Their involvement in planning, data analysis, and status reviews must be assured. This problem also occurs with supervision if they are not carefully brought into the effort.

Systems changes are inherently difficult to implement. Changing the entire manufacturing system is a huge job and the changes at the factory floor level will force changes at the enterprise. Companies spend freely for new manufacturing processes but not for new manufacturing systems. It is easier to justify new hardware for the old manufacturing system than to rearrange and modify the old hardware into a new manufacturing system (linked cells). However, anyone with capital can buy the newest equipment, often creating another island of automation.

Converting to linked cells will free up additional capacity (setup time saved) and capital (funds not tied up in inventory), but such conversions are long term projects and will require expenditure of funds for equipment modifications and employee training in quality, maintenance, of machine tools, setup reduction, problem solving, and so forth.

The top management of the company may lose interest if they don't see quick results or get involved in some new fad which dissipates support for lean manufacturing. Generally, few of the companies have long range plans and programs that support concepts like lean manufacturing. Top management tends to manage for quarterly results and immediate fixes. Therefore it is critical that the steps outlined here are followed for a successful implementation of lean manufacturing.

Decision making is choosing among the alternatives in the face of uncertainty. The greater the uncertainty, the more likely that the "do-nothing" alternatives will be selected. The fear of the unknown can lead to uncertainty. In addition, many managers harbor faulty criteria for decision making. Decisions should be based on the ability of the company to compete (quality, reliability, unit cost, delivery time, flexibility for product or volume change) rather than on price alone. Getting the internal customers (production workers) involved in the decision making process is a significant change and critical to getting them committed to the new standardized work methods required for lean manufacturing.

■ 43.9 SUMMARY: USE IM,C NOT CIM

There are many who believed that the only way in which manufacturing companies can compete is to automate, robotize, and computerize. This is the CIM approach. In a nutshell, the CIM concept tries to achieve integration through computerization and automation. Lean manufacturing is a different approach that might be called IM,C (integrate the manufacturing system, then computerize and automate). The development of manufacturing and assembly cells is just the first step in the manufacturing systems design. Lean manufacturing should be implemented prior to any efforts to computerize the system. While costs of these systems are difficult to obtain, the lean cell approach is significantly less costly than the CIM approach.

We hear a lot of talk today about continuous improvement. Continuous improvement requires the continuous redesign of the manufacturing system. We hear a lot of talk about technology transfer (TT). Technology transfer happens when buyers share their experience in L-CMS with their suppliers on a one to one basis (step 8).

The lean factory is based on a different design for the manufacturing system in which the sources of variation in time and quality are attacked and the delays and defects in the system removed. In summary, the next generation of American factories (the factories with a future) will be designed with manufacturing and assembly cells linked together with a pull system for material and information control. In this L-CMS, downstream process will dictate upstream production rates. The L-CMS design simplifies the manufacturing system, integrates the critical control functions before applying technology (automation, robotization, and computerization), avoids risks, makes automation easier to implement, and most importantly, provides a system that fosters continuous improvement of the work (and the workplace) by the people who do the work.

■ KEY WORDS

8-4-8-4 scheme
andon
autonomation
computer integrated
   manufacturing (CIM)
computer-aided design (CAD)
computer-aided
   manufacturing (CAM)
decoupler
interim cell

inventory control
linked-cell manufacturing
   system (L-CMS)
machine tool
make one, check one, move
   one on (MO-CO-MOO)
manufacturing system design
   (MSD)
mass production system
pokayoke

production control
production ordering kanban
   (POK)
quality control
six sigma capability
single sourcing (sole source)
single-cycle automatic
stock-on-hand (SOH)
takt time
technology transfer (TT)

throughput time (TPT)
Toyota Production System
   (TPS)
WLK Withdrawal Kanban
work-in-process (WIP)
Yo-i-don

■ REVIEW QUESTIONS

1. If the first step is to form single-piece flow cells, what will be the consequence of all the hidden variations that existed in the job shop design?
2. How will you prepare the workforce to deal with the conversion to the lean shop?
3. What can be done before cell formation that will minimize the variation in cycle time?
4. Who has the best ideas about how to improve work in the cells?
5. Who should be tasked with sustaining and maintaining the standardized work instructions?
6. What common element exists in implementing both integrated quality control and integrated preventative maintenance?
7. Describe the difference between leveling and balancing.
8. Describe the difference between sequencing and synchronizing.
9. What is a Kanban system? Explain how it works in a manufacturing setting. Also, describe how the Kanban system can be used to lower inventory levels.
10. Many companies build and customize their equipment. What are the advantages of this strategy? What are alternative strategies if a company lacks the technology to build their own equipment?

11. Give a detailed definition of a manufacturing system. Be sure to include all the elements of a manufacturing system.
12. List and define the functional requirements of the new integrated manufacturing system design, the lean shop.
13. What is the objective in manned Linked-Cell Manufacturing System Design? Describe the basis for maintaining this objective.
14. Define and give examples of internal elements and external elements of setup operations.
15. What are the advantages of integrated quality control? What is the fundamental difference between how quality control is implemented in a lean system and how quality control is implemented in other systems?
16. The originators of control charts believed that the variability in a process was inherent. What did they mean by that?
17. How is production control accomplished in lean manufacturing? How do the parts know where to go and how does the manufacturing system know when to make which part?
18. What is build-schedule stability for a vendor?
19. Name a pokayoke device that was built into your car.
20. If the cycle time for the paint cell is 60 seconds, how long will the painted parts in Figure 43-14 have to dry?

■ PROBLEMS

1. Try this exercise in SMED. Study how long it takes to change a tire on a car. What are the work elements of the current method? Did you videotape the changeover? Separate the elements into internal and external. Now study the tire change process again. How much time did you take out of the setup? What has NASCAR done to reduce tire change to a minimum (20 seconds)?

2. Prepare for class a three-minute presentation on one of the seven tools of quality control.
3. The automobile has about 5,000 parts. How many parts does a typical mountain bike have? Which of these parts would be synchronized in delivery to final assembly?
4. A top-level manager at a manufacturing company just read a journal article on lean manufacturing and the Kanban system.

Currently, his manufacturing company does not operate on a lean system. However, the manager was so influenced by the article that he thinks his company can benefit from the ideas he read about, and he wants to take immediate action to implement a Kanban system. He asks you, one of the engineers, if the Kan-

ban can be implemented successfully. What is your reply to him? Do you think the manager's current plan will be successful? If not, develop a successful plan of action, listing the steps required to achieve lean production. Be sure to warn the manager of the common reasons attempts at lean production fail.

www.wiley.com/college/degarmo

*C*hapter 43   CASE STUDY

Automobile Water Pump Impeller

Figure CS-43 is a water pump impeller used by a major automotive parts manufacturer. The outer diameter of the component is 2.75 inches and the height is specified at 0.75 ± 0.005 inches. The I.D. of the center hole is 0.625 inches. The dimensions and positioning of the six blades is important from the viewpoint of vibration and balance. Smooth finish is desirable for good fluid flow. The operating temperature range has been estimated as $-60°F$ to $+300°F$, and the mechanical property requirements have been given as 30,000 psi minimum tensile strength and a minimum elongation of only 0.5%. Since there is no metal-to-metal contact, high wear resistance is not necessary. The contact fluid should be a water/antifreeze mixture with corrosion-resistant additives. Production volume should be high, so a low total cost (material plus manufacturing) is a prime objective.

1. Based on the size and shape of the product, describe at least three reasonable ways in which the component could be produced. For each method, briefly discuss its relative pros and cons.

2. What types of engineering materials might be able to meet the desired requirements? What would be the pros and cons of each general family?

3. For each of the shape generation methods in part 1, select an appropriate material from the alternatives discussed in part 2. (*Note:* Casting alloys should be matched with casting processes, high machinability alloys would be favored for cutting applications, etc.)

4. Which of the above alternatives do you feel would be the "best" solution to the problem? Did you consider making the part in a manufacturing cell? For this system, outline the specific steps that would be necessary to produce the part from reasonable starting material.

5. For your proposed solution in part 4, would any additional heat treatment or surface treatment be required? If so, what would you recommend?

6. Could this be a possible application for molded nylon, other polymer, or composite material? If so, what would be the candidate material? What would be the major differences in properties, performance, and manufacturing process?

FIGURE CS-43

SELECTED REFERENCES FOR ADDITIONAL STUDY

HANDBOOKS AND GENERAL REFERENCES

ASM Metals Reference Book. 3rd Ed., ASM, Materials Park, OH (1993).

ASM Engineered Materials Reference Book, 2nd Ed., ASM, Materials Park, OH (1993).

Tool and Manufacturing Engineers Handbook, 4th Ed., Society of Manufacturing Engineers, Dearborn, MI.
Vol. 1, Machining (1983)
Vol. 2, Forming (1984)
Vol. 3, Materials, Finishing and Coating (1985)
Vol. 4, Quality Control and Assembly (1987)
Vol. 5, Manufacturing Management (1987)
Vol. 6, Design for Manufacturability (1992)
Vol. 7, Continuous Improvement (1994)
Vol. 8, Plastic Part Manufacturing (1996)
Vol. 9, Material and Part Handling in Manufacturing (1997)

ASME Handbook, McGraw-Hill, New York.
Vol. 1, Metals Engineering: Design
Vol. 2, Metals Properties
Vol. 3, Engineering Tables
Vol. 4, Metals Engineering: Processes

ASM Engineered Materials Handbook Series, ASM, Materials Park, OH.
Vol. 1, Composites (1987)
Vol. 2, Engineering Plastics (1988)
Vol. 3, Adhesives and Sealants (1990)
Vol. 4, Ceramics and Glasses (1991)

ASM Engineered Materials Handbook, Desk Ed., ASM, Materials Park, OH (1995).

ASTM Standards (multiple volumes, published annually), ASTM, Philadelphia, PA.

CRC Handbook of Materials Science, C. T. Lynch (ed.), CRC Press, FL (1971).
Vol. 1, General Properties
Vol. 2, Metals, Composites and Refractory Materials

Manufacturing Processes Reference Guide, R. H. Todd, D. K. Allen, and F. Alting, Industrial Press (1994).

Materials Handbook, 13th Ed., George S. Brady and Henry R. Clauser, McGraw-Hill (1995).

Materials and Processes—Part A: Materials; Part B: Processes— 3rd Ed., James F. Young and Robert S. Shane (eds.), Marcel Dekker (1985).

Metals Handbook, Desk Ed., 2nd Ed., ASM, Materials Park, OH (1998).

Metals Handbook, 9th Ed., ASM, Materials Park, OH.
Vol. 1, Properties amid Selection: Irons and Steels (1978)
Vol. 2, Properties and Selection: Nonferrous Alloys and Pure Metals (1979)
Vol. 3, Properties and Selection: Tool Materials and Special-Purpose Metals (1980)
Vol. 4, Heat Treating (1981)
Vol. 5, Surface Cleaning, Finishing, and Coating (1982)
Vol. 6, Welding, Brazing and Soldering (1983)
Vol. 7, Powder Metallurgy (1984)
Vol. 8, Mechanical Testing (1985)
Vol. 9, Metallography and Microstructures (1985)
Vol. 10, Materials Characterization (1986)
Vol. 11, Failure Analysis and Prevention (1986)
Vol. 12, Fractography (1987)
Vol. 13, Corrosion (1987)
Vol. 14, Forming and Forging (1988)
Vol. 15, Casting (1988)
Vol. 16, Machining (1989)
Vol. 17, Nondestructive Evaluation & Quality Control (1989)

Metals Handbook, 10th Ed., ASM, Materials Park, OH.
Vol. 1, Properties and Selection: Irons, Steels and High-Performance Alloys (1990)
Vol. 2, Properties and Selection: Nonferrous Alloys and Special—Purpose Materials (1991)
Vol. 3, Alloy Phase Diagrams (1992)
Vol. 4, Heat Treating (1991)
Vol. 5, Surface Engineering (1994)
Vol. 6, Welding, Brazing and Soldering (1993)
Vol. 7, Powder Metal Technologies and Applications (1998)
Vol. 8, Mechanical Testing and Evaluation (2000)
Vol. 11, Failure Analysis and Preventions (2002)
Vol. 18, Friction, Lubrication and Wear Technology (1992)
Vol. 19, Fatigue & Fracture (1996)
Vol. 20, Materials Selection and Design (1997)
Vol. 21, Composites (2001)

Military Standardization Handbook—MIL - HDB K-5C - "Metallic Materials and Elements for Aerospace Vehicle Structures," U.S. Department of Defense (1976).

Production Processes: The Productivity Handbook, 5th Ed., Roeer W. Bolz (ed.), Industrial Press (1981).

Production Handbook, 4th Ed., John A. White (ed.), Wiley (1987).

SAE Handbook, Part 1—Materials (issued annually). Society of Automotive Engineers, Warrendale, PA.

Smithells Metals Reference Book, 6th Ed., F. A. Brandes, Butterworths (1983).

Source Book on Industrial Alloy and Engineering Data, ASM, Materials Park, OH (1978).

Source Book on Materials Selection, Vols. 1 and 2, ASNI, Materials Park, OH (1977).

Troubleshooting Manufacturing Processes, 4th Ed., SME (1988).

Woldman's Engineering Alloys, 9th Ed., J. Frick, ASM, Materials Park, OH (2000).

BASIC AND GENERAL TEXTBOOKS IN MATERIALS AND PROCESSES

21st Century Manufacturing, Paul Kenneth Wright, Prentice Hall (2001).

Elements of Materials Science and Engineering, 6th Ed., L. H. VanVlack, Addison Wesley (1989).

Engineering Materials, 6th Ed., Kenneth Budinski and Michael Budinski, Prentice Hall (1999).

Fundamentals of Modern Manufacturing, Mikell P. Groover, Prentice Hall (1996).

Introduction to Materials Science for Engineers, 5th. Ed., James F. Shackelford, Prentice-Hall (2000).

Introduction of Manufacturing Processes, 3rd Ed., John A. Schey, McGraw-Hill (2000).

Introduction to Manufacturing Processes and Materials, Robert C. Creese, Marcel Dekker (1999).

Manufacturing Engineering and Technology, 4th Ed., Serope Kalpakjian and Steven R. Schmid, Prentice-Hall (2001).

Manufacturing Processes and Systems, 9th Ed., Phillip F. Ostwald and Jairo Muñoz, Wiley (1997).

Manufacturing Engineering Processes, Leo Alting, Marcel Dekker, New York (1982).

Manufacturing Engineering, Kenneth C. Ludema, R. M. Caddell, and A. C. Atkins, Prentice Hall (1987).

Manufacturing Processes and Equipment, Jiri Tlusty, Prentice Hall, Upper Saddle River, NJ (1999).

Manufacturing Processes and Systems, 9th Ed., Phillip F. Ostwald and Jairo Muñoz, Wiley (1997).

Manufacturing Processes and Materials, 4th Ed., George F. Schrader and Ahmad K. Elshennawy, Society of Manufacturing Engineers (2000).

Manufacturing Science, A. Ghosh and A. K. Mallik, Ellis Harwood Ltd. (1986).

Materials Science and Engineering, C. F. Carter and D. E. Paul, ASM (1991).

Materials Science and Engineering Materials: An Introduction, 4th Ed., W. D. Callister, Wiley (1997).

Mechanical Metallurgy, 3rd Ed., G. E. Dieter, McGraw-Hill, New York (1986).

Modern Manufacturing Process Engineering, B. W. Niebel, A. B. Draper, and R. A. Wysk, McGraw-Hill (1989).

Physical Metallurgy Principles, 3rd Ed., R. E. Reed-Hill and R. Abbaschian, PWS-Kent (1992).

Principles of Materials Science and Engineering, 4th Ed., W. F. Smith, McGraw-Hill (1996).

Principles of Engineering Manufacture, 3rd Ed., Stewart C. Black, Vic Chiles, A. J. Lissaman, and S. J. Martin, (1982).

Processes and Design for Manufacturing, 2nd Ed., Sherif D. El Wakil, PWS Publishing Company (1998).

Processes and Materials of Manufacture, 4th Ed., Roy A. Lindherg, Allyn and Bacon (1990).

Structures and Properties of Engineering Alloys, 2nd Ed., W. F. Smith, McGraw-Hill New York (1993).

The Mechanical Behavior of Materials, T. H. Courtney, McGraw-Hill (1990).

The Science and Engineering of Materials, 3rd Ed., Donald R. Askeland, PWS (1994).

The Science and Design of Engineering Materials, 2nd Ed., J. P. Schaffer et al., McGraw-Hill (1999).

FERROUS METALS

ASM Specialty Handbook - Carbon and Alloy Steels, ASM (1996).

ASM Specialty Handbook - Cast Irons, ASM (1996).

ASM Specialty Handbook - Stainless Steels, ASM (1994).

Cast Iron: Physical and Engineering Properties, H. T. Angus, Butterworths (1976).

Design Guidelines for the Selection and Use of Stainless Steel, Designers Handbook Series, Specialty Steel Industry of the United States, Washington, DC (1992).

Engineering Properties of Steel, ASM, Materials Park, OH (1982).

Handbook of Stainless Steels, Donald Peckner and I. M. Bernstein (eds.), McGraw-Hill (1977).

Introduction to Stainless Steels, 3rd Ed., ASM (1999).

Metallurgy and Heat Treatment of Tool Steels, Robert Wilson, McGraw-Hill (1975).

Modern Steels and Their Properties, (various editions) Bethlehem Steel Corporation, Bethlehem, PA.

Properties and Selection of Tool Materials, ASM, Materials Park, OH (1975).

Source Book on Stainless Steels, ASM, Materials Park, OH (1976).

Stainless Steel, R. A. Lula, ASM, Metals Park, OH (1985).

Steel Selection: A Guide for Improving Performance and Profits, R. F. Kern and M. E. Suess, Wiley-Interscience (1979).

Steels: Metallurgy & Applications, D. T. Llewellyn, Butterworth Heinemann (1992).

Steels: Heat Treatment and Processing Principles, G. Krauss, ASM, Materials Park, OH (1990).

The Making, Shaping and Treating of Steel, 10th Ed., Association of Iron and Steel Engineers (1985).

Tool Steels, 5th Ed., G. A. Roberts, G. Krauss and R. Kennedy, ASM, Materials Park, OH (1998).

Worldwide Guide to Equivalent Irons and Steels, 4th Ed., ASM, Materials Park, OH (2000).

NONFERROUS METALS

Aluminum Alloys: Structure and Properties, V. F. Mondolfo, Butterworths (1975).

Aluminum, ASM, Materials Park, OH (1967).
Vol. 1, Properties, Physical Metallurgy and Phase Diagrams
Vol. 2, Design and Application
Vol. 3, Fabrication and Finishing

Aluminum Standards and Data, The Aluminum Association, Washington, DC (bi-annual updates).

Aluminum: Properties and Physical Metallurgy, John E. Hatch (ed.), ASM, Materials Park, OH (1984).

ASM Specialty Handbook - Aluminum and Aluminum Alloys, ASM (1993).

ASM Specialty Handbook - Copper and Copper Alloys, ASM (2001).

ASM Specialty Handbook - Magnesium and Magnesium Alloys, ASM (1999).

ASM Specialty Handbook - Nickel, Cobalt, and Their Alloys, ASM (2000).

Engineering Properties of Zinc Alloys, International Lead Zinc Research Organization, New York (1980).

Introduction to Aluminum Alloys and Tempers, J. G. Kaufman, ASM (2000).

Materials Properties Handbook: Titanium Alloys, R. Boyer, E. W. Collings, and G. Welsch, ASM, Materials Park, OH (1994).

Properties of Aluminum Alloys: Tensile, Creep and Fatigue Data at High and Low Temperatures, J. G. Kaufman (ed.), ASM and Aluminum Association (1999).

Source Book on Copper and Copper Alloys, ASM, Materials Park, OH (1979).

Source Book on Selection and Fabrication of Aluminum Alloys, ASM, Materials Park, OH (1978).

Standards Handbook: Copper, Brass, and Bronze (7 volumes) Copper Development Association.

Superalloys: A Technical Guide, 2nd Ed., ASM (2002).

Titanium and Titanium Alloys Source Book, ASM, Materials Park, OH (1982).

Titanium: A Technical Guide, 2nd Ed., M. Donachie (ed.), ASM, Materials Park, OH (2000).

Titanium Alloys Handbook, Metals and Ceramics Information Center, Battelle Columbus Laboratories, Columbus, OH (1972).

Worldwide Guide to Equivalent Nonferrous Metals and Alloys, 4th Ed., ASM, Materials Park, OH (2001).

PLASTICS, COMPOSITES, AND CERAMICS

Advanced Polymer Composites: Principles and Applications, B. J. Jang. ASM, Materials Park, OH (1993).

Composite Materials Handbook, 2nd Ed., M. M. Schwartz, McGraw-Hill, New York (1995).

Engineering Plastics and Composites, 2nd Ed., W. A. Woishnis (ed.), ASM, Materials Park, OH (1993).

Engineers Guide to Composite Materials, John W. Weeton (ed.), ASM, Materials Park, OH (1986).

Fabrication of Composite Materials—Source Book, M. M. Schwartz (ed.), ASM, Materials Park, OH (1985).

Fundamentals of Metal Matrix Composites, S. Suresh (ed.), Butterworth-Heinemann (1993).

Handbook of Composites, George Lubin (ed.), Van Nostrand Reinhold, New York (1982).

Handbook of Plastics, Elastomers and Composites, 2nd Ed., C. A. Harper (ed.), McGraw-Hill (1992).

Handbook of Thermoset Plastics, S. H. Goodman, Noyes Publications (1986).

Modern Plastics Encyclopedia, published annually by *Modern Plastics Magazine*, a McGraw-Hill publication.

Plastic Blow Molding Handbook, N. C. Lee (ed.), Van Nostrand Reinhold (1991).

Plastic Part Technology, E. A. Muccio, ASM, Materials Park, OH (1991).

Plastic Materials and Processes, S. Schwartz and S. H. Goldman, Van Nostrand Reinhold, New York (1982).

Plastics Product Design Engineering Handbook, 2nd Ed., Sidney Levy and J. H. DuBois, Chapman & Hall (1984).

Plastics Processing Technology, E. A. Muccio, ASM, Materials Park, OH (1994).

Plastics Process Engineering, James L. Throne, Marcel Dekker, New York (1979).

Plastics' Technology Handbook, 2nd Ed., M. Chanda and S. K. Roy (eds.), Marcel Dekker (1992).

Structural Ceramics, M. Schwartz, McGraw-Hill (1994).

ELEVATED TEMPERATURE APPLICATIONS

ASM Specialty Handbook - Heat-Resistant Materials, ASM (1997).

Engineer's Guide to High Temperature Materials, F. J. Clauss, Addison Wesley, Reading, MA (1969).

High Temperature Property Data: Ferrous Alloys, M. F. Rothman (ed.), ASM, Materials Park, OH (1987).

Properties of Refractory Metals, W. D. Wilkinson, Gordon and Breach (1969).

Source Book on Materials for Elevated Temperature Applications, ASM, Materials Park, OH (1979).

ELECTRONIC AND MAGNETIC APPLICATIONS

ASM Ready Reference: Electrical and Magnetic Properties of Materials, ASM (2000).

Electronic Materials Handbook, Vol. 1, Packaging, ASM (1989).

Electronic Materials & Processes Handbook, 2nd Ed., C. A. Harper and R. M. Sampson, McGraw-Hill (1994).

Electronic Packaging and Interconnection Handbook, C. A. Harper, McGraw-Hill, New York (2000).

Fundamentals of Semiconductor Processing Technology, B. El-Kareh Kluwer Academic Publishers, Boston (1995).

Guidelines for Laboratory Design: Health and Safety Considerations, L. J. Diberardinis, G. T. Gatwood, J. S. Baum, E. Groden, M. W. First, and A. K. Seth, Wiley, New York (1993).

Integrated Circuit Fabrication Technology, D. J. Elliott, McGraw-Hill, New York (1989).

Soldering in Electronics Assembly, M. Judd and K. Brindley, Reed International Books, Boston (1992).

Solid State Technology, 35, 37–39, DI. L. Flamm (1992).

The Magnetic Properties of Materials, J. E. Thompson, CRC Press, Cleveland, OH (1968).

The Science and Engineering of Microelectronic, S. A. Campbell, Oxford University Press (2001).

DESIGN

Application of Fracture Mechanics—for Selection of Metallic Structural Materials, J. E. Campbell, et al., ASM, Materials Park, OH (1982).

Atlas of Stress-Strain Curves, H. E. Boyer (ed.), ASM, Materials Park, OH (1986).

Atlas of Fatigue Curves, Howard E. Boyer (ed.), ASM, Materials Park, OH (1986).

Atlas of Stress Corrosion and Corrosion Fatigue Curves, A. J. McEvily, ASM, Materials Park, OH (1990).

Axiomatic Design, Nam P. Suh, Oxford University Press (2001).

Concurrent Design of Products and Processes, J. L. Nevins and Dan Whitney, McGraw-Hill (1989).

Engineering Design: A Materials and Processing Approach, 3rd Ed., George Dieter, McGraw-Hill (2000).

Fatigue Data Book: Light Structural Alloys, ASM, Materials Park, OH (1995).

Fracture Mechanics: Fundamentals and Applications, T. L. Anderson, CRC Press (1991).

Fundamentals of Tool Design, 3rd Ed., David T. Reid (Cd.), Society of Manufacturing Engineers (1991).

Handbook of Product Design for Manufacturing—A Practical Guide for Low-Cost Production, James G. Bralla (ed.), McGraw-Hill, New York (1986).

The Principles of Design, Nam P. Suh, Oxford University Press (1990).

The Principles of Material Selection for Engineering Design, Pat L. Mangonon, Prentice Hall (1999).

Tool Design, 3rd Ed., C. Donaldson, G. LeCain, and V. C. Gould, Glencoe (1993).

CASTING

Cast Metals Handbook, 4th Ed., American Foundryman's Society (1957).

Iron Castings Handbook, Charles F. Walton, Iron Castings Society (1981).

Gray and Ductile Iron Casting Handbook, Gray and Ductile Iron Founder's Society, Cleveland, OH (1971).

Steel Castings Handbook, 6th Ed., Steel Founder's Society of America, OH (1995).

Foundry Technology Sourcebook, ASM and AFS, Materials Park, OH (1982).

Investment Casting Handbook—1980, Investment Casting Institute, Chicago, IL (1979).

Principles of Metal Casting, 2nd Ed., Heine, Loper and Rosenthal, McGraw-Hill (1967).

The NFFS (Non-Ferrous Founders Society) Guide to Aluminum Casting Design: Sand and Permanent Mold, NFFS (1994).

Copper, Brass and Bronze Castings: Their Structures, Properties, and Applications, Non-Ferrous Founder's Society, Cleveland, OH (1961)—2 volumes.

Copper Alloy Pressure Die Casting, A. A. Machonis (ed.), International Copper Research Assoc., New York (1975).

WELDING AND JOINING

Adhesives Technology Handbook, Arthur H. Landrock, Noyes Publications (1985).

Adhesives in Modern Manufacturing, Society of Manufacturing Engineers Dearborn, MI (1970).

Brazing Handbook, American Welding Society (1991).

Ceramic Joining, M. M. Schwartz, ASM, Materials Park, OH (1990).

Design of Welded Structures, James F. Lincoln Foundation.

Design of Weldments, James F. Lincoln Foundation.

Handbook of Adhesives, 2nd Ed., Irving Skiest (ed.), Van Nostrand Reinhold (1977).

The Basics of Soldering, A. Rahn, Wiley (1993).

Joining of Composite Matrix Materials, M. M. Schwartz, ASM, Materials Park, OH (1994).

Metals Joining Manual, M. M. Schwartz, McGraw-Hill (1979).

Modern Welding Technology, 2nd Ed., H. B. Cary, Prentice Hall, NJ (1989).

Principles of Soldering and Brazing, G. Humpston and D. M. Jacobson, ASM, Materials Park, OH (1993).

Soldering Manual, 2nd Ed., American Welding Society (1978).

Solders and Soldering, 3rd Ed., H. H. Manko, McGraw-Hill (1994).

Source Book on Electron Beam and Laser Welding, ASM, Materials Park, OH (1980).

Source Book on Brazing and Brazing Technology, ASM, Materials Park, OH (1980).

Standard Handbook of Fastening and Joining, 2nd Ed., R. O. Parmley, McGraw-Hill (1994).

Welding Encyclopedia, L. B. MacKenzie, 16th Ed., revised and re-edited by T. B. Jefferson, Monticello Books, Morton Grove, IL (1968).

Welding Handbook, 8th Ed., American Welding Society, New York (1987).

FABRICATION AND FORMING

Extrusion, K. Laue and H. Stenger, ASM, Materials Park, OH (1981).

Finite-Element Plasticity and Metalforming Analysts, G. W. Rowe, C. E. N. Sturgess, P. Hartley, and I. Pillingen, Cambridge University Press (1991).

Forging Industry Handbook, 3rd Ed., T. G. Bryer (ed.), Forging Industry Association, Cleveland, OH and ASM (1981).

Forging Equipment, Materials and Practice (MCIC-HB-03), Metals and Ceramics Information Center, Battelle Columbus Labs.

Forging Materials and Practices, A. M. Sabroff et al., Reinhold, New York (1968).

Handbook of Metalforming Processes, B. Avitzur, Wiley-Interscience, New York (1983).

Handbook of Metal Forming, Kurt Lange (ed.), McGraw-Hill, New York (1985).

Handbook of Fabrication Processes, O. D. Lascoe, ASM, Materials Park, OH (1988).

Mechanics of Plastic Deformation in Metal Processing, E. G. Thomsen, C. T. Yang, and S. Kobayashi, Macmillan (1965).

Mechanics of Sheet Metal Forming: Material Behavior and Deformation Analysis, D. Koistinen, Plenum (1977).

Metal Forming: Fundamentals and Applications, T. Altan, S-I Oh, and H. Gegal, ASM, Materials Park, OH (1983).

Metalforming and the Finite-Element Method, S. Kohavashi, S. Oh, and T. Altan, Oxford University Press (1989).

Metalworking Science and Engineering, Edward M. Mielnik, McGraw-Hill (1991).

Principles of Industrial Metalworking Processes, G. W. Rowe, Arnold, London (1977).

Source Book on Cold Forming, ASM, Materials Park, OH (1975).

Techniques of Pressworking Sheet Metal, 2nd Ed., D. F. Eary and E. A. Reed, Prentice Hall (1974).

POWDER METALLURGY

Handbook of Powder Metallurgy, H. H. Hausner, Chemical Publishing Co. (1973).

Introduction to Powder Metallurgy, J. S. Hirschorn, American Powder Metallurgy Institute, New York (1969).

Injection Molding of Metals and Ceramics, Randall M. German and Animesh Bose, Metal Powder Industries Federation (1997).

P/M Materials Standards and Specifications—MPIF Standard 35, Metal Powder Industries Federation (1981).

Powder Metallurgy Science, 2nd Ed., R. M. German, Metal Powder Industries Federation (1994).

Powder Metallurgy Processing: New Techniques and Analyses, H. A. Kuhn and A. Lawley, Academic Press (1978).

Powder Metallurgy Design Manual, 2nd Ed., Metal Powder Industries Federation (1995).

Powder Metallurgy, Principles and Applications, F. V. Lenel, Metal Powder Industries Federation, Princeton, NJ (1980).

Powder Metallurgy Equipment Manual—3, Samuel Bradbury (ed.), MPIF (1986).

Powder Injection Molding, R. M. German, Metal Powder Industries Federation (1990).

Source Book on Powder Metallurgy, ASM, Materials Park, OH (1979).

MACHINING

Analysis of Material Removal Processes, W. R. DeVries, Springer-Verlag (1992).

Application of Metal Cutting Theory, Fryderyk E. Gorczyca, Industrial Press, Inc. (1987).

Design of Cutting Tools, A. Bhattacharyya and Inyong Ham, Society of Manufacturing Engineers.

Fundamentals of Metal Machining and Machine Tools, G. Boothroyd and Winston Knight, Marcell Dekker (1989).

High Speed Machining, ASME PED Vol. 12 (1984).

Machining of Plastics, Akira Kobayashi, Krieger Pub. Co. (1981).

Machining Data Handbook, 3rd Ed., (2 volumes) Machinability Data Center, Cincinnati, OH (1980).

Machining—Theory and Practice, ASM, Materials Park, OH (1950).

Machining Source Book, ASME International, Metals Park, OH (1988).

Manufacturing Analysis, Nathan H. Cook, Addison-Wesley (1966).

Materials Issues in Machining and The Physics of Machining Processes, Robin Stevenson and David A. Stephenson (eds.), A publication of ASME and TMS (1992).

Mechanics of Machining, P. L. B. Oxlev, Ellis Harwood Ltd. (1989).

Metal Cutting, 4th Ed., Edward M. Trent, Paul K. Wright, Butterworth Heineman (2000).

Metal Cutting Principles, M. C. Shaw, Oxford University Press (1984).

The Machining of Metals, E. J. A. Armarego and R. H. Brown, Prentice Hall (1969).

HEAT TREATMENT

Atlas of Isothermal and Cooling Transformation Diagrams, ASM, Materials Park, OH (1977).

Atlas of Isothermal and Cooling Transformation Diagrams for Engineering Steels, ASM, Materials Park, OH (1980).

Atlas of Time-Temperature Diagrams for Nonferrous Alloys, G. VanderVoort (ed.), ASM, Materials Park, OH (1991).

Atlas of Time-Temperature Diagrams for Irons and Steels, G. VanderVoort (ed.), ASM, Materials Park, OH (1991).

Handbook of Quenchants and Quenching Technology, G. E. Totten, C. E. Bates, and N. A. Clinton, ASM, Materials Park, OH (1992).

Heat Treater's Guide: Practices and Procedures for Irons and Steels, 2nd Ed., ASM, Materials Park, OH (1995).

Heat Treater's Guide: Practices and Procedures for Nonferrous Alloys, ASM (1996).

Heat Treatment, Structure and Properties of Nonferrous Alloys, Charles R. Brooks, ASM, Materials Park, OH (1982).

Practical Heat Treating, Howard E. Boyer, ASM, Materials Park, OH (1984).

Practical Induction Heat Treating, R. E. Haimbaugh, ASM (2001).

Principles of Heat Treatment of Steel, G. Krauss, ASM, Materials Park, OH (1980).

Principles of Heat Treatment, M. A. Grossman and E. C. Bain, ASM, Materials Park, OH (1964).

Principles of the Heat Treatment Plain Carbon and Low Alloy Steels, C. R. Brooks, ASM, Materials Park, OH (1996).

Source Book on Heat Treating, ASM, Materials Park, OH (1975).
Vol. 1, Materials and Processes
Vol. 2, Production and Engineering Practices

Steel and Its Treatment, Bofors Handbook, K. E. Thelning, Butterworths (1975).

Steel Heat Treatment Handbook, G. E. Totten and M. A. H. Howes, Marcel Dekker (1997).

Surface Hardening of Steels: Understanding the Basics, J. R. Davis (ed.), ASM (2002).

The Heat Treating Source Book, ASM, Materials Park, OH (1986).

SURFACES AND FINISHES

Metal Finishing: Guidebook Directory, Metals and Plastics Publications, Inc., Westwood, NJ (published annually).

Surface Preparation and Finishes for Metals, J. A. Murphy (ed.), Society for Manufacturing Engineers, Dearborn, Mich. (1971).

Basic Metal Finishing, J. A. Von Frauenhofer, Chemical Publishing (1976).

Surface Finishing Systems, George Rudski, ASM, Materials Park, OH (1983).

Electroplating Engineering Handbook, 4th Ed., Lawrence Durney (ed.), Van Nostrand-Reinhold (1984).

Properties of Electroplated Metals and Alloys, 2nd Ed., Wm. H. Safranek, Amer. Electroplaters and Finishers Soc. (1986).

Metallizing of Plastics: Handbook of Theory and Practice, R. Suchentruck, Finishing Publications Ltd and ASM (1993).

The Surface Treatment and Finishing of Aluminum and Its Alloys, 6th Ed., S. Wernick, R. Pinner and P. B. Sheasby, Finishing Publications Ltd. and ASM (2001).

Surface Engineering for Corrosion and Wear Resistance, J. R. Davis (ed.), IOM Communications and ASM (2001).

Coatings and Coating Processes for Metals, J. H. Lindsay (ed.), ASM (1998).

CORROSION

Handbook of Corrosion Data, 2nd Ed., B. Craig and D. Anderson (eds.), ASM, Materials Park, OH (1995).

NACE Corrosion Engineer's Reference Book, 2nd Ed., R. S. Treseder (ed.), National Association of Corrosion Engineers (NACE) (1991).

Corrosion Resistance Tables, 4th Ed., P. A. Schweitzer, Marcel Dekker (1995).

An Introduction to Metallic Corrosion, 3rd Ed., Ulick R. Evans, Edward Arnold Ltd. and ASM (1981).

Corrosion and Its Control—An Introduction to the Subject, Atkinson and Van Droffelaar, National Association of Corrosion Engineers, Houston, TX (1982).

Corrosion Engineering, 3rd Ed., M. G. Fontana, McGraw-Hill, New York (1986).

Corrosion and Corrosion Control, 3rd Ed., H. H. Uhlig and R. Winston Revie, Wiley-Interscience, New York (1985).

Corrosion, 2nd Ed.—Vol. 1: Corrosion of Metals and Alloys and Vol. 2: Corrosion Control, L. L. Shrier (ed.).

Material Selection for Corrosion Control, S. I. Chawla and R. K. Gupta, ASM, Materials Park, OH (1993).

Corrosion Prevention by Protective Coatings, Charles G. Munger, National Association of Corrosion Engineers (1984).

Fundamentals of Electrochemical Corrosion, E. E. Stansbury and R. A. Buchanan, ASM (2000).

Corrosion: Understanding the Basics, J. R. Davis (ed.), ASM (2000).

NDT, METROLOGY, QUALITY CONTROL, AND TAGUCHI METHODS

Creating Quality, William J. Kolarik, McGraw-Hill (1995).

Guide to Quality Control, Kaoru Ishikawa, Asian Prod. Organization (1984).

Handbook of Dimensional Measurement, 2nd Ed., F. T. Farago, Industrial Press (1982).

Handbook of Industrial Metrology, Prentice Hall.

Hardness Testing, 2nd Ed., H. Chandler (ed.), ASM (1999).

Introduction to Quality Engineering, Genichi Taguchi, Kraus Int. Pub. (1986).

Introduction to Statistical Quality Control, 2nd Ed., D. C. Montgomery, Wiley (1981).

ISO System of Limits and Fits, General Tolerances and Deviations, American National Standards Institute.

Metallography: Principles and Practice, George VanderVoort, McGraw-Hill (1984).

Nondestructive Testing Handbook, 2nd Ed. (multiple volumes), American Society for Nondestructive Testing (1982–1985).

Nondestructive Testing, Louis Cartz, ASM, Materials Park, OH (1995).

Practical Non-Destructive Testing, 2nd Ed., B. Raj, T. Jaykumar, and M. Thavasimuthu, Narosa Publishing and ASM (2002).

Process Quality Control, 2nd Ed., E. R. Ott and E. G. Schilling, McGraw-Hill (1990).

Quality Engineering in Production Systems, G. Taguchi, E. A. ElSayed, and T. Hsiang, McGraw-Hill (1989).

Quality Control Handbook, 4th Ed., J. M. Juran and F. M. Gigna, McGraw-Hill (1988).

Quality, Productivity, and Competitive Position, W. Edwards Deming, MIT Press, Boston (1982).

Quality Control Source Book, ASM, Materials Park, OH (1982).

Quality Planning and Analysis, 2nd Ed., Juran and Gryan, McGraw-Hill (1988).

Statistical Quality Design and Control, R. E. De Vor, T. Chang, and J. W. Sutherland, Macmillan (1992).

Statistical Quality Control, 6th Ed., Eugene L. Grant and Richard S. Leavenworth, McGraw-Hill (1988).

Taguchi Techniques for Quality Engineering, P. J. Russ, McGraw-Hill (1988).

The Testing of Engineering Materials, 4th Ed., H. E Davis, G. E. Troxell, and G. F. W. Hauck, McGraw-Hill, New York (1982).

FAILURE ANALYSIS AND PRODUCT LIABILITY

Analysis of Metallurgical Failures, V. J. Conangelo and F. A. Heiser, Wiley-Interscience (1974).

Case Histories in Failure Analysis, ASM, Materials Park, OH (1979).

Engineering Aspects of Product Liability, V. J. Conangelo and P. A. Thornton, ASM, Materials Park, OH (1981).

Failure Analysis: Case Histories, F. K. Naumann, ASM, Materials Park, OH (1983).

Failure Analysis: The British Engine Technical Reports, F. K. Hutchings and P. M. Unterweiser, ASM, Materials Park, OH (1981).

Failure of Materials in Mechanical Design, J. A. Collins, Wiley-Interscience (1981).

Fractography in Failure Analysis, American Society for Testing and Materials, Philadelphia, PA (1978).

Handbook of Case Histories in Failure Analysis: Vols. 1 and 2, K. A. Esaklul (ed.), ASM, Materials Park, OH (1992–93).

Metal Failures: Mechanisms, Analysis, Prevention, A. J. McEvily (ed.), John Wiley & Sons, New York (2002).

Metallography in Failure Analysis, J. L. McCall and P. M. French, Plenum (1977).

Metallurgical Failure Analysis, C. R. Brooks and A. Choudhury, McGraw-Hill (1993).

Product Liability and the Reasonably Safe Product, A. S. Weinstein et al., Wiley-Interscience, New York (1978).

Residual Stresses and Fatigue in Metals, J. O. Almen and P. H. Black, McGraw-Hill, New York (1963).

Source Book on Failure Analysis, ASM, Materials Park, OH (1975).

Stress Corrosion Cracking: Materials Performance and Evaluation, Russell Jones (ed.), ASM, Materials Park, OH (1992).

Tool and Die Failures Source Book, ASM, Materials Park, OH (1982).

Understanding How Components Fail, 2nd Ed., Donald J. Wulpi, ASM, Materials Park, OH (1999).

Why Metals Fail, R. D. Barer and B. F. Peters, Gordon and Breach, New York (1970).

MANUFACTURING SYSTEMS ANALYSIS

Computational Intelligence in Design and Manufacturing, Andrew Kusiak, Wiley (2000).

Fundamentals of Systems Engineering, C. Jotin Kristy and Jamshid Mohammadi, Prentice-Hall (2001).

Manufacturing Systems Design and Analysis, 2nd Ed., B. Wu, Chapman and Hall (1994).

Manufacturing Systems Engineering, S. B. Gershwin, PTR, Prentice Hall (1994).

Modeling and Analysis of Manufacturing Systems, R. G. Askin and C. R. Standridge, Wiley (1993).

Performance Modeling of Automated Manufacturing Systems, N. Viswanadham and Y. Narahari, Prentice Hall (1992).

Stochastic Models of Manufacturing Systems, J. A. Buzacott and J. C. Shanthikumar, Prentice Hall (1993).

Systems Analysis and Modeling, Donald W. Boyd, Academic Press (2001).

MANUFACTURING ENGINEERING

Concurrent Design of Products and Processes, J. L. Nevins and D. E. Whitney, McGraw-Hill (1989).

Expert Systems Applications in Engineering and Manufacturing, A. B. Badiv, Prentice Hall (1992).

Intelligent Manufacturing Systems, A. Kusiak, Prentice Hall (1990).

Manufacturing Engineering, 2nd Ed., D. T. Koenig, Taylor and Francis (1994).

Manufacturing Engineering, 2nd Ed., J. P. Tanner, Marcel Dekker (1991).

Manufacturing Intelligence, P. K. Wright and D. A. Bourne, Addison Wesley (1988).

Manufacturing Systems Engineering, 2nd Ed., K. Hitomi, Taylor and Francis Ltd. (1996).

ROBOTICS

Assembly with Robots, T. Owen, Prentice Hall (1985).

Handbook of Industrial Robotics, Nof (ed.), Wiley (1999).

Industrial Robotics, M. Groover et al., McGraw-Hill (1986).

Industrial Robots and Robotics, E. Kafrissen and M. Stephens, Reston (1984).

Industrial Robots: Computer Interfacing and Control, W. E. Snyder, Prentice Hall (1985).

Introduction to Robot Technology, P. Coiffet and M. Chizouze, McGraw-Hill (1993).

Robotics: Control Sensing, Vision and Intelligence, K. S. Fu, R. C. Gonzalez, C. S. Gonzalez, and C. S. G. Lee, McGraw-Hill (1987).

Robotics for Engineers, Y. Koren, McGraw-Hill (1985).

Robotics and Manufacturing Automation, 2nd Ed., C. R. Asfahl, Wiley (1992).

AUTOMATION, CNC, FMS AND CIM

An Introduction to CNC Machining and Programming, D. Gibbs and T. M. Crandall, Industrial Press (1991).

Anatomy of Automation, G. H. Amber and P. S. Amber, Prentice Hall (1962).

Automatic Assembly, C. Boothroyd, C. Poli, and L. F. March, Marcel Dekker (1982).

Automation, Production Systems, and Computer-Integrated Manufacturing, 2nd Ed., Mikell P. Groover, Prentice Hall (2001).

CIM Technology—Fundamentals and Applications, Russell Biekert, The Goodheart-Willcox Company (1998).

Computer Control of Machining and Processes, J. G. Bolhinger and N. A. Duffie, Addison Wesley (1988).

Computer Integrated Manufacturing, Vol. III, R. U. Ayres, W. Haywood, and I. Tchijov (eds.), Chapman and Hall (1992).

Computer Integrated Manufacturing, Vol. II, R. U. Ayres, W. Hywood, M. E. Merchant, J. Rant, and H.-J. Warnecke (eds.), Chapman and Hall (1992).

Computer Integrated Manufacturing, Vol. IV, R. U. Ayres, R. Dobrinsky, W. Haywood, Kuno, and Zuscovitch (eds.), Chapman and Hall (1992).

Computer Integrated Manufacturing, Vol. 1, R. U. Ayres, Chapman and Hall (1991).

Computer-Aided Process Planning, Hsu-Pin Wang and Jian-Kang Li, Elsevier (1991).

Computer-Integrated Design and Manufacturing, D. D. Bedworth, M. R. Henderson, and P. M. Wolfe, McGraw-Hill (1991).

Flexible Manufacturing Systems Handbook, Noyes Publishing (1984).

Handbook of Manufacturing Automation and Integration, J. Stark (ed.), Auerbach (1989).

Introduction to Computer Numerical Control, J. V. Valentino and J. Goldenberg, Regents/Prentice Hall (1993).

Microcomputer Applications in Manufacturing, A. G. Ulsoy and W. R. DeVries, Wiley (1989).

NC Machine Programming and Software Design, C-H Chang and M. A. Melkanoff, Prentice Hall (1989).

Numerical Control and Computer Aided Manufacturing, R. S. Pressman and I. J. E. Williams, Wiley (1977).

Principles of Numerical Control, J. L. Childs, Industrial Press (1982).

The Design and Operation of FMS-Flexible Manufacturing Systems, P. G. Ranky, IFS Publications Ltd. (UK) and North-Holland Publishing Company, New York (1983).

COST ESTIMATING

Cost Estimator's Reference Manual, R. D. Steward and R. W. Wyskida, Wiley (1987).

Engineering Cost Estimating, 3rd Ed., P. E. Ostwald, Prentice Hall (1992).

Manufacturing Cost Estimating Guide, P. R. Ostwald, American Machinist (1982).

Realistic Cost Estimating for Manufacturing, 2nd Ed., W. Winchell, Society of Manufacturing Engineers (1989).

NTM/LASER MACHINING SOURCES

Advance Machining Technology Handbook, James Brown, McGraw-Hill, New York (1998).

Advanced Methods of Machining, J. A. McGeough, Chapman and Hall, New York (1988).

Laser Material Processing, W. M. Steen, Springer, London, England (1998).

Laser Machining, G. Chryssolouris, Springer-Verlag (1991).

Machining Data Handbook, 3rd Ed., Vol. 2, Machinability Data Center, Metcut Research Associates, Inc., Cincinnati, OH (1980).

Non-Traditional Machining Handbook, Carl Sommer, Advance Publishing, Inc., Houston, TX (2000).

Nontraditional Machining Processes, 2nd Ed., E. J. Weller (ed.), Society of Manufacturing Engineers, Dearborn, MI (1984).

Nontraditional Manufacturing Processes, G. F. Benedict, Marcel Dekker, New York (1987).

The EDM Handbook, E. B. Guitrau and Hanser Gardner, Cincinnati, OH (1997).

RAPID PROTOTYPING

"The World of Rapid Prototyping," Proceedings of the Fourth International Conf Desktop Manufacturing, T. Wohlers, Management Roundtable, Boston, MA, Sept. 24-25, San Jose, CA (1992).

Automated Fabrication: Improving Productivity in Manufacturing, M. Burns, Prentice Hall, Englewood Cliffs, NJ (1993).

Rapid Prototyping and Other RP&M Technologies: From Rapid Prototyping to Rapid Tooling, P. F. Jacobs, SME, Dearborn, MI (1996).

Rapid Automated Prototyping: An Introduction, Lamont Wood, Industrial Press, Inc., New York (1993).

Rapid Manufacturing: The Technologies and Applications of Rapid Prototyping and Rapid Tooling, D. T. Pham and S. S. Dimov, Springer-Verlag, London, UK (2001).

Rapid Tooling: Technologies and Industrial Applications, Peter D. Hilton and P. F. Jacobs, Marcel Dekker, New York (2000).

Solid Freeform Manufacturing, D. Kocran, Elsevier, Amsterdam (1993).

LEAN MANUFACTURING

A Revolution in Manufacturing: The SMED System, Shigeo Shingo, Productivity Press (1985).

A Study of The Toyota Production System, Shigco Shingo, Productivity Press (1989).

Design, Analysis and Control and Manufacturing Cells, PED-Vol. 53, J. T. Black, B. C. Jiang, and G. J. Wicns, ASME (1991).

Flexible Manufacturing Cells and Systems, William W. Luggen, Prentice Hall (1991).

Handbook of Cellular Manufacturing Systems, Shahrukh A. Irani (ed.), Wiley (1999).

Introduction to TPM: Total Productive Maintenance, Seiichi Nakajima, Productivity Press (1988).

JIT Factory Revolution, H. Hirano and J. T. Black, Productivity Press (1988).

Japanese Manufacturing Techniques: Nine Hidden Lessons in Simplicity, Richard J. Schonberger (1982). Free Press.

Just-in-Time Manufacturing, C. A. Voss, IFS Publications Ltd. (UK), Springer-Verlag (1987).

Kaizen for Quick Changeover, Kenichi Sekine and Keisuke Arai (translated by Bruce Talbot), Productivity Press (1992).

Kanban: Just-In-Time at Toyota, translation by D. J. Lu, Japan Management Association, Productivity Press (1986).

Kawasaki U.S.A.: A Case Study, Robert Hall, APICS (1982).

One-Piece Flow: Cell Design for Transforming the Production Process, Kenichi Sekine, Productivity Press (1990).

Reinventing The Factory II: Managing the World Class Factory, Roy L. Harmon, Free Press (1992).

The Design of the Factory with A Future, J. T. Black, McGraw-Hill (1991).

The Shift to JIT: How People Make the Difference, Ichiro Majima, Productivity Press (1992).

The New Manufacturing Challenge, Kioyoshi Suzaki, The Free Press (1987).

The Machine That Changed The World, James P. Womack, Daniel T. Jones, and Daniel Roos, Harper Perennial (1991).

Toyota Production System: Beyond Large-Scale Production, Taiichi Ohno, Productivity Press (1988).

Toyota Production System—An Integrated Approach to Just-In-Time, 3rd Ed., Yasuhiro Monden (1998).

Toyota Production System, Yasuhiro Monden, Industrial Engineering and Management Press (1993).

TPM, Introduction to TPM: Total Productive Maintenance, S. Nakajima, Productivity Press (1988).

World Class Manufacturing, Richard J. Schonberger, The Free Press (1986).

World Class Manufacturing Casebook: Implementing JIT and TOC, Richard J. Schonberger, The Free Press (1987).

Zero Inventories, Robert W. Hall, Dow Jones-Irwin (1983).

Experimental Design, Walter T. Federer, MacMillan (1955).

Flexible Manufacturing Cells and Systems, William W. Luggen, Prentice Hall (1991).

Guide to Quality Control, Asian Productivity Organization (1972).

Japan Quality Control Circles, Asian Productivity Organization (1972).

Management for Quality Improvement—The Seven New Tools, Shigeru Mizuno (ed.), Productivity Press (1979).

Practical Experimental Designs, 2nd Ed., Van Nostrand Reinhold (1989).

Quality Engineering in Production Systems, Genichi Taguchi, Elsayed A. Edsayed, and Thomas C. Hsiang, McGraw-Hill (1989).

Reliability and Maintainability Management, Balbir S. Dhillon and Hans Reiche, Von Nostrand Reinhold (1985).

Quality Control and Industrial Statistics, 5th Ed., Acheson J. Duncan, Irwin McGraw-Hill (1986).

Quality Control for Managers and Engineers, Elwood G. Kirkpatrick, Wiley (1970).

Statistical Quality Control, 6th Ed., Eugene L. Grant and Richard S. Leavenworth, McGraw-Hill (1988).

Total Quality Management—Text, Cases and Readings, Joel E. Ross, St. Lucie Press (1993).

Zero Quality Control: Source Inspection and the Poka-Yoke System, Shigeo Shingo, Productivity Press (1986).

INDEX

HGVS	Human-Guided Vehicle System *(fork-lift with driver)*	PAW	Plasma Arc Welding *(PAC = Cutting) (PAM = Machining)*
HIP	Hot Isostatic Pressing	PCB	Printed Circuit Board
IGES	Initial Graphics Exchange System	PD	Pitch Diameter
IMPSs	Integrated Manufacturing Production Systems	PDES	Product Design Exchange Specification
I/O	Input/Output	PLC	Programmable Logic Controller
IOCS	Input/Output Control System	POK	Production Ordering Kanban
JIT	Just-In-Time	PROM	Programmable Read-Only Memory
LAN	Local Area Network	PS	Production System
LASER	Light Amplification by Stimulated Emission of Radiation	P/M	Powder Metallurgy
		PVD	Physical Vapor Deposition
LBM	Laser Beam Machining *(LBW = Welding) (LBC = Cutting)*	QC	Quality Control
		QMS	Quality Management System
L-CMS	Linked-Cell Manufacturing System	RAM	Random Access Memory
LED	Light Emitting Diode	RIM	Reaction Injection Molding
LP	Lean Production	ROM	Read-Only Memory
LSI	Large Scale Integration	SAW	Submerged Arc Welding
MAP	Manufacturing Automation Protocol	SCA	Single Cycle Automatic
MCU	Machine Control Unit	SMAW	Shielded Metal Arc Welding
MDI	Manual Data Input	SPC	Statistical Process Control
MIG	Metal–Inert Gas	SPF	Single Piece Flow
MPS	Manufacturing Production System	SQC	Statistical Quality Control
mrp	Material Requirements Planning	TCM	Thermochemical Machining
MRPII	Manufacturing Resources Planning	TIR	Total Indicator Readout
MSD	Manufacturing System Design	TPS	Toyota Production System
NC	Numerically Control	TQC	Total Quality Control
NDT	NonDestructive Testing *(NDE = Evaluation) (NDI = Inspection)*	USM	Ultrasonic Machining *(USW = Welding)*
		VA	Value Analysis
OCR	Optical Character Recognition	WAN	Wide Area Network
OM	Orthogonal Machining	WIP	Work-In-Progress (or Process)
OPM	Orthogonal Plate Machining	WJM	Water Jet Machining
OS	Operating System	WLK	Withdrawl Kanban
OTT	Orthogonal Tube Turning	YAG	Yttrium-Aluminum Garnet